Advanced Structured Materials

Volume 116

Common engineering materials reach in many applications their limits and new developments are required to fulfil increasing demands on engineering materials. The performance of materials can be increased by combining different materials to achieve better properties than a single constituent or by shaping the material or constituents in a specific structure. The interaction between material and structure may arise on different length scales, such as micro-, meso- or macroscale, and offers possible applications in quite diverse fields.

This book series addresses the fundamental relationship between materials and their structure on the overall properties (e.g. mechanical, thermal, chemical or magnetic etc) and applications.

The topics of *Advanced Structured Materials* include but are not limited to

- classical fibre-reinforced composites (e.g. glass, carbon or Aramid reinforced plastics)
- metal matrix composites (MMCs)
- micro porous composites
- micro channel materials
- multilayered materials
- cellular materials (e.g., metallic or polymer foams, sponges, hollow sphere structures)
- porous materials
- truss structures
- nanocomposite materials
- biomaterials
- nanoporous metals
- concrete
- coated materials
- smart materials

Advanced Structured Materials is indexed in Google Scholar and Scopus.

More information about this series at http://www.springer.com/series/8611

Loutfy H. Madkour

Nanoelectronic Materials

Fundamentals and Applications

Springer

Loutfy H. Madkour
Faculty of Science, Chemistry Department
Al Baha University
Baljarashi, Saudi Arabia

ISSN 1869-8433 ISSN 1869-8441 (electronic)
Advanced Structured Materials
ISBN 978-3-030-21620-7 ISBN 978-3-030-21621-4 (eBook)
https://doi.org/10.1007/978-3-030-21621-4

This Springer imprint is published by the registered company Springer Nature Switzerland AG
The registered company address is: Gewerbestrasse 11, 6330 Cham, Switzerland

Preface

Nanoscience and nanotechnology have become highly popular in the last few years. The nanoobjects of importance include quantum dots and nanocrystals of metals, semiconductors, oxides and other materials as well as one-dimensional nanostructures such as nanotubes and nanowires. Synthesis, characterization and applications of these nanomaterials are being explored widely. Typical of the novel applications of nanomaterials that are emerging include single molecule electronics based on nanotubes and nanowires, nanocatalysis and biological sensors using nanocrystals or nanotubes. One-dimensional nanostructures of materials have received great attention since the discovery of the carbon nanotubes.

Nanomedicine involves utilization of nanotechnology for the benefit of human health and well being. The use of nanotechnology in various sectors of therapeutics has revolutionized the field of medicine where nanoparticles of dimensions ranging between 1–100 nm are designed and used for diagnostics, therapeutics and as biomedical tools for research. It is now possible to provide therapy at a molecular level with the help of these tools, thus treating the disease and assisting in study of the pathogenesis of disease.

Nanotechnology is an inherently interdisciplinary field that has generated significant scientific and engineering interest in recent years. Nanomaterials can be found everywhere in nature and have been part of the environment since our planet was created about 4.5 billion years ago.

The impact of nanotechnology in all areas of science and technology is evident. Nanotechnology increases the strengths of many materials and devices, as well as enhances efficiencies of monitoring devices, remediation of environmental pollution and renewable energy production. While these are considered as the positive effect of nanotechnology, there are certain negative impacts of nanotechnology on environment in many ways, such as increased toxicological pollution on the environment due to the uncertain shape, size and chemical compositions of some of the nanotechnology products (or nanomaterials). It can be vital to understand the risks of using nanomaterials, and cost of the resulting damage. It is required to conduct a risk assessment and full life-cycle analysis for nanotechnology products at all stages of products to understand the hazards of nanoproducts and the resultant knowledge that can then be used to predict the possible positive and negative impacts of the nanoscale products. Choosing right, less toxic materials (e.g. graphene) will make huge impacts on the environment. This can be very useful for the training and protection of students, as well as scientists, engineers, policymakers and regulators working in the field.

Nanotechnology refers to an emerging field of science that includes synthesis and development of various nanomaterials. Presently, different metallic nanomaterials are being produced using copper, zinc, titanium, magnesium, gold, alginate and silver. Due to their incredible properties, production of nanomaterials has been constantly evolving over the last few years for manifold applications in chemical surface processes, chemical industry, environmental pollution monitoring, agriculture, smart materials, sensors, nanoscale biostructures, energy capture and storage, magnets, fabrication, electronics, optical and biomedical fields. Nanostructured materials (NSMs) are divided into nanoparticles (NPs), nanotubes (NTs), nanocomposites (NCMs) and nanowires (NWs). Nanoparticle technologies have great potentials, being able to convert poorly soluble, poorly absorbed and labile biologically active substance into promising deliverable substances. Nanoparticles are applied as delivery systems, e.g. for drugs or bioactive food ingredients. They are designed to target drugs to specific organs or to increase the bioavailability of bioactive food ingredients that may have a health impact. Nanomedicine has tremendous prospects for the improvement of the diagnosis and treatment of human diseases. Use of microbes in biosynthesis of nanoparticles is an environmentally acceptable procedure. Nanomedicine is the application of nanotechnology (the engineering of tiny machines) to the prevention and treatment of disease in the human body. Nanomaterials can impart antibacterial and anti-odour functionality on human skin in powder, gel, stick or spray underarm products. It has also antimicrobial and anti-irritant properties. This discipline is in its infancy. It has the potential to change medical science dramatically in the twenty-first century.

The expanding availability of a variety of nanostructures with properties in the nanometer size range has sparked widespread interest in their use in biotechnological systems, including the field of environmental remediation. Nanomaterials can be used as catalysts, adsorbents, membranes, water disinfectants and additives to increase catalytic activity and capability due to their high specific surface areas and nanosize effects. Thus, nanomaterials appear promising for new effective environmental technologies. Definitely, nanotechnology applications for site remediation and wastewater treatment are currently in research and development stages, and innovations are underway. The synthesis of metallic nanoparticles has been intensively developed not only due to its fundamental scientific interest but also for many technological applications. The use of microorganisms in the synthesis of nanoparticles is a relatively new eco-friendly and promising area of research with considerable potential for expansion. On the other hand, chemical synthesis occurs generally under extreme conditions (e.g. pH, temperature) and chemicals used may have associated environmental and human health impacts. The use of microorganisms during the biosynthesis of metallic nanoparticles and their unique properties that make them good candidates for many applications, including in biotechnology should be considered. Thorough cost-benefit analyses are essential to further evaluate the applicability of nanotechnology. Economic analyses must take into consideration the synthesis of nanomaterials, the benefit of application as well as the cost associated with the potential environmental impacts. Lowcost nanomaterials should be explored for potential environmental applications. Nanotechnology based water treatment technologies will only be able to compete with conventional treatment if the cost of nanomaterials as well as the systems utilizing nanomaterials becomes comparable. Applications of nanomaterials are wide field intervention in industry, environment, medicine and food fields, will save time and effort, and will be relied upon very dramatically. Environmental impact can be improved by utilizing nanostructure particulates in corrosion inhibition, coating and eliminating the requirement of toxic solvents. The application of nanotechnology in the corrosion protection of metals has recently gained momentum and is of real promise.

This book in detail explains the classification of environmental nanomaterials and how to deal with their formation, diffusion, environmental fate and impacts, and our exposure to them. This book is an overview of current research worldwide on the classification, use and all types of metallic nanomaterials and their unique properties that make them good candidates for many applications, including nanotechnology and nanobiotechnology. Thus, the aim of this book is to summarize the fundamentals and technical approaches in processing and behaviour of nanostructured materials to provide the readers systematic, comprehensive and brief information in the challenging field of nanomaterials NMs and nanotechnology. Therefore, this book is a general introduction for all branches of academic science, nanoscience, nanochemistry, physical science, nanotechnology, nano-biotechnology, nanomedicine and in general materials science.

Finally, we would like to thank Editor, all reviewers, and the Editorial Board Members for their invaluable contributions.

- Correspondence: **Loutfy H. Madkour**, Professor of Physical Chemistry and Nano Science at *Chemistry Department, Faculty of Science, Al Baha University, Saudi Arabia (KSA).*
- Tel. +966 533899075; fax: +966 77247272;
- E-mail: loutfy_madkour@yahoo.com; lha.madkour@gmail.com; lmadkour@bu.edu.sa.

Baljarashi, Saudi Arabia Loutfy H. Madkour

About This Book

Nanostructured materials (NSM) are divided into nanoparticles (NPs), nanotubes (NTs), nanocomposites (NCMs) and nanowires (NWs). Nanoparticle technologies have great potentials, being able to convert poorly soluble, poorly absorbed and labile biologically active substance into promising deliverable substances. The book includes 17 chapters and gives emphasis to current classification and future trends in vision of life sciences for processing, techniques, synthesis methods and advanced applications for nanoelectronic materials and nanodevices in industries and research laboratory. The nanoobjects of importance include quantum dots and nanocrystals of metals, semiconductors, oxides and other materials as well as one-dimensional nanostructures such as NTs and NWs. Nanoelectronics is certainly among the most vibrant research fields, which is highly pushed by technology industries. The numerical calculation of the electronic properties and in particular the quantum transport properties of devices at the nanoscale is considered. Principles of computational simulations devices and characterization of nanoelectronic materials are discussed. The book precisely has explained the synthesis techniques for the preparation of low-dimensional nanomaterials including 0D (quantum dot), 1D (Nanowire, Nanotube) and 2D (Thin films, few layer) and their potential applications in nanoelectronic systems. Allotropic forms of elemental carbon nanomaterials (fullerene, nanodiamond, graphite, carbon nanotubes CNT, graphene) and two-dimensional transition metal dichalcogenides (2D TMDCs) with their future applications are discussed. This book has been concerned with the size effects and going from bulk materials to nanomaterials, electronic properties of nanoscale devices, different classes of nanomaterials from microelectronics to nanoelectronics into molecular electronics. Further, the book demonstrates the structural stability, physical, chemical, magnetic, optical, electrical, thermal, electronic, mechanical properties of the nanomaterials, physicochemical properties of NPs and gives aspects of their applications. Characterization, fabrication techniques from lab-scale to mass-production and functionality of nanomaterials by chemical or defect engineering has been studied. Environmental impact of nanotechnology and novel applications of NMs, NWs, NTs, NPs and nanodevices in mechanical industries, the environment, energy harvesting, clean energy, manufacturing materials, electronics, transistors, health and/or medical therapy are explained. Interfacing biology systems with nanoelectronics and examples for nanoelectronic—cell interfaces of the advanced medicine therapeutics applications has been discussed.

Summary

Today, the feature size of integrated circuits are of the order of 100 nm. In the near future, this order will most likely approach 10 nm and lead to yet more powerful computers. Nanomaterials NMs surreptitiously enter the environment through water, soil, and air during various human activities.

Nanoelectronics is certainly among the most vibrant research fields, which is highly pushed by technology industries. The numerical calculation of the electronic properties and in particular the quantum transport properties of devices at the nanoscale is considered. Principles of computational simulations devices and characterization of nanoelectronic materials are discussed. This book has been concerned with the size effects and going from bulk materials to nanomaterials, electronic properties of nanoscale devices, different classes of nanomaterials from microelectronics to nanoelectronics into molecular electronics. The electronic characteristics such as thermal, chemical and mechanical stability of nanomaterials are investigated. Characterization, fabrication techniques from lab-scale to mass-production and functionality of nanomaterials by chemical or defect engineering has been studied. The comprehensive description of the family of all nanomaterials and their applications is considered. The book precisely discusses about various preparation and synthesis techniques for low-dimensional nanomaterials including 0D (quantum dot), 1D (Nanowire, Nanotube) and 2D (Thin films, few layer) and their potential applications in nanoelectronic systems. Allotropic forms of elemental carbon nanomaterials (fullerene, nanodiamond, graphite, carbon nanotubes CNT, graphene) and two-dimensional transition metal dichalcogenides (2D TMDCs) with their applications are discussed. These informations give a clear vision for synthesis and applications of nanomaterials among the researchers. It is necessary to handle systems comprising millions of atoms, and this will require new efficient algorithms for the most time-consuming stages of the calculations. Further, the book demonstrates the structural stability, physical, chemical, magnetic, optical, electrical, thermal, electronic, mechanical properties of the nanomaterials, physicochemical properties of NPs and gives emphasis to applications of them. Applications and novel applications of NMs, NWs, NTs, NPs and nanodevices in mechanical industries, the environment, energy harvesting, clean energy, manufacturing materials, electronics, transistors, health and medical therapy are discussed through the chapters of the book. Fundamentals of nanoelectronics in biological system and medicine therapeutics applications are studied.

Besides many positive effects of NMs in industrial and medical applications, there are certain toxicities and negative impacts associated with using NPs and other NMs on environment in many ways. It can be vital to understand the risks of using NMs, and cost of the resulting damage. It is required to conduct a risk assessment and full life-cycle analysis for NMs at all stages of products to understand the hazards of nanoproducts and the resultant knowledge that can then be used to predict the possible positive and negative impacts of the nanoscale products.

Generally, the book gives emphasis to current and future trends in vision of life sciences for synthesis processing and novel applications of nanoelectronic materials and nanodevices in industries and research laboratory.

Our recommendation for future work is that different reaction parameters such as temperature, pressure, time and pH can play important role in controlling the shape and morphology of the NMs and NPs, so that should be optimized for achieving specific characteristic product. Besides this, for good implications and properties study specific characterization techniques should be used. More importantly, environmental issues should be taken into account before using these materials for any applications, especially in case of heavy metals, which are prone to environmental hazards and can also affect the livings as well.

Keywords Nanotechnology · NMs · NSs · NSMs, NPs · NWs · NCMs · SiNWs · CPNWs · NTs · CNTs · DWNTs · CNT · 0D · 1D · 2D · 2D TMDCs · 3-D · TMDCs · DFT · CVD · CVC · GCP · HWCVD · Graphene · Fullerene · NanoES · NW-FET · NW-TFT · PVD · PV · QDs · SET · SiGe NWs · SSDs · SSPD · VLS · VS · VSS · SLS · NEMS · MEMS · FET · ISFET · EN · EDL · SNSPD · SPR · TMD-NDs · SiNWs · SiONWs · CMOS · CP NWs · ENIAC · HGNPs · HOPG · HS · LUMO · HOMO

A Look Ahead

Our book is clearly oriented to give a broad introduction to the field of nanomaterials to the topics, classification and most basic concepts of nanoscience. Book involves 17 chapters, which can satisfy many useful criteria: Development in Nanotechnology; Electronic properties of nanoscale devices; Classes of nanomaterials; Microelectronics, nanoelectronics and molecular electronics; Characterization and fabrication techniques; Classification of nanoelectronic materials; Synthesis techniques of NMs; 0D (quantum dot), 1D (Nanowire, Nanotube) and 2D (Thin films, few layer); Properties and physicochemical properties of NPs; Applications and Novel applications of NMs; Certain toxicities of NPs; Current and future trends for NMs and nano devices; Fundamentals of nanoelectronics in biological system; Vision for Life Sciences.

Nanomaterials are all around us, which exist in nature at our life environment.

We begin by the periodical various stages of development in nanotechnology. Our study of nanostructured materials (NSM) begins with giving a brief of some definitions: nanotechnology, nanoscale, nanoscience, nanochemistry, nanoparticles, nanowires, nanorodes, nanofibers and nanotubes with referring to the fundamental issues in nanomaterials.

During this book, we answered many questions such as:

- What is different at the nanoscale?
- What is new about nanoscience?
- What are nanomaterials NMs?
- What are the fundamental issues in nanomaterials?
- Where are nanomaterials found?
- What nanomaterials exist in nature?
- What is the importance of NSMs in our life?
- Why is there so much interest in nanomaterials?
- What is at nanoscale in nanomaterials?
- Why and what is graphene?
- Are pure low-dimensional systems interesting and worth pursuing?
- Are nanotechnology products available?

- What are sensors?
- How Artificial Intelligence (AI) and Nanotechnology do They Work Together?
- What are the recent advances in nanoelectronic materials?
- What are the novel applications of NMs?

Principles of computational simulations devices and characterization of nanoelectronic materials are explained.

We describe the role of (NSM) in our modern society, by studying the nature, unique physical properties and its types by giving examples of nanostructured materials.

Next, we build up our vocabulary in learning the role of surfaces and interfaces, which is the study of a nanostructure, with the diameter of the order of a nanometer (10^{-9} meters). It can also be defined as the ratio of the length to width being greater than 1000.

We study in detail allotropic forms of elemental carbon nanomaterials (graphite, diamond, fullerene, carbon nanotubes and graphene), by giving their industrial applications.

Two-dimentional transition metal dichalcogenides (2D TMDCs), synthesis methods and properties of 2D nanostructured materials are investigated.

Next, we examine the classification of nanomaterials according to Gleiter's classification and dimensionality depending on the number of reduced dimensions, which usually refers to the number of degrees of freedom in the momentum from 3D to 0D.

Nanowires (NWs) are divided as metallic, semi-conducting, insulating and molecular, which involves repeating organic or inorganic molecular units.

We learn in detail the physics and ways of processing nanostructured materials, either by bottom-up and/or top-down approaches as techniques for elaboration and synthesis of nanostructured materials and nanowires.

We continue our study of physical properties (morphology, lattice parameter and the phase changes), electronic, magnetic, optical, electrical and mechanical (elasticity, plasticity, dislocations, hardening, twinning, toughness and ductility) properties of nanomaterials.

We then extend our discussion to the backbone in environmental impact of nanotechnology research involving classical and kinetic Monte-Carlo methods.

We then become acquainted with some present, future and novel applications of nanomaterials and nanodevices of nanotechnology for the chemical industry, environmental pollution monitoring, corrosion protection of metals and the human health fields, which can be precisely controlled during synthesis, chemical composition, length, diameter and doping/electronic properties.

We present a vision for Life Sciences at nanomedicine through interfaces between nanoelectronic and biological systems.

We give conclusion about nanomaterials and nanotechnology, which has been playing an increasingly important role in supporting innovative technological advances.

With the increasing use of nanomaterials for commercial and industrial purposes, the debate becomes whether the numerous benefits of nanoparticles can overcome the economic costs, environmental impacts and unknown risks resulting from their use.

The risk assessment of such nanomaterials requires evaluation of their mobility, reactivity, environmental toxicity and stability. An understanding of the interactions between nanoparticles and biological systems is of significant interest.

To date, few studies have been conducted on the toxic and environmental effects that result from direct and indirect exposure to nanoparticles, and there are no clear standards to determine their effects. Lack of technical information in this regard has provided an appropriate context for supporters and opponents of nanoparticles to present contradictory and ill-considered results. Such an uncertain atmosphere has caused increased concerns about the effects of nanoparticles. Therefore, adequate studies to determine the exact, real risks of the use of nanomaterials are required. The information resulting from these studies can be useful in minimizing the environmental hazards that could arise from the use of nanomaterials. Thus, this book briefly explains the classification of environmental nanomaterials and how to deal with their formation, diffusion, environmental fate and impacts, and our exposure to them.

Generally, the book gives emphasis to current and future trends in vision of life sciences for synthesis processing and novel applications of nanoelectronic materials and nanodevices in industries and research laboratory.

In general, my book has been satisfying the following criteria:

(1) Summarize the fundamentals and technical approaches in processing and behaviour of nanostructured materials to provide the readers systematic, comprehensive and brief information in the challenging field of nanomaterials NMs and nanotechnology. Gives a clear vision for the types, classification, synthesis, properties, applications and novel applications of all types of nanomaterials among the researchers.

(2) The book has discussed the interfaces between biology systems and nanoelectronics devices by giving some essential applications examples.

(3) The book demonstrates the structural stability, electronic and mechanical properties of the nanomaterials and gives emphasis to applications of them. These information give a clear vision for synthesis and applications of nanomaterials among the researchers.

(4) The book discussed various preparation techniques for low dimensional nano materials including 0D (quantum dot), 1D (Nanowire, Nanotube) and 2D (Thin films, few layer) and their potential application in nanoelectronic system. The book deals with the different techniques of synthesis of quantum

well, quantum wire and quantum dot materials. Mechanism of the various nano-materials.

(5) The book gives the structural stability, electronic and mechanical properties of these materials and gives emphasis to applications of nanomaterials.

(6) Allotropic forms of elemental carbon nanomaterials (fullerene, nanodiamond, graphite, carbon nanotubes CNT, graphene) and two-dimensional transition metal dichalcogenides (2D TMDCs) with their applications are discussed in details. It gives examples of nanomaterials with various morphologies.

(7) The book gives the numerical calculation of the electronic properties and in particular the quantum transport properties of devices at the nanoscale.

(8) This book has been concerned with the size effects and going from bulk materials to nanomaterials, electronic properties of nanoscale devices, different classes of nanomaterials and their electronic characteristics, thermal, chemical and mechanical stability of nanomaterials. Characterization and fabrication techniques from lab-scale to mass-production and functionality of nanomaterials by chemical or defect engineering.

(9) The book demonstrates the structural stability, physical, chemical, magnetic, optical, electrical, thermal, electronic, mechanical properties of nanomaterials and physicochemical properties of NPs.

(10) The book gives emphasis to present and future novel applications of nanomaterials and nanodevices in industries and research laboratory.

(11) The book demonstrates and gives a clear vision for synthesis and applications of nanomaterials among the researchers. It is necessary to handle systems comprising millions of atoms, and this will require new efficient algorithms for the most time-consuming stages of the calculations.

(12) The book discusses the advantages and disadvantages of using nanomaterials; there are certain toxicities which are associated with NPs and other nanomaterials and basic knowledge is required for these toxic effects to encounter them properly.

(13) The book has answered about almost many essential and important questions throughout its 17 chapters contained for all various classes of nanomaterials.

(14) Generally, the book gives emphasis to current and future trends to vision for life sciences of synthesis processing and novel applications of nanomaterials and nanodevices in industries and research laboratory. It is considered actually as an essential reference for Nanoelectronic Materials: Fundamentals and Applications.

Contents

About the Author

Prof. LOUTFY H. MADKOUR: (Egyptian) has been a Professor of Physical Chemistry and Nano Science at *Chemistry Department, Faculty of Science and Arts, Al Baha University, Baljarashi, 65635, Saudi Arabia (KSA)*, since 2012 till now.

Scopus Author ID: Author ID: 57201889680
http://orcid.org/0000-0002-3101-8356
https://www.mendeley.com/profiles/prof-loutfy-h-madkour/
https://scholar.google.com/citations?hl=en&user=A378_tsAAAAJ&view_op=list_works&sortby=pubdate
https://www.researchgate.net/profile/Loutfy_Madkour
https://hindawi.academia.edu/LoutfyMadkour

He received his B.Sc., M.Sc. and Ph.D. from the Cairo, Minia and Tanta Universities of the AR Egypt, respectively in Physical Chemistry. He worked as a lecturer in chemistry at the Tanta University since 1982 and as a professor of physical chemistry in 1999. He has conducted a series of studies in the field of physical chemistry, electrochemistry, corrosion science, density functional theory, molecular dynamic simulation, quantum, theoretical chemistry, chemical equalization principles, nanoscience, nanotechnology, nanomedicine, electrometallurgy, electroanalytical chemistry, analytical chemistry, polarography, electrolytic extraction of heavy metals from natural ores and deposits, electrochemical thermodynamics and environmental chemistry. His earlier research accomplishments include Biosynthesis of Metallic Nanoparticles (MNPs)s and toxicology studies for Pharmacological applications in medicine and therapy. He has published 150 peer-reviewed original research articles, 11 review articles, and 4 books in the area of physical chemistry, practical and applied chemistry, corrosion science, nanoscience and nanomedicine. He is serving in different positions in Egypt, Kuwait, Yemen and Saudi Arabia.

He is appointed as the prestigious *Editorial Board member* of several international journals such as: International Journal of Industrial Chemistry (IJIC) Published by the SPRINGER; International Journal of Ground Sediment & Water; E-Cronicon Chemistry (EC Chemistry); BAOJ Chemistry; Global Drugs and Therapeutics (GDT); Chronicles of Pharmaceutical Science (ISSN 2572-7761) journal; Journal of Targeted Drug Delivery. UNIQUE PUB INTERNATIONAL UPI Journal of Pharmaceutical, Medical and Health Sciences; and Global Journal of Nanomedicine (GJN) ISSN: 2573-2374; Research and Reviews in Computational Chemistry Journal and Journal of Pharmaceutical sciences-current research (JPSCR) in the area of Pharmaceutical, Pharmacology, Journal of Pharmacology & Pharmaceutical Research (JPPR); Archives of Pharmacy & Pharmacology Research journal- APPR and Drug Designing & Development (NAPDD) website; the Pharmaceutical Sciences & Analytical Research Journal (PSARJ); Nanotechnology & Applications (NTA); the Clinical Pharmacology and Toxicology Research; The Laboratory Medicine Journal; Journal Drug Design Development & Therapy MEDCRAVEONLINE.COM MOJDDT; Journal of Pharmacy and Drug Development. E Scientific Publishers and ADVANCES AND APPLICATIONS OF PHARMACY Journal published online by MDPI quarterly. He has joined as an Editorial Board member in Journal of Clinical and Medical Research; International Journal of Environmental Chemistry, Science PG Publishing Group and LOJ Pharmacology & Clinical Research (LOJPCR) website.

He is a Reviewer for many international Elsevier and Springer journals. He is a member for many interested international societies, including American Association for the Advancement of Science (AAAS), European Desalination Society (EDS), Egyptian Chemical Society (ECS), Egyptian Corrosion Bulletin Society and American Chemical Society (ACS).

E-mail address: loutfy_madkour@yahoo.com; lha.madkour@gmail.com; lmadkour@bu.edu.sa

* *Corresponding author*. Tel. +966 533899075; fax: +966 77247272 *Saudi Arabia (KSA)*

Editorial Board Member by Author

[1] International Journal of Industrial Chemistry (IJIC) Published by the Springer. http://www.springer.com/chemistry/industrial+chemistry+and+chemical+engineering/journal/40090?detailsPage=editorialBoard
[2] International Journal of Ground Sediment & Water.
http://www.directoryofscience.com/site/4549486
http://ijgsw.comze.com/
[3] Research and Reviews in Computational Chemistry.
http://www.rrcchemistry.com/index.php/rrcc/announcement/view/1
[4] Journal of Pharmaceutical Sciences - Current Research (JPSCR).
https://vagusinprosysonline.org/journals/journal-details/editorial-board.php?id=36
[5] Journal of Pharmacology & Pharmaceutical Research (JPPR).
http://researchopenworld.com/category/journal-of-pharmacology-pharmaceutical-research/
[6] E-Cronicon Chemistry (EC Chemistry).
https://www.ecronicon.com/chemistry-editorial-panel.php
https://www.ecronicon.com/chemistry.php
[7] BAOJ Chemistry.
http://bioaccent.org/chemistry/index.php
http://bioaccent.org/chemistry/editorialboard.php
[8] Global Drugs and Therapeutics (GDT).
http://www.oatext.com/Global-Drugs-and-Therapeutics-GDT.php#Editorial_Board
https://oatext.com/Global-Drugs-and-Therapeutics-GDT.php#Editorial_Board
[9] Chronicles of Pharmaceutical Science (ISSN 2572-7761) journal.
https://scientiaricerca.com/cops-eb.php
[10] Journal of Targeted Drug Delivery.
http://sciaeon.org/targeted-drug-delivery/editorial-board
[11] He is appointed as the prestigious editorial board member of UPI Journal of Pharmaceutical, Medical and Health Sciences.
https://uniquepubinternational.com/wp-content/uploads/2018/01/Dr.-Loutfy-H.-Madkour.pdf
[12] He has joined as an Editorial Board member in Global Journal of Nanomedicine (GJN) ISSN: 2573-2374. Published by Insights in Mining Science & Technology (IMST).
https://juniperpublishers.com/gjn/editorialboard.php
[13] He is appointed as an Editorial panel member for Archives of Pharmacy & Pharmacology Research journal - APPR.
https://irispublishers.com/appr/editorialboard.php
[14] He is joined as the editorial board member/reviewer of Science Journal of Chemistry.
http://www.journalchemistry.org/editorialboard

[15] He has recognized under “Quarterly Franklin Membership” (Membership ID#EI15469) in international, peer-reviewed, refereed journal, London Journal of Research in Science: Natural and Formal (LJRS).
support@journalspress.com
1210th, Arlington Business Park, Theale, United Kingdom.
[16] He has joined as an Editorial Board Member of journal entitled “Chemical & Pharmaceutical Research”.
http://www.jocpr.com/editorial-board.html
[17] He has joined as an Editorial Board member in Drug Designing & Development (NAPDD) website.
https://juniperpublishers.com/napdd/editorialboard.php
[18] He has uploaded as an Editorial Board member for the Pharmaceutical Sciences & Analytical Research Journal (PSARJ).
https://chembiopublishers.com/PSARJ/editorial-board.php
[19] Editor for Book: Corrosion Science: Theoretical and Practical Applications: Pub Date: December 2017 Hard ISBN: 9781771886024.
Published by Apple Academic Press, Inc. is an independent international publisher focusing on academic and professional books in STEM and other fields. With a focus on relevant content as well as first-class production, Apple Academic Press is dedicated to publishing cutting-edge, informative books written and edited by internationally renowned experts in their fields.
http://appleacademicpress.com/corrosion-science-theoretical-and-practical-applications-/9781771886024#bios
http://appleacademicpress.com/category.php?id=19
[20] He has joined as an Editorial Board member in Nanotechnology & Applications Journal (NTA).
http://www.scivisionpub.com/journals/editorialboard-nanotechnology-applications
[21] He has uploaded as an Editorial Board member for the Clinical Pharmacology and Toxicology Research.
https://www.pulsus.com/clinical-pharmacology-toxicology-research/editorial-board.html
[22] He has joined as an Editorial Board member in The Laboratory Medicine Journal.
https://makperiodicallibrary.com/thelaboratorymedicine/editorial-board/
[23] He has joined as an Honorable Editor for MOJ Drug Design Development & Therapy Journal Drug Design Development & Therapy.
https://medcraveonline.com/MOJDDT/editorial-board
[24] He has uploaded as an Editorial Board member for the Journal of Pharmacy and Drug Development. E Scientific Publishers. Is an Open Access Publication that aims to publish Scholarly Articles pertaining to research of Pharmacy and drug products?
https://escientificpublishers.com/editors/journal-of-pharmacy-and-drug-development

[25] He has joined as an Editorial Board member in Advances and Applications of Pharmacy Journal published online by MDPI quarterly.
https://researchnovelty.com/aap.php#
Website: www.researchnovelty.com
[26] He has joined as an Editorial Board member in Journal of Clinical and Medical Research.
https://www.redelve.com/editorialmember.php
[27] He has been appointed as the editorial board member of International Journal of Environmental Chemistry, Science PG Publishing Group.
http://www.sciencepublishinggroup.com/journal/editorialboard?journalid=292
[28] He has been uploaded as the Editorial Board member in LOJ Pharmacology & Clinical Research (LOJPCR) website.
http://www.lupinepublishers.com/pharmacology-clinical-research-journal/editorial-committee.php

Recent Published Research Articles by Author

[1] Loutfy H. Madkour. Cytotoxic Mechanisms of Heavy Metals Exposure Induced Oxidative DNA Damage and the Site of Reactive Oxygen Species (ROS) Production. Global Drugs and Therapeutics (2019) (in press).

[2] Loutfy H. Madkour. Nanoelectronics Applications of Biosensors in Macromolecules Living Organisms Cells. International Journal of Research Studies in Science, Engineering and Technology, Volume 6, Issue 4 (2019), pp. 23–43. http://ijrsset.org/pdfs/v6-i4/4.pdf

[3] Loutfy H. Madkour. Nanoparticles as Targeted Drug Co-Delivery in Cancer Therapeutics. Chronicles of Pharmaceutical Science Journal, Volume 3, Issue 2 (2019), pp. 800–804.
https://scientiaricerca.com/srcops/SRCOPS-03-00081.php

[4] Loutfy H. Madkour. Ecotoxicology-Nanotoxicology and Reactive Oxygen Species (ROS) Stress Combination of Free Radicals and Nanoparticles towards Antioxidant Defense Therapeutics. Journal of Targeted Drug Delivery, J Target Drug Deliv. Volume 3(1) (2019), pp. 1–58.

[5] Yue Xu, Shengtao Zhang, Wenpo Li, Lei Guo, Shenying Xu, Li Feng, Loutfy H. Madkour. Experimental and theoretical investigations of some pyrazolo-pyrimidine derivatives as corrosion inhibitors on copper in sulfuric acid solution. Applied Surface Science, Volume 459, 30 November (2018), pp. 612–620.
https://doi.org/10.1016/j.apsusc.2018.08.037
https://www.sciencedirect.com/science/article/pii/S0169433218321561

[6] Lei Guo, Min Wu, Savaş Kaya, Meihang Chen, and Loutfy H. Madkour. Influence of the alkyl chain length of alkyltriazoles on the corrosion inhibition of iron: A DFTB study. AIP Conference Proceedings 1995, 020015 (2018).
https://doi.org/10.1063/1.5048746
https://aip.scitation.org/doi/abs/10.1063/1.5048746
https://aip.scitation.org/doi/pdf/10.1063/1.5048746

[7] Li Feng, Shengtao Zhang, Yujie Qiang, Yue Xu, Lei Guo, Loutfy H. Madkour, and Shijin Chen. Experimental and Theoretical Investigation of Thiazolyl Blue as a Corrosion Inhibitor for Copper in Neutral Sodium Chloride Solution. Materials (Basel) (2018) Jun; 11(6): 1042. Published online (2018) Jun 19.
https://doi.org/10.3390/ma11061042
https://www.ncbi.nlm.nih.gov/pmc/articles/PMC6025645/pdf/materials-11-01042.pdf

[8] Y. Xu, S. Zhang, L. Guo, B. Tan, C. Liao, Y. Zhou, and Loutfy H. Madkour. Halogen-substituted pyrazolo-pyrimidine derivatives as corrosion inhibitors for copper in sulfuric acid solution. Int. J. Corros. Scale Inhib. (2018), 7(2) 236–249.
https://www.researchgate.net/publication/325644146_Halogen-substituted_pyrazolopyrimidine_derivatives_as_corrosion_inhibitors_for_copper_in_sulfuric_acid_solution

[9] Loutfy H. Madkour, Savaş Kaya, Lei Guo, Cemal Kaya. Quantum chemical calculations, molecular dynamic (MD) simulations and experimental studies of using some azo dyes as corrosion inhibitors for iron. Part 2: Bis–azo dye derivatives. Journal of Molecular Structure, 1163 (2018), 397–417.
https://doi.org/10.1016/j.molstruc.2018.03.013
https://www.sciencedirect.com/science/article/pii/S0022286018302965?dgcid=STMJ_73515_AUTH_SERV_PPUB_V38

[10] Loutfy H. Madkour, Savaş Kaya, Ime Bassey Obot. Computational, Monte Carlo simulation and experimental studies of some arylazotriazoles (AATR) and their copper complexes in corrosion inhibition process. Journal of Molecular Liquids, 260 (2018) 351–374.
https://doi.org/10.1016/j.molliq.2018.01.055
https://www.sciencedirect.com/science/article/pii/S0167732217335651#bg0005

[11] Lei Guo, Ying Gao, Yongxian Xu, Renhui Zhang, Loutfy H. Madkour, and Yingchang Yang. Understanding the corrosion behavior of amorphous multiple-layer carbon coating. Advances in Materials, Machinery, Electronics II: Proceedings of the 2nd International Conference on Advances in Materials, Machinery, Electronics (AMME 2018), Volume number: 1955 Published: Apr 18, 2018. AIP Conference Proceedings 1955, 020001 (2018).
https://doi.org/10.1063/1.5033573
https://aip.scitation.org/doi/pdf/10.1063/1.5033573

[12] Loutfy H. Madkour. Applications of gold nanoparticles in medicine and therapy. Pharmacy & Pharmacology International Journal (2018); 6(3) 157–174.
https://doi.org/10.15406/ppij.2018.06.00172
http://medcraveonline.com/PPIJ/PPIJ-06-00172.pdf

[13] Loutfy H. Madkour. Toxic Effects of Environmental Heavy Metals on Cardiovascular Pathophysiology and Heart Health Function: Chelation Therapeutics. UPI Journal of Pharmaceutical, Medical and Health Sciences (UPI–JPMHS) (2018); 1(1) 19–62.
https://uniquepubinternational.com/wp-content/uploads/2018/03/UPI-JPMHS-2018-7.pdf

[14] Loutfy H. Madkour. Biogenic–Biosynthesis Metallic Nanoparticles (MNPs) for Pharmacological, Biomedical and Environmental Nanobiotechnological Applications. Chronicles of Pharmaceutical Science Journal, Volume 2, Issue 1 (2018), pp. 384–444.
https://scientiaricerca.com/srcops/SRCOPS-02-00038.php

[15] Loutfy H. Madkour. Ecofriendly green biosynthesized of metallic nanoparticles: Bio-reduction mechanism, characterization and pharmaceutical applications in biotechnology industry. Global Drugs and Therapeutics (2018), Volume 3 (1): 1–11.
http://www.oatext.com/ecofriendly-green-biosynthesized-of-metallic-nanoparticles-bio-reduction-mechanism-characterization-and-pharmaceutical-applications-in-biotechnology-industry.php

[16] Loutfy H. Madkour. *Review Article*: Advanced AuNMs as nanomedicine's central goals capable of active targeting in both imaging and therapy in biomolecules. Global Drugs and Therapeutics (2017), Volume 2(6): 1–12.
http://www.oatext.com/advanced-aunms-as-nanomedicines-central-goals-capable-of-active-targeting-in-both-imaging-and-therapy-in-biomolecules.php

[17] Loutfy H. Madkour. Biotechnology of Nucleic Acids Medicines as Gene Therapeutics and Their Drug Complexes. Chronicles of Pharmaceutical Science Journal, Volume 1, Issue 4 (2017), pp. 204–253.
https://scientiaricerca.com/srcops/pdf/SRCOPS-01-00023.pdf

[18] Loutfy H. Madkour. Advanced AuNMs as Nanomedicine's central Goals Capable of Active Targeting in Both Imaging and Therapy in Biomolecules. Bio Accent Online BAOJ Nanotechnology, (2017), Volume 3, Issue 1, 015, 1–18.
https://bioaccent.org/nanotechnology/nanotechnology15.pdf

[19] Loutfy H. Madkour. Vision for life sciences: interfaces between nanoelectronic and biological systems. Global Drugs and Therapeutics, 2(4), (2017), 1–4.
https://doi.org/10.15761/gdt.1000126
https://oatext.com/Vision-for-life-sciences-interfaces-between-nanoelectronic-and-biological-systems.php

[20] Loutfy H. Madkour, Savaş Kaya, Cemal Kaya, Lei Guo. Quantum chemical calculations, molecular dynamics simulation and experimental studies of using some azo dyes as corrosion inhibitors for iron. Part 1: Mono-azo dye derivatives. Journal of the Taiwan Institute of Chemical Engineers, 68, (2016), 461–480.
84993118310
http://www.sciencedirect.com/science/article/pii/S1876107016303522

[21] Loutfy H. Madkour, I. H. Elshamy. Experimental and computational studies on the inhibition performances of benzimidazole and its derivatives for the corrosion of copper in nitric acid. International Journal of Industrial Chemistry (IJIC), Volume 7, Number 2 (2016) 7, 195–221.
84971325185
http://link.springer.com/article/10.1007/s40090-015-0070-8

[22] Savaş Kaya, Cemal Kaya, Lei Guo, Fatma Kandemirli, Burak Tüzün, İlkay Uğurlu, Loutfy H. Madkour, Murat Saraçoğlu. Quantum chemical and molecular dynamics simulation studies on inhibition performances of some thiazole and thiadiazole derivatives against corrosion of iron. Journal of Molecular Liquids, 219, (2016), 497–504.
84962828467
http://www.sciencedirect.com/science/article/pii/S016773221531254X

[23] Loutfy H. Madkour, S. K. Elroby. Inhibitive properties, thermodynamic, kinetics and quantum chemical calculations of polydentate Schiff base compounds as corrosion inhibitors for iron in acidic and alkaline media. International Journal of Industrial Chemistry (IJIC), Volume 6, Number 3 (2015) 6: 165–184.
84940640561
http://link.springer.com/article/10.1007/s40090-015-0039-7

[24] Loutfy H. Madkour. Electro-Thermal and Semiconductivity Behaviour of Natural Sintered Complex Carbonate Ore for Thermo-Technological Applications. Hindawi Publishing Corporation. Journal of Geochemistry Volume (2014), Article ID 451782, 10 pages.
https://www.hindawi.com/archive/2014/451782/

[25] Loutfy H. Madkour, S. K. Elroby. Correlation between corrosion inhibitive effect and quantum molecular structure of Schiff bases for iron in acidic and alkaline media. Standard Scientific Research and Essays, Vol 2(13): 680–704, December Special Issue (2014) (ISSN: 2310-7502).
http://standardresearchjournals.org/journals/SSRE/Pdf/2014/december_special_issue/Madkour%20and%20Elroby.pdf

[26] Loutfy H. Madkour, U. A. Zinhome. Inhibition, kinetic and thermodynamic effects of new Azo derivatives on iron corrosion in acidic and alkaline solutions. Standard Scientific Research and Essays, Vol 2(13): 705–724, December Special Issue (2014) (ISSN: 2310-7502).
http://standardresearchjournals.org/journals/SSRE/Pdf/2014/december_special_issue/Madkour%20and%20Zinhome.pdf

[27] Loutfy H. Madkour. Electro-Thermal and Semiconductivity Behaviour of Natural Sintered Complex Carbonate Ore for Thermo-Technological Applications. International Research Journal of Geology and Mining (IRJGM) (2276-6618), Vol. 4(1) pp. 37–50, January (2014).
https://doi.org/10.14303/irjgm.2013.034
Copyright © (2014) International Research Journals.
http://www.interesjournals.org/irjgm/january-2014-vol-4-issue-1/electro-thermal-and-semiconductivity-behaviour-of-natural-sintered-complex-carbonate-ore-for-thermo-technological-applications

[28] Loutfy H. Madkour, S. K. Elroby. Aminic nitrogen- bearing polydentate Schiff base compounds as corrosion inhibitors for iron in acidic and alkaline media: A combined experimental and DFT studies. Journal of Corrosion Science and Engineering, Volume 17, (2014).
84892389669

[29] Loutfy H. Madkour, U. A. Zinhome. Adsorption and corrosion inhibitive properties, thermodynamic and quantum chemical studies of Polydentate Schiff base Compounds (PSCs) on iron electrode in acidic and alkaline media. Merit Research Journal of Engineering, Pure and Applied Sciences, Vol. 1(1) pp. 01–27, October (2013).
http://www.meritresearchjournals.org/epas/index.htm

[30] Loutfy H. Madkour, U. A. Zinhome. Kinetic-thermodynamic studies of substituted mono- and bis- azo dyes as corrosion inhibitors for iron in nitric acid and sodium hydroxide solutions. Journal of Corrosion Science and Engineering, Volume 15, (2012).
84870359337

Recent Published Books by Author

[1] Book Title: Practical and Applied Chemistry
Published By: The Public Authority for Applied Education and Training, Kuwait, first edition (1993), 5 Chapters, and 565 large size pages.

[2] Book Title: Corrosion Science: Theoretical and Practical Applications
Pub Date: December (2017)
Hard ISBN: 9781771886024
By (authors): Savaş Kaya, Ime Bassey Obot, Ph.D., and Loutfy H. Madkour, Ph.D.
Published by: Apple Academic Press Inc. is an independent international publisher focusing on academic and professional books in STEM and other fields. With a focus on relevant content as well as first-class production, Apple Academic Press is dedicated to publishing cutting-edge, informative books written and edited by internationally renowned experts in their fields.
http://appleacademicpress.com/corrosion-science-theoretical-and-practical-applications-/9781771886024#bios
http://appleacademicpress.com/category.php?id=19

[3] Book Title: Toxic Effects of Environmental Heavy Metals on Cardiovascular Pathophysiology and Heart Health Function: Chelation Therapeutics
By (author): Loutfy H. Madkour
Published by: LAP Lambert Academic Publishing (2018-10-15)
Book Details:
Publishing house: LAP LAMBERT Academic Publishing
Website: https://www.lap-publishing.com/
ISBN-13: 978-613-9-92290-1
ISBN-10: 6139922909
EAN: 9786139922901
Book language: English
Published on: 2018-10-15
Keywords: Heavy metals, Cardiovascular Disease, oxidative stress, ROS, EDTA, Curcumin chelation.
https://www.morebooks.shop/store/gb/book/toxic-effects-of-environmental-heavy-metals-on-cardiovascular-pathophy/isbn/978-613-9-92290-1

[4] Book Title: Nanoelectronic Materials
Book Subtitle: Fundamentals and Applications
By (author): Loutfy H. Madkour
Series Title: Advanced Structured Materials
Series Volume: 116
Copyright: 2019
Publisher: Springer International Publishing
Copyright Holder: Springer Nature Switzerland AG
eBook ISBN: 978-3-030-21621-4
DOI: 10.1007/978-3-030-21621-4
Hardcover ISBN: 978-3-030-21620-7
Series ISSN: 1869-8433

Edition Number: 1
Number of Pages: XLIII, 777
Number of Illustrations: 122 b/w illustrations, 494 illustrations in colour
Topics: Nanotechnology

[5] Book Title: Nucleic Acids as Gene Anticancer Drug Delivery Therapy
By (author): Loutfy H. Madkour
ISBN: 9780128197776
Publishing house: ELS USA (Elsevier Inc-US) (2020) (in press).

[6] Book Title: Heavy Metals and Nanoparticles–Induced Reactive Oxygen Species (ROS): Antioxidants Defenses Therapy
By (author): Loutfy H. Madkour
Publishing house: ELS USA (Elsevier Inc-US) (2020) (under publication).

Abbreviations

0-D	Zero-dimensional (quantum dot: all dimensions at the nanoscale)
1-D	One-dimensional (e.g. surface films, nanowire, nanotube: two dimensions at the nanoscale and one dimension at the macroscale)
2-D	Two-dimensional (e.g. strands or fibers, thin films, few layer: one dimension at the nanoscale and two dimensions at the macroscale)
2D TMDCs	Two-dimensional transition metal dichalcogenides
3-D	Three-dimensional space (e.g. Particles: No dimensions at the nanoscale, all dimensions at the macroscale)
AES	Auger electron spectroscopy
AFM	Atomic force microscopy
AI	Artificial Intelligence
ALD	Atomic layer deposition
AMBIO	Advanced Nanostructured Surfaces for the Control of Biofouling
ANNs	Artificial neural networks
ANTT	Active nanotube transistor
a-Si	Amorphous silicon
ATP	Adenosine triphosphate
AuNPs	Gold nanoparticles
BET	Brunauer–Emmett–Teller
BHJ	Bulk heterojunction
C_{60} and C_{70}	Fullerene with the diameter of 7.114 and 7.648 nm
C_{82} or C_{84}	Endohedral fullerenes (larger fullerenes)
CBE	Chemical beam epitaxy
CBM	Conduction band minimum
C_e	Electron specific heat
CFL	Compact fluorescent lamp
CHEM-FET	Chemical field effect transistor
CIGS	Copper indium gallium (di) selenide
CMOS	Complementary metal oxide semiconductor
CNT	Carbon nanotubes

c_p	Phonon specific heat
CP NWs	Conducting polymer nanowires
CRT	Cathode ray tube
CS	Coconut shell
CSD	Cell-surface display
CVD	Chemical vapor deposition
DC	Direct Current
DCR	Dark count rate
DDU	Density gradient ultracentrifugation
DFT	Density functional theory
DNA	Deoxyribonucleic acid
DOS	density of states
DPN	Dip-pen nanolithography
DRIE	Deep reactive-ion etching
DRS	UV/vis-diffuse reflectance spectrometer
DS	Dodecyl sulfate
DWNTs	Double-walled carbon nanotubes
EBID	Electron beam induced deposition
ECAP	Equal-channel angular pressing
ECMs	Extracellular matrices
EDAX	Energy Dispersive X-Ray Spectroscopy
EDL	Electric double layer
EDX	Energy dispersive X-ray
EELS	Electron energy loss spectroscopy
EFM	Electrostatic force microscopy
EMC	Ensemble Monte Carlo model
EN	Engineered nanomaterials
ENA	Electrochemical noise analysis
ENIAC	Electronic Numerical Integrator and Computer
EOT	Equivalent oxide thickness
EQE	External quantum efficiency
ETE	Energy transfer efficiency
FAS	Field assisted sintering
FEL	Free Electron Laser
Fe_2O_3	Maghemite
Fe_3O_4	Magnetite
FCWLEDs	Flip-chip white light-emitting diodes
FEDs	Field Emission Displays
FET	Field-effect transistors
FIB	Focused ion beam
FQM	Fluorescence Quenching Microscopy
FRET	Fluorescence resonance energy transfer
FWHM	Full width at half maximum
GaAs	Gallium arsenide
GBs	Grain boundaries

GCP	Gas Condensation Processing
GO	Graphene oxide
GPa	Gigapascals
H	Hexagonal
HAGO	Hydrogenation assisted graphene origami
hBN	Hexagonal boron nitride
HDTVs	High definition televisions
HGNPs	Hollow gold NPs
HIP	High isostatically pressing
HJBT	Heterojunction bipolar transistors
HK	High-k materials
HOMO	Highest occupied molecular orbital
HOPG	Highly oriented pyrolytic graphite
HRS	High resistance state
HRTEM	High resolution transmission electron microscopy
HS	Humic substances
HSZ	High silica zeolites
HWCVD	Hot Wire filament chemical vapor deposition
IC	Integrated circuit level
ICMA	International Conference on Mechatronics and Automation
ICP	Inductively coupled plasmaoptical emission spectrometry
IEA	International Energy Agency
IGC	Inert gas condensation
IM	Irresistible Materials Ltd.
IQE	Internal quantum efficiency
IR	Infrared
I_r	Return current
ISFET	Ion-selective field effect transistor
ITO	Indium tin oxide
κ	Thermal conductivity
L	Channel length (L) of the transistor
LCD	Liquid-crystal displays
LDL	Low density lipoprotein
LEDs	Light-emitting diodes
LEED	Low energy electron diffraction
L_{litho}	Lithographic limit
LOT	Lateral olfactory tract
LPCVD	Performed at low pressure chemical vapor deposition
L_{phys}	Physical limit
LPNE	Lithographically patterned nanowire electrodeposition
LRS	Low resistance state
LSPR	Localized Surface Plasmon resonance
LSZ	Low-silica zeolite
LUMO	Lowest unoccupied molecular orbital
MA	Mechanical attrition

MBE	Molecular beam epitaxy
μc-Si	Microcrystalline silicon
MD	Molecular Dynamic Simulations
MEAs	Metal electrodes and arrays
MEMS	Microelectronic-mechanical systems
mfp	Mean free paths
MFCs	Mass flow controllers
Mg-Ni	Magnesium-nickel
MIT	Massachusetts Institute of Technology
MNPs	Magnetic nanoparticles
MOCVD	Metal-organic chemical vapor deposition
MODFET	Modulation-doped field effect transistor
MOEMS	Micro-optoelectronic- mechanical systems
MOFs	Metal organic frameworks
MOMBE	Metallorganic molecular beam epitaxy
MoS_2	Molybdenum disulfide
$MoSe_2$	Molybdenum diselenide
MOSFET	Metal-oxide semiconductor field-effect transistor
MPs	Microparticles
MQWs	Multiple quantum wells
MRI	Magnetic resonance imaging
MSM-UPDs	Metal-semiconductor-metal ultraviolet photo detectors
MVs	Mono-vacancies
MWCNTs	Multi-walled carbon nanotubes
MWNTs	Multiwalled nanotubes
NanoES	Nanoelectronic scaffolds
NBs	Nanobelts
Nc	Nanocrystalline materials
NCMs	Nanocomposites
NEGF	Nonequilibrium Green's function
NEMS	Nanoelectromechanical system
nm	Nanometer
NMs	Nanomaterials
NNI	National Nanotechnology Initiative
NOR	NOT-OR
NPs	Nanoparticles
NS-SC	Nanostructured semiconducting
NSs	Nanostructures
NSMs	Nanostructured materials
NT	Nanotechnology
NTs	Nanotubes
NVM	Nonvolatile memory
NWs	Nanowires
NW-FET	Nanowire-based field-effect transistor
NW-TFT	Nanowire-based thin-film transistor

OFETs	Organic field-effect transistors
OLED	Organic light emitting diodes
OPVs	Organic photovoltaic devices
PANA	Polymerizing anthranilic acid
PANA-Fe	Polyanthranilic acid iron Nano composite
PANA-Mg	Polyanthranilic acid-magnesium Nanocomposite
PANA-Zn	Polyanthranilic acid-Zinc Nano composite
PAW	Projector augmented wave
PC	Phosphatide-choline vesicles
PCA	Principal component analysis
PCE	Power conversion efficiency
PCBM	Phenyl C61-butyric acid methyl ester
PEC	Photo electrochemical
PECS	Pulsed electric current sintering
PECVD	Plasma-Enhanced chemical vapor deposition
PEG	Polyethylene glycol
PEN	Poly (ethylene naphthalate)
PEO	Polyethylene oxide
PET	Poly (ethylene terephthalate)
PD	Phage display
PGM	Platinum group metals
PL	Photoluminescence
PLA	Polylactic acid
PMMA	Poly(methyl methacrylate)
pMOSFET	p-type metal-oxide semiconductor field-effect transistor
PMTs	Photomultipliers
Pn	Pico Newton's
PNCA	Plasma Nanoscience Center Australia
PNP	polymer nanoparticle
PNRSNSPD	photon-number-resolving superconducting nanowire single-photon detector
POM	Polarized optical microscopy
PSA	Prostate specific antigen
PSPR	Propagating Surface Plasmon resonance
$P(t)$	Timedependent power of radiation
PV	Photovoltaic
PVD	Physical vapor deposition
PZT	Lead zirconium titania
QDs	Quantum dots
QKD	Quantum key distribution
R	Rhombohedral
RBAI	Root bark extract of *Azadirachta indica*
RES	Reticulum endothelial system
RET	Resonant energy transfer
RF	Radio-Frequency

RHEED	Reflection high-energy electron diffraction
RNA	Ribonucleic acid
Rpo	The coating resistance
RRAM	Resistive random access memory
R_{sheet}	Sheet resistance
RTA	Rapid thermal annealing
RTN	Random telegraph noise
SAED	Selected area electron diffraction
SAXS	Small-angle X-ray scattering
SBU	Secondary builing units
SDA	Structure-directing agent
SEI	Solid electrolyte interface
SEM	Scanning Electron Microscope analysis
SET	Single-Electron Transistor
SHG	Second harmonic generation
SHIBL	Scanning He^+ ion beam lithography
SI	International System of units
SiC	Silicon carbide
SiGe	Silicon-germanium
Si_3N_4	Silicon nitride
SiNWs	Silicon nanowires
SiONWs	Silica nanowires
SKPM	Scanning Kelvin probe microscopy
SLS	Solid-Liquid-Solid
SNAP	Superconducting nanowire avalanche photodetector
S/N	Signal-to-noise ratio
SnO_2	Tin dioxide
SNOM	Scanning near-field optical microscopy
SNSPD	Superconducting nanowire single-photon detector
SPAD	Single photon avalanche photodiodes
SPD	Single-photon detector
SPF	Sun protection factor
SPM	Electrical scanning probe microscopy techniques
SPR	Surface Plasmon resonance
SPS	Spark plasma sintering
SPW	Surface Plasmon wave
SRAM	Static random access memory
SSDs	Self-switching diodes
SSPD	Superconducting single-photon detector
STEM	Scanning transmission electron microscopy
SSM	Simulation of styrene molecules
STM	Scanning tunneling microscopy
STS	Scanning tunneling spectroscopy
SWCNTs	Single-walled carbon nanotubes
SWNTs	Single-walled carbon nanotubes

T	Tetragonal
τ	Recovery time
T_0	Substrate temperature
TCO	Transparent conductive oxide
TCSPC	Time-correlated single-photon counting
TDDFT	Time dependent DFT
T_e	Electron subsystem\
T_{es}	Phonon escape time
TEM	Transmission electron microscopy
TEOS	Tetraethyl orthosilicate
TERS	Tip-Enhanced Raman Spectroscopy
TFBGs	Tilted fiber Bragg gratings
TFTs	Thin film transistors
THG	Third harmonic generation
TiO_2	Titanium dioxide
TMDs	Transition metal dichalcogenides
TMD-NDs	Transition-metal dichalcogenide nanodots
TNT	Tri-nitro toluene
T_p	Phonon subsystem
TPR	Transient photoimpedence response
UFM	Ultrasonic force microscopy
UHV	Ultrahigh vacuum
ULCI	Ultra-large integrated circuits
UV	Ultraviolet
UV–Vis	Ultraviolet–visible
VASP	Vienna ab initio simulation package
VBM	VALENCE band maximum
VDS	Drain-source voltage
vdW	Van der Waals
VLS	Vapor—liquid—solid
VS	Vapor-Solid
VSS	Vapor-solid-solid
WLEDs	White light-emitting diodes
WO_3	Tungsten oxide
WS_2	Tungsten disulfide
WSe_2	Tungsten diselenide
XPS	X–ray photoelectron spectroscopy
XRD	X–ray diffraction
ZnO	Zinc oxide

Chapter 1
Introduction to Nanotechnology (NT) and Nanomaterials (NMs)

Abstract Nanotechnology (NT) is the complex interdisciplinary science including nanoscience, nanochemistry, nanophysics, nanomaterials, nanoelectronics, nanometrology, nanobionics, etc. Nanotechnology is a relatively new branch of science that has found a wide range of applications that range from energy production to industrial production processes to biomedical applications. Nanomaterials (NMs) can be engineered to possess unique composition and functionalities, which can provide novel tools and techniques. In this chapter, we aim to discuss the key of many concepts from nanoscale, and nanostructures. Periodical development in nanotechnology and areas of study within nanotechnology have been discussed. The fundamental issues in nanomaterials are considered. We will attempt to correlate the improvements in the material properties that are achieved due to the fine microstructures arising from the size of the grains and/or dimensionality.

Keywords Nanotechnology · Nanoscience · Nanochemistry · MEMS · MOEMS · Nanoparticles · Nanowires · Nanotubes · Nanorodes · Nanofibers

Nanotechnology deals with materials in the size of 0.1–100 nm; however it is also inherent that these materials should display different properties such as electrical conductance chemical reactivity, magnetism, optical effects and physical strength, from bulk materials as a result of their small size. Nanotechnology works on matter at dimensions in the nanometer scale length (1–100 nm), and thus can be used for a broad range of applications and the creation of various types of nano materials and nano devices.

L. H. Madkour, *Nanoelectronic Materials*, Advanced Structured Materials 116, https://doi.org/10.1007/978-3-030-21621-4_1

1.1 Nanotechnology Debate

Nano: Greek prefix which means dwarf. The term nanotechnology comes from the combination of two words: the Greek numerical prefix nano referring to a billionth and the word technology. As an outcome, Nanotechnology or Nanoscaled Technology [1–7] is generally considered to be at a size below 0. 1 μm or 100 nm (a nanometer is one billionth of a meter, 10^{-9} m). Nanotechnology, shortened to "nanotech", is the study of the controlling of matter on an atomic and molecular scale. Generally nanotechnology deals with structures of the size 100 nm or smaller in at least one dimension, and involves developing materials or devices within that size. Nanotechnology range is very wide from the smallest particles that cannot be seen by naked eye to the food you ate and clothes you wear.

Definition

Nanotechnology can be defined as:

1. Research and technology development at the atomic, molecular or macromolecular levels, in the length scale of approximately 1–100 nm
2. Creating and using structures, devices and systems that have novel properties and functions because of their small and/or intermediate size
3. Ability to control or manipulate on the atomic scale.

History of Nanotechnology

The development in the field of nanotechnology started in 1958 and the various stages of development have been summarized in Table 1.1.

Nanotechnology is one of the leading scientific fields today since it combines knowledge from the fields of Physics, Chemistry, Biology, Medicine, Informatics, and Engineering. It is an emerging technological field with great potential to lead in great breakthroughs that can be applied in real life. Novel nano and biomaterials, and nanodevices are fabricated and controlled by nanotechnology tools and techniques, which investigate and tune the properties, responses, and functions of living and non-living matter, at sizes below 100 nm. The application and use of nanomaterials in electronic and mechanical devices, in optical and magnetic components, quantum computing, tissue engineering, and other biotechnologies, with smallest features, widths well below 100 nm, are the economically most important parts of the nanotechnology nowadays and presumably in the near future. The number of nanoproducts is rapidly growing since more and more nanoengineered materials are reaching the global market. The continuous revolution in nanotechnology will result in the fabrication of nanomaterials with properties and functionalities which are going to have positive changes in the lives of our citizens, be it in health, environment, electronics or any other field. In the energy generation challenge where the conventional fuel resources cannot remain the dominant energy source, taking into account the increasing consumption demand and the CO_2 emissions alternative renewable energy sources based on new technologies have to be promoted. Innovative solar cell technologies that utilize nanostructured materials and

Table 1.1 Periodical development in nanotechnology

Year	Development in nanotechnology
3.5 Mrd. Years	Procaryontic cells with nano machines
400 B.C.	Demokrit: Reasoning about atoms and matter
1905 A.D.	Albert Einstein: Calculate molecular diameter
1931	Max Knoll and Ernst Ruska: Electron microscope
1959	Richard Feynman: "There's Plenty of Room at the Bottom"
1968	Alfred Y. Cho and John Arthur (Bell Labs): MBE (atomic layer growth)
1974	The term nanotechnology was used by Taniguchi for the first time. Norio Taniguchi: Nanotechnology for fabrication methods below 1 μm
1981	IBM Scanning Tunneling Microscope. Gerd Binning & Heinrich Rohrer: Nobel Prize for Scanning Tunneling Microscope
1985	"Bucky Ball". Robert F. Curl, Harold W. Kroto, Richard Smalley: Buckminster fullerenes (Bucky balls)
1986	First book on nanotechnology Engines of Creation published by K. Eric Drexler, Atomic Force Microscope. K. Eric Drexler: Engines of Creation
1989	IBM logo was made with individual atoms. M. Eigler: Writing with a STM tool
1991	S. Iijima discovered Carbon Nano tube for the first time. Sumio Iigima (NEC): Carbon nanotubes
1993	Warren Robinett, R. Stanley Williams: Combination of SEM and VR (virtual reality system)
1998	Cees Dekker et al.: Carbon nano tube transistor
1999	James M. Tour and Mark A. Read: Single molecule switch. 1st nano medicine book by R. Freitas "Nano medicine" was published
2000	For the first time National Nanotechnology Initiative was launched. Eigler et al.: Construction of quantum corrals and quantum mirrors
2001	For developing theory of nanometer-scale electronic devices and for synthesis and characterization of carbon nanotubes and nano wires, Feynman Prize in Nanotechnology was awarded. Florian Bamberg: Soldering of nanotubes with e-beam
2002	Feynman Prize in Nanotechnology was awarded for using DNA to enable the self-assembly of new structures and for advancing our ability to model molecular machine systems
2003	Feynman Prize in Nanotechnology was awarded for modeling the molecular and electronic structures of new materials and for integrating single molecule biological motors with nano-scale silicon devices
2004	First policy conference on advanced nanotech was held. First center for nano mechanical systems was established, Feynman Prize in Nanotechnology was warded for designing stable protein structures and for constructing a novel enzyme with an altered function. Intel launches the Pentium IV "Prescott" processor based on 90 nm technology
2006	Yi Lu et al.: Smart nanomaterials with DNA molecules
2005–2010	3D Nano systems like robotics, 3D networking and active nano products that change their state during use were prepared
2011	Era of molecular nano technology started

composite systems such as organic photovoltaics offer great technological potential due to their attractive properties such as the potential of large-scale and low-cost roll-to-roll manufacturing processes The advances in nanomaterials necessitate parallel progress of the nanometrology tools and techniques to characterize and manipulate nanostructures. Revolutionary new approaches in nanometrology will be required in the near future and the existing ones will have to be improved in terms of better resolution and sensitivity for elements and molecular species.

Applications of Nanotechnology

The different fields that find potential applications of nanotechnology are as follows:

a. Health and Medicine
b. Electronics
c. Transportation
d. Energy and Environment
e. Space exploration

Nanotechnology [8–13] is truly a multidisciplinary field that includes individuals from chemistry, physics, biology, materials science, and engineering all working in a collaborative manner to better understand and apply knowledge of objects that meet the scale classification. Advancements in bioengineering, instrumentation, materials science, and manufacturing diminish the distinguishing characteristics of science and technological disciplines. Rapid discovery, development, and advancement have resulted in relationships among science, technology, and society becoming increasingly stronger. Progress in science and engineering at the nanoscale are critical for national security, prosperity of the economy, and enhancement of life quality for society as a whole. The National Science Foundation has identified that a major outcome goal of the United States is the development of "a diverse, competitive, and globally engaged U.S. workforce of scientists, engineers, technologists and well-prepared citizens. It has been projected that approximately two million workers will be required worldwide during the next decade. Many science, engineering, and technology disciplines converge at the nanoscale, because they consist of many common principles and tools of investigation. By 2015, it is predicted that nano-related goods and services will be a market in excess of one trillion dollars. Thus, nano businesses will become the fastest growing industry in history, outranking the telecommunications and information technology industries combined.

Nanoscience and nanotechnology are the study and application of extremely small things and can be used across all the other science fields, such as chemistry, biology, physics, materials science, and engineering. There is obviously no denying that nanosciences and nanotechnologies focus on the study of nano-sized objects. However, there is currently no consensus on what can be considered as nano and what is excluded from this field. The definition of nanosciences and nanotechnologies varies according to contexts and authors.

Nanotechnology is not simply working at ever smaller dimensions; rather, working at the nanoscale enables scientists to utilize the unique physical, chemical, mechanical, and optical properties of materials that naturally occur at that scale. Nanotechnology therefore is *the complex interdisciplinary science* including: Nanochemistry, Nanophysics, Nanomaterials, Nanoelectronics, Nanobionics and Nanometrology. The biotechnology sequence consists of DNA technology, gene detection, enzyme replacement, cell culture, and associated nanotechnologies. A major portion of the biotechnology sequence is dedicated to the study of nanotechnology.

Nanotechnology as we now know it began more than 30 years ago, when tools to image and measure at the nanoscale became available. Around the turn of the century, government research managers in the United States and other countries observed that physicists, biologists, chemists, electrical engineers, optical engineers, and materials scientists were working on interconnected, multidisciplinary issues emerging at the nanoscale. In 2000, the U.S. National Nanotechnology Initiative (NNI) was created to help these researchers benefit from each other's insights, accelerate technology development, and foster commercialization across disciplines.

In reality the Nanotechnology NT have been arisen in early 80-th, when the scanning tunneling microscopy, the atomic force and other probe microscopes were invented. These have given the opportunity to realize *the main concept of NT* formulated by Richard Feynman, namely, to assemble artificially the nanowork-pieces and nanodevices from single atoms and molecules.

Huge advantage of Pentium-4 over IBM-360 have been achieved by a miniaturizing of integrated circuits and fabricating of microchips containing ca. $\sim 10^9$ units/cm^2 of ~ 200 nm in size. And this is not a limit; the size of individual units may be decreased at least on the orders of magnitudes. With regard to nanoworld, a natural question has arisen "*where are its boundaries*?" Formally it is restricted by size of nanoparticles, $d < 100$ nm. Physically it is determined by a variety of size effects. Decrease in size results in the particles physical chemical properties changing and, consequently, the properties of nano-materials are changed dramatically and sometime cordially. The size effects may be divided into two types, the internal and external ones, as well as the classical and quantum effects. Internal or *intrinsic size effects* are determined as a change of the properties peculiar to particles (the lattice parameters, melting temperature, hardness, band gap, luminescence, diffusion coefficients, chemical activity, sorption, etc.) irrespective of external disturbances. *External size effects* arise inevitably and always in the processes of interaction between different physical fields and matters under decreasing of their building units (the particles, grains, domains) down to a crucial value, when this size becomes to be comparable with a length of physical phenomena (the free length of electrons, phonons, coherent length, screening length, irradiative wave length, etc.). In turn *the classical size effects* appear to become apparent in variation of lattice parameters, hardness, plasticity, thermal conductivity, diffusion, etc. *The quantum size effects* manifest themselves in a blue shift of luminescence, in the rise of peculiar low dimensional quantum states, in the quantization of electro conductivity in magnetic field, in the oscillation of the superconductivity critical temperature, magneto resistance and other physical characteristics, in the generation of hypersound, etc. Hence, studying the size effects in novel nanostructured materials activated by different external fields one can hope for the discovery of novel effects and phenomena and for the development of novel nanotechnology on this base.

There are many different points of view about the nanotechnology. These differences start with the definition of nanotechnology. Some define it as any activity that involves manipulating materials between one nanometer and 100 nm in size. However the original definition of nanotechnology involved building machines at the molecular scale and involves the manipulation of materials on an atomic (about two-tenths of a nanometer) scale. To give you an idea of how small that is, it would take eight hundred 100 nm particles side by side to match the width of a human hair.

Nanotechnology (NT) is proposed presently to define as the complex of fundamental and engineering sciences [14–19] that integrates a chemistry, physics and biology of nanostructures with a materials science, electronics, and processes technologies focused on a comprehensive research of nanostructures, on a development of atomistic physical chemical processes, self- and automatic-assembling of nanomaterials and workpieces using complex probe microscopes combined with other tools, resulted in a fabrication and manufacturing of nanodevices, nanomachines, ultra-low integrated circuits, microopto-electro-mechanical systems, nanobiorobots, etc.

The debate continues with varying opinions about exactly what nanotechnology can achieve. Some researchers believe nanotechnology can be used to significantly extend the human lifespan or produce replicator-like devices that can create almost anything from simple raw materials. Others see nanotechnology only as a tool to help us do what we do now, but faster or better. The third major area of debate concerns the timeframe of nanotechnology-related advances. Will nanotechnology have a significant impact on our day-to-day lives in a decade or two, or will many of these promised advances take considerably longer to become realities?

Now that you have an idea of how small a scale nanotechnologists work with, consider the challenge they face. Think about how difficult it is for many of us to insert thread through the eye of a needle. Such an image helps you imagine the problem scientists have working with nanoparticles that can be as much as one millionth the size of the thread. Only through the use of powerful microscopes can they hope to 'see' and manipulate these nano-sized particles.

Nanotechnology is highly interdisciplinary, which requires collaboration between different sciences and may present further challenges for environmental scientists and engineers. Nanotechnology refers to the creation and utilization of materials whose constituents exist at the nanoscale; and, by convention, be up to 100 nm in size. Nanotechnology explores electrical, optical, and magnetic activity as well as structural behavior at the molecular and submolecular level. It has the potential to revolutionize a series of medical and biotechnology tools and procedures so that they are portable, cheaper, safer, and easier to administer. Finally, the development of specific guidance for the safety evaluation of nanotechnology products is strongly recommended.

Nanotechnology (NT) is the complex interdisciplinary science including [20–26]:

1.1.1. Nanochemistry (nanocolloid, sol-gel and quantum chemistry) destined for self-assembling and synthesis of nanoparticles as well as for research of their intrinsic size effects.

1.1.2. Nanophysics (quantum physics, spintronics, and photonics) destined for artificial assembling and fabrication of nanostructures as well as for research of their external size effects.

1.1.3. Nanomaterials *science* (nanopowder technology, nanoceramics compounds, nanotribology, nanosintering and other nanoprocesses) destined for research, development and production of novel nanostructured architectures, functional nanomaterials and smart nanocomponents with unique properties.

1.1.4. Nanoelectronics, *optoelectronics and nanoengineering* destined for development of novel technological processes, nanomotors, nanoactuators, nanodevices, micro-optoelectronic- mechanical systems (MEMS, MOEMS), ultra-large integrated circuits (ULCI), nanorobots, etc.

1.1.5. Nanobionics destined for development of novel biomachine complexes, such as nanobiochips, nanobiorobots, etc.

1.1.6. Nanometrology, *nanodevice building and nano-hand-craft* destined for development of special nanotools, instrumentations, information and computational systems for support of NT itself.

The association of these sciences in nanotechnology reflects both their inherent interconnection around the nanoobjects and *the change in technology paradigm*, namely, the nanomaterial, nanodevice or nanosystem seems to be fabricated by the automatic artificial assembling or self-assembling from molecules or clusters *in whole*, in situ, *in place*, in the single technological process incorporating them then in microdevices, rather than by aggregating of different components as now.

1.1.7. Nanoscience: is the science of objects with typical sizes of 1–100 nm. *Nanoscience* study on fundamental relationships between physical properties and material dimension on the nanometer scale. Nanoscience is not physics, chemistry, engineering or biology. It is all of them. S. M. Lindsay, Introduction to Nanoscience, Oxford University Press (2009).

Biomaterial devices have been tailored at a micro and nanoscale to be used in medicine. The devices can be produced with different shapes (i.e. particles, fibers, rods, discs) for drug delivery of a variety of molecules. A dendrimer is morphologically characterized by a branched structure grown from one or more cores. The size of these NPs is easily controlled by the number of generations that are allowed to grow over these cores. Dendrimers present difficulties regarding drug incorporation and release, being their synthesis quite time-consuming [27] as shown in Fig. 1.1.

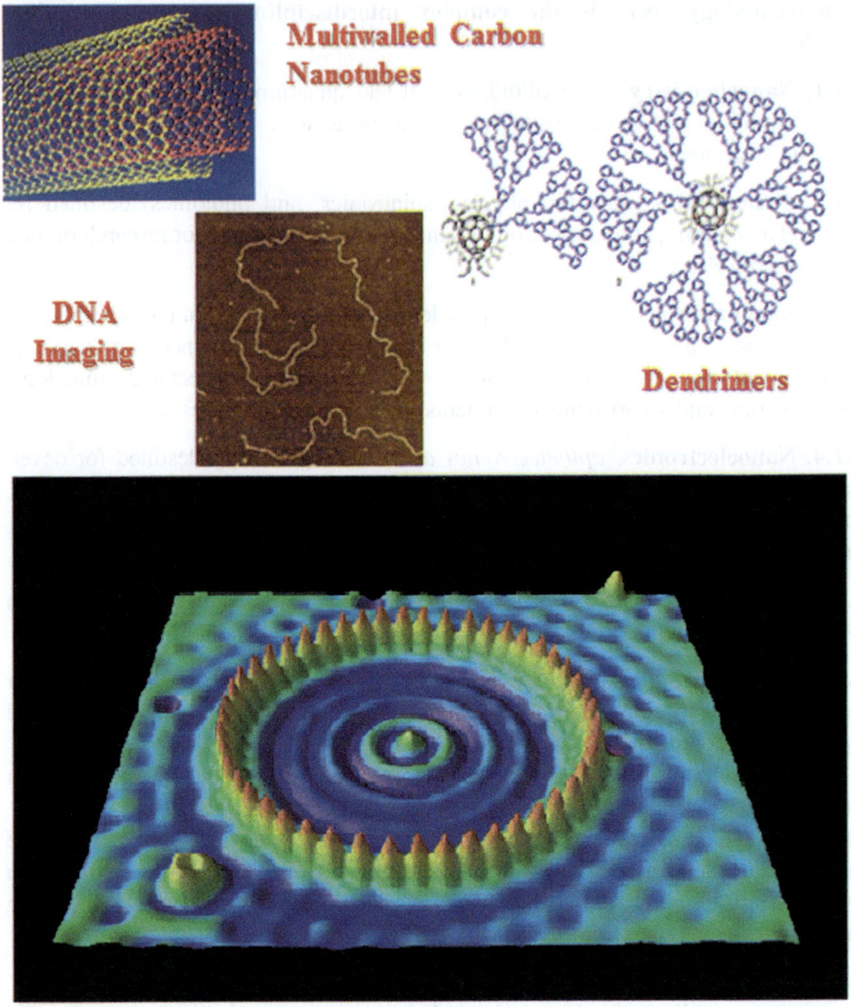

Fig. 1.1 Different shapes and types of nanomaterials

Microparticles (MPs), defined as particles whose size ranges from 1 to 1000 μm have been studied [28]. They can be administered by different routes, such as intranasally [29], subcutaneously [30] or orally [31]. Additionally, they have high specific surface area, when comparing with macroparticles, which ensures high quantity of functional groups for chemical reactions [32]. Stable suspensions containing MPs characterized by narrow size dispersion and presenting constant size along time, can be obtained [32]. Gellan-gum [33], chitosan [34–36] or dextran [37] are just examples of a broad range of biomaterials used to produce these MPs. MPs can be tailored to avoid clearance mechanisms from the body, by increasing for example their hydrophilicity [38]. Moreover, they can be targeted to specific anatomical sites of the human body [39]. When used as drug delivery devices, MPs are able to entrap efficiently active agents, taking into account the hydrophobic or hydrophilic drug character, protecting the drug from undesired premature degradation and interactions with the biological environment [40] (Fig. 1.2).

Nanoscience is about the phenomena that occur in systems with nanometer dimensions (Fig. 1.3).

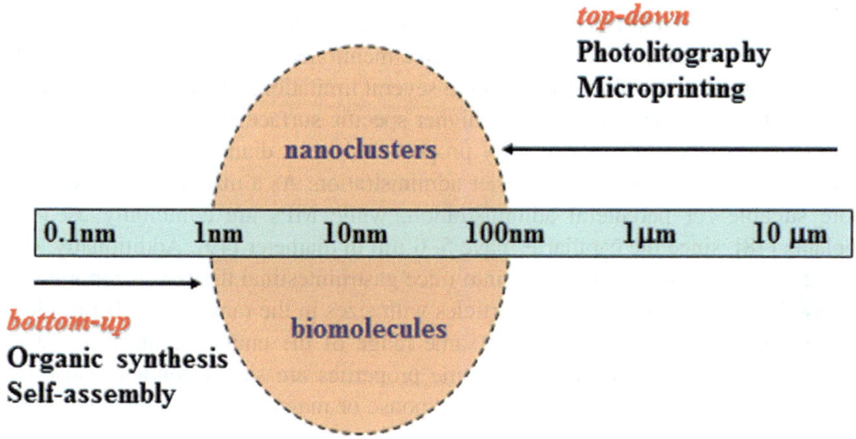

Fig. 1.2 Nanoclusters and biomolecules

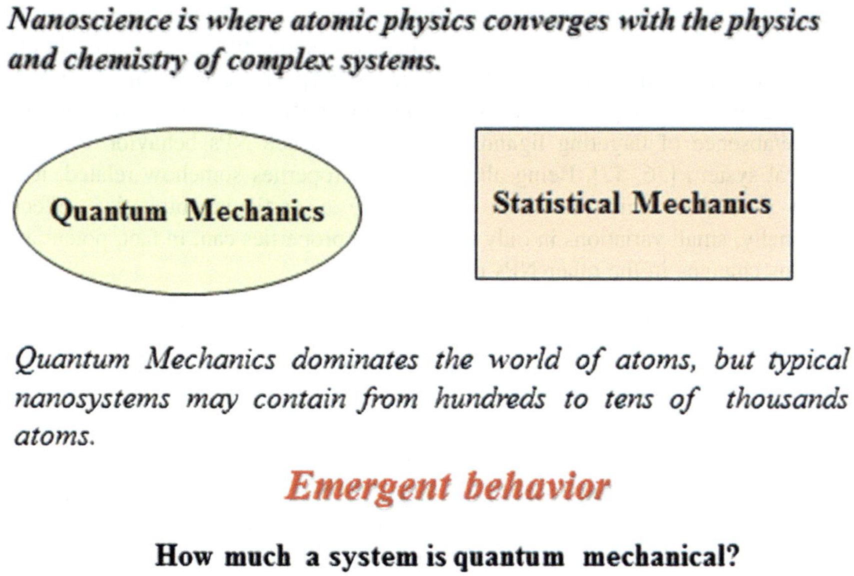

Fig. 1.3 Correlations between quantum and statistical mechanics

1.1.8. Nanochemistry: is related with the production and the reactions of nanoparticles and their compounds. It is concerned with unique properties associated with assemblies of atoms or molecules on a scale between that of the individual building blocks and the bulk material (from 1 to 1000 nm).

1.1.9. Nanoparticles: are particles between 1 and 100 nm in size.

1.1.10. Nanowires: A nanoscale rod of semiconducting material used in miniature transistors and some laser applications.

Nanoparticles (NPs) are regarded as nanoentities whose size ranges from 10 to 1000 nm [41]. Besides sharing all the above-mentioned MPs advantages [42–47], NPs also have the capability to overcome several limitations of MPs. In fact, due to their smaller size, they present much higher specific surface area, because the total surface area of a particle is inversely proportional to its diameter [32]. Moreover, this reduction in size enables an easier administration. As a matter of fact, NPs are more suitable for parenteral administration, while MPs are commonly used as implants [48], since the capillaries have 5–6 μm in diameter [49]. Additionally, the uptake efficiency of nanostructures into mice gastrointestinal tissue was reported to be 15–250 times higher than that particles with sizes in the range of 1–10 μm [50]. Furthermore, NPs have sizes in the same range of the entities controlling basic cellular functions [51]. In addition, some properties are size-dependent and only found in the nanoscale, such as optical response or magnetism [51], allowing their use also as diagnosis/imaging devices.

NPs properties

The properties of the NPs influence their behavior in vivo. Particularly, morphological properties like shape and size, can influence NPs circulation and targeting within the body [52, 53]. These properties are also responsible for variations in the degradation rate of NPs and drug release kinetics [54]. The shape and size of NPs are also responsible for specific cell signaling [55]. NPs surface properties and the presence/absence of targeting ligands can also influence NPs behavior within a biological system [56, 57]. Being all these NPs properties somehow related, it is difficult to define which of them will ensure a particular biological effect. Additionally, small variations in only one of these properties can, in fact, potentiate enormous changes in the other NPs performance.

Size and polydispersity Index

Ideally, NPs should be in circulation until they reach the target anatomical site. A schematic representation of the distribution of particles with different sizes after intravenous administration is presented on Fig. 1.4.

Studies have reported that MPs larger than 7 μm are filtered mechanically, being entrapped in the capillary network of the lungs [58, 59]. Particles with diameters in the range of 0.1–7 μm are detected by the reticulum endothelial system (RES) in the liver or spleen, being phagocytized by Kupffer cells or by spleen macrophages, respectively [58, 60, 61]. If the particle diameter is lower than 100 nm, it will remain in the blood vessels within fenestrae of the endothelial lining [62], since the

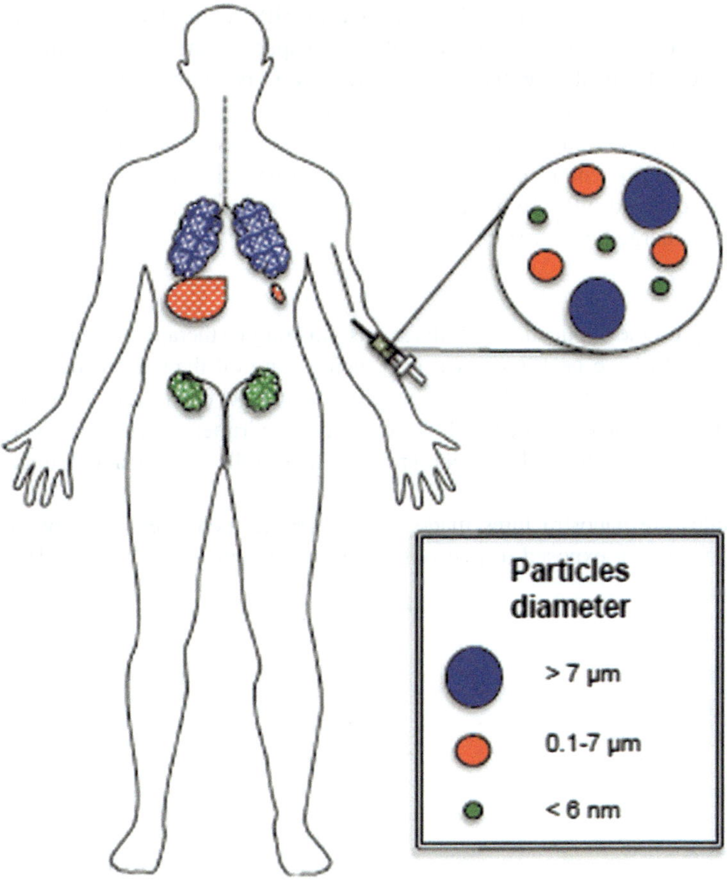

Fig. 1.4 Expected body distribution of intravenously administration of particles with different sizes

possibility of being uptaken by macrophages or recognized by opsonins is reduced. The smallest NPs, with diameter inferior to 6 nm, suffer from glomerular filtration in the kidneys [63–65].

Shape

Recent reports have suggested a significant role of particle shape (sphere, ring, disc) in the circulation of NPs, distribution within the body, cellular uptake and general in vivo behavior [66, 67]. The influence of the shape in the NPs transport within the human body was already evaluated in several works [62, 68, 69].

Surface properties

Distinct surface properties, such as hydrophobicity and surface charge, have been used to characterize NPs [70, 71]. Hydrophobicity has huge relevance since it influences NPs clearance from the body due to RES action. Indeed, a decrease in the hydrophobicity leads to reduction of the nonspecific interactions with proteins. NPs surface electrical potential, also known as zeta potential, is the potential of a particle or molecule due to its charge in a certain medium and gives the tendency for the particles to undergo aggregation [72]. In fact, as the absolute value of this surface electric potential increases, the repulsion between NPs is intensified [72]. Positively charged NPs are more nonspecifically internalized than their neutral or negatively charged counterparts [47, 73].

Types of NPs

Depending on the application (i.e. diagnosis, imaging or therapy), different types of NPs (Fig. 1.5) have been proposed [74], being some of them used for more than one aim. In this section, NPs types are divided in two main groups: organic and inorganic NPs. The first group includes micelles, dendrimers, liposomes, hybrid and compact polymeric NPs. The second group includes fullerenes, quantum dots, silica and gold NPs.

Micelles are nanostructures made of amphiphilic molecules, like polymers or lipids [75]. When exposed to aqueous environments, they hide their hydrophobic

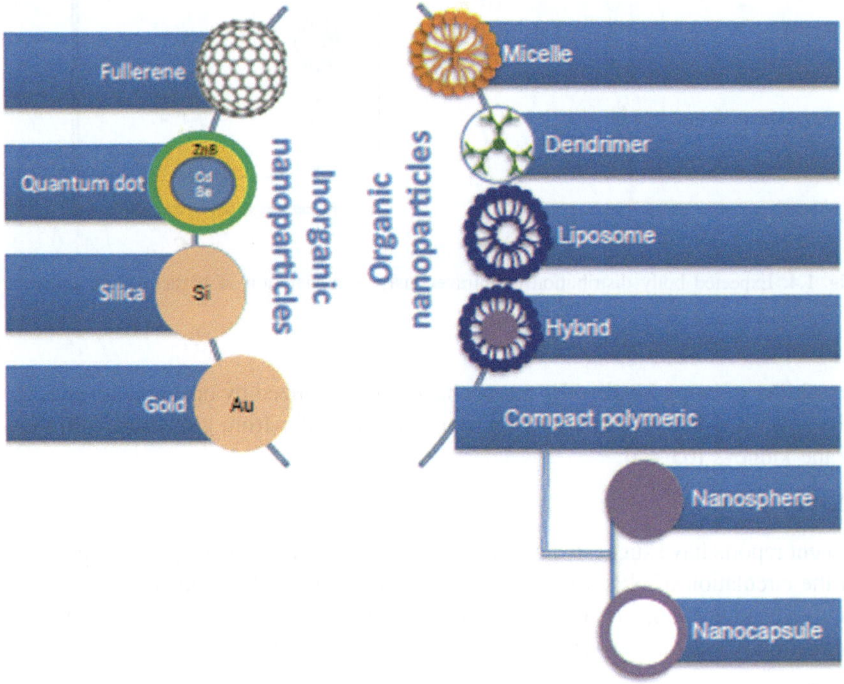

Fig. 1.5 Main types of NPs used in various applications; image not at scale

groups inside the structure and expose the hydrophilic groups. On the other hand, when in lipid-rich environments, their structure may organizes in a reverse way [75].

Physics and chemistry of nanostructures or nanophysics and nanochemistry are relatively new areas of science arisen in last decade of past century after discovery of fullerenes and nanotubes. It is introduction into more extent interdisciplinary integrated modern science known now as nanotechnology rapidly developing. At this stage of growing knowledge authors have shortly outlined the subject and classifications of nanostructures, interesting milestones, main principles, methods, techniques, as well as general directions of future perspective research to be a guideline in a set of modern research. Main mechanisms of chemical processes affected formation of nanostructured materials and their properties are clearly expressed, in particular, a dielectric permittivity as a principal characteristic of electric, magnetic, acoustic, optic transparency, superconducting, and other properties of nanoceramics and nanometals. The peculiar properties of nanostructures are emphasized to be result of size effects, external and internal, classical and quantum ones that arise in zero-dimensional quantum dots, one-dimensional wires, and two-dimensional layers. Numerous applications are considered including microlasers, photonic crystals, probe microscopy, left-handed materials with negative refraction index, etc. Novel idea is advanced that new discovery of novel fundamental laws, phenomena and applied effects are possible only in artificially fabricated nanostructures with new effect theoretically predicted and designed in advance. Content of the course covers the types, classification and peculiarities of nanostructures, size effects, synthesis and growth, fullerenes, nanotubes, microlasers, photonics, scanning probe microscopy, nanomanipulation, etc.

Nanoscience primarily deals with synthesis, characterization, exploration, and exploitation of nanostructured materials. These materials are characterized by at least one dimension in the nanometer range. A nanometer (nm) is one billionth of a meter, or 10^{-9} m. One nanometer is approximately the length equivalent to 10 hydrogen or 5 silicon atoms aligned in a line. The processing, structure and properties of materials with grain size in the tens to several hundreds of nanometer range are research areas of considerable interest over the past years. A revolution in materials science and engineering is taking place as researchers find ways to pattern and characterize materials at the nanometer length scale. New materials with outstanding electrical, optical, magnetic and mechanical properties are rapidly being developed for use in information technology, bioengineering, and energy and environmental applications. On Nano scale, some physical and chemical material properties can differ significantly from those of the bulk structured materials of the same composition; for example, the theoretical strength of nanomaterials can be reached or *quantum effects* may appear; crystals in the nanometer scale have a low melting point (the difference can be as large as 1000 °C) and reduced lattice constants, since the number of surface atoms or ions becomes a significant fraction of the total number of atoms or ions and *the surface energy* plays a significant role in the thermal stability. Therefore, many material properties must now be revisited in light of the fact that a considerable increase in surface-to-volume ratio is associated with the reduction in material size to the nanoscale, often having a prominent effect on material performance. Historically, fundamental material properties such as elastic modulus have been characterized in bulk specimens using macroscopic, and more recently microscopic, techniques. However, as nanofabrication advances continue, these bulk properties are no longer sufficient to predict performance when devices are fabricated with small critical dimensions. Although nanotechnology is a new area of research, nanomaterials are known to be used for many industrial applications. For example, the Chinese used gold nanoparticles as an inorganic dye to introduce red color into their ceramic porcelains more than thousand years ago. Roman glass artifacts contained metal nanoparticles, which provided beautiful colours. In medivial times, nanoparticles were used for decoration of cathedral windows.

What new about nanoscience?

is the combination of our ability to see and *manipulate matter* on the nanoscale and our understanding of atomic scale interactions? Advances in the materials processing along with the precipitous rise in the sophistication of routine, commonly available tools capable for characterization of materials with force, displacement and spatial resolutions as small as pico Newtons (Pn = 10^{-12} N), nanometer (nm = 10^{-9} m) and Angstrom (A = 10^{-10} m), respectively, have provided unprecedented opportunities to probe the structure and mechanical response of materials on nanoscale. In addition, major improvements in computer support have allowed *the simulations of material structures* and behavior with a degree of accuracy unimaginable as recently as a decade ago. Although study on materials in the nanometer scale can be traced back for centuries, the current fever of nanotechnology is at least partly driven by the ever *shrinking of devices* in the semiconductor industry.

In place of the traditional processes of thermo mechanical treatment (the rolling, cutting, welding, soldering, molding, etc.) and of microelectronics processes (the chemical and physical vapor depositions, lithography, etc.) the novel atomistic nanotechnology processes (the nanomanipulation, artificial- and self-assembling, nanolithography, membrane templating synthesis, sol-gel synthesis, molecular-beam epitaxy, etc.) are expected to will come.

Living in macroworld human come into controllable tunable contact with nanoworld mainly by means of a tip of probe microscope, so the contact "tip-surface" is the contact of macroworld with nanoworld.

Therefore the key problem of nowadays nanotechnology is a comprehensive research of atomistic mechanisms of *the nanocontact phenomena* (adhesion, indentation, friction, wear, etc.) in dependence on a type of interatomic inter-molecular bonds, type and structure of contact materials, size of tip and nanostructure, value of load, width of gap, environment, temperature, external electric and magnetic fields, frequency and intensity of electromagnetic waves, and so on. These researches have to be expressed in development of *the techniques for tunable manipulation, characterization, control, and position assembling of nanostructures*, particulary, seizure, gripping, restraining, turning, moving, breaking, reset and adhesion of a molecular building block onto prescribed place. Such operations at atomic and molecular level are just the ones which become to be principal for nanotechnology. It should be emphasized that NT has not intended to replace the existence micro technologies, but to stay in close connection with them to complement them in the deeper study and advanced control of nanoworld. Atoms, molecules, clusters, fullerenes, supramolecular structures, their crystals, nanotubes, nanowires, nanorodes, their arrays and photonic crystals serve as *NT objectives.*

Fullerenes and atomic clusters

are the smallest zero-dimensional (0D) nanostructures called quantum dots possessing the properties inherent for nanomaterial rather than for single atom. Note that for fullerenes it should mean not only the buckyball C60, but the multitude of another carbon Cn and noncarbon clusters and metcarbes Me@Cn. Presently a number of experimental nanodevices was developed on this base, e.g. the switchers, diodes, transistors, amplifiers, sensors, optical filters, solar cells, magneto optical recorders, etc.

Nanotubes, nanorodes, nanowires, nanofibers

manifest more advanced and promising properties as being the 1D quantum wires nanoscopic in diameter but microscopic in length. Their unique properties stem from capability of the ring and cylindrical types of acoustic and electromagnetic waves to propagate that makes them a unique nanolaboratory for research of quantum resonance phenomena. All stated above also concern to noncarbon 1D nanowires and nanotubes based on boron-nitride, oxides, chalcogenides, dichalcogenides, chalogenides, and some other III–V and II–VI compounds possessing of the most manifold physical-chemical characteristics. Reduced two-dimensional *2D heterostructures*, nanolayers and nanodisks as being the well-known 2D quantum wells are believed to migrate from micro- to nanoelectronics. In addition the 2D arrays of nanowires and nanotubes ordered in 2D forest arrays or 2D crystals seem to be novel and very perspective core of NT. Their unique properties have to be determined by new principles of electromagnetic waves propagation based on the Bragg diffraction law rather than on the total internal reflection. They are the quantum and, in the same time, the macroscopic 2D crystals in which the various quantum states and resonance effects are expected. Actually such resonance states can be recognized as the novel state of matter, research of which appears to become the advanced direction in nanophysics. On this base the waveguides, laser emitting diodes, infrared sensors and other nanodevices have already been developed.

Design and assembling of such artificial media, search for new unusual effects and phenomena, as well as development of the up-to-date nanodevices on their base seems to be the most promising way in the nearest NT development. The example is the discovery of “left” matter or meta materials, in which unconventional inverse refraction law, inverse Doppler and inverse Cherenkov effects were observed. In nanomaterials science the structure-form engineering will put in the forefront in addition to the impurity engineering. Material becomes not a raw or a pig but it is forming at once as a nano-workpiece. Note that advantage of nanomaterials is hoped to proclaim itself just at developing of nanodevices, the electronic gnat for example, rather than in the large scale industry.

By *peculiarity of the nanoworld* is the cancellation of distinctions between the living and inorganic matter. The exchange of substance being the indication of life manifests itself on the supramolecular level rather than a molecular one. Proteins, membranes, and nuclein acids refer to giant natural nanostructures built in result of self-assembling. The analogy opens a fantastic opportunity for nanomaterials and nanodevices fabrication by such biomimicry. Artificial growth of pearls inside mussels, as well as ordering of non-equilibrium defects into 2D nanostructures on a surface of semiconductors under the ion bombardment and implantation are the examples.

Principal question is "what are the peculiar features inherent to nowadays nanotechnology taking into consideration that atomic and molecular physics, chemical synthesis technologies, microelectronics, etc., were existed before NT era?" *The novelty* includes:

- the artificial manipulating by nanoobjects and manual or automatic assembling of the nanodevices designed beforehand using a "bottom-up" approach;
- the deliberate meddling in processes mechanisms with the comprehensive control of a chemical self-assembling at molecular level;
- the invention, design and production of nanodevices of submicrometer size followed by their integration into micro-, mezzo-, -and macro-systems. Entering into NT it should be warn of some *illusions and problems*.

Firstly, decrease in particles size *is restricted from below* because it does not always result into improvement of the properties. For instance, the optimal size of disperse inclusions in oxide ceramics ca. ~10–20 μk was shown to exist at which the optimal combination of hardness and durability is achieved.

Secondly, with particles size decrease the processes of *thermal instability* and phase transitions were shown to take place resulting in nondurability of nanosystems. For instance, the well-known words IBM, NANO, and corals drown on substrate by atomic force microscopy were turn out to be unstable due to fast surface diffusion of building atoms. Since the covalent bonded semiconductors and ceramics preferably appear to be stable and durable, the nanomaterials for NT are thought to be nonmetallic.

Thirdly, a cosmic irradiation and radiation background are capable atoms to knockout from nanostructures leading in *degradation* of their properties and in worsening of nanodevice operation.

Fourthly, a thermal noise and vibrations will be significant circumstances influenced the properties and characteristics of nanodevices. In particular, it limits certainty of probe microscope position, which must never be less than a half-amplitude of thermal vibrations.

Fifthly, even negligible concentration of inherent impurities and *irremovable contamination* enable to destroy the assembling processes, so a super-high-purity feed reagents and clean-room processes are required.

Concluding, all physical discoveries in vacuum have been already made except further discovery of the vacuum itself. Novel discoveries, laws, phenomena,

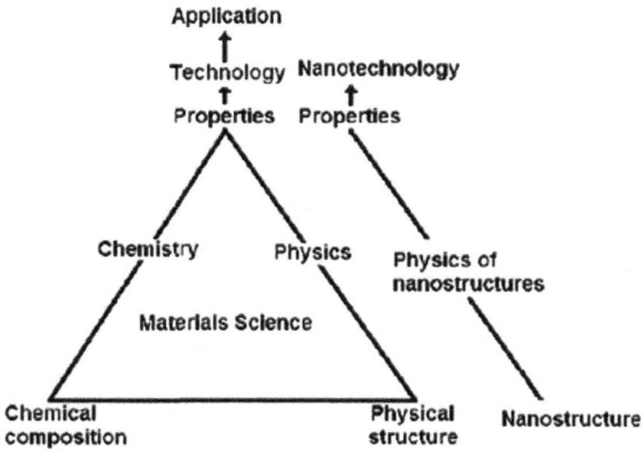

Fig. 1.6 Fundamental triad of materials science

technical decisions, solutions, and inventions will be possibly made only in special designed and assembled artificial nanostructures to be fabricated by future materials science.

Materials science concept is shown in Fig. 1.6 illustrating the inherent interconnection between the composition, structure, properties, technology and applications.

Nanotechnology in pharmacy and medicine is going to have a major impact on the survival of the human race. The unique optical, catalytic, electronic and physical properties (melting point) of metallic nanoparticles have made them potential candidates in the field of nanotechnology. The synthesis of metallic nanoparticles is being carried out by various methods. Method of synthesis is one of the important factors, which largely influences their biological effectiveness. Moreover, conventional physical and chemical processes involve the use of expensive chemicals and these methods are non-ecofriendly.

Are nanotechnology products available?

Yes, nanotechnology is becoming ubiquitous in our daily lives and has found its way into many commercial products, for example, strong, lightweight materials for better fuel economy; targeted drug delivery for safer and more effective cancer treatments; clean, accessible drinking water around the world; superfast computers with vast amounts of storage; self-cleaning surfaces; wearable health monitors; more efficient solar panels; safer food through packaging and monitoring; regrowth of skin, bone, and nerve cells for better medical outcomes; smart windows that lighten or darken to conserve energy; and nanotechnology-enabled concrete that dries more quickly and has sensors to detect stress or corrosion at the nanoscale in roads, bridges, and buildings. By some estimates, revenue from the sale of nanotechnology-enabled products made in the United States has grown more than six-fold from 2009 through 2016 and is projected to exceed $500 billion in 2016.

Table 1.2 highlights the targeted areas of study within nanotechnology, the proportional emphasis dedicated to the specific unit, and the content of each area.

Table 1.2 Areas of study within nanotechnology

Nano topics	Proportion (%)	Content
Microscale and nanoscale	15	"Nano" and nanometer, English expression, magnitude as a number and exponent, relative size
Carbon forms	8	Physical properties, oxidation state, compounds, carbon form uses
Modeling and simulation	15	Data structuring, process structuring with choice of 3DS applications, Flash applications, Truespace applications
Water purification	8	Contaminants, microfiltration, porous nanoparticles, nanotubes
Nanorobotics	6	Biosensors, control design, controlled manipulation, bio-uses
Biomedical applications	12	Drug delivery, bio-machines, biosensors
Future application	12	Fuel cells, solar cells, nanofibre armour
Risks and benefits	12	Security, health, environment
Nano ethics	12	Public safety, regulation, standards

Animation courses greatly contribute to the nanotechnology modeling and simulation requirements in the Emerging Issues in Technology course. An animation still of an atomic force microscope used to manipulate atoms. Nanomanipulation, or the ability to move atoms individually and arrive at predetermined arrangements, is an important first step in realizing the potential of nanotechnology.

Finally, all the opinions about what nanotechnology can help us achieve echo with ethical challenges. If nanotechnology helps us to increase our lifespans or produce manufactured goods from inexpensive raw materials, what is the moral imperative about making such technology available to all? Is there sufficient understanding or regulation of nanotech based materials to minimize possible harm to us or our environment?

1.2 Nanomaterials (NMs)

1.2.1 What Are the Fundamental Issues in Nanomaterials?

The term "nanomaterial" refers to nanoscale materials, or materials that contain nanoscale structures internally or on their surfaces. These can include engineered (or man-made) nanometer-scale objects such as nanoparticles, nanotubes, and nanofilms, as well as naturally occurring nanoparticles such as volcanic ash, sea spray, and smoke.

The fundamental issues in this domain of nanomaterials are:

(1) ability to control the scale (size) of the system,
(2) ability to obtain the required composition—not just the average composition—but details such as defects, concentration gradients, etc.,
(3) ability to control the modulation dimensionality,
(4) during the assembly of the nano-sized building blocks, one should be able to control the extent of the interaction between the building blocks as well as the architecture of the material itself.

More specifically the following issues have to be considered for the future development of nanomaterials:

- Development of synthesis and/or fabrication methods for raw materials (powders) as well as for the nanostructured materials.
- Better understanding of the influence of the size of building blocks in nano structured materials as well as the influence of microstructure on the physical, chemical and mechanical properties of this material.
- Better understanding of the influence of interfaces on the properties of nanostructured material.
- Development of concepts for nanostructured materials and in particular their elaboration.
- Investigation of catalytic applications of mono- and plurimetallic nanomaterials
- Transfer of developed technologies into industrial applications including the development of the industrial scale of synthesis methods of nanomaterials and nanostructured systems.

Nanomaterials are cornerstones of nanoscience and nanotechnology. Nanostructure science and technology is a broad and interdisciplinary area of research and development activity that has been growing explosively worldwide in the past few years. It has the potential for revolutionizing the ways in which materials and products are created and the range and nature of functionalities that can be accessed. It is already having a significant commercial impact, which will assuredly increase in the future.

- What can nanomaterials be?
- What is a dimension?
- Does size mean a lot?
- What is a potential well?

Material is not a dull bar, blank, block, pig, but it is the immense word, the Universe, the media in which new physical laws may be discovered. Actually, there are 100 pure natural elements in the Periodical Table on base of which 10,000 XY binary, 1,000,000 XYZ ternary, 100,000,000 quaternary, etc., compounds may theoretically exist accounting the chemical composition. This abundance in many times increases with account of physical structure including nanostructures. However only 500,000 compounds are known presently to exist in modern crystallography database. Hence, the abundance of novel undiscovered compounds with new unique properties is very huge forming the challenging frontier of research for future nanotechnology.

At present time, we meet NT in child age. The announcement of grand projects, such as biochips and nanobiorobots for medicine, smart dust for space research, etc., have become as motivation for its intense development, that may influence upon a civilization development. In USA, EC, Japan, Russia and other leading countries the great funds were released for NT projects. The perspectives of NT at the beginning of 21 century looks very optimistic, since a severe reality is capable to darken these somewhat naive prospects. However, in any case, the development of NT is unavoidable and it is doomed to success.

Due to their incredible properties, nanomaterials (NMs) and nanoparticles (NPs) have become significant in many fields in recent years such as energy, health care, environment, agriculture etc. Nanoparticle technologies have great potentials, being able to convert poorly soluble, poorly absorbed and labile biologically active substance into promising deliverable substances.

"Nano" is the prefix word originating from the Latin word '*nanus*' meaning literally 'dwarf' and according to an International System of units (SI) it is used to indicate a reduction factor of 10^9 times. So nanosized materials are typically measured in terms of nanometers (1 nm is 10^{-9} m). Nanoscience can be defined as a study of the phenomenon and size reduction of materials at atomic or molecular scales. Pharmaceutical nanotechnology embraces applications of nanoscience to pharmacy as nanomaterial's and as devices like drug delivery, diagnostics, imaging and biosensors. The term nanomedicine refers to the application of nanotechnology to diagnosis, monitoring, control and treatment of diseases. Biologists are embracing nanotechnology as the engineering and manipulation of entities in 1–100 nm range and exploiting its potential to develop new therapeutics and diagnostics. Due to the advent of many modern and advanced analytical tools and capabilities to measure particle size in nanometres, particulate drug delivery systems research and development has been moving from macro to nano. It is an extremely large field ranging from in vivo and in vitro diagnostics to therapy, including targeted delivery and regenerative medicine. Nanostructured materials brought about by the creation of functional materials with desired physicochemical properties.

Historically, India was probably the first country to maintain records of useful drugs. *Charaka Samhita* written by Acharya Charaka, the great Ayurvedic scholar of 1500 BC and later treatises describe the medicinal properties of various metals like gold, lead, mercury and silver. The metals used in Ayurvedic system of medicine includes copper, gold, iron, lead, mercury, silver and zinc. The term metallic nanoparticles (MNPs) are used to describe nanosized metals with dimensions (length, width or thickness) with size ranges from 1 to 100 nm. The existence of MNPs in solution was first recognized by Faraday in 1857 and a qualitative explanation of their colour was given by Mie in 1908. The current research in the field of MNPs is largely studied due to their special properties. Especially, from the last two decades, considerable attention has been devoted to the synthesis of MNPs because of their unusual properties and potential applications in optical, electronic, catalytic, magnetic materials and in medicine.

Scientists have been studying and working with nanoparticles NPs for centuries, but the effectiveness of their work has been hampered by their inability to see the structure of nanoparticles. In recent decades the development of microscopes capable of displaying particles as small as atoms has allowed scientists to see what they are working with. The following illustration titled “The Scale of Things”, created by the U. S. Department of Energy, provides a comparison of various objects to help you begin to envision exactly how small a nanometer is. The chart starts with objects that can be seen by the unaided eye, such as an ant, at the top of the chart, and progresses to objects about a nanometer or less in size, such as the ATP molecule used in humans to store energy from food.

1.2.2 Nano Scale and Nanostructures

Nanoscale particles have an average particle size smaller than $\sim$100 nm (Fig. 1.7).

The nano scale is the place where the properties of most common things are determined just above the scale of an atom. Nano scale objects have at least one dimension (height, length, depth) that measures between 1 and 999 nm

Nanoscale materials are defined as a set of substances where at least one dimension is less than approximately 100 nm. A nanometer is one millionth of a millimeter—approximately 100,000 times smaller than the diameter of a human hair. Nanomaterials are of interest because at this scale unique optical, magnetic, electrical, and other properties emerge. These emergent properties have the potential for great impacts in electronics, medicine, and other fields.

Nanomaterials are important because they can be used in many domains giving strategic assets to those who use and develop them. This point is, for example, pointed out by Eiji Kobayashi, from Panasonic, who says: “Those who control materials control technology”. Scientists are now able to create a wide variety of building blocks that they assemble to make nanostructured materials. Using nanoscale building blocks allows us to have high surface areas, many interfaces and, in the case of nanostructures, size confinement (Fig. 1.8).

Today, nanochemists work in biomedical chemistry, polymer chemistry, product synthesis, and a host of other areas. They use a wide variety of methods to prepare and assemble “little pieces of matter” with novel electronic, magnetic, optical, chemical, and mechanical behaviors that can be attributed to their nanometer-scale size. Because almost all of the atoms in a nanoparticle are on the surface, these particles have chemical and physical properties that differ from those of individual molecules or larger aggregates of atoms or molecules. Nanoparticles have already made an impact on the consumer market, in the form of sunscreen lotions that absorb ultraviolet light without giving you a white nose and clear coatings for cars that protect the bright paint colors underneath without detracting from their appearance (Fig. 1.9).

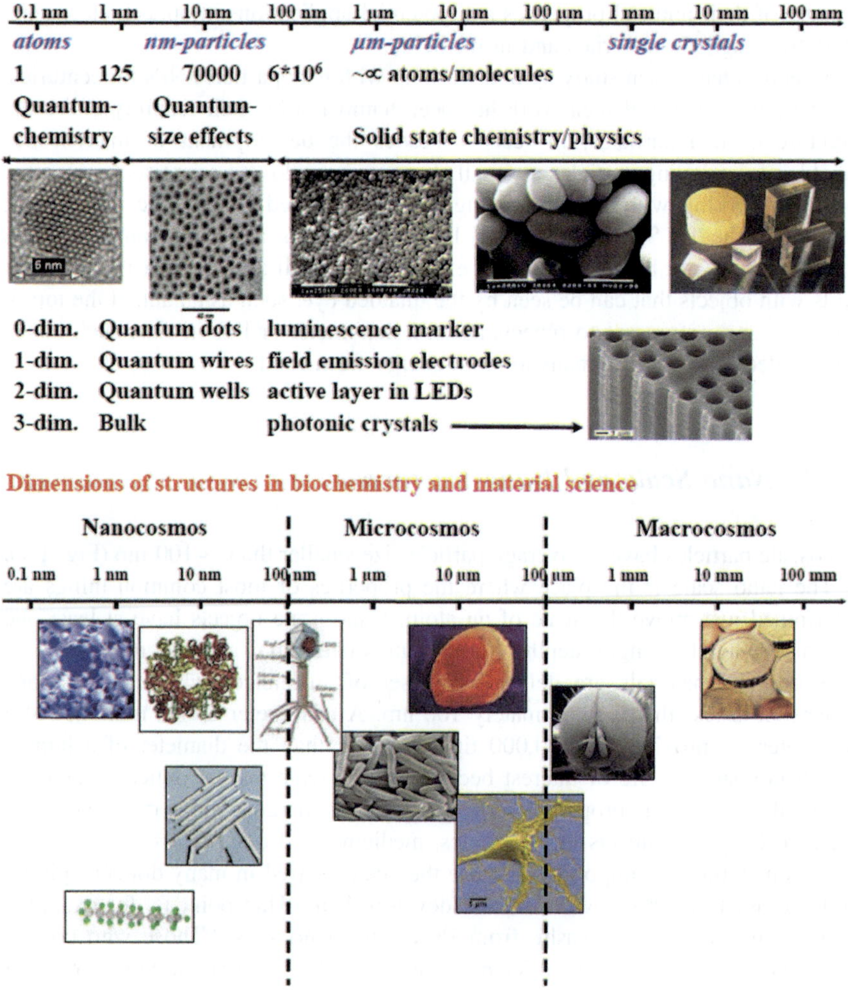

Fig. 1.7 Dimensions of structures and quantum dots

The nanostructuration of materials may lead to several changes of the properties of materials:

- A global change of the properties can be obtained. This is, for example, observed by including nanoparticles in a polymer material. Because there is a huge interface area between the nanoparticles and the polymer, an interaction between the two components takes place which modifies the properties of the bulk material. This happens for tires where carbon black is used to improve their properties, particularly their durability. This is also the case in cements. Concrete is a building material extensively used around the world. It is a low cost and easy to manufacture material with good structural properties. However,

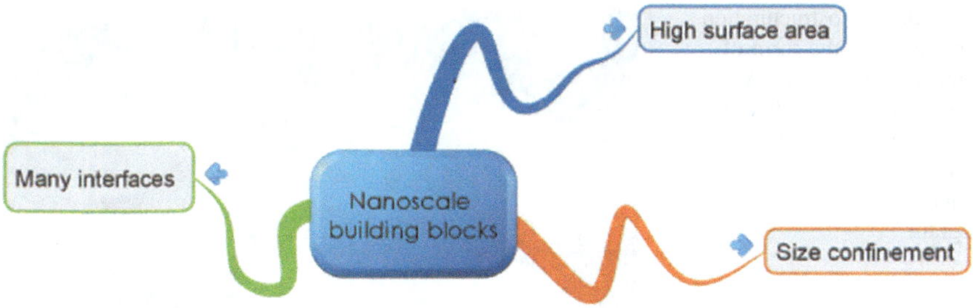

Fig. 1.8 Properties of nanoscale building blocks which will have tremendous consequences for nanostructured materials

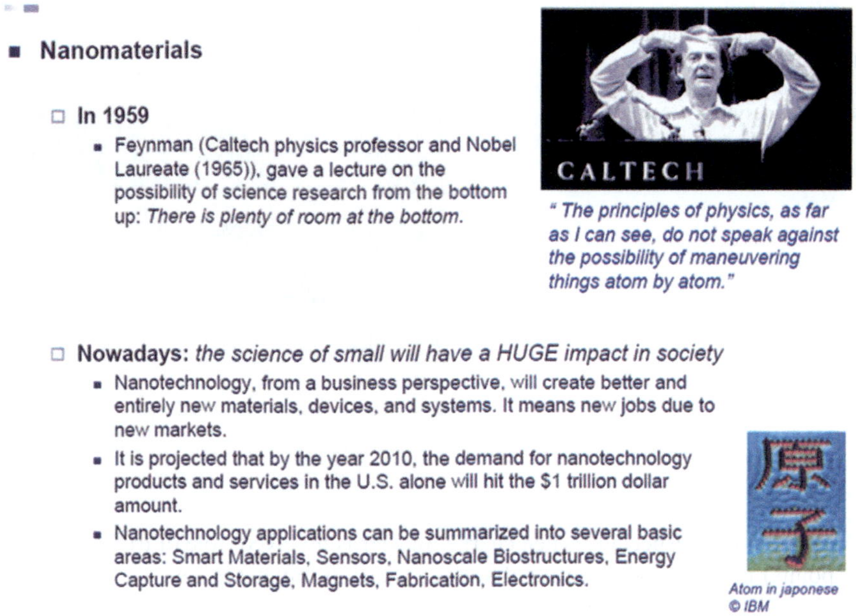

Fig. 1.9 Development in nanotechnology and nanomaterials

it has a low resistance to tensile forces and is porous. A considerable improvement of its properties is obtained by adding silica smoke containing silica nanoparticles. Concrete with silica smoke is significantly less porous and less permeable to water, which is for example valuable when this material is in contact with sea water.

- Nanoparticles can also be used to introduce specific optical, magnetic, and catalytic properties. In many cases, it is the characteristics of the nanoparticles which are important and not the interaction with the inert matrix housing them, but interfaces between nanoparticles and host matrix or support can play an

Fig. 1.10 Nanostructuration of platinum in the form of nanoparticles. Clef CEA n○59. Courtesy of CEA. The results have been obtained by the CEA in the framework of a Challenge innovation program

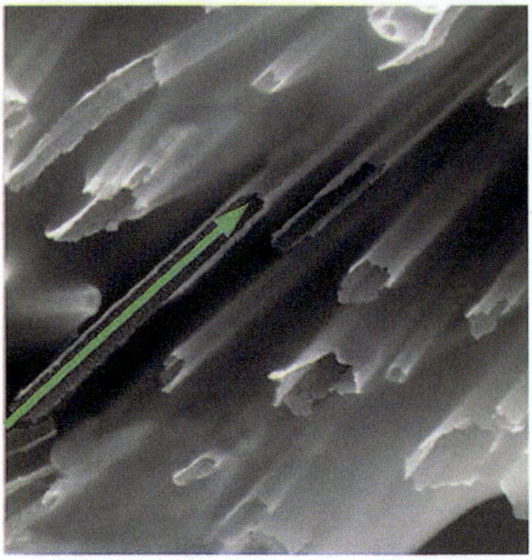

Fig. 1.11 Nanostructuration of platinum in the form of porous nanotubes. Clef CEA n○59. Courtesy of CEA

essential role, especially when charge transfer is involved: role of triple lines (or points) in nanocatalysts, role of surface states in luminescence.

Figures 1.10 and 1.11 show examples of the nanostructuration of platinum. In Fig. 1.10, the nanostructuration is done in the form of nanoparticles. The length of the horizontal figure is about 800 nm.

In Fig. 1.11, the nanostructuration of platinum is produced in the form of porous nanotubes which could be used in catalysis. The green arrow indicates the oriented porous catalytic nanostructure. The length of the horizontal figure corresponds to about 7 μm.

1.2.3 Nanostructured Materials

Nanostructured materials are materials with structures in the nanoscale range (1–100 nm). The properties of the nanostructured materials depend on the size and nature of their microstructure. If the characteristic length scale of the microstructure is comparable to the lengths associated with fundamental physical phenomena, large changes in the properties of the material compared to a non-nanostructured material can be observed. The whole structure—like for many materials—can be in or out of thermodynamic equilibrium. Using crystallites of nanometer size of elements like gold or sodium chloride, for instance, can produce a great variety of nanostructuration depending on the crystallographic orientation, the chemical composition of the mixture and the possibility to have non-equilibrium structures which provides certain property advantages.

Nanoparticles are of course often used as building blocks to make nanostructured materials. They can be of various natures: fullerenes, nanotubes, nanocrystallites, nanofibers, etc. If the simplest building blocks of nanostructured nanomaterials are nanoparticles, more complicated elementary structures can be used as well, as in nanocomposites, for example (Fig. 1.12).

Nano-intermediates are the building blocks of nanostructured materials. Nanoparticles are the simplest objects we may think of from which to build nanomaterials. However, it is not always easy to align nanoparticles according to a given template by self-assembly. There are only few examples (mainly sulphides or selenides) combining with success soft templates, self-alignment of nanoparticles by dipolar interactions and oriented attachment resulting in 1D structures or hierarchical structures. Hard nanotemplates can be designed using different techniques such as electron lithography, for example, or anodization as for alumina membranes. Soft nanotemplates (such as mesophases and micellar systems) may also be

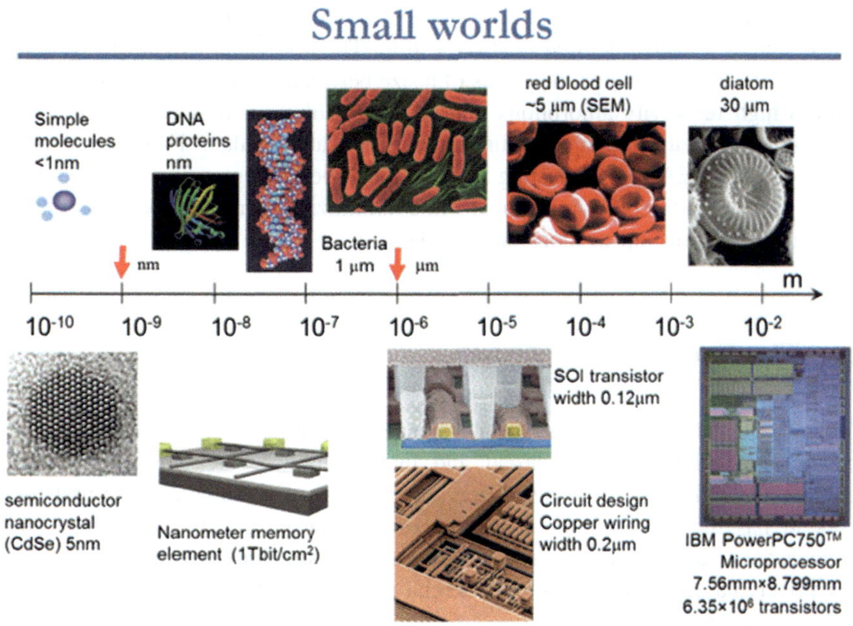

Fig. 1.12 Scale of things

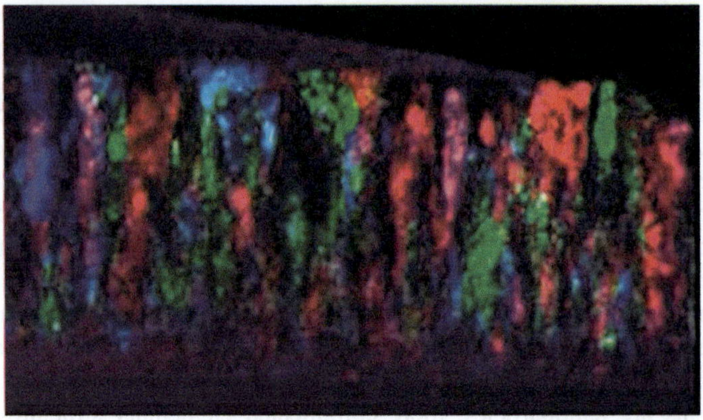

Fig. 1.13 Transverse view of a thin film of FeHf(N,O) nanostructured material. The ferro-magnetic grains have a size between 5 and 10 nm, finely dispersed across a high-resistivity amorphous matrix. This gives the material a high permeability and a high resistivity. From clef CEA n○59. Courtesy of CEA

used in the synthesis of nanoparticles with controlled size and shape, or mesoporous materials.

Films with a thickness less than 100 nm, supramolecular assemblies, dispersions of nanoparticles, etc. are nano-intermediates currently used to improve different technologies such as solar cells, batteries, catalysts, and drug delivery systems. Example of a nanostructured film. An illustration is shown in Fig. 1.13.

The picture shows a transverse section observed with a transmission electron microscope of a nanostructured FeHf(NO) film. The horizontal length of Fig. 1.13 is about 750 nm. Ferromagnetic grains of a size between 5 and 10 nm are dispersed inside a high resistivity amorphous matrix. The resulting nanostructured film has both high resistivity and high saturation magnetization value. These properties, obtained at the CEA, are interesting because they allow integrating this material in circuits close to inductive components, while minimizing parasitic capacities at high frequency operation (up to more than 2 GHz).

1.3 The Nanoworld

Nanomaterials are cornerstones of nanoscience and nanotechnology. Nanostructure science and technology is a broad and interdisciplinary area of research and development activity that has been growing explosively worldwide in the past few years. It has the potential for revolutionizing the ways in which materials and products are created and the range and nature of functionalities that can be accessed. It is already having a significant commercial impact, which will assuredly increase in the future (Fig. 1.14).

- The nanoworld
  - It ranges from 1 nm to ~100 nm.
  - Nanomaterials have at least one dimension in the nanometric range.

- Nanoscience has already many applications:
  - Electronics, Mechanics, Optics…
  - Biomedicine…

- One simple obvious characteristic of the nanomaterials is related to their surfaces or interfaces.
  - A large percentage of atoms are on the surface or in grain boundaries.
  - 2000 atoms in a 2nm gold cluster, ~50% on the surface.
  - The surface energy γ is the work needed to create a surface. The lowest values corresponds to the denser surfaces.
  - The size of the nanograins is a parameter that should be included into classical description of bulk materials when they are for nanomaterials.

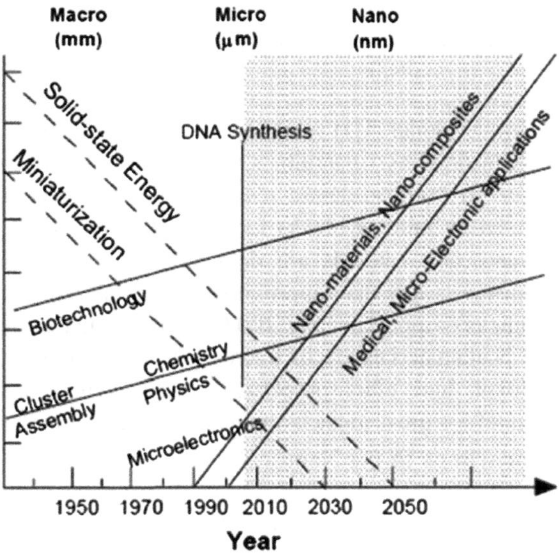

Fig. 1.14 Evolution of science, technology, and the future

Nanomaterials have attracted tremendous interest due to their noticeable performance in electronics, optics, and photonics. Nanomaterials are typically classified into three groups: 0-dimensional, 1-dimensional, and 2-dimensional. 0-dimensional nanostructures, referred to as quantum dots or nanoparticles with an aspect ratio near unity, have been extensively used in biological applications. 2-dimensional nanomaterials, such as thin films, have also been widely used as optical coatings, corrosion protection, and semiconductor thin film devices.

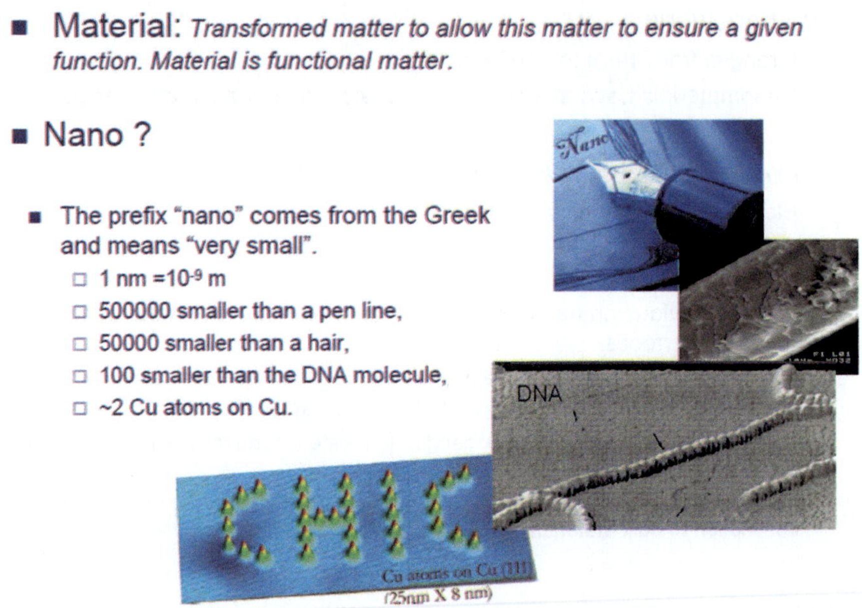

Fig. 1.15 Nanomaterial is functional matter

One-dimensional (1D) semiconductor nanostructures such as nanowires, nanorods (short nanowires), nanofibers, nanobelts, and nanotubes have been of intense interest in both academic research and industrial applications because of their potential as building blocks for other structures (Fig. 1.15).

Nanoscale materials are defined as a set of substances where at least one dimension is less than approximately 100 nm. A nanometer is one millionth of a millimeter—approximately 100,000 times smaller than the diameter of a human hair. Nanomaterials are of interest because at this scale unique optical, magnetic, electrical, and other properties emerge. These emergent properties have the potential for great impacts in electronics, medicine, and other fields.

Comparisons to the Nanoscale:

- One meter = One billion (1,000,000,000) nanometers
- One inch = 25,400,000 nm
- A human hair = approximately 50,000 nm in diameter
- Bacteria (single-celled microorganisms) = from a few hundred to 1,000 nm across
- 10 hydrogen atoms in a line = 1 nm (Figs. 1.16 and 1.17).

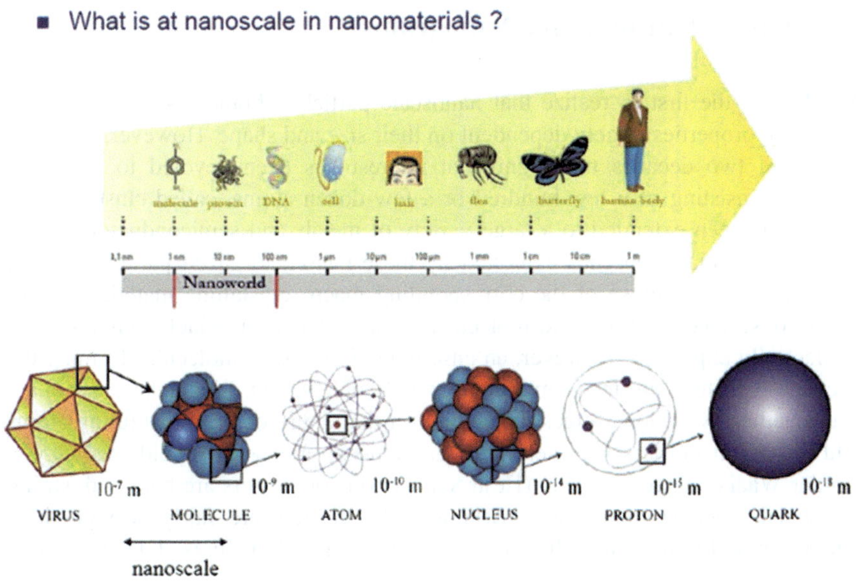

Fig. 1.16 Nanoscale in nanomaterials

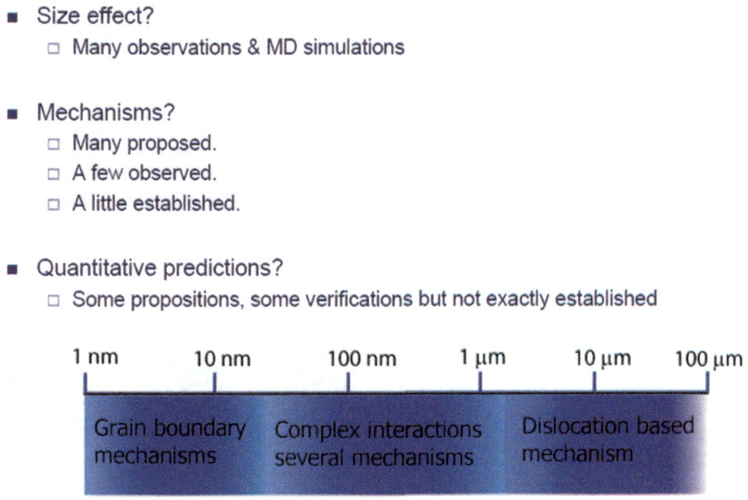

Fig. 1.17 Size effect, mechanisms and quantitative predictions

1.4 Atoms, Clusters and Nanograins

Oswald was the first to realize that nanoscale particles should display novel and interesting properties largely dependent on their size and shape. However, it is only in the last two decades that significant interest has been devoted to inorganic particles consisting of a few hundred or a few dozen atoms, called clusters. This interest has been extended to a large variety of metals and semiconductors and is due to the special properties exhibited by these nanometer sized particles, which differ greatly from those of the corresponding macro crystalline material. Matter that is constituted of atoms and molecules as such, has been widely classified and satisfactorily explained. However, an ensemble of atoms, or molecules forming the so-called 'Clusters' are far from being properly understood. Elemental clusters are held together by various forces depending on the nature of the constituting atoms hold elemental clusters together: Inert gas clusters are weakly held together by van-der-Waals interactions, e.g. (He)n Semiconductor clusters are held with strong directional covalent bonds, e.g. (Si)n Metallic clusters are fairly strongly held together by delocalized non-directional bonding, e.g. (Na)n (Figs. 1.18 and 1.19).

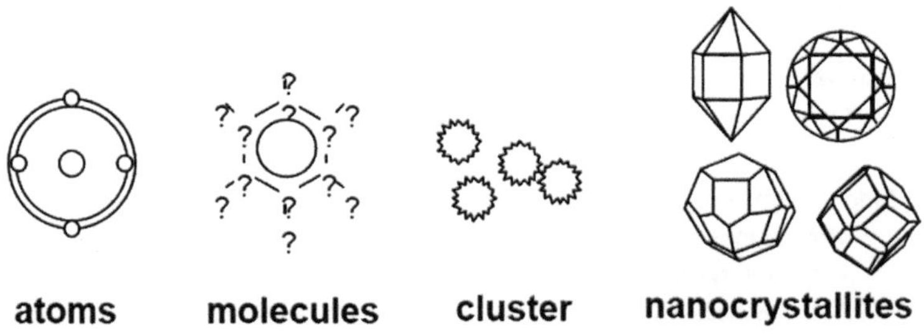

Fig. 1.18 Schematic representation of various states of matter

No. Of Shells	1	2	3	4	5
No. of atoms	M_{13}	M_{55}	M_{147}	M_{309}	M_{561}
Percentage of atoms	92%	76%	63%	52%	45%

Fig. 1.19 Idealized representation of hexagonal close packed full-shell 'magic number' clusters. Note that as the number of atoms increases, the percentage of surface atoms decreases

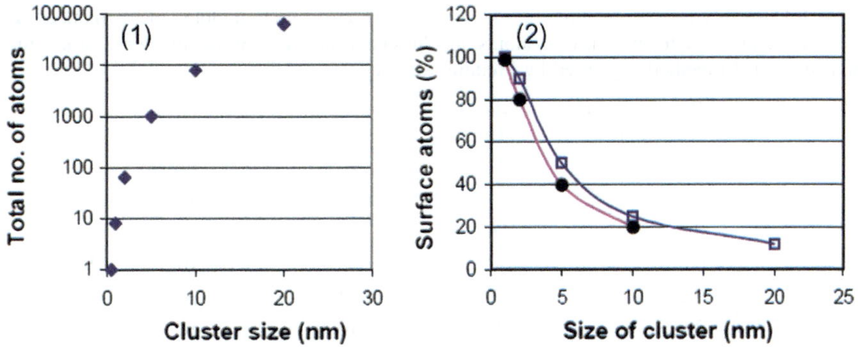

Fig. 1.20 (1) Total number of atoms with size of the cluster. (2) Number of surface atoms for a hypothetical model sphere of diameter 0.5 nm and density 1000 Kg m^{-3} with a mass of 6.5 10^{-26} kg occupying a volume of about 6.5×10^{-29} m^3 with a geometrical cross-section of 2×10^{-19} m^2 (in terms of atomic mass the sphere is considered to have a mass of 40 amu, where 1 amu = 1.67×10^{-27} kg). Calculated by **a** considering dense structures (Square), *ix* and **b** method suggested by Preining (dark circle)

Either elemental clusters or a mixture of clusters of different elements constitute the vast expanding field of materials sciences called ‘nanomaterials’. One has to be clear right at this stage that clusters are not a fifth state of matter, as sometimes believed, but they are simply intermediate between atoms on one hand, and solid or liquid state of matter on the other, with widely varying physical and chemical properties. Depending on the number of atoms forming the cluster determines the percentage of atoms that are exposed on the surface of the cluster. An example of such an ensemble of metal atoms show the decreasing number of surface atoms with increasing size of the cluster as shown in Fig. 1.20. When an ensemble of atoms add up to form a few nanometer sized clusters, they form what we call ‘nanoparticles’, since only a few atoms forming clusters are called ‘molecular clusters’. Agglomeration of a few atoms have been studied in great details by physicists working with molecular beams. Today, the mystery related to larger ensemble of atoms (in other words ‘nanomaterial’) are getting clearer due to active research being carried over across the world over the last decade or two (Table 1.3).

In the table, we see that the smaller particles contain only a few atoms, practically all at the surface. As the particle size increases from 1 to 10 nm, cross-section increases by a factor of 100 and the mass number of molecules by a factor of 1000. Meanwhile, the proportion of molecules at the surface falls from 100% to just 25%. For particles of 20 nm size, a little more than 10% of the atoms are on the surface. Of course, this is an idealized hypothetical case. If particles are formed by macromolecules (that are larger than the present example), number of molecules per particle will decrease and their surface fraction increase. The electronic properties

Table 1.3 Idealized representation of the variation of cross section, total mass, number of molecules and the effective surface atoms in clusters. *Note* (a) considering dense structures (Square), and (b) method suggested by Preining (dark circle)

Size (nm)	Cross section (10^{18} m^2)	Mass (10^{25} kg)	No. of molecules	Fraction of molecules at surface (%)	
				a	b
0,5	0.2	0.65	1		
1	0.8	5.2	8	100	99
2	3.2	42	64	90	80
5	20	650	1000	50	40
10	80	5200	8000	25	20
20	320	42,000	64,000	12	

Table 1.4 Particle size, surface area and surface energy of $CaCO_3$ (The surface energy of bulk $CaCO_3$ (calcite) is 0.23 J m^{-2})

Size (nm)	Surface area (m^2 mol^{-1})	Surface energy (J mol^{-1})
1	1.11×10^9	2.55×10^4
2	5.07×10^8	1.17×10^4
5	2.21×10^8	5.09×10^3
10	1.11×10^8	2.55×10^3
20	5.07×10^7	1.17×10^3
10^2	1.11×10^7	2.55×10^2
10^3(1 μm)	1.11×10^6	2.55×10^1

of this ensemble of atoms or molecules will be the result of their mutual interactions so that the overall chemical behaviour of the particles will be entirely different from the individual atoms or molecules that they are constituted of. They will also be different from their macroscopic bulk state of the substance in question under the same conditions of temperature and pressure (Table 1.4).

The idea of tailoring properly designed atoms into agglomerates has brought in new fundamental work in the search for novel materials with uncharacteristic properties. Among various types of nanomaterials, cluster assembled materials represent an original class of nanostructured solids with specific structures and properties. In terms of structure, they could be classified in between amorphous and crystalline materials. In fact, in such materials the short-range order is controlled by the grain size and no long-range order exists due to the random stacking of nanograins characteristic of cluster assembled materials. In terms of properties, they are generally controlled by the intrinsic properties of the nanograins themselves and by the interactions between adjacent grains. Cluster assembled films are formed by the deposition of these clusters onto a solid substrate and are generally highly porous with densities as low as about one half of the corresponding bulk materials densities and both the characteristic nanostructured morphology and a possible memory effect of the original free cluster structures are at the origin of their specific properties. From recent developments in the cluster source technologies (thermal, laser vaporization and sputtering), it is now possible to produce intense cluster beams of any materials, even the most refractory or complex systems (bimetallic, oxides and so on), for a wide range of size from a few atoms to a few thousands of atoms.

1.5 What Is Different at the Nanoscale?

Material is functional matter. Each of the different sized arrangement of gold atoms absorbs and reflects light differently based on its energy levels, which are determined by size and bonding arrangement. This is true for many materials when the particles have a size that is less than 100 nm in at least one dimension.

1. ***Below a certain length scale (that depends on interaction strengths) systems must be described using quantum mechanics.***

Ex. quantum dots, nanocatalysts, electronic transport through nanowires amd thin films

2. ***Many processes depend on the number of available energy states per unit energy. This quantity varies with the dimensionality of the system.***

3. ***The effective concentration of reactants that are confined in nanostructures may be very high.***

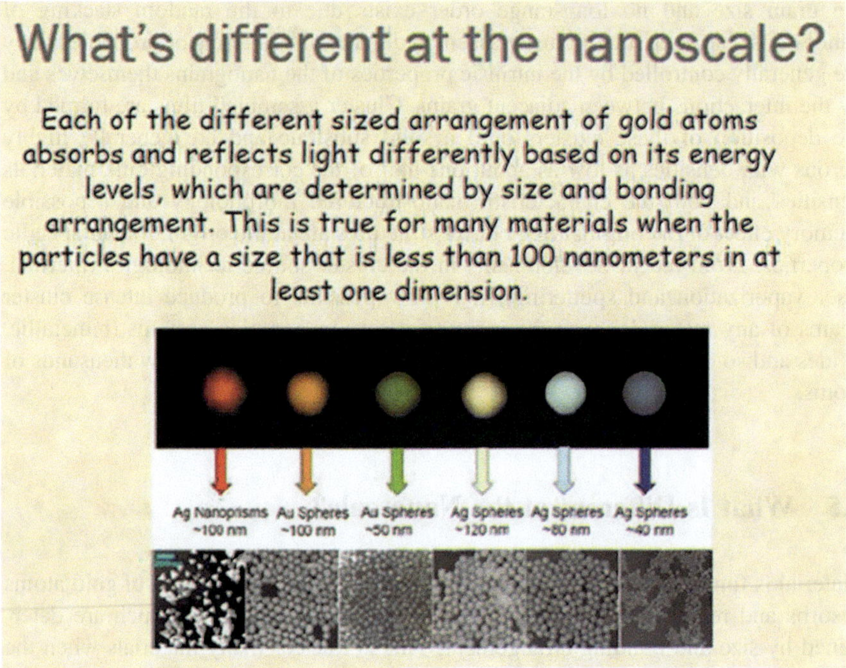

Material: *Material is functional matter.*

- *Grains in polycrystals*

Usually, crystalline materials are in the micrometric range, even being made of atoms with a nanometric size.

This micrometric range refers to the size of the grains in polycrystals.

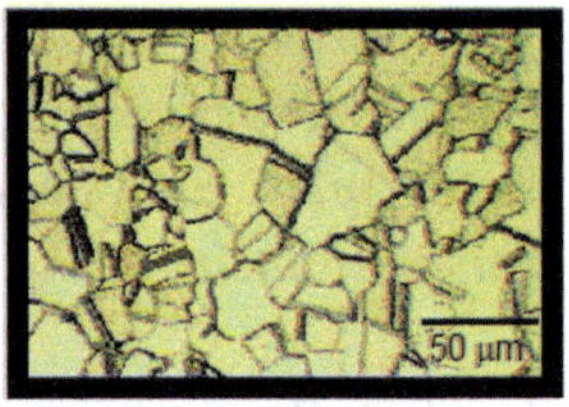

Grains in bulk gold

How many atoms in a 10 µm gold grain ?

2.5 10^{14} atoms assuming a spherical shape

- Nanoclusters

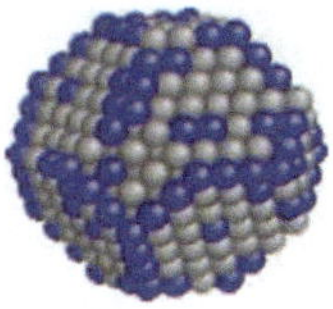

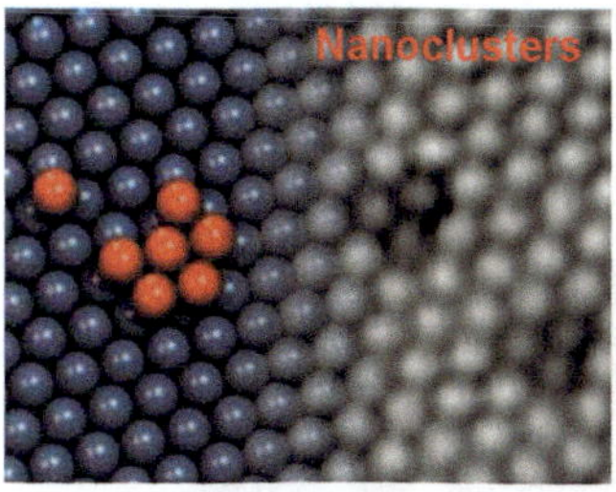

Deposition of atoms or molecules on a clean surface at low temperature.

The size ranges from 0.1 nm to a few nanometers.

Self organization
Specific properties → Molecular engineering

- Nanoclusters

Array of 25 Fe atoms dots
Laboratory of Nanostructures at Surfaces, EPFL

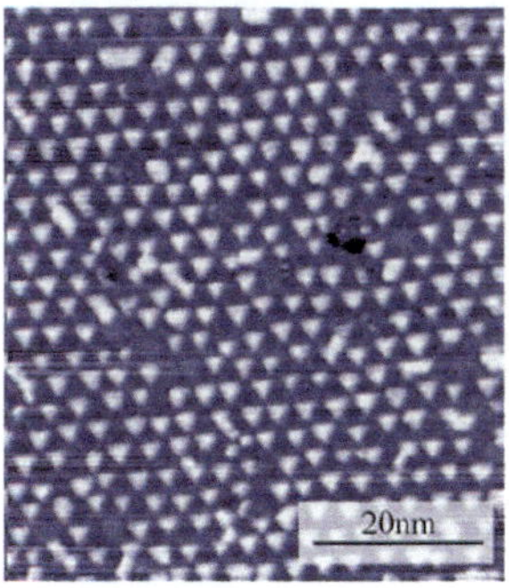

0.3ML MgO/Ag(100)
270nm x 200nm, T=50K
Laboratoire de Physique des Surfaces , EPFL

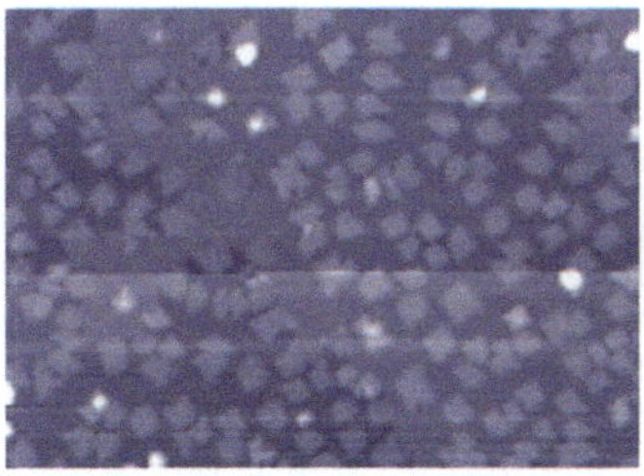

- Nanoclusters

Silver dendrite grown at 130 K on a platinum (111) surface
Laboratory of Nanoscale Science
Max-Planck-Institute for Solid State

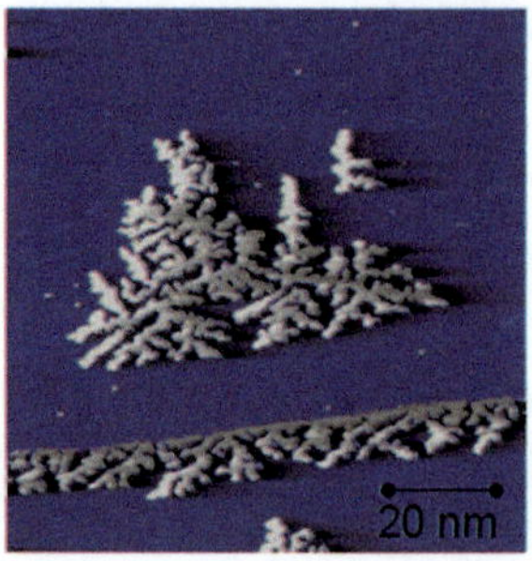

$C_{42}H_{28}$ (Rubrena)/Au(111)
60nm x 60nm, T=5K
Laboratoire de Physique des Surfaces , EPFL

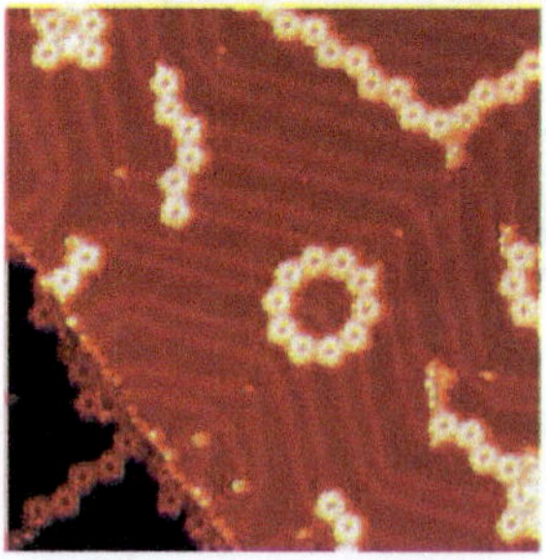

- From nanoclusters to nanomaterials
  - 1D: nanotubes
  - 2D: multilayers
  - 3D: embedded clusters

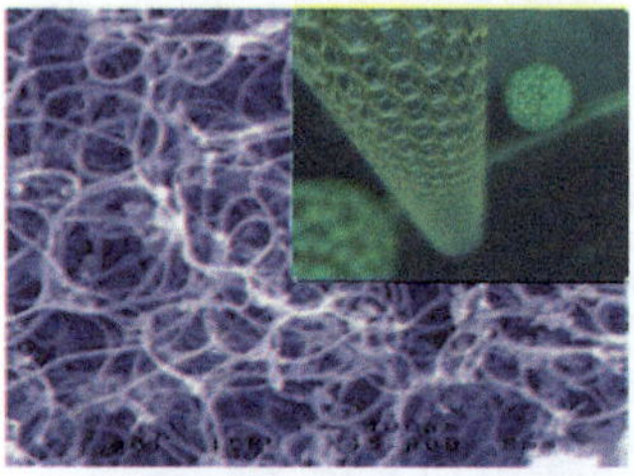

- **SEM pictures of carbon nanotubes**
diameters in the range 1-10 nm

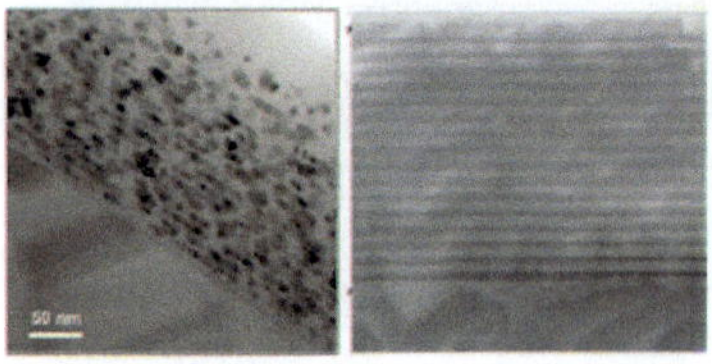

- *Multilayers of Fe-Pt-C on Si with a period of a few nm.*
D. Babonneau, LMP, Poitiers

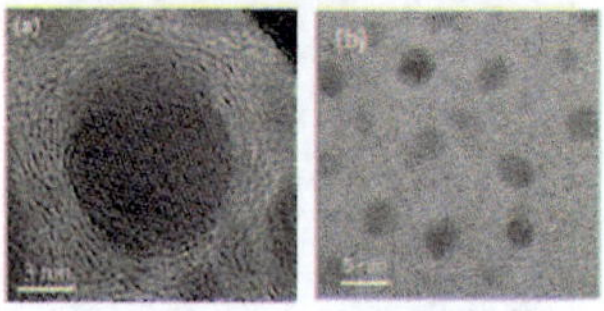

- *HRTEM of embedded silver clusters in (a) C and (b) BN.*
D. Babonneau, LMP, Poitiers

- From nanoclusters to nanomaterials
  - 3D: nanograins in thin films

TEM cross-sectional image and SAED pattern of 300nm **FePt thin film**

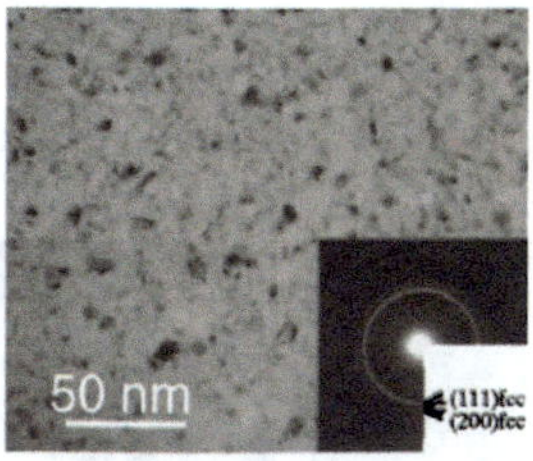

10nm FePt thin film

Nanomaterials: *at least one dimension at the nanometric range*

- Nanostructures (NSs)
  - A structure with a **d** at least less or equal to a critical one $\mathbf{d_c}$.
  - $\mathbf{d \leq d_c} \approx 100$ nm.
  - The value of $\mathbf{d_c}$ is dictated by a critical characteristic of some physical phenomena *(free path length of electrons, phonons, length of de Broglie wave, length of external electromagnetic and acoustical waves, correlation length, penetration length, diffusion length…)* giving rise to the size effects.
  - Nanostructures (NSs) are characterized by a form and a dimensionality **k**.

- Nanostructured Materials (NSMs)
  - Low dimensional materials comprising of building units of a submicron or nanoscale size at least in one direction and exhibiting size effects.
  - Nanostructured materials are characterized by a form, a dimensionality and by a composition.
- Nanocomposites (NCMs)
  - Nanocomposite materials (two *or more* component materials) cover a vast range of ceramic, metallic and ceramometallic materials in the form of bulk materials and thin films.

- The elementary "blocks"
  - All NSs can be built from elementary units (blocks) having low dimensionality 0D, 1D, and 2D.
  - *The 3D units are excluded because they can't be used to build low dimensional NSs except 3D matrix. 3D structures can be considered as NSMs if they involve the 0D, 1D, 2D NSs (Gleiter's classification of NSMs).*

- The 0D units (*3 dimensions in the nanometric range*)
  - Molecules, clusters, fullerenes, rings, metcarbs (metal carbides), particles, powders, grains...

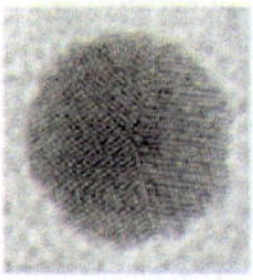

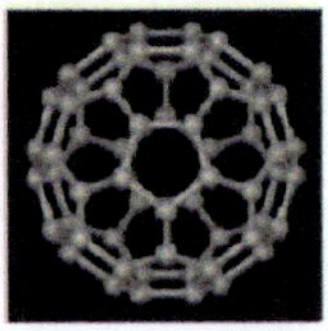

- The 1D units (*2 dimensions in the nanometric range*)
  - Nanotubes, fibers, filaments, whiskers, spirals, belts, springs, columns, needles,

Nanowires and nanoribbons of GaN
S. Biswas et al, Mat. Res. Bull 42, 2007

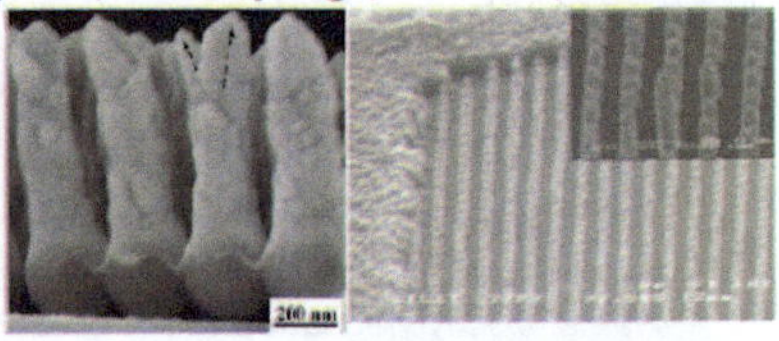

Ta and CdSe Nanopillars, CM Zhou et al, Thin Sol. Films 515 (2006) L.K. Teh et al., Thin Solid Films 515 (2007)

ZNO nanoneedles, R. Wahab et al, Applied Surf. Science 2007

- The 2D units (*1 dimension in the nanometric range*)
  - Layers

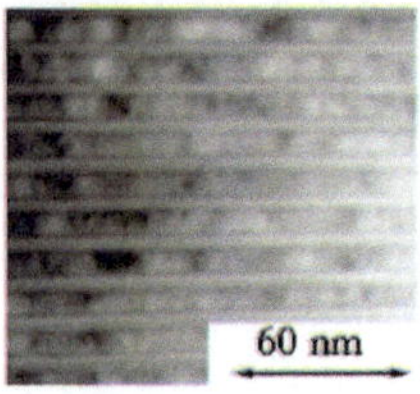

nc-Si/a-SiO2 superlattices, D.J. Lockwood, Journal of Crystal Growth 237–239 (2002) 1898

Future's Recommendations

- Cutting-edge paper-based battery is powered by electron-harvesting bacteria
- The best vaporizers for flower and concentrates
- A new way to 'freeze' water could help transform organ preservation
- Regular paints and plastics will soon be able to 'heal' like skin
- There's a new way to 3D print graphene, the strongest material on Earth.

1.6 History of Nanomaterials

Nanomaterials are interesting in many respects. Some have been used (or developed) over a considerable time and have been manufactured by chance without any understanding of the microscopic structure of matter. This is for example the case of the "Lycurgus cup", a glass cup with a mythological scene made probably in Rome around the 4th century AD. It is a vessel known as a cage-cup which was made by blowing or casting a thick glass blank. The glass was then cut and ground away leaving a decorative cage at the surface. The cup is made with a dichroic glass probably synthesized by accident. A change of color of the cup is observed when it is illuminated from inside: the opaque green cup becomes glowing translucent red. This phenomenon is caused by gold and silver nanoparticles which were formed during the fabrication of glass. The technique was further developed during the medieval age to make stained glass windows. The so-called "Ruby glass" it is still used today in order to make cadmium free intensely red glass.

The history of nanomaterials began immediately after the big bang when Nanostructures were formed in the early meteorites. Nature later evolved many other Nanostructures like seashells, skeletons etc. Nanoscaled smoke particles were formed during the use of fire by early humans. The scientific story of nanomaterials however began much later. One of the first scientific report is the colloidal gold particles synthesized by Michael Faraday as early as 1857. Nanostructured catalysts have also been investigated for over 70 years. By the early 1940s, precipitated and fumed silica nanoparticles were being manufactured and sold in USA and Germany as substitutes for ultrafine carbon black for rubber reinforcements.

Nanosized amorphous silica particles have found large-scale applications in many every-day consumer products, ranging from non-diary coffee creamer to automobile tires, optical fibers and catalyst supports. In the 1960s and 1970s metallic nanopowders for magnetic recording tapes were developed. In 1976, for the first time, nanocrystals produced by the now popular inert- gas evaporation technique was published by Granqvist and Buhrman. Recently it has been found that the Maya blue paint is a nanostructured hybrid material. The origin of its color and its resistance to acids and biocorrosion are still not understood but studies of authentic samples from Jaina Island show that the material is made of needle-shaped palygorskite (clay) crystals that form a superlattice with a period of 1.4 nm, with intercalates of amorphous silicate substrate containing inclusions of metal (Mg) nanoparticles. The beautiful tone of the blue color is obtained only when both these nanoparticles and the superlattice are present, as has been shown by the fabrication of synthetic samples.

• ***1981 Invention of Scanning Tunneling Microscopy***

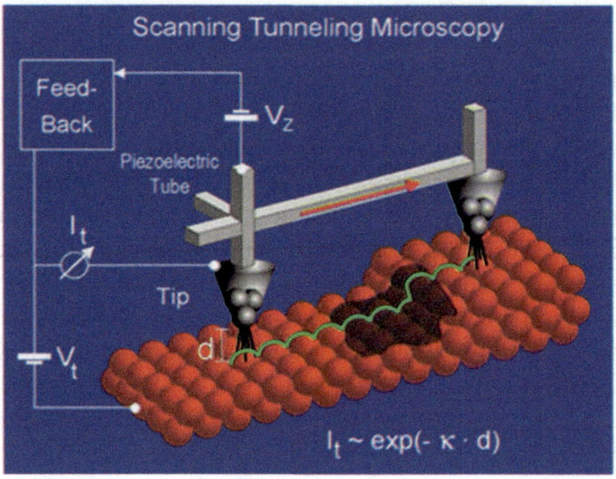

• G. Binning, H. Rohrer, C. Berger and E. Weibel
Surface studies by Scanning Tunneling Microscopy,
Phys. Rev. Lett. 49, 57-61 (1982)

• 1986 – First International Conference on STM
•Santiago de Compostela, Spain, 14-18 July

• 1986 – Nobel Prize to G. Binning and H. Rohrer

• 1988 – STM Imaging of DNA and biological structures

•1986 Invention of Atomic Force Microscopy

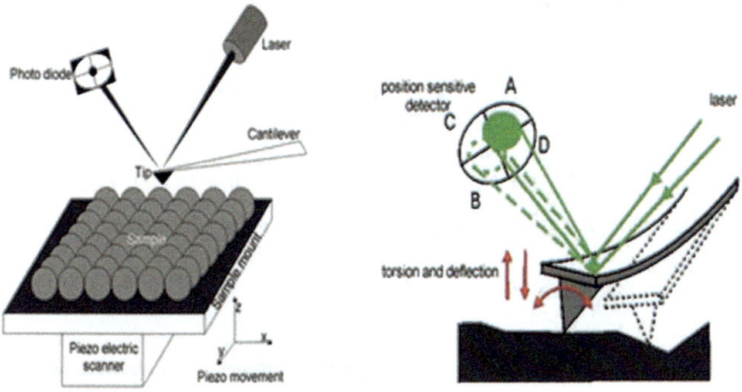

G. Binning, C.F. Quate and C. Berger
Atomic Force Microscopy, Phys. Rev. Lett. 56, 930-933 (1986)

Electron Beam Lithography

- *"We can reverse the lens of an electron microscope in order to demagnify as well as magnify……This, when you demagnify it 25,000x, it is still 80Å in diameter – 32 atoms across."*

Current e-beam technology allows features as small as 10 nm to be written.

Fresnel Lens made by EBL for focusing X-rays (submicron patterning)

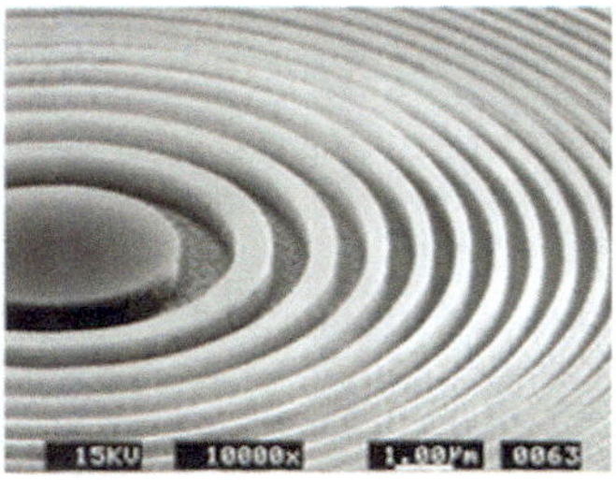

C. David, Paul Scherrer Institut

Today nanophase engineering expands in a rapidly growing number of structural and functional materials, both inorganic and organic, allowing to manipulate mechanical, catalytic, electric, magnetic, optical and electronic functions. The production of nanophase or cluster-assembled materials is usually based upon the creation of separated small clusters which then are fused into a bulk-like material or on their embedding into compact liquid or solid matrix materials. E.g. nanophase silicon, which differs from normal silicon in physical and electronic properties, could be applied to macroscopic semiconductor processes to create new devices. For instance, when ordinary glass is doped with quantized semiconductor "colloids," it becomes a high performance optical medium with potential applications in optical computing.

Stamping Nanostructures

- *"We would just have to press the same metal plate again into the plastic and we would have another copy."*

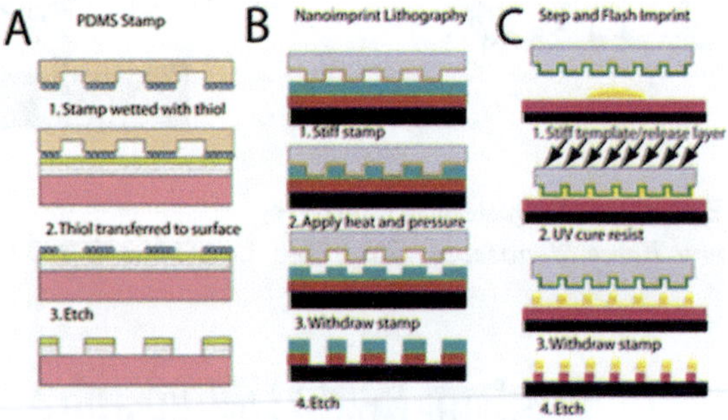

PDMS Stamp technology

Ion Beam Milling

- *"A source of ions, sent through the lens in reverse, could be focused to a very small spot."*

FIB = Focused Ion Beam

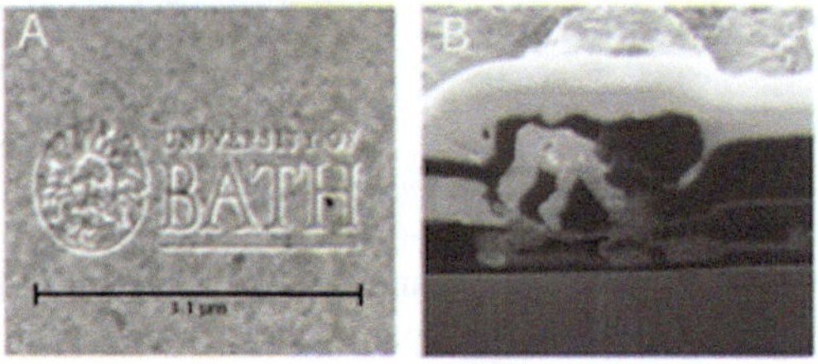

(Courtesy of FEI Inc.)

- Nanoparticles and nanostructured materials (NSM)
- Grain size on the order of 10^{-9} m or nm

- Extremely large specific surface area
- Manifest fascinating and useful properties
- Structural and non-structural applications
- Stronger, more ductile materials
- Chemically very active materials.

The continued decrease in device dimensions has followed the well-known *Moore's law* predicted in 1965 and illustrated in Fig. 1.21. The trend line illustrates the fact that the transistor size has decreased by a factor of 2 every 18 months since 1950.

There are many *nanoscale electronic devices* available now: tunneling junctions; devices with negative differential electrically configurable switches; carbon nanotube transistor; and single molecular transistor; ultrahigh density nanowires lattices and circuits with metal and semiconductor nanowires; etc. Devices have also been connected together to form circuits capable of performing single functions such as basic memory and logic function. Computer architecture based on nanoelectronics (also known as nanocomputers) has also been intensively studied. Various processing techniques have been applied in the fabrication of nanoelectronics such as focused ion beam (FIB), electron beam lithography, and imprint lithography. Major obstacles preventing the development of such devices include addressing nanometer-sized objects such as nanoparticles and molecules, molecular vibrations, robustness and the poor electrical conductivity. Certainly, nanomaterials play an important role not only in semiconductor—based electronics. *Nano-sized materials* currently are used in numerous industries, e.g., carbon black particles make rubber tires wear resistant; nanofibers are used for insulation and reinforcement of composites; iron oxide creates the magnetic material used in disk drives and audio-video tapes; nano-zinc oxides and titania are used as sunblocks for UV rays; etc. Nanoscale particles and nanothin layers of materials are being used, among other things, to make products lighter, stronger or more conductive. Some of the products on the market using nanotechnology are: magnetic recording tapes; computer hard drivers; bumpers on cars; solid–state compasses; protective and

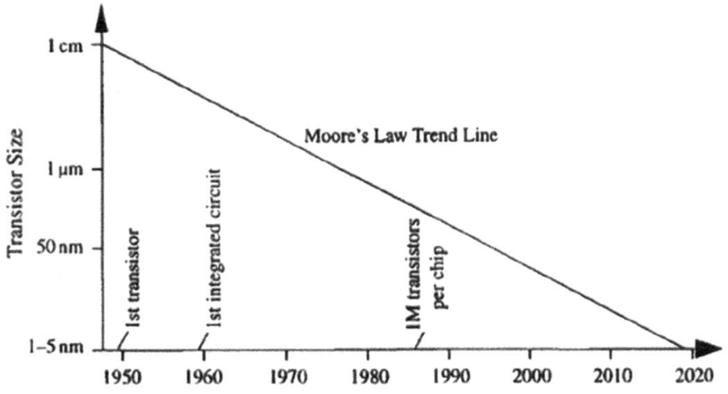

Fig. 1.21 "Moore's Law" plot of transistor size versus year

glare–reducing coatings for eyeglasses and windows; automobile catalytic converters; metal–cutting tools; dental bonding agents; longer–lasting tennis ball; burn and wound dressing; ink; etc. Promising applications of nanotechnology in medicine and/or biology have attracted a lot of attention and have become a fast growing field. One of the attractive applications in nanomedicine is the creation of nanoscale devices for improved therapy and diagnostics. Such nanoscale devices or nanorobots serve as vehicles for delivery of therapeutic agents, detectors or guardians against early disease and perhaps repair of metabolic or genetic defects. For applications in medicine, the major challenge is "miniaturization": new instruments to analyze tissues literally down to the molecular level, sensors smaller than a cell allowing looking at ongoing functions, and small machines that literally circulate within a human body pursuing pathogens and neutralizing chemical toxins. Researchers expect to develop *new commercial* ***applications for nanotechnology*** for the next several years. They include: advanced drug – delivering systems, including implantable devices that automatically administer drugs and sense drug levels; medical diagnostic tools, such as cancer–tagger mechanisms and "lab-on-a-chip" diagnostics for physicians; cooling chips or wafer to replace compressors in cars, refrigerators, air conditioners and other devices, using no chemicals or moving parts; sensors for airborne chemicals or other toxins; solar fuel cells and portable power to provide inexpensive, clean energy; etc.

In the following chapters we will review the various developments that have been revolutionizing the synthesis and application of nanomaterials. We will attempt to correlate the improvements in the material properties that are achieved due to the fine microstructures arising from the size of the grains and/or dimensionality.

References

1. Allhoff, F., Lin, P., Moore, D.: What is Nanotechnology and Why Does it Matter?: From Science to Ethics, p. x, 293. Wiley-Blackwell, Chichester, UK, Malden, MA (2010)
2. Viswanathan, B.: Nanomaterials. Narosa Publishing House
3. Zhang, J.Z.: Optical Properties and Spectroscopy of Nanomaterials. World Scientific Publishing Co. Pte. Ltd
4. Sajanlal, P.R., Sreeprasad, T.S., Samal, A.K., Pradeep, T.: Anisotropic nanomaterials: structure, growth, assembly, and functions. Nano Rev. **2** (2011)
5. Lue, J.T.: Physical properties of nanomaterials. Encyclopedia of Nanoscience and Nanotechnology, vol. X, pp. 1–46
6. Nanomaterials—An Introduction to Synthesis, Properties and Application. Environ. Eng. Manag. J. **7**(6), 865–870 (2008)
7. Law, M., Goldberger, J., Yang, P.: Annu. Rev. Mater. Sci. **34**, 83 (2004)
8. Wu, Y., Yang, P.: J. Am. Chem. Soc. **123**, 3165 (2001)
9. Wu, Y., Yan, H., Huang, M., Messer, B., Song, J., Yang, P.: Chem. Euro. J. **8**, 1260 (2002)
10. Huang, M., Mao, S., Feick, H., Yan, H., Wu, Y., Kind, H., Weber, E., Russo, R., Yang, P.: Science **292**, 1897 (2001)
11. Kuykendall, T., Pauzauskie, P.J., Zhang, Y., Goldberger, J., Sirbuly, D., Denlinger, J., Yang, P.: Nat. Mater. **3**, 528 (2004)

12. Wu, Y., Fan, R., Yang, P.: Nano Lett. **2**, 83 (2002)
13. Choi, H., Johnson, J., He, R., Lee, S., Saykally, R., Yang, P.: J. Phys. Chem. B **107**, 8721 (2003)
14. He, R., Law, M., Fan, R., Kim, F., Yang, P.: Nano Lett. **2**, 1109 (2002)
15. Goldberger, J., He, R., Lee, S., Zhang, Y., Yan, H., Choi, H., Yang, P.: Nature **422**, 599 (2003)
16. Messer, B., Song, J.H., Yang, P.: J. Am. Chem. Soc. **122**, 10232 (2000)
17. Yang, P.: Nature **425**, 243 (2003)
18. Tao, A., Kim, F., Hess, C., Goldberger, J., He, R., Sun, Y., Xia, Y., Yang, P.: Nano Lett. **3**, 1229 (2003)
19. Lin, Y.-M., Dresselhaus, M.S.: Phys. Rev. B **68**, 075304 (2003)
20. Li, D., Wu, Y., Fan, R., Yang, P., Majumdar, A.: Appl. Phys. Lett. **83**, 3186 (2003)
21. Li, D., Wu, Y., Kim, P., Shi, L., Yang, P., Majumdar, A.: Appl. Phys. Lett. **83**, 2934 (2003)
22. Johnson, J., Choi, H.J., Knutsen, K.P., Schaller, R.D., Yang, P., Saykally, R.J.: Nat. Mater. 101 (2002)
23. Johnson, J., Yan, H., Yang, P., Saykally, R.: J. Phys. Chem. B **107**, 8816 (2003)
24. Johnson, J., Yan, H., Schaller, R., Saykally, R., Yang, P.: J. Phys. Chem. B **105**, 11387 (2001)
25. Kind, H., Yan, H., Messer, B., Law, M., Yang, P.: Adv. Mater. **14**, 158 (2002)
26. Law, M., Sirbuly, D., Johnson, J., Goldberger, J., Saykally, R., Yang, P.: Science **305**, 1269 (2004)
27. Adair, J., Parette, M., Erhan, I., Kester, M.: Nanoparticulate alternatives for drug delivery. ACS Nano **4**(9), 4967–4970 (2010)
28. Lockman, P., Mumper, R., Khan, M., Allen, D.: Nanoparticle technology for drug delivery across the blood-brain barrier. Drug Dev. Ind. Pharm. **28**(1), 1–13 (2002)
29. Takenaga, M., Serizawa, Y., Azechi, Y., et al.: Microparticle resins as a potential nasal drug delivery system for insulin. J. Control. Release **52**(1–2), 81–87 (1998)
30. Hilbert, A., Fritzsche, U., Kissel, T.: Biodegradable microspheres containing influenza A vaccine: immune response in mice. Vaccine **17**(9–10), 1065–1073 (1999)
31. Bodmeier, R., Chen, H., Paeratakul, O.: A novel approach to the oral delivery of micro- or nanoparticles. Pharm. Res. **6**(5), 413–417 (1989)
32. Kawaguchi, H.: Functional polymer microspheres. Prog. Polym. Sci. **25**(8), 1171–1210 (2000)
33. Pereira, D., Silva-Correia, J., Caridade, S., et al.: Development of gellan gum-based microparticles/hydrogel matrices for application in the intervertebral disc regeneration. Tissue Eng. Part C **17**(10), 961–972 (2011)
34. Custódio, C., Cerqueira, M., Marques, A., Reis, R., Mano, J.: Cell selective chitosan microparticles as injectable cell carriers for tissue regeneration. Biomaterials **43**, 23–31 (2015)
35. Costa, R., Custódio, C., Arias, F., Rodríguez-Cabello, J., Mano, J.: Nanostructured and thermoresponsive recombinant biopolymer-based microcapsules for the delivery of active molecules. Nanomed. Nanotechnol. Biol. Med. **9**(7), 895–902 (2013)
36. Custódio, C., Santo, V., Oliveira, M., Gomes, M., Reis, R., Mano, J.: Functionalized microparticles producing scaffolds in combination with cells. Adv. Funct. Mater. **24**(10), 1391–1400 (2014)
37. Lima, A., Puga, A., Mano, J., Concheiro, A., Alvarez-Lorenzo, C.: Free and copolymerized γ-cyclodextrins regulate the performance of dexamethasone-loaded dextran microspheres for bone regeneration. J. Mater. Chem. B **2**(30), 4943–4956 (2014)
38. Simón-Yarza, T., Formiga, F., Tamayo, E., Pelacho, B., Prosper, F., Blanco-Prieto, M.: PEGylated-PLGA microparticles containing VEGF for long term drug delivery. Int. J. Pharm. **440**(1), 13–18 (2013)
39. Manocha, M., Pal, P., Chitralekha, K., et al.: Enhanced mucosal and systemic immune response with intranasal immunization of mice with HIV peptides entrapped in PLG microparticles in combination with Ulex Europaeus-I lectin as M cell target. Vaccine **23**(48–49), 5599–5617 (2005)

40. Wischke, C., Schwendeman, S.: Principles of encapsulating hydrophobic drugs in PLA/PLGA microparticles. Int. J. Pharm. **364**(2), 298–327 (2008)
41. Hamidi, M., Azadi, A., Rafiei, P.: Hydrogel nanoparticles in drug delivery. Adv. Drug Deliv. Rev. **60**(15), 1638–1649 (2008)
42. Grislain, L., Couvreur, P., Lenaerts, V., Roland, M., Deprez-Decampeneere, D., Speiser, P.: Pharmacokinetics and distribution of a biodegradable drug-carrier. Int. J. Pharm. **15**(3), 335–345 (1983)
43. Dubernet, C., Couvreur, P.: Nanoparticles in cancer therapy and diagnosis. Adv. Drug Deliv. Rev. **54**(Supplement), 631–651 (2002)
44. Muller, R., Maaben, S., Weyhers, H., Mehnert, W.: Phagocytic uptake and cytotoxicity of solid lipid nanoparticles (SLN) sterically stabilized with Poloxamine 908 and Poloxamer 407. J. Drug Target. **4**(3), 161–170 (1992)
45. Lamprecht, A., Ubrich, N., Yamamoto, H., Afer, U., Takeuchi, H.: Biodegradable nanoparticles for targeted drug delivery in treatment of inflammatory bowel disease. J. Pharmacol. Exp. Ther. **299**(2), 775–781 (2001)
46. Subia, B., Kundu, S.: Drug loading and release on tumor cells using silk fibroin-albumin nanoparticles as carriers. Nanotechnology **24**(3), 035103 (2013)
47. Alexis, F., Pridgen, E., Molnar, L., Farokhzad, O.: Factors affecting the clearance and biodistribution of polymeric nanoparticles. J. Pharm. Sci. **5**(4), 505–515 (2008)
48. Danhier, F., Ansorena, E., Silva, J., Coco, R., Le Breton, A., Préat, V.: PLGA-based nanoparticles: an overview of biomedical applications. J. Control. Release **161**(2), 505–522 (2012)
49. Singh, R., Lillard, J.: Nanoparticle-based targeted drug delivery. Exp. Mol. Pathol. **86**(3), 215–223 (2009)
50. Desai, M., Labhasetwar, V., Amidon, G., Levy, R.: Gastrointestinal uptake of biodegradable microparticles: effect of particle size. Pharm. Res. **13**(12), 1838–1845 (1996)
51. Ratner, B., Hoffman, A., Schoen, F., Lemons, J.: Biomaterials Science: An Introduction to Materials in Medicine. Academic Press, Third Edit. Canada (2014)
52. Moghimi, S., Hunter, A., Murray, J.: Long-circulating and target-specific nanoparticles: theory to practice. Pharmacol. Rev. **53**(2), 283–318 (2001)
53. Patil, V., Campbell, C., Yun, Y., Slack, S., Goetz, D.: Particle diameter influences adhesion under flow. Biophys. J. **80**(4), 1733–1743 (2001)
54. Dunne, M., Corrigan, O., Ramtoola, Z.: Influence of particle size and dissolution conditions on the degradation properties of polylactide-co-glycolide particles. Biomaterials **21**(16), 1659–1668 (2000)
55. Jiang, W., Kim, B., Rutka, J., Chan, W.: Nanoparticle-mediated cellular response is size-dependent. Nat. Nanotechnol. **3**(3), 145–150 (2008)
56. Xiao, K., Li, Y., Luo, J., et al.: The effect of surface charge on in vivo biodistribution of PEG-oligocholic acid based micellar nanoparticles. Biomaterials **32**(13), 3435–3446 (2012)
57. Shah, N., Vercellotti, G., White, J., Fegan, A., Wagner, C., Bischof, J.: Blood-nanoparticle interactions and in vivo biodistribution: impact of surface PEG and ligand properties. Mol. Pharm. **9**(8), 2146–2155 (2012)
58. Illum, L., Davis, S., Wilson, C., Thomas, N., Frier, M., Hardy, J.: Blood clearance and organ deposition of intravenously administered colloidal particles. The effects of particle size, nature and shape. Int. J. Pharm. **12**(2–3), 135–146 (1982)
59. Kanke, M., Simmons, G., Weiss, D., Bivins, B., Deluca, P.: Clearance of 141Ce-labeled microspheres from blood and distribution in specific organs following intravenous and intraarterial administration in beagle dogs. J. Pharm. Sci. **69**(7), 755–762 (1980)
60. Ikada, Y.: Phagocytosis of polymer microspheres by macrophages. Adv. Polym. Sci. **94**, 107–141 (1990)
61. Moghimi, S., Hedeman, H., Muir, I., Illum, L., Davis, S.: An investigation of the filtration capacity and the fate of large filtered sterically-stabilized microspheres in rat spleen. Biochim. Biophys. Acta **1157**(3), 233–240 (1993)

62. Stolnik, S., Illum, L., Davis, S.: Long circulating microparticulate drug carriers. Adv. Drug Deliv. Rev. **16**(95), 195–214 (1995)
63. Longmire, M., Choyke, P., Kobayashi, H.: Clearance properties of nano-sized particles and molecules as imaging agents: considerations and caveats. Nanomedicine **3**(5), 703–717 (2012)
64. Choi, H., Liu, W., Misra, P., et al.: Renal clearance of quantum dots. Nat. Biotechnol. **25**(10), 1165–1170 (2007)
65. Vinogradov, S., Bronich, T., Kabanov, A.: Nanosized cationic hydrogels for drug delivery: preparation, properties and interactions with cells. Adv. Drug Deliv. Rev. **54**(1), 135–147 (2002)
66. Muñoz-Mármol, M., Crespo, J., Fritts, M., Maojo, V.: Towards the taxonomic categorization and recognition of nanoparticle shapes. Nanomedicine **11**(2), 457–465 (2015)
67. Truong, N., Whittaker, M., Mak, C., Davis, T.: The importance of nanoparticle shape in cancer drug delivery. Expert Opin. Drug Deliv. **12**(1), 1–14 (2014)
68. Champion, J., Katare, Y., Mitragotri, S.: Particle shape: a new design parameter for micro- and nanoscale drug delivery carriers. J. Control. Release **121**(1–2), 3–9 (2014)
69. Geng, Y., Dalhaimer, P., Cai, S., et al.: Shape effects of filaments versus spherical particles in flow and drug delivery. Nat. Nanotechnol. **2**(5), 249–256 (2007)
70. Choi, C., Chae, S., Nah, J.-W.: Thermosensitive poly (Nisopropylacrylamide)- b-poly (3-caprolactone) nanoparticles for efficient drug delivery system. Polymer (Guildf). **47**(13), 4571–4580 (2006)
71. Jiang, J., Oberdorster, G., Biswas, P.: Characterization of size, surface charge, and agglomeration state of nanoparticle dispersions for toxicological studies. J. Nanoparticle Res. **11**(1), 77–89 (2009)
72. Wang, A., Pu, K., Dong, B., et al.: Role of surface charge and oxidative stress in cytotoxicity and genotoxicity of graphene oxide towards human lung fibroblast cells. J. Appl. Toxicol. **33** (10), 1156–1164 (2013)
73. Thiele, L., Merkle, H., Walter, E.: Phagocytosis and phagosomal fate of surface-modified microparticles in dendritic cells and macrophages. Pharm. Res. **20**(2), 221–228 (2003)
74. Parveen, S., Misra, R., Sahoo, S.: Nanoparticles: a boon to drug delivery, therapeutics, diagnostics and imaging. Nanomed. Nanotechnol. Biol. Med. **8**(2), 147–166 (2012)
75. Orive, G., Anitua, E., Pedraz, J., Emerich, D.: Biomaterials for promoting brain protection, repair and regeneration. Nat. Rev. Neurosci. **10**(9), 682–692 (2009)

Chapter 2
Principles of Computational Simulations Devices and Characterization of Nanoelectronic Materials

Quantum mechanical effects can arise and the classical Ohm's law of electronic transport breaks down. On the other hand, such quantum effects are not necessarily harmful if we have the basic understanding of them, as they can be used to design radically new types of electronic components. The development and manufacturing of such components at the scale of nanometers is the key aspect and idea of nanotechnology.

Besides technological prospects, nano-scale devices also offer a convenient platform to explore the fundamental physics of electrons in solids and molecules. In this ongoing journey of understanding the nature of materials, researchers around the world have developed many theories, among which the density functional theory (DFT) has emerged as one of the most powerful tools. This theory, which in its modern version dates back to the seminal work by Kohn and his collaborators in the mid-1960s, has been the most prominent contributor to the rising field of computational materials science. In this field, the basic properties of atomic configurations are investigated from first principles by using numerical simulations, which are often quite expensive. Consequently, such computational approaches depend equally on both state-of-the-art computer hardware and efficient numerical algorithms in order to be successful.

One of the most active areas of computational materials research in recent years, yet also one of the numerically most costly, is the combination of DFT with modern theories of quantum transport in the attempt to study electronic transport through nano-scale devices under external bias. Among the variety of schemes developed, a special place is occupied by the application of the nonequilibrium Green's function (NEGF) method within the DFT approach. This particular approach is by far the most widely used method, and, with some appropriate approximations and optimized algorithms, NEGF-DFT has become a cost-effective first-principles method and has reproduced a variety of groundstate properties within a few percentage of error compared with experimental data. However, for the vast majority of

L. H. Madkour, *Nanoelectronic Materials*, Advanced Structured Materials 116, https://doi.org/10.1007/978-3-030-21621-4_2

interesting cases, the sizes of the nanoscale systems considered in calculations have been much smaller than realistic experimental configurations. Usually the number of atoms is no more than a few hundred. The reason for this is in general a lack of sufficient computational resources and/or efficiently implemented numerical algorithms.

The computational aspects of modeling electronic transport in nano-scale devices. From a physical starting point, an accurate quantum mechanical model of these devices should have the ability to capture the following fundamental effects:

- Electron wave interference.
- Quantum mechanical tunneling.
- Discrete energy levels due to confinement in 2D and 3D device geometries.
- Scattering processes (electron-phonon, electron-electron).

The first three effects can be accommodated most simply in the Landauer-Büttiker transport theory in conjunction with the solution of the electronic Schrödinger's equation. The fourth effect requires, in general, the NEGF formalism, or equivalent theories, to account for energy, momentum and phase relaxation due to inelastic scattering processes.

2.1 Charged Particle Single Nanometre Manufacturing

Methods for nanostructuring—the fabrication of structures with sizes on the nanoscale—are required for the fabrication of nextgeneration nanoelectronics, such as quantum-based devices, for nanophotonics, nanobiotechnology, nanomaterials and nanoelectro-mechanical systems. Until now, the leading method for scaled-up fabrication of nanostructures has been optical lithography, combined with pattern transfer techniques including plasma etching. Despite its success, optical lithography is reaching its resolution limits and new structuring techniques are required. Among several candidates, molecular self-assembly and self-organization of structures represent the so-called bottom-up approach. Nanoindentation, thermal scanning probe lithography, local oxidation lithography, dip-pen lithography, extreme UV lithography or X-ray lithography are the leading top-down methods. In this review we focus on another topdown generic technology, namely nanostructuring by charged particle beams used to expose a resist, which can be used as a mask for pattern transfer etching and metal deposition etc.

Nanolithography using charged particle beams can be divided into two approaches: either the particle beam is generated far from the sample and scanned over it using electric or magnetic focusing and deflection optics or the beam is generated at a field-emission tip located in close proximity to the sample and the tip itself is scanned to generate the lithographic pattern. The first approach, termed scanned beam technology, comprises electron and ion beam lithographies and electron/ion beam induced deposition. It has its origins in the scanning electron microscope and, more recently, the scanning ion microscope. Scanning probe lithography, the second approach, also stems from microscopy in the form of scanning tunneling microscopy and atomic force microscopy; the corresponding lithography techniques include scanning tunneling lithography and field emission scanning probe lithography.

Since the introduction of the integrated circuit in the late 1950s, the number of individual transistors that can be placed upon a single integrated circuit chip has approximately quadrupled every three years (Moore's law—the number of transistors on a single integrated circuit chip roughly doubles every 18 months [1]). Remarkably, this projection has remained valid since the early 1960s. This impressive, and longstanding, rate of technological progress is poised to significantly decrease in the near future, as the ultimate limits of conventional transistor technology are being reached. Aside from the atomic nature of materials, the continual scaling of transistors is being challenged by many so-called "short-channel" effects which arise in sub-50 nm devices and compromise device functionality. With the next generation of state-of-the-art commercial devices achieving 22 nm, it is believed that without a major revitalization Moore's law is likely nearing its end. In 1980, Hewlett-Packard produced a single-chip microprocessor containing approximately half million devices in its 1 cm^2 area. This chip was produced with transistors having a nominal 1.25 mm gate length and was considered a remarkable step forward. In contrast, by 2007, the printed gate length of production microprocessor transistors was 48 nm and the physical gate length was closer to 25 nm. Research devices have been demonstrated down to 10 nm gate length or less. Clearly, current integrated circuit manufacturing is truly a nanoscale technology. For a 25 nm gate length Si device, the number of atoms spanning the channel is on the order of a 100 or less. It has been concluded that one can easily downsize the transistor to a gate length of 30 nm if macroscopic transport theory continued to hold. There were predictions [www.itrs.net] of scaling down to 10 nm gate length before a serious "brick wall" is encountered and 15 nm gate lengths are scheduled for production by 2010. Laboratory MOSFET devices with gate lengths down to 15 nm have been reported by Intel [2] and AMD [3] which exhibit excellent I-V characteristics, and 6 nm gate length p-channel transistors have been reported by IBM [4].

The (slow) death of Moore's law has been the main driver of progress in today's information and communications society. One of the possible routes by which the current technology can be improved, is called "Beyond Moore" and it includes radically new system architectures whose building blocks may be atomic and molecular devices, and also the substitution of Boolean logic by more efficient algorithms, possibly based on quantum computation. Very clever designs are pushing the limits of today's CMOS technology even further, since as Gordon Moore put it, "No Exponential is forever, but we can delay 'forever'". But there is no doubt that in the end the venerable CMOS micro- and even "nano-electronics" technology must give way to radically new approaches, that will be based in quantum physics and chemistry, like graphene electronics, or molecular electronics [5, 6] or spinotronics.

2.2 Exotic Effects and Potential

Even though molecular electronics is a relatively new field the advances in this area of research have been significant Molecules have a series of advantages over the traditional silicon based lithographic techniques. Besides the obvious huge increase in packing density of electronic components, they can be synthesized easily and with low cost in many cases; they can be grown in three dimensions as opposed to the typical 2D semiconductor lithographic layout; they can self-assemble; they are expected to reduce heat and noise production and, most importantly, they can display a whole new world of properties and functionalities.

Recent growing interest in making electronic devices using molecules comprising of Carbon chains or benzene rings aims at replacing standard semiconductors with organic materials. They have the advantages of being produced with low-temperature low-cost chemical methods, instead of expensive high-temperature solid-state growth (e.g. molecular beam epitaxy) and patterning (lithography) techniques. In addition the endless possibilities of chemical synthesis, end-group and side-chain engineering give good expectation for new concept devices. From the electronic side negative differential resistance [7] and rectification [8] have already been proved at the molecular level and fully functional molecular transistors [9], memories and logic gates [10, 11] have been demonstrated suggesting a possible road-map to the post-silicon era. They should produce future generations of computers, together with magnetic data storage devices exceeding the Terabit/in^2 storage limit. The readout of such high-density data storage media will be achieved using nanoscale devices with magnetic atomic point contacts [12, 13]. Since the seminal work of Reed et al. [14], several experiments have been performed on a variety of molecules [7–9, 15–20]. Although the feasibility of single molecule devices has been shown, a number of issues have been raised by these experiments. Probably the most important of them is the nature of the contact between the molecule and the surface of the electrodes. One of the fundamental aspects of molecular electronics is that the bonding between a molecule and the current/voltage probes can dramatically change the current flowing through a device compare to conventional inorganic heterostructures. The bonding site and the bond angle usually depend on the layer density (coverage), and these can be further tuned by changing the anchoring groups. Ultimately we can say that molecules present great potential but also introduce new complications, especially from the fabrication point of view. Greater control and characterization is necessary when designing devices.

2.3 Preliminary Concepts: Elements from Solid State Physics

In order to make the reader minimally familiar with the important concepts and approximations that underly the numerical calculation of electronic transport, we will begin this thesis from the point of view of solid state physicists. The goal is to convey in few words how the stationary quantum mechanical properties of nano-scale structures can be found from the self-consistent solution of an eigenvalue problem and a Poisson problem. In addition, we also wish to demonstrate that the solution can be obtained in a numerical calculation on a standard computer, in a matter of minutes, if the structures are relatively small or bulk-like. We refer the reader to the excellent text books by Ashcroft and Mermin [21], Martin [22], and Datta [23] for more detailed presentations.

2.4 Computing Electronic Transport

As the miniaturization of devices begins to reveal the atomic nature of materials, where chemical bonding and quantum effects are important, one must resort to a parameter-free theory for predictions. Alongside of large experimental investigations and explorations on nanoscale devices, equally large efforts have been devoted to develop efficient computational methods for evaluating conduction characteristics of nanoscale devices. Simultaneously or many times way ahead of experimental progress, there have been theoretical/computational works predicting and analysing the physics behind the transport of electrons through molecular junctions. This is a great theoretical challenge since advanced quantum transport algorithms must be combined with state of the art electronic structure methods. Modem theory of quantum transport has developed a range of methods for calculating transport in nanoscale conductors. Various methods which are used for computing study of electronic transport can be divided in following categories.

2.4.1 Electronic Structure Calculations

Nano-scale devices usually consist of thousands of atoms which, in turn, contains equally many nuclei and possibly a much larger number of electrons. Nuclei and electrons interact with each other and mutually among themselves. These interactions establishes the structure and electronic configuration of the system and must be properly described by the many-particle Schrödinger's equation

$$\hat{H}\Psi(\mathbf{r}_1,\ldots,\mathbf{r}_N;\{\mathbf{R}_m\}) = E\Psi(\mathbf{r}_1,\ldots,\mathbf{r}_N;\{\mathbf{R}_m\}), \tag{2.1}$$

where r_i are the positions of the N electrons, $\{\mathbf{R}m\}$ is the set of nuclei coordinates which is assumed fixed in space (Born-Oppenheimer approximation), and the many-particle Hamiltonian operator is defined as

$$\hat{H} = -\frac{1}{2}\sum_i \nabla_i^2 + \sum_{i<j}\frac{1}{|\mathbf{r}_i - \mathbf{r}_j|} + \sum_i V_{\text{ext}}(\mathbf{r}_i;\{\mathbf{R}_m\}), \tag{2.2}$$

composed of terms for the kinetic energy, the electron-electron interaction, and an external potential, which in this description accommodates the influence of the electron-nucleus and nucleus-nucleus interactions. The nuclei equilibrium positions can be found by minimizing the total energy E with respect to $\{\mathbf{R}m\}$.

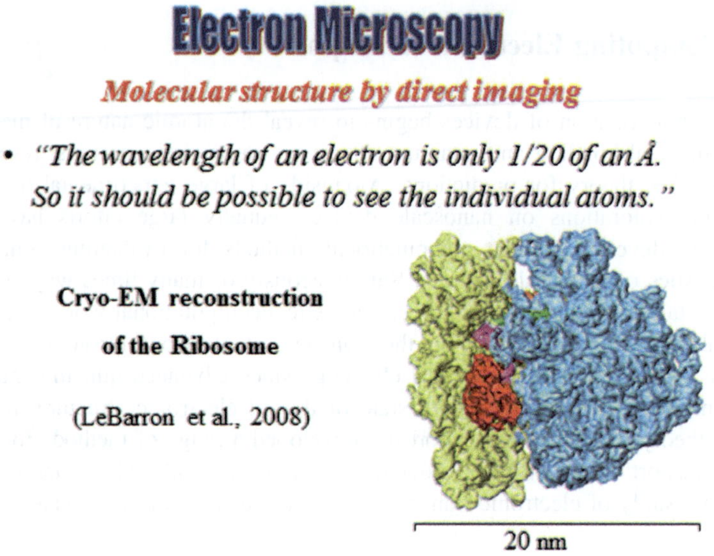

2.4.2 Density-Functional Theory

Methods based on Landauer-type scattering formalism in combination with ground state DFT [24–32]. They are based upon the assumption that, regardless of the details of a possible transient, a steady state is always achieved. These methods are widely used for ballistic as well as non equilibrium conductance calculations.

An unfortunate aspect of the above formulation is that the many-particle wave function Ψ provides the complete knowledge of the stationary properties of the system, but is numerically infeasible to compute for more than a few particles [33]. It is therefore more practical to consider the one-body electron density,

$$n(\mathbf{r}) = N \int d\mathbf{r}_2 \cdots \int d\mathbf{r}_N |\Psi(\mathbf{r}, \mathbf{r}_2, \ldots, \mathbf{r}_N; \{\mathbf{R}_m\})|^2, \tag{2.3}$$

as the central quantity of interest. The number of degrees of freedom is then reduced from $3N$ to 3 if Ψ is not explicitly evaluated. Moreover, the Hohenberg-Kohn theorems [34] prove that all ground state properties of the many-particle system can be calculated from the unique ground state density n(r) that corresponds to the non-degenerate ground state energy $E_0 = \min E[\mathrm{n}(\mathrm{r})]$.

2.4.3 Another Three Alternate Approaches Are

(a) Use of Kohn-Sham effective single-particle version of a 'master equation' formulation of transport [35–37].
(b) With the use of time dependent DFT (TDDFT), one obtains the current by calculating the time evolution of a system consisting of a molecule coupled to two finite metallic contacts and turning on a potential step, resulting in two different chemical potentials [38].
(c) A method that uses large finite leads, and watches a capacitance discharge [39].

Methods described above essentially begin from a static distribution, apply some change, and allow the system to evolve to a steady, but non-equilibrium distribution. For non-interacting electrons in the weak bias limit, all agree, both with each other and the standard approach, but likely disagree in the general case of interacting electrons in finite bias. Under certain limiting conditions, such as adiabatic approximations to TDDFT, and local approximations to ground-state DFT, they yield the same results.

A schematic drawing of a molecular junction is shown in Fig. 2.1, where one or more molecules are sandwiched between two semi-infinite electrodes. Mainly, the junction used in molecular electronics has three parts:

(1) The electrodes or reservoirs,
(2) The backbone of the molecule, and
(3) The interfaces between lead and the molecule.

These three parts contribute to the total transmission of electrons through the junction. Changing any physical or chemical aspect of any part is manifested in

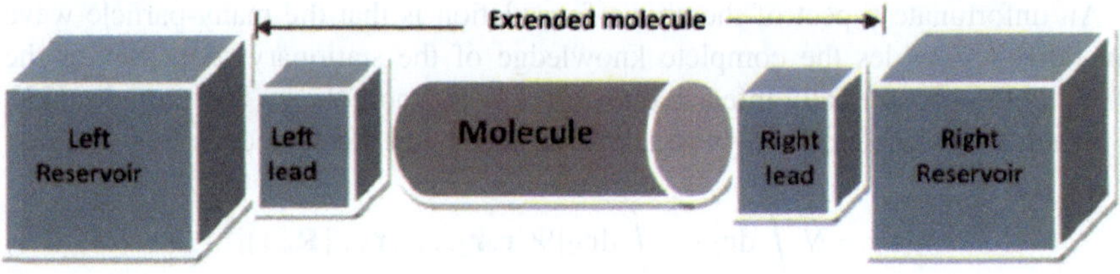

Fig. 2.1 Schematic diagram used for electronic transport calculation of nanostructures

drastic change in behaviour of electron transport through the junction. There have been large number of first principle investigations on various aspects of a molecular devices such as the material dependent effects; structures and relaxation of electrodes [40–43], effect of electrode work function [44], Fermi level alignment [45, 46], interface/contact dependence [47], effects of roughness of interface [48, 49], tilt-angle and temperature-dependence [50], effects of functional, anchoring and side groups [51, 52], importance of geometry over chemistry [53], effects of molecular symmetry [54], molecular conformation [55] and molecular length dependence [56–60].

The conductance properties of atomic chains of different length [61–63], atomic clusters of different sizes [64, 65], cage structures having different number and types of impurities [66], graphene sheets [67, 68], passivated and non-passivated nanoribbons [69–71], of different width and length, SW, DW and MW nanotubes [72, 73] of different diameter and chirality, nanotubes with impurities and defects [74], benzene [75, 76], molecules made by number of phenyl rings [57] and having different groups attached on different sites of it [51], polyynes [77, 78], alkane and alkenes [56, 79–81] are very well studied, exploring new phenomenon and searching for science behind them and giving answers to many queries. There still remain many unanswered questions and there are also newly aroused questions from recent discoveries that should be addressed properly.

For a very small or no external bias voltage applied to the electrodes, one can assume the electrode-scatterer-electrode system in equilibrium. This linear regime is useful to study the adsorption induced modifications in electronic structure of electrode-molecule-electrode system which are more dominant than field induced modification. Due to nanometric size of molecules used as the scatterer in nanodevices, the length of the scatterer is essentially smaller than the mean free path of the electron. Therefore, the electrons incident on one electrode-molecule junction can be considered to reach at the other junction without experiencing any scattering within the scatterer. This type of electron transport is known as ballistic transport.

Molecular Beam Epitaxy (MBE)

Making materials from atomic layers

- "So, you simply evaporate until you have a block of stuff which has the elements...... What could we do with layered materials with just the right layers?"

Alternate layers of HgTe and HgCdTe

T. Aoki, M. Takeguchi, P. Boieriu, R. Singh, C. Grein, Y. Chang, S. Sivananthan and D. J. Smith, *Microstructural characterization of HgTe/HgCdTe superlattices* J. Cryst, Growth, 2004, 271, 29-36,

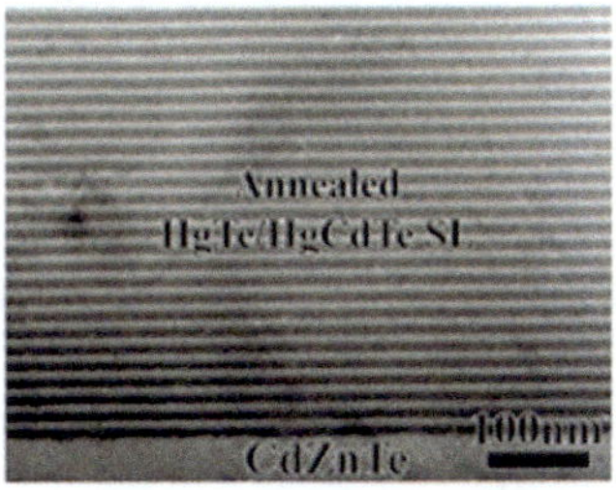

Atomic scale synthesis by "pushing atoms"

- *"We can arrange atoms the way we want."*

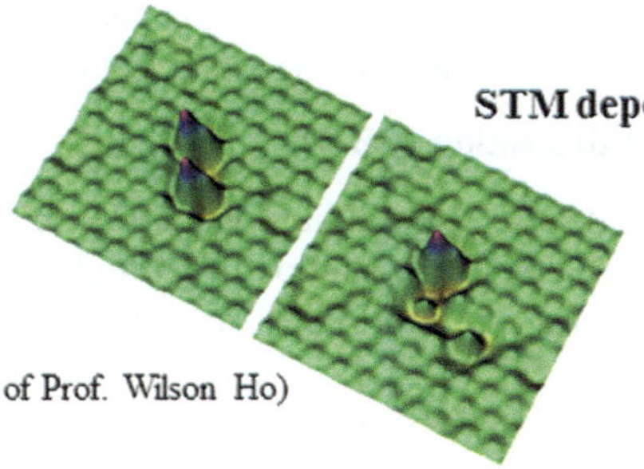

STM deposition

(Courtesy of Prof. Wilson Ho)

Single Molecule Chemistry

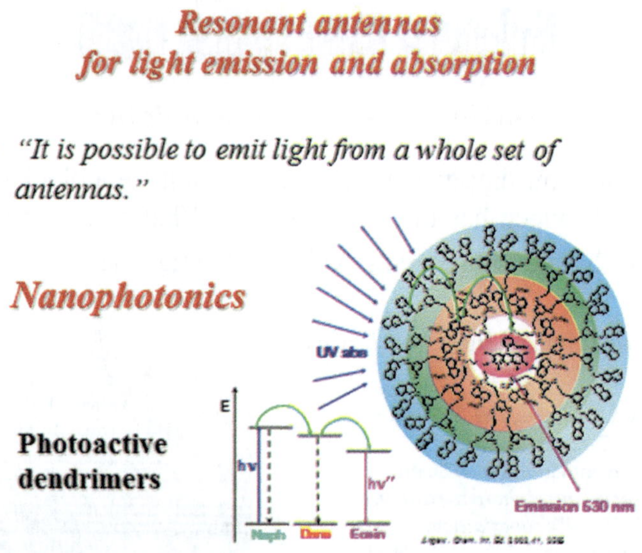

The computational aspects of modeling electronic transport in nano-scale devices. From a physical starting point, an accurate quantum mechanical model of these devices should have the ability to capture the following fundamental effects [23]:

- Electron wave interference.
- Quantum mechanical tunneling.
- Discrete energy levels due to confinement in 2D and 3D device geometries.
- Scattering processes (electron-phonon, electron-electron).

Electron beam induced deposition (EBID).

Figure 2.2 shows schematically the evolution of EBID into a controlled nanopatterning technique since the first demonstration of its high resolution, followed by simulations and experiments on thin films to study the ultimate resolution achievable.

This has ultimately resulted in the successful fabrication of dense lines on a bulk substrate, demonstrating the evolution of EBID into a well-controlled nanopatterning technique and a good candidate for lithography. One of the major considerations in this application is the purity of the deposited material. This has been reported in detail in the literature, with the demonstration of both post processing and in situ techniques for achieving, inter alia, high-purity metallic deposits [82–84] and will not be discussed further here.

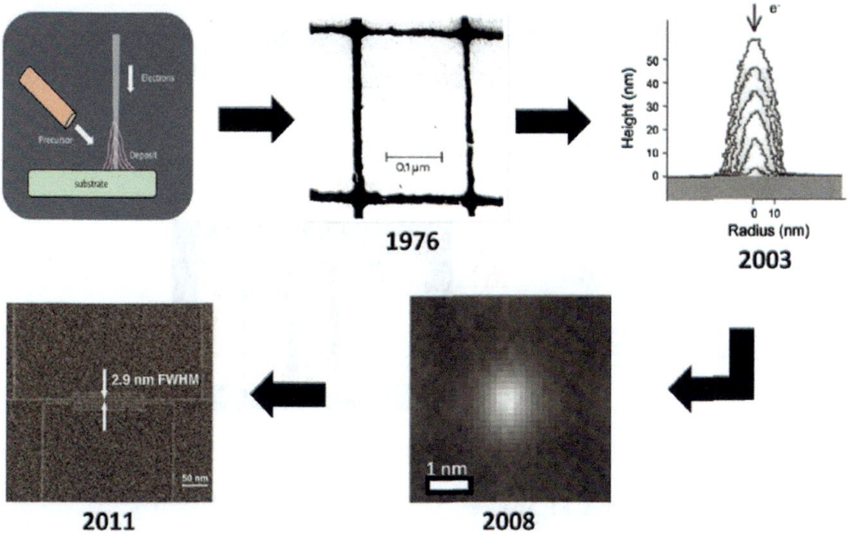

Fig. 2.2 Schematic representation of the evolution of high-resolution electron beam induced deposition as a nanopatterning technique since its first demonstration in 1976. (Clockwise order) Schematic of EBID, 8 nm lines patterned by Broers et al. on a thin membrane [85]; simulation of dot grown on a thin membrane using a zero-diameter electron beam demonstrating the relevance of deposit-generated SEs in EBID [86]; experimental demonstration of dots with average diameter of ca. 1 nm patterned on a thin membrane by EBID [87]; direct patterning of 3 nm dense lines on bulk Si/SiO_2 by EBID [88]. Reprinted with permission from [85] and [86], copyright 2003, 2013 AIP Publishing; reprinted with permission from [88], copyright 2011 American Vacuum Society

Quantum Computing

- *"We could use, not just circuits, but some systems involving quantized energy levels, or the interaction of quantized spins."*

Electron spin valves have become the dominant readout device in the disk drives.

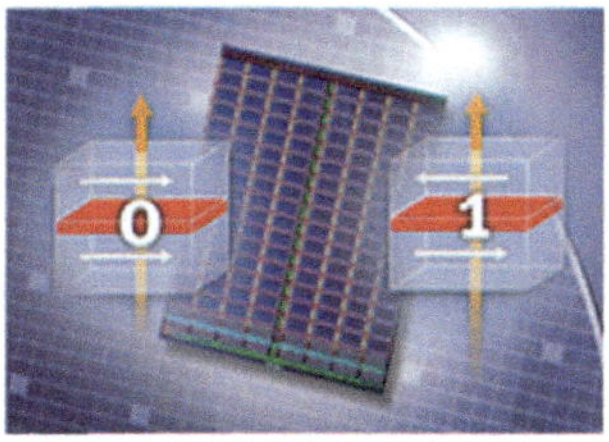

Spintronics

Quantum Mechanics

- Particle in a box $E_1 = \dfrac{\hbar^2 \pi^2}{2mL^2}$

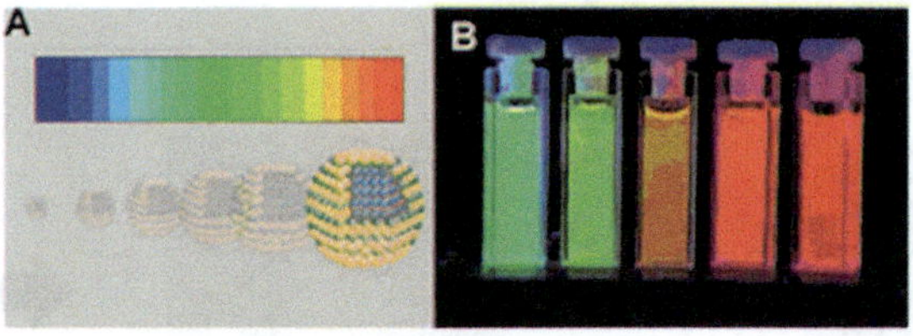

Quantum dots

(Courtesy of Dylan M. Spencer)

The first three effects can be accommodated most simply in the Landauer-Büttiker transport theory in conjunction with the solution of the electronic Schrödinger's equation. The fourth effect requires, in general, the NEGF formalism, or equivalent theories, to account for energy, momentum and phase relaxation due to inelastic scattering processes.

Fluctuations

- *Fluctuations play a large role in small systems simply because they are relatively larger in smaller systems.*

Fluctuations scale as $\sqrt{N}/N$ with respect to the mean energy

But $\sqrt{N}/N \to 1$ in small systems

- **Complexity** is a rapidly increasing function of N:

$$\propto \frac{A^N}{N}$$

- **Adequate complexity and fluctuation.**

  The critical size scale where fluctuations are big enough and the system is complex enough is indeed the nanoscale.

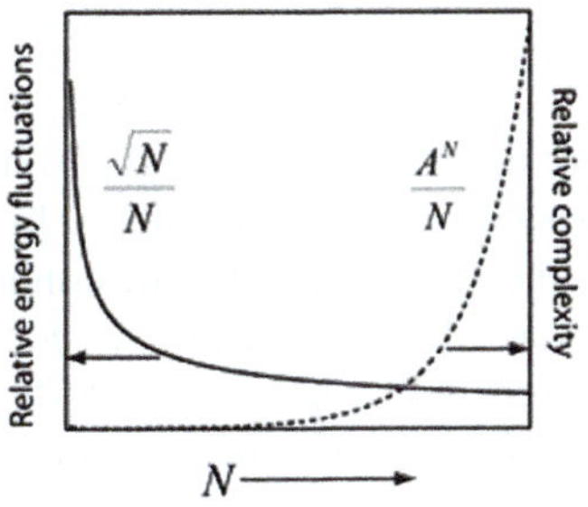

However, if we are interested in electronic transport at low temperatures and apply only a small voltage across the device, the effects of inelastic scattering at the nano-scale can be neglected.

Scattering and range depend upon a combination of electron scattering and nuclear scattering. The latter is of particular importance for heavy ions like Ga^+ and causes damage to the resist through sputtering and loss of substrate crystallinity. The situation for Ga^+ ions is further complicated by the property of Ga^+ as a p-type dopant of silicon. Scattering and range can be calculated using Monte Carlo simulation codes with typical results as shown in Fig. 2.3 [89].

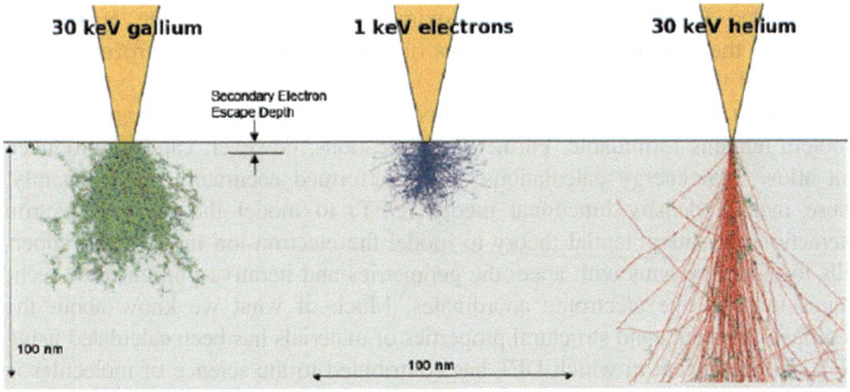

Fig. 2.3 Electron and ion beam substrate penetration—Monte Carlo simulations of charged particle paths showing forward and back scattering. Reproduced with permission through Creative Commons Attribution (CC BY) from [89], 2007 AIP Conference Series

Table 2.1 Comparison of beam diameters and scattering effects for different charged particle lithography methods

Method	Ga^+ SIBL	EBL	He^+ SHIBL
Probe size (nm)	>1	>1	<0.5
Scattering effects	No proximity effect	Proximity effect	Low proximity effect

The advantages of SHIBL using He ions over SIBL using Ga ions is clear from the reduced lateral spread of the beam in the top 10–20 nm of the surface where the ultrathin resist layer is located for nanolithography. In practice, the effect of the resist can be ignored and the lateral spread is determined by the wafer substrate alone. The resist layer is exposed not by the primary ions but by the electrons they generate, which have lower energies compatible with breaking the chemical bonds of the resist—the so-called "δ-rays". The main points of comparison are summarized in Table 2.1.

2.5 Basics of DFT and Methodology

In quantum mechanics we learn that all information we can possibly have about a given system is contained in the system's wave function. Prediction of the electronic and geometric structure of atoms, molecules and solids requires calculation of the quantum-mechanical total energy and subsequent minimization of that energy with respect to the electronic and nuclear coordinates. Because of the large difference in mass between the electrons and nuclei and the fact that the forces on the particles are the same, the electrons respond essentially instantaneously to the motion of the nuclei. Thus the nuclei can be treated adiabatically, leading to a separation of electronic and nuclear coordinates in the many-body wave function—the so-called Born-Oppenheimer approximation. This reduces the many-body problem to the solution of the dynamics of the electrons in some frozen-in configuration of the nuclei.

Even with the introduction of Born-Oppenheimer approximation, the many-body problem remains formidable. Further simplifications, however, can be introduced that allow total energy calculations to be performed accurately and efficiently. These include density functional theory (DFT) to model the electron-electron interactions, pseudopotential theory to model the electron-ion interactions, supercells to model systems with aperiodic geometries and iterative minimization techniques to relax the electronic coordinates. Much of what we know about the electrical, magnetic, and structural properties of materials has been calculated using DFT, and the extent to which DFT has contributed to the science of molecules is reflected by the 1998 Nobel Prize in Chemistry, which was awarded to Kohn [90], the founding father of DFT, and Pople [91], who was instrumental in implementing

DFT in computational chemistry. This approach of DFT forms the basis of the large majority of electronic-structure calculations in physics and chemistry [92–102].

For system with more than one electron (i.e., a many-body problem) Schrödinger's equation becomes

$$\left[\sum_{i}^{N}\left(-\frac{\hbar^2\nabla_i^2}{2m}+v(\mathbf{r_i})\right)+\sum_{i<j}U\left(\mathbf{r_i},\mathbf{r_j}\right)\right]\Psi(\mathbf{r}_1,\mathbf{r}_2,\ldots,\mathbf{r}_\mathrm{N})=E\Psi(\mathbf{r}_1,\mathbf{r}_2,\ldots,\mathbf{r}_\mathrm{N}) \tag{2.4}$$

where N is the number of electrons and $U(\mathrm{r_i}, \mathrm{r_j})$ is the electron-electron interaction. For a Coulomb system one has

$$\hat{U}=\sum_{i<j}U\left(\mathbf{r_i},\mathbf{r_j}\right)=\sum_{i<j}\frac{q^2}{\left|\mathbf{r_i}-\mathbf{r_j}\right|} \tag{2.5}$$

Note that this is the same operator for any system of particles interacting via the Coulomb interaction, just as the kinetic energy operator

$$\hat{T}=-\frac{\hbar^2}{2m}\sum_i\nabla_i^2 \tag{2.6}$$

is the same for any nonrelativistic system. Whether our system is an atom, a molecule, or a solid thus depends only on the potential $v(\mathrm{r_i})$. For a molecule or solid e.g.,

$$\hat{V}=\sum\nolimits_i v(\mathbf{r_i})=\sum_{ik}\frac{Q_k q}{|\mathbf{r}_i-\mathbf{R}_k|} \tag{2.7}$$

where the sum on k extends over all nuclei in the system, each with charge $Qk = Zke$ and position Rfc. It is only the spatial arrangement of the Rk (together with the corresponding boundary conditions) that distinguishes, fundamentally, a molecule from a solid. Similarly, it is only through the term U that the (essentially simple) single-body quantum mechanics of differs from the extremely complex many-body problem. These properties are built into DFT in a very fundamental way.

Fueled by the demand for higher performance electronics, the size of devices has been shrinking exponentially for the last five decades. Because of this forecast, a wealth of alternative nano-scale electronic devices based on novel materials (e.g., quantum dots, molecules, nanowires, graphene, topological insulators) and designs (e.g., tunneling transistors, phase-change devices, nanoelectromechanical systems) have been proposed. Although these novel systems may present new and interesting characteristics, both the experimental realization and theoretical treatment of such nano-scale devices presents substantial difficulties. Experimentally, an exceptional degree of control is required to systematically build and characterize nano-scale devices, where atomic imperfections and variability are inevitable and can play a crucial role.

From a theoretical viewpoint, whether considering current generation transistors or emerging novel devices, classical and semi-classical transport theories often can no longer adequately describe devices in which the carriers are phase-coherent quantum mechanical entities. Moreover, for nano-scale systems, the particular atomic structure and chemical environment are important factors that influence the electronic properties and can easily fluctuate from one device to another. Thus, parameter-based theories are clearly at a disadvantage when considering advanced materials in conjunction with surfaces, interfaces and low-dimensional systems, since many system-dependent parameters must be predetermined. Furthermore, attempts to adapt existing semiclassical theories in order to describe ever smaller devices have led to a dramatic increase in the number of parameters, which must be fit to experimental data. Such a process is expensive and undesirable, as the theory loses its predictive power. Hence, there is an urgent need for a practical modeling tool that doesn't rely on phenomenological parameters. The development and engineering of nanoelectronic devices could greatly benefit from an atomistic parameter-free quantum theory that is capable of accurately predicting transport characteristics.

Such a formalism was developed by Taylor et al. [103] in 2001, in which density functional theory (DFT) was combined with nonequilibrium Green's functions (NEGF). DFT is a powerful, practical and widely used parameter-free theory which describes the mechanical and electronic properties of materials from their constituent atoms. It successfully converts the many-body wavefunction problem into a single particle effective Schrödinger-type equation, via a mean-field approach [104].

DFT is commonly applied to two cases, namely isolated systems (i.e., molecules) and periodic systems (i.e., bulk solids). However, an electronic device is neither isolated nor periodic, since it is connected to the outside world via metallic electrodes. Moreover, since we are concerned with current flow, one must consider the nonequilibrium physics of this problem.

NEGF is a fully quantum mechanical formalism that describes electronic/spin transport, and its population statistics, under nonequilibrium conditions (i.e., finite applied voltage) [23, 105]. This technique assumes that the device is connected to left and right semi-infinite leads, each with their own respective chemical potential. The infinite-sized electrodes are effectively folded into the finite-sized device Green's function via lead self-energies. Scattering can be included, but without such interactions conduction results from phase-coherent ballistic transport.

The groundbreaking work of Taylor et al. [103] established how to combine both DFT and NEGF in order to simulate the nonequilibrium quantum transport properties of nano-scale devices from first principles. Roughly speaking, DFT provides

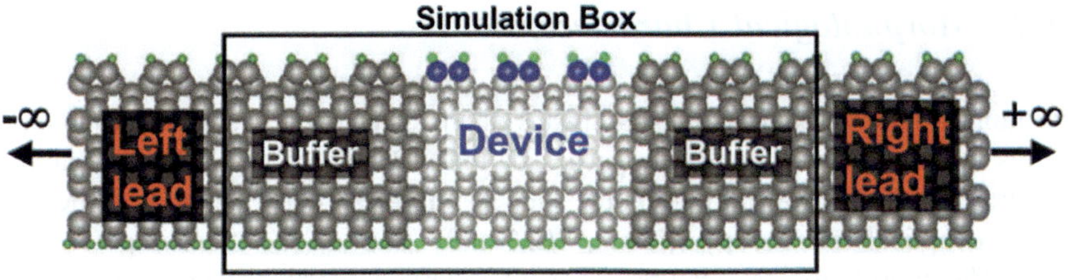

Fig. 2.4 Diagram of a simulated nano-scale device, which includes a device region and left/right semi-infinite electrodes. A portion of the leads, called "buffer layers", enter the simulation box in order to guarantee a smooth matching of the potential and charge density at the edges. The simulation box is repeated periodically in the plane perpendicular to transport

the device Hamiltonian and density of states, while NEGF describes the nonequilibrium statistics of the electronic states. Using the so-called "screening approximation" DFT can be used to handle open device boundaries with real-space numerical techniques. Figure 2.4 shows a schematic illustration of a simulated device within the NEGF-DFT formalism. The NEGF-DFT method is currently the state-of-the-art and has been successfully implemented by many groups [106–111] and applied to numerous transport problems, such as: conduction through a C60 molecule [112], a carbon atomic wire [113], ethylbenzene molecular wires on a Si substrate [114], alkanethiol molecular wires [115], current density of B- and P-doped carbon nanotubes [116], electron-phonon coupling in molecular devices [117, 118], magnetoresistance of molecular tunnel junctions [110], the magnetic tunnel junction Fe/MgO/Fe [119], STM simulation of styrene molecules on Si [120], molecular spintronics [121], the effect of lead-molecule geometry [122], thermopower of molecular junctions [111] and the effect surface roughness on Cu interconnects [123].

2.6 Characterization of Nanomaterials

A range of experimental techniques can be used to determine the structure, number of layers, chemical composition, and electrical, optical, and mechanical properties of 2D materials. These include optical, scanning probe, and electron microscopy, electronic transport measurement, electrochemistry, Raman and PL spectroscopy, and UV-vis spectroscopy. Different characterization techniques have been practiced for the analysis of various physicochemical properties of NPs. These include techniques such as X-ray diffraction (XRD), X-ray photoelectron spectroscopy (XPS), infrared (IR), SEM, TEM, Brunauer–Emmett–Teller (BET), and particle size analysis.

2.6.1 Morphological Characterizations

The morphological features of NPs always attain great interest since morphology always influences most of the properties of the NPs. There are different characterization techniques for morphological studies, but microscopic techniques such as polarized optical microscopy (POM), SEM and TEM are the most important of these. The morphological features of ZnO modified metal organic frameworks (MOFs) were studied through SEM technique, which indicates the ZnO NPs dispersion and morphologies of MOFs at different reaction conditions (Fig. 2.5) [124].

Similarly, TEM is based on electron transmittance principle, so it can provide information of the bulk material from very low to higher magnification.

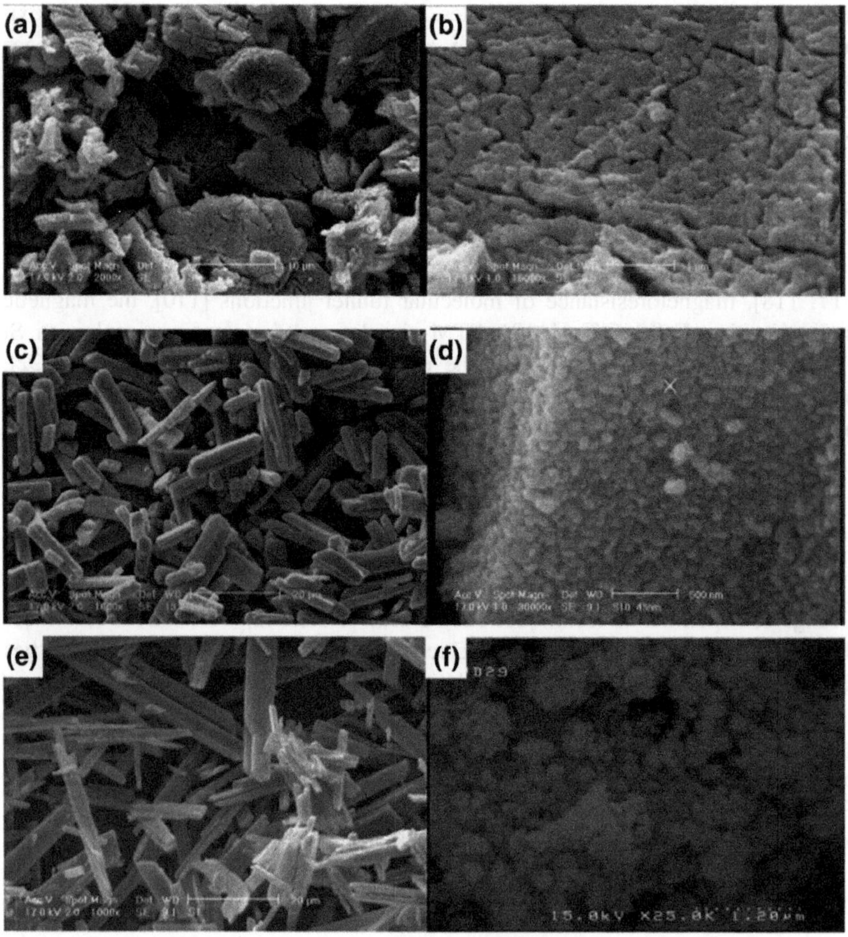

Fig. 2.5 SEM images of ZnO modified MOFs at different temperatures [124]

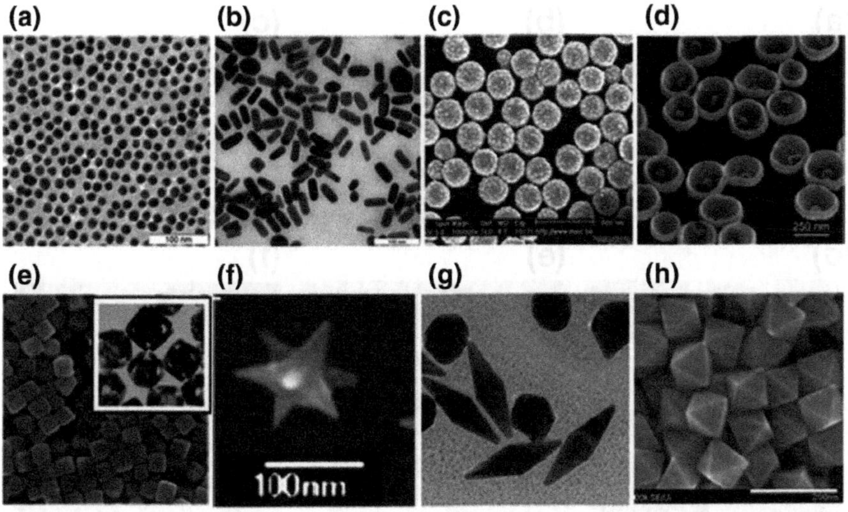

Fig. 2.6 TEM images of different form of gold NPs, synthesized by different techniques [125–127]

The different morphologies of gold NPs are studied via this technique. Figure 2.6 provides some TEM micrographs showing various morphologies of gold NPs, prepared via different methods [125–127].

TEM also provides essential information about two or more layer materials, such as the quadrupolar hollow shell structure of Co_3O_4 NPs observed through TEM. These NPs founded to be exceptionally active as anode in Li-ion batteries (Fig. 2.7). Porous multishell structure induces shorter Li^+ diffusion path length with adequate annulled space to buffer the volume expansion, good cycling performance, greater rate capacity, and specific capacity as well [128].

2.6.2 Structural Characterizations

The structural characteristics are of the primary importance to study the composition and nature of bonding materials. It provides diverse information about the bulk properties of the subject material. XRD, energy dispersive X-ray (EDX), XPS, IR, Raman, BET, and Zieta size analyzer are the common techniques used to study structural properties of NPs.

XRD is one of the most important characterization techniques to reveal the structural properties of NPs. It gives enough information about the crystallinity and phase of NPs. It also provides rough idea about the particle size through Debye Scherer formula [129–131]. This technique worked well in both single and multiphase NPs identification [132]. Nevertheless, in the case of smaller NPs having

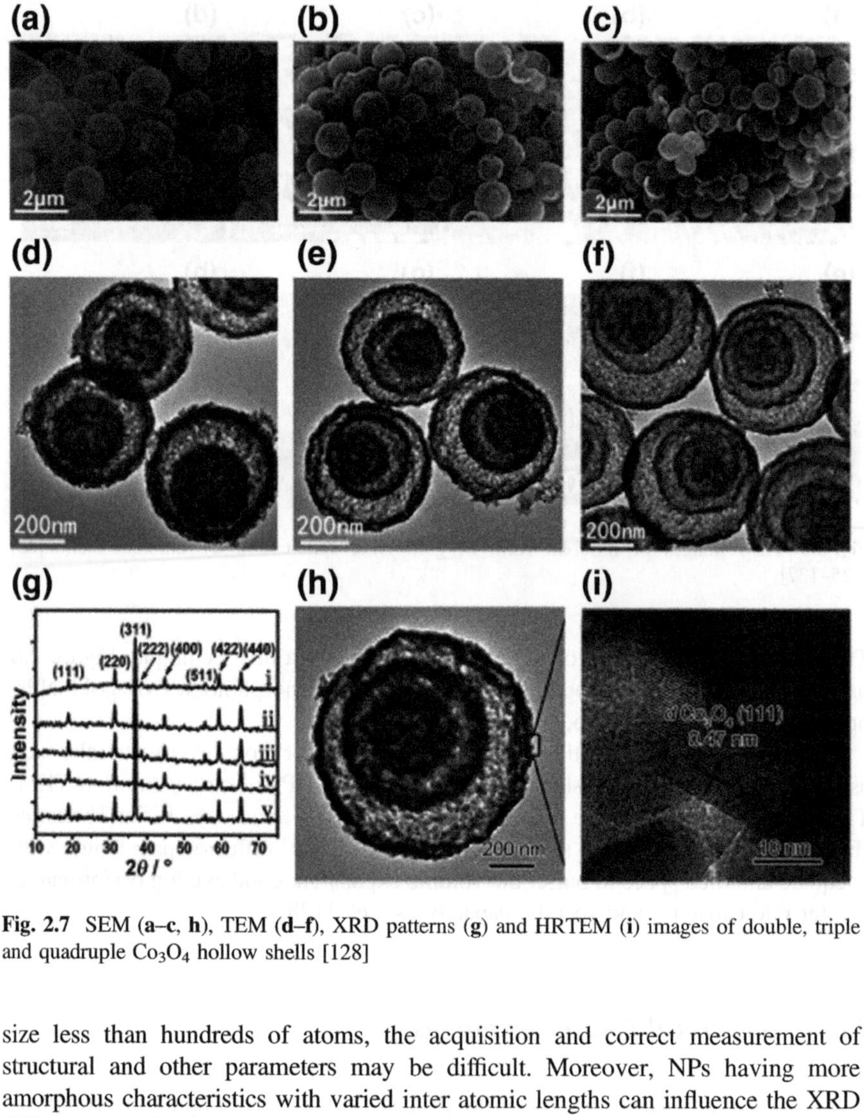

Fig. 2.7 SEM (**a–c**, **h**), TEM (**d–f**), XRD patterns (**g**) and HRTEM (**i**) images of double, triple and quadruple Co_3O_4 hollow shells [128]

size less than hundreds of atoms, the acquisition and correct measurement of structural and other parameters may be difficult. Moreover, NPs having more amorphous characteristics with varied inter atomic lengths can influence the XRD diffractogram. In that case, proper comparison of the diffractograms of bimetallic NPs with those of the corresponding monometallic NPs and their physical mixtures is required to obtain accurate information. Comparison of computer simulated structural model of bimetallic NPs with observed XRD spectra is the best way to get good contrast [133]. EDX, which is normally fixed with field emission scanning electron miscopy (FE-SEM) or TEM device is widely used to know about the elemental composition with a rough idea of % wt. The electron beam focused over a single NP by SEM or TEM through the program functions, to acquire the insight information from the NP under observation. NP comprises of constituent elements and each of them emits characteristics energy X-rays by electron beam irradiation.

The intensity of specific X-ray is directly proportional to the concentration of the explicit element in the particle. This technique is widely used by researchers to give support to SEM and other techniques for the confirmation of their elements in prepared materials [134, 135].

XPS is considered to be the most sensitive technique and it is widely used to determine the exact elemental ratio and exact bonding nature of the elements in NPs materials. It is surface sensitive technique and can be used in depth profiling studies to know the overall composition and the compositional variation with depth. XPS is based on the basic spectroscopic principles and typical XPS spectrum is composed of the number of electrons on Y-axis plot versus the binding energy (eV) of the electrons on X-axis. Each element has their own fingerprint binding energy value and thus gives specific set of XPS peaks.

Vibrational characterization of nanoparticles is normally studied via FT-IR and Raman spectroscopies. These techniques are the most developed and feasible as compared to other elemental analytical methods. The most important range for NPs is the fingerprint region, which provides signature information about the material. In one study, functionalization of Pt NPs (1.7 nm mean size) and its interaction with Alumina substrate studied via FT-IR and XPS technique. FT-IR confirms the functionalization as it showed the signature vibrational peaks of carboxylated C–O 2033 cm^{-1}, respectively in addition to a broader O–H peak at 3280 cm^{-1}. The degree of functionalization was revealed from the red shift values of FT-IR bands (Fig. 2.8) [136].

More recently surface enhanced Raman spectroscopy (SERS) is evolving as vibrational conformational tool due to its signal enhanced capability via SPR phenomenon [137]. One study reported SERS technique to study the vibrational properties with phonons modes in nanostructured and quantum dots NPS of TiO_2, ZnO and PbS. They concluded that the enhanced spectra can be attributed to the plasmonic resonances in semiconductor systems [138].

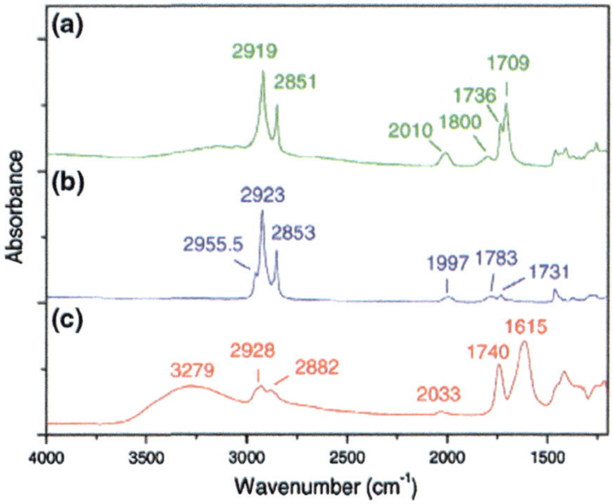

Fig. 2.8 FTIR spectra of platinum (1.7 nm) **a** extracted from polyol, **b** dodecanethiol coated Pt, and **c** MUDA coated Pt [136]

2.6.3 Particle Size and Surface Area Characterization

Different techniques can be used to estimate the size of the NPs. These include SEM, TEM, XRD, AFM, and dynamic light scattering (DLS). SEM, TEM, XRD and AFM can give better idea about the particle size [139], but the zeta potential size analyzer/DLS can be used to find the NPs size at extremely low level. Gross et al. detected the particle size and concentration of different sized NPs in suspensions of polymer and protein samples and provided an overview on the effect of experimental and data evaluation parameters [140].

Large surface area of nanomaterials offers great room for various applications and BET is the best technique to determine the surface area of NPs materials. This technique is based on adsorption and desorption principle and Brunauer–Emmett–Teller (BET) theorem. Normally nitrogen gas is used for this purpose. BET produces four types of isotherm specifically, which are labeled as Type-I, Type-II, Type-III and Type-IV [141]. The fresh $7Cu_3Ce/ZSM$-5 showed typical features of Type-I isotherm obtain from nitrogen adsorption/desorption. It was discovered that N_2 adsorption volume is progressively increased with relative pressure until certain limit signifying the availability of pores. The BET specific surface area for this material was 133–144 m^2/g, while the total pore volume was 0.112–0.185 cm^3/g. But after sulphidation process, the BET surface area reduced to 110 m^2/mg and the pore volume decreased to 0.096 cm^3/g, respectively [142].

2.6.4 Optical Characterizations

Optical properties are of great concerned in photocatalytic applications and therefore, photo-chemists acquired good knowledge of this technique to reveal the mechanism of their photochemical processes. These characterizations are based on the famous beer-lambert law and basic light principles [143]. These techniques give information about the absorption, reflectance, luminescence and phosphorescence properties of NPs. It is widely known that NPs especially metallic and semiconductor NPs possess different colors and therefore, best harmonized for photo-related applications. So, it is always interesting to know the value of absorption and reflectance of these materials to understand the basic mechanism for each application. Ultraviolet-visible (UV-Vis), photoluminescence (PL) and the null ellipsometer are the well-known optical instruments, which can be used to study the optical properties of NPs materials.

The UV/vis-diffuse reflectance spectrometer (DRS) is a fully equipped device which can be used to measure the optical absorption, transmittance and reflectance. The former two are supplementary to each other while the latter (DRS) is a special technique use for sold samples mostly. The method is exceptionally acceptable for the determination of bandgaps of NPs and other nanomaterials. Bandgap of materials is very important to conclude about the photoactivity and conductance of the material. The carbon nanodot-carbon nitride (C_3N_4) was found to be a metal free water splitting photocatalyst. The photo ability of this material is directly correlated to the bandgap value of 2.74–2.77 eV, which was calculated using UV-Vis spectroscopy [144, 145]. Similarly, this technique also use to see the absorption shift in case of doping, composite formation or heterostructure NPs materials. Peng et al. synthesis MMT, $LaFeO_3$ and $LaFeO_3$/MMT nanocomposites and studied variation in their electromagnetic radiations absorption through UV-vis DRS to reconnoiter their optical characteristics. The strong red shift observed in case of nanocomposite as compared to pristine MMT and $LaFeO_3$ NPs. $LaFeO_3$ and $LaFeO_3$/MMT displayed rather broad absorption band from 400 to 620 nm, showing decrease in their bandgap. This property makes these catalysts considerable for solar light driven photocatalysis [146].

In addition to UV, PL also considers valuable technique to study the optical properties of the photoactive NPs and other nanomaterials. This technique offers additional information about the absorption or emission capacity of the materials and their effect on the overall excitation time of photoexcitons. Thus, it provides significant information about the charge recombination and half-life of the excited materials in their conductance band, which are useful for all photo related and imaging applications. The PL spectrum can be recorded as emission or absorbance depending on the nature of study. Figure 2.9 shows a typical PL spectrum of pristine and modified ZnO NPs. It is evident from this figure that pristine ZnO NPs show high PL intensity as compared to CdS modified ZnO NPs. The gold embedded CdS/Au/ZnO composite shows the lowest intensity. This quenching from pure ZnO to CdS/Au/ZnO can be attributed to the decrease in the rate of charge recombination and larger lifetime of photoexcitons in the latter case [147]. In addition, this technique is successfully used to determine the thickness of layer

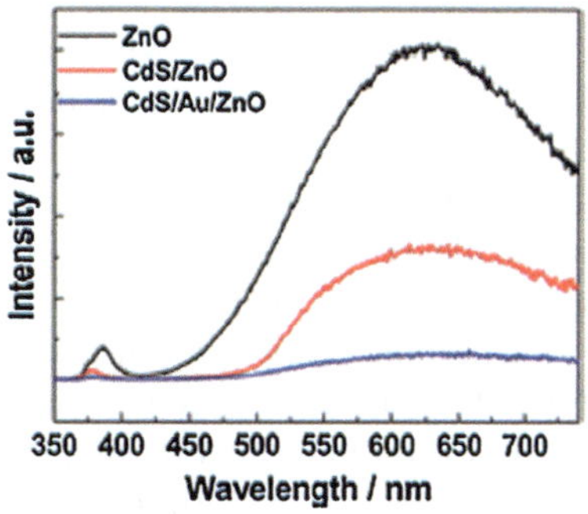

Fig. 2.9 Photoluminescence (PL) spectra of pristine ZnO, CdS/ZnO, and CdS/Au/ZnO measured with 270 nm excitation wavelength at normal temperature [147]

[148], doping quantity of [149, 150] material and defects/oxygen vacancies determination [151] of NPs.

Similarly, Wan et al. determined the values of refractive index and extinction coefficient for hollow gold NPs (HGNPs) via spectroscopic ellipsometry. They prepared a series of HG-NPs, with different morphologies and plasmonic properties and the optical constants were calculated. The values were compared with the optical constant values of solid gold NPs, which gave good indication to use these materials in chemical sensing applications due to their sensitive nature as revealed from ellipsometric values [152].

2.7 Characterization Techniques

Examining the vibrational and molecular structure of 2D nanostructured materials is integrally interesting due to their size. On the other hand, various approaches have been established to facilitate their description and credentials. In this section, we will discuss the basic principles of selected characterization techniques and their use for analyzing single- and multi-layer-thick materials. In addition, we will also highlight recently developed techniques, which have the potential to explain structural and spectroscopic information about the layers.

2.7.1 Microscopy Techniques for 2D Materials

The optical properties of 2D materials are heavily affected by their physical structure, which can be studied using various microscopy techniques. Mainly atomic force microscopy (AFM), scanning electron microscopy (SEM), ultrasonic force microscopy (UFM), and transmission electron microscopy (TEM) is used for morphological and topographical analysis.

2.7.1.1 Atomic Force Microscopy

The AFM uses a sharp tip with a radius of nanometer magnitude, which is attached to a cantilever. When the tip is brought close to the sample surface, in the case of 2D layers electrostatic forces act on the tip, which makes the cantilever bend according to Hooke's law. This force is detected by the deflection of a laser against the cantilever, into a photodiode detector. An electronic feedback detection system repositions a piezo stage where the sample is held, which moves the sample up and down, keeping the tip at a constant height above the surface. Many researchers have used AFM to measure the thickness of 2D layers.

AFM is an influential technique to identify layer thickness (Fig. 2.10b, c) with 5% preciseness [153–156]. However, inconsistencies originate from dissimilarities

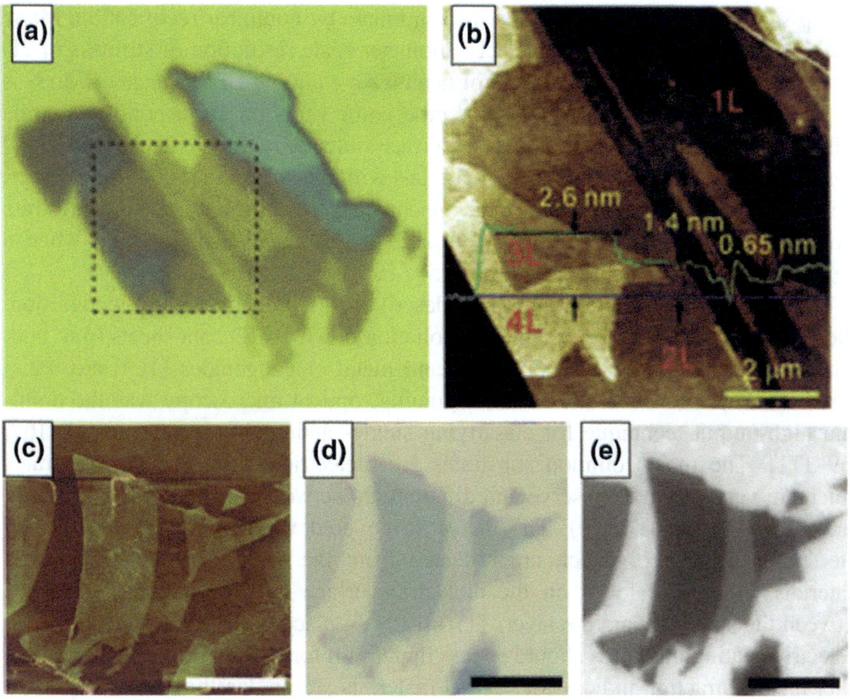

Fig. 2.10 **a** Optical micrograph of MoS_2 layer [160] **b** AFM of MoS_2 [160], **c** AFM of graphene [156], **d** optical micrograph of graphene [156], **e** FQM image of graphene using PVP/fluorescein [156]. Scale bars are 10 lm. Reprinted by permission of ACS Publishers

in the tip interactions with substrate and the layered materials [157]. For example, a single layer thickness could be known by evaluating the height between the first two layers, instead of measuring the height between the substrate and a first layer because the tip-layer interactions are constant in the former. To achieve the perfect height profile of a single layer and control these differences, it is essential to optimize the tip surface distance from any hysteretic pieces [157].

2.7.1.2 Ultrasonic Force Microscopy

UFM can also be used to measure the stiffness of materials (for example, mechanical coupling to the substrate) on the nanometer scale [158, 159]. For this technique, the sample is vibrated at small vibrational amplitude (0.5–2 nm) and high frequency (2–10 MHz), greater than the AFM cantilever resonance frequencies. The restoring force of the sample is then modified by the stiffness of the material, and this can be used to detect below-surface features such as cavities, subsurface interfaces and sample-substrate interfaces. UFM can be measured concurrently with AFM, detected on a separate channel where ultrasonic vibration at the tip-sample junction is converted into an image by nonlinear rectification. Many researchers have used UFM to detect nanometer scale resolution of stiffness of 2D structures of few atomic monolayers of layers. An example UFM image of MoS_2 is given in Fig. 2.10a [160, 161]. In this case color variation correspond to sample stiffness. This is used to detect below surface defects, regions of suspension and measure mechanical coupling of 2D materials with the substrate [160–163]. Others, used this technique to the show weak mechanical bonding between the substrate and the layers for free standing sheets and strong mechanical bonding when a dielectric capping layer is added [153, 154, 164].

The primary task in the mono-layer description is preparation and recognition. The simplest technique is mechanical exfoliation in which 2D nanosheets from bulk materials such as graphene [165–168], and metal chalcogenides [169] are transferred to a suitable substrate. In the beginning, optical microscopy was the dominant high-output technique for classifying single- and/or few-layer material [167, 169–171]. The most common substrates are dielectric-coated SiO_2/Si substrates which are very useful to observe and differentiate between single and few layers in nanoscale (Fig. 2.10a, d, e). The dielectric covered wafer color depends on an interference influence, which are reflected from the two surfaces. The layered materials (single- or few-) on the dielectric surface produce a color difference between the substrate and the layers [170, 172]. In place of ideal color contrast, the dielectric coating thickness should be less than 5 nm for the perfect measure. On the other hand, refraction indices for several materials are unidentified, it is frequently essential to exfoliate onto a various substrates having various dielectric thicknesses in order to obtain the ideal thickness.

Imaging mono layers on different substrates free from interference methods has also been discovered. Some TMDs (MoS_2, WS_2) have direct band gaps when it is mono layers, enabling their uninterrupted visualization using fluorescence microscopy [159]. On the other hand, when single layers changes to multiple layers, these TMD materials band gaps convert to indirect, thus their fluorescence will be significantly decreased. Fluorescence Quenching Microscopy (FQM) can be used for an organic fluorophore whose fluorescence is quenched with mono- or multi-layer sheets (Fig. 2.10e). Various layer structured materials characteristic either low enough band gaps or the suitable band alignment to reduce nearby fluorophores using a fluorescence resonance energy transfer mechanism (FRET). Mono layers of any 2D materials can be identified on a substrate by observing the decrease in fluorescence intensity of a thin fluorophore layer that was spin-coated on the top [161, 163, 173].

2.7.1.3 Electron Microscopy Techniques

Diffraction limits the resolution of an optical microscope to around 1 μm. Electrons have a small de Broglie wavelength, and therefore higher spatial resolution can be achieved. Frequently used microscopy methods are SEM and TEM. A focused beam of electrons is produced thermoionically in a tungsten/LaB_6/field ion gun cathode and raster scanned across the surface of the sample. Secondary electrons are emitted from the sample due to inelastic scattering within a few nanometers to a couple of micrometer from the surface [153]. Electrons are collected by a scintillator grid, which converts the intensity into an image. The resolution of the SEM is normally <50 nm [164]. Unlike SEM, where secondary electrons are emitted from the sample surface, electrons in TEM are transmitted through the entire sample; usually after the sample has been thinned to nm thickness. The electrons interact with the atoms during transmission and deflection can be used to produce an image [154]. TEM has a higher resolution compared to SEM as high voltage (200–300 kV) is used to generate electrons in TEM which has a shorter wavelength and thus enhance the resolution, but requires samples to be extremely thin (<200 nm) which is a disadvantage for many systems [164].

These techniques can provide comprehensive evidence on the crystalline nature, interlayer stacking relationships, layer sizes, and elemental configuration (Fig. 2.11a, b) [153, 155, 169, 174–176]. For graphene, selected area electron diffraction (SAED) may differentiate a monolayer or multilayers when the (110) to (100) diffraction intensity ratio is smaller or larger than one. This aberration in diffraction intensity could also be distinguished within the monolayer or multilayer of any 2D materials when there is a variation in interlayer recording. Dark-field microscopy is the recent development for identifying the layer numbers and stacking order of the layered material [177, 178]. Scanning transmission electron microscopy (STEM), coupled with electron energy loss spectroscopy (EELS) could be very useful for imaging an individual or separate atoms in a single layer. In this technique, a sub-angstrom electron beam is formed across the material and it is scattered as soon as it strikes a particular atom. A STEM image of oxygen and carbon substitutions in a boron nitride single layer has been shown in Fig. 2.11c. In STEM, beam damage can be reduced by imaging at low accelerating voltages (<60 keV).

Another important probe based technique is Scanning Tunneling Microscopy (STM) which can be used to evaluate the topographic and electronic structure of single layered materials. This technique could control single atom at exact points in order to characterize the 2D layer materials that are detached from the substrate (Fig. 2.11d) [179]. For Examples, STM studies of MoS_2 nanoislands [180] and BN [181], as well as characterization of the unusual cyclotron quantization in graphene [182, 183]. At present concentration has been shifted to the adsorbates and defect studies, containing metal atoms on graphene [184, 185] and hexagonal BN [186, 187] and molecules on MoS_2 [188]. Epitaxially grown layers are the most commonly studied samples in STM because of the alignment challenge in a specific part of the substrate with the distant STM tip. Near-field tip-enhanced Raman spectroscopy (TERS) is nothing but silver or gold coated tip in AFM or STM which can improve the local Raman spectra. This technique has established to be a great combined image and spectroscopic technique, for example, it is very useful to find defects and grain boundaries of TMD layers [189].

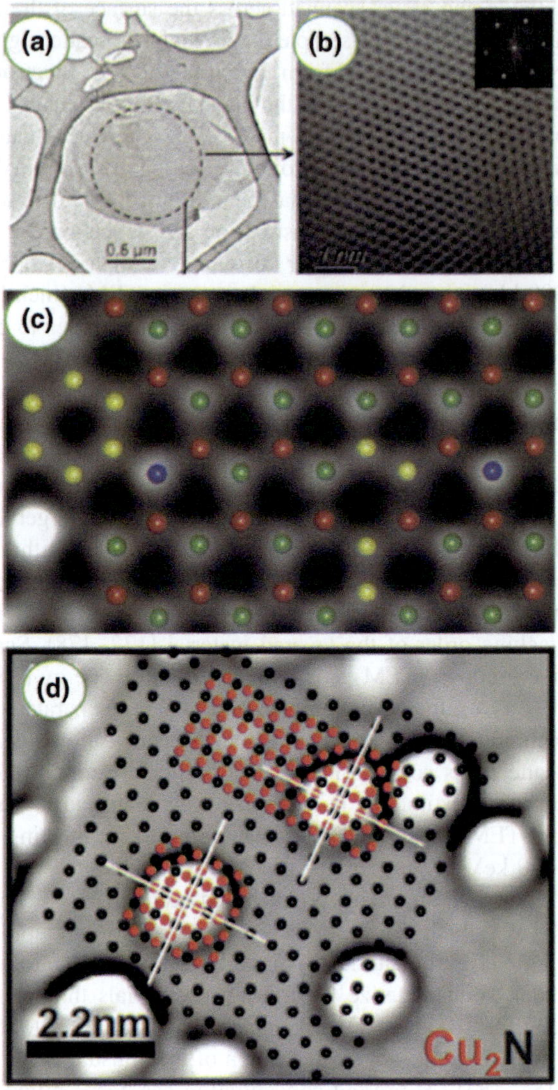

Fig. 2.11 **a** and **b** Low and high resolution TEM image of a 2D WS_2 crystal [176], **c** Annular dark field-STEM image of a single BN layer and part of DFT simulation observed substitutional impurities overlaid on the corresponding part of experimental image. (yellow, green, blue, and red colors are corresponds to C, N, O, and B, respectively) [178], **d** STM image of a Cu2 N island with adsorbed Co atoms on Cu site (top center) and N site (bottom left) (0.1 V, 0.25 nA) [179]. Reprinted by permission of ACS Publishers, Nature Publishing Group, and AIP Publishing LLC

2.7.2 Raman Spectroscopy

When a sample is irradiated, despite of being absorbed by the sample and then reemitted, a small proportion of the incoming light is scattered at a wavelength that is shifted from the original laser wavelength. Raman Spectroscopy is based on an inelastic scattering of monochromatic light. There are two kinds of inelastic scattering. Stokes (output energy is smaller than input energy) and anti-stokes scattering (output energy larger than input energy). The difference between the output and

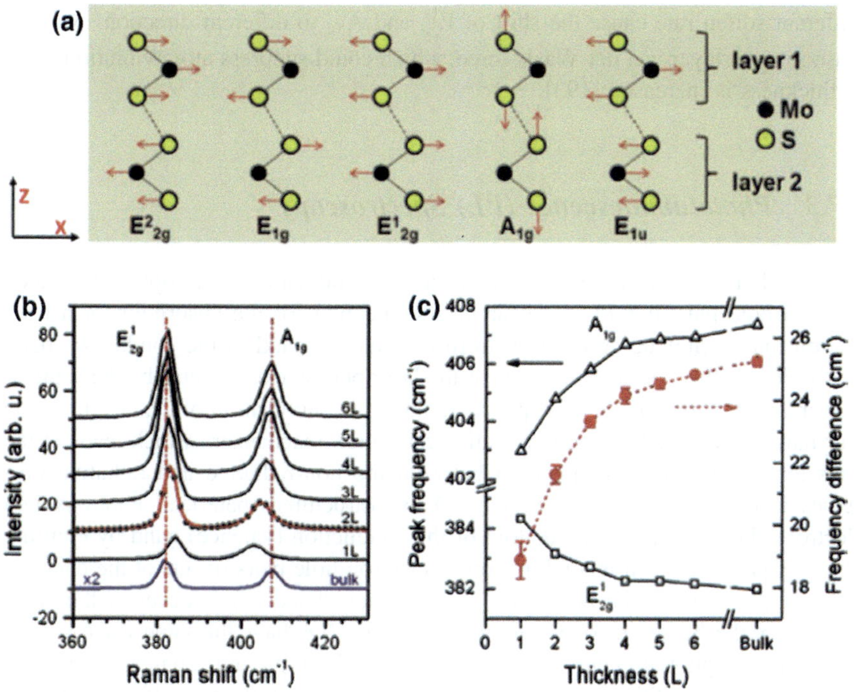

Fig. 2.12 **a** Vibrational modes of MoS_2 (Z direction is out of plane), **b** thickness dependent MoS_2 Raman spectra, and **c** frequency difference between E^1_{2g} and A1g mode. Reprinted from Ref. [160] with permission of ACS Publishers

input light provides the information of vibrational modes, which are materials fingerprints.

The vibrational modes of MoS_2 are shown in Fig. 2.12a. Among them, the two strongest Raman peaks are E^1_{2g} and A_{1g} modes. E_{1g} mode cannot be seen due to the selection rules [190] and E^1_{2g} mode is also unable to be observed because of the limited rejection of Rayleigh scattered radiation [191]. E_{1u} mode is not Raman active.

Raman spectra of MoS_2 are thickness dependent [160]. As thickness increases from single-layer to bulk, E^1_{2g} vibration makes red shifts and A_{1g} vibration makes blue shifts (Fig. 2.12b, c). The E^1_{2g} peak is at 382 cm^{-1} and the A_{1g} peak is at 407 cm^{-1} for bulk MoS_2. The E^1_{2g} peak blue shifts to 385 cm^{-1} and A_{1g} peak red shits to 403 cm^{-1} in single-layer MoS_2. Therefore, the Raman peak frequencies of E^1_{2g} and A_{1g} peaks can be used to determine the thickness of MoS_2.

The Raman shift of MoS_2 is the result of several effects. One of the reasons of the Raman peak shift of MoS_2 is the thermal effect of the laser [192]. Increasing the laser power will soften the bond, so the Raman spectra will shift to the red side.

Different soften rate cause the shift of E^1_{2g} and A_{1g} to different direction. Another reason is interlayer van der Waals force, which could suppress atom vibration more as thickness is increased [193].

2.7.3 Photoluminescence (PL) Spectroscopy

PL is a photon emission process due to the recombination of an optically created electron hole pair. For PL to occur in a semiconductor the absorption of a laser photon must stimulate an electron from the valence band to the conduction band [194]. For electron promotion to occur the laser photon energy must be the same or higher than the material band-gap. The radiative recombination of the electron-hole pair may be preceded by a non-radiative relaxation mechanism, which leads to the emission of phonons. Competition between the non-radiative and radiative processes is important for most of the 2D nanostructured materials. Photo-excited electrons (holes) relax to the bottom of the conduction (valence) band by emitting phonons [195]. Optical recombination of electron-hole pairs occurs at the energies close to the edge of the band-gap [196]. If these band edges occur at the same position of the Brillouin zone in momentum space (i.e. have the same wave vector) emission of a photon occurs with conservation of energy and momentum without further need for additional processes such as emission of phonons. If they occur at different positions (i.e. have different wave vectors), then emission of a particle becomes a second order process and requires additional emission of phonon in order to conserve momentum [197]. This is therefore a slower process and produces a lower emission rate of photons. The first order process is characteristic to direct band-gap semiconductors and is found in the case of most of TMD layers.

In an atom, electrons assemble in discrete energy levels and therefore produce a discrete florescence emission spectrum with line widths determined by the uncertainty principle [198]. In semiconductors the electrons and holes relax into the bottom of the valence and conduction band respectively, forming a thermal distribution before recombining [194]. The line width of emission is related to this thermal distribution KBT, where T is temperature and KB is the Boltzmann constant, along with the density of states, carrier density and inhomogeneity of the crystal. If confinement becomes great enough, quantum effects can occur. In the case of quantum wells and 2D materials such as MoS_2, the electron is unrestricted in movement in two directions, and confined in the third axis. This modifies the density of states to a step like structure with each step corresponding to a confined energy state [195]. On the other hand, localized strain on the atoms in a crystal lattice effect their relative positions leading to a disordered band-edge. Radiative recombination of electron-hole pairs will therefore have a slightly different energy at different positions in the crystal. There is an evidence that non-radiative processes are dominant in the recombination of electron-hole pair in 2D materials. It has been found that competition between radiative and non-radiative processes affect the behavior of PL when varying temperature and incident power. There is an evidence to support previous results showing a low quantum efficiency in 2D layers [199], finding that PL emission does not saturate at high pumping power. Time resolved PL is also used to measure the temporal characteristics of light emission as a result of sample excitation with a pulsed laser. Typically, the PL decays in the nanosecond range in structures where the electron-hole pairs are tightly bound [200]. Instead of continuous beam of photons, 2D layers can be excited by very short laser pulses. The PL spectrum is then recorded as a function of time from the arrival of the excitation pulse. Resolution of the detector plays an important role in these experiments.

2.7.4 X-Ray Diffraction

X-ray diffraction (XRD) is a non-destructive analytical characterization method to determine the crystal phase and structure. An X-ray diffractometer generates an X-ray beam striking a material as a function of incident and scattered angle, polarization, and wavelength or energy. Both the unit cell size and geometry are resolved from the angular positions of the X-ray diffraction peaks.

Wide angle XRD and small-angle X-ray scattering (SAXS) are highly used to collect the information about the material structure and component, thickness of the sheet, and the placement of stacked nanosheets. Inter-layer stacking peaks occur only at low angles which can be measured by SAXS [175]. By pretending this pattern, sheet thickness, the stacking alignment, and distance between two layers or sheets can be measured. For example, wet colloidal and dried aggregates of layers always show different XRD peaks: only layer registry may be concerned for wet colloidal aggregates, however the inner level lattice planes are retained, proposing detached nanosheets (Fig. 2.13) [201, 202]. On the other hand, for dried nanosheets, low-angle pattern shows a well-organized nanosheets arrangement along the stacking axis, and similarly, the basal reflections disappears in the pattern shows random alignments for the group of nanosheets [202]. The layer orientation on substrates helps the thin axis alignment. Specific peak broadening under the layer orientation is the indication of the thin axis [203]. For additional information, the XRD pattern of nanosheets could be demonstrated by the Lorentz polarization factor, the structural influence, the temperature influence and the Laue interference function [153, 201]. Structural parameters for monolayer and multilayers could be computed assuming the unit cell structure, atomic positions, and orientation to a substrate [153]. This modeling results could help to find the nanosheets important units such as the spacing between the nanosheets, and if the nanosheets are rearranged. Inner layer XRD information could be very useful to detect hkl reflections of the nanosheets 2D unit cell [153, 155]. This hkl reflection measurement is a valuable technique to detect absent peaks from well-ordered nanosheets on a substrate. By modifications of these data unit cell parameters could be extracted and compared to bulk nanosheets. For example, XRD investigation of Germanane

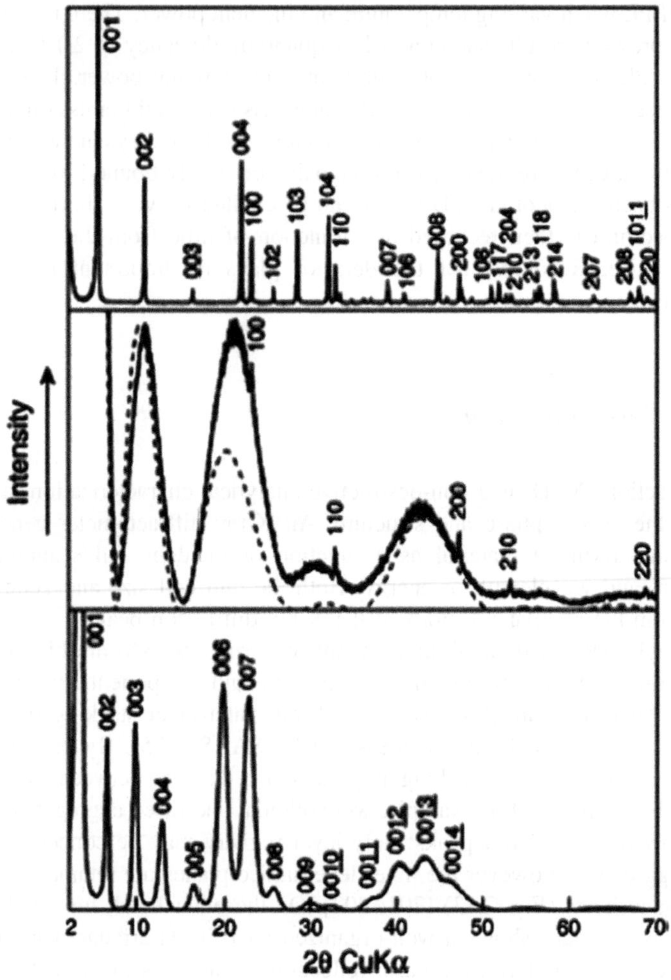

Fig. 2.13 XRD patterns of nanosheets **a** bulk niobate, **b** wet colloidal aggregates of nanosheets as solid line: and calculated for a single layer as broken line, and **c** dried aggregates of nanosheets. Reprinted from Ref. [201] with permission of Elsevier Publishers

(GeH) showed that it could be fitted to a 2H unit cell (2 GeH layers per hexagonal unit cell spacing) with $a = 3.380$ A and $c = 11.04$ A (5.5 A per layer). In the a-direction, the hydrogen-terminated germanane is slightly contracted compared to the bulk unit cell parameters of $a = 3.987$, $c = 30.0582$, (6 layer stacking, $c/6 = 5.097$ A) but expanded in the c-direction due to the replacement of Ca^{2+} with 2 Ge–H bonds between each layer [204].

The reflections from the basal plane will be absent in the case of single mono layer, whereas a peak from the basal plane will appear in the XRD pattern for stacked nanosheets. The intensity of the reflection depends on the number of sheets stacked. A classic example of this is MoS_2 [205]. (100) reflection appears in the XRD pattern when single layer MoS_2 is present with no (110) and (002) reflections. In case of loose MoS_2 agglomerates, (110) reflection also appears due to having no sign of (002) reflection. Whereas few layers stacked MoS_2 shows all the three (100), (110) and (002) reflections due to the appropriate orientation in between MoS_2 nanosheets. Therefore, information regarding single and few layer nanosheets can be easily obtained using XRD.

2.7.5 Characterization Possibilities

The progress and execution of novel characterization methods which can non-destructively probe and quickly investigate the structural and physicochemical properties of 2D layer materials is essential to improve the field. Most of the 2D layers are used for electronic application, it is important to measure the electronic properties and it is mostly controlled by minority defects. Aberration corrected STEM and STM is most powerful and incredibly valuable techniques which can easily identify minor defects, edge states and other species present in the material. The development of imaging techniques with spectroscopic combination such as Tip-Enhanced Raman Spectroscopy (TERS) can considerably assist in identifying and characterizing the single layers and also it can minimize the substrate signals.

References

1. Moore, G.E.: Cramming more components onto integrated circuits. Electronics 114–117 (1965)
2. Chau, R., Doyle, D., Doczy, M. et al.: In: Proceedings of the 61st Device Research Conference, pp. 123–126, Salt Lake City, 23–25 June 2003
3. Doris, B., et al.: In: Technical Digest of the IEEE International Device Meeting, p. 937 (2001)
4. Yu, B., et al.: In: Technical Digest of the IEEE International Device Meeting, p. 267 (2002)
5. Nitzan, A., Ratner, M.A.: Science **300**, 1384 (2003)
6. Aviram, A., Ratner, M.A.: Chem. Phys. Lett. **52**, 9071 (1974)
7. Chen, J., Reed, M.A., Rawlett, A.M., Tour, J.M.: Science **286**, 1550 (1999)
8. Yao, Z., Postman, H.W.C., Balents, L., Dekker, C.: Nature (London) **402**, 273 (1999)
9. Tans, S.J., Verschueren, A.R.M., Dekker, C.: Nature (London) **393**, 49 (1998)
10. Collier, C., Wong, E., Belohradsky, M., Raymo, F., Stoddart, J., Kuekes, P., Williams, R., Heath, J.: Science **285**, 391 (1999)
11. Huang, Y., Duan, X., Cui, Y., Lauhon, L., Kim, K.-H., Lieber, C.: Science **294**, 1313 (2001)
12. Garcia, N., Munoz, M., Zhao, Y.W.: Phys. Rev. Lett. **82**, 2923 (1999)
13. Versluijs, J.J., Bari, M.A., Coey, J.M.D.: Phys. Rev. Lett. **87**, 26601 (2001)
14. Reed, M.A., Zhou, C., Muller, C.J., Burgin, T.P., Tour, J.M.: Science **278**, 252 (1997)
15. Venkataraman, L., Klare, J.E., Nuckolls, C., Hybertsen, M.S., Steigerwald, M.L.: Nature **442**, 904 (2006)
16. Chen, J., Reed, M.A.: Chem. Phys. **281**, 127 (2002)

17. Wang, W., Lee, T., Reed, M.A.: Phys. Rev. B **68**, 035416 (2003)
18. Xiao, X.Y., Xu, B.Q., Tao, N.J.: Nano Lett. **4**, 267 (2004)
19. Hybertsen, M.S., Venkataraman, L., Klare, J.E., Whalley, A.C., Steigerwald, M., Nuckolls, C.: J. Phys. Condens. Matter **20**, 374115 (2008)
20. Teramae, Y., Horiguchi, K., Hashimoto, S., Tsutsui, M., Kurokawa, S., Sakai, A.: Appl. Phys. Lett. **93**, 083121 (2008)
21. Ashcroft, N.W., Mermin, D.N.: Solid State Phys. (1976). Brooks Cole
22. Martin, R.M.: Electronic Structure. Cambridge University Press, Cambridge, UK (2004)
23. Datta, S.: Quantum Transport: Atom to Transistor. Cambridge University Press, Cambridge, UK (2005)
24. Datta, S.: Electronic Transport in Mesoscopic Systems. Cambridge University Press, Cambridge, UK (1995)
25. Tada, T., Kondo, M., Yoshizawa, K.: J. Chem. Phys. **121**, 8050 (2004)
26. Lang, N.D.: Phys. Rev. B **52**, 5335 (1995)
27. Ventra, M.D., Pantelides, S.T., Lang, N.D.: Phys. Rev. Lett. **84**, 979 (2000)
28. Rocha, A.R., Sanvito, S.: Phys. Rev. B **70**, 94406 (2004)
29. Zhang, C., Zhang, X.-G., Krstic, P.S., Cheng, H., Butler, W.H., MacLaren, J.M.: Phys. Rev. B **69**, 134406 (2004)
30. Evers, F., Weigend, F., Koentopp, M.: Phys. Rev. B **69**, 235411 (2004)
31. Choi, H.J., Ihm, J.: Phys. Rev. B **59**, 2267 (1999)
32. Kong, L., Chelikowslcy, J.R., Neaton, J.B., Louie, S.G.: Phys. Rev. B **76**, 235422 (2007)
33. Sørensen, H.H.B.: Correlations in many-body systems with the stochastic variational method. Master's thesis, Aarhus University, Aarhus, Denmark (2005)
34. Hohenberg, P., Kohn, W.: Inhomogeneous electron gas. Phys. Rev. **136**, 864–871 (1964)
35. Burke, K., Car, R., Gebauer, R.: Phys. Rev. Lett. **94**, 146803 (2005)
36. Beenaklcer, C.: Phys. Rev. B **44**, 1646 (1991)
37. Gebauer, R., Car, R.: Phys. Rev. B **70**, 125324 (2004)
38. Stefanucci, G., Almbladh, C.-O.: Phys. Rev. B **69**, 195318 (2004)
39. Ventra, M.D., Todorov, T.N.: J. Phys. Condens. Matter **16**, 8025 (2004)
40. Ke, S.H., Baranger, H.U., Yang, W.: J. Chem. Phys. **123**, 114701 (2005)
41. Muller, K.H.: Phys. Rev. B **73**, 045403 (2006)
42. Qian, Z., Hou, S., Ning, J., Li, R., Shen, Z., Zhao, X., Xue, Z.: J. Chem. Phy. **126**, 084705 (2007)
43. Feng, X., Bengone, O., Alouani, M., Lebegue, S., Rungger, I., Sanvito, S.: Phys. Rev. B **79**, 174414 (2009)
44. Rauba, J.M.C., Strange, M., Thygesen, K.S.: Phys. Rev. B **78**, 165116 (2008)
45. Garcia-Suarez, V.M., Lambert, C.J.: Phys. Rev. B **78**(235412), 9 (2008)
46. Stadler, R., Jacobsen, K.W.: Phys. Rev. B **74**, 161405R (2006)
47. Zhang, Z., Yang, Z., Yuan, J., Qiu, M.: J. Chem. Phys. **128**, 044711 (2008)
48. Fadlallah, M.M., Schuster, C., Schwingenschlogl, U., Wunderlich, T., Sanvito, S.: J. Phys. Condens. Matter **21**, 315001 (2009)
49. Xue, Y., Ratner, M.A.: Phys. Rev. B **68**, 115407 (2003)
50. Pauly, F., Viljas, J.K., Cuevas, J.C., Schon, G.: Phys. Rev. B **77**, 155312 (2008)
51. Mowbray, D.J., Jones, G., Thygesen, K.S.: J. Chem. Phys. **128**, 111103 (2008)
52. Zhang, Z.H., Yang, Z., Yuan, J.H., Zhang, H., Ding, X.Q., Qiu, M.: J. Chem. Phys. **129**, 094702 (2008)
53. Finch, C.M., Sirichantaropass, S., Bailey, S.W., Grace, I.M., Garcia-Suarez, V.M., Lambert, C.J.: J. Phys. Condens. Matter **20**, 022203 (2008)
54. Martin, S., Manrique, D.Z., Garcia-Suarez, V.M., Haiss, W., Higgins, S.J., Lambert, C.J., Nichols, R.J.: Nanotechnology **20**, 125203 (2009)
55. George, C.B., Ratner, M.A., Lambert, J.B.: J. Phys. Chem. A **113**, 3876 (2009)
56. Zhou, Y.X., Jiang, F., Chen, H., Note, R., Mizuseki, H., Kawazoe, Y.: J. Chem. Phys. **128**, 044704 (2008)
57. Crljen, Z., Grigoriev, A., Wendin, G., Stokbro, K.: Phys. Rev. B **71**, 165316 (2005)

58. Zhou, Y.X., Jiang, F., Chen, H., Note, R., Mizuseki, H., Kawazoe, Y.: Phys. Rev. B **75**, 245407 (2007)
59. Pauly, F., Viljas, J.K., Cuevas, J.C.: Phys. Rev. B **78**, 035315 (2008)
60. Lang, N.D.: Phys. Rev. Lett. **97**, 1357 (1997)
61. Tongay, S., Dag, S., Durgun, E., Senger, R.T., Ciraci, S.: J. Phys. Condens. Matter **17**, 3823–3836 (2005)
62. Cheraghchi, H., Esfarjani, K.: Phys. Rev. B **78**, 085123 (2008)
63. Ambavale, S.K., Sharma, A.C.: J. Comput. Theor. Nanosci. **6**, 1549–1555 (2009)
64. Roland, C., Meunier, V., Larade, B., Guo, H.: Phys. Rev. B **66**, 035332 (2002)
65. Dai, Z.X., Shi, X.Q., Zheng, X.X.H., Wang, X.L., Zeng, Z.: Phys. Rev. B **75**, 155402 (2007)
66. Kong, L., Chelikowsky, J.R.: Phys. Rev. B **77**, 073401 (2008)
67. Li, G.Q., Cai, J., Denng, J.K., Rocha, A.R., Sanvito, S.: Appl. Phys. Lett. **92**, 163104 (2008)
68. Jayasekera, T., Mintmire, J.W.: Nanotechnology **18**, 424033 (2007)
69. Andriotis, A.N., Richter, E, Menon, M.: Appl. Phys. Lett. **91**, 152105 (2007)
70. Sun, L., Li, Y., Li, Z., Li, Q., Zhou, Z., Chen, Z., Yang, J., Hou, J.G.: J. Chem. Phys. **129**, 174114 (2008)
71. Yamamoto, M., Takane, Y., Wakabayashi, K.: Phys. Rev. B **79**, 125421 (2009)
72. Stojkovic, D., Zhang, P., Crespi, V.H.: Phys. Rev. Lett. **87**, 125502 (2001)
73. Zhu, J.-L., Xu, F.-F., Jia, Y.-F.: Phys. Rev. B **74**, 155430 (2006)
74. Li, Y.-F., Li, B.-R., Zhang, H.-L.: J. Phys. Condens. Matter **20**, 415207 (2008)
75. Wheeled, W.D., Dahnovsky, Y.: J. Chem. Phys. **129**, 154112 (2008)
76. Morari, C., Rignanese, G.-M., Melinte, S.: Phys. Rev. B **76**, 115428 (2007)
77. Crljen, Z., Baranovic, G.: Phys. Rev. Lett. **98**, 116801 (2007)
78. Evans, J.S., Cheng, C.-L., Van Voorhis, T.: Phys. Rev. B **78**, 165108 (2008)
79. Dalgleish, H., Kirczenow, G.: Phys. Rev. B **73**, 235436 (2006)
80. Engelkes, V.B., Beebe, J.M., Frisbie, C.D.: J. Am. Chem. Soc. **126**, 14287 (2004)
81. Batista, R.J.C., Ordejon, P., Chacham, H., Artacho, E.: Phys. Rev. B **75**, 041402(R) (2007)
82. Botman, A., Mulders, J.J.L., Hagen, C.W.: Nanotechnology **20**, 372001 (2009). https://doi.org/10.1088/0957-4484/20/37/372001
83. Mehendale, S., Mulders, J.J.L., Trompenaars, P.H.F.: Nanotechnology **24**, 145303 (2013). https://doi.org/10.1088/0957-4484/24/14/145303
84. Sychugov, I., Nakayama, Y., Mitsuishi, K.: J. Phys. Chem. C **113**, 21516–21519 (2009). https://doi.org/10.1021/jp9079684
85. Broers, A.N., Molzen, W.W., Cuomo, J.J., Wittels, N.D.: Appl. Phys. Lett. **29**, 596–598 (1976). https://doi.org/10.1063/1.89155
86. Silvis-Cividjian, N., Hagen, C.W., Kruit, P.: J. Appl. Phys. **98**, 084905 (2005). https://doi.org/10.1063/1.2085307
87. van Dorp, W.F., Hagen, C.W., Crozier, P.A., Kruit, P.: Nanotechnology **19**, 225305 (2008). https://doi.org/10.1088/0957-4484/19/22/225305
88. van Oven, J.C., Berwald, F., Berggren, K.K., Kruit, P., Hagen, C.W.: J. Vac. Sci. Technol. B Nanotechnol. Microelectron. Mater. Process. Meas. Phenom. 29, 06F305 (2011). https://doi.org/10.1116/1.3640743
89. Notte, J., Ward, B., Economou, N., Hill, R., Percival, R., Farkas, L., McVey, S., Seiler, D. G., Diebold, A.C., McDonald, R., Garner, C.M., Herr, D., Khosla, R.P., Secula, E.M.: AIP Conf. Proc. **489**, 931 (2007). https://doi.org/10.1063/1.2799423
90. Kohn, W.: Rev. Mod. Phys. **71**, 1253 (1999)
91. Pople, J.A.: Rev. Mod. Phys. **71**, 1267 (1999)
92. Hohenberg, P., Kohn, W.: Phys. Rev. **136**, B864 (1964)
93. Kohn, W., Sham, L.J.: Phys. Rev. **140**, A1133 (1965)
94. Parr, R.G., Yang, W.: Density Functional Theory of Atoms and Molecules. Oxford press, New York (1989)
95. Burke, K.: The ABC of DFT (2007). http://chem.ps.uci.edu/Tderon/dft/book/
96. Kuno, M.: Introduction to Nanoscience and Nanotechnology: A Workbook
97. Maizari, N.: Ab initio molecular dynamics for metallic systems (thesis) (1996)

98. Payne, M.C., Teter, M.P., Allan, D.C., Arias, T.A., Joannopoulos, J.D.: Rev. Mod. Phys. **64**, 1045 (1992)
99. Hirose, K., Ono, T., Fujimoto, Y., Tsukamoto, S.: First-Principles Calculations in Real-Space Formalism, Electronic Configurations and Transport Properties of Nanostructures. Imperial College Press (2005)
100. Rocha, A.R.: Theoretical and computational aspects of electronic transport at the nanoscale (thesis) (2007)
101. Chen, F., Hihath, J., Huang, Z., Li, X., Tao, N.J.: Measurement of single-molecule conductance. Annu. Rev. Phys. Chem. **58**, 535–564 (2007)
102. Capelle, K.: A bird's-eye view of density-functional theory. arXiv:condmat/0211443v5 [cond-mat.mtrl-sci] (2006)
103. Taylor, J., Guo, H., Wang, J.: Ab initio modeling of quantum transport properties of molecular electronic devices. Phys. Rev. B **63**(24), 245407 (2001)
104. Kohn, W., Sham, L.J.: Self-consistent equations including exchange and correlation effects. Phys. Rev. **140**(4A), A1133–A1138 (1965)
105. Haug, H., Jauho, A.-P.: Quantum Kinetics in Transport and Optics of Semiconductors. Springer Series in Solid State Sciences. Springer, Berlin (1998)
106. Brandbyge, M., Mozos, J.-L., Ordejón, P., Taylor, J., Stokbro, K: Density-functional method for nonequilibrium electron transport. Phys. Rev. B **65**, 165401 (2002)
107. Xue, Yongqiang, Datta, Supriyo, Ratner, Mark A.: First-principles based matrix green's function approach to molecular electronic devices: general formalism. Chem. Phys. **281**(23), 151–170 (2002)
108. Faleev, S.V., Léonard, F., Stewart, D.A., van Schilfgaarde, M.: Ab initio tight-binding LMTO method for nonequilibrium electron transport in nanosystems. Phys. Rev. B **71**(19), 195422 (2005)
109. Rocha, A.R., Garcia-suarez, V.M., Bailey, S.W., Lambert, C.J., Ferrer, J., Sanvito, S.: Towards molecular spintronics. Nat. Mater. **4**, 335–339 (2005)
110. Waldron, D., Haney, P., Larade, B., MacDonald, A., Guo, H.: Nonlinear spin current and magnetoresistance of molecular tunnel junctions. Phys. Rev. Lett. **96**, 166804 (2006)
111. Ke, S.-H., Yang, W., Curtarolo, S., Baranger, H.U.: Thermopower of molecular junctions: an ab initio study. Nano Lett. **9**(3), 1011–1014 (2009)
112. Taylor, J., Guo, H., Wang, J:. Ab initio modeling of open systems: charge transfer, electron conduction, and molecular switching of a *C*60 device. Phys. Rev. B **63**(12), 121104 (2001)
113. Larade, B., Taylor, J., Mehrez, H., Guo, H.: Conductance, I-V curves, and negative differential resistance of carbon atomic wires. Phys. Rev. B **64**, 075420 (2001)
114. Smeu, M., Wolkow, R.A., Guo, H.: Conduction pathway of π-stacked ethylbenzene molecular wires on si(100). J. Am. Chem. Soc. **131**(31), 11019–11026 (2009). PMID: 19603787
115. Kaun, C.-C., Guo, H.: Resistance of alkanethiol molecular wires. Nano Lett. **3**(11), 1521–1525 (2003)
116. Liu, Y., Guo, H.: Current distribution in b- and n-doped carbon nanotubes. Phys. Rev. B **69**, 115401 (2004)
117. Sergueev, N., Roubtsov, D., Guo, H.: *Ab Initio* analysis of electron-phonon coupling in molecular devices. Phys. Rev. Lett. **95**, 146803 (2005)
118. Sergueev, N., Demkov, A.A., Guo, H.: Inelastic resonant tunneling in C60 molecular junctions. Phys. Rev. B **75**, 233418 (2007)
119. Waldron, D., Liu, L., Guo, H.: Ab initio simulation of magnetic tunnel junctions. Nanotechnology **18**(42), 424026 (2007)
120. Bevan, K.H., Zahid, F., Kienle, D., Guo, H.: First-principles analysis of the STM image heights of styrene on si(100). Phys. Rev. B **76**, 045325 (2007)
121. Ning, Z., Zhu, Y., Wang, J., Guo, H.: Quantitative analysis of nonequilibrium spin injection into molecular tunnel junctions. Phys. Rev. Lett. **100**, 056803 (2008)

122. Basch, H., Cohen, R., Ratner, M.A.: Interface geometry and molecular junction conductance: geometric fluctuation and stochastic switching. Nano Lett. **5**(9), 1668–1675 (2005). PMID: 16159203
123. Ke, Y., Zahid, F., Timoshevskii, V., Xia, K., Gall, D., Guo, H.: Resistivity of thin Cu films with surface roughness. Phys. Rev. B **79**, 155406 (2009)
124. Mirzadeh, E., Akhbari, K.: Synthesis of nanomaterials with desirable morphologies from metal-organic frameworks for various applications. CrystEngComm **18**, 7410–7424 (2016). https://doi.org/10.1039/C6CE01076H
125. Khlebtsov, N., Dykman, L.: Biodistribution and toxicity of engineered gold nanoparticles: a review of in vitro and in vivo studies. Chem. Soc. Rev. **40**, 1647–1671 (2011). http://dx.doi.org/10.1039/C0CS00018C
126. Khlebtsov, N., Dykman, L.: Plasmonic nanoparticles. 37–85 (2010). http://dx.doi.org/10.1201/9781439806296-c2
127. Khlebtsov, N.G., Dykman, L.A.: Optical properties and biomedical applications of plasmonic nanoparticles. J. Quant. Spectrosc. Radiat. Transf. **111**, 1–35 (2010). http://dx.doi.org/10.1016/j.jqsrt.2009.07.012
128. Wang, J., Yang, N., Tang, H., Dong, Z., Jin, Q., Yang, M., Kisailus, D., Zhao, H., Tang, Z., Wang, D.: Accurate control of multishelled Co_3O_4 hollow microspheres as high-performance anode materials in lithium-ion batteries. Angew. Chemie Int. Ed. **52**, 6417–6420 (2013). https://doi.org/10.1002/anie.201301622
129. Khan, I., Ali, S., Mansha, M., Qurashi, A.: Sonochemical assisted hydrothermal synthesis of pseudo-flower shaped Bismuth vanadate ($BiVO_4$) and their solar-driven water splitting application. Ultrason. Sonochem. **36**, 386–392 (2017). http://dx.doi.org/10.1016/j.ultsonch.2016.12.014
130. Khan, I., Ibrahim, A.A.M., Sohail, M., Qurashi, A.: Sonochemical assisted synthesis of RGO/ZnO nanowire arrays for photoelectrochemical water splitting. Ultrason. Sonochem. **37**, 669–675 (2017). https://doi.org/10.1016/j.ultsonch.2017.02.029
131. Ullah, H., Khan, I., Yamani, Z.H., Qurashi, A.: Sonochemicaldriven ultrafast facile synthesis of SnO_2 nanoparticles: growth mechanism structural electrical and hydrogen gas sensing properties. Ultrason. Sonochem. **34**, 484–490 (2017). https://doi.org/10.1016/j.ultsonch.2016.06.025
132. Emery, A.A., Saal, J.E., Kirklin, S., Hegde, V.I., Wolverton, C.: High-throughput computational screening of perovskites for thermochemical water splitting applications. Chem. Mater. **28** (2016). http://dx.doi.org/10.1021/acs.chemmater.6b01182
133. Ingham, B.: X-ray scattering characterisation of nanoparticles. Crystallogr. Rev. **21**, 229–303 (2015). http://dx.doi.org/10.1080/0889311X.2015.1024114
134. Avasare, V., Zhang, Z., Avasare, D., Khan, I., Qurashi, A.: Room-temperature synthesis of TiO_2 nanospheres and their solar driven photoelectrochemical hydrogen production. Int. J. Energy Res. **39**, 1714–1719 (2015). https://doi.org/10.1002/er.3372
135. Iqbal, N., Khan, I., Yamani, Z.H., Qurashi, A.: Sonochemical assisted solvothermal synthesis of gallium oxynitride nanosheets and their solar-driven photoelectrochemical water-splitting applications. Sci. Rep. **6**, 32319 (2016). https://doi.org/10.1038/srep32319
136. Dablemont, C., Lang, P., Mangeney, C., Piquemal, J.-Y., Petkov, V., Herbst, F., Viau, G.: FTIR and XPS study of Pt nanoparticle functionalization and interaction with alumina. Langmuir **24**, 5832–5841 (2008). https://doi.org/10.1021/la7028643
137. Muehlethaler, C., Leona, M., Lombardi, J.R.: Review of surface enhanced Raman scattering applications in forensic science. Anal. Chem. **88**, 152–169 (2016). http://dx.doi.org/10.1021/acs.analchem.5b04131
138. Ma, S., Livingstone, R., Zhao, B., Lombardi, J.R.: Enhanced Raman spectroscopy of nanostructured semiconductor phonon modes. J. Phys. Chem. Lett. **2**, 671–674 (2011). http://dx.doi.org/10.1021/jz2001562
139. Kestens, V., Roebben, G., Herrmann, J., Jämting, A., Coleman, V., Minelli, C., Clifford, C., De Temmerman, P.-J., Mast, J., Junjie, L., Babick, F., Cölfen, H., Emons, H.: Challenges in

the size analysis of a silica nanoparticle mixture as candidate certified reference material. J. Nanopart. Res. **18**, 171 (2016). http://dx.doi.org/10.1007/s11051-016-3474-2
140. Gross, J., Sayle, S., Karow, A.R., Bakowsky, U., Garidel, P.: Nanoparticle tracking analysis of particle size and concentration detection in suspensions of polymer and protein samples: influence of experimental and data evaluation parameters. Eur. J. Pharm. Biopharm. **104**, 30–41 (2016). http://dx.doi.org/10.1016/j.ejpb.2016.04.013
141. Fagerlund, G.: Determination of specific surface by the BET method. Matériaux Constr. **6**, 239–245 (1973). http://dx.doi.org/10.1007/BF02479039
142. Liu, D., Zhou, W., Wu, J.: CuO–CeO_2/ZSM–5 composites for reactive adsorption of hydrogen sulphide at high temperature. Can. J. Chem. Eng. **94**, 2276–2281 (2016). https://doi.org/10.1002/cjce.22613
143. Swinehart, D.F.: The Beer-Lambert law. J. Chem. Educ. **39**, 333 (1962). https://doi.org/10.1021/ed039p333
144. Liu, D., Li, C., Zhou, F., Zhang, T., Zhang, H., Li, X., Duan, G., Cai, W., Li, Y.: Rapid synthesis of monodisperse Au nanospheres through a laser irradiation-induced shape conversion, self-assembly and their electromagnetic coupling SERS enhancement. Sci. Rep. **5**, 7686 (2015). https://doi.org/10.1038/srep07686
145. Liu, J., Liu, Y., Liu, N., Han, Y., Zhang, X., Huang, H., Lifshitz, Y., Lee, S.-T., Zhong, J., Kang, Z.: Metal-free efficient photocatalyst for stable visible water splitting via a two–electron pathway. Science **80**(347), 970–974 (2015). https://doi.org/10.1126/science.aaa3145
146. Peng, K., Fu, L., Yang, H., Ouyang, J.: Perovskite $LaFeO_3$/montmorillonite nanocomposites: synthesis, interface characteristics and enhanced photocatalytic activity. Sci. Rep. **6**, 19723 (2016). https://doi.org/10.1038/srep19723
147. Yu, Z.B., Xie, Y.P., Liu, G., Lu, G.Q., Ma, X.L., Cheng, H.-M.: Self-assembled CdS/Au/ZnO heterostructure induced by surface polar charges for efficient photocatalytic hydrogen evolution. J. Mater. Chem. A **1**, 2773 (2013). http://dx.doi.org/10.1039/c3ta01476b
148. Lin, G., Zhang, Q., Lin, X., Zhao, D., Jia, R., Gao, N., Zuo, Z., Xu, X., Liu, D.: Enhanced photoluminescence of gallium phosphide by surface plasmon resonances of metallic nanoparticles. RSC Adv. **5**, 48275–48280 (2015). http://dx.doi.org/10.1039/C5RA07368E
149. Gupta, K., Singh, R.P., Pandey, A., Pandey, A.: Photocatalytic antibacterial performance of TiO_2 and Ag-doped TiO_2 against S. aureus. P. aeruginosa and E. coli. Beilstein J. Nanotechnol. **4**, 345–351 (2013). https://doi.org/10.3762/bjnano.4.40
150. Pal, M., Pal, U., Jiménez, J.M.G.Y., Pérez-Rodríguez, F.: Effects of crystallization and dopant concentration on the emission behavior of TiO_2: Eu nanophosphors. Nanoscale Res. Lett. **7**, 1 (2012). http://dx.doi.org/10.1186/1556–276X–7–1
151. Torchynska, T.V., El Filali, B., Ballardo Rodríguez, I.C., Shcherbyna, L.: Defect related emission of ZnO and ZnO Cu nanocrystals prepared by electrochemical method. Phys. Status Solidi **13**, 594–597 (2016). http://dx.doi.org/10.1002/pssc.201510277
152. Wan, D., Chen, H.-L., Lin, Y.-S., Chuang, S.-Y., Shieh, J., Chen, S.-H.: Using spectroscopic ellipsometry to characterize and apply the optical constants of hollow gold nanoparticles. ACS Nano **3**, 960–970 (2009). http://dx.doi.org/10.1021/nn8009008
153. Fukuda, K., Akatsuka, K., Ebina, Y., Ma, R., Takada, K., Nakai, I., et al.: Exfoliated nanosheet crystallite of cesium tungstate with 2D pyrochlore structure: synthesis, characterization, and photochromic properties. ACS Nano **2**, 1689–1695 (2008)
154. Osada, M., Takanashi, G., Li, B.W., Akatsuka, K., Ebina, Y., Ono, K., et al.: Controlled polarizability of one-nanometer-thick oxide nanosheets for tailored, high-j nanodielectrics. Adv. Funct. Mater. **21**, 3482–3487 (2011)
155. Aksit, M., Hoselton, B.C., Kim, H.J., Ha, D.-H., Robinson, R.D.: Synthesis and properties of electrically conductive, ductile, extremely long ($\sim$50 μm) nanosheets of $K_xCoO_2{\cdot}yH_2O$. ACS Appl. Mater. Interf. **5**, 8998–9007 (2013)
156. Kim, J., Cote, L.J., Kim, F., Huang, J.: Visualizing graphene based sheets by fluorescence quenching microscopy. J. Am. Chem. Soc. **132**, 260–267 (2009)

157. Nemes-Incze, P., Osvath, Z., Kamaras, K., Biro, L.: Anomalies in thickness measurements of graphene and few layer graphite crystals by tapping mode atomic force microscopy. Carbon **46**, 1435–1442 (2008)
158. Wang, X., Song, J., Liu, J., Wang, Z.L.: Direct-current nanogenerator driven by ultrasonic waves. Science **316**, 102–105 (2007)
159. Splendiani, A., Sun, L., Zhang, Y., Li, T., Kim, J., Chim, C.-Y., et al.: Emerging photoluminescence in monolayer MoS_2. Nano Lett. **10**, 1271–1275 (2010)
160. Lee, C., Yan, H., Brus, L.E., Heinz, T.F., Hone, J., Ryu, S.: Anomalous lattice vibrations of single-and few-layer MoS_2. ACS Nano **4**, 2695–2700 (2010)
161. Kim, J., Cote, L.J., Kim, F., Yuan, W., Shull, K.R., Huang, J.: Graphene oxide sheets at interfaces. J. Am. Chem. Soc. **132**, 8180–8186 (2010)
162. Zheng, Q., Ip, W.H., Lin, X., Yousefi, N., Yeung, K.K., Li, Z., et al.: Transparent conductive films consisting of ultralarge graphene sheets produced by Langmuir-Blodgett assembly. ACS Nano **5**, 6039–6051 (2011)
163. Tan, A.T., Kim, J., Huang, J.K., Li, L.J., Huang, J.: Seeing two-dimensional sheets on arbitrary substrates by fluorescence quenching microscopy. Small **9**, 3253–3258 (2013)
164. Akatsuka, K., Takanashi, G., Ebina, Y., Haga, M.-A., Sasaki, T.: Electronic band structure of exfoliated titanium-and/or niobium based oxide nanosheets probed by electrochemical and photoelectrochemical measurements. J. Phys. Chem. C **116**, 12426–12433 (2012)
165. Novoselov, K., Geim, A.K., Morozov, S., Jiang, D., Grigorieva, M.K.I., Dubonos, S., et al.: Two-dimensional gas of massless Dirac fermions in graphene. Nature **438**, 197–200 (2005)
166. Novoselov, K.S., Geim, A.K., Morozov, S., Jiang, D., Zhang, Y., Dubonos, S., et al.: Electric field effect in atomically thin carbon films. Science **306**, 666–669 (2004)
167. Novoselov, K., Jiang, D., Schedin, F., Booth, T., Khotkevich, V., Morozov, S., et al.: Two-dimensional atomic crystals. Proc. Natl. Acad. Sci. U.S.A. **102**, 10451–10453 (2005)
168. Lui, C.H., Liu, L., Mak, K.F., Flynn, G.W., Heinz, T.F.: Ultraflat graphene. Nature **462**, 339–341 (2009)
169. Late, D.J., Liu, B., Matte, H., Rao, C., Dravid, V.P.: Rapid characterization of ultrathin layers of chalcogenides on SiO_2/Si substrates. Adv. Funct. Mater. **22**, 1894–1905 (2012)
170. Benameur, M., Radisavljevic, B., Heron, J., Sahoo, S., Berger, H., Kis, A.: Visibility of dichalcogenide nanolayers. Nanotechnology **22**, 125706 (2011)
171. Sterrer, M., Risse, T., Pozzoni, U.M., Giordano, L., Heyde, M., Rust, H.-P., et al.: Control of the charge state of metal atoms on thin MgO films. Phys. Rev. Lett. **98**, 096107 (2007)
172. Blake, P., Hill, E., Neto, A.C., Novoselov, K., Jiang, D., Yang, R., et al.: Making graphene visible. Appl. Phys. Lett. **91**, 063124 (2007)
173. Kim, J., Cote, L.J., Huang, J.: Two dimensional soft material: new faces of graphene oxide. Acc. Chem. Res. **45**, 1356–1364 (2012)
174. Coleman, J.N., Lotya, M., O'Neill, A., Bergin, S.D., King, P.J., Khan, U., et al.: Two-dimensional nanosheets produced by liquid exfoliation of layered materials. Science **331**, 568–571 (2011)
175. Schliehe, C., Juarez, B.H., Pelletier, M., Jander, S., Greshnykh, D., Nagel, M., et al.: Ultrathin PbS sheets by two-dimensional oriented attachment. Science **329**, 550–553 (2010)
176. Gutierrez, H.R., Perea-Lopez, N., Elias, A.L., Berkdemir, A., Wang, B., Lv, R., et al.: Extraordinary room-temperature photoluminescence in triangular WS_2 monolayers. Nano Lett. **13**, 3447–3454 (2012)
177. Hong, J., Park, M.K., Lee, E.J., Lee, D., Hwang, D.S., Ryu, S.: Origin of new broad Raman D and G peaks in annealed graphene. Sci. Rep. **3** (2013)
178. Krivanek, O.L., Chisholm, M.F., Nicolosi, V., Pennycook, T.J., Corbin, G.J., Dellby, N., et al.: Atom-by-atom structural and chemical analysis by annular dark-field electron microscopy. Nature **464**, 571–574 (2010)
179. Choi, T., Ruggiero, C., Gupta, J.: Tunneling spectroscopy of ultrathin insulating Cu_2N films, and single Co adatoms. J. Vacuum Sci. Technol. B Microelectron. Nanomet. Struct. **27**, 887–890 (2009)

180. Helveg, S., Lauritsen, J.V., Lagsgaard, E., Stensgaard, I., Norskov, J.K., Clausen, B., et al.: Atomic-scale structure of single-layer MoS_2 nanoclusters. Phys. Rev. Lett. **84**, 951 (2000)
181. Auwarter, W., Kreutz, T., Greber, T., Osterwalder, J.: XPD and STM investigation of hexagonal boron nitride on Ni(111). Surf. Sci. **429**, 229–236 (1999)
182. Li, G., Andrei, E.Y.: Observation of Landau levels of Dirac fermions in graphite. Nat. Phys. **3**, 623–627 (2007)
183. Miller, D.L., Kubista, K.D., Rutter, G.M., Ruan, M., de Heer, W.A., First, P.N., et al.: Observing the quantization of zero mass carriers in graphene. Science **324**, 924–927 (2009)
184. Brar, V.W., Decker, R., Solowan, H.-M., Wang, Y., Maserati, L., Chan, K.T., et al.: Gate-controlled ionization and screening of cobalt adatoms on a graphene surface. Nat. Phys. **7**, 43–47 (2011)
185. Gyamfi, M., Eelbo, T., Waśniowska, M., Wiesendanger, R.: Fe adatoms on graphene/Ru (0001): adsorption site and local electronic properties. Phys. Rev. B **84**, 113403 (2011)
186. Preobrajenski, A., Ng, M.L., Vinogradov, N., Vinogradov, A., Lundgren, E., Mikkelsen, A., et al.: Impact of oxygen coadsorption on intercalation of cobalt under the h-BN nanomesh. Nano Lett. **9**, 2780–2787 (2009)
187. Natterer, F.D., Patthey, F., Brune, H.: Ring state for single transition metal atoms on boron nitride on Rh(111). Phys. Rev. Lett. **109**, 066101 (2012)
188. Besenbacher, F., Lauritsen, J., Linderoth, T., Lagsgaard, E., Vang, R., Wendt, S.: Atomic-scale surface science phenomena studied by scanning tunneling microscopy. Surf. Sci. **603**, 1315–1327 (2009)
189. Sanchez, S.I., Small, M.W., Sivaramakrishnan, S., Wen, J.-G., Zuo, J.-M., Nuzzo, R.G.: Visualizing materials chemistry at atomic resolution. Anal. Chem. **82**, 2599–2607 (2010)
190. Verble, J., Wieting, T.: Lattice mode degeneracy in MoS_2 and other layer compounds. Phys. Rev. Lett. **25**, 362 (1970)
191. Verble, J., Wietling, T., Reed, P.: Rigid-layer lattice vibrations and van der Waals bonding in hexagonal MoS_2. Solid State Commun. **11**, 941–944 (1972)
192. Najmaei, S., Liu, Z., Ajayan, P., Lou, J.: Thermal effects on the characteristic Raman spectrum of molybdenum disulfide (MoS_2) of varying thicknesses. Appl. Phys. Lett. **100**, 013106 (2012)
193. Li, H., Zhang, Q., Yap, C.C.R., Tay, B.K., Edwin, T.H.T., Olivier, A., et al.: From bulk to monolayer MoS_2: evolution of Raman scattering. Adv. Funct. Mater. **22**, 1385–1390 (2012)
194. Perkowitz, S.: Optical characterization of semiconductors: infrared, Raman, and photoluminescence spectroscopy (1993). Elsevier
195. Fox, M.: Optical Properties of Solids. Oxford Master Series in Condensed Matter Physics. Oxford University Press, Oxford (2001)
196. Sze, S.M., Ng, K.K.: Physics of semiconductor devices. Wiley (2006)
197. Grundmann, M.: The physics of semiconductors, vol. 19, p. 37. Springer, Berlin (2006)
198. Foot, C.J.: Atomic Physics. Oxford University Press (2004)
199. Wang, Q.H., Kalantar-Zadeh, K., Kis, A., Coleman, J.N., Strano, M.S.: Electronics and optoelectronics of two-dimensional transition metal dichalcogenides. Nat. Nanotechnol. **7**, 699–712 (2012)
200. Couto Jr., O., Sercombe, D., Puebla, J., Otubo, L., Luxmoore, I., Sich, M., et al.: Effect of a GaAsP shell on the optical properties of selfcatalyzed gaAs nanowires grown on silicon. Nano Lett. **12**, 5269–5274 (2012)
201. Ebina, Y., Sasaki, T., Watanabe, M.: Study on exfoliation of layered perovskite-type niobates. Solid State Ionics **151**, 177–182 (2002)
202. Bizeto, M.A., Shiguihara, A.L., Constantino, V.R.: Layered niobate nanosheets: building blocks for advanced materials assembly. J. Mater. Chem. **19**, 2512–2525 (2009)
203. Ding, Z., Bux, S.K., King, D.J., Chang, F.L., Chen, T.-H., Huang, S.-C., et al.: Lithium intercalation and exfoliation of layered bismuth selenide and bismuth telluride. J. Mater. Chem. **19**, 2588–2592 (2009)

204. Bianco, E., Butler, S., Jiang, S., Restrepo, O.D., Windl, W., Goldberger, J.E.: Stability and exfoliation of germanane: a germanium graphane analogue. ACS Nano **7**, 4414–4421 (2013)
205. Peng, Y., Meng, Z., Zhong, C., Lu, J., Yu, W., Jia, Y., et al.: Hydrothermal synthesis and characterization of single-molecular-layer MoS_2 and $MoSe_2$. Chem. Lett. **30**, 772–773 (2001)

Chapter 3
Where Are Nanomaterials (Nms) Found?

Some nanomaterials (NMs) occur naturally, but of particular interest are engineered nanomaterials (EN), which are designed for, and already being used in many commercial products and processes. They can be found in such things as sunscreens, cosmetics, sporting goods, stain-resistant clothing, tires, electronics, as well as many other everyday items, and are used in medicine for purposes of diagnosis, imaging and drug delivery.

3.1 Nanoparticles Are All Around Us

You are looking at an electron micrograph of gold nanoparticles; a snap shot of tiny gold crystals that are 1/10,000th the diameter of a human hair. Nanoparticles just these like may soon transform every aspect of our day-to-day lives, from the size of our personal electronic devices to the way diagnose and treat cancer; all part of the promised nanotechnology revolution. The very word "nanotechnology" seems to suggest something alien; something that belongs far in the future or in the realm of our favorite sci-fi movies. However, *nanoparticles*, one of the "building blocks" of nanotechnology are all around us right now, and have been all around us throughout human history. They were with us when human beings began making their first tools, and they are present in products we buy at the grocery store every day. They largely flew under the radar until electron microscopes become commonplace several decades ago, but now, the more we turn our microscopes on everyday objects, the more nanoparticles we seem to find. Even the most seemingly mundane objects can give rise to nanoparticles; detecting them is simply a matter of being able to look closely enough to see them (no simple matter for such small materials). You could find nanoparticles in your jewelry box or the drawer with your family's fanciest silverware. I have to see this first hand while I was working in the

L. H. Madkour, *Nanoelectronic Materials*, Advanced Structured Materials 116, https://doi.org/10.1007/978-3-030-21621-4_3

Hutchison lab at the University of Oregon several years ago. Some of my colleagues were trying to understand why silver nanoparticles change size and shape so rapidly, even when they are just left in storage on the shelf. Because they saw such rapid changes in the size and shape of silver nanoparticles, they thought to look and see if large every day pieces of silver and copper (Sterling silver forks, earrings, and wires) might give off nanoparticles. To test this, they simply left the fork (or any of the other items) on an electron microscopy grid for several hours, then took the fork away, and had a peek at what it had left behind. Surprisingly, they found that the silver and copper items had left silver and copper nanoparticles behind all over the grid; a most elegant demonstration that human beings can come into contact with a variety of nanoparticles, even in our own homes. Forks and earrings are merely the tip of the iceberg, though. Wherever we go during our day-to-day routine, we can encounter nanoparticles (both synthetic and natural).

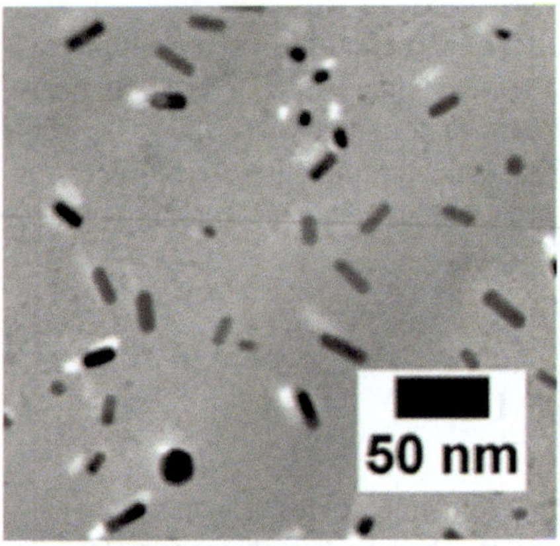

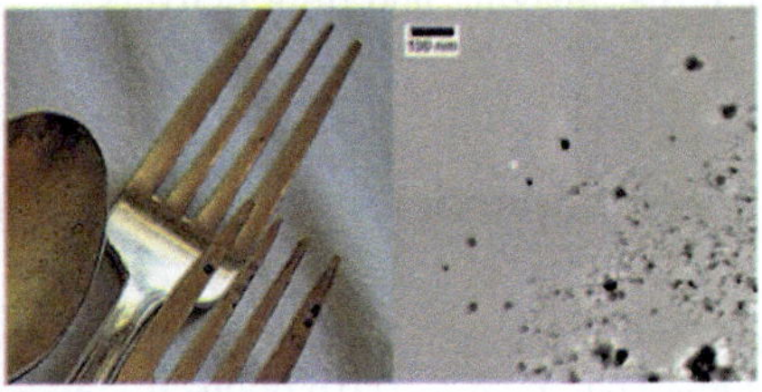

Engineered nanomaterials are resources designed at the molecular (nanometre) level to take advantage of their small size and novel properties, which are generally not seen in their conventional, bulk counterparts. The two main reasons why materials at the nano scale can have different properties are increased relative surface area and new quantum effects. Nanomaterials have a much greater surface area to volume ratio than their conventional forms, which can lead to greater chemical reactivity and affect their strength. Also at the nano scale, quantum effects can become much more important in determining the materials properties and characteristics, leading to novel optical, electrical and magnetic behaviors.

Nanomaterials are already in commercial use, with some having been available for several years or decades. The range of commercial products available today is very broad, including stain-resistant and wrinkle-free textiles, cosmetics, sunscreens, electronics, paints and varnishes. Nanocoatings and nanocomposites are finding uses in diverse consumer products, such as windows, sports equipment, bicycles and automobiles. There are novel UV-blocking coatings on glass bottles, which protect beverages from damage by sunlight, and longer-lasting tennis balls using butyl rubber/nano-clay composites. Nanoscale titanium dioxide, for instance, is finding applications in cosmetics, sun-block creams and self-cleaning windows, and nanoscale silica is being used as filler in a range of products, including cosmetics and dental fillings.

An experiment you could do at home (if you have your very own electron microscope)! Sterling silver forks release small quantities of silver nanoparticles into the surrounding environment. Fork image source.

It seems weird to say, but (as the fork study shows) nanoparticles are actually a fairly common type of material in many different environments, and they can pass by almost undetected unless you are looking for them. In fact, many kinds of physical and chemical processes (both human activities and natural processes) produce nanoparticles. Naturally occurring nanoparticles can be found in volcanic ash, ocean spray, fine sand and dust, and even biological matter (e.g. viruses). Synthetic nanoparticles are equally, if not more diverse than their naturally occurring counterparts.

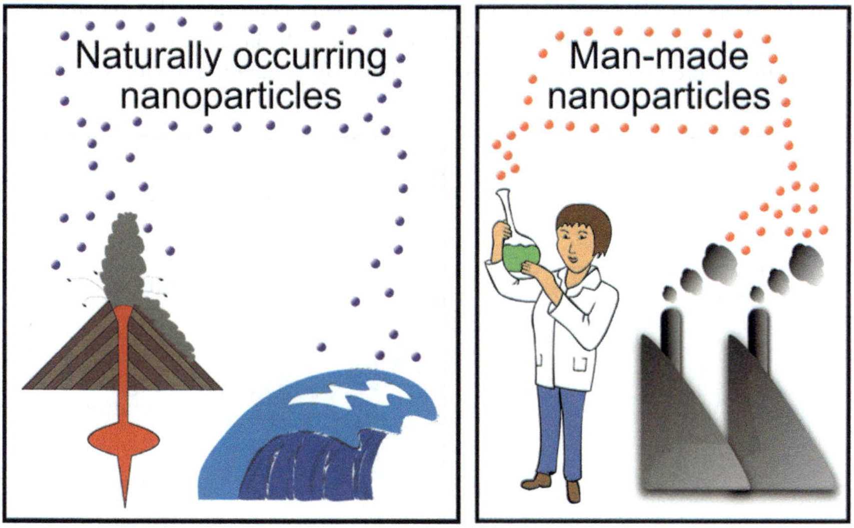

Synthetic nanoparticles (sometimes-called *anthropogenic* nanoparticles) fall into two general categories: "incidental" and "engineered" nanoparticles. Incidental nanoparticles are the byproducts of human activities, generally have poorly controlled sizes and shapes, and may be made of a hodge-podge of different elements. Many of the processes that generate incidental nanoparticles are common every day activities: running diesel engines, large-scale mining, and even starting a fire.

A few sources of incidental nanoparticles

Engineered nanoparticles on the other hand, have been specifically designed and deliberately synthesized by human beings. Not surprisingly, they have very precisely controlled sizes, shapes, and compositions. They may even contain "layers" with different chemical compositions (e.g. a core made out of gold, covered in a shell of silica, and coated with specifically chosen antibodies). Although engineered nanoparticles get more sophisticated with each passing year, simple engineered nanoparticles can be created by relatively simple chemical reactions that have been within the scope of chemists and alchemists for many centuries. This means that long before people could "see" a nanoparticle through an electron microscope, human beings were both deliberately and accidentally generating a wide variety of these materials.

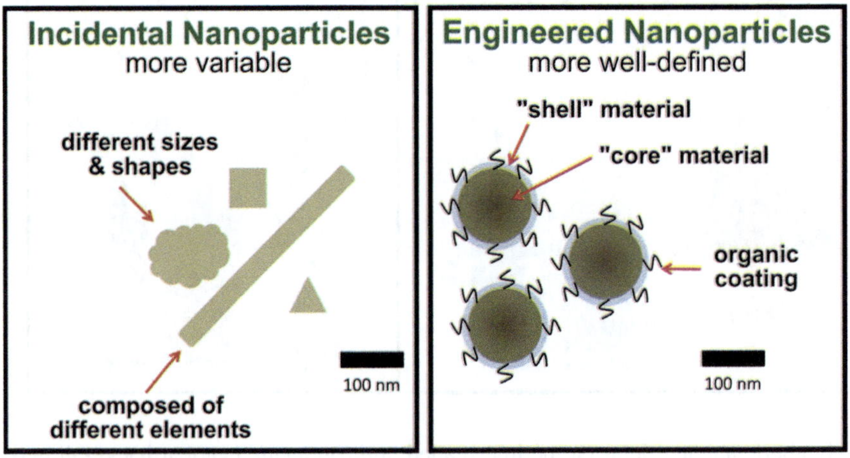

There are two types of synthetic nanoparticles: "incidental" (byproducts of other human activities) and engineered (made to order). It is hard to say when human beings started making incidental nanoparticles, but probably as soon as people started taming fire—you can find small nano-scale particles of soot in the embers and smoke. Certainly, by the Bronze Age, incidental copper nanoparticles would have been prevalent in human civilizations. The earliest engineered nanoparticles are often attributed to the ancient Romans, Egyptians, and Chinese. Despite not having any idea about the scientific implications of what they were making, they were successfully able to prepare nanoparticle solutions of gold and other precious metals with reasonably precise control over particle size and composition. In fact, some of the methods used in these ancient cultures to make nanoparticles were very similar to the way many metal nanoparticles are still synthesized in laboratories today. These antiquarian gold nanoparticle solutions were vibrantly colored (generally reds and purples) and could be impregnated into glass to make stained glass and jewelry. In addition, many people ingested these colloidal gold and silver solutions as health tonics; specifically to treat high fevers and syphilis. Though it's unlikely that drinking metal nanoparticles has any truly positive aspects for your health, some people still swear by this folk remedy, with often unappetizing, though not necessarily *lethal*, results.

Roman colloidal gold. (L) The Lycurgus cup. Gold and silver nanoparticles in the glass make for some fantastic and unique color effects. (R) A solution of gold nanoparticles in water. Some people still drink this as a health tonic. I would not recommend it. Cup image source.

As human society has advanced technologically, it is likely that exposure to all kinds of synthetic nanoparticles has dramatically increased and will continue to increase. The advent of the industrial revolution provided an opportunity for human beings to be exposed to the byproducts of mining and fossil fuel burning on a scale never before seen. Indeed, the prevalence of black lung in miners for the past several hundred years can be linked to the respiration of ultrafine (*aka* nano) particles, and even today, people living near high-traffic areas/large-scale manufacturing operations often experience a much higher incidence of chronic respiratory problems that are beginning to be linked with nanoparticles found in diesel exhaust. Ever since researchers began to consider nanotechnology to be a unique research field, however, we have also seen a sharp rise in consumer products that deliberately include synthetic nanoparticles. These nano-enabled products range from cosmetics to sporting goods and the nanoparticles they contain serve a variety of functions: everything from acting as UV-blocking agents to making plastics that are extremely light, but stronger than steel. Some of these nanoparticle-enabled products actually made it onto the market before people recognized that their benefits were specifically related to the nanoparticles that they contained (e.g. silver impregnated wound dressings). The Woodrow Wilson Institute supports a database of ordinary consumer products that contain nanoparticles, and they now list more than 800 registered nanoparticle-enabled products available to consumers.

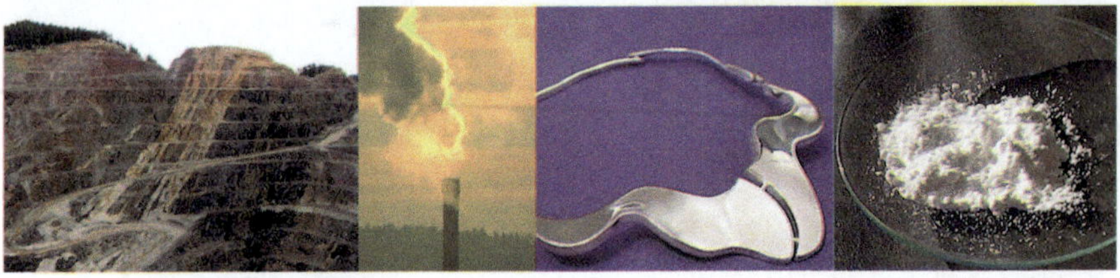

Where can we find nanoparticles in our everyday world? (L-R) Industrial processes like large-scale mining and fossil fuel burning can release nanoparticles, your best silverware and jewelry can give off small quantities of silver or copper nanoparticles, powdered titanium dioxide, like you might find in sunscreen.

Perhaps the easiest place to find nanoparticles in your very own home is in health care products and cosmetics. A number of companies have, for at least the past decade, sold sun block that contains combinations of zinc oxide (ZnO) and titanium dioxide (TiO_2) nanoparticles because these materials absorb UV-light very strongly, preventing your skin from feeling the unwanted effects of ultraviolet exposure. The smaller these particles are, the easier it is to keep them evenly mixed in the liquid portion of the sunblock. This means you are less likely to notice the chalky bits of semiconductor on your skin and (most importantly, of course!) the less likely you are to look like you are covered in chalk. Therefore, using TiO_2 and ZnO nanoparticles is an excellent way to ensure comfortable and uniform sun protection.

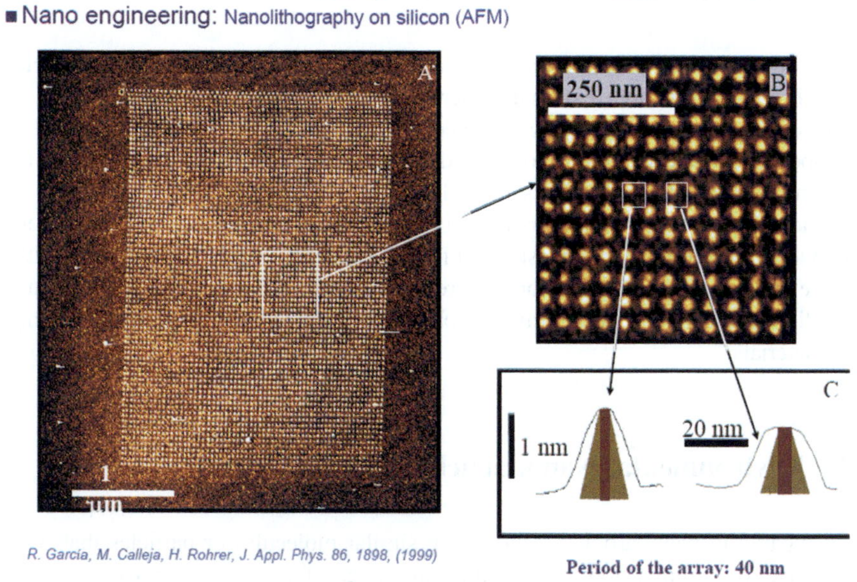

Clearly, it's not that hard to find nanoparticles anywhere we go in our everyday lives. We inhale them in tiny amounts as we go about our day-to-day activities, we buy products that use nanoparticles to protect our health, and even objects as mundane as a fork leave behind nanoparticles as they sit unused in drawers. Until recently, we have not really noticed how often we are exposed to nanoparticles, because they can be extremely difficult to detect (unless you are specifically looking for them with the right equipment). One question this raises (and I wish we had time to explore it in detail today) is: "Is this ever-increasing level of nanoparticle exposure good for us?" Unfortunately, as Stephanie previously pointed out, we have no easy answer to that. As it happens, some of the nanoparticles that you are exposed to everyday that can be harmful to you (particularly the nano-sized particles that can be found in certain types of pollution) but others have already become our go-to weapons to prevent disease (whether that disease is skin cancer or bacterial infection). So maybe the amount of nanoparticle exposure is not the most important question to ask. Maybe it would be more useful, instead, to ask: "Can we come up with some guidelines that allow us to 'spot' which nanoparticles benefit human health and which nanoparticles will potentially do the most harm? And can we tailor our exposure to both accordingly?" We are working on answers, so stay tuned!

3.2 What Nanomaterials Exist in Nature?

According to the European Commission, a 'nanomaterial' is defined as any particle that exhibits one or more external dimensions in the size range 1–100 nm. As a result of their spatial constraint in the electronic properties, nanoparticles exhibit a high specific surface area and a potential difference in their behavior as compared to their bulk counterparts when present at this extremely small scale [1].

Nanosized particles, whether they originate from natural or manufactured materials, must therefore be investigated for their potential impact on human health and the environment. A vast amount of research has been devoted to investigating naturally occurring the structural and chemical varieties of naturally occurring nanomaterials.

3.3 Environmental Nanoparticles and Colloids

As compared to nanoparticles, colloids are similar molecules or particles that will instead exhibit a dimension that is within the range of 1 nm to 1 μm (μm). When present in nature, environmental colloids and nanoparticles can exist in a variety of different compositions and conformations.

As a result of biological decaying and chemical weathering processes, numerous environmental colloids can arise. These nanoparticles are principally oxides and oxyhydroxides of iron (Fe), manganese (Mn), aluminum (Al) and aluminosilicates [1].

3.4 Humic Substances

Humic substances (HS), which are chemically extracted fractions of total natural organic carbon, are the most studied environmental colloid, as their trace metal binding ability [1]. As a dispersed material, HS can also aggregate through charge and static stabilization to form larger structures that can increase in sizes greater than that nanorange.

3.5 Volcanic Ashes

Erosion and volcanic eruption are commonly referenced origins of natural nanoparticles. The ashes that are released during volcanic eruptions can reach temperatures that exceed 1,400 °C [2]. As these ashes reach the atmosphere and neighboring water supplies, their chemical interaction with the environment can result in the deposition of a wide range of nanoparticles that can have negative health effects.

These particles typically range in size from 100 to 200 nm in diameter and will be readily suspended in the air. As a result of this suspension within the atmosphere, the inhalation of nanoparticles that originate from volcanic ashes can lead to serious respiratory disorders following the deposition of these particles in the upper, tracheobronchial and alveolar regions of the respiratory tract.

3.6 Desert Sources of Nanoparticles

Dust storms appear to be the most important source of nanoparticles in the atmosphere, as it has been estimated that approximately 50% of aerosols in the troposphere are minerals that originate from deserts. The chemical composition of nanoparticles that originate from deserts has been determined to exhibit high silicon (Si) concentrations, as well as some traces of Al, calcium (Ca) and Fe.

In fact, a recent study conducted in Xian, China determined that dust transport from the Gobi desert contained high a carbon and nitrogen concentration, as well as sulfate, nitrate and ammonium ions present within the wind carried sample.

3.7 Biological Nanoparticles

A number of biological processes are associated with the production of nanoparticles. For example, bacteria such as *Shewanella* and *Lactobacillus* species are responsible for the fermentation of milk proteins but are also capable of reducing selenite to elemental selenium nanoparticles [2]. These biological origins of nanoparticles are important for future endeavors involved in the removal of environmental contaminants, such as heavy metals, organic and inorganic pollutants, as natural nanoparticles derived from plants, fungi, and bacteria are environmentally friendly and economical alternatives to traditional chemical methods.

Researchers have also been investigating the potential of harmless microorganisms, such as *Saccharomyces cerevisiae* and *Staphylococcus carnosus* to generate homogenous selenium nanoparticles for application in food supplements and even antimicrobial agents [2]. This area of nanotechnology research is particularly promising in the agriculture industry, as a method of enriching the soil with selenium for fortified food products while simultaneously providing plants with a natural defense system is a promising future against harmful pathogens that exhibit destructive tendencies against crops.

References

1. Hartland, A., Lead, J.R., Slaveykova, V.I., O'Carroll, D., ValsamiJones, E.: The environmental significance of natural nanoparticles. Nat. Educ. Knowl. **4**(8), 7 (2013)
2. Griffin, S., Masood, M.I., Nasim, M.J., Sarfraz, M., Ebokaiwe, A.P., Schafer, K., Keck, C.M., Jacob, C.: Natural nanoparticles: a particular matter inspired by nature. Antioxidants (2017). https://doi.org/10.3390/antiox7010003

Chapter 4
Benefits of Nanomaterials and Nanowire Geometry

Where are we heading in terms of nanomaterial synthesis? Two stages will be followed as nanotechnology develops: the "nanobulk" stage and the "nanoworld" stage. They are conceptually different and reflect the two major paradigms of nanofabrication.

4.1 The Nanobulk Stage (10–15 Years)

In the nanobulk stage, which has already begun and that is expected to further develop for 10–15 years, the benefits of nanomaterials will be exploited at the macroscale. In this stage, a nanomaterial is classified by its action at the human scale with novel properties determined by its nanoscopic internal structure.

To clarify this, let us review a few realistic examples of bulk nanomaterials:

1. Cosmetics containing nanoparticles are be found on the market today;
2. In healthcare, nanoparticle suspensions or solutions can be injected into living tissues to help diagnostics or drug delivery;
3. Energy: large-scale nanostructured materials can enhance every type of surface or interface-based chemical reaction, such as are important in energy generation, conversion and storage;
4. Automotive/infrastructure: composite materials containing nano-objects with extraordinary mechanical properties can result in outstanding performance enabling for instance the application of novel coatings with unprecedented resistances.

L. H. Madkour, *Nanoelectronic Materials*, Advanced Structured Materials 116, https://doi.org/10.1007/978-3-030-21621-4_4

4.2 Advances of Nanomaterials (NMs)

- Tunable/tailorable chemical, mechanical and physical properties due to extremely fine grain size (1–100 nm)
- Superior formability and potential superplasticity
- Possess high strength, toughness and ductility
- Possess enhanced activity (extremely large specific surface area)
- Reduced thermal conductivity
- Dispersions can be utilized to further increase strength
- Reduced energy costs.

From the implementation point of view, it is clear that the most important figure of merit for the above-mentioned applications is yield. If nanomaterials are to be implemented in macroscale applications, cost-effective mass-production is essential. Important parameters will come into play such as uniformity, purity, toxicology and stability. Today, applied research will try to target these requirements, with the aim of using relatively cheap and environmentally-friendly synthesis facilities.

However, solving specific issues and unveiling performance-limiting factors will call for a deep understanding of the chemical and physical processes involved in the fabrication. This is where basic research and characterisation techniques come in. Even if the ultimate aim is producing trillions of nominally-equivalent nanocomponents, the focus is put on the study of the synthesis of an individual nanocomponent, to understand how it is formed and how we can better control the process. What can occur during synthesis must be understood, including why two components may be nominally equivalent but not fully equivalent.

4.3 The Nanoworld Stage (15–40 Years)

This paradigm no longer considers nanocomponents as all nominally-equivalent, but aims to take advantage of the natural or induced diversity between nanoblocks. Different nanounits fabricated by the same technique may diversify to have different functions. Treating nanocomponents no longer as a mass but as individuals is obviously more challenging. It implies that the synthesis should have a very precise and independent degree of control over the structure of every single nano-object and/or the location where it is assembled or grown.

Technologically, the nanoworld stage is similar to the nanobulk one. However, basic and applied research now merge. The fabrication will no longer target mass production nor cheap strategies. The goal is to fabricate an "intelligent" nanoworld, without necessarily the driver of low cost. Precision will be more important than yield. Therefore, the facilities that only played a characterisation and fundamental role for the nanobulk could become the only possible way towards applied

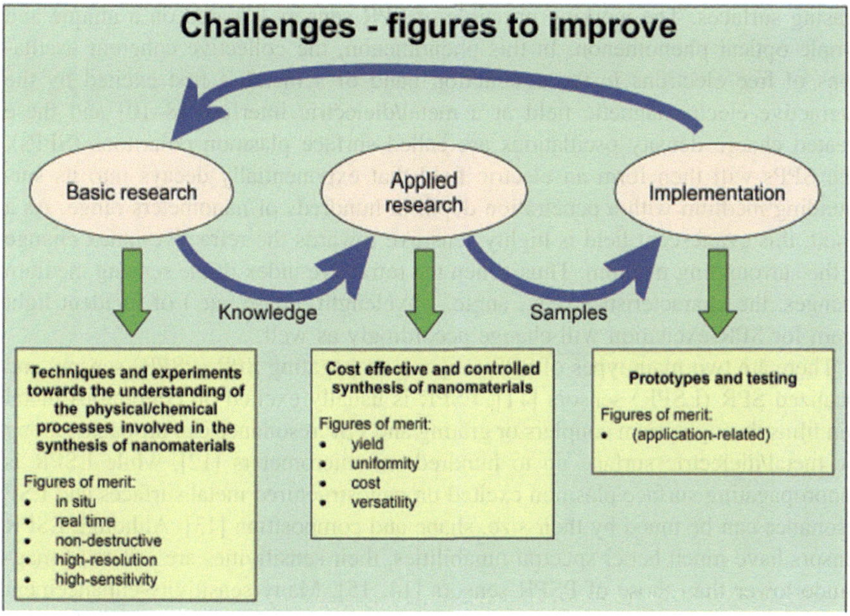

Fig. 4.1 Roadmap for improvements of nanotechnology

synthesis with nanoscale deterministic control. A good example is 3D lithography that uses high-brilliance synchrotron light to guide the process during manufacturing.

The development framework for the nanobulk stage can be summarised by Fig. 4.1 which represents the traditional mode of innovation.

4.4 NMs Enhanced Surface Plasmon Resonance for Biological and Chemical Sensing Applications

The main challenge for all electrical, mechanical and optical sensors is to detect low molecular weight (less than 400 Da) chemical and biological analytes under extremely dilute conditions. Surface plasmon resonance sensors are the most commonly used optical sensors due to their unique ability for real-time monitoring the molecular binding events.

Surface plasmon resonance (SPR) sensors are powerful tools for real-time monitoring interactions of various biological and chemical analytes [1–5]. Their unique abilities for characterizing and quantifying low molecular weight molecules have led to their wide applications in theranostics, pharmaceutics, food safety, environmental monitoring and homeland security [6, 7]. SPR sensors are optical refractometers that can measure refractive-index changes of the medium at the SPR sensing surfaces. The working principle of SPR sensors is based on a unique and simple optical phenomenon. In this phenomenon, the collective coherent oscillations of free electrons in the conduction band of a metal is first excited by the interactive electromagnetic field at a metal/dielectric interface [8–10] and these created charge density oscillations are called surface plasmon polaritons (SPPs). The SPPs will then form an electric field that exponentially decays into its surrounding medium with a penetration depth in hundreds of nanometers range. As a result, this evanescent field is highly sensitive towards the refractive-index change of the surrounding medium. Thus, when the refractive index of the sensing medium changes, the characteristics (e.g., angle, wavelength, phase, etc.) of incident light beam for SPR excitation will change accordingly as well.

There are two main types of SPR sensors: propagating SPR (PSPR) sensors and localized SPR (LSPR) sensors [11]. PSPR is usually excited on continuous metal thin films through prism couplers or grating and PSP resonance can propagate along the metal/dielectric surface up to hundreds of micrometers [12], while LSPR is nonpropagating surface plasmon excited on nanostructured metal surfaces and LSP resonance can be tuned by their size, shape and composition [13]. Although LSPR sensors have much better spectral tunabilities, their sensitivities are orders of magnitude lower than those of PSPR sensors [14, 15]. Many sensitivity-enhancement methods have been proposed and reported as shown in Scheme 4.1, most of which

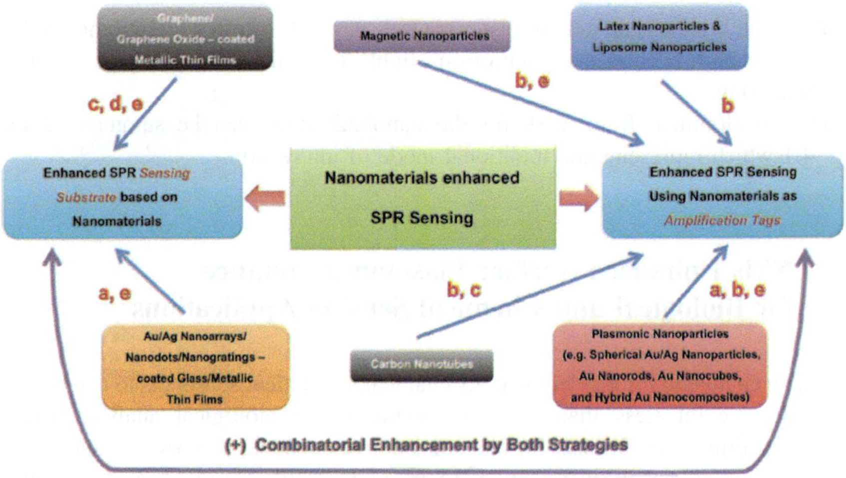

Scheme 4.1 Various types of nanomaterials for enhanced SPR sensing. Enhanced mechanisms: **a** electric field enhancement by coupling the LSPR excited on the surface of nanomaterials with the surface plasmon wave (SPW) excited on the sensing film; **b** large surface mass loading of the nanomaterials leading to large perturbations on the sensing surface; **c** charge transfer from the nanomaterial surface to the metallic sensing film surface that would induce larger evanescent field enhancement thereby magnifying the SPR signals; **d** enhanced adsorption efficiency due to pi-stacking force between target analyte and the nanomaterial surface; **e** catalytic activity of functionalized nanomaterials to further trigger secondary signal amplification—optional

are based on nanomaterials including metallic nanoparticles (e.g., Au and Ag nanoparticles), magnetic nanoparticles, carbon-based nanostructures (e.g., graphene), latex nanoparticles and liposome nanoparticles [12, 13, 16–19]. This scheme will serve as a useful platform for the researchers to have a bird'seye view of the current state-of-art of using various nanomaterials for enhanced SPR sensing applications ranging from medicine to environmental monitoring. More importantly, this scheme will allow one to select the most appropriate enhanced mechanisms for engineering the specific type of nanomaterial-based enhanced SPR sensing system for detecting the desired biological or chemical molecules in different solvents. It is worth noting that the two main strategies as introduced in Scheme 4.1, (i) enhanced SPR sensing substrate based on nanomaterials and (ii) enhanced SPR sensing using nanomaterials as amplification tags can be further combined to achieve a greater SPR sensing performance.

Latex and liposome nanoparticles might not be the most popular enhanced nanomaterials to be used for SPR sensing as they can only work as the amplification tags in terms of their high refractive index that results from their large particle size [20], but one can use them as a powerful platform to create hybrid colloidal nanoparticles as "effective" amplification tags that will be very useful for ultra-sensitive SPR detection. For example, hybrid colloidal particles such as Au NPs-coated polystyrene microspheres [21], Ag NPs-coated polystyrene beads [22], and Au NPs-loaded phosphatidyl- choline (PC) vesicles [23, 24] have recently been reported and they can be easily incorporated with various SPR sensing strategies for screening desired biomolecules such as pathogens. More importantly, we can foresee that these hybrid nanoparticles can be functionalized with targeting molecules for further improving the sensitivity of PSPR sensors. Nevertheless, we want to emphasize that the preparation of these nanoparticles are not straightforward and a skillful nanotechnologist is required to synthesize the optimized particles formulation for SPR detection since there are many different synthesis protocols available in the literatures.

Noble metallic nanoparticles, especially gold nanoparticles (Au NPs), are commonly used nanomaterials for improving the sensing performance of PSPR sensors [12, 14, 16]. These particles are also called plasmonic nanoparticles. They exhibit strong absorption in the visible and near-infrared wavelength regions. Consequently, a large electric field due to LSPR will be excited on the surface of the particles. It has been demonstrated that the SPR sensitivity is closely related to the excited electric field: the larger the electric field, the more sensitive of SPR sensors to the change of its surrounding medium [16]. Coupling the LSPR with surface plasmon waves generated from metal thin film on a conventional PSPR sensor is found to be an efficient way to obtain a larger field enhancement thereby resulting in a higher sensitivity. Also, the ease of functionalizing the metallic nanoparticle's surface with various functional molecules allows one to tailor the surface coverage and binding efficiency of target molecules on the surface of the particle for specifically interacting with the sensing film [13]. Recently, functionalized magnetic nanoparticles (MNPs) such as Fe_3O_4 nanoparticles have been employed for signal amplification of PSPR sensors as well [18, 25–28]. Compared to plasmonic nanoparticles, the main advantages of using MNPs are that the production cost of the MNPs is lower and the step for initial receptor immobilization onto the SPR sensing film is not needed [18]. Moreover, the MNPs can be attracted and manipulated by externally applied magnetic fields and form an "aggregate" layer with a strong refractive index contrast on the metallic sensing film that will lead to a notable SPR signal when sensing trace amounts of biological and chemical samples [18, 25]. Besides having such improved sensitivities, these nanoparticle-enhanced PSPR sensors also retain the advantages of conventional PSPR sensors such as flexibility in quantitative and kinetic analyses of biomolecules that are a challenging to perform using the complementary counterpart of the optical sensing tools such as fluorescence sensors [29–31].

In addition to the above-mentioned zero/one-dimensional nanoparticles as SPR signal amplification tags, two-dimensional nanostructures have recently been developed and integrated into the SPR sensing films for both sensitivity enhancement and preserving the label-free feature of the SPR sensors [14, 32–34].

Nanosized amorphous silica particles have found large-scale applications in many every-day consumer products, ranging from non-diary coffee creamer to automobile tires, optical fibers and catalyst supports. In the 1960s and 1970s metallic nanopowders for magnetic recording tapes were developed. In 1976, for the first time, nanocrystals produced by the now popular inert-gas evaporation technique was published by Granqvist and Buhrman. Recently it has been found that the Maya blue paint is a nanostructured hybrid material. The origin of its color and its resistance to acids and biocorrosion are still not understood but studies of authentic samples from Jaina Island show that the material is made of needle-shaped palygorskite (clay) crystals that form a superlattice with a period of 1.4 nm, with intercalates of amorphous silicate substrate containing inclusions of metal (Mg) nanoparticles. The beautiful tone of the blue color is obtained only when both these nanoparticles and the superlattice are present, as has been shown by the fabrication of synthetic samples.

Today nanophase engineering expands in a rapidly growing number of structural and functional materials, both inorganic and organic, allowing manipulating mechanical, catalytic, electric, magnetic, optical and electronic functions. The production of nanophase or cluster-assembled materials is usually based upon the creation of separated small clusters, which then are fused into a bulk-like material or on their embedding into compact liquid or solid matrix materials. e.g. Nano phase silicon, which differs from normal silicon in physical and electronic properties, could be applied to macroscopic semiconductor processes to create new devices. For instance, when ordinary glass is doped with quantized semiconductor "colloids," it becomes a high performance optical medium with potential applications in optical computing.

The nanowire geometry provides potential advantages over planar wafer based or thin-film solar cells in every step of the photoconversion process. These advantages include reduced reflection, extreme light trapping, improved band gap tuning, facile strain relaxation, and increased defect tolerance. These benefits are not expected to increase the maximum efficiency above standard limits; instead,

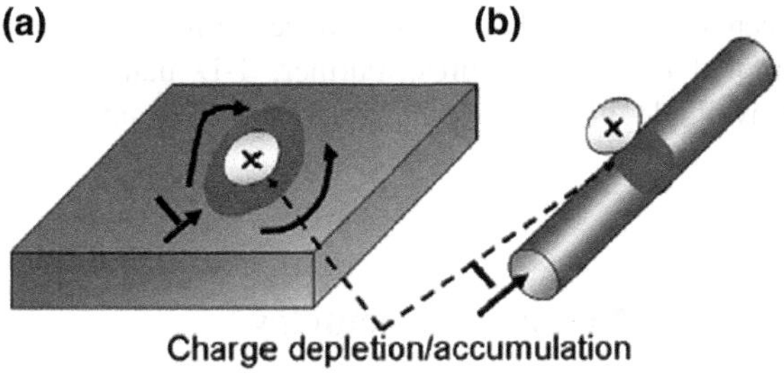

Fig. 4.2 The major advantage of 1-D nanostructures (**b**) over 2-D thin film (**a**). Binding to 1-D nanowire leads to depletion or accumulation in the "bulk" of the nanowire as opposed to only the surface in 2-D thin-film case

they reduce the quantity and quality of material necessary to approach those limits, allowing for substantial cost reductions. Additionally, nanowires provide opportunities to fabricate complex single-crystalline semiconductor devices directly on low-cost substrates and electrodes such as aluminum foil, stainless steel, and conductive glass, addressing another major cost in current photovoltaic technology.

One dimensional (1-D) nanostructures such as nanowires, nanotubes, nanobelts and nanosprings have become the focus of intensive research due to their unique properties and their potential for fabrication into high density nanoscale devices including sensors, electronics, and optoelectronics. Further, 1-D nanostructures can be used for both efficient transport of electrons and optical excitation, and these two factors make them critical to the function and integration of nanoscale devices. In fact, 1-D systems are the smallest dimension structures that can be used for efficient transport of electrons and are thus critical to the function and integration of these nanoscale devices. Because of their high surface-to-volume ratio and tunable electron transport properties due to quantum confinement effect, their electrical properties are strongly influenced by minor perturbations (Fig. 4.2). Compared to 2-D thin films where binding to the surface leads to depletion or accumulation of charge carriers only on the surface of a planar device (Fig. 4.2a), the charge accumulation or depletion in the 1-D nanostructure takes place in the "bulk" of the structure (Fig. 4.2b) thus giving rise to large changes in the electrical properties that potentially enables the detection of a single molecule. 1-D nanostructures thus avoid the reduction in signal intensities that are inherent in 2-D thin films as a result lateral current shunting. This property of the 1-D nanostructures provides a sensing modality for label-free and direct electrical readout when the nanostructure is used as a semiconducting channel of a chemiresistor or field-effect transistor. Such label-free and direct detection is particularly desirable for rapid and real-time monitoring of receptor—ligand interaction with a receptor-modified nanostructure, particularly when the receptor is a biomolecule such as antibody, DNA, and proteins. This is critical for clinical diagnosis and biowarfare agent's detection applications. Additionally, the sizes of biological macromolecules, such as proteins and nucleic acids are comparable to nanoscale building blocks. Therefore, any interaction between such molecules should induce significant changes in the electrical properties of 1-D nanostructures. Further, 1-D nanostructures offer new capabilities not available in larger scale devices (for example, study of single molecule properties) [1-D].

4.5 Benefits of the Nanowire Geometry

In any solar cell, there are many sources of loss that must be minimized. The steps required to convert light to electricity are photon absorption, exciton creation, exciton separation to free carriers, and carrier collection by the electrodes [35]. Using nanowires instead of wafers or thin films provides opportunities to minimize losses in each step at lower costs. The potential cost benefits come primarily from lowering the purity standard and the amount of semiconductor material needed to obtain high efficiencies; increasing the defect tolerance; and enabling new single-crystalline materials to be used without expensive, lattice-matched substrates. The major benefits associated with each photoconversion step are depicted latter in details in Chap. 15; however has been described below.

4.5.1 Absorption

Absorption losses can be divided into two categories: reflection, in which not all of the light is able to enter the semiconductor, and transmission, in which the sample either is too thin or does not have sufficient light trapping to quench all the photons. Antireflection coatings address the first loss, whereas light-trapping schemes affect the second loss. Although both antireflection and light-trapping schemes improve absorption, it is important to differentiate between the two mechanisms to understand how photocurrent can be increased for different solar cell structures.

Reflection at an interface arises from a difference in refractive index between two media [36]. In the case in which there is no antireflection coating, most inorganic semiconductors will reflect between 10 and 50% of the light in the range of 400–2,000 nm, where most of the sun's power lies. For example, a silicon solar cell without an antireflection coating will lose more than 30% of its power to reflection [37]. One common strategy to reduce reflection is to add one or more coatings that have a refractive index intermediate between that of the semiconductor and that of air. Using an appropriate coating thickness and refractive index leads to destructive interference between reflected light and incident light, eliminating reflection at both interfaces. Unfortunately, this technique is completely effective only for a single wavelength at normal incidence, so there will still be some reflectance over most of the spectrum. Double- and triple-layer coatings can spectrally broaden the response, but the ideal coating would have a continuously graded refractive index. Tapered nanowire arrays (nanocones) with tips much smaller than the wavelength of light

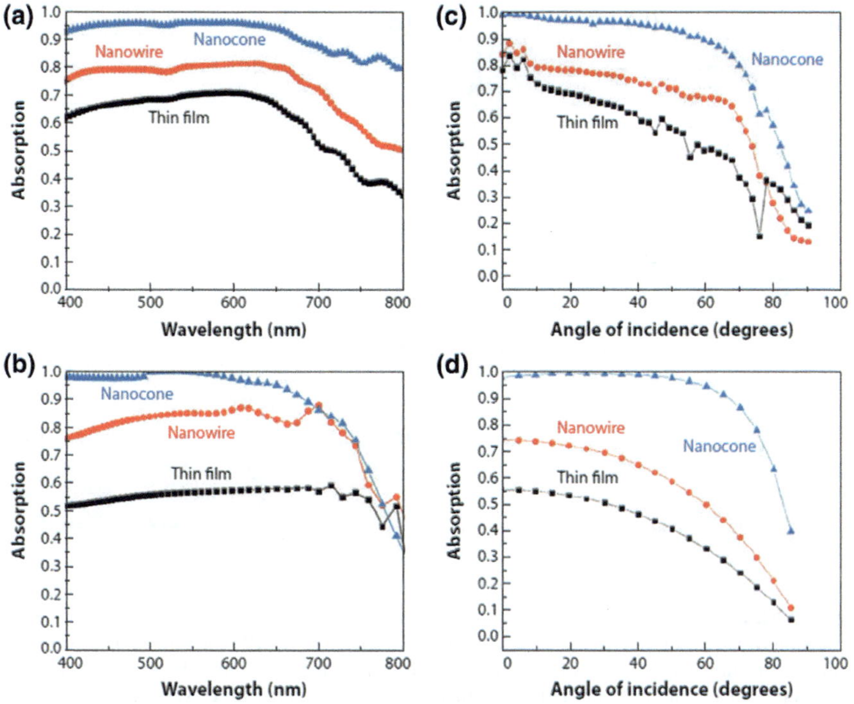

Fig. 4.3 Antireflection properties of nanowires and nanocones. **a** Measured and **b** simulated values of absorption on samples with amorphous Si:H (a-Si:H) thin films, nanowire arrays, and nanocone arrays over a large range of wavelengths at normal incidence. **c** Measured and **d** calculated values of absorption on samples with a-Si:H thin films, nanowire arrays, and nanocone arrays over different angles of incidence (at wavelength $\lambda = 488$ nm). Adapted with permission from Reference [38]. Copyright © 2009, American Chemical Society

act as an effective medium, with a gradual change in refractive index given by the weighted average of the material and air. Figure 4.3 shows that this approach leads to greatly enhanced absorption compared with planar samples over a broad wavelength and angle range [38].

Cao et al. [39–42] studied these leaky-mode resonances, which resemble a whispering gallery mode in a microdisk cavity, in greater detail by measuring the photocurrent and scattering spectrum of single nanowires lying flat on a substrate (Fig. 4.4). They found a redshift in the scattering and absorption efficiency peaks with increasing diameter and matched the experimental data to Lorentz-Mie scattering calculations and finite difference time domain simulations. The simulated field plots showed distinct resonant modes at spectral peaks, as expected. The absorption efficiency decreased as the incident angle moved away from the normal direction, again confirming that these leaky-mode resonances are the dominant light-trapping effect in single nanowires or arrays of horizontal nanowires.

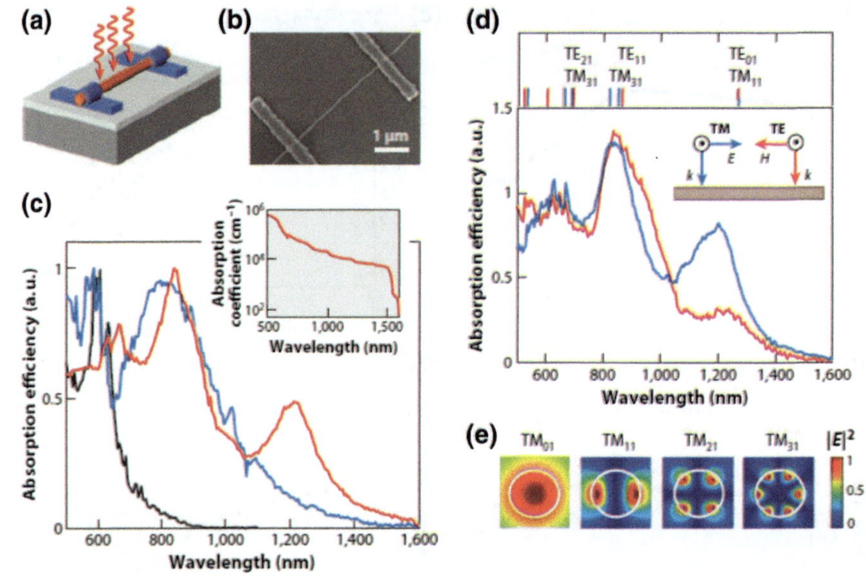

Fig. 4.4 Measurement of light absorption in individual germanium nanowire devices and correlation of the absorption peaks with leaky-mode resonances (LMRs). **a** Schematic illustration of the germanium nanowire device used for photocurrent measurements. **b** Scanning electron microscopy image of a 25-nm-radius germanium nanowire device. **c** Measured spectra of absorption efficiency *Q*abs for unpolarized light taken from individual germanium nanowires with radii of 10 nm (*black*), 25 nm (*blue*), and 110 nm (*red*). The spectra are normalized to their maximum absorption efficiency to highlight the tunability of the nanowire absorption properties. (*Inset*) Absorption coefficient of bulk germanium as a function of wavelength. **d** *Q*abs spectra of a 110-nm-radius germanium nanowire taken using linearly polarized transverse-electric (TE; *red*) or transverse-magnetic (TM; *blue*) light. The spectra are normalized to an internal quantum efficiency of four, which is assumed to yield the best fit between the calculated and experimental results. The inset illustrates the illumination geometry for the TE (the electric field of the light polarized perpendicular to the axis of the wire) and TM (the electric field of the light polarized parallel to the axis of the wire) polarizations. The red/blue lines at the top indicate locations of all LMRs of the 110-nm-radius germanium nanowire in this spectral region. Only some of the modes are labeled for visualization convenience. *E*, *H*, and *k* are the electric field, magnetic field, and *k* vector of the incoming light, respectively. **e** The configuration of the electric field intensity for typical TM leaky modes. The white circle denotes the nanowire/air interface. Adapted from Reference [40]

Finally, nanowires can show increased absorption via resonant energy transfer from dyes or quantum dots located near the semiconductor surface. Because of the larger surface area compared with planar samples, this energy transfer can substantially affect absorption (Fig. 4.5).

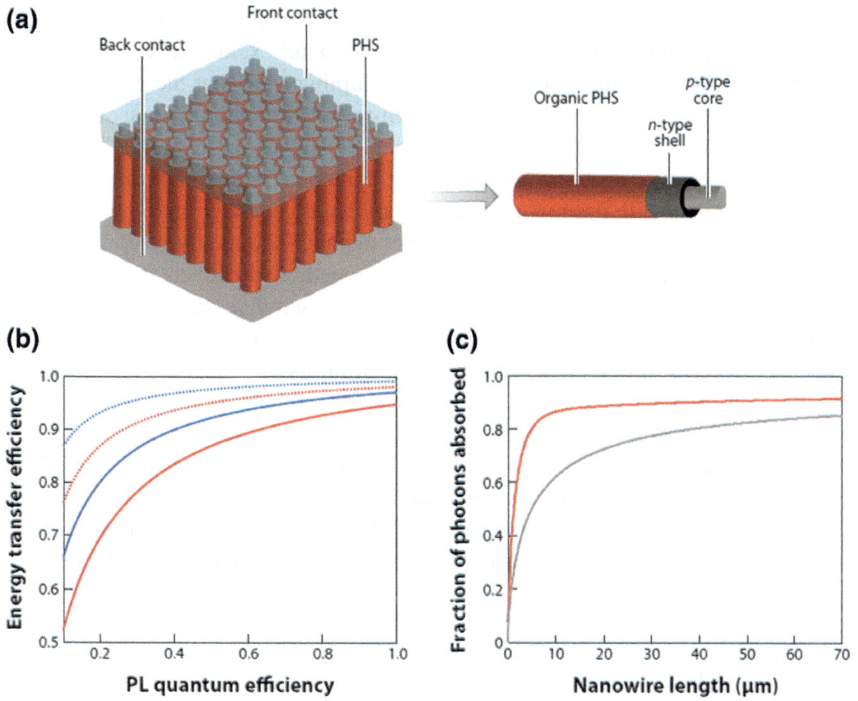

Fig. 4.5 Increasing light absorption in nanowires through resonant energy transfer from photon-harvesting shells (PHS). **a** Schematic of a nanowire solar cell with a PHS. Organic molecules are chemically attached to the surface of the nanowires. The organic shell absorbs light and transfers the energy to the silicon nanowires via resonant energy transfer. The silicon nanowires have radial *p-n* junctions that selectively separate and transport the charges to the respective electrodes (shown in the isolated nanowire to the *right*). **b** Energy transfer efficiency (ETE) as a function of dye photoluminescence (PL) efficiency for dye A (*blue*) and dye B (*red*). The solid and dashed lines are for x values of 2.7 nm and 1.7 nm, respectively, where x is the dye separation distance from the nanowire surface. The much weaker variation of the dashed curves compared with the solid curves shows that it is important to carefully control the dye-silicon separation to achieve a high ETE for arbitrary dyes. **c** Fractions of photons above the band gap of silicon that are absorbed with bare silicon nanowires (*gray*) and with dyes A and B attached to the nanowires (*red*) as a function of nanowire length. The nanowires are 50 nm in diameter, with a center-to-center distance of 70 nm. Dyes A and B are assumed to have equal surface coverage with a molar absorptivity of $2 \times 105\ M^{-1}\ cm^{-1}$ and surface concentration on the wires of $7.5 \times 10^{13}\ cm^{-2}$. Adapted with permission from Reference [43]. Copyright © 2009, American Institute of Physics

4.5.2 Exciton Formation

After light is absorbed in the solar cell, the next loss mechanism is carrier relaxation. If the exciton binding energy is weak compared with the available thermal energy at room temperature (e.g., in bulk silicon), the excited carriers will relax to the band edge as free carriers. If the exciton binding energy is strong (as in most organic semiconductors), the carriers will form a bound exciton state. This carrier relaxation normally occurs by carrier-phonon coupling such that the energy is lost as heat. This is the single greatest loss mechanism in every photovoltaic device and typically represents between 30 and 40% of the incident solar energy [44]. The simplest way to reduce heat losses is to choose a material with an optimal band gap or several materials with optimal complementary band gaps for single-junction and multi-junction solar cells, respectively. One method commonly used to adjust material band gaps is to alloy multiple materials together. However, some common alloys, such as indium gallium nitride, show a miscibility gap at intermediate concentrations caused by large lattice mismatch strain, which makes it difficult to synthesize uniform single-crystalline films at arbitrary compositions. Nanowires close this miscibility gap by providing improved strain relaxation, opening up a broader range of absorption energies than is possible with thin-film materials [45]. Additional tuning can come from reducing the nanowire diameter to near or below the Bohr radius of the material, when quantum confinement begins to cause an increase in band gap. This is practically useful only in materials with large Bohr radii and relatively small starting band gaps, such as germanium and lead selenide.

4.5.3 Charge Separation

Nanowires also provide an opportunity to use new charge separation mechanisms. Wu et al., [46] used density functional theory to show that tapered nanowires that exhibit different degrees of quantum confinement along the length can separate electrons and holes without any dopants. Figure 4.6 shows the spatial separation between the highest occupied molecular orbital (HOMO) and the lowest unoccupied molecular orbital (LUMO) wave functions in a tapered quantum silicon nanowire, leading to a spontaneous potential gradient.

4.5.4 Carrier Collection

Once carriers have been separated, they must diffuse to the contacts to be extracted for useful work. The nanowire geometry is especially beneficial for this step because it enables rapid radial charge separation and efficient carrier collection through band conduction [47]. Law et al. [48] demonstrated such a scheme, using zinc oxide nanowires in a dye-sensitized solar cell. They showed that the carrier collection efficiency is much higher for the nanowire case compared with the standard nanoparticle case due to faster band conduction rather than a trap-limited diffusion transport mechanism. A much shorter zinc oxide nanowire network also showed improved charge transport in inverted polymer bulk heterojunction solar cells compared with controls on planar zinc oxide films manifested in higher *FF*

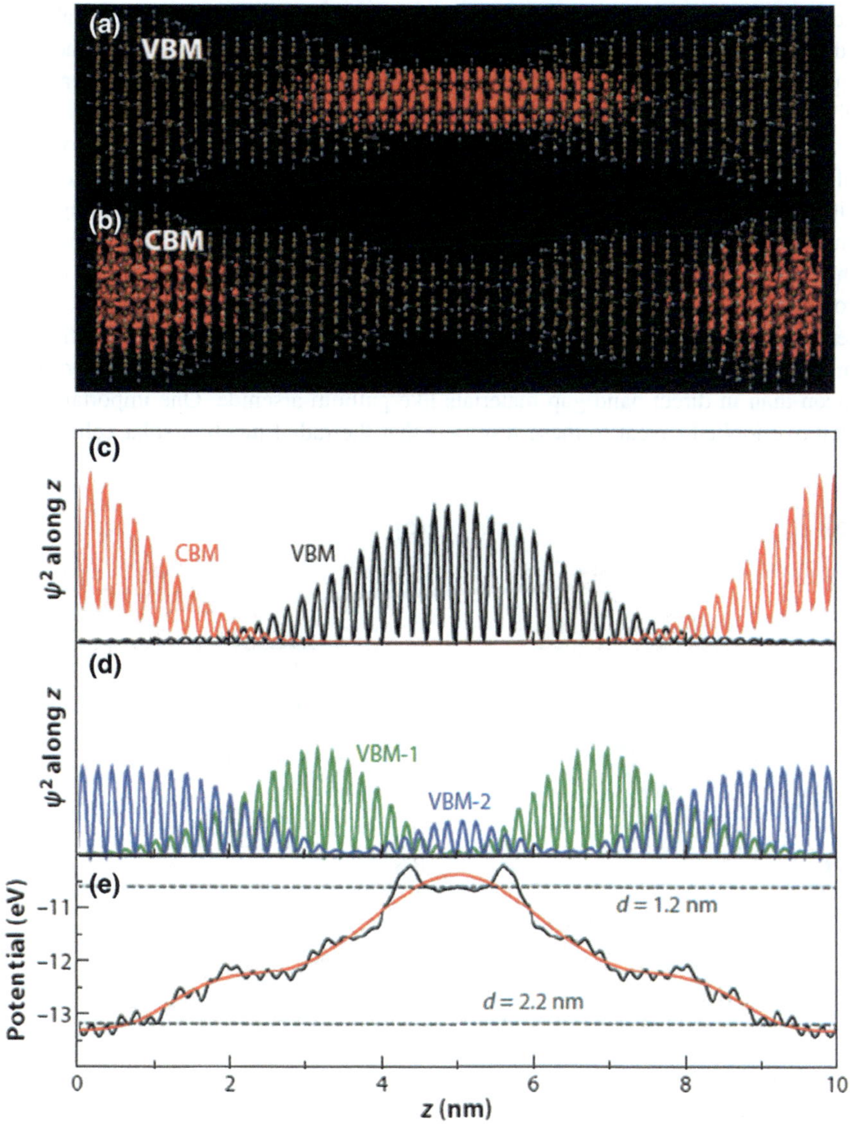

Fig. 4.6 Charge separation using quantum confinement. Panels **a** and **b** show isosurfaces (*red*) of the square of the wave functions of the valence band maximum (VBM) and conduction band minimum (CBM) in the tapered nanowires at values of 10% of their respective maximum. Panels **c** and **d** show the charge density distribution of the VBM, CBM, VBM-1, and VBM-2 states along the wire axis (integrated over the *xy* plane). Panel **e** shows the planar-averaged total potential $V(z)$ along the wire axis for the tapered nanowires (*solid black*) and the straight-edged nanowires (*dashed black*). The red curve is a polynomial fit to the black solid curve. Adapted with permission from Reference [46]. Copyright © 2008, American Physical Society

and overall efficiency [49, 50]. Considering the light-trapping discussion above, ordered nanowire arrays may be able to provide additional gains in efficiency by improving absorption in both dye-sensitized and bulk heterojunction polymer solar cells.

Another advantage of the radial junction is the decreased sensitivity to bulk defects. Kayes et al. [47] predicted that silicon nano- or microwires with radial *p-n* junctions should show only a slight efficiency degradation with decreasing lifetime until the minority carrier diffusion length approaches the nanowire radius, whereas planar *p-n* junction cells show severe degradation once the diffusion length drops below the film thickness. Figure 4.7 shows that this effect is strongest when there is a large mismatch between the absorption and minority carrier diffusion lengths, which is more likely to occur in indirect band gap semiconductors like silicon than in direct band gap materials like gallium arsenide. One important and often overlooked caveat to these results is that the radial junction solar cells show

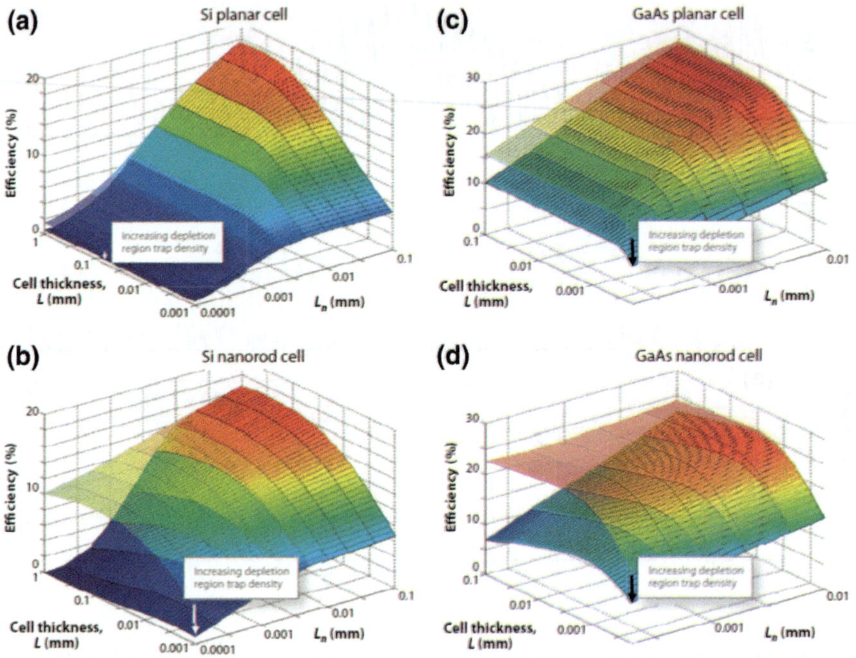

Fig. 4.7 Efficiency versus cell thickness L and minority electron diffusion length Ln for **a** a conventional planar *p-n* junction silicon cell, **b** a radial *p-n* junction nanorod silicon cell, **c** a conventional planar *p-n* junction gallium arsenide (GaAs) cell, and **d** a radial *p-n* junction nanorod GaAs cell. In all cases the top surface shown in the plot has a depletion region trap density fixed at 1014 cm^{-3} so that the carrier lifetimes $\tau\ n0$, $\tau\ p0 = 1$ μs, whereas the bottom surface has a depletion region trap density equal to the trap density in the quasi-neutral region at each value of Ln. In the radial *p-n* junction nanorod case, the cell radius R is set equal to Ln, a condition that is near optimal. Adapted with permission from Reference [47]. Copyright © 2005, American Institute of Physics

an efficiency advantage primarily when the defects are concentrated in the quasi-neutral region of the device. Verifying the benefits of the radial geometry requires some control over the minority carrier diffusion length or carrier extraction length in nanowire cells.

Single-nanowire solar cell measurements showed an efficiency increase of approximately a factor of five for radial versus axial *p-n* junctions, due primarily to an improved photocurrent (Fig. 4.8) [51, 52]. It is tempting to assume that the increased efficiency (and photocurrent) stems from improved minority carrier collection in the radial geometry.

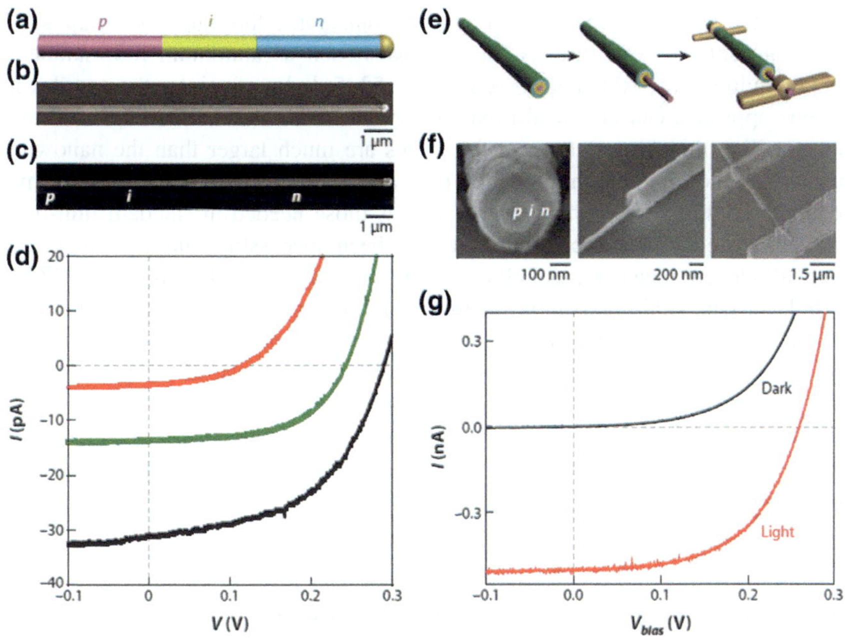

Fig. 4.8 Axial and radial *p-i-n* silicon nanowire solar cells. **a** Schematic of an axially modulated *p-i-n* silicon nanowire. Growth is mediated with a gold catalyst (gold-colored hemisphere) in a sequence beginning with the *p* and *i* regions and ending with the *n* region. **b** Scanning electron microscopy (SEM) image of a uniform *p-i-n* axial silicon nanowire with a gold catalyst (d = 250 nm) visible on the right end. **c** SEM image of a *p-i-n* axial nanowire after selective wet etching. The *p*- and *i*-type silicon regions have faster etching rates than the *n*-type region. **d** Light *I-V* characteristics for the *i* length = 0-, 2-, and 4-μm devices (*red*, *green*, and *black*, respectively); the illumination intensity is 100 mW cm^{-2}, air-mass 1.5 global (AM 1.5) spectrum. **e** Schematics of radial *p-i-n* junction device fabrication. (*Left*) Pink, yellow, cyan, and green layers correspond to the *p* core, *i* shell, *n* shell, and plasma-enhanced chemical vapor deposition–coated silicon dioxide, respectively. (*Middle*) Selective etching to expose the *p* core. (*Right*) Metal contacts deposited on the *p* core and *n* shell. **f** SEM images corresponding to the schematics in panel **e**. **g** Dark and light *I-V* curves of a typical radial *p-i-n* junction device. Panels **a–d**. Adapted with permission from Reference [52]. Copyright © 2008, American Chemical Society. Panels **e–g** adapted with permission from Reference [51]. Copyright © 2007, Macmillan

4.5.5 Cost

All the above benefits of the nanowire geometry promise to lower the cost of solar cells by reducing the amount of material needed (lower reflection, superior light trapping), by expanding the list of available materials (band gap tuning, new charge separation mechanisms), or by relaxing the materials purification requirements (enhanced charge separation in low-lifetime materials). But nanowires may also offer an enormous cost benefit for the fabrication steps of high-performance multijunction solar cells by removing the need for an expensive lattice-matched substrate. Many materials relevant to solar cells, including zinc oxide, germanium, silicon, indium gallium nitride, and cadmium sulfide, can be made into single-crystalline nanowires on non-epitaxial substrates through bottom-up synthesis, with some grown directly on electrodes such as aluminum foil, stainless steel, or indium tin oxide-coated glass [48, 53–56]. In principle, the top-down synthetic approach can also lead to single-crystalline nanowires starting with any polycrystalline thin film as long as the grains are much larger than the nanowire diameter and the film is highly textured (i.e., all the grains are oriented in the same direction). These requirements are the same as those needed in standard thin-film solar cells to reduce recombination and have been successfully implemented in a variety of relevant materials, including silicon, cadmium telluride, CIGS, and CZTS [57–60]. Surface strain relaxation facilitates epitaxial growth of subsequent layers onto nanowires, whereas the radial geometry increases the defect tolerance, which may allow for solution-processed polycrystalline layers instead of vapor phase epitaxy. The promise of bringing multijunction solar cell fabrication costs down to near traditional thin-film costs with only a moderate efficiency loss compared with world record cells is one of the most exciting possibilities for nanowire photovoltaics.

4.6 Disadvantages of Nanomaterials (NMs)

(i) Instability of the particles—Retaining the active metal nanoparticles is highly challenging, as the kinetics associated with nanomaterials is rapid. In order to retain nanosize of particles, they are encapsulated in some other matrix. Nanomaterials are thermodynamically metastable and lie in the region of high-energy local-minima. Hence they are prone to attack and undergo transformation. These include poor corrosion resistance, high solubility, and phase change of nanomaterials. This leads to deterioration in properties and retaining the structure becomes challenging.

(ii) Fine metal particles act as strong explosives owing to their high surface area coming in direct contact with oxygen. Their exothermic combustion can easily cause explosion.

(iii) Impurity—Because nanoparticles are highly reactive, they inherently interact with impurities as well. In addition, encapsulation of nanoparticles becomes necessary when they are synthesized in a solution (chemical route). The stabilization of nanoparticles occurs because of a non-reactive species engulfing the reactive nano-entities. Thereby, these secondary impurities become a part of the synthesized nanoparticles, and synthesis of pure nanoparticles becomes highly difficult. Formation of oxides, nitrides, etc. can also get aggravated from the impure environment/surrounding while synthesizing nanoparticles. Hence retaining high purity in nanoparticles can become a challenge hard to overcome.

(iv) Biologically harmful—Nanomaterials are usually considered harmful as they become transparent to the cell-dermis. Toxicity of nanomaterials also appears predominant owing to their high surface area and enhanced surface activity. Nanomaterials have shown to cause irritation, and have indicated to be carcinogenic. If inhaled, their low mass entraps them inside lungs, and in no way they can be expelled out of body. Their interaction with liver/blood could also prove to be harmful (though this aspect is still being debated on) (Fig. 4.9).

(v) Difficulty in synthesis, isolation and application—It is extremely hard to retain the size of nanoparticles once they are synthesized in a solution. Hence, the nanomaterials have to be encapsulated in a bigger and stable molecule/material. Hence free nanoparticles are hard to be utilized in isolation, and they have to be interacted for intended use via secondary means of exposure. Grain growth is inherently present in nanomaterials during their processing. The finer grains tend to merge and become bigger and stable grains at high temperatures and times of processing.

(vi) Recycling and disposal—There are no hard-and-fast safe disposal policies evolved for nanomaterials. Issues of their toxicity are still under question,

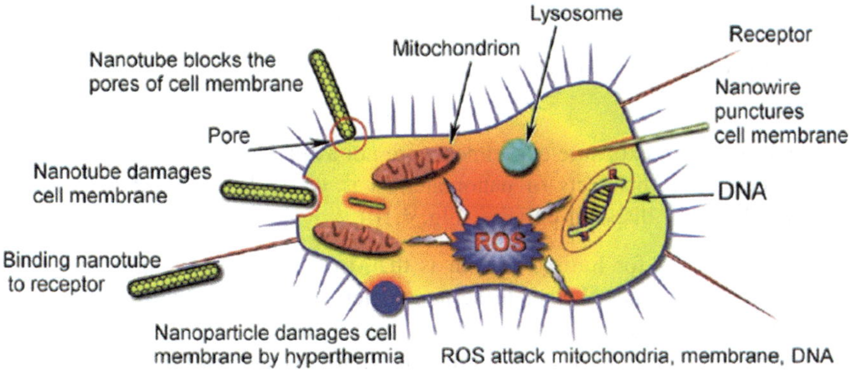

Fig. 4.9 Possible harmful effects when nanomaterials interact with cells

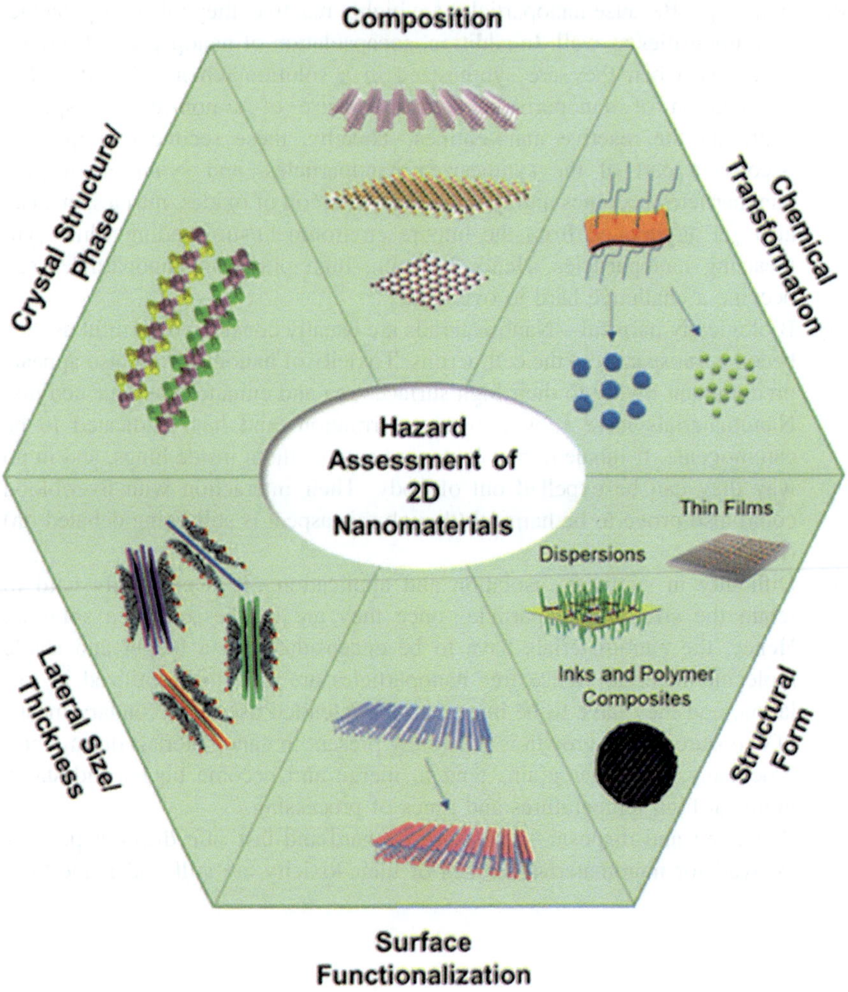

Fig. 4.10 Hazard assessment of nanomaterials

and results of exposure experiments are not available. Hence the uncertainty associated with affects of nanomaterials is yet to be assessed in order to develop their disposal policies (Fig. 4.10).

What this means for future research is that, in some cases, it might be more important to understand whether these nanomateriale aggregates are toxic to bacteria and simple organisms, rather than whether the original "as synthesized" nanoparticles would be hazardous. Furthermore, it appears that the chemical connection between the capping agent and the nanoparticle crystal surface may be an important factor in determining whether the nanoparticle breaks down when it enters the environment, or whether the original NP remains intact. The Center for

Sustainable Nanotechnology has already begun to consider the changeable nature of nanoparticles released into the environment, as we design new experiments to study all their interactions.

References

1. Prasad, P.N.: Introduction to Biophotonics. Wiley-Interscience, New York (2003)
2. Homola, J., Yee, S.S., Gauglitz, G.: Sens. Actuators B **54**, 3–15 (1999)
3. Hoa, X.D., Kirk, A.G., Tabrizian, M.: Biosens. Bioelectron. **23**, 151–160 (2007)
4. Shalabney, A., Abdulhalim, I.: Laser Photon. Rev. **5**, 571–606 (2011)
5. Bedford, E.E., Spadavecchia, J., Pradier, C.M., Gu, F.X.: Macromol. Biosci. **12**, 724–739 (2012)
6. Situ, C., Buijs, J., Mooney, M.H., Elliott, C.T.: Trac-Trends Anal. Chem. **29**, 1305–1315 (2010)
7. Szunerits, S., Maalouli, N., Wijaya, E., Vilcot, J.P., Boukherroub, R.: Anal. Bioanal. Chem. **405**, 1435–1443 (2013)
8. Raether, H.: Surface Plasmons on Smooth and Rough Surfaces and on Gratings. Springer, Berlin (1988)
9. Ho, H.P., Lam, W.W.: Sens. Actuators, B **96**, 554–559 (2003)
10. Ho, H.P., Law, W.C., Wu, S.Y., Liu, X.H., Wong, S.P., Lin, C.L., Kong, S.K.: Sens. Actuators, B **114**, 80–84 (2006)
11. Stewart, M.E., Anderton, C.R., Thompson, L.B., Maria, J., Gray, S.K., Rogers, J.A., Nuzzo, R.G.: Chem. Rev. **108**, 494–521 (2008)
12. Zeng, S.W., Yu, X., Law, W.C., Zhang, Y., Hu, R., Dinh, X.Q., Ho, H.P., Yong, K.T.: Sens. Actuators, B **176**, 1128–1133 (2013)
13. Zeng, S.W., Yong, K.T., Roy, I., Dinh, X.Q., Yu, X., Luan, F.: Plasmonics **6**, 491–506 (2011)
14. Kabashin, A.V., Evans, P., Pastkovsky, S., Hendren, W., Wurtz, G.A., Atkinson, R., Pollard, R., Podolskiy, V.A., Zayats, A.V.: Nat. Mater. **8**, 867–871 (2009)
15. Law, W.C., Markowicz, P., Yong, K.T., Roy, I., Baev, A., Patskovsky, S., Kabashin, A.V., Ho, H.P., Prasad, P.N.: Biosens. Bioelectron. **23**, 627–632 (2007)
16. Lee, K.S., Lee, M., Byun, K.M., Lee, I.S.: J. Mater. Chem. **21**, 5156–5162 (2011)
17. Wu, L., Chu, H.S., Koh, W.S., Li, E.P.: Opt. Express **18**, 14395–14400 (2010)
18. Severs, A.H., Schasfoort, R.B.M.: Biosens. Bioelectron. **8**, 365–370 (1993)
19. Wink, T., van Zuilen, S.J., Bult, A., van Bennekom, W.P.: Anal. Chem. **70**, 827–832 (1998)
20. Liu, Q.T., Boyd, B.J.: Analyst **138**, 391–409 (2013)
21. Han, D., Hwang, H., Park, J.K.: Appl. Phys. Lett. **102**, 054105 (2013)
22. Zhang, N., Yu, X., Hu, J., Xue, F., Ding, E.: RSC Adv. **3**, 13740–13747 (2013)
23. Rasch, M.R., Rossinyol, E., Hueso, J.L., Goodfellow, B.W., Arbiol, J., Korgel, B.A.: Nano Lett. **10**, 3733–3739 (2010)
24. Zeng, S., Baillargeat, D., Ho, H.P., Yong, K.T.: Nanomaterials enhanced surface plasmon resonance for biological and chemical sensing applications. Chem. Soc. Rev. **43**, 3426–3452 (2014)
25. Wang, J.L., Munir, A., Zhu, Z.Z., Zhou, H.S.: Anal. Chem. **82**, 6782–6789 (2010)
26. Wang, J.A., Sun, Y., Wang, L.Y., Zhu, X.N., Zhang, H.Q., Song, D.Q.: Colloids Surf. B **81**, 600–606 (2010)
27. Wang, J.L., Zhu, Z.Z., Munir, A., Zhou, H.S.: Talanta **84**, 783–788 (2011)
28. Wang, L.Y., Sun, Y., Wang, J., Wang, J.A., Yu, A.M., Zhang, H.Q., Song, D.Q.: Colloids Surf. B **84**, 484–490 (2011)
29. Yu, F., Persson, B., Lofas, S., Knoll, W.: J. Am. Chem. Soc. **126**, 8902–8903 (2004)

30. Fan, X.D., White, I.M., Shopova, S.I., Zhu, H.Y., Suter, J.D., Sun, Y.Z.: Anal. Chim. Acta **620**, 8–26 (2008)
31. Granqvist, N., Hanning, A., Eng, L., Tuppurainen, J., Viitala, T.: Sensors **13**, 15348–15363 (2013)
32. Xia, L.P., Yin, S.Y., Gao, H.T., Deng, Q.L., Du, C.L.: Plasmonics **6**, 245–250 (2011)
33. Salihoglu, O., Balci, S., Kocabas, C.: Appl. Phys. Lett. **100**, 213110 (2012)
34. Wang, L., Zhu, C.Z., Han, L., Jin, L.H., Zhou, M., Dong, S.J.: Chem. Commun. **47**, 7794–7796 (2011)
35. Garnett, E.C., Brongersma, M.L., Cui, Y., McGehee, M.D.: Nanowire solar cells. Annu. Rev. Mater. Res. **41**, 269–295 (2011)
36. Hecht, E.: Optics, p. 698. Pearson, San Francisco (2002)
37. Green, M.A.: Solar Cells: Operating Principles, Technology and System Applications, 274 pp. University New South Wales, Kensington, NSW, Australia (1982)
38. Zhu, J., Yu, Z., Burkhard, G.F., Hsu, C., Connor, S.T., et al.: Optical absorption enhancement in amorphous silicon nanowire and nanocone arrays. Nano Lett. **9**(1), 279–282 (2009)
39. Cao, L., Park, J., Fan, P., Clemens, B., Brongersma, M.L.: Resonant germanium nanoantenna photodetectors. Nano Lett. **10**(4), 1229–1233 (2010)
40. Cao, L., White, J.S., Park, J., Schuller, J.A., Clemens, B.M., Brongersma, M.L.: Engineering light absorption in semiconductor nanowire devices. Nat. Mater. **8**(8), 643–647 (2009)
41. Cao, L., Fan, P., Vasudev, A.P., White, J.S., Yu, Z., et al.: Semiconductor nanowire optical antenna solar absorbers. Nano Lett. **10**(2), 439–445 (2010)
42. Cao, L., Fan, P., Barnard, E.S., Brown, A.M., Brongersma, M.L.: Tuning the color of silicon nanostructures. Nano Lett. **10**(7), 2649–2654 (2010)
43. Peters, C.H., Guichard, A.R., Hryciw, A.C., Brongersma, M.L., McGehee, M.D.: Energy transfer in nanowire solar cells with photon-harvesting shells. J. Appl. Phys. **105**(12), 124509 (2009)
44. Ginley, D., Green, M., Collins, R.: Solar energy conversion toward 1 terawatt. MRS Bull. **33** (355), 373 (2008)
45. Kuykendall, T., Ulrich, P., Aloni, S., Yang, P.: Complete composition tunability of InGaN nanowires using a combinatorial approach. Nat. Mater. **6**(12), 951–956 (2007)
46. Wu, Z., Neaton, J.B., Grossman, J.C.: Quantum confinement and electronic properties of tapered silicon nanowires. Phys. Rev. Lett. **100**, 246804 (2008)
47. Kayes, B.M., Atwater, H.A., Lewis, N.S.: Comparison of the device physics principles of planar and radial *p-n* junction nanorod solar cells. J. Appl. Phys. **97**(11), 114302 (2005)
48. Law, M., Greene, L.E., Johnson, J.C., Saykally, R., Yang, P.: Nanowire dye-sensitized solar cells. Nat. Mater. **4**(6), 455–459 (2005)
49. Takanezawa, K., Tajima, K., Hashimoto, K.: Efficiency enhancement of polymer photovoltaic devices hybridized with ZnO nanorod arrays by the introduction of a vanadium oxide buffer layer. Appl. Phys. Lett. **93**(6), 063308 (2008)
50. Takanezawa, K., Hirota, K., Wei, Q., Tajima, K., Hashimoto, K.: Efficient charge collection with ZnO nanorod array in hybrid photovoltaic devices. J. Phys. Chem. C **111**(19), 7218–7223 (2007)
51. Tian, B., Zheng, X., Kempa, T.J., Fang, Y., Yu, N., et al.: Coaxial silicon nanowires as solar cells and nanoelectronic power sources. Nature **449**(7164), 885–889 (2007)
52. Kempa, T.J., Tian, B., Kim, D.R., Hu, J., Zheng, X., Lieber, C.M.: Single and tandem axial *p-i-n* nanowire photovoltaic devices. Nano Lett. **8**(10), 3456–3460 (2008)
53. Fan, Z., Razavi, H., Do, J., Moriwaki, A., Ergen, O., et al.: Three-dimensional nanopillar-array photovoltaics on low-cost and flexible substrates. Nat. Mater. **8**(8), 648–653 (2009)
54. Dong, Y., Tian, B., Kempa, T.J., Lieber, C.M.: Coaxial group III-nitride nanowire photovoltaics. Nano Lett. **9**(5), 2183–2187 (2009)
55. Fan, Z., Kapadia, R., Leu, P.W., Zhang, X., Chueh, Y., et al.: Ordered arrays of dual-diameter nanopillars for maximized optical absorption. Nano Lett. **10**(10), 3823–3827 (2010)

56. Chan, C.K., Peng, H., Liu, G., McIlwrath, K., Zhang, X.F., et al.: High-performance lithium battery anodes using silicon nanowires. Nat. Nanotechnol. **3**(1), 31–35 (2008)
57. Wu, X.: High-efficiency polycrystalline CdTe thin-film solar cells. Sol. Energy **77**(6), 803–814 (2004)
58. Ramanathan, K., Contreras, M.A., Perkins, C.L., Asher, S., Hasoon, F.S., et al.: Properties of 19.2% efficiency ZnO/CdS/CuInGaSe2 thin-film solar cells. Prog. Photovolt. Res. Appl. **11** (4), 225–230 (2003)
59. Hibberd, C.J., Chassaing, E., Liu, W., Mitzi, D.B., Lincot, D., Tiwari, A.N.: Non-vacuum methods for formation of Cu(In, Ga)(Se, S)2 thin film photovoltaic absorbers. Prog. Photovolt. Res. Appl. **18**(6), 434–452 (2009)
60. Green, M., Basore, P., Chang, N., Clugston, D., Egan, R., et al.: Crystalline silicon on glass (CSG) thin-film solar cell modules. Sol. Energy **77**(6), 857–863 (2004)

Chapter 5
Why So Much Interest in Nanomaterials (NMs)?

These materials have created a high interest in recent years by virtue of their unusual mechanical, electrical, optical and magnetic properties. Some examples are given below:

- Nanophase ceramics are of particular interest because they are more ductile at elevated temperatures as compared to the coarse-grained ceramics.
- Nanostructured semiconductors are known to show various non-linear optical properties. Semiconductor Q-particles also show quantum confinement effects which may lead to special properties, like the luminescence in silicon powders and silicon germanium quantum dots as infrared optoelectronic devices. Nanostructured semiconductors are used as window layers in solar cells.
- Nanosized metallic powders have been used for the production of gas tight materials, dense parts and porous coatings. Cold welding properties combined with the ductility make them suitable for metal-metal bonding especially in the electronic industry.
- Single nanosized magnetic particles are mono-domains and one expects that also in magnetic nanophase materials the grains correspond with domains, while boundaries on the contrary to disordered walls. Very small particles have special atomic structures with discrete electronic states, which give rise to special properties in addition to the superparamagnetism behaviour. Magnetic nanocomposites have been used for mechanical force transfer (ferrofluids), for high density information storage and magnetic refrigeration.
- Nanostructured metal clusters and colloids of mono- or plurimetallic composition have a special impact in catalytic applications. They may serve as precursors for new type of heterogeneous catalysts (Cortex-catalysts) and have been shown to offer substantial advantages concerning activity, selectivity and lifetime in chemical transformations and electrocatalysis (fuel cells). Enantioselective catalysis was also achieved using chiral modifiers on the surface of nanoscale metal particles.

L. H. Madkour, *Nanoelectronic Materials*, Advanced Structured Materials 116, https://doi.org/10.1007/978-3-030-21621-4_5

- Nanostructured metal-oxide thin films are receiving a growing attention for the realization of gas sensors (NO_x, CO, CO_2, CH_4 and aromatic hydrocarbons) with enhanced sensitivity and selectivity. Nanostructured metal-oxide (MnO_2) finds application for rechargeable batteries for cars or consumer goods. Nanocrystalline silicon films for highly transparent contacts in thin film solar cell and nano-structured titanium oxide porous films for its high transmission and significant surface area enhancement leading to strong absorption in dye sensitized solar cells.
- Polymer based composites with a high content of inorganic particles leading to a high dielectric constant are interesting materials for photonic band gap structure.

5.1 Recent Advances of Nanostructured Materials

- New Technologies
- New Products
- New Markets
- Financial Rewards
- Process Capabilities
- Production Facilities
- Analytical Capabilities
- Variety of Products
- Stability and Profitability.

Thus, nanomaterials have better properties than the bulk substances such as:

- Mechanical strength
- Thermal stability
- Catalytic activity
- Electrical conductivity
- Magnetic properties
- Optical properties.

Figure 5.1 representing a comprehensive review of nanomaterials-based enhanced surface plasmon resonance sensing of chemical and biological analytes.

5.2 New Properties Can Be Created

- Inert materials become catalysts.
- Insulators become conductors.
- Stable materials become combustible.

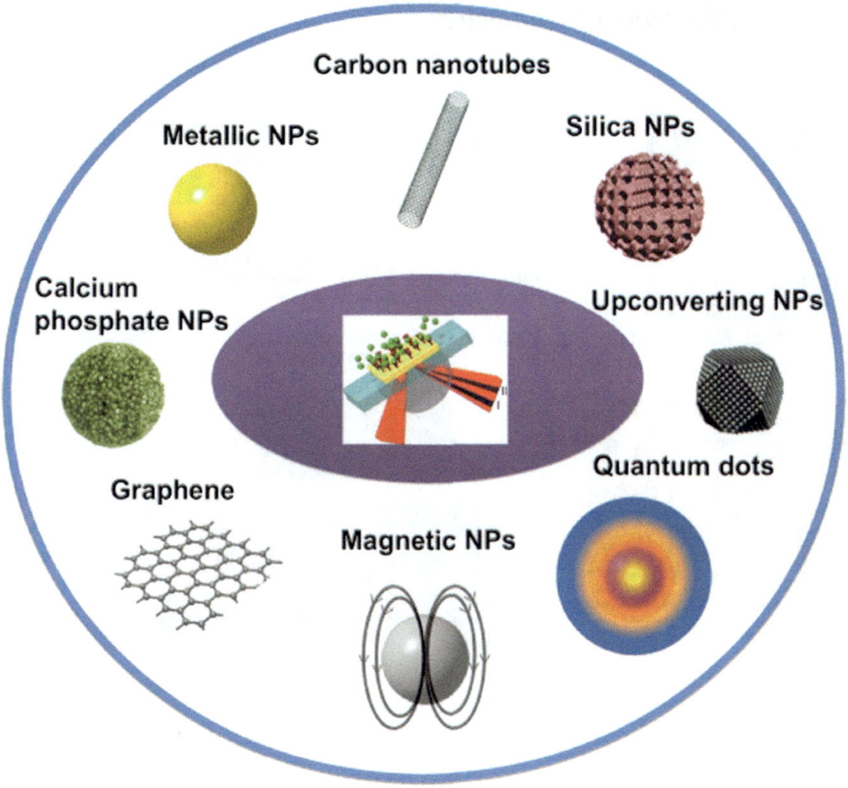

Fig. 5.1 Representing a comprehensive review of nanomaterials-based enhanced surface plasmon resonance sensing of chemical and biological analytes

- A rather inert material Au may become an efficient and selective catalyst when the size is of a few nm. Some examples are given below:
  The Center for Sustainable Nanotechnology has already begun to consider the changeable nature of nanoparticles released into the environment, as we design new experiments to study their biological interactions.

5.3 Some Present and Future Applications of Nanomaterials

Here we list some of the present and future applications of nanomaterials that has been reported in recent literature (Table 5.1).

5.3.1 *Applications of Nanowires*

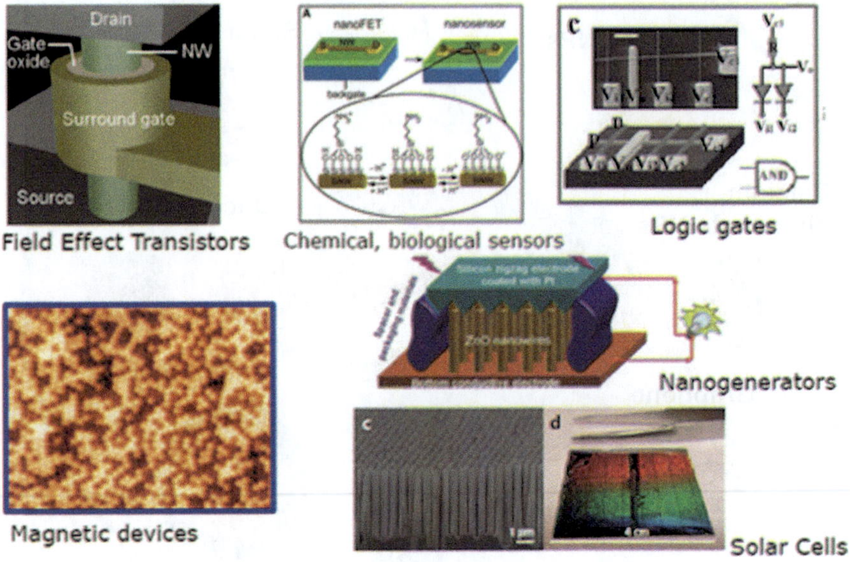

Table 5.1 Some typical properties of nano-structured materials and possibilities of future applications

Bulk	
Single magnetic domain	Magnetic recording
Small mean free path of electrons in a solid	Special conductors
Size smaller than wavelength	Light or heat absorption, scattering
High and selective optical absorption of metal particles	Colours, filters, solar absorbers, photovoltaics, photographic material, phototropic material
Formation of ultra fine pores due to superfine agglomeration of particles	Molecular filters
Uniform mixture of different kinds of superfine particles	R&D of mew materials
Grain size too small for stable dislocation	High strength and hardness of metallic materials
Surface/Interface	
Large specific surface area	Catalysis, sensors
Large surface area, small heat capacity	Heat-exchange materials Combustion Catalysts

(continued)

Table 5.1 (continued)

Lower sintering temperature	Sintering accelerators
Specific interface area large boundary area	Nano-Structured materials
Superplastic behaviour of ceramics	Ductile ceramics
Cluster coating and metallization	Special resistors, temperature sensors
Multi-shell particles	Chemical activity of catalysts Taitored optical elements

5.4 Engineered Nanoparticles Change Shape in Soil and Groundwater

Engineered nanoparticles (NPs) are now components of thousands of consumer products, including personal care products, sporting goods, solar panels, and next-generation batteries. In fact, demand for electric car batteries (built using lithium-nickel-cobalt nanosheets) might cause lithium prices to skyrocket in the coming decade, and completely change the way in which the element lithium is recovered and processed. Engineered NPs pose a poorly-understood (but potentially meaningful) threat to human and environmental health, because they have unique size-dependent properties and reactivity, not found in molecular chemicals or bulk materials (like metals or semiconductors). As with any chemical, the risk that engineered nanomaterials pose to human or environmental health is related to both their hazards (possible toxicity) and their bioavailability (Fig. 5.2).

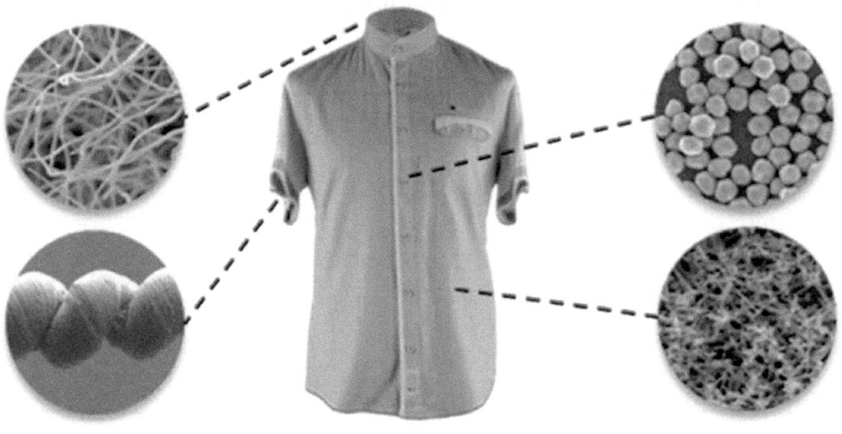

Fig. 5.2 Fabrics containing nanosilver are one of the many uses for nanomaterials in commercial products (image from Yetisen et al. [1], used with permission from the American Chemical Society)

But how do we know what happens to NPs when they get into the environment?

Bioavailability is a single concept unifying various important factors that determine how easily chemical contaminants might be taken up by organisms [2]. For example, how quickly nanoparticles travel through ecosystems, whether their chemical composition changes in the environment, and whether organisms that encounter them will consume them, all contribute to bioavailability. The easier it is for nanoparticles to move around in soil or ground water, the more likely any nanomaterial contamination would be to become widespread. Alternatively, if nanomaterials are quickly trapped by algae, biofilms, or terrestrial plants, they might be ingested in large quantities, and quickly biomagnified up the food web. Biomagnification is a significant concern for accurately predicting human exposure to other contaminants, like heavy metals or persistent organic pollutants.

One of the Center for Sustainable Nanotechnology's primary goals is to connect nanoparticle properties like size, shape, and composition to the way NPs interact with biological systems. A typical engineered nanoparticle consists of a metal, diamond, or semiconductor crystalline core, which is then coated with small organic molecules called "capping agents." Typical capping agents include small molecules like citric acid (the molecule responsible for the sour taste in lemons), ascorbic acid (Vitamin C), or surfactants. You can learn more about the composition and synthesis of engineered NPs in this blog post. Designing effective experiments that relate NP properties to their bioavailability is more challenging than it first appears, though, because the size and composition of nanomaterials often change when they enter a new environment, be it groundwater, soil, or blood (Fig. 5.3).

When nanoparticles are made in the research lab, their properties and composition can be precisely defined. For instance, the size of the nanoparticle's core can be determined with an electron microscope. The coverage of the organic molecules on the particle's surface can be determined using other chemical instrumentation,

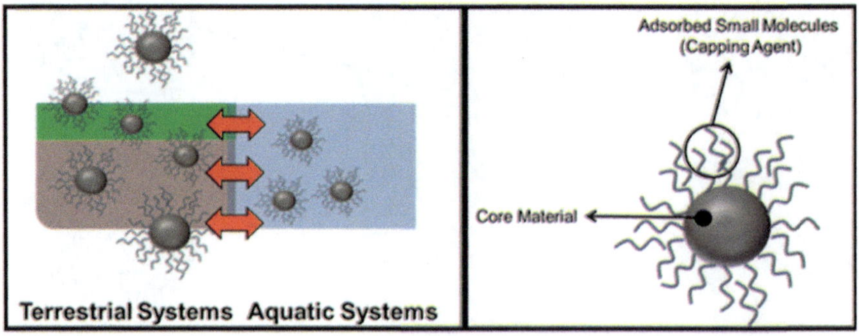

Fig. 5.3 Understanding engineered NPs environmental transport, such as how engineered NPs might move from terrestrial ecosystems to aquatic ecosystems (Left) is essential to predicting routes of NP environmental exposure. A diagram of the structure of a typical engineered NP (Right) (images by Sam Lohse)

like thermogravimetric analysis or ^{1}H-NMR spectroscopy [3]. The size, composition, and type of molecules on the particle's surface constitute a "synthetic identity" for the nanoparticle, similar to a finger print. As long as the nanoparticle stays in the research lab, dissolved in highly purified (deionized) water, this synthetic identity remains more or less intact. Once the NPs enter the real world (for example, maybe one day they will be injected into a patient's blood stream), the NPs may experience many changes: they may stick to proteins present in blood, high salt concentrations may cause the particles to clump together, or the particles may break down into the individual metal ions or atoms that make up the core. As a result, the "biological identity" of a nanoparticle can be entirely different from their original "synthetic identity."

Both these "physiochemical transformations" of the NPs and their odd size (somewhere between a typical protein and a very large virus), scientists predict that most engineered nanomaterials probably move through ecosystems quite differently than typical chemical contaminants, like TCE or PAHs. Many nanoparticles are similar in size to the kinds of particles and large molecules that make up soil, algae, and bacterial biofilms, while molecules like TCE and PAH are much smaller.

Several simple laboratory models to imitate soil, algae biofilms, and ground water. Then introduced some nanomaterials to our simulated soil and biofilm matrices, and measured how they moved using some basic chemical instruments (such as absorbance spectroscopy) [4] (Fig. 5.4).

A final challenge in this kind of study is that there are so many different kinds of nanoparticles! We began this study by looking at gold nanoparticles (check out the

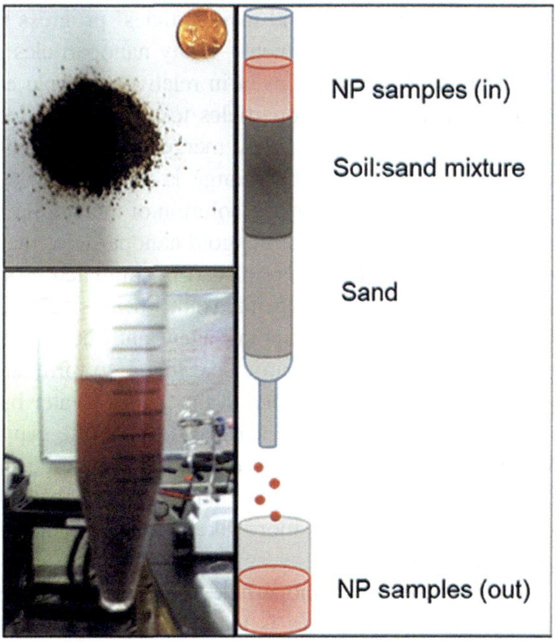

Fig. 5.4 The standard experimental setup used in NP-soil interaction studies. Clockwise, from upper left: a sample of Lockwood Farms soil, soil columns used to determine the extent to which NPs were retained in soil, and a gold nanoparticle solution layered on top of a soil sample (image adapted from Lohse et al. (2017) [4], used with permission of the American Chemical Society)

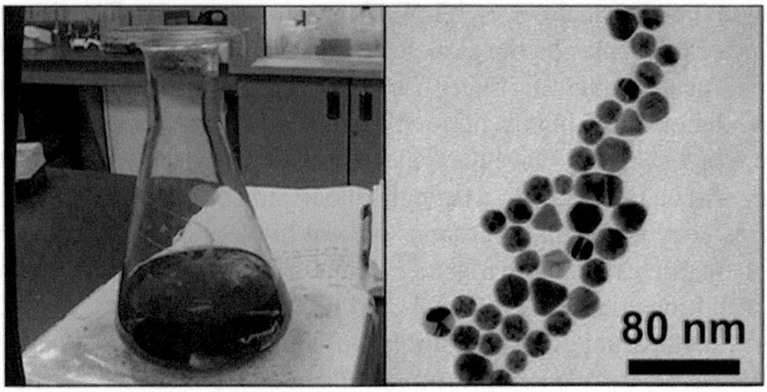

Fig. 5.5 Gold nanoparticles typically produce bright red solutions if they are dissolved in water (left). Red solutions indicate that the particles remain distinct and well-separated, as seen in the electron microscope image (right) (images by Sam Lohse)

image below). Why? Gold nanoparticles are particularly convenient engineered nanoparticles to study. They are easy to prepare in different shapes and sizes, easy to track through soil with simple laboratory instruments, and they are typically considered very stable. For instance, unlike iron, which rusts easily, gold does not tarnish. In addition, unlike silver, which quickly leaches silver ions into water, gold doesn't dissolve easily (Fig. 5.5).

Going into experiments, the gold nanoparticles would remain intact in the soil and simulated ground water, and that the nanoparticles would move as individual particles, with their "synthetic identity" intact through our soil and algae columns. The ability to track the gold nanoparticles' progress through our soil samples was key to a successful experiment—many nanoparticles made of non-precious metals cannot be easily detected, even in relatively simple environmental systems.

Some of the gold nanoparticles tested seemed decidedly unstable. When gold nanoparticles break down, they merge into larger gold crystals through a process called aggregation, and the change is very obvious, even to the naked eye. The vibrant red or brown color of a solution of intact gold nanoparticles gives way to an (ugly) grey-blue. Several of the gold nanoparticles that we tested aggregated almost immediately when they were placed in simulated ground water. We found that the stability of the particle depended not on its size and shape, but on the chemical connection between the capping agent and the crystal surface. For instance, gold nanoparticles coated with citric acid merge to form larger nanowire structures quite quickly, because calcium ions in the ground water break the citric acid molecules loose from the nanoparticle surface. Loss of the capping agent opened up surface sites where the individual gold crystals could merge and form irregular wire-shaped structures. In contrast, gold nanoparticles that were wrapped in polymers (like polyacrylate) were more likely to retain their synthetic identity in soil or ground water (Fig. 5.6).

Another way that nanoparticles can change their synthetic identity when they enter soil or groundwater, is to become entangled in naturally occurring organic

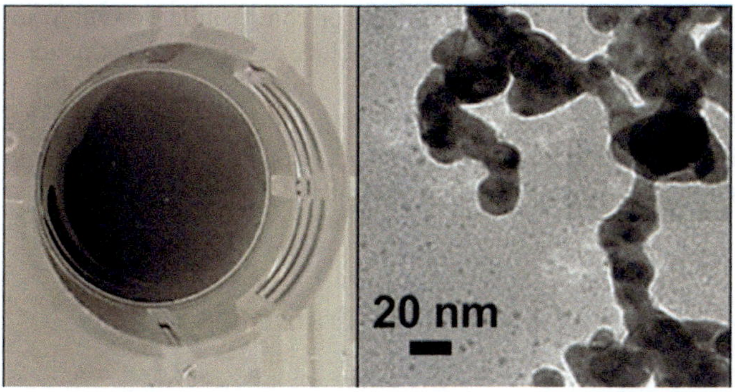

Fig. 5.6 If certain gold nanoparticles are exposed to solutions containing calcium cations, they will aggregate irreversibly. This can be seen in the solution's color, which becomes pale blue-grey (left) or in electron micrograph images, which show gold NPs aggregating into extended nanowire structures (right) (images by Sam Lohse)

polymers (like alginic acid, an important building block of algae biofilms). We found that if alginic acid was present in our ground water samples, alginic acid polymer strands would stick to all the gold nanoparticles that we tested, winding the nanoparticles into large algae-nanoparticle capsules, which could actually be seen with the naked eye. Figure 5.7 shows both the large algae-NP capsules (the red-brown flecks visible in the left-hand picture) and electron microscope images of individual gold nanorods trapped in nanoscale polymer capsules. These algae polymer capsules change the effective size of the nanoparticles in ground water and may cause engineered NPs to move through the environment in the same way as clumps of algae.

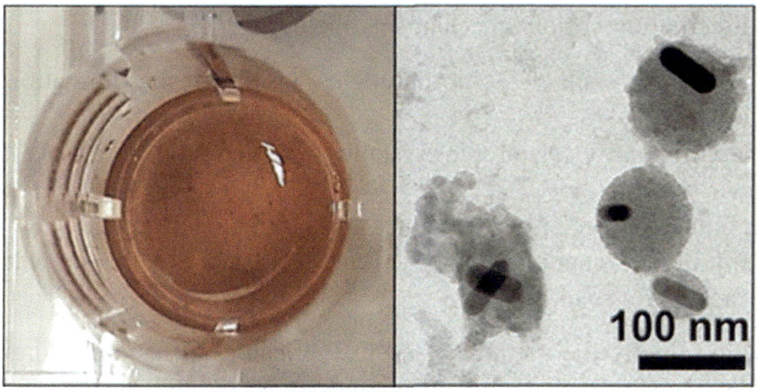

Fig. 5.7 Gold nanoparticles captured by alginic acid. (Left) Algae films form macroscale capsules around gold nanoparticles in water, which are visible to the naked eye. (Right) Electron microscopy image of gold nanoparticles surrounded by smaller alginic acid capsules (images by Sam Lohse)

Engineered NPs can change size, shape, or composition once they are released into the environment. These changes may include aggregating into larger crystals or becoming trapped in natural organic matter (such as algae) present in the environment. The nanoparticles' new "biological identity" may make them less mobile in the environment, or more likely to "piggy back" along environmental transport pathways taken by algae, silicates, or aluminates.

5.5 Applications of Field-Effect Transistors (FET)

In this part we review recent advances in biosensors based on one-dimensional (1-D) nanostructure field-effect transistors (FET). Specifically, we address the fabrication, functionalization, assembly/alignment and sensing applications of FET based on carbon nanotubes, silicon nanowires and conducting polymer nanowires. The advantages and disadvantages of various fabrication, functionalization, and assembling procedures of these nanosensors are reviewed and discussed. We evaluate how they have been used for detection of various biological molecules and how such devices have enabled the achievement of high sensitivity and selectivity with low detection limits. Finally, we conclude by highlighting some of the challenges researchers face in the 1-D nanostructures research arena and also predict the direction toward which future research in this area might be directed.

5.6 Fabrication of 1-D Nanostructures

A good method for generating 1-D nanostructures should enable simultaneous control of the dimensions, properties, and morphology. In general, 1-D nanostructures are synthesized by promoting the crystallization of solid-state structures along one direction by various mechanisms including:

- use of templates with 1-D morphologies to direct the formation of 1-D nanostructures
- use of the intrinsically anisotropic crystallographic structure of a solid to achieve 1-D growth
- use of a liquid/solid interface to reduce the symmetry of a seed
- use of appropriate capping agents to kinetically control the growth rates of various facets of a seed.

We describe briefly the methods used to fabricate 1-D nanostructures such as silicon nanowires (SiNWs), carbon nanotubes (CNTs), and conducting polymer nanowires (CP NWs).

5.6.1 *Carbon Nanotubes (CNTs)*

Carbon nanotubes are carbon sheets rolled to form either a seamless cylinder, known as single-walled carbon nanotubes (SWCNTs), or many cylinders stacked one inside the other, known as multi-walled carbon nanotubes (MWCNTs). The lengths of these tubes generally range from several hundred nanometers (nm) to several micrometers (μm), though tubes in the millimetre range have also been reported. Their diameter is dependent on the number of walls and is usually >100 nm.

5.6.2 *Silicon Nanowires (SiNWs)*

The vapor–liquid–solid (VLS) method, involves the use of liquid–metal solvents with low solubility for Si and other elemental semiconductor materials. This method has been very successful in producing SiNWs in large quantities and at low temperatures. The vapor–liquid–solid (VLS) method is the most successful in generating large quantities of nanowires with single crystalline structures. Another method that has been used to synthesize high purity SiNWs is the laser ablation technique. Thermal methods have been applied to the synthesis of SiNWs. For example, thermal evaporation of a mixture of Si and SiO_2 yielded SiNWs that consisted of a polycrystalline Si core with a high density of defects and a silicon oxide shell. Highly oriented, long SiNWs have been obtained in large yields on flat silicon substrates by the thermal evaporation of SiO. Ultrafine SiNWs of diameters between 1 and 5 nm, sheathed with a SiO_2 outer layer of 10–20 nm, were synthesized by oxide-assisted growth via the disproportionation of thermally evaporated SiO using a zeolite template.

The zeolite restricted the growth of the nanowires laterally and supplied the oxide to form the outer sheath (Fig. 5.8).

Single-crystal silicon nanowires with diameters approaching molecular dimensions were synthesized using gold [5] nanocluster-catalyzed 1D growth. High-resolution transmission electron microscopy studies show that silicon nanowires grown with silane reactant in hydrogen are single crystal with little or no visible amorphous oxide down to diameters as small as 3 nm. Structural characterization of a large number of samples shows that the smallest-diameter nanowires grow primarily along the {110} direction, whereas larger nanowires grow along the {111} direction. In addition, cross-sectional transmission electron microscopy was used to address the importance of surface energetics in determining the growth direction of the smallest nanowires (Fig. 5.9).

Generally speaking, it is necessary to enhance the field in the tunnel oxide and possibly at the same time reduce the field in the top oxide. Nanowires and to a

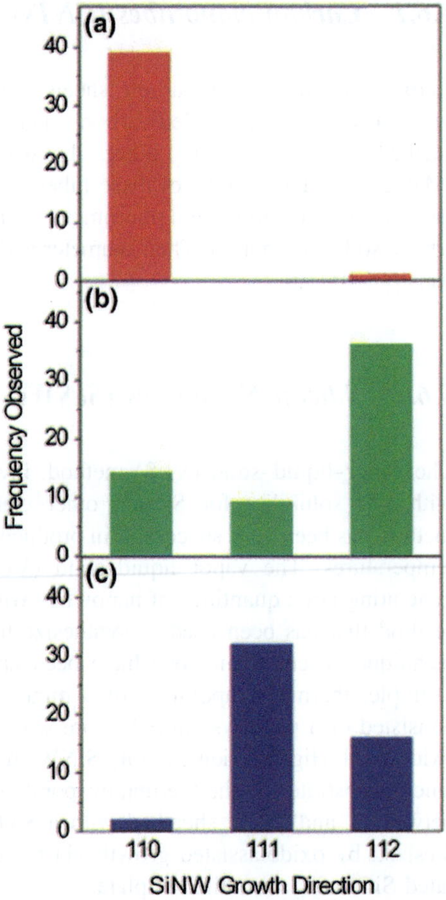

Fig. 5.8 Histograms of the growth directions for SiNWs with diameters from **a** 3 to 10 nm, **b** 10 to 20 nm, and **c** 20 to 30 nm

lower extent tri-gates [6] offer a unique opportunity here, since the geometry will automatically generate a higher field at the bottom oxide and the extension of the field increase can be controlled by the diameter of the nanowire. Figure 5.10 illustrates that situation. Therefore, nanowire devices have been extensively explored for charge trapping devices.

The ability to prepare well-defined molecular-scale single-crystal silicon nanowires opens up new opportunities for both fundamental studies and nanodevice applications.

5.6.3 SiONWs Are Interest in SNOM and Integrated Optics

Silica nanowires (SiONWs) using an excimer laser ablation method. The SiONWs emit stable and high brightness blue light at energies of 2.65 and 3.0 eV. The

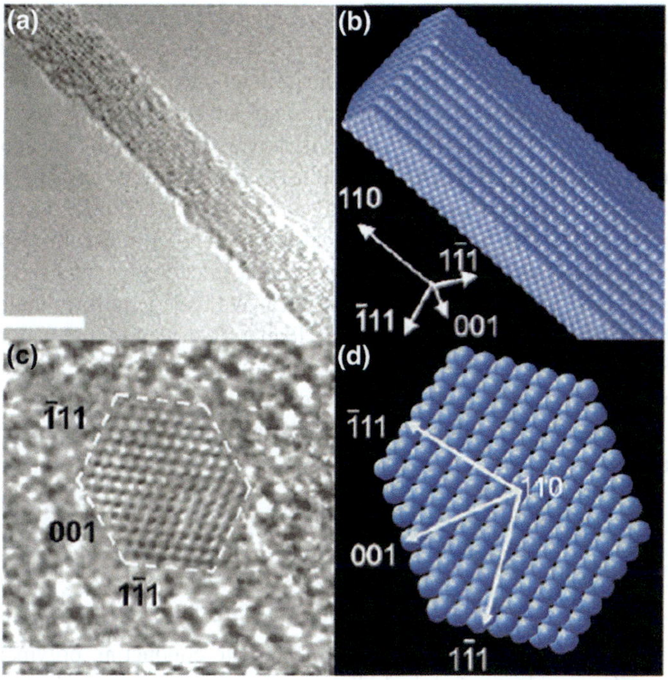

Fig. 5.9 **a** TEM images of 3.8-nm SiNWs grown along the (110) direction, **c** HRTEM cross-sectional image, and equilibrium shapes for the **b** NW and the **d** NW cross sections predicted by Wulff construction. The scale bars are 5 nm. SiNWs dispersed in epoxy were sectioned with a microtome to slices ca. 50 nm thick and were then transferred to copper TEM grids [5]

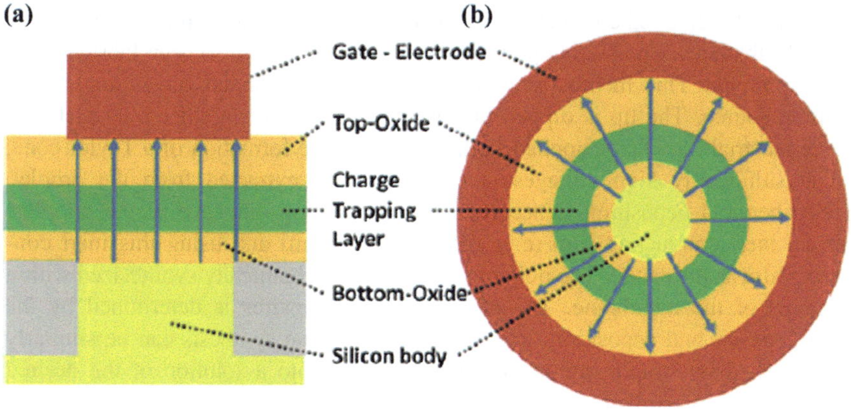

Fig. 5.10 Schematic cut in width direction of a planar (**a**) and a nanowire based (**b**) charge trapping memory cell. The arrows indicate the field lines in the device in programing or erase operation. In the nanowire device the field lines are denser in the bottom oxide. Therefore the exchange of carriers between the charge trapping layer and the silicon transistor body is enhanced and the charge exchange between the charge trapping layer and the gate-electrode is reduced

intensity of the emission is two orders of magnitude higher than that of porous silicon. SiONWs have potential applications in high-resolution optical heads of scanning near-field optical microscope or nanointerconnections in future integrated optical devices [7]. Nano-scale optical wires are of scientific and technological interest in the fields of localization of light, lower dimensional waveguide, and scanning near-field optical microsocpy (SNOM). With the development of science towards mesoscopic scale and of technological advances in the integrated optics, it is important to synthesize nano-scale optical wires that can meet the demands of further applications. Thus, nanowires of amorphous silica were synthesized in large scale using the laser ablation approach. The SiONWs have a uniform diameter and long length. The growth of SiONWs is most likely controlled by the vapor–liquid–solid (VLS) mechanism similar to that in Si nanowire growth. An extraordinarily intensive blue light emission is observed which could be attributed to defect centers of oxygen deficiency in the wires. The SiONWs are probably of interest in SNOM and integrated optics.

5.6.4 Conducting Polymer Nanowires (CP NWs)

Among the methods that have been used to fabricate conducting polymers on the nanometer scale are electrochemical dip-pen lithography, mechanical stretching [8, 9], electrospinning [10, 11] and template-directed electrochemical synthesis [12]. Dip-pen nanolithography (DPN) is a scanning probe nanopatterning technique in which an AFM tip is used to deliver molecules to a surface via a solvent meniscus which naturally forms in the ambient atmosphere. Doped polyaniline and polypyrrole lines down to 310 nm and 290 nm widths, respectively, were formed using ionically charged conducting polymers as the "ink" for writing on oppositely charged substrates. Conducting polymers as nanofibers have also been fabricated by electrospinning. This method uses a microfabricated scanned tip as an electrospinning source. The tip is dipped in a polymer solution to gather a droplet as a source material. A voltage applied to the tip causes the formation of a Taylor cone, and at sufficiently high voltages, a polymer jet is extracted from the droplet. Electrochemical deposition is accomplished by coating one face of the template with an inert conducting film (e.g. gold and platinum) and using this inert conducting film as the anode. The polymer is then electrochemically synthesized within the pores of the membrane. The length of the nanowires is determined by the current density and deposition time. Chemical template synthesis can be similarly accomplished by simply immersing the membrane into a solution of the desired monomer and its oxidizing agent. The diameter of nanowire is determined by the pore diameter of template. Figure 5.11 depicts a schematic of the template-based

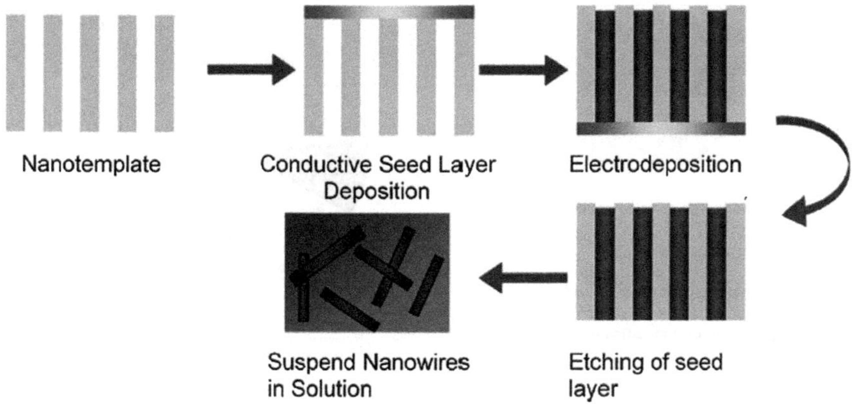

Fig. 5.11 Schematic of template-based synthesis of conducting polymer nanowires

synthesis process. Conducting polymers show preferential deposition along the walls of the polycarbonate membrane resulting in nanotubule structures due to solvophobic interactions. These tubules close up as the deposition time is increased and eventually results in nanowires. Following nanowires electrodeposition, the conducting film used for electrochemical polymerization and the template are then subsequently dissolved using appropriate acids or bases. Organic solvents may be used to dissolve polymer templates. It has been shown that template- directed conducting polymer nanotubes and nanowires have higher electronic conductivities than bulk samples of the same material. Furthermore, the mechanism of electronic conduction in these template-directed nanostructures can be varied at will by changing their diameter.

Direct electrochemical synthesis of large arrays of uniform and oriented nanowires of conducting polymers with a diameter of less than 100 nm on a variety of substances such as Pt, Si, Au and carbon without using a supporting template was recently reported. A three-step electrochemical deposition procedure was utilized. In the first step, a large current density was used to create the nucleation sites on the substrate. The initial stage is followed by continued polymerization with reduced current density.

Economical Relevance

World market for nanotechnology in percent is shown in Figs. 5.12a, b and c.

Finally the global metal oxide nanoparticles (NPs) market predict that, the market is projected to reach $16,919.9 metric tons by 2021 as shown in diagram Fig. 5.13.

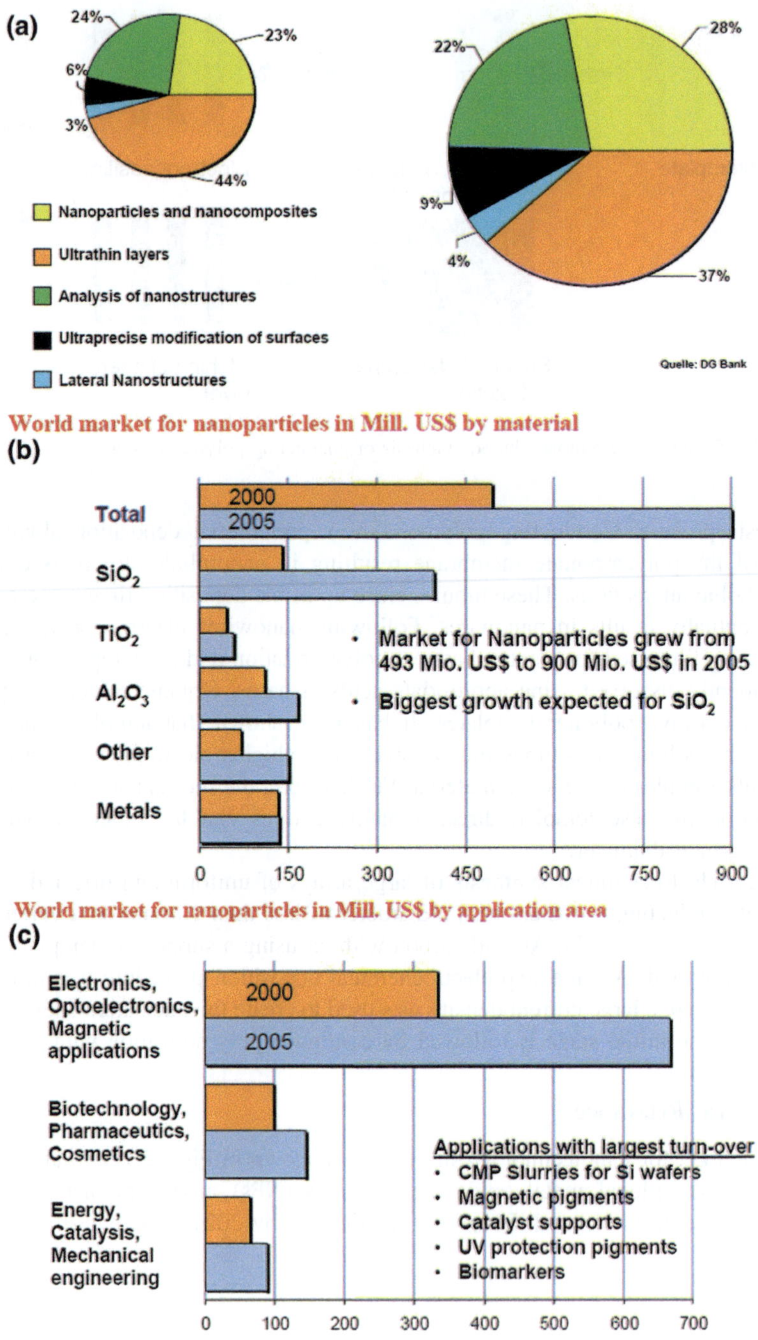

Fig. 5.12 **a**, **b**, **c** World market for nanotechnology in percent

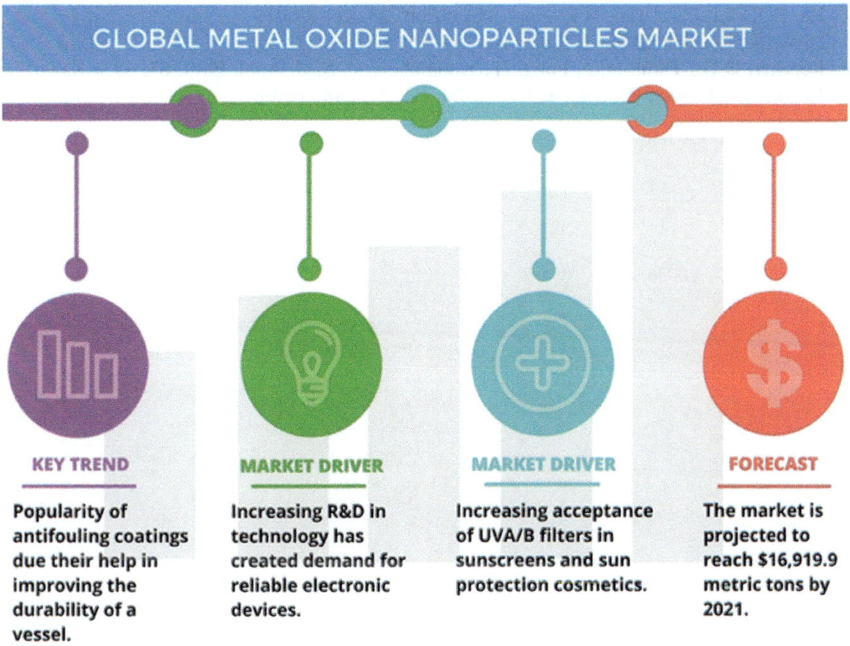

Fig. 5.13 Global metal oxide nanoparticles (NPs) market

References

1. Yetisen, et al.: Nanotechnology in textiles. ACS Nano **10**(3), 3042–3068 (2016). https://doi.org/10.1021/acsnano.5b08176
2. Semple, K.T., et al.: Environ. Science Technol. **38**(12), (2004). https://doi.org/10.1021/es040548w
3. Thermogravimetric Analysis (TGA): frequently asked questions. Accessed 20 Feb 2018
4. Lohse, S.E., Abadeer, N.S., Zoloty, M., White, J.C., Newman, L.A., Murphy, C.J.: Nanomaterial probes in the environment: gold nanoparticle soil retention and environmental stability as a function of surface chemistry. ACS Sustain. Chem. Eng. **5**(12), 11451–11458 (2017). https://doi.org/10.1021/acssuschemeng.7b02622
5. Yue, W., Cui, Y., Huynh, L., Barrelet, C.J., Bell, D.C., Lieber, C.M.: Controlled growth and structures of molecular-scale silicon nanowires. Nano Lett. **4**(3), 433–436 (2004)
6. Specht, M., Kömmling, R., Hofmann, F., Klandzievski, V., Dreeskornfeld, L., Weber, W., Kretz, J., Landgraf, E., Schulz, T., Hartwich, J., Rösner, W., Städele, M., Luyken, R.J., Reisinger, H., Graham, A., Hartmann, E., Risch, L.: Novel dual bit tri-gate charge-trapping memory devices. IEEE Electron Device Lett. **25**, 810 (2004)
7. Yu, D.P., Hang, Q.L., Ding, Y., Zhang, H.Z., Bai, Z.G., Wang, J.J., Zou, Y.H., Qian, W., Xiong, G.C., Feng, S.Q.: Amorphous silica nanowires: intensive blue light emitters. Appl. Phys. Lett. **73**(21), 23 (1998)
8. Wanekaya, A.K., Chen, W., Myung, N.V., Mulchandani, A.: Nanowire-based electrochemical biosensors. Electroanalysis **18**(6), 533–550 (2006)

9. He, H.X., Li, C.Z., Tao, N.J.: Appl. Phys. Lett. **78**, 811 (2001)
10. Doshi, J., Reneker, D.H.: J. Electrostat. **35**, 151 (1999)
11. Reneker, D.H., Yarin, A.L., Fong, H., Koombhongse, S.: J. Appl. Phys. **87**, 4531 (2000)
12. Jerome, C., Jerome, R.: Angew. Chem. Int. Ed. **37**, 2488 (1998)

Chapter 6
Examples of Nanomaterials with Various Morphologies

As the world is advancing, there are many nanomaterials that being produces and engineered. Nanomaterials is also a new step in the evolution of understanding and utilization of materials. Example of nanomaterials are carbon nanotube, nanoparticle, metal rubber, quantum dots, nanopores and many more.

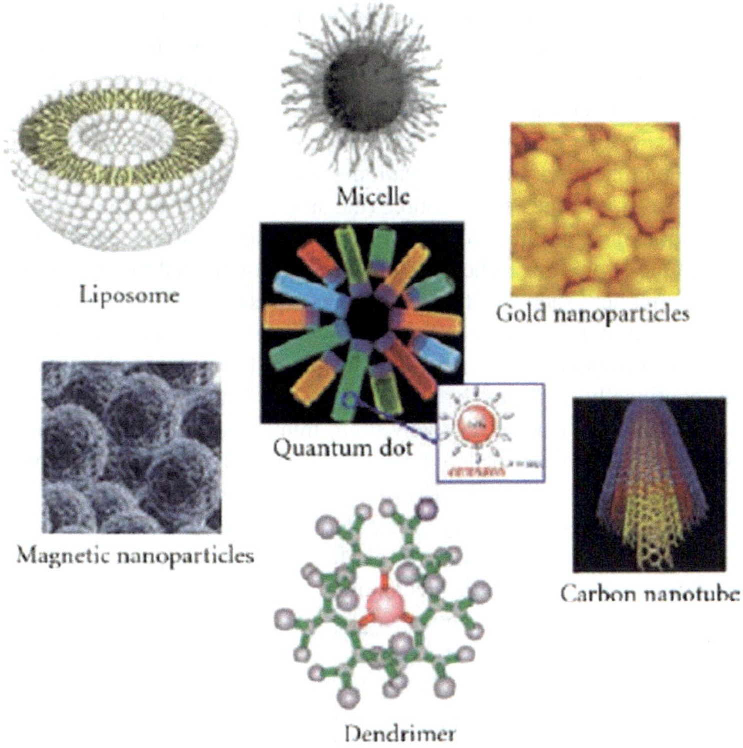

L. H. Madkour, *Nanoelectronic Materials*, Advanced Structured Materials 116, https://doi.org/10.1007/978-3-030-21621-4_6

6.1 Carbon Nanotubes (CNTs)

Carbon nanotubes, first observed in 1991 [1], occur as multiwalled nanotubes (MWNTs) and single-walled nanotubes (SWNTs) [2]. Primary methods for CNTs synthesis include arc-discharge [1, 3], laser ablation [4], gas-phase catalytic growth from carbon monoxide [5], and chemical vapor deposition (CVD) from hydrocarbons [6–8]. Because impurities in the form of catalyst particles, amorphous carbon, and nontubular fullerenes are also produced during CNTs synthesis, subsequent purification steps are required to separate the tubes. The gas-phase processes tend to produce nanotubes with fewer impurities and are more amenable to large-scale processing. The arc-discharge technique generally involves the use of two high purity graphite rods as the anode and cathode. The rods are brought together under a helium atmosphere and a voltage is applied until a stable arc is achieved. The exact process variables depend on the size of the graphite rods. As the anode is consumed, a constant gap between the anode and cathode is maintained by adjusting the position of the anode. The material deposits on the cathode to form a buildup consisting of an outside shell of fused material and a softer fibrous core containing nanotubes and other carbon particles. To create SWNTs, the electrodes are doped with a small amount of catalyst particles [2, 3, 9–11]. Initially used for the production of fullerenes, laser ablation technique has, over the years, been improved to allow the production of SWNTs [4, 12, 13]. In this technique, a graphite target held in a controlled atmosphere oven at temperatures near 1200 °C is vaporized by a laser and the vapors condensed over a water-cooled target. To produce SWNTs, the graphite target is doped with a catalyst [4]. Both the arc-discharge and the laser ablation techniques produce a limited volume of sample in relation to the size of the carbon source (the anode in arc-discharge and the target in laser ablation) and require subsequent purification steps to separate the tubes from undesirable byproducts. These limitations have motivated the development of gas-phase techniques, such as chemical vapor deposition (CVD), in which nanotubes are formed by the decomposition of a hydrocarbon or carbon monoxide gas. The gas-phase techniques are amenable to continuous processes since the carbon source is continually replaced by flowing gas. In addition, the purity of the as-produced nanotubes can be relatively high, minimizing subsequent purification steps. Nikolaev et al. described the gas-phase growth of SWNTs with carbon monoxide as the carbon source [5]. They reported that the highest yields of SWNTs occurred at the high temperature and pressure (1200 °C, 10 atm). Smalley and his co-workers have refined the process to produce large quantities of SWNTs with increased purity. Other gas-phase techniques utilize hydrocarbon gases as the carbon source for production of both single and multi walled carbon nanotubes [14–17].

As a result of their good physical and chemical properties such as the strongest and stiffest materials yet discovered in terms of tensile strength and elastic modulus respectively and good electrical and thermal conductor, carbon nanotubes have potential use in a variety of industrial and biomedical applications including:

- Catalysts
- Biosensors
- Composite materials with improved mechanical and electrical properties, and
- Drug carriers.

6.2 Nanoparticles

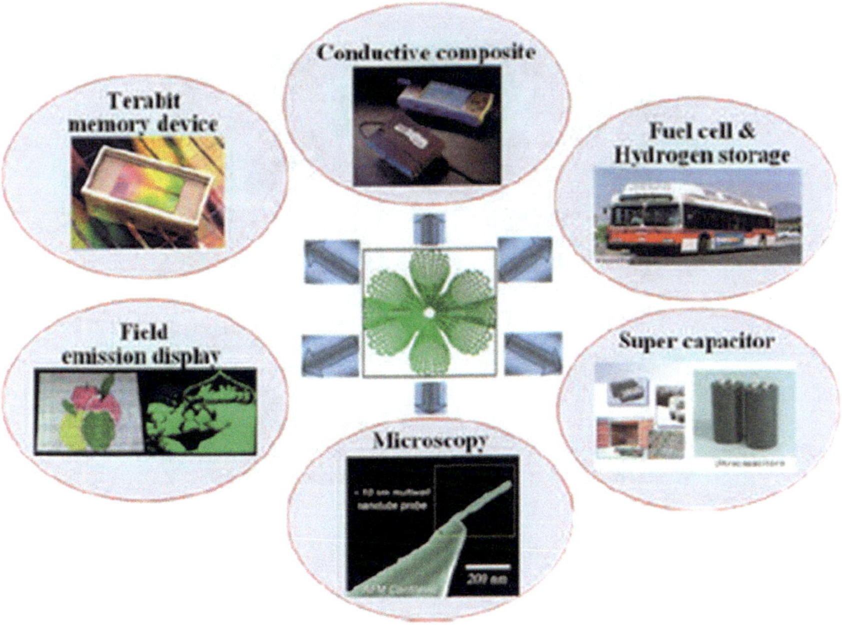

In nanotechnology, a nanoparticle is defined as a small object that behaves as a whole unit in terms of its transport and properties. It is further classified according to size: in terms of diameter. Particles in the nano range are being investigated for a number of applications, like: wear resistance, scratch protection, blocking of light. A good example of this last application is the blocking of ultraviolet light by zinc oxide particles for cosmetics. The blocking power is directly related to particle size, size distribution, particle loading and the dispersion.

6.2.1 Classification of NPs

NPs are broadly divided into various categories depending on their morphology, size and chemical properties. Based on physical and chemical characteristics, some of the wellknown classes of NPs are given as below.

6.2.1.1 Carbon-Based NPs

Fullerenes and carbon nanotubes (CNTs) represent two major classes of carbon-based NPs. Fullerenes contain nanomaterial that are made of globular hollow cage such as allotropic forms of carbon. They have created noteworthy commercial interest due to their electrical conductivity, high strength, structure, electron affinity, and versatility (as discussed in details latter in Sect. 7.2).

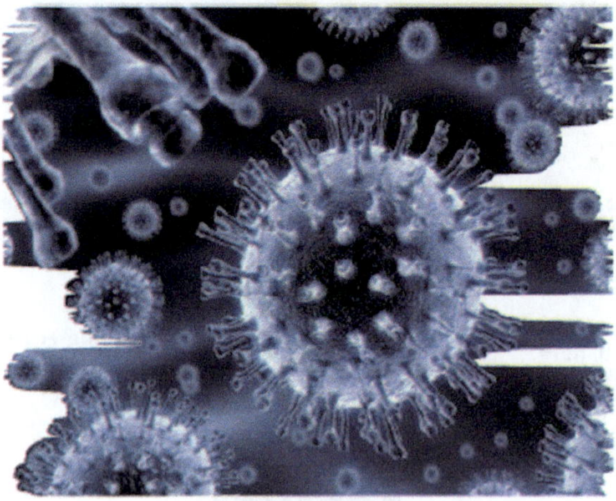

6.2.1.2 Metallic Nano Particles

Metal NPs are purely made of the metals precursors. Due to well-known localized surface plasmon resonance (LSPR) characteristics, these NPs possess unique optoelectrical properties. NPs of the alkali and noble metals i.e. Cu, Ag and Au have a broad absorption band in the visible zone of the electromagnetic solar spectrum. The facet, size and shape controlled synthesis of metal NPs is important in present day cutting-edge materials [18]. Due to their advanced optical properties, metal NPs find applications in many research areas. Gold NPs coating is widely used for the sampling of SEM, to enhance the electronic stream, which helps in obtaining high quality SEM images (Fig. 6.1).

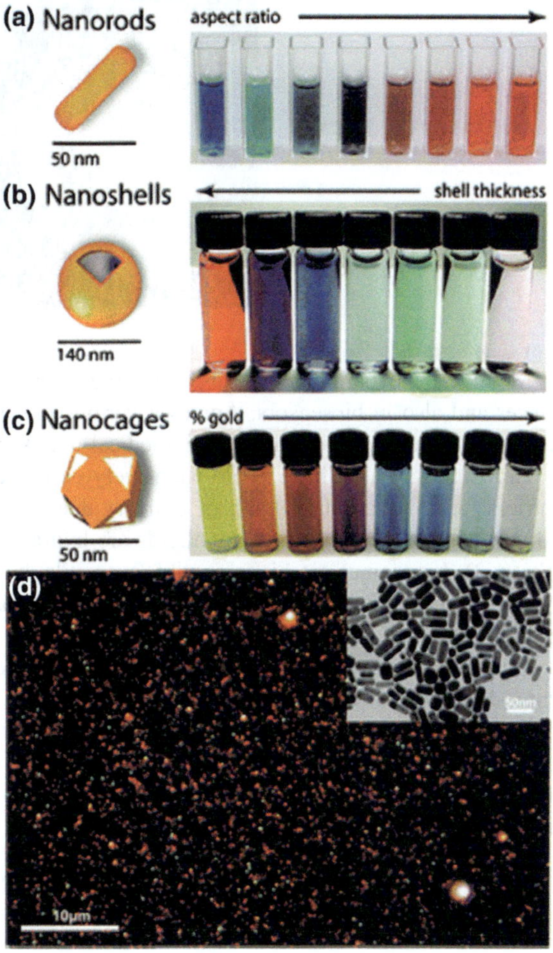

Fig. 6.1 Color dependence of Au NPs on size and shape [18]

Fig. 6.2 Surface functionalized gold nano particles

Metallic nano particles have used in drug delivery, especially in treatment of cancer and also in biosensors. Amongst various metals, silver and gold nano particles are of prime importance for biomedical use (Fig. 6.2).

Liposomes: These have been extensively explored and most developed nano carriers for novel and targeted drug delivery due to their small size, these are 50–200 nm in size. When dry phospholipids are hydrated, closed vesicles are formed (Fig. 6.3). Liposomes are biocompatible, versatile and have good entrapment efficiency. It finds application as long circulatory and in passive and active delivery of gene, protein and peptide.

Dendrimers: Dendrimers are hyper branched, tree-like structures. It contains three different regions: core moiety, branching units, and closely packed surface (Fig. 6.4). It has globular structure and encloses internal cavities. Its size is less than 10 nm. These are used for long circulatory, controlled delivery of bioactive material, targeted delivery of bioactive particles to macrophages and liver targeted delivery.

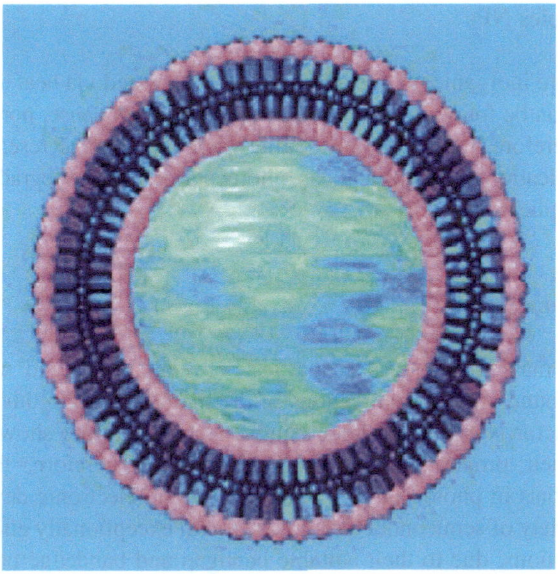

Fig. 6.3 Structure of liposomes

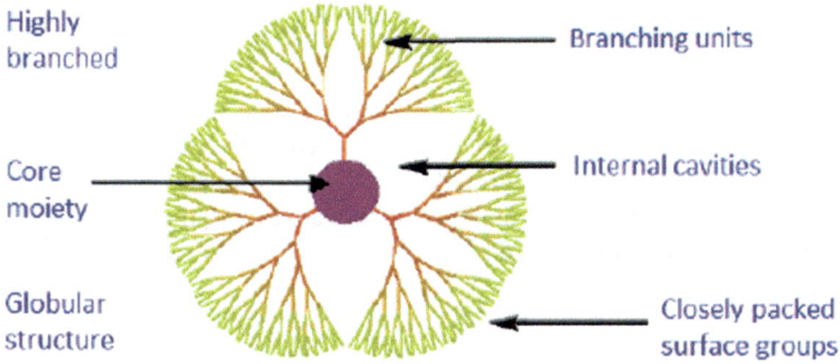

Fig. 6.4 Schematic representation of a dendrimers showing core, branches, and surface

There are many other applications, which are deeply discussed in applications section of this book.

6.2.1.3 Ceramics NPs

Ceramics NPs are inorganic nonmetallic solids, synthesized via heat and successive cooling. They can be found in amorphous, polycrystalline, dense, porous or hollow forms [19]. Therefore, these NPs are getting great attention of researchers due to their use in applications such as catalysis, photocatalysis, photodegradation of dyes, and imaging applications [20].

6.2.1.4 Semiconductor NPs

Semiconductor materials possess properties between metals and nonmetals and therefore they found various applications in the literature due to this property [21, 22]. Semiconductor NPs possess wide bandgaps and therefore showed significant alteration in their properties with bandgap tuning. Therefore, they are very important materials in photocatalysis, photo optics and electronic devices [23]. As an example, variety of semiconductor NPs are found exceptionally efficient in water splitting applications, due to their suitable bandgap and bandedge positions [24].

6.2.1.5 Polymeric NPs

These are normally organic based NPs and in the literature a special term polymer nanoparticle (PNP) collective used for it. They are mostly nanospheres or nanocapsular shaped [25]. The former are matrix particles whose overall mass is generally solid and the other molecules are adsorbed at the outer boundary of the spherical surface. In the latter case the solid mass is encapsulated within the particle completely [26]. The PNPs are readily functionalize and thus find bundles of applications in the literature [27, 28].

6.2.1.6 Lipid-Based NPs

These NPs contain lipid moieties and effectively using in many biomedical applications. Generally, a lipid NP is characteristically spherical with diameter ranging from 10 to 1000 nm. Like polymeric NPs, lipid NPs possess a solid core made of lipid and a matrix contains soluble lipophilic molecules. Surfactants or emulsifiers stabilized the external core of these NPs [29]. Lipid nanotechnology [30] is a special field, which focus the designing and synthesis of lipid NPs for various applications such as drug carriers and delivery [31] and RNA release in cancer therapy [32].

6.3 Other Application Examples of Nanoparticles are

- Filter media. The active component is an alumina (AlOOH) fiber 2 nm in diameter. The nano alumina fiber is highly electropositive, and will attract and retain particles.
- Silver has long been known to inhibit oxygen exchange, effectively killing bacteria. Nanosilver has been demonstrated to enhance that effect, and is currently being used in wound dressings for critical-burn victims.
- Other nanoparticles can be used as pharmaceutical carriers [33] to specifically deliver disease-fighting drugs to afflicted body areas, thus minimizing secondary effects.
- In fuel cells perovskites are being discussed as electrode materials for use as cathodes. Specifically the high surface area and the high ionic conductivity of the material is of interest.

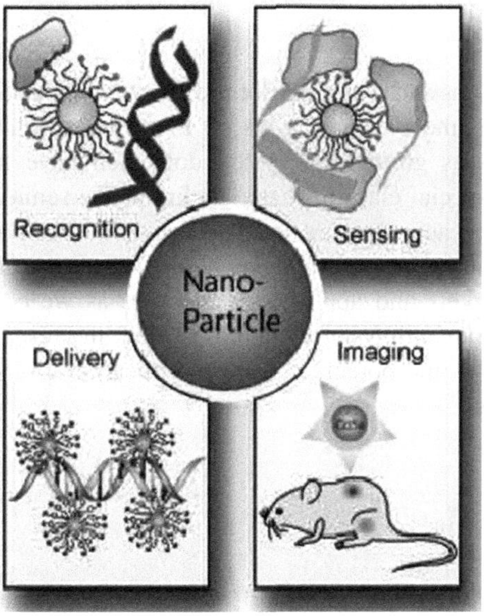

6.4 Quantum Dots

A semiconductor whose excitons are confined in all three spatial dimensions. Basically a single electron trapped.

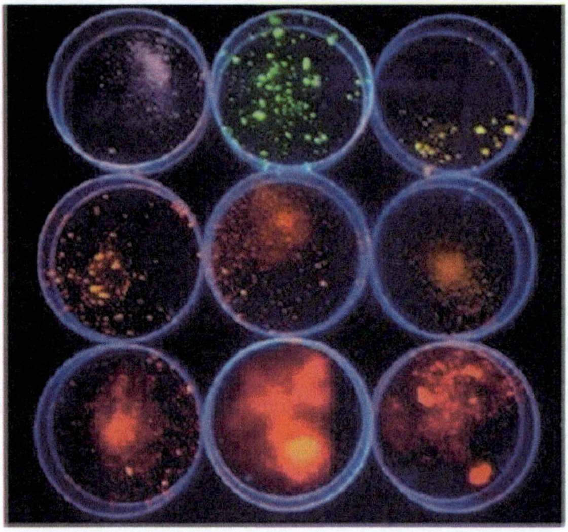

Inside a cage of atoms and when the dot is exposed to a pulse of light at the right time and wavelength, the electron is raised to an exited state. Thus when the electron fall back to its groundstate, these dots then have the ability to glow. Quantum dots are a special class of materials known as semiconductor. Quantum dots are very special because they are so small that 4,000,000 dots take up 2 cm. At these small sizes quantum dots enable never before seen applications to science and technology. New markets and applications emerges as we continue to innovate, works and master the quantum dot technology. Now the researchers are trying to harness quantum dots full potential. Many application [34] used is based on quantum dots such as:

- Used for LED
- Memory storage
- Solar cells and
- Disease.

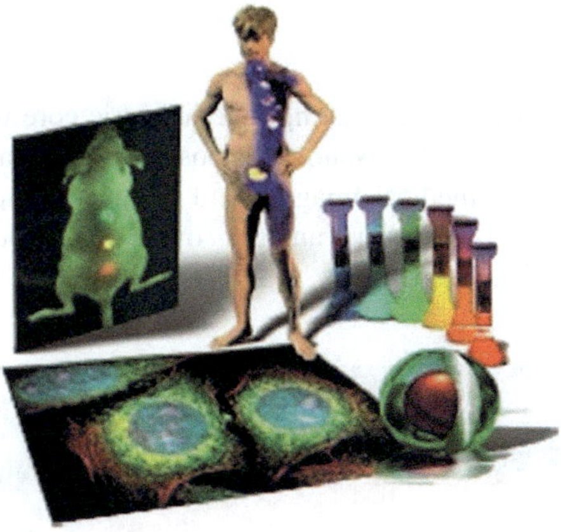

In the near future there will be a foldable wide screen TV, or a truly foldable phone, right in your pocket, instead of paint, you have walls which are screens that can display movies or tv shows, walls that can change colours to your mood, and help you get to sleep, and a newspaper that actually only on a single surface.

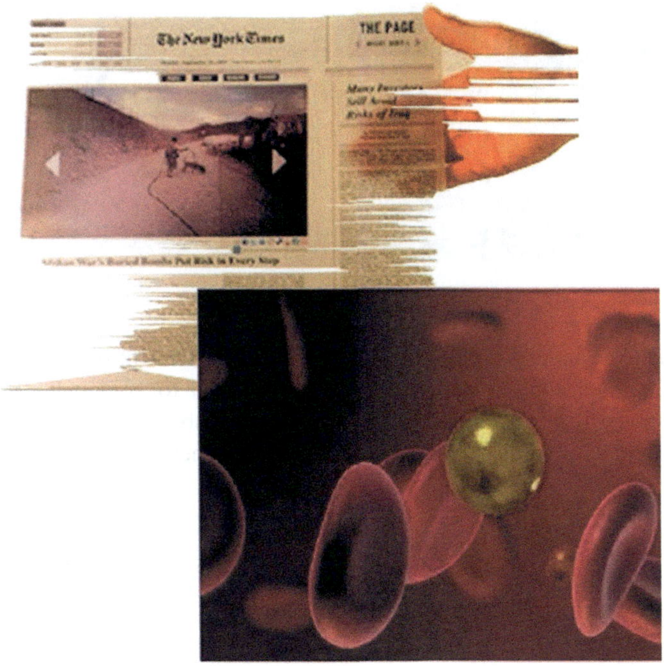

6.5 Nanoshell

A type of spherical nanoparticle consisting of a dielectric core which is covered by a thin metallic shell (usually gold). Nanoshells possess highly favorable optical and chemical properties for biomedical imaging and therapeutic applications [35]. In addition, nanoshells offer other advantages over conventional organic dyes including improved optical properties.

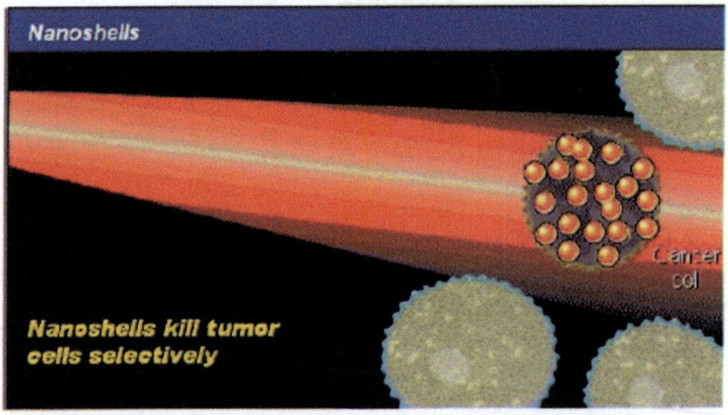

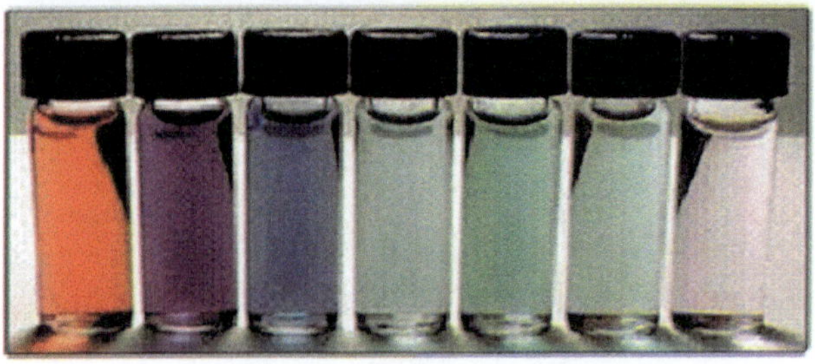

Nanoshells designed to absorb various wavelengths of light (the six vials on the right), including infrared (vial at far right) compared to gold colloid (far left). Nanoshell properties is chemically inert and water-soluble, can be used as efficient catalysts, forms stable dispersions and also non-magnetic; therefore they can be useful in biological applications.

6.6 Metal Rubber

A self-assembled nanocomposite material that combines the high electrical conductivity of metals with the low mechanical modulus of elastomers. This metal rubber also flexible, indestructible material that can be heated, frozen, washed or doused with jet fuel, and still retain its electricity-conducting properties.

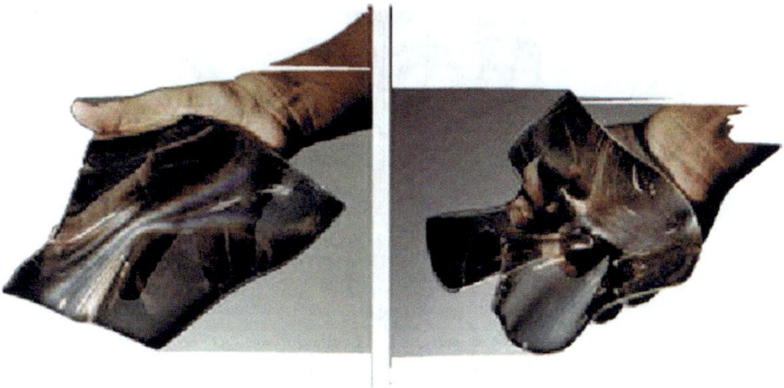

A process called "self-assembly" is used by chemists and engineers to make this rubber metal. The material is repeatedly dipped into positively charged and negatively charged solutions. The positive and negative charges bond, forming layers that conduct electricity. Metal rubber is used for:

- Bendy, electrically charged aircraft wings
- Body armor
- Durable electronic sensors
- Artificial muscles
- Wearable computers and
- Abuse-resistant, flexible circuits, like cell phones.

6.7 Nanopores

Nanoporous materials area class of compounds that contain a regular ordered framework possessing pores on the nanoscale. The collection includes mainly natural or synthetic zeolites. There are many types of nanopores such as biological nanopores, solid-state nanopores, and hybrid nanopores.

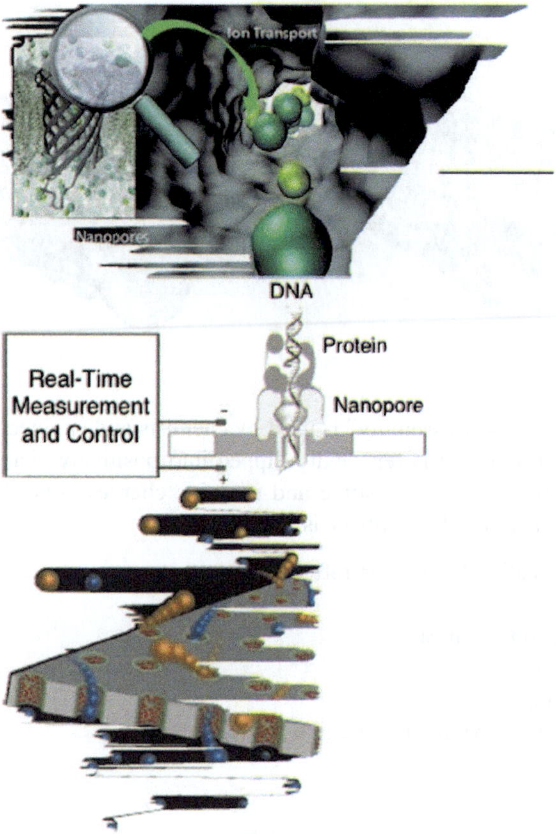

Nanopore analysis [36] is an emerging technique that involves using a voltage to drive molecules through a nanoscale pore in a membrane between two electrolytes, and monitoring how the ionic current through the nanopore changes as single molecules pass through it. Using this approach, many application can be used such as DNA sequencing, transport of various ions, detect target molecules at very low concentrations from very small sample volumes, simultaneously screen panels of biomarkers or genes and eliminate cumbersome amplification and conversion steps.

6.8 Nanoparticles with Different Morphologies

Nanomaterials (gold, carbon, metals, metal oxides and alloys) with variety of morphologies (shapes) are depicted in Fig. 6.5. Nanoparticles are being used for diverse purposes, from medical treatments, using in various branches of industry production such as solar and oxide fuel batteries for energy storage, to wide incorporation into diverse materials of everyday use such as cosmetics or clothes, optical devices, catalytic, bactericidal, electronic, sensor technology, biological labelling and treatment of some cancers. Due to their exceptional properties including antibacterial activity, high resistance to oxidation and high thermal conductivity, nanoparticles have attracted considerable attention in recent years. Nanoparticles can be synthesized chemically or biologically. Metallic nanoparticles that have immense applications in industries are of different types, namely, Gold, Silver, Alloy, magnetic etc. This part aims to present an overview of nanoparticles, with special reference to their mechanism of biosynthesis and types.

Silver: Silver nanoparticles have proved to be most effective because of its good antimicrobial efficacy against bacteria, viruses and other eukaryotic microorganisms. They are undoubtedly the most widely used nanomaterials among all, thereby being used as antimicrobial agents, in textile industries, for water treatment, sunscreen lotions etc. Studies have already reported the successful biosynthesis of silver nanoparticles by plants such as *Azadirachta indica*, *Capsicum annuum* and *Carica papaya*.

Gold: Gold nanoparticles (AuNPs) are used in immunochemical studies for identification of protein interactions. They are used as lab tracer in DNA fingerprinting to detect presence of DNA in a sample. They are also used for detection of aminoglycoside antibiotics like streptomycin, gentamycin and neomycin. Gold nanorods are being used to detect cancer stem cells, beneficial for cancer diagnosis and for identification of different classes of bacteria.

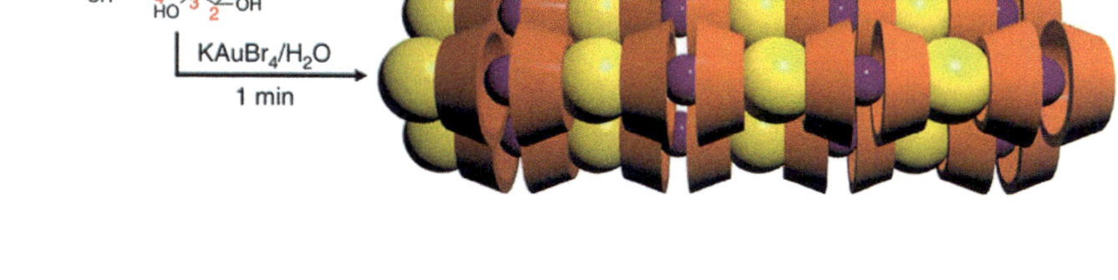

Alloy: Alloy nanoparticles exhibit structural properties that are different from their bulk samples. Since Ag has the highest electrical conductivity among metal fillers and, unlike many other metals, their oxides have relatively better conductivity, Ag flakes are most widely used. Bimetallic alloy nanoparticles properties are influenced by both metals and show more advantages over ordinary metallic NPs.

Magnetic: Magnetic nanoparticles like Fe_3O_4 (magnetite) and Fe_2O_3 (maghemite) are known to be biocompatible. They have been actively investigated for targeted cancer treatment (magnetic hyperthermia), stem cell sorting and manipulation, guided drug delivery, gene therapy, DNA analysis, and magnetic resonance imaging (MRI).

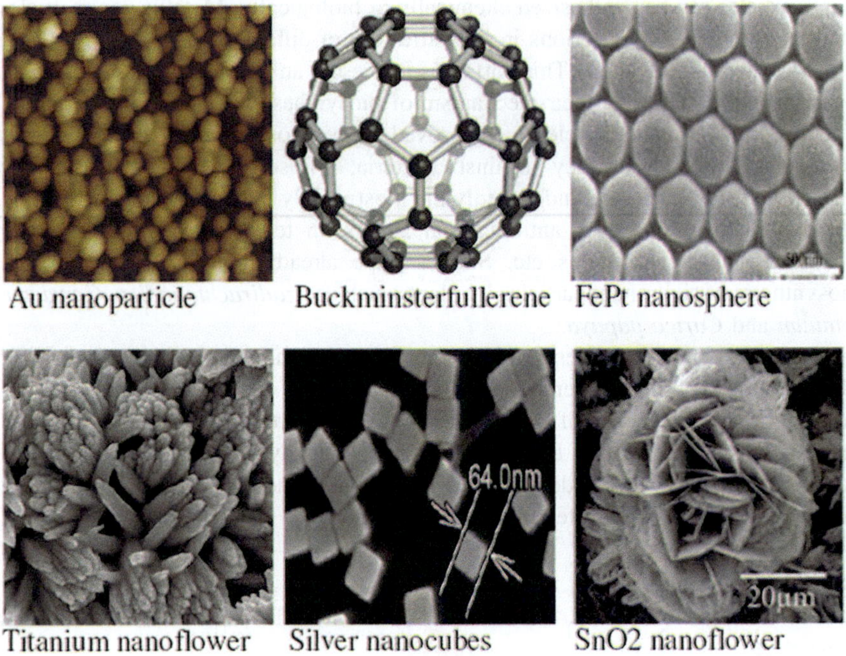

Tungstic acid gel

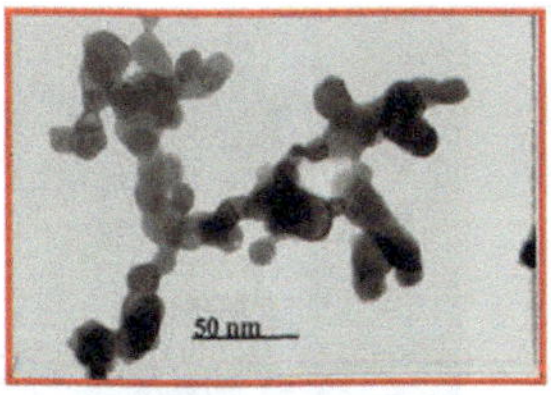

Elemental tungsten

W-Mo Solid Solution

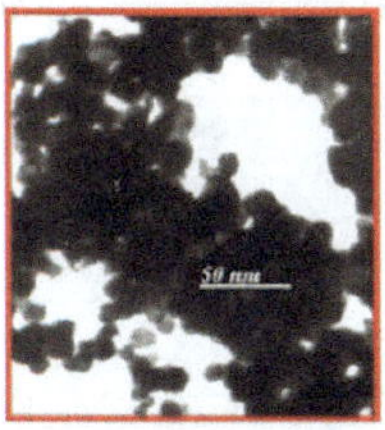

W-Ta Solid Solution

W-Re alloy

W-Ni alloy

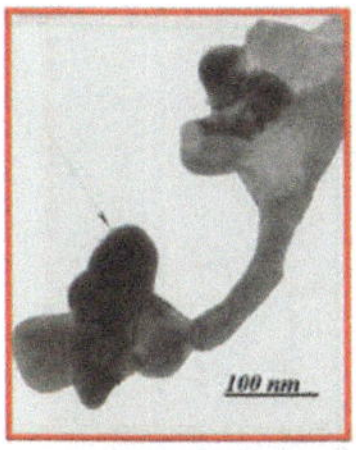

Cu-coated W

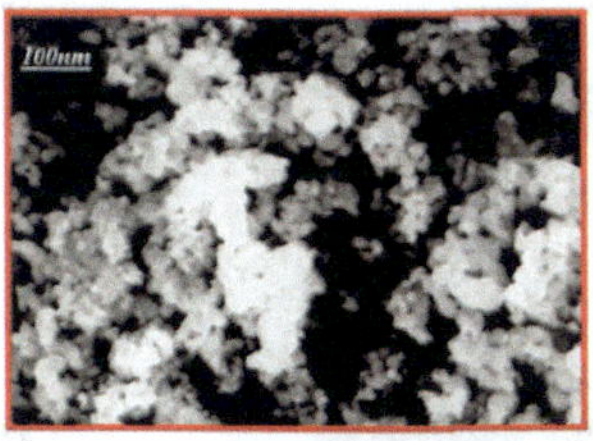

WC-Co

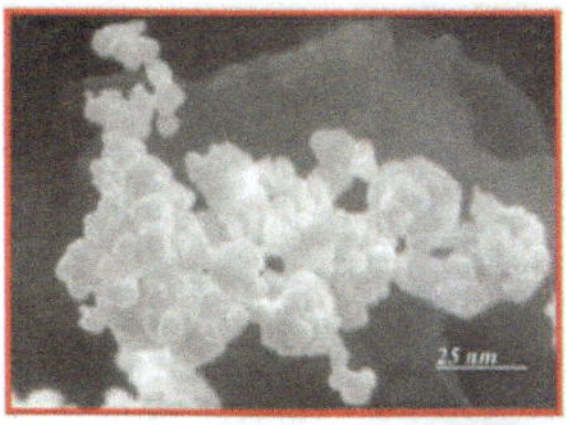

AlN

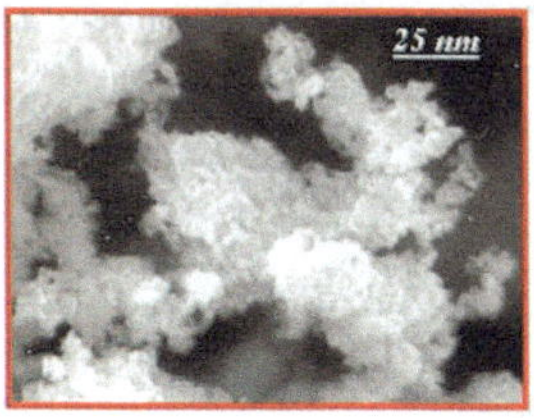

B_4C

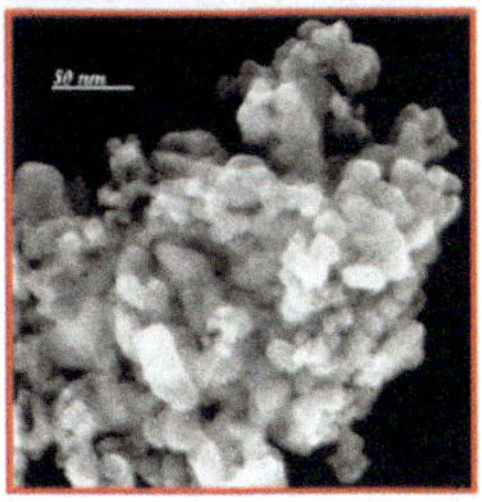

TiB_2

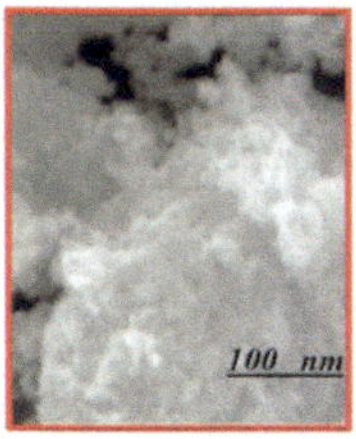

SiC

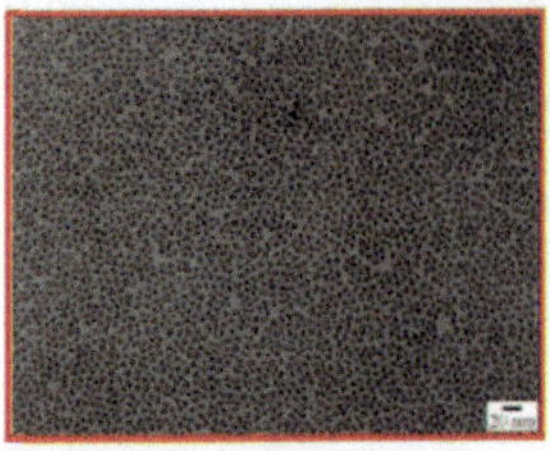

Fe

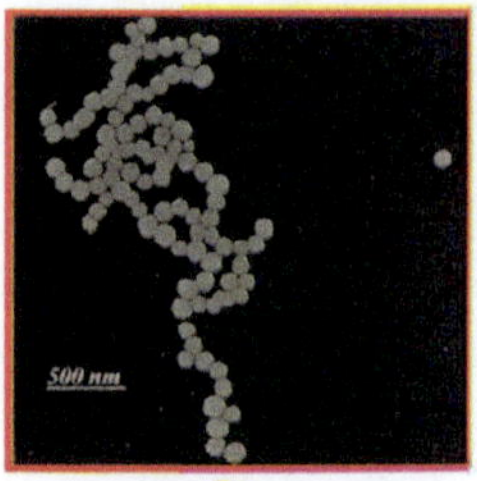

Co

Co

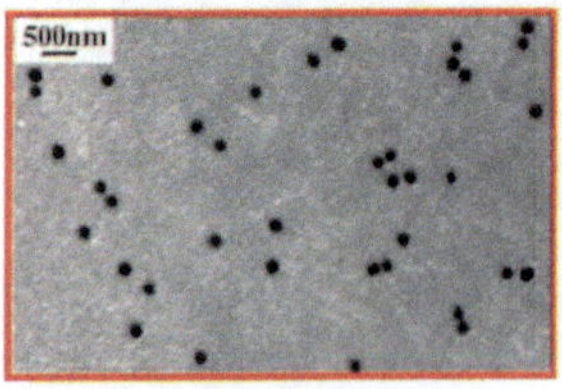

Fe-Co Alloy

CoPt Alloy

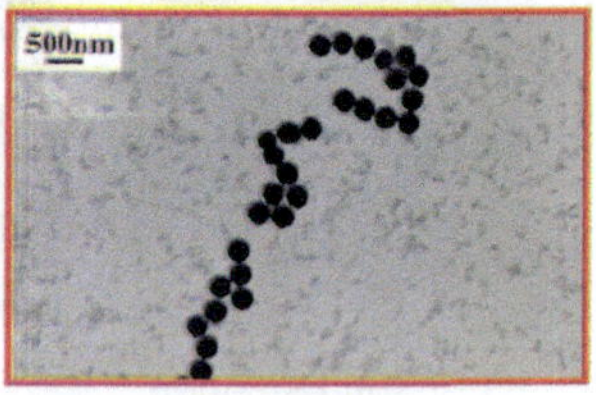

Fe-Co Alloy

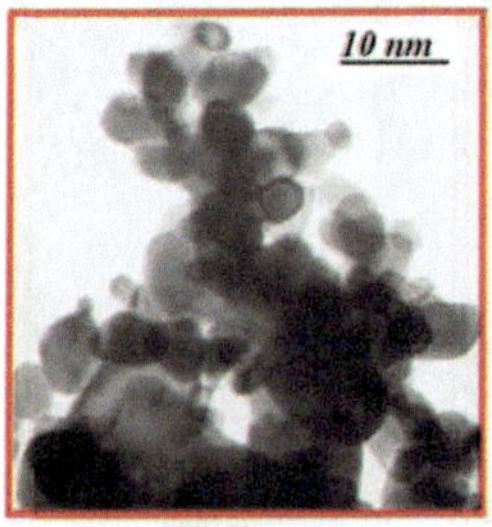

V_2O_5

TiO_2

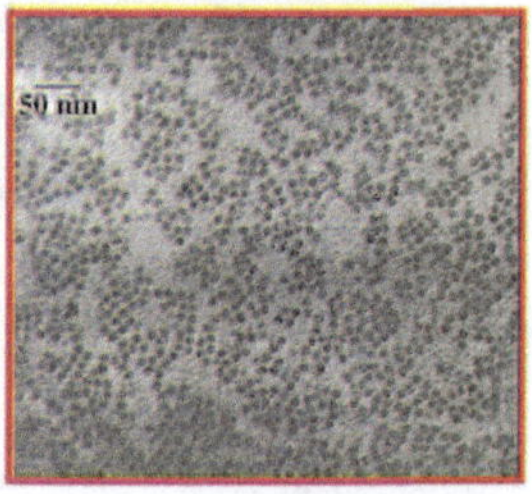

γ-Fe_2O_3

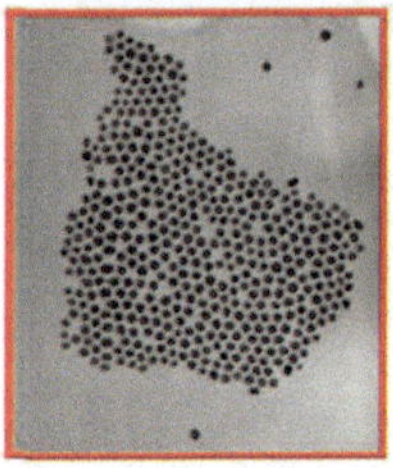

Nano Gold (8-10 nm)

Talc

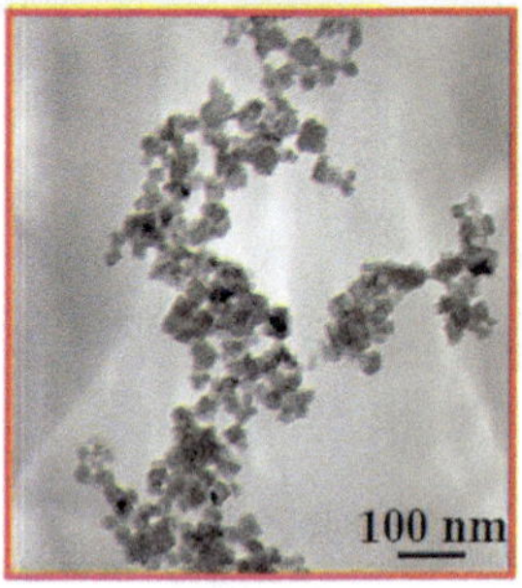

$CaCO_3$

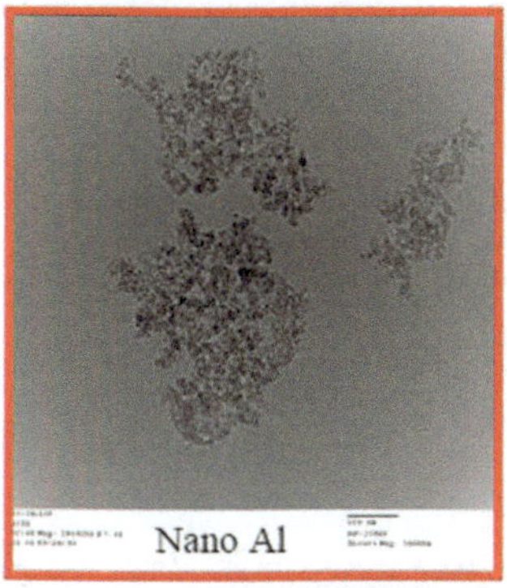
Nano Al

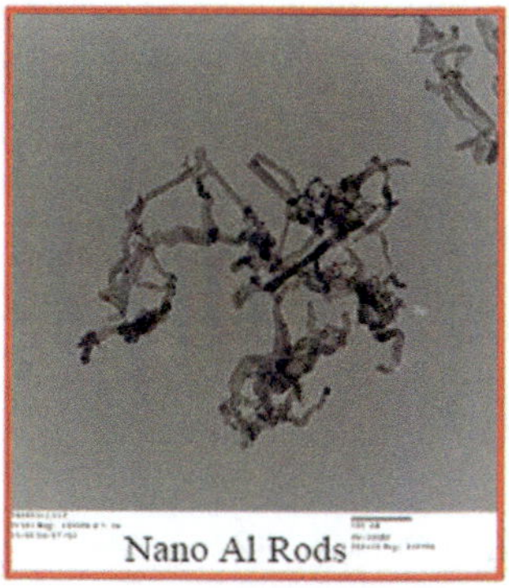
Nano Al Rods

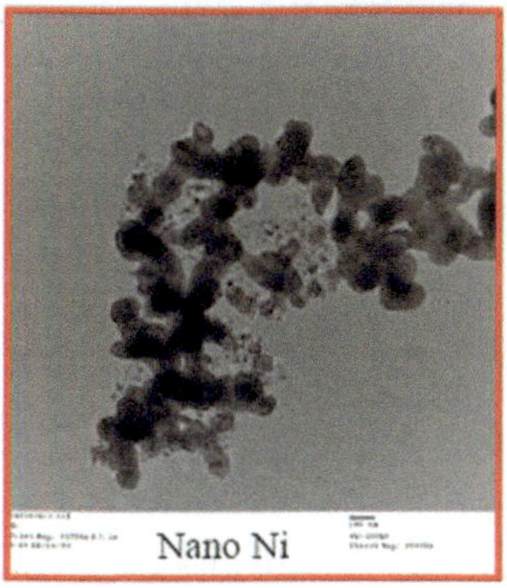
Nano Ni

Nano Re

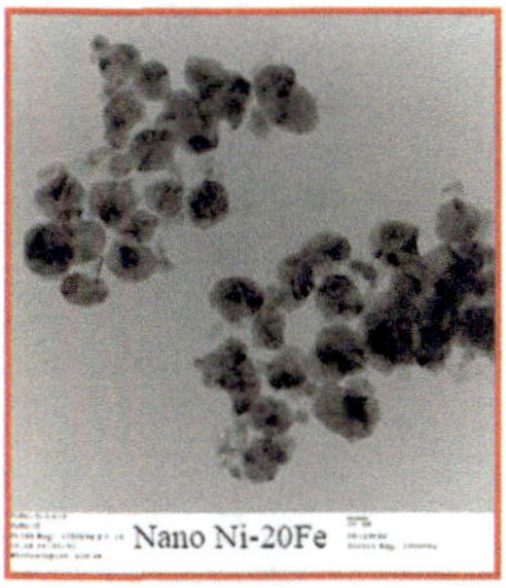
Nano Ni-20Fe

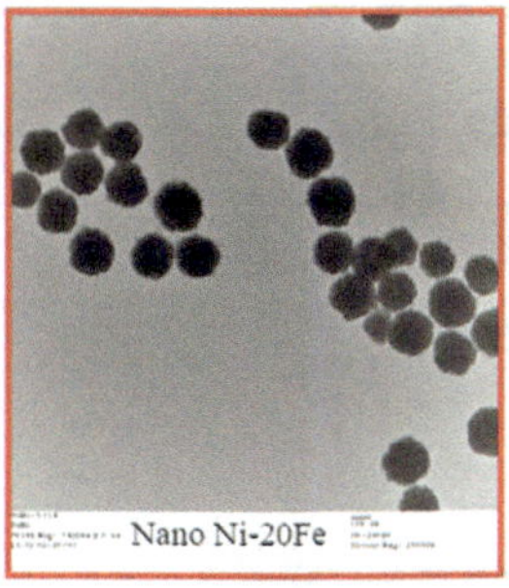
Nano Ni-20Fe

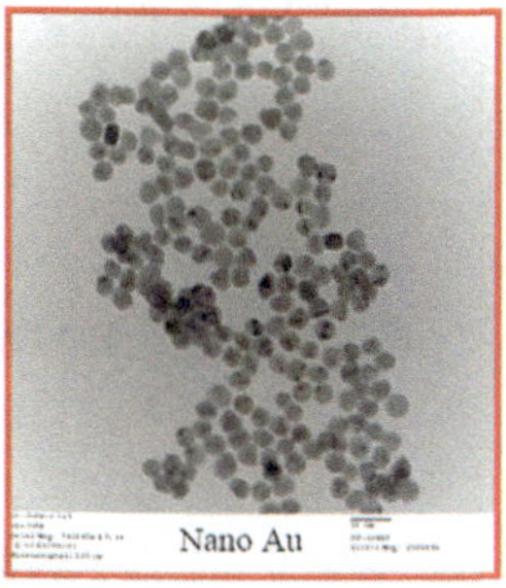
Nano Au

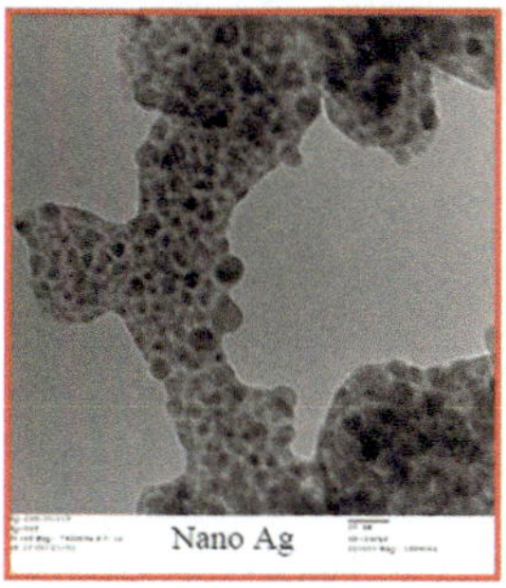
Nano Ag

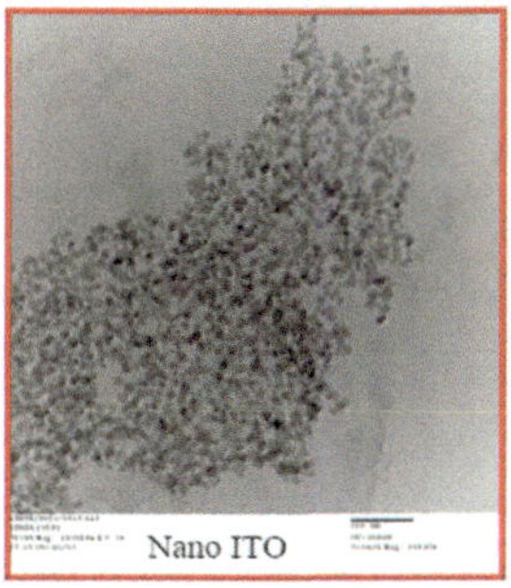
Nano ITO

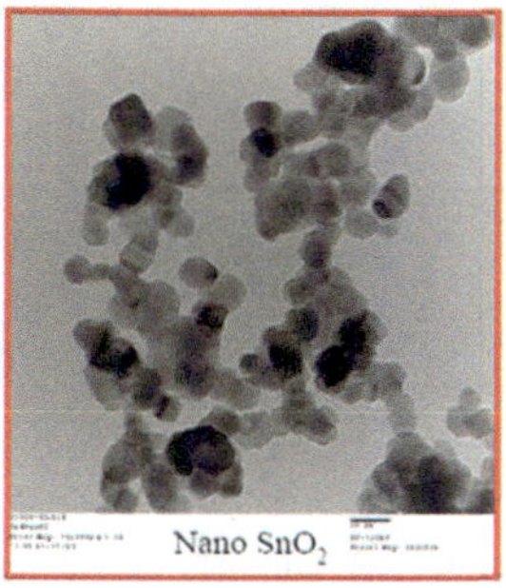
Nano SnO_2

Fig. 6.5 A summary of different shapes that have been achieved for various metal nanocrystals with a variety of morphologies. Reproduced with permission from [37]. Copyright 2008, Wiley

6.8.1 Example of a Phase Contrast

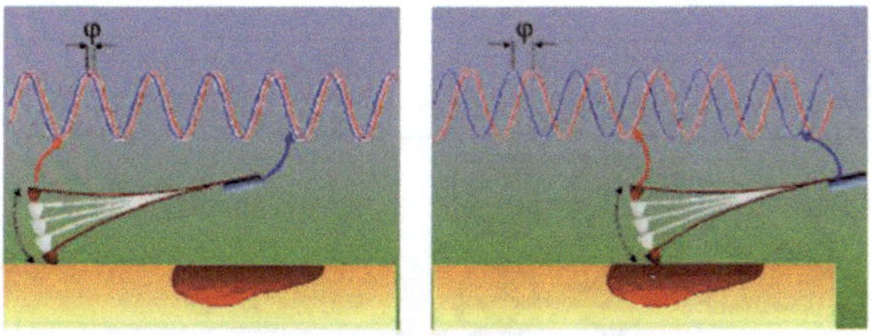

$$\sin(\varphi) = \frac{\omega A(\omega)}{\omega_0 A(\omega_0)} + \frac{Q E_{ps}}{\pi k A_0 A(\omega)}$$

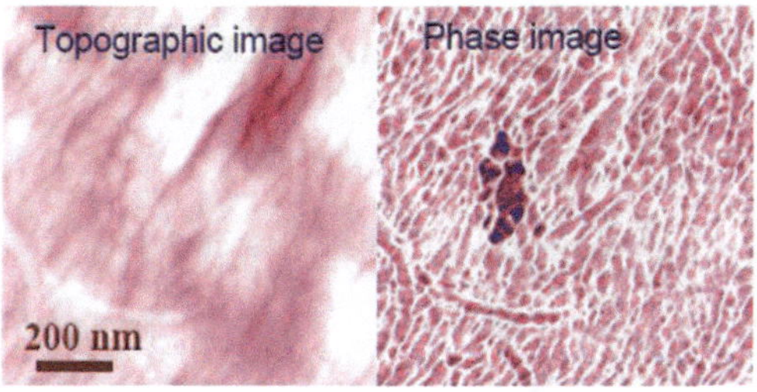

6.8.2 Summary of Different Shapes for Various Metal Nanocrystals

Structures	Shapes	Schematic drawings	Metals
single-crystal	perfect/truncated cube[a]		Pd, Ag, Au, Pt, Cu, Rh, Bi, Fe
	perfect/truncated octahedron[a]	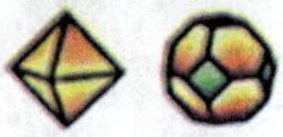	Pd, Ag, Au, Pt
	perfect/truncated tetrahedron[a]		Ag, Au, Pt, Rh
	rectangular bar		Pd, Ag, Pt
	octagonal rod		Pd, Au, Fe, Co, Ni
	rectangular or octagonal wire		Pb, In, Sn, Sb, Fe, Co
singly twinned	right bipyramid		Pd, Ag
	beam		Ag
multiply twinned	decahedron[a]		Pd, Ag, Au
	icosahedron[a]		Pd, Au
	five-fold twinned pentagonal rod		Pd, Ag, Au, Cu
	five-fold twinned pentagonal wire	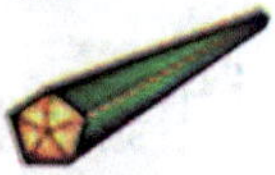	Ag, Au, Cu
	triangular/hexagonal plate	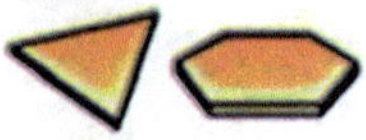	Pd, Ag, Au, Cu, Pb, Bi, Co, Ni
	disc		Sn, Co

References

1. Iijima, S.: Nature **354**, 56 (1991)
2. Iijima, S., Ichihashi, T.: Nature **363**, 603 (1993)
3. Journet, C., Maser, W.K., Bernier, P., Loiseau, A., Lamy de la Chappelle, M., Lefrant, S., Deniard, P., Lee, R., Fischer, J.E.: Nature **388**, 756 (1997)
4. Rinzler, A.G., Liu, J., Dai, H., Nikolaev, P., Huffman, C.B., Rodriguez-Macias, F.J., Boul, P. J., Lu, A.H., Heymann, D., Colbert, D.T., Lee, R.S., Fischer, J.E., Rao, A.M., Eklund, P.C., Smalley, R.E.: Appl. Phys. A **67**, 29 (1998)
5. Nikolaev, P., Bronikowski, M.J., Bradley, R.K., Rohmund, F., Colbert, D.T., Smith, K.A., Smalley, R.E.: Chem. Phys. Lett. **313**, 91 (1999)
6. Ren, Z.F., Huang, Z.P., Xu, J.W., Wang, D.Z., Wen, J.G., Wang, J.H.: Appl. Phys. Lett. **75**, 1086 (1999)
7. Ren, Z.F., Huang, Z.P., Xu, J.W., Wang, J.H., Bush, P., Siegal, M.P., Provencio, P.N.: Science **282**, 1105 (1998)
8. Huang, Z.P., Xu, J.W., Ren, Z.F., Wang, J.H., Siegal, M.P., Provencio, P.N.: Appl. Phys. Lett. **73**, 3845 (1998)
9. Bethune, D.S., Kiang, C.H., Devries, M.S., Gorman, G., Savoy, R., Vazquez, J.: Nature **363**, 605 (1993)
10. Shi, Z., Lian, Y., Liao, F.H., Zhou, X., Gu, Z., Zhang, Y.: J. Phys. Chem. Solids **61**, 1031 (2000)
11. Saito, Y., Nishikubo, K., Kawabata, K., Matsumoto, T.: J. Appl. Phys. **80**, 3062 (1996)
12. Thess, A., Lee, R., Nikolaev, P., Dai, H.J., Petit, P., Robert, J.: Science **273**, 483 (1996)
13. Zhang, Y., Iijima, S.: Appl. Phys. Lett. **75**, 3087 (1999)
14. Ge, M., Sattler, K.: Appl. Phys. Lett. **64**, 710 (1994)
15. Che, G., Lakshmi, B.B., Martin, C.R., Fisher, E.R., Ruoff, R.S.: Chem. Mater. **10**, 260 (1998)
16. Li, W.Z., Xie, S.S., Qian, L.X., Chang, B.H., Zou, B.S., Zho, W.Y.: Science **274**, 1701 (1999)
17. Zhang, X.X., Li, Z.Q., Wen, G.H., Fung, K.K., Chen, J., Li, Y.: Chem. Phys. Lett. **333**, 509 (2001)
18. Dreaden, E.C., Alkilany, A.M., Huang, X., Murphy, C.J., El-Sayed, M.A.: The golden age: gold nanoparticles for biomedicine. Chem. Soc. Rev. **41**, 2740–2779 (2012). https://doi.org/10.1039/C1CS15237H
19. Sigmund, W., Yuh, J., Park, H., Maneeratana, V., Pyrgiotakis, G., Daga, A., Taylor, J., Nino, J.C.: Processing and structure relationships in electrospinning of ceramic fiber systems. J. Am. Ceram. Soc. **89**, 395–407 (2006). https://doi.org/10.1111/j.1551-2916.2005.00807.x
20. Thomas, S., Harshita, B.S.P., Mishra, P., Talegaonkar, S.: Ceramic nanoparticles: fabrication methods and applications in drug delivery. Curr. Pharm. Des. **21**, 6165–6188 (2015). https://doi.org/10.2174/1381612821666151027153246
21. Ali, S., Khan, I., Khan, S.A., Sohail, M., Ahmed, R., Rehman, A., Ur Ansari, M.S., Morsy, M.A.: Electrocatalytic performance of Ni@Pt core–shell nanoparticles supported on carbon nanotubes for methanol oxidation reaction. J. Electroanal. Chem. **795**, 17–25 (2017). https://doi.org/10.1016/j.jelechem.2017.04.040
22. Khan, I., Abdalla, A., Qurashi, A.: Synthesis of hierarchical WO_3 and Bi_2O_3/WO_3 nanocomposite for solar-driven water splitting applications. Int. J. Hydrogen Energy **42**, 3431–3439 (2017). https://doi.org/10.1016/j.ijhydene.2016.11.105
23. Sun, S.: Monodisperse FePt nanoparticles and ferromagnetic FePt nanocrystal superlattices. Science **80**(287), 1989–1992 (2000). https://doi.org/10.1126/science.287.5460.1989
24. Hisatomi, T., Kubota, J., Domen, K.: Recent advances in semiconductors for photocatalytic and photoelectrochemical water splitting. Chem. Soc. Rev. **43**, 7520–7535 (2014). https://doi.org/10.1039/c3cs60378d

25. Mansha, M., Khan, I., Ullah, N., Qurashi, A.: Synthesis, characterization and visible-light-driven photoelectrochemical hydrogen evolution reaction of carbazole-containing conjugated polymers. J. Hydrogen Energy, Int (2017). https://doi.org/10.1016/j.ijhydene.2017.02.053
26. Rao, J.P., Geckeler, K.E.: Polymer nanoparticles: preparation techniques and size-control parameters. Prog. Polym. Sci. **36**, 887–913 (2011). https://doi.org/10.1016/j.progpolymsci.2011.01.001
27. Abd Ellah, N.H., Abouelmagd, S.A.: Surface functionalization of polymeric nanoparticles for tumor drug delivery: approaches and challenges. Expert Opin. Drug Deliv. 1–14 (2016). http://dx.doi.org/10.1080/17425247.2016.1213238
28. Abouelmagd, S.A., Meng, F., Kim, B.-K., Hyun, H., Yeo, Y.: Tannic acid-mediated surface functionalization of polymeric nanoparticles. ACS Biomater. Sci. Eng. 6b00497 (2016). https://doi.org/10.1021/acsbiomaterials.6b004
29. Rawat, M.K., Jain, A., Singh, S., Mehnert, W., Thunemann, A.F., Souto, E.B., Mehta, A., Vyas, S.P.: Studies on binary lipid matrix based solid lipid nanoparticles of repaglinide: in vitro and in vivo evaluation. J. Pharm. Sci. **100**, 2366–2378 (2011). https://doi.org/10.1002/jps.22435
30. Mashaghi, S., Jadidi, T., Koenderink, G., Mashaghi, A., Lipid nanotechnology. Int. J. Mol. Sci. **14**, 4242–4282 (2013). https://doi.org/10.3390/ijms14024242
31. Puri, A., Loomis, K., Smith, B., Lee, J.-H., Yavlovich, A., Heldman, E., Blumenthal, R.: Lipid-based nanoparticles as pharmaceutical drug carriers: from concepts to clinic. Crit. Rev. Ther. Drug Carrier Syst. **26**, 523–580 (2009)
32. Gujrati, M., Malamas, A., Shin, T., Jin, E., Sun, Y., Lu, Z.-R.: Multifunctional cationic lipid-based nanoparticles facilitate endosomal escape and reduction-triggered cytosolic siRNA release. Mol. Pharm. **11**, 2734–2744 (2014). https://doi.org/10.1021/mp400787s
33. Nanoparticles, Gunter Schmidt Edition. Wiley (2004)
34. Lewis, L.N.: Chem. Rev. (Washington, D.C.) **93**, 2693 (1993)
35. Iqbal, S., Bashir, R.: Nanopores: Sensing and Fundamental Biological Interactions. Springer (2011)
36. Brongersma, M.L.: Nanoshells: gifts in a gold wrapper. Nat. Mater. **2**, 296–297 (2003)
37. Xia, Y., Xiong, Y., Lim, B., Skrabalak, S.E.: Shape-controlled synthesis of metal nanocrystals: simple chemistry meets complex physics. Angew. Chem. Int. Ed. Engl. **48**, 60–103 (2009)

Chapter 7
Carbon Nanomaterials and Two-Dimensional Transition Metal Dichalcogenides (2D TMDCs)

Carbon nanostructures are a leading material in the nanotechnology field. The discovery and research of carbon materials has considerably contributed to the advancement of modern day science and technology. After the discovery of fullerene and single walled carbon nanotube (CNT), which are zero-dimensional and one-dimensional carbon nanomaterials respectively, the researchers have tried to isolate 2D graphitic material or to make 1D nano-ribbons from 2D crystals. The efforts have started to pay off since 2004 with the first isolation and electrical characterization of graphene transistor reported by Geim and Novoselov [1]. Most of these nanostructures are compiled exclusively of sp^2 hybridized carbon atoms establishing several structures, from the soccer ball like fullerene (C_{60}) to the single atom thick plane of carbon called graphene, to the "rolled up", tubular sheets of graphene: carbon nanotubes as shown in Fig. 7.1 [1–4]. The physicochemical properties of these graphene like structures, even though their being composed of sp^2 carbon, are as different as their atomic structure. From the earliest theoretical investigation into the electronic properties of graphene and graphite, it has become clearer that graphite has some unusual properties [5]. Later, many more properties of graphite, graphene and CNT have come to light, for example the high charge carrier mobility of graphene and CNT [6–9].

2D materials in nano and micro form with 1D restricted that are in close contact with each other to form boundaries. 2D nanosheets and quantum features in their optical and electronic properties can be prepared both uniform and equivalent in thickness and size by several approaches presently used in material synthesis. It was not until 2004 that physicists were able to first isolate individual graphene planes using adhesive tape to obtain flakes that exhibited unique electronic properties [10]. The discovery has led to a great interest in the science community to study the large scale production of 2D materials. Van der Waals materials can be chemically and mechanically exfoliated into single or few layers which was shown in the last decade [11]. This preliminary work has motivated many researchers to obtain and distinguish these layers to unveil new properties. In the field of graphene research, several methods are reported for preparing, detecting, transferring, characterizing

L. H. Madkour, *Nanoelectronic Materials*, Advanced Structured Materials 116, https://doi.org/10.1007/978-3-030-21621-4_7

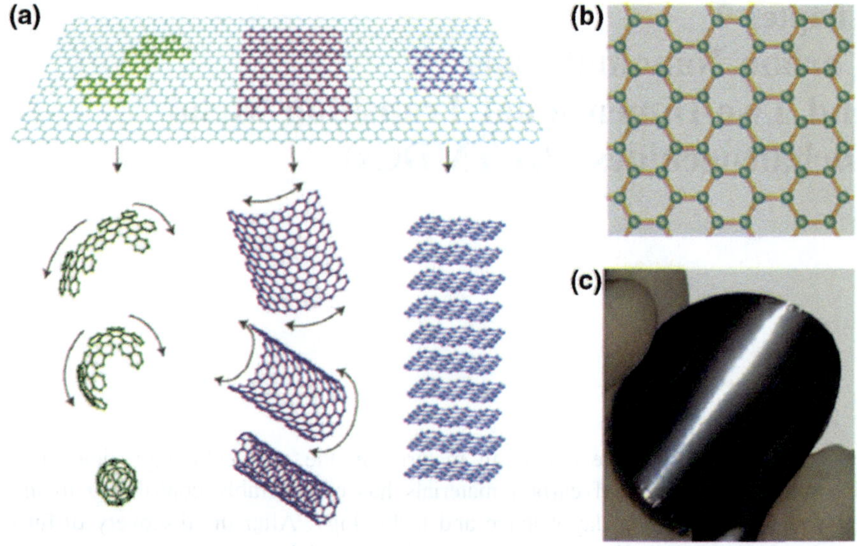

Fig. 7.1 **a** A C_{60} fullerene molecule, a carbon nanotube and graphite are in the order from left to right. Graphene, a single sheet of graphite, can be considered as building block of all these carbon structure (image produced from Ref. [2]). **b** Honey-comb lattice structure. **c** Dark shiny appearance of graphene (**b**, **c** images produced from Ref. [3])

and controlling the properties of 2D layered materials [12, 13]. Additionally, new synthetic techniques including solvothermal, solution based and epitaxial methodologies have allowed the possible ways to produce new single- or few layers and van der Waals materials. These well-known techniques have allowed the field of 2D materials beyond graphene to establish rapidly. Soon after these discoveries, many graphene's "cousins" had been suggested that 2D nanosheets of groups IV and II–VI elements, such as silicone [14, 15] and germanene [16, 17] might demonstrate similar properties as graphene. In addition, these 2D semiconductor materials are very useful structural agents that can be modified and combined into composites for an extensive application.

Physicochemical properties and applications of graphene based materials and reduced graphene oxide had been reported earlier [18–21]. After the discovery of graphene, many researchers would not realize that more than a dozen 2D materials can be isolated and studied in less than 10 years. Progress in graphene research had a spillover effect by way of leading to an interest in other 2D materials like for example metal nitrides and carbides. 2D nanostructures are one of the greatest widely studied materials because of the unique physical singularities that happen when heat transport and charge is confirmed to a plane. Due to the unique properties, 2D nanostructures are anticipated to have an important influence on a huge diversity of applications, extending from high performance sensors, electronics to gas separation or storage, catalysis, inert coatings and support membranes, etc. Right now, the 2D materials family includes not just carbon material but also transition metal dichalcogenides (TMDs), and layered metal oxides. One of the most promising applications of 2D materials is in electronic devices [22, 23]. With respect to the electrical properties, one can fabricate a new generation of superconductors, metallic materials, semimetals, semiconductors, insulators from the 2D materials.

Even though graphene is the last one to be isolated in the carbon materials family, it serves as the building block for other family members with different dimensionalities. In particular, single layers of TMDs have attracted notable attention because of their diverse properties and natural abundance.

Despite the similarity in the chemical formula, MX_2, where typically M is transition metal of groups 4–10 (typically Mo, Nb, W, Ni, V, or Re) [24, 25] and X is a chalcogen (typically Se, Te, or S) [26–28]. While the bonding of the tri-layers are covalent, neighboring sheets are bonded via Van der Waals interactions to grow 3D crystals. There are currently 40 different types of combination of TMDs with different chalcogen atoms. Based on the co-ordination and oxidation state of the metal atoms, TMDs ranging from insulators such as HfS_2, semiconductors such as MoS_2, semimetal such as $TiSe_2$, to true metals such as $NbSe_2$, which can even exhibit superconductivity at low temperature [29–31]. Besides, a solid attention in group-IV 2D oxides has newly developed. These different types of nanosheets whether oxides or TMDs, have been studied experimentally or computationally to understand their electronic structure and compared with graphene [13].

Table 7.1 lists all the current members in the 2D layered materials family [11]. However, stability is a critical issue. The blue shaded materials are constant under ambient environments (room temperature in air) for monolayers. Perhaps green shaded materials are stable in air, but those may be steady only in inert atmosphere are shaded pink. Gray shading means monolayer has been exfoliated, but no further information has yet been provided other than atomic force microscope (AFM). Still, there is a complete periodic table of crystalline materials individually having various properties such as mechanical, transport and electronic properties, and the probability to make single- or several-atom polyhedral dense 2D sheets from any

Table 7.1 2D materials family (table produced from Ref. [29])

<table>
<tr><td>Graphene family</td><td colspan="2">Graphene</td><td>hBN “white graphene”</td><td>BCN</td><td>Fluorographene</td><td>Graphene oxide</td></tr>
<tr><td>2D chalcogenides</td><td colspan="2">MoS_2, WS_2, $MoSe_2$, WSe_2</td><td colspan="2">Semiconducting dichalcogenides: $MoTe_2$, WTe_2, ZrS_2, $ZrSe_2$, and so on</td><td colspan="2">Metallic dichalcogenides: $NbSe_2$, NbS_2, TaS_2, TiS_2, $NiSe_2$ and so on
Layered Semiconductors: GaSe, GaTe, InSe, Bi_2Se_3 and so on</td></tr>
<tr><td rowspan="2">2D oxides</td><td>Micas, BSCCO</td><td>MoO_3, WO_3</td><td colspan="2" rowspan="2">Perovskite-type: $LaNb_2O_7$, $(Ca,Sr)_2Nb_3O_{10}$, $Bi_4Ti_3O_{12}$, $Ca_2Ta_2TiO_{10}$ and so on</td><td colspan="2">Hydroxides: $Ni(OH)_2$, $Eu(OH)_2$ and so on</td></tr>
<tr><td>Layered Cu oxides</td><td>TiO_2, MnO_2, V_2O_5, TaO_3, RuO_2 and so on</td><td colspan="2">others</td></tr>
</table>

material remains. Furthermore, this layered 2D materials exhibits a combination of excellent electronic [32], mechanical [33], optical [34], and thermal properties [35], which may substitute silicon in electronics, photonics, and nanoelectromechanical system (NEMS) as well in this century or next. This is most researchers dream about 2D materials into real life applications.

7.1 Classification of 2D Materials

In order to accomplish the reliable synthesis, detailed understanding of the properties and induction in a multitude of applications, it is essential to understand the 2D materials structure composed of single atom thick or single-polyhedral-thick layer. Therefore, 2D materials are categorized in three classes.

7.1.1 Layered van der Waals Solids

These types of materials are the front-runner and the most common material to exfoliate mono and few layered nanosheets. These materials exhibit strong in-plane covalent or ionic bonding and weak out of the plane van der Waal or hydrogen bonding. The weak nature of van der Waal bonding ($\sim$40–70 meV) and surface tension (60–90 mJ/m^2) in the range of common solvents used for graphite exfoliation facilitate the exfoliation of bulk parent crystals into nanosheets either by micromechanical exfoliation or liquid exfoliation [36, 37]. Once exfoliated these nanosheets possess a high surface area and aspect ratio. These exfoliated nanosheets are usually few lm in lateral size and less than 1 nm in thickness. Transition metal dichalcogenides (TMDs) (MX_2 (X–M–X layer) where M = Mo, Bi, Ta, V, Ti, Nb, W, Hf, Zr and X = S, Se Te) are a well-studied system for layering van der Waals solids, especially MoS_2, $MoSe_2$ and WS_2. Currently over 40 different TMDs have been reported in layered form [38]. Exotic properties of single and few layer derivatives in term of direct and indirect band gap, Dirac point, electrochemical and mechanical behavior have emerged them as 21st century material having potential in numerous applications such as semiconductor [22], hydrogen production [39, 40], thermo-electric devices [41] etc. In TMDs, transition metal occupies octahedral coordinates (for d0, d3, and some d1 metals) and trigonal prismatic coordinates (for d1 and d2 metals) in IT and 2H crystal structure, respectively. These coordinates share edges with nearest neighbors within each layer and form hexagonal honeycomb like structure. *h*-BN [42], SiC [43], vanadium oxide [44] and Sb_2Te_3 [45] are some other member of this class.

7.1.2 Layered Ionic Solids

Layered ionic compounds constitute another important class of layered 2D materials that have been gaining popularity. These types of solids consist of a charged polyhedral layer sandwiched between hydroxide or halide layers by electrostatic forces. Single and few layered materials can be easily exfoliated by ion intercalation or ion exchange liquid exfoliation method. Adsorption of bulkier foreign molecules and exchange of parent ions with heavier ions increases the spacing between the basal planes and weakens the basal plane bonding. Exfoliated europium hydroxide is a classic example of layered ionic solids [46]. Perovskite type oxides such as $La_{0.90}Eu_{0.05}Nb_2O_7$ [47], $K_{1.5}Eu_{0.5}Ta_3O_{10}$ [48], $KLnNb_2O_7$, $K_2Ln_2Ti_3O_{10}$, and $RbLnTa_2O_7$ (Ln: Lanthanide Ion) [49], $KCa_2Nb_3O_{10}$ [50, 51], metal hydroxides such as $Eu(OH)_{2.5}(DS)_{0.5}$ [46], $Co_{2/3}Fe_{1/3}(OH)_2^{1/3+}$ [52] are some examples of layered materials exfoliated from the ion-exchange method. The measured thickness of monolayers is usually higher than the crystallographic thickness due to the adsorption of solvent and bulkier exchanged ions.

7.1.3 Surface Assisted Nonlayered Solids

This class of atomically thin layered material is artificially synthesized on a substrate via chemical vapor deposition and epitaxial growth. Silicene, an analogue of graphene, is a perfect example of surface assisted nonlayered solids. Low buckled hexagonal structure of Si atoms has been deposited on Ag(111) [53], Ir(111) [54] and a thin film of ZrB_2 on silicon wafer [55]. Unlike graphene, instability of silicene at the ambient condition has limited its potential in electronic and optoelectronic applications. Silicene has been discussed in detail Al_2O_3/SiO_2 [56], $Cu_2N/Cu(100)$ [57], MgO/Mo(001) [58], TiO_2 [59]; and Ge/Ag(100) and Ag(111) [60–64] nanosheets are grown for application in energy storage, field effect transistor (FET), controlling the charge density of nanoparticles. In Fig. 7.2, we show the glimpse of atomic arrangements in different classes of 2D materials (MoS_2, $La_{0.90}Eu_{0.05}Nb_2O_7$, silicene and MnO_2).

7.2 2D Materials, Their Properties, and Applications

As we know that the bulk layered TMDs have strong intralayer covalent M–X bonds and weak interlayer van der Waals interactions, which may be cleaved to single- or few-layer TMDs. Whether these materials are able to be exfoliated effortlessly depends on their surface energies. Surface energy is the energy required to create new surfaces when the layered crystals are peeled; peeling would introduce two new surfaces. The surface energy of MoS_2 and WS_2 (65–75 mJ m^{-2}

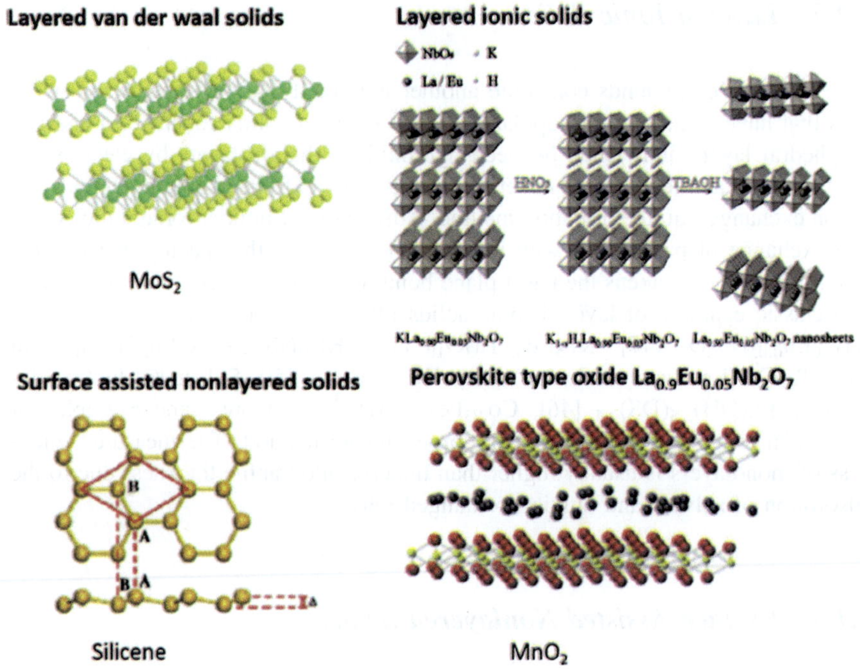

Fig. 7.2 Atomic arrangement in different classes of layered 2D materials. MoS_2 and MnO_2 from Ref. [65], layered ionic solids from Ref. [66] and silicene from Ref. [14]. Reprinted by permission of the American Association for the Advancement of Science and ACS Publishers

[36, 67]) are similar to the surface energy of graphene (65–120 mJ m^{-2} [68, 69]) making them potentially cleavable. Theoretical results also reveal the stability of single-layer TMD (summarized in Fig. 7.3) [70]. Among them, chromium (Cr), molybdenum (Mo) and tungsten (W) based layered TMDs are semiconductors. However, CrS_2 and $CrSe_2$ are not stable [71] and $CrTe_2$ has a small band gap [72] with triclinic structure [73]. Thus, only Mo and W dichalcogenides meet the requirements to form semiconductors. Furthermore, Mo and W dichalcogenides have sizable band gaps around 1–2 eV. The results from both experiments (marked by "E") and theoretic calculations (marked by "T") indicate that single- or few-layer Mo and W dichalcogenides are satisfactory for electronic devises. In addition, the abundance of Mo and W dichalcogenides also makes them possible for industrial application with a low cost.

Two dimensional materials have been investigated for over 50 years but since the isolation of graphene [74] in 2004 there has been growing interest in the field. There have been significant improvements in the identification [75] and production [76] of layered materials which opens up a range of applications including heterostructure components [77] and specialised electronic devices [78]. There are several reasons for the growing number of devices being engineered using these materials but chief amongst them is the control of electronic properties [79],

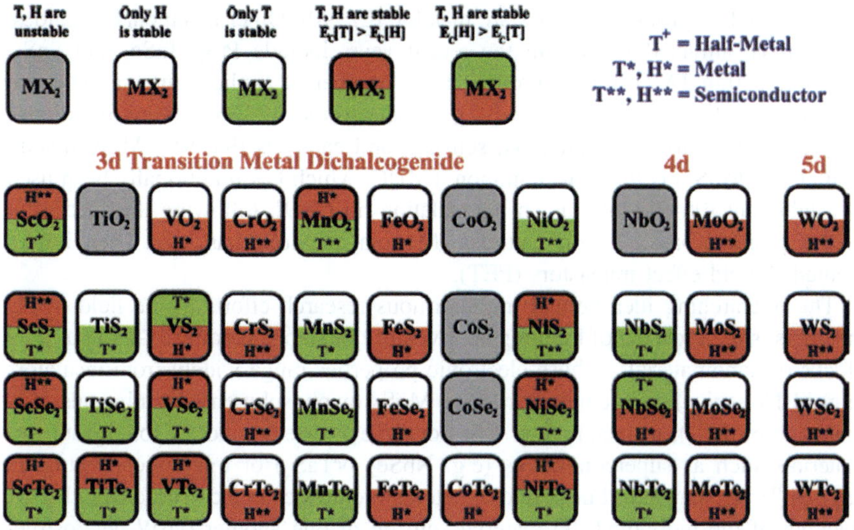

Fig. 7.3 Summary of the calculation results of stability of MX_2 (M = transition metal atoms, X = chalcogen atoms) compounds. (The lower-lying structure in each box (H or T) is the most stable phase and the upper-lying structure is the less stable phase. White color means only one phase is stable. Orange color indicates 1H phase and Green color indicates 1T phase. Gray color means neither 1T nor 1H phase is stable. "+" means half-metal; "*" means metal and "**" means semiconductor.) Reprinted from Ref. [70] with permission of ACS Publishers

confinement effects [80] and increased surface areas [81], all due to reduced dimensionality. Very often the thin film structures lead to flexible devices which have beneficial optical properties, including transparency [82, 83]. The change from bulk graphite to single layer graphene for example shows the extent to which dimensionality plays a role in electronic behaviour. Graphite is semi-metallic with a mobility of roughly 3,000 cm^2/V s at 77 K [84], whereas in graphene even at room temperature the mobility has been found to be around 10,000 cm^2/V s [85]. Much of this is due to the difference in electronic behaviour, whilst both are semi-metallic; in single layer graphene the band structure close to the Fermi level is linear and the electrons are described as massless Dirac fermions [86]. This has in turn led to some relatively unusual phenomena including a room temperature quantum Hall effect [85] and a minimum level of conductance [86]. This has been matched by mechanical properties that outperform steel [87] and thermal transport properties better than diamond [88].

TMDCs are a large group of crystals with a generic chemical formula of MX_2 (M = transition metal, X = S, Se, or Te), about two thirds of which are known to form layered structures [38]. Significant scientific knowledge of bulk TMDCs has been accumulated in 1960–80s, especially for the group 6 TMDCs, where M = Mo or W, and, following the success of graphene, their mono- and few-layer forms were soon produced [89]. Typically, group 4 or 6 TMDCs are semiconductors or insulators, while group 5 TMDCs are metallic [22, 90]. Other metal chalcogenides, which have attracted interest in the recent years include PtS_2, InSe, and SnS_2. Semiconducting TMDCs hold promise for a number of applications ranging from photovoltaics [91], photodetection [92], optoelectronics [22], light-emitting diodes [93], to energy storage/conversion, sensing, and catalysis [94, 95]. Molybdenum disulphide (MoS_2) is the most common TMDC, which has traditionally been used as a lubricant [96–98], and as an intercalation host [99], following the successful use of graphite-based Li-ion batteries. Monolayer MoS_2 has soon been implemented in field-effect transistors (FET).

The overarching idea behind the enormous research efforts in the field of 2D materials is the prospect of building an inventory of stable, atomically-thin building blocks of nanomaterials, whose electronic properties range widely from insulators (e.g. hBNor mica), semiconductors (e.g. MoS_2 or phosphorene), semimetals (e.g. graphene or TiS_2), to metals (e.g. VSe_2 or $PtTe_2$), or even more exotic groups of materials such as superconductors (e.g. $NbSe_2$ orTaS_2) or topological insulators (e.g. Bi_2Te_3 or stanene). Furthermore, heterostructures can be constructed either by stacking of the individual 2D materials on top of one another (vdW heterostructures), or through synthesis of different 2D materials in the same plane (lateral heterostructures), which opens a much wider world of artificial nanomaterials with diverse functionalities packed into a thickness of a few nm [100, 101].

The field of 2D materials is an ever-expanding research area, and the search for other 2D materials beyond graphene is not just limited to TMDs. New 2D materials such as silicene and phosphorene are strong contenders in the rapidly emerging realms of 2D materials [102, 103]. Several theoretical studies address the fundamental properties of these new 2D materials; however, experimental perspectives are still in their infancy due to stability issues.

Two-dimensional (2D) materials have, within just one decade, reshaped many disciplines of modern science and technology, both through intensive fundamental research and early commercial applications. Recent advances in atomically thin two-dimensional transition metal dichalcogenides (2D TMDs) have led to a variety of promising technologies for nanoelectronics, photonics, sensing, energy storage, and opto-electronics, to name a few. These nanomaterials are essential for the next generation of devices in tuneable electrochemistry, sensing, and photovoltaics.

This chapter is focus on group transition metal dichalcogenides (TMDCs) and other emerging 2D materials beyond graphene, including their heterostructures. This chapter reviews the recent progress in 2D materials beyond graphene and includes mainly transition metal dichalcogenides (TMDs) (e.g. MoS_2, WS_2, $MoSe_2$, and WSe_2). The preparation and applications of the heterostructures from various 2D layered materials are explained. These materials are finding niche applications for next-generation electronics and optoelectronics devices relying on ultimate atomic thicknesses. Albeit several challenges in developing scalable and defect-free TMDs on desired substrates, new growth techniques compatible with traditional and unconventional substrates have been developed to meet the ever-increasing demand of high quality and controllability for practical applications.

This chapter highlights the recent advances in the synthesis of large-scale and defect-free 2D TMDs. Moreover, we focus on the recent progress in electronic, opto-electronic, and electrochemical properties of newly studied TMDs with rational designs and new structures for potential applications in electronics, sensors, and energy storages. Additionally, we discuss recent breakthroughs in the newest families of 2D materials like silicene and phosphorene. The wide range of interesting properties and potential for use in emergent technologies suggest TMDs are likely to remain an important research area for years to come.

By use of density functional theory (DFT) this thesis will address some of the key areas of interest at this stage in the characterisation of single layer graphene, silicene, germanene and bilayer graphene and silicene.

1. Identifying the various geometric configurations of these structures and comparing and contrasting the energy differences between them. This along with the electronic band occupancy, i.e. bonding or anti-bonding states, and vibrational properties will identify the stable forms of these materials.
2. The analysis of electronic properties focuses on identifying any linear band structure and subsequent Fermi velocity; the presence of which would signify that many of the phenomena that are a product of this can be expected to arise. These are well known and outlined here in the case of graphene and can then be expected to be found in other materials where the electronic properties are similar.
3. The vibrational properties are investigated as a route to identification; where a combination of infrared and Raman spectroscopy techniques can be utilised, along with the calculated frequency of active modes, to establish distinct differences between materials that would provide a 'finger print'. This has been used as a highly successful non-invasive method of identifying graphene compared with graphite and could be extended to other two dimensional materials.
4. Where possible the electron-phonon coupling properties will be calculated, which is known to impact on carrier mobility and exciton relaxation. This will be based on calculations of the dynamical matrix elements.
5. There is also a fundamental issue with the current assumptions applied to the study of electron-phonon interactions; namely the Born-Oppenheimer approximation which is now known to be violated in graphene due to the pulsation time of vibrational modes and it is possible to perform a similar analysis on the materials.

The atomic structure of graphene was confirmed as being a honeycomb lattice of free standing 2D sheets. Although exhibiting a hexagonal symmetry the two carbon atoms in the unit cell, A and B, are each described as being part of a sublattice triangular periodicity. Figure 7.4 shows the real space geometry and the Fourier transformed k space where the primitive unit cell is now given by the first Brillouin zone in reciprocal space.

A full description of the Fourier transformation is given as well as a derivation of the band structure of the π and π^* orbitals following a tight binding method. Reciprocal space is also known as k-space or momentum space and it allows the

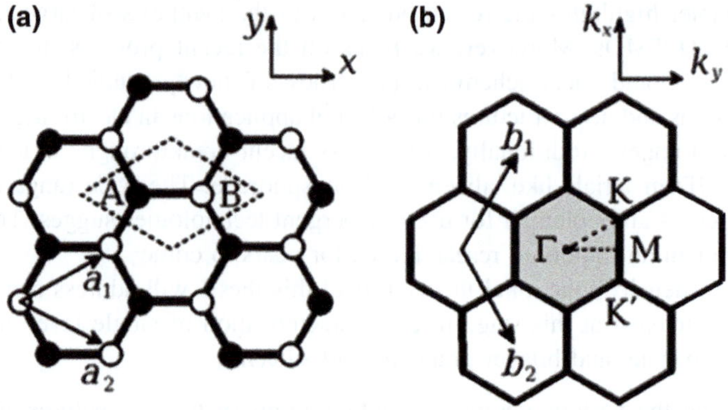

Fig. 7.4 Adapted from Ref. [104], the structure of graphene in **a** real space and **b** reciprocal space. The two carbon atoms of the primitive unit cell, A and B, correspond to a triangular sublattice structure and the first Brillouin zone is given by the shaded hexagon with high symmetry points Γ, K and M. a1 and a2 are the shortest lattice vectors between equivalent atomic points in the graphene sheet, vectors b1 and b2 are the Fourier transformation of the lattice constants in momentum space

representation of periodic systems in a reduced formalism. A primitive unit cell of graphene is given by the blue kite shape in Fig. 7.5 and is a real space geometry that if repeated periodically along each blue edge can reproduce the entire two dimensional material; for example Fig. 7.5 shows a 3 × 3 supercell of graphene.

Figure 7.5 shows the lattice vectors ($\vec{a_1}$) and ($\vec{a_2}$); which signify two of the vectors between equivalent points in the graphene sheet i.e. the primitive vectors of the hexagonal lattice. The lattice vectors can be expressed in Cartesian coordinates as:

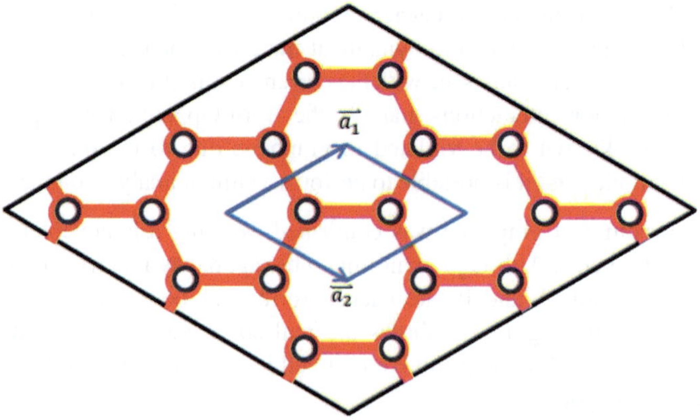

Fig. 7.5 Representation of the graphene lattice with the lattice vectors ($\vec{a_1}$) and ($\vec{a_2}$) given along with ($\vec{a_1}$) and ($\vec{a_2}$) vectors that complete the unit cell. The magnitude of $|\vec{a_1}| = |\vec{a_2}| = a$ which is the lattice constant for graphene = 2.46 Å

$$\vec{a_1} = a\ \cos\ 30\hat{x} + a\ \sin\ 30\hat{y} = \frac{\sqrt{3}}{2} a\hat{x} + \frac{a}{2}\hat{y}$$

$$\vec{a_2} = a\ \cos\ 30\hat{x} + a\ \sin\ 30\hat{y} = \frac{\sqrt{3}}{2} a\hat{x} - \frac{a}{2}\hat{y}$$

where a is the lattice constant = 2.46 Å for graphene and ^^ are unit vectors. We can then apply standard rules for the transformation to reciprocal lattice vectors $\vec{b_1}$ and $\vec{b_2}$.

The heterostructures formed by alien semiconductors play an important role in modern semiconductor industry, and they have been widely used as an essential building block for electronic devices. A heterojunction can be formed by interfacing two different semiconductors, where the electronic band structure near the interface will be changed according to electrostatics. The semiconductor heterojunctions have been applied in many solidstate devices, such as solar cells, photo detectors, semiconductor lasers, and light-emitting diodes (LEDs). Since the great influence of the semiconductor junction to our daily life, Nobel Prize in physics was awarded to Zhores I. Alferov, Herbert Kroemer and Jack S. Kilby for their contribution toward the development of semiconductor heterostructures [105, 106] and to Isamu Akasaki, Hiroshi Amano, and Shuji Nakamura for the invention of blue LEDs [107, 108].

The great success of graphene has been followed by an equally impressive surge for the development of other 2D materials that can form atomic sheets with extraordinary properties. Interestingly, the 2D library grows every year and feature more than 150 exotic layered materials that can be easily split into a subnanometer-thick materials [37, 109, 110]. These include 2D TMDs (e.g. molybdenum disulfide (MoS_2), molybdenum diselenide ($MoSe_2$), tungsten disulfide (WS_2), and tungsten diselenide (WSe_2)), hexagonal boron nitride (h-BN), borophene (2D boron), silicene (2D silicon), germanene (2D germanium), and MXenes (2D carbides/nitrides) [14, 111–117]. Figure 7.6 is a year-wise publication list of 2D materials that show the increasing trend of studying TMDs. Depending on their chemical compositions and structural configurations, atomically thin 2D materials can be categorized as metallic, semi-metallic, semiconducting, insulating, or superconducting. The first graphene descendants that sparked intense research activity are TMDs, which are almost as thin, transparent and flexible as graphene [118, 119]. Unlike graphene, many 2D TMDs are semiconductor in nature and possess huge potential to be made into ultra-small and low power transistors that are more efficient than state-of-the-art silicon based transistors fighting to cope with ever-shrinking devices [120, 121]. Besides sharing the similarities of a band gap in the visible-near IR range, high carrier mobility, and on/off ratio with ubiquitous silicon, TMDs can be deposited onto flexible substrates and survive the stress and strain compliance of flexible supports [122, 123].

2D TMDs exhibit unique electrical and optical properties that evolve from the quantum confinement and surface effects that arise during the transition of an indirect bandgap to a direct bandgap when bulk materials are scaled down to monolayers. This tunable bandgap in TMDs is accompanied by a strong

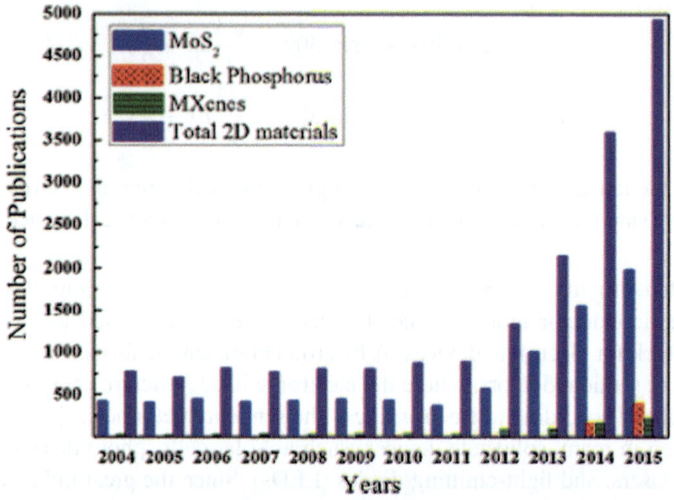

Fig. 7.6 Year-wise publication plots for 2D TMDs including MoS_2, $MoSe_2$, black phosphorus, MXenes, and total 2D TMDs in the period of 2005–2016 (searched by SciFinder Scholar (https://scifinder.cas.org), American Chemical Society database (https://www.acs.org/content/acs/en.html), August 10th, 2016)

photoluminescence (PL) and large exciton binding energy, making them promising candidate for a variety of opto-electronic devices, including solar cells, photo-detectors, light-emitting diodes, and photo-transistors [24, 124–127]. For example, unique properties of MoS_2 include direct bandgap ($\sim$1.8 eV), good mobility ($\sim$700 $cm^2\ V^{-1}\ s^{-1}$), high current on/off ratio of $\sim 10^7$–10^8, large optical absorption ($\sim 10^7\ m^{-1}$ in the visible range) and a giant PL arising from the direct bandgap (1.8 eV) in monolayer; thus, it has been studied widely for electronics and optoelectronics applications [128].

7.3 Crystal Structure of 2D Materials

Two-dimensional (2D) materials are crystalline solids with a high ratio between their lateral size ($\sim$1–10,000 μm) and thickness (<1 nm) [129]. The layered structure of the corresponding bulk crystals, which facilitates their exfoliation to 2D layers, originates from the strong intralayer covalent bonding and comparatively weaker interlayer van der Waals (vdW) interaction with typical binding energies of 40–70 meV [37]. The electronic transport and therefore conductivity in layered materials is much more efficient through the layers (i.e. in the direction perpendicular to the vertical c-axis) than between the layers, typically by 3 or 4 orders of magnitude [130]. This is due to the strong charge carrier localisation within the individual layers and simultaneous transport of charge carriers through multiple layers [131]. The basal plane of 2D and layered materials (perpendicular to the

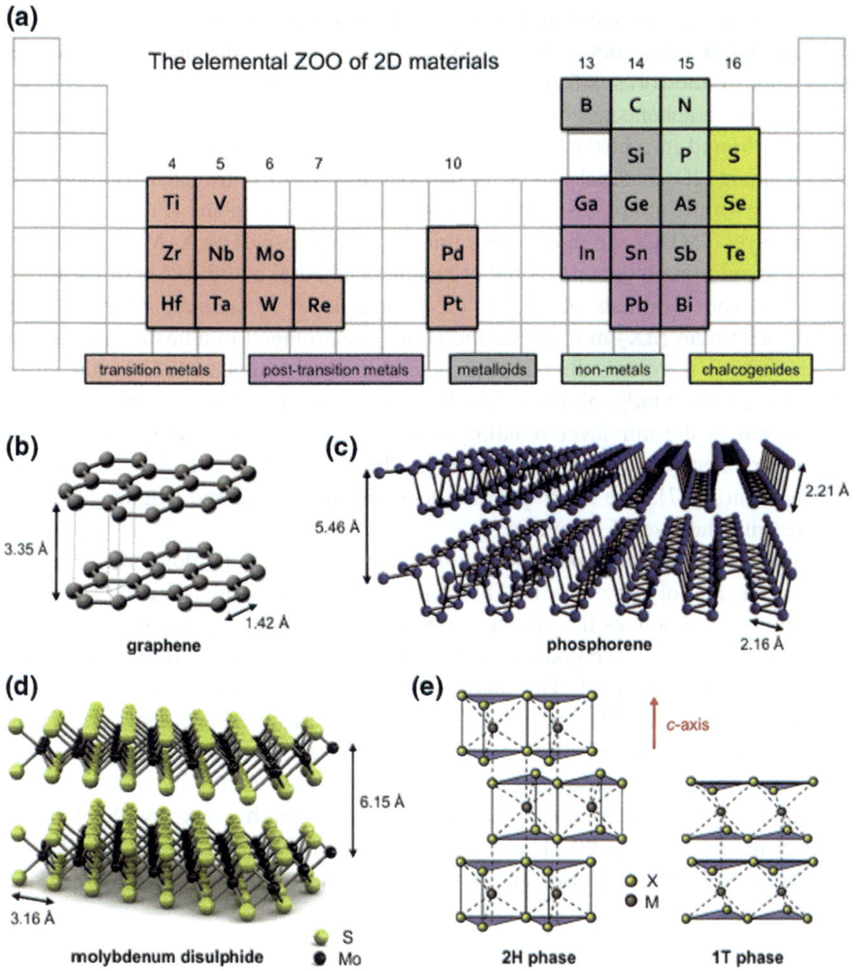

Fig. 7.7 Two-dimensional ZOO. **a** Simplified periodic table highlighting the elements, which form the most common layered and 2D materials. **b**, **c**, **d** Crystal structures of graphene, phosphorene, and MoS_2 bilayers, respectively. **e** Schematic of the two most common TMDC phases, 2H and 1T. The lattice parameters in **b**, **c**, and **d** are taken from [139], [140], and [90], respectively. Figures adapted with permission from: **b** [141] © 2011 American Physical Society. **c** [142] © 2000 John Wiley and Sons. **d** [143] © 2011 Nature Publishing Group. **e** [22] © 2012 Nature Publishing Group

c-axis) is well defined and chemically stable surface due to the absence of dangling bonds, unlike the edge plane (parallel to the c-axis), which is usually terminated by various surface groups [94]. The discovery of graphene and its unique properties in 2004 [10] launched a scientific revolution due to the fact that 2D materials were long thought to be thermodynamically unstable [132]. A large number of other layered materials with diverse properties based on the elements highlighted in

Fig. 7.7a were gradually exfoliated in 2D. These include hexagonal boron nitride (hBN), metal and non-metal chalcogenides, oxides, hydroxides and halides, covalent organic frameworks, silicates, and perovskites [11, 12, 38, 89, 133–135].

Monolayer graphene and hBN are the only true 2D crystals where all the atoms, covalently bound in a hexagonal lattice, occupy the same plane (Fig. 7.7b). Other single-element 2D materials such as phosphorene form corrugated (or buckled) structures with the atoms distributed across more than one plane (Fig. 7.7c).

Monolayer TMDCs are three atoms thick and comprise of an X–M–X sandwich (Fig. 7.7d and e) with either trigonal prismatic octahedral coordination of the metal atom [90]. The d orbitals of M and the *p* orbitals of X facilitate their covalent bonding within an MX_2 monolayer, the atoms are arranged in a hexagonal pattern when viewed in the c-axis direction, and the individual layers are held by the weak vdW forces. Importantly, all the accessible orbitals of M and X at the basal surface are involved in the intralayer bonding, leaving only the high-energy antibonding orbitals for interlayer or external bonding, which leads to a complete absence of dangling bonds [97]. The interlayer distances and intralayer bond lengths generally increase with the size of M and X.

The metal coordination and stacking order between the individual layers defines the phase or the polytype of a TMDC (Fig. 7.7e). Common phases are 1T, 2H, or 3R, where 1, 2, 3 defines the number of X–M–X sandwiches per unit cell in the c axis direction and T (tetragonal), H (hexagonal), R (rhombohedral) denotes the crystal symmetry [90, 136]. The phase determines the properties of TMDCs: i.e. MoS_2 exist either as 2H, which is a thermodynamically stable semiconducting phase with a bulk band gap of ~ 1.2 eV, or as 1T, which is a metastable metallic phase. The 2H and 1T phases can be transformed to one another by simple gliding of the atomic planes, demonstrated by an in situ electron microscopy [137], or by chemical modification [138]. The coordination of the M atoms is a trigonal prismatic in the 1H and 3R phases and octahedral in the 1T phase (Fig. 7.7e) [90].

7.4 Electronic, Optical, and Mechanical Properties of 2D Materials

Transition metal dichalcogenides (TMDs) provide a wide range properties such as semiconductors, half metallic magnetism, superconductivity, or charge density waves and applications have included photovoltaics, super capacitors transistors, battery electrodes, and magneto optic devices. Layered TMDs have a hexagonal structure with one transition metal atom and two chalcogens, which form three layers: an upper and lower sublattice of chalcogen atoms and a middle layer of transition metal atoms (Fig. 7.8).

The electronic behaviour of TMD layers depends on the coordination and oxidation state of the metal atoms, for instance they can be semiconducting for Mo and W, or metallic for Nb or Re. Heterostructures composed of TMD layers provide the

Fig. 7.8 Atomic structure of a typical transition metal dichalcogenides

opportunity to combine a range of electronic behaviours into a small scale design but by far the most studied TMD is molybdenum disulphide, MoS_2. From an electronic perspective MoS_2 has generated considerable interest due to its absorption in the solar spectral region and subsequent applications from photovoltaic to photocatalytic materials. Perhaps most interestingly bulk MoS_2 has an indirect band gap of 1.29 eV whereas a single layer has a direct band gap of 1.9 eV. The reason for this is the characteristics of the d-electron orbitals which make up the conduction and valence bands; or more specifically the valence band maximum.

Importantly, the properties of 2D materials can be tuned "on-demand" by external stimuli such as electric field, strain, illumination, or functionalisation. In the following sections we will review the fundamental properties of 2D materials in the context of electrochemistry and condensed matter physics.

Optoelectronic properties of 2D materials are dominated by quasiparticles such as excitons and trions, i.e. bound complexes of photogenerated holes and electrons. Their binding energies are typically 1 or 2 orders of magnitude larger than in conventional 3D semiconductors, which results in their longer lifetimes and diffusion paths prior to their recombination, making them relevant and observable at room temperature [144].

Two-dimensional (2D) materials commonly possess unique optical bandgap structures, extremely strong light–matter interactions, and large specific surface area. Graphene, hexagonal boron nitride (h-BN) and some transition metal dichalcogenides (TMDs) have emerged as promising building blocks for novel nanoelectronics, providing a full range of materials types, including large band gap insulators [145, 146], semiconductors [147, 148], and semimetals [149, 150]. 2D semiconductors exhibit high carrier mobility, high on-off current ratio and excellent bendability that suit for future low-power consumption and flexible electronics [151–155]. These materials are laterally composed by strong covalent bonds, which provide great in-plane stability. On the other hand, the weak van der Waals (vdW) interlayer force allows us to isolate 2D monolayer and restack them into arbitrary stacking heterojunctions without the need to consider the atomic commensurability as in their bulk counterparts. Hence, a new research field of heterojunctions formed by 2D layered materials has emerged [156–158]. The vertically stacked 2D layers can be formed by mechanical stacking, serving as a quick and convenient way of forming heterostructures. Besides, the chemical vapor deposition (CVD) method is a growing category for fabricating 2D heterojunction in the past few years. CVD process not only provides large-area 2D material for mechanical assembling. It is also possible to grow various stacking structures directly. The synthesized heterojunction can provide a much cleaner interface for fundamental research and hence better device performance. Furthermore, the development of this methodology has shown the possibility to synthesize lateral 2D heterojunction, which goes beyond the limit of the mechanical stacking method. 2D lateral heterojunction has opened a new realm in material science and device applications.

Many scientific and technologically interesting properties emerge when the dimensionality of a material is controlled, for example the broader bands observed in bulk materials become confined quantum states when reduced to quantum dot dimensions ($\approx$20 nm^3) [159] and mimic quantized atomic states. Two dimensional, especially layered, materials could potentially be combined more easily than bulk compounds, creating hybrid systems with shared electronic and mechanical properties. The reduced dimensionality also greatly increases the active surface area and as we shall discuss there are significant technological advantages from this. What will be primarily outlined here though is that the electronic and phonon properties can be profoundly affected by dimensionality; the case of carbon (i.e. diamond, graphite and graphene) shows how drastic this can be.

The sp^2 carbon family of materials comes in a range of geometries: graphite (three dimensional, 3D), graphene (two dimensional, 2D), carbon nanotubes (one dimensional, 1D) and fullerenes (zero dimensional, 0D). Electronic properties for these materials vary significantly: graphite is a semi-metal [160], although the resistance is much larger perpendicular to the layer planes [161], graphene is a zero-gap semi-conductor or semimetal [74] with massless carriers, carbon nanotubes are either metallic or semi-conductors depending on their chirality [162] and fullerenes exhibit quantum dot characteristics [163]. Of course carbon also forms in an sp^3 bonded crystal, diamond, which is a wide band gap semi-conductor/insulator [164] but the majority of the change in electronic properties in this instance is due to tetrahedral geometry and sp3 hybridization which eliminates the loosely bound/ delocalised electrons. Other than the control of intrinsic electronic properties the reduced dimensionality also presents the opportunity to take advantage of the high aspect ratios, such as gas sensors [165, 166], and heterostructure designs which combine several desired properties together, such as field effect transistors [77, 167, 168]. Applications for graphene range from inclusion in high-speed electronics [169] and optical devices [170], energy generation and storage [170–173] including hydrogen storage [174, 175], hybrid materials [176, 177], chemical sensors [178, 179], DNA sequencing [81, 180, 181], and as a support substrate for transmission electron microscopy [182]. However graphene alone is unable to fulfil many application needs due to the absence of a band gap, which makes it unusable in logic gate systems, where sensitive device on/off ratios our paramount [183]. There have been many attempts to overcome this shortfall by using bilayer graphene where an induced and controllable band gap is possible [184–188]. Another significant limitation of the use of graphene in high speed applications is that the production method significantly affects the quality of the graphene sheet and subsequently its electronic properties. The issues surrounding production and quality of the produced material is not isolated to the case of graphene. There are a wide range of two dimensional materials where their production methods cause significant damage and/or functionalization of the desired layered material [189].

Since the experimental isolation of graphene there has been renewed interest in a range of 2D compounds; these include transition metal dichalcogenides [190], layered transition oxides [191], hexagonal boron nitride (hBN) [190, 192, 193] and layered topological insulators [194, 195]; the majority of these layered materials form in a stacked structure in their bulk 3-D crystals. The advantage of using 2D materials, which naturally form in a stacked 3D structure, is that the bonding between layers is based upon van der Waals interactions, which is significantly weaker than covalent bonding. Having a range of 2D materials has the advantage of also providing a range of electronic properties to choose from including: insulator to metal behaviour, optical absorption/emission [196] or transparency [197] and more exotic properties such as electronic properties controlled by the number of layers (e.g. topological insulators) [198, 199] superconductivity [200], and thermoelectricity [201]. In addition to the 2D materials that naturally form in a layered structure there is growing interest in two dimensional silicon and germanium; although energetically preferring bulk sp3 bonded structures, they have stable layered configurations known as silicene and germanene. What follows will be a more detailed discussion of several key two dimensional materials starting with the most well-known, graphene, before considering the latest results for silicene and germanene and comparing their properties, production methods and technological issues with other layered structures.

7.5 2D van der Waals Heterostructures

Early in 2011, Yandong Ma's group has calculated that the binding energy of per C atom binding to MoS_2 is −23 meV and the forming interlayer spacing between graphene and MoS_2 is 3.32 Å. Due to the variation of on-site energy induced by MoS_2, band structure of graphene could be largely preserved in this hybrid structure

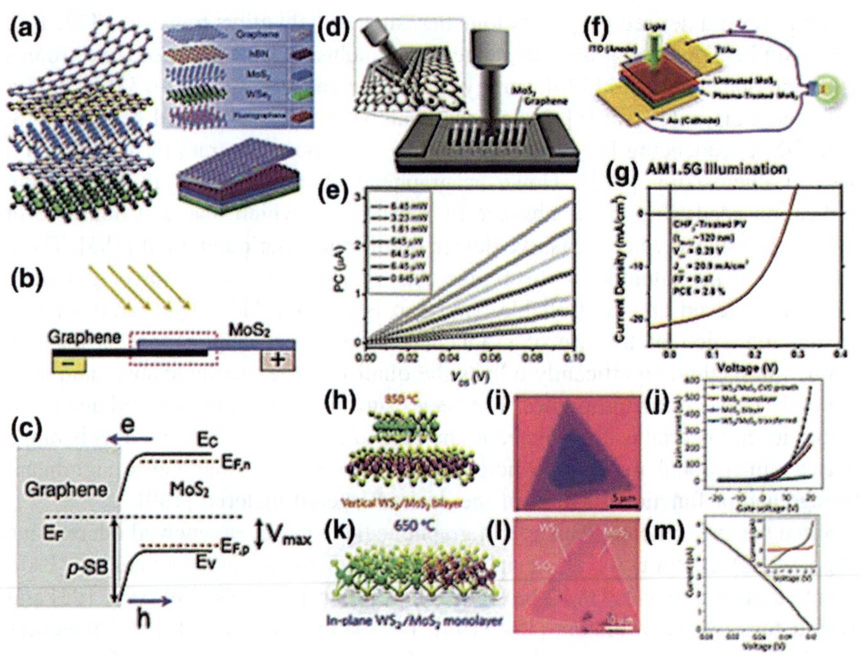

Fig. 7.9 2D van der Waals heterostructures. **a** Schematic illustrations of Lego-like van der Waals heterostructures design and construction based on 2D layered materials. **b** MoS_2/graphene interface and Schottky barrier solar cell. M1 and M2 are low and high work function metals respectively. **c** Band alignment at a MoS_2/graphene interface. p-SB: hole Schottky barrier; Ef: fermi energy; Ec: conduction band bottom; Ev: valence band top; EF,n and EF,p: quasi-Fermi levels for electrons and holes under illumination; Vmax: maximum Voc. **d** Schematic diagram of a graphene-MoS_2 hybrid phototransistor under light irradiation. **e** Photocurrent of the graphene-MoS_2 transistor exhibited in **d** for different optical power as a function of drain-source voltage (VDS). **f** Schematic illustration of the plasma-treated MoS_2 photovoltaic device and **g** I–V characteristics measured under illumination of AM1.5 (100 mW/cm^2). **h** Schematic and **i** optical images of the vertically stacked WS_2/MoS_2 heterostructures. **j** Typical plot of gating voltage versus source/drain current of a CVD-grown WS_2/MoS_2 bilayer, a mechanically transferred WS_2/MoS_2 bilayer, a MoS_2 bilayer and monolayer MoS_2. **k** Schematic and **l** optical images of the WS_2/MoS_2 in-plane heterojunctions. **m** Photovoltaic effect of the in-plane heterojunction. Inset is the typical IeV curve of the junction with (black) and without (red) illumination. Panel **a** reprinted with permission from Ref. [12]. Copyright 2013, Rights Managed by Nature Publishing Group. Panel **b**, **c** reprinted with permission from Ref. [91]. Copyright 2013 American Chemical Society. Panel **d**, **e** reprinted with permission from Ref. [204]. Copyright 2014 WILEY-VCH Verlag GmbH & Co. KGaA, Weinheim. Panel **f**, **g** reprinted with permission from Ref. [14]. Copyright 2014 American Chemical Society. Panel **h–m** reprinted with permission from Ref. [206]. Copyright 2014, Rights Managed by Nature Publishing Group

while introducing a small bandgap of 2 meV which was almost negligible [202]. Further analysis indicated that this band gap was tunable by varying the interlayer spacing, highlighting the prospect in designing of devices with tunable bandgap and high electron mobility simultaneously. In 2011, A. K. Geim proposed "van der

Waals heterostructures" on Nature (Fig. 7.9a) [11], showing a landscape for future development of 2D hybrid structures. Many challenges were accomplished afterwards, pushing devices into real practical applications. J. C. Grossman and his coworkers confirmed the feasibility by studying the performance of 1 nm-thick solar cell based on MoS_2/graphene through first principles calculations (Fig. 7.9b, c) [91]. First, MoS_2 like TMDCs monolayers could absorb 5–10% incident sunlight within 1 nm in thickness, exceeding that of traditional semiconductors (GaAs and Si) more than one order of magnitude. In addition, A type-II Schottky junction (Fig. 7.9c) within 1 nm which would greatly facilitate separation and transport of carriers in the stacking interfaces was constructed, exporting a high power conversion efficiency (PCE) up to $\sim 1\%$. Moreover, MoS_2/graphene solar cell demonstrated a power density of 0.25–2.5 mW/kg, which were higher by approximately 1–3 orders of magnitude than the best existing ultrathin solar cells [91]. Further experimental observation proved the ultrafast interfacial charge transfer in TMDCs stacking structures, ensuring the effective charge collection and utilization in later circuits [203], opening up the development for light detection and harvesting in atomically thin devices.

Two primary approaches have been adopted to construct devices based on 2D van der Waals heterostructures. One was through layer-by-layer stacking and another was the direct CVD preparation as mentioned above in Part 2. As of layer-by-layer stacking, numerous achievements have been reported. For examples, Hua Xu's group presented a high responsivity 2D graphene-MoS_2 hybrid phototransistor to be continuously tuned from 0 to 10^4 mA/W by the gate voltage [204] (Fig. 7.9d, e). In 2014, Wenjing Zhang's group demonstrated a photodetector with an extremely high photoresponsivity of 10^7 A/W based on graphene-MoS_2, which could achieve a photogain greater than 10^8 [205]. For direct CVD preparation, one breakthrough was proposed by Kathleen M. McCreary and his coworkers, which realized continuous and uniform MoS_2 single-layer films growth on large-area graphene. Recently, P. M. Ajayan's group reported the preparation of both vertically stacked (Fig. 7.9h, i) and in-plane interconnected WS_2/MoS_2 heterostructures (Fig. 7.9k, l) through a convenient one-step CVD growth [206]. The vertically stacked layers built a typical type-II band structure, achieving an on/off ratio of up to 10^6 together with the field-effect mobility as high as 15–34 $cm^2/(V\ s)$, which exceeded the performance of a mechanically transferred WS_2/MoS_2 bilayer (Fig. 7.9j) [206]. For the in-plane seamless p-n heterostructures, due to the strong enhancement of localized photoluminescence effect, an open circuit voltage (V_{oc}) of 0.12 V with an short circuit current (I_{sc}) of 5.7 pA could be generated, which exhibited great potential to be used as atomically thin solar cells (Fig. 7.9m) [206]. Sunjin Wi's group demonstrated MoS_2 based solar cell exhibiting a superior PCE up 2.8% with short circuit current density (J_{sc}) of 20.9 mA/cm^2 (Fig. 7.9f, g) [207].

van der Waals (vdW) gaps between each neighboring layer and large specific surface area due to sheet-like structures are distinct features that make 2D TMDs highly attractive for capacitive energy storage (e.g. supercapacitors and batteries) and sensing applications [208, 209]. The large surface-to-volume ratio bestows 2D TMDs based sensors with improved sensitivity, selectivity and low power consumption. Unlike digital sensors, TMDs based sensors do not have physical gates for selectively reacting to the targeted gas molecules or biomolecules [210, 211]. MoS_2-based FET devices have potential applications in gas, chemical, and bio-sensors. Another aspect of the weakly bonded 2D TMDs atomic layers is that they can be easily isolated and stacked with other TMDs to construct a wide range of vdW heterostructures without the limitation of lattice matching [11, 212]. Stacking together one-atom-thick sheets of dissimilar TMDs, for example, vertically stacked heterostructures allows for the realization of unique functions and superior properties that cannot be obtained otherwise. By exploiting such novel properties in these vdW heterostructures as band alignment, tunneling transports, and strong interlayer coupling, several new electronic/opto-electronic devices such as tunneling transistors, barristers, photodetectors, LEDs and flexible electronics can be fabricated [213, 214]. Figure 7.10 depicts the diverse devices constructed from the 2D TMDs by using their unique physical, chemical, and opto-electronic properties [215–220].

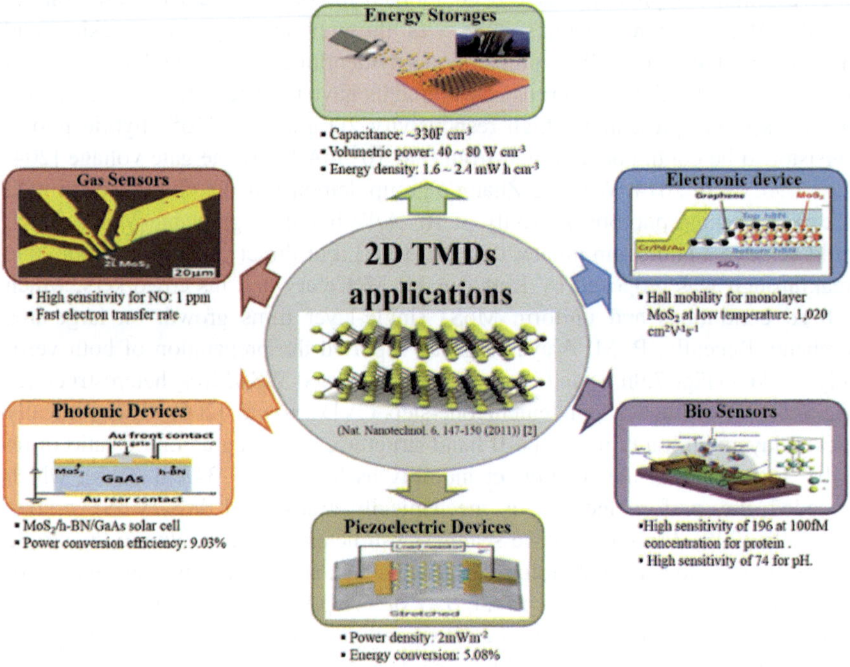

Fig. 7.10 Electronic, opto-electronic and energy devices based on 2D transition metal dichalcogenides (TMDs). (Reprinted with permission from Ref. [215]. 2015, Nature Publishing Group; Ref. [216]. 2014, Nature Publishing Group; Ref. [217]. 2015, Royal Society of Chemistry; Ref. [218]. 2012, John Wiley and Sons; Ref. [219]. 2015, Nature Publishing Group; and Ref. [220]. 2014, American Chemical Society.)

Optical visualisation and imaging of 2D materials on suitable substrates still remains one of the most popular ways of rapid identification of thin crystals. The wavelength-dependent contrast between monolayer graphene and an SiO_2/Si substrate reaches 10% for green light and 90/280 nm thick SiO_2 [221].

7.6 Fabrication of 2D Heterostructures

Owing to the layer structures of 2D materials, the formation of heterostructure can be in vertical or lateral fashion. Depending on the architecture of the heterostructures, the fabrication methods vary. Here, we sort the fabrication processes into two categories: one is the mechanical stacking method and the other is the directly synthesis.

7.6.1 Heterostructures by Manual Stacking

Thanks to the invention of exfoliation [74] and transfer [171, 222] methods of layered materials, the mechanically or chemically exfoliated 2D flakes can be manually stacked to form 2D heterostructures, where the interlayer vdW force holds the heterostructure securely. The 2D flakes can be mechanically/chemically exfoliated from bulk materials or isolated from the synthetic (grown) 2D layers on substrates. The stacking order and the interface of the heterojunction are critical for their electrical and optical properties. The possibility to form various stacking lattice orientation provides the heterojunction interface with tunable physical properties depending on the interaction strength of the two layered materials. For example, the carrier mobility of graphene on SiO_2 is normally limited by the scattering effect from charge impurities, substrate surface roughness and SiO_2 optical phonons. Dean and co-workers have demonstrated a graphene device on h-BN interface layers by using mechanical tacking method (Fig. 7.11a) [222]. The atomic flat and nearly free charge trapping of h-BN layers serves as excellent substrate for graphene. The carrier mobility of graphene devices fabricated on h-BN substrates is nearly an order higher than the devices on amorphous SiO_2 substrates. The graphene layers on h-BN show reduced roughness, less doping and improved chemical stability, demonstrating the critical role of the interfaces. On the other hand, Hsu and co-workers perform the second harmonic generation (SHG) measurement to demonstrate the effect of the twisting angle on stacking TMD bilayers [223]. The SHG from the twisted bilayers is a coherent superposition of the second harmonic fields from the individual stacking layers and the phase difference is only determined by the stacking angle. For the interlayer coupling, Fang and co-workers have investigated the optical transition properties on MoS_2/WSe_2 stacking heterojunction (Fig. 7.11b) [224]. The band alignment and the interlayer coupling at the interface result in a Stokes shift of ~ 100 meV, where the

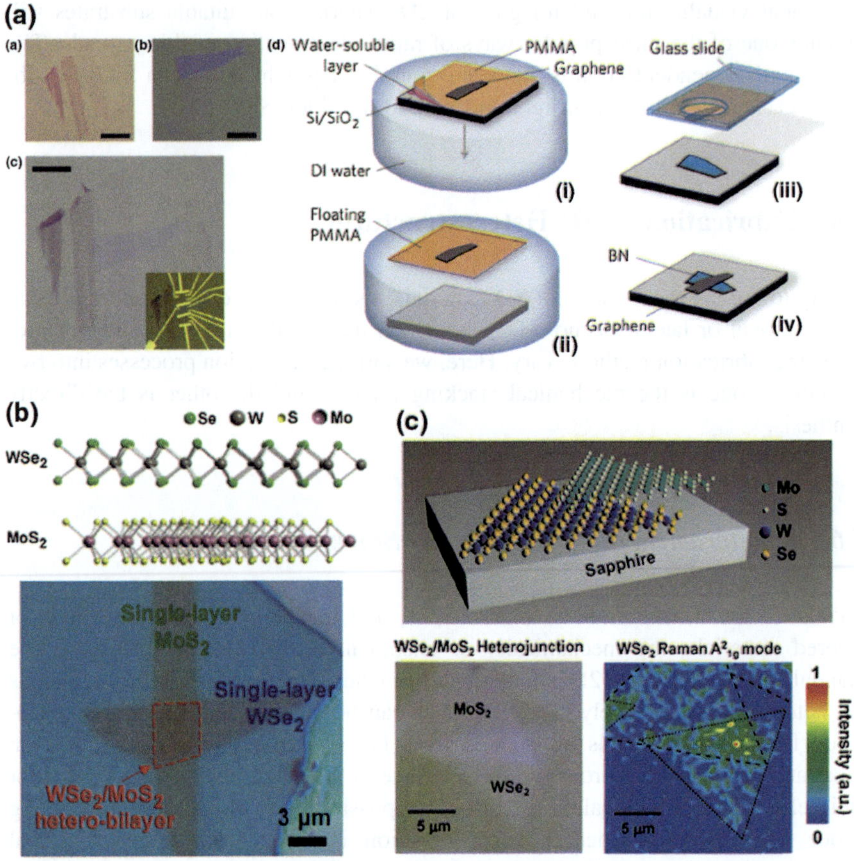

Fig. 7.11 Mechanical stacking process for vertical heterostructures. **a** Schematic illustration of the transfer and stacking process and the optical image of the graphene/h-BN devices (scale bars, 10 mm) [222]. Reproduced with permission from Ref. [222]. Copyright 2010, Nature Publishing Group. **b** Schematic illustration and optical microscope image of the stacked WSe_2/MoS_2 heterojunction [224]. Reproduced with permission from Ref. [224]. Copyright 2014, National Academy of Sciences. **c** Schematic illustration and optical microscope image of the CVD MoS_2/WSe_2 stacking heterojunction. The spatial mapping of the Raman intensity for WSe_2 A_{1g}^2 peak shows the interlayer coupled area [225]. Reproduced with permission from Ref. [225]. Copyright 2014, American Chemical Society

direct absorption and indirect emission were observed on the heterojunction. Chiu and co-workers have fabricated vertical MoS_2/WSe_2 heterojunction by manually stacking CVD-grown MoS_2 and WSe_2 monolayers and demonstrated that the interlayer coupling can be enhanced by thermal treatment (Fig. 7.11c) [225]. Through the study on interlayer coupling with optical transition and the band alignment with scanning tunneling spectroscopy (STS) and X-ray photoemission spectroscopy (XPS), the band alignment and the exciton binding energy of the heterojunction can be determined. The mechanical stacking method provides a simplest way to construct arbitrary 2D heterostructures. However, the quality of the interface can be easily affected by the trapping of solvents or chemicals used for transfer, which still remains as an issue to be solved.

7.6.2 Direct Synthesis of 2D Heterostructures

Although the exfoliation technique can produce high quality 2D crystals for fundamental study, it still remains challenging to control the location, layer number and interface of the produced heterojunction, which hinders the practical fabrication. Very recently, the CVD technique shows great promise for the production of large domain 2D building blocks for TMD heterostructures. It also allows the direct synthesis of various 2D heterojunctions with vertically stacked or laterally stitched interface.

The CVD is one of the most effective methods to achieve large area growth of atomically thin 2D TMDs for the successful device applications. The simplest form of CVD to grow 2D TMDs is the co-evaporation of metal oxides and chalcogen precursors that lead to vapor phase reaction followed by the formation of a stable 2D TMD over a suitable substrate. The growth mechanism of CVD method differs in each synthesis process as the materials forming process also depends on (1) properties of substrate, (2) temperature and (3) atomic gas flux.

$$MoO_3\ (S) + S(g) \rightarrow MoO_{3-x}\ (S) + SO_x\ (g)$$
$$MoO_{3-x}\ (S) + S(g) \rightarrow MoS_2\ (S)$$

Another facile method for growing large area and continuous TMDs is using the 'two-step method, depositing transition metal thin film (e.g. Mo, W, Nb, etc.) on substrate (usually Si/SiO_2) followed by thermal reaction with chalcogen (S, Se, Te) vapor. The following reaction occurs to form a stable 2D TMD during the CVD process at high temperatures (300–700 °C) and inert atmosphere.

$$M\ (s) + 2S\ (g) \rightarrow MS_2\ (s)$$

Song et al. [226] demonstrated the wafer scale growth of WS_2 using the ALD growth of tungsten oxide (WO_3) with subsequent conversion via H2S annealing. Figure 7.12a shows the ALD growth steps for the synthesis of WS_2 nanosheets. The number of MoS_2 layers can be controlled effectively by tuning the number of ALD cycles for MoO_3 growth. The camera image of large-area (approximately 13 cm in length) mono-, bi-, and tetralayer WS_2 nanosheets on SiO_2 substrate is shown in Fig. 7.12b. Monolayer WS_2 FETs fabricated in the top-gate geometry showed n-type conduction with an electron mobility of $\sim 3.9\ cm^2\ V^{-1}\ s^{-1}$ (Fig. 7.12c).

Here, we introduce the vertical and lateral heterojunctions fabricated by CVD method separately.

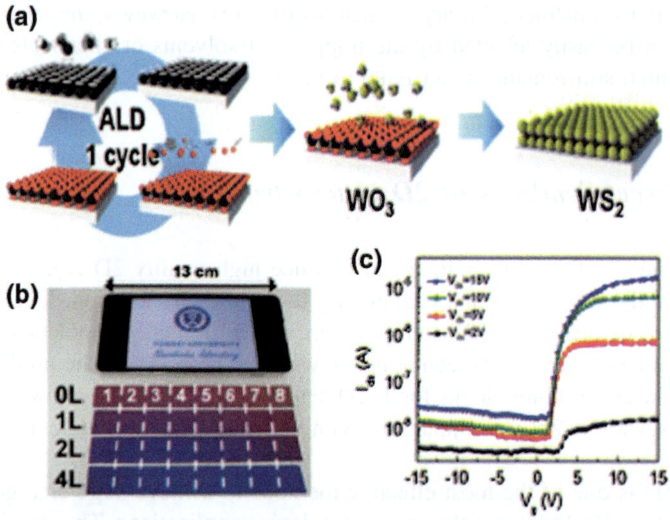

Fig. 7.12 **a** Schematic of the ALD process for the synthesis of large area and thickness controlled WS_2 films and **b** demonstration of the large area (approx. 13 cm) mono-, bi-, and tetralayer WS_2 fabricated on Si/SiO_2 substrates. The growth area is about the size of a cellular phone display screen. **c** Transfer characteristics of single layer WS_2 FET exhibiting a n-type conduction behavior. (Reprinted with permission from Ref. [226]. 2013, American Chemical Society.)

Vertically stacked 2D heterojunctions

With the CVD methods, the vertically stacked heterojunction can be obtained by growing one 2D material on another. The direct growth in principle results in clean heterojunction interfaces. Since there exists only vdW interaction between 2D layers, the mechanism is likely through the vdW epitaxy, where the lattice mismatch is the key challenge to overcome. The early experiment starts from graphene and h-BN which share similar lattice constant. Yang and co-workers have reported a plasma-assisted deposition method for the growth of single domain graphene on h-BN substrate (Fig. 7.13a) [227]. Graphene registers on the h-BN lattice with a preferred orientation and the size of graphene is only restricted by the area of underlying h-BN. Furthermore, Shi and co-workers have obtained a vertically stacked MoS_2/graphene structure via thermal decomposition of ammonium thiomolybdate precursors on graphene surfaces (Fig. 7.13b) [228]. Lin and co-workers have further demonstrated the direct growth of MoS_2, WSe_2, and h-BN on graphene through CVD methods, where the underlying graphene was grown epitaxially on SiC (Fig. 7.13c) [229]. It has been revealed that the morphology of the graphene layer strongly affects with the growth and the properties of top heterostructures, where strain, wrinkling, and defects on the surface of graphene provides the nucleation centers for upper layer material growth. Another research interest is focused on TMD heterojunctions formation. Gong and co-workers have reported a one-step vapor phase growth method for the direct formation of vertical

WS_2/MoS_2 heterostructures [1]. The WS_2 can grow on top of MoS_2 with the 2H stacking structure. On the other hand, Lin and co-workers present the direct synthesis of MoS_2/WSe_2/graphene and WSe_2/MoS_2/graphene heterostructures ex situ by oxide powder vaporization and metal-organic chemical vapor deposition (MOCVD) methods as shown in Fig. 7.13d [230]. The electrical transport of the heterostructure exhibits a sharp negative differential resistance at room temperature resulting from the special resonant tunneling of charge carriers at interface. This observation has not been found in mechanically stacked junctions. These direct syntheses are in principle applicable to other materials which opens new ways for the fabrication of various heterojunctions. However, the thermodynamically dominated process might restrict the controllability on stacking order, twisting angle or domain size.

Laterally stitched 2D heterojunctions

Since the 2D materials are covalently bonded, the lateral heterojunction can only be formed by direct growth. Levendorf and coworkers have demonstrated the lateral stitched graphene/h-BN heterojunction by growing h-BN on patterned graphene with CVD method, which results in a continuous 2D heterojunction film (Fig. 7.14a) [231]. $MoSe_2/WSe_2$ by physical vapor transport method with mixed sources [232] and WS_2/MoS_2 by one-spot CVD growth with multi-sources (Fig. 7.14b) [1]. Li and co-workers have recently demonstrated a two-step CVD approach for epitaxial growth of WSe_2/MoS_2 lateral junction, where WSe_2 is firstly synthesized through van der Waals epitaxy on substrate followed by the edge epitaxy of MoS_2 along the W growth front (Fig. 7.14c) [233].

7.7 2D Heterostructures and Their Applications

2D TMD materials are considered attractive for diverse applications including electronics, photonics, sensing, and energy devices. These applications are inspired by the unique properties of layered materials such as thin atomic profile that represents the ideal conditions for maximum electrostatic efficiency, mechanical strength, tunable electronic structure, optical transparency, and sensor sensitivity [234]. Of particular interest for applications is flexible nanotechnology, which is considered for potentially ubiquitous electronics and energy devices that can benefit from the range of outstanding properties afforded by 2D materials. Flexible technology comprises a wide array of scalable large-area devices including thin film transistors (TFTs), displays, sensors, transducers, solar cells and energy storage on mechanically compliant substrates. The recent advances in 2D heterostructures indicate a broad range of applications such as field effect/tunneling transistors, biosensors, light emitting diodes, light detectors, photovoltaic and energy storage devices. The 2D heterostructures and applications reported in the recent five years are summarized and several applications are discussed in the following paragraphs.

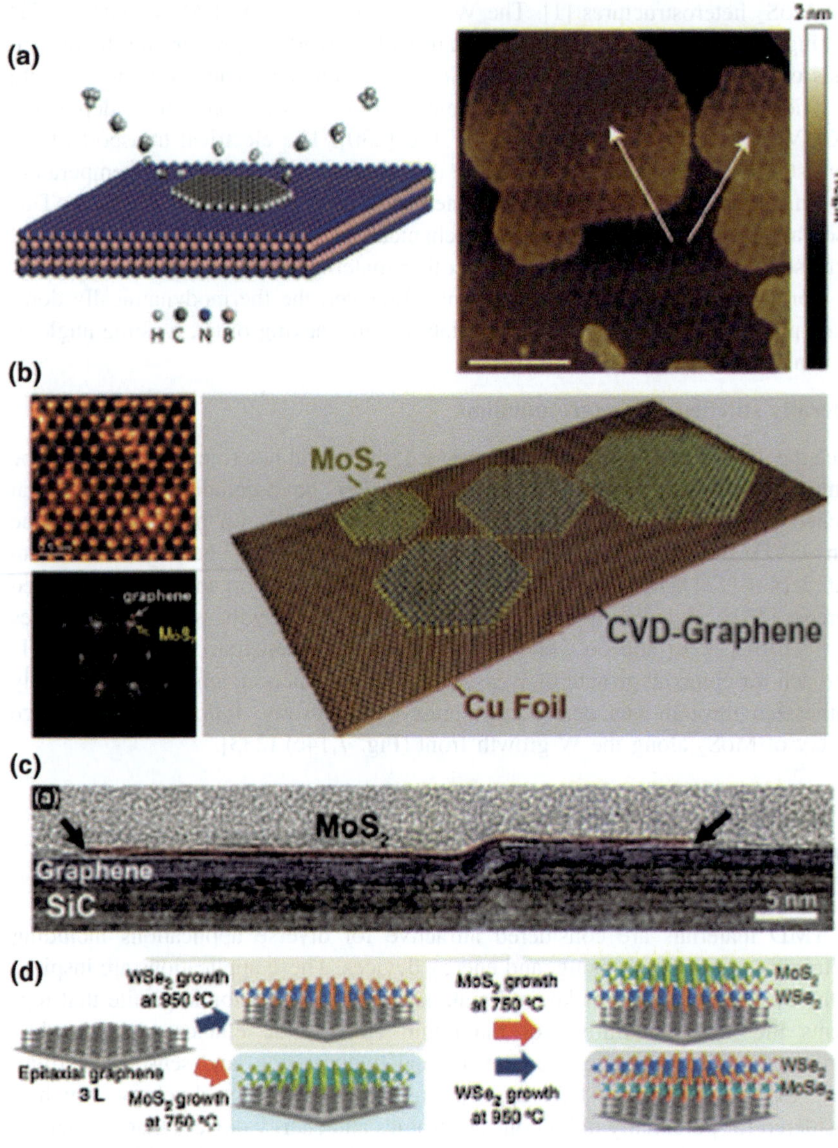

Fig. 7.13 Growth of 2D vertical heterostructures. **a** Schematic illustration of the plasma-assisted deposition method for graphene grown on h-BN and the Morie' pattern of the heterostructure. The white arrows indicate the graphene layer (scale bars 100 nm) [227]. Reproduced with permission from Ref. [227]. Copyright 2013, Nature Publishing Group. **b** The TEM image and the schematic illustration of the MoS_2 grown on graphene [228]. Reproduced with permission from Ref. [228]. Copyright 2012, American Chemical Society. **c** Cross-sectional HRTEM of MoS_2/graphene stacked heterojunction on a SiC substrate step edge [229]. Reproduced with permission from Ref. [229] Copyright 2014, American Chemical Society. **d** Schematic illustration of the MoS_2/WSe_2/graphene and WSe_2/MoS_2/graphene vertical heterostructures formed by oxide powder vaporization and MOCVD approaches [230]. Reproduced with permission from Ref. [230]. Copyright 2015, Nature Publishing Group

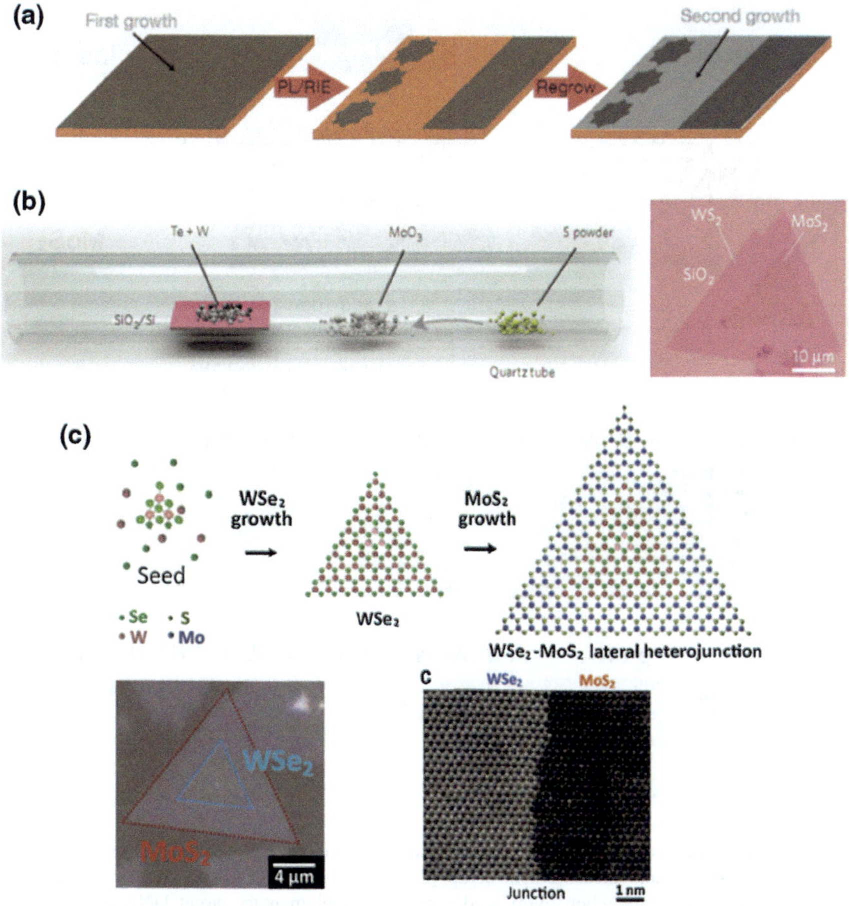

Fig. 7.14 In-plane epitaxial growth of 2D lateral heterostructures. **a** Schematic illustration for formation of graphene/h-BN lateral heterojunctions using CVD growth method [231]. Reproduced with permission from Ref. [231]. Copyright 2012, Nature Publishing Group. **b** Schematic illustration of the one-pot synthesis setup and optical image of the WS_2/MoS_2 lateral heterojunction [1]. Reproduced with permission from Ref. [1]. Copyright 2014, Nature Publishing Group. **c** Schematic illustration of the two-step growth of the monolayer WSe_2/MoS_2 lateral heterojunction, and the optical image of the junction. The right picture shows the corresponding High-resolution STEM images taken at the interface [233]. Reproduced with permission from Ref. [233]. Copyright 2015 American Association for the Advancement of Science

7.7.1 Biosensor

The large aspect ratio of 2D materials and their ultrasensitivity to environmental changes make them suitable for possible applications in bio-sensing [150, 235]. Previous works have demonstrated the DNA differentiation ability in MoS_2 through

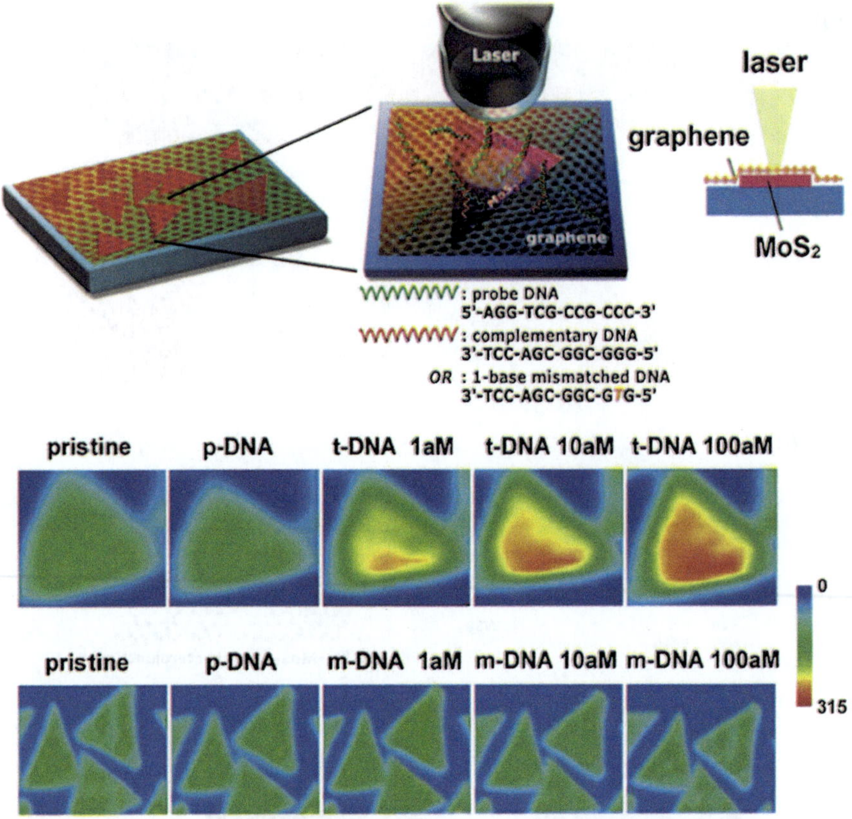

Fig. 7.15 Heterostructures for biosensor. Schematic illustration of the graphene/MoS_2 stacking heterostructure sensor and measurement setup [238]. Bottom images show the MoS_2 photoluminescence peak mappings when hybridized with the complementary target DNA (up) and the mismatched DNA (down) at different concentrations. Reproduced with permission from Ref. [238]. Copyright 2014, WILEY-VCH Verlag GmbH & Co. KGaA, Weinheim

its optical and electro-chemical characterization [236, 237]. However, MoS_2 is also sensitive with the presence of moisture and oxygen, which limits their applications in aqueous solutions. In order to increase the stability and sensitivity, Loan and co-workers have demonstrated a graphene/MoS_2 heterostructure for label-free selective detection of DNA hybridization through the photoluminescence (PL) intensity changing of MoS_2 (Fig. 7.15) [238]. The top graphene layer not only avoids the contact of MoS_2 with moisture and oxygen but also hosts the DNA owing to its biocompatibility [238, 239]. The detection ability of the graphene/MoS_2 sensor can down to 1 atto mole (10^{-18} M) with few minutes of fast response time, demonstrating the potential of ultrasensitive DNA detection using 2D hetero-structures.

7.7.2 *Solar Cells (Photovoltaic)*

A solar cell comprises an interface (junction) of two materials with opposite carrier types, which allows efficient carriers separation of the photogenerated electron–hole pairs at the interface to induce the photovoltaic (PV) effect. Although the interface can be easily constructed by forming a Schottky or p–n junction in 2D hetero-junctions, the key question is still whether it is possible to achieve enough solar energy conversion within such a thin layer. Bernardi and co-workers have studied the PV effects of 2D heterojunction theoretically, including the Schottky barrier between MoS_2 and graphene and the MoS_2/WS_2 stacking heterostructure [240]. The simulation shows that even the atomically thin active layer can produce power conversion efficiencies (PCE) up to $\sim 1\%$ or a power density of up to 2.5 MW/kg. Compared with the existing ultrathin solar cells, the 2D hetero-structure solar cell could achieve 1–3 orders of magnitude higher efficiencies.

Furchi and co-workers have revealed the PV effects on a stacking heterojunction diode made of n-type MoS_2 and p-type WSe_2 monolayers (Fig. 7.16a) [241]. By adjusting the electrical doping level through a back gate tuning, the 2D hetero-diode shows the PCE about 0.2% and the external quantum efficiency (EQE) about 1.5%. However, the low in-plane mobility of the TMD layer results in a short carrier diffusion time comparable to the interlayer electron–hole recombination lifetime, which leads to the de-creased photoresponsivity. In order to overcome the carrier transport issue in this kind of stacking device, Lee and co-workers demonstrate a MoS_2/WSe_2 stacked junction with graphene electrodes directly on the top and the bottom of the heterostructure [242]. The carrier extraction rate is enhanced through the direct charge transfer to graphene semi-metal layer and the photo response has increased about factor of 5. Another similar TMD p–n heterojunction is formed by using p-MoS_2/n-MoS_2 junction. Wi and co-workers demonstrate the PV device with a vertically stacking indium tin oxide (ITO)/n-MoS_2/plasma-doped MoS_2/Au heterostructure (Fig. 7.16b) [243]. This stacking device exhibits a high short-loop photocurrent density of 20.9 mA/cm^2 (under AM1.5 illumination), PCE of 2.8% and high EQE values from 37 to 78% for the wavelengths between 300 and 700 nm. Tsai and co-workers further utilize CVD MoS_2 monolayers with p-Si to form a large scale type II stacking p–n heterojunction, where the built-in electric field near the interface can facilitate photogenerated carrier separation [244]. This large scale heterojunction device can achieve a PCE of 5.23%, and the successful combination of MoS_2 with Si implies the possibility for integrating 2D materials into the Si manufacturing process. Another stacking photovoltaic device is through the formation of simple Schottky barrier with metal (semi-metal)/semiconductor junction to establish the built-in electric field at the interface. Shanmugam and co-workers have reported the graphene/WS_2 Schottky barrier solar cell by using WS_2 as a photoactive layer, which shows a good photon absorption ability in the visible spectral range and a photoelectric conversion efficiency about 3.3% [245].

Besides the vertically stacked heterojunction, Gong and co-workers have reported the lateral p–n junction with WS_2/MoS_2 heterostructure, which shows a clear PV

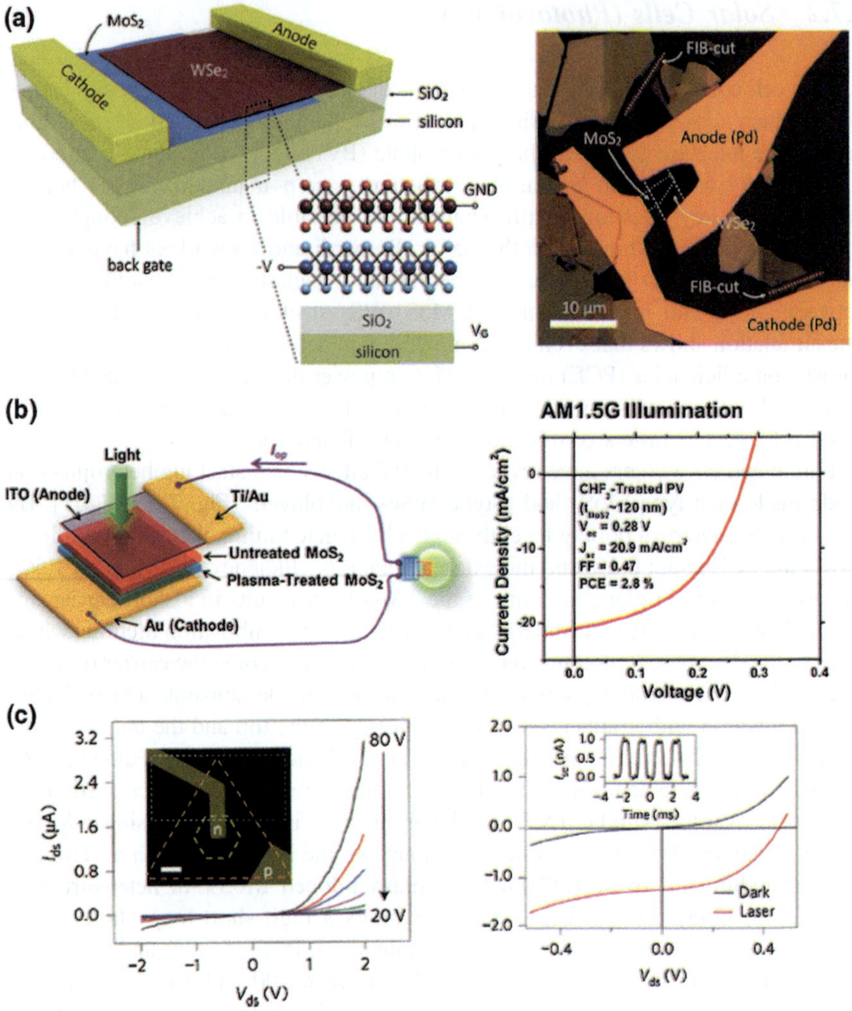

Fig. 7.16 Heterostructures for photovoltaic devices. **a** Schematic illustration and optical microscope image of the MoS_2/WSe_2 hetero-diode [241]. **b** Schematic illustration of the p-MoS_2/n-MoS_2 photovoltaic device and the current–voltage characteristics under AM1.5 G solar simulator illumination [243]. Reproduced with permission from Refs. [241, 243]. Copyright 2014, American Chemical Society. **c** The optical image of the lateral WSe_2/WS_2 heterojunction p–n diode and its I–V characterization under different gate voltage (left) and under laser illumination (right). Scale bar, 2 mm [246]. Reproduced with permission from Ref. [246]. Copyright 2014, Nature Publishing Group

effect [206]. An open-loop voltage of 0.12 V and close-loop current of 5.7 pA is measured without further external electrical field tuning. Duan and co-workers forms a lateral heterojunction p–n junction with WSe_2 and WS_2 (Fig. 7.16c) [246]. The lateral diode shows an open-circuit voltage of ~0.47 V and short-circuit current of

~1.2 nA. The EQE of the photon-to-electron conversion is determined as ~9.9%, and internal quantum efficiency of ~43%. Li and co-workers further demonstrate a controlled-edge epitaxy lateral WSe_2–MoS_2 mono-layer heterojunction. The intrinsic p–n junction exhibits open-circuit voltage of 0.22 V and close-circuit current of 7.7 pA under white light illustration, and a PCE at least 0.2% [247]. The smaller photo active area and the monolayered thickness in these in-plane solar cells might restrict their solar absorption. The low PCE issue may be resolved by introducing metallic nanoparticles or photonic structures on it.

7.7.3 Field Effect Transistors (FET)

Although graphene shows potential in realizing 2D high speed bipolar field effect transistor (FET) devices due to its high carrier mobility and ambipolar effect, the zero band gap and Klein tunneling nature limit its application for FET due to low ON/OFF current ratios. One approach for realizing high performance 2D FETs is to use TMD as the channel material. The proper bandgap and reasonable mobility of the TMD materials make them more promising than graphene. Roy and co-workers have demonstrated an n-type MoS_2 FET with all components formed by 2D materials, including MoS_2 as the active channel material, h-BN as the dielectric layer for top electrode isolation, and graphene as the semi-metal contacts (Fig. 7.17a) [248]. The high mobility and tunable Fermi-level properties of graphene allow it to from an Ohmic contact with TMD materials. The h-BN layers can act as an isolating layer with its large bandgap, and it has atomically smooth surfaces without charge trapping, making it a great interface or substrate for 2D materials [222]. The MoS_2 n-type FET exhibits an ON/OFF ratio larger than 106, and an electron mobility of ~33 cm^2/V s. Likewise, a p-type FET has been demonstrated using WSe_2 as the active channel [153]. Since WSe_2 is a slightly p-doped material, it is easy to tune the transport carrier types by extra doping or by selection of the contact metal to serve for n-type or p-type FETs [153, 249, 250]. Furthermore, the complementary metal oxide semiconductor (CMOS) has been reported by using two different channel materials [251] or materials with different doping levels [250, 252, 253]. Yu and co-workers have demonstrated the high performance CMOS by using WSe_2 as the main active channel and tune the carrier channel through chemical doping (Fig. 7.17b) [253]. This WSe_2 CMOS exhibits full logic swing, ideal voltage transfer characteristic, high noise margin and large voltage gain up to 38 in inverter operation, promising for low power electronic devices.

Vertical transport through 2D heterojunctions is another type of 2D FET. Britnell and co-workers have reported a tunneling FET with the vertically stacked heterostructure composed of graphene and thin h-BN or MoS_2 (Fig. 7.17c) [254]. The basic principle is based on the quantum tunneling, where the graphene electrodes are separated by thin insolation barriers (h-BN or MoS_2) and the tunneling is controlled by tuning the density of states in graphene and adjoined barrier by an external voltage. The tunneling FET exhibits an ON/OFF ratio of ~50 and ~10^4 by using BN and MoS_2 as barrier, respectively. The transit time is expected to be

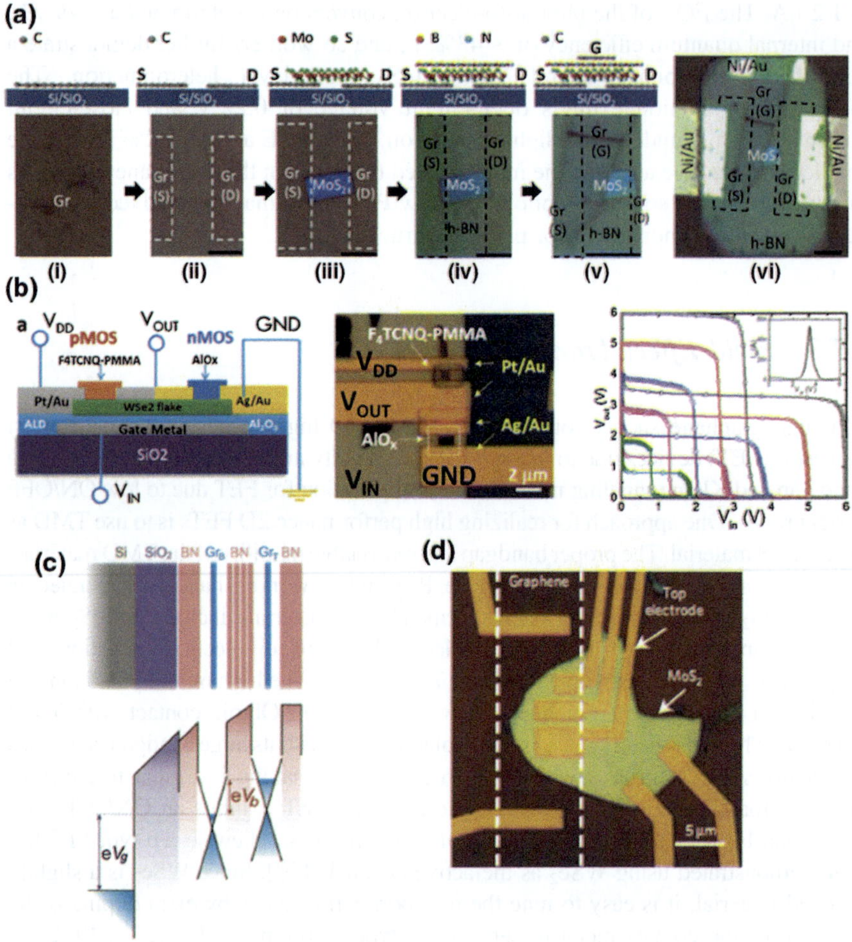

Fig. 7.17 Heterostructures for field-effect transistors. **a** Schematic illustration of the fabrication process and the related optical images of the all-2D FETs [248]. Reproduced with permission from Ref. [248]. Copyright 2014, American Chemical Society. **b** Schematic illustration and optical image of the WSe_2 CMOS structure. The right figure shows its voltage transfer characteristics of the logic inverter with Vdd supply from 2 to 6 V [253]. Reproduced with permission from Ref. [253]. Copyright 2015, American Chemical Society. **c** Schematic illustration of the graphene field-effect tunneling transistor and the corresponding band structure under operation [254]. Reproduced with permission from Ref. [254]. Copyright 2015 American Association for the Advancement of Science. **d** Optical image of the graphene/MoS_2/metal vertical transistor device [256]. Reproduced with permission from Ref. [256]. Copyright 2013, Nature Publishing Group

much faster than planar FET by tunneling mechanism, which is potential for high-speed operation. Sharing the same idea, Georgiou and co-workers have reported a graphene vertical FET by using WS_2 layers as barrier. The small band gap of WS_2 not only allows for tunneling transport but also for thermionic transport by electrically tuning the Fermi-level of graphene. Meanwhile, it exhibits a high

ON current and better current modulation with an ON/OFF current ratio up to 10^6 [255]. This device can further operate stably on transparent and flexible substrates with similar performance on SiO_2 substrates. Another type of vertical FET is simply with the structure of graphene/TMD/metal, where proper metal layer pro-vides an Ohmic contact with TMD to facility carrier transport and the current modulation is completed by electrical tuning through Schottky barrier between graphene and TMD layers. Yu and co-workers have reported a vertical FET with a graphene/few-layer MoS_2/metal heterostructure, which exhibits an ON/OFF ratio larger than 10^3 and high current density of 5000 A/cm^2 (Fig. 7.17d) [256]. Moriya and co-workers further improved the current modulation to larger than 10^5 and current density up to 10^4 A/cm^2 with a similar structure but better interface fabrication [257]. The advantages of this type of vertical FETs are the large current density and the small device scale, providing high potential for future high-density integration circuits. Besides the above commercial tunneling FETs, Sarkar and co-workers have recently revealed a band-to-band tunnel FET with the vertical stacking heterojunction of p-type Ge and n-type MoS_2, where its subthreshold swing (SS) 31.1 mV/decades is significantly lower than the thermionic limitation [258]. Their results open a venue for future high-performance and low power electronics.

7.7.4 *Photodetector*

The zero band gap and high mobility properties of graphene can provide broadband absorption and fast response time in photo-electronics, but the zero gap nature and weak optical absorption limit the photoresponsivity. The monolayer TMD materials have a direct band gap and thus a high absorption coefficient. However, the low mobility of TMDs will reduce the response time and extraction rate. Therefore, the development in related heterostructures is trying to combine both advantages to enhance the photo-responses. Britnell and co-workers have demonstrated an efficient photodetector with graphene/TMDs/graphene stacking heterostructures (Fig. 7.18a) [259]. The TMDs layers serve as a light absorber and the graphene electrodes can provide good carrier separation by tuning its Fermi level through doping. The combined heterostructure (graphene/WS_2/graphene) for PV devices shows photo-responsivity above 0.1 A/W, corresponding to the EQE up to 30%. Furthermore, Yu and co-workers provide an extra gate voltage control in graphene/MoS_2/graphene heterostructure to adjust the Schottky barrier between graphene and MoS_2, where the external electrical field can enhance the photo-carrier generation, separation and transport processes in this vertical stacking device (Fig. 7.18b) [260]. It has been demonstrated that the EQE can achieve 55% and 85% of internal quantum efficiency (IQE) by proper photocurrent modulation. Lee and co-workers have extended the range of these vertical stacking heterostructure through inducing p–n junction into the photovoltaic device, where the MoS_2/WSe_2 p–n junction is sandwiched by graphene electrodes [242]. The p–n junction can facilitate photo-excited electron–hole separation and the EQE has been shown to be dependent on the thickness of p–n junction. Cheng

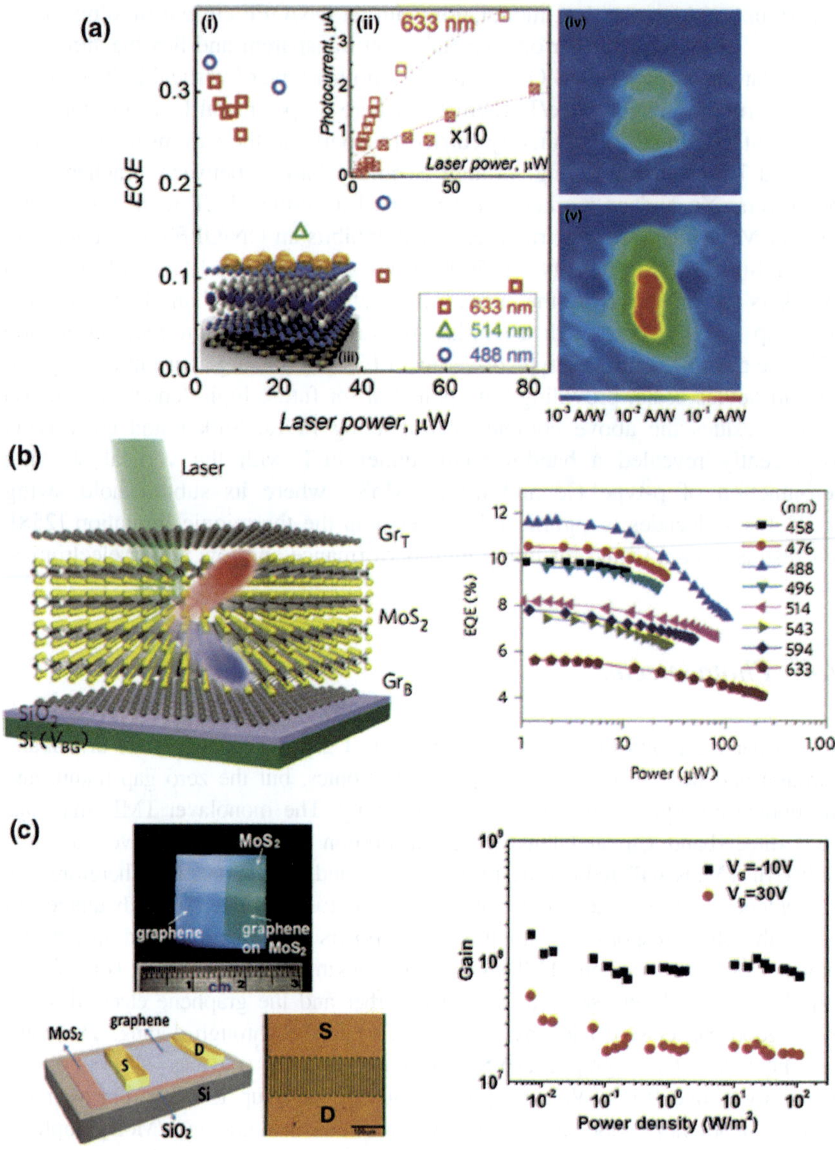

Fig. 7.18 Heterostructure for photodetector. **a** External quantum efficiency (EQE) of the graphene/WS_2/graphene heterostructure at different wavelengths and powers. Right image shows the photocurrent maps of h-BN/Gr/MoS_2/Gr devices without (up) and with (down) spattering of gold nanoparticles for plasmonic enhancement [259]. Reproduced with permission from Ref. [259]. Copyright 2013, American Association for the Advancement of Science. **b** Schematic illustration of the side view of the graphene/MoS_2/graphene stacking heterostructure device, and the EQE at different laser powers [260]

and co-workers have fabricated a vertical p–n junction by monolayer CVD WSe_2 and multilayer exfoliated MoS_2 [261]. The device shows photoresponses on entire overlapping area with a maximum EQE up to 12%. Similarly, Deng and co-workers report a p–n diode based on a p-type black phosphorus/n-type monolayer MoS_2 stacked p–n heterojunction [262]. The black phosphorus is an intrinsic p-type semiconductor with a high carrier mobility value ($\sim$10,000 cm^2/V s) and a small direct band gap about 0.3 eV in monolayer, suitable for broadband photodetectors. This black phosphorus/ MoS_2 heterojunction exhibits the photo-responsivity of 418 mA/W and EQE of 0.3% under 633 nm laser illumination, potentially useful for broadband photodetectors. It is noteworthy that the electroluminescence peaks have been observed at various junctions including vertically stacked p–n junction [261, 263], electrical field induced junction [264–267], and double heterojunction [268], suggesting the possibility of using TMD for LED devices.

Besides above mentioned heterostructures, the simple graphene/TMD heterojunction can provide sensitive photodetection by properly tuning the electrical potential at the interface. Zhang and co-workers have constructed a large area graphene/MoS_2 heterojunction with monolayer CVD graphene and MoS_2, where the electrodes are contacted only on top graphene layer and the MoS_2 is covered by graphene (Fig. 7.18c) [269]. The photo-excited electrons from MoS_2 will transport to graphene and the holes are trapped in MoS_2 based on the tunable build-in potential at the interface. At the same time, the high mobility in graphene will facilitate the electron extraction, but the holes will be trapped in MoS_2 resulting in a multiple recirculation of electrons in graphene. The recirculation phenomenon leads to a photogain higher than 10^8 and a photoresponsivity larger than 10^7 A/W. On the other hand, Roy and co-workers shows that the graphene/multi-layer MoS_2 heterojunction not only provides a high photoresponsivity near 5×10^8 A/W but also exhibits persistent photoconductivity through electric field tuning, which can serve as gate-tunable photodetectors and optical switches [270].

7.7.5 Thermoelectric Devices

On the basis of the layered stacking structure, the heat dissipation in 2D heterostructures is limited by the thermal transport in vertical direction [271]. The temperature gradient can be easily built in few-layer 2D materials or vertical heterostructures, potential for applications in thermoelectrics. Chen and co-workers have reported a Seebeck study on a graphene/h-BN/graphene heterostructure [272]. Through applying a temperature gradient on top and bottom graphene layers and measuring the thermo-electric voltage, the heterostructure shows a Seebeck coefficient about −99 μV/K. Together with the electrical and thermal conductance, the power factor (S^2G) of 1.51×10^{-15} W/K^2 and the thermoelectric figure of merit $ZT = 1.05 \times 10^{-6}$ can be derived. Although the thermoelectric energy conversion efficiency is still low in this device, the research area is at an early stage and it is anticipated that progresses can be made in the near future.

7.8 Fullerenes Molecules

Fullerenes, a carbon allotrope, also called as "bucky balls" were discovered in 1985. The Buckminster fullerene is the most common form of fullerene measuring about 7 Å in diameter with 60 carbon atoms arranged in a shape known as truncated icosahedrons. It resembles a soccer ball with 20 hexagons and 12 pentagons and is highly symmetrical.

Fullerenes are being investigated for drug transport of antiviral drugs, antibiotics and anticancer agents. Fullerenes can also be used as free radical scavengers due to presence of high number of conjugated double bonds in the core structure. These are found to have a protective activity against mitochondrial injury induced by free radicals. However, fullerenes can also generate reactive oxygen species during photosensitization. This property can be used in cancer therapy.

Fullerenes have the potential to stimulate host immune response and production of fullerene specific antibodies. Animal studies with C60 fullerene conjugated with thyroglobulin have produced a C60 specific immunological response which can be detected by ELISA with IgG specific antibodies.

This can be used to design methods of estimation of fullerene levels in the body when used for therapeutic or diagnostic purposes. On intravenous injection, these get distributed to various parts of the body and get excreted unchanged through the kidney. Soluble derivates of fullerenes are more biocompatible compared to insoluble forms of fullerenes and have low toxic potential even at higher doses. Further, the degree of purification of fullerene determines its cost and highly purified fullerenes are expensive, restricting its application in medical field.

Types of fullerenes:

1. Alkali doped fullerenes are structures with alkali metal atoms in between fullerenes contributing valence electrons to neighbouring fullerenes. They occur because of the electronegative nature of the fullerenes.
2. Endohedral fullerenes have another atom enclosed inside the buckyball. If a metallic atom is enclosed, these are called as metallofullerenes. Due to the small size of C_{60} fullerene, it is difficult to synthesize endohedral C_{60} fullerenes. However, larger fullerenes such as C_{82} or C_{84} fullerenes are used for synthesizing endohedral fullerenes. Endohedral metallofullerenes can be used for diagnostic purposes as radio contrast media in magnetic resonance imaging and other imaging procedures. Since the radioactive metal is enclosed within the buckyball, these are less toxic and safer. This method can also be employed for imaging organs as radioactive tracers.
3. Exohedral fullerenes also called as fullerene derivatives, are synthesized by chemical reaction between the fullerene and other chemical groups. These are also called as functionalized fullerenes.

   Such fullerenes can be used as photosensitizers in photodynamic therapy for malignancies. These generate reactive oxygen species when stimulated by light and kills the target cells. This method is now also being investigated for antimicrobial property as these cause cell membrane disruption especially in Gram positive bacteria and mycobacterium.

4. Heterofullerenes are fullerene compounds where one or more carbon atoms are replaced by other atoms like nitrogen or boron.

The Advantages of Carbon can be summarizes as:

- Carbon is a basic element of life.
- Carbon is special because of its ability to bond too many elements in many different ways.
- It is the sixth most abundant element in the universe.
- The most known types of carbon materials: diamond; graphite; fullerenes; and carbon nanotubes.

Allotropic forms of elemental carbon

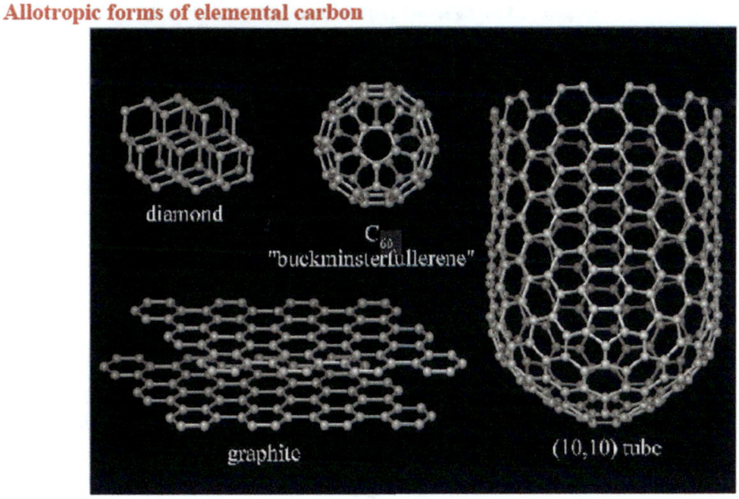

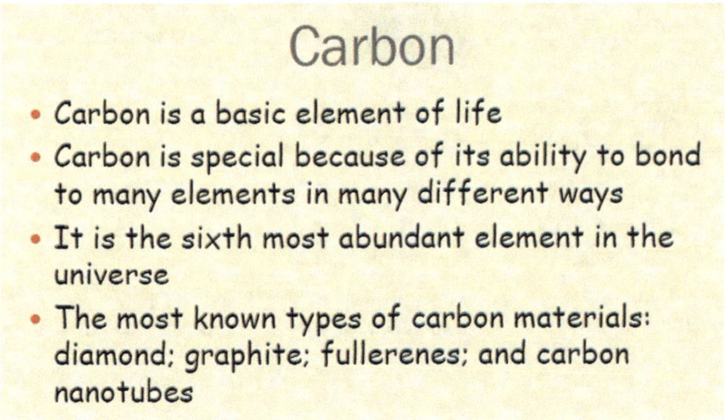

Size scale

Bohr radius = 0.5292Å ≈ 0.05 nm

C atom (VdW radius)=0.17 nm

In a 1nm line: 3 C atoms

In a 1nm·1nm surface: 9 C atoms

In a 1nm·1nm·1nm cube: 27 C atoms

In a 1m·1m·1m cube: $2.7 \cdot 10^{28}$ C atoms

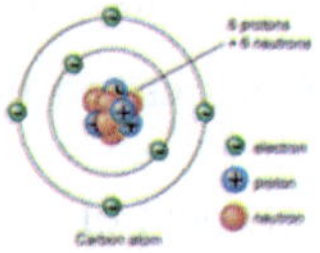

Typical nanosystems may contain from hundreds to tens of thousands of atoms.

Graphite: $2.3 \cdot 10^3$ Kg·m^{-3} = $1.15 \cdot 10^{29}$ C atoms·m^{-3}

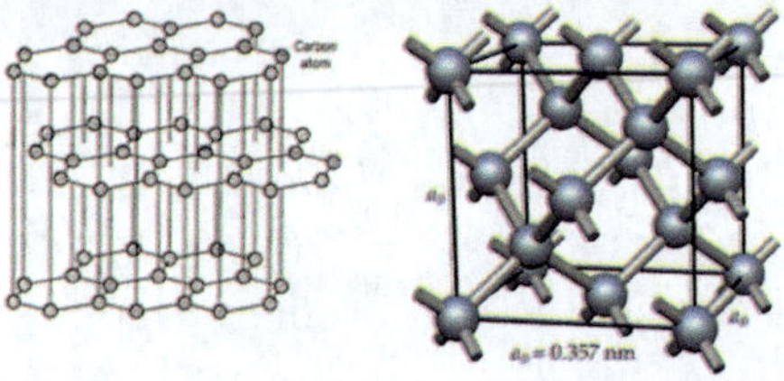

Diamond: $3.5 \cdot 10^3$ Kg·m^{-3} = $1.76 \cdot 10^{29}$ C atoms·m^{-3}

Carbon materials

2s and 2p electrons available for bonding

Diamond and graphite are **two allotropes** of carbon: pure forms of the same element that differ in structure.

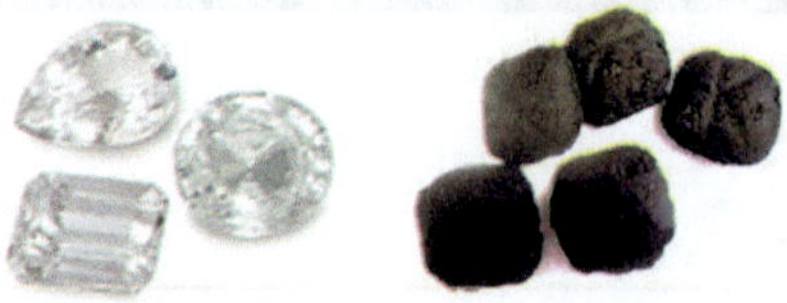

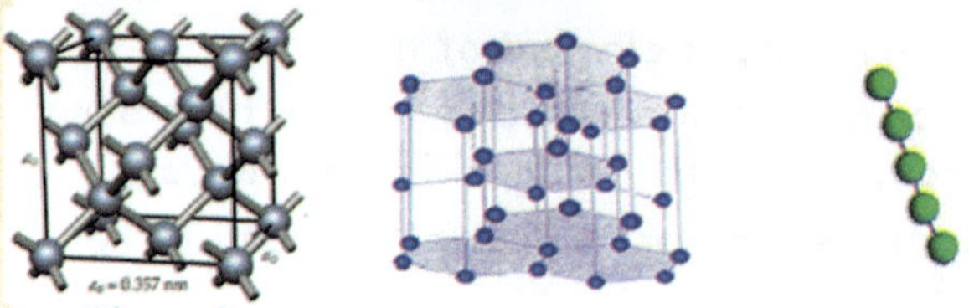

Diamond	Graphite	Chain
sp^3 (3D) 1332 cm^{-1}	sp^2 (2D) 1582 cm^{-1}	sp^1 (1D) 1855 cm^{-1}

Mainly, Carbon based molecular devices, which have aromatic rings or alkanes and structures like fullerene, nanotubes, graphene, nanoribbon and graphane are widely considered, as they are characterized by an amazingly large variety of structures and the properties that originate from the combination of *sp*, *sp2* and *sp3* hybridisations of Carbon atoms. The main parameters that affect the transport properties of two probe molecular devices are related to:

(i) Change in geometry/chemistry of molecules.
(ii) Change in geometry of contact between molecule and electrode.
(iii) Change in chemistry/geometry/dimensions of electrodes.

7.9 Diamond Molecules

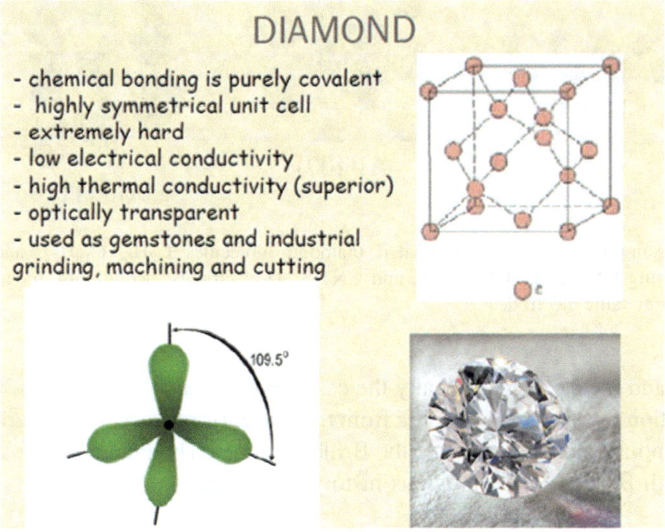

We here consider the relatively small example systems shown in Fig. 7.19, corresponding to three different hydrogenated diamond molecules, three different semi-conducting carbon nanotubes, both of growing sizes, and three different metals. In all cases, we represent the computational objects in a minimal single-ζ (SZ) basis, except for the gold electrodes, where single-ζ-polarized (DZP) is used. We apply periodic boundary conditions by taking advantage of the supercell method. For the molecules, this means that the simulation box has vacuum regions big enough in all dimensions to neglect interaction between repeated images. For

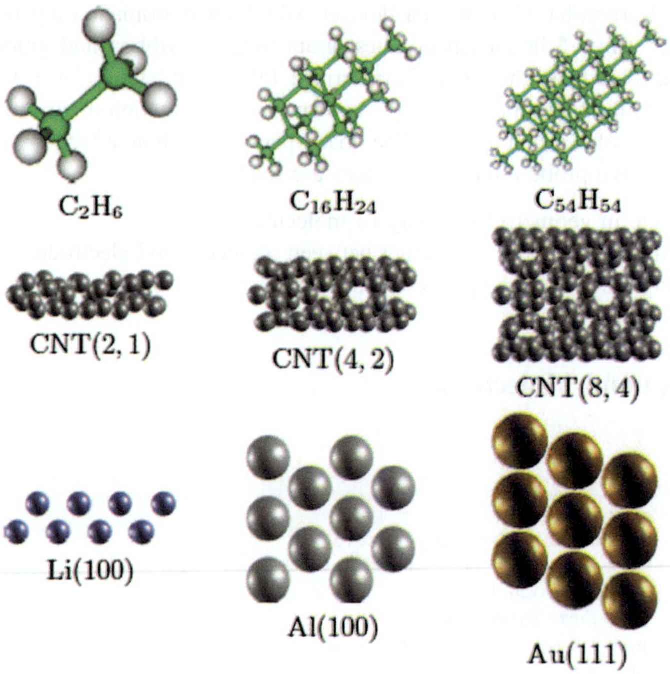

Fig. 7.19 Example systems: hydrogenated diamond molecules C_2H_6, $C_{16}H_{24}$, and $C_{54}H_{54}$; semi-conducting CNT(2, 1), CNT(4, 2), and CNT(8, 4) electrodes; bcc-lithium, fcc-aluminium, and fcc-gold metallic electrodes

the CNTs and metals, this is equally the case for the $\hat{x}$ and $\hat{y}$ dimensions, while for the $\hat{z}$ direction the system is periodic from the outset (for an accurate description we use 100 k-points along $\hat{z}$ to sample the Brillouin zone). The Pulay mixing scheme is applied with $\beta = 0.1$ and 6 previous history steps taken into account.

7.10 Carbon Nanotubes (Carbon-Based NPs)

Carbon nanotubes discovered in 1991 are tubular structures like a sheet of graphite rolled into a cylinder capped at one or both ends by a buckyball. Nanotubes can be single walled carbon nanotube (SWCNT) or multiwalled carbon nanotube (MWCNT) in concentric fashion. Single walled nanotube has an internal diameter of 1–2 nm and multiwalled nanotube has a diameter of 2–25 nm with 0.36 nm distance between layers of MWCNT.

Carbon Nano Tubes: Structures and properties

Carbon Nano Tubes: Physical properties

Average diameter of Single Wall Nano Tubes (SWNTs)	**1.2 - 1.4 nm**
Lattice parameter	
(10,10) Armchair	**16.78 Å**
(17,0) Zigzag	**16.52 Å**
Density	
(10,10) Armchair	**1.33 g/cm^3**
(17,0) Zigzag	**1.34 g/cm^3**
(12,6) Chiral	**1.40 g/cm^3**
Optical band gap	
For (n,m): n-m is divisible by 3 (metallic)	**0 eV**
For (n,m): n-m is not divisible by 3 (semi-cond.)	**~0.5 eV**
Electrical transport	
Conductance quantization	**(12.9 kΩ)$^{-1}$**
Resistivity	**10^{-4} Ωcm**
Maximum current density	**10^{13} A/m^2**
Thermal transport	
Thermal conductivity	**~ 2000 W/m·K**
Elastic behavior	
Young's modulus (SWNTs)	**~ 1 TPa**
Young's modulus (MWNTs)	**1.28 TPa**
Maximum tensile strength	**~ 100 GPa**

These vary in their length ranging from 1 μm to a few micrometers. These are characterized by greater strength and stability hence can be used as stable drug carriers. Cell specificity can be achieved by conjugating antibodies to carbon nanotubes with fluorescent or radiolabelling. Entry of nanotubes into the cell may be mediated by endocytosis or by insertion through the cell membrane.

Carbon Nano Tubes: Structures and properties

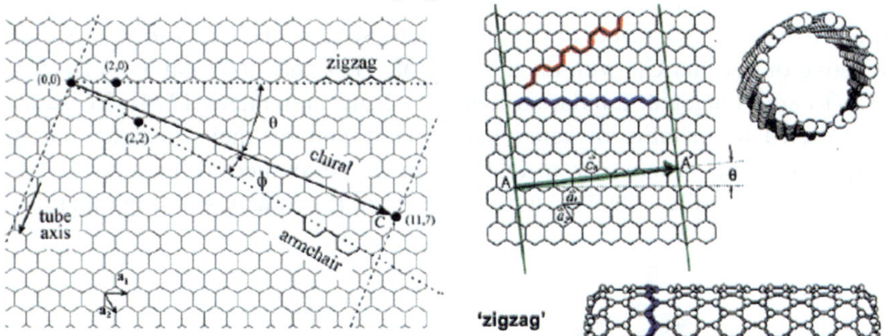

n and m determine the chirality and thus conductance, density, lattice structure, etc.

Metallic	Semiconducting
(n,n) armchair	(n,0) zigzag
(n,m) where n-m = 3x	(n,m) where n-m ≠ 3x
	band gap ~ 0.5 eV

Carbon nanotubes can be made more soluble by incorporation of carboxylic or ammonium groups to their structure and can be used for the transport of peptides, nucleic acids and other drug molecules. Indium-111 radionuclide labelled carbon nanotubes are being investigated for killing cancer cells selectively. Fullerenes and carbon nanotubes (CNTs) represent two major classes of carbon-based NPs. Fullerenes contain nanomaterial that are made of globular hollow cage such as allotropic forms of carbon. They have created noteworthy commercial interest due to their electrical conductivity, high strength, structure, electron affinity, and versatility [273].

Carbon Nano Tubes: Structures and properties

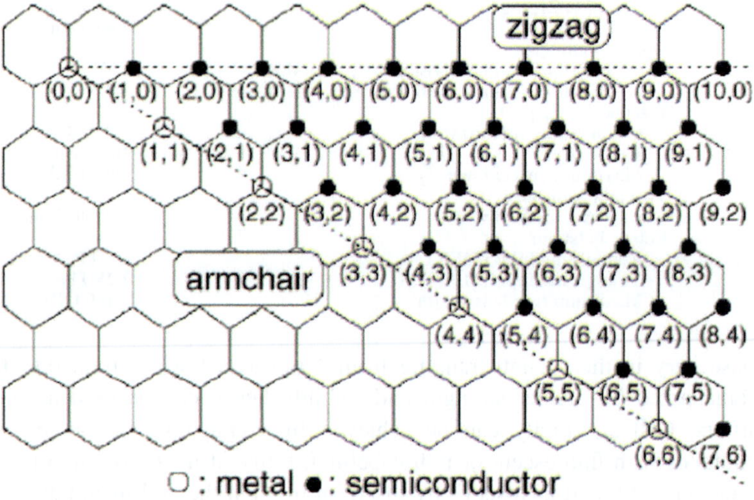

These materials possess arranged pentagonal and hexagonal carbon units, while each carbon is sp^2 hybridized. Figure 7.20 shows some of the well-known fullerenes consisting of C_{60} and C_{70} with the diameter of 7.114 and 7.648 nm, respectively.

CNTs are elongated, tubular structure, 1–2 nm in diameter [274]. These can be predicted as metallic or semiconducting reliant on their diameter telicity [275]. These are structurally resembling to graphite sheet rolling upon itself (Fig. 7.21). The rolled sheets can be single, double or many walls and therefore

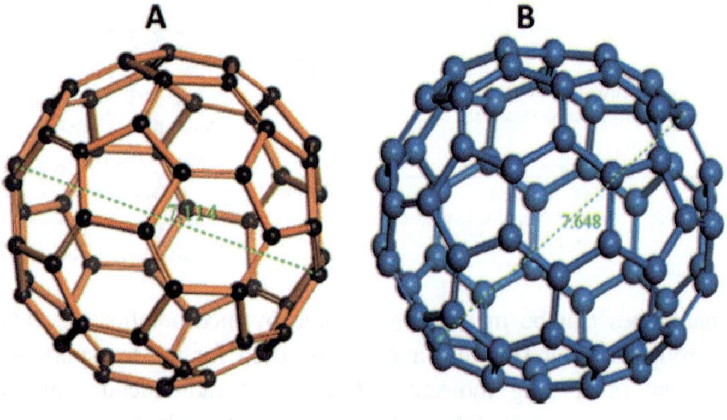

Fig. 7.20 Different form of fullerenes/buck balls **a** C_{60} and **b** C_{70}

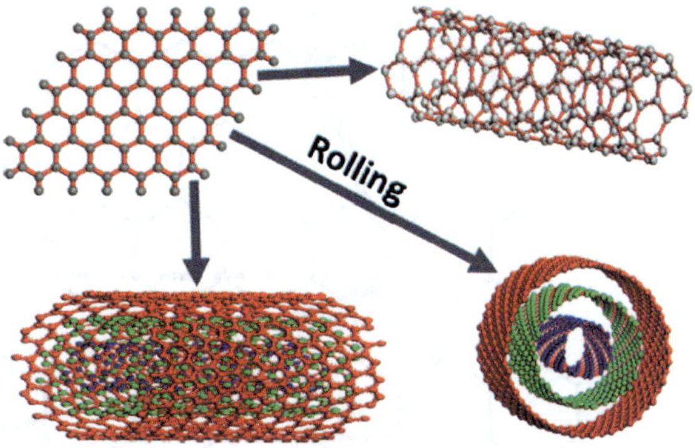

Fig. 7.21 Rolling of graphite layer into single-walled and multi-walled CNTs

they named as single-walled (SWNTs), double-walled (DWNTs) or multi-walled carbon nanotubes (MWNTs), respectively. They are widely synthesized by deposition of carbon precursors especially the atomic carbons, vaporized from graphite by laser or by electric arc on to metal particles. Lately, they have been synthesized via chemical vapor deposition (CVD) technique [276]. Due to their unique physical, chemical and mechanical characteristics, these materials are not only used in pristine form but also in nanocomposites for many commercial applications such as fillers [277, 278], efficient gas adsorbents for environmental remediation [279], and as support medium for different inorganic and organic catalysts [280] (Fig. 7.22).

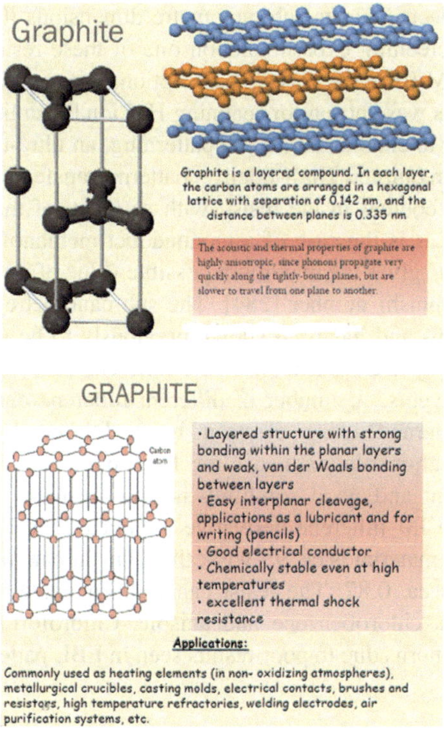

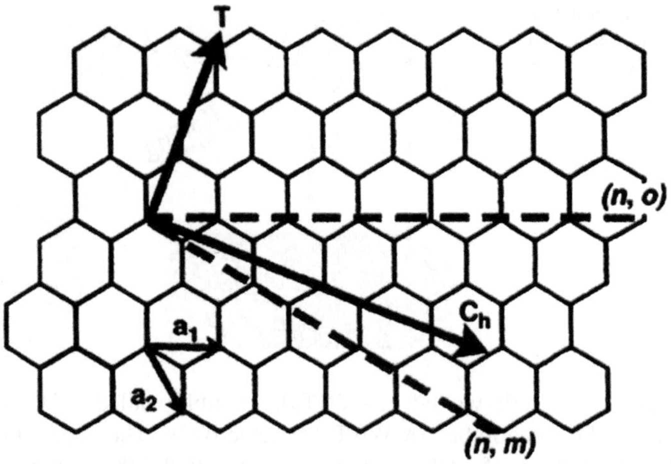

Fig. 7.22 Graphitic sheet showing the basis vectors a_1 and a_2 of the two-dimensional unit cell, the axis vector *T* about which the sheet is rolled to generate the armchair structure nanotube, and the circumferential vector C_h at right angle to *T*. Other orientations of *T* on the sheet generate the zigzag and chiral structures

Novel fullerene formulations from Irresistible Materials Ltd (IM) and Birmingham University, UK, which have demonstrated etch selectivity with respect to Si of 9:1 in fluorine plasma chemistry [281]. These resists are polymerized from the C_{60} molecule, selected for its sub-nanometre dimensions (0.7 nm) and its stability [282]. Resist screening experiments on one of these resists with the experimental codename HM-01 have revealed exceptional plasma etch resistance and stability [283] and this was chosen for scanning He^+ ion beam lithography (SHIBL) experiments. In order to enable sub-10 nm patterning, an ultra-thin resist film, with small molecules is required. High-fidelity pattern transfer via plasma etching requires high carbon content in the resist, with as many of the carbons in a ring structure as possible, as is the case of a monoadduct methanofullerene derivative. The fullerene molecule gives the maximum possible value of ring parameter and the minimum possible Ohnishi number [284]. The sub-nanometre fullerene molecule forms amorphous films and has been shown previously to be suitable for electron beam lithography [285]. Derivatisation of the fullerene renders it soluble in common spin-coating solvents. A number of different fullerene derivatives have been spin-coated and patterned using electron beam lithography [285, 286]. An ultrathin-film, ultrahigh-resolution resist for He^+ ion beam exposure with high overall carbon content and good solubility is now available from IM Ltd. The structural class of the IM fullerene derivative is indicated in Fig. 7.23 [287]. The carbon content of the material is ca. 95 wt%; the Ohnishi parameter is ca. 1.26, and the ring parameter is ca. 0.87. The resist can be formulated in various solvents, including chloroform, chlorobenzene and anisole. Chloroform was rejected as a solvent for SHIBL patterns due to poor results seen in EBL patterning and due to its

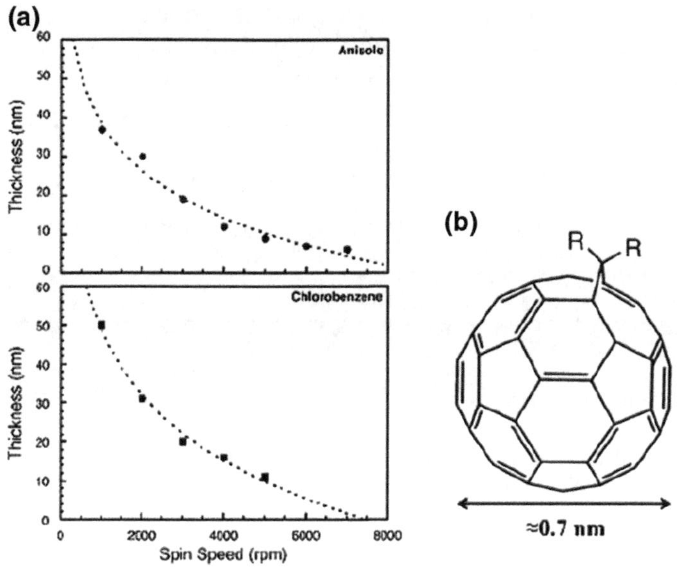

Fig. 7.23 **a** Spin-coating results for fullerene resists HM-01A and HM-01C using anisole and chlorobenzene solvents, respectively. **b** Schematic of the monoadduct methanofullerene molecule used in the HM resist series [287]

hazardous nature, and development has been focused on two formulations: HM-01A using anisole solvent and HM-01C using chlorobenzene. The spin-coatability of the two formulations is shown in Fig. 7.23a.

Figure 7.24 shows 8 nm isolated lines written with a line dose of 0.08 nC cm^{-1} in a 10 nm layer of HM-01A negative tone fullerene resist [288]. (Many novel resists require physical vapour deposition but HM-01A and HM-01C have the advantage of being spin-coatable.)

A comparison of the sensitivity of the HM resists for He^+ ion beam lithography at a beam energy of 30 keV and that for electron beam lithography at the same beam energy is shown in Fig. 7.25 [288].

The HM resists are up to 500-times more sensitive in SHIBL than in EBL at the same beam energy. This is markedly better than HSQ for which the corresponding figure is 4.4-times. The contrast in SHIBL is higher for the anisole version HM-01A.

The results of SHIBL in 10 nm thick HM-01A fullerene resist at 30 keV He^+ ion beam energy are shown in Fig. 7.26 for dense features [288]. Figure 7.26a shows dense single-pixel features exposed at a line dose of 0.09 nC cm^{-1}. The SE contrast measurement reveals continuous lines with 8.5 nm line width and 17 nm pitch. Figure 7.26b shows that 6 nm lines on 12 nm pitch were resolvable but the lines were broken at a line dose of 0.04 nC cm^{-1}, which is equivalent to 25 ions per nanometre, i.e., a signal-to-noise ratio of 5:1. The line discontinuity is caused by shot noise producing missing pixels. Thus 6 nm represents the current limit of 1:1 dense features using SHIBL to expose HM-01A fullerene resist.

Fig. 7.24 Isolated dose-optimised SHIBL experiments on HM01 fullerene resist: 8 nm wide sparse exposed using 30 keV He^+ ions [288]

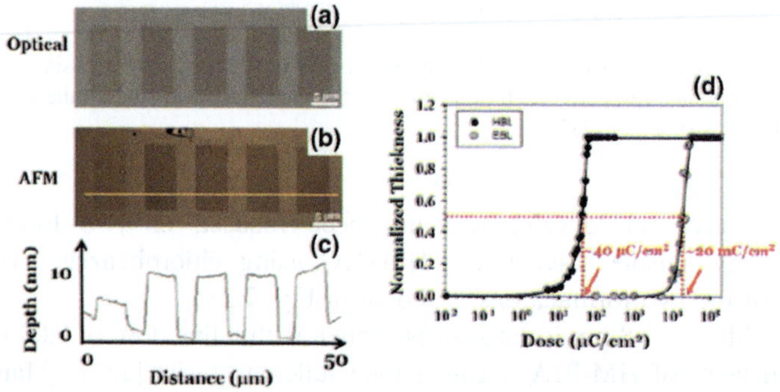

Fig. 7.25 Results of SHIBL at 30 keV beam energy: **a** optical micrograph, **b** AFM image, **c** line profile (HM-01 fullerene resist thickness ca. 10 nm). He^+ doses left to right: 37.9, 56.5, 85.0, 128.0, and 192.0 μC cm^{-2}, **d** comparison of the dose response curves for HM-01 in SHIBL and EBL. Sensitivities are 40 μC cm^{-2} and 20 mC cm^{-2}, respectively, revealing a 500-fold increased sensitivity in SHIBL [288]

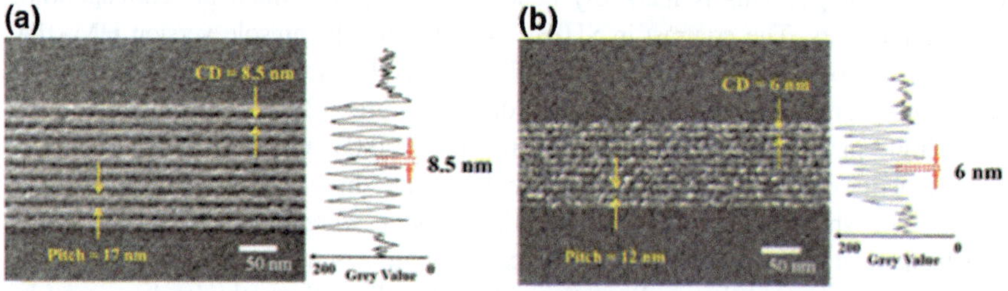

Fig. 7.26 HIM image of dense (1:1) single-pixel features exposed at **a** 0.09 nC cm^{-1} and **b** at 0.04 nC cm^{-1} in ca. 10 nm thick HM-01A fullerene resist, with line scans of secondary electron intensity providing measurements of linewidth [288]

7.11 Graphene Background

Graphene has definitely been the single biggest source of the interest and is the most widely used and studied two dimensional material but two of the main draw backs are the lack of a band gap and the need to redesign existing devices to incorporate carbon based electronics and traditional Si-based device architectures. Whilst overcoming these issues continues to be the subject of intense attention alternatives to single layer graphene have become the subject of growing focus, for instance bilayer graphene, where a band gap can be introduced, and silicene and germanene which can be more easily incorporated with the existing electronic technology. Several areas of interest emerge when considering bilayer graphene and silicon/germanium based alternatives and this book will cover several of those relating to electronic and vibrational properties. For bilayer graphene the main consideration will be the effect of non-covalent bonding between layers on the electronic behaviour and the vibrational modes. In silicene and germanene the change in mass and electron density is the largest concern, where more loosely bound outer electrons could have profound repercussions.

Graphene is now being viewed as a strategic resource in most of the industrial economies of the world. We believe that Imagine Intelligent Materials IM is positioned at the forefront of developing commercially viable graphene solutions. Imagine IM is the leading Australian developer of graphene-based coatings for industrial textiles and fibres. It was founded in 2014 by a group of scientists led by Chris Gilbey and Phil Aitchison, with a vision to create disruptive products and solutions that use graphene. Imagine IM has developed a licensing and certification model to ensure that participants in the supply chain are required to meet a set of standards of materials quality. Imagine IM is the first company in the world to develop conductive geomaterials using functionalised graphene.

Graphene stands out amongst other 2D materials due to its unique electronic band structure and the associated record carrier mobility, thermal conductivity, and high transparency. These properties make it a serious candidate for the replacement of indium tinoxide (ITO), a transparent conductive material, successfully used in liquid crystal displays, touch panels, or light-emitting diodes, whose performance is compromised by its low-flexibility, conflicting opposing relationship between its optical transparency and conductivity, and low abundance of indium [289]. However, many other 2D materials have what graphene does not: a band gap of 1–2 eV, which is well matched to the solar spectrum and is com-parable to the current industrial standards in photovoltaics: Si (1.1 eV), GaAs (1.4 eV), or CdTe (1.5 eV) [290]. 2D crystals of high-κ dielectrics, i.e. highly insulating materials such as hBN, mica, or perovskites, can be used as tunnelling barriers in capacitors and gate insulators [133]. The future generations of nanoelectronics, beyond the silicon-based era, need to address the ever-increasing drive for the reduction in size of the circuit-elements in digital electronics without compromising their quality. Utilisation of the sub-nanometre thickness and high quality of atomically-thin 2D materials in electronic circuits offers an effective electrostatic control of conductivity. Furthermore, the lack of surface groups or dangling bonds in 2D materials reduces charge carrier scattering otherwise common in bulk materials. The high surface-to-volume ratio of 2D materials is directly exploitable in electrochemistry [291] and results in their extreme sensitivity to the environment applicable in molecular sensors [218]. The vanishingly small thick-ness of 2D materials should further lead to a reduction in volume, mass, and ultimately the cost of devices in many applications such as transparent solar cells, conductive displays, or protective light-absorbing layers. Other applications such spintronics, valleytronics, photonics, or plasmonics, exploiting more exotic properties of 2D materials, were also proposed or realised [22, 37, 292–295].

What is Graphene?

The simplest way to describe graphene is that it is a single, thin layer of graphite—the soft, flaky material used in pencil lead. Graphite is an allotrope of the element carbon, meaning it possesses the same atoms but they're arranged in a different way, giving the material different properties. For example, both diamond and graphite are forms of carbon, yet they have wildly different natures. Diamonds are incredibly strong, while graphite is brittle. Graphene's atoms are arranged in a hexagonal arrangement as shown in Fig. 7.27.

Interestingly, when graphene is isolated from graphite it takes on some miraculous properties. It is a mere one atom thick, the first two-dimensional material ever discovered. Despite this, graphene is also one of the strongest materials in the known universe. With a tensile strength of 130 GPa (gigapascals), it is more than 100 times stronger than steel.

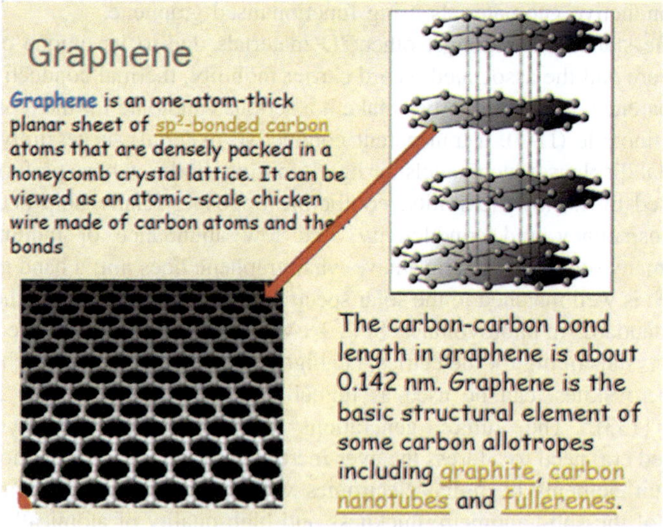

Fig. 7.27 Graphene's atoms are arranged in a honeycomb pattern Alex LMX/Shutterstock

Graphene's incredible strength despite being so thin is already enough to make it amazing, however, its unique properties do not end there. It is also flexible, transparent, highly conductive, and seemingly impermeable to most gases and liquids. It almost seems as though there is no area in which graphene does not excel.

Typical configurations for mono-vacancies (MVs) in graphene, MoS_2 and phosphorene are shown in Fig. 7.28. It is seen from Fig. 7.28a that there is only one chemical type of MV in graphene. For MoS_2, there are two types of MVs: Mo vacancy (VM) and S vacancy (VS), as shown in Fig. 7.28b. For phosphorene, as shown in Fig. 7.28c, there are two types of MV with the same degenerate energy level [296]. An important quantity in discussing vacancies is their formation energy (E_f), which essentially dictates their thermodynamic equilibrium concentration. The value of E_f for MV in phosphorene is 1.65 eV, significantly smaller than that of graphene of 7.57 eV [296]. Since the defect population depends exponentially on E_f, the equilibrium concentration of MVs in phosphorene should be exceedingly larger than that of graphene under the same conditions. The much lower value of E_f in phosphorene is associated with the inherently softer P–P bonds compared with the much stronger C–C bond, and also the curvature effect related to the structural buckling. For MoS_2, the value of E_f for V_s in monolayer MoS_2 (from 1.22 to 2.25 eV) is smaller than or comparable to that in phosphorene [297, 298]. This may

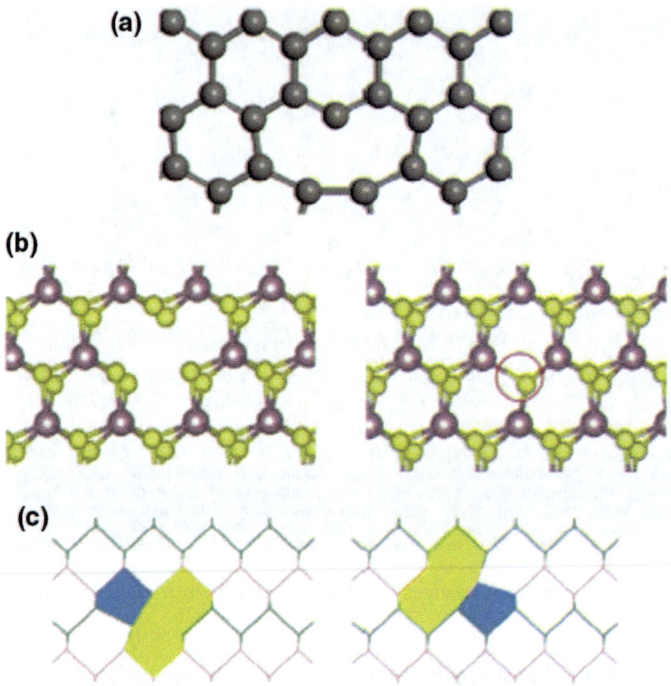

Fig. 7.28 Vacancy configurations in 2D materials: **a** graphene, **b** MoS_2 (Mo purple and S yellow), and **c** phosphorene. (For interpretation of the references to color in this figure legend, the reader is referred to the web version of this article.)

explain the high Vs concentration often observed experimentally. Here, the variability of the chemical potential (elemental rich and poor conditions) defines the upper and lower bounds of E_f.

The mobility of MV is determined by their diffusion energy barrier, E_b. The sequence of E_b for MV diffusion in the above 2D materials follows: phosphorene (0.40 eV) < graphene (1.39 eV) < MoS_2 (V_S, 2.27 eV) [296]. Therefore, phosphorene has the lowest value of E_b, which is only one third of that of graphene. According to the Arrhenius formula, the hopping rate (v) can be calculated by $v = v\text{s}\ \exp(-E_b/kT)$, where vs is the characteristic frequency, normally around 1013 Hz, k is the Boltzmann constant and T is the temperature. At room temperature, the rate-limiting process of the migration of MV in phosphorene is estimated to be amazingly rapid, around 16 orders of magnitude faster than that in graphene, signifying a significantly greater vacancy activity [296].

History of Graphene: A Roll of Tape, and a Dream

Graphite has been a known quantity long time (humans have been using it since the Neolithic era). Its atomic structure is well documented, and for a long time, scientists pondered whether single layers of graphite could be isolated. Until recently, however, graphene was merely a theory, as scientists were unsure if it would ever be possible to slice graphite down to a single, atom-thin sheet. The first isolated sample of graphene was discovered in 2004 by Andre Geim and Konstantin Novoselov at the University of Manchester. One might expect that they isolated the fabled substance using some massive, expensive piece of machinery, but the tool they used was amusingly simple: a roll of scotch tape.

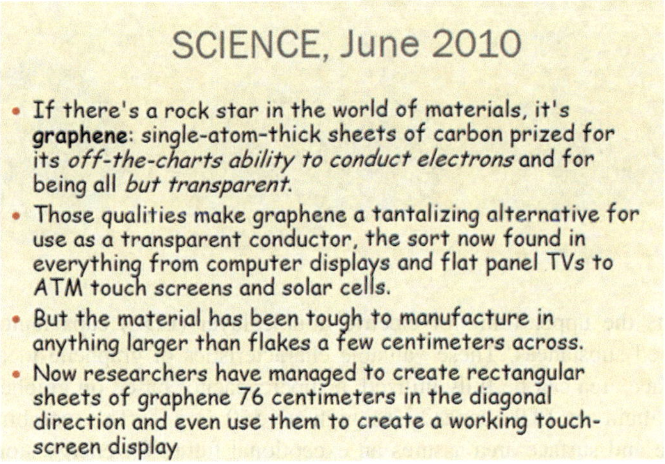

When using tape to polish a large block of graphite, the researchers noticed exceptionally thin flakes on the tape. Continuing to peel layer and layer from the flakes of graphite, they eventually produced a sample as thin as possible. They had found graphene. The discovery was so bizarre, the scientific world was skeptical at first. The popular journal *Nature* even rejected their paper on the experiment twice. Eventually, their research was published, and in 2010 Geim and Novoselov were awarded the Nobel Prize in Physics for their discovery.

Why Graphene?

One of the numerous sought after properties of graphene is its high surface area (up to 2630 $m^2\ g^{-1}$). This is of notable interest for supercapacitors. Researchers from Tsinghua University, Beijing, and Arizona State University recently detailed the intrinsic capacitance of single layer graphene to be 21-μF cm^{-2}.

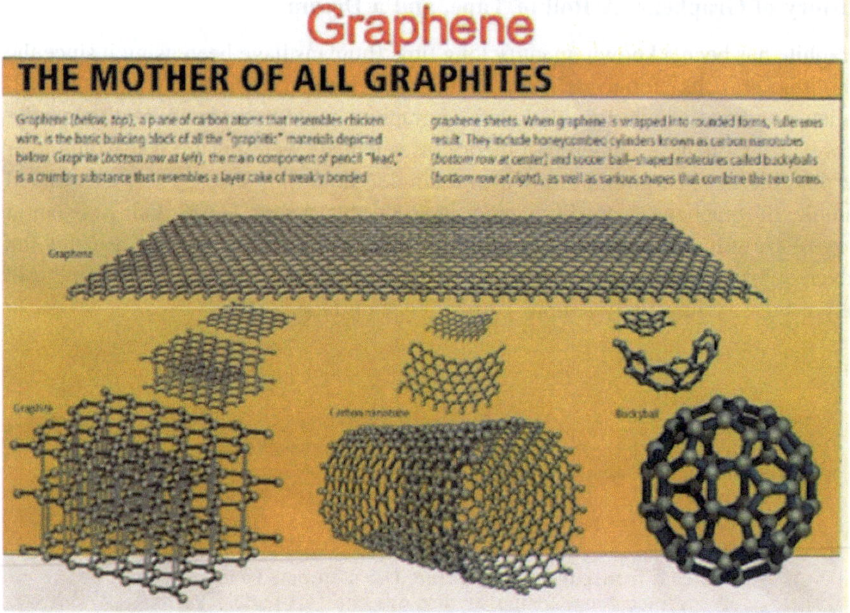

This sets the upper limit for electric double layer (EDL) capacitance for all carbon based substances. These valuable characteristics of graphene mean, if the whole surface area can be fully utilized, a supercapacitor based on graphene could in theory attain an EDL capacity as high as 550 F g^{-1}. This combination of capacitance and surface area assures an exceptional future for energy storage.

The performance of graphene supercapacitors has been outstanding since the initial early studies. Recently, a researcher at Australia's Swinburne University has created a battery that can be used repeatedly and charges up in seconds. Dr. Han Lin, of the Centre for Micro-Photonics, has developed a supercapacitor that not only overcomes the barriers of long charging times and costs less than a traditional lithium ion battery over its lifetime.

Researchers from UC Santa Cruz and Lawrence Livermore National University recently doubled the performance of 3D printed graphene based supercapacitors. Their method entails sandwiching perchlorate ion and lithium ion between layers of graphene in aerogel electrodes.

This allowed the researchers to strengthen the 3D printed graphene aerogel. Yat Li, UCSC professor and corresponding author said that they, "exfoliated the stacked graphene layers and functionalizing their surface, without damaging structural integrity."

Graphene is also useful for a wide scope of batteries as well as supercapacitors:

- Lithium-ion batteries
- Metal-air
- Redox flow
- Lithium-sulfur.

First principles calculations reveal that lithium ions can be stored on both sides of graphene to form C_3Li. Graphene can accordingly deliver a theoretical capacity of 744 mAh g^{-1}, which is about double the capacity of conventional graphite electrodes. This was shown in a recent example featured in the Journal of Materials Chemistry.

7.12 Potential Applications of Graphene

The 2D nature of graphene makes the chemical functionalization of graphene a promising approach to modulating the graphene properties. For example, hydrogenation of graphene involves bonding atomic hydrogen to the carbon atoms in graphene [299]. Such a reaction changes the hybridization of the C–C bonds in graphene from sp2 into sp3. As a result, the 2D atomic structure of pristine graphene is distorted [300]. It has been shown that suitable single-sided hydrogenation of graphene can lead to folding of graphene in a programmable fashion. This feature can be leveraged to achieve the hydrogenation assisted graphene origami (HAGO), in which initially planar, suitably patterned graphene can selfassemble into three dimensional nanoscale objects of desirable shapes (Fig. 7.29) [301]. A unique feature of the HAGO process is that the resulting nanostructure can be modulated by an external electric field, enabling programmable opening and closing of the nano-objects, a desirable feature to achieve molecular mass manipulation, storage and delivery. For example, MD simulations have demonstrated HAGO-enabled nanocages for controllable uptake and release of nanoparticles as well as ultra-high density of hydrogen storage.

Recent experiments showed that monolayer graphene can be patterned via optical lithography into desirable shapes, such as a spiral spring, a kirigami pyramid, a cantilever or a nanomesh [302, 303]. Such a patterned graphene structure is planar as fabricated, but can deflect and twist out of the plane when stretched. The out-of-plane deformation can accommodate huge in-plane elongation without substantial strain in a suitably patterned structure [304]. As a result, the patterned graphene achieves significant elastic stretchability [302, 303, 305], way beyond the elastic limit of pristine graphene (Fig. 7.30). For example, systematic coarse-grained simulations revealed that a suitably patterned graphene nanomesh can be made extremely compliant with nearly zero stiffness up to about 20% elongation and then remain highly compliant up to about 50% elongation (Fig. 7.30b).

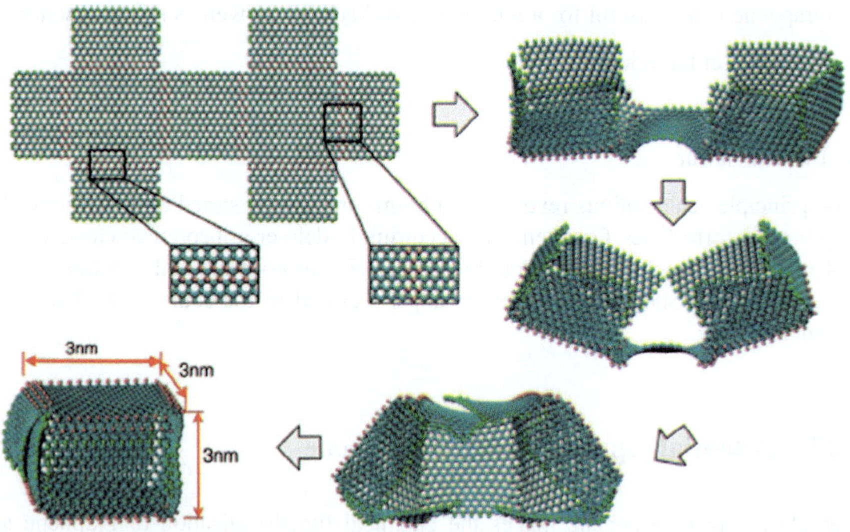

Fig. 7.29 A monolayer graphene patterned into a double-cross shape and suitably hydrogenated (insets) can spontaneously fold and form a graphene nanocage. *Source* Figure adapted from [301]

These features of graphene kirigami are desirable for graphene-based functional devices such as epidermal electronics and sensing prosthesis [303] as well as stretchable and tunable quantum dot arrays [306].

Graphene as nanomaterial is used for our largest focus industries that produce products such as cement and asphalt concretes, composites, protective coatings. As well as energy generation and storage businesses. Fullerenes or graphene, which are of nanosize, have even been found in space while they have been synthesized by man only recently. Wood is one of the most common natural nanomaterials. It has a hierarchical scale structure. At the largest scale, wood contains soft fibers with a diameter of about 20–30 μm and a length typically between 2 and 5 mm. At an intermediate hierarchical scale, nanofibers are present with a diameter less than 100 nm and a length greater than 1 μm. The smallest scale contains crystallites with a width less than 5 nm and a length less than 300 nm. It turns out that mechanical properties improve as the size of the structure decreases. For example, the elasticity is multiplied by almost 12 and the strength by 100 as one goes from softwood structure to wood nanocrystals. Nanocellulosic materials are also found in nature. These can be collagen fibrils originating from animal sources; nanofibers originating from wood, plants, crops or bacteria; crystals or whiskers deriving from wood, plants and crops. Cellulose nanocrystals or whiskers can be isolated by mechanical and chemical processes for subsequent use in polymer nanocomposite materials. Nature has developed several ways to build macroscopic systems from nanobricks, like seashell or bones, for

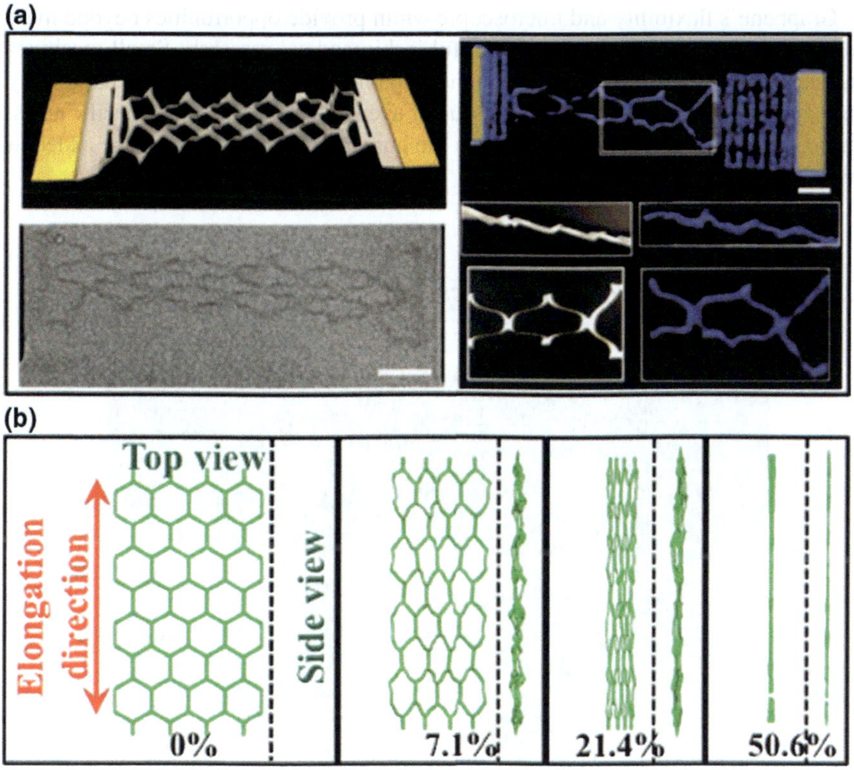

Fig. 7.30 **a** (Left) Paper and graphene in-plane kirigami springs, respectively. (Right) Three-dimensional reconstruction of the deformed graphene spring under large elongation. **b** Deformation sequence of a graphene nanomesh subject to an elongation beyond 50%. *Source* Figures adapted from: **a** [302] and **b** [303]

example. The synthesis is carried out using nanocrystals at moderate temperature and pressure. The biomineralization process involves calcium carbonate or phosphate. Biomimetic strategies are now frequently used.

If graphene had merely one of its many superlative traits, it would be the subject of intense research into potential uses. Begin so remarkable in so many ways, graphene has inspired scientists to think of a wide range of uses for the material, in fields as varied as consumer tech and environmental science.

In addition to its powerful electrical properties, graphene is also highly flexible and transparent. This makes it perfect for use in portable electronics. Smart phones and tablets could become much more durable using graphene, and perhaps could even be folded up like paper. Wearable electronic devices have been growing in popularity recently. With graphene, these devices could be made even more useful, designed to fit snugly around limbs and bending to accommodate exercise.

Graphene's flexibility and microscopic width provide opportunities beyond mere consumer devices. It could also be useful in biomedical research. Small machines and sensors could be made with graphene, capable of moving easily and harmlessly through the human body, analyzing tissue or even delivering drugs to specific areas. Carbon is already a crucial ingredient in the human body; a little graphene added in might not hurt.

7.12.1 Solar Cells/Photovoltaics

Graphene is both highly conductive and transparent. As such, it has great potential as a material in solar cells. Typically, solar cells use silicon, which produces a charge when a photon hits the materials, knocking loose a free electron. Silicon only releases one electron per photon that hits it. Research has indicated that graphene can release multiple electrons for each photon that hits it. As such, graphene could be far better at converting solar energy, with a projected 60% efficiency compared to the roughly 25% efficiency that current silicon cells are capable of.

Unfortunately, this is all theoretical. Current graphene cells are not yet up to par with silicon cells. Thankfully, research into graphene solar cells is ongoing, and the efficiency of such cells is increasing. Before long, cheaper, more powerful graphene cells could produce a massive surge in renewable energy.

In less grandiose fields, graphene's photovoltaic properties also mean that it could be used to develop better image sensors for devices such as cameras.

7.12.2 Semiconductors

Due to its high conductivity, graphene could be used in semiconductors to greatly increase the speed at which information travels. Recently the Department of Energy conducted tests which demonstrated that semiconductive polymers conduct electricity much faster when placed atop a layer of graphene than a layer of silicon. This holds true even if the polymer is thicker. A polymer 50-nm thick, when placed on top of a graphene layer, conducted a charge better than a 10 nm thick layer of the polymer. This flew in the face of previous wisdom which held that the thinner a polymer is, the better it can conduct charge. Yet another example of graphene's remarkable properties.

The biggest obstacle to graphene's use in electronics is its lack of a band gap, the gap between valence and conduction bands in a material that, when crossed, allows for a flow of electrical current. The band gap is what allows semiconductive materials such as silicon to function as transistors; they can switch between insulating or conducting an electric current, depending on whether their electrons are pushed across the band gap or not. Because graphene has no band gap, electrical charges simply flow along, as in any metal.

New research conducted by the Georgia Institute of Technology may have changed this, however. The researchers claim to have produced graphene samples with a band gap of 0.5 electronvolts, which they say should be enough for graphene to function as a semiconductor. If true, this could lead to graphene supplanting silicon as the semiconductor of choice, enabling much faster electronics to be made.

7.12.3 Water Filtration

Graphene's tight atomic bonds make it impermeable for nearly all gasses and liquids. Curiously, water molecules are an exception. Because water can evaporate through graphene while most other gasses and liquids cannot, graphene could be an exceptional tool for filtration. Researchers at the University of Manchester tested graphene's permeability with alcohol, and were able to distill very strong samples of spirits, as only the water in the samples was able to pass through the graphene.

Of course, graphene's use as a filter has potential beyond distilling stronger spirits. Graphene could also be immensely helpful in purifying water of toxins. In a study published by The Royal Society of Chemistry, researchers showed that oxidized graphene could even pull in radioactive materials such as uranium and plutonium present in water, leaving the liquid free of contaminants. The implications of this study are massive. Some of the biggest environmental hazards in history, including nuclear waste and chemical runoff, could be cleansed from water sources thanks to graphene.

As overpopulation continues to be one of the world's most pressing environmental concerns, maintaining clean water supplies will only become more important. Indeed, water scarcity afflicts more than a billion people worldwide, a number that will only continue to rise given current trends. Graphene filters have immense potential to improve water purification, increasing the amount of fresh water available. In fact, Lockheed Martin recently developed a graphene filter called "Perforene," which the company claims could revolutionize the desalination process.

Current desalination plants use a method called reverse osmosis to filter salt out of seawater. Reverse osmosis uses pressure to move water through a membrane. In order to produce large amounts of drinkable water, the pressure involved requires enormous amounts of energy. A Lockheed Martin engineer claims their Perforene filters could reduce the energy requirements a hundred times less than that of other filters.

7.12.4 Superconductivity

Not long after scientists at Cambridge demonstrated that graphene can act as a superconductor (a material with no electrical resistance) when paired with praseodymium cerium copper oxide, researchers at MIT recently discovered another astounding property: It can apparently function as a superconductor alone, in the right configuration. The researchers stacked two slices of graphene, but offset them by an angle of 1.1 degrees. According to a report published in Nature, "Physicist Pablo Jarillo-Herrero at the Massachusetts Institute of Technology (MIT) in Cambridge and his team weren't looking for superconductivity when they set up their experiment. Instead, they were exploring how the orientation dubbed the magic angle might affect graphene."

What they discovered is that, when they ran electricity through the offkilter graphene stack, it functioned as a superconductor. This simple process of applying electricity makes graphene easier to study than a similar class of superconductors, cuprates, although those materials display superconductivity at much higher temperatures. Most materials that display superconductivity only do so near a temperature of absolute zero. Some so-called "high temperature superconductors" can display superconductivity at temperatures around 133 K (−140 °C), which is relatively high; hydrogen sulfide, under enough pressure, displays the property at a miraculous −70 °C!

The graphene arrangement had to be cooled to 1.7 degrees above absolute zero, however the researchers consider its behavior similar to that of cuprates, and so they hope that it will be a much easier material for studying unconventional superconductivity, which is still an area of great disagreement among physicists. Because superconductivity typically only happens at such low temperatures, superconductors are only used in costly machinery like MRI machines, but scientists hope to one day find a superconductor that works at room temperature, which would bring down costs by removing the need for cooling units.

7.12.5 The Latest Developments Graphene Supercapacitors

Interest in graphene for energy storage applications has hugely increased over the last ten years. Almost 500 papers were published on the subject in the first half of this year alone, with many more certain to happen. In 2008, one of the first graphene based ultracapacitors (also known as supercapacitors) was announced and interest has been increasing ever since.

Published in NanoLetters by Meryl D. Stoller and colleagues, the 2008 work illustrated the performance of chemically modified graphene in an ultracapacitor cell. The author stated, the strong performance and high electrical conductivity over a wide span of voltage scan rates showed the "exciting potential for high performance, electrical energy storage devices based on this new class of carbon material."

Carbon nanotubes, first observed in 1991, occur as multiwalled nanotubes (MWNTs) and single walled nanotubes (SWNTs). Primary methods for CNTs synthesis include arc-discharge, laser ablation, gas-phase catalytic growth from carbon monoxide, and chemical vapor deposition (CVD) from hydrocarbons. Because impurities in the form of catalyst particles, amorphous carbon, and non-tubular fullerenes are also produced during CNTs synthesis, subsequent purification steps are required to separate the tubes. The gas-phase processes tend to produce nanotubes with fewer impurities and are more amenable to large-scale processing (Fig. 7.31).

The properties of graphene, carbon sheets that are only one atom thick, have caused researchers and companies to consider using this material in several fields.

Graphene include using graphene sheets as electrodes in ultracapacitors, which will have as much storage capacity as batteries but will be able to recharge in minutes, attaching strands of DNA to graphene to form sensors for rapid disease diagnostics, replacing indium in flat screen TVs and making high strength composite materials. A catalyst made from graphene doped with cobalt can be used to produce hydrogen from water. Graphene can replace indium-based electrodes in organic light emitting diodes (OLED). These diodes are used in electronic device display screens which require low power consumption. The use of graphene instead of indium not only reduces the cost but eliminates the use of metals in the OLED, which may make devices easier to recycle.

Adding graphene to epoxy composites may result in stronger/stiffer components than epoxy composites using a similar weight of carbon nanotubes. Graphene appears to bond better to the polymers in the epoxy, allowing a more effective coupling of the graphene into the structure of the composite. This property could result in the manufacture of components with high strength to weight ratio for such uses as windmill blades or aircraft components.

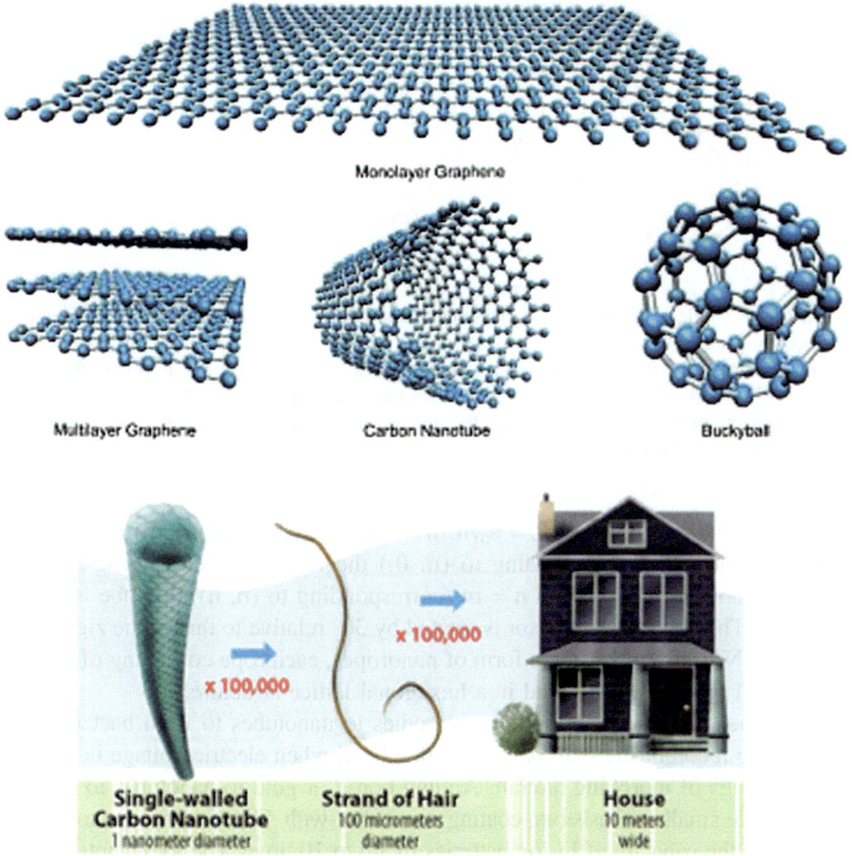

Fig. 7.31 Nanomaterial (for example: carbon nanotube)

7.13 Applications of Carbon Nanotubes (CNTs)

Iijima in the form of multiple coaxial carbon fullerene shells (multi-wall nanotubes, MWNTs) discovered nanotubes. Later, in 1993 single fullerene shells (singlewall nanotubes, **SWNTs**) were synthesized using transition metal catalysts. A nanotube can be simply described as a sheet of graphite (or graphene) coaxially rolled to create a cylindrical surface (as shown in Fig. 7.32a).

In this way the 2D hexagonal lattice of grapheme is mapped onto a cylinder of radius **R**. The mapping can be realized with different helicities resulting in different nanotubes. Each nanotube is characterized by a set of two integers **(n, m)** indicating the components of the chiral vector $\mathbf{C} = \mathbf{na_1} + \mathbf{ma_2}$ in terms of the 2D hexagonal Bravais lattice vectors of graphene, $\mathbf{a_1}$ and $\mathbf{a_2}$, as illustrated in Fig. 7.32b. The chiral vector is a circumferential vector and the tube is obtained by folding the graphene such that the two ends of **C** are coincident. The radius of the tube is given in terms

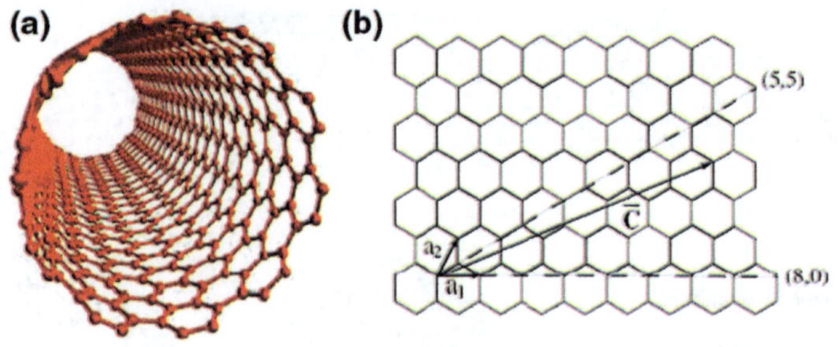

Fig. 7.32 **a** A single-wall nanotube is graphene wrapped on a cylinder surface. **b** Nanotubes are described by a set of two integers (**n**, **m**) which indicate the graphite lattice vector components. A chiral vector can be defined as $\mathbf{C} = \mathbf{na_1} + \mathbf{ma_2}$. Tubes are called 'zigzag' if either one of the integers is zero (**n**, **0**) or called 'armchair' if both integers are equal (**n**, **n**) [K.S. Ciraci, A. Buldum, I.P. Batra, Quantum effects in electrical and thermal transport through nanowires, 2001, Journal of Physics Condensed Matter]

of (**n**, **m**) through the relation $R = a_0\sqrt{n}^2m^2 + nm/2\pi$, where $|\mathbf{a_1}| = |\mathbf{a_2}| = \mathbf{a_0}$. When **C** involves only $\mathbf{a_1}$ (corresponding to (**n**, **0**)) the tube is called 'zigzag', and if **C** involves both $\mathbf{a_1}$ and $\mathbf{a_2}$ with **n** = **m** (corresponding to (**n**, **n**)) the tube is called 'armchair'. The chiral (**n**, **n**) vector is rotated by 30° relative to that of the zigzag (**n**, **0**) tube. SWNTs are found in the form of nanoropes, each rope consisting of up to a few hundred nanotubes arranged in a hexagonal lattice structure.

Carbon nanotubes include adding antibodies to nanotubes to form bacteria sensors, making a composite with nanotubes that bend when electric voltage is applied bend the wings of morphing aircraft. Adding boron or gold to nanotubes to trap oil spills, include smaller transistors, coating nanotubes with silicon to make anodes the can increase the capacity of Li-ion batteries by up to 10 times. Carbon nanotubes are used in strong, lightweight car parts and sporting equipment, as well as composites that protect airplanes from lightning strikes and keep computer circuits cool.

The conventional amorphous carbon was the most commonly used material of electronic conductors due to its high conductivity. Unfortunately, amorphous carbon coating was found to be prone to increase the unfavorable metal ions in Nickle-rich layered materials which have attracted extensive attentions as the cathode material for Li-ion batteries owing to its low cost, low toxicity, excellent structure stability and long cycling life. The uniform distribution of graphene nanodots with the size of 5 nm on the $LiNi_{0.8}Co_{0.15}Al_{0.0}5O_2$ particle surface can dramatically improve the electronic conductivity. Furthermore, the moderate graphene nanodots coating can provide abundant lithium ions transportation pathway for the enhanced rate performance. As a result, the graphene nanodots modified $LiNi_{0.8}Co_{0.15}Al_{0.05}O_2$ with 0.5 wt% coating amount exhibits a high discharge capacity of 150 mAh g^{-1} at the high rate of 5 C. Generally carbon or design a graphene nanodots work provides new strategy for realizing the promise of high-energy and high-power lithium-ion batteries and in all industrial applications of Nanoelectronic Materials.

Fig. 7.33 A scanning electron microscope photo of hollow carbon nanofiber-encapsulated sulfur tubes, at the heart of a new battery design

The properties of *carbon nanotubes* have caused researchers and companies to consider using them in several fields, in electronics, energy, healthcare and environment.

7.13.1 Carbon Nanotubes and Electronics

Building transistors from carbon nanotubes enables minimum transistor dimensions of a few nanometers and the development of techniques to manufacture integrated circuits built with nanotube transistors (Fig. 7.33).

7.13.2 Carbon Nanotubes and Energy

Carbon nanotubes in 3-D structured electrodes. silicon coated carbon nanotubes in anodes for Li-ion batteries. They are predicting that the use of silicon can increase the capacity of Li-ion batteries by up to 10 times. However silicon expands during a batteries discharge cycle, which can damage silicon based anodes. By depositing silicon on nanotubes aligned parallel to each other the researchers hope to prevent damage to the anode when the silicon expands.

7.13.3 Carbon Nanotubes in Healthcare

Sensor that uses nanotubes and gold nanoparticles to detect proteins that indicate the presence of oral cancer.

7.13.4 *Carbon Nanotubes and the Environment*

Carbon nanotubes are being developed to clean up oil spills. Researchers have found that adding boron atoms during the growth of carbon nanotubes causes the nanotubes to grow into a sponge like material that can absorb many times it's weight in oil. These nanotube sponges are made to be magnetic, which should make retrieval of them easier once they are filled with oil.

Carbon nanotubes can be used as the pores in membranes to run reverse osmosis desalination plants. Water molecules pass through the smoother walls of carbon nanotubes more easily than through other types of nanopores, which requires less power. Other researchers are using carbon nanotubes to develop small, inexpensive water purification devices needed in developing countries.

Sensors using carbon nanotube detection elements are capable of detecting a range of chemical vapors. These sensors work by reacting to the changes in the resistance of a carbon nanotube in the presence of a chemical vapor.

7.14 The Future of Graphene Research

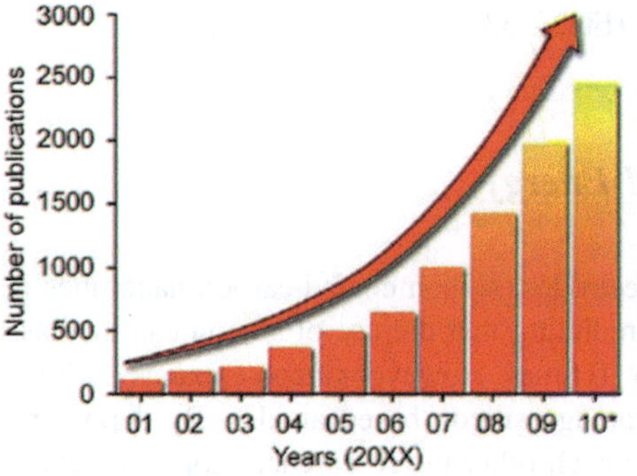

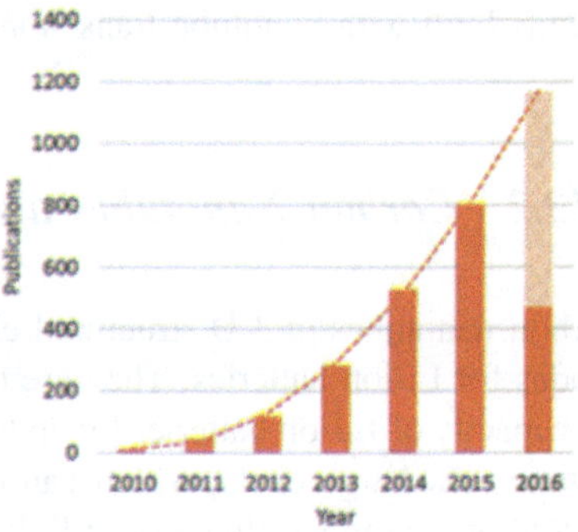

Given graphene's seemingly endless list of strengths, one would expect to see it everywhere. Why, then, has graphene not been widely adopted? As with most things, it comes down to money. Graphene is still extremely expensive to produce in large quantities, limiting its use in any product that would demand mass production. Moreover, when large sheets of graphene are produced, there is increased risk of tiny fissures and other flaws appearing in the material. No matter how incredible a scientific discovery may be, economics will always decide success.

Production issues aside, graphene research is by no means slowing down. Research laboratories the world over—including the University of Manchester, where graphene was first discovered—are continually filing patents for new methods of creating and using graphene. The European Union in 2013 approved funding for a flagship program that will fund graphene research for use in electronics. Perhaps more importantly, many major tech companies in Asia are conducting research on Graphene, including mobile giant Samsung. As the EU tries to steady its footing in the face of explosive economic growth in Asia, graphene could be an important battlefield in international politics in the coming years. Again, empires often rise and fall based on their control of resources.

Despite its current limitations, researchers are trying out all sorts of uses for graphene, from lightbulbs to body armor.

Revolutions don't happen overnight. Silicon was discovered in the mid 19th century, but it took nearly a century before silicon semiconductors paved the way for the rise of computers. Might graphene, with its almost mythical qualities, be the resource that drives the next era of human history?

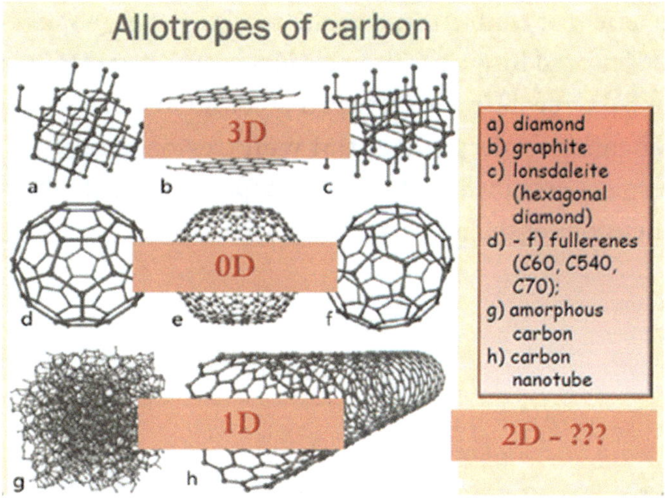

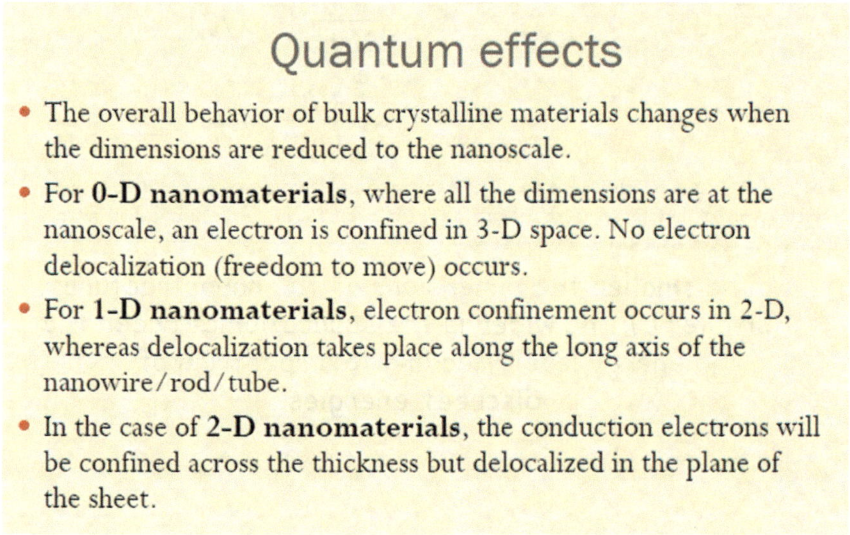

Electrons confinement

- For **0-D nanomaterials** the electrons are fully confined.
- For **3-D nanomaterials** the electrons are fully delocalized.
- In **1-D and 2-D nanomaterials**, electron confinement and delocalization coexist.
- The effect of confinement on the resulting energy states can be calculated by quantum mechanics, as the **"particle in the box"** problem. An electron is considered to exist inside of an infinitely deep **potential well** (region of negative energies), from which it cannot escape and is confined by the dimensions of the nanostructure.

Energies

(0-D) $$E_n = \left[\frac{\pi^2\hbar^2}{2mL^2}\right]\left(n_x^2 + n_y^2 + n_z^2\right)$$

(1-D) $$E_n = \left[\frac{\pi^2\hbar^2}{2mL^2}\right]\left(n_x^2 + n_y^2\right)$$

(2-D) $$E_n = \left[\frac{\pi^2\hbar^2}{2mL^2}\right]\left(n_x^2\right)$$

where $h^{-} \equiv h/2\pi$, h is Planck's constant, m is the mass of the electron, L is the width (confinement) of the infinitely deep potential well, and nx, ny, and nz are the principal quantum numbers in the three dimensions x, y, and z.

The **smaller** the dimensions of the nanostructure (smaller L), the **wider** is the separation between the energy levels, leading to a spectrum of **discreet energies**.

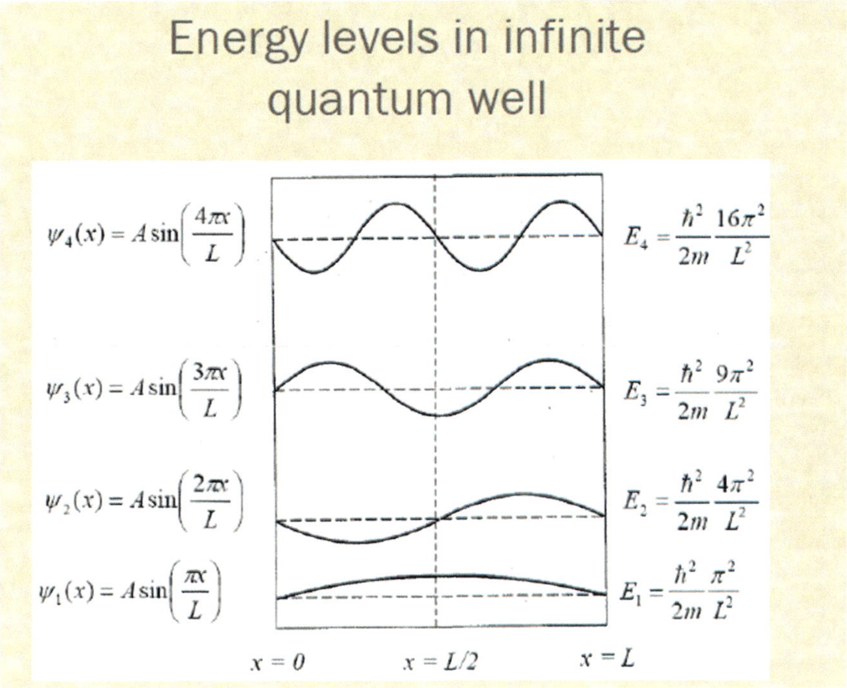

Energy levels in infinite quantum well
$\psi_4(x) = A\sin\left(\frac{4\pi x}{L}\right)$
$E_4 = \frac{\hbar^2}{2m}\frac{16\pi^2}{L^2}$
$\psi_3(x) = A\sin\left(\frac{3\pi x}{L}\right)$
$E_3 = \frac{\hbar^2}{2m}\frac{9\pi^2}{L^2}$
$\psi_2(x) = A\sin\left(\frac{2\pi x}{L}\right)$
$E_2 = \frac{\hbar^2}{2m}\frac{4\pi^2}{L^2}$
$\psi_1(x) = A\sin\left(\frac{\pi x}{L}\right)$
$E_1 = \frac{\hbar^2}{2m}\frac{\pi^2}{L^2}$
$x = 0$
$x = L/2$
$x = L$

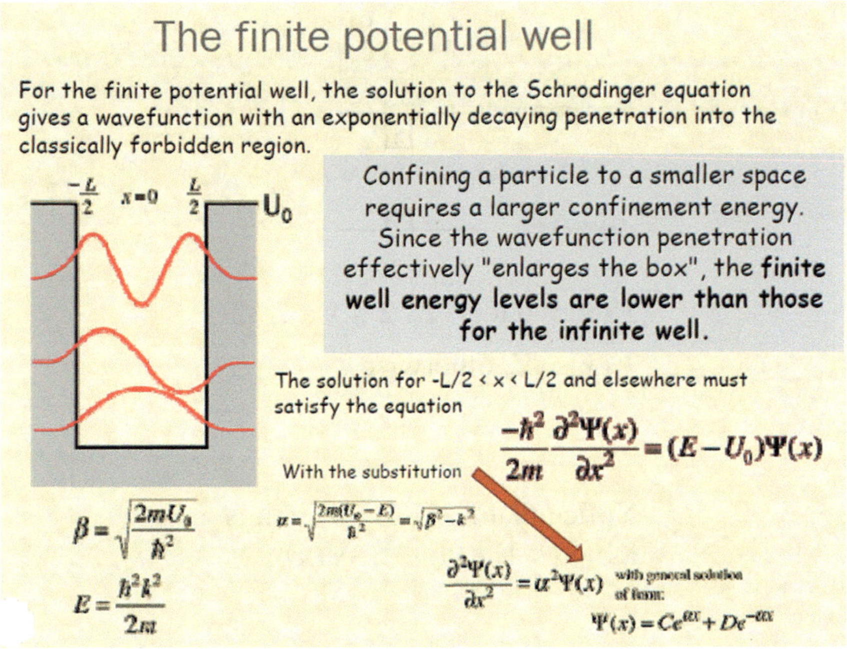

The finite potential well
For the finite potential well, the solution to the Schrodinger equation gives a wavefunction with an exponentially decaying penetration into the classically forbidden region.
$-\frac{L}{2}$ $x=0$ $\frac{L}{2}$ U_0
Confining a particle to a smaller space requires a larger confinement energy. Since the wavefunction penetration effectively "enlarges the box", the finite well energy levels are lower than those for the infinite well.
The solution for -L/2 < x < L/2 and elsewhere must satisfy the equation
$\frac{-\hbar^2}{2m}\frac{\partial^2\Psi(x)}{\partial x^2} = (E - U_0)\Psi(x)$
With the substitution
$\alpha = \sqrt{\frac{2m(U_0 - E)}{\hbar^2}} = \sqrt{\beta^2 - k^2}$
$\frac{\partial^2\Psi(x)}{\partial x^2} = \alpha^2\Psi(x)$
with general solution of form:
$\Psi(x) = Ce^{\alpha x} + De^{-\alpha x}$
$\beta = \sqrt{\frac{2mU_0}{\hbar^2}}$
$E = \frac{\hbar^2 k^2}{2m}$

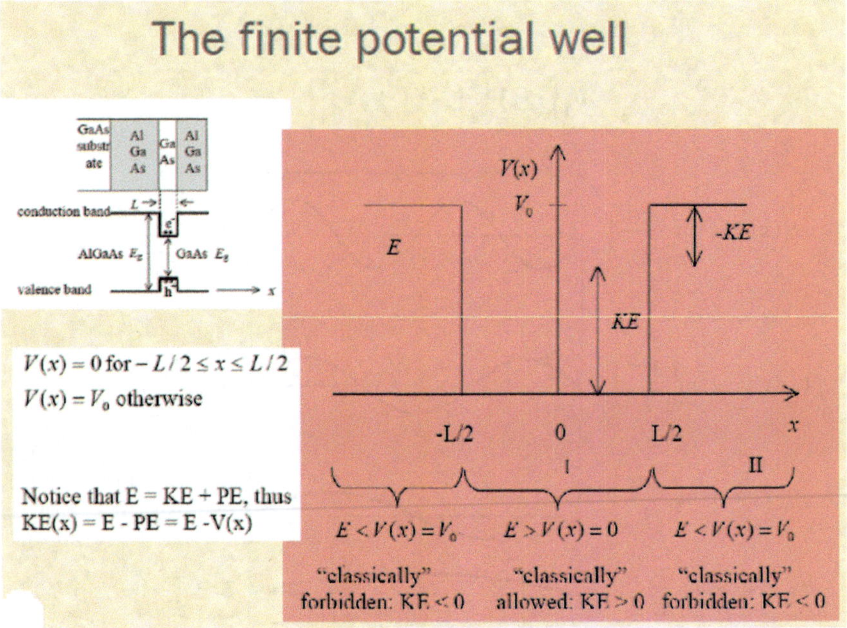

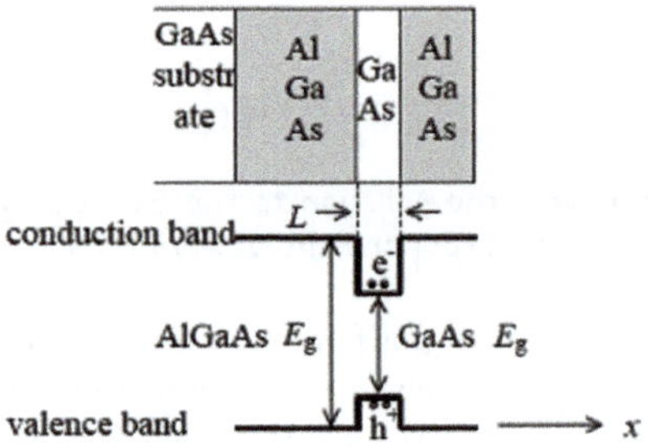

$$V(x) = 0 \text{ for } -L/2 \le x \le L/2$$
$$V(x) = V_0 \text{ otherwise}$$

Notice that E = KE + PE, thus
KE(x) = E - PE = E -V(x)

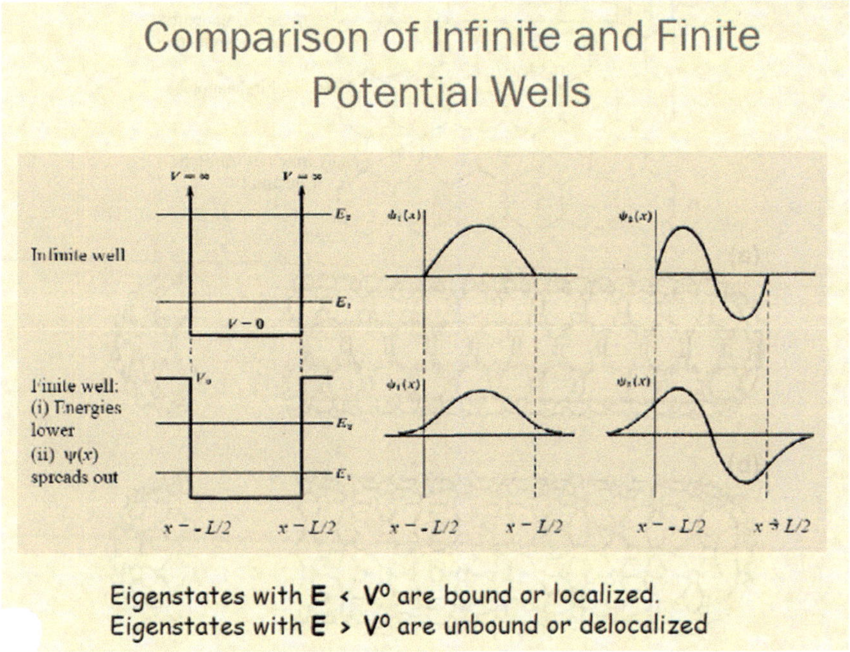
Comparison of Infinite and Finite Potential Wells
Infinite well
Finite well:
(i) Energies lower
(ii) ψ(x) spreads out
Eigenstates with E < V⁰ are bound or localized.
Eigenstates with E > V⁰ are unbound or delocalized

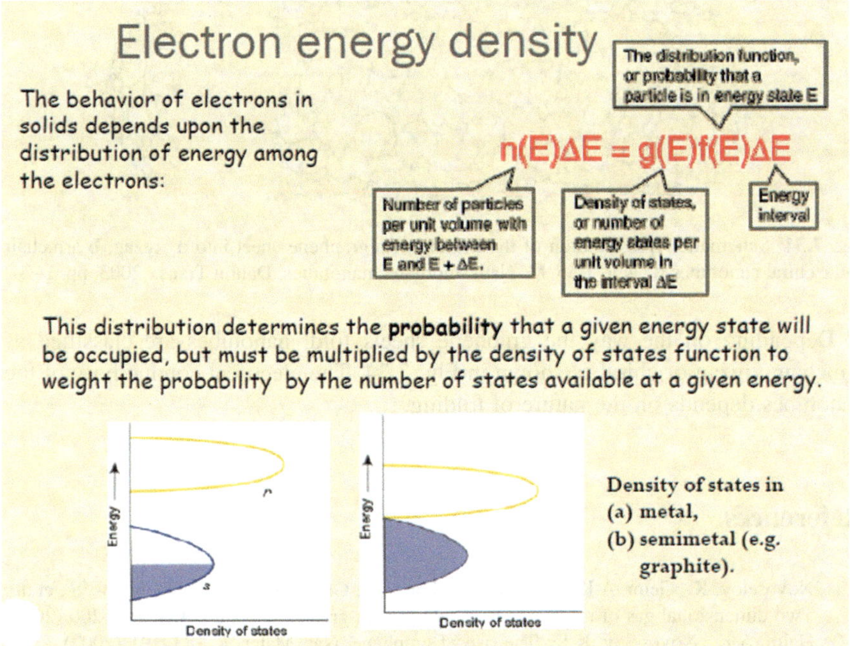
Electron energy density
The behavior of electrons in solids depends upon the distribution of energy among the electrons:
The distribution function, or probability that a particle is in energy state E
n(E)ΔE = g(E)f(E)ΔE
Number of particles per unit volume with energy between E and E + ΔE.
Density of states, or number of energy states per unit volume in the interval ΔE
Energy interval
This distribution determines the probability that a given energy state will be occupied, but must be multiplied by the density of states function to weight the probability by the number of states available at a given energy.
Energy
Density of states
Energy
Density of states
Density of states in
(a) metal,
(b) semimetal (e.g. graphite).

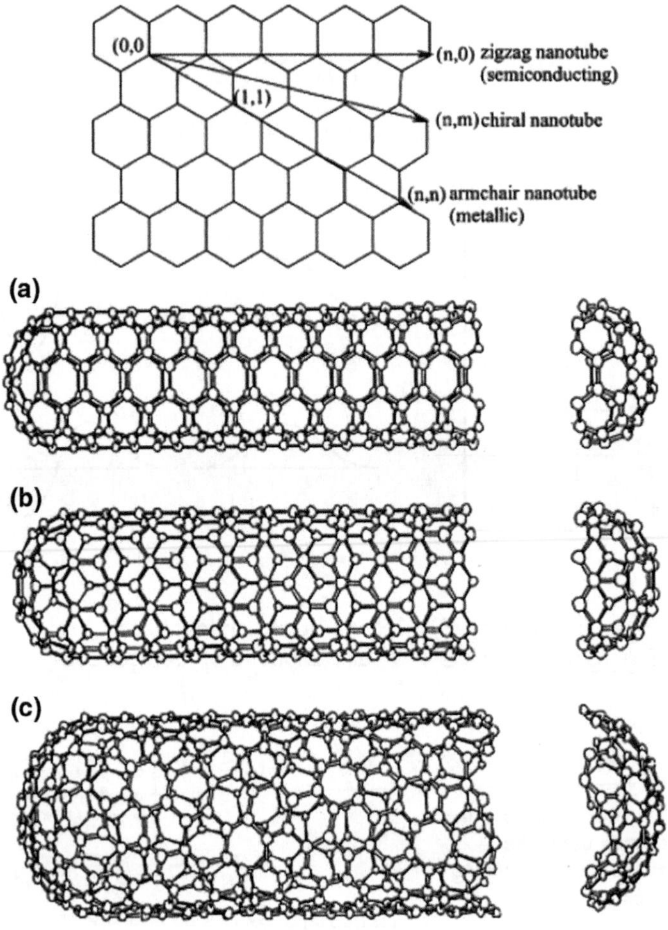

Fig. 7.34 Schematic representation of the folding of a graphene sheet into **a** zigzag, **b** armchair and **c** chiral nanotubes [C.N.R. Rao, M. Nath, Inorganic nanotubes, Dalton Trans., 2003, pp. 1–24]

Depending on the way the graphene sheets fold, nanotubes are classified as armchair, zigzag or chiral as shown in Fig. 7.34. The electrical conductivity of the nanotubes depends on the nature of folding.

References

1. Novoselov, K., Geim, A.K., Morozov, S., Jiang, D., Grigorieva, M.K.I., Dubonos, S., et al.: Two dimensional gas of massless Dirac fermions in graphene. Nature **438**, 197–200 (2005)
2. Geim, A.K., Novoselov, K.S.: The rise of graphene. Nat. Mater. **6**, 183–191 (2007)
3. Li, D., Kaner, R.B.: Materials science. Graphene-based materials. Science **320**, 1170–1171 (2008)

4. Li, D., Kaner, R.B.: Graphene-based materials. Nat. Nanotechnol. **3**, 101 (2008)
5. Wallace, P.R.: The band theory of graphite. Phys. Rev. **71**, 622 (1947)
6. Zou, J., Liu, J., Karakoti, A.S., Kumar, A., Joung, D., Li, Q., et al.: Ultralight multiwalled carbon nanotube aerogel. ACS Nano **4**, 7293–7302 (2010)
7. Du, X., Skachko, I., Barker, A., Andrei, E.Y.: Approaching ballistic transport in suspended graphene. Nat. Nanotechnol. **3**, 491–495 (2008)
8. Fuhrer, M.S., Lau, C.N., MacDonald, A.H.: Graphene: materially better carbon. MRS Bull. **35**, 289–295 (2010)
9. Durkop, T., Getty, S., Cobas, E., Fuhrer, M.: Extraordinary mobility in semiconducting carbon nanotubes. Nano Lett. **4**, 35–39 (2004)
10. Novoselov, K.S., Geim, A.K., Morozov, S., Jiang, D., Zhang, Y., Dubonos, S., et al.: Electric field effect in atomically thin carbon films. Science **306**, 666–669 (2004)
11. Geim, A., Grigorieva, I.: Van der Waals heterostructures. Nature **499**, 419–425 (2013)
12. Mas-Balleste, R., Gomez-Navarro, C., Gomez-Herrero, J., Zamora, F.: 2D materials: to graphene and beyond. Nanoscale **3**, 20–30 (2011)
13. Xu, M., Liang, T., Shi, M., Chen, H.: Graphene-like two-dimensional materials. Chem. Rev. **113**, 3766–3798 (2013)
14. Ni, Z., Liu, Q., Tang, K., Zheng, J., Zhou, J., Qin, R., et al.: Tunable bandgap in silicene and germanene. Nano Lett. **12**, 113–118 (2011)
15. Tabert, C.J., Nicol, E.J.: AC/DC spin and valley Hall effects in silicene and germanene. Phys. Rev. B **87**, 235426 (2013)
16. Wei, W., Dai, Y., Huang, B., Jacob, T.: Many-body effects in silicene, silicane, germanene and germanane. Phys. Chem. Chem. Phys. **15**, 8789–8794 (2013)
17. Cai, Y., Chuu, C.-P., Wei, C., Chou, M.: Stability and electronic properties of two-dimensional silicene and germanene on graphene. Phys. Rev. B **88**, 245408 (2013)
18. Singh, V., Joung, D., Zhai, L., Das, S., Khondaker, S.I., Seal, S.: Graphene based materials: past, present and future. Prog. Mater. Sci. **56**, 1178–1271 (2011)
19. Yang, N., Zhai, J., Wang, D., Chen, Y., Jiang, L.: Two-dimensional graphene bridges enhanced photoinduced charge transport in dye-sensitized solar cells. ACS Nano **4**, 887–894 (2010)
20. Ghosh, S., Sarker, B.K., Chunder, A., Zhai, L., Khondaker, S.I.: Position dependent photodetector from large area reduced graphene oxide thin films. Appl. Phys. Lett. **96**, 163109 (2010)
21. Chunder, A., Pal, T., Khondaker, S.I., Zhai, L.: Reduced graphene oxide/copper phthalocyanine composite and its optoelectrical properties. J. Phys. Chem. C **114**, 15129–15135 (2010)
22. Wang, Q.H., Kalantar-Zadeh, K., Kis, A., Coleman, J.N., Strano, M.S.: Electronics and optoelectronics of two-dimensional transition metal dichalcogenides. Nat. Nanotechnol. **7**, 699–712 (2012)
23. Wilson, J., Di Salvo, F., Mahajan, S.: Charge-density waves in metallic, layered, transition-metal dichalcogenides. Phys. Rev. Lett. **32**, 882 (1974)
24. Mak, K.F., Lee, C., Hone, J., Shan, J., Heinz, T.F.: Atomically thin MoS_2: a new direct-gap semiconductor. Phys. Rev. Lett. **105**, 136805 (2010)
25. Canadell, E., LeBeuze, A., El Khalifa, M.A., Chevrel, R., Whangbo, M.H.: Origin of metal clustering in transition-metal chalcogenide layers MX2 (M = Nb, Ta, Mo, Re; X = S, Se). J. Am. Chem. Soc. **111**, 3778–3782 (1989)
26. Jin, S., Lukowski, M.A., Daniel, A.S., English, C.R., Meng, F., Forticaux, A., et al.: Highly active hydrogen evolution catalysis from metallic WS_2 nanosheets. Energy Environ. Sci. (2014)
27. Andriotis, A.N., Menon, M.: Tunable magnetic properties of transition metal doped MoS_2. Phys. Rev. B **90**, 125304 (2014)
28. He, J., Hummer, K., Franchini, C.: Stacking effects on the electronic and optical properties of bilayer transition metal dichalcogenides MoS_2, $MoSe_2$, WS_2, and WSe_2. Phys. Rev. B **89**, 075409 (2014)

29. Huang, Y., Peng, C., Chen, R., Huang, Y., Ho, C.: Transport properties in semiconducting NbS_2 nanoflakes. Appl. Phys. Lett. **105**, 093106 (2014)
30. Moore, D.B., Beekman, M., Disch, S., Zschack, P., Hausler, I., Neumann, W., et al.: Synthesis, structure, and properties of turbostratically disordered $(PbSe)_{1.18}(TiSe_2)_2$. Chem. Mater. **25**, 2404–2409 (2013)
31. Jeong, S., Yoo, D., Jang, J.-t., Kim, M., Cheon, J.: Well-defined colloidal 2-D layered transition-metal chalcogenide nanocrystals via generalized synthetic protocols. J. Am. Chem. Soc. **134**, 18233–18236 (2012)
32. Eda, G., Fujita, T., Yamaguchi, H., Voiry, D., Chen, M., Chhowalla, M.: Coherent atomic and electronic heterostructures of single-layer MoS_2. ACS Nano **6**, 7311–7317 (2012)
33. Li, H., Lu, G., Wang, Y., Yin, Z., Cong, C., He, Q., et al.: Mechanical exfoliation and characterization of single- and few-layer nanosheets of WSe_2, TaS_2, and $TaSe_2$. Small **9**, 1974–1981 (2013)
34. Li, H., Lu, G., Yin, Z., He, Q., Li, H., Zhang, Q., et al.: Optical identification of single- and few-layer MoS_2 sheets. Small **8**, 682–686 (2012)
35. Tongay, S., Zhou, J., Ataca, C., Lo, K., Matthews, T.S., Li, J., et al.: Thermally driven crossover from indirect toward direct bandgap in 2D semiconductors: $MoSe_2$ versus MoS_2. Nano Lett. **12**, 5576–5580 (2012)
36. Coleman, J.N., Lotya, M., O'Neill, A., Bergin, S.D., King, P.J., Khan, U., et al.: Two-dimensional nanosheets produced by liquid exfoliation of layered materials. Science **331**, 568–571 (2011)
37. Butler, S.Z., Hollen, S.M., Cao, L., Cui, Y., Gupta, J.A., Gutierrez, H.R., et al.: Progress, challenges, and opportunities in two-dimensional materials beyond graphene. ACS Nano **7**, 2898–2926 (2013)
38. Wilson, J.A., Yoffe, A.D.: Transition metal dichalcogenides discussion and interpretation of observed optical, electrical, and structural properties. Adv. Phys. **18**, 193–335 (1969)
39. Li, Y., Wang, H., Xie, L., Liang, Y., Hong, G., Dai, H.: MoS_2 nanoparticles grown on graphene: an advanced catalyst for the hydrogen evolution reaction. J. Am. Chem. Soc. **133**, 7296–7299 (2011)
40. Wang, T., Liu, L., Zhu, Z., Papakonstantinou, P., Hu, J., Liu, H., et al.: Enhanced electrocatalytic activity for hydrogen evolution reaction from self-assembled monodispersed molybdenum sulfide nanoparticles on an Au electrode. Energy Environ. Sci. **6**, 625–633 (2013)
41. Poudel, B., Hao, Q., Ma, Y., Lan, Y., Minnich, A., Yu, B., et al.: High-thermoelectric performance of nanostructured bismuth antimony telluride bulk alloys. Science **320**, 634–638 (2008)
42. Taha-Tijerina, J., Narayanan, T.N., Gao, G., Rohde, M., Tsentalovich, D.A., Pasquali, M., et al.: Electrically insulating thermal nanooils using 2D fillers. ACS Nano **6**, 1214–1220 (2012)
43. Lin, S.S.: Light-emitting two-dimensional ultrathin silicon carbide. J. Phys. Chem. C **116**, 3951–3955 (2012)
44. Yang, S., Gong, Y., Liu, Z., Zhan, L., Hashim, D.P., Ma, L., et al.: Bottom-up approach toward single-crystalline VO2-graphene ribbons as cathodes for ultrafast lithium storage. Nano Lett. **13**, 1596–1601 (2013)
45. Kong, D., Dang, W., Cha, J.J., Li, H., Meister, S., Peng, H., et al.: Few-layer nanoplates of Bi_2Se_3 and Bi_2Te_3 with highly tunable chemical potential. Nano Lett. **10**, 2245–2250 (2010)
46. Hu, L., Ma, R., Ozawa, T.C., Sasaki, T.: Exfoliation of layered europium hydroxide into unilamellar nanosheets. Chem. Asian J. **5**, 248–251 (2010)
47. Ozawa, T.C., Fukuda, K., Akatsuka, K., Ebina, Y., Sasaki, T.: Preparation and characterization of the Eu^{3+} doped perovskite nanosheet phosphor: $La_{0.9}Eu_{0.05}Nb_2O_7$. Chem. Mater. **19**, 6575–6580 (2007)
48. Ozawa, T.C., Fukuda, K., Akatsuka, K., Ebina, Y., Sasaki, T., Kurashima, K., et al.: $(K_{1.5}Eu_{0.5})Ta_3O_{10}$: a far-red luminescent nanosheet phosphor with the double perovskite structure. J. Phys. Chem. C **112**, 17115–17120 (2008)

49. Ida, S., Ogata, C., Eguchi, M., Youngblood, W.J., Mallouk, T.E., Matsumoto, Y.: Photoluminescence of perovskite nanosheets prepared by exfoliation of layered oxides, $K_2Ln_2Ti_3O_{10}$, $KLnNb_2O_7$, and $RbLnTa_2O_7$ (Ln: lanthanide ion). J. Am. Chem. Soc. **130**, 7052–7059 (2008)
50. Ebina, Y., Sasaki, T., Harada, M., Watanabe, M.: Restacked perovskite nanosheets and their Pt-loaded materials as photocatalysts. Chem. Mater. **14**, 4390–4395 (2002)
51. Osada, M., Akatsuka, K., Ebina, Y., Funakubo, H., Ono, K., Takada, K., et al.: Robust high-j response in molecularly thin perovskite nanosheets. ACS Nano **4**, 5225–5232 (2010)
52. Ma, R., Liu, Z., Takada, K., Iyi, N., Bando, Y., Sasaki, T.: Synthesis and exfoliation of Co^{2+}–Fe^{3+} layered double hydroxides: an innovative topochemical approach. J. Am. Chem. Soc. **129**, 5257–5263 (2007)
53. Lalmi, B., Oughaddou, H., Enriquez, H., Kara, A., Vizzini, S., Ealet, B., et al.: Epitaxial growth of a silicene sheet. Appl. Phys. Lett. **97**, 223109 (2010)
54. Meng, L., Wang, Y., Zhang, L., Du, S., Wu, R., Li, L., et al.: Buckled silicene formation on Ir(111). Nano Lett. **13**, 685–690 (2013)
55. Fleurence, A., Friedlein, R., Ozaki, T., Kawai, H., Wang, Y., Yamada-Takamura, Y.: Experimental evidence for epitaxial silicene on diboride thin films. Phys. Rev. Lett. **108**, 245501 (2012)
56. Jung, H., Park, J., Oh, I.-K., Choi, T., Lee, S., Hong, J., et al.: Fabrication of transferable Al_2O_3 nanosheet by atomic layer deposition for graphene FET. ACS Appl. Mater. Interfaces **6**, 2764–2769 (2014)
57. Ruggiero, C.D., Badal, M., Choi, T., Gohlke, D., Stroud, D., Gupta, J.A.: Emergence of surface states in nanoscale Cu_2N islands. Phys. Rev. B **83**, 245430 (2011)
58. Sterrer, M., Risse, T., Pozzoni, U.M., Giordano, L., Heyde, M., Rust, H.-P., et al.: Control of the charge state of metal atoms on thin MgO films. Phys. Rev. Lett. **98**, 096107 (2007)
59. Hao, B., Yan, Y., Wang, X., Chen, G.: Synthesis of anatase TiO_2 nanosheets with enhanced pseudocapacitive contribution for fast lithium storage. ACS Appl. Mater. Interfaces **5**, 6285–6291 (2013)
60. Oughaddou, H., Aufray, B., Biberian, J.P., Hoarau, J.Y.: Growth mode and dissolution kinetics of germanium thin films on Ag (001) surface: an AES–LEED investigation. Surf. Sci. **429**, 320–326 (1999)
61. Oughaddou, H., Gay, J.M., Aufray, B., Lapena, L., Lay, G.L., Bunk, O., et al.: Ge tetramer structure of the $p(2\sqrt{2} \times 4\sqrt{2})R45°$ surface reconstruction of Ge/Ag(001): a surface X-ray diffraction and STM study. Phys. Rev. B **61**, 5692 (2000)
62. Golias, E., Xenogiannopoulou, E., Tsoutsou, D., Tsipas, P., Giamini, S.A., Dimoulas, A.: Surface electronic bands of submonolayer Ge on Ag(111). Phys. Rev. B **88**, 075403 (2013)
63. Oughaddou, H., Mayne, A., Aufray, B., Biberian, J.P., Lay, G.L., Ealet, B., et al.: Germanium adsorption on Ag(111): an AES-LEED and STM study. J. Nanosci. Nanotechnol. **7**, 3189–3192 (2007)
64. Cullis, A.G., Booker, G.R.: The epitaxial growth of silicon and germanium films on (111) silicon surfaces using UHV sublimation and evaporation techniques. J. Cryst. Growth **9**, 132–138 (1971)
65. Nicolosi, V., Chhowalla, M., Kanatzidis, M.G., Strano, M.S., Coleman, J.N.: Liquid exfoliation of layered materials. Science **340**, 1226419 (2013)
66. Ozawa, T.C., Fukuda, K., Akatsuka, K., Ebina, Y., Sasaki, T.: Preparation and characterization of the Eu^{3+} doped perovskite nanosheet phosphor: $La_{0.90}Eu_{0.05}Nb_2O_7$. Chem. Mater. **19**, 6575–6580 (2007)
67. Cunningham, G., Lotya, M., Cucinotta, C.S., Sanvito, S., Bergin, S.D., Menzel, R., et al.: Solvent exfoliation of transition metal dichalcogenides: dispersibility of exfoliated nanosheets varies only weakly between compounds. ACS Nano **6**, 3468–3480 (2012)
68. Hernandez, Y., Nicolosi, V., Lotya, M., Blighe, F.M., Sun, Z., De, S., et al.: High-yield production of graphene by liquid-phase exfoliation of graphite. Nat. Nanotechnol. **3**, 563–568 (2008)

69. Diaz, E., Ordonez, S., Vega, A.: Adsorption of volatile organic compounds onto carbon nanotubes, carbon nanofibers, and high-surface-area graphites. J. Colloid Interface Sci. **305**, 7–16 (2007)
70. Ataca, C., Sahin, H., Ciraci, S.: Stable, single-layer MX_2 transition-metal oxides and dichalcogenides in a honeycomb-like structure. J. Phys. Chem. C **116**, 8983–8999 (2012)
71. Fang, C., Van Bruggen, C., De Groot, R., Wiegers, G., Haas, C.: The electronic structure of the metastable layer compound. J. Phys.: Condens. Matter **9**, 10173 (1997)
72. Zhang, J.H., Birdwhistell, T.L., O'Connor, C.J.: Magnetic and electrical properties of a new chromium telluride phase: $CrTe_2$. Solid State Commun. **74**, 443–446 (1990)
73. Koneshova, T., Babitsyna, A., Emel'yanova, T.: Cu–Te and Cr–Te Phase Diagrams (2001)
74. Novoselov, K.S., et al.: Electric field effect in atomically thin carbon films. Science **306** (5696), 666–669 (2004)
75. Balendhran, S., et al.: Atomically thin layers of MoS_2 via a two-step thermal evaporation-exfoliation method. Nanoscale **4**(2), 461–466 (2012)
76. Nicolosi, V., et al.: Liquid exfoliation of layered materials. Science **340**, 6139 (2013)
77. Wang, J., et al.: Epitaxial $BiFeO_3$ multiferroic thin film heterostructures. Science **299**(5613), 1719–1722 (2003)
78. Zixu, Z., et al.: Graphene geometric diodes for terahertz rectennas. J. Phys. D Appl. Phys. **46** (18), 185101 (2013)
79. O'Hare, A., Kusmartsev, F.V., Kugel, K.I.: A stable, "flat" form of two-dimensional crystals: could graphene, silicene, germanene be minigap semiconductors? Nano Lett. **12**(2), 1045–1052 (2012)
80. Kuc, A., Zibouche, N., Heine, T.: Influence of quantum confinement on the electronic structure of the transition metal sulfide TS2. Phys. Rev. B **83**(24), 245213 (2011)
81. Xu, M., et al.: Unique synthesis of graphene-based materials for clean energy and biological sensing applications. Chin. Sci. Bull. **57**(23), 3000–3009 (2012)
82. Zhu, W., et al.: Graphene radio frequency devices on flexible substrate. Appl. Phys. Lett. **102**(23), 233102 (2013)
83. Bae, S., et al.: Roll-to-roll production of 30-inch graphene films for transparent electrodes. Nat. Nanotechnol. **5**(8), 574–578 (2010)
84. Klein, C.A., Straub, W.D.: Carrier densities and mobilities in pyrolytic graphite. Phys. Rev. **123**(5), 1581–1583 (1961)
85. Novoselov, K.S., et al.: Room-temperature quantum hall effect in graphene. Science **315** (5817), 1379 (2007)
86. Novoselov, K.S., et al.: Two-dimensional gas of massless Dirac fermions in graphene. Nature **438**(7065), 197–200 (2005)
87. Stankovich, S., et al.: Graphene-based composite materials. Nature **442**(7100), 282–286 (2006)
88. Balandin, A.A., et al.: Superior thermal conductivity of single-layer graphene. Nano Lett. **8** (3), 902–907 (2008)
89. Novoselov, K.S., Jiang, D., Schedin, F., Booth, T.J., Khotkevich, V.V., Morozov, S.V., Geim, A.K.: Two-dimensional atomic crystals. Proc. Natl. Acad. Sci. USA **102**, 10451–10453 (2005)
90. Mattheiss, L.F.: Band structures of transition-metal-dichalcogenide layer compounds. Phys. Rev. B **8**, 3719–3740 (1973)
91. Bernardi, M., Palummo, M., Grossman, J.C.: Extraordinary sunlight absorption and one nanometer thick photovoltaics using two-dimensional monolayer materials. Nano Lett. **13**, 3664–3670 (2013)
92. Lee, H.S., Min, S.-W., Chang, Y.-G., Park, M.K., Nam, T., Kim, H., Kim, J.H., Ryu, S., Im, S.: MoS_2 nanosheet phototransistors with thickness-modulated optical energy gap. Nano Lett. **12**, 3695–3700 (2012)
93. Withers, F., Del Pozo-Zamudio, O., Mishchenko, A., Rooney, A.P., Gholinia, A., Watanabe, K., Taniguchi, T., Haigh, S.J., Geim, A.K., Tartakovskii, A.I., Novoselov, K.S.:

Light-emitting diodes by band-structure engineering in van der Waals heterostructures. Nat. Mater. **14**, 301–306 (2015)
94. Chhowalla, M., Shin, H.S., Eda, G., Li, L.J., Loh, K.P., Zhang, H.: The chemistry of two-dimensional layered transition metal dichalcogenide nanosheets. Nat. Chem. **5**, 263–275 (2013)
95. Chia, X., Eng, A.Y.S., Ambrosi, A., Tan, S.M., Pumera, M.: Electrochemistry of nanostructured layered transition-metal dichalcogenides. Chem. Rev. (Washington, DC, US) **115**, 11941–11966 (2015)
96. Windom, B.C., Sawyer, W.G., Hahn, D.W.: A Raman spectroscopic study of MoS_2 and MoO_3: applications to tribological systems. Tribol. Lett. **42**, 301–310 (2011)
97. Fleischauer, P.D., Lince, J.R., Bertrand, P.A., Bauer, R.: Electronic structure and lubrication properties of MoS_2: a qualitative molecular orbital approach. Langmuir **5**, 1009–1015 (1989)
98. Salomon, G., De Gee, A.W.J., Zaat, J.H.: Mechano-chemical factors in MoS_2-film lubrication. Wear **7**, 87–101 (1964)
99. Azhagurajan, M., Kajita, T., Itoh, T., Kim, Y.-G., Itaya, K.: In situ visualization of lithium ion intercalation into MoS_2 single crystals using differential optical microscopy with atomic layer resolution. J. Am. Chem. Soc. **138**, 3355–3361 (2016)
100. Novoselov, K.S., Mishchenko, A., Carvalho, A., Castro Neto, A.H.: 2D materials and van der Waals heterostructures. Science **353**, aac9439 (2016)
101. Ajayan, P., Kim, P., Banerjee, K.: Two-dimensional van der Waals materials. Phys. Today **69**, 38–44 (2016)
102. Oughaddou, H., et al.: Prog. Surf. Sci. **90**, 46 (2015)
103. Liu, H., et al.: ACS Nano **8**, 4033 (2014)
104. Abergel, D.S.L., et al.: Properties of graphene: a theoretical perspective. Adv. Phys. **59**(4), 261–482 (2010)
105. Alferov, Z.I.: Rev. Mod. Phys. **73**(3), 767 (2001)
106. Kroemer, H.: Rev. Mod. Phys. **73**(3), 783 (2001)
107. Isamu, A., et al.: J. Lumin. **48–49**(Part 2), 666 (1991)
108. Amano, H., et al.: Jpn. J. Appl. Phys. **28**(12A), L2112 (1989)
109. Duan, X., et al.: Chem. Soc. Rev. **44**, 8859 (2015)
110. Jariwala, D., et al.: ACS Nano **8**, 1102 (2014)
111. Mahatha, S., et al.: J. Phys.: Condens. Matter **24**, 475504 (2012)
112. Hwang, W.S., et al.: Appl. Phys. Lett. **101**, 013107 (2012)
113. Vogt, P., et al.: Phys. Rev. Lett. **108**, 155501 (2012)
114. Pakdel, A., et al.: Mater. Today **15**, 256 (2012)
115. Song, L., et al.: Phys. Rev. B **86**, 075429 (2012)
116. Ci, L., et al.: Nat. Mater. **9**, 430 (2010)
117. Liu, Z., et al.: Nano Lett. **11**, 2032 (2011)
118. Zhang, H.: ACS Nano **9**, 9451 (2015)
119. Das, S., et al.: Crit. Rev. Solid State Mater. Sci. **39**, 231 (2014)
120. Wang, X.-R., et al.: Chin. Phys. B **22**, 098505 (2013)
121. Das, S., et al.: Annu. Rev. Mater. Res. **45**, 1 (2015)
122. Cheng, R., et al.: Nat. Commun. **5**, 5143 (2014)
123. Akinwande, D., et al.: Nat. Commun. **5**, 5678 (2014)
124. Bao, W., et al.: Appl. Phys. Lett. **102**, 042104 (2013)
125. Frey, G., et al.: Phys. Rev. B **57**, 6666 (1998)
126. Islam, M.R., et al.: Nanoscale **6**, 10033 (2014)
127. Choudhary, N., et al.: J. Phys.: Condens. Matter **28**, 364002 (2016)
128. Fuhrer, M.S., Hone, J.: Nat. Nanotechnol. **8**, 146 (2013)
129. Novoselov, K.S., Fal'ko, V.I., Colombo, L., Gellert, P.R., Schwab, M.G., Kim, K.: A road map for graphene. Nature **490**, 192–200 (2012)
130. Tributsch, H., Bennett, J.C.: Electrochemistry and photochemistry of MoS_2 layer crystals. I. J. Electroanal. Chem. **81**, 97–111 (1977)

131. Fivaz, R., Mooser, E.: Mobility of charge carriers in semiconducting layerstructures. Phys. Rev. **163**, 743–755 (1967)
132. Geim, A.K.: Nobel lecture: random walk to graphene. Rev. Mod. Phys. **83**, 851–862 (2011)
133. Osada, M., Sasaki, T.: Two-dimensional dielectric nanosheets: novel nanoelectronics from nanocrystal building blocks. Adv. Mater. **24**, 210–228 (2012)
134. Miró, P., Audiffred, M., Heine, T.: An atlas of two-dimensional materials. Chem. Soc. Rev. **43**, 6537–6554 (2014)
135. Wang, F., Wang, Z., Wang, Q., Wang, F., Yin, L., Xu, K., Huang, Y., He, J.: Synthesis, properties and applications of 2D non-graphene materials. Nanotechnology **26**, 292001 (2015)
136. Bromley, R.A., Murray, R.B., Yoffe, A.D.: The band structures of some transition metal dichalcogenides. III. Group VIA: trigonal prism materials. J. Phys. C: Solid State Phys. **5**, 759–778 (1972)
137. Lin, Y.-C., Dumcenco, D.O., Huang, Y.-S., Suenaga, K.: Atomic mechanism of the semiconducting-to-metallic phase transition in single-layered MoS_2. Nat. Nanotechnol. **9**, 391–396 (2014)
138. Kappera, R., Voiry, D., Yalcin, S.E., Branch, B., Gupta, G., Mohite, A.D., Chhowalla, M.: Phase-engineered low-resistance contacts for ultrathin MoS_2 transistors. Nat. Mater. **13**, 1128–1134 (2014)
139. Slonczewski, J.C., Weiss, P.R.: Band structure of graphite. Phys. Rev. **109**, 272–279 (1958)
140. Castellanos-Gomez, A., Vicarelli, L., Prada, E., Island, J.O., Narasimha-Acharya, K.L., Blanter, S.I., Groenendijk, D.J., Buscema, M., Steele, G.A., Alvarez, J.V., Zandbergen, H. W., Palacios, J.J., van der Zant, H.S.J.: Isolation and characterization of few-layer black phosphorus. 2D Mater. **1**, 025001 (2014)
141. Novoselov, K.S.: Nobel lecture: graphene: materials in the flatland. Rev. Mod. Phys. **83**, 837–849 (2011)
142. Guo, Z., Zhang, H., Lu, S., Wang, Z., Tang, S., Shao, J., Sun, Z., Xie, H., Wang, H., Yu, X.-F., Chu, P.K.: From black phosphorus to phosphorene: basic solvent exfoliation, evolution of Raman scattering, and applications to ultrafast photonics. Adv. Funct. Mater. **25**, 6996–7002 (2015)
143. Radisavljevic, B., Radenovic, A., Brivio, J., Giacometti, V., Kis, A.: Single-layer MoS_2 transistors. Nat. Nanotechnol. **6**, 147–150 (2011)
144. Zhu, X., Monahan, N.R., Gong, Z., Zhu, H., Williams, K.W., Nelson, C.A.: Charge transfer excitons at van der Waals interfaces. J. Am. Chem. Soc. **137**, 8313–8320 (2015)
145. Kim, K.K., et al.: Nano Lett. **12**(1), 161 (2011)
146. Shi, Y., et al.: Nano Lett. **10**(10), 4134 (2010)
147. C.-H. Chen, et al.: 2D Mater. **1**(3), 034001 (2014)
148. Liu, K.-K., et al.: Nano Lett. **12**(3), 1538 (2012)
149. Li, X., et al.: Science **324**(5932), 1312 (2009)
150. Chen, C.-H., et al.: Small **8**(1), 43 (2012)
151. Radisavljevic, B, et al.: Nat. Nanotechnol. **6**(3), 147 (2011)
152. He, Q., et al.: Small **8**(19), 2994 (2012)
153. Fang, H., et al.: Nano Lett. **12**(7), 3788 (2012)
154. Huang, J.-K., et al.: ACS Nano **8**(1), 923 (2013)
155. Pu, J., et al.: Nano Lett. **12**(8), 4013 (2012)
156. Geim, A.K., Grigorieva, I.V.: Nature **499**(7459), 419 (2013)
157. Wang, H., et al.: Nanoscale **6**(21), 12250 (2014)
158. Buscema, M., et al.: Chem. Soc. Rev. **44**(11), 3691 (2015)
159. Park, N.-M., Kim, T.-S., Park, S.-J.: Band gap engineering of amorphous silicon quantum dots for light-emitting diodes. Appl. Phys. Lett. **78**(17), 2575 (2001)
160. Partoens, B., Peeters, F.M.: From graphene to graphite: electronic structure around the K point. Phys. Rev. B **74**(7), 075404 (2006)
161. Wallace, P.R.: The band theory of graphite. Phys. Rev. **71**(9), 622–634 (1947)

162. Wilder, J.W.G., et al.: Electronic structure of atomically resolved carbon nanotubes. Nature **391**(6662), 59–62 (1998)
163. Golden, M.S., et al.: The electronic structure of fullerenes and fullerene compounds from high-energy spectroscopy. J. Phys.: Condens. Matter **7**(43), 8219 (1995)
164. Herman, F.: Electronic structure of the diamond crystal. Phys. Rev. **88**(5), 1210–1211 (1952)
165. Bauer, L.A., Birenbaum, N.S., Meyer, G.J.: Biological applications of high aspect ratio nanoparticles. J. Mater. Chem. **14**(4), 517–526 (2004)
166. Yoriya, S., et al.: Initial studies on the hydrogen gas sensing properties of highly-ordered high aspect ratio TiO_2 nanotube-arrays 20 μm to 222 μm in length. Sens. Lett. **4**(3), 334–339 (2006)
167. Bimberg, D., Grundmann, M., Ledentsov, N.N.: Quantum Dot Heterostructures, vol. 471973882. Wiley, Chichester (1999)
168. Xiang, J., et al.: Ge/Si nanowire heterostructures as high-performance field-effect transistors. Nature **441**(7092), 489–493 (2006)
169. Lin, Y.-M., et al.: 100-GHz transistors from wafer-scale epitaxial graphene. Science **327** (5966), 662 (2010)
170. Liu, M., et al.: A graphene-based broadband optical modulator. Nature **474**(7349), 64–67 (2011)
171. Kim, K.S., et al.: Large-scale pattern growth of graphene films for stretchable transparent electrodes. Nature **457**(7230), 706–710 (2009)
172. Kim, Y.J., et al.: Chemically modified multiwalled carbon nanotubes as an additive for supercapacitors. Small **2**(3), 339–345 (2006)
173. Zhu, Y., et al.: Carbon-based supercapacitors produced by activation of graphene. Science **332**(6037), 1537–1541 (2011)
174. Dimitrakakis, G.K., Tylianakis, E., Froudakis, G.E.: Pillared graphene: a new 3-D network nanostructure for enhanced hydrogen storage. Nano Lett. **8**(10), 3166–3170 (2008)
175. Henwood, D., Carey, J.D.: Ab initio investigation of molecular hydrogen physisorption on graphene and carbon nanotubes. Phys. Rev. B **75**(24), 245413 (2007)
176. El-Kady, M.F., et al.: Laser scribing of high-performance and flexible graphene-based electrochemical capacitors. Science **335**(6074), 1326–1330 (2012)
177. Yang, X., et al.: Graphene uniformly decorated with gold nanodots: in situ synthesis, enhanced dispersibility and applications. J. Mater. Chem. **21**(22), 8096–8103 (2011)
178. Schedin, F., et al.: Detection of individual gas molecules adsorbed on graphene. Nat. Mater. **6**(9), 652–655 (2007)
179. Deng, M., et al.: Electrochemical deposition of polypyrrole/graphene oxide composite on microelectrodes towards tuning the electrochemical properties of neural probes. Sens. Actuators B: Chem. **158**(1), 176–184 (2011)
180. Xu, M., Fujita, D., Hanagata, N.: Perspectives and challenges of emerging single-molecule DNA sequencing technologies. Small **5**(23), 2638–2649 (2009)
181. Garaj, S., et al.: Graphene as a subnanometre trans-electrode membrane. Nature **467**(7312), 190–193 (2010)
182. Wilson, N.R., et al.: Graphene oxide: structural analysis and application as a highly transparent support for electron microscopy. ACS Nano **3**(9), 2547–2556 (2009)
183. Kim, K., et al.: A role for graphene in silicon-based semiconductor devices. Nature **479** (7373), 338–344 (2011)
184. Castro, E., et al.: Biased bilayer graphene: semiconductor with a gap tunable by the electric field effect. Phys. Rev. Lett. **99**(21), 216802 (2007)
185. McCann, E.: Asymmetry gap in the electronic band structure of bilayer graphene. Phys. Rev. B **74**(16), 161403 (2006)
186. Ohta, T., et al.: Controlling the electronic structure of bilayer graphene. Science (New York, N.Y.) **313**(5789), 951–954 (2006)
187. Samuels, A.J., Carey, J.D.: Molecular doping and band-gap opening of bilayer graphene. ACS Nano **7**(3), 2790–2799 (2013)

188. Zhang, Y., et al.: Direct observation of a widely tunable bandgap in bilayer graphene. Nature **459**(7248), 820–823 (2009)
189. Xu, M., et al.: Graphene-like two-dimensional materials. Chem. Rev. **113**(5), 3766–3798 (2013)
190. Novoselov, K.S., et al.: Two-dimensional atomic crystals. Proc. Natl. Acad. Sci. USA **102** (30), 10451–10453 (2005)
191. Osada, M., Sasaki, T.: Two-dimensional dielectric nanosheets: novel nanoelectronics from nanocrystal building blocks. Adv. Mater. **24**(2), 210–228 (2012)
192. Giovannetti, G., et al.: Substrate-induced band gap in graphene on hexagonal boron nitride: ab initio density functional calculations. Phys. Rev. B **76**(7), 073103 (2007)
193. Sławioska, J., Zasada, I., Klusek, Z.: Energy gap tuning in graphene on hexagonal boron nitride bilayer system. Phys. Rev. B **81**(15), 155433 (2010)
194. Zhang, H., et al.: Topological insulators in Bi_2Se_3, Bi_2Te_3 and Sb_2Te_3 with a single Dirac cone on the surface. Nat. Phys. **5**(6), 438–442 (2009)
195. Tang, H., et al.: Two-dimensional transport-induced linear magneto-resistance in topological insulator Bi_2Se_3 nanoribbons. ACS Nano **5**(9), 7510–7516 (2011)
196. Gourmelon, E., et al.: MS2 (M = W, Mo) photosensitive thin films for solar cells. Sol. Energy Mater. Sol. Cells **46**(2), 115–121 (1997)
197. Eda, G., et al.: Transparent and conducting electrodes for organic electronics from reduced graphene oxide. Appl. Phys. Lett. **92**(23), 233305 (2008)
198. Motohiko, E.: A topological insulator and helical zero mode in silicene under an inhomogeneous electric field. New J. Phys. **14**(3), 033003 (2012)
199. Shahil, K.M.F., et al.: Crystal symmetry breaking in few-quintuple Bi2Te3 films: applications in nanometrology of topological insulators. Appl. Phys. Lett. **96**(15), 153103 (2010)
200. Gamble, F.R., Silbernagel, B.G.: Anisotropy of the proton spin–lattice relaxation time in the superconducting intercalation complex TaS_2 (NH_3): structural and bonding implications. J. Chem. Phys. **63**(6), 2544–2552 (1975)
201. Tang, X., et al.: Preparation and thermoelectric transport properties of high performance p-type Bi_2Te_3 with layered nanostructure. Appl. Phys. Lett. **90**(1), 012102 (2007)
202. Ma, Y., Dai, Y., Guo, M., Niu, C., Huang, B.: Graphene adhesion on MoS_2 monolayer: an ab initio study. Nanoscale **3**, 3883–3887 (2011)
203. Hong, X., Kim, J., Shi, S.F., Zhang, Y., Jin, C., Sun, Y., et al.: Ultrafast charge transfer in atomically thin MoS_2/WS_2 heterostructures. Nat. Nanotechnol. **9**, 682–686 (2014)
204. Xu, H., Wu, J., Feng, Q., Mao, N., Wang, C., Zhang, J.: High responsivity and gate tunable graphene-MoS_2 hybrid phototransistor. Small **10**, 2300–2306 (2014)
205. Zhang, W., Chuu, C.P., Huang, J.K., Chen, C.H., Tsai, M.L., Chang, Y.H., et al.: Ultrahigh-gain photodetectors based on atomically thin graphene-MoS_2 heterostructures. Sci. Rep. **4**(3826), 1–8 (2014)
206. Gong, Y., Lin, J., Wang, X., Shi, G., Lei, S., Lin, Z., et al.: Vertical and in-plane heterostructures from WS_2/MoS_2 monolayers. Nat. Mater. **13**, 1135–1142 (2014)
207. Wi, S., Kim, H., Chen, M., Nam, H., Guo, L.J., Meyhofer, E., et al.: Enhancement of photovoltaic response in multilayer MoS_2 induced by plasma doping. ACS Nano **8**, 5270–5281 (2014)
208. Choudhary, N., et al.: J. Mater. Res. **31**, 824 (2016)
209. Varghese, S.S., et al.: Electronics **4**, 651 (2015)
210. Lee, J., et al.: Sci. Rep. **4**, 7352 (2014)
211. Perera-Lopez, N., et al.: 2D Mater. **1**, 011004 (2014)
212. Choudhary, N., et al.: Sci. Rep. **6**, 25456 (2016)
213. Lee, C.-H., et al.: Nat. Nanotechnol. **9**, 676 (2014)
214. Mak, K.F., Shan, J.: Nat. Photonics **10**, 216 (2016)
215. Lin, S., et al.: Sci. Rep. **5**, 15103 (2015)
216. Wu, W., et al.: Nature **514**, 470 (2014)
217. Choudhary, N., et al.: J. Mater. Chem. A **3**, 24049 (2015)

218. Li, H., et al.: Small **8**, 63 (2012)
219. Cui, S., et al.: Nat. Commun. **6**, 8632 (2015)
220. Sarkar, D., et al.: ACS Nano **8**, 3992 (2014)
221. Blake, P., Hill, E.W., Castro Neto, A.H., Novoselov, K.S., Jiang, D., Yang, R., Booth, T.J., Geim, A.K.: Making graphene visible. Appl. Phys. Lett. **91**, 063124 (2007)
222. Dean, C.R., et al.: Nat. Nanotechnol. **5**(10), 722 (2010)
223. Hsu, W.-T., et al.: ACS Nano **8**(3), 2951 (2014)
224. Fang, H., et al.: Proc. Natl. Acad. Sci. USA **111**(17), 6198 (2014)
225. Chiu, M.-H., et al.: ACS Nano **8**(9), 9649 (2014)
226. Song, J.-G., et al.: ACS Nano **7**, 11333 (2013)
227. Zhou, X., et al.: J. Am. Chem. Soc. **137**, 7994 (2015)
228. Keum, D.H., et al.: Nat. Phys. **11**, 482 (2015)
229. Ali, M.N., et al.: Nature **514**, 205 (2014)
230. Lee, S.Y., et al.: ACS Nano **9**, 9034 (2015)
231. Han, M.Y., et al.: Phys. Rev. Lett. **98**, 206805 (2007)
232. Del Pozo-Zamudio, O., et al.: 2D Mater. **2**, 035010 (2015)
233. Tongay, S., et al.: Nat. Commun. **5**, 3252 (2014)
234. Kang, J., et al.: Proc. SPIE **9083**, 908305 (2014)
235. Chen, C.-H., et al.: Nanomed. Nanotechnol. Biol. Med. **9**(5), 600 (2013)
236. Zhu, C., et al.: J. Am. Chem. Soc. **135**(16), 5998 (2013)
237. Loo, A.H., et al.: Nanoscale **6**(20), 11971 (2014)
238. Loan, P.T.K., et al.: Adv. Mater. **26**(28), 4838 (2014)
239. Mak, K.F., et al.: Nat. Mater. **12**(3), 207 (2013)
240. Bernardi, M., et al.: Nano Lett. **13**(8), 3664 (2013)
241. Furchi, M.M., et al.: Nano Lett. **14**(8), 4785 (2014)
242. Lee, C.-H., et al.: Nat. Nanotechnol. **9**(9), 676 (2014)
243. Wi, S., et al.: ACS Nano **8**(5), 5270 (2014)
244. Tsai, M.-L., et al.: ACS Nano **8**(8), 8317 (2014)
245. Shanmugam, M., et al.: Nanoscale **6**(21), 12682 (2014)
246. Duan, X., et al.: Nat. Nanotechnol. **9**, 1024 (2014)
247. Li, M.-Y., et al.: Science **349**(6247), 524 (2015)
248. Roy, T., et al.: ACS Nano **8**(6), 6259 (2014)
249. Fang, H., et al.: Nano Lett. **13**(5), 1991 (2013)
250. Tosun, M., et al.: ACS Nano **8**(5), 4948 (2014)
251. Liu, H., et al.: ACS Nano **8**(4), 4033 (2014)
252. Das, S., Roelofs, A. In: 2014 72nd Annual Device Research Conference (DRC), p. 185 (2014)
253. Yu, L., et al.: Nano Lett. **15**(8), 4928 (2015)
254. Britnell, L., et al.: Science **335**(6071), 947 (2012)
255. Georgiou, T., et al.: Nat. Nanotechnol. **8**(2), 100 (2013)
256. Yu, W.J., et al.: Nat. Mater. **12**(3), 246 (2013)
257. Moriya, R., et al.: Appl. Phys. Lett. **105**(8), 083119 (2014)
258. Sarkar, D., et al.: Nature **526**(7571), 91 (2015)
259. Britnell, L., et al.: Science **340**(6138), 1311 (2013)
260. Yu, W.J., et al.: Nat. Nanotechnol. **8**(12), 952 (2013)
261. Cheng, R., et al.: Nano Lett. **14**(10), 5590 (2014)
262. Deng, Y., et al.: ACS Nano **8**(8), 8292 (2014)
263. Lopez-Sanchez, O., et al.: ACS Nano **8**(3), 3042 (2014)
264. Pospischil, A., et al.: Nat. Nano **9**(4), 257 (2014)
265. Baugher, B.W.H., et al.: Nat. Nano **9**(4), 262 (2014)
266. Ross, J.S., et al.: Nat. Nano **9**(4), 268 (2014)
267. Zhang, Y.-J., et al.: Science **344**(6185), 725 (2014)
268. Withers, F., et al.: Nat. Mater. **14**(3), 301 (2015)
269. Zhang, W., et al.: Sci. Rep. **4**, 3826 (2014)

270. Roy, K., et al.: Nat. Nanotechnol. **8**(11), 826 (2013)
271. Pop, E., et al.: MRS Bull. **37**(12), 1273 (2012)
272. Chen, C.-C., et al.: Nano Res. **8**(2), 666 (2015)
273. Astefanei, A., Núñez, O., Galceran, M.T.: Characterisation and determination of fullerenes: a critical review. Anal. Chim. Acta **882**, 1–21 (2015). http://dx.doi.org/10.1016/j.aca.2015.03.025
274. Ibrahim, K.S.: Carbon nanotubes—properties and applications: a review. Carbon Lett. **14**, 131–144 (2013). https://doi.org/10.5714/CL.2013.14.3.131
275. Aqel, A., El-Nour, K.M.M.A., Ammar, R.A.A., Al-Warthan, A.: Carbon nanotubes, science and technology part (I) structure, synthesis and characterisation. Arab. J. Chem. **5**, 1–23 (2012). https://doi.org/10.1016/j.arabjc.2010.08.022
276. Elliott, J.A., Shibuta, Y., Amara, H., Bichara, C., Neyts, E.C.: Atomistic modelling of CVD synthesis of carbon nanotubes and graphene. Nanoscale **5**, 6662 (2013). https://doi.org/10.1039/c3nr01925j
277. Saeed, K., Khan, I.: Preparation and characterization of single-walled carbon nanotube/nylon 6,6 nanocomposites. Instrum. Sci. Technol. **44**, 435–444 (2016). http://dx.doi.org/10.1080/10739149.2015.1127256
278. Saeed, K., Khan, I.: Preparation and properties of single-walled carbon nanotubes/poly (butylene terephthalate) nanocomposites. Iran. Polym. J. **23**, 53–58 (2014). https://doi.org/10.1007/s13726-013-0199-2
279. Ngoy, J.M., Wagner, N., Riboldi, L., Bolland, O.: A CO_2 capture technology using multi-walled carbon nanotubes with polyaspartamide surfactant. Energy Procedia **63**, 2230–2248 (2014). https://doi.org/10.1016/j.egypro.2014.11.242
280. Mabena, L.F., Sinha Ray, S., Mhlanga, S.D., Coville, N.J.: Nitrogen-doped carbon nanotubes as a metal catalyst support. Appl. Nanosci. **1**, 67–77 (2011). https://doi.org/10.1007/s13204-011-0013-4
281. Manyam, J., Manickam, M., Preece, J.A., Palmer, R.E., Robinson, A.P.G.: Proc. SPIE **7972**, 79722N (2011). https://doi.org/10.1117/12.879469
282. Kroto, H.W., Heath, J.R., O'Brien, S.C., Curl, R.F., Smalley, R.E.: Nature **318**, 162–163 (1985). https://doi.org/10.1038/318162a0
283. de Marneffe, J.F., Cooke, M., Goodyear, A., Braithwaite, N.S.J., Sutton, Y., Bowden, M., Altimarano-Sanchez, E., Zotovich, A., El Otell, Z., Chan, B.T., Knoll, A., Rawlings, C., Duerig, U., Spieser, M., Kaestner, M., Neuber, C., Rangelow, I.: Advanced etching for nano-devices and 2D materials. In: Proceedings of the 42nd International Conference on Micro and Nano Engineering, SNM-1-52016 (2016). https://www.researchgate.net/publication/308764935
284. Gogolides, E., Argitis, P., Couladouros, E.A., Vidali, V.P., Vasilopoulou, M., Cordoyiannis, G., Diakoumakos, C.D., Tserepi, A.: J. Vac. Sci. Technol. B: Microelectron. Nanometer. Struct. Process. Meas. Phenom. **21**, 141 (2003). https://doi.org/10.1116/1.1535930
285. Chen, X., Palmer, R.E., Robinson, A.P.G.: Nanotechnology **19**, 275308 (2008). https://doi.org/10.1088/0957-4484/19/27/275308
286. Tada, T., Kanayama, T.: Jpn. J. Appl. Phys. **35**(Part 2), L63–L65 (1996). https://doi.org/10.1143/jjap.35.l63
287. Gibbons, F.P., Robinson, A.P.G., Palmer, R.E., Manickam, M., Preece, J.A.: Small **2**, 1003–1006 (2006). https://doi.org/10.1002/smll.200500443
288. Shi, X., Prewett, P., Huq, E., Bagnall, D.M., Robinson, A.P.G., Boden, S.A.: Microelectron. Eng. **155**, 74–78 (2016). https://doi.org/10.1016/j.mee.2016.02.045
289. Kim, H., Gilmore, C.M., Piqué, A., Horwitz, J.S., Mattoussi, H., Murata, H., Kafafi, Z.H., Chrisey, D.B.: Electrical, optical, and structural properties of indium–tin–oxide thin films for organic light-emitting devices. J. Appl. Phys. **86**, 6451–6461 (1999)
290. Haynes, W.M. (ed.): CRC Handbook of Chemistry and Physics. CRC Press (2016). http://www.hbcponline.com
291. Pumera, M., Sofer, Z., Ambrosi, A.: Layered transition metal dichalcogenides for electrochemical energy generation and storage. J. Mater. Chem. A **2**, 8981–8987 (2014)

292. Koskinen, P., Fampiou, I., Ramasubramaniam, A.: Density-functional tight-binding simulations of curvature-controlled layer decoupling and band-gap tuning in bilayer MoS_2. Phys. Rev. Lett. **112**, 186802 (2014)
293. Lee, H.S., Luong, D.H., Kim, M.S., Jin, Y., Kim, H., Yun, S., Lee, Y.H.: Reconfigurable exciton-plasmon interconversion for nanophotonic circuits. Nat. Commun. **7**, 13663 (2016)
294. Mak, K.F., He, K., Shan, J., Heinz, T.F.: Control of valley polarization inmonolayer MoS_2 by optical helicity. Nat. Nanotechnol. **7**, 494–498 (2012)
295. Novoselov, K.S., Mishchenko, A., Carvalho, A., Castro Neto, A.H.: 2D materialsand van der Waals heterostructures. Science **353**, aac9439 (2016)
296. Cai, Y., et al.: Highly itinerant atomic vacancies in phosphorene. J. Am. Chem. Soc. **138**, 10199–10206 (2016)
297. Zou, X., Liu, Y., Yakobson, B.I.: Predicting dislocations and grain boundaries in two-dimensional metal-disulfides from the first principles. Nano Lett. **13**, 253–258 (2013)
298. Zhou, W., et al.: Intrinsic structural defects in monolayer molybdenum disulphide. Nano Lett. **13**, 2615–2622 (2013)
299. Elias, D., et al.: Control of graphene's properties by reversible hydrogenation: evidence for graphane. Science **323**, 610–613 (2009)
300. Zhu, S., Li, T.: Hydrogenation enabled scrolling of graphene. J. Phys. D Appl. Phys. **46**, 075301 (2013)
301. Zhu, S., Li, T.: Hydrogenation-assisted graphene origami and its application in programmable molecular mass uptake, storage, and release. ACS Nano **8**, 2864–2872 (2014)
302. Blees, M.K., et al.: Graphene kirigami. Nature **524**, 204–207 (2015)
303. Zhu, S., Huang, Y., Li, T.: Extremely compliant and highly stretchable patterned graphene. Appl. Phys. Lett. **104**, 173103 (2014)
304. Li, T., et al.: Compliant thin film patterns of stiff materials as platforms for stretchable electronics. J. Mater. Res. **20**, 3274–3277 (2005)
305. Qi, Z., Campbell, D.K., Park, H.S.: Atomistic simulations of tension-induced large deformation and stretchability in graphene kirigami. Phys. Rev. B **90**, 245437 (2014)
306. Bahamon, D., et al.: Graphene kirigami as a platform for stretchable and tunable quantum dot arrays. Phys. Rev. B **93**, 235408 (2016)

Chapter 8
Nanoelectronics and Role of Surfaces Interfaces

Electronic and optoelectronic devices [1–23] impact many areas of society, from simple household appliances and multimedia systems to communications, computing, and medical instruments. Given the demand for ever more compact and powerful systems, there is growing interest in the development of nanoscale devices that could enable new functions and/or greatly enhanced performance. Semiconductor nanowires are emerging as a powerful class of materials that, through controlled growth and organization, are opening up substantial opportunities for novel nanoscale photonic and electronic devices.

8.1 The Development of Microelectronics

The number of atoms and the switching energy necessary for operating one bit are reduced by several powers of ten as shown in Fig. 8.1.

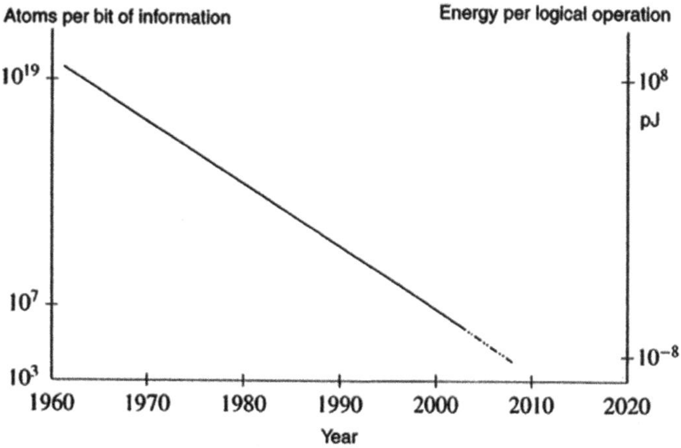

Fig. 8.1 The development of integrated circuits shows changes in the orders of magnitude

L. H. Madkour, *Nanoelectronic Materials*, Advanced Structured Materials 116, https://doi.org/10.1007/978-3-030-21621-4_8

8.2 The Region of Nanostructures

See Fig. 8.2.

- Structure, Phase changes

  - Large nanoclusters have the same structure as the bulk with **somewhat different lattice parameters** (in general clusters are slightly contracted). Smaller clusters might have structures that differ from the bulk materials.
  - **Structural magic numbers** occur when an exact amount of atoms is needed for a specific structure, to minimize the energy (jellium model).
  - The **general tendency** is **to form structures where the total surface energy** (and internal energy) **is minimized**.

  - Another effect caused by a high surface to volume ratio in clusters is **a substantially lower melting temperature** as compared to bulk materials.

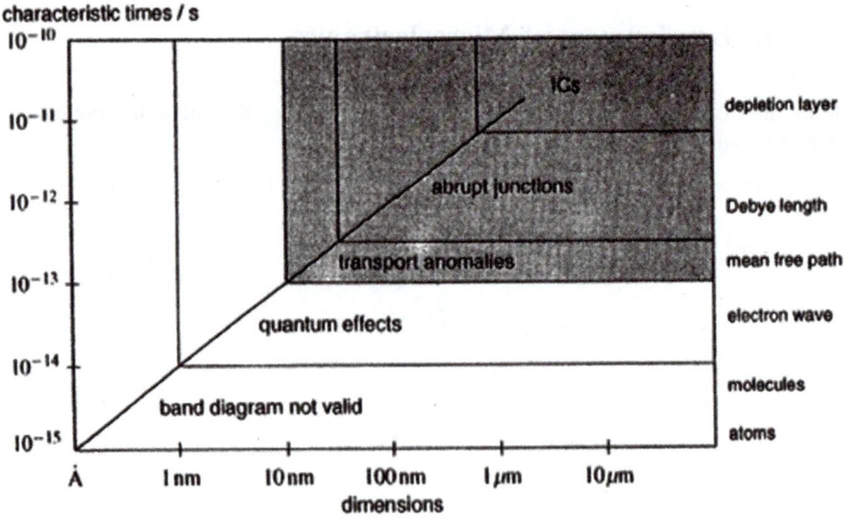

Fig. 8.2 Characteristic times and structures for semiconductor devices. The grey area concerns the present integrated circuits

8.3 Crystal Structure and Dense Planes

- The crystal structure is a unique arrangement of atoms in a crystal. It is composed of a unit cell with a set of atoms at each vertex. This is periodically repeated in 3D in a lattice.
- The various possible structures are defined by their symmetries.
- These symmetries and structure play a role in determining many of its properties.

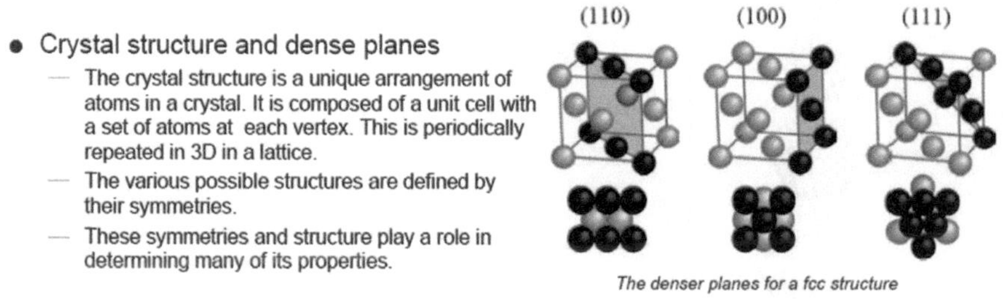

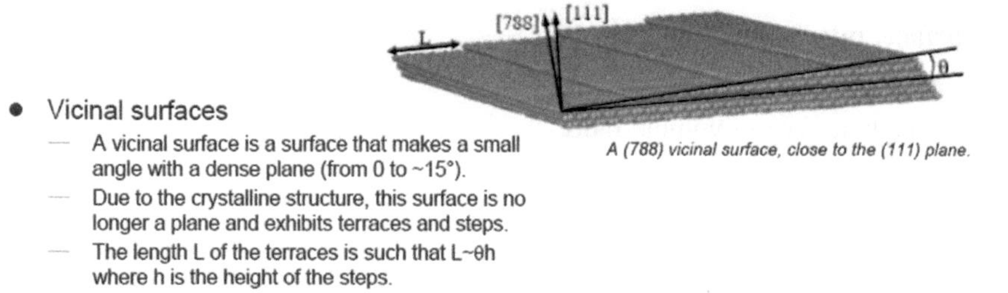

- **Surface**
  - The atoms on the surface play a dominant role for the physical and chemical properties that are controlled by surface exchanges.
  - Contrary to classical materials, the proportion of atoms on the surface of the nanoclusters becomes non negligible.

- surface
  - The atoms on the surface play a dominant role for the physical and chemical properties that are controlled by surface exchanges
  - Contrary to classical materials, the proportion of atoms on the surface of the nanoclusters becomes non negligible.

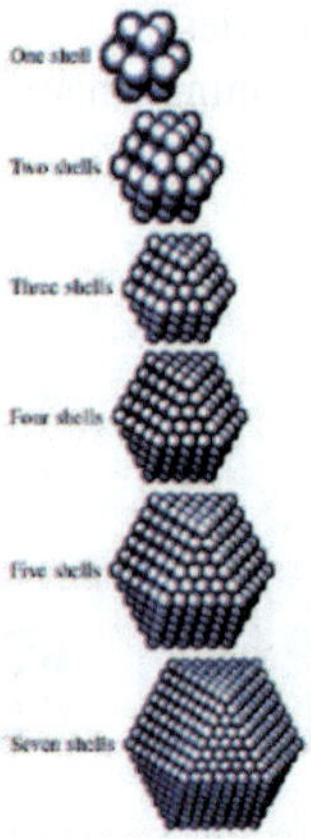

Total number of atoms	Percentage of surface atoms
13	92 %
55	76 %
147	63 %
309	52 %
561	45 %
1415	35%

- **Surface, interfaces**
  - **Nanoclusters**:
    a high surface-to-volume ratio.

- Surface, interfaces
  - Nanoclusters:
    a high surface-to-volume ratio.

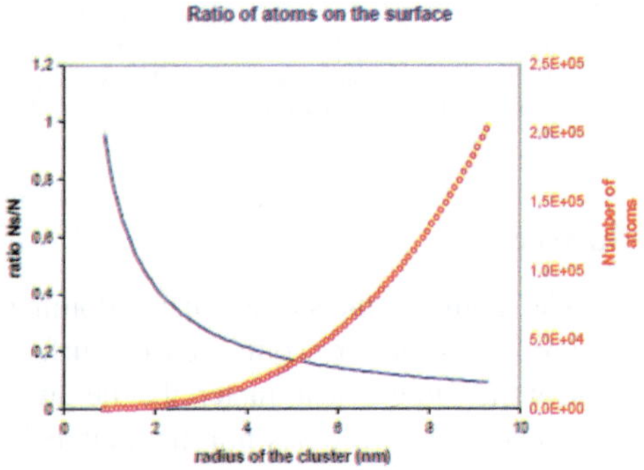

  - Nanomaterials:
    a high volume of atoms in grain boundaries and triple junctions.

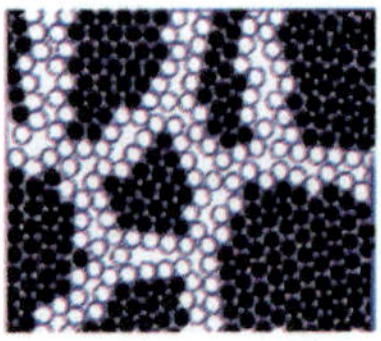

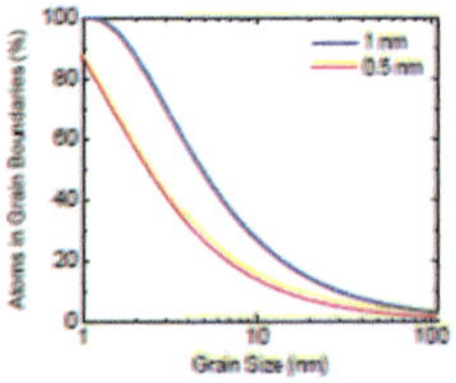

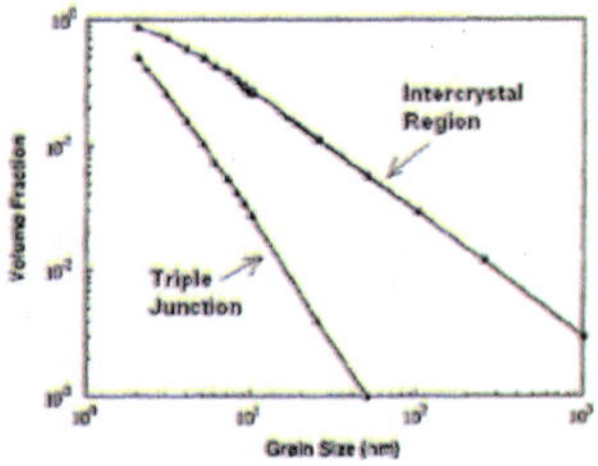

- **Nanomaterials**:
  a high volume of atoms in grain boundaries and triple junctions.

8.4 The Surface Energy γ

- in J m^{-2}
- $dW = \gamma\, dS$ is the work needed to increase the surface by dS. In this transformation, the atomic surface concentration is kept constant.
- If the surface is increased while keeping the same number of atoms (the concentration is decreased), the work needed to increase the surface by dS is now: $dW = \sigma_{ij}\, dS$ where σ_{ij} is the stress tensor.
- $\sigma_{ij} = \gamma\delta_{ij} + \frac{\partial\gamma}{\partial u_{ij}}$ where u_{ij} is the strain tensor.

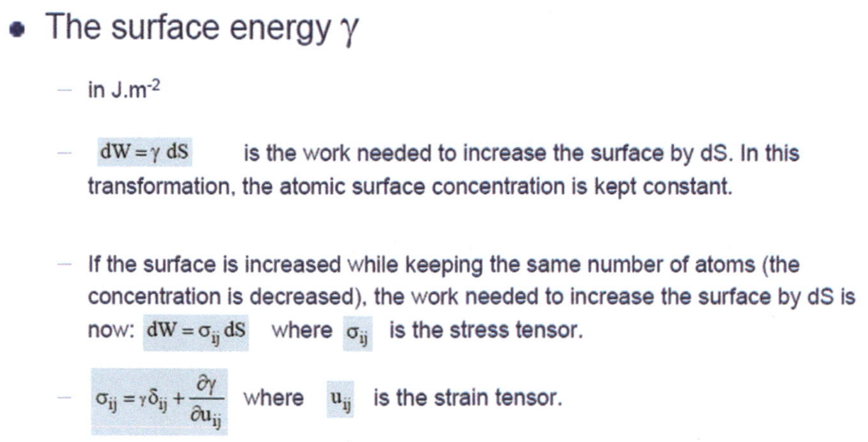

- To create a surface needs an energy dW. This energy depends on the orientation of the surface.
- The surface energy depends on the number of cut bonds between nearest neighbor and next nearest neighbor interactions.

"Square" lattice

$\tan\theta = m/q$

$m = 0$ and $q = 1$ for the Σ plane.

$m = 1$ and $q = 3$ for the Σ′ plane.

- The surface energy
  - To create a surface needs an energy dW. This energy depends on the orientation of the surface.
  - The surface energy depends on the number of cut bonds between nearest neighbor and next nearest neighbor interactions.

"Square" lattice

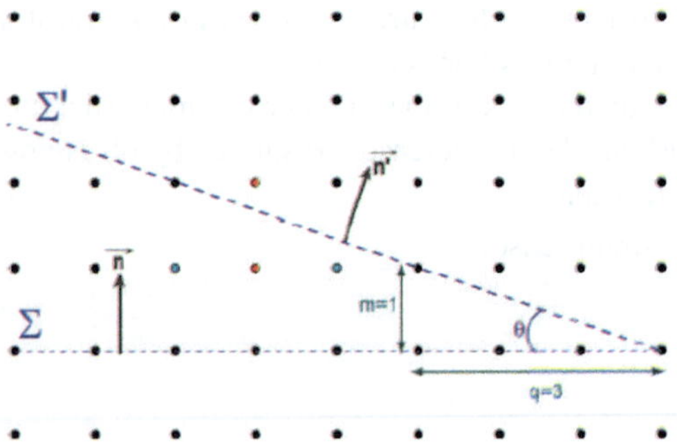

$$\tan\theta = \frac{m}{q}$$

m=0 and q=1 for the Σ plane.
m=1 and q=3 for the Σ' plane.

- The number of nearest neighbor cut bonds:
- $N_{1\Sigma} = 1$ and $N_{1\Sigma'} = 4$
- $N_1 = q + m$
- The number of next nearest neighbor cut bonds:
- $N_{2\Sigma} = 2$ and $N_{2\Sigma'} = 6$
- $N_2 = 2q$

- The surface energy
  - The number of nearest neighbor cut bonds:
    - $N_{1\Sigma} = 1$ and $N_{1\Sigma'} = 4$
    - $N_1 = q+m$
  - The number of next nearest neighbor cut bonds:
    - $N_{2\Sigma} = 2$ and $N_{2\Sigma}' = 6$
    - $N_2 = 2q$

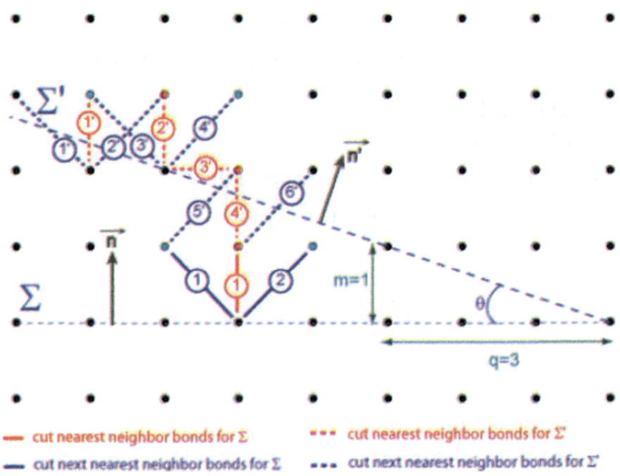

- It is clear that the creation of a dense surface cut less bonds per unit surface than the creation of a less dense surface.
- Thus, the surface energy of a given surface is higher than the surface energy of the denser surface.

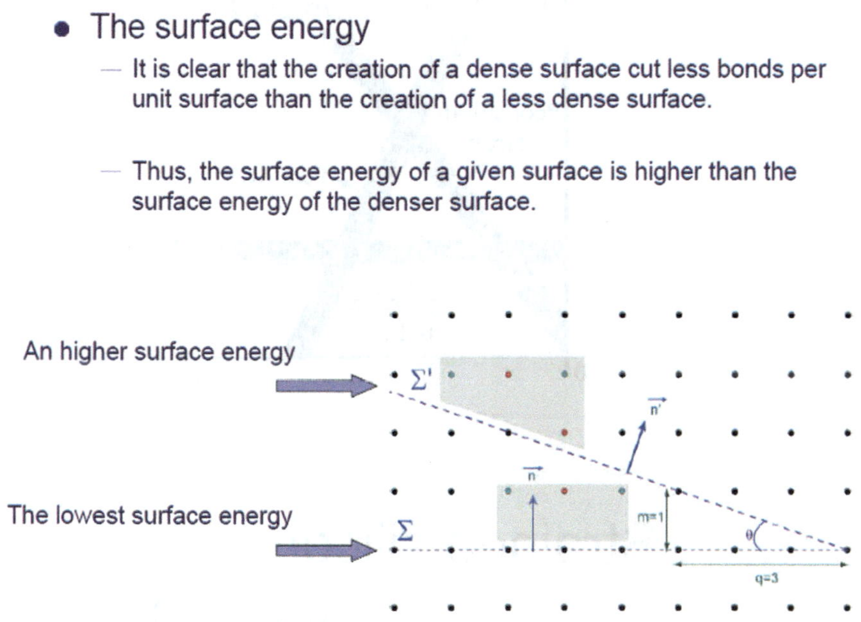

8.5 Transistor Scaling

Nodes with feature size less than 100 nm can be referred to as nanotechnology. By 2011, the gate length is expected to be at or below 10 nm. Transistor scaling will be enabled by integration of emerging nanotechnology options on to the Si platform (Figs. 8.3, 8.4, 8.5 and 8.6).

- Third condition is gate oxide reliability $T_{ox} = V_{dd} / E_{ox}^{max}$

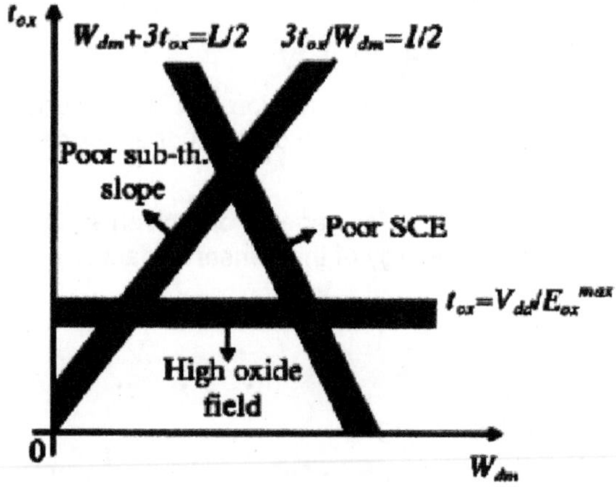

Strained silicon

- •The mobility of the carriers increases with increase in strain
- •Two types of strain : biaxial strain and uniaxial strain
- •The size of silicon small than compound Silicon structure introduces the biaxial strain
- •Uniaxial:Placing at the edge of the source or drain.
- •The Ion of device improves 15% to 30% with out scaling
- •90nm and 65 nm technologies used strained silicon technology
- •Next generation technology (45nm)uses High –k and metal gate with strain silicon

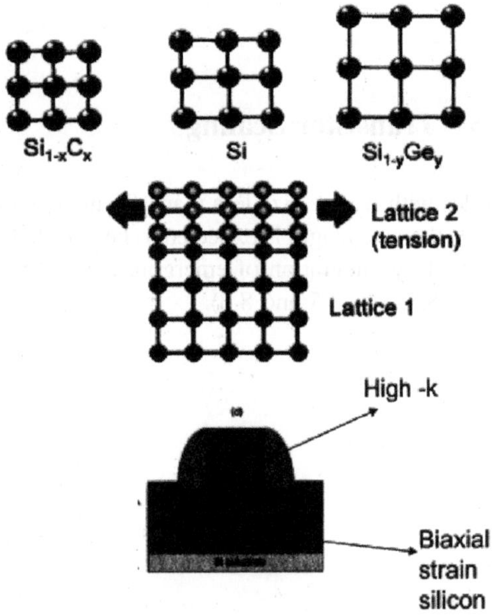

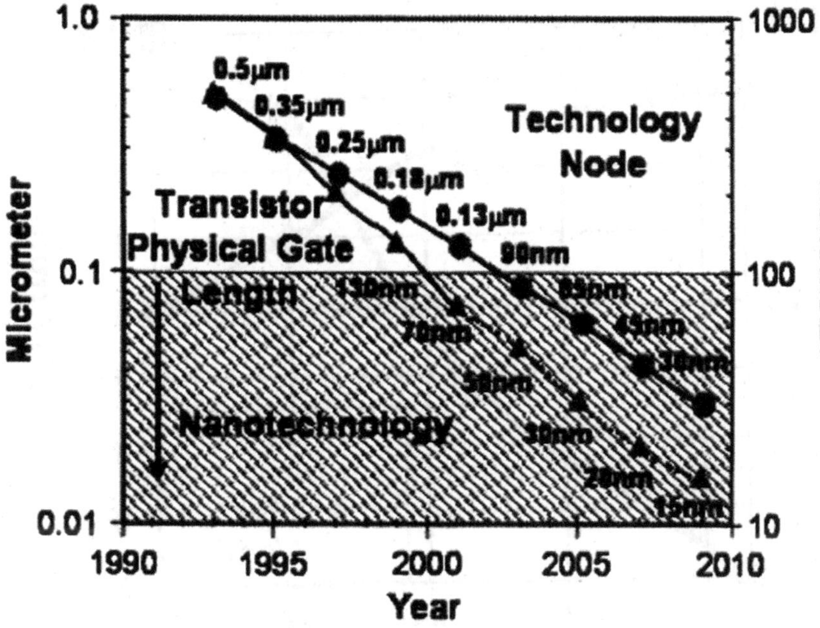

Fig. 8.3 Scaling of transistor size (physical gate length) with technology node to sustain Moore's Law

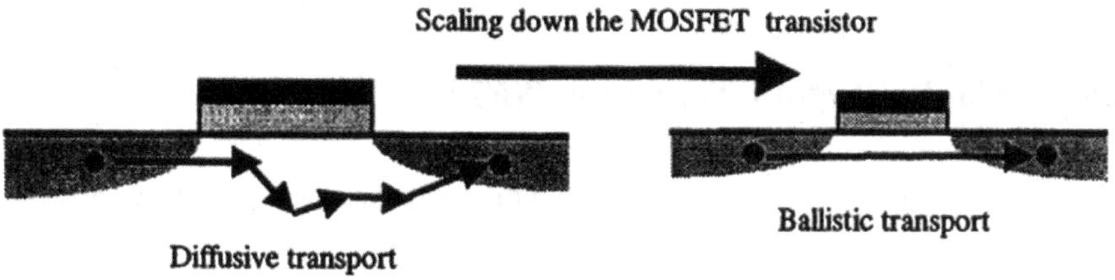

Fig. 8.4 By scaling down the MOSFET transistor the diffusive transport transforms into ballistic transport

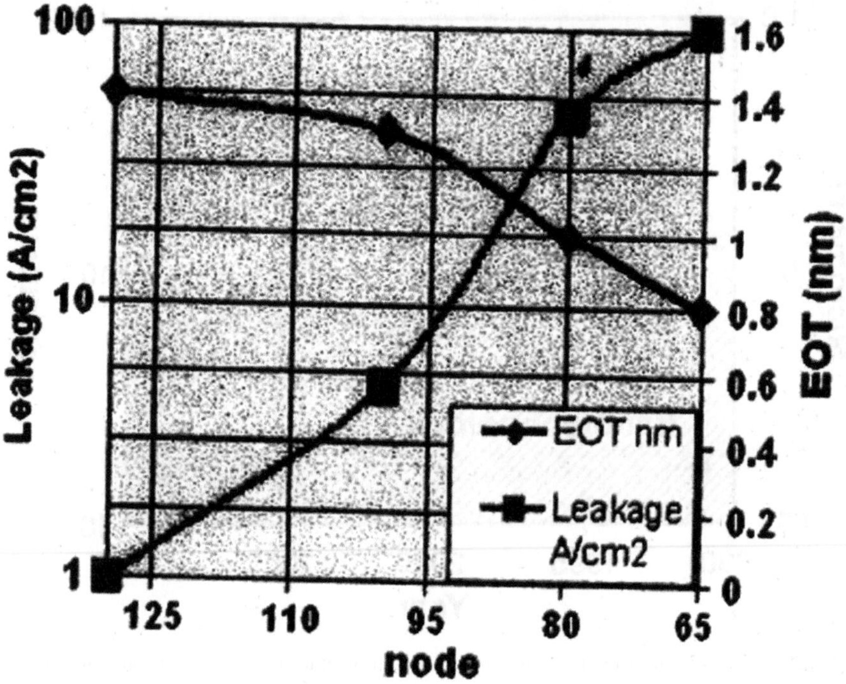

Fig. 8.5 Gate leakage current and equivalent oxide thickness (EOT) versus technology node given in nm

8.5.1 Single-Electron Transistor (SET)

The SET combines the Coulomb Block with the single Electron Box (Figs. 8.7, 8.8 and 8.9).

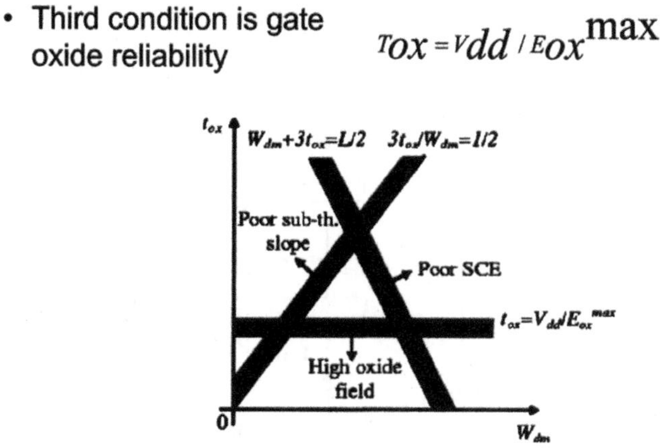

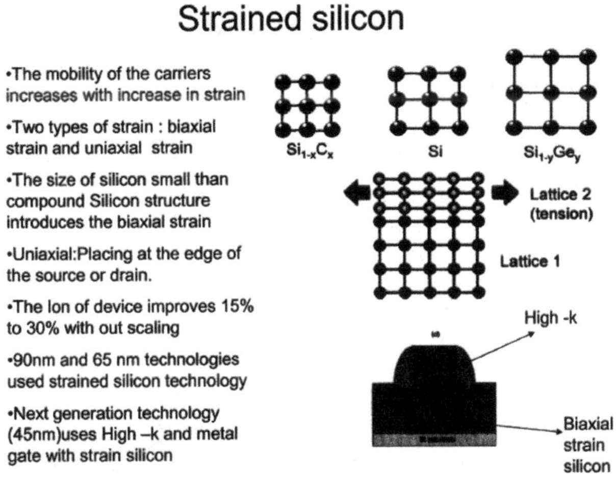

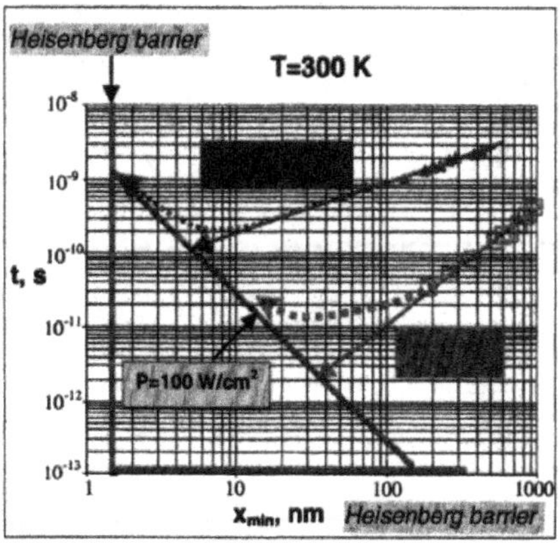

Fig. 8.6 Possible inflections of ITRS scaling vector for memory and logic devices by approaching limits of heat removal

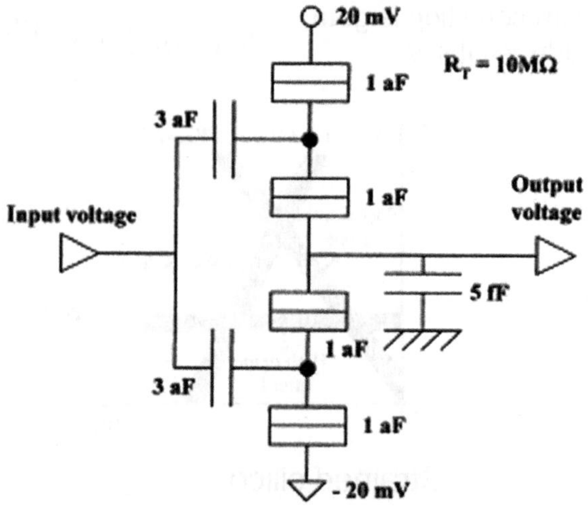

Fig. 8.7 Simple SET inverter

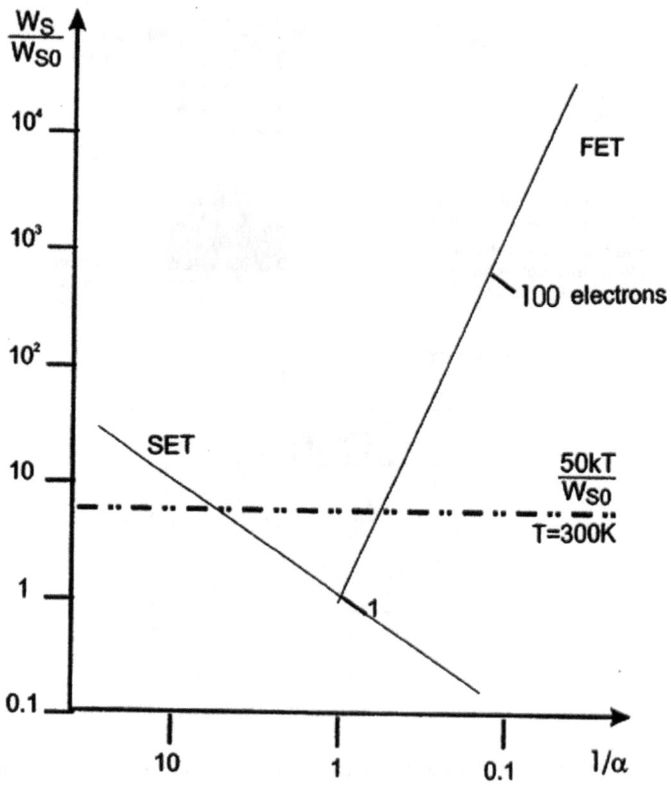

Fig. 8.8 The power-delay diagram reveals the impacts of scaling on the SET and FET

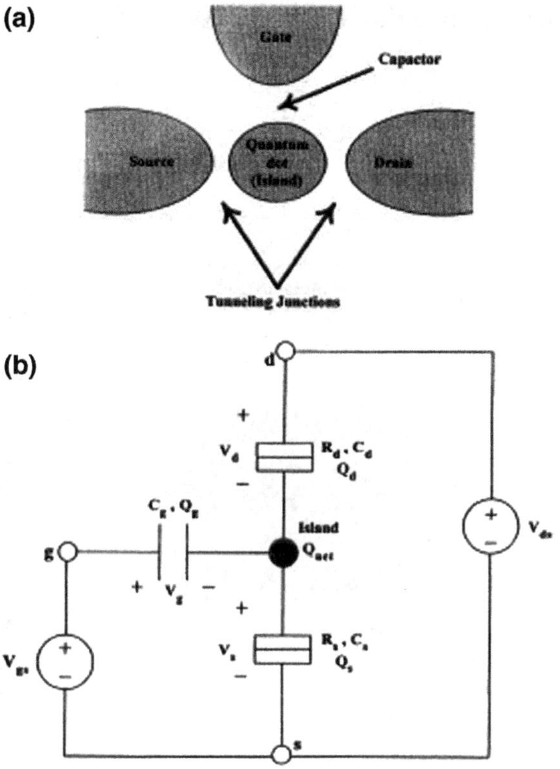

Fig. 8.9 Single electron transistor **a** physical structure and **b** circuit for analysis

8.6 Molecular Electronics

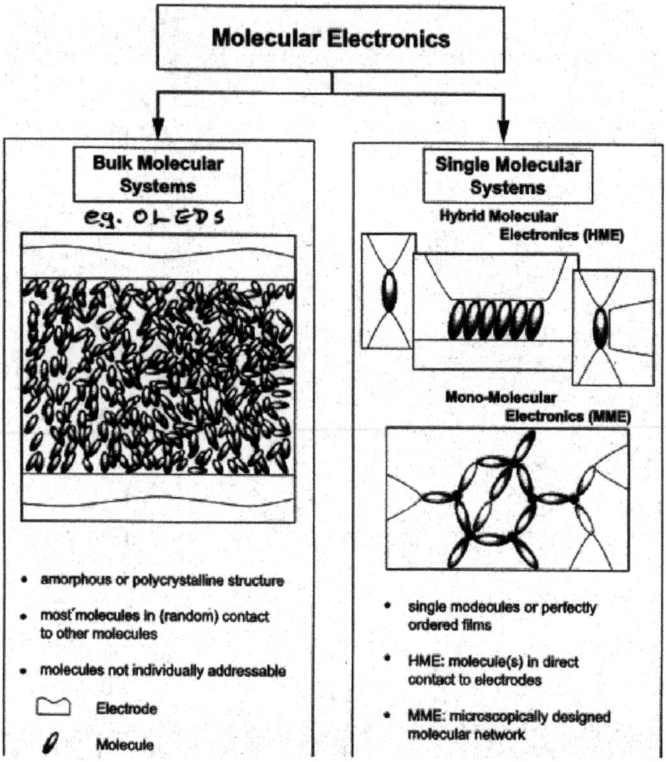

8.7 Multi Walled Carbon Nanotubes (CNTs)

If the (0, 0) point is folded to any of the points along the armchair line, we get a metallic nanotube. However, if it is folded along the zigzag line or in the chair area, which is the area between the zigzag and armchair, we get either semiconductor or metallic tube, depending on the n and m rules as shown in Fig. 8.13 (Figs. 8.10, 8.11, 8.12).

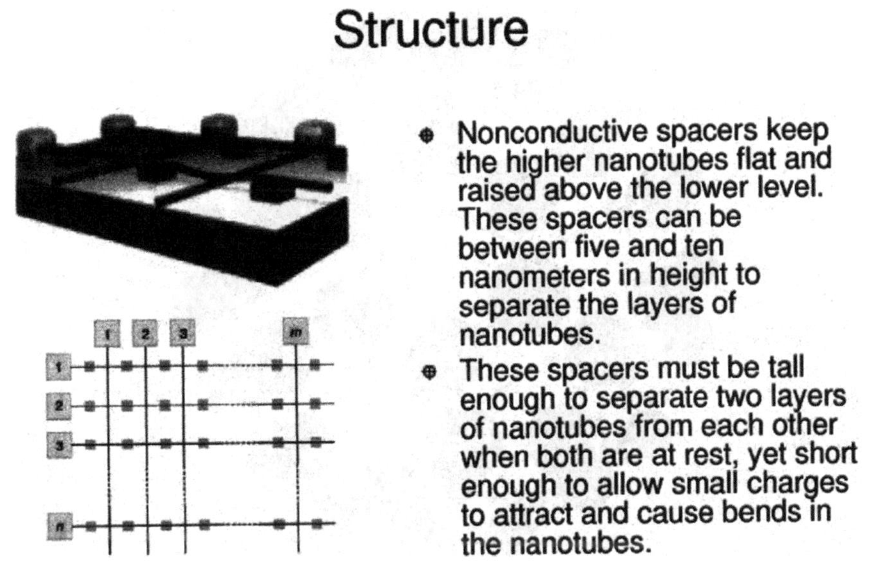

Source Thomas Rueckes et al., "Carbon Nanotube Based Nonvolatile Random Access Memory for Molecular Computing" Science, Vol 289, 7 July 2000

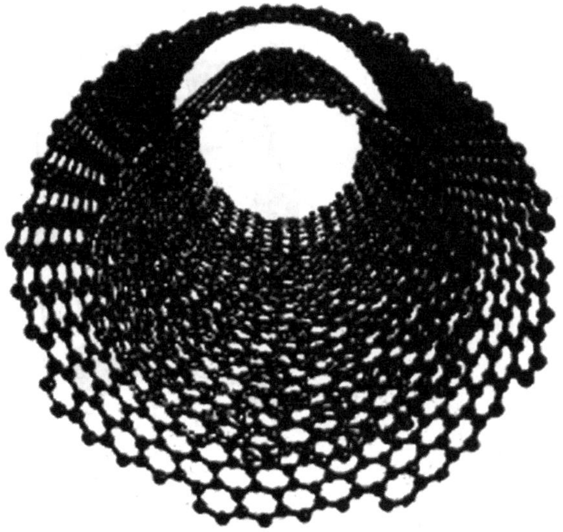

Fig. 8.10 Illustration of a nested nanotube in which one tube is inside the another

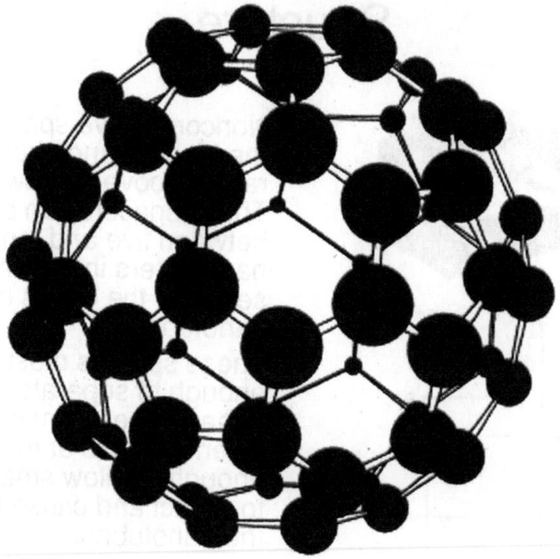

Fig. 8.11 Structure of the C_{60} fullerene molecule

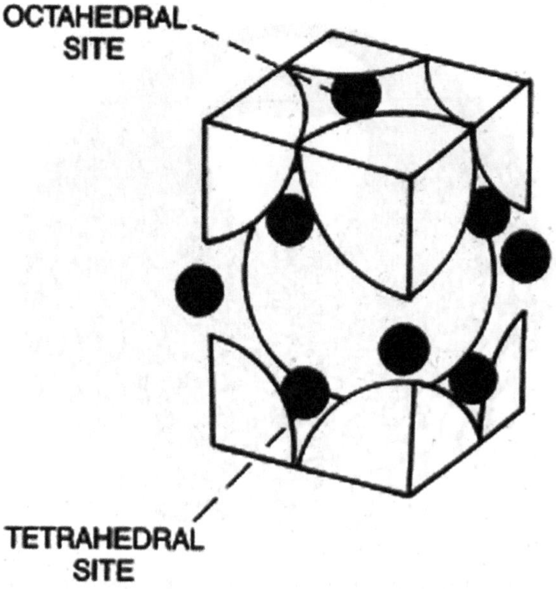

Fig. 8.12 Crystal lattice unit cell of C_{60} molecules (large spheres) doped with alkali atoms (dark circles) [From F. J. Owens and C. P. Poole, Jr., The New Superconductors, Plenum, (1998)]

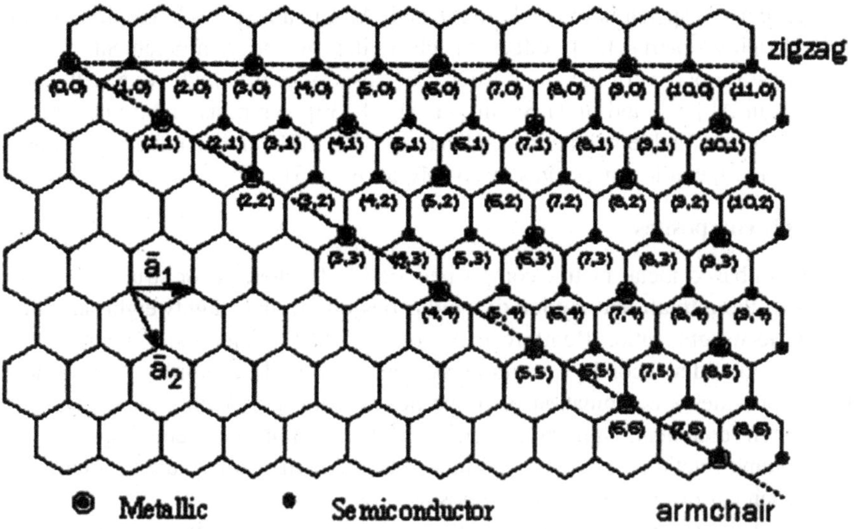

Fig. 8.13 Metallic or semiconducting depending on which point the (0, 0) point is rolled to

Summary

- **The nanoworld**
  - It ranges from 1 nm to ~100 nm.
  - Nanomaterials have at least one dimension in the nanometric range.
- **Nanoscience has already many applications**:
  - Electronics, Mechanics, Optics…
  - Biomedicine…
- **One simple obvious characteristic of the nanomaterials is related to their surfaces or interfaces.**
  - A large percentage of atoms are on the surface or in grain boundaries.
  - 2000 atoms in a 2 nm gold cluster, ~50% on the surface.
  - The surface energy γ is the work needed to create a surface. The lowest values corresponds to the denser surfaces.
  - The size of the nanograins is a parameter that should be included into classical description of bulk materials when they are for nanomaterials.

During the past few years, studies have been successfully achieved a very uniform distribution of the reinforcement phases in different types of matrices through the solid-state powder processing technique of MA [24–33]. These include homogeneous dispersion of lead in Al and Al–Cu alloys [24], graphite in an Al-6061 alloy matrix [25], effect of clustering of the reinforcement on the mechanical properties of the composites [26], dispersion of a high volume fraction of Al_2O_3 in Al [27], and of Ti_5Si_3 in γ-TiAl [28–30], synthesis of $MoSi_2$ + Si_3N_4 composites for hightemperature applications [31, 32], and synthesis of amorphous + intermetallic composites in Al–Mg alloys [33].

Al–Al_2O_3 composites

Aluminum-based metal matrix composites are ideal materials for structural applications in the aircraft and automotive industries due to their lightweight and high strength-to-weight ratio. Reinforcement of the ductile aluminum matrix with stronger and stiffer second-phase reinforcements like oxides, carbides, borides, and nitrides provides a combination of properties of both the metallic matrix and the ceramic reinforcement. Uniform dispersion of the fine reinforcements, especially at large volume fractions, and a fine grain size of the matrix contribute to improving the mechanical properties of the composite.

Most of the study carried out to date on discontinuously reinforced aluminum was primarily concerned with dispersing reinforcements on the micrometer scale (a few to a couple of hundred micrometers). Even though studies were carried out on nanometer-sized reinforcements, the volume fraction of the reinforcements was very small. Thus, there is a gap in understanding the structure and properties of high volume fraction composites containing nanometersized reinforcements.

Microstructural observations in new results suggested that the distribution of Al_2O_3 particles in the Al matrix was uniform after 20 h of milling. This was true in almost all the cases investigated. Figure 8.14 shows that a very uniform dispersion of the fine 50-nm Al_2O_3 particles was achieved even when the volume fraction of the reinforcement was 50%. This corresponds to the finest particle size and largest volume fraction, the maximum value achieved so far in any nanocomposite. Similar uniform distributions were obtained in the larger size and smaller volume fractions of the reinforcement also. Since the uniform distribution of the Al_2O_3 particles in the matrix may not be detected at this resolution, X-ray elemental mapping was conducted in the SEM.

Figure 8.15 clearly shows that a uniform distribution of the reinforcement phase was obtained as revealed by the uniform distribution of the spots in the aluminum and oxygen dot maps [27].

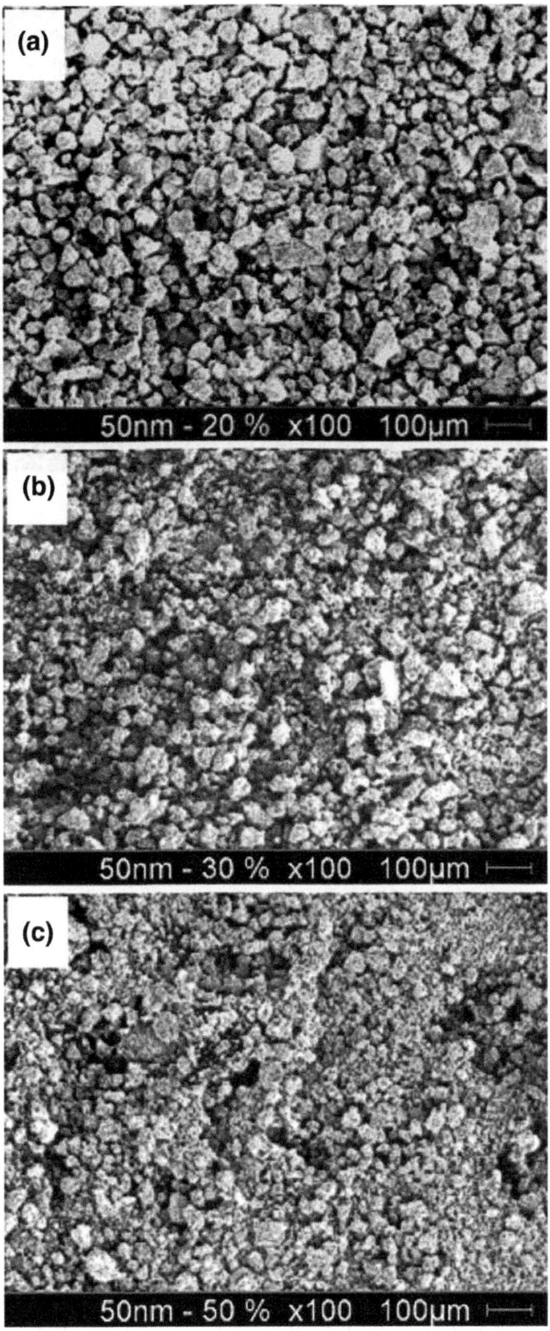

Fig. 8.14 SEM micrographs of Al–Al_2O_3 (50 nm) powders milled to the steady-state condition (20 h) showing a uniform distribution of Al_2O_3 in the Al matrix. **a** 20 vol.%, **b** 30 vol.%, and **c** 50 vol.% Al_2O_3

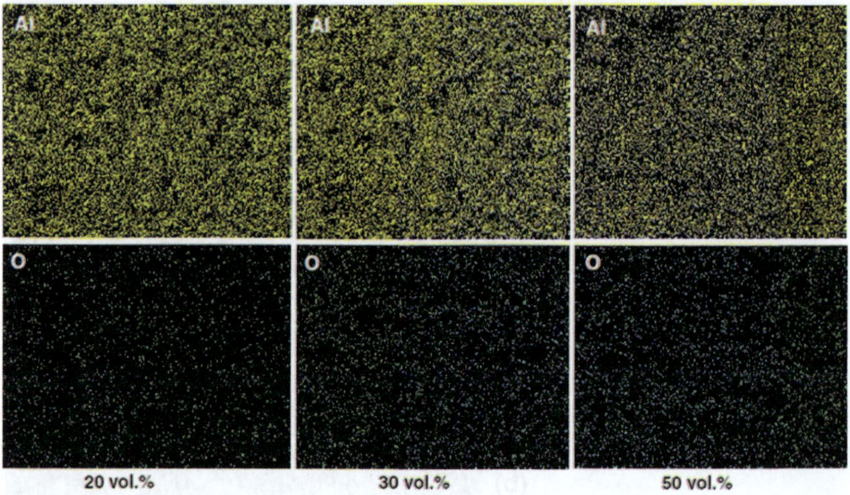

Fig. 8.15 X-ray maps showing distribution of Aluminum (top) and Oxygen (bottom) atoms in the Al–Al_2O_3 (50 nm) powders milled for 20 h

References

1. Qian, F., et al.: Nano Lett. **5**, 2287 (2005)
2. Hayden, O., et al.: Adv. Mater. **17**, 701 (2005)
3. Lu, W., et al.: Proc. Natl. Acad. Sci. U.S.A. **102**, 10046 (2005)
4. Xiang, J., et al.: Nature **441**, 489 (2006)
5. Li, Y., et al.: Nano Lett. **6**, 1468 (2006)
6. Jin, S., et al.: Nano Lett. **4**, 915 (2004)
7. Bryllert, T., et al.: IEEE Electron Device Lett. **27**, 323 (2006)
8. Doh, Y., et al.: Science **309**, 272 (2005)
9. Zhong, Z., et al.: Nano Lett. **5**, 1143 (2005)
10. Tilke, A., et al.: J. Appl. Phys. **89**, 8159 (2001)
11. McAlpine, M.C., et al.: Nano Lett. **3**, 443 (2003)
12. Friedman, R.S., et al.: Nature **434**, 1085 (2005)
13. Cui, Y., et al.: Science **293**, 1289 (2001)
14. Hahm, J., Lieber, C.M.: Nano Lett. **4**, 51 (2004)
15. Patolsky, F., et al.: Proc. Natl. Acad. Sci. U.S.A. **101**, 14017 (2004)
16. Wang, W.U., et al.: Proc. Natl. Acad. Sci. U.S.A. **102**, 3208 (2005)
17. Zheng, G., et al.: Nat. Biotechnol. **23**, 1294 (2005)
18. Patolsky, F., Lieber, C.M.: Mater. Today **8**(4), 20 (2005)
19. Huang, Y., et al.: Science **294**, 1313 (2001)
20. Zhong, Z., et al.: Science **302**, 1377 (2003)
21. Huang, Y., et al.: Small **1**, 142 (2005)
22. Duan, X.F., et al.: Nature **421**, 241 (2003)
23. Hayden, O., et al.: Nat. Mater. **5**, 352 (2006)
24. Kim, H.M., Kim, T.S., Suryanarayana, C., Chun, B.S.: Mater. Sci. Eng. A **287**, 59 (2000)
25. Son, H.T., Kim, T.S., Suryanarayana, C., Chun, B.S.: Mater. Sci. Eng. A **348**, 163 (2003)
26. Hong, S.J., Kim, H.M., Huh, D., Suryanarayana, C., Chun, B.S.: Mater. Sci. Eng. A **347**, 198 (2003)

27. Prabhu, B., Suryanarayana, C., An, L., Vaidyanathan, R.: Mater. Sci. Eng. A **425**, 192 (2006)
28. Klassen, T., Bohn, R., Suryanarayana, C., Fanta, G., Bormann, R.: In: Shaw, L., Suryanarayana, C., Mishra, R.S. (eds.) Processing and properties of structural nanomaterials, p. 93. PA, TMS, Warrendale (2003)
29. Klassen, T., Suryanarayana, C., Bormann, R.: Scripta Mater. **59**, 455 (2008)
30. Suryanarayana, C., Klassen, T., Ivanov, E.: Synthesis of nanocomposites and amorphous alloys by mechanical alloying. J. Mater. Sci. **46**, 6301–6315 (2011)
31. Mohan, P., Suryanarayana, C., Desai, V.: In: Bandyopadhyay, S., et al. (eds.) Nanomaterials: synthesis, characterisation, and application, p. 171. Tata McGraw-Hill Publ. Co., Ltd., New Delhi, India (2004)
32. Suryanarayana, C.: Mater. Sci. Eng. A **479**, 23 (2008)
33. Al-Aqeeli, N., Mendoza-Suarez, G., Suryanarayana, C., Drew, R.A.L.: Mater. Sci. Eng. A **480**, 392 (2008)

Chapter 9
Classification of Nanostructured Materials

Nano materials can be classified dimension wise into following categories:

- Nano rods, nano wires have dimension less than 100 nm.
- Tubes, fibers, platelets have dimensions less than 100 nm.
- Particles, quantum dots, hollow spheres have 0 or 3 Dimensions <100 nm.

On the basis of phase composition, nano materials in different phases can be classified as,

- The nano material is called single phase solids. Crystalline, amorphous particles and layers are included in this class.
- Matrix composites, coated particles are included in multi-phase solids.
- Multi-phase systems of nano material include colloids, aero gels, Ferro fluids, etc.

There are several ways to classify nanomaterials [1–8]. This depends very much on the applications and philosophy which are adopted. We shall just present here two extreme classifications which can be found in the literature. In the first one, the focus is done on the macroscopic dimensions of the material. Nanomaterials are then defined as materials containing structures having at least one dimension less than about 100 nm. According to this definition, a thin film (with a thickness less than 100 nm) is considered as a nanomaterial (Fig. 9.1) since one of the dimensions is nanometric. Thin films have been used for a long time, before there was any interest in "nanotechnology". If two dimensions are "nano", we have nanofibers, nanowires etc., and if all three dimensions are "nano" we have quantum dots. In this classification this no reference to the structure of the material.

Another classification, which might be more relevant as far as the nanostructure of a material is concerned, is to consider the dimensionality of the nanoscale component with which the material is made. A nanostructure is said to have one dimension, for example, if it has a length larger than 100 nm in one direction only. According to this definition, a nanoparticle is considered to have zero dimension (it

L. H. Madkour, *Nanoelectronic Materials*, Advanced Structured Materials 116, https://doi.org/10.1007/978-3-030-21621-4_9

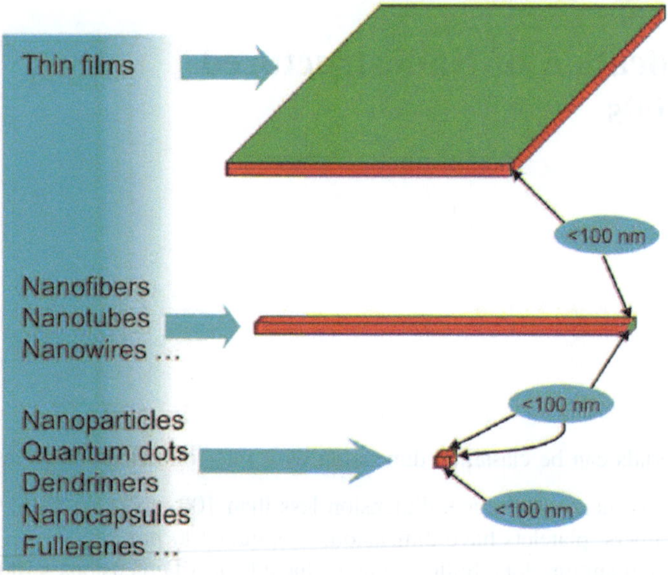

Fig. 9.1 Classification where 1D, 2D and 3D objects have one, two or three dimensions in the nanometer range

has no dimension with a length larger than 100 nm). A wire or a fiber is a 1-dimensional object and a thin film a 2-dimension nanostructure. To summarize, 0D nanomaterials contain spheres or clusters which are considered as point-like particles. 1D nanomaterials contain nanofibers, wires, rods, etc., 2D use films, plates, multilayers, or networks. 3D nanomaterials are nanophase materials consisting of equiaxed nanometer-sized grains. This is shown schematically in Fig. 9.2.

Bulk nanomaterials are larger objects made from structures having well-identified domains with an average size less than 100 nm, for instance the grain size in ceramics. For comparison, the diameter of a human hair is about 100,000 nm and the size of a single particle of smoke is around 1,000 nm. In this case a nanoparticle is considered to have zero dimension, a wire, 1 dimension There are several ways to classify nanostructured nanomaterials, but this depends, to some extent, on the applications under consideration. As sketched in Fig. 9.2, a nanomaterial can be classified according to the dimension of the nanostructure used to make it: zero dimensional (0D), one dimensional (1D), two dimensional (2D) and three dimensional (3D).

Nanomaterials have extremely small size [9–14] which having at least one dimension 100 nm or less. Nanomaterials can be nanoscale in one dimension (e.g. surface films), two dimensions (e.g. strands or fibres), or three dimensions (e.g. particles). They can exist in single, fused, aggregated or agglomerated forms with spherical, tubular, and irregular shapes. Common types of nanomaterials include nanotubes, dendrimers, quantum dots and fullerenes [15–19]. Nanomaterials have

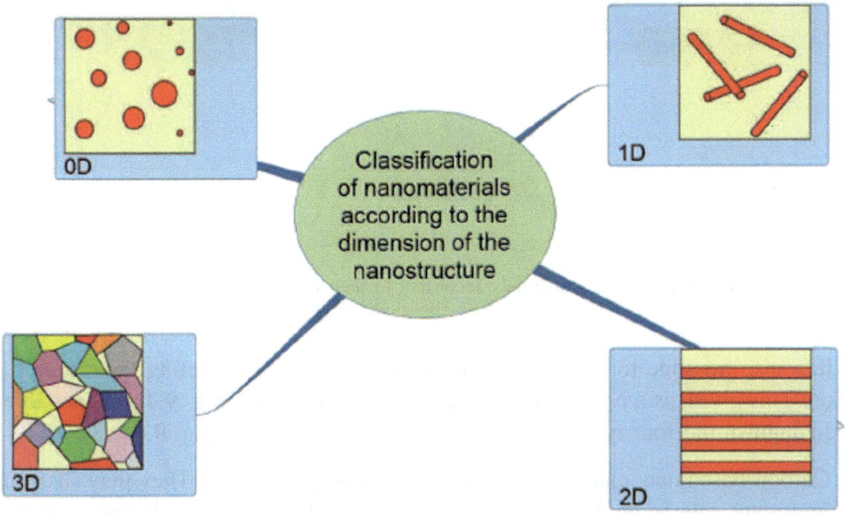

Fig. 9.2 Classification of materials according to the dimensionality of the nanostructures. 0D, 1D, 2D and 3D nanostructured materials are sketched in this figure. In the 2D illustration, we have 2D layers embedded in the bulk and in the 3D crystallites or grains

applications in the field of nano technology, and displays different physical chemical characteristics from normal chemicals (i.e., silver nano, carbon nanotube, fullerene, photocatalyst, carbon nano, silica).

According to Siegel, Nanostructured materials are classified as Zero dimensional, one dimensional, two dimensional, three dimensional nanostructures.

Nanomaterials are materials that are characterized by an ultra-fine grain size (<50 nm) or by a dimensionality limited to 50 nm. Nanomaterials can be created with various modulation dimensionalities as defined by Richard W. Siegel: zero (atomic clusters, filaments and cluster assemblies), one (multilayers), and two (ultrafine-grained overlayers or buried layers), and three (nanophase materials consisting of equiaxed nanometer sized grains) as shown in the Fig. 9.3.

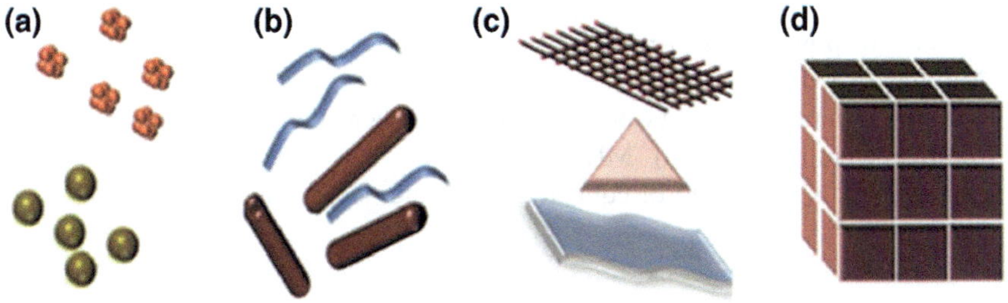

Fig. 9.3 Classification of nanomaterials. **a** 0D spheres and clusters, **b** 1D nanofibers, wires, and rods, **c** 2D films, plates, and networks, **d** 3D nanomaterials

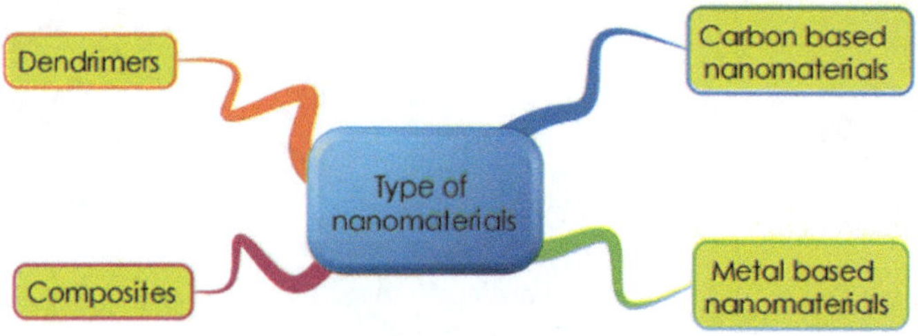

Fig. 9.4 Different types of nanomaterials according to their composition

It is also possible to classify nanomaterials in families reflecting their composition. This gives, as a possibility, the classification shown in Fig. 9.4. It is possible to distinguish the four types of nanomaterials as indicated in Fig. 9.4.

- Carbon-based nanomaterials are composed mostly of carbon. They play such an important role in applications as well as in the historical development of the nanotechnology domain. This classification includes fullerenes, carbon nanotubes, graphene and the like.
- Metal-based nanomaterials are materials made of metallic nanoparticles like gold, silver, metal oxides, etc. For example, titanium dioxide (TiO_2) nanoparticles are extensively used in applications such as paint, sunscreen, and toothpaste.
- Dendrimers are nanosized polymers built from branched units. They can be functionalized at the surface and can hide molecules in their cavities. A direct application of dendrimers is for drug delivery.
- Composite nanomaterials contain a mixture of simple nanoparticles or compounds [20–28] such as nanosized clays within a bulk material. The nanoparticles give better physical, mechanical, and/or chemical properties to the initial bulk material.

Nanomaterials consisting of nanometer sized crystallites or grains and interfaces may be classified according to their chemical composition and shape (dimensionality), as discussed above. According to the shape of the crystallites or grains we can broadly classify nanomaterials into four categories:

1. Clusters or powders (MD = 0)
2. Multilayers (MD = 1)
3. Ultrafine grained overyears or buried layers (where the layer thickness or the rod-diameters are <50 nm) (MD = 2)
4. Nanomaterials composed of equiaxed nanometer-sized grains (MD = 3).

9.1 Glitter's Classification of Nanostructured Materials (NSM)

The materials and/or devices sintered by means of the controlled manipulation of their microstructure on the atomic level may be divided into three categories [29–34].

The first category comprises materials and/or devices with reduced dimensions and/or dimensionality in the form of isolated, substrate-supported or embedded nanometer-sized particles, thin wires or thin films. The techniques that are most frequently used to produce this type of microstructure are chemical vapor deposition (CVD), physical vapor deposition (PVD), various aerosol techniques, and precipitation from the vapor, supersaturated liquids or solids. Well-known examples of technological applications of materials the properties of which depend on this type of microstructure are catalysts and semiconductor devices utilizing single or multilayer quantum well structures.

The second category comprises materials and/or devices in which the nanometersized microstructure is limited to a thin (nanometer-sized) surface region of a bulk material. PVD, CVD, ion implantation and laser beam treatments are the most widely applied procedures to modify the chemical composition and/or atomic structure of solid surfaces on a nanometer scale. Surfaces with enhanced corrosion resistance, hardness, wear resistance or protective coatings are examples taken from today's technology in which the properties of a thin surface layer are improved by means of creating a nanometer-sized microstructure in a thin surface region. For example, patterns in the form of an array of nanometer-sized islands (e.g. quantum dots) connected by thin (nanometer scale) wires. Patterns of this type may be synthesized by lithography, by means of local probes (e.g. the tip of a tunneling microscope, near-field methods, focused electron or ion beams) and/or surface precipitation processes. Such kind of processes and/or devices are expected to play a key role in the production of the next generation of electronic devices such as highly integrated circuits, terabit memories, single electron transistors, quantum computers, etc.

The third category comprises bulk solids with a nanometer-scale microstructure. Those are solids in which the chemical composition, the atomic arrangement and/or the size of the building blocks (e.g. crystallites or atomic/molecular groups) forming the solid varies on a length scale of a few nanometers throughout the bulk.

One of the basic results of the materials science is the insight that most properties of solids depend on the microstructure [35–41]. A reduction in the spatial dimension, or confinement of particles or quasi-particles in a particular crystallographic direction within a structure generally leads to changes in physical properties of the system in that direction. Hence the another classification of nanostructured materials and systems essentially depends on the number of dimensions which lie within the nanometer range: (a) 3D-systems confined in three dimensions, e.g. structures typically composed of consolidated equiaxed crystallites; (b) 2D-systems confined in two dimensions, e.g. filamentary structures where the length is substantially greater than the cross-sectional dimensions; (c) 1D-systems confined in one

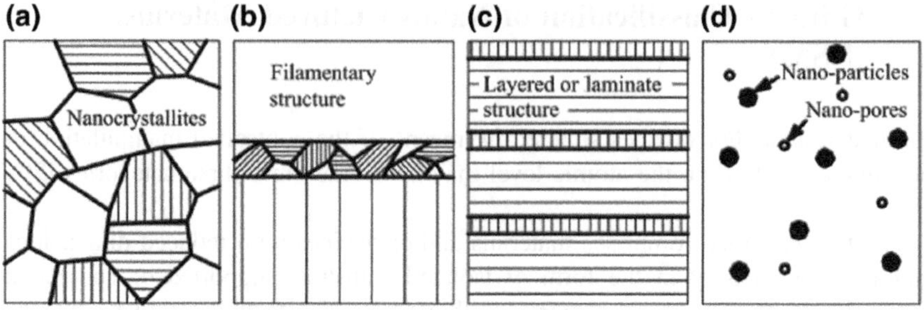

Fig. 9.5 Schematic classification of nano-materials: **a** three-dimensional structures; **b** two-dimensional; **c** one-dimensional; and **d** zero-dimensional structures

dimension, e.g. layered or laminate structures; (d) 0D-zero-dimensional structures, e.g. nano-pores and nano-particles, Figs. 9.5 and 9.6.

The most well-known example of the correlation between the atomic structure and the properties of a bulk material is probably the spectacular variation in the hardness of carbon when it transforms from diamond to graphite. Comparable variations have been noted if the atomic structure of a solid deviates far from equilibrium or if its size is reduced to a few interatomic spacing. An example of the latter case is the change in color of CdS crystals if their size is reduced to a few nano-meters.

Fig. 9.6 Definition of nanomaterial's following Siege

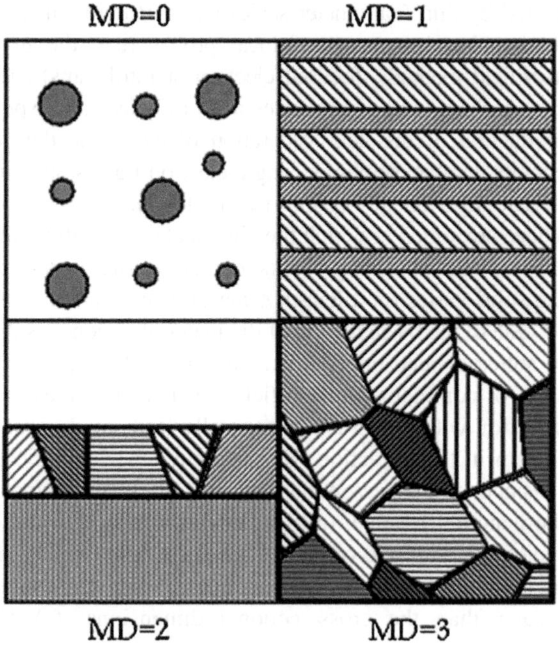

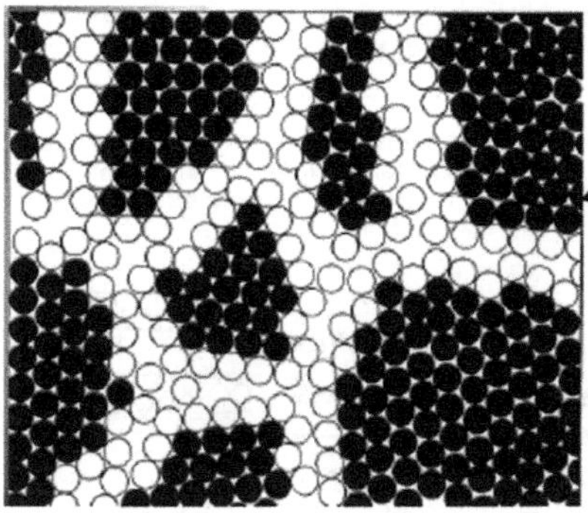

Fig. 9.7 Schematic model of a nanostructured material (adapted from Gleiter H., Acta Mater., 2000, vol. 48 [1])

Three-dimensional structures or bulk materials with a nanometer-sized microstructure are assembled of nanometer-sized building blocks or grains that are mostly crystallites. The schematic model of the nanostructured material is shown in Fig. 9.7. These building blocks may differ in their atomic structure, their crystallographic orientation and/or their chemical composition. If the building blocks are crystallites, incoherent or coherent interfaces may be formed between them, depending on the atomic structure, the crystallographic orientation and/or the chemical composition of adjacent crystallites. In other words, materials assembled of nanometer-sized building blocks are microstructurally heterogeneous consisting of the building blocks (e.g. crystallites) and the regions between adjacent building blocks (e.g. grain boundaries). It is this inherently heterogeneous structure on a nanometer scale that is crucial for many of their properties and distinguishes them from glasses, gels, etc. that are micro structurally homogeneous.

Nanostructured materials (NSMs) as a subject of nanotechnology are low dimensional materials comprising building units of a submicron or nanoscale size at least in one direction and exhibiting size effects. Development of any science needs in classification. First classification scheme of NSMs was proposed by H. Gleiter in 1995 and further was extended by V. Pokropivny and V. Skorokhod in 2005. In recent years the hundreds of new NSMs and abundance of novel nanostructures (NSs) have been obtained so the need in their classification is ripen.

Crystalline forms and chemical composition was assumed by Gleiter as a basis of a classification scheme of NSMs where both intercrystalline grain boundaries and crystallites were regarded as building blocks (Fig. 9.8). However this scheme seems to be incomplete because of zero- and one-dimensional (0D, 1D) structures such as fullerenes and nanotubes were not considered. Therefore, in this scheme there are actually 3 classes and 4 types in each of them rather than 12 classes.

Figure 9.8 shows classification schema for nanomaterials according to their chemical composition and the dimensionality (shape) of the crystallites (structural elements) forming the nanomaterial. The boundary regions of the first and second

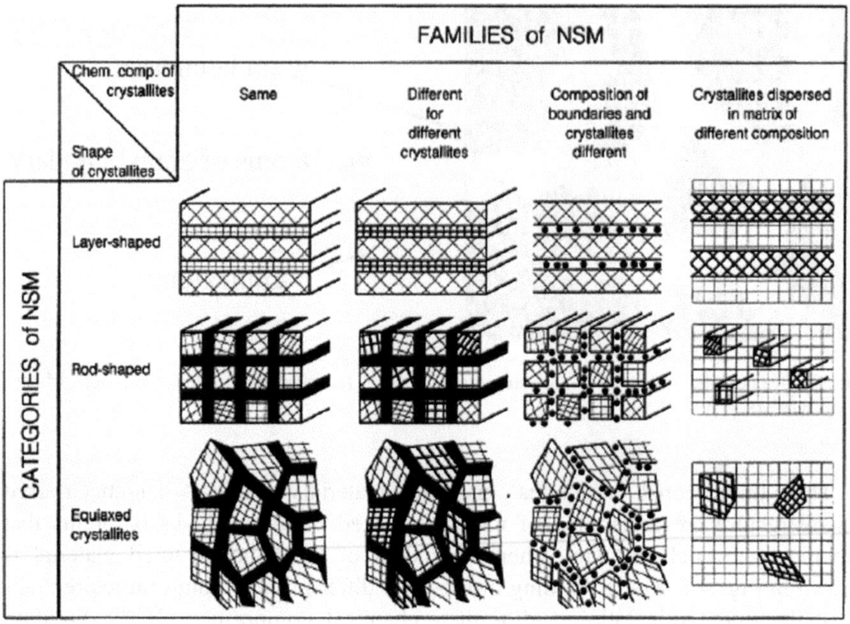

Fig. 9.8 Gleiter's classification schema for NSM according to their chemical composition and the dimensionality (shape) of the crystallites (structural elements) forming the NSM

family of nanomaterials are indicated in black to emphasize the different atomic arrangements in the crystallites and in the boundaries. The chemical composition of the (black) boundary regions and the crystallites is identical in the first family. In the second family, the (black) boundaries are the regions where two crystals of different chemical composition are joined together causing a steep concentration gradient.

The latter three categories can be further grouped into four families as shown in Fig. 9.8.

- In the most simple case (first family in the Fig. 9.8), all grains and interfacial regions have the same chemical composition. E.g. Semi crystalline polymers (consisting of stacked lamellae separated by non-crystalline region), multilayers of thin film crystallites separated by an amorphous layer (a-Si:N:H/nc Si) etc.
- As the second case, we classify materials with different chemical composition of grains. Possibly quantum well structures are the best example of this family.
- In the third family includes all materials that have a different chemical composition of its forming matter (including different interfaces) e.g. Ceramic of alumina with Ga in its interface.
- The fourth family includes all nanomaterials formed by nanometer sized grains (layers, rods or equiaxed crystallites) dispersed in a matrix of different chemical composition. Precipitation hardened alloys typically belong to this family. E.g.

Nanometer sized Ni_3Al precipitates dispersed in a nickel matrix generated by annealing a supersaturated Ni–Al solution—are an example of such alloys. Most high-temperature materials used in modern jet engines are based on precipitation-hardened Ni_3Al/Ni alloys.

9.2 Classification of Nanomaterials by Dimensionality

Nanostructures (NSs) should be separated from NSMs because the former (NSs) are characterized by a form and dimensionality while the last (NSMs) by a composition in addition. Hence NSs should be classified accurately upon one of these sign, namely, dimensionality, as being the general natural attribute, integrated a size and shape or form. Abundance of forms for bulk 3D materials is infinite. Under transition into nanoworld an atomic difference between some shapes can be neglected regarding these forms as the same due to their low dimension. Hence one can conclude that a number of NS-classes becomes to be finite. This brings up the problem of modern NSs classification [42–49].

Under a nanostructure we understand the structure the one size of which d at least is less or equal to a critical one d^*, $d \leq d^* \approx 102$ nm. The value of d^* have not certain meanings because it is dictated by a critical characteristic of some physical phenomena (free path length of electrons, phonons, length of de Broglie wave, length of external electromagnetic and acoustical waves, correlation length, penetration length, diffusion length, etc.) giving rise to the size effects. We constitute our classification of NSs on their dimensionality. It may be one of the four, namely, 0D, 1D, 2D or 3D. All NSs can be built from elementary units (blocks) having low dimensionality 0D, 1D, and 2D. The 3D units are excluded because they can't be used to build low dimensional NSs except 3D matrix. However 3D structures can be considered as NSMs if they involve the 0D, 1D, 2D NSs. This is just the case that Gleiter considered in his classification of NSMs. Let us introduce the notation of NSs:

$$k\,\mathrm{D}\,l\;mn$$

where k is a dimensionality of NS as a whole, while the integers l, m, n denotes the dimensionality of the NS's building units of different types. Each integer l, m, n refers to different type unit, so the number of these integers must be equal to the number of the different constituting units. From the definition of NSs the condition leads, namely, $k \geq l, m, n$, and $k, l, m, n = 0, 1, 2, 3$.

It follows from this conditions that restricted number of NSs classes exists, namely, 3 sorts of elementary units (0D, 1D, 2D), 9 single classes of $k\mathrm{D}l$ type built of 1 sort units, 19 binary classes of $k\mathrm{D}lm$ type built of 2 sort units, and variety of ternary, quaternary, etc., classes. Restricting the classification by 5 main ternary structures of *kDlmn* type built of 3 sort units, we obtain in the result 3 + 9 + 19 + 5 = 36 classes of NSs shown in Fig. 9.9. Classification is based on the number of dimensions, which are not confined to the nanoscale range (<100 nm).

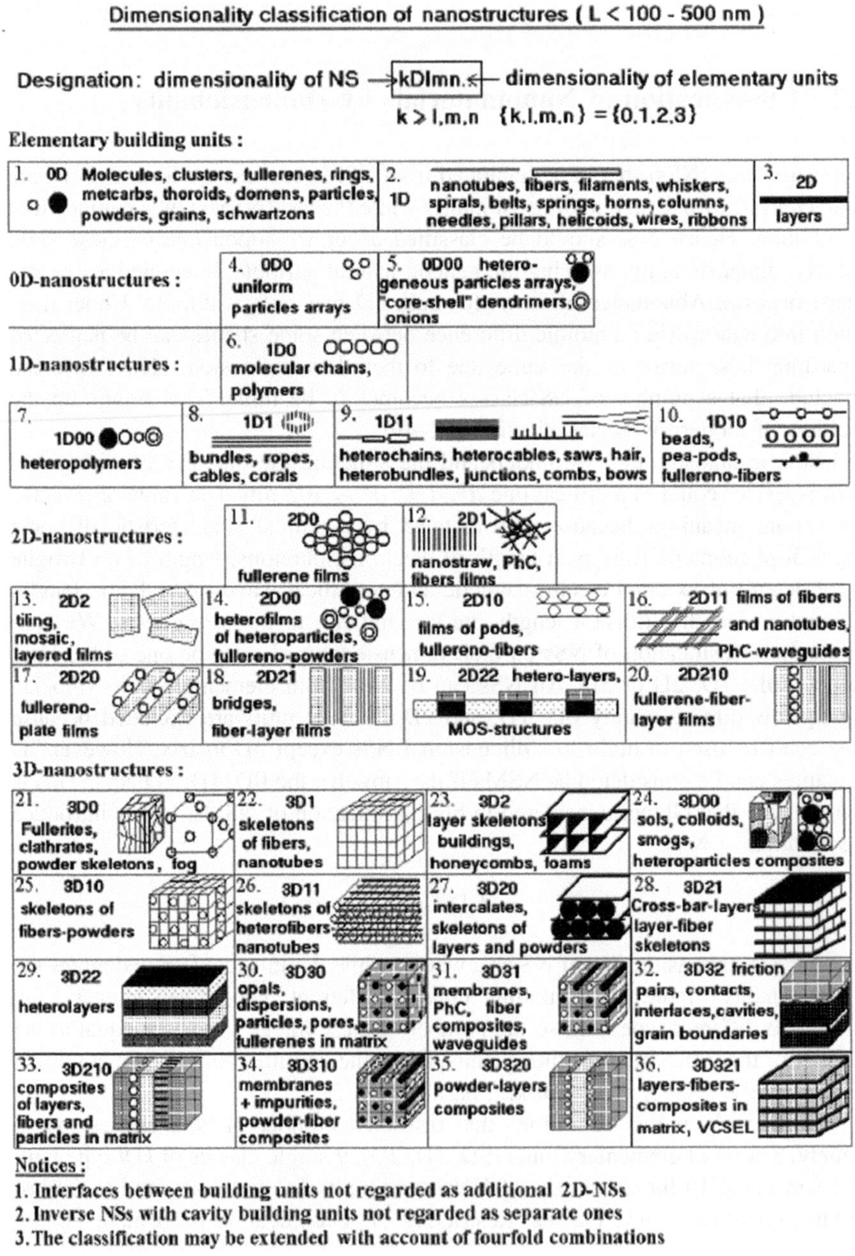

Fig. 9.9 Dimensionality classification of nanostructures

(1) zero-dimensional (0-D),
(2) one-dimensional (1-D),
(3) two-dimensional (2-D), and
(4) three-dimensional (3-D).

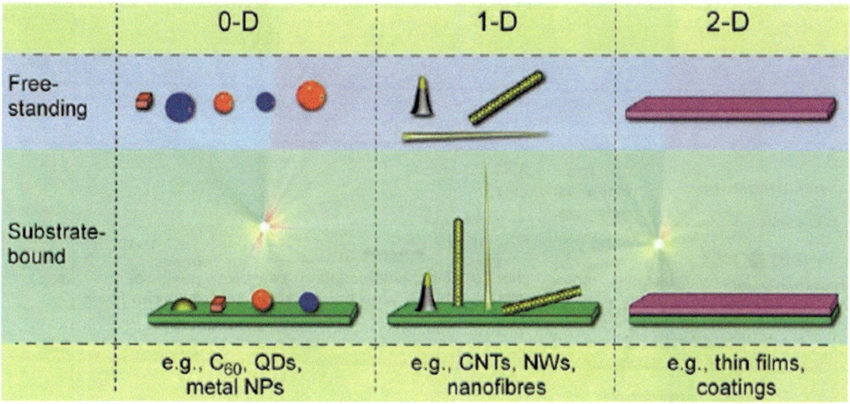

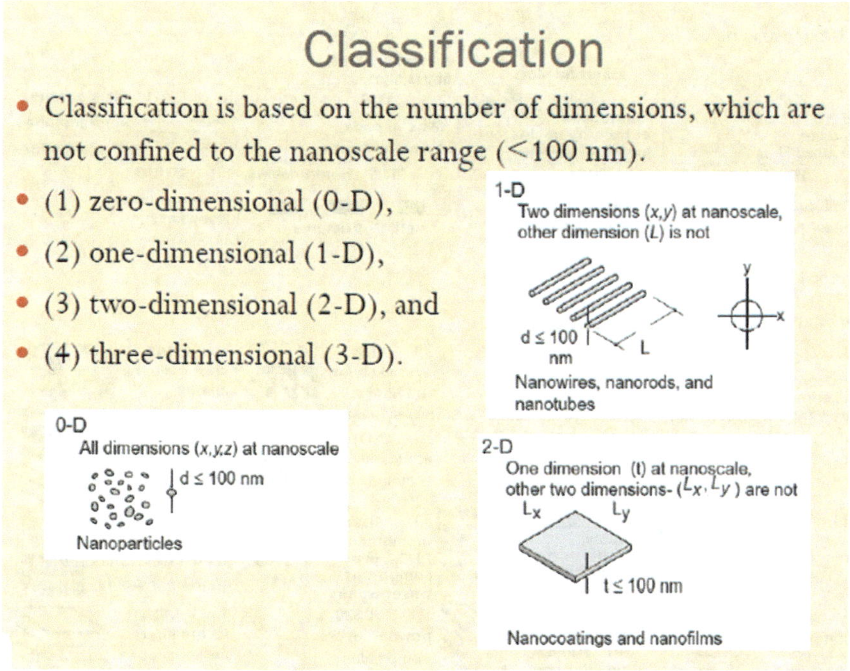

□ The 0Dℓm structures

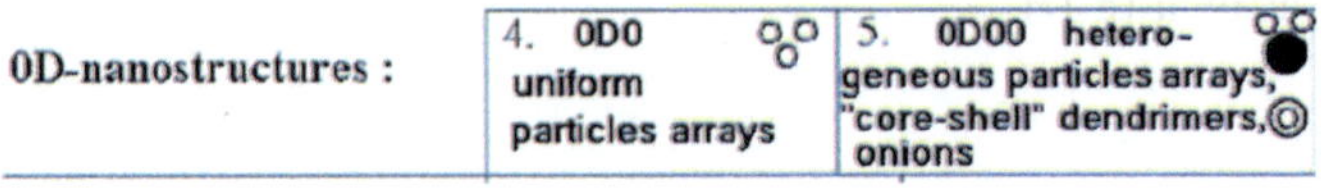

□ The 1Dℓm structures

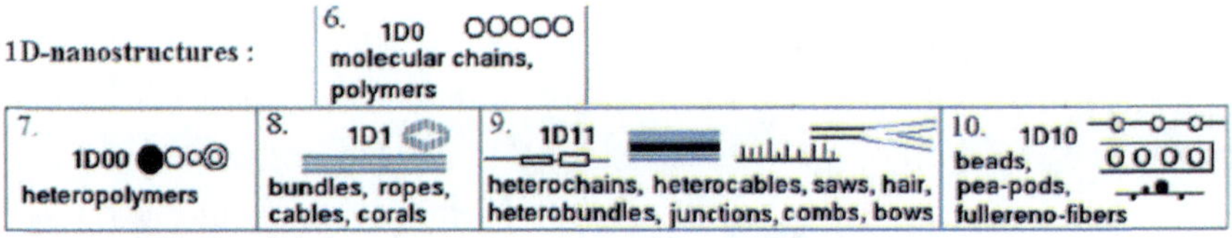

□ The 2Dℓmn structures

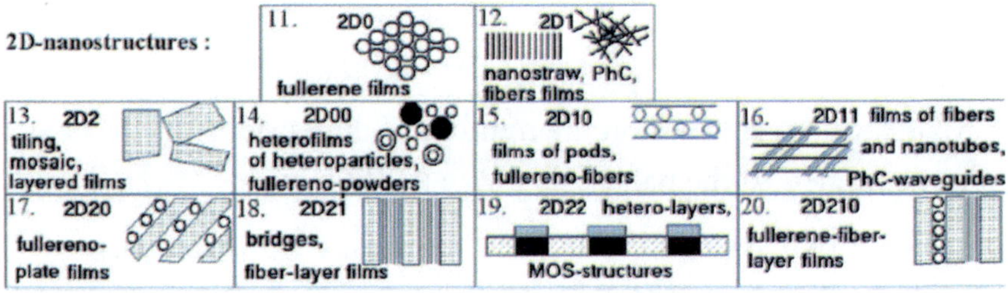

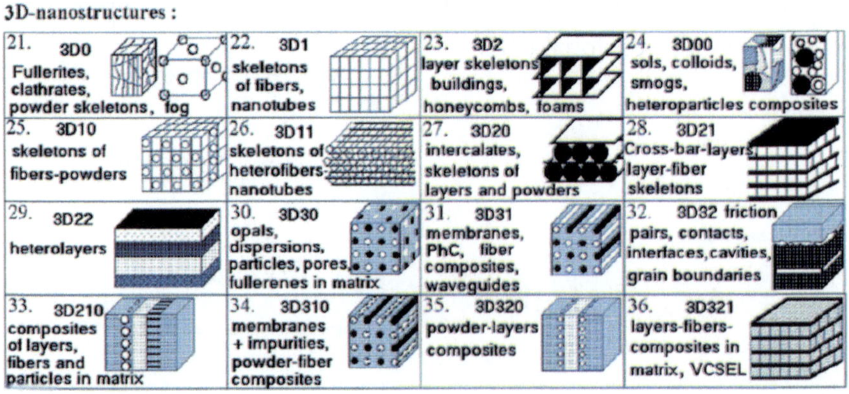

All kinds of NSs known in the literature [50–59] belong to one of these classes. However some of classes still remain to be poor demonstrated although there is the predictive ability of the suggested classification. On this basis the combined classification of NSMs can be further developed with account of the secondary signs, in particular, the type and composition of materials, such as polymers, metals, dielectrics, semiconductors, ceramics (carbides, nitrides, borides, oxides, etc.), cermets, etc.

9.3 Some Classifications Definitions

9.3.1 Nanostructures (NSs)

- A structure with a **d** at least less or equal to a critical one $\mathbf{d_c}$.
- $\mathbf{d} \leq \mathbf{d_c} \approx 100$ nm.
- The value of $\mathbf{d_c}$ is dictated by a critical characteristic of some physical phenomena (*free path length of electrons, phonons, length of de Broglie wave, length of external electromagnetic and acoustical waves, correlation length, penetration length, diffusion length...*) giving rise to the size effects.
- Nanostructures (NSs) are characterized by a form and a dimensionality **k**.

9.3.2 Nanostructured Materials (NSMs)

- Low dimensional materials comprising of building units of a submicron or nanoscale size at least in one direction and exhibiting size effects.
- Nanostructured materials are characterized by a form, a dimensionality and by a composition.

9.3.3 Nanocomposites (NCMs)

- Nanocomposite materials (two *or more* component materials) cover a vast range of ceramic, metallic and ceramometallic materials in the form of bulk materials and thin films.

9.4 Elementary Building Units (Nanostructures)

- The elementary "blocks"
- All NSs can be built from elementary units (blocks) having low dimensionality 0D, 1D, and 2D.
- *The 3D units are excluded because they can't be used to build low dimensional NSs except 3D matrix. 3D structures can be considered as NSMs if they involve the 0D, 1D, 2D NSs (Gleiter's classification of NSMs).*
- The 0D units (*3 dimensions in the nanometric range*)
- Molecules, clusters, fullerenes, rings, metcarbs (metal carbides), particles, powders, grains…
- **The 1D units (*2 dimensions in the nanometric range*)**
- Nanotubes, fibers, filaments, whiskers, spirals, belts, springs, columns, needles.

9.5 Quantum Confinement from 3D to 0D

Zero-dimensional nanomaterials

Materials wherein all the dimensions are measured within the nanoscale (no dimensions, or 0-D, are larger than 100 nm).

The most common representation of zero-dimensional nanomaterials are nanoparticles.

- **Nanoparticles can**:
  - Be amorphous or crystalline
  - Be single crystalline or polycrystalline
  - Be composed of single or multi-chemical elements
  - Exhibit various shapes and forms
  - Exist individually or incorporated in a matrix
  - Be metallic, ceramic, or polymeric.

Zero-dimensional nanomaterials

- Materials wherein all the dimensions are measured within the nanoscale (no dimensions, or 0-D, are larger than 100 nm).
- The most common representation of zero-dimensional nanomaterials are **nanoparticles.**
  - **Nanoparticles** can:
- Be amorphous or crystalline
- Be single crystalline or polycrystalline
- Be composed of single or multi-chemical elements
- Exhibit various shapes and forms
- Exist individually or incorporated in a matrix
- Be metallic, ceramic, or polymeric

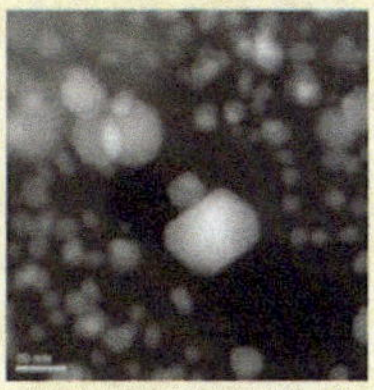

One-dimensional nanomaterials

- One dimension that is outside the nanoscale.
- This leads to needle like-shaped nanomaterials.
- 1-D materials include **nanotubes, nanorods, and nanowires**.
  - **1-D nanomaterials** can be
- Amorphous or crystalline
- Single crystalline or polycrystalline
- Chemically pure or impure
- Standalone materials or embedded in within another medium
- Metallic, ceramic, or polymeric

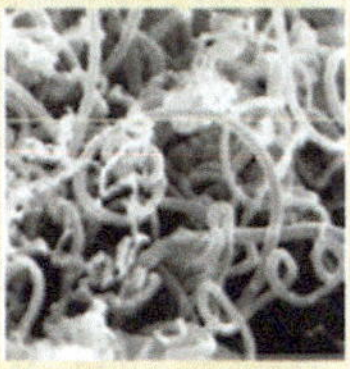

Two-dimensional nanomaterials

- Two of the dimensions are not confined to the nanoscale.
- 2-D nanomaterials exhibit plate-like shapes.
- Two-dimensional nanomaterials include **nanofilms, nanolayers, and nanocoatings.**

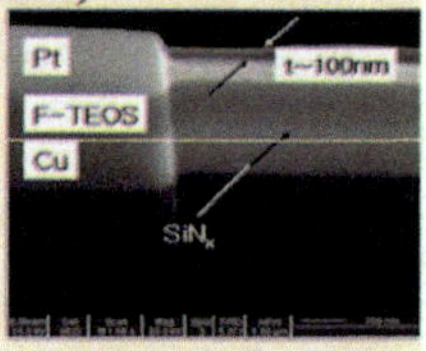

- **2-D nanomaterials** can be:
- Amorphous or crystalline
- Made up of various chemical compositions
- Used as a single layer or as multilayer structures
- Deposited on a substrate
- Integrated in a surrounding matrix material
- Metallic, ceramic, or polymeric

Three-dimensional nanomaterials

- **Bulk** nanomaterials are materials that are not confined to the nanoscale in any dimension. These materials are thus characterized by having three arbitrarily dimensions above 100 nm.
- Materials possess a nanocrystalline structure or involve the presence of features at the nanoscale.
- In terms of nanocrystalline structure, bulk nanomaterials can be composed of a multiple **arrangement of nanosize crystals**, most typically in different orientations.
- With respect to the presence of features at the nanoscale, **3-D nanomaterials** can contain dispersions of nanoparticles, bundles of nanowires, and nanotubes as well as multinanolayers.

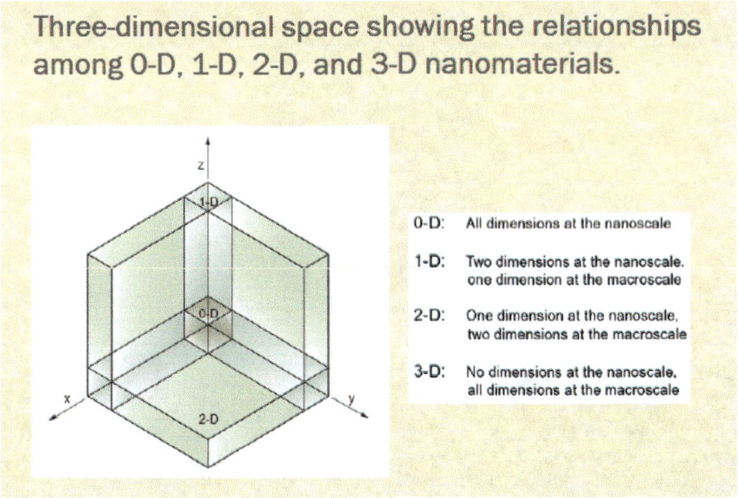

Two-dimensional (2D) gold nanoparticles can possess novel physical and chemical properties, which will greatly expand the utility of gold nanoparticles in a wide variety of applications ranging from catalysis to biomedicine. However, colloidal synthesis of such particles generally requires sophisticated synthetic techniques to carefully guide anisotropic growth. Here we report that 2D hyper-branched gold nanoparticles in the lateral size range of about 50, 120 nm can be synthesized selectively on a 2D immiscible oil/water interface in a few minutes at room temperature without structure-directing agents. An oleic acid/water interface can provide diffusion-controlled growth conditions, leading to the structural evolution of a smaller gold nucleus to 2D nanodendrimer and nanourchin at the interface. Simulations based on the phase field crystal model match well with experimental observations on the 2D branching of the nucleus, which occurs at the early stage of growth. Branching results in higher surface area and stronger near-field enhancement of 2D gold nanoparticles. This interfacial synthesis can be scaled up by creating an emulsion and the recovery of oleic acid is also achievable by centrifugation.

A small nucleus grows on an oleic acid/water interface into a nanodendrimer, and further structural evolution from dendrimer to urchin takes place while the thicknesses of these nanoparticles remains nearly unchanged, confirming that the nanoparticles on the interface are 2D (Fig. 9.10a). The morphological transition to nanourchin is completed within a few minutes. The experimental data for the 2D branching of the nucleus at the early stage of growth is consistent with simulations based on the phase field crystal model.

In addition to the aforementioned advantage of having a fast reaction time, no requirement for structural directing agents and room temperature synthesis, the interfacial synthesis of 2D gold nanoparticles has more practical implications (Fig. 9.10b). The synthesis can be scaled up by creating an emulsion, allowing the

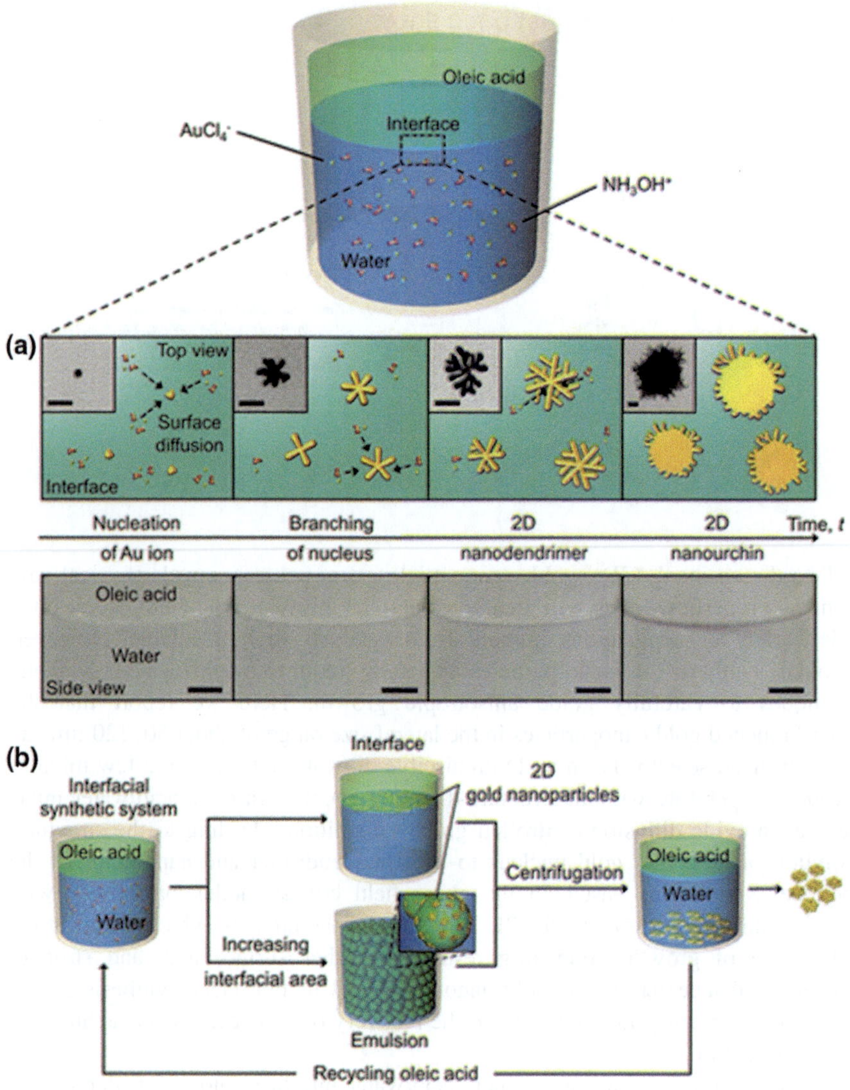

Fig. 9.10 Schematic illustrations and an experimental demonstration for our oil/water interfacial synthesis of two-dimensional (2D) gold nanoparticles. **a** Schematic representation and time-resolved optical images of an oleic acid/water interface for the synthesis of 2D gold nanoparticles. **b** Schematic illustration of the interfacial synthesis for recycling of the biocompatible oleic acid and scaling up (creating an emulsion). Scale bars are 25 nm (transmission electron microscopy (TEM) images in (**a**)) and 5 mm (optical images in (**a**)), respectively

production of a large amount of colloidal particles in a short time. Oleic acid is biocompatible and can be easily recovered due to the different densities of oil (i.e., oleic acid) and water after the interfacial synthesis. 2D hyper-branched gold nanoparticles can show a large surface area that is available to neighboring molecules without diffusion limitation, strong near-field enhancement and tunable far-field response. The size can be as small as ca. 50 nm, which is well suited for in vivo applications.

Three-3-D dimensional nanomaterials

Bulk nanomaterials are materials that are not confined to the nanoscale in any dimension. These materials are thus characterized by having three arbitrarily dimensions above 100 nm.

Materials possess a nanocrystalline structure or involve the presence of features at the nanoscale.

In terms of nanocrystalline structure, bulk nanomaterials can be composed of a multiple arrangement of nanosize crystals, most typically in different orientations.

With respect to the presence of features at the nanoscale, 3-D nanomaterials can contain dispersions of nanoparticles, bundles of nanowires, and nanotubes as well as multinanolayers.

Three-dimensional space showing the relationships among 0-D, 1-D, 2-D, and 3-D nanomaterials (Fig. 9.11).

Summary of 2-D and 3-D crystalline structures

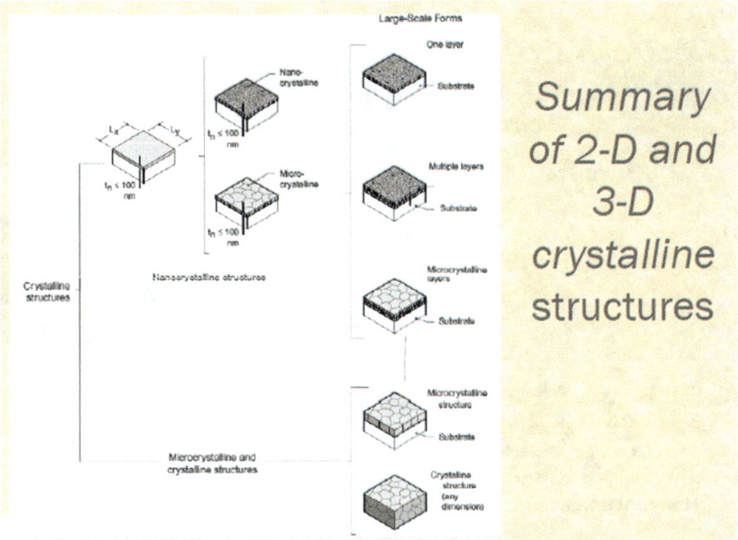

9.5.1 *Physical and Chemical Nature of Nanoparticles*

However, this scaling trend is almost reached to its physical limitation and is being faced with inevitable problems so called ‘short channel effects’ (SCEs) that mainly arisen from the weakened gate controllability onto the channel [60]. Furthermore, the fabrication optimization in such extremely small geometrical volume is much challenging since the tiny statistical deviation arisen from the device process will lead to the serious performance variability even though it associates with few atoms in the channel [61, 62].

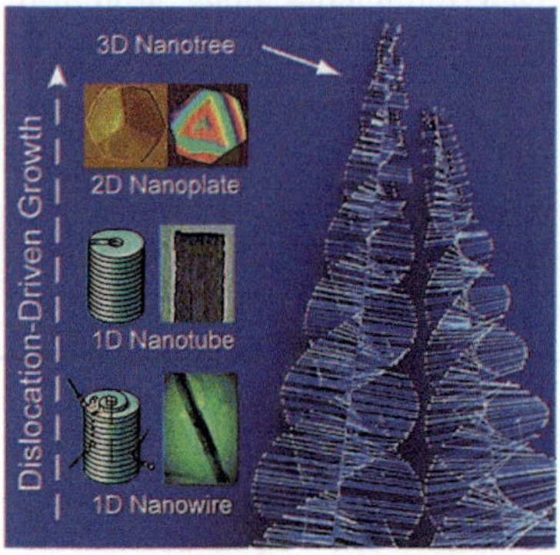

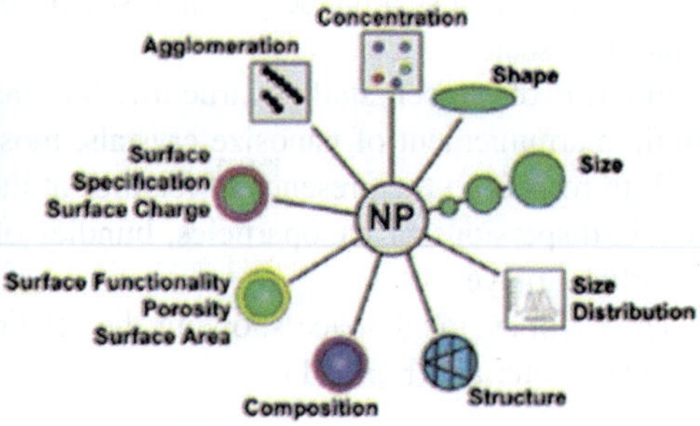

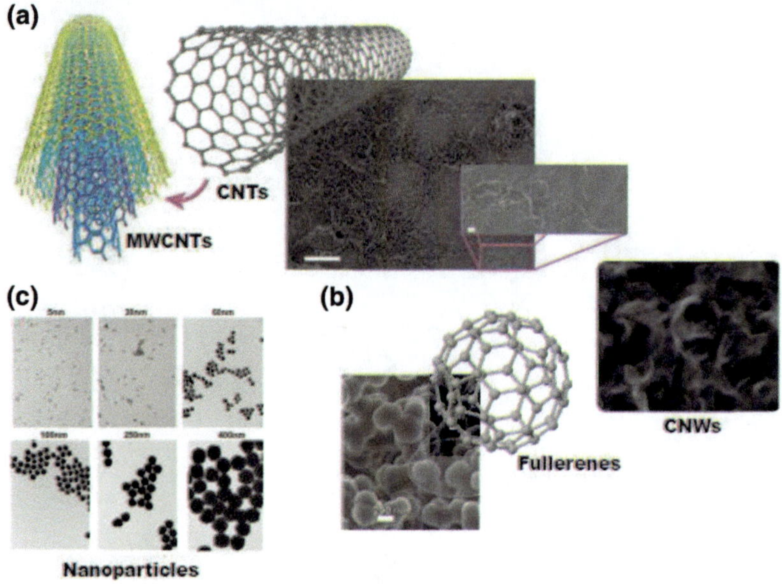

Fig. 9.11 Typical nanomaterial paradigms used for enhancing the performance of sensing configurations, like electrodes for bio-electrochemical applications. CNTs and MWCNTs. **a** fullerenes and CNWs **b** and nanoparticles of various sizes (source: Sigma-Aldrich) **c** are widely used examples

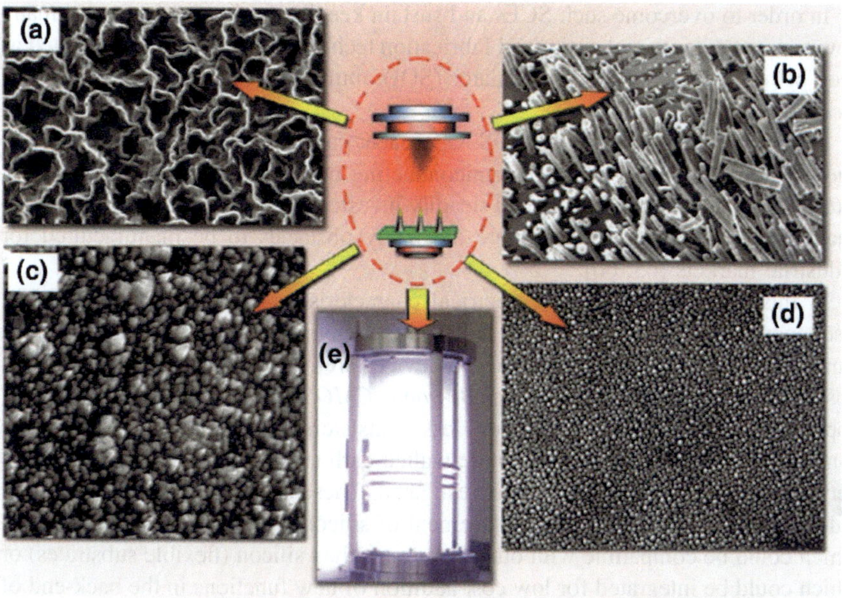

Photos as Examples of nanomaterials fabricated in the Plasma Nanoscience Center Australia (PNCA).

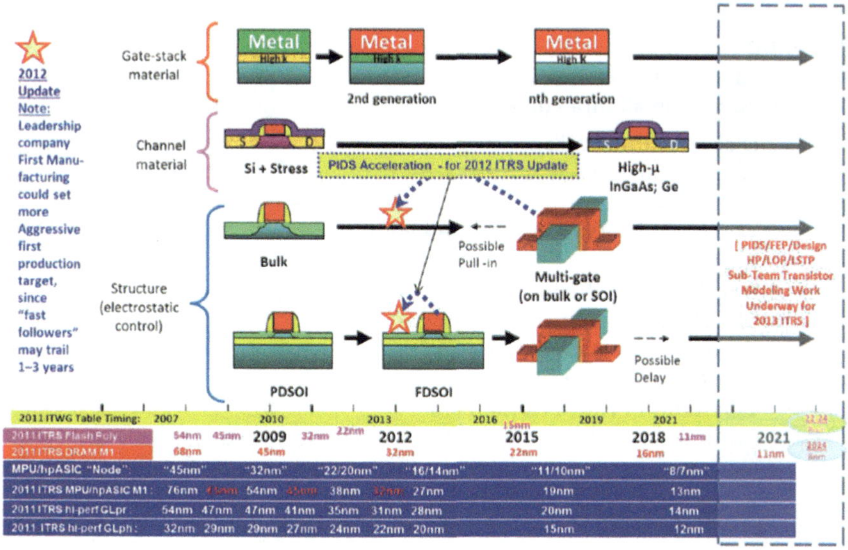

2012 ITRS "Equivalent Scaling" process technologies timing Overall Roadmap Technology Characteristics (ORTC) microprocessor unit (MPU)/High-performance ASIC half pitch and gated length trends and timing, and industry "Nodes". [http://www.itrs.net/].

In order to overcome such SCEs and sustain keeping such a scaling trend, many new concept devices and optimized fabrication technologies have been suggested as seen below such as silicon on insulator (SOI), multi-gate structure, tunneling FET etc. [60, 63, 64].

In addition, as a "Beyond CMOS" approach, carbon based materials (graphene and car-bon nanotubes (CNTs)), semiconducting nanowires, and topological insulators have been widely studied and steadily invested as a possible alternative active channel material, even though there are many obstacles to be manufactured for industrial markets [65, 66].

In this dissertation, various electrical characteristics based on I-V, C-V, low frequency noise (LFN), and low temperature environment including numerical simulation for top-down and bottom-up devices are respectively investigated. As a possible emerging candidate for "*Be-yond CMOS*" and "*more than Moore*" approaches, heavily doped junction less transistors (JLTs), quasi-1D tin-oxide (SnO_2) nanowire FETs and -2D single-walled carbon nanotube (SWCNT) random network thin film transistors were investigated. These are not only focusing on the reduction of SCEs but also the replacement of silicon based rigid channel material which could be compatible with other substrates than silicon (flexible substrates) or which could be integrated for low cost addition of new functions in the back-end of line of CMOS.

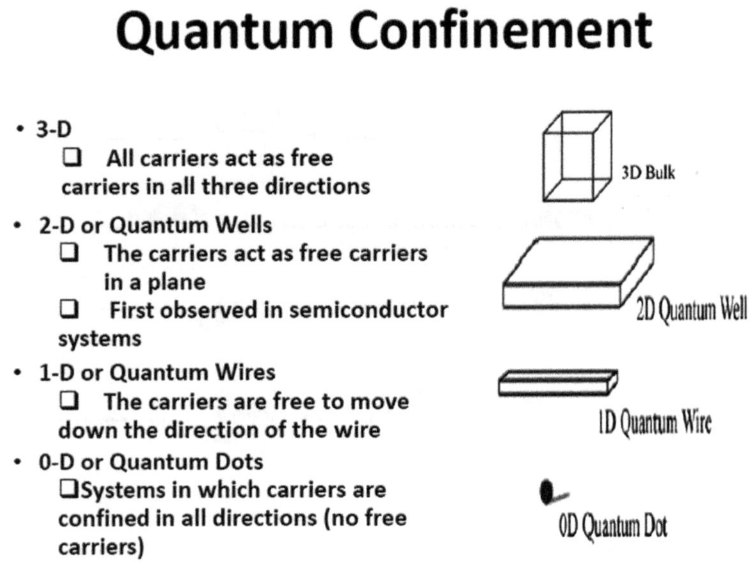

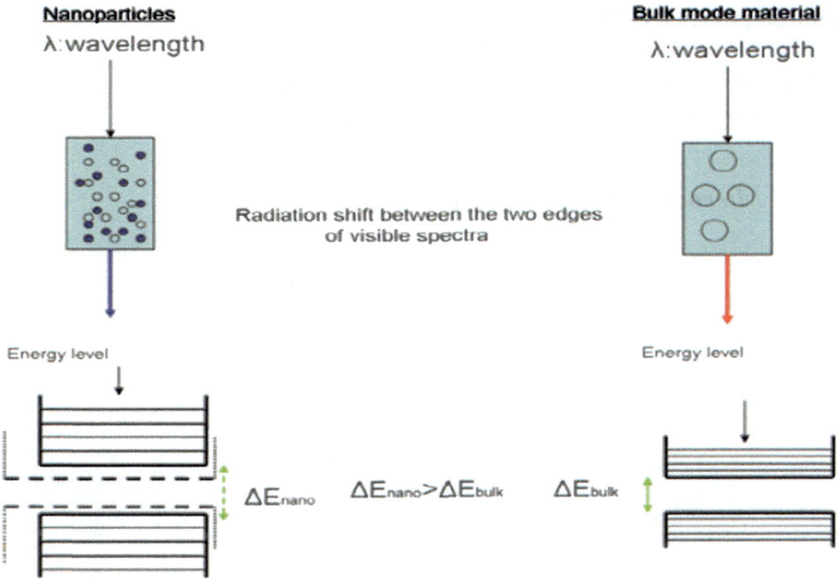

Quantum confinement is responsible for the increase of energy difference between energy states and band gap. $\Delta E_{nano} > \Delta E_{bulk}$

9.6 Matrix-Reinforced and Layered Nanocomposites

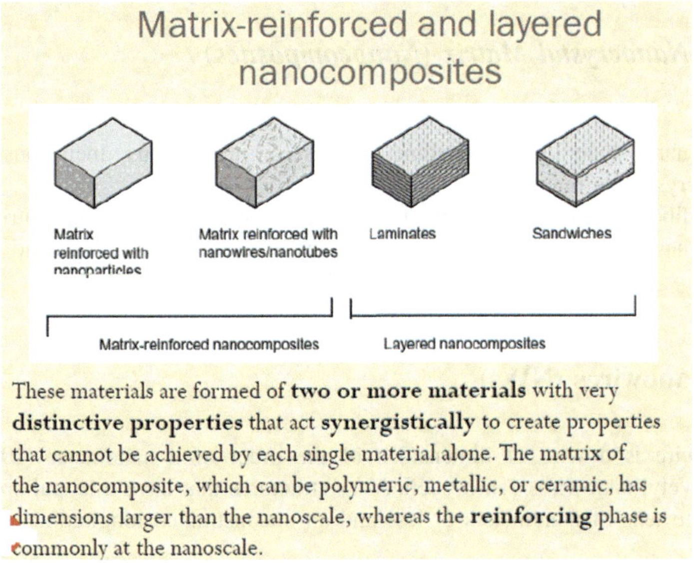

Nanocomposites (NCMs)

- Nanocomposites can be classified according to their structures.
- Two main groups of nanocomposites are distinguished:

9.6.1 *Microcrystal Matrix (Micro-Nano Type)*

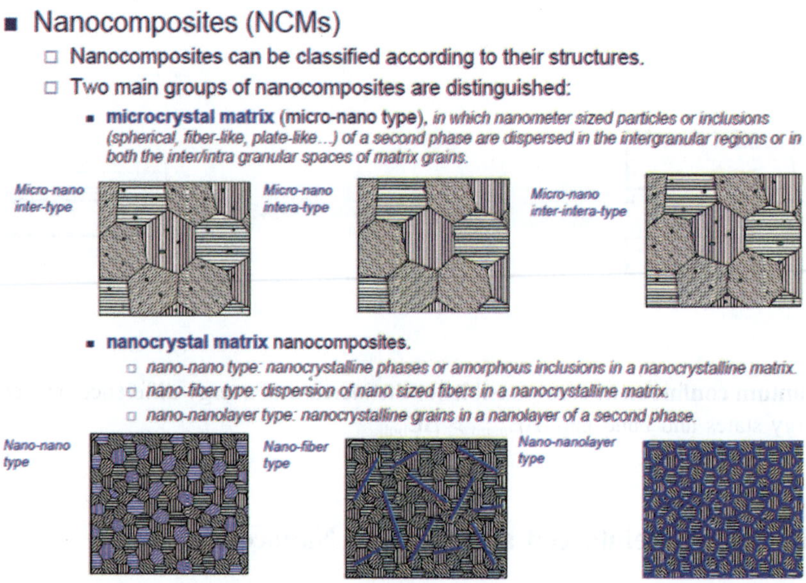

9.6.2 *Nanocrystal Matrix (Nanocomposites)*

- nano-nano type: nanocrystalline phases or amorphous inclusions in a nanocrystalline matrix.
- nano-fiber type: dispersion of nanosized fibers in a nanocrystallinematrix.
- nano-nanolayer type: nanocrystalline grains in a nanolayer of a second phase.

9.7 Nanowires (NWs)

A nanowire is a wire of dimensions of the order of a nanometer (10^{-9} m). Alternatively, nanowires can be defined as structures that have a lateral size constrained to tens of nanometers or less and an unconstrained longitudinal size. At these scales, quantum mechanical effects are important—hence such wires are also known as "quantum wires". A **nanowire** is a nanostructure, with the diameter of the order of a nanometer (10^{-9} m). It can also be defined as the ratio of the length to width being greater than 1000. Alternatively, nanowires can be defined as structures that have a thickness or diameter or a lateral size constrained to tens of nanometers or less and an unconstrained longitudinal size. They represent the smallest dimension for efficient transport of electrons and excitons, and thus will be used as interconnects and critical devices in nanoelectronics and nano-optoelectronics. There is no restriction on how wide they can grow, but cannot grow more than a few nanometers in height. Have nanometer size in one of the dimension which produces quantum confinement in material and changes its properties. At these scales, quantum mechanical effects are important—hence such wires are also known as "quantum wires".

1D nanostructures(Quantum Wires)

- 1D nanostructures represent the smallest dimension structure that can efficiently transport electrical carriers.

- 1D nanostructures can also exhibit such device functions that can be exploited as both **wiring** and **device elements** in future architectures for functional nanosystems.

- In this regard, two material classes:
  I. **(NTs) carbon nanotubes**
  II. **(NWs) nanowires**

  have shown particular promise

A nanowire is an extremely thin wire with a diameter on the order of a few nanometers (nm) or less, where 1 nm = 10^{-9} m. Two processes in nanotechnology by which nanowires can be manufactured are suspension and deposition. Semiconductor nanowires have recently emerged as a new class of materials with significant potential to reveal new fundamental physics and to propel new applications in quantum electronic and optoelectronic devices. Semiconductor nanowires show exceptional promise as nanostructured materials for exploring physics in reduced dimensions and in complex geometries, as well as in one-dimensional nanowire devices. They are compatible with existing semiconductor technologies and can be tailored into unique axial and radial heterostructures.

A suspended nanowire is held up by the ends in an evacuated chamber, and then is chemically etched or bombarded with high-speed atom s or molecule s to reduce its diameter. Another method involves indenting the surface of a wire in the center of a suspended span, raising the temperature, and then stretching the wire while it is near its melting point. A deposited nanowire is fabricated on a surface consisting of some non-conducting substance such as plastic or glass. The process is similar to that by which semiconductor chips are grown, except that the result is a linear (one-dimensional) structure rather than a flat (two-dimensional) or solid (three-dimensional) structure.

Fields expected to benefit from nanotechnology include water purification, sanitation, agriculture, alternative energy (particularly photovoltaics), home and business construction, computer manufacturing, communications, and medicine.

In 1965, engineer Gordon Moore predicted that the number of transistors on an **integrated circuit**—a precursor to the microprocessor—would double approximately every two years. Today, we call this prediction **Moore's Law**, though it's not really a scientific law at all. Moore's Law is more of a **self-fulfilling prophecy** about the computer industry. Microprocessor manufacturers strive to meet the prediction, because if they don't, their competitors will [source: Intel].

Nanowires and nanotubes, slender structures that are only a few billionths of a meter in diameter but many thousands or millions of times longer, have become hot materials in recent years. They exist in many forms—made of metals, semiconductors, insulators and organic compounds—and are being studied for use in electronics, energy conversion, optics and chemical sensing, among other fields.

The initial discovery of carbon nanotubes—tiny tubes of pure carbon, essentially sheets of graphene rolled up unto a cylinder—is generally credited to a paper published in 1991 by the Japanese physicist Sumio Ijima (although some forms of carbon nanotubes had been observed earlier). Almost immediately, there was an explosion of interest in this exotic form of a commonplace material. Nanowires—solid crystalline fibers, rather than hollow tubes—gained similar prominence a few years later.

Due to their extreme slenderness, both nanotubes and nanowires are essentially one-dimensional. "They are quasi-one-dimensional materials," says MIT associate professor of materials science and engineering Silvija Gradečak: "Two of their dimensions are on the nanometer scale." This one-dimensionality confers distinctive electrical and optical properties.

For one thing, it means that the electrons and photons within these nanowires experience "quantum confinement effects," Gradečak says. And yet, unlike other materials that produce such quantum effects, such as quantum dots, nanowires' length makes it possible for them to connect with other macroscopic devices and the outside world.

The structure of a nanowire is so simple that there's no room for defects, and electrons pass through unimpeded, Gradečak explains. This sidesteps a major problem with typical crystalline semiconductors, such as those made from a wafer of silicon: There are always defects in those structures, and those defects interfere with the passage of electrons.

Made of a variety of materials, nanowires can be "grown" on many different substrates through a vapor deposition process. Tiny beads of molten gold or other metals are deposited on a surface; the nanowire material, in vapor, is then absorbed by the molten gold, ultimately growing from the bottom of that bead as a skinny column of the material. By selecting the size of the metal bead, it is possible to precisely control the size of the resulting nanowire.

In addition, materials that don't ordinarily mix easily can be grown together in nanowire form. For example, layers of silicon and germanium, two widely used semiconductors, "are very difficult to grow together in thin films," Gradečak says. "But in nanowires, they can be grown without any problems." Moreover, the equipment needed for this kind of vapor deposition is widely used in the semiconductor industry, and can easily be adapted for the production of nanowires.

While nanowires' and nanotubes' diameters are negligible, their length can extend for hundreds of micrometers, even reaching lengths visible to the unaided eye. No other known material can produce such extreme length-to-diameter ratios: millions of times longer than they are wide.

Because of this, the wires have an extremely high ratio of surface area to volume. That makes them very good as detectors, because all that surface area can be treated to bind with specific chemical or biological molecules. The electrical signal generated by that binding can then easily be transmitted along the wire.

Similarly, nanowires' shape can be used to produce narrow-beam lasers or light-emitting diodes (LEDs), Gradečak says. These tiny light sources might someday find applications within photonic chips, for example—chips in which information is carried by light, instead of the electric charges that relay information in today's electronics.

Compared to solid nanowires, nanotubes have a more complex structure: essentially one-atom-thick sheets of pure carbon, with the atoms arranged in a pattern that resembles chicken wire. They behave in many ways as one-dimensional materials, but are actually hollow tubes, like a long, nanometer-scale drinking straw.

The properties of carbon nanotubes can vary greatly depending on how they are rolled up, a property called chirality. (It's similar to the difference between forming a paper tube by rolling a sheet of paper lengthwise versus on the diagonal: The different alignments of fibers in the paper produce different strength in the resulting tubes.) In the case of carbon nanotubes, chirality can determine whether the tubes behave as metals or as semiconductors.

But unlike the precise manufacturing control that is possible with nanowires, so far methods for making nanotubes produce a random mix of types, which must be sorted to make use of one particular kind. Besides single-walled nanotubes, they also exist in double-walled and multi-walled forms.

In addition to their useful electronic and optical properties, carbon nanotubes are exceptionally strong, and are used as reinforcing fibers in advanced composite materials. "In any application where one-dimensionality is important, both carbon nanotubes and nanowires would provide benefits," Gradečak says. Many different types of nanowires exist, including metallic (e.g., Ni, Pt, Au), semiconducting (e.g., InP, Si, GaN, etc.), and insulating (e.g., SiO_2, TiO_2). Molecular nanowires are composed of repeating molecular units either organic or inorganic. Thus, **nanowires** represent the smallest dimension for efficient transport of electrons and excitons, and thus will be used as interconnects and critical devices in nano electronics and nano-opto electronics.

Increased surface area

- Diameter-dependent bandgap
- Increased surface scattering for electrons and phonons
- Controlled synthesis
- Diameter, length, composition
- Electronic structure
- Quantum confinement
- Present in some, absent in others
- Unique magnetic and electronic properties.

9.7.1 Unique Applications of Nanowires

Semiconductor nanowires (NWs) have been studied extensively for over two decades for their novel electronic, photonic, thermal, electrochemical and mechanical properties.

Nanowires are promising materials for many novel applications for their unique geometry and unique physical properties such as:

- Electrical
- Magnetic
- Optical
- Mechanical

Important applications of nanowires in environment:

- using silver chloride nanowires as a photocatalysis to decompose organic molecules in polluted water.
- using an electrified filter composed of silver nanowires, carbon nanotubes and cotton to kill bacteria in water.
- using nanowire mats to absorb oil spill.
- using nanowires made an alloy of iron and nickel to creat dense memory devices in electronics.
- using silver nanowires embedded in polymer to make conductive layers that can flex, without damaging the conductor.
- Sensors using zinc oxide nano-wire detection element, capable of detecting a range of chemical vapors.

9.7.2 *Different Types of Nanowires*

Two material classes:

- **(NTs) carbon nanotubes**
- **(NWs) nanowires**
- Can be either electrically conductive or semi conductive, depending on their magnetic felicity.
- Exhibit electrical conductivity as high as copper.
- Have thermal conductivity as high as diamond.
- Strength100 times greater than steel at one-sixth the weight and high strain to failure.

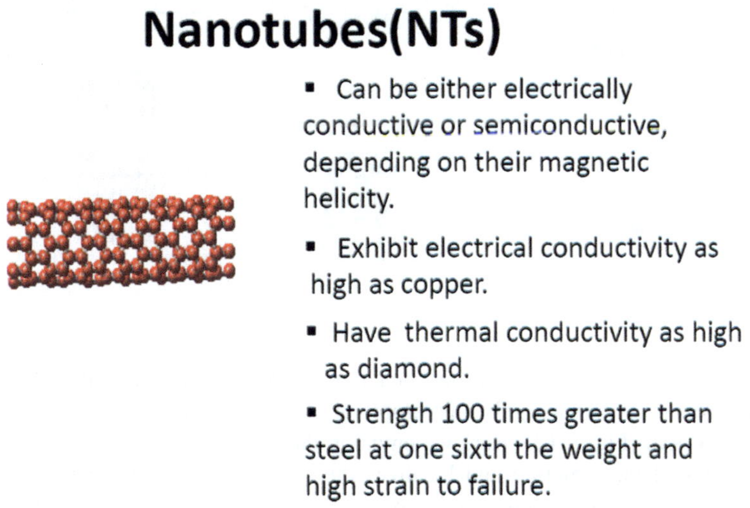

9.7.3 *Basic Growth Mechanism*

Growth mechanism of nanotubes (NTs)

- The way in which nanotubes are formed is not exactly known.
- More than one mechanism might be operative during the formation of CNTs.

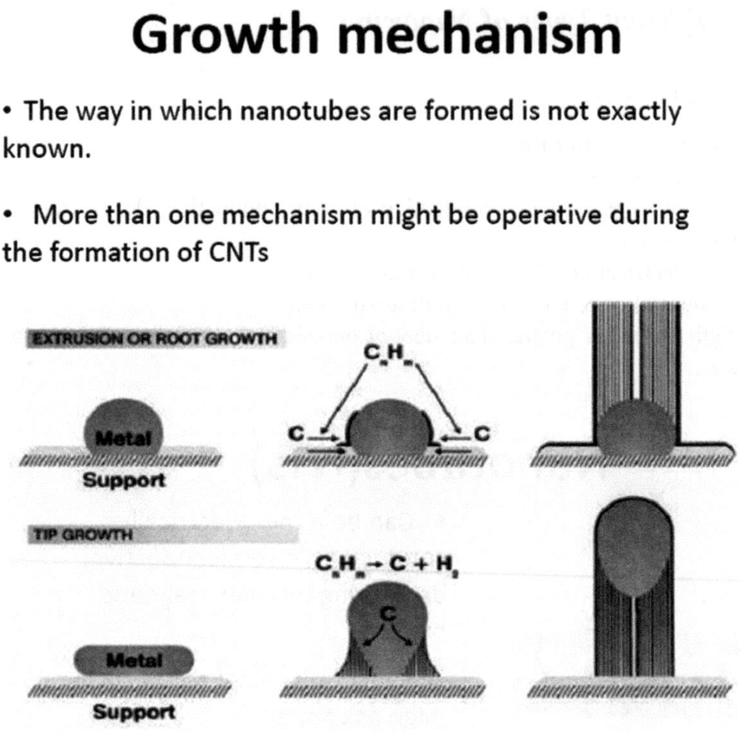

From the outset we believed that a catalytic approach, in which the catalyst is used to direct growth in a highly anisotropic or 1D manner, would enable us to meet this overall goal (Fig. 9.12). Here we envision the catalyst as a cluster that defines the diameter of the structure and localizes the reactant at the end of the growing nanowire much like a catalyst in living polymerization. An important feature of this idea is that it readily provides the intellectual underpinning needed for the prediction of good catalysts and synthesis conditions. First, one uses equilibrium phase diagrams to choose a catalyst that can form a liquid alloy with the nanowire material of interest. The phase diagram is then used to choose a specific composition (catalyst: nanowire material) and synthesis temperature so that there is a coexistence of liquid alloy and solid nanowire material. The liquid catalyst alloy cluster serves as a preferential site for absorption of reactant (i.e., there is a much higher sticking probability on liquid vs solid surfaces) and, when supersaturated, the nucleation site for crystallization. Preferential 1D growth occurs in the presence of reactant as long as the catalyst remains liquid.

Within this framework one can readily imagine synthesizing nanowires of many different materials and diameters, if catalyst clusters of nanometer dimensions are available. This raises an important point, because equilibrium thermodynamics can be used to define the minimum radius, r_{min}, of a liquid metal cluster as:

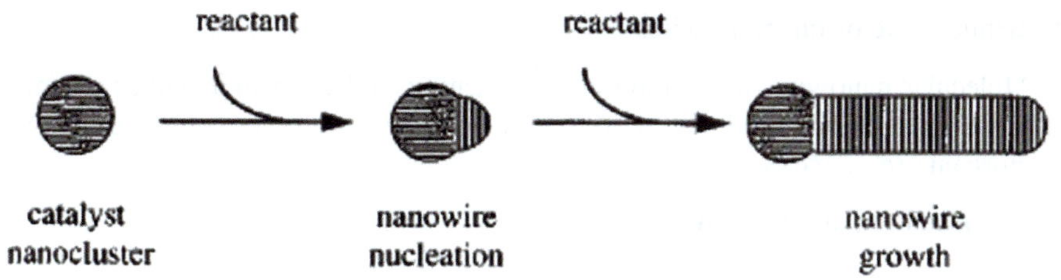

Fig. 9.12 Schematic diagram illustrating the catalytic synthesis of nanowires. Reactant material, which is preferentially absorbed on the catalyst cluster, is added to the growing nanowire at the catalyst-nanowire interface

$$r_{\min} = 2\sigma_{LV} V_L / RT \ln \sigma$$

where σ_{LV} is the liquid-vapor surface free energy, V_L is the molar volume, R and T have the usual meaning, and σ is the vapor phase supersaturation [50, 51].

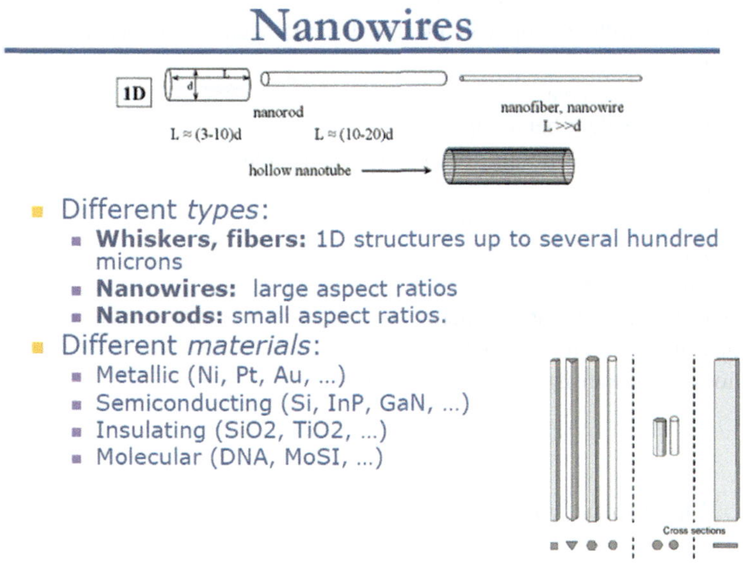

Many different types of nanowires exist, including metallic (e.g., Ni, Pt, Au), superconducting (e.g., YBCO), semiconducting (e.g., InP, Si, GaN, etc.), and insulating (e.g., SiO_2, TiO_2).

- Semiconductor nanowires (Silicon nanowires).
- Oxide nanowires.
- Multi-segment nanowires.

- Semiconductor quantum wires.

Molecular nanowires are composed of repeating molecular units either organic or inorganic are composed of repeating molecular units either organic (e.g. DNA) or inorganic (e.g. $Mo_6S_{9-x}I_x$).

1D Nanostructures (Quantum Wires)

- 1D nanostructures represent the smallest dimension structure that can efficiently transport electrical carriers.
- 1D nanostructures can also exhibit such device functions that can be exploited as both **wiring** and **device elements** in future architectures for functional nanosystems.

1D nanostructures represent the smallest dimension structure that can efficiently transport electrical carriers (e−, h+):

- Can act as both nanoscale devices and wiring to link components into extremely small circuits.
- Unique density of electronic states:
- Significantly different optical, electrical, and magnetic properties from bulk counterparts.
- Increased surface area
- Diameter-dependent band gap
- Increased surface scattering for electrons and phonons
- Controlled synthesis
- Diameter, length, composition
- Electronic structure
- Quantum confinement
- Present in some, absent in others
- Unique magnetic and electronic properties
- Millions more transistors per microprocessor.

9.7.4 Why Study Nanowires?

- Central importance of nanoscale wires in integrated Nano systems
- Fundamental scientific questions in 1-dimensional systems
- Synthetic challenge of controlling structure and composition on many length scales
- New/novel materials can make revolutionary vs. evolutionary changes in science and technology!

- **Nanowires: A General and Predictable Approach**

- Breaking symmetry for 1D growth. Nanoscale wires can be prepared rationally by exploiting a catalyst to direct preferentially the addition of reactant.
- The key issue for controlled nanowire growth is the generation of nanometer scale 'catalyst' clusters.
- The growth process begins when the catalyst becomes supersaturated with reactant, and terminates when the nanowires pass out of the hot reaction zone.

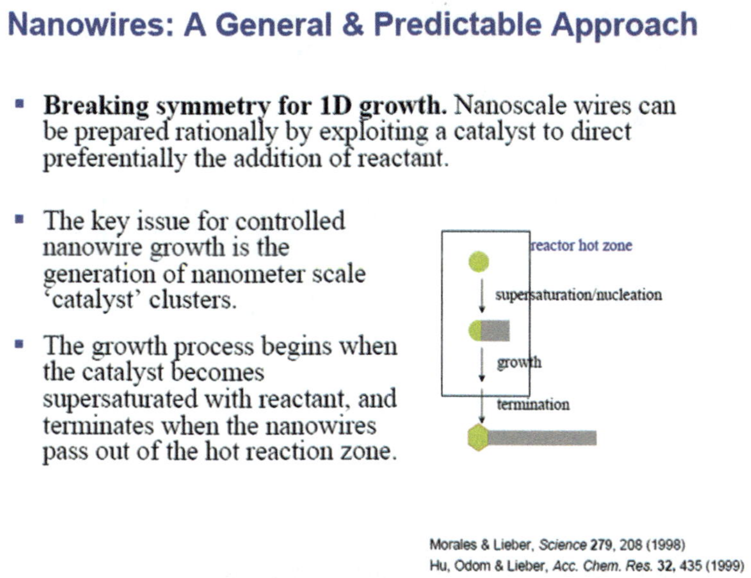

- 1D nanostructures represent the smallest dimension structure that can efficiently transport electrical carriers.
- 1D nanostructures can also exhibit such device functions that can be exploited as both **wiring** and **device elements** in future architectures for functional nanosystems.
- In this regard, two material classes:

I. **(NTs) carbon nanotubes**

II. **(NWs) nanowires**

Classification according to the number of reduced dimensions (usually refers to the number of degrees of freedom in the momentum).

From 3D to 0D

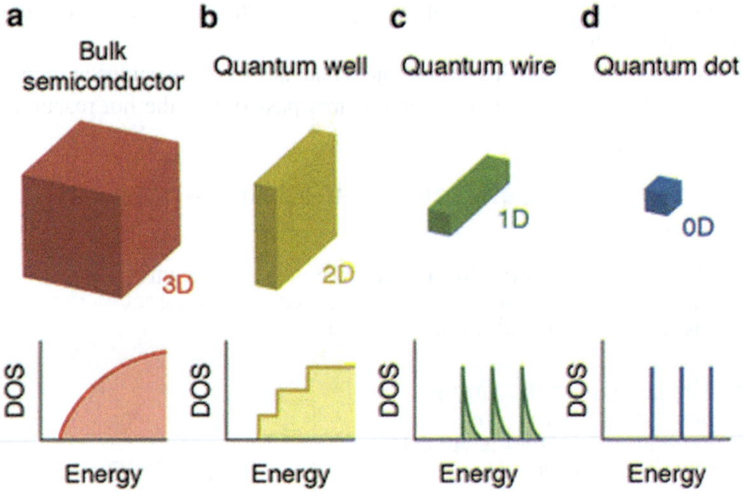

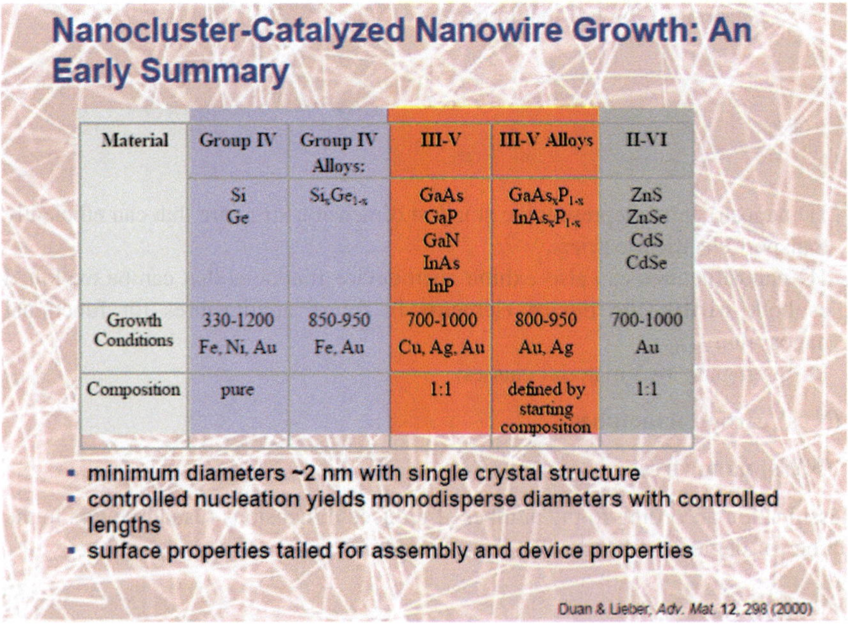

Nanocluster-Catalyzed Nanowire Growth: An Early Summary

Material	Group IV	Group IV Alloys:	III-V	III-V Alloys	II-VI
	Si Ge	Si_xGe_{1-x}	GaAs GaP GaN InAs InP	$GaAs_xP_{1-x}$ $InAs_xP_{1-x}$	ZnS ZnSe CdS CdSe
Growth Conditions	330-1200 Fe, Ni, Au	850-950 Fe, Au	700-1000 Cu, Ag, Au	800-950 Au, Ag	700-1000 Au
Composition	pure		1:1	defined by starting composition	1:1

- minimum diameters ~2 nm with single crystal structure
- controlled nucleation yields monodisperse diameters with controlled lengths
- surface properties tailed for assembly and device properties

Duan & Lieber, *Adv. Mat.* 12, 298 (2000)

From 3D to 0D

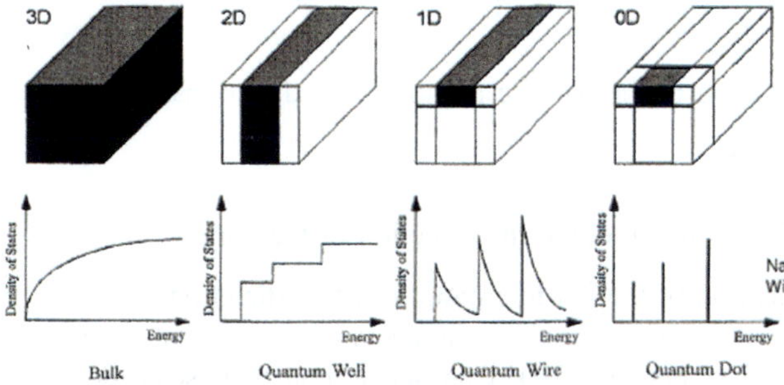

- **0D: quantum dots**
- **1D: Nanowires**
- **2D: superlattices, heterostructures**

Nanoscale Materials in Chemistry, Wiley, 2001

- Classification according to the number of reduced dimensions (usually refers to the number of degrees of freedom in the momentum).
- 3 D or bulk system: no wuantization of the particle motion.
- 2D system: quantization of the particle motion in one direction.
- 1D system: quantization of the particle motion in two directions.
- 0D system: quantization of the particle motion in all directions.

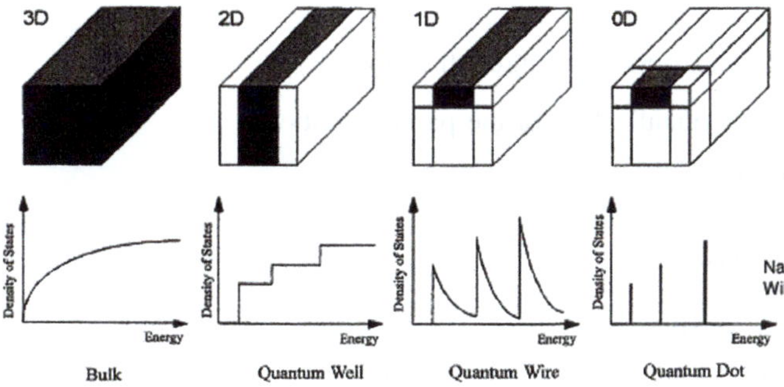

- **0D: quantum dots**
- **1D: Nanowires**
- **2D: superlattices, heterostructures**

Nanoscale Materials in Chemistry, Wiley, 2001

$$\frac{dn}{dE} \propto E^{\frac{1}{2}}$$

$$\frac{dn}{dE} = \frac{Am}{2\pi\hbar^2}$$

Constant for each electronic band

$$\frac{dn}{dE} \propto E^{-\frac{1}{2}}$$

At each atomic level, the DOS decreases as the reciprocal of the square root of energy.

Sharp levels corresponding to the eigenstates of the system

0-D or Quantum Dots

- Systems in which carriers are confined in all directions

  (no free carriers)
  0D system: quantization of the particle motion in all directions.

1-D or Quantum Wires

- The carriers are free to move down the direction of the wire.

  1D system: quantization of the particle motion in two directions.
  1D nanostructures represent the smallest dimension structure that can efficiently

Transport electrical carriers (e−, h+): Can act as both nanoscale devices and wiring to link components into extremely small circuits.

Unique density of electronic states: significantly different optical, electrical, and magnetic properties from bulk counterparts.

2-D or Quantum Wells

- The carriers act as free carriers in a plane
- First observed in semiconductor systems.

  2D system: quantization of the particle motion in one direction.

3-D

- All carriers act as free carriers in all three directions

  3 D or bulk system: no quantization of the particle motion.

Nanowires 1D structures

- Diameter: 1–100 nm (10^{-9} m)
- Length: microns (10^{-6} m)
- Typical aspect ratios of 1000 or more.
- Crystal structures close to that of the bulk material
- Promising framework for applying the "bottomup" approach for the design of nanostructures.

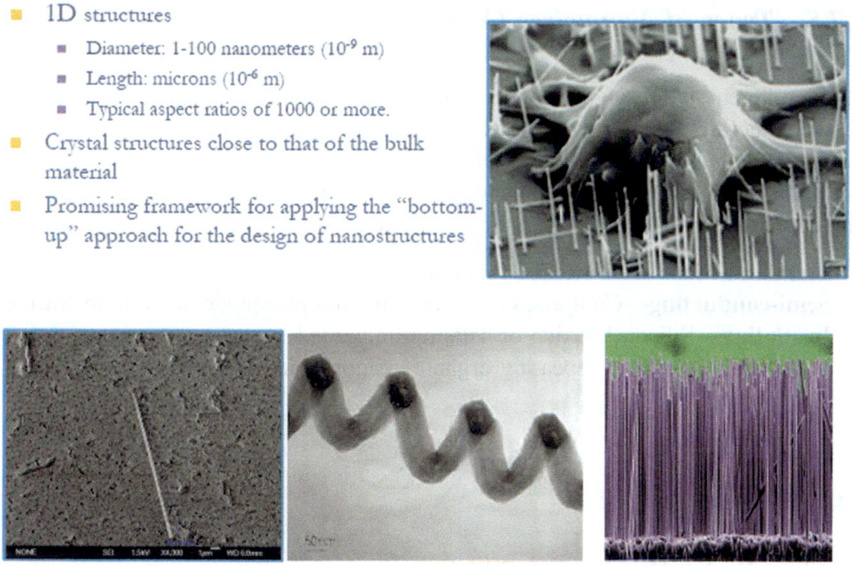

Typical nanowires exhibit aspect ratios (length-to-width ratio) of 1000 or more [67–70]. As such they are often referred to as one-dimensional (1-D) materials. Nanowires have many interesting properties that are not seen in bulk or 3-D (three-dimensional) materials. This is because electrons in nanowires are quantum confined laterally and thus occupy energy levels that are different from the traditional continuum of energy levels or bands found in bulk materials.

Peculiar features of this quantum confinement exhibited by certain nanowires manifest themselves in discrete values of the electrical conductance. Such discrete values arise from a quantum mechanical restraint on the number of electrons that can travel through the wire at the nanometer scale. These discrete values are often referred to as the quantum of conductance and are integer multiples of

$$\frac{2e^2}{h} \simeq 77.41\ \mu\text{S}$$

They are inverse of the well-known resistance unit h/e^2, which is roughly equal to 25812.8 Ω, and referred to as the von Klitzing constant R_K (after Klaus von Klitzing, the discoverer of exact quantization). Since 1990, a fixed conventional value R_{K-90} is accepted.

9.7.5 Types of Nanowires (NWs)

Many different types of nanowires exist, including metallic (e.g., Ni, Pt, Au), superconducting (e.g., YBCO), semiconducting (e.g., InP, Si, GaN, etc.), and insulating (e.g., SiO_2, TiO_2). Molecular nanowires are composed of repeating molecular units either organic or inorganic are composed of repeating molecular units either organic (e.g. DNA) or inorganic (e.g. $Mo_6S_{9-x}I_x$).

- **Metallic**—Made from Nickel, Platinum or Gold
- **Semi-conducting**—Comprises of Silicon, Indium phosphide or Gallium Nitride
- **Insulating**—Silicon Dioxide or Titanium dioxide
- **Molecular**—Involves repeating organic or inorganic molecular units.

References

1. Gleiter, H.: Acta Mater. **48**, 1–29 (2000)
2. Rao, C.N.R., Nath, M.: Inorganic nanotubes. Dalton Trans. 1–24 (2003)
3. Ikkala, O., Brinke, G.T.: Functional materials based on self-assembly of polymeric supramolecules. Science **295**(5564), 2407–2409 (2002). New Series
4. Van Swygenhoven, H., Caro, A., Farkas, D.: A molecular dynamics study of polycrystalline FCC metals at the nanoscale Grain boundary structure and its influence on plastic deformation. Mater. Sci. Eng. A (2001)
5. Braginsky, L., Shklover, V., Hofmann, H., Bowen, P.: High-temperature thermal conductivity of porous Al_2O_3 nanostructures. Phys. Rev. B **70**, 134201 (2004)
6. Braginsky, L., Lukzen, N., Shklover, V., Hofmann, H.: Phys. Rev. B **66**, 134203 (2002)
7. Wang, X., et al.: Nature **437**, 121 (2005)
8. Odom, T., et al.: J. Phys. Chem. B **104**, 2794 (2000)
9. Ouyang, M., et al.: Acc. Chem. Res. **35**, 1018 (2002)
10. Ouyang, M., et al.: Annu. Rev. Phys. Chem. **53**, 201 (2002)
11. Yao, Z., et al.: In: Dresselhaus, M.S., et al. (eds.) Carbon Nanotubes: Synthesis, Structure, Properties and Applications, p. 147. Springer, New York (2000)
12. Liu, J., et al.: MRS Bull. **29**(4), 244 (2004)
13. McEuen, P.L., Park, J.Y.: MRS Bull. **29**(4), 272 (2004)
14. Morales, A.M., Lieber, C.M.: Science **279**, 208 (1998)
15. Levitt, A.P. (ed.): Whisker Technology. Wiley, New York (1970)
16. Wagner, R.S., Ellis, W.C.: Appl. Phys. Lett. **4**, 89 (1964)
17. Duan, X.F., Lieber, C.M.: Adv. Mater. **12**, 298 (2000)
18. Cui, Y., et al.: J. Phys. Chem. B **104**, 5213 (2000)
19. Duan, X.F., Lieber, C.M.: J. Am. Chem. Soc. **122**, 188 (2000)
20. Duan, X.F., et al.: Appl. Phys. Lett. **76**, 1116 (2000)
21. Gudiksen, M., Lieber, C.M.: J. Am. Chem. Soc. **122**, 8801 (2000)
22. Wang, J., et al.: Science **293**, 1455 (2001)
23. Duan, X.F., et al.: Nature **409**, 66 (2001)
24. Gudiksen, M., et al.: J. Phys. Chem. B **105**, 4062 (2001)
25. Huang, Y., et al.: Nano Lett. **2**, 101 (2002)
26. Cui, Y., et al.: Nano Lett. **3**, 149 (2003)
27. Zhong, Z., et al.: Nano Lett. **3**, 343 (2003)
28. Barrelet, C.J., et al.: J. Am. Chem. Soc. **125**, 11498 (2003)

29. Greytek, A.B., et al.: Appl. Phys. Lett. **84**, 4176 (2004)
30. Wu, Y., et al.: Nano Lett. **4**, 433 (2004)
31. Zheng, G., et al.: Adv. Mater. **16**, 1890 (2004)
32. Radovanovic, P.V., et al.: Nano Lett. **5**, 1407 (2005)
33. Gudiksen, M., et al.: Nature **415**, 617 (2002)
34. Wu, Y., et al.: Nature **430**, 61 (2004)
35. Yang, C., et al.: Science **310**, 1304 (2005)
36. Lauhon, L., et al.: Nature **420**, 57 (2002)
37. Qian, F., et al.: Nano Lett. **4**, 1975 (2004)
38. Qian, F., et al.: Nano Lett. **5**, 2287 (2005)
39. Hayden, O., et al.: Adv. Mater. **17**, 701 (2005)
40. Lu, W., et al.: Proc. Natl. Acad. Sci. U.S.A. **102**, 10046 (2005)
41. Xiang, J., et al.: Nature **441**, 489 (2006)
42. Li, Y., et al.: Nano Lett. **6**, 1468 (2006)
43. Jin, S., et al.: Nano Lett. **4**, 915 (2004)
44. Bryllert, T., et al.: IEEE Electron Device Lett. **27**, 323 (2006)
45. Doh, Y., et al.: Science **309**, 272 (2005)
46. Zhong, Z., et al.: Nano Lett. **5**, 1143 (2005)
47. Tilke, A., et al.: J. Appl. Phys. **89**, 8159 (2001)
48. McAlpine, M.C., et al.: Nano Lett. **3**, 443 (2003)
49. Friedman, R.S., et al.: Nature **434**, 1085 (2005)
50. Cui, Y., et al.: Science **293**, 1289 (2001)
51. Hahm, J., Lieber, C.M.: Nano Lett. **4**, 51 (2004)
52. Patolsky, F., et al.: Proc. Natl. Acad. Sci. U.S.A. **101**, 14017 (2004)
53. Wang, W.U., et al.: Proc. Natl. Acad. Sci. U.S.A. **102**, 3208 (2005)
54. Zheng, G., et al.: Nat. Biotechnol. **23**, 1294 (2005)
55. Patolsky, F., Lieber, C.M.: Mater. Today **8**(4), 20 (2005)
56. Huang, Y., et al.: Science **294**, 1313 (2001)
57. Zhong, Z., et al.: Science **302**, 1377 (2003)
58. Huang, Y., et al.: Small **1**, 142 (2005)
59. Duan, X.F., et al.: Nature **421**, 241 (2003)
60. Sze, S.M., Ng, K.K.: Physics of semiconductor devices. Wiley (2006)
61. Liu, T.J.K., Chang, L.: Transistor scaling to the limit. In: Huff, H. (eds.) Into the Nano Era. Springer, Berlin, Heidelberg, pp. 191–223 (2009)
62. Ghibaudo, G.: Evaluation of variability performance of junctionless and conventional Trigate transistors, Solid-State Electron. **75**, 13–15 (2012)
63. Kim, D.M.: Introductory quantum mechanics for semiconductor nanotechnology. Wiley (2010)
64. von Haartman, M., Östling, M.: Low-frequency noise in advanced MOS devices. Springer (2010)
65. Ellinger, F.: Radio frequency integrated circuits and technologies. Springer (2008)
66. Shulaker, M.M., Hills, G., Patil, N., Wei, H., Chen, H.-Y., Wong, H.S.P., Mitra, S.: Carbon nanotube computer, Nature **501**(7468), 526–530 (2013)
67. Hayden, O., et al.: Nat. Mater. **5**, 352 (2006)
68. Mukai, T.: IEEE J. Sel. Top. Quantum Electron. **8**, 264 (2002)
69. Wagner, R.S., Ellis, W.C.: Vapor-solid-growth mechanism of single crystal growth. Appl. Phys. Lett. **4**, 89 (1964)
70. Wagner, R.S.: VLS mechanism of crystal growth. In: Levitt, A.P. (ed.) Whisker Technology. Wiley, New York (1970)

Chapter 10
Processing of Nanomaterials (NMs)

Nanoparticles can be synthesized chemically or biologically. A huge body of literature is now devoted to nanomaterial synthesis, and excellent reviews describe the growth mechanisms [1–4]. Many adverse effects have been associated with chemical synthesis methods due to the presence of some toxic chemical absorbed on the surface. Eco friendly alternatives to Chemical and physical methods are Biological ways of nanoparticles synthesis using microorganisms, enzymes, fungus, and plants or plant extracts. The development of these eco-friendly methods for the synthesis of nanoparticles is evolving into an important branch of nanotechnology especially silver nanoparticles, which have many applications.

Approaches to Nanoscale Structures

The two major approaches [5] to get nano materials are—one is the bottom up and the other is top down approach. Bottom up produce components which are made of single molecules, and covalent forces hold them together that are far stronger than the forces that hold together macro-scale components. Enormous amount of information could be stored in devices build from the bottom up. For example, use of AFM, liquid phase techniques based on inverse micelles, sol-gel processing, and chemical vapor deposition (CVD), laser pyrolysis and molecular self-assembly use bottom up approach for nano scale material manufacturing.

Top manufacturing involves the construction of parts through methods such as cutting, carving and molding and due to our limitations in these processes highly advanced nano devices are yet to be manufactured. Laser ablation, milling, nano-lithography, hydrothermal technique, physical vapor deposition and electrochemical method (electroplating) uses top down approach for nano-scale material manufacturing.

L. H. Madkour, *Nanoelectronic Materials*, Advanced
Structured Materials 116, https://doi.org/10.1007/978-3-030-21621-4_10

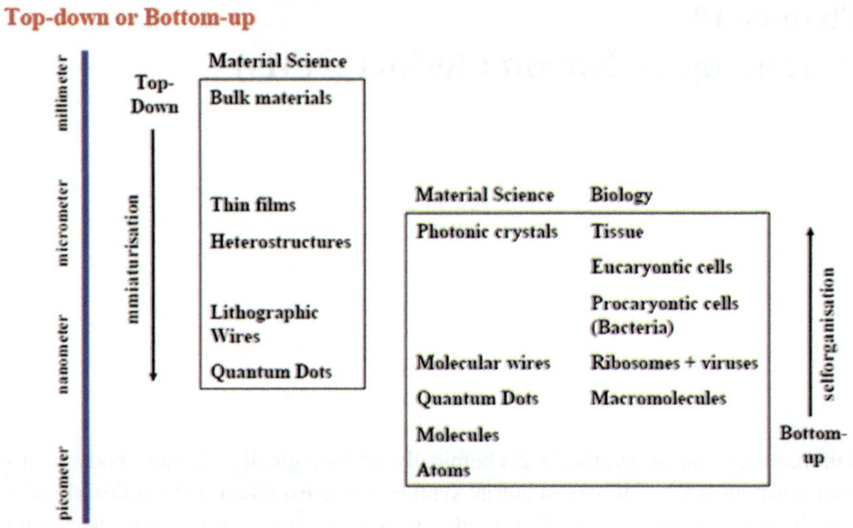

Every element of periodic table can be utilized in nanotechnology depending upon the target material which someone is going to fabricate range from nano medicine and goes up to nano concrete via nano electronics. Nanotechnology provides us the chance to synthesize nano scale building blocks with control on size, composition etc. Materials manufacturing will be revolutionized by further assembling into larger structures with designed properties. Without machining, metals, polymers, ceramics etc. can be manufactured at exact shape.

Methods to achieve nanostructures

Zero dimensional structures → nanoparticles, quantum dots
- **Precipitation in confined structures**
- **Physical treatment of targets**

One dimensional structures → tubes, wires
- **Electro deposition**
- **Chemical Vapour Deposition**

Two dimensional structures → wells, layers
- **Chemical Vapour Deposition**
- **Photolithography**

Three dimensional structures → photonic crystals, superlattices
- **Replacement reactions**
- **Crystallisation of nanocrystals**
- **DNA based assembly**

Driving force for particle growth

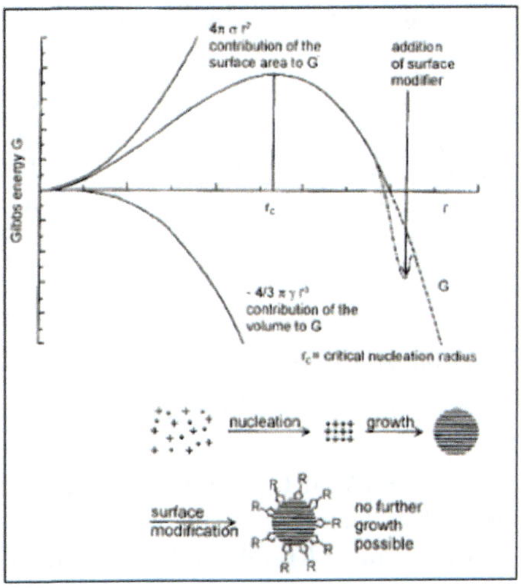

Gibbs energy minimization by reducing the surface/volume ratio

Surface energy increases quadratically, but volume energy decreases cubically

Very small particles (r_c) are highly reactive, due to their high chemical potential

<u>Comment</u>
5 nm particles ~ 50% surface atoms
1 μm particles ~ 0.6% surface atoms

Stabilisation of small colloidal particles

a) electrostatic **b) thermodynamic** **c) static**

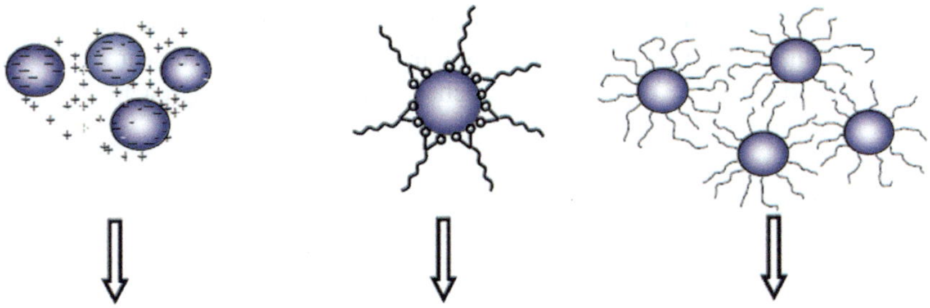

Surface charging | **Surface complexation (modification)** | **Adsorption of polymer molecules by the surface**

Thermordynamic stabilisation of nanoparticles

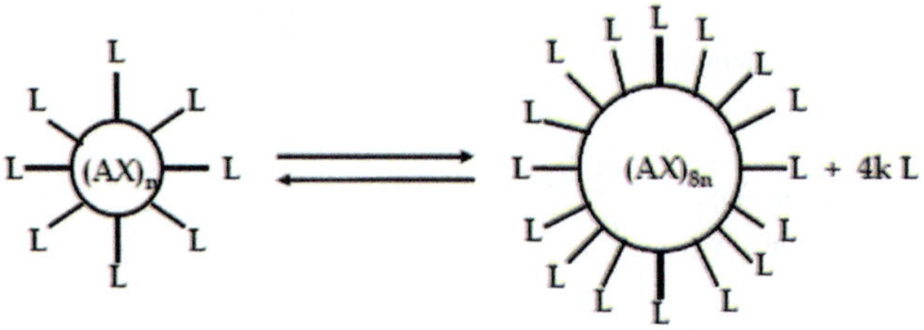

Ligandmoleküle fest an der Oberfläche gebunden
Energie-Bilanz bei der Verdopplung des Teilchenradius:

$(AX)_nL_k \rightarrow (AX)_{8n}L_{4k}$+ 4k L (free Ligands) ⇒ Cleavage of the metal-ligand bond required for particle growth

- **Strong metal-ligand bond** small cluster
- **Weak metal-ligand bond** large cluster

⇒ Cluster size scales with metal-ligand bond strength

Top-Down and Bottom-Up Approaches

The synthesis of nanomaterials is key to the future success of this new technology and in principle, the approaches to the synthesis of nanomaterials can be divided into two main classes: Top-down approaches and bottom-up approaches.

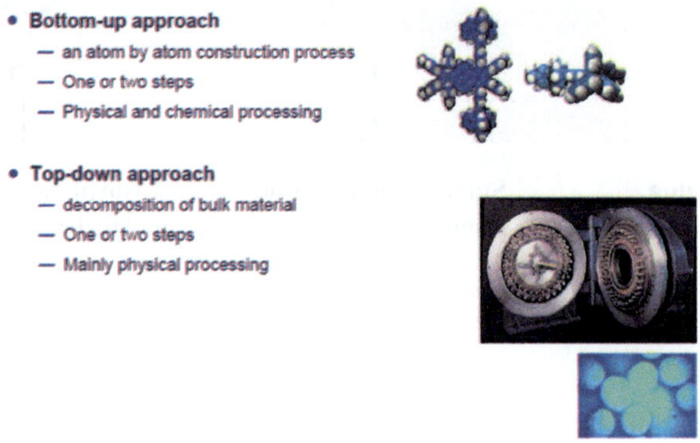

Nanomaterial's deal with very fine structures: a nanometer is a billionth of a meter. This indeed allows us to think in both the 'bottom up' or the 'top down' approaches (Fig. 10.1) to synthesize nanomaterials, i.e. either to assemble atoms together or to dis-assemble (break, or dissociate) bulk solids into finer pieces until they are constituted of only a few atoms. This domain is a pure example of interdisciplinary work encompassing physics, chemistry, and engineering up to medicine.

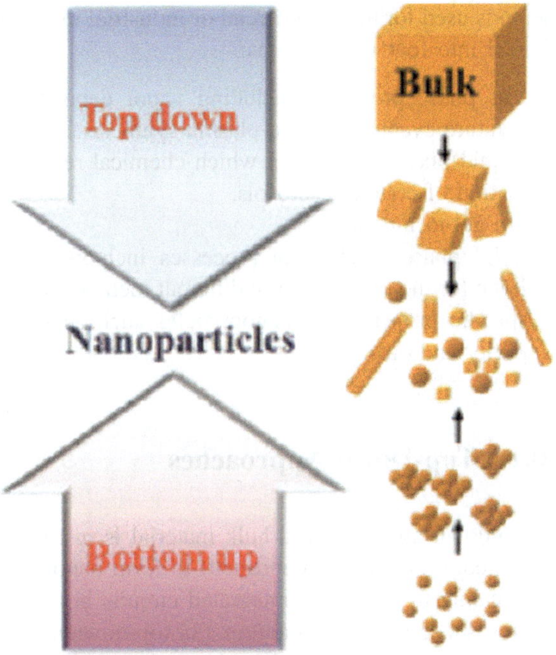

Fig. 10.1 Schematic illustration of the preparative methods of nanoparticles

There are two general approaches to the synthesis of nanomaterials and the fabrication of nanostructures: one is *the bottom-up approach*, that is the miniaturization of the components, as articulated by Feynman in his famous lecture in 1959 stated that "there is plenty of room at the bottom"; and *the top-down approach* of *the self-assembly* of molecular components, where each nanostructured component becomes part of a superstructure. Attrition or milling is a typical top-down method in making nanoparticles, whereas the colloidal dispersion is a good example of bottom-up approach in the synthesis of nanoparticles. Bottom-up, or self-assembly, approaches to nanofabrication use chemical or physical forces operating at the nanoscale to assemble basic units into larger structures. Bottom-up approaches seek to have smaller components arrange themselves into more complex assemblies, while top-down approaches seek to create nanoscale devices by using larger, externally-controlled ones to direct their assembly. Lithography may be considered as a hybrid approach, since the growth of thin films is bottom-up whereas etching is top-down, while nanolithography and nanomanipulation are commonly a bottom-up approach. The purpose of this chapter is to introduce the methods of the production of ultra-fine particles, which form the building blocks of nanostructured materials. Most of the time, nanoparticles are embedded into a matrix to obtain desirable properties of material, or nanoparticles themselves are processed into a film or a coating. Nanoparticles can be synthesized in many different ways, both in chemical reactions and physical processes. Most common methods used for the commercial or industrial manufacture of nanoparticles may be divided into four main groups:

- Gas phase processes including vapor deposition, flame pyrolysis, high temperature evaporation and plasma synthesis.
- Liquid phase methods in which chemical reactions in solvents lead to the formation of colloids, aerosols.
- Sol-gel technique.
- Solid phase mechanical processes including grinding, milling and alloying. Even for the same material different methods are often used in order to optimize specific properties of nanoparticles such as size, size distribution, symmetry, purity and others.

10.1 Top-Down Approaches

In top-down approaches, a bulk material is restructured (i.e. partially dismantled, machined, processed or deposited) to form nanomaterials (Fig. 10.2). The aggressive scaling of electronic integrated circuits in recent years can be considered the greatest success of this paradigm. For top-down methods, the challenges increase as devices size is reduced and as the desired component designs become larger and more complex. Also, the top-down assembly of nanocomponents over large areas is difficult and expensive.

Patterned chemical etching is a top-down or hybrid top-down/bottom-up approach that involves a lithography step followed by an etch step. Figure 10.3

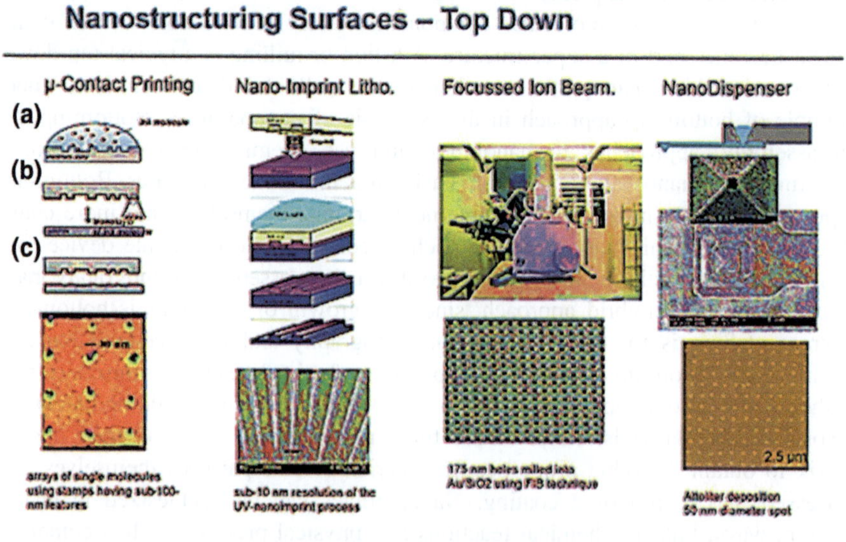

Fig. 10.2 Top-down nanosynthesis methods. Image courtesy of CSEM–International Conference on Mechatronics and Automation (ICMA), Beijing, China, August 2011

shows the fabrication steps for one possible hybrid scheme and images of the resulting nanowire array. The initial lithography step is either a top-down approach such as e-beam and nanoimprint or a spontaneous bottomup approach such as nanosphere assembly, anodic alumina, and block copolymer formation [6–10]. The patterned substrate is dissolved with a directional, substrate-selective etch such as deep reactive-ion etching (DRIE) or metal-assisted chemical etching [10–13]. The nanosphere assembly patterning approach is particularly useful, as it provides a method for controlling the nanowire diameter and pitch independently simply by changing the sphere diameter and etch time prior to the substrate etch step [12]. Silica and polymer spheres can be synthesized with a wide range of diameters (typically 100–1,000 nm) and can be assembled over large areas using Langmuir-Blodgett, dip-coating, and roll-coating methods [12, 14–16]. Anodic alumina and block copolymer templates can access much smaller pattern dimensions (typically 10–100 nm) and have been used to fabricate nanowires with diameters of $\sim$10 nm [9, 13]. The patterned chemical etching technique has the advantage that the starting wafer or thin film sets the doping level and material composition, which allows for more precise control over the material parameters and simplifies the material characterization.

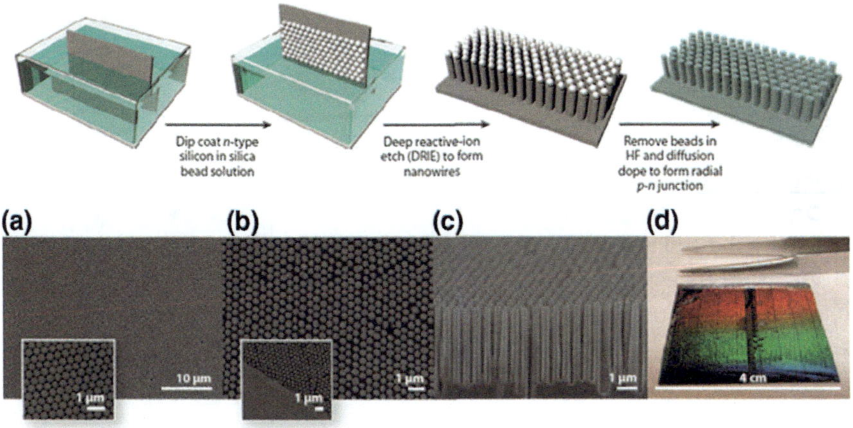

Fig. 10.3 Nanowire solar cell fabrication. (*Top row*) Schematic of the fabrication process. HF denotes hydrofluoric acid. **a** Scanning electron microscopy (SEM) image of a close-packed monolayer of silica beads assembled on a silicon wafer using dip coating. **b** Plan-view SEM image of a completed ordered silicon nanowire radial *p-n* junction array solar cell made by bead assembly and deep reactive-ion etching. The inset shows the edge of a top contact finger, demonstrating that the metal completely fills in between the nanowires. **c** Tilted cross-sectional SEM of the solar cell in panel **b**. **d** Tilted optical image of 36 silicon nanowire radial *p-n* junction solar cell arrays from panels **a**–**c**. The color dispersion demonstrates the excellent periodicity present over the entire substrate. Adapted with permission from Ref. [14]. Copyright_c 2010, American Chemical Society

10.1.1 *Ball Milling: Mechanical Crushing of Solids into Nanocrystallites*

Top-down: Mechanical crushing of solids into nanocrystallites

L

Advantages

- Inexpensive
- Large scale process
- Old well-established process
- Down to 2 – 20 nm possible

Disadvantages

- Irregular nanoparticles
- Introduction of defects
- Introduction of impurities from balls and milling additives

A ball mill being used as part of a gold mining operation in Peru

Top-down: Mechanical crushing of solids into nanocrystallites

L

Milling process

- Dry
- Dry + solid additive (salt)
- Wet (slurry)

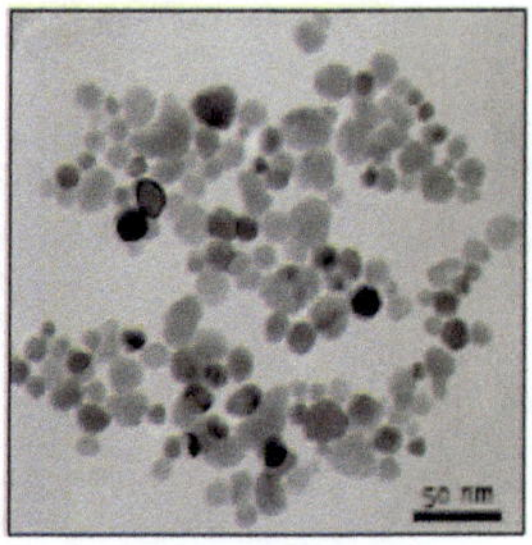

ZnO nanoscale particles from a milling process

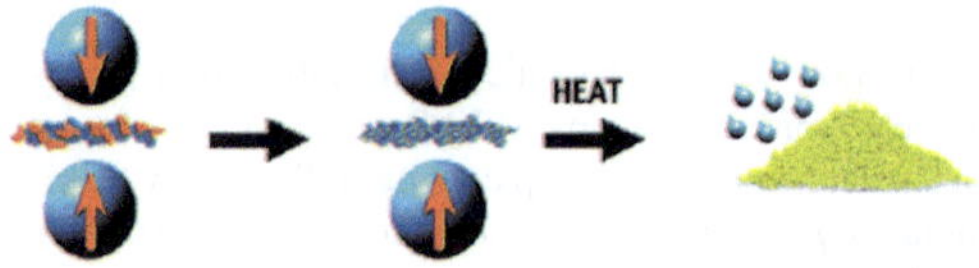

Dry milling salt added washed product

10.1.2 Photolithography

Top-down: Photochemical manufacturing of micro- and nanostructures for integrated circuits

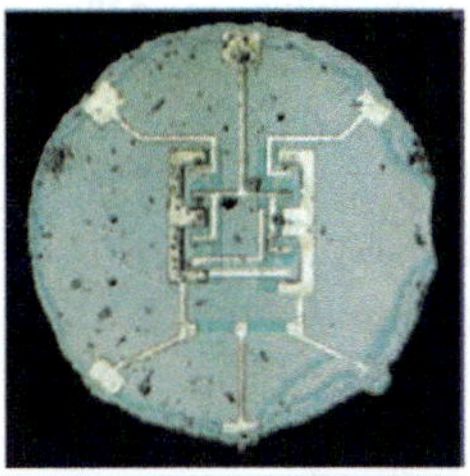

History of integrated circuit development

1947	**Invention of the transistor**
1961	**First integrated circuit**
1965	**Moore's law published**
1975	**Intel 8080 chip: 4500 transistors**
1981	**„640 kByte ought to be enough for anybody" (Bill Gates)**
1993	**Pentium I: 3.1 Mio. transistors**
1997	**Pentium II: 7.5 Mio. transistors 250 nm line width**
2000	**180 nm line width**
2002	**130 nm structures**
2005	**45 nm technology**

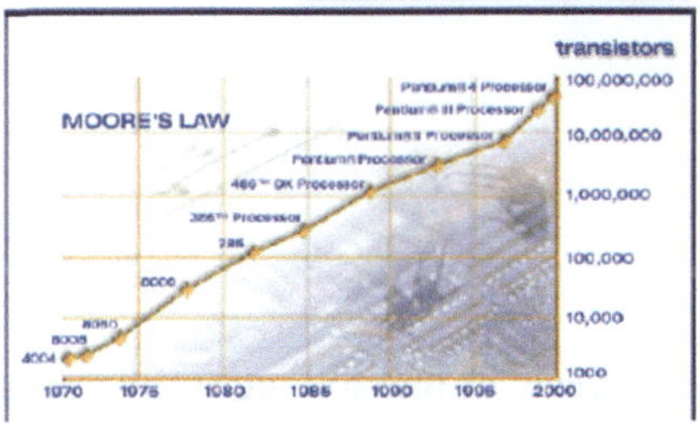

Process

1. **Photoresist layer deposition by spin coating**
2. **Exposure to UV Radiation**
3. **Removal of photoresist**
4. **Etching of unprotected areas**
5. **Doping by ion implantation**
6. **Repeat cycle to deposit metal connector**

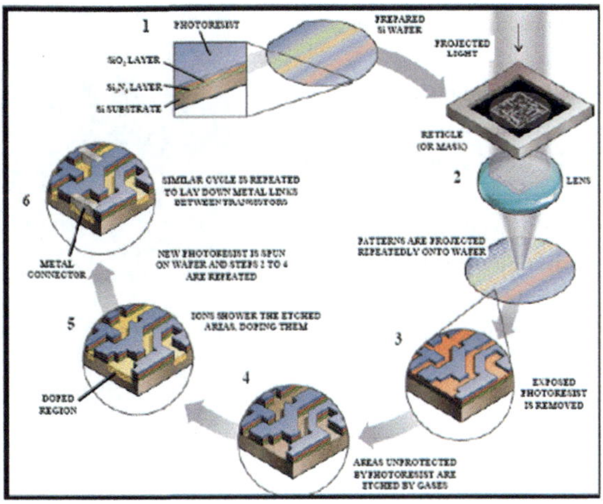

Evolution of the lithography process

Stepper	λ [nm]	Radiation source
Visible g-line	436	high-pressure Hg
Visible i-line	365	high-pressure Hg
High deep UV	248	KrF Excimer LASER
Deep UV	193	ArF Excimer LASER
Deep UV	157	F_2 Excimer LASER
Extreme UV	13.7	
Electron beam		

Schematic process flow

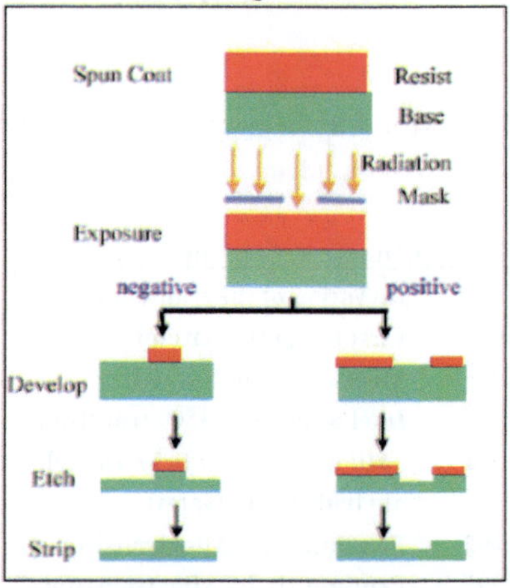

Limitations of photolithography and miniaturisation of transistors

Technical limitations

- Wavelength of radiation source
- Focussing of radiation

Fundamental limitation

- Typically a semiconductor of 1000 nm³ comprises one free electron
- Extrapolation of the present development shows that "single electron transistors" will be reality at ~ 2020

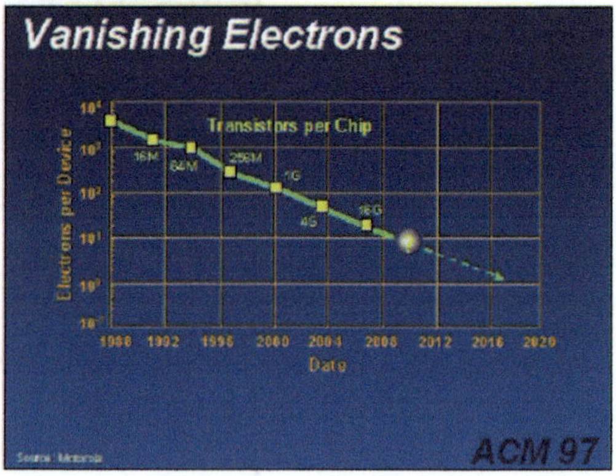

10.1.3 Gas Phase Processes

Top-down: LASER-Ablation for the synthesis of Carbon Nano Tubes (CNTs)

The vaporization of a target at a fixed temperature by a continuous CO_2 laser beam ($\lambda = 10.6\ \mu m$) is shown in the bottom figure.
The power can be varied from 100 W to 1600 W.
The temperature of the target is measured with an optical pyrometer.
These measurements are used to regulate the laser power to maintain a constant vaporization temperature.
The gas, heated by the contact with the target, acts as a local furnace and creates an extended hot zone, making an external furnace unnecessary.
The gas is extracted through a silica pipe, and the solid products formed are carried away by the gas flow through the pipe and then collected on a filter.

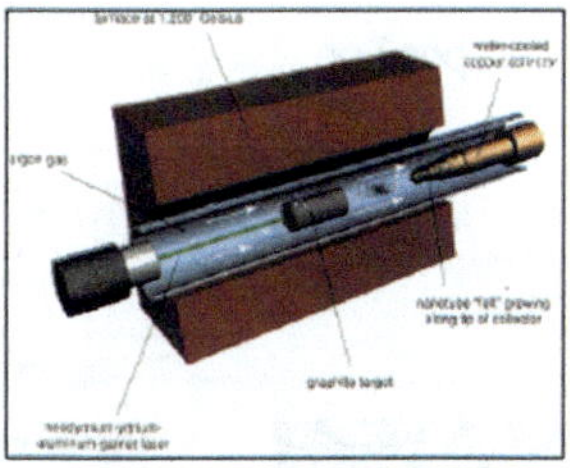

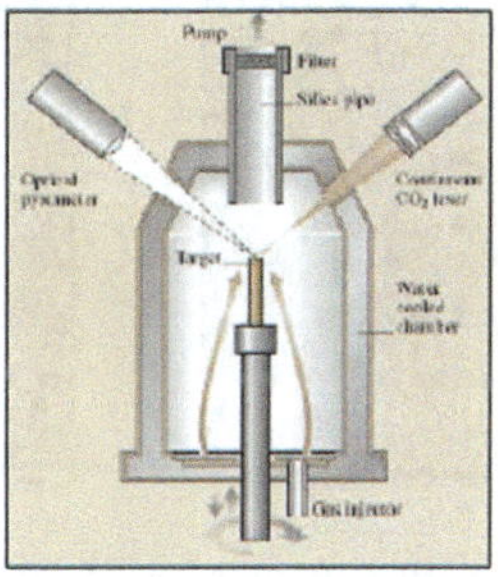

10.2 Bottom-Up Approach

In bottom-up approaches nanomaterials are assembled from basic building blocks, such as molecules or nanoclusters (Fig. 10.4). The basic building blocks, in general, are nanoscale objects with suitable properties that can be grown from elemental precursors. The concept of the bottom-up paradigm is that the complexity of nanoscale components should reside in their self-assembled internal structure, requiring as little intervention as possible in their fabrication from the macroscopic world.

A typical example of bottom-up processing is given in Fig. 10.5 for nanocomposite magnets from individual high-magnetization and high-coercivity nanoparticles.

The assembling critically depends on availability of anisotropic (single crystal) hard magnetic nanoparticles. Anisotropic nanoparticles produced via surfactant-assisted high energy ball milling satisfy the major requirements for this application.

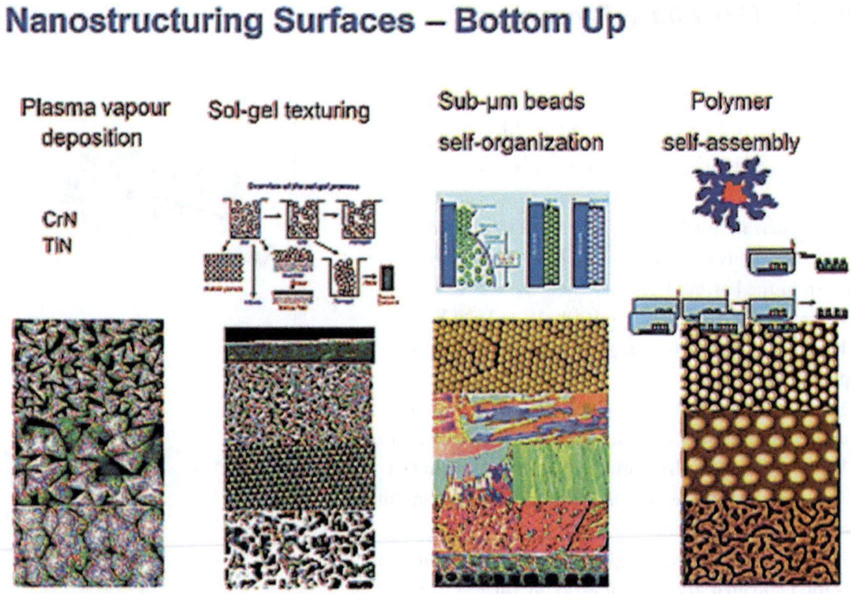

Fig. 10.4 Bottom-up methods for nanosynthesis. Image courtesy of CSEM—International Conference on Mechatronics and Automation (ICMA), Beijing, China, August 2011

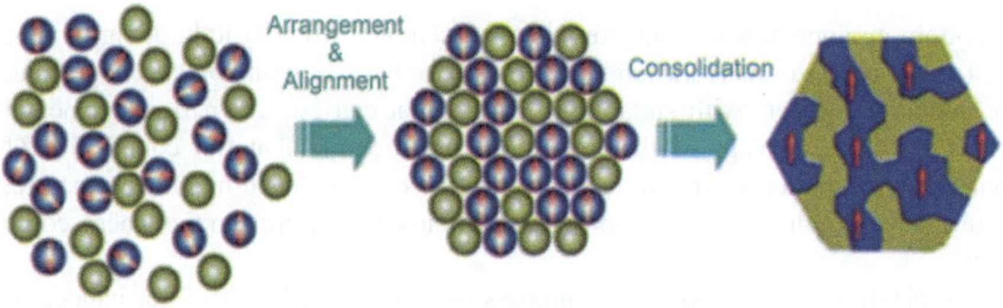

Fig. 10.5 Bottom-up fabrication of nanocomposite magnets. Image courtesy of P. Perlo Torino e-district, IFEVS—GENNESYS Barcelona conference 2010

10.2.1 Gas Phase Processes

Bottom-up: Chemical Vapour Deposition (CVD)

Advantages of CVD

- **Ease of control of layer thickness**
- **Good layer homogeneity**
- **„Universal" process**

CVD parameters

- **Volatility of precursor**
- **Ease of decomposition & volatility of fragments**
- **Relative concentration**
- **Catalyst on target surface (e.g. Ni or Co)**
- **Crystallographic arrangement of surface**
- **Process temperature**
- **Gas pressure**
- **…..**

Bottom-up: Chemical Vapour Deposition (CVD)

Chemical process for depositing thin solid films of various materials onto a substrate

Deposition material: **Diamond, metals, oxides, nitrides, borides, carbides …**
Substrate: **Si, SiO_2, Si_3N_4, Al_2O_3 …**

Application areas

- **Coating of tools (W, Diamond)**
- **Metal contacts on semiconductors (Cu, Ag)**
- **Transparent conductors onto glass (ITO)**
- **Insulation layers (SiO_2)**
- **Composite materials (Al)**
- **Preparation of semiconductors**
  - **(Al,In,Ga)N**
  - **(Al,In,Ga)P**
  - **Ga(As,P)**

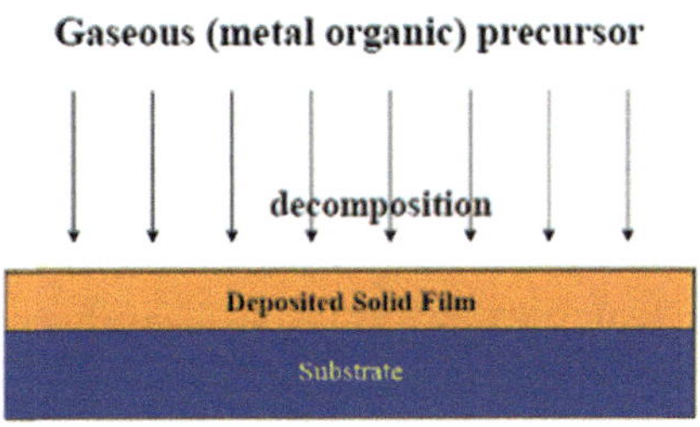

Bottom-up: Chemical Vapour Deposition (CVD) of Platinum layers

Precursor: $[Pt^{II}(hfac)_2]^0$ **hfac = hexafluoroacetylacetonate**

Reductive thermal decomposition:
$[Pt^{II}(hfac)_2]^0 + H_2 \rightarrow Pt + 2\ H_2fac$

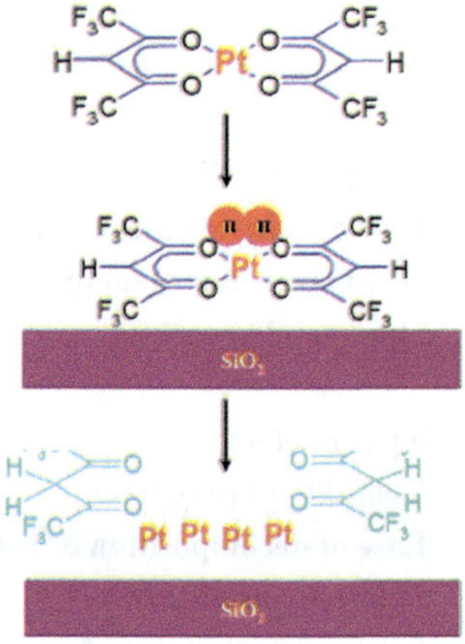

Result

Bottom-up: Other CVD Methods

Decomposition due to	Method
LASER beam	Laser assisted CVD
(Microwave) plasma	Plasma enhanced CVD
High temperature	Thermal CVD, fluid bed CVD

.....

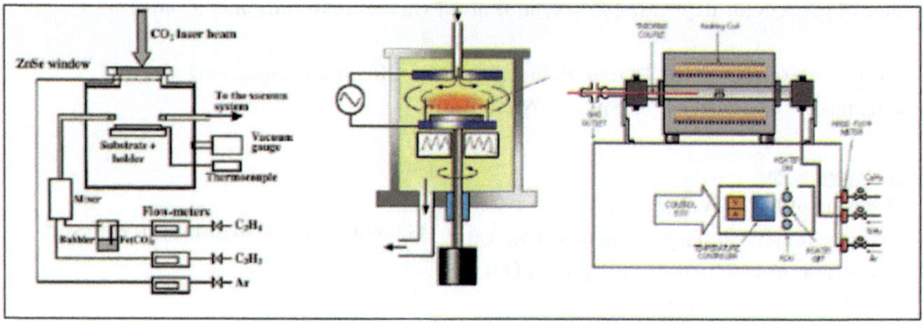

Bottom-up: CNT growth synthesis by DC Plasma Chemical Vapour Deposition

Bottom-up: CNT growth synthesis by DC Plasma Chemical Vapour Deposition

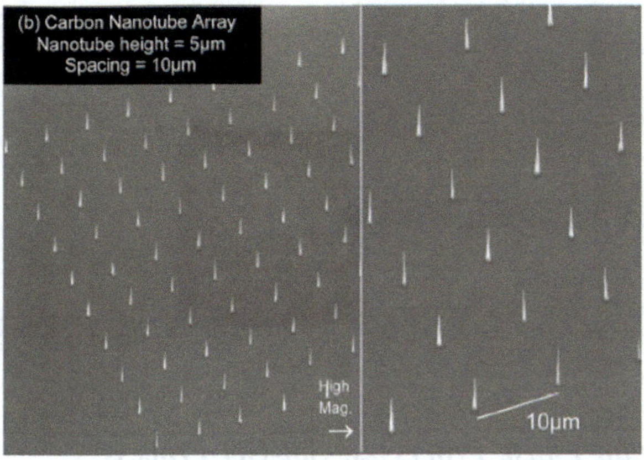

W. Milne, J. Robertson, K. Teo, Cambridge U. UK
(FP5-EU-Program CARDECOM Partner)

Bottom-up: CNT growth synthesis by DC Plasma CVD

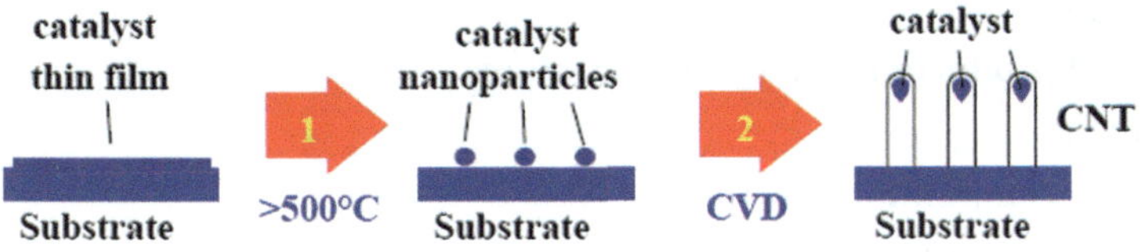

1. Deposition of a catalyst film (Ni, Fe, Co)

2. Formation of catalyst nanoparticles at 500° - 900°C (growth temp.): Catalyst film breaks into nanoparticles

3. Plasma CVD: C-carrier (CH_4, C_2H_2) provides C for tube growth, etchant (NH_3, N_2, H_2) removes unwanted amorphous carbon

Bottom-up: CNTs grown by DC Plasma CVD

Source: Peter K. Bachmann
Philips Research Laboratories Aachen

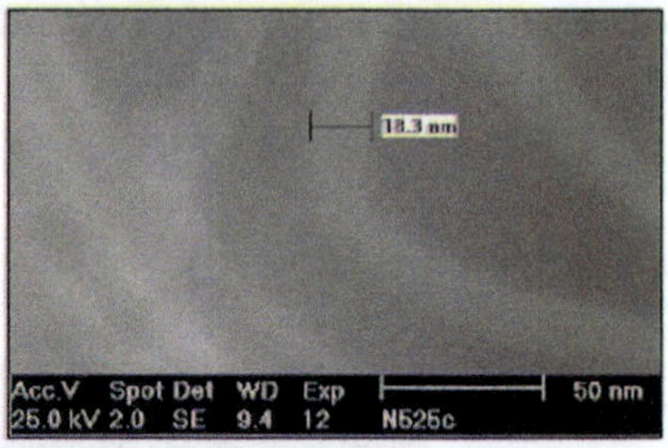

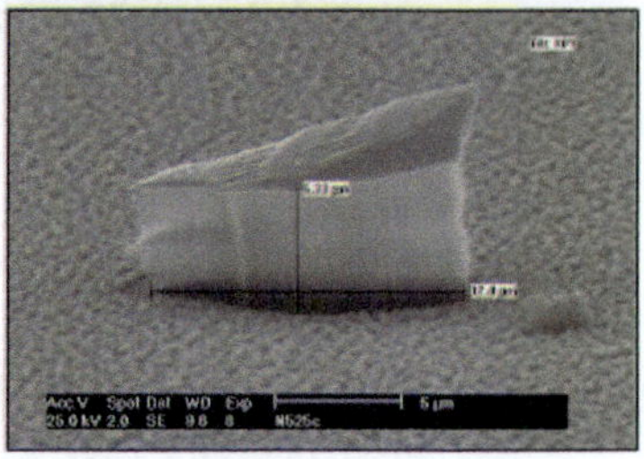

10.2.2 Liquid Phase Processes: Sol-Gel Process

Bottom-up: Sol-Gel Process

Advantages

- **Ease of production of large area coatings**
- **Scalable**
- **Precise composition control**
- **Low temperature synthesis**
- **High homogeneity**
- **Tunable layer composition**

Disadvantages

- **Sensitivity for atmosphere condition**
- **Cost of raw materials**
- **Use of toxic solvent system**

Bottom-up: Sol-Gel Process

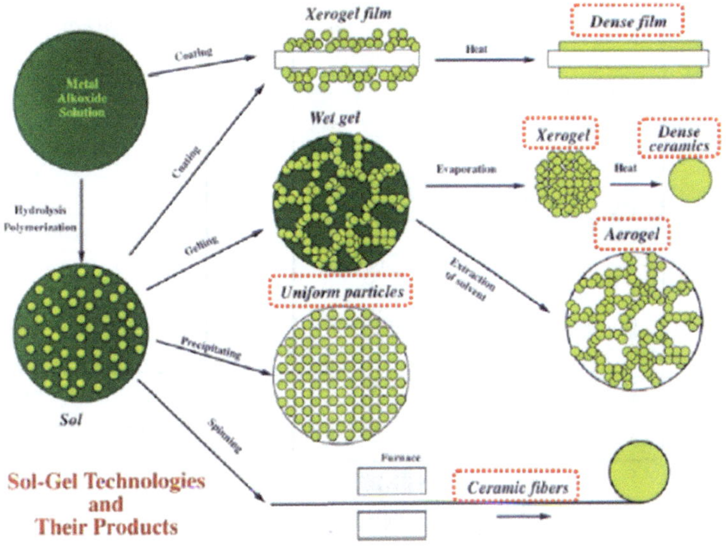

Bottom-up: Sol-Gel Process

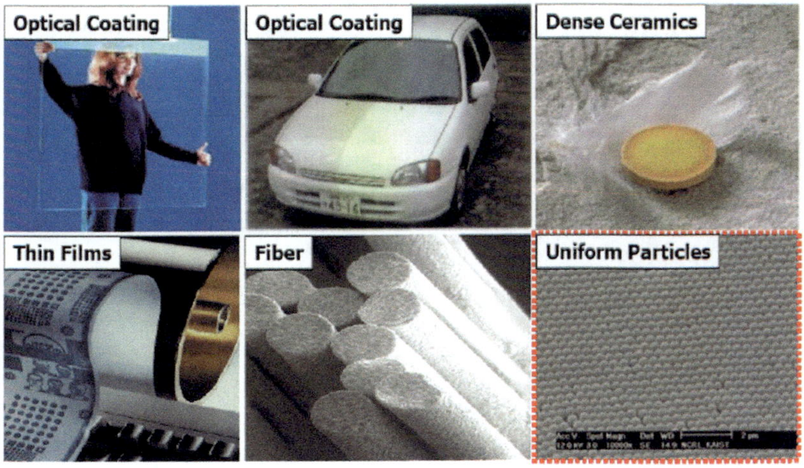

Bottom-up: Sol-Gel Process

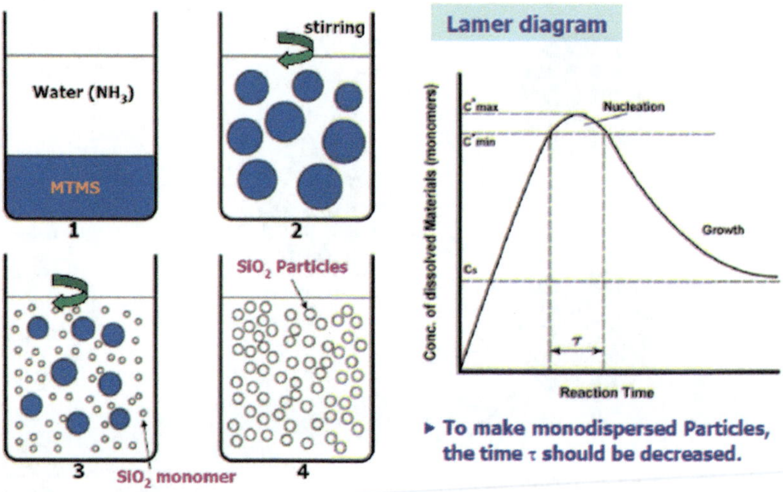

MTMS = Methyltrimethoxysilane

Bottom-up: Sol-Gel Process

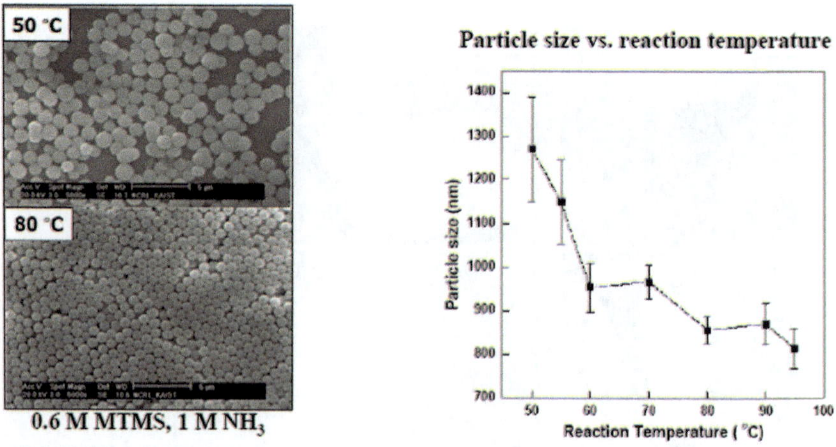

0.6 M MTMS, 1 M NH_3

Temperature↑ ⇒ Reaction rate↑ ⇒ Nucleation time↓ ⇒ Size Distribution↓

10.2.3 *Liquid Phase Processes: Synthesis of Metal Nanoparticles*

Bottom-up: Sol-Gel Process „Stöber Process“

Photonic crystal out of SiO_2 spheres (Opal)

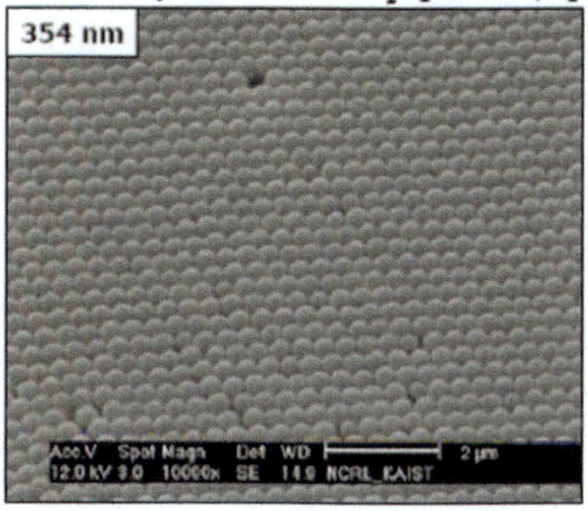

Reflection spectra = $f(d_{111})$

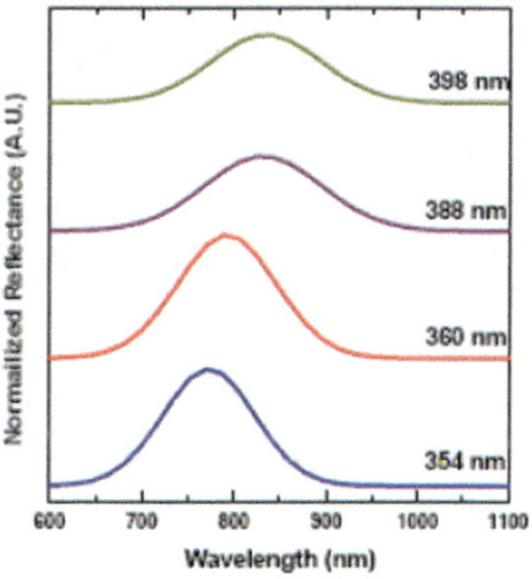

Bragg's law: $\lambda_{Bragg} = 2n_{eff} \cdot d_{111}$

with $n_{eff} = \sqrt{n_{SiO_2}{}^2 \cdot f + n_{air}{}^2 \cdot (1-f)}$

and $d_{111} = (2/3)^{1/2} \cdot D$, f = filling ratio

Bottom-up: Sol-Gel Process „Stöber Process“

$H_2O/EtOH = 0.2$, $0.7\ M\ NH_3$, $22°C$

TEOS Concentration↑ ⇒ Particle size↑

Particle size vs. TEOS concentration

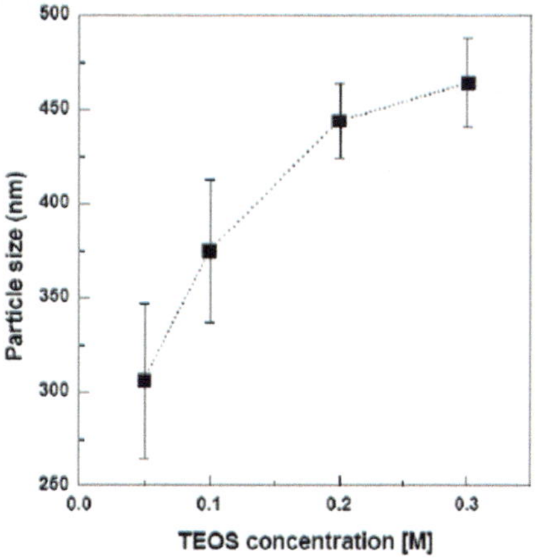

Bottom-up: Synthesis of metal nanoparticles

Application of metal nanoparticles

- **Electronic materials, e.g. Ni**
- **Magnetic materials, e.g. Fe**
- **Catalysts, e.g. Pt**
- **Explosives, e.g. Al**
- **Powder metallurgy, e.g. Cr**
- **Photographic films, e.g. Ag**

Synthesis approach: Reduction of metal salts

(a) In organic solvents (non noble metals)

$Co^{2+} + 2\ BH_4^- \rightarrow Co + H_2 + B_2H_6$ in Dyglyme

$AlCl_3 + 3\ K \rightarrow Al + 3\ KCl$ in Xylene

(b) In water (noble metals, i.e. Rh, Ir, Pd, Pt, Cu, Ag, Au)

$6\ HAuCl_4 + HOC(COOH)(CH_2COOH)_2 + 24\ OH^- \rightarrow 6\ Au + 24\ Cl^- + 6\ CO_2 + 19\ H_2O$

Bottom-up: Synthesis of metal nanoparticles

Example: Manufacturing of multi-layer ceramic capacitors (MLCCs)

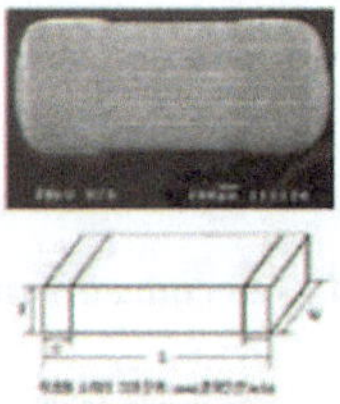

Commercial Ni nanoscale powders

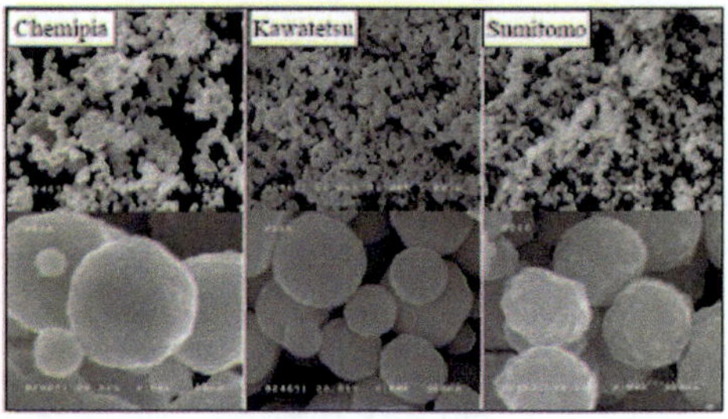

Ongoing improvement of small high capacitance MLCCs require a continuous reduction of

- **dielectric layer thickness**
- **internal Ni electrode layer thickness (80 ± 20 nm Ni particles)**

Bottom-up: Synthesis of metal nanoparticles

Reduction of Ni^{2+} in solution

$2\,Ni^{2+} + N_2H_4 + 4\,OH^- \rightarrow 2\,Ni^0 + N_2 + 4\,H_2O$ **Hydrazine = Reducing agent**

Result

Particles with Random shape and size and a broad broad size distribution

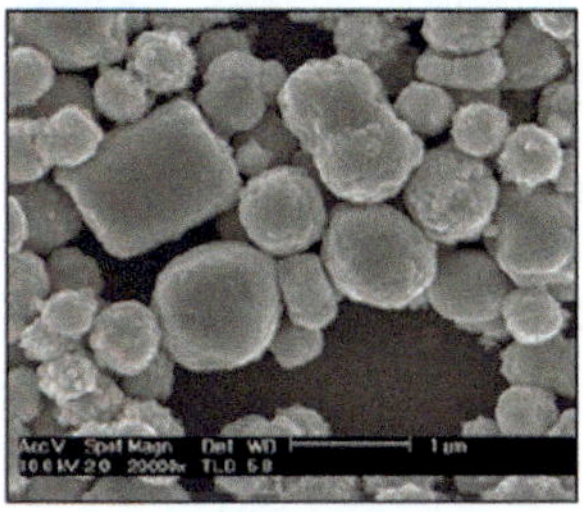

Advanced precipitation processes

- **Application of an additive (dispersant, complex agent)**
- **Heterogeneous nucleation (seed)**

Bottom-up: Synthesis of metal nanoparticles

Guideline for the reducing agent

Metal	E^0 [V] vs. SHE	Reducing agent	Conditions
Au, Pd, Pt, Ag	> + 0.7	organic acids, ROH, polyols	> 70°C
Rh, Ir, Hg		aldehyds, sugars	< 50°C
		hydrazine, H_2SO_3, H_3PO_2	ambient temp.
		$NaBH_4$, boranes, hydrated e^-	ambient temp.
Cu, Re, Ru	< + 0.7 > 0.0	polyols	> 120°C
		aldehydes, sugars (Fehling sol.!)	> 70°C
		hydrazine, hydrogen	< 70°C
		$NaBH_4$	ambient temp.
Cd, Co, Ni, Fe	< 0.0 > - 0.6	polyols	>180°C
In, Sn, Mo, W		hydrazine, hydroxylamine	70 – 100°C
		$NaBH_4$, boranes, hydrated e^-	ambient temp.
Cr, Mn, Ta, Y	< - 0.6	$NaBH_4$, boranes	T, P > ambient
		hydrated e^-	ambient temp.

Bottom-up: Synthesis of metal nanoparticles

How to control metal particle size?

- **Processing parameter**
  - **Temperature, pressure, concentration, pH, ultrasound application, …**
- **Seed (heterogeneous nucleation)**
  - **Type and concentration of applied seed**
- **Complex agent**
  - **Complex stability**

Redox potential of Ag^+ as function of the presence of a complexing agents

Redox system	-log K_B	E^0 [V] vs. SHE
$Ag^+ + e^- \rightarrow Ag^0$	-	+ 0.799
$[Ag(NH_3)_2]^+ + e^- \rightarrow Ag^0 + 2\ NH_3$	7.2	+ 0.38
$[Ag(SO_3)_2]^{3-} + e^- \rightarrow Ag^0 + 2\ SO_3^{2-}$	8.7	+ 0.29
$[Ag(S_2O_3)_2]^{3-} + e^- \rightarrow Ag^0 + 2\ S_2O_3^{2-}$	13.4	+ 0.01
$[AgI_4]^{3-} + e^- \rightarrow Ag^0 + 4\ I^-$	15.0	- 0.09
$[Ag(CN)_3]^{2-} + e^- \rightarrow Ag^0 + 3\ CN^-$	22.2	- 0.51

Bottom-up: Synthesis of metal nanoparticles

Heterogeneous nucleation

(a) $Ag^+ + e^- \rightarrow Ag^0$(seed)

(b) Ag^0(seed) + $Ni^{2+} + 2e^- \rightarrow Ni^0$ (nanoscale particle)

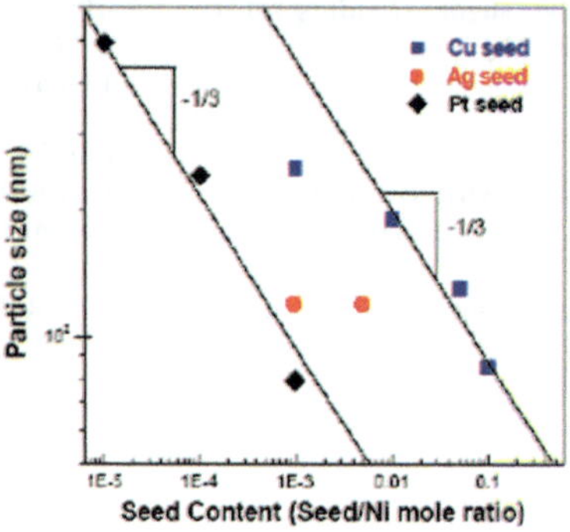

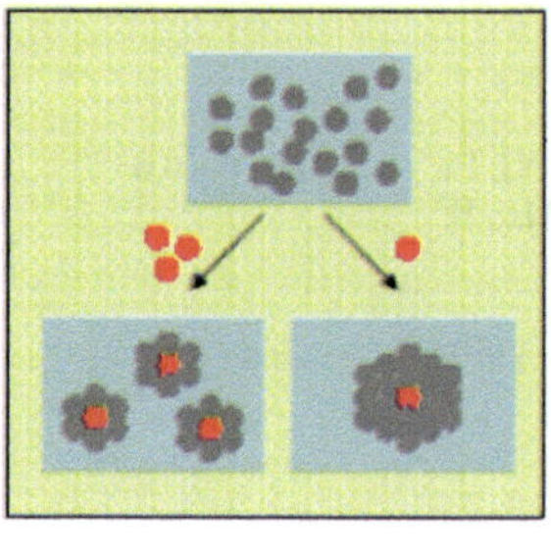

$$d_m = A \cdot [Seed]^{-1/3}$$

Bottom-up: Synthesis of metal nanoparticles

Heterogeneous nucleation

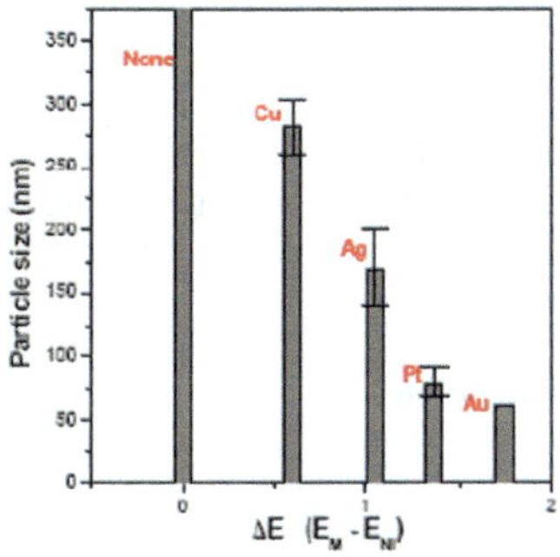

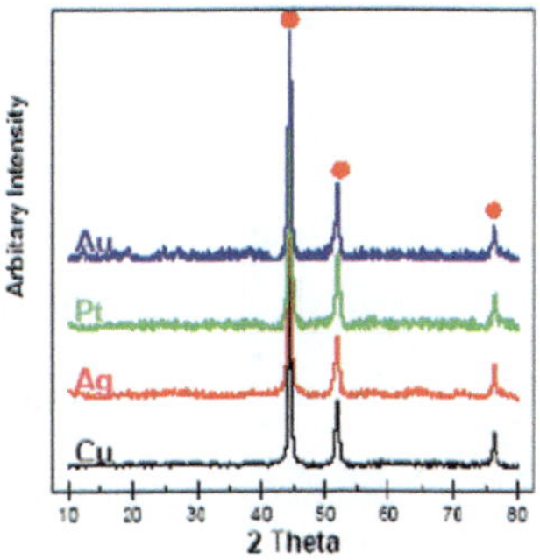

Seed	E^0 [V] vs. SHE	Ni particle size [nm]
Cu	+0.34	282
Ag	+0.80	170
Pt	+1.18	90
Au	+1.50	60

Bottom-up: Synthesis of metal nanoparticles

Formation mechanism of seed-mediated growth for Ag and Au rods

(a) Ag- or Au-salt + $NaBH_4$ + citrate → seed

(b) seed + metal salt + ascorbic acid + CTAB → nanorods
(CTAB = Cetyltrimethylammonium bromide)

Chemical	Role
$NaBH_4$	strong reducing agent
Citrate	capping agent
Ascorbic acid	weak reducing agent
CTAB	rod like template

Decreasing the seed concentration increases the aspect ratio and the color (absorption edge) of nanorods

Aspect ratio increases from 1:1 to 1:10 (left graphs)

Ag nanorods

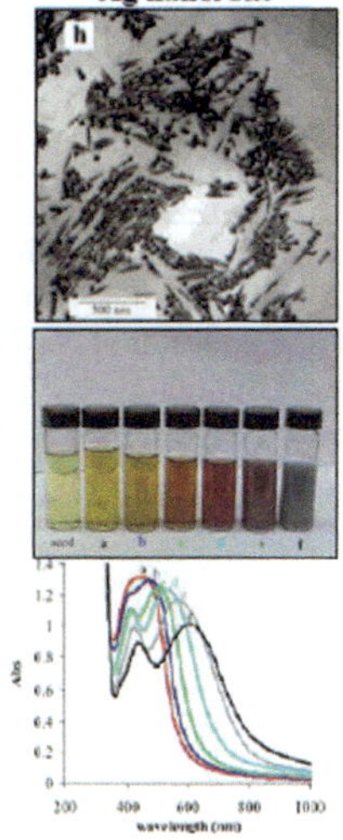

Bottom-up: Synthesis of metal nanoparticles

Polyol method

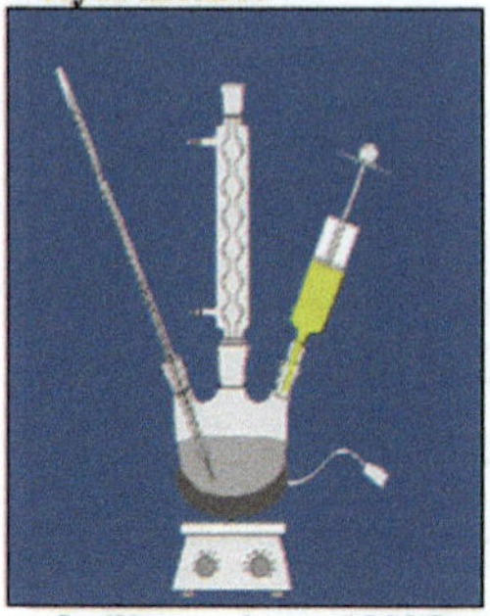

boiling point > 120°C

e.g. ethylen glycol

<u>Synthesis pathway</u>

1. Dissolve soluble metal salts, e.g. acetates in pure alcohol (reducing agent) with a rather high boiling point

2. Boil for several hours

3. Separate by centrifugation

4. Wash by alcohol

→ Au, Pt, Pd, Ag, Rh, Hg, Ir, Cu, Re, Ru, Cd, Co, Ni, Fe, In, Sn, Mo, W, ...

Bottom-up: Synthesis of phosphate nanoparticles

<u>Synthesis in alcohols (van Veggel)</u>

1. **$LnCl_3$ and trisalkylphosphate in methanol**
2. **Addition of H_3PO_4**
3. **Addition of trioctylamine to deprotonate phosphoric acid**

⇒ **Formation of $LnPO_4$ nanoscale particles with alkyl phosphate capping**

<u>Synthesis in highly boiling polyols (Haase)</u>

1. **$Ln(ac)_3$ in ethylen glycol**
2. **Addition of trioctylphosphinoxide**
3. **Addition of Na_2HPO_4**
4. **Boiling**

⇒ **Formation of $LnPO_4$ nanoscale particles with TOPO capping**

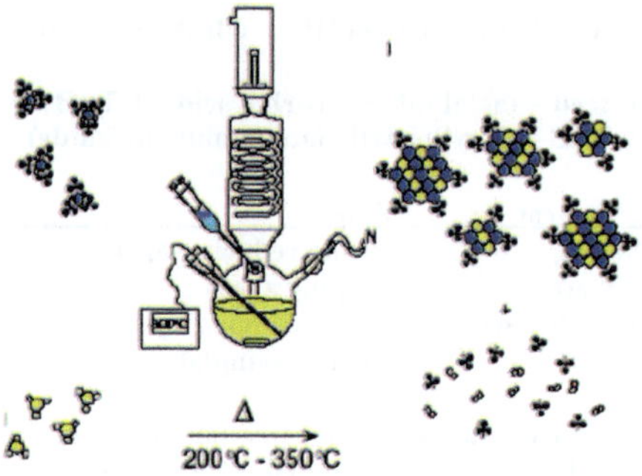

Bottom-up: Synthesis of phosphate nanoparticles

Synthesis in water (Merikhi, Bachmann, Jüstel et al.)

1. **$Ln(ac)_3$ in acidic water**
2. **Addition of complexing agent**
3. **Addition of an excess of citric acid**
4. **Addition of $H_2PO_4^-$**
5. **Enhance pH to 9–10 and temper at 80–90°C**

⇒ **Formation of $LnPO_4$ nanoscale particles with citric acid capping**

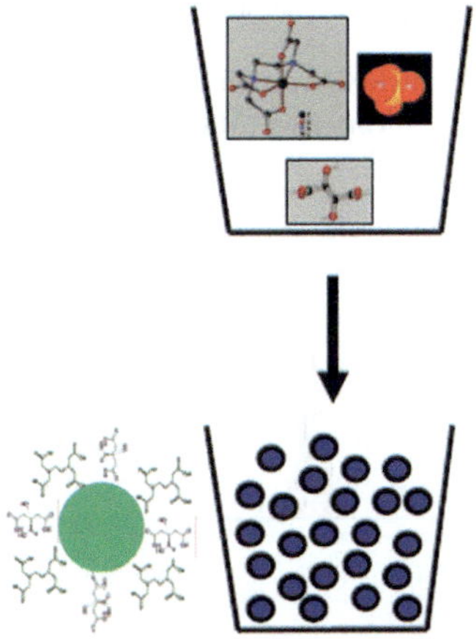

Bottom-up: Synthesis of phosphate nanoparticles

Example: Colloidal YPO_4:Tb^{3+} nanoscale particles in aequous solution

Particle size distribution by dynamic light scattering (Malvern Nanosizer)

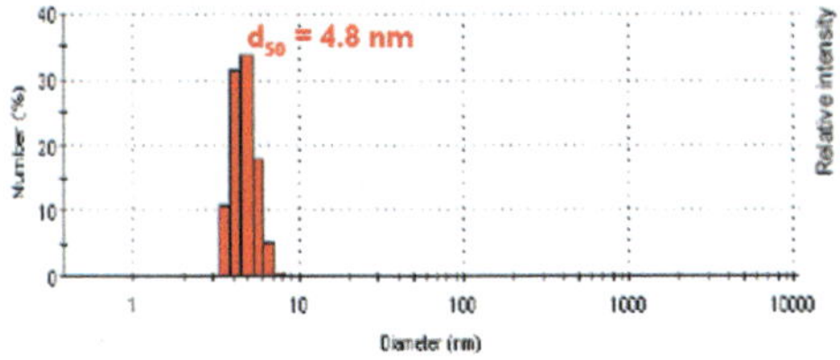

Emission and excitation spectrum

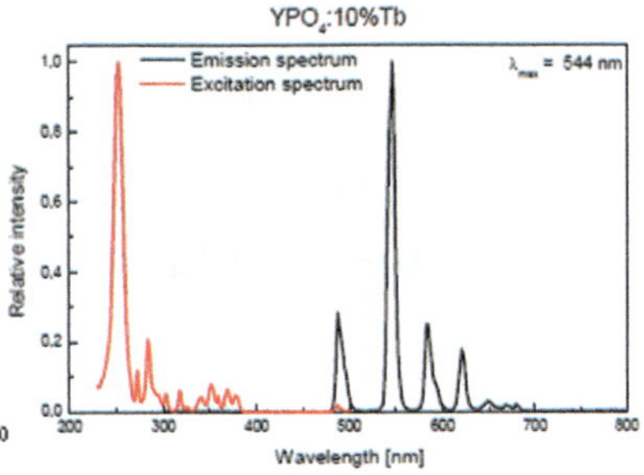

Bottom-up: Synthesis of phosphate nanoparticles

Example: Colloidal $YPO_4:Eu^{3+}$ nanoscale particles in aequous solution

Particle size distribution by dynamic light scattering (Malvern Nanosizer)

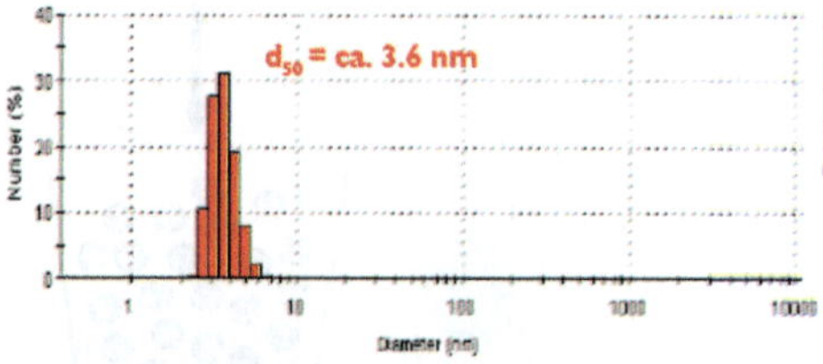

Emission and excitation spectrum

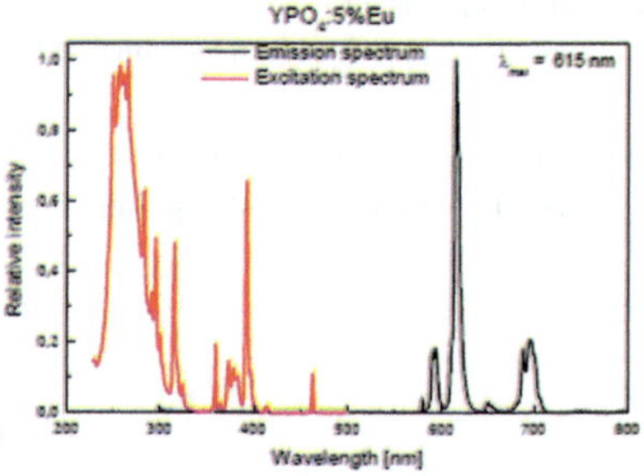

Bottom-up: Synthesis of phosphate nanoparticles

YPO_4:Pr and $LuPO_4$:Pr as 235 nm emitter

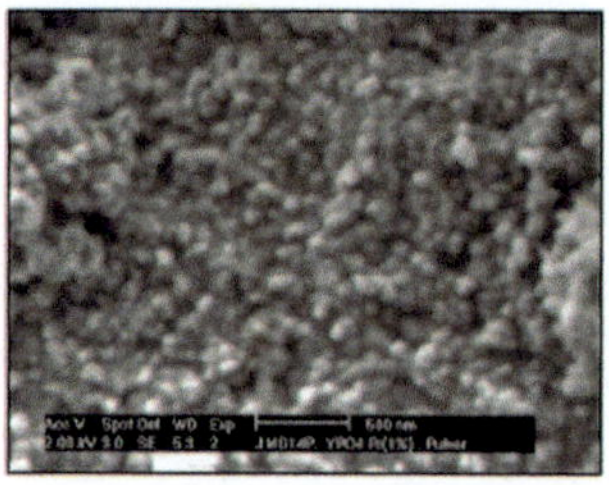

YPO_4:Pr (30 nm particles)

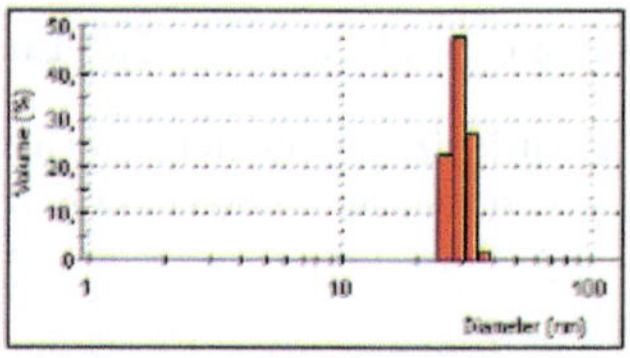

$LuPO_4$:Pr (5 nm particles)

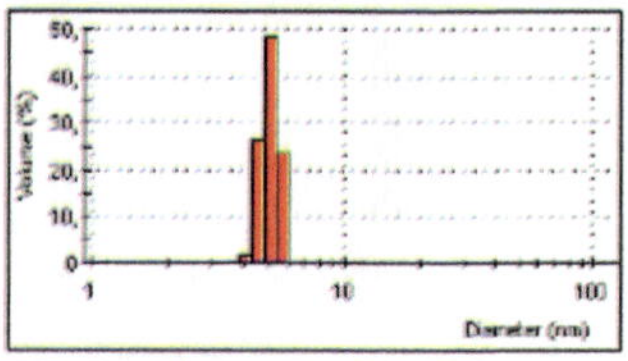

Further examples: $GdPO_4$:Pr, $LaPO_4$:Pr, YPO_4:Ln (Ln = Ce, Gd, Nd)

Bottom-up: Synthesis of phosphate nanoparticles using rod like template

YPO_4:Bi Nanorods | **YPO_4:Ce Nanorods**

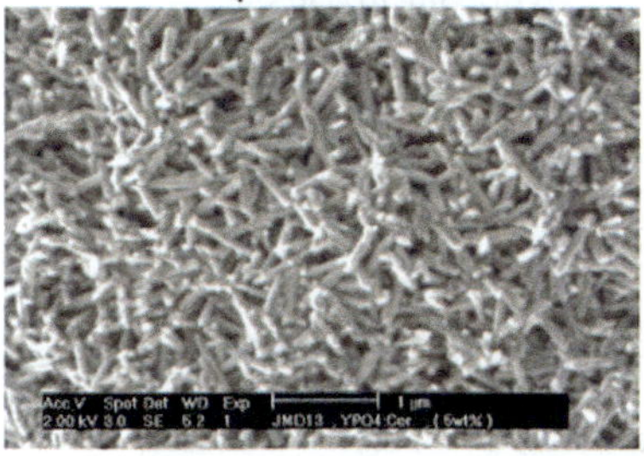

Application areas

- **Precursor for phosphor syntheses**
- **Precoatings of lamp glass**
- **Phosphors for UV emitting discharge lamps**

10.2.4 Material Synthesis

Synthesis pathways to Y_2O_3:Eu powder

1. **Ceramic method**

$$Y_2O_3 + Eu_2O_3 \rightarrow (Y_{1-x}Eu_x)_2O_3$$

2. **Precipitation methods**

(a) **Oxalate route (non-homogeneous precipitation)**

$$2(1\text{-}x)\ Y^{3+} + 2x\ Eu^{3+} + 3\ C_2O_4^{2-} \rightarrow (Y_{1-x}Eu_x)_2(C_2O_4)_3$$

$$(Y_{1-x}Eu_x)_2(C_2O_4)_3 \rightarrow (Y_{1-x}Eu_x)_2O_3 + 3\ CO_2 + 3\ CO$$

(b) **Hydroxide route (homogeneous precipitation)**

$$H_2N\text{-}CO\text{-}NH_2 + H_2O \rightarrow 2\ NH_3 + CO_2$$

$$NH_3 + H_2O \rightleftharpoons NH_4^+ + OH^-$$

$$(1\text{-}x)\ Y^{3+} + x\ Eu^{3+} + 3\ OH^- \rightarrow (Y_{1-x}Eu_x)OH_3$$

$$2\ (Y_{1-x}Eu_x)OH_3 \rightarrow (Y_{1-x}Eu_x)_2O_3 + 3\ H_2O$$

Synthesis of Y_2O_3:Eu nanoscale powder

Hydroxide precursor

Y_2O_3:Eu phosphor

Emission and excitation spectrum

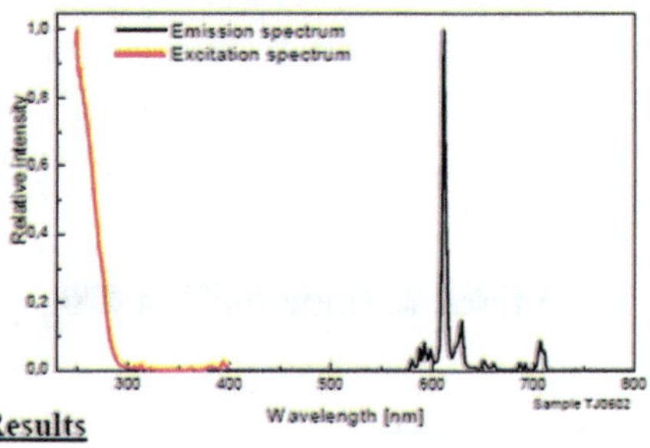

<u>Results</u>

- **200 nm particles**
- **Narrow particle size distribution**
- **Homogeneous particle morphology**
- **Quantum efficiency > 65%**

Synthesis of $(Y,Gd)_2O_3$:Eu nanoscale powder

Homogeneous particle morphology

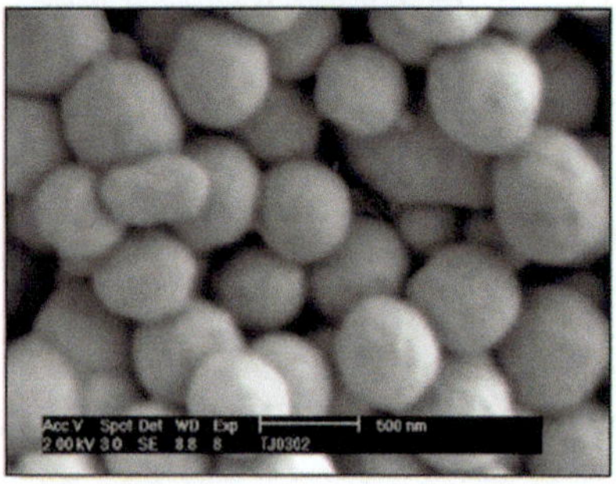

⇒ homogeneous layer morphology

- **$(Y,Gd)_2O_3$ particle diameter ~ 200–500 nm**
- **Thin and dense layer, e.g. as a reflector layer**

Synthesis of homogeneous (Y,Gd)BO_3:RE powder for PDPs

(a) Synthesis of a insoluble precursor $LnB(OH)_4CO_3$
$\Rightarrow$ 0.6 – 7.0 μm particles

(b) Conversion into the phosphor by annealing

- Blue — $LnBO_3$:Ce, $LnBO_3$:Tm
- Green — $LnBO_3$:Tb, $LnBO_3$:Er
- Red — $LnBO_3$:Eu, $LnBO_3$:Sm

"Precursor route"

$LnBO_3$:Eu (solid state synthesis)

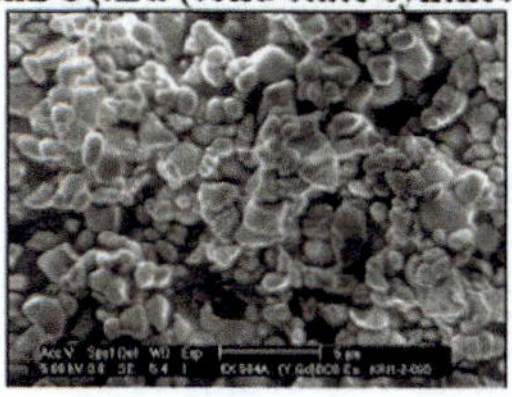

Precursor $LnB(OH)_4CO_3$

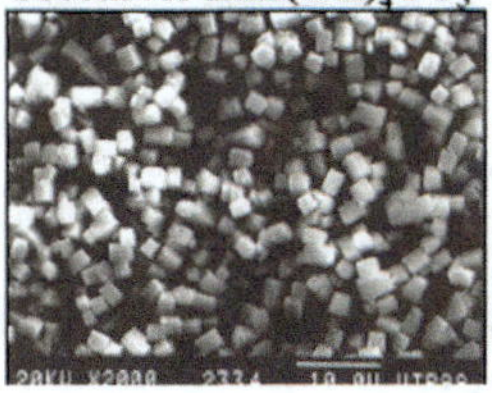

Phosphor $LnBO_3$

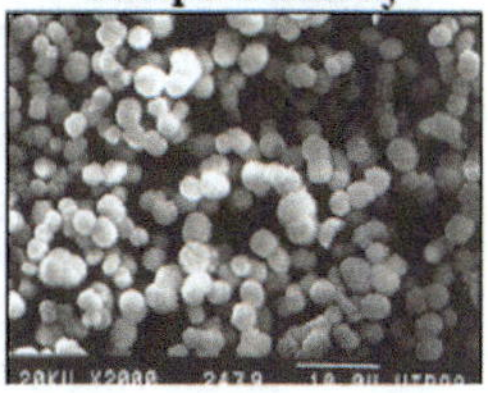

Synthesis of optically transparent ceramics

- Optical transmission: $I/I_0 = (1-R)^2 \exp(-\mu x)$
  R: reflectivity, μ: absorption coefficient, x: sample thickness
- $\mu = a + S_{im} + S_{op}$
  a: electronic absorption, S_{im}: impurity scattering, S_{op}: optical anisotropy scattering
- Requirements for transparency: optically isotropic host lattice, no pores or second phases

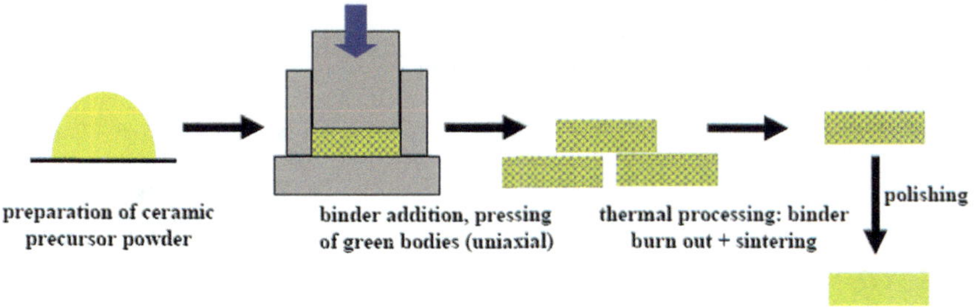

Synthesis of $Y_3Al_5O_{12}$:Ce (cubic) powder for ceramic processing

1. **Ceramic method**
   $3\,Y_2O_3 + 5\,Al_2O_3$ + Flux (BaF_2) $\rightarrow 2\,Y_3Al_5O_{12}$ $\Rightarrow$ µm-particles

2. **Homogeneous precipitation of hydroxycarbonates + surfactants, e.g. sodium dodecylsulfate**
   $\Rightarrow$ Reduction in particle size and improvement of homogeneity $\Rightarrow$ nm-particles

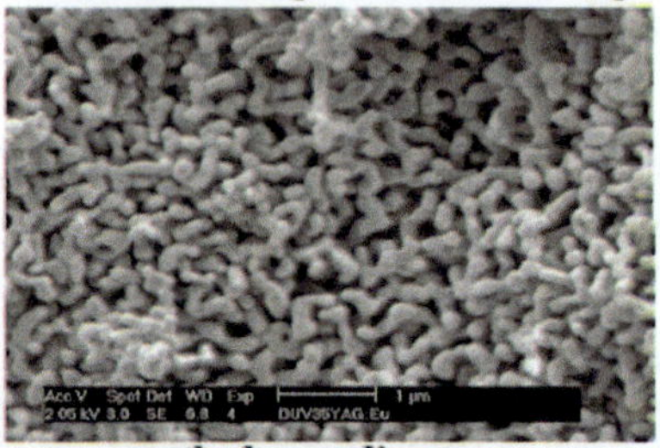

method according to Konoshima Chemical

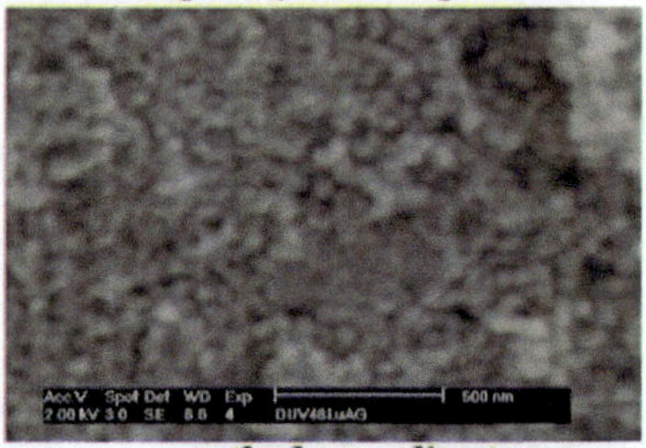

method according to D. Uhlich, Diploma thesis (2004)

Synthesis of transparent $Y_3Al_5O_{12}$:Ce „Yellow phosphor for LEDs“

1. **Precipitation of nanopowder**
   $Al(t\text{-}BuO)_3 + Y(NO_3)_3$ in ethyl acetate
2. **Sintering: 1000°C in air**
3. **Pellets pressing (PVA binder) in SiC pit**
4. **Microwave treatment (400 W, 35 min.) + polishing**

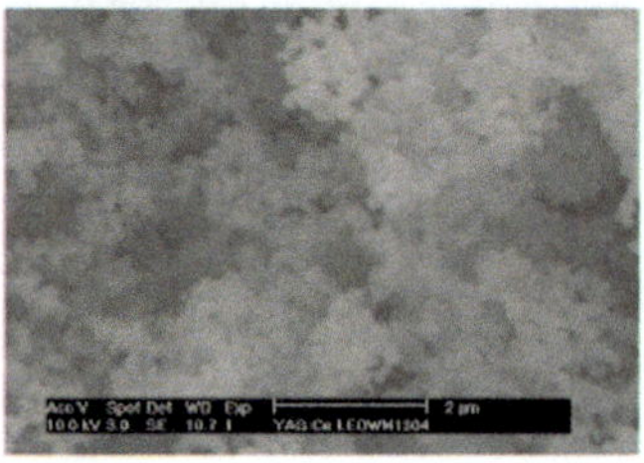

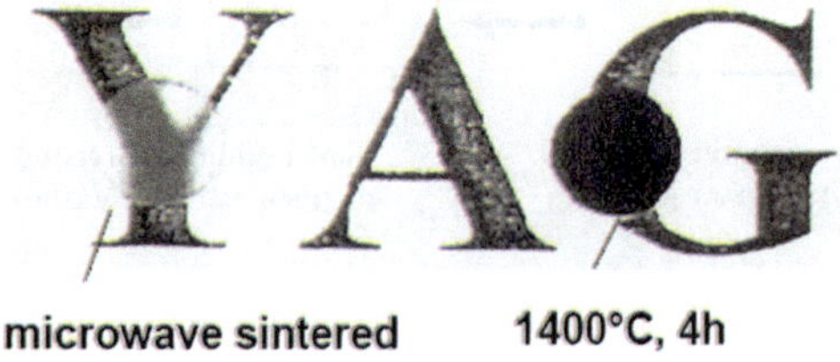

10.3 Two Approaches with the Same Goal

The practical applications of the two nanosynthesis approaches are demonstrated in Fig. 10.6.

Top-down strategies already exist today and will continue to play an important role despite their intrinsic limitations. Bottom-up methods will drive the nanobulk stage and ultimately take over certain aspects of top-down processing via intelligent self-assembly. The history of nanosynthesis methods for electronic materials is shown in Fig. 10.7 as a function of a time showing a crossing point around 2010.

- **An atom by atom construction**
  - It starts with atoms, ions, or molecules, to form nanostructured "building blocks", e.g. crystallites, or nanoscale clusters, which are then assembled into final bulk materials.
- **One step processes**
  - Electrodeposition: *from chemical compounds in vapor liquid phase.*
  - Chemical vapor deposition (CVD): *chemical compounds are used either in vapor or liquid phase.*

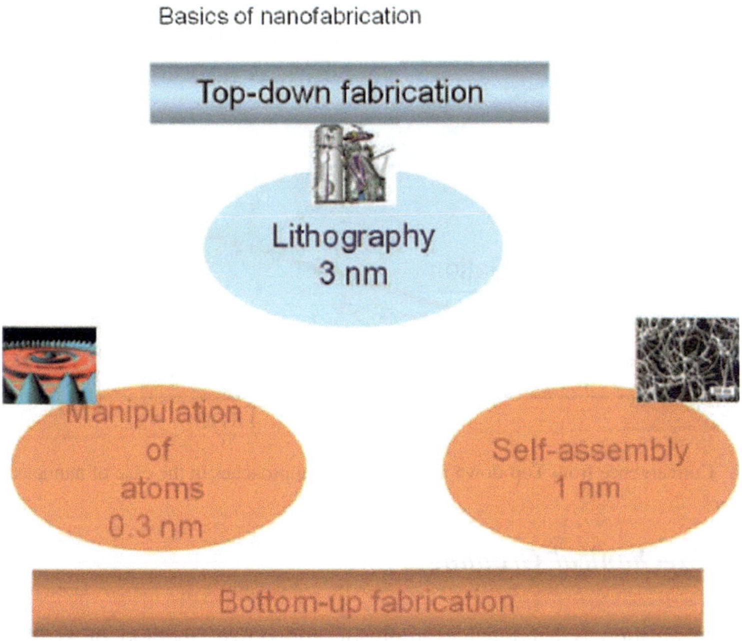

Fig. 10.6 Schematic position of top-down and bottom-up fabrication mechanisms for nano-electronics. Image courtesy of IMEC Louvain

  - Physical vapor deposition (PVD): *when it is from the vapor phase, material is extracted from a source either by heating or ion bombardment.*
  - Sol-Gel processes: *from colloidal solutions. It allows preparing free-standing nanoparticles or nanocrystals directly in solution.*

- **Two-step processes**
  - Nanoparticle's production by CVD, PVD and chemical routes + compaction.

10.4 Methods for Creating Nanostructures

There are many different ways of creating nanostructures: of course, macro-molecules or nanoparticles or buckyballs or nanotubes and so on can be synthesized artificially [17–21] for certain specific materials. They can also be arranged by methods based on equilibrium or near-equilibrium thermodynamics such as methods of self-organization and self-assembly (sometimes also called bio-mimetic processes). Using these methods, synthesized materials can be arranged into useful shapes so that finally the material can be applied to a certain application.

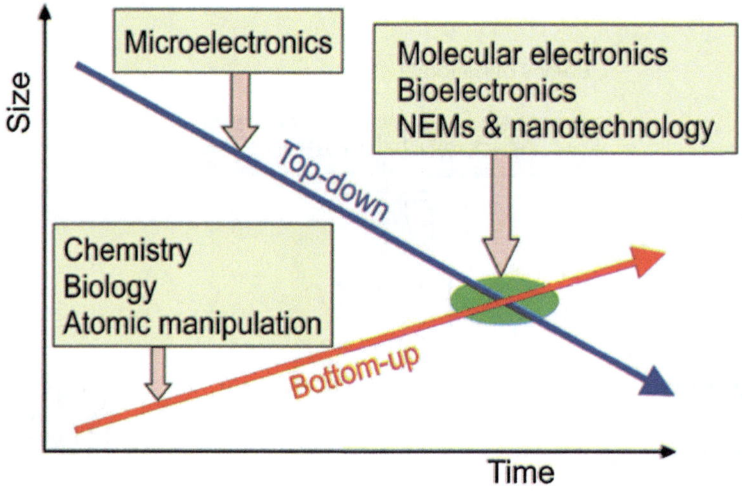

Fig. 10.7 Convergence from Top-down and Bottom-up approaches in the case of nanoelectronics

10.4.1 Mechanical Grinding

Mechanical attrition is a typical example of 'top down' method of synthesis of nanomaterials, where the material is prepared not by cluster assembly but by the structural decomposition of coarser-grained structures as the result of severe plastic deformation. This has become a popular method to make nanocrystalline materials because of its simplicity, the relatively inexpensive equipment needed, and the applicability to essentially the synthesis of all classes of materials. The major advantage often quoted is the possibility for easily scaling up to tonnage quantities of material for various applications. Similarly, the serious problems that are usually cited are;

1. contamination from milling media and/or atmosphere, and
2. to consolidate the powder product without coarsening the nanocrystalline microstructure.

In fact, the contamination problem is often given as a reason to dismiss the method, at least for some materials. Here we will review the mechanisms presently believed responsible for formation of Nano crystalline structures by mechanical attrition of single-phase powders, mechanical alloying of dissimilar powders, and mechanical crystallization of amorphous materials. The two important problems of contamination and powder consolidation will be briefly considered (Fig. 10.8).

Mechanical milling is typically achieved using high-energy shaker, planetary ball, or tumbler mills. The energy transferred to the powder from refractory or steel balls depends on the rotational (vibrational) speed, size and number of the balls, ratio of the ball to powder mass, the time of milling and the milling atmosphere. Nanoparticles are produced by the shear action during grinding. Milling in cryogenic liquids can greatly increase the brittleness of the powders influencing the

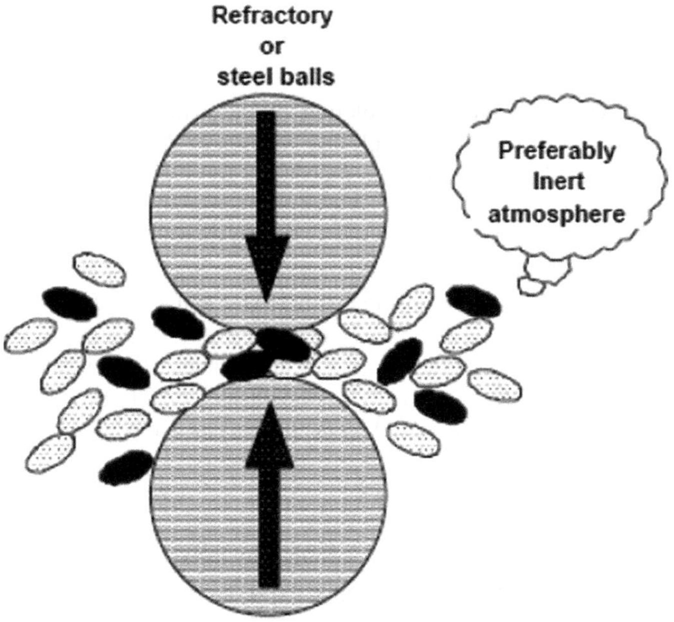

Fig. 10.8 Schematic representation of the principle of mechanical milling

fracture process. As with any process that produces fine particles, an adequate step to prevent oxidation is necessary. Hence, this process is very restrictive for the production of non-oxide materials since then it requires that the milling take place in an inert atmosphere and that the powder particles be handled in an appropriate vacuum system or glove box. This method of synthesis is suitable for producing amorphous or nanocrystalline alloy particles, elemental or compound powders. If the mechanical milling imparts sufficient energy to the constituent powders, a homogeneous alloy can be formed. Based on the energy of the milling process and thermodynamic properties of the constituents the alloy can be rendered amorphous by this processing.

10.4.2 Wet Chemical Synthesis of Nanomaterials

In principle, we can classify the wet chemical synthesis of nanomaterials into two broad groups:

1. The top down method: where single crystals are etched in an aqueous solution for producing nanomaterials, For example, the synthesis of porous silicon by electrochemical etching.
2. The bottom up method: consisting of sol-gel method, precipitation etc. where materials containing the desired precursors are mixed in a controlled fashion to form a colloidal solution.

10.4.2.1 Sol-Gel Process

The sol-gel process, involves the evolution of inorganic networks through the formation of a colloidal suspension (**sol**) and gelation of the sol to form a network in a continuous liquid phase (**gel**). The precursors for synthesizing these colloids consist usually of a metal or metalloid element surrounded by various reactive ligands. The starting material is processed to form a dispersible oxide and forms a sol in contact with water or dilute acid. Removal of the liquid from the sol yields the gel, and the sol/gel transition controls the particle size and shape. Calcination of the gel produces the oxide.

Sol-gel processing refers to the hydrolysis and condensation of alkoxide-based precursors such as $Si(OEt)_4$ (tetraethyl orthosilicate, or TEOS). The reactions involved in the sol-gel chemistry based on the hydrolysis and condensation of metal alkoxides.

M(OR)z can be described as follows:

$$\mathrm{MOR + H_2O \rightarrow MOH + ROH}\ (\text{hydrolysis})$$

$$\mathrm{MOH + ROM \rightarrow M\text{-}O\text{-}M + ROH}\ (\text{condensation})$$

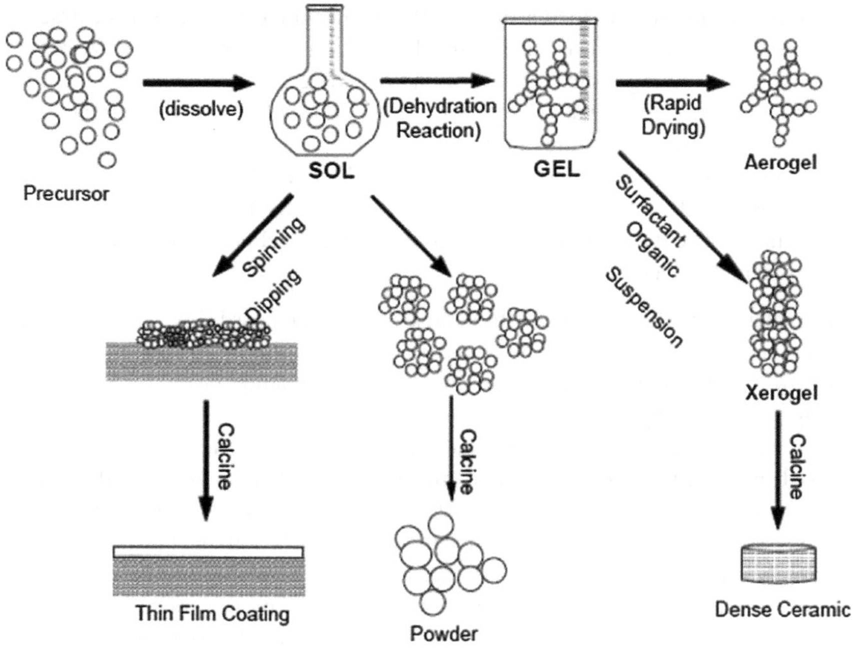

Fig. 10.9 Schematic representation of sol-gel process of synthesis of nanomaterials

Sol-gel method of synthesizing nanomaterials is very popular amongst chemists and is widely employed to prepare oxide materials. The sol-gel process can be characterized by a series of distinct steps (Fig. 10.9).

1. Formation of different stable solutions of the alkoxide or solvated metal precursor.
2. Gelation resulting from the formation of an oxide- or alcohol-bridged network (the gel) by a polycondensation reaction that results in a dramatic increase in the viscosity of the solution.
3. Aging of the gel (Syneresis), during which the polycondensation reactions continue until the gel transforms into a solid mass, accompanied by contraction of the gel network and expulsion of solvent from gel pores. Ostwald ripening (also referred to as coarsening, is the phenomenon by which smaller particles are consumed by larger particles during the growth process) and phase transformations may occur concurrently with syneresis. The aging process of gels can exceed 7 days and is critical to the prevention of cracks in gels that have been cast.
4. Drying of the gel, when water and other volatile liquids are removed from the gel network. This process is complicated due to fundamental changes in the structure of the gel. The drying process has itself been broken into four distinct steps: (i) the constant rate period, (ii) the critical point, (iii) the falling rate period, (iv) the second falling rate period. If isolated by thermal evaporation, the

resulting monolith is termed a *xerogel*. If the solvent (such as water) is extracted under supercritical or near super critical conditions, the product is an *aerogel*.

5. Dehydration, during which surface-bound M–OH groups are removed, there by stabilizing the gel against rehydration. This is normally achieved by calcining the monolith at temperatures up to 8000 °C.
6. Densification and decomposition of the gels at high temperatures (T > 8000 °C). The pores of the gel network are collapsed, and remaining organic species are volatilized. The typical steps that are involved in sol-gel processing are shown in the schematic diagram below.

The interest in this synthesis method arises due to the possibility of synthesizing nonmetallic inorganic materials like glasses, glass ceramics or ceramic materials at very low temperatures compared to the high temperature process required by melting glass or firing ceramics.

The major difficulties to overcome in developing a successful bottom-up approach is controlling the growth of the particles and then stopping the newly formed particles from agglomerating. Other technical issues are ensuring the reactions are complete so that no unwanted reactant is left on the product and completely removing any growth aids that may have been used in the process. In addition, production rates of nano powders are very low by this process. The main advantage is one can get monosized nano particles by any bottom up approach.

10.4.3 Gas Phase Synthesis of Nanomaterials

Gas phase deposition can be used to fabricate thin films, multilayers, nanotubes, monofilaments or nanometer-sized particles. The general techniques can be classified broadly as either physical vapor deposition (PVD) or chemical vapor deposition (CVD). PVD involves the conversion of solid material into a gaseous phase by physical processes; this material is then cooled and re-deposited on a substrate with perhaps some modification, such as reaction with a gas. Examples of PVD conversion processes include thermal evaporation (such as resistive or electron beam heating or even flame synthesis), laser ablation or pulsed laser deposition (where a short nanosecond pulse from a laser is focused onto the surface of a bulk target), spark erosion and sputtering (the removal of a target material by bombardment with atoms or ions).

Most nanoparticle synthesis methods in the gas phase are based on homogeneous nucleation of a supersaturated vapor and subsequent particle growth by condensation, coagulation and capture. The gas-phase synthesis methods are of increasing interest because they allow elegant way to control process parameters in order to be able to produce size, shape and chemical composition controlled nanostructures. Before we discuss a few selected pathways for gas-phase formation of nanomaterials, some general aspects of gas-phase synthesis needs to be discussed. In conventional chemical vapour deposition (CVD) synthesis, gaseous products either are allowed to react homogeneously or heterogeneously depending on a particular application.

1. In homogeneous CVD, particles form in the gas phase and diffuse towards a cold surface due to thermophoretic forces, and can either be scrapped of from the cold surface to give nano-powders, or deposited onto a substrate to yield what is called '*particulate films*'.
2. In heterogeneous CVD, the solid is formed on the substrate surface, which catalysis the reaction and a dense film is formed.

In order to form nanomaterials several modified CVD methods have been developed. Gas phase processes have inherent advantages, some of which are noted here:

- An excellent control of size, shape, crystallinity and chemical composition
- Highly pure materials can be obtained
- Multicomponent systems are relatively easy to form
- Easy control of the reaction mechanisms.

Most of the synthesis routes are based on the production of small clusters that can aggregate to form nano particles (condensation). Condensation occurs only when the vapour is supersaturated and in these processes homogeneous nucleation in the gas phase is utilized to form particles. This can be achieved both by physical and chemical methods.

In general, vapor forms within an aerosol reactor at elevated temperatures. The precursor material in the form of as solid, liquid or gas is introduced into the reactor where it is heated and mixed with a carrier gas. The supersaturated vapor is produced by cooling or by chemical decomposition reaction or by some combination of these. The most straightforward method of achieving super saturation is to evaporate a solid into a background gas. By including a reactive gas such as oxygen, oxides or other compounds of the evaporated material can be produced.

The nucleation process is initiated by the formation of very small nucleus from the molecular phase. These nuclei subsequently grow by surface growth mechanisms (heterogeneous condensation, surface reaction) and by collision and coagulation. Further collisions can result in the formation of loosely bound agglomerates or chain like, dendritic forms. The most common heating or evaporation process are: the flame pyrolysis, furnace flow reactors, laser induced pyrolysis, laser vaporization, thermal plasma, microwave plasma, sputtering, and laser ablation.

10.4.3.1 Furnace

The simplest fashion to produce nanoparticles is by heating the desired material in a heat resistant crucible containing the desired material. This method is appropriate only for materials that have a high vapour pressure at the heated temperatures that can be as high as 2000 °C. Energy is normally introduced into the precursor by arc

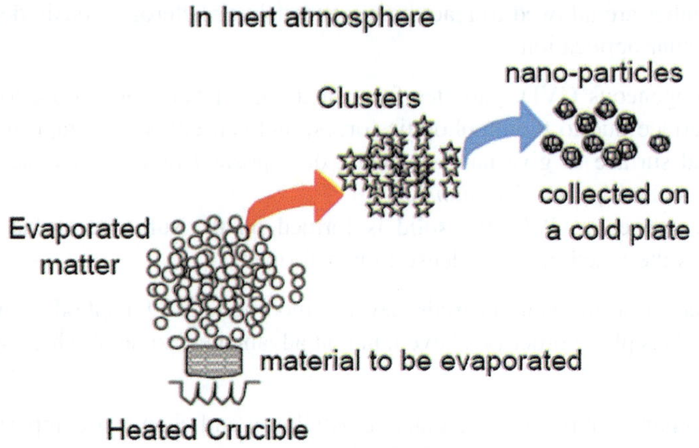

Fig. 10.10 Schematic representation of gas phase process of synthesis of single phase nanomaterials from a heated crucible

heating, electron beam heating or Joule heating. The atoms are evaporated into an atmosphere, which is either inert (e.g. He) or reactive (so as to form a compound). To carry out reactive synthesis, materials with very low vapour pressure have to be fed into the furnace in the form of a suitable precursor such as organometallics, which decompose in the furnace to produce a condensable material. The hot atoms of the evaporated matter lose energy by collision with the atoms of the cold gas and undergo condensation into small clusters via homogeneous nucleation. In case a compound is being synthesized, these precursors react in the gas phase and form a compound with the material that is separately injected in the reaction chamber. The clusters would continue to grow if they remain in the supersaturated region. To control their size, they need to be rapidly removed from the supersaturated environment by a carrier gas. The cluster size and its distribution are controlled by only three parameters:

(1) The rate of evaporation (energy input),
(2) The rate of condensation (energy removal), and
(3) The rate of gas flow (cluster removal) (Fig. 10.10).

Because of its inherent simplicity, it is possible to scale up this process from laboratory (mg/day) to industrial scales (tons/day).

10.4.3.2 Flame Assisted Ultrasonic Spray Pyrolysis

In this process, precursors are nebulized and then unwanted components are burnt in a flame to get the required material, e.g. ZrO_2 has been obtained by this method from a precursor of $Zr(CH_3\ CH_2\ CH_2O)_4$. Flame hydrolysis that is a variant of this process is used for the manufacture of fused silica. In the process, silicon

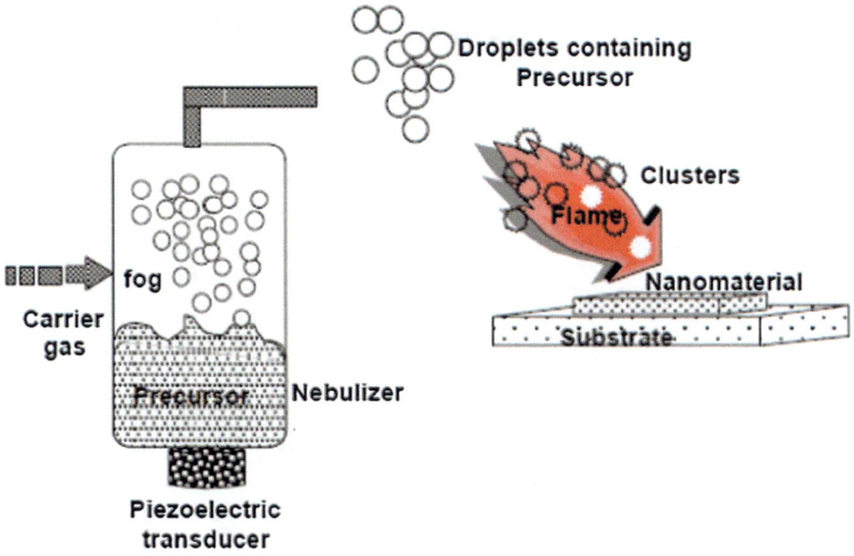

Fig. 10.11 Flame assisted ultrasonic spray pyrolysis

tetrachloride is heated in an oxy-hydrogen flame to give a highly dispersed silica. The resulting white amorphous powder consists of spherical particles with sizes in the range 7–40 nm. The combustion flame synthesis, in which the burning of a gas mixture, e.g. acetylene and oxygen or hydrogen and oxygen, supplies the energy to initiate the pyrolysis of precursor compounds, is widely used for the industrial production of powders in large quantities, such as carbon black, fumed silica and titanium dioxide. However, since the gas pressure during the reaction is high, highly agglomerated powders are produced which is disadvantageous for subsequent processing. The basic idea of low-pressure combustion flame synthesis is to extend the pressure range to the pressures used in gas phase synthesis and thus to reduce or avoid the agglomeration. Aerosol scientists to study particle formation in the flame have extensively used low-pressure flames (Fig. 10.11).

A key for the formation of nanoparticles with narrow size distributions is the exact control of the flame in order to obtain a flat flame front. Under these conditions the thermal history, i.e. time and temperature, of each particle formed is identical and narrow distributions result. However, due to the oxidative atmosphere in the flame, this synthesis process is limited to the formation of oxides in the reactor zone.

10.4.3.3 Gas Condensation Processing (GCP)

In this technique, a metallic or inorganic material, e.g. a suboxide, is vaporised using thermal evaporation sources such as crucibles, electron beam evaporation

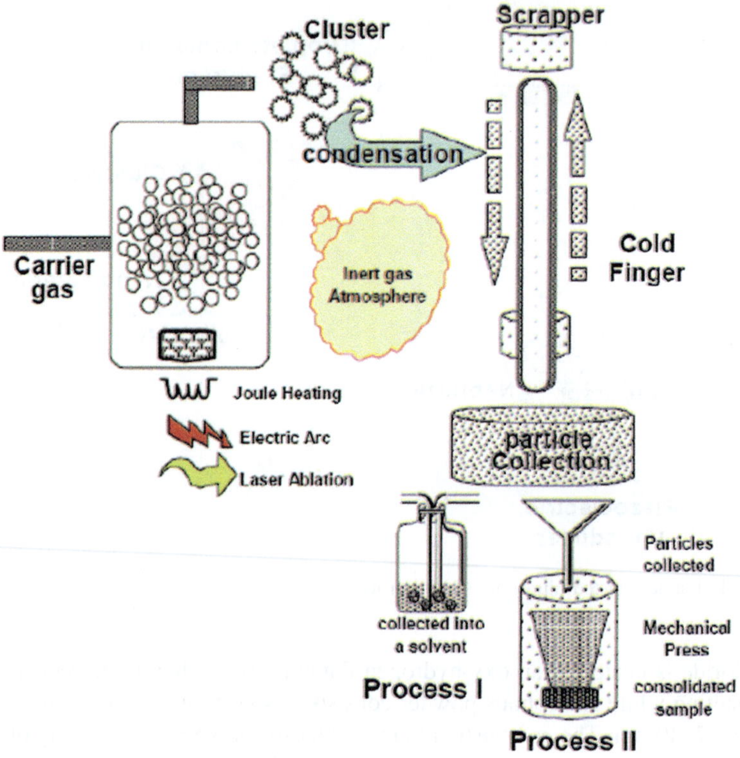

Fig. 10.12 Schematic representation of typical set-up for gas condensation synthesis of nanomaterials followed by consolidation in a mechanical press or collection in an appropriate solvent media

devices or sputtering sources in an atmosphere of 1–50 mbar He (or another inert gas like Ar, Ne, Kr). Cluster form in the vicinity of the source by homogenous nucleation in the gas phase and grow by coalescence and incorporation of atoms from the gas phase (Fig. 10.12).

The cluster or particle size depends critically on the residence time of the particles in the growth system and can be influenced by the gas pressure, the kind of inert gas, i.e. He, Ar or Kr, and on the evaporation rate/vapour pressure of the evaporating material. With increasing gas pressure, vapour pressure and mass of the inert gas used the average particle size of the nanoparticles increases. Lognormal size distributions have been found experimentally and have been explained theoretically by the growth mechanisms of the particles. Even in more complex processes such as the low pressure combustion flame synthesis where a number of chemical reactions are involved the size distributions are determined to be lognormal.

Originally, a rotating cylindrical device cooled with liquid nitrogen was employed for the particle collection: the nanoparticles in the size range from 2 to 50 nm are extracted from the gas flow by thermophoretic forces and deposited loosely on the surface of the collection device as a powder of low density and no agglomeration. Subsequently, the nanoparticles are removed from the surface of the cylinder by means of a scraper in the form of a metallic plate. In addition to this cold finger device several techniques known from aerosol science have now been implemented for the use in gas condensation systems such as corona discharge, etc. These methods allow for the continuous operation of the collection device and are better suited for larger scale synthesis of nanopowders. However, these methods can only be used in a system designed for gas flow, i.e. a dynamic vacuum is generated by means of both continuous pumping and gas inlet via mass flow controller. A major advantage over convectional gas flow is the improved control of the particle sizes. It has been found that the particle size distributions in gas flow systems, which are also lognormal, are shifted towards smaller average values with an appreciable reduction of the standard deviation of the distribution. Depending on the flow rate of the He-gas, particle sizes are reduced by 80% and standard deviations by 18%.

The synthesis of nanocrystalline pure metals is relatively straightforward as long as evaporation can be done from refractory metal crucibles (W, Ta or Mo). If metals with high melting points or metals which react with the crucibles, are to be prepared, sputtering, i.e. for W and Zr, or laser or electron beam evaporation has to be used. Synthesis of alloys or intermetallic compounds by thermal evaporation can only be done in the exceptional cases that the vapour pressures of the elements are similar. As an alternative, sputtering from an alloy or mixed target can be employed. Composite materials such as Cu/Bi or W/Ga have been synthesized by simultaneous evaporation from two separate crucibles onto a rotating collection device. It has been found that excellent intermixing on the scale of the particle size can be obtained.

However, control of the composition of the elements has been difficult and reproducibility is poor. Nanocrystalline oxide powders are formed by controlled postoxidation of primary nanoparticles of a pure metal (e.g. Ti to TiO_2) or a suboxide (e.g. ZrO to ZrO_2). Although the gas condensation method including the variations have been widely employed to prepared a variety of metallic and ceramic materials, quantities have so far been limited to a laboratory scale. The quantities of metals are below 1 g/day, while quantities of oxides can be as high as 20 g/day for simple oxides such as CeO_2 or ZrO_2. These quantities are sufficient for materials testing but not for industrial production. However, it should be mentioned that the scale-up of the gas condensation method for industrial production of nanocrystalline oxides by a company called nanophase technologies has been successful.

10.4.3.4 Chemical Vapour Condensation (CVC)

As shown schematically in Figure, the evaporative source used in GPC is replaced by a hot wall reactor in the Chemical Vapour Condensation or the CVC process. Depending on the processing parameters nucleation of nanoparticles is observed during chemical vapour deposition (CVC) of thin films and poses a major problem in obtaining good film qualities. The original idea of the novel CVC process which is schematically shown below where, it was intended to adjust the parameter field during the synthesis in order to suppress film formation and enhance homogeneous nucleation of particles in the gas flow. It is readily found that the residence time of the precursor in the reactor determines if films or particles are formed. In a certain range of residence time both particle and film formation can be obtained.

Adjusting the residence time of the precursor molecules by changing the gas flow rate, the pressure difference between the precursor delivery system and the main chamber occurs. Then the temperature of the hot wall reactor results in the fertile production of nanosized particles of metals and ceramics instead of thin films as in CVD processing. In the simplest form a metal organic precursor is introduced into the hot zone of the reactor using mass flow controller. Besides the increased quantities in this continuous process compared to GPC has been demonstrated that a wider range of ceramics including nitrides and carbides can be synthesized. Additionally, more complex oxides such as $BaTiO_3$ or composite structures can be formed as well. Appropriate precursor compounds can be readily found in the CVD literature. The extension to production of nanoparticles requires the determination of a modified parameter field in order to promote particle formation instead of film formation. In addition to the formation of single phase nanoparticles by CVC of a single precursor the reactor allows the synthesis of:

1. Mixtures of nanoparticles of two phases or doped nanoparticles by supplying two precursors at the front end of the reactor, and
2. Coated nanoparticles, i.e., n-ZrO_2 coated with n-Al_2O_3 or vice versa, by supplying a second precursor at a second stage of the reactor. In this case nanoparticles which have been formed by homogeneous nucleation are coated by heterogeneous nucleation in a second stage of the reactor (Fig. 10.13).

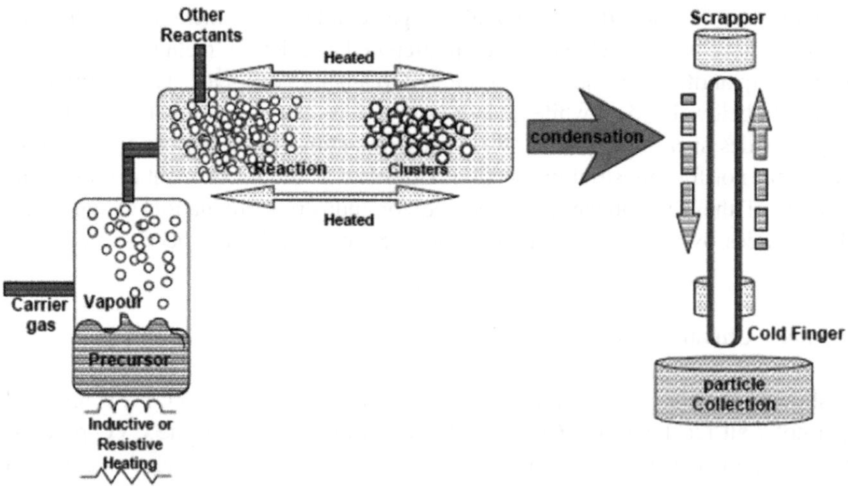

Fig. 10.13 A schematic of a typical CVC reactor

Because CVC processing is continuous, the production capabilities are much larger than in GPC processing. Quantities in excess of 20 g/h have been readily produced with a small scale laboratory reactor. A further expansion can be envisaged by simply enlarging the diameter of the hot wall reactor and the mass flow through the reactor.

10.4.4 Sputtered Plasma Processing

In this method is yet again a variation of the gas-condensation method excepting the fact that the source material is a sputtering target and this target is sputtered using rare gases and the constituents are allowed to agglomerate to produce nanomaterial. Both dc (direct current) and rf (radio-frequency) sputtering has been used to synthesize nanoparticles. Again reactive sputtering or multitarget sputtering has been used to make alloys and/or oxides, carbides, nitrides of materials. This method is specifically suitable for the preparation of ultrapure and non-agglomerated nanoparticles of metal.

10.4.4.1 Microwave Plasma Processing

This technique is similar to the previously discussed CVC method but employs plasma instead of high temperature for decomposition of the metal organic precursors. The method uses microwave plasma in a 50 mm diameter reaction vessel made of quartz placed in a cavity connected to a microwave generator. A precursor such as a chloride compound is introduced into the front end of the reactor. Generally, the microwave cavity is designed as a single mode cavity using the TE10 mode in a WR975 waveguide with a frequency of 0.915 GHz. The major advantage of the plasma assisted pyrolysis in contrast to the thermal activation is the low temperature reaction which reduces the tendency for agglomeration of the primary particles. This is also true in the case of plasma-CVD processes. Additionally, it has been shown that by introducing another precursor into a second reaction zone of the tubular reactor, e.g. by splitting the microwave guide tubes, the primary particles can be coated with a second phase. For example, it has been demonstrated that ZrO_2 nanoparticles can be coated by Al_2O_3. In this case the inner ZrO_2 core is crystalline, while the Al_2O_3 coating is amorphous. The reaction sequence can be reversed with the result that an amorphous Al_2O_3 core is coated with crystalline ZrO_2. While the formation of the primary particles occurs by homogeneous nucleation, it can be easily estimated using gas reaction kinetics that the coating on the primary particles grows heterogeneously and that homogeneous nucleation of nanoparticles originating from the second compound has a very low probability. A schematic representation of the particle growth in plasma's is given below.

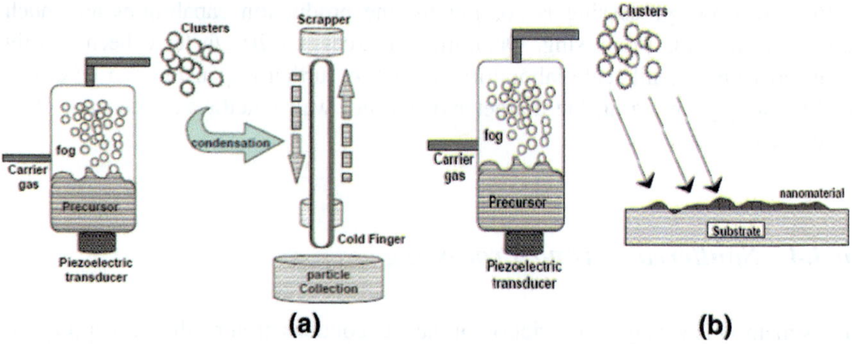

Fig. 10.14 Schematic representation of **a** nanoparticle, and **b** particulate film formation

10.4.5 Particle Precipitation Aided

In another variation of this process, colloidal clusters of materials are used to prepare nanoparticles. The CVD reaction conditions are so set that particles form by condensation in the gas phase and collect onto a substrate, which is kept under a different condition that allows heterogeneous nucleation. By this method both nanoparticles and particulate films can be prepared. An example of this method has been used to form nanomaterials e.g. SnO_2, by a method called pyrosol deposition process, where clusters of tin hydroxide are transformed into small aerosol droplets, following which they are reacted onto a heated glass substrate (Fig. 10.14).

10.4.6 Laser Ablation

Laser ablation has been extensively used for the preparation of nanoparticles and particulate films. In this process, a laser beam is used as the primary excitation source of ablation for generating clusters directly from a solid sample in a wide variety of applications. The small dimensions of the particles and the possibility to form thick films make this method quite an efficient tool for the production of ceramic particles and coatings and also an ablation source for analytical applications such as the coupling to induced coupled plasma emission spectrometry, ICP, the formation of the nanoparticles has been explained following a liquefaction process which generates an aerosol, followed by the cooling/solidification of the droplets which results in the formation of fog. The general dynamics of both the aerosol and the fog favours the aggregation process and micrometer-sized fractal-like particles are formed. The laser spark atomizer can be used to produce highly mesoporous thick films and the porosity can be modified by the carrier gas flow rate. ZrO_2 and SnO_2 nanoparticulate thick films were also synthesized successfully using this process with quite identical microstructure. Synthesis of other materials such as lithium manganate, silicon and carbon has also been carried out by this technique.

References

1. Hochbaum, A.I., Yang, P.: Semiconductor nanowires for energy conversion. Chem. Rev. **110**(1), 527–546 (2010)
2. Law, M., Goldberger, J., Yang, P.: Semiconductor nanowires and nanotubes. Annu. Rev. Mater. Res. **34**(1), 83–122 (2004)
3. Schmidt, V., Wittemann, J.V., Gösele, U.: Growth, thermodynamics, and electrical properties of silicon nanowires. Chem. Rev. **110**(1), 361–388 (2010)
4. Fan, H.J., Werner, P., Zacharias, M.: Semiconductor nanowires: from self-organization to patterned growth. Small **2**(6), 700–717 (2006)
5. Khan, Y.: The great partition: the making of India and Pakistan (2007)
6. Fan, Z., Razavi, H., Do, J., Moriwaki, A., Ergen, O., et al.: Three-dimensional nanopillar-array photovoltaics on low-cost and flexible substrates. Nature Mater. **8**(8), 648–653 (2009)
7. Schmid, H., Björk, M.T., Knoch, J., Riel, H., Riess, W., et al.: Patterned epitaxial vapor-liquid-solid growth of silicon nanowires on Si(111) using silane. J. Appl. Phys. **103**(2), 024304 (2008)
8. Mårtensson, T., Carlberg, P., Borgström, M., Montelius, L., Seifert, W., Samuelson, L.: Nanowire arrays defined by nanoimprint lithography. Nano Lett. **4**(4), 699–702 (2004)
9. Zschech, D., Kim, D.H., Milenin, A.P., Scholz, R., Hillebrand, R., et al.: Ordered arrays of <100>-oriented silicon nanorods by CMOS-compatible block copolymer lithography. Nano Lett. **7**(6), 1516–1520 (2007)
10. Huang, Z., Fang, H., Zhu, J.: Fabrication of silicon nanowire arrays with controlled diameter, length, and density. Adv. Mater. **19**(5), 744–748 (2007)
11. Garnett, E.C., Yang, P.: Silicon nanowire radial *p-n* junction solar cells. J. Am. Chem. Soc. **130**(29), 9224–9225 (2008)
12. Hsu, C., Connor, S.T., Tang, M.X., Cui, Y.: Wafer-scale silicon nanopillars and nanocones by Langmuir–Blodgett assembly and etching. Appl. Phys. Lett. **93**(13), 133109 (2008)
13. Huang, Z., Zhang, X., Reiche, M., Liu, L., Lee, W., et al.: Extended arrays of vertically aligned sub-10 nm diameter [100] Si nanowires by metal-assisted chemical etching. Nano Lett. **8**(9), 3046–3051 (2008)
14. Garnett, E., Yang, P.: Light trapping in silicon nanowire solar cells. Nano Lett. **10**(3), 1082–1087 (2010)
15. Bogush, G.H., Tracy, M.A., Zukoski IV, C.F.: Preparation of monodisperse silica particles: control of size and mass fraction. J. Non-Cryst. Solids **104**, 95–106 (1988)
16. Jeong, S., Hu, L., Lee, H.R., Garnett, E., Choi, J.W., Cui, Y.: Fast and scalable printing of large area monolayer nanoparticles for nanotexturing applications. Nano Lett. **10**, 2989–2994 (2010)
17. Viswanathan, B.: Nanomaterials. Narosa Publishing House
18. Zhang, J.Z.: Optical properties and spectroscopy of nanomaterials. World Scientific Publishing Co. Pte. Ltd
19. Sajanlal, P.R., Sreeprasad, T.S., Samal, A.K., Pradeep, T.: Anisotropic nanomaterials: structure, growth, assembly, and functions. Nano Rev. **2** (2011)
20. Lue, J.T.: Physical Properties of Nanomaterials. In: Encyclopedia of nanoscience and nanotechnology, vol. X, pp. 1–46
21. Vollath, D.: Book review nanomaterials—an introduction to synthesis, properties and application. Environ. Eng. Manage. J. **7**(6), 865–870 (2008). WILEY-VCH Verlag GmbH & Co. KGaA, Weinheim, Germany. ISBN: 978-3-527-31531-4

Chapter 11
Techniques for Elaboration of Nanomaterials

Nanotechnology has received much interest nowadays. It is defined as techniques, methods or processes to fabricate nanoscale structures with size less than 100 nm. The delivery of a compound via skin offers many advantages including being pain-free, ease of administration, avoidance of hepatic firstpass metabolism and low cost. However, the major barrier of this route is the stratum corneum, which is the outermost layer of the epidermis. Thus effective nanodelivery systems are essential and no wonder why these systems reveal significant roles in delivery of cosmetics, drugs and biological products via topical routes these days. There are a variety of nanodelivery systems for topical route. However, this presentation covers polymeric nanoparticles and amphiphilic-association structures, which are the systems of the author's particular interest.

Polymeric nanoparticles: The polymeric nanoparticles are solid particles prepared using biodegradable and biocompatible polymers or monomers having size in the nanometer range. The preparation of the nanoparticles can be divided into two techniques: the dispersion of preformed polymers and the polymerization of monomers [1]. Depending on the preparation techniques, templates as well as conditions used such as stirring speed, pH, ionic strength, type and concentration of surfactant/stabilizer and temperature etc., the nanoparticles with desired properties can be prepared. The advantages of nanoparticles for topical delivery include: (1) physically stable compared to liposomes; (2) an encapsulated active compound can be stabilized; (3) site-specific targeting can be achieved and thus low dose of the active compound is needed; (4) the release of the compound can be controlled and (5) their small size (<50 nm) enables them to penetrate through skin.

Amphiphilic-association structures: Amphiphilic-association structures including micelles, microemulsions, and lyotropic liquid crystals are formed by self-assembly of amphiphilic molecules. The formation of the surfactant aggregates with different shapes can be determined by the critical packing parameter (CPP),

L. H. Madkour, *Nanoelectronic Materials*, Advanced
Structured Materials 116, https://doi.org/10.1007/978-3-030-21621-4_11

which is defined in terms of area of hydrophilic head group per hydrophobic chain volume as following equation [2]:

$$CPP = v/a_o l$$

where v is the volume of hydrocarbon chain, a_o is the area of hydrophilic head group, and l is the hydrocarbon chain length of a surfactant. When CPP is less than 1, surfactant film generally bends toward polar core. Oppositely, when it is more than 1, surfactant film bends toward non-polar core.

However, the lamellar structure is obtained when CPP is equal to 1. The advantages of amphiphilic-association structures are being thermodynamically stable, ease of preparation and low production costs [3]. Interestingly these delivering systems can be incorporated in both hydrophilic and hydrophobic active compounds. Thus they are candidate systems for herbal extracts containing a mixture of complex compounds, which is known to be challenging to efficiently formulate.

Vapor phase deposition can be used to fabricate thin films, multilayers, nanotubes, nanofilaments or nanometer-sized particles. The general techniques can be classified broadly as either physical vapor deposition (PVD) or chemical vapor deposition (CVD). PVD involves the conversion of solid material into a gaseous phase by physical processes; this material is then cooled and re-deposited on a substrate with perhaps some modification, such as reaction with a gas. Examples of PVD conversion processes include thermal evaporation (such as resistive or electron beam heating or even flame synthesis), laser ablation or pulsed laser deposition (where a short nanosecond pulse from a laser is focused onto the surface of a bulk target), spark erosion and sputtering (the removal of a target material by bombardment with atoms or ions). Most nanoparticle synthesis methods in the gas phase are based on homogeneous nucleation of a supersaturated vapor and subsequent particle growth by condensation, coagulation and capture. In general, vapor forms within an aerosol reactor at elevated temperatures. The precursor material in the form of as solid, liquid or gas is introduced into the reactor where it is heated and mixed with a carrier gas. The supersaturated vapor is produced by cooling or by chemical decomposition reaction or by some combination of these. The most straightforward method of achieving super saturation is to evaporate a solid into a background gas. By including a reactive gas such as oxygen, oxides or other compounds of the evaporated material can be produced. The nucleation process [4–9] is initiated by the formation of very small nucleus from the molecular phase. These nuclei subsequently grow by surface growth mechanisms (heterogeneous condensation, surface reaction) and by collision and coagulation. Further collisions can result in the formation of loosely bound agglomerates or chain like, dendritic forms. The most common heating or evaporation process are: the flame pyrolysis, furnace flow reactors, laser induced pyrolysis, laser vaporization, thermal plasma, microwave plasma, sputtering, and laser ablation.

11.1 Vapor-Phase Synthesis

- **CVD**
  - Volatile precursors react and/or decompose on the substrate surface to produce the desired deposit.
  - A variety of CVD processes which differ by which chemical reactions are initiated and conditions of processing. Examples:

    LPCVD → performed at low pressure, which improves deposit uniformity and composition.
    PECVD → Plasma-Enhanced which boost the chemical reaction rates of the precursors.
    HWCVD → hot filament (Hot Wire) is used to chemically decompose the source gases.

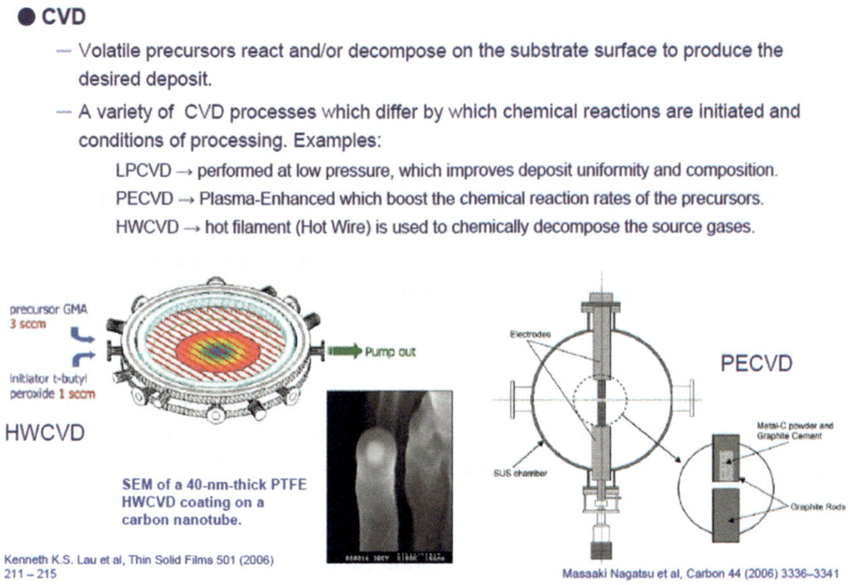

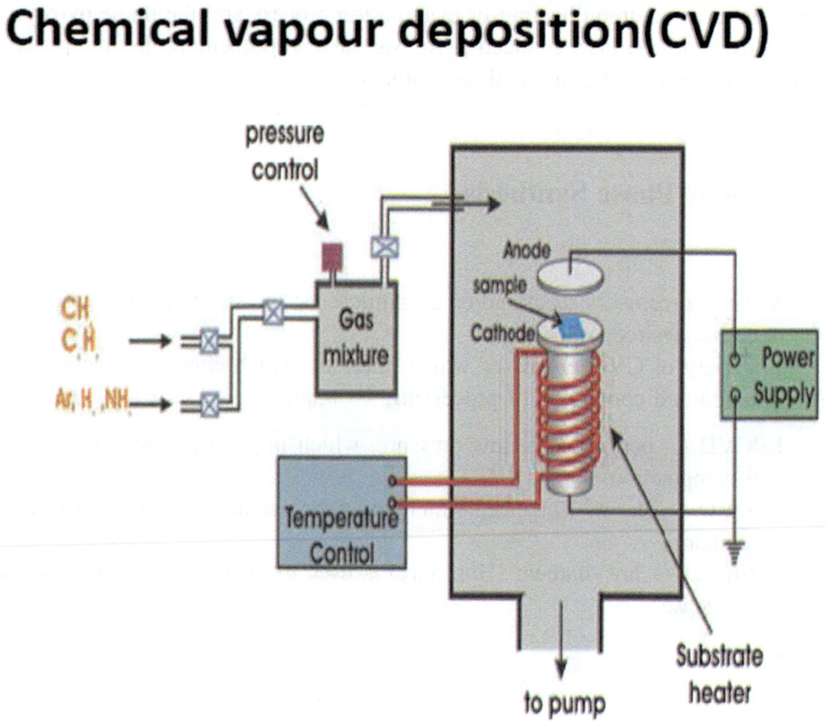

- **PVD**
  - Material to be deposited is placed in an energetic environment (like submitted to ion or electron beam) and particles of material escape its surface and travel to the cooled substrate surface to form a solid layer.
  - Processing is done under vacuum to allow a sufficient mean free path of the particles.
  - Examples of PVD:

    Thermal evaporation: using an electrical wire to melt the material. Molecular beam epitaxy is a particular sophisticated form of thermal evaporation.
    Electron beam evaporation: melting is done by the electron beam.
    Sputtering by an ion beam: Rare gas Ions sputter the material target. Deposition can be assisted using another "reactive beam".
    Pulsed laser deposition: it is an ablation process by pulses of focused laser which vaporizes the material.

- **PVD**
  - Material to be deposited is placed in an energetic environment (like submitted to ion or electron beam) and particles of material escape its surface and travel to the cooled substrate surface to form a solid layer.
  - Processing is done under vacuum to allow a sufficient mean free path of the particles.
  - Examples of PVD:

    Thermal evaporation: *using an electrical wire to melt the material. Molecular beam epitaxy is a particular sophisticated form of thermal evaporation.*

    Electron beam evaporation: *melting is done by the electron beam.*

    Sputtering by an ion beam: *Rare gas Ions sputter the material target. Deposition can be assisted using another "reactive beam".*

    Pulsed laser deposition: *it is an ablation process by pulses of focused laser which vaporizes the material.*

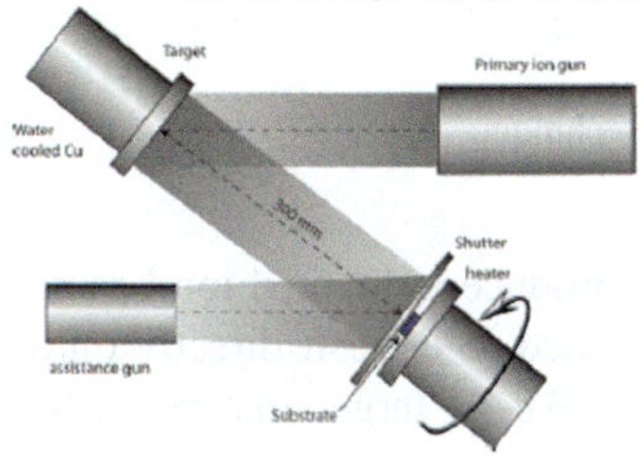

Reactive Dual Ion Beam Sputtering (RDIBS)

TiN/Ag multilayers prepared by RDIBS

Tse et al, Surface and Coatings Technology 180 –181 (2004) 470–477

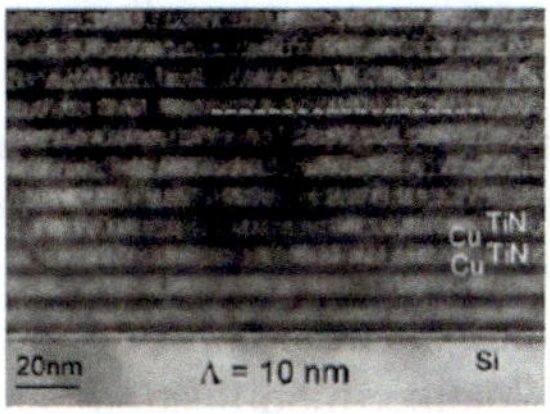

11.1.1 Gas-Vapor Deposition

Chemical Vapor Deposition (CVD) methods are well known in a semiconductors industry. In CVD process, vapor is formed in a reaction chamber by pyrolysis, reduction, oxidation or nitridation, and then deposited on the surface. Areas of growth are controlled by patterning processes like photolithography or photomasking (deposition patterns are etched on to the surface layers of the wafers).

The most important application of CVD methods is the synthesis of carbon nanotubes where CVD is considered to offer one of the most effective routes for scaling up to industrial production. Many other nanoparticles are synthesized by CVD as well.

11.1.2 Plasma-Based Synthesis

Plasma spraying of materials onto substrates to form protective coatings is widely used industrial practice. The use of plasma (i.e., ionized gas) during vapor deposition allows accessing to different chemical and physical processes and obtaining final materials of high-purity. There are several different types of *plasma deposition* reactor for plasmaassisted PVD and CVD. In plasma reactors temperatures of the order of 10,000 °C can be achieved, causing evaporation or initiating chemical reactions. The main types of plasma used are Direct Current (DC) plasma jet, DC art plasma and Radio-Frequency (RF) induction plasma. DC glow discharge involves the ionization of gas atoms by electrons emitted from a heated filament.

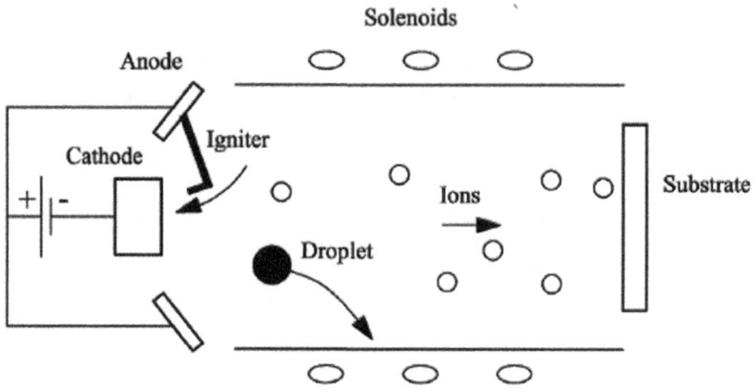

Fig. 11.1 Schematic diagram of a vacuum arc deposition technique

The gas ions in the plasma are then accelerated to produce a directed ion beam. If the gas is a reactive precursor gas, this ion beams used to deposit directly onto a substrate. If an inert gas is used, the ion beam strikes a target material which sputters neutral atoms onto a neighboring substrate.

Another modification is a *magnetron sputtering*. In sputtering methods material is vaporized from a solid surface by bombardment with ions of inert gas from sputter sources like an ion gun or hollow cathode plasma sputter. Plasma is created by the application of a large DC potential between two parallel plates. A static magnetic field is applied near a sputtering target and confines the plasma to the vicinity of the target. Ions from the high-density plasma sputter material, predominantly in the form of neutral atoms, from the target onto a substrate. One of the greatest benefits of the magnetrons is high deposition rate (about 1 μm/min) that makes the method to be industrially viable. Moreover, multiple targets can be rotated so as to produce a multilayered coating on the substrate.

Nowadays, *vacuum arc deposition* is well-established process for producing of thin films and nanoparticles. This technique involves the initiation of an arc by contacting a cathode made of a target material. An igniter is attached to an anode in order to generate a low-voltage, high-current self-sustaining arc. The arc ejects ions and material droplets from a small area on the cathode. Further, the ions are accelerated towards a substrate while any large droplets are filtered out before deposition (Fig. 11.1).

One of the outstanding strides in plasma processing for nanoparticles synthesis is the developed process of the vapor condensation. The principle of this method is illustrated in Fig. 11.2. The precursor material is put into the working chamber with a stable arc. The chamber is filled by reactive gas that becomes ionized; then molecular clusters are formed and cooled to produce nanoparticles. In the plasma-assisted PVD processes the vapor phases originate from a solid target. Instead, plasma-enhanced CVD employs gas phase precursors that are dissociated to form molecular fragments which condense to form thin films or nanoparticles. The dissociation temperatures required for CVD tend to be much lower than for

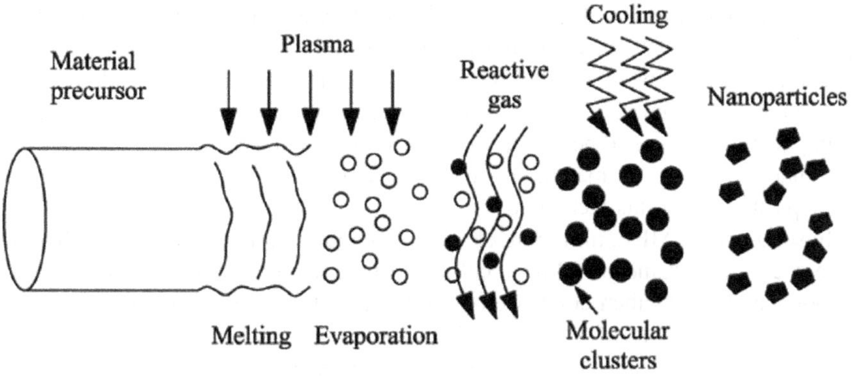

Fig. 11.2 Principle of the vapor condensation process

conventional CVD processes due to the high energy of the plasma, and this may be of importance for deposition on sensitive substrates such as semiconductors and polymers.

11.1.3 Molecular Beam Epitaxy

A molecular beam epitaxy (MBE) machine is essentially an ultra-high-precision, ultra clean evaporator, combined with a set of in situ tools, such as Auger electron spectroscopy (AES) and/or reflection high-energy electron diffraction (RHEED), for characterization of the deposited layers during growth. The reactor consists of an ultrahigh-vacuum chamber (typically better than 5×10^{-14} atm pressure) of approximately 1.5 m diameter (Fig. 11.3). The most important aspect of MBE is

Fig. 11.3 Schematic diagram of a molecular beam epitaxy thin film deposition system (adapted from *Nanoscale Science and Technology*, Eds. R.W. Kelsall, I.W. Hamley, M. Geoghegan, Wiley Ltd., 2005)

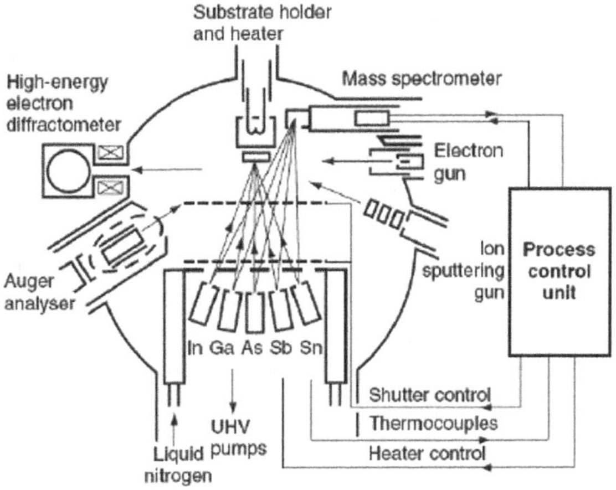

the slow deposition rate (1–300 nm per minute), which allows the films to grow epitaxialy on a heated substrate under UHV conditions.

The sources can be either solid or gaseous and an MBE machine will typically have an array of multiple sources, which can be shuttered to allow layered, alternating heterostructures to be produced. Semiconductor quantum wells, superlattices and quantum wires and metallic or magnetic multilayers for spin valve structures are deposited using this technique.

In solid-source MBE, ultra-pure elements such as gallium and arsenic are heated in separate quasi-Knudsen effusion cells until they begin to slowly evaporate. The evaporated elements then condense on the wafer, where they may react with each other. In the example of gallium and arsenic, single-crystal gallium arsenide is formed. The term "beam" simply means that evaporated atoms do not interact with each other or any other vacuum chamber gases until they reach the wafer, due to the long mean free paths of the beams. The substrate is rotated to ensure even growth over its surface. By operating mechanical shutters in front of the cells, it is possible to control which semiconductor or metal is deposited. For example, opening the Ga and As cell shutters results in the growth of GaAs. Shutting the Ga cell and opening the Al cell switches the growth to AlAs. As the shutters can be switched rapidly, in comparison to the rate at which material is deposited, it is possible to grow very thin layers exhibiting very sharp interfaces. Other effusion cells contain elements required for doping, and it is possible to monitor the growth by observing the electron diffraction pattern produced by the surface. MBE can also be performed using gaseous sources, and this is often termed chemical beam epitaxy (CBE). When the sources are metal-organic compounds, the process is known as *metalorganic MBE (MOMBE).*

11.1.4 Inert Gas Condensation

Gas condensation, as a technique for producing nanoparticles, refers to the formation of nanoparticles in the gas phase, i.e., condensing atoms and molecules in the vapor phase. *The inert gas condensation* (*IGC*) process is one of the most known and well established procedures for the production of nanopowders. These powders are widely used for electrically conductive adhesives and polymers, which find application, among other things, for the surface mounting technique in electronics. Figure 11.4 gives a schematic overview of the IGC procedure. Here a material, often a metal, is evaporated from a heated metallic source into a chamber which has been previously evacuated to about 10^{-7} torr and backfilled with inert gas to a low-pressure. The metal vapor cools through collisions with the inert gas atoms, becomes supersaturated and then nucleates homogeneously; the particle size

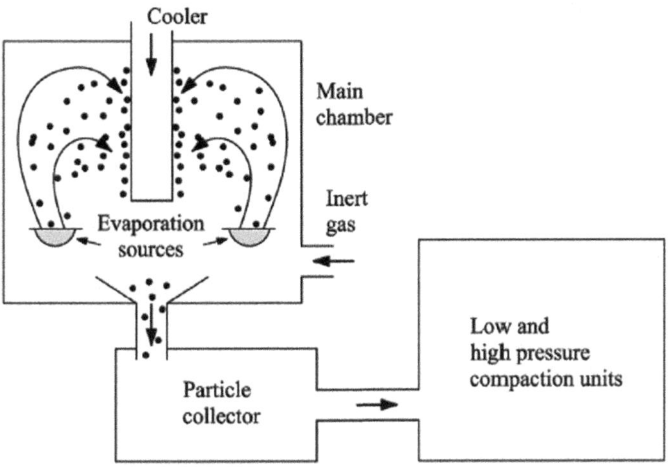

Fig. 11.4 Schematic diagram of an inert gas condensation apparatus

is usually in the range 1–100 nm and can be controlled by varying the inert gas pressure. Ultimately, the particles are collected and may be compacted to produce a dense nanomaterial.

11.1.5 Flame Pyrolysis

Flame pyrolysis is widely used in production of carbon black, fumed silica (SiO_2), ultrafine TiO_2 and many other materials. In this process flame heat is used to initiate the chemical reactions. The disadvantage of this method is that it usually yields agglomerated particles. Furnace flow reactors (Fig. 11.5) are the simplest systems used to produce saturated vapor for substances having a large vapor pressure at intermediate temperatures. In these systems a crucible containing the source material is placed in a heated flow of inert carrier gas. Materials with low vapor pressure can be fed in as suitable precursors.

In the *laser pyrolysis* technique IR laser is used to rapidly heat a flowing reacting gas. The source molecules are heated selectively by absorption of the laser energy whereas the carrier gas is not. Heating leads to decomposition of the precursors and super saturation is created resulting in nanoparticle formation.

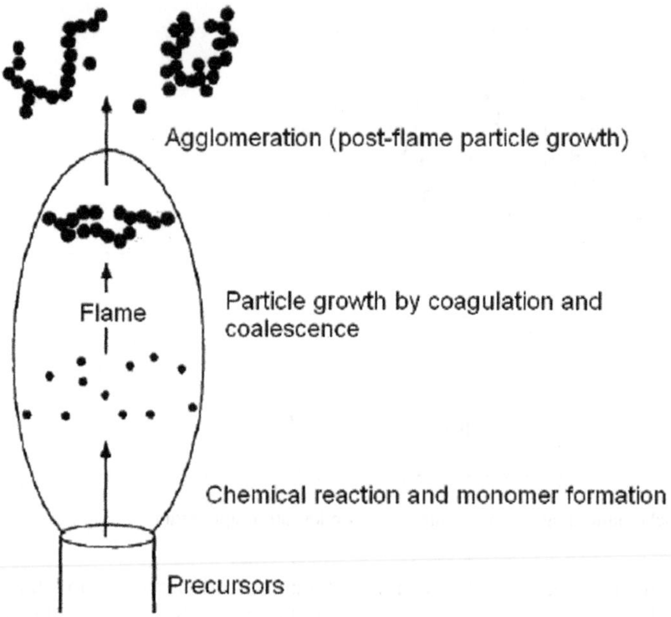

Fig. 11.5 Schematic diagram of a flame pyrolysis

11.2 Liquid Phase Synthesis

Precipitating nanoparticles [10–15] from a solution of chemical compounds can be classified into five major categories: (1) colloidal methods; (2) sol–gel processing; (3) water–oil microemulsions method; (4) hydrothermal synthesis; and (5) polyol method. Some of these techniques of nanoparticles producing are discussed below.

11.2.1 Colloidal Methods

Colloidal methods are simple and well established wet chemistry precipitation processes in which solutions of the different ions are mixed under controlled temperature and pressure to form insoluble precipitates. For metal nanoparticles the basic principles of colloidal preparation were known since antiquity. E.g. gold colloids used for high quality red and purple stained glass from medieval times to date. However proper scientific investigations of colloidal preparation methods started only in 1857 when Faraday has published results of his experiments with gold. He prepared gold colloids by reduction of $HAuCl_4$ with phosphorus. Today, colloidal processes are widely used to produce such nanomaterials like metals, metal oxides, organics, and pharmaceuticals. An important sub-set of colloidal methods are *son chemistry methods*, in which acoustic cavitation is used to control the process. By applying the ultrasonic radiation to the precursor solution the chemical reactions is initiated. The ultrasound leads to creation, growth and rapid collapse of small bubbles, which act like nucleation centers. The growth of nucleus ends after bubble collapse. Nanoparticles produced by wet chemical methods can remain in liquid suspension for further use or may be collected by filtering or by spray drying to produce a dry powder.

11.2.2 Solution Precipitation

This method relies on the precipitation of nanometer-sized particles within a continuous fluid solvent. An inorganic metal salt, such as chloride, nitride and so on, is dissolved in water. Metal cations exist in the form of metal hydrate species, for example, $Al(H_2O)^{3+}$ or $Fe(H_2O_6)^{3+}$. These hydrates are added with basic solutions, such as NaOH or Na_4OH. The hydrolyzed species condense and then washed, filtered, dried and calcined in order to obtain the final product. Described method is relatively simple and widely used for production of single and multi-components oxide nanopowders using optimized reactions and reaction conditions. Moreover, subsequent processing of colloids can include additional colloidal precipitation on particle surfaces to produce a core–shell nanoparticle structure, deposition on substrates to produce quantum dots, self-assembly on substrates as ordered 2D and even 3D arrays, and finally embedding in other media to form nanocomposites. One problem inherent in many colloidal methods is aging of colloid solutions, e.g. the particles can increase their size as a function of time [16, 17].

11.2.3 Electrodeposition

- The electrodeposition consists of different steps
  - Stabilization of the particles in the bath
  - Transport of the particles from the solution toward the electrode
  - Adhesion of the particles to the substrate.

- Surfactants are added to industrial plating bath to prevent particle aggregation.
- The efficiency (rate) of the electrode position is determined by the flow intensity, electrode polarization and ionic strength of the solution.
- Electrophoresis is the movement of an electrically charged substance under the influence of an electric field. This movement is due to the Lorentz force.

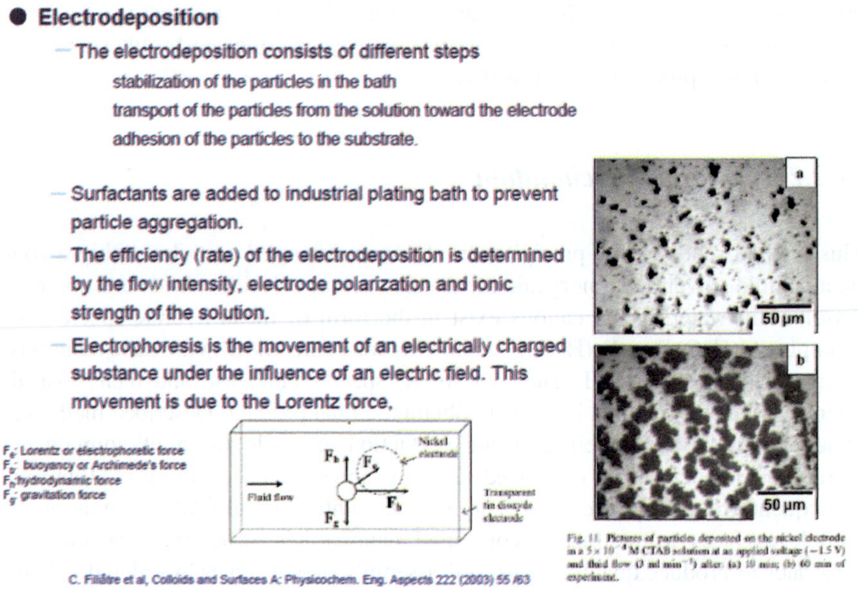

Fe: Lorentz or electrophoretic force, Fb: buoyancy or Archimedes' force, Fh: hydrodynamic force, Fg: gravitation force.
C. Filiâtre et al., Colloids and Surfaces A: Physicochem. Eng. Aspects 222 (2003) 55/63.

The principle of electrodeposition is inducing chemical reactions in an aqueous electrolyte solution with the help of applied voltage, e.g. this is the process of using electrical current to coat an electrically conductive object with a relatively thin layer of metal. This method is relevant to deposition of nanostructured materials include metal oxides and chalcogenides. The process of electrodepositing may be either anodic or cathodic. In an anodic process, a metal anode is electrochemically oxidized in the presence of other ions in solution, which then react together and deposit on the anode. Meanwhile in a cathodic process, components are deposited onto the cathode from solution precursors. Electrodeposition is relatively cheap and can be performed at low temperatures which will minimize interdiffusion of materials in the case of a multilayered thin film preparation. The film thickness can be controlled by monitoring the amount of charge delivered, whereas the deposition rate can be followed by the variation of the current with time. The composition and defect chemistry can be controlled by the magnitude of the applied potential, which can be used to deposit non-equilibrium phases. Pulsing or cycling the applied current or potential in a solution containing a mixture of precursors allows the production of a multilayered material. The potential during the pulse will determine the species deposited whilst the thickness of individual layers is determined by the charge passed. Alternatively, the substrate can be transferred periodically from one electrolytic cell to another. The final films can range in thickness from a few nanometers to tens of microns and can be deposited onto large specimen areas of complex shape, making the process highly suitable for industrial use. Electrodeposition can also be performed within a nanoporous membrane which serves to act as a template for growth; for example, anodized alumina has cylindrical nanopores of uniform dimensions and electrodeposition within this membrane can produce nanocylinders. Deposition on planar substrates can also limit nanocrystal growth and produce ordered arrays; if the growth is epitaxial then any strain due to lattice mismatch between the nanocrystal and the substrate can be growth-limiting. Furthermore, it is possible to modify the surface of substrates (e.g., by STM or AFM) to produce arrays of defects which can act as nucleation sites for the electrodeposition of nanocrystals.

Electrodeposition has three main attributes that make it so well suited for nano-, bio and micro technologies: (1)—It can be used to grow functional material through complex 3D masks; (2)—It can be performed near room temperature from water based electrolytes; (3)—It can be scaled down to the deposition of a few atoms or up to large dimensions.

11.3 Sol–Gel Technique

- A transition from a liquid phase (colloidal solution → "sol") into a solid phase (→ "gel") phase.
- Organometallic precursors (TMOS, TEOS, Zr(IV)-Propoxide, Ti(IV)-Butoxide …) submitted to hydrolysis and condensation reactions:
  - $M{-}O{-}R + H_2O \rightarrow M{-}OH + R{-}OH$ (hydrolysis)
  - $M{-}OH + HO{-}M \rightarrow M{-}O{-}M + H_2O$ (water condensation)
  - $M{-}O{-}R + HO{-}M \rightarrow M{-}O{-}M + R{-}OH$ (alcohol condensation).

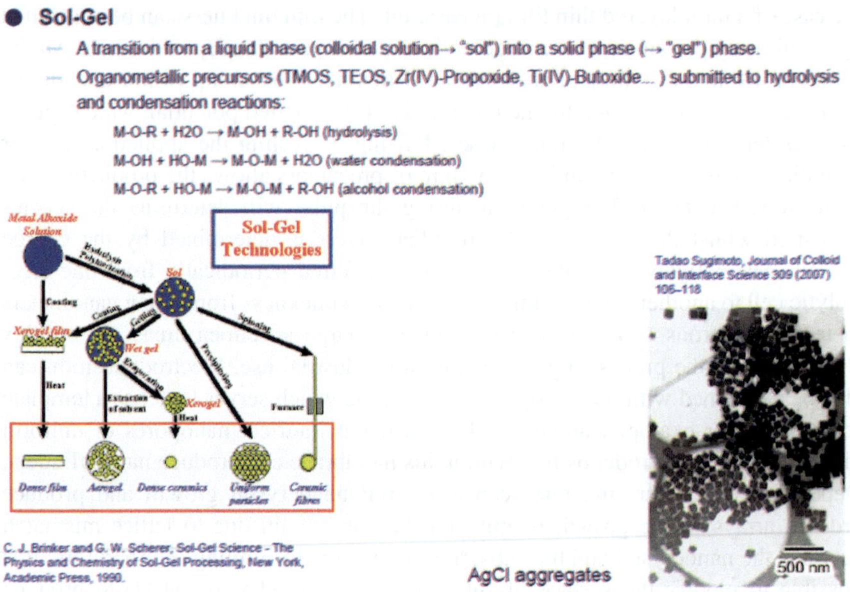

AgCl aggregates Tadao Sugimoto, Journal of Colloid and Interface Science 309 (2007) 106–118.

Sol–gel technology is a well-established colloidal chemistry technology, which offers possibility to produce various materials with novel, predefined properties in a simple process and at relatively low process cost. *The sol* is a name of a colloidal solution made of solid particles few hundred nm in diameter, suspended in a liquid phase. *The gel* can be considered as a solid macromolecule immersed in a solvent. So, in general terms, the sol–gel process consists in the chemical transformation of a liquid (the sol) into a gel state and with subsequent post-treatment and transition into solid oxide material. Sol–gel process has been used for glass coating since the 1939 and today it is a common technology for fabrication of ultra-fine powders, monolithic ceramics and glasses, ceramic fibers, inorganic membranes, aerogels and other types of materials. *Sol–gel technique* is one of the most popular solutions processing for nanoparticles (mostly oxides) production. This methods involve a set of chemical reactions which irreversibly convert a homogeneous solution of molecular reactant precursors (a sol) into a three-dimensional polymer (a gel) forming an elastic solid filling the same volume as the solution.

11.3.1 Sol–Gel Process

Figure 11.6 gives an insight to the process. The hydrolysis of metal alkoxides involves nucleophilic reaction with water:

$$M(OR)_y + xH_2O \rightarrow M(OR)_{y-x}(OH) + xROH$$

Condensation occurs when either hydrolyzed species react with each other and release a water molecule, or a hydrolyzed species react with an unhydrolyzed species and release an alcohol molecule. Processing involves forming the gel followed by drying to remove the solvent. Gels can be cast and molded to form a micro porous pre-form and dried to produce a monolithic bulk material (e.g., a xerogel or an aerogel) that can be used to form filters and membranes. They can also be spin coated or dipped to produce thin (typically 50–500 nm) films on substrates. These films are used for electronic thin film devices, for wear, chemical or oxidation protection, as well as for their optical properties (e.g., antireflection). Alternately, fibers can be drawn from the gel; e.g., silica fibers for light transmission. The interconnected nanoscale porosity in the dried gel can be filled via incorporation of a second material using techniques such as liquid infiltration or chemical reaction in order to obtain the nanocomposites.

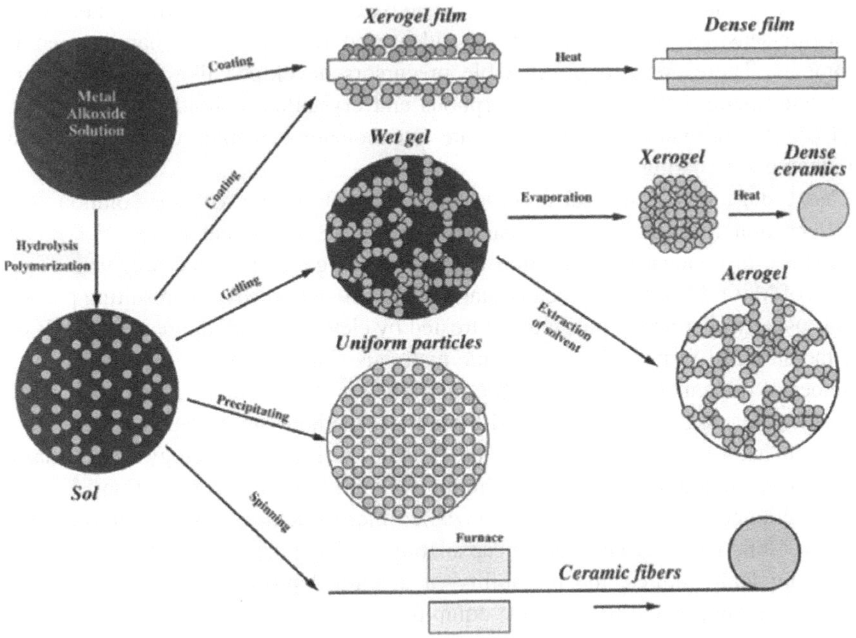

Fig. 11.6 Sol–gel technology and its final products

Diphase gels use the initial gel host for the precipitation of a second phase by sol–gel routes. In another variant, organic material can be incorporated as a monomer within an inorganic gel host; the monomers can be subsequently polymerized to form a hybrid material. If a dense rather than a nanoporous material is desired, drying is followed by sintering at higher temperatures. The high surface area leads to rapid densification, which can be accompanied by significant grain growth if temperatures are too high.

The main benefits of sol–gel processing are the high purity and uniform nanostructure achievable at low temperatures. First stage of the sol–gel process is a preparation of precursor solution. Precursors play a key role in sol–gel technology directly affecting the porosity, refractive index, hardness and other performance characteristics of the resultant material. Precursor can be inorganic, but more often metal organic precursors are used. Common metal organic precursors for the sol–gel process are metal alkoxides ($M(OR)_z$), where R stands for an alkyl group (C_xH_{2x+1}). In a typical sol–gel process, the precursor is subjected to a series of hydrolysis and polymerization reactions to form a colloidal suspension (the sol). For example, in the case of metal alkoxide, it is dissolved in alcohol and then the water is added under acidic, neutral or basic conditions. Addition of the water leads to hydrolysis in which alkoxide ligand is replaced with a hydroxyl liquid:

$$M(OR)_z + H_2O \rightarrow M(OR)_{z-1}(OH) + ROH$$

The chemical reactivity of metal alkoxides is related to the R—the larger the R, the slower the hydrolysis of metal alkoxides. Metal alkoxides are very moisture-sensitive (except silicon alkoxides) and require special handling environments. There are also more stable precursors developed like e.g. metal carboxylate, metal dialkylamides, amorphous and crystalline colloidal sol solutions, and organic/inorganic hybrids. They are not moisture-sensitive, easy to use, and produce good coatings.

Next stage of the sol–gel process after preparation of precursor solution is a condensation reaction in which particles condense in the gel phase. For metal alkoxides, in condensation reaction the hydroxyl ligands produce polymers composed of M–O–M bonds. In the last stage of the sol–gel process the resulting porous gel is usually chemically purified and treated by elevated temperatures, or by UV or IR radiation, to form high purity oxide materials. The gel can be modified with a number of dopants to produce unique properties in the resultant glass unattainable by other means. Materials produced by sol–gel method can be either totally inorganic in nature or both inorganic and organic. Sol–gel technology offers many advantages, including excellent stoichiometry control of precursor solutions, ease of compositional modifications, customizable microstructure, ease of introducing various functional groups or encapsulating sensing elements, relatively low annealing temperatures, the possibility of coating deposition on large-area substrates, and simple and inexpensive equipment.

11.3.2 Sol–Gel Coating Processes

The most common application of sol–gel process is fabrication of various coatings and films, and it is reasonable to consider common wet coating technologies. *Dip Coating* is illustrated in Fig. 11.7. In the dip coating process the substrate is immersed into a sol and then withdrawn with a well-defined speed under controlled temperature and atmospheric conditions. The sol left on substrate forms a film with thickness mainly defined by the withdrawal speed, the solid content and the viscosity of the liquid. Next stage is a gelation (densification) of the layer by solvent evaporation and finally annealing to obtain the oxide coating.

More recently, an *angle-dependent dip coating* process has been developed (Fig. 11.8). In this case, the coating thickness is dependent also on the angle between the substrate and the liquid surface, so different layer thickness can be obtained on the top and bottom side of the substrate.

Spin Coating is illustrated in Fig. 11.9. Spin coating is used for making a thin coating on relatively flat substrates or objects. The material to be made into coating is dissolved or dispersed into a solvent, and then deposited onto the surface and spun off to leave a uniform layer for subsequent processing stages and ultimate use. General stages in spin coating include deposition of the coating fluid onto the

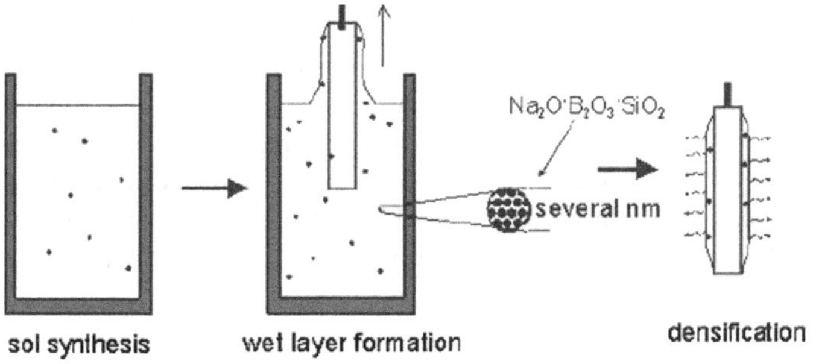

Fig. 11.7 Example of formation of sodium borosilicate coating on glass by dip coating (H. Schmidt, M. Mennig, Wet Coating Technologies for Glass, Tutorial, November 2000)

Fig. 11.8 Schematic of angle dependent dip coating

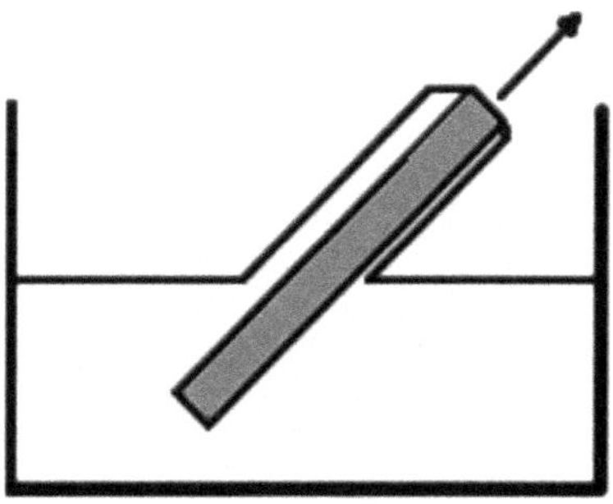

Fig. 11.9 Stages of the spin coating process: deposition of the sol, spin up, spin off and gelation by solvent evaporation

substrate, aggressive fluid expulsion from the substrate surface by the rotational motion, gradual fluid thinning, and coating thinning by solvent evaporation.

The coating thickness is inversely proportional to the square root of the rotation speed and also depends on the coating solution properties like viscosity and composition. *Flow coating* processes is illustrated in Fig. 11.10. In the flow coating process the liquid coating system is poured over the substrate to be coated as shown schematically on the figure below.

The coating thickness depends on the angle of inclination of the substrate, the coating liquid viscosity and the solvent evaporation rate. The advantage of the flow-coating process is that non-planar large substrates can be coated rather easily. As a variation of this process, the spinning of the substrate after coating may be helpful in order to obtain more homogenous coatings. If no spinning process is employed, the coating thickness increases from the top to the bottom of the substrate.

Capillary coating is illustrated in Fig. 11.11. Capillary or laminar flow coating process combines the high optical quality of the dip coating process with the

Fig. 11.10 Scheme of the flow-coating process

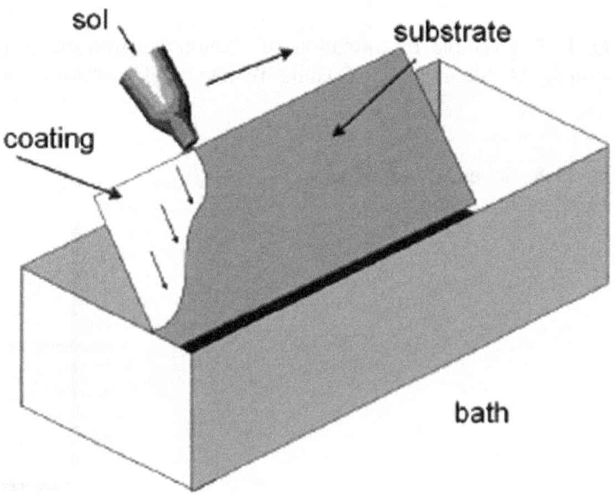

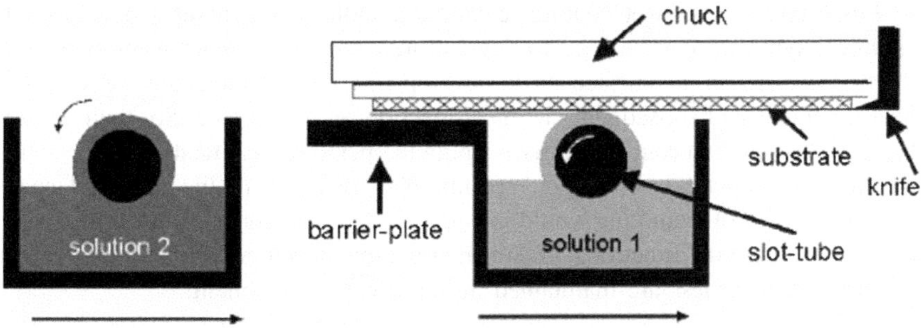

Fig. 11.11 Schematics of the capillary coating process (B.T. Chen, Polymer Engineering and Science 23 (1983) 399–403)

advantage that all the coating liquid can be exploited. In this process the tubular dispense unit is moved under the substrate surface without physical contact. A spontaneous meniscus is created between the top of the slot tube (or porous cylinder) and the substrate surface, and achieving laminar deposition conditions a coating is deposited with high uniformity. Multilayer coatings can be fabricated by using two dispense lines one after the other.

Roll coating. Roll coating is a process by which a thin liquid film is formed on a continuously moving web or substrate by using one or more rotating rolls. It is of importance that in order to avoid structures in the surface cutted roles have to be used, and coating thickness and viscosity of the liquid have to be adapted very thoroughly. Using cutted roles, the amount of liquid transported onto the glass surface is defined by the voids cutted out of the role. After the deposition, the parts have to coagulate and to form a homogeneous film. For this reason, the wetting behavior of the glass against the liquid has to be perfect and the drying speed has to be adapted to the film forming velocity. Therefore, temperature and atmosphere have to be controlled perfectly.

Spray coating techniques. Spray coating is used e.g. for coating irregularly shaped glass forms like pressed glass parts, lamps or container glass (cold end coating). Using special automatic flat spraying equipment it is possible to prepare glass like coatings (coloured coatings and electrochromic WO_3 coatings) with thickness in the range between 100 and 220 nm. The preparation of optical coatings by spraying offers several advantages compared to the dip coating technique, since it is several times faster, the waste of coating sols is much smaller, coating sols with rather short pot lives can be used and the coating step is suitable for establishing an in-line process.

There are also processes like the pyrosol-process, where very fine droplets are produced and sprayed onto the surface. The coating material does not hit the surface in form of liquid droplets but more or less in form of dried small particles in the nanometer range. Due to the high reactivity of these particles when reaching the hot surface, a continuous and very homogeneous glass film can be formed.

Post-treatment. In wet coating techniques, molecular structures developed by chemical synthesis can be used to develop new properties either when preserving these structures on the surface, or to develop new desired molecular structures by heat treatment and subsequent chemical reaction on the surface. So, there are two basic routes: the first route comprises a high temperature treatment after the coating step in order to get "glass-like" or "ceramic-like" materials on the glass surface and the second type of technique would include a low-temperature UV- or infrared type of curing, where the functional chemical structures developed in the liquid coating material, more or less are maintained during this post-treatment.

11.3.3 Reverse Micelles as Nanoreactors

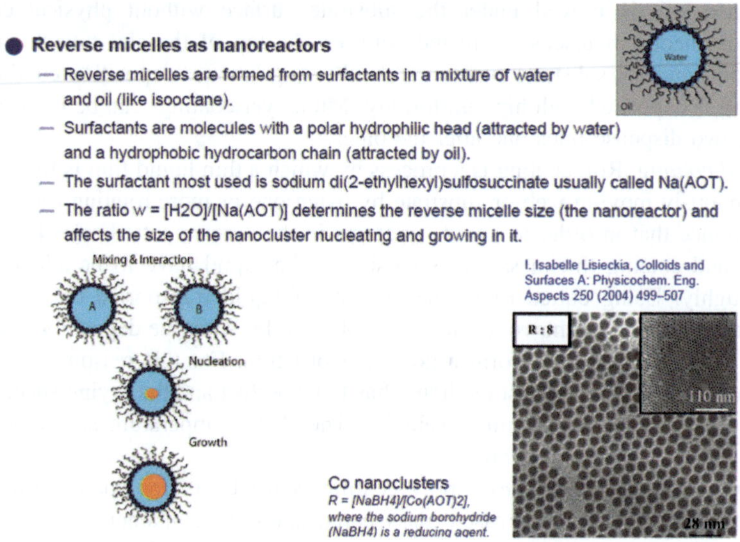

- Reverse micelles are formed from surfactants in a mixture of water and oil (like isooctane).
- Surfactants are molecules with a polar hydrophilic head (attracted by water) and a hydrophobic hydrocarbon chain (attracted by oil).
- The surfactant most used is sodium di (2-ethylhexyl) sulfosuccinate usually called Na(AOT).
- The ratio w = $[H_2O]/[Na(AOT)]$ determines the reverse micelle size (the nanoreactor) and affects the size of the nanocluster nucleating and growing in it.

11.3.4 Sol–Gel Applications

The applications for sol gel-derived products are numerous. Sol–gel technologies are used in various applications including optics, electronics, energy, space, sensors and separation technology. Sol–gel coatings are extensively used for such diverse applications as protective and optical coatings, anti-reflection coatings, passivation and planarization layers, sensors, high or low dielectric constant films, inorganic membranes, electro-optic and nonlinear optical films, electrochromics, semiconducting anti-static coatings, superconducting films, strengthening layers and ferroelectrics. Cast into a mold, and with further drying and heat-treatment, dense ceramic or glass particles with novel properties can be obtained in the forms that cannot be created by any other method. With the viscosity of a sol adjusted into a proper range, different fibers and needles can be drawn which can be used e.g. for fiber optic sensors or thermal insulation. Ultra-fine and uniform ceramic powders can be formed by precipitation. These powders of single- and multicomponent compositions can be made in submicron particle size for dental and biomedical applications. Composite powders have been patented for use as agrochemicals and herbicides. Also powder abrasives, used in a variety of finishing operations, are made using a sol–gel type process. Another important application of sol–gel processing is to carry out zeolite synthesis. Other elements (metals, metal oxides) can be easily incorporated into the final product and the silicalite sol formed by this method is very stable. If the liquid in a wet gel is removed under a supercritical condition, a highly porous and extremely low-density material called aerogel is obtained. Drying the gel by means of low temperature treatments (25–100 C), it is possible to obtain porous solid matrices called xerogels. Other products fabricated with this process include various ceramic membranes for microfiltration, ultrafiltration, nanofiltration, evaporation and reverse osmosis.

11.4 Solid-State Phase Synthesis

One of the nanofabrication processes of major industrial importance is the high-energy ball milling, also known as mechanical attrition or mechanical alloying. In contrast to previous three groups, where particles were produced in down up processes, mechanical methods based on top down (size reduction) processes like attrition and disintegration of larger particles. When particle size lies in nanometer range, terms ultrafine grinding or nanosizing are often used for the process. Mechanical attrition methods give very high production rates (up to tones per hour) and are widely used for industrial production of clay, coal and metal powders. The production process often involves wet milling in milling chamber with rotating perforated plates. Wet milling results in suspension that needs to be stabilized by adjustment of the pH to prevent particle recombination due to increasing particle–particle interactions.

11.4.1 Mechanical Milling, Attrition and Alloying

In the 1970s, the method of *mechanical attrition* (*MA*) of powder particles followed by high temperature sintering was developed as an industrial process to successfully produce new alloys and phase mixtures. For example, this powder metallurgical process allows the preparation of alloys and composites, which cannot be synthesized via conventional casting routes. In the 1980s, the method of *high-energy milling* gained a lot of attention as a nonequilibrium solid-state process resulting in materials with nanoscale microstructures. Typical objectives of the milling include particle size reduction, solid-state alloying, mixing or blending, and particle shape changes. A variety of ball mills have been developed for different purposes, among them there are tumbler-, attrition-, shaker-, vibratory-, and planetary mills, etc. However, the basic principal of mechanical attrition is similar for all processes and can be illustrated as shown in Fig. 11.12. Coarse-grained materials in the form of powders are crushed mechanically in rotating drums by hard steel or tungsten carbide balls, usually under controlled atmospheric conditions to prevent unwanted reactions such as oxidation. This repeated deformation can cause large reductions in grain size via the formation and organization of grain boundaries within the powder particles.

High-energy milling forces can be obtained by using high frequencies and small amplitudes of vibration. Ball mills are highly energetic, and reactions can take place by one order of magnitude faster compared with other types of mills. Since the kinetic energy of the balls is a function of their mass and velocity, dense materials (steel or tungsten carbide) are preferable to ceramic balls. A conventional planetary ball mill consists of a rotating horizontal drum half-filled with small steel balls. As the drum rotates the balls drop on the powders that are being ground; the rate of

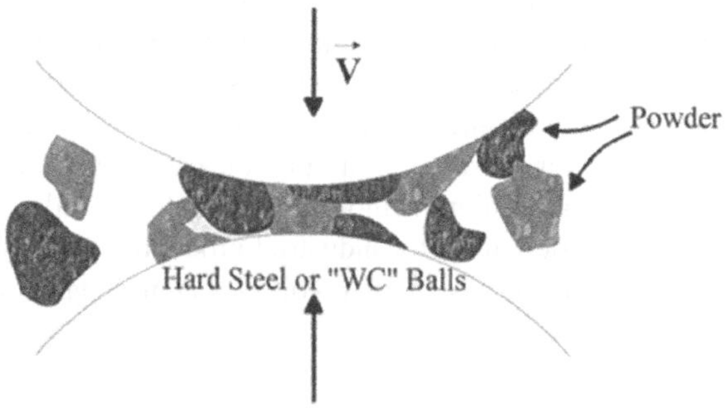

Fig. 11.12 Schematic representation of the process of mechanical attrition

grinding increases with the speed of rotation. When speeds are too high, however, the centrifugal force acting on the flying balls exceeds the force of gravity, and the balls are pinned to the wall of the drum. As a result, the grinding action stops. An attractor consists of a vertical drum with a series of impellers inside it. Set progressively at right angles to each other, the impellers energize the ball charge, causing powder size reduction because of impact between balls, between balls and container wall, and between balls, agitator shaft, and impellers. Some size reduction appears to take place by interparticle collisions and by ball sliding. A powerful motor rotates the impellers, which in turn agitate the balls in the drum.

Due to the continuous severe plastic deformation, a continuous refinement of the internal structure of the powder particles to nanometer scale occurs during high-energy mechanical attrition. The increase in temperature during this process is modest and is generally estimated to be in between 100 and 200 °C. The collision time roughly corresponds to about 2 μs. The rate of structural refinement is dependent on the mechanical energy input and the work hardening of the material. The process of grain refinement starts from the localizing of shear bands that indicates an increase in dislocations density. Hall–Petch relationship gives the value of the yield stress, σ, required to deform a polycrystalline material by dislocation movement:

$$\sigma = \sigma o + kd^{-1/2},$$

where d is the average grain size; σo and k are material constants. σo may be treated as a friction stress below which dislocations are immobile if there are no grain boundaries in material. At a certain level of stress, the dislocations create small angle boundaries that separate newly developed sub-grains of nano- or/and micro-meter size. Continuing mill leads to high angle boundaries formation and separation of nano-grains. An extrapolation to nanocrystalline dimensions shows that very high stresses are required to maintain plastic deformation. Experimental values for k and σo are typically of about 0.5 M $Nm^{-3/2}$ and 50 MPa, respectively. For a grain size of 10 nm, the minimum yield stress is of the order of 5 GPa corresponding to 15% of the theoretical shear stress, which sets a limit to the grain size reduction achieved by plastic deformation during ball milling. Therefore, the reduction of grain size to a few nanometers is limited by the stresses applied during ball milling as long as no dramatic elastic softening of the crystal lattice occurs. Further energy storage by mechanical deformation is only possible by an alternative mechanism. Grain boundary sliding has been observed in many cases at high temperatures leading to superplastic behavior. Alternatively, grain boundary sliding can also be achieved at a very small grain size and low temperature by diffusional flow of atoms along the intercrystalline interfaces, which allows the synthesis of ductile ceramics. This provides a mechanism for the self-organization and rotation of the grains, thus increasing the energy of the grain boundaries proportional to

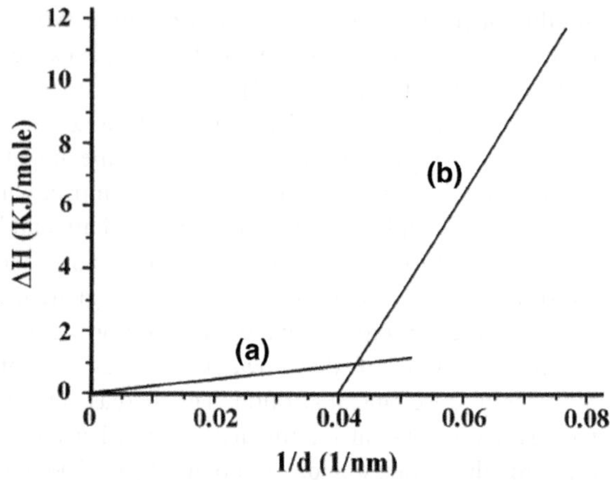

Fig. 11.13 The stored enthalpy *H* as function of reciprocal grain size 1/*d* of Fe at different levels of mechanical attrition

their misorientation angle and excess volume. This energetic microstructure-property relationship is schematically represented in Fig. 11.13. Here the stored enthalpy, ΔH, in attrited Fe powder is shown as a function of average reciprocal grain size, $1/d$, since $1/d$ scales also with the volume density of grain boundaries in the nanocrystalline material. Two different regimes can be clearly distinguished: (a)—the stored enthalpy shows only a weak grain size dependence typical for dislocation controlled deformation processes for small grain size decrease at the early stages of mechanical attrition; and (b)—an energy storage becomes more efficient when the average domain size is reduced below $d^* = 30$–40 nm. The critical grain size, d^*, corresponds to the size of nanograins which are formed within the shear bands. Therefore, for domain with size $d < d^*$ deformation is controlled by the properties of the low and, later, high angle grain boundaries which are developing in stages (b).

Milling can be used to induce chemical reactions through ball—powder colliding interactions. *Mechanical alloying* is a high-energy ball-milling process in which elemental blended powders are continuously welded and fractured to achieve alloying at the atomic level. Different components can be mechanically alloyed together by cold welding to produce nanostructured alloys. A nanometer dispersion of one phase in another can also be achieved. Microstructures and phases produced in this way can often be thermodynamically metastable. For example, recently the synthesis of ferrite compounds with a spinel structure (such as Fe_3O_4, $CoFeO_4$) by ball-milling aqueous solutions of metal chlorides and NaOH was reported. Existing factors influencing the mechanical alloying/milling processes include milling time, charge ratio, milling environment, and the internal mechanics specific to each mill. Cryomilling reduces oxygen contamination from the atmosphere and minimizes the heat generated during milling; therefore, fracturing is favored over welding, especially in milling of ductile materials. For example, the grain size of nickel has been sufficiently refined when milling is conducted in liquid nitrogen instead of methanol. Despite of its advances, mechanical attrition process for nanoparticles synthesis has some shortcomings that limit the application of this technique. One of the greatest problems of milling is high level of impurities or surface contaminations. Another problem is lack of control of the particle size distribution, and inability to tailor precisely shape and size of particles in the nanometers range.

11.4.2 Severe Plastic Deformation

Generally any form of mechanical deformation under shear conditions and high strain rates can lead to the formation of nanostructures, since energy pumped into crystalline structures results in lattice defects creation. The severe plastic deformation that occurs during machining, cold rolling, drawing, cyclic deformation or sliding wear has also been reported to form nanostructured material. *Severe plastic deformation (SPD)* processing, in which materials are subjected to the imposition of very large strains without the introduction of concomitant changes in cross-sectional dimensions of the samples, is one of the most successful top-down approaches. Materials produced by SPD techniques have grain sizes in the range of 50–1000 nm. However, they have sub-grain structures, which are often much smaller than 100 nm. Now there are several SPD processing available: *equal-channel angular pressing (ECAP), high-pressure torsion, accumulative roll-bonding, repetitive corrugation and friction stir processing.* Torsion straining under high pressure and equal channel angular pressing are the most well-known methods of providing large plastic deformations and formation of nanostructures. Figure 11.14 schematically shows the principles of SPD techniques [18–20].

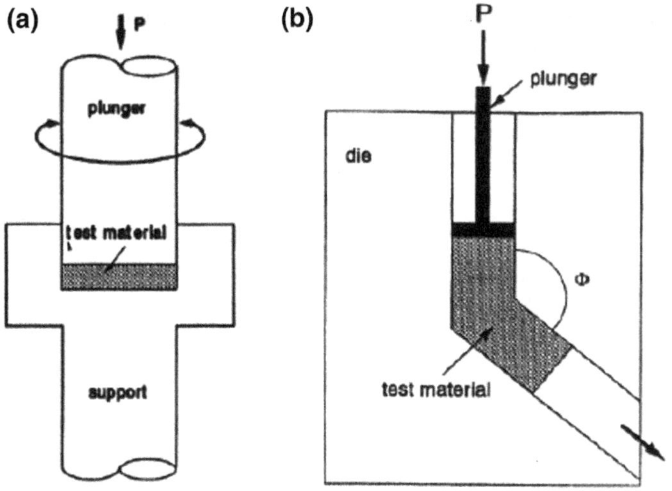

Fig. 11.14 Schematics of SPD methods: **a**—torsion straining; **b**—equal-channel angular (ECA) pressing

A method of torsion straining under high pressure can be used for fabrication of disk type samples (Fig. 11.14a). An ingot is held between anvils and strained in torsion under the applied pressure (P) of several GPa. A lower holder rotates and surface friction forces deform the ingot by shear. Due to the specific geometric shape of the sample, the main volume of the material is strained in conditions of quasi-hydrostatic compression under the applied pressure and the pressure of sample outer layers. As a result, in spite of large strain values, the deformed sample is not destroyed. Severe torsion straining can be successfully used not only for the refinement of a microstructure but also for the consolidation of powders. During torsion straining at room temperature, high pressure can provide a rather high density that may be close to 100% in the processed disk sample. During ECA pressing a billet of material is multiple pressed through a special die (Fig. 11.14b). The angle of two channels intersection is usually 900, however, in the case of a hard-to-deform material, the angle may be changed. Moreover, for processing of hardly deformed materials, ECA pressing can be conducted at elevated temperatures. During ECAP the direction and number of billet passes through the channels are very important parameter for microstructure development. Up to now, there is no common view on mechanisms of grains refinement. One of the proposed mechanisms is concentration of dislocations in cell walls and their absent inside cells in materials with high stacking fault energy (for ex., Cu and Ni). Further increase in the strain results in a decrease in a cell size and an increase in cell misorientations due to dislocations movement. That can lead to intensification of rotation modes of deformation within the whole sample. In materials with relatively low stacking fault energy, as, for ex., Ni–Cr alloy, the microstructure refinement is due to shear bands formation. On the basis of the experimental data the following model of evolution of the materials defect structure during SPD is proposed, Fig. 11.15. The main idea is transformation of a cellular structure (Fig. 11.15a) to a granular one. This stage is outlined by decrease in dislocation density in the grain (cell) boundaries caused by partial annihilation of dislocations of different signs when the dislocation density achieves some critical value (Fig. 11.15b). As a result,

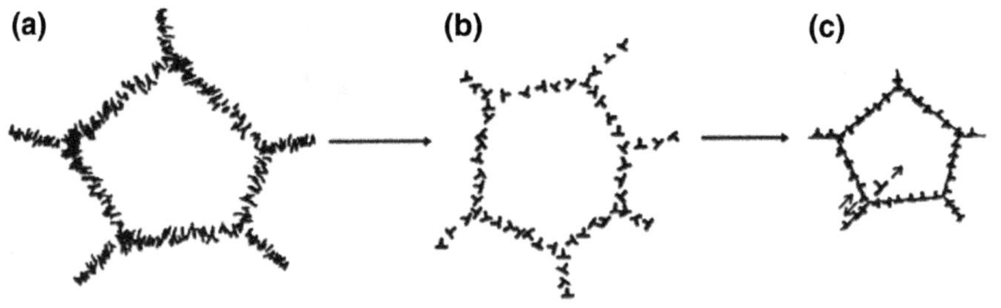

Fig. 11.15 Schematic model of dislocation structure evolution at different stages during severe plastic deformation (adapted from R.Z. Valiev, R.K. Islamgaliev, I. Alexandrov. *Bulk nanostructured materials from severe plastic deformation*, Progress in Mat. Sci., 2000, v. 45, 103–189)

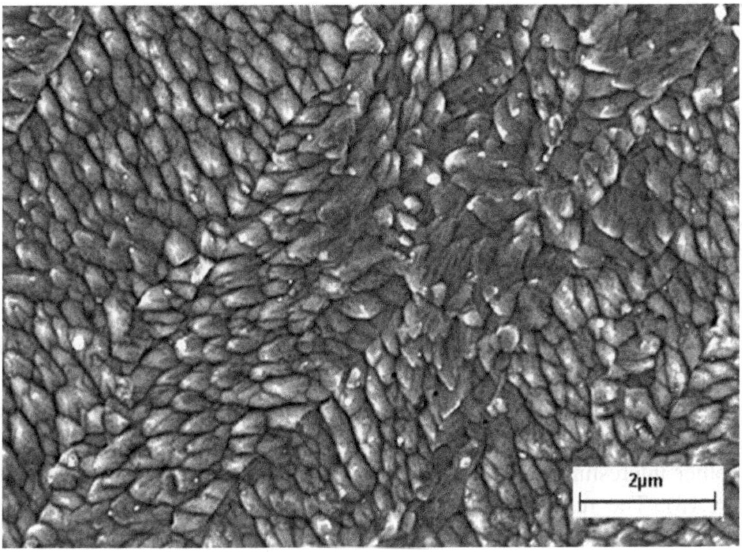

Fig. 11.16 Shear bands in ultra-fine grained copper produced by SPD method

there is an excess of dislocations with similar sign (Fig. 11.15c). Remaining dislocations further increase misorientation in the case of dislocations with Burgers vectors directed perpendicular to the boundary; at the same time, long range stress fields are connected with glide dislocations which can also lead to sliding of grains along grain boundaries, i.e. reveal rotation deformation modes which are indicated earlier for large plastic straining.

Most of the grain boundaries in nanocrystalline materials (nc) are at non-equilibrium state. Non-equilibrium grain boundaries in NSM are characterized by excess energy and long range elastic stresses due to the presence of a high density of extrinsic defects in their structure. These stresses result in significant distortions and dilatations of the crystal lattice near grain boundaries which are revealed experimentally by TEM and Xray methods. In turn, atomic displacements in near boundary regions change the dynamics of lattice vibrations and, as a result, change such fundamental properties as elastic modula, Debye and Curie temperatures and others. By now the nanostructures are obtained for a number of pure metals, alloys, steels and intermetallic compounds, as well as for metal—matrix composites and semiconductors. Microstructure of nano-copper produced with the help of SPD is shown in Fig. 11.16.

11.5 Other Methods

Here are some more noteworthy methods [21–30] among other existing methods. Fujitsu Laboratories Ltd. developed a novel method for fabrication of size and alignment controlled oxide dots on a semiconductor substrate shown in Figs. 11.10, 11.11, 11.12, 11.13, 11.14, 11.15, 11.16, 11.17. In this method, the AFM tip is moved near the semiconductor surface and a voltage is applied. As a result, water molecules in the atmosphere decompose into Hydrogen (H^+) and Hydroxide (OH^-) ions, and OH^- ions oxidize the substrate under the tip. The diameter of the oxide dots can be controlled by adjusting the durations of the voltage applications. In the next step oxide dots are removed with an etching solution or by ultrasonic cleaning in water, forming holes on the substrate. Finally, the substrate is placed into a molecular beam epitaxial (MBE) chamber, and quantum dots are fabricated using a process similar to self-assembled growth.

Another interesting method is fabrication of semiconductor tubes by rolling up (Fig. 11.18). The release of thin semiconductor layers from a substrate surface results in the formation of a novel class of free-standing nano- and micro objects. The right hand side shows a thin strained bilayer that rolls up into a tube as soon as the sacrificial buffer layer is selectively etched away. The technique allows accurately positioning of the novel semiconductor nanostructures and therefore constitutes a powerful method to combine top-down and bottom-up approach in nanotechnology.

A lot of efforts are focused on development of the new and on improvement of the existing methods, to make the fabrication of nanoparticles more and more simple, cheaper and easier to control.

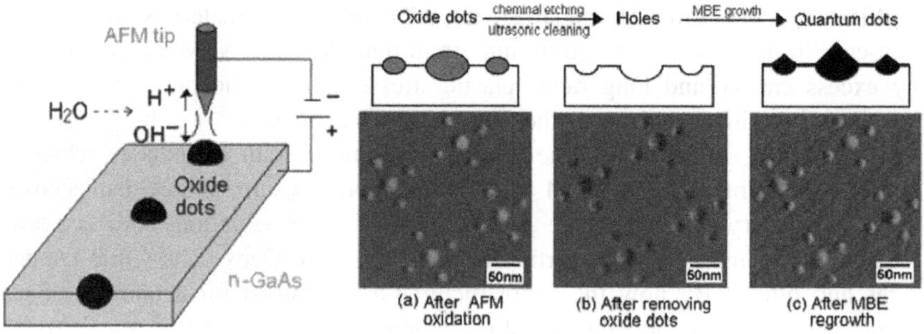

Fig. 11.17 Quantum dot fabrication and array of quantum dots

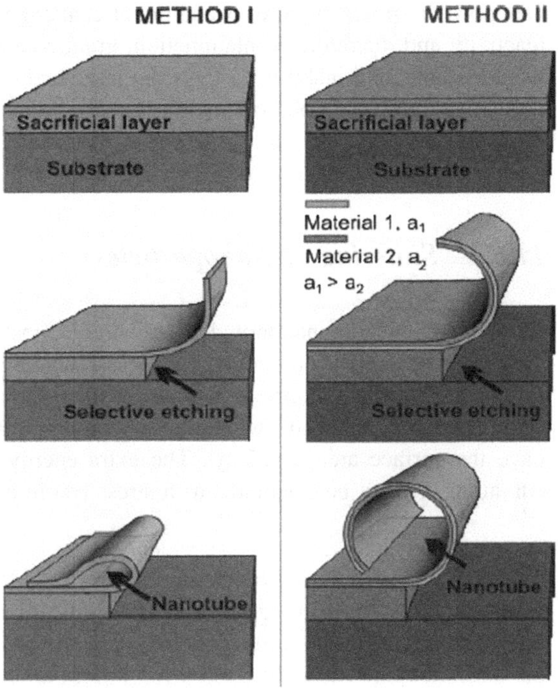

Fig. 11.18 Rolling up technique

11.6 Consolidation of Nanopowders

To take advantage of the unique properties of bulk nanocrystalline materials, the nanometer range powders have to be densified into parts of certain properties, geometry, and size. The key to the nanopowder consolidation process is to achieve densification with minimal microstructural coarsening and/or undesirable microstructural transformations. In addition, the fully dense specimen must be of sufficient size for reliable testing of final properties or a useful final product. Attempts to produce and densify nanopowders started as early as 1968. These efforts were related to sintering MgO to achieve superplastic behavior. In the 80s, when nanopowder production was initiated on a larger scale, attention was directed to nanopowder processing. However, densification commonly resulted in either grain coarsening, or unacceptably small specimen size and insufficient bonding. Special precautions must be taken to reduce the interparticle friction and heating; as well as particles activity to minimize the danger of explosion. The early 90s emphasized the need in development of the reproducible processing methods for manufacturing nanopowders into sizable parts that retain nanometer features. The past decade has brought significant advances in the practice and theory of nano-sintering, which consequently resulted in the production of fully dense parts with nanometer grain size. The densification process for conventional powders is well known, both theoretically and practically. However, the densification of nanopowders poses significant additional challenges. Powder agglomeration, high reactivity and, therefore, contamination, grain coarsening, and ultimate loss of the nano-features, and inability to fabricate large and dense parts are among the main problems. Outlined problems may adversely influence the overall quality of final product.

11.6.1 Sintering of Nanoparticles

Thermodynamically, nanopowders are highly unstable. The sintering process is driven by the tendency to reduce the excessively large surface area per unit volume. For example, 1 kg of copper spherical particles of 5 nm in radius has a total surface area of around 75,000 m^2, while the same amount of particles of radius 50 μm will have the surface area of 15 m^2. The extra energy of a surface with a radius of curvature, *R*, may be calculated as a stress (σ) in a Laplace equation:

$$\sigma = \gamma/R$$

where γ is the surface energy. In nanomaterials, this sintering stress may reach very high values. For instance, the sintering stress may be as large as 300 MPa in 10 nm particles compared to only 3 MPa for 1 μm particles, if γ has a typical value of 1.5 J/m^2.

Usually, the surface energy is assumed to be isotropic. For nanocrystals with significant surface area, the anisotropy problem becomes much more critical. First, sintering starts at a lower temperature. In this case, the effects of surface energy anisotropy are more pronounced. The highly anisotropic behavior of nanoparticle surfaces may be responsible for the crystallographic alignment, which has been often observed. The first step of powder consolidation is *compaction* of nanopowders. To produce a so-called green body, process goes at relatively low or moderate temperatures. Most sintering defects may be related to the microstructure of such green body. Inhomogeneities in density, packing, and particle size in green compact will limit the final sintered density. Such an example is the crack generation in ceramics upon sintering inhomogeneous cold compacts. Generally, nanocrystalline powders are more sensible to defects in green compacts compared to conventional powders. Due to Hall–Petch relationship between yield stress and particle size, the cold compaction of nanoparticles requires stresses in gigapascal range. That means the new non-conventional compaction routes are necessary. On the nano-scale, mechanical friction becomes substantial due to many interparticle contacts. These forces are a result of mechanical, electrostatic, van der Waals, and surface adsorption phenomena that are much more significant when particle size decreases down to nanometers. Friction limits particles motion and rearrangement and may cause a lower density green body formation. Rearrangement of the particles may be facilitated by the use of lubricants or water, often in combination with ultrasonic agitation or centrifuging to obtain an acceptable powder packing prior to pressing and/or sintering. Recently, osmotic compaction has been applied to ceramic powder compaction based on the osmotic chemical potential action. Equal or greater green densities as compared to physical pressure applications have been achieved with no mechanical breakage that external forces may produce.

Sintering process starts when powder is packed together and heated to high temperatures, typically about 2 $T_{melt}/3$. At this stage, diffusion becomes significant. The densification process consists of solid particle bonding or neck formation followed by continuous closing of pores from a largely open porosity to essentially a pore-free body. Multiple mechanisms are involved throughout sintering process, namely, evaporation condensation, surface diffusion, grain boundary diffusion, bulk diffusion, viscous flow, and plastic deformation. Each transport process exhibits a particular dependence on the particle/grain size and defect density. The highest sensitivity on particle size is that of surface and grain boundary diffusion. Although simultaneous mechanisms participate, the common sintering models attribute a predominant mass transport path to a specific sintering stage. For instance, surface diffusion is considered the principle mechanism during the initial stage when the main event is the neck formation. For nanoparticles with excessive surface area, highly curved surfaces, and reduced diffusion distances, surface diffusion is expected to be extremely rapid in early sintering stages. The general relationship between sintering parameters may be expressed as follows:

$$\frac{d\rho}{dt} \sim \frac{1}{d^n} \exp\left(-\frac{Q}{RT}\right)$$

where n is a constant, ρ is the density, Q is the activation energy for sintering and d is the mean powder particle diameter. The n is usually about 3 and Q is considered to be equal to the activation energy for grain boundary diffusion. Ideally, it would be desired to start sintering with producing a green body with a larger number of initial point contacts, smaller pores in a high density compact, and a uniform pore distribution. That would allow shortening the sintering time and using lower sintering temperatures might. The sintering of both metals and ceramics starts at temperatures of 0.2–0.4 T_{melt} while sintering of conventional powders starts at 0.5–0.8 T_{melt}. The lower temperatures for minimizing grain growth may hinder good intergranular bonding, thus compromising the expected high mechanical strength and ductility. Such low sintering temperatures may also interfere with the thermochemical oxide reduction on particle surfaces necessary for subsequent sinter bonding. The most recent efforts have been very fruitful in overcoming some of these problems (e.g., agglomeration and grain size control). This has been accomplished by major improvements in the nanopowder synthesis methods and understanding of the densification processes such as pore effects in nanosintering. The most common problem is the elimination of large pores that originate from the green compact. This elimination requires high temperatures upon subsequent sintering, thereby, promoting unwanted grain growth and losing the desired nanosize features. Similar to equation given above for sintering parameters, the pores fraction may be estimated by:

$$\frac{1}{\rho(t)(1-\rho(t))}\cdot\frac{d\rho}{dt}\sim\frac{1}{r_p(t)}\cdot\frac{1}{d^n}\exp\left(-\frac{Q}{RT}\right)$$

where $\rho(t)$ and $rp(t)$ are the instantaneous density and pore radius, respectively. This equation predicts that the highest densification rate occurs for the finest pore size. This relationship has two implications. First, the pore size, in addition to grain size, should be controlled during sintering. Fast sintering kinetics result from fine pores' size. Second, the densification rate is dictated by the instantaneous pore size, not only the initial pore size. Therefore, to maintain a fast sintering rate, the pore should remain small even in late sintering stages. A small pore size throughout the sintering process is also critical in controlling the final grain size. For these purposes, a small and uniform pore population is desired in the green compact. Most often, such a pore distribution is associated with a high green density in non- or weakly agglomerated powders. The success in nanopowder consolidation is intimately related to the control of the competition between densification and coarsening. The driving force for densification is largely due to the excess of surface energy. As sintering progresses and grain boundaries are created, the driving force, Δp, for mass transport may be expressed as a Gibbs-Thomson effect:

$$\Delta p=\frac{2\gamma_b}{d}+\frac{2\gamma}{r}$$

where γb is the grain boundary energy, d is the average grain size, γ is the surface free energy, and r is the radius of curvature of pore surfaces. The first term is the grain coarsening tendency, while the second is the sintering driving force. When pores are small, sintering predominates or the pores control grain growth. When particles bond and grain sizes are small, the driving force for coarsening becomes important. The key to nanopowder consolidation is to control this coarsening which inevitably competes with sintering. As a surface controlled process, sintering is critically dependent on particle surface conditions, e.g. particle contamination. Particles produced by MA techniques are more contaminated than those fabricated by, for example, IGC. Oxides, nitrides, and other compounds picked-up from the milling media are often found in consolidated parts made of attrition-milled nanopowders. However, these compounds may prevent grain coarsening when present as fine dispersions. For instance, when the inherent surface SiO_2 is present on Si_3N_4 particles, sintering is enhanced, but creep strength and toughness are impaired during Si_3N_4 part service. For nano metals production, the oxygen contaminations may cause problems because metal oxides tend to form a thin film on the surface and inhibit sintering. To solve some problems in nano materials sintering, a number of non-conventional consolidation methods have been applied to nanopowder densification.

11.6.2 Non-conventional Processing

To overcome the problem of grain growth, unconventional sintering and densification techniques have been proposed. These include the use of grain growth inhibitors, high pressure densification, shock densification, magnetic pulse compaction, microwave heating and field assisted methods. However, while some success was attained by these methods, the results fall short of the ideal goal of having a fast, reproducible technique which can produce a large number of high density samples (e.g., >98% relative density) with a grain size below 20 nm. The ideal method should also be able to use agglomerated powders, which are the typical products of most wet chemistry methods developed in the last few years for the synthesis of pure and doped nanopowders. Therefore, the main purpose in using the non-conventional methods is to enhance densification, thus, reducing the sintering temperature or time with the ultimate benefit of preserving final fine grain sizes.

11.6.2.1 Microwave Sintering

Microwave heating offers many advantages over conventional heating. *Microwave sintering* has a rapid processing time, two to fifty times faster than conventional heating. There is also an acceleration of sintering and diffusion in the material because of high electrical fields; thus densification can occur at lower temperatures. A typical microwave sintering apparatus operates at a 2.45 GHz frequency with power output in the range of 1–6 kW. The sintering chamber consists of ceramic insulation housing (batch system) or an alumina tube insulated with ceramic insulation from outside, Fig. 11.19. The primary function of the insulation is to preserve the heat generated in the work piece. The temperatures are monitored by optical pyrometers, IR sensors and/or sheathed thermocouples placed close to the

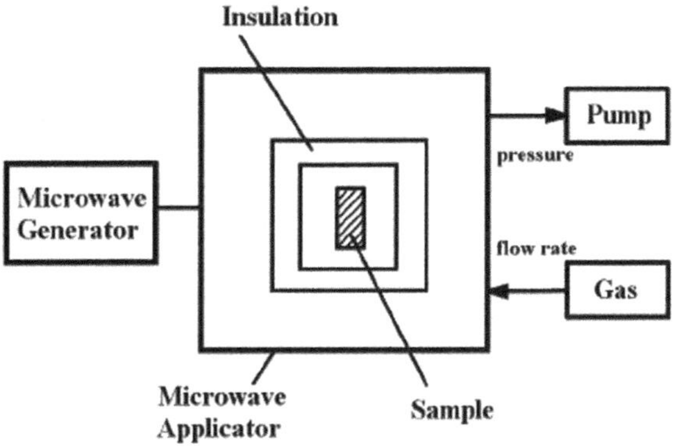

Fig. 11.19 Schematic diagram of microwave heating technique

surface of the sample. The system is equipped with appropriate technique to provide the desired sintering atmosphere, such as H_2, N_2, Ar, etc., and is capable of achieving temperatures up to 1600 °C.

11.6.2.2 Field-Assisted Sintering (FAS)

One of the most interesting and promising methods of powder consolidation is *field-assisted sintering* (*FAS*). Electric field assisted sintering is an emerging technology for the fabrication of metals, ceramics and their composites starting from powders. Field assisted sintering (FAS) often referred as spark plasma sintering (SPS) or pulsed electric current sintering (PECS). All these methods are essentially identical in the application of a pulsed discharge and subsequent or simultaneous resistance sintering. The technique is similar to traditional hot-pressing, but in this case the sample is heated by a high-intensity, low-voltage pulsed DC electric current flowing directly through the sample (if electrically conducting) and through a die, typically made out of graphite. This makes possible faster heating rates (up to 1000 °C/min) than in traditional hotpress systems. In addition to providing Joule heating, high intensity electric currents have also been shown to produce significant modification in the reactivity in several solid-state systems. Kinetically, activated sintering is a result of lowering the activation energies for densification. The schematic of FAS techniques is shown in Fig. 11.20.

The equipment consists of a mechanical device capable of uniaxial pressure application and electrical components to apply the pulsed and/or steady DC current. The loose powders are directly loaded into a punch and die unit without any additives. The machines are equipped with chambers for vacuum or controlled environment. The pressure used in the FAS method is generally limited by the compressive strength of the material used for the dies, which for typical

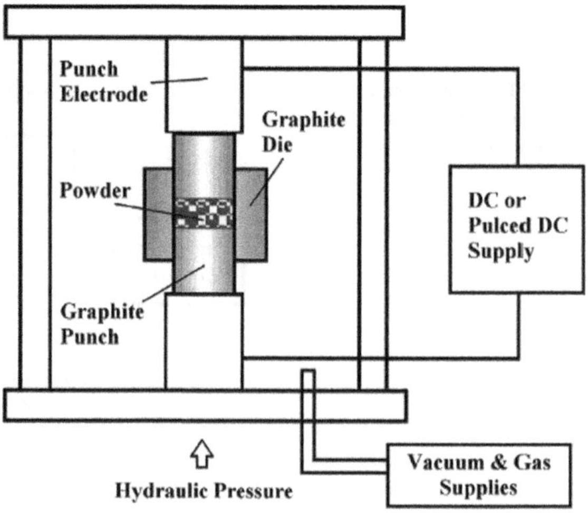

Fig. 11.20 Schematic diagram of a FAS technique

high-density graphite is about 140 MPa. With such a pressure, combined with an appropriate thermal cycle, fully-dense ceramics with grain size between 50 and 100 nm can be obtained. The pressure may be constant throughout the sintering cycle or changed in different densification stages. The consolidation process consists of two stages: (1) an initial activation through the application of a pulsed voltage, and (2) the subsequent heating and densification by using DC current. The sintering cycles are very short. Typically less than 10 min for the full densification of both conductive and non-conductive materials. FAS for 3 min at 9980 C of mechanically synthesized Fe-85%Fe_3C achieved 99% density with a final grain size of 45 nm. This is to be compared to high isostatically pressing (HIP) of the same powders at 12,980 C for 60 min, with the same density value but final grain size of 87 nm. The pure α-Al_2O_3 is consolidated to 99% density in less than 10 min at 18,460 C. The conventional sintering of such powders reaches a similar density at 20,460 C in 3 h. It was postulated that electrical current application creates favorable conditions for the removal of impurities and the activation of the powder particle surfaces. This activation explains the high densities obtained in ceramics without additives and direct grain-to-grain contact at atom scale. The pulsed current promotes electrical discharges at powder particle surfaces, thus activating them for subsequent bonding. Usually, high densities and fine grain sizes characterize FAS sintered materials. For instance, in TiN nanopowders, coarsening during FAS resulted in a final grain size that is at least 1 order of magnitude less than in the conventional sintering of the same powders. The limited coarsening observed in FAS sintering is attributed to a very short time at high temperatures and also reflects minimal coarsening during the initial and intermediate stages of sintering. Generally, significant densification with the little coarsening is noticed in specimens during heating phase. For example, the enhancement of the sintering behavior is observed in WC-Co powders during the heating before melting for a conventional liquid—phase sintering. The use of FAS in pre-melting state induced 98% density while only 70% may be achieved with conventional sintering of the same powder. This enhanced low-temperature sintering is attributed to the surface activation and faster heating rate in FAS as compared to those in conventional sintering. All FAS results indicate accelerated densification with minimum grain coarsening while achieving a good metallurgical grain-to-grain bonding. The latter may be partly explained by the ability to remove the oxides and impurities from the particle surfaces. In general the remaining oxides on powder particles are known to cause consolidation difficulties and low mechanical properties (e.g. ductility and fracture toughness) of the sintered part. These properties are critically dependent on the metallurgical bonding to take full advantage of the intrinsic strength of the material. Sintering at high temperatures in vacuum or reducing atmospheres usually provides good antiparticle bonding due to oxide decomposition. In powders with more stable oxides direct grain contact may be achieved either by mechanical removal of the surface oxides or by physical activation of the particle surface prior to final densification. In field sintering oxide removal and subsequent good interparticle bonding may be attributed to phenomena ranging from resistance heating to thermal and electrical breakdown of insulating films and discharge or arcing. As in any sintering process field activated densification starts with a highly porous body. The initial pressure application proceeds with neck formation. The neck formation is due to the geometric amplification of the pressure on the interparticle point contacts, but as the necks grow, the local pressure at the necks is substantially reduced. At this stage, the pulsed current is applied and a current path is established in metal or conductive ceramic powders. The goal is to achieve a uniform current path rather than local channels that may concentrate all passing current. Another notable feature of FAS consolidation is the enhancement of either phase transformations in single-phase ceramics (e.g., anatase to rutile in TiO_2) or in reactions of single components to form compounds (for example, the formation of Al_2TiO_2 from Al_2O_3 and TiO_2 components).

11.6.2.3 Shockwave Consolidation

Dynamic or shockwave consolidation proceeds with the passage of a large-amplitude compressive stress generated by plate impact or explosion without any external heating. Dynamic consolidation commonly uses explosives or high velocity impact to generate a shockwave. During the travel of the wave front, a pressure level exceeding 1 GPa is applied on a green compact confined in an enclosed container. The wave pressure imposes plastic deformation on the container, which results in an ultra-rapid consolidation of the powders into bulk materials. The high pressure is exerted for a very short duration, in the order of a microsecond, resulting in densification at extremely high strain rates (10^7–10^8 s^{-1}). Rapid processing induces particle surface heating, as compared to bulk heating for conventional processes, and allows interfacial melting while maintaining relatively low temperature inside the particles and providing densification by plastic yielding for both metals and ceramics. Localized heating due to particle interfriction enables good interparticle bonding. In nanosize powders, the heat may transfer throughout the entire particle, thus providing an advantage over coarser materials where the heating is only superficial. Best results are achieved when high temperatures are reached before the shockwave passes. If particles are heated, they may deform rather than fracture when the stress is applied. This very short, high-temperature exposure provides the best means to retain fine grain size or non-equilibrium conditions such as amorphous structures, or supersaturated solid solutions. The major drawback is the difficult coordination of these short stress and heat application events which may result in specimen cracking. Shockwave consolidation has been applied to consolidation of both ceramic and metal nanoparticles. For instance, fully dense specimens with grain size of 20 nm were obtained in ball milled Fe–N solid solution. In the Ti–Si system, shock consolidation yielded 30–40 nm grain size of crystalline $TiSi_2$ and Ti_5Si_3 phases. Only limited grain coarsening took place upon subsequent annealing at 800 °C for one hour. Full densification was also reported in mechanically alloyed TiAl specimens with final grain sizes of 15 nm. These results are compared with HIP ping that provided full consolidation at 13,480 C, 207 MPa, 2 h, but grain sizes were about 100 nm.

References

1. Rao, J.P., Geckeler, K.E.: Polymer nanoparticles: preparation techniques and size-control parameters. Prog. Polym. Sci. **36**(7), 887–913 (2011)
2. Israelachvili, J.N., Mitchell, D.J., Ninham, B.W.: Theory of selfassembly of hydrocarbon amphiphiles into micelles and bilayers. J. Chem. Soc. Faraday Trans. 2 Mol. Chem. Phys. **72**, 1525–1568 (1976)
3. Pitaksuteepong, T.: Nanotechnology: effective topical delivery systems. Asian J. Pharm. Sci. **11**, 16–17 (2016)
4. Viswanathan, B.: Nanomaterials. Narosa Publishing House
5. Zhong, Z.J.: Optical Properties and Spectroscopy of Nanomaterials. World Scientific Publishing Co. Pte. Ltd
6. Sajanlal, P.R., Sreeprasad, T.S., Samal, A.K., Pradeep, T.: Anisotropic nanomaterials: structure, growth, assembly, and functions. Nano Rev. **2** (2011)
7. Lue, J.T.: Physical properties of nanomaterials. Encycl. Nanosci. Nanotechnol. **X**, 1–46
8. Nanomaterials: an introduction to synthesis, properties and application. Environ. Eng. Manag. J. **7**(6), 865–870 (2008)
9. Wang, Z.L.: Nanowires and Nanobelts, Vol. I: Metal and Semiconductor Nanowires, and Vol. II: Nanowire and Nanobelt of Functional Oxide. Kluwer Academic Publisher, Norwell, MA (2003)
10. Service, R.F.: Science **276**, 895 (1997)
11. Kong, X.Y., Wang, Z.L.: Nano Lett. **3**, 1625 (2003)
12. Arnold, M.S., Avouris, Ph, Pan, Z.W., Wang, Z.L.: J. Phys. Chem. B **107**, 659 (2003)
13. Huang, M.H., Mao, S., Feick, H., Yan, H., Wu, Y., Kind, H., Weber, E., Russo, R., Yang, P.: Science **2001**, 292 (1897)
14. Liu, C., Zapien, J.A., Yao, Y., Meng, X., Lee, C.S., Fan, S., Lifshitz, Y., Lee, S.T.: Adv. Mater. **15**, 838 (2003)
15. Bai, X.D., Wang, E.G., Gao, P.X., Wang, Z.L.: Nano Lett. **3**, 1147 (2003)
16. Ng, H.T., Chen, B., Li, J., Han, J., Meyyappan, M., Wu, J., Li, X., Haller, E.E.: Appl. Phys. Lett. **82**, 2023 (2003)
17. Zhao, Q.X., Willander, M., Morjan, R.R., Hu, Q.H., Campbell, E.E.B.: Appl. Phys. Lett. **83**, 165 (2003)
18. Gao, P.X., Ding, Y., Wang, Z.L.: Nano Lett. **3**, 1315 (2003)
19. Park, W.I., Kim, D.H., Jung, S.W., Yi, G.C.: Appl. Phys. Lett. **80**, 4232 (2002)
20. Harnack, O., Pacholski, C., Weller, H., Yasuda, A., Wessels, J.M.: Nano Lett. **3**, 1097 (2003)
21. Kempa, K., Kimball, B., Rybczynski, J., Huang, Z.P., Wu, P.F., Steeves, D., Sennett, M., Giersig, M., Rao, D.V.G.L.N., Carnahan, D.L., Wang, D.Z., Lao, J.Y., Li, W.Z., Ren, Z.: F. Nano Lett. **3**, 13 (2003)
22. Huang, Z.P., Carnahan, D.L., Rybczynski, J., Giersig, M., Senett, M., Wang, D.Z., Wen, J.G., Kempa, K., Ren, Z.F.: Appl. Phys. Lett. **82**, 460 (2003)
23. Wang, X.D., Gao, P.X., Li, J., Summers, C.J., Wang, Z.L.: Adv. Mater. **14**, 1732 (2002)
24. Johnson, J.C., Yan, H., Yang, P., Saykally, R.J.: J. Phys. Chem. B **107**, 8816 (2003)
25. Birefringence swap at the transition to hyperbolic dispersion in metamaterials. Phys. Rev. B **85**, 165408 (2012)
26. Rational design of sub-parts per million specific gas sensors array based on metal nanoparticles decorated nanowire enhancement-mode transistors. Nano Lett. **13**, 3287 (2013)
27. Nanowire-Composite based flexible thermoelectric nanogenerators and self-powered temperature sensors. Nano Res. **5**, 888 (2012)
28. Nature **451**, 809–813 (14 February 2008)
29. Science **316**, 102 (2007)
30. Adv. Funct. Mat. **23**, 2445–2449 (2013)

Chapter 12
Synthesis Methods For 2D Nanostructured Materials, Nanoparticles (NPs), Nanotubes (NTs) and Nanowires (NWs)

12.1 Synthesis Methods for 2D Materials

Two-dimensional (2D) materials have, within just one decade, reshaped many disciplines of modern science and technology, both through intensive fundamental research and early commercial applications. Two-dimensional (2D) materials are crystalline solids with a high ratio between their lateral size ($\sim$ 1–10,000 μm) and thickness (<1 nm). 2D materials systems offer unique properties not available in current bulk materials. For example, the band structure of 2D graphene enables world-record carrier mobility. When isolated to monolayer form, the molybdenite mineral becomes a direct gap semiconductor. It absorbs and emits light rather efficiently owing to much enhanced light-matter interaction in 2D. It also exhibits exciting spin to valley degrees of freedom (often referred as valleytronics), possesses 100% quantum efficiency, and the many-body interactions in the mineral help to stabilize the excitons at room temperatures. Similarly, 2D pictinides possess high temperature superconductivity while 2D black phosphorus exhibits negative Poisson's ratio. Owing to their thinness, they can be electrically biased and operated at low voltages. Moreover, the mechanical properties of 2D materials also enable flexible/stretchable electronic applications. Indeed, if we think back carefully, the very existence of 2D materials should come as a surprise: They are so thin, delicate, exposed, and potentially chemically active! Working with 2D materials is analogous to working with Saran Wrap. How often do you get frustrated with this notso-cooperative polymeric sheet? Now imagine the difficulty when material itself is only one unit cell thick…

L. H. Madkour, *Nanoelectronic Materials*, Advanced Structured Materials 116, https://doi.org/10.1007/978-3-030-21621-4_12

12.1.1 Micromechanical Exfoliation Using Scotch Tape

The layered structure of the corresponding bulk crystals, which facilitates their exfoliation to 2D layers, originates fromthe strong intralayer covalent bonding and comparatively weaker interlayer van der Waals (vdW) interaction with typical bindingenergies of 40–70 meV. Atomically and few layers thick sheets of 2D layered inorganic materials can be obtained by micromechanical exfoliation. This technique is similar to the method used by Grim and his coworkers to separate single layer graphene from highly oriented pyrolytic graphite (HOPG) [1]. Commercially available inorganic layered material is rubbed/peeled off against scotch tape on photoresist surface. This process leaves a variety of flakes including monolayers on the photoresist surface due to the cleavage of weak van der Waal forces between two layers. These monolayers are washed with acetone and transferred on oxidized silicon wafer. This method has been employed to obtain monolayers of WS_2 [2], MoS_2 [3, 4], h-BH [4, 5], $NbSe_2$ [4], Bi_2 $Sr_2CaCu_2O_x$ [4] from the bulk. However, this process is relatively simple, fast and cost effective but it has certain disadvantages. As we know monolayers are second building block of layered materials, therefore, to understand the fundamental/intrinsic properties of materials, separation of monolayers from thicker flakes is necessary. The monolayer yield of this process is extremely low; hence, this method is only suitable for laboratory scale studies and cannot be utilized for large-scale production for high end technological application. On the other hand, since this process does not utilize any chemicals and only rely on the shear force applied during the peeling process, structural integrity, and high crystallinity are maintained. After the cleavage of inter-planer van der Waal bonds, optical microscopy (phase contrast due to single and few layers) and AFM (thickness measurement in tapping mode) are employed at their best to screen the monolayers from thicker structures. Characterization of 2D nanostructured materials is discussed in detail in Sects. 2.6 and 2.7. Single and few layers of exfoliated 2D layers are stable at ambient conditions for days to months similar to single graphene layer as evident from the time dependent high resolution transmission electron microscopy (HRTEM) and scanning tunneling microscopy (STM) [4, 6].

12.1.2 Liquid Exfoliation

Mechanical exfoliation of 2D layered materials is a semiquantitative low yield method, therefore it was necessary to develop facile methods for high quality and large quantity of single and few layer of 2D materials. As an alternative to mechanical exfoliation, liquid exfoliation is a reliable method to produce single and few layers 2D sheets at bulk scale. This method has been extensively used to obtain mono and few layers nanosheets of MoS_2, WS_2, BN, $NbSe_2$, $NiTe_2$, $MoTe_2$, $MoSe_2$, $TaSe_2$, Bi_2Te_3 and so on [7–10]. The liquid exfoliation method has four different forms: (i) oxidation, (ii) intercalation (iii) ion exchange and (iv) ultrasonic cleavage [11].

Oxidative liquid exfoliation is an established technique for the synthesis of graphene oxide. Graphite flakes are treated with sulfuric acid or nitric acid or mixture of sulfuric and nitric acid; and potassium chlorate/potassium permanganate to passivate the surface of individual graphite layers with the addition of OH or epoxide functional groups and their subsequent dispersion in polar solvent and sonication results in the exfoliation of graphene oxide. This method is popularly known as a Hummers method (sulfuric acid) or modified Hummers method (mixture of acids) [12, 13]. It is to be noted that extreme care should be taken while selecting the suitable oxidizer and solvent for different layered material. Layered materials having low reduction potential are best suited for this oxidative liquid exfoliation.

Weak out-of-plane bonding in the layered materials and high surface area results in the opening for molecules to be adsorbed on the basal planes and from inclusion complex. The Intercalation of organic or ionic species causes the weakening of interlayer adhesion due to the increase in the interlayer spacing and energy barrier required for the exfoliation. Similar to graphite, this method has been mainly applied for TMDs. On the other hand, ionic species intercalate and exchange ions present on the basal planes by charge transfer and weakening of interlayer adhesion. Ion exchange liquid exfoliation has been widely used to exfoliate oxide nanosheets for optoelectronic applications [14]. In the ion exchange exfoliation process, layered europium ($Eu(OH)_{2.5}Cl_{0.5}{\cdot}0.9H_2O$) first react with dodecyl sulfate (DS) by replacing Cl^- ions in ion exchange process followed by ultrasonication or shear mixing to separate the positively charged sheets [15]. Similar to europium hydroxide, TiO_2 layered crystals are tends to be negatively charged and their charge neutrality is maintained by Cs^+ ions. Therefore, negatively charged oxides are treated with bulkier cations such as tetrabutylammonium. This method is widely known as proton exchange and has used in the exfoliation of perovskite type oxides.

Ultrasonic cleavage of weak out-of-plane bonding is another liquid exfoliation method to obtain mono and few layer nanosheets. Bulk layered material is dispersed in suitable solvent and ultrasonicated for 1–3 h. These high energy ultrasonic waves create cavitation bubbles in the solvent [11]. Sudden burst of cavitation bubbles generates high energy via release of pressure followed by an exfoliation of layers in the solvent. The resultant dispersion is centrifuged to separate large thickness crystals. Ultrasonication time and centrifugation rate are some of the critical parameters which dictate the quality of the resultant solution. Mono and few layers nanosheets remain suspended in the supernatant solution. Selection of optimum solvent is key to tap the full potential of this technique. Detailed studies have been carried out for [7] more than 25 solvent on the exfoliation nature of a wide range of 2D material based on the yield, crystallinity, thickness and size of the resultant layers. Solvents having surface tension 35–40 mJ/m^2 maximize the exfoliation rate examined using UV–Vis spectroscopy. Use of solvent having surface tension similar to surface energy of layered 2D material not only minimize the energetic cost of exfoliation but also prevents the restacking of nanosheets [7, 11, 16]. Earlier comprehensive study [7] has acted as a catalyst for the synthesis of 2D layered materials with yield of up to 30 mg/mL [11, 16–18]. NMP, DMF, DMSO, IPA, cyclohexyl-pyrrolidinone, Ndodecyl- pyrrolidone, NVP are used as common solvents for ultrasonic exfoliation of h-BN, MoS_2, WS_2, $MoTe_2$, VS_2 [7, 10, 19–23]. Aqueous solution of surfactant or polymer has also been utilized to exfoliate nanosheets that are electrostatically or sterically stabilized. Nanosheets obtained from ultrasonication are highly crystalline and micrometer in size as confirmed from HRTEM. However, it is also reported that, the in-plane cleavage of layered materials in gadolinium hydroxide with an interlayer of Cl^- anion results in flakes of ~ 100 nm size [24]. Schematic representation of the various liquid exfoliation processes is presented in Fig. 12.1.

Density gradient ultracentrifugation (DDU) of 2D material containing solution further allows the separation of single layers from few layers. This technique was first successfully applied to separate single layer graphene oxide using sodium cholate as intercalating agent (surfactant) from the solution [25]. Buoyant density difference of mono and few layer nanosheets separate single layer, few layers and bulk crystals. Despite the large scale production ability of this method, mechanical exfoliation remains the preferred choice for layered 2D materials due to the less destructive nature as well as mono layers of size up to 10 lm have been peeled off on a variety of substrates [26, 27].

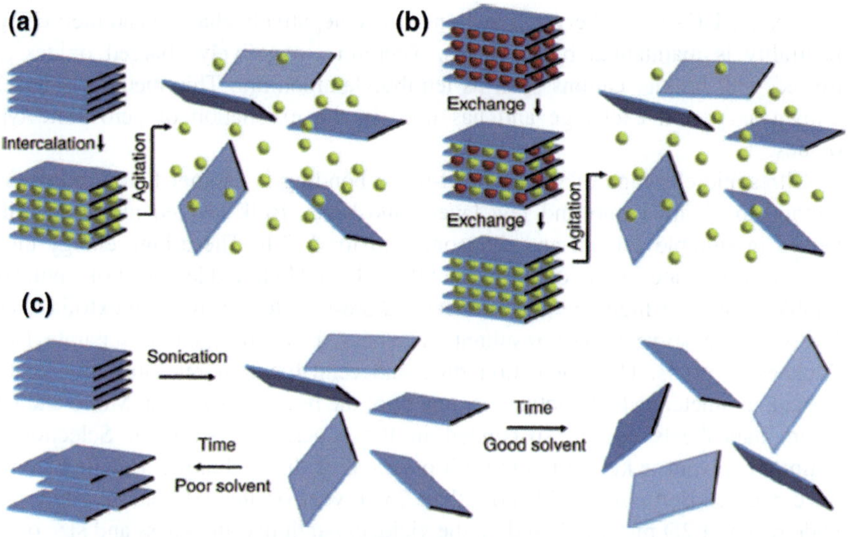

Fig. 12.1 Schematic representation of variation in liquid exfoliation process: **a** intercalation, **b** ion exchange and **c** ultrasonic exfoliation. Reprinted from Ref. [11] with permission of the American Association for the Advancement of Science Publisher

12.1.3 Chemical Vapor Deposition (CVD)

The CVD technique has enabled the synthesis of large area and uniform thickness 2D layer of metal and insulating surfaces for the large-scale device fabrication including electronic and flexible optoelectronic devices. Recently MoS_2 nanosheets have been grown on insulating oxidized silicon substrateusing two step bottom-up approaches [28]. First, a Mo thin layer was deposited on oxidized silicon substrate by e-beam evaporation. Second, elemental sulfur was heated just above the melting point to generate vapor further reacting with the Mo thin film at 750 °C and form 2D MoS_2. However, this method does not provide atom thick layer of MoS_2 (1–3 layers). Moreover, 0.72 nm thick MoS_2 single layers has been synthesized in one step bottom-up approach using MoO_3 and elemental S at 650 °C [29]. Before the decomposition of precursor, SiO_2/Si substrate was first coated with GO to serve as a growth template for MoS_2 layer. In another attempt to obtain large size MoS_2 sheet, SiO_2/Si substrate was dip-coated with ammonium thiomolybdate [$(NH_4)_2MoS_4$] dissolved in DMF and annealed at 500 °C (Fig. 12.2) [30].

During the second step, annealed substrate was heated to 1000 °C and allow to react with sulfur vapor. During the thermolysis process, $(NH_4)_2MoS_4$ forms MoO_3 which later converted into MoS_2 after reacting with S vapor at temperature above 800 °C. Another research group [31] has studied the growth and grain boundary structure of CVD grown MoS_2 layer on SiO_2/Si substrate. They have shown the evidence of grain coalescence during growth of MoS_2 grains results in the formation of an array of dislocation of 5–7-fold rings. Additionally, the growth of one grain above another grain without any chemical bonding have also been observed.

Besides MoS_2, h-BN has been extensively prepared and studied using CVD on transaction metal surfaces such as Pt, Ni, Ru, Pd, Rh, etc. for well-ordered layer structure BN by thermal decomposition of Btrichloroborazine $(ClBNH)_3$ or borazine $(BN)_3H_6$ [32–36]. Extreme conditions, including high temperature (>700 °C) and ultra-high vacuum are required for this CVD process. It is worth mentioning here that a lattice mismatch between the metal surface and BN result in the formation of two different superstructures, i.e. types A and B [32]. In case of minor lattice mismatch, type A structure develops during the CVD process, whereas for

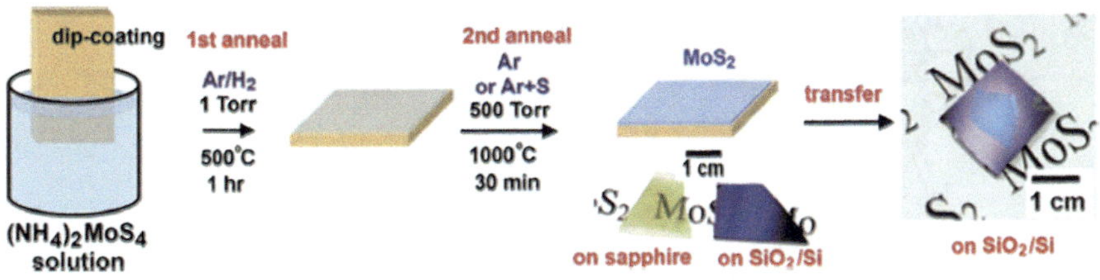

Fig. 12.2 Schematic representation of dip and anneal technique for large area MoS_2 nanosheet synthesis. Insulator substrate was first dip coated with Mo precursor followed by annealing under optimized conditions. Resultant product is uniform large surface area MoS_2 nanosheet. Reprinted from Ref. [30] with permission of ACS Publisher

large mismatch such as in the case of Pt(111) (mismatch ~11%) both types A and B structure form as observed from low energy electron diffraction (LEED) pattern (Fig. 12.3a). For type A structure, B atoms prefer to stay at hcp or fcc sites and N atoms at the top of the substrate atoms (Fig. 12.3b). While in type B structure, B and N atoms are at on top of the metal surface atoms and on the bridge sites (Fig. 12.3c).

12.1.4 Van der Waal Epitaxial Growth on Substrate

Van der Waal epitaxial is an another form of CVD method. In contrast to CVD where substrate surface acts as a catalyst, substrate often acts as a seed crystal. This

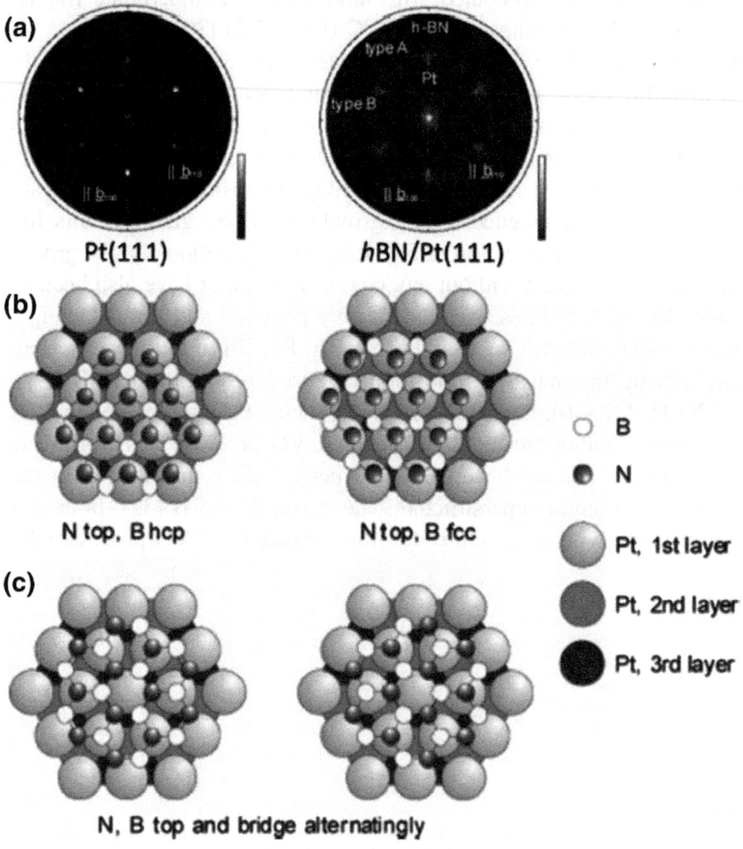

Fig. 12.3 **a** LEED pattern of Pt(111) surface and h-BN/Pt(111) surface showing both types A and B structure. Atomic model of **b** type A and **c** type B structure in h-BN layer. Reprinted from Ref. [32] with permission of ACS Publisher

method has been employed to fabricate atom thick, large aspect ratio several 2D material such as hexagonal Si [37], h-BN [32], GaSe [38, 39], MoS_2 [40], etc.

In situ deposition of Si on Ag(001) by direct current heating of Si wafer under ultrahigh vacuum condition develop two distinct superstructures in Si stripe examined using XFRD, STM and LEED [41]. A $p(3 \times 3)$ superstructure was observed during the first stage growth of Si nanostructure up to single monolayer followed by a complex superstructure denoted as $p(7 \times 4)$. Both the structures were oriented along the [110] or [–110] direction of Ag substrate. The unit cell of $p(3 \times 3)$ consists of two tilted dimers arranged in a rectangular fashion to make one Si tetramer. Figure 12.4a and b shows the filled state STM image and atomic model reconstruction of $p(3 \times 3)$ superstructure on Ag(001). In the unit cell, side distances of the tetramer measured by SXRD are 0.234 and 0.40 nm whereas the height distance between Si atoms are 0.13 nm. Further growth of monolayer develops 2 nm wide stripes having $p(7 \times 4)$ superstructure on $p(3 \times 3)$ structure containing defects and no long-range ordering were observed (Fig. 12.4c). The width of the stripe is equivalent to $7a_{Ag}$ and two joined hexagons equivalent to $4a_{Ag}$. Detailed atomic reconstruction model of $p(7 \times 4)$ superstructure is shown in Fig. 12.4d. It is worth to mention here that these self-organized Si structure produce minor distortion in the Ag substrate. On Ag(110) substrate, 1.6 nm width and

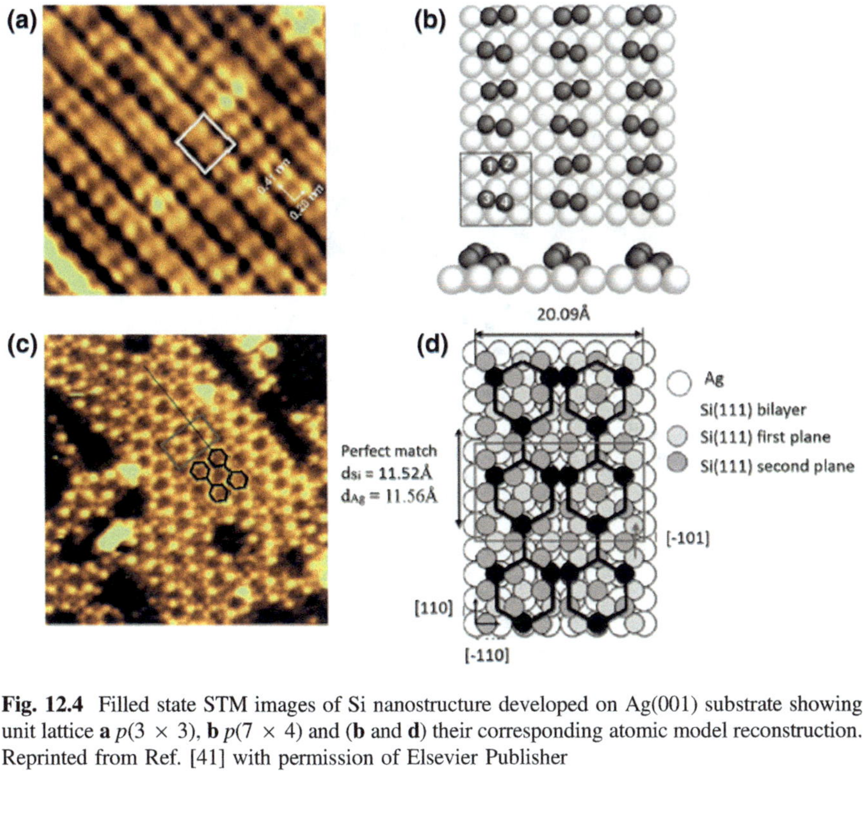

Fig. 12.4 Filled state STM images of Si nanostructure developed on Ag(001) substrate showing unit lattice **a** $p(3 \times 3)$, **b** $p(7 \times 4)$ and (**b** and **d**) their corresponding atomic model reconstruction. Reprinted from Ref. [41] with permission of Elsevier Publisher

0.2 nm thick Si nanoribbons with honeycomb structure were synthesized by direct current deposition [42, 43]. On the other hand, Ge tetramers structure $p(2\sqrt{2} \times 4\sqrt{2})$ $R45°$ on Ag(001) produces a strong distortion in Ag substrate due to high solubility in Ag [44]. Germanium monolayer have also been tried to synthesize on Ag(110) and Ag(111) substrates. On Ag(110) substrate, 0.5 Ge monolayer coverage was obtained which has c(4 × 2) superstructure examined from LEED [45, 46]. This c (4 × 2) superstructure consists of two tetramers per unit cell [45, 46]. However, for Ag(111) substrate, different superstructures have been observed. Initial 1/3 Ge monolayers show a $(\sqrt{3} \times \sqrt{3})$R30° that convert into $c(\sqrt{3} \times 7)$ superstructure on 1 monolayer Ge coverage [45, 47]. This $c(\sqrt{3} \times 7)$ consists of four Ge tetramers per unit cell.

Further changing the substrate to Ag(111) (Fig. 12.5a), precise control of substrate temperature (250 °C) and Si flux resulted in the synthesis of single monolayer of silicene where Si atoms are arranged in a honeycomb lattice examined by STM (Fig. 12.5b) [37]. A height difference of 0.02 nm in line profile of the STM image gives the evidence of Si atoms being present in two different sub-latticessuggesting the possible formation of sp^2–sp^3 hybridization. Peak-to-peak and valley-to-valley analysis measures the Si–Si atom distance to be 0.19 ± 0.01 nm which is

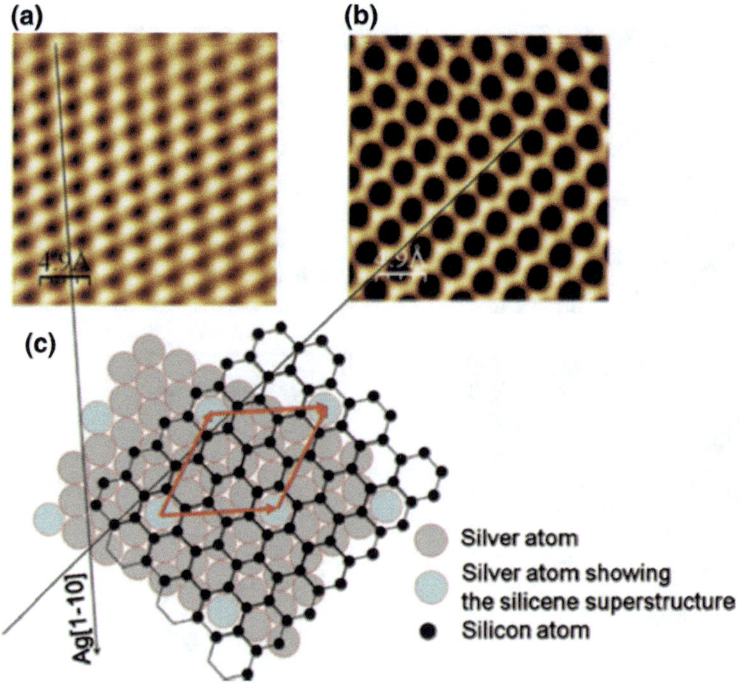

Fig. 12.5 Filled state STM image of **a** Ag(111) substrate and **b** hexagonal silicene. **c** Orientation relationship of Si with Ag(111) surface showing unit cell. Reprinted from Ref. [37] with permission of AIP Publishing LLC

significantly lower than the bulk Si (0.235 nm). LEED analysis characterized the silicene monolayer consists of $(2\sqrt{3} \times 2\sqrt{3})R30°$ superstructure having long range periodicity (Fig. 12.5c) [37]. This discrepancy in the Si–Si distance from the silicene structure has been explained by the misinterpretation of pure Ag(111) surface [48].

Recently, researchers have also fabricated buckled silicene having $(\sqrt{3} \times \sqrt{3})$ super lattice structure consider with $(\sqrt{7} \times \sqrt{7})$ superstructure on Ir(111) substrate obtained from LEED [49]. These findings were further supported by DFT based first principles calculations performed using the Vienna ab initio simulation package (VASP). The projector augmented wave (PAW) potentials and generalized gradient approximation were utilized in the first principle calculation. Similar to the Ag(111) substrate silicene, height difference has been observed between Si adatom layers. However, the Si–Si nearest neighbor distance is measured to be 0.72 nm. The relaxed lowest energy configuration from DFT reveal that a single unit cell is made of six Si atoms present in different chemical environments. One Si atom is directly on top of an iridium atom; two atoms at the hcp hollow and fcc hollow sites of the Ir(111) lattice; and the rest are at the bridge sites which result in the buckling in the silicene monolayer.

Besides the substrate orientation, deposition temperature also causes the change in the orientation of Si nanostructure. Four different orientations i.e. (4×4), $(\sqrt{13} \times \sqrt{13})$, $(2\sqrt{3} \times 2\sqrt{3})$ and $(\sqrt{7} \times \sqrt{7})$ have been reported for Si nanostructures on Ag(111) substrate at various temperature [50–53]. Moreover, epitaxial growth of MoS_2 has been reported on CVD grown graphene on Cu foil which act as a template where dangling bonds and intrinsic weak van der Waal forces assist in the growth process [40]. Use of CVD grown graphene–Cu template also able to lower the deposition temperature to 400 °C.

As discussed in the above examples, van der Waal epitaxial growth of thin layers allow the fabrication of large sheets of novel 2D material which is not possible through liquid exfoliation and hydrothermal methods. However, substrate chemistry and orientation are critical to obtain the desired structure and properties. Harsh temperature, controlled atmosphere, and high cost limit the wide use of van der Waal epitaxy in future commercial application.

12.1.5 Hydrothermal Synthesis

Crystallization of substance at high temperature and at high vapor pressure from aqueous or organic solution is primarily defined as hydrothermal method. It is particularly suitable for precursor having best solubility and stability of growth phase at high temperature and high pressure. Materials having high vapor pressure near the melting point can also be synthesized by hydrothermal method. This method has been employed for the synthesis of layered 2D materials. Chemical reaction between ammonium molybdate $[(NH_4)_6Mo_7O_{24}{\cdot}4H_2O]$ and sulfur and/or selenium in hydrazine monohydrate solution at 150–180 °C for 48 h results in the synthesis of single monolayer MoS_2 and $MoSe_2$ [54]. The presence of single layer was confirmed by the absence of (002) diffraction peak in the XRD spectrum. It is important to mention that no restacking tendencies were observed in both monolayers samples without the use of surfactant. MoS_2 flakes of 3–10 layers were reported in the literature on GO in a one-step solvothermal reaction of $(NH_4)_2MoS_4$ and hydrazine in DMF solution for hydrogen evolution [55]. Inspired by the finding of single and few MoS_2 layers, use of hydrothermal method has been extended to obtain other 2D layered materials.

Recently the synthesis of groups IV and V transaction metal chalcogenides crystals from metal chloride in oleylamine is reported [56]. CS_2, elemental S and Se were used as the primary chalcogen source. Poor crystallinity and irregular shaped crystal were observed in the case of elemental S due to the formation of highly reactive radicals, whereas the use of CS_2 generate intermediate H_2S which further reacts with the metal precursor and form metal disulfides crystals of (001) layers of varying lateral size (Fig. 12.6a, b). Growth of layered nanocrystal in (001) plane is attributed to the low surface energy in (001) plane compared to the (100) and (010) plane (Fig. 12.6c). Contrary to S, use of Se as chalcogen source gives better layered crystals due to its high stability compared to CSe_2 [57]. This article also demonstrates the effect of radicals on TiS_2 and $TiSe_2$ nanocrystals with time. Reaction of TiS_2 crystal at 300 °C with S and AIBN degraded the crystal due to the radical generation of highly reactive nature, whereas mild degradation was observed for $TiSe_2$ crystal (Fig. 12.6d–f). On the other hand, the reaction of radicals does not degrade the closed pack zinc blend structure compounds suggesting that radicals react only on the edges and penetrate between the open spaces of two layers to react with these layered crystals. Therefore, the selection of low reactive radical generating chalcogen is needed to synthesize such layered nanocrystals.

Facile large scale production of nanosheets from the hydrothermal method also makes it an attractive choice for industrial incubation. Recently, h-BN of ~4 nm thickness were prepared on the gram scale from boric anhydride, Zn powder and $N_2H_4{\cdot}2HCl$ at 500 °C [58]. Similar results were obtained from the reaction of boric acid and urea at 900 °C [59]. Besides TMDs, this method has also been extensively utilized to prepare nanosheets of metal oxides and hydroxides. Hydrothermal reduction of V_2O_5 in the presence of graphene oxide (GO) result in the formation of VO_2 nanoribbons of thickness ~10 nm, width 50–500 nm and length ~10 μm on graphene oxide nanosheets for the cathode material in lithium ion batteries [60]. Synthesis of Eu doped yttrium hydroxide nanosheets and nanotubes have been reported via hydrothermal method using triethylamine as alkaline and complexing agent [61]. HRTEM, XPS and STM characterization have revealed high crystallinity and composition control on such crystal grown by a hydrothermal method. These synthesized 2D materials can be resuspended in the desired solvent and deposited to prepare nanocomposites, nanodevices and future nanotechnology applications.

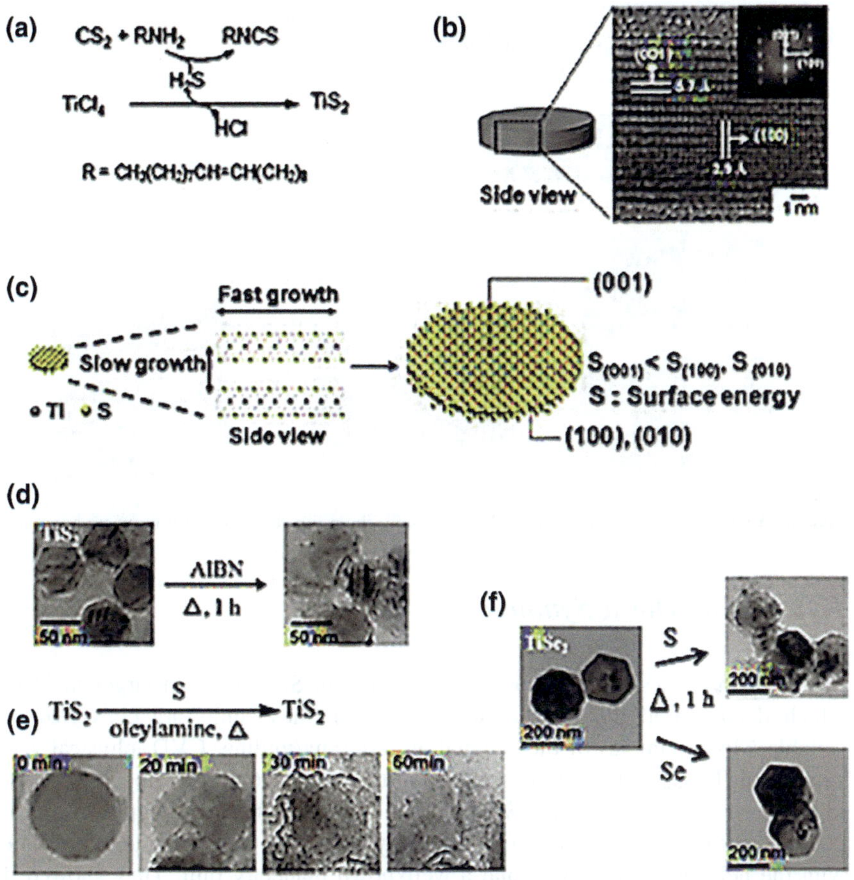

Fig. 12.6 **a** Chemical reaction for the synthesis of TiS_2 layered nanocrystal. **b** HRTEM image of TiS2 nanocrystal from reaction (**a**). **c** Schematic representation of growth model of ToS_2 nanocrystal based on surface energy. **d** TEM image showing the radical effect on crystal while reaction with AIBN at 300 °C. **e** Morphological change in TiS_2 nanocrystal with time when react with elemental S. **f** TEM images of $TiSe_2$ nanocrystal change when react with elemental S and Se. Reprinted from Ref. [56] with permission of ACS Publisher

12.2 Synthesis Methods of Nanoparticles NPs

Various methods can be employed for the synthesis of NPs, but these methods are broadly divided into two main classes i.e. (1) Bottom-up approach and (2) Top-down approach [62] as shown in Scheme 12.1 [63]. These approaches further divide into various subclasses based on the operation, reaction condition and adopted protocols.

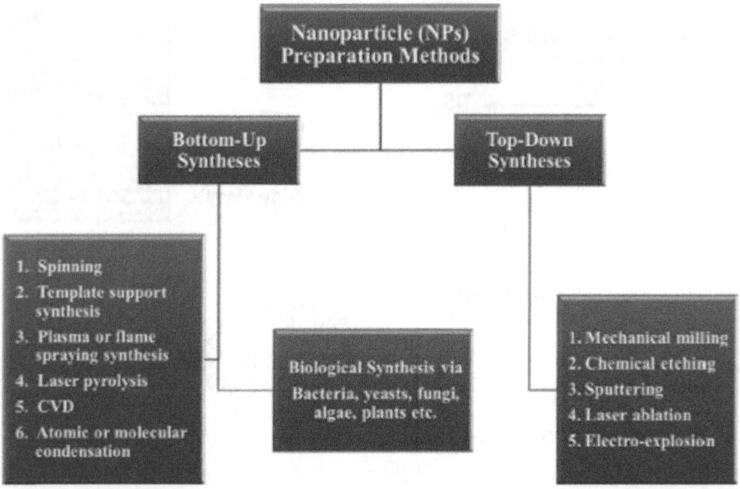

Scheme 12.1 Typical synthetic methods for NPs for the **a** top-down and **b** bottom-up approaches

12.2.1 Top-Down Syntheses

In this method, destructive approach is employed. Starting from larger molecule, which decomposed into smaller units and then these units are converted into suitable NPs. Examples of this method are grinding/milling, CVD, physical vapor deposition (PVD) and other decomposition techniques [63]. This approach is used to synthesized coconut shell (CS) NPs. The milling method was employed for this purpose and the raw CS powders were finely milled for different interval of times, with the help of ceramic balls and a wellknown planetary mill. They showed the effect of milling time on the overall size of the NPs through different characterization techniques. It was determined that with the time increases the NPs crystallite size decreases, as calculated by Scherer equation. They also realized that with each hour increment the brownish color faded away due to size decrease of the NPs. The SEM results were also in an agreement with the Xray pattern, which also indicated the particle size decreases with time [64].

One study revealed the spherical magnetite NPs synthesis from natural iron oxide (Fe_2O_3) ore by top-down destructive approach with a particle size varies from $\sim$20 to $\sim$50 nm in the presence of organic oleic acid [65]. A simple top-down route was employed to synthesize colloidal carbon spherical particles with control size. The synthesis technique was based on the continuous chemical adsorption of polyoxometalates (POM) on the carbon interfacial surface. Adsorption made the carbon black aggregates into relatively smaller spherical particles, with high dispersion capacity and narrow size distribution as shown in Fig. 12.7 [66]. It also revealed from the micrographs, that the size of the carbon particles become smaller with sonication time. A series of transition-metal

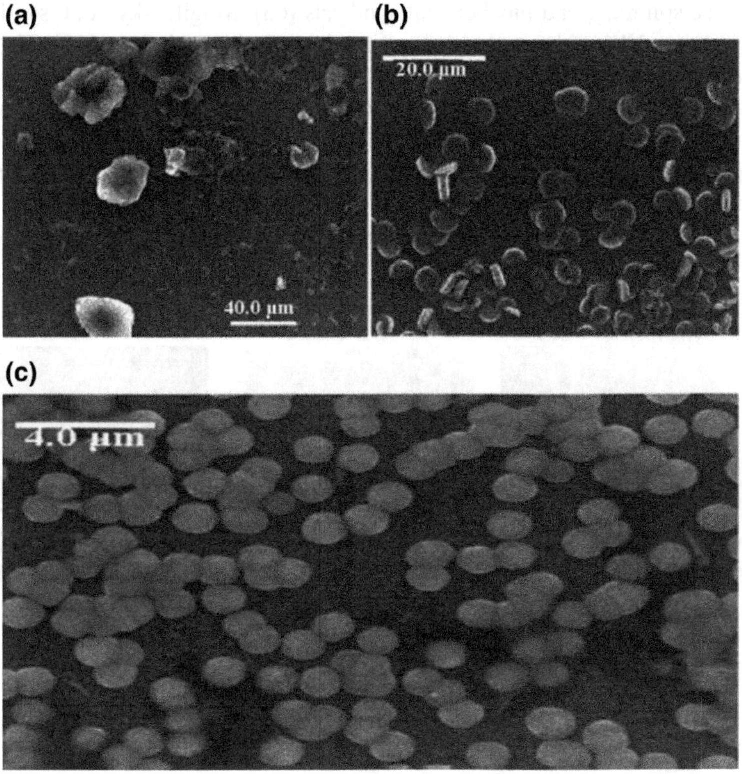

Fig. 12.7 SEM images of **a** the untreated carbon black, **b** and **c** 10 min and 1 h ultrasonication in POM solution [66]

dichalcogenide nanodots (TMD-NDs) were synthesized by combination of grinding and sonication top-down techniques from their bulk crystals. It was revealed that almost all the TMD-NDs with sizes <10 nm show an excellent dispersion due to narrow size distribution [67]. Lately, highly photoactive active Co_3O_4 NPs were prepared via top down laser fragmentation, which is a top-down process. The powerful laser irradiations generate well-uniform NPs having good oxygen vacancies [68]. The average size of the Co_3O_4 was determined to be in the range of 5.8 ± 1.1 nm.

12.2.2 Bottom-Up Syntheses

This approach is employed in reverse as NPs are formed from relatively simpler substances, therefore this approach is also called building up approach. Examples of this case are sedimentation and reduction techniques. It includes sol gel, green synthesis, spinning, and biochemical synthesis [63]. Mogilevsky et al. synthesized TiO_2 anatase NPs with graphene domains through this technique [69]. They used alizarin and titanium isopropoxide precursors to synthesize the photoactive composite for photocatalytic degradation of methylene blue. Alizarin was selected as it offers strong binding capacity with TiO_2 through their axial hydroxyl terminal groups. The anatase form was confirmed by XRD pattern. The SEM images taken for different samples with reaction scheme are provided in Scheme 12.2. SEM indicates that with temperature elevation, the size of NPs also increases [69].

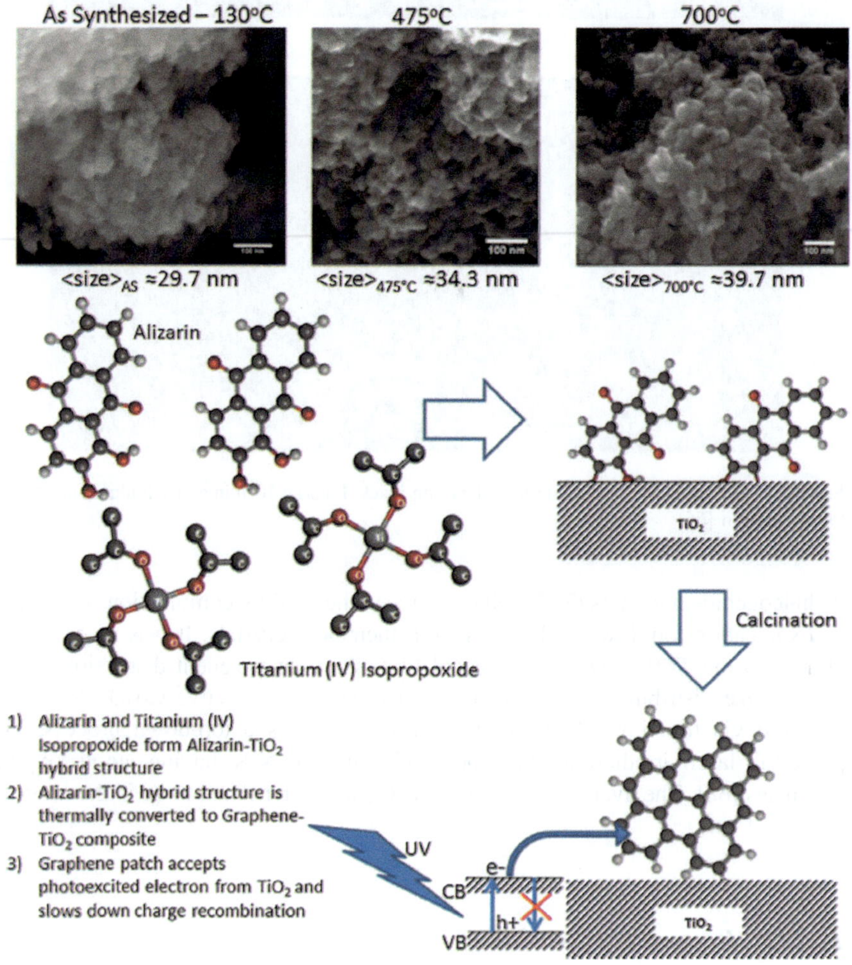

Scheme 12.2 Synthesis of TiO_2 via bottom-up technique. SEM images showing the TiO_2 NPs [69]

Well-uniform spherical shaped Au nanospheres with monocrystalline have been synthesized via laser irradiation top-down technique [70, 71]. Liu et al. selectively transform the octahedra morphology to spherical shape by controlling the laser treatment time and other reaction parameters. Figure 12.8 provides the SEM and

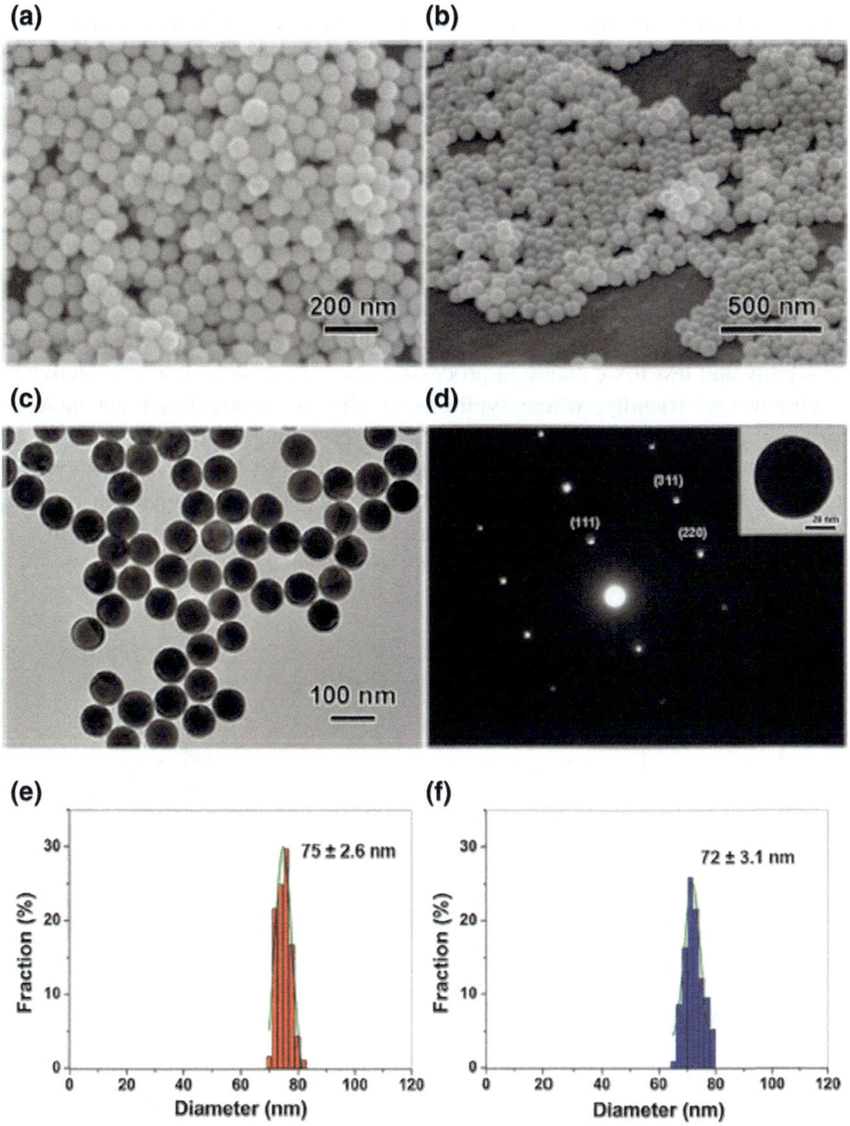

Fig. 12.8 SEM for Au nanospheres **a** top view, **b** tilted view, **c** TEM image of Au nanospheres, **d** SAED pattern (inset: TEM of single Au particle), **e** and **f** size distribution spectra of spherical and octahedral Au NPs [70, 71]

TEM of the prepared Au nanospheres, which showed average diameter of 75 ± 2.6 nm of Au nanospheres (red column Fig. 12.8e) and 72 ± 3.1 in edge length of Au octahedra per particle (blue column Fig. 12.8f).

More recently, solvent-exchange method is used to achieve limit sized low density lipoprotein (LDL) NPs for medical cancer drug delivery purpose by Needham et al. In this method nucleation is the bottom approach followed by growth which is the up approach. The LDL NPs were obtained without using phospholipid and possessed high hydrophobicity, which is essential for drug delivery applications [72].

The monodispersed spherical bismuth (Bi) NPs were synthesized by both top-down and bottom-up approaches [62]. These NPs have excellent colloidal properties. In the bottom-up approach bismuth acetate was boiled within ethylene glycol, while in top-down approach the bismuth was converted into molten form and then the molten drop was emulsified within the boiled diethylene glycol to produce the NPs. The size of the NPs obtained by both methods was varied from 100 to 500 nm [62]. The details of this study are provided in Scheme 12.3.

Green and biogenic bottom-up synthesis attracting many researchers due to the feasibility and less toxic nature of processes. These processes are cost-effective and environmental friendly, where synthesis of NPs is accomplished via biological systems such as using plant extracts. Bacteria, yeast, fungi, Aloe vera, tamarind and even human cells are used for the synthesis of NPs. Au NPs have been synthesis from the biomass of wheat and oat [73] and using the microorganism and plant extracts as reducing agent [74]. The merits and demerits of various top-down and bottom-up techniques with general remarks were registered [75].

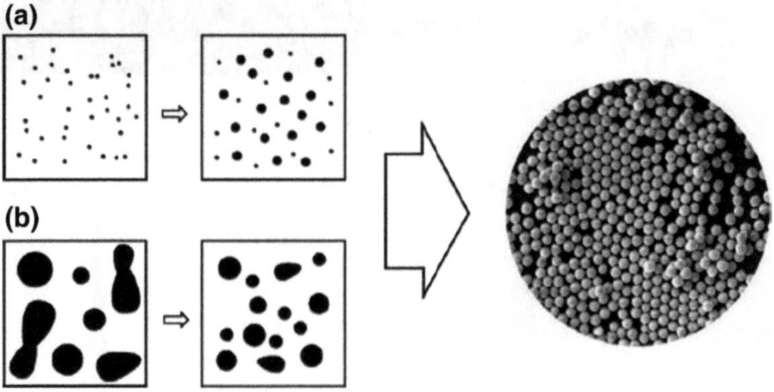

Scheme 12.3 **a** Bottom-up approach: a molecular precursor is disintegrated to simpler metal atoms that grow into colloids. **b** Topdown approach: large drops of a metal broken into smaller drops [62]

12.3 Synthesis Methods of Nanotubes (NTs)

12.3.1 Arc Discharge

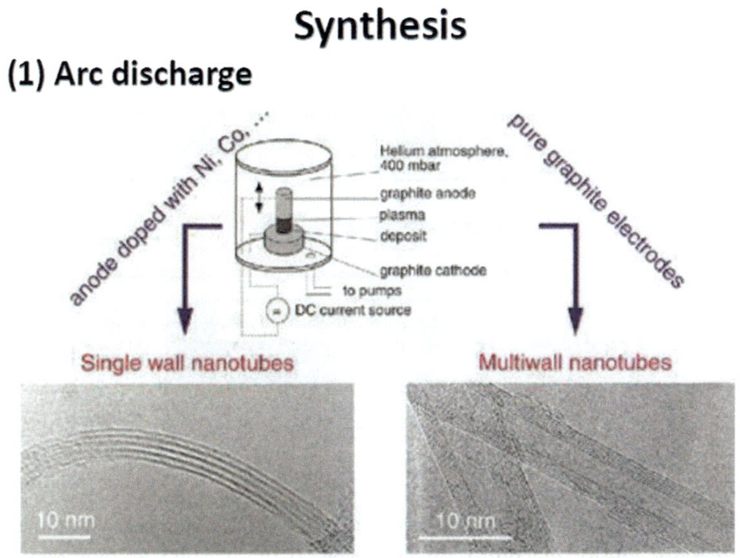

(a) Inert Gas: Argon with lower diffusion coefficient and thermal conductivity smaller diameter (1.2 nm). (b) Optical Plasma Control: distance between anode and cathode is increases, anode vaporization increases, the nanotubes diameter ranges from 1.27 to 1.37 nm. (c) Catalyst: By changing metal catalyst, the nanotubes with a diameter of 0.6 to 1.2 nm are produced. Catalysts used are Co and Mo. (d) Open Air Synthesis with Welding Arc Torch: Steelbased electrode, torch arc aimed at the edge of target and soot is deposited on substrate.

Production of MWNTs

(a) Synthesis in Liquid Nitrogen: Arc-discharge generated in liquid nitrogen. Yield is about 70% of reaction product.
(b) Magnetic Field Synthesis: Arc-discharge is controlled by a magnetic field around the arc plasma. Extremely pure graphite rods (purity > 99.999%) are used as electrodes. Highly pure MWNTs (purity > 95%) are obtained.
(c) Plasma Rotating Arc Discharge: The centrifugal force caused by the rotation generates turbulence Plasma volume and the plasma temperature raises. The yield can be increased up to 90% after purification if the rotation speed (5000 rpm)

12.3.2 Laser Ablation for Production of SWNTs

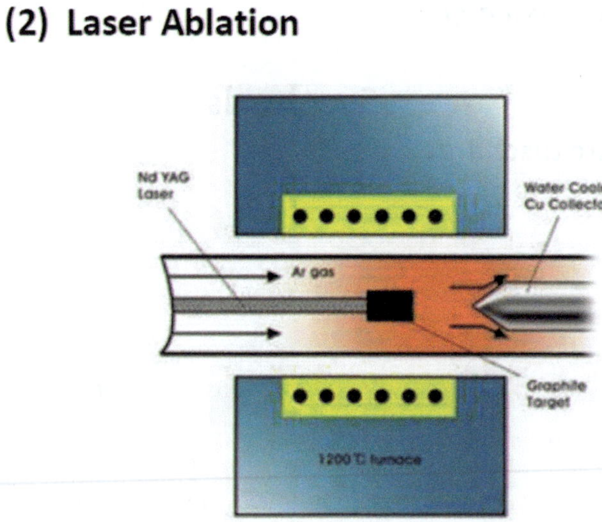

(a) Ultra-Fast Pulses from a Free Electron Laser (FEL) Method. The pulse and the repetition rate of the pulse are increased. If this system is upgraded with full power a yield of 45 gm/h can be achieved.
(b) Continuous Wave Laser-Powder Method: Instead of Nd laser, CO_2 laser is used in an argon stream. Thermal conductivity losses are significantly reduced. It is more economical in comparison with Nd (Neodymium) laser system and yield is 5gm/h.

12.3.3 Chemical Vapour Deposition (CVD)

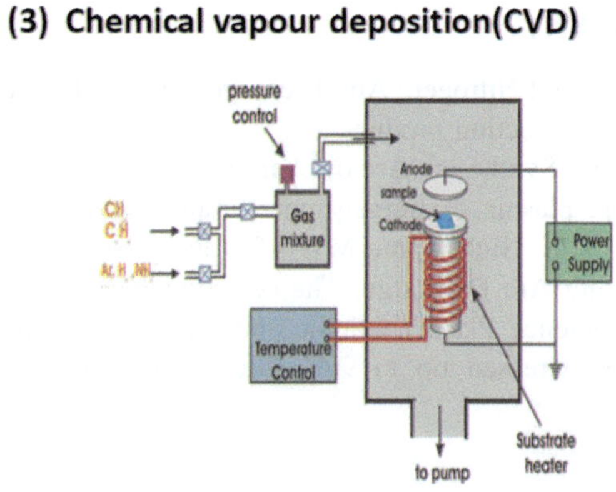

12.3.4 Flame Synthesis

(4) Flame synthesis

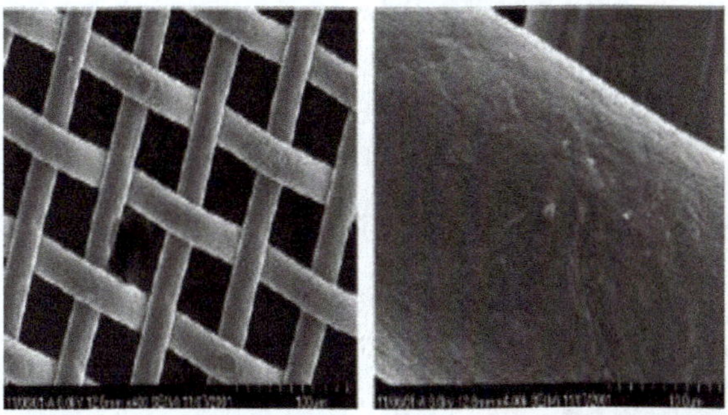

Meshes on which the metal catalyst is coated, used in flame synthesis

Types of Carbon Nanotubes

(1) A "chiral" carbon nanotube.
(2) A "zig-zag" carbon nanotube.
(3) An "armchair" carbon nanotube.

Types of carbon nanotubes

(1) A "chiral" carbon nanotube.

(2) A "zig-zag" carbon nanotube

(3) An "armchair" carbon nanotube.

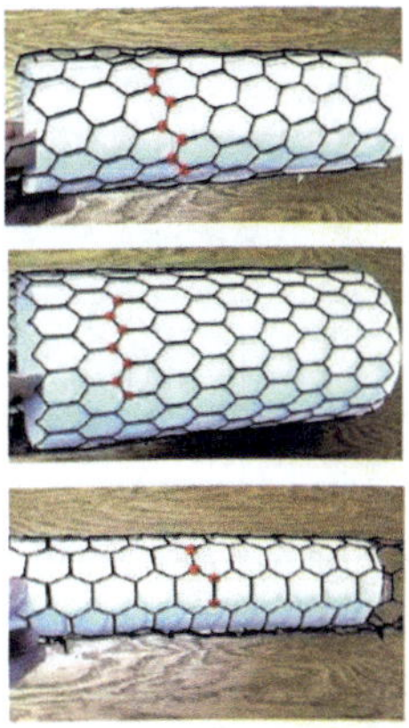

Nanotube Manipulation

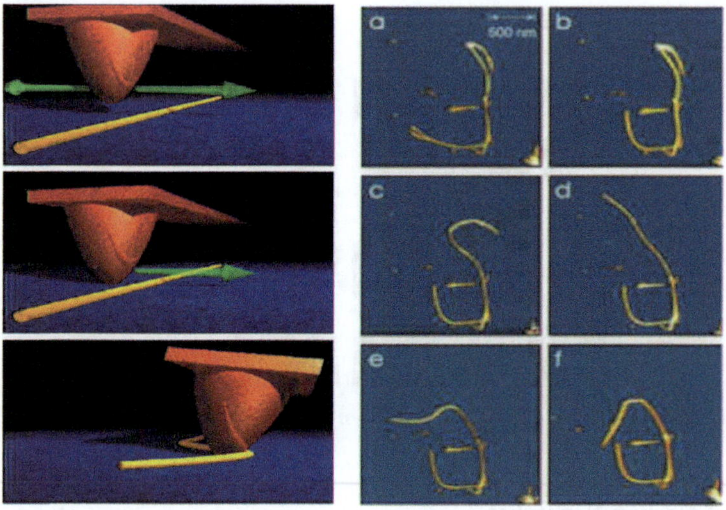

Atomic force microscope (AFM) is used

The Greek letter "theta" from a 2.5 micron long nanotubeover a silicon substrate is formed.

Ni Nanotubes

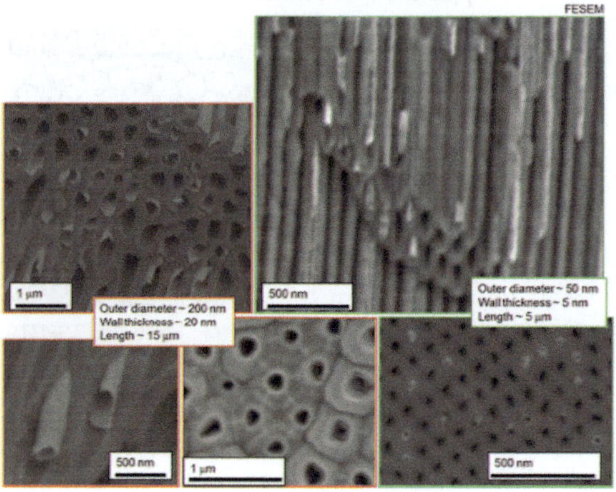

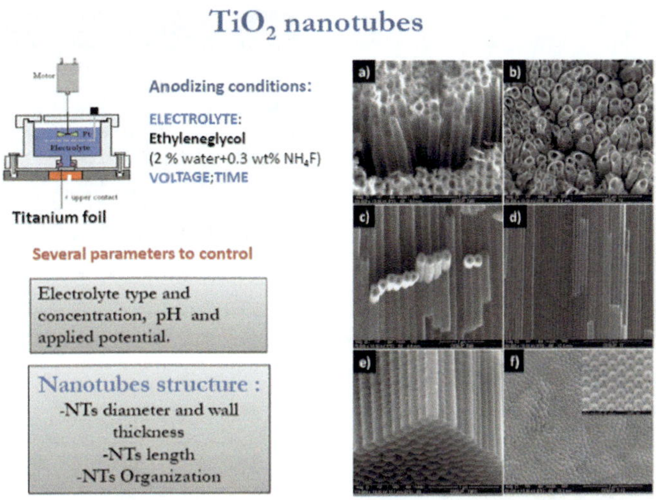

12.4 Synthesis Methods of Nanowires NWs

"Nanowires represent the smallest dimension for efficient transport of electrons and excitons, and thus will be used as interconnects and critical devices in nanoelectronic and nano-optoelectronics".

Types of Nanowires

- **Metallic**—Made from Nickel, Platinum or Gold
- **Semi-conducting**—Comprises of Silicon, Indium phosphide or Gallium Nitride
- **Insulating**—Silicon Dioxide or Titanium dioxide
- **Molecular**—Involves repeating organic or inorganic molecular units.

- 1D nanostructures represent the smallest dimension structure that can efficiently transport electrical carriers (e-, h+):
  - Can act as both nanoscale devices and wiring to link components into extremely small circuits.
- Unique density of electronic states:
  - significantly different optical, electrical, and magnetic properties from bulk counterparts.

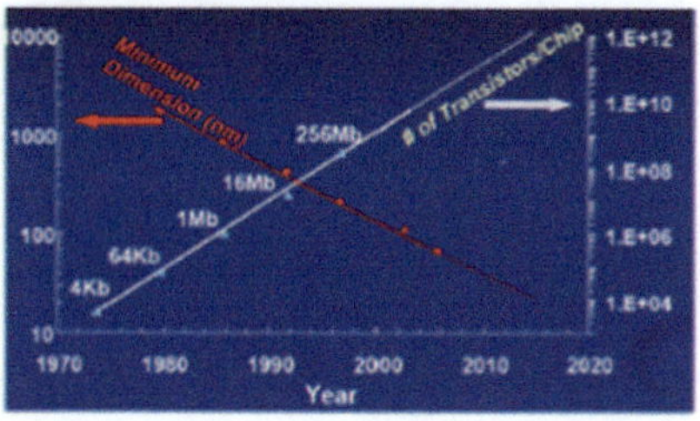

- Increased surface area
- Diameter-dependent bandgap
- Increased surface scattering for electrons and phonons
- Controlled synthesis
  - Diameter, length, composition
  - Electronic structure

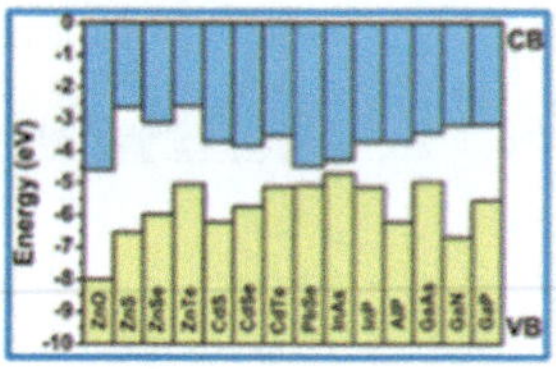

- Quantum confinement
  - Present in some, absent in others
  - Unique magnetic & electronic properties
- Millions more transistors per microprocessor

Nanowires: synthesis

- Development of 1D structures for long hindered by difficulties associated with its syntesis and fabrication (dimension control, morphology, chemical composition).
- Can be fabricated using a number of advanced nanolithography techniques:
  - e-beam lithography
  - Focused Ion Beam
  - X-ray, extreme-UV lithography
- Unconventional methods might provide alternative approach in terms of cost, throughput and high volume production
- Different stratagies use individual building blocks to fabricate increasingly complex structures
  - Challenging processing
  - Low device-to-device reproducibility
  - Alignment is the critical first step for developing devices

- **Lithography**
- **Spontaneous growth:**
  - Evaporation condensation
  - Dissolution condensation
  - Vapor-Liquid-Solid growth (VLS)
- **Template-based synthesis:**
  - Electrochemical deposition
  - Electrophoretic deposition
  - Colloid dispersion, melt, or solution filling
  - Conversion with chemical reaction

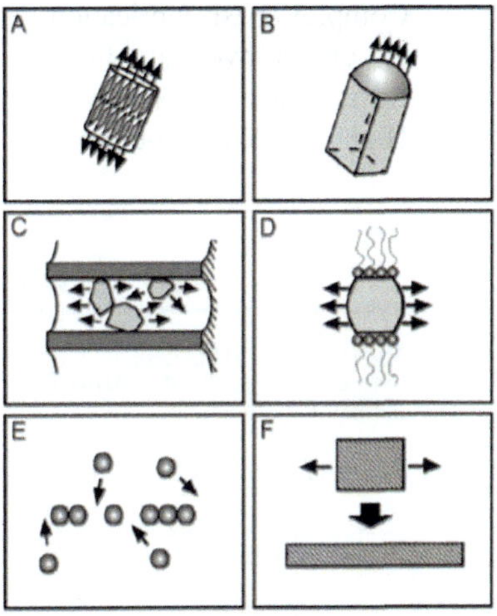

Adv. Mater. 15, 353 (2003)

12.4.1 Lithography (Top-Down)

- Connects high proportion of the nanowires to the electrodes, without the necessity to link the two components individually.
- It is the process of transferring geometric shapes on a mask to the surface of a silicon wafer
- Top-down approaches used to reduce the size of a thin film into a 1D structure
  - e-beam lithography

    vacuum environment
    direct write systems (software masks)
    slow writting over large areas
    very high system cost

  - Focused ion beam

    Resists are more sensitive than electron beam resists
    Site-specific analysis
    Deposition
    Ablation of materials
    Expensive, slow

  - X-ray lithography

    Air environment
    Very small wavelength (<14Å)

Complex mask fabrication
No optics involved—limited to 1:1 shadow printing
Resists have low sensivity
High cost X-ray sources (e.g. Synchrotron).

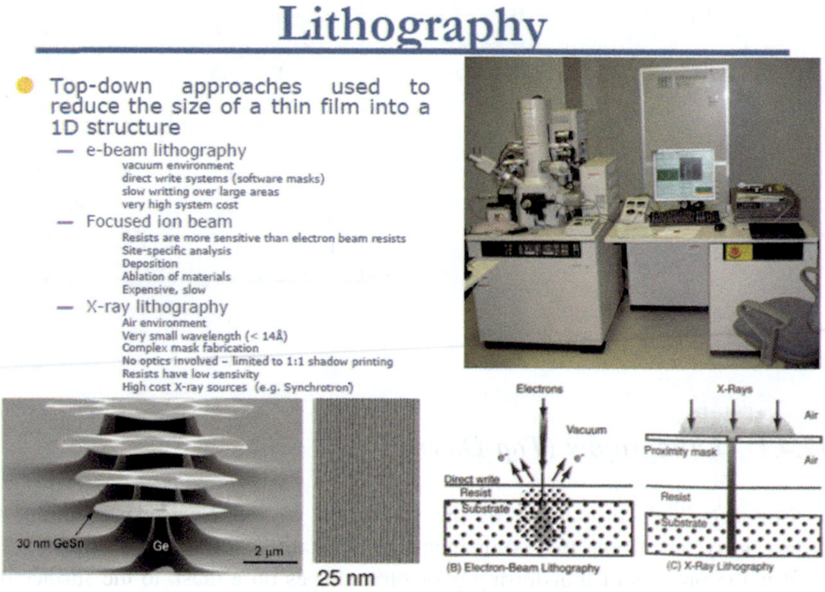

Highly Focused Ion Beam Sputters Material From a Selected Domain on a Sample Surface.

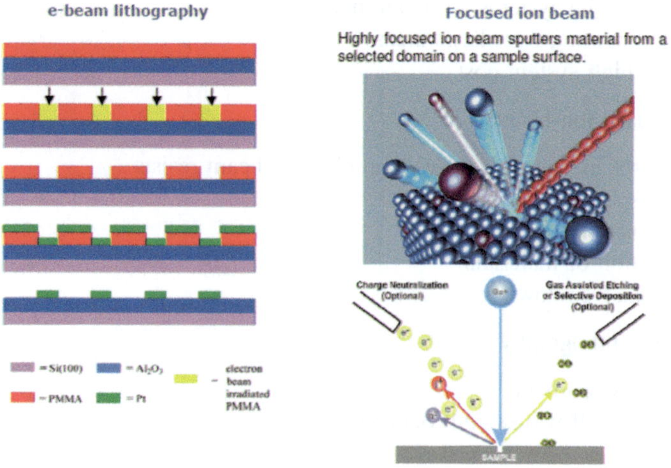

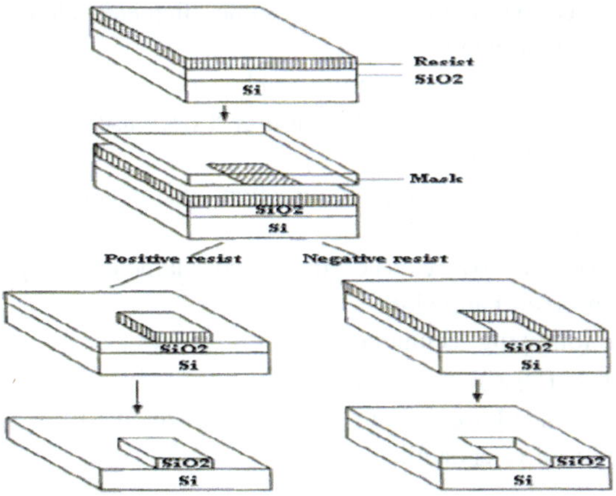

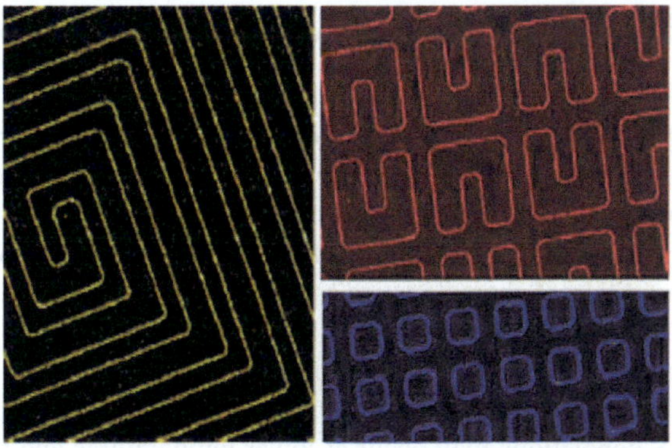

Using the lithographically patterned nanowire electrodeposition (LPNE) method, photolithography is used to define the position and two-dimensional trajectory of nanowires [5 nm (h) and 11 nm (w)].

12.4.2 Spontaneous Growth

General Idea:

- Anisotropic growth is required.
- Crystal growth proceeds along one direction, where as there is no growth along other direction.

- Uniformly sized nanowires (i.e. the same diameter along the longitudinal direction of a given nanowire).
- A growth driven by the reduction of Gibbs free energy or chemical potential.
- This can be from either recrystallization or a decrease in supersaturation.
- Anisotropic growth is required.
- Crystal growth proceeds along one direction, where as there is no growth along other direction.
- Uniformly sized nanowires (i.e. the same diameter along the longitudinal direction of a given nanowire).
- Involves the following techniques:
  - Evaporation (Dissolution) condensation
  - Vapor (or solid)-Liquid-Solid growth (VLS or SLS)
  - Stress induced re-crystallization

Nanowire Growth:

The bottom-up growth of nanowires may be achieved by a variety of methods, such as solution-phase or template-directed deposition. This work will, however, exclusively focus on the vapor-based growth of nanowires and specifically via the vapor-liquid-solid (VLS) process. The VLS process was first demonstrated in the 1960s by Wagner with the growth of silicon whiskers (wire-like structures 100's of nm to μm in diameter). The name for the VLS process was coined to describe the step-by-step phase changes of the reactant molecules as they proceed through the entire crystal growth process. It is important to note that the VLS growth process possesses many different characteristics as compared to plasma or thermal chemical vapor deposition (CVD) processes. Whereas traditional CVD relies upon the thermal (or plasma) activated decomposition of gaseous reactants, the VLS process utilizes a metal catalyst to facilitate the deposition of the gaseous reactants. Typically, VLS depositions may occur at lower temperatures and produce products with a greater crystalline quality than low-temperature thermal or plasma-CVD processes. As a generality in VLS growth, the diameter of the metal catalyst particle roughly determines the diameter of the grown nanowire. For example, if large catalyst particles are used (i.e. micron sized metal particles), then large, micron-sized whisker structures will be obtained from VLS growth, as was first established by Wagner in the 1960s. Metal nanoparticles may be deposited directly onto a growth substrate. The diameter and density of the deposited nanoparticles directly influences these same properties of the grown nanowire forests. If growth in only certain areas is desired, the nanoparticles may be patterned via means such as photolithography.

Spontaneous growth

- A growth driven by the reduction of Gibbs free energy or chemical potential.
- This can be from either recrystallization or a decrease in supersaturation.
- Anisotropic growth is required
- Crystal growth proceeds along one direction, where as there is no growth along other direction.
- Uniformly sized nanowires (i.e. the same diameter along the longitudinal direction of a given nanowire)

12.4.2.1 Evaporation (Dissolution) Condensation

Nanowires and nanorods grown by evaporation-condensation are generally single crystals with very few imperfections.

- Different facets in the crystal have different growth rate
- Presence of imperfections in specific crystal directions
- Preferential accumulation of impurities ion specific facets
- Formation of nanowires or nanorods is due to the anisotropic growth.

Advantages of Vapor-Solid (VS) Technique

- Simple and accessible
- Vapor species are generated (e.g. by evaporation) and transported and condensated onto a substrate placed in a zone with temperature lower than that of the sourcematerial
- Nanowires grown by this method are commonly single crystals with few imperfections
- The formation of nanowires is due to anisotropic growth
- Different facets in a crystal have different growth rates
- No control on the direction of growth of nanowire using this method.

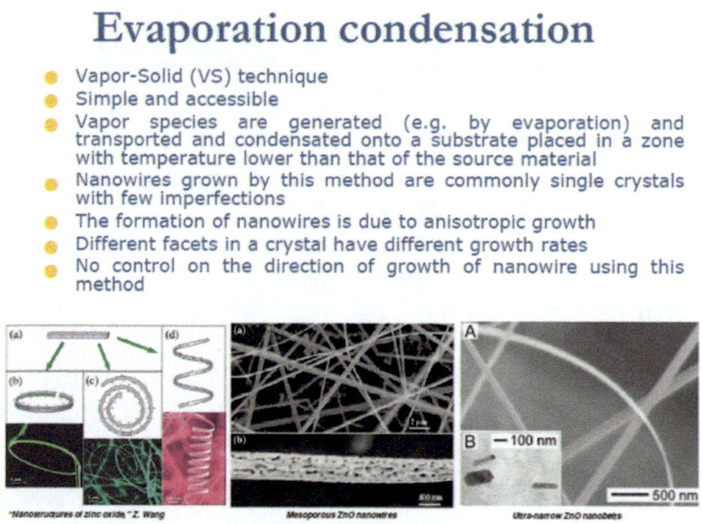

http://www.materialstoday.com/pdfs_7_6/zhang.pdf

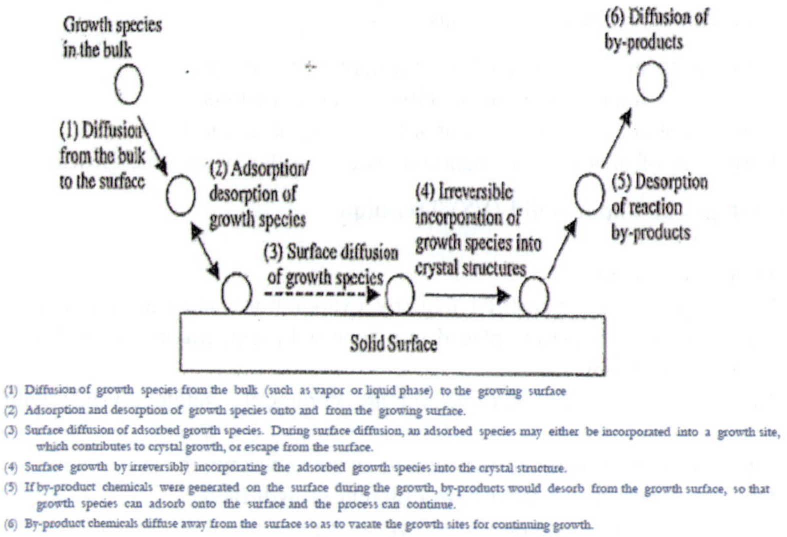

(1) Diffusion of growth species from the bulk (such as vapor or liquid phase) to the growing surface.
(2) Adsorption and desorption of growth species onto and from the growing surface.

(3) Surface diffusion of adsorbed growth species. During surface diffusion, an adsorbed species may either be incorporated into a growth site, which contributes to crystal growth, or escape from the surface.
(4) Surface growth by irreversibly incorporating the adsorbed growth species into the crystal structure.
(5) If by-product chemicals were generated on the surface during the growth, by-products would desorb from the growth surface, so that growth species can adsorb onto the surface and the process can continue.
(6) By-product chemicals diffuse away from the surface so as to vacate the growth sites for continuing growth

$$J = \frac{\alpha \sigma P_0}{\sqrt{2\pi mkT}}$$

where α is the accommodation coefficient, σ = is the supersaturation of the growth species Pois the equilibrium vapor at temperature T, m is the mass of the growth species and k is the Boltzmann constant. J = (atoms $cm^{-2}s^{-1}$).

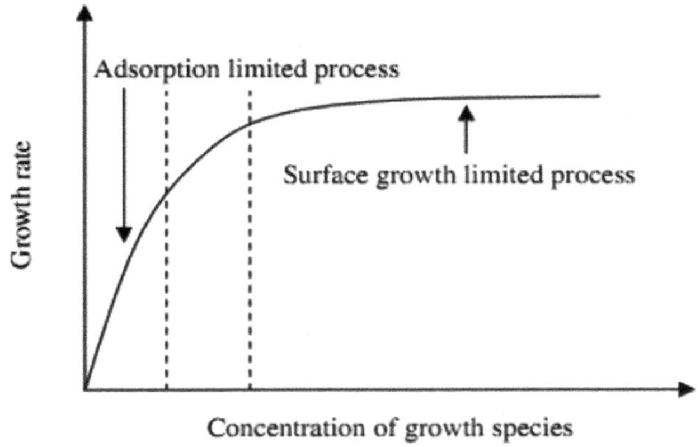

At low concentration, growth is diffusion limited and thus increases linear with increasing reactant concentration. At high concentration, surface reaction is the limit step and thus the growth rate becomes independent of reactant concentration.

12.4.2.2 Vapor (or Solid)-Liquid-Solid Growth (VLS or SLS)

A common technique for creating a nanowire is vapor-liquid-solid method (VLS). This process can produce crystalline nanowires of some semiconductor materials. It uses a source material from either laser ablated particles or a feed gas such as silane. VLS synthesis requires a catalyst. For nanowires, the best catalysts are liquid metal (such as gold) nanoclusters, which can either be self-assembled from a thin film by rewetting, or purchased in colloidal form and deposited on a substrate. The source enters these nanoclusters and begins to saturate them. On reaching super saturation, the source solidifies and grows outward from the nanocluster. Simply turning off the source can adjust the final length of the nanowire. Switching sources while still in the growth phase can create compound nanowires with super-lattices of alternating materials. A single-step vapour phase reaction at elevated temperature synthesizes inorganic nanowires such as $Mo_6S_{9-x}I_x$. From another point of view, such nanowires are cluster **polymers**:

Solution-Phase Synthesis

Solution-phase synthesis refers to techniques that grow nanowires in solution. They can produce nanowires of many types of materials. Solution-phase synthesis has the advantage that it can produce very large quantities, compared to other methods. In one technique, the polyol synthesis, ethylene glycol is both solvent and reducing agent. This technique is particularly versatile at producing nanowires of lead, platinum, and silver. The supercritical fluid-liquid-solid growth method can be used to synthesize semiconductor nanowires, e.g., Si and Ge. By using metal nanocrystals as seeds, Si and Ge organometallic precursors are fed into a reactor filled with a supercritical organic solvent, such as toluene. Thrombolysis results in degradation of the precursor, allowing release of Si or Ge, and dissolution into the metal nanocrystals. As more of the semiconductor solute is added from the supercritical phase (due to a concentration gradient), a solid crystallite precipitates, and a nanowire grows uniaxial from the nanocrystal seed.

Non-catalytic Growth

Nanowires can be also grown without the help of catalysts, which gives an advantage of pure nanowires and minimizes the number of technological steps. The simplest methods to obtain metal oxide nanowires use ordinary heating of the metals, e.g. metal wire heated with battery, by Joule heating in air can be easily done at home. The vast majority of nanowire-formation mechanisms are explained through the use of catalytic nanoparticles, which drive the nanowire growth and are either added intentionally or generated during the growth. However the mechanisms for catalyst-free growth of nanowires (or whiskers) were known from 1950s. Spontaneous nanowire formation by non-catalytic methods were explained by the dislocation present in specific directions or the growth anisotropy of various crystal faces. More recently, after microscopy advancement, the nanowire growth driven by screw dislocations or twin boundaries were demonstrated. The picture on the right shows a single atomic layer growth on the tip of CuO nanowire, observed by in situ TEM microscopy during the non-catalytic synthesis of nanowire. The VLS growth of a nanowire can be summarized into three steps: (I) liquid alloy droplet formation, (II) nanowire nucleation, and (III) nanowire elongation (i.e. growth).

- Catalyst forms a liquid droplet by itself.
- Acts as a trap for growth species.
- The growth species is evaporated first and then diffuses and dissolves into a liquid droplet.
- It precipitates at the interface between the substrate and the liquid.

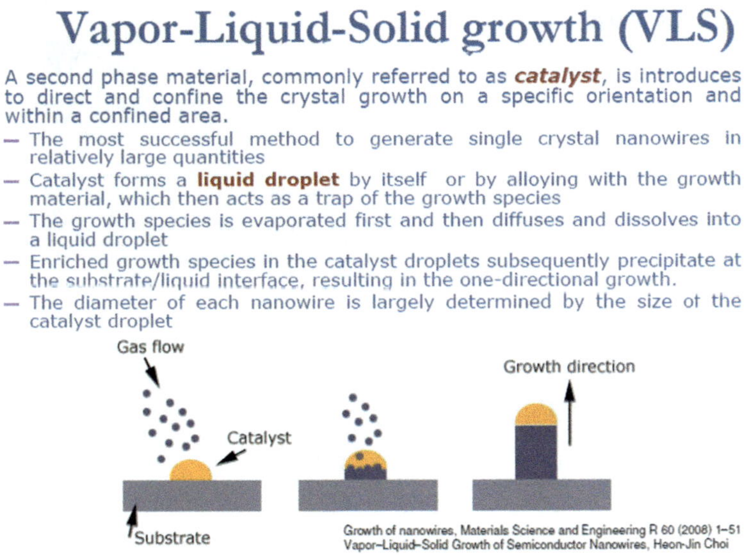

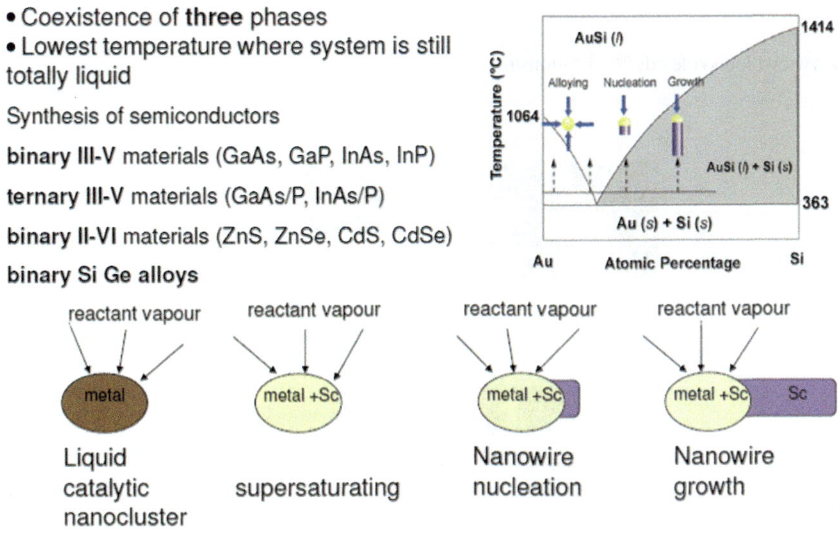

- Coexistence of **three** phases
- Lowest temperature where system is still totally liquid

Synthesis of semiconductors

binary III-V materials (GaAs, GaP, InAs, InP)

ternary III-V materials (GaAs/P, InAs/P)

binary II-VI materials (ZnS, ZnSe, CdS, CdSe)

binary Si Ge alloys

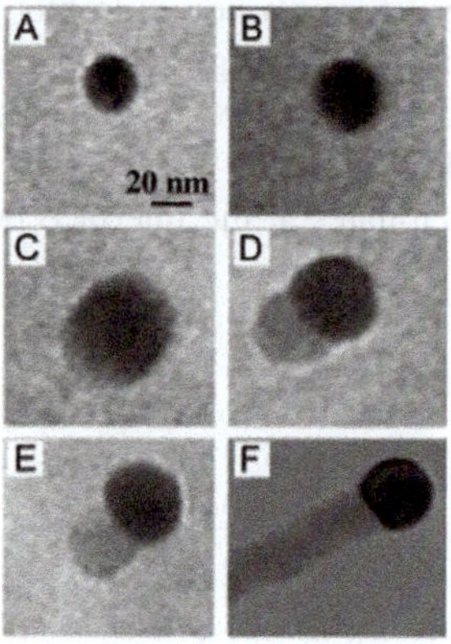

The birth of a Ge nanowire on a Au nanocluster, as observed using in-situ TEM. It is clearly seen that the Au nanocluster started to melt after the formation of Ge–Au alloy, and this was followed by an increase in the liquid droplet size during the Ge vapor condensation process. When the droplet was supersaturated with the Ge component, a Ge nanowire grew out of this droplet of Au–Ge alloy and became longer as time elapsed

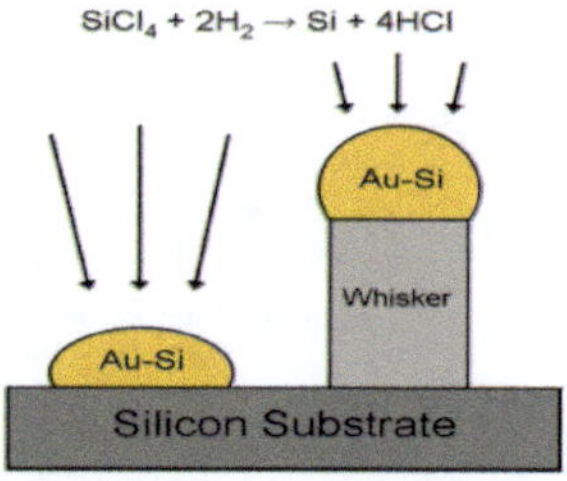

Si whisker growth from the reaction of $SiCl_4$ and H_2 vapor phases. This reaction is catalyzed by gold-silicon droplet deposited on the wafer surface prior to whisker growth.

www.cvd.louisville.edu/Publications/recentpublications/proceedings_mrs_fall2001.pdf

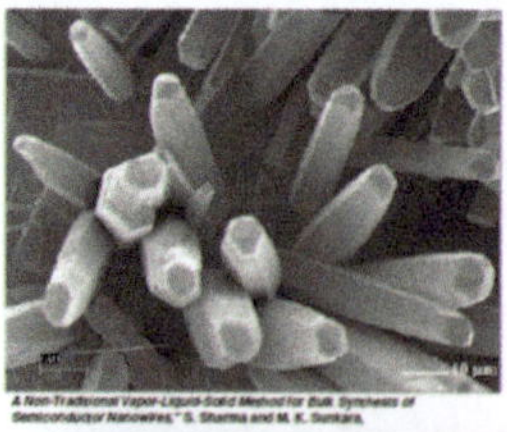

A Non-Traditional Vapor-Liquid-Solid Method for Bulk Synthesis of Semiconductor Nanowires," S. Sharma and M. K. Sunkara.

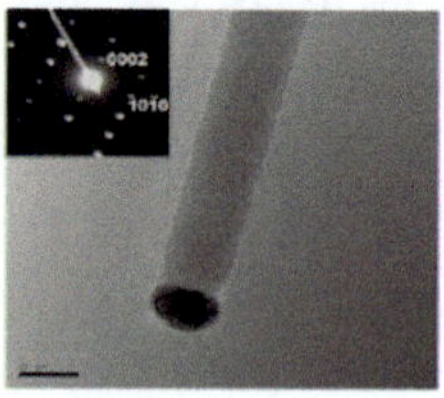

TEM of a single crystal ZnO nanorod

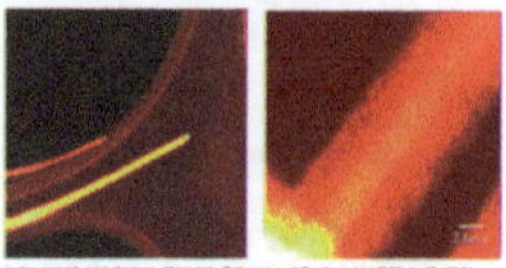

ZnO nanowire growth and devices, Materials Science and Engineering R47) 1–47 (2004

Z-contrast scanning transmission electron microscopy image of a (Zn,Mg)O nanorod with a Ag catalyst particle at the rod tip.

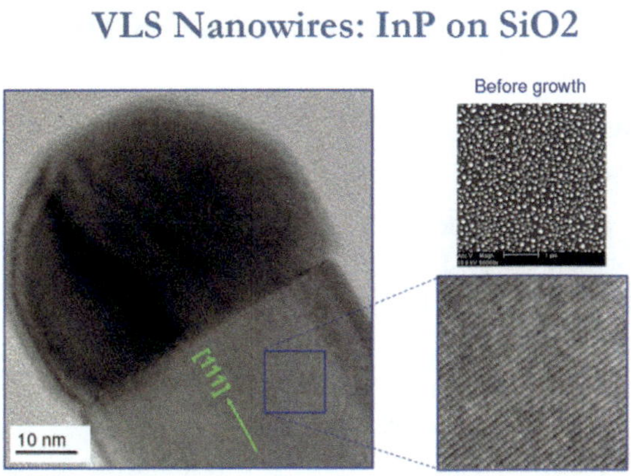

VLS Nanowires: Heterostructures

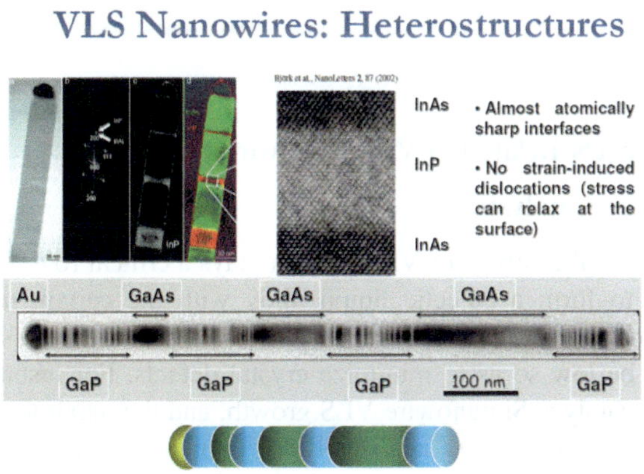

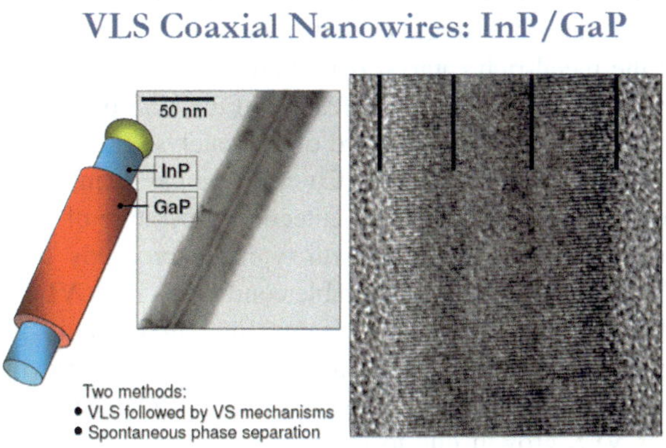

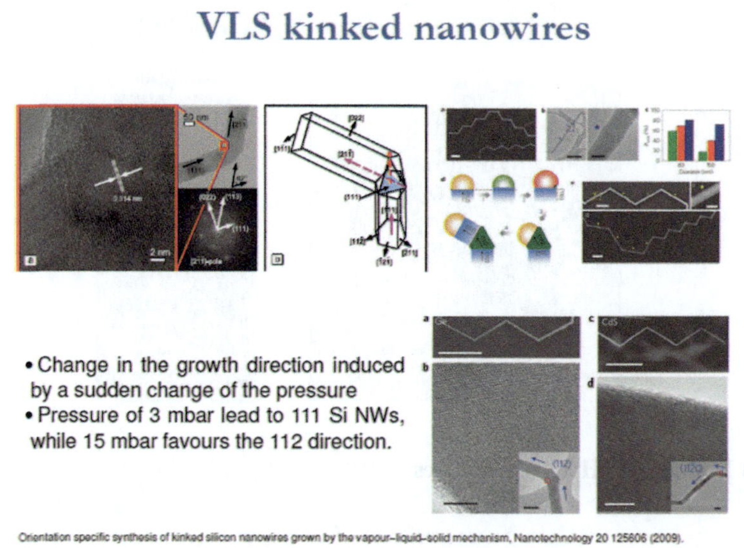

Important Aspects Related to VLS Nanowire Growth

1. **Nanoparticle Catalysts**

The choice of metal catalyst for VLS growth plays a critical role, the metal catalyst must be able to form a eutectic liquid alloy with the certain nanowire growth species. Moreover, the solid solubility of the catalyst in the final single-crystalline material must be low so as to minimize crystal defects. For instance, Au is commonly used to catalyze Si nanowire VLS growth, and the maximum solid solubility of Au in Si is just $\sim 2 \times 10^{-4}$ at.% at 1280 °C, making it an appropriate choice.

Another important consideration is whether the metal catalyst can alternatively perform nanowire growth via the vapor-solid-solid (VSS) process. VSS growth proceeds similarly to the VLS process, a catalytic metal nanoparticle acts as a growth site for a nanowire single-crystal. Unlike the VLS process, VSS growth proceeds with the metal-rich catalyst particle remaining in a solid state throughout the entire growth process. One notable feature of the VSS process is that it can take place far below the eutectic temperature of the catalyst and crystal material alloy. For instance, Si nanowire growth with Cu catalysts has been demonstrated via the VSS mechanism at $\sim$400 °C while the eutectic temperature of the system is $\sim$800 °C. Moreover, the VSS nanowire growth typically proceeds 10–100× slower as compared to VLS growth under comparable conditions. Both VLS and VSS growth have been observed occurring at the same time in a single growth chamber; therefore in order to maintain good control over nanowire growth it necessary to properly control and choose the appropriate growth mechanism. Au is often chosen as the alloy for VLS growth due to the fact that it possesses a low eutectic temperature in most alloys that it forms, suppressing the likelihood of VSS growth. The density and morphology of the metal nanoparticles are a critical determinate of nanowire growth. Since VLS grown nanowires roughly take on the diameter of the metal catalyst nanoparticle, the presence of a wide distribution of nanoparticle sizes will produce nanowires possessing the same large size distribution.

2. **Nanoscale Eutectic Temperature**

For nanoscale systems (i.e. metal nanoparticles or thin films), it has been widely noted that an alloy's eutectic temperature is typically suppressed from the well-known bulk values. This effect contributes to the reduction of the minimal temperature necessary for nanowire VLS growth.

3. **Nucleation of Nanoparticles**

Although ideally in VLS growth there should be a 100% nucleation rate of each catalyst nanoparticle to initiate the growth of an individual nanowire, this does not always happen under practical conditions. Under higher growth temperatures and higher partial pressures of the growth reactants, more nucleation events occurred.

4. **Nanowire Sidewall Deposition**

In the idealized VLS nanowire growth mechanism, deposition of material on the grown nanowire sidewalls is assumed not to occur. However, under certain conditions this may in fact take place, leading to the growth of tapered nanowires. Tapered nanowires feature a wide base and narrow top (the end where axial growth ceases). This morphology is due to the fact that sidewall deposition is typically homogenous and occurs at a uniform rate on a nanowire; thus, the oldest grown segments of the nanowire experience the most deposition and the newly grown sections of a nanowire will experience the least amount of total sidewall deposition. Various strategies for reducing or eliminating sidewall deposition have been demonstrated. Two parameters known to play a large role in uncatalyzed deposition are growth temperature and the partial pressures of reactants. Increasing either parameter will increase the amount of thermal decomposition of material species on grown nanowires and the substrate itself. To carry out proper nanowire growth, these two parameters must be properly adjusted to address the two important needs of: (1) high yield nanowire nucleation, and (2) the prevention of sidewall deposition.

Ge nanowire growth provides a good example of the rational choice of the growth temperature in order to address these two simultaneous needs. It has been established that Ge nanowires grown from GeH_4 are particularly susceptible to unanalyzed sidewall deposition if grown near the bulk eutectic temperatures on Ge and Au ($\sim$360 °C). A successful method to achieve both high nucleation rates and uniform nanowire diameters is through the use of a two-step growth process. A higher temperature is used initially in the VLS growth process in order to cause the nucleation of a large percentage of the alloy nanoparticle droplets. Then, once high yield nucleation is achieved, the temperature is ramped down to a lower level to prevent sidewall deposition while simultaneously providing sufficient nanowire axial elongation growth rates.

Another strategy involves the use of an appropriate carrier gas to functionalize the grown nanowire surfaces with the correct functionalization group, which prevents sidewall deposition. This was demonstrated with the use of H_2 as a carrier gas in order to H- terminate silicon nanowires during VLS growth. Additionally, the use of local substrate heaters (as opposed to hot-wall chambers) has been shown to reduce the amount of unwanted thermal decomposition of CVD reactants and thus reduce nanowire sidewall deposition.

5. **Nanowire Growth Rates**

Another essential parameter for nanowire growth is the rate of axial elongation of the nanowire. Depending on deposition conditions, this rate can be drastically altered. For instance, the CVD reactant Si_2H_6 was found to produce nanowires at growth rates over 100× as great as compared to SiH_4. It was reasoned that Si_2H_6 more readily decomposes as compared to SiH_4. Besides the choice of reactant species, other parameters which play a critical role are the temperature and partial pressure of the reactance species. Increasing either of these parameters will also increase the nanowire growth rate.

Specifically, it has been shown that the amount of supersaturation of the crystal growth material in the liquid alloy nanoparticle is directly influenced by the partial pressure of the gaseous reactants. This amount of supersaturation is what directly effects nanowire growth rates. Thus, it has been observed that typically a higher partial pressure of the reactants leads to higher growth rates.

6. **Nanowire Heterostructures**

Another benefit that the bottom-up growth of nanowires provides is that different materials may be incorporated into a single nanowire in a heterogeneous configuration during growth. A schematic of the possible configurations of the heterogeneous nanowire material structure is shown in Fig. 12.9. The base case (starting point) to heterogeneous nanowire growth is the VLS growth of a single material nanowire (Fig. 12.9a) [76]. The VLS growth process has already been extensively covered earlier in this chapter. By modifying the reactant gases during VLS growth, axial or radial material nanowire heterostructures may be obtained. An axial heterostructure (Fig. 12.9b), where the material varies along the long axis of the nanowire, can be achieved by varying reactants during the axial elongation phase of VLS growth.

Radial heterostructures (Fig. 12.9c), where the material composition varies symmetrically along the radial cross-section of a nanowire, are synthesized by performing sidewall (i.e. homogeneous) deposition of a new material after axial elongation of the nanowire is ceased. A two-material radial heterostructure, as depicted in Fig. 12.9c, is typically termed a core-shell nanowire, where the core comprises of the starting nanowire material, and the shell is made up of the later, homogeneously deposited exterior material. The limit to the amount of shells capable of being deposited is in theory unlimited, but the practical limit is determined by the specific material system being deposited.

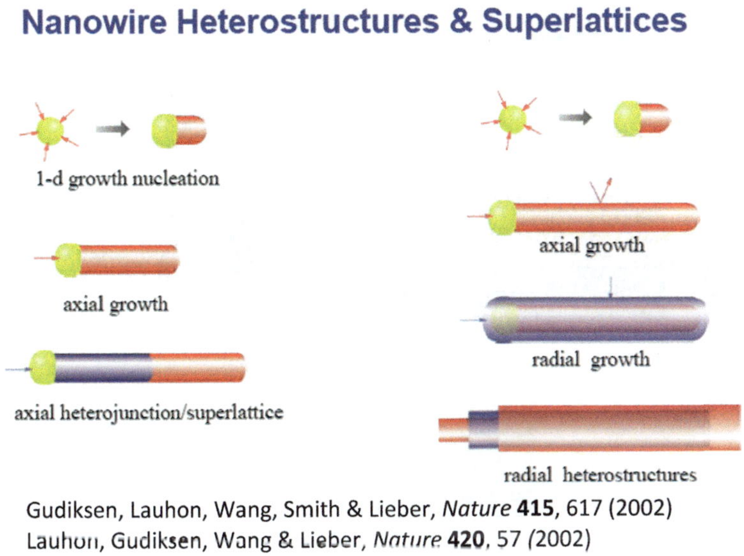

Fig. 12.9 Nanowire Heterostructure Configurations Description given in text. Arrow denotes preferential site for reactant adsorption. Reprinted with permission from. Copyright 2003, Materials Research Society; Gudiksen, Lauhon, Wang, Smith & Lieber, *Nature* **415**, 617 (2002); Lauhon, Gudiksen, Wang & Lieber, *Nature* **420**, 57 (2002)

Nanowire Growth Setups

CVD Nanowire Growth

One of the most common setups used to carry out the VLS growth of nanowires is in a low-pressure CVD, i.e. LPCVD, apparatus (Fig. 12.10). Typically, the setup comprises of a two-ended quartz tube placed in a ceramic-encased resistive element furnace. One end of the tube (the inlet side) is hooked up to mass flow controllers (MFCs) which regulate the input flow of compressed process gases. Besides the main process gas (i.e. SiH_4 or GeH_4), there may be also be dopant gases or dilutant gases (i.e. H_2 or Ar) fed into the growth chamber. The other end of the tube (outlet side) is hooked up to a vacuum pump via an adjustable vacuum throttle valve. The nanowire growth substrates are placed in the center of the tube furnace where temperature uniformity is the highest.

The nanowire growth is typically run at low to medium vacuum ($\sim$1 to 760 torr) and this is controlled by properly adjusting the inlet flow of the gases as well as the strength of the vacuum pump via its throttle valve connection to the outlet of the tube. Similar to conventional LPCVD systems, the major reason for low-pressure depositions is to increase uniformity across the deposition substrates [77]. Additionally it is important to be able to adjust the process pressures for VLS nanowire growth, since these parameters directly affect all aspects of nanowire growth (as detailed earlier).

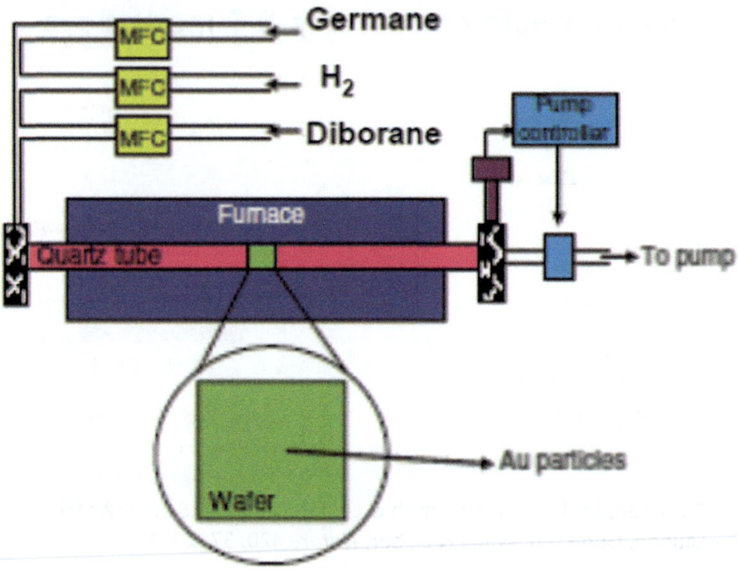

Fig. 12.10 CVD setup for Ge nanowire growth. Reprinted by permission from Macmillan Publishers Ltd: Nature Protocols copyright 2004

Vapor-Transport Metal Oxide Nanowire Growth

An alternative method for nanowire growth via the VLS method relies upon the vapor transport of evaporated source materials from a solid source (typically in powder form). This method is schematically similar to the CVD growth method outlined earlier, with the one difference being that the source materials are supplied via thermal evaporation rather than compressed gas sources. A typical setup for this process is depicted in Fig. 12.11. As before with CVD systems, the setup comprises of a two-ended quartz tube placed in a ceramic-encased resistive element furnace.

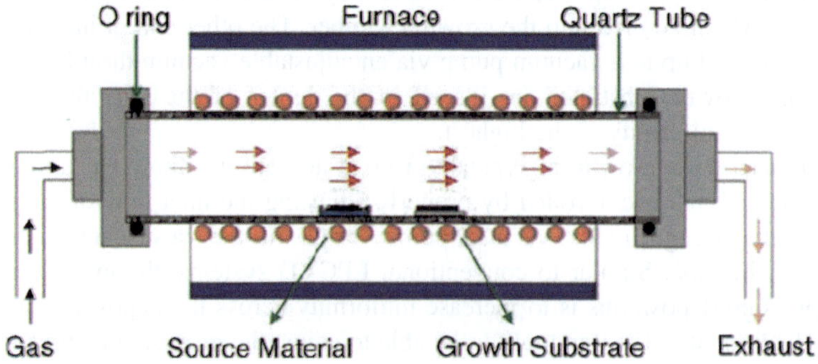

Fig. 12.11 Vapor-transport apparatus for nanowire growth. Reprinted from [78] with permission from Elsevier

One end of the tube (the inlet side) is hooked up to mass flow controllers (MFCs) which regulate the input flow the compressed carrier/process gases. The other end of the tube (outlet/exhaust side) is hooked up either to an exhaust or a vacuum pump depending on whether atmospheric pressure or vacuum is desired. Inside the quartz, tube is first placed the source material and then downstream is placed the growth substrate.

Regardless of the process pressure of the tube (either atmospheric or vacuum), a carrier gas is supplied at the inlet end of the tube via a MFC in order to direct the transport of the evaporated source species to the downstream growth substrate. This carrier gas is typically either Ar or N_2; the carrier gas species and flow rate are important parameters for nanowire growth [79]. Near the middle of the tube lies the source material. During process, this area of the furnace is heated up to a high temperature so as to create a partial pressure of the species by thermal evaporation. This partial pressure is determined by the evaporation rate and vapor pressure (for a certain temperature) of the source material. Downstream of the source material is placed the growth substrate. The growth substrate is typically heated either to the same temperature or to a lower one as compared to the source material (depending on process requirements). Additionally, the distance between the source material and growth substrate is an important parameter, which affects the amount of material species that reach the growth substrate.

The previously described vapor transport growth using evaporated source materials method can grow a wide variety of nanowires, including even Si or Ge nanowires, however its use in the growth of metal oxide nanowires will be explored in detail. Metal oxide semiconductor nanowires; such as SnO_2, ZnO, or In_2O_3; are a class of materials that readily lend themselves to be grown by the vapor transport growth method. In order to synthesize oxide nanowires with this method, two options are available: either the semiconductor oxide material may be evaporated directly (i.e. SnO_2, ZnO, or In_2O_3 powder), or a pure metal powder (i.e. Sn, Zn, or In) may be evaporated in the presence of O_2 in the growth tube. The gaseous O_2 may be supplied either via a MFC after pumping out residual atmospheric O_2 from the process tube; or instead, trace amounts of atmospheric O_2 that either leak into the tube or which are residual quantities may be utilized. This amount of O_2 is a critical parameter of nanowire growth, it not only influences the volatility of the source material (due to competitive oxidation of the metallic atoms) but also influences the yield, morphology, and growth rate of the nanowires.

Metal Oxide Nanowire Vapor-Phase Growth

Metal oxide nanowires (i.e. SnO_2, ZnO, or In_2O_3) may be grown using vapor-phase, metal catalyzed methods which are widely assumed to be similar to the previously detailed VLS method for Si and Ge nanowires. However, the exact growth mechanism of metal oxide nanowires is still under much debate in the literature. The main uncertainty lies in discerning the role of the metal catalyst during growth. The following four separate theories on metal oxide nanowire growth will be presented: (1) VLS growth, (2) VSS growth, (3) self-catalytic VLS growth, and (4) VS growth.

1. **VLS Growth of Metal Oxide Nanowires**

The theorized VLS growth process for a ZnO nanowire will now be described. ZnO is chosen arbitrarily for this discussion, the process is applicable to any metal oxide nanowire. It is proposed that the VLS growth proceeds from a binary alloy droplet of the vapor phase Zn material and the chosen catalyst (i.e. Au). Upon sufficient super saturation of the droplet, Zn precipitates from the droplet at the liquid-solid interface of the droplet and the ZnO nanowire. Oxygen is claimed not to be present in the alloy droplet (experimental data is cited in [80]), and the makeup of the Zn/Au alloy droplet is hypothesized to follow Wagner's original VLS theory of a binary eutectic alloy droplet. Gaseous oxygen atoms are next presumed to diffuse towards the Zn metal interfacial region (but not through the Au-Zn droplet bulk) and then carry out the oxidation of the Zn atoms into ZnO and thus elongate the nanowire [80–82]. It is important to note that the authors derived the above theory from experimental ZnO nanowire growth that took place near the melting temperature of Zn ($\sim$420 °C) [81–83]. This theory however does not properly explain (seemingly-VLS) ZnO nanowire growth at temperatures far above Zn's melting temperature, i.e. NW growth has been demonstrated at $\sim$800 °C [84]. At such high temperatures, any precipitated Zn metal would quickly re liquefy which invalidates the previous model. A less widely cited proposal may explain such high temperature growths. It hypothesizes that oxygen diffuses into the alloy droplet during growth, causing supersaturating of the metal oxide inside the droplet, thus bringing about precipitation of the metal oxide and nanowire growth [85] (Fig. 12.12).

In addition to ZnO, Au-catalyzed VLS growth has been used to grow many other metal oxide nanowires, including SnO_2 and In_2O_3 nanowires [87]. The VLS growth first described by Wagner and Ellis [88] the growth occurs via phase changes that are mediated through a catalyst particle. That means that the material to be grown,

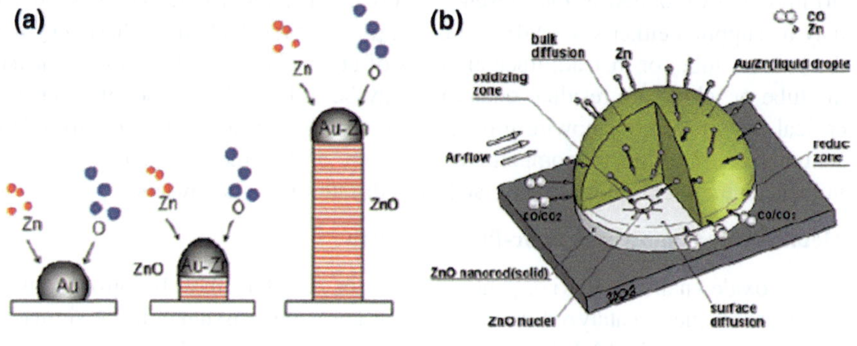

Fig. 12.12 Model for metal oxide nanowire VLS growth. **a** A proposed model for the VLS growth of a ZnO nanowire. The vapor phase Zn atoms are presumed to form an alloy with the Au catalyst nanoparticle. Nanowire growth proceeds via the precipitation of Zn from the droplet, which then forms ZnO by oxidation by oxygen atoms present in the chamber. Reprinted from [86] with permission from Elsevier. **b** A more detailed schematic of this model. Oxidative species (i.e. CO/CO_2 in this figure) oxidize the precipitating Zn atoms at the liquid-solid interface of the catalytic droplet and growing nanowire. Reprinted from [80] with permission from Elsevier

in this case silicon, is delivered in the gas phase, either molecularly or in the form of a gas compound that is introduced into the growth chamber. Figure 12.13 shows a schematic sketch illustrating the widely adopted VLS growth. In the case of Si, molecular Si precursors can be evaporated either by Si effusion cells, or pulsed vapor deposition. Silicon gas precursors include mono-silane (SiH_4), trichlorosilane ($SiHCl_3$) or higher order silanes. The catalyst particle acts as a collector of silicon. In the case of employing gas precursors, the catalytic nature leads to lowering of the dissociation energies of the gas. Consequently, a high concentration of silicon atoms is found at the catalyst particle′s surface, leading to diffusion into the cluster.

Nucleation of Si occurs at the footprint of the Au–Si catalyst layer by layer. Consequently, a silicon monolith or nanowire is formed. Figure 12.14 shows the

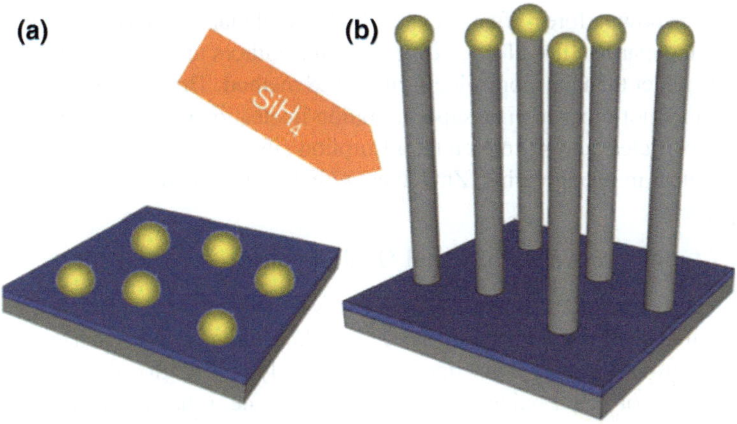

Fig. 12.13 Schematic view of VLS growth of silicon nanowires. **a** Gold particles formed on the growth substrate. **b** VLS growth using silane as silicon precursor

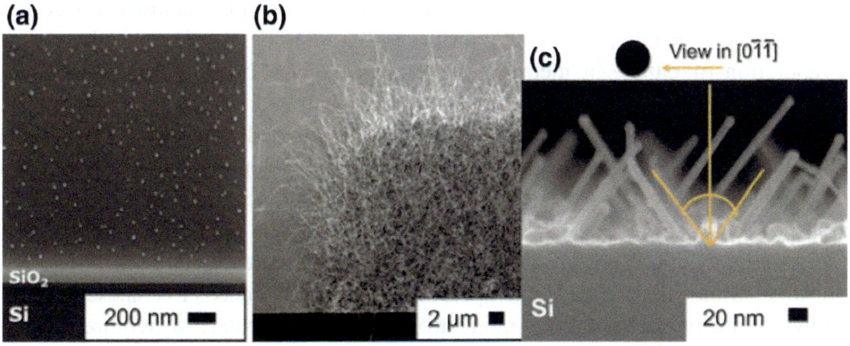

Fig. 12.14 Nanowires grown by the VLS process. **a** Shows the catalytic gold particles formed by sub–nanometer sputter deposition of Gold and a subsequent coalescence with a combination of a thermal anneal in hydrogen atmosphere at 450 °C for 300 s and a subsequent plasma treatment [89]. **b** Shows nanowires grown on an oxidized silicon substrate and c shows a close-up of nanowires grown on a (100) oriented single crystalline silicon substrate

starting gold catalysts in Fig. 12.14a, silicon nanowires grown on an oxidized silicon wafer in Fig. 12.14b and a close up view of {110} oriented silicon nanowires lattice matched grown on a crystalline silicon (100) wafer. An important implication of the VLS mechanism is that the size of the Au cluster defines the diameter of the nanowire without the tedious need of further lithographic means. In addition, the catalyst position on the substrate dictates the nanowire placement.

2. **VSS Growth of Metal Oxide Nanowires**

An alternative explanation for metal nanoparticle catalyzed ZnO nanowire growth relies upon the vapor-solid-solid (VSS) growth mechanism. It is important to note that this theory was derived from experimental measurements on ZnO nanowire growths that were conducted at ~ 850 °C and does consistently explain growth at this high temperature [90]. The basis for their theory is attributed to XRD measurements taken on a ZnO nanowire forest after growth, which reveal that no Zn was present in the Au catalyst nanoparticle tips. This finding leads the authors to suggest that VSS growth occurred rather than the more widely cited VLS method. The authors propose that the growth does not proceed via a supersaturation-nucleation process, but rather that the Au nanoparticles act as preferential adsorption sites for Zn species which then are incorporated in the growing ZnO nanowire by surface diffusion along the Au nanoparticle.

3. **Self-Catalytic Growth of Metal Oxide Nanowires**

In the previous two methods, a separate catalyst metal (i.e. Au) was used in order to catalyze the growth of metal oxide nanowires. An alternative approach to metal oxide nanowire growth instead relies upon the self-catalytic NW growth properties of the metal species of metal oxide nanowires. For instance, it has been shown that ZnO nanowires have been selectively grown from Zn isolated metal islands without the use of any additional metal catalyst by simple oxidation annealing [91, 92]. Since in both references, the reaction takes place above the melting point of Zn (~ 500 °C), it is hypothesized that Zn liquid originating on the Zn islands act as a catalyst to ZnO nanowire growth [93]. The vapor-phase growth of ZnO nanowires has been demonstrated on completely bare surfaces (i.e. SiO_2 or alumina) by the thermal evaporation of either Zn metal in the presence of O_2 or ZnO powder [86, 94]. Moreover, a similar synthesis has also been demonstrated with SnO_2 nanowires [95, 96]. The mechanism for such nanowire growth is explained by two hypotheses: either self-catalytic VLS or vapor-solid (VS) growth. For both methods, source materials are evaporated at high temperature (usually in a tube furnace) into a vapor state and transported via carrier gas.

In self-catalytic VLS growth, the vapor phase species reaches the deposition substrate while still in the high temperature zone, and proceeds to coalesce onto the growth substrates to form liquid droplets. In particular, a requirement for self-catalytic VLS growth is that the deposition substrate temperature be above the melting point of the metal species to allow for liquid droplet formation. An example of this growth process for a SnO_x nanobelt is shown in Fig. 12.15. One theory proposed to explain the self-catalytic growth process is that liquid metal droplets are

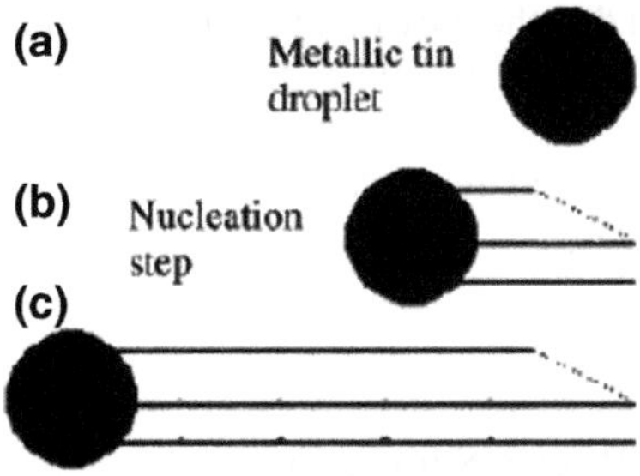

Fig. 12.15 Model for Self-Catalytic VLS growth of a SnO_x Nanobelt. Sn droplet formation via the coagulation of vapor Sn atoms, **b** nucleation of the SnO_x nanobelt after the Sn droplet becomes supersaturated with SnO_x, **c** axial elongation of the nanobelt via precipitation of supersaturated SnO_x, similar to the VLS mechanism. Adapted with permission from [97]. Copyright 2006 American Chemical Society

formed which then catalyze nanowire growth as a result of the supersaturating of metal oxide in the droplet [97]. An alternative theory surmises that instead of metal oxide super saturation in the droplet, gaseous oxygen reacts at the surface of the droplet (not in the bulk of droplet) to form the metal oxide nanowire [91]. Importantly, Au-catalyzed VLS growth and self-catalytic VLS growth may proceed simultaneously. Branched SnO_2 nanowires have been grown from Au catalysts due to the self-catalysis of branched nanowire formation from adsorbed Sn droplets on the nanowire surface [85].

4. **Vapor-Solid Growth of Metal Oxide Nanowires**

An alternative hypothesis to the self-catalytic formation of metal oxide nanowires is the VS growth process, whereby metal oxide vapor directly deposits onto the growth substrate in solid form. As a distinction to the VLS process, there is no intermediate liquid phase present. Crystal nucleation occurs directly as vapor metal oxide species condense (and solidify) onto a growth substrate. Nanowire growth proceeds due to the fact the nanowire growth front presents a thermodynamically favorable place for additional vapor phase species to solidify. Notably, the VS process is able to cohesively explain the self-catalytic growth of metal oxide [98] and even metal [99] nanowires on deposition substrates held at temperatures lower than the melting point of the metal species. Furthermore, the VS process has also been deduced to operate at temperatures above the metal melting point [100] because metal oxides typically possess melting temperatures greater than the metal element itself. A disadvantage of the VS process is that it is difficult to control the morphology of the grown nanowires (i.e. diameter) since no separate metal catalyst is required.

12.4.2.3 Stress Induced Re-Crystallization

- Application of pressure on solids at elevated temperatures is known to result in the growth of whiskers or nanowires with diameters as small as 50 nm.

- It was demonstrated that the growth rate of tin whiskers increased proportionally with the applied pressure and could be four orders of magnitude when a pressure of 7500 psi was applied.
- This technique is not widely explored in the recent studies on the growth of nanorods and nanowires.

12.4.3 Template-Based Synthesis

General Idea:

- Use in fabrication of nanorods, nanowires, and Nanotubes of polymers, metals, semiconductors, and oxides.
- Some porous membrane with nano-size channels (pores) are used as templates to conduct the growing of nanowires.
- Pore size ranging from 10 to 100 mm can be achieved.

Involves the following techniques:

- Electrochemical deposition
- Electrophoretic deposition
- Colloid dispersion, melt or solution filling
- Conversion with chemical reaction.

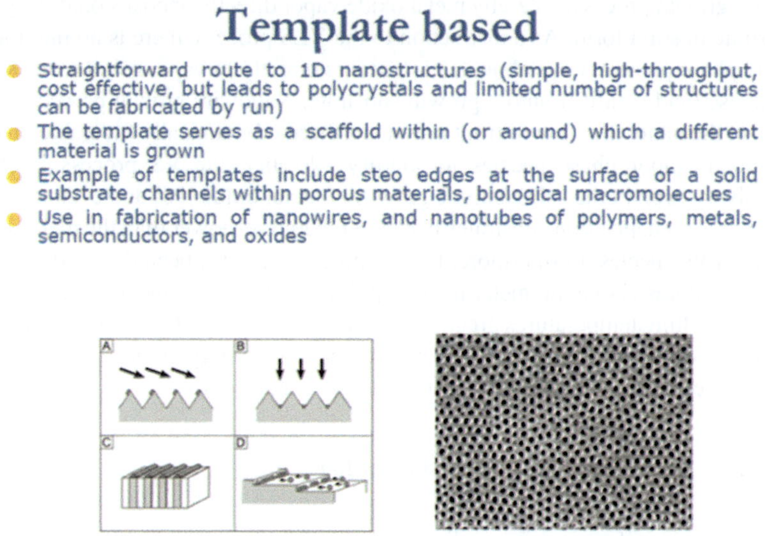

DNA Based Templates

- Nanowires are formed on the outer surface of the templates
- Diameter of the nanowires is not restricted by the template sizes and can be controlled by adjusting the amount of materials deposited on the templates DNA is an excellent choice as a template to fabricate nanowires because its diameter is ~2 nm and its length and sequence can be precisely controlled
- Fix a DNA strand between two electrical contacts
- Exposed to a solution containing some ions
- Ions bind to DNA and are then form some nanoparticles decorating along the DNA chain.

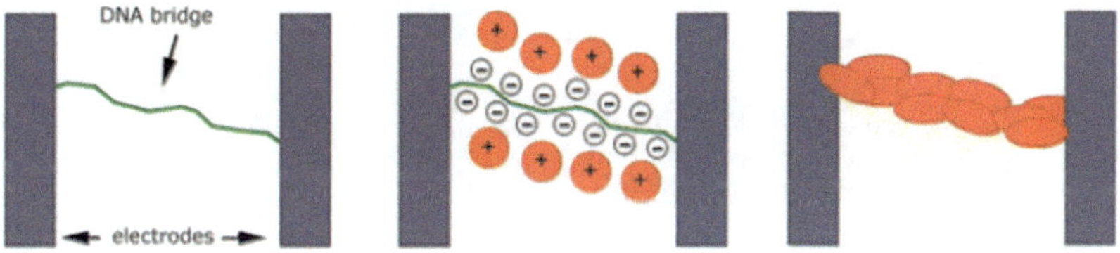

Surface Step-Edge Templates

- Relief structures (natural or patterned) on the surface of a substrate can be used as templates to grow nanowires.
- The method takes the advantage of the fact that deposition of many materials on the surface often starts preferentially at defect sites, such as surface step-edges.
- The problem is that these nanowires cannot be easily removed from the surface on which they are deposited.

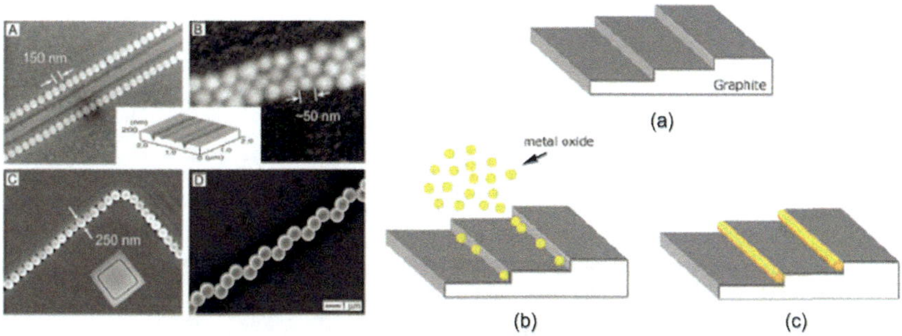

MAGNETISM IN ONE-DIMENSIONAL METAL SYSTEMS, PIETRO GAMBARDELLA

Nanoporous Alumina Templates

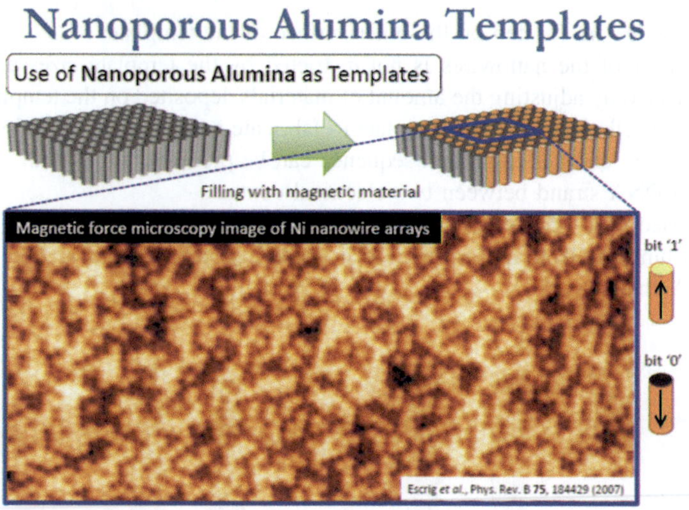

Anodization Process

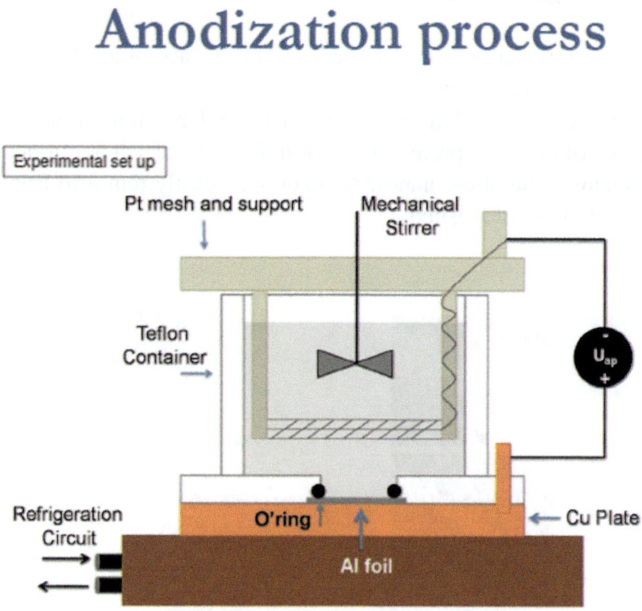

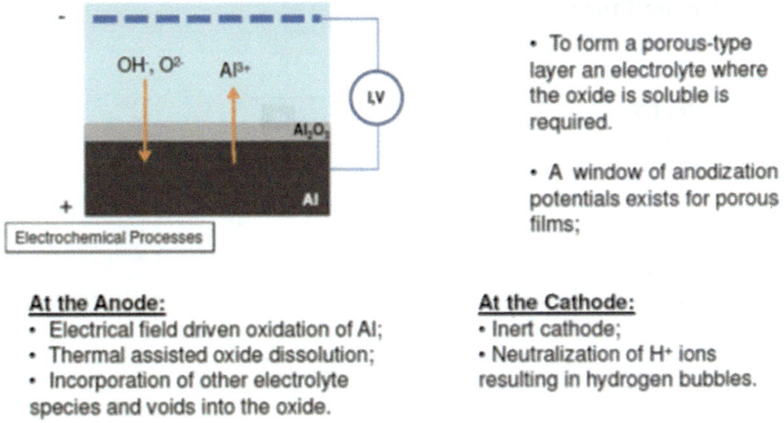

At the Anode:

- Electrical field driven oxidation of Al;
- Thermal assisted oxide dissolution;
- Incorporation of other electrolyte species and voids into the oxide.

$$2Al(s) + 3H_2O(l) \rightarrow Al_2O_3(s) + 6H^+(aq) + 6e^-$$

$$Al_2O_3(s) + nH_2O(l) \rightarrow Al^{3+}(aq) + O^{2-}(aq) + OH^-(aq) + H^+(aq)$$

At the Cathode:

- Inert cathode;
- Neutralization of H^+ ions resulting in hydrogen bubbles.

$$6H^+(aq) + 6e^- \rightarrow 2H_2(g)$$

In Anodization Process

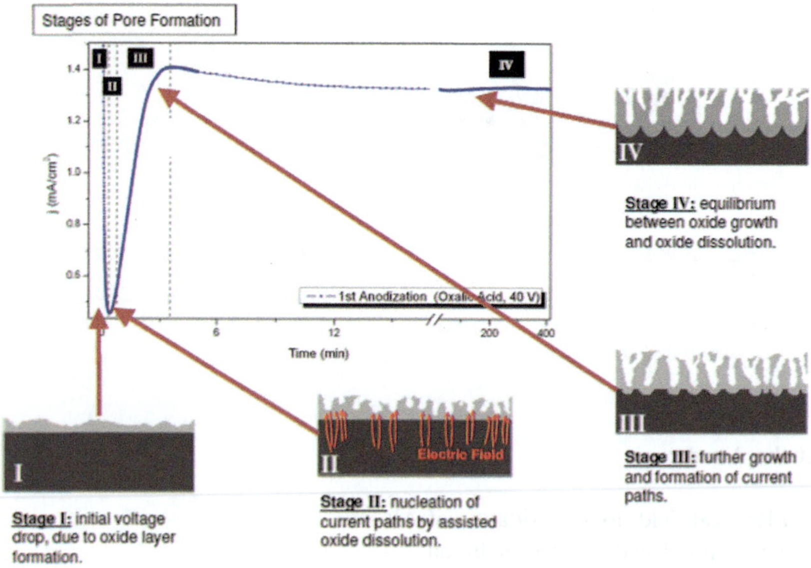

Two-Step Anodization

First Anodization: competition among the nucleated current paths (pores). The difference Δ emphasizes the distinct number of current paths grown.

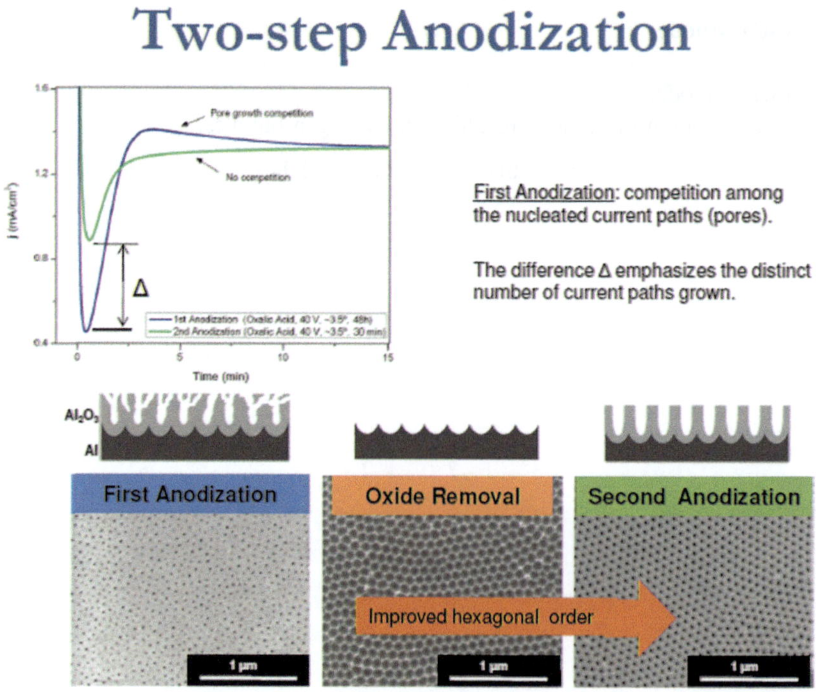

Nanoporous Alumina Templates

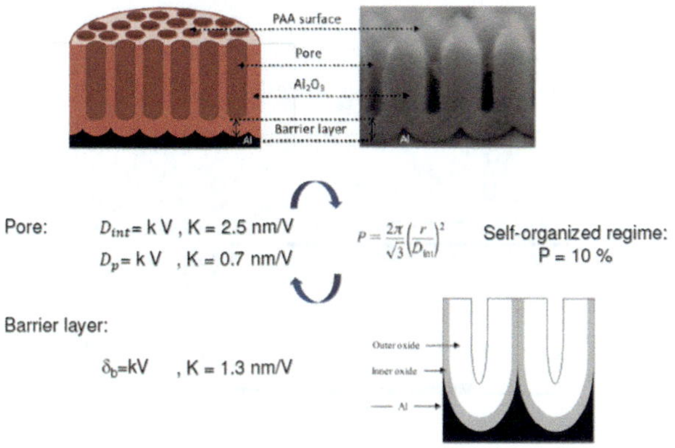

Alumina templates preparation

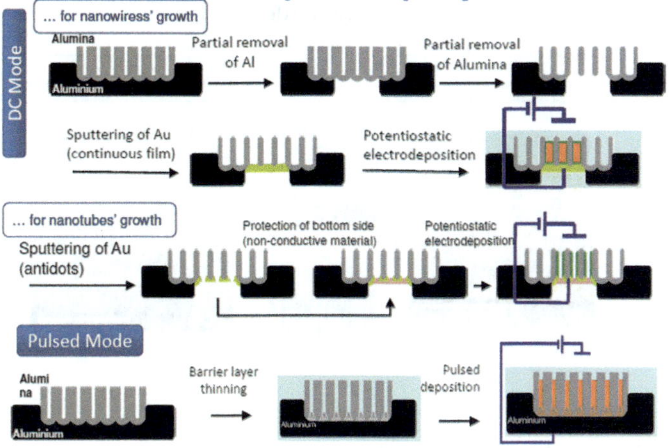

Filling methods

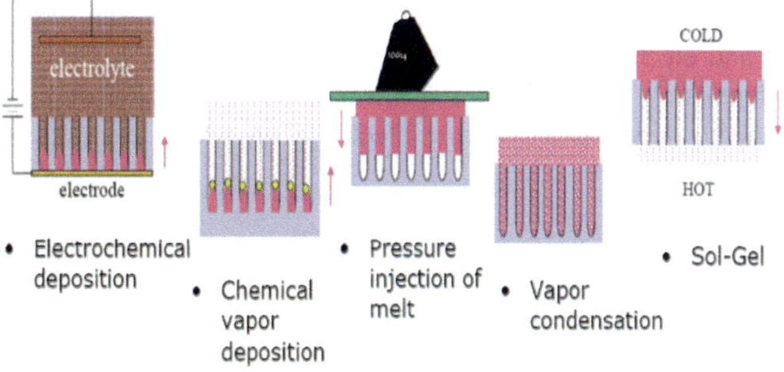

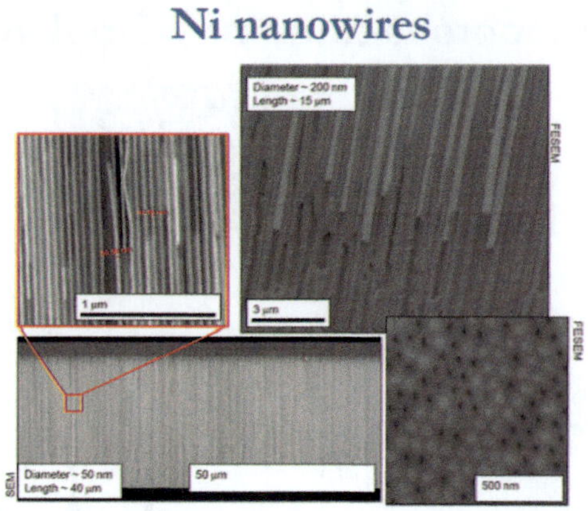

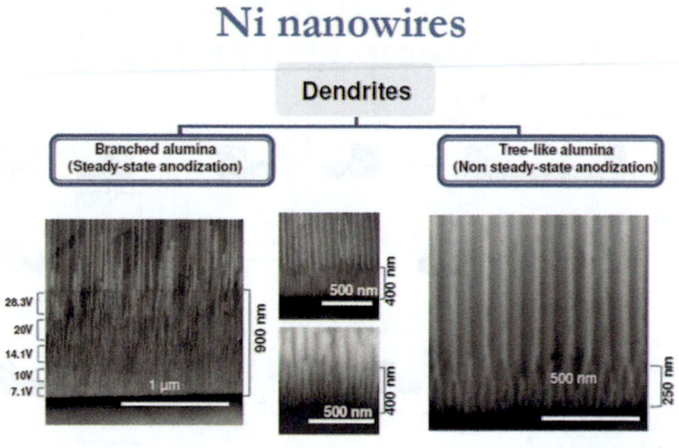

SEM images in tree-like alumina and branched alumina.

12.4.3.1 Electrochemical Deposition

- This method can be understood as a special *electrolysis* resulting in the deposition of solid material on an electrode
- Only applicable to electrically conductive materials: metals, alloys, semiconductors, and electrical conductive polymers
- Electrodeposition often requires a *metal film* on one side of the freestanding membrane to serve as a working electrode on which electrodeposition takes place

- If dissolve away the host solid material, free-standing nanowires are obtained.
- The diameter of the nanowires is determined by the geometrical constraint of the pores
- Electrodeposited nanowires tend to be dense, continuous, and highly crystalline in contrast to other deposition methods
- the ability to control the aspect ratio of the metal nanowires by monitoring the total amount of passed charge.

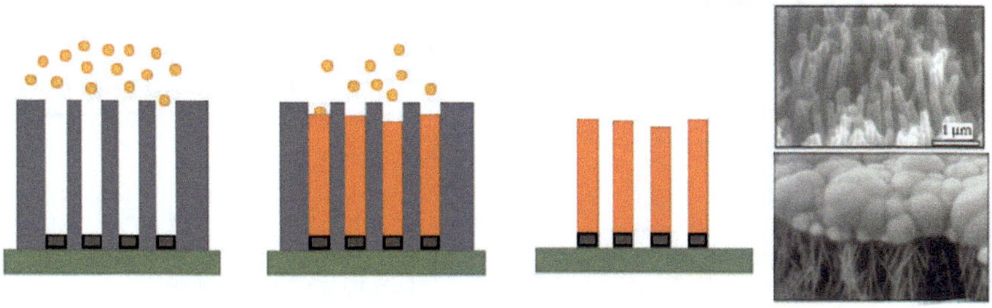

- This is a self-propagating process
- Only applicable to electrically conductive materials (metals, alloys, semiconductors, and electrical conductive Polymers)
- Use prefabricated cylindrical nanopores in a solid material as templates.
- If dissolve away the host solid material, freestanding nanowires are obtained.

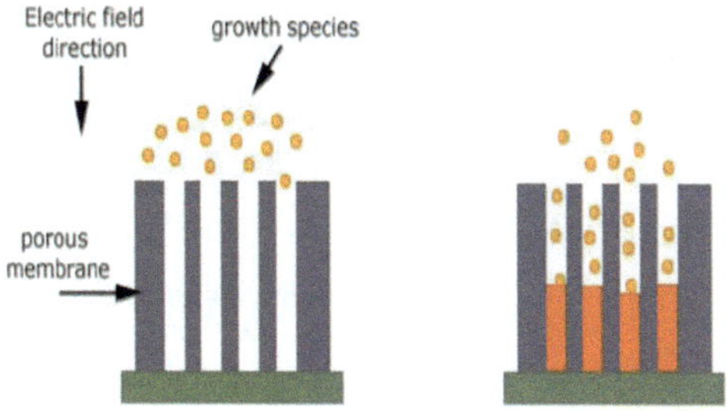

Electrodeposition

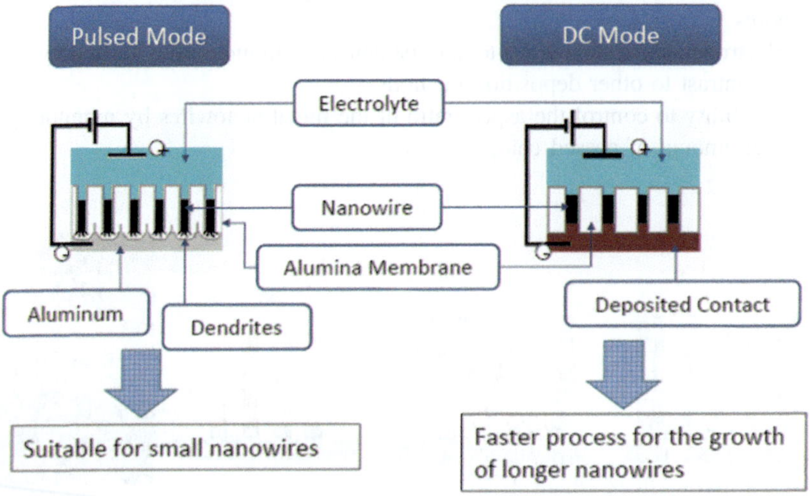

Alumina Barrier Layer

A non-conductive barrier-layer exists at the pores bottom: $\delta b \approx \alpha V$ where, α between 1.2 and 1.4 nm/V. To electrodeposit metallic nanowires, the electrical current has to flow, thus $t_{barrier}$ must be thinned down.

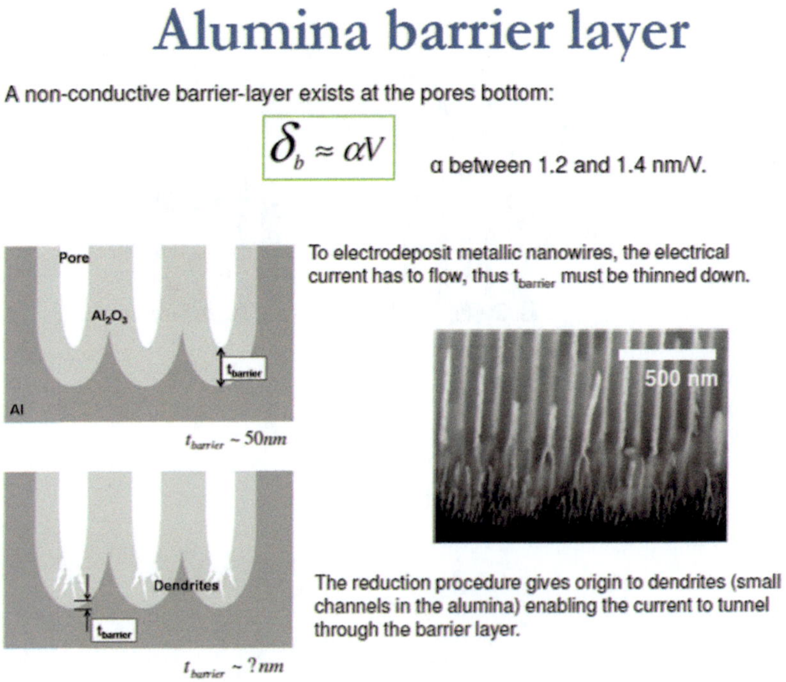

Barrier Layer Reduction

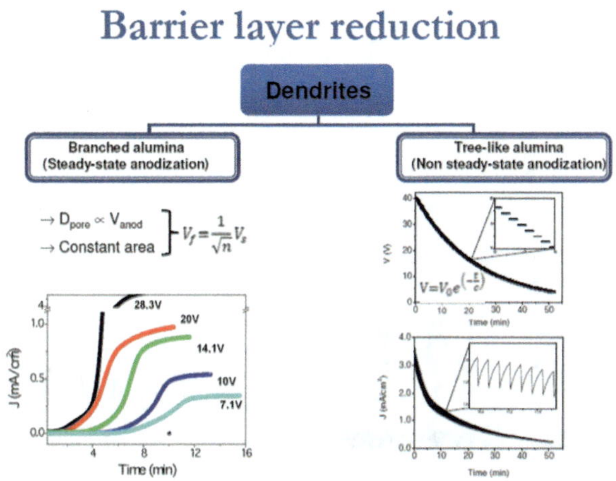

Pulsed Electrode Position

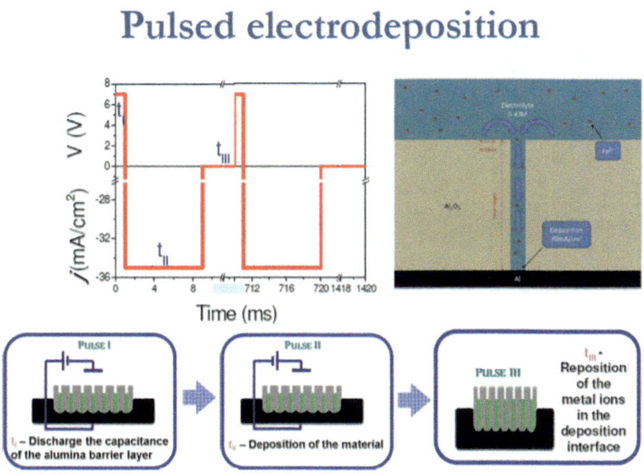

Branched Pore Filling

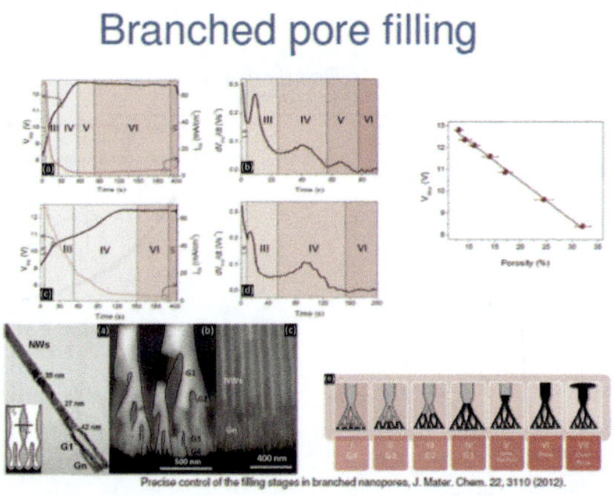

Tuning Pore Filling Controlling δ_b

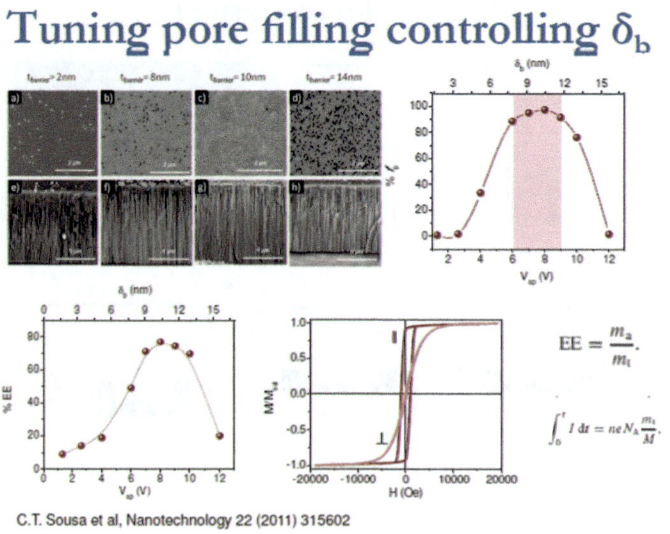

Electrodeposition Parameters Optimization

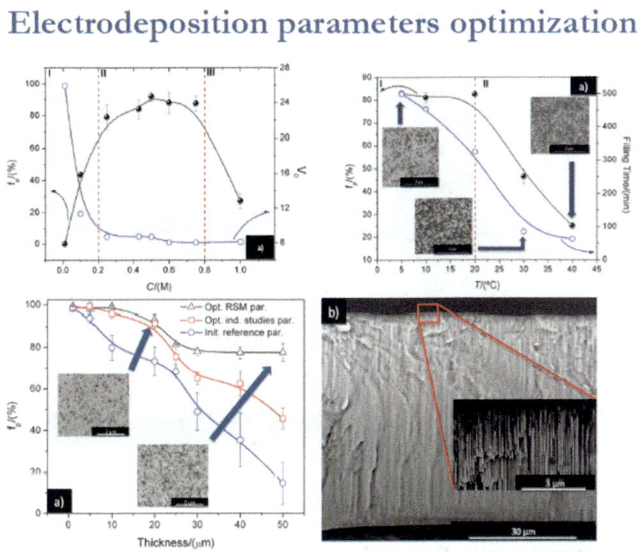

Modulated Nanowires

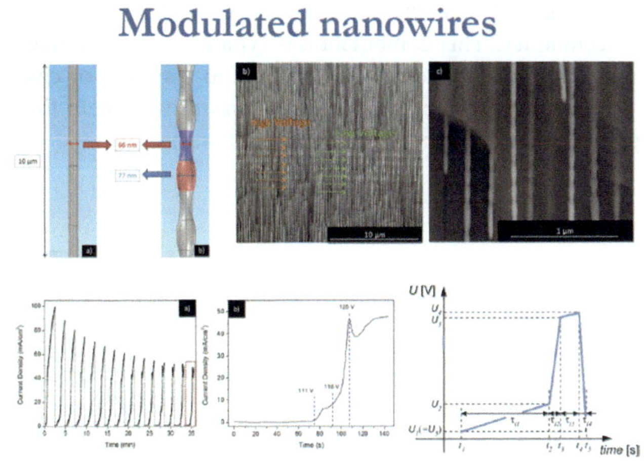

Segmented Nanowires

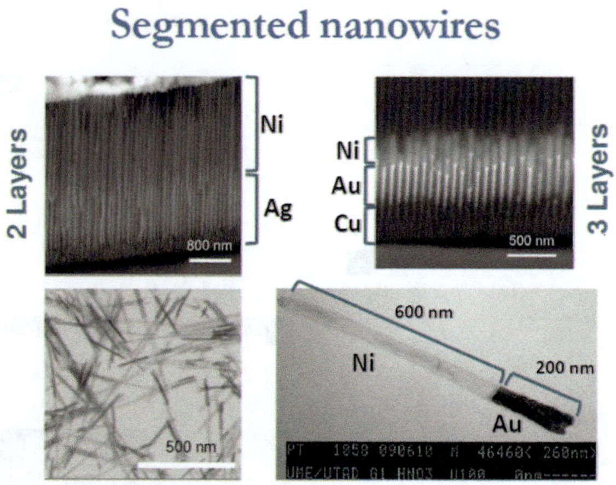

12.4.3.2 Electrophoretic Deposition (Electrophoresis)

- The deposit need not be electrically conductive
- Particularly suitable for oxide nanowires: SiO_2, TiO_2, Bi_2O_3, etc.
- Over the surface of nanoparticles develops an electrical charge via some chemical techniques. This combination is typically called counter-ion
- Upon application of an external electric filed to a system of charged nanosize particle system, the particles are set in motion in response to the electric filed
- This type of motion is referred to as electrophoresis.

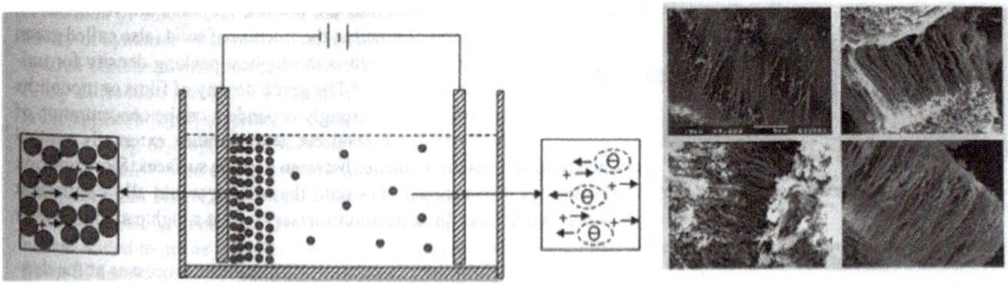

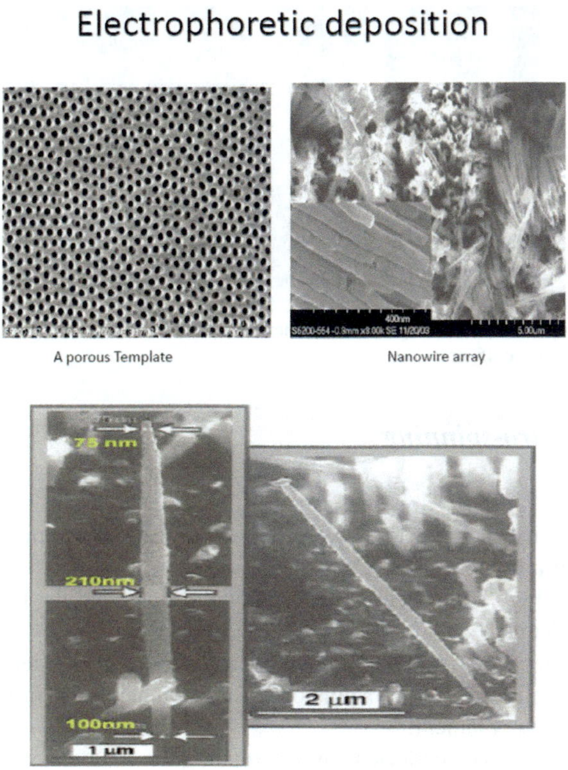

Nanowires grown in a 80 nm template membrane after dissolution of the membrane.

Advantages of Electrophoretic Deposition

- The ability to create highly conductive nanowires because electro-deposition relies on electron transfer, which is the fastest along the highest conductive path.
- Electro-deposited nanowires tend to be dense continuous and highly crystalline in contrast toother deposition methods.
- The ability to control the aspect ratio of the metal nanowiresby monitoring the total amount of passed charge.
- Differs from electrochemical deposition in several aspects.
- The deposit does not need to be electrically conductive.
- Particular method for obtaining oxide nanowires like SiO_2, TiO_2, Bi_2O_3, etc.

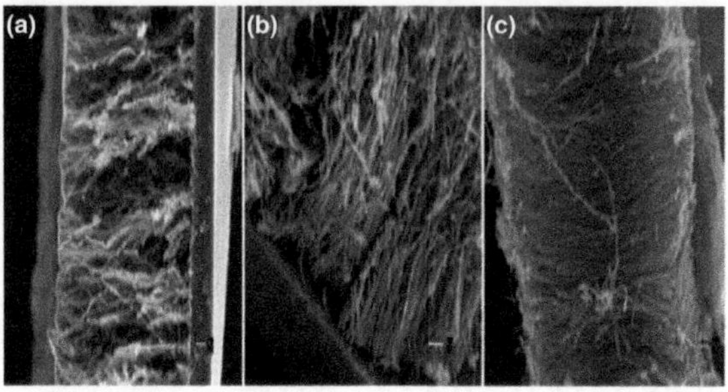

Different sizes of TiO nanorods grown in a membrane of electrophoretic deposition Diameters: **a** 180 nm, **b** 90 nm, **c** 45 nm

12.4.4 Electro-spinning

- Uses an electrical charge to draw very fine (typically on the micro or nanoscale) fibres from a liquid.
- Sufficiently high voltages applied to a liquid droplet and the body of the liquid becomes charged.
- When the electrostatic repelling force overcomes the surface tension force of the polymer solution, the liquid spills out of the spinneret and forms an extremely fine continuous filament.
- These filaments are collected onto a rotating or stationary collector with an electrode where they accumulate and bond together to form nanofiber fabric (Figs. 12.16).

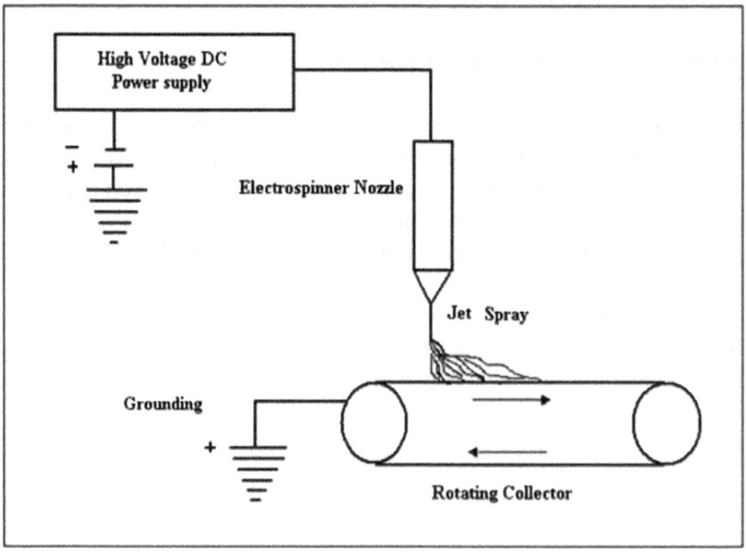

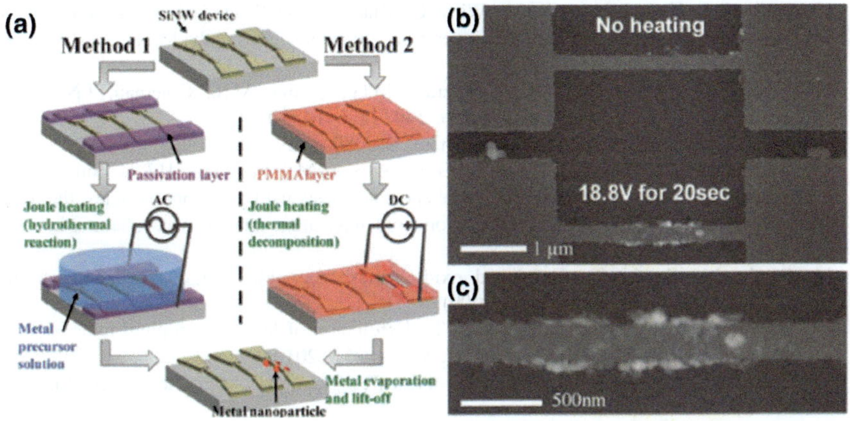

Fig. 12.16 **a** Schematic description of surface modification by self-heating of a nanowire: in method 1, nanoparticles are formed by a hydrothermal reaction via Joule heating of a SiNW. In method 2, a metal thin film is locally deposited on a SiNW after PMMA decomposition, metal evaporation and lift-off. SEM images of the nanoparticle-decorated SiNW via Joule heating in a liquid metal precursor environment; **b** Pd nanoparticles selectively coated on the heated SiNW; and **c** high magnification SEM image of Pd nanoparticles on the heated SiNW. This figure is composed of figures taken from [101], reprinted with permission from the Royal Society of Chemistry ©

References

1. Novoselov, K.S., Geim, A.K., Morozov, S.V., Jiang, D., Zhang, Y., Dubonos, S.V., et al.: Electric field effect in atomically thin carbon films. Science **306**, 666–669 (2004)
2. Thripuranthaka, M., Late, D.J.: Temperature dependent phonon shifts in single-layer WS_2. ACS Appl. Mater. Interface **6**, 1158–1163 (2014)
3. Miremadi, B.K., Morrison, S.R.: High activity catalyst from exfoliated MoS_2. J. Catal. **103**, 334–345 (1987)
4. Novoselov, K.S., Jiang, D., Schedin, F., Booth, T.J., Khotkevich, V.V., Morozov, S.V., et al.: Two-dimensional atomic crystals. Proc. Natl. Acad. Sci. U.S.A. **102**, 10451–10453 (2005)
5. Pacile, D., Meyer, J.C., Girit, C.O., Zettl, A.: The two-dimensional phase of boron nitride: few-atomic-layer sheets and suspended membranes. Appl. Phys. Lett. **92**, 133107 (2008)
6. Padova, P.D., Ottaviani, C., Quaresima, C., Olivieri, B., Imperatori, P., Salomon, E., et al.: 24 h stability of thick multilayer silicene in air. 2D Mater **1**, 021003 (2014)
7. Coleman, J.N., Lotya, M., O'Neill, A., Bergin, S.D., King, P.J., Khan, U., et al.: Two-dimensional nanosheets produced by liquid exfoliation of layered materials. Science **331**, 568–571 (2011)
8. Wang, Q.H., Kalantar-Zadeh, K., Kis, A., Coleman, J.N., Strano, M.S.: Electronics and optoelectronics of two-dimensional transition metal dichalcogenides. Nat. Nanotechnol. **7**, 699–712 (2012)
9. Zhou, K.-G., Mao, N.-N., Wang, H.-X., Peng, Y., Zhang, H.-L.: A mixed-solvent strategy for efficient exfoliation of inorganic graphene analogues. Angew. Chem. Int. Ed. **50**, 10839–10842 (2011)

10. Smith, R.J., King, P.J., Lotya, M., Wirtz, C., Khan, U., De, S., et al.: Large-scale exfoliation of inorganic layered compounds in aqueous surfactant solutions. Adv. Mater. **23**, 3944–3948 (2011)
11. Nicolosi, V., Chhowalla, M., Kanatzidis, M.G., Strano, M.S., Coleman, J.N.: Liquid exfoliation of layered materials. Science **340**, 1226419 (2013)
12. Hummers, W., Offeman, R.: J. Am. Chem. Soc. **80**, 1339 (1958)
13. Goncalves, G., Marques, P.A.A.P., Granadeiro, C.M., Nogueira, H.I.S., Singh, M.K., Gracio, J.: Surface modification of graphene nanosheets with gold nanoparticles: the role of oxygen moieties at graphene surface on gold nucleation and growth. Chem. Mater. **21**, 4796–4802 (2009)
14. Osada, M., Sasaki, T.: Exfoliated oxide nanosheets: new solution to nanoelectronics. J. Mater. Chem. **19**, 2503–2511 (2009)
15. Hu, L., Ma, R., Ozawa, T.C., Sasaki, T.: Exfoliation of layered europium hydroxide into unilamellar nanosheets. Chem. Asian J. **5**, 248–251 (2010)
16. Coleman, J.N.: Liquid exfoliation of defect-free graphene. Acc. Chem. Res. **46**, 14–22 (2013)
17. Khan, U., O'Neill, A., Lotya, M., De, S., Coleman, J.N.: High-concentration solvent exfoliation of graphene. Small **6**, 864–871 (2010)
18. Hernandez, Y., Lotya, M., Rickard, D., Bergin, S.D., Coleman, J.N.: Measurement of multicomponent solubility parameters for graphene facilitates solvent discovery. Langmuir **26**, 3208–3213 (2010)
19. Zhi, C.Y., Bando, Y., Tang, C.C., Kuwahara, H., Golberg, D.: Large-scale fabrication of boron nitride nanosheets and their utilization in polymeric composites with improved thermal and mechanical properties. Adv. Mater. **21**, 2889–2893 (2009)
20. Tang, Q., Zhou, Z.: Graphene-analogous low-dimensional materials. Prog. Mater Sci. **58**, 1244–1315 (2013)
21. Feng, J., Peng, L., Wu, C., Sun, X., Hu, S., Lin, C., et al.: Giant moisture responsiveness of VS_2 ultrathin nanosheets for novel touchless positioning interface. Adv. Mater. **24**, 1969–1974 (2012)
22. Matte, H.S.S.R., Plowman, B., Datta, R., Rao, C.N.R.: Graphene analogues of layered metal selenides. Dalton Trans. **40**, 10322–10325 (2011)
23. O'Neill, A., Khan, U., Coleman, J.N.: Preparation of high concentration dispersions of exfoliated MoS_2 with increased flake size. Chem. Mater. **24**, 2414–2421 (2012)
24. Lee, B.-I., Lee, K.S., Lee, J.H., Lee, I.S., Byeon, S.-H.: Synthesis of colloidal aqueous suspensions of a layered gadolinium hydroxide: a potential MRI contrast agent. Dalton Trans. 2490–2495 (2009)
25. Green, A.A., Hersam, M.C.: Solution phase production of graphene with controlled thickness via density differentiation. Nano Lett. **9**, 4031–4036 (2009)
26. Butler, S.Z., Hollen, S.M., Cao, L., Cui, Y., Gupta, J.A., Gutierrez, H.R., et al.: Progress, challenges, and opportunities in two dimensional materials beyond graphene. ACS Nano **7**, 2898–2926 (2013)
27. Mas-Balleste, R., Gomez-Navarro, C., Gomez-Herrero, J., Zamora, F.: 2D materials: to graphene and beyond. Nanoscale **3**, 20–30 (2011)
28. Zhan, Y., Liu, Z., Najmaei, S., Ajayan, P.M., Lou, J.: Large-area vapor-phase growth and characterization of MoS_2 atomic layers on a SiO_2 substrate. Small **8**, 966–971 (2012)
29. Lee, Y.-H., Zhang, X.-Q., Zhang, W., Chang, M.-T., Lin, C.-T., Chang, K.-D., et al.: Synthesis of large-area MoS_2 atomic layers with chemical vapor deposition. Adv. Mater. **24**, 2320–2325 (2012)
30. Liu, K.-K., Zhang, W., Lee, Y.-H., Lin, Y.-C., Chang, M.-T., Su, C.-Y., et al.: Growth of large-area and highly crystalline MoS_2 thin layers on insulating substrates. Nano Lett. **12**, 1538–1544 (2012)
31. Najmaei, S., Liu, Z., Zhou, W., Zou, X., Shi, G., Lei, S., et al.: Vapour phase growth and grain boundary structure of molybdenum disulphide atomic layers. Nat. Mater. **12**, 754–759 (2013)

32. Muller, F., Stowe, K., Sachdev, H.: Symmetry versus commensurability: epitaxial growth of hexagonal boron nitride on Pt (111) from B-trichloroborazine $(ClBNH)_3$. Chem. Mater. **17**, 3464–3467 (2005)
33. Rokuta, E., Hasegawa, Y., Itoh, A., Yamashita, K., Tanaka, T., Otani, S., et al.: Vibrational spectra of the monolayer films of hexagonal boron nitride and graphite on faceted Ni(755). Surf. Sci. **427–428**, 97–101 (1999)
34. Auwarter, W., Suter, H.U., Sachdev, H., Greber, T.: Synthesis of one monolayer of hexagonal boron nitride on Ni(111) from Btrichloroborazine $(ClBNH)_3$. Chem. Mater. **16**, 343–345 (2004)
35. Corso, M., Auwarter, W., Muntwiler, M., Tamai, A., Greber, T., Osterwalder, Jr.: Boron nitride nanomesh. Science **303**, 217–220 (2004)
36. Nagashima, A., Tejima, N., Gamou, Y., Kawai, T., Oshima, C.: Electronic states of monolayer hexagonal boron nitride formed on the metal surfaces. Surf. Sci. **357–358**, 307–311 (1996)
37. Lalmi, B., Oughaddou, H., Enriquez, H., Kara, A., Vizzini, S., Ealet, B., et al.: Epitaxial growth of a silicene sheet. Appl. Phys. Lett. **97**, 223109 (2010)
38. Zheng, Y., Koebel, A., Petroff, J.F., Boulliard, J.C., Capelle, B., Eddrief, M.J.: Cryst Growth **162**, 135–141 (1996)
39. Palmer, J.E., Saitoh, T., Yodo, T., Tamura, M.: J. Appl. Phys. **74**, 7221 (1993)
40. Shi, Y., Zhou, W., Lu, A.-Y., Fang, W., Lee, Y.-H., Hsu, A.L., et al.: Van der Waals epitaxy of MoS2 layers using graphene as growth templates. Nano Lett. **12**, 2784–2791 (2012)
41. Le´andri, C., Oughaddou, H., Aufray, B., Gay, J.M., Lay, G.L., Ranguis, A., et al.: Growth of Si nanostructures on Ag(001). Surf. Sci. **601**, 262–267 (2007)
42. Aufray, B., Kara, A., Vizzini, S., Oughaddou, H., Leandri, C., Ealet, B., et al.: Graphene-like silicon nanoribbons on Ag(110): a possible formation of silicene. Appl. Phys. Lett. **96**, 183102 (2010)
43. Padova, P.D., Kubo, O., Olivieri, B., Quaresima, C., Nakayama, T., Aono, M., et al.: Multilayer silicene nanoribbons. Nano Lett. **12**, 5500–5503 (2012)
44. Oughaddou, H., Gay, J.M., Aufray, B., Lapena, L., Lay, G.L., Bunk, O., et al.: Ge tetramer structure of the $p(2\sqrt{2} \times 4\sqrt{2})R45°$ surface reconstruction of Ge/Ag(001): a surface X-ray diffraction and STM study. Phys Rev B **61**, 5692 (2000)
45. Kara, A., Enriquez, H., Seitsonen, A.P., Voon, L.C.L.Y., Vizzini, S., Aufray, B., et al.: A review on silicene–new candidate for electronics. Surf. Sci. Rep. **67**, 1–18 (2012)
46. Leandri, C., Oughaddou, H., Gay, J., Aufray, B., Le Lay, G., Biberian, J., et al.: Self-assembled germanium nano-clusters on silver (110). Surf. Sci. **573**, L369–L374 (2004)
47. Meunier, I., Gay, J.-M., Lapena, L., Aufray, B., Oughaddou, H., Landemark, E., et al.: Atomic structure of the SbCu surface alloy: a surface X-ray diffraction study. Surf. Sci. **422**, 42–49 (1999)
48. Lay, G.L., Padova, P.D., Resta, A., Bruhn, T., Vogt, P.: Strained? Escibs. Epitaxial silicene: can it be strongly strained? J. Phys. D Appl. Phys. **45**, 392001 (2012)
49. Meng, L., Wang, Y., Zhang, L., Du, S., Wu, R., Li, L., et al.: Buckled silicene formation on Ir(111). Nano Lett. **13**, 685–690 (2013)
50. Chiappe, D., Grazianetti, C., Tallarida, G., Fanciulli, M., Molle, A.: Local electronic properties of corrugated silicene phases. Adv. Mater. **24**, 5088–5093 (2012)
51. Feng, B., Ding, Z., Meng, S., Yao, Y., He, X., Cheng, P., et al.: Evidence of silicene in honeycomb structures of silicon on Ag(111). Nano Lett. **12**, 3507–3511 (2012)
52. Lin, C.L., Arafune, R., Kawahara, K., Tsukahara, N., Minamitani, E., Kim, Y., et al.: Structure of silicene grown on Ag(111). Appl. Phys. Exp. **5**, 045802 (2012)
53. Jamgotchian, H., Colignon, Y., Hamzaoui, N., Ealet, B., Hoarau, J.Y., Aufray, B., et al.: Growth of silicene layers on Ag(111): unexpected effect of the substrate temperature. J. Phys. Condens. Matter **24**, 172001 (2012)
54. Peng, Y., Meng, Z., Zhong, C., Lu, J., Yu, W., Jia, Y., et al.: Hydrothermal synthesis and characterization of single-molecular-layer MoS_2 and $MoSe_2$. Chem. Lett. **30**, 772–773 (2001)

55. Li, Y., Wang, H., Xie, L., Liang, Y., Hong, G., Dai, H.: MoS_2 nanoparticles grown on graphene: an advanced catalyst for the hydrogen evolution reaction. J. Am. Chem. Soc. **133**, 7296–7299 (2011)
56. Jeong, S., Yoo, D., J-t, Jang, Kim, M., Cheon, J.: Well-defined colloidal 2-D layered transition-metal chalcogenide nanocrystals via generalized synthetic protocols. J. Am. Chem. Soc. **134**, 18233–18236 (2012)
57. Wudl, F.: Process for producing chalcogen containing compounds. In: USPTO, editor. USA: AT & T Bell Laboratories; 31 July 1984
58. Wang, L., Sun, C., Xu, L., Qian, Y.: Convenient synthesis and applications of gram scale boron nitride nanosheets. Catal. Sci. Technol. **1**, 1119–1123 (2011)
59. Nag, A., Raidongia, K., Hembram, K.P., Datta, R., Waghmare, U.V., Rao, C.: Graphene analogues of BN: novel synthesis and properties. ACS Nano **4**, 1539–1544 (2010)
60. Yang, S., Gong, Y., Liu, Z., Zhan, L., Hashim, D.P., Ma, L., et al.: Bottom-up approach toward single crystalline VO_2–graphene ribbons as cathodes for ultrafast lithium storage. Nano Lett. **13**, 1596–1601 (2013)
61. Zhang, N., Liu, X., Yi, R., Shi, R., Gao, G., Qiu, G.: Selective and controlled synthesis of single-crystalline yttrium hydroxide/oxide nanosheets and nanotubes. J. Phys. Chem. C **112**, 17788–17795 (2008)
62. Wang, Y., Xia, Y.: Bottom-up and top-down approaches to the synthesis of monodispersed spherical colloids of low melting-point metals. Nano Lett. **4**, 2047–2050 (2004). https://doi.org/10.1021/nl048689j
63. Iravani, S.: Green synthesis of metal nanoparticles using plants. Green Chem. **13**, 2638 (2011). https://doi.org/10.1039/c1gc15386b
64. Bello, S.A., Agunsoye, J.O., Hassan, S.B.: Synthesis of coconut shell nanoparticles via a top down approach: assessment of milling duration on the particle sizes and morphologies of coconut shell nanoparticles. Mater. Lett. (2015). http://dx.doi.org/10.1016/j.matlet.2015.07.063
65. Priyadarshana, G., Kottegoda, N., Senaratne, A., de Alwis, A., Karunaratne, V., Priyadarshana, G., Kottegoda, N., Senaratne, A., de Alwis, A., Karunaratne, V.: Synthesis of magnetite nanoparticles by top-down approach from a high purity ore. J. Nanomater. **2015**, 1–8 (2015). https://doi.org/10.1155/2015/317312
66. Garrigue, P., Delville, M.-H., Labrugère, C., Cloutet, E., Kulesza, P.J., Morand, J.P., Kuhn, A.: Top-down approach for the preparation of colloidal carbon nanoparticles. Chem. Mater. **16**, 2984–2986 (2004). https://doi.org/10.1021/cm049685i
67. Zhang, X., Lai, Z., Liu, Z., Tan, C., Huang, Y., Li, B., Zhao, M., Xie, L., Huang, W., Zhang, H.: A facile and universal top-down method for preparation of monodisperse transition-metal dichalcogenide nanodots. Angew. Chemie Int. Ed. **54**, 5425–5428 (2015). https://doi.org/10.1002/anie.201501071
68. Zhou, Y., Dong, C.-K., Han, L., Yang, J., Du, X.-W.: Topdown preparation of active cobalt oxide catalyst. ACS Catal. **6**, 6699–6703 (2016). https://doi.org/10.1021/acscatal.6b02416
69. Mogilevsky, G., Hartman, O., Emmons, E.D., Balboa, A., DeCoste, J.B., Schindler, B.J., Iordanov, I., Karwacki, C.J.: Bottom-up synthesis of anatase nanoparticles with graphene domains. ACS Appl. Mater. Interfaces **6**, 10638–10648 (2014). https://doi.org/10.1021/am502322y
70. Liu, D., Li, C., Zhou, F., Zhang, T., Zhang, H., Li, X., Duan, G., Cai, W., Li, Y.: Rapid synthesis of monodisperse Au nanospheres through a laser irradiation -induced shape conversion, self-assembly and their electromagnetic coupling SERS enhancement. Sci. Rep. **5**, 7686 (2015). https://doi.org/10.1038/srep07686
71. Liu, J., Liu, Y., Liu, N., Han, Y., Zhang, X., Huang, H., Lifshitz, Y., Lee, S.-T., Zhong, J., Kang, Z.: Metal-free efficient photocatalyst for stable visible water splitting via a two-electron pathway. Science **80**(347), 970–974 (2015). https://doi.org/10.1126/science.aaa3145
72. Needham, D., Arslanagic, A., Glud, K., Hervella, P., Karimi, L., Hoeilund-Carlsen, P.-F., Kinoshita, K., Mollenhauer, J., Parra, E., Utoft, A., Walke, P.: Bottom up design of

nanoparticles for anti-cancer diapeutics: "put the drug in the cancer's food". J. Drug Target 1–21 (2016). http://dx.doi.org/10.1080/1061186X.2016.1238092
73. Parveen, K., Banse, V., Ledwani, L.: Green synthesis of nanoparticles: their advantages and disadvantages. Acta Nat. 20048 (2016). http://dx.doi.org/10.1063/1.4945168
74. Ahmed, S., Annu, S., Yudha, S.S.: Biosynthesis of gold nanoparticles: a green approach. J. Photochem. Photobiol. B Biol. **161**, 141–153 (2016). https://doi.org/10.1016/j.jphotobiol.2016.04.034
75. Biswas, A., Bayer, I.S., Biris, A.S., Wang, T., Dervishi, E., Faupel, F.: Advances in top-down and bottom-up surface nanofabrication: techniques, applications & future prospects. Adv. Coll. Interface Sci. **170**, 2–27 (2012). https://doi.org/10.1016/j.cis.2011.11.001
76. Lieber, C.M.: Nanoscale science and technology: building a big future from small things. MRS Bull. **28**, 486–491 (2003)
77. Plummer, J.D., Deal, M.D., Griffin, P.B.: Silicon VLSI technology: fundamentals, practice, and modelling. Prentice Hall (2000)
78. Lu, J.G., Chang, P., Fan, Z.: Quasi-one-dimensional metal oxide materials—synthesis, properties and applications. Mat. Sci. Eng. R-Rep. **52**, 49–91 (2006)
79. Wang, Z.L.: Functional oxide nanobelts: materials, properties and potential applications in nanosystems and biotechnology. Annu. Rev. Phys. Chem. **55**, 159–196 (2004)
80. Hejazi, S.R., Hosseini, H.R.M., Ghamsari, M.S.: The role of reactants and droplet interfaces on nucleation and growth of ZnO nanorods synthesized by vapor–liquid-solid (VLS) mechanism. J. Alloy. Compd. **455**, 353–357 (2008)
81. Park, J., Choi, H.-H., Siebein, K., Singh, R.K.: Two-step evaporation process for formation of aligned zinc oxide nanowires. J. Cryst. Growth **258**, 342–348 (2003)
82. Chen, J.Y., Pan, C.J., Tsao, F.C., Kuo, C.H., Chi, G.C., Pong, B.J., Chang, C.Y., Norton, D. P., Pearton, S.J.: Characterization of ZnO nanowires grown on Si (100) with and without Au catalyst. Vacuum **84**, 1076–1079 (2009)
83. Huang, M.H., Wu, Y., Feick, H., Tran, N., Weber, E., Yang, P.: Catalytic growth of zinc oxide nanowires by vapor transport. Adv. Mater. **13**, 113–116 (2001)
84. Li, S.Y., Lee, C.Y., Tseng, T.Y.: Copper-catalyzed ZnO nanowires on silicon (1 0 0) grownby vapor-liquid-solid process. J. Cryst. Growth **247**, 357–362 (2003)
85. Wang, J.X., Liu, D.F., Yan, X.Q., Yuan, H.J., Ci, L.J., Zhou, Z.P., Gao, Y., Song, L., Liu, L. F., Zhou, W.Y., Wang, G., Xie, S.S.: Growth of SnO_2 nanowires with uniform branched structures. Solid State Commun. **130**, 89–94 (2004)
86. Chen, J.Y., Pan, C.J., Tsao, F.C., Kuo, C.H., Chi, G.C., Pong, B.J., Chang, C.Y., NortonS, D.P., Pearton, J.: Characterization of ZnO nanowires grown on Si (100) with and without Au catalyst. Vacuum **83**, 1076–1079 (2009)
87. Johnson, M.C., Aloni,S., McCready, D.E., Bourret-Courchesne, E.D.: Controlled vapor–liquid–solid growth of indium, gallium, and tin oxide nanowires via chemical vapor
88. Wagner, R.S., Ellis, W.C.: Vapor–liquid–solid mechanism of single crystal growth. Appl. Phys. Lett. **4**, 89–90 (1964)
89. Weber, W., et al.: Silicon nanowires: catalytic growth and electrical characterization. Phys. Stat. Sol. (b) **243**, 3340–33451–6 (2006)
90. Kirkham, M., Wang, X., Wang, Z.L., Snyder, R.L.: Solid Au nanoparticles as a catalyst for growing aligned ZnO nanowires: a new understanding of the vapour–liquid–solid process. Nanotechnology **18**, 365304 (2007)
91. Law, J.B.K., Boothroyd, C.B., Thong, J.T.L.: Site-specific growth of ZnO nanowires from patterned Zn via compatible semiconductor processing. J. Cryst. Growth **310**, 2485–2492 (2008)
92. Fan, H.J., Scholz, R., Kolb, F.M., Zacharias, M.: Two-dimensional dendritic ZnO nanowiresfrom oxidation of Zn microcrystals. Appl. Phys. Lett. **85**, 4142–4144 (2004)
93. Dang, H.Y., Wang, J., Fan, S.S.: The synthesis of metal oxide nanowires by directly heating metal samples in appropriate oxygen atmospheres. Nanotechnology 738 (2003)

94. Wan, Q., Lin, C.L., Yu, X.B., Wang, T.H.: Room-temperature hydrogen storage characteristics of ZnO nanowires. Appl. Phys. Lett. **84**, 124–126 (2004)
95. Chen, Y.Q., Jiang, J., Wang, B., Hou, J.G.: Synthesis of tin-doped indium oxide nanowires by self-catalytic VLS growth. J. Phys. D Appl. Phys. 3319 (2004)
96. Chen, Y., Cui, X., Zhang, K., Pan, D., Zhang, S., Wang, B., Hou, J.G.: Bulk quantity synthesis and self-catalytic VLS growth of SnO_2 nanowires by lower-temperature evaporation. Chem. Phys. Lett. **369**, 16–20 (2003)
97. Orlandi, M.O., Leite, E.R., Aguiar, R., Bettini, J., Longo, E.: Growth of SnO nanobelts and dendrites by a self-catalytic VLS process. J. Phys. Chem. B **110**, 6621–6625 (2006)
98. Dai, Z.R., Pan, Z.W., Wang, Z.L.: Novel nanostructures of functional oxides synthesized by thermal evaporation. Adv. Func. Mater. **13**, 9–24 (2003)
99. Hsu, Y.-J., Lu, S.-Y.: Vapour-solid growth of Sn nanowires: growth mechanism and superconductivity. J. Phys. Chem. B **109**, 4398–4403 (2005)
100. Pan, Z.W., Dai, Z.R., Wang, Z.L.: Nanobelts of semiconducting oxides. Science **291**, 1947–1949 (2001)
101. Yun, J., Jin, C.Y., Ahn, J.H., Jeon, S., Park, I.: A self-heated silicon nanowire array: selective surface modification with catalytic nanoparticles by nanoscale Joule heating and its gas sensing applications. Nanoscale **5**, 6851–6856 (2013)

Chapter 13
Chemistry and Physics for Nanostructures Semiconductivity

Materials that show peculiar properties are the basis of modern devices. The contemporary technology stimulated great interest on searching for functional devices based on new materials.

During the last 5 years the potential for applications of semiconductor nanowires has grown rapidly via the development of methods for catalytically induced nanowire growth using the, so-called vapor–liquid–solid (VLS) growth mode. The VLS method offers a high degree of control of parameters such as position, diameter, length and composition, including the realization of atomically abrupt heterostructure interfaces inside a nanowire. Progress in semiconductor physics and electronics for the last 30–40 years has been strongly influenced by the development of advanced techniques for epitaxial growth of planar, ultrathin layers and the related formation of atomically sharp heterostructure interfaces between semiconductors of different compositions. In this way, ideal two-dimensional structures can be formed which have been used for the basic studies of transport in 2D-electron gas systems and the realization of quantum devices such as quantum well lasers and heterostructure electronic devices, e.g. HEMTs and resonant tunneling diodes. During the last 5 years, important progress has been made in the growth of ideal1D structures, such as carbon nanotubes and semiconductor nanowires, but it is only during the last few years that heterostructures in 1D systems have become possible to master, investigate and explore for applications.

Dimensionality plays a critical role in determining the properties of materials due to, for example, the different ways that electrons interact in three-dimensional, two-dimensional (2D), and one-dimensional (1D) structures [1–5]. The study of dimensionality has a long history in chemistry and physics, although this has been primarily with the prefix "quasi" added to the description of materials; that is, quasi-1D solids, including square-planar platinum chain and metal trichalcogenide compounds [2, 6], and quasi-2D layered solids, such as metal dichalcogenides and copper oxide superconductors [3–5, 7, 8]. The anisotropy inherent in quasi-1D and -2D systems is central to the unique properties and phases that these materials exhibit, although the small but finite interactions between 1D chains or 2D layers in

L. H. Madkour, *Nanoelectronic Materials*, Advanced
Structured Materials 116, https://doi.org/10.1007/978-3-030-21621-4_13

bulk materials have made it difficult to address the interesting properties expected for the pure low-dimensional systems.

Are Pure Low-Dimensional Systems Interesting and Worth Pursuing?
We believe that the answer to this question is an unqualified yes from the standpoints of both fundamental science and technology. One-dimensional nanostructures are of both fundamental and technological interest. They not only exhibit interesting electronic and optical properties associated with their low dimensionality and the quantum confinement effect, but they also represent critical components in potential nanoscale devices. In addition to the recent extensive studies on "single-component" nanowires, of increasing importance is incorporating different interfaces and controlling doping profiles within individual single-crystalline nanowires. Epitaxial growth plays a significant role in fabricating such nanowire heterostructures.

One needs to look no further than past studies of the 2D electron gas in semiconductor heterostructures, which have produced remarkably rich and often unexpected results [9, 10], and electron tunneling through 0D quantum dots, which have led to the concepts of the artificial atom and the creation of single electron transistors [11–15]. In these cases, low dimensional systems were realized by creating discrete 2D and 0D nanostructures. 1D nanostructures, such as nanowires and nanotubes, are expected to be at least as interesting and important as 2D and 0D systems [16, 17]. 1D systems are the smallest dimension structures that can be used for efficient transport of electrons and optical excitations, and are thus expected to be critical to the function and integration of nanoscale devices. However, little is known about the nature of, for example, localization that could preclude transport through 1D systems. In addition, 1D systems should exhibit density of states singularities, can have energetically discrete molecular like states extending over large linear distances, and may show more exotic phenomena, such as the spin-charge separation predicted for a Luttinger liquid [1, 2]. There are also many applications where 1D nanostructures could be exploited, including nanoelectronics, superstrong and tough composites, functional nanostructured materials, and novel probe microscopy tips [16–29].

To address these fascinating fundamental scientific issues and potential applications requires answers to two questions at the heart of condensed matter chemistry and physics research: (1) How can atoms or other building blocks be rationally assembled into structures with nanometer-sized diameters but much longer lengths? (2) What are the intrinsic properties of these quantum wires and how do these properties depend, for example, on diameter and structure?

An effective approach is demonstrated for growing large-area, hexagonally patterned, aligned ZnO nanorods. The synthesis uses a catalyst template produced by a self-assembled monolayer of submicron spheres and guided vapor–liquid–solid (VLS) growth on a single crystal alumina substrate. The ZnO nanorods have uniform shape and length, align vertically on the substrate, and are distributed according to the pattern defined by the catalyst template. The nanorods grow along [0001] with side surfaces defined by {2110}. This approach opens the possibility of creating patterned one-dimensional nanostructures for applications as sensor arrays, piezoelectric antenna arrays, optoelectronic devices, and interconnects.

Helical crystalline silicon carbide nanowires covered with a silicon oxide sheath (SiC/SiO_2) have been synthesized by a chemical vapor deposition technique. The SiC core typically has diameters of 10–40 nm with a helical periodicity of 40–80 nm and is covered by a uniform layer of 30–60 nm thick amorphous SiO_2. A screw-dislocation-driven growth process is proposed for the formation of this novel structure based on detailed structural characterizations. The helical nanostructures may find applications as building blocks in nanomechanical or nanoelectronic devices. The screw-dislocation-induced growth mechanism suggests that similar helical nanostructures of a wide range of materials may be synthesized.

13.1 Conductivity of Nanowires NWs

The radial NW concept also offers substantial opportunities for NW optoelectronics since the required n- and p-type active materials can be incorporated as the core and shell, which enables carrier injection or collection over a much larger area than possible in crossed NW devices and axial NW heterostructures.

We have shown that semiconductor NWs offer many opportunities for the assembly of nanoscale electronic and optoelectronic devices and arrays by the bottom-up paradigm. Central to our progress in the field and future efforts at realizing applications has been the rational control of key NW parameters during growth, including chemical composition, structure, size, morphology, and doping, since it is these parameters that determine predictable device function. The examples described here illustrate how, with increasing control over the parameters of the basic NW building blocks from homogenous doped materials to increasingly complex axial and radial heterostructures, it has been possible to demonstrate key advantages of NWs in electronics and photonics compared with conventional technologies. Integration strategies have also been developed that remove the constraints of lithography facing conventional top-down technologies today. Looking into the future, we believe that continued advances in our capability to control the structural/compositional complexity of NWs during growth, which correspondingly determines the functional complexity of the building blocks, together with advances in organizing them into larger integrated arrays, will lead to increasingly unique nanoelectronic and optoelectronic circuits and systems that will create the technologies of the future.

Several physical reason predict that the conductivity of a nanowire will be much less than that of the corresponding bulk material. First, there is scattering from the wire boundaries, whose effect will be very significant whenever the wire width is below the free electron mean free path of the bulk material. In copper, for example, the mean free path is 40 nm. Copper nanowires less than 40 nm wide will shorten the mean free path to the wire width. Nanowires also show other peculiar electrical properties due to their size. Unlike single wall carbon nanotubes, whose motion of electrons can fall under the regime of ballistic transport (meaning the electrons can travel freely from one electrode to the other), nanowire conductivity is strongly influenced by edge effects. The edge effects come from atoms that lay at the nanowire surface and are not fully bonded to neighboring atoms like the atoms within the bulk of the nanowire. The unbonded atoms are often a source of defects within the nanowire, and may cause the nanowire to conduct electricity more poorly than the bulk material. As a nanowire shrinks in size, the surface atoms become more numerous compared to the atoms within the nanowire, and edge effects become more important. Furthermore, the conductivity can undergo a quantization in energy: i.e. the energy of the electrons going through a nanowire can assume only discrete values, which are multiples of the conductance quantum $G = 2e^2/h$ (where e is the charge of the electron and h is the Planck constant. See also the Quantum Hall effect). The conductivity is hence described as the sum of the transport by separate *channels* of different quantized energy levels. The thinner the wire is, the smaller the number of channels available to the transport of electrons. This quantization has been demonstrated by measuring the conductivity of a nanowire suspended between two electrodes while pulling it: as its diameter reduces, its conductivity decreases in a stepwise fashion and the plateaus correspond to multiples of G. The quantization of conductivity is more pronounced in semiconductors like Si or GaAs than in metals, due to their lower electron density and lower effective mass. It can be observed in 25 nm wide silicon fins, and results in increased threshold voltage. In practical terms, this means that a MOSFET with such nanoscale silicon fins, when used in digital applications, will need a higher gate (control) voltage to switch the transistor on.

One-dimensional (1D) materials such as various kinds of nanowires and nanotubes have attracted considerable attention due to their potential application in electronic and energy conversion devices [30–33]. In contrast with the extensive studies on electron transport, investigation of phonon transport in 1D nanostructures [34–38] has been initiated only recently. When crystalline solids are confined to the nanometer range, phonon transport within them can be significantly altered due to various effects, namely (i) increased boundary scattering; (ii) changes in phonon dispersion relation; and (iii) quantization of phonon transport. For example, theoretical studies [36, 37] have suggested that, as the diameter of a Si nanowire becomes smaller than 20 nm, the phonon dispersion relation could be modified due to phonon confinement, such that the phonon group velocities would be significantly less than the bulk value. Molecular dynamics simulations [38] have shown that, for wires of nanometer diameter, the thermal conductivities could be two orders of magnitude smaller than that of bulk silicon. However, to the best of our knowledge, no systematic experimental results have been reported on the thermal conductivities of Si nanowires. It is, therefore, important to experimentally validate these theoretical predictions to understand the underlying physics. In addition, a deeper understanding of thermal transport in nanostructures also has practical implications [39] in the design and performance of modern microelectronic devices that have sub-100 nm features and recently proposed nanowire-based thermoelectric devices. It was found that [40] Si nanowire thermal conductivity is much lower than the corresponding bulk value, which can be explained by increased phonon boundary scattering.

Semiconductors are the most important materials in our modern technology-based world. In particular, in the last two decades, semiconductor nanostructures have become one of the most rapidly developing area within the nanoscience research community. Among the different semiconductor nanostructures, SiGe nanowires, that can be manipulated through size reduction, geometry variation, and alloying, are considered one of the key developments for next generation technologies, due to their ease of processing, unique properties, and compatibility with the existent silicon based microelectronics industry.

13.2 Welding Nanowires

To incorporate nanowire technology into industrial applications, researchers in 2008 developed a method of welding nanowires together: a sacrificial metal nanowire is placed adjacent to the ends of the pieces to be joined (using the manipulators of a scanning electron microscope); then an electric current is applied, which fuses the wire ends. The technique fuses wires as small as 10 nm. For nanowires with diameters less than 10 nm, existing welding techniques, which require precise control of the heating mechanism and which may introduce the possibility of damage, will not be practical. Recently scientists discovered that single-crystalline ultrathin gold nanowires with diameters $\sim$3–10 nm can be "cold-welded" together within seconds by mechanical contact alone, and under remarkably low applied pressures (unlike macro- and micro-scale cold welding process). High-resolution transmission electron microscopy and in situ measurements reveal that the welds are nearly perfect, with the same crystal orientation, strength and electrical conductivity as the rest of the nanowire. The high quality of the welds is attributed to the nanoscale sample dimensions, oriented-attachment mechanisms and mechanically assisted fast surface diffusion. Nanowire welds were also demonstrated between gold and silver, and silver nanowires (with diameters $\sim$5–15 nm) at near room temperature, indicating that this technique may be generally applicable for ultrathin metallic nanowires. Combined with other nano- and microfabrication technologies, cold welding is anticipated to have potential applications in the future bottom-up assembly of metallic one-dimensional nanostructures.

In the next section we have thoroughly discussed the major advances in the synthesis of SiGe nanowires, their experimental and theoretical characterization and the progresses done in device applications.

13.3 Silicon-Germanium Nanowires SiGe NWs

Despite of the amount of literature devoted to SiGe NWs, a unified description of the theoretical concepts and experimental results on the main properties of SiGe NWs is still lacking.

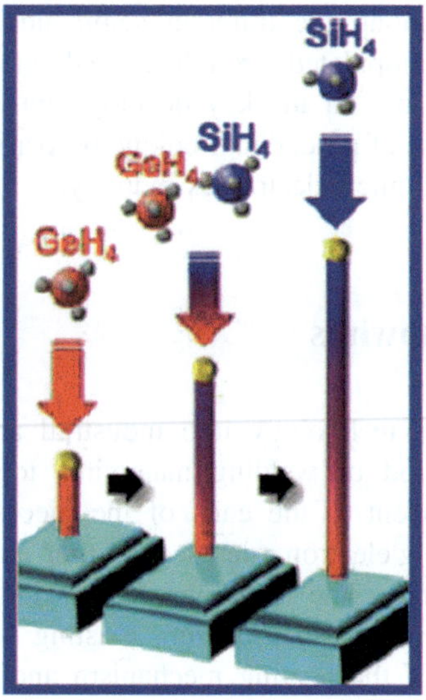

The trend predicted by Moore's law sets forth strict requirements on the electronic properties of materials that cannot always be satisfied by conventional semiconductors. Thus, the last twenty years and, in particular, the first decade of this century have witnessed a progressive interest in materials and devices with reduced size and dimensionality. Nanostructures and nanomaterials have been considered the key development for next generation technology, due their ease of processing, new properties and compatibility with the existent silicon (Si) microelectronics. Great progresses have been made, in particular regarding the use of Si nanostructures in microelectronics, photonics, and photovoltaics [41–45].

However, the scaling down of semiconductor structures to nanometer sizes alone does not satisfy the need of new unique properties in a variety of applications. On the other hand, the manipulation through alloying together with the variation of the size provides a natural platform for the development and creation of materials that present the desired properties but, at the same time, can be easily integrated into the existent Si technology. A meaningful example of this possibility is given by the use of silicon-germanium (SiGe) nanostructures.

The idea of combining the chemical and physical properties of Si and Ge represents a well-established strategy to engineer innovative materials and devices. In particular SiGe systems have garnered much attention as promising technology building blocks able to improve some of the properties of Si, producing enhanced versions of Si microlectronic devices, still maintaining its cheap fabrication processes. Indeed the smaller band gap of Ge compared to that of Si and the possibility to strain Si, due to the lattice mismatch of the two materials, offered the possibility to engineer the electronic band gap thereby increasing the devices inherent switching speeds due to the higher Ge mobility. The real strength of SiGe systems for electronics applications lies in the possibility to integrate analog, radio frequency, and digital on a single chip using existing complementary metal oxide semiconductor (CMOS) fabrications technologies, which is not possible with any other technologies (like, e.g., devices based on GaAs). Furthermore, SiGe materials are actively investigated in the field of silicon photonics and optoelectronics with the hope for the future to maintain the versatility and cheap costs of silicon technology to create complete optoelectronic systems. Thus, SiGe systems have been experiencing tremendous development, starting from more than four decades ago, as high quality SiGe thin layers and superlattices became the subjects of an intense research and development. Historically [46] the point of departure was the theoretical suggestion by Gnutzmann and Clausecker [47] that, due to zone folding effects in superlattices, it would be possible to obtain quasi-direct band gap semiconductors starting from the indirect band gap ones. The first experimental attempt to do this can be ascribed to Kasper et al. [48], who using an ultrahigh vacuum (UHV) evaporation technique at a substrate temperature of 750 °C obtained heterostructures of medium quality, due to the large lattice mismatch. Improvements were made, in the 80s, by Bean et al. [49] using molecular beam epitaxy (MBE) and through lowering of the substrate growth temperature and by the Abstreiter's group, which, using the concept of strain symmetrization and growing an intermediate SiGe alloy buffer layer, was able to grown thick pseudomorphic multilayer structures [50]. In the 90s the first observations of band edge photoluminescence on fully strained SiGe alloys and Si/SiGe multiple quantum wells (MQWs) were thus reported [51, 52] greatly exciting the entire field. Since then, major progresses in chemical synthesis, materials processing, device engineering, characterization and modeling have enabled the fabrication of devices with improved performance and reproducibility. As a result of these efforts, SiGe layered heterostructures boosted important technological breakthrough in solid-state electronics [53], for example as heterojunction bipolar transistors (HJBT) [54], heterojunction internal photoemission photodetector [55], metal-on-insulator field effect transistor (MOSFET) with high performance [56], and modulation-doped field effect transistor (MODFET) [57, 58]. Many other devices, such as optical modulators, waveguides and detectors, resonant tunneling diodes, lasers, and quantum devices have been envisaged and proposed. Reviews of these exciting results for SiGe heterostructures were given in the last two decades by Schäffler [59] and Whall and Parker [53, 60] and recently by Paul [61] and Lee et al. [62].

In the new millennium nano definitely enters into play. Nanostructures are low-dimensional confined systems, where quantum confinement can act in three spatial directions; therefore, one has two-dimensional (quantum wells), one dimensional (quantum wires or nanowires), and zero-dimensional (quantum dots or nanocrystals) confined structures. Thus, concerning SiGe nanostructures a huge amount of literature has been devoted to Si/SiGe quantum well structures [63, 64] and to the spontaneous formation, selforganization, and properties of SiGe quantum dots [64, 65].

13.4 Growth Techniques, Morphology, and Structural Properties of SiGe NWs

Semiconductor nanowires are commonly grown through the vapor–liquid–solid (VLS) mechanism, which was first proposed in the 1960s by Wagner and Ellis [66] for the growth of Si whiskers. In a nutshell, the method consists in using a catalytic nanoparticle, typically Au, to promote the decomposition of an appropriate gaseous precursor. Let us see in more detail the case of the standard growth of a Si NW. First an Au nanoparticle is deposited onto a Si substrate. Au and Si react and form an Au–Si alloy droplet. Next, SiH_4 in vapor phase, the Si precursor, is inserted in the growth chamber and starts to be adsorbed in the Au–Si droplet. When supersaturation is reached, Si atoms precipitate and the nanowire starts its nucleation.

Many different flavors of this general recipe have been reported and some of the ingredients can change (precursor gas and catalytic nanoparticle), but the process always relies on the catalyzed decomposition of a precursor on a liquid metal droplet, which yields a growth rate which is orders of magnitude higher in one direction than in the others. The main distinctive element among different approaches to VLS is how the precursor is supplied and reach the catalytic droplet. (i) In chemical vapor deposition (CVD) a volatile gaseous precursor-SiH_4, Si_2H_6, or $SiCl_4$, to stick to the case of Si NWs—reaches the Au particle and decompose. (ii) In molecular beam epitaxy (MBE), a solid ultrapure Si source is heated until it starts to sublimate; then, a gaseous beam of Si atoms is directed to the substrate, where it reacts with the Au particle. (iii) In laser ablation, used to produce the first truly nanoscopic Si [67] and, as we shall see briefly, SiGe alloy NWs [68], a high-power pulsed laser is used to ablate material from a mixed Si–Au target; the Si ablated from the target collides with inert gas molecules present in the furnace and the atoms condense to liquid nanodroplets that contain both Si and the catalyst material. Other popular choices include oxide assisted growth, based in the evaporation of SiO [69], solution-based techniques, where growth takes place in liquid media [70], and metal-assisted etching [71].

13.4.1 Alloyed Nanowires

The observation that the eutectic temperature of Au–Si and Au–Ge are practically identical suggested immediately the possibility of growing NWs that contained both species. Roughly speaking, if Si and Ge precursors are incorporated in the catalytic particle and therein decompose until supersaturation at the same temperature, it is in principle enough to add simultaneously both precursors in the growth chamber. This was first accomplished by Duan and Lieber [68] in 2000 by laser-assisted catalytic growth [72].

Significant progress in this direction was reported by Lew et al. [73] where SiGe alloy NWs were grown by CVD from SiH_4 and GeH_4. Initially, at 275 °C no growth was observed, while subeutectic growth of Ge NWs has been previously reported. SiGe NWs growth was observed upon raising the temperature to 350 °C. Wires with uniform diameter along all their length were obtained for growth temperature of up to 400 °C. In this narrow window T is high enough to promote VLS growth but also low enough to prevent a noncatalyzed deposition of Ge on the wire sidewalls. Indeed, above 400 °C tapering was observed and it increased approximately linearly with temperature.

Recently, the importance of obtaining a uniform distribution of atoms into alloyed wires has been clearly pointed out by Ma et al. [74]. By combining mechanical properties measurements and atomistic simulations, they demonstrated that, in alloyed SiGe NWs grown with the VLS method, the presence of interstitial Ge atoms can strongly affect the mechanical response of the wire, modifying its fracture strength and fracture strain. In particular, they used a nanomanipulator in a focused ion beam (FIB) system combined to a scanning electron microscope (SEM) to perform tensile experiments. In a first step, they measured the fracture strength and strain in SiGe NWs finding a nonlinear dependence of these quantities on the Ge content. Then, they employed a rapid thermal annealing (RTA) process to remove incorporated interstitial atoms and performed the same type of mechanical investigation. Surprisingly, after the RTA process, the fracture strength and strain were found to have a linear increasing relationship that is proportional to the Ge concentration. This pointed out the central role of interstitial defects on the structural stability of the wires.

The effect of dopant incorporation, necessary for device design, on the morphology, growth rate, and alloy composition has been studied by Givan et al. [75]. Diborane (B_2H_6) and phosphine (PH_3) were used to achieve p- and n-type doping, respectively. *p* doping perturbs minimally the morphology and growth kinetic of the $Si_{1-x}Ge_x$ NW: no tapering is observed, the growth rate is minimally affected and the Ge concentration exhibits only a slight decrease. On the other hand, B doping appears to be more critical, making more difficult to obtain smooth nanowires with a controllable composition. The presence of diborane favors the decomposition of silane and silicon is more easily incorporated in the nanowire. While this fact can lead to an increase of the growth rate of up to the 35%, the Ge/Si ratio is halved for all the doping level considered. Additionally, B doping results in a noticeable tapering (0.5–1 nm/μm), which increases with the doping level.

13.4.2 Axial Heterostructures

Two-dimensional heterojunctions between semiconductors have been the privileged playground for the study of many fundamental properties in mesoscopic physics as well as the basic building block for several electronic and optoelectronic devices.

In two-dimensional interfaces, however, requirements on the lattice match of the two materials are quite stringent to obtain dislocation-free junctions. One-dimensional systems allow envisaging a broader class of possible heterostructures, because the lattice misfit can be more easily accommodated due to the new boundary conditions; that is, the stress can be released by radial expansion. Si and Ge have a lattice mismatch of $\sim$4.2%. This rather large value, besides making the growth of defect-free 2D-interfaces challenging [76, 77], is responsible for sophisticated growth mechanism as the well-known Stranski-Krastanov growth, used to grow Ge quantum dots on Si substrates.

The first successful attempt at growing an axially modulated SiGe one-dimensional nanostructure was reported by Wu et al. [78, 79]. They used an ingenious combination of laser ablation and CVD growth to synthesized Si/SiGe superlattice NWs (Fig. 13.1). The process consists in carrying out a conventional VLS-CVD growth of a Si NW, while a pulsed laser is periodically turned on and off. When the laser is on, pulsed ablation of a Ge target provides the Ge vapor for the SiGe alloyed segment. In this way the superlattice periodicity is simply controlled through the on/off frequency of the laser (5 s on, 25 s off in the experiment of Wu et al. [78]).

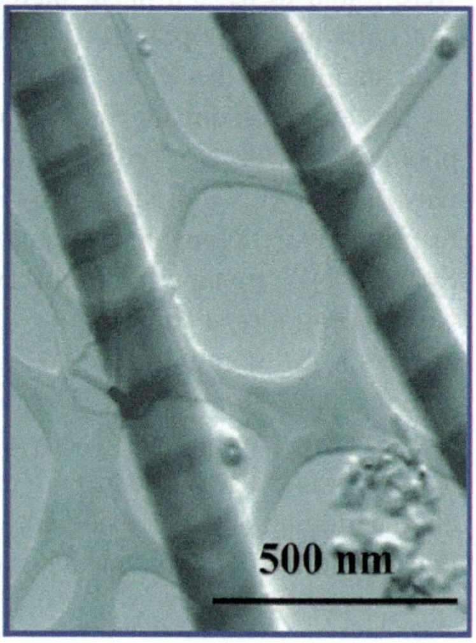

Fig. 13.1 STEM image of two heterostructured Si/SiGe nanowires in bright field mode. The scale bar is 500 nm. Reprinted with permission from Ref. [78]. Copyright 2002 American Chemical Society

13.4.3 Radial Heterostructures

An important class of advanced devices are based on core–shell nanowires. Radial heterostructures are in principle more challenging than axial ones because they require the ability of controlling axial versus radial growth at the same time that the inlet gas precursors are switched. SiGe and GeSi core–shell NWs were first reported by Lauhon et al. [80], where axial VLS growth of a Ge or a Si NW was followed by uncatalyzed, homogeneous vapor-phase deposition on the nanowire sidewalls of the complementary chemical species (Fig. 13.2). With this approach they demonstrated Ge core–p-Si shell, as well as Si core–Ge shell NWs. All of the structures were monocrystalline, though Si shells, initially amorphous, needed a 600 °C thermal annealing to crystallize.

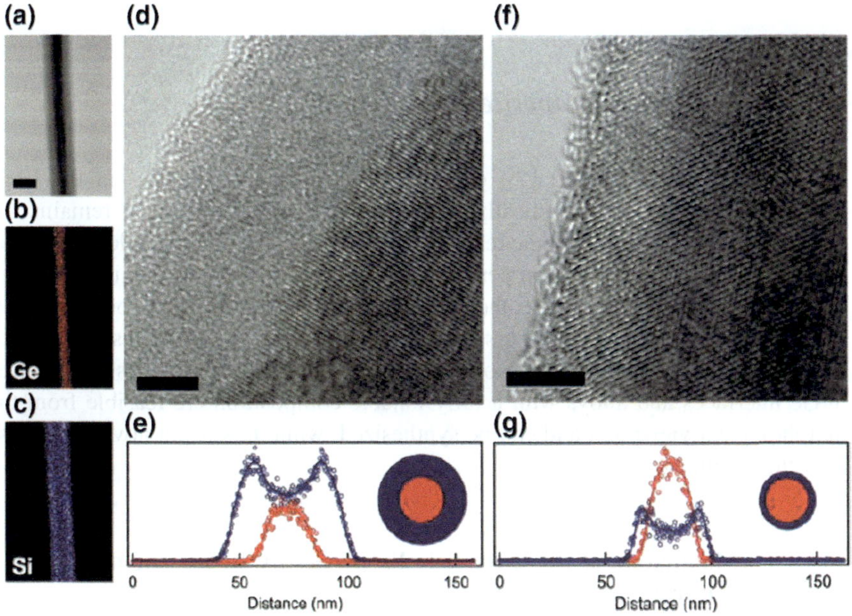

Fig. 13.2 **a** Bright-field image of an unannealed Ge-core/Si-shell NW with an amorphous p-Si shell. Scale bar is 50 nm. **b, c** Scanning TEM elemental maps of Ge and Si concentrations of the nanowire shown in panel **a**. **d** High-resolution TEM image of a nanowire grown under the same conditions as the wire in panel **a–c**. Scale bar is 5 nm. **e** Elemental mapping cross-section showing the Ge (red circles) and Si (blue circles) concentrations. **f** High-resolution TEM image of annealed Ge-core/Si-shell NW with a crystalline *p*-Si shell. Scale bar is 5 nm. **g** Elemental mapping cross-section that gives a 5-nm shell thickness with a sharp interface consistent with the TEM image in panel **f**. Reprinted with permission from Ref. [80]. Copyright 2002 Nature Publishing Group

13.5 Chemical and Physical Properties of Nanowires

The chemical and physical properties of pure Si and Ge nanowires have been extensively reviewed elsewhere, covering growth techniques, experimental measurements, and theoretical calculations [43, 81–85].

Recent years have witnessed an enormous progress in the methods of synthesis of SiGe nanowires, making them very attractive for their high mobility and band gap engineering, in the field of photovoltaics, nanoelectronics, optoelectronics and nanomedicine. Nevertheless, their use in real device applications is still limited due to the extreme difficulty in the control of the different growth parameters, and in the understanding of their relation with the desired structural, electronic, transport, and optical properties. For this reason a large part of the works that we are reviewing in this section aims at clarifying which is the relation between growth parameters, shape, homogeneity, crystallinity of SiGe nanowires, and their chemical and physical properties.

13.5.1 Electronic Properties

Silicon has been the material of choice for semiconductor devices in the latest decades and, although periodically other materials are proposed, it remains the undisputed king of the electronic realm. Sometimes, however, we would like silicon to have just slightly different properties. A simple route to this fine-tuning is adding some amount of germanium to the recipe, creating alloy or more complex interfaces. While this can also be pursued with bulk materials, nanowires offer much more flexibility to achieve such modulation of the electronic properties. Defect-free Si/Ge interfaces and alloys with a fully tunable composition are feasible from the morphological viewpoint and can be synthesized, while this is not always the case with their bulk counterparts.

13.5.1.1 Modulation of the Electronic Properties by Composition Control

Motivated by the long years of experience on SiGe bulk and thin-film alloys, the electronic, transport, and optical properties of one-dimensional $Si_{1-x}Ge_x$ alloy NWs have been first investigated with the idea to use them as possible active elements in electronic and optoelectronic devices. Indeed, several groups have demonstrated the kinetic control of the catalytic decomposition of the chemical precursors during chemical vapor synthesis, to grow single-crystalline random $Si_{1-x}Ge_x$ alloy nanowires, against the general tendency toward phase segregation of the corresponding bulk crystal.

13.5.1.2 Interfaces at Work: Strain, Band-Offset, and Carrier Gases

As we have, SiGe alloy nanowires, it is possible to grow SiGe NW heterojunctions with axial or radial core–shell heterointerfaces. Rather than from composition, as in alloy NWs, the rich physics of these systems stems from the nature of the interface, namely the strain of the mismatched lattice and the band-offset of the two materials. Any attempt of heteroepitaxy involving bulk Si and Ge irremediably ends up with the formation of dislocations that develops as a direct consequence of the strain induced by the 4% lattice mismatch. Actually, this strain can have an important role and is the driving force of the well-known but nontrivial Stranski-Krastanow method that leads to the nucleation of Ge quantum dots on Si. Nevertheless, in axial Si/Ge NWs, the strain induced by the lattice mismatch is partially relieved at the interface by lateral displacement of atoms, and the resulting defect-free interfaces can be exploited in the engineering of efficient nanodevices. These effects were quantified with meticulous photoluminescence (PL) measurements by Chang et al. [86], who characterized the Si/Ge heterojunctions of a crystalline, dislocation-free nanowire.

The present strong interest toward devices based on core/shell SiGe nanowires started after the first observation of hole gas accumulation by Lu et al. [87] in 2005. Following the approach described in Ref. [80], these authors reported the growth of undoped core/shell nanowires with sharp Ge/Si interface and without dislocations and discussed their transport properties in top- and back-gated FET devices, prepared with lithographically patterned Ni source/drain electrodes. Their room temperature electrical transport measurements in a backgated device 10-nm core diameter NW with S/D separation of 10 nm, clearly showed a *p*-type depletion mode behavior of the current *I* versus the source-drain applied voltage V_{SD}. In a more recent work Lauhon and collaborators [88] reported the direct detection of hole accumulation in Ge-core/Si-shell nanowires by the analysis of Raman spectra.

13.5.1.3 Doped Nanowires

The operation of electronic devices is based on the interaction of adjacent regions of a semiconductor with different, tailor-made conductive properties. The way to design such regions is doping, the controlled addition of impurities able to tune the electronic properties of the host semiconductor. SiGe NWs are no exception and impurity doping is routinely used to modulate their physical and chemical properties to design functional devices. B and P are the most common impurities used to p- and n type dope SiGe NWs. Doping is normally carried out in situ, by the catalytic decomposition of B_2H_4 or PH_3 gas precursors that are mixed with silane and/or germane at growth time. Several groups have demonstrated the controlled coflow of the different precursor gases and the successful achievement of n and *p*-type doping. Already in their pioneering work, Lieber and co-workers reported core–shell SiGe heterostructures in which composition and/or doping were modulated at the nanoscale [80]. From the X-ray diffraction (XRD) and micro-Raman

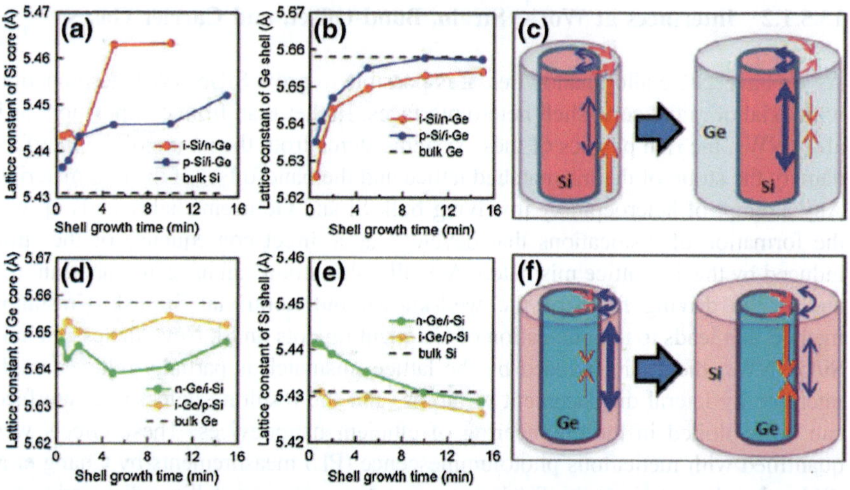

Fig. 13.3 Lattice constant versus shell growth time of **a** the Si core in the i-Si/n-Ge and p-Si/i-Ge core–shell NWs, **b** the Ge shell in the i-Si/n-Ge and p-Si/i-Ge core–shell NWs, **d** the Ge core in the n-Ge/i-Si and i-Ge/p-Si core–shell NWs, and **e** the Si shell in the n-Ge/i-Si and i-Ge/p-Si core–shell NWs. Panels **c** and **f** illustrate the stress induced by the formation of Si/Ge and Ge/Si core–shell NWs respectively. Reprinted with permission from Ref. [89]. Copyright 2012 American Chemical Society

spectra it was found that, Ge (Si)-shell applies tensile (compressive) stress to the Si (Ge)-core, whereas Si (Ge)-core applies compressive (tensile) stress on the Ge (Si)-shell. This is well summarized in Fig. 13.3.

13.5.2 Thermal and Thermoelectric Properties

Due to the next future energy challenge, thermoelectric materials able to convert heat to electrical energy and vice versa are the object of increasing interest by researchers and industries. As we will see in the following, in the last years SiGe nanowires started to play a prominent role among the different systems under investigation to enhance the efficiencies and reduce the costs of the present thermoelectric devices.

As envisaged by several authors, a promising strategy to greatly enhance ZT is to scale down the material size, taking advantage of confinement effects: these should increase S and σ which depend on the density of states (DOS) and its energy derivative near the Fermi energy and at the same time reduce the thermal conductivity κ, due to an increased phonons scattering rate. In this regard, silicon nanowires [90, 91] manifest promising thermoelectric properties, having a thermal conductivity 100 times smaller than bulk silicon which is a very poor thermoelectric material. Then, in the last years, many researchers have focused on SiGe nanowires with the idea to use the advantages of alloy and quantum confinement effects.

13.6 Theoretical Modeling

13.6.1 Electronic Structure

The confinement quantizes the kinetic energy of the particle, influencing its eigenstate energies as the quantum number n, the effective mass m^* and the width of the potential well d (clearly $d = 2R$, where R is the wire radius) are changed:

$$E_n = \frac{\hbar^2 n^2 \pi^2}{2m^* d^2}$$

13.6.1.1 Quantum Confinement Effect and Band Offset

The physical origin of band offset is related to discontinuities occurring in the valence bands and in the conduction bands when two semiconductors are joined to form an heterojunction [92, 93]. When Si and Ge form an abrupt interface, their band structure line-up in order to accommodate their band gap differences. This process induces offsets both in conduction and valence bands (an example of valence band offsets is shown in Fig. 13.4).

13.6.1.2 Size Effects

The ability to control the electronic and optical response of nanowires by changing their size is of crucial importance for nanodevices. For this reason in the case of pure Si and Ge NWs a lot of studies have been focused on the analysis of the dependence of the energy band gap on the wire's diameter.

13.6.1.3 Alloying and Interface Effects

The most innovative degree of freedom in SiGe NWs properties with respect to the pure NWs is the possibility to change the electronic structure by varying the relative composition of the wire.

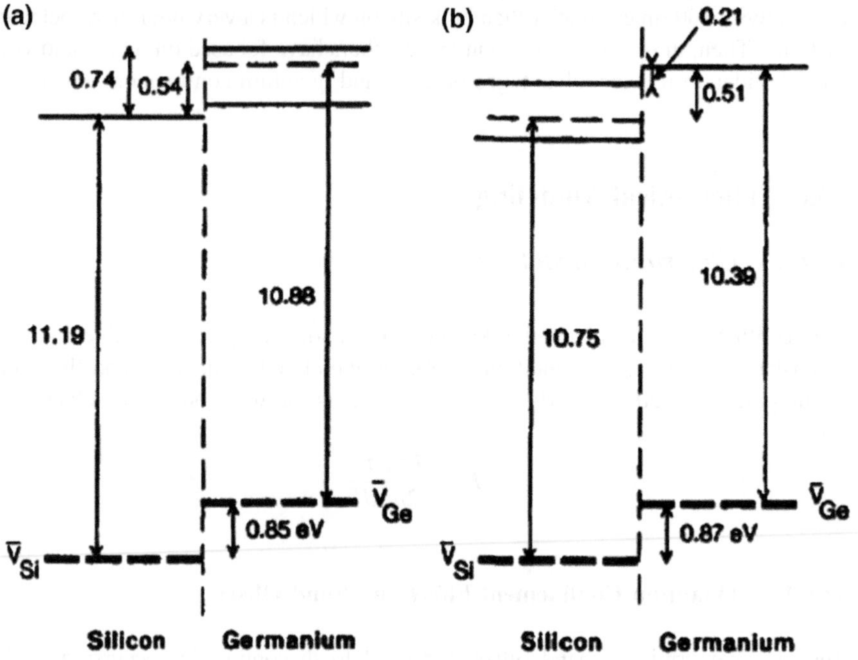

Fig. 13.4 Calculated band-offsets of the Si/Ge interface. Left panel: a∥ = 5.43 Å (cubic Si, strained Ge); right panel: a∥ = 5.65 Å (cubic Ge, strained Si) ∇Si and ∇Ge are the average potentials for Si and Gevalence bands, while dashed lines are the weighted averages of band splitting due to strain. Reprinted with permission from Ref. [94]. Copyright 1985 American Vacuum Society

13.6.1.4 Strain Effects

For bulk SiGe alloys and for SiGe superlattices it has been demonstrated that strain effects, altering the intrinsic interatomic distances, modify the energy levels [94–96].

13.6.1.5 Addition of Impurities

The understanding of doping at the nanoscale represents one of the main topic for the design and development of SiGe nanowire-based devices. The intentional addition of impurities has the general purpose of tuning the number of free charges into the semiconductor, hence controlling the wire conductivity. In Si, Ge or SiGe bulk this goal is fulfilled by incorporating very small amounts of selected additives, usually boron or phosphorus atoms, to them (generally no more than a few parts per million). These impurities increase the number of electrons or holes and therefore the mobility of the material.

13.6.1.6 Electronic Transport

The electrical conductivity of different types of SiGe NWs has been mostly studied in structures that have been proposed for thermoelectric applications. These works are much less abundant than those purely focused on thermal transport, due to the implicit assumption that the electrical conductivity can ultimately be tuned by adjusting the concentration of dopants. SiGe superlattices are amenable to electrical conductivity modulation, in principle, but it is only with nanowires that defect-free interfaces can be obtained, thanks to the lateral release of the strain deriving from the lattice mismatch. Shi et al. [97] performed DFT calculations of a few representative SiGe superlattices where the thickness of the Ge segment was varied, calculating the ballistic conductance with non-equilibrium Green's functions (NEGF) within the Landauer formalism.

13.6.1.7 Optical Properties

The possibility of design the electronic structure of SiGe NWs by playing with intrinsic and extrinsic parameters offers also a useful and straightforward tool to modulate their optical response.

13.6.2 Phonons and Thermal Conductivity

13.6.2.1 Breakdown of Fourier's Law at Nanoscale

Usually, in semiconductors the heat conduction is mainly due to the contribution of phonons while the influence of electrons and photons in the process is negligible. The standard macroscopic approach to describe heat transport is the wellknown Fourier's law, which states that the heat flux density in the material is proportional to temperature gradient as reported in the following equation:

$$J = -k\nabla T$$

The quantity k is the lattice thermal conductivity of the material and it is an intrinsic property. The above equation is a diffusive equation in which the thermal conductivity, which is a function of the temperature, is independent of the size and geometry [98–100].

13.6.2.2 Numerical Simulations of Thermal Properties

The great interest in the thermal properties of SiGe NWs is related to the possibility of engineering phonon dispersion and thermal conductivity by playing on some parameters of the system. In the last decades a lot of theoretical and experimental works, mainly focused on the nanostructuring of existing thermoelectric materials, have shown a strong reduction of thermal conductivity (if compared with bulk values) when the dimensions of the system are at the nanometer scale.

During axial elongation, whereby the resulting axial junctions can yield controlled nanoscale device function without the need for lithography. A representative example is a GaAs/GaP compositionally modulated axial heterostructures (Fig. 13.5a) [101]. Since GaAs is a direct bandgap semiconductor and GaP has an

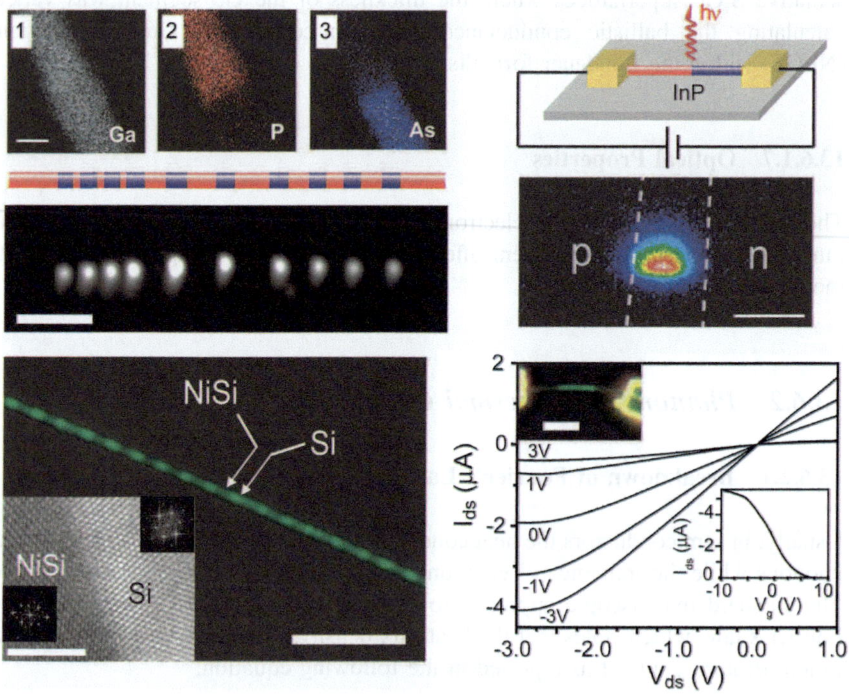

Fig. 13.5 Axial NW heterostructures. **a**, top 1–3, TEM elemental mapping of a single GaAs/GaP nanowire heterojunction, showing the spatial distribution of Ga (gray), *P* (red), and As (blue) at the junction. Scale bar, 20 nm. **a**, bottom, Schematic and photoluminescence (PL) image of a 21-layer, (GaP/GaAs)10GaP, nanowire superlattice. The ten bright regions correspond to GaAs (blue, direct bandgap) regions, while the dark segments are from the GaP (red, indirect bandgap) regions. **b** Schematic of a modulation-doped InP NW LED and image of the emission from the device. Dashed white lines indicate the edges of the electrodes. Scale bar, 3 μm. **c** Dark-field optical image of a single NiSi/Si NW superlattice heterostructure. The bright green segments correspond to Si and the dark segments to NiSi; scale bar is 10 μm. Inset shows a high-resolution TEM image of the atomically abrupt interface between the NiSi and Si; scale bar is 5 nm. **d** Ids-Vds curves of a NiSi/p-Si/NiSi heterojunction NWFET fabricated using a 30 nm diameter p-Si NW; upper inset is a dark-field optical image of the same device showing that the contacts are made to the metallic NiSi regions only. Scale bar, 3 μm. Lower inset is the I_{ds}-V_g obtained with $V_{ds} = -3$ V (Reprinted with permission from Ref. [37, 38]. © 2002 and 2004 Nature Publishing Group, respectively)

indirect bandgap, these NW heterostructures can be patterned synthetically and emit light as nanoscale barcodes. In addition, *p-n* junctions formed within individual NWs can also be prepared in a similar way. Forward biased n-InP/p-InP single NW devices function as nanoscale LEDs with light emission at the *p-n* interface as shown in Fig. 13.5b.

We have taken this key concept of composition modulation to define functional devices in several other directions relevant to electronic and optoelectronic devices. First, we demonstrated the selective transformation of Si NWs into metallic NiSi NWs and NiSi/Si NW heterostructures by thermal annealing as-made SiNWs with Ni (Fig. 13.5c) [102]. Significantly, this method yielded the first example of atomically sharp metal-semiconductor interfaces between single metallic (NiSi) and semiconductor (Si) nanowires. In these heterostructures, Si NWFET source-drain contacts are defined by the metallic NiSi NW regions, which function as excellent ohmic contacts at room temperature (Fig. 13.5d), and thus provide an integrated solution for nanoscale contacts and interconnects.

References

1. Voit, J.: One-dimensional fermi liquids. Rep. Prog. Phys. **57**, 977 (1994)
2. Devreese, J.T., Evrard, R.P., van Doren, V.E. (eds.): Highly Conducting One-Dimensional Solids. Plenum, New York (1979)
3. Lieber, C.M., Wu, X.L.: Scanning tunneling microscopy studies of low-dimensional materials: probing the effects of chemical substitutions at the atomic level. Acc. Chem. Res. **24**(4), 223 (1991)
4. Dai, H., Lieber, C.M.: Scanning tunneling microscopy studies of low-dimensional materials: charge density wave pinning and melting in two dimensions. Annu. Rev. Phys. Chem. **44**, 237 (1993)
5. Lieber, C.M., Liu, J., Sheehan, P.E.: Understanding and manipulating inorganic materials using scanning probe microscopes. Angew. Chem., Int. Ed. Engl. **35**, 686 (1996)
6. Rouxel, J. (ed.): Crystal Chemistry and Properties of Materials with Quasi-One-Dimensional Structures. Reidel, Boston (1986)
7. Wilson, J.A., DiSalvo, F.J., Mahajan, S.: Charge density waves and superlattices in the metallic layered transition metal dichalcogenides. Adv. Phys. **24**, 117 (1975)
8. Batlogg, B.A.: Cuprate superconductors: science beyond high Tc. Solid State Commun. **107**, 639 (1998)
9. Das-Sarma, S., Pinczuk, A. (eds.): Perspectives in Quantum Hall Effects. Wiley, New York (1997)
10. Stormer, H.L.: Fractional quantum hall effect today. Solid State Commun. **107**, 617 (1998)
11. Grabert, H., Devoret, M.H. (eds.): Single Charge Tunneling. Plenum, New York (1991)
12. Dvoret, M.H., Esteve, D., Urbina, C.: Single-electron transfer in metallic nanostructures. Nature **360**, 547 (1992)
13. Ashoori, R.C.: Electrons in artificial atoms. Nature **379**, 413 (1996)
14. Alivisatos, A.P.: Semiconductor clusters, nanocrystals, and quantum dots. Science **271**, 933 (1996)
15. Klein, D.L., Roth, R., Lim, A.K.L., Alivisatos, A.P., McEuen, P.L.: A single-electron transistor made from a cadmium selenide nanocrystal. Nature **389**, 699 (1997)

16. Lieber, C.M., Morales, A.M., Sheehan, P.E., Wong, E.W., Yang, P.: One-dimensional nanostructures: rational synthesis, novel properties and applications. In: Proceedings of the Robert A. Welch Foundation 40th Conference on Chemical Research: Chemistry on the Nanometer Scale. Welch Foundation, Houston (1997)
17. Lieber, C.M.: One-dimensional nanostructures: chemistry, physics & applications. Solid State Commun. **107**, 607 (1998)
18. Bockrath, M., Cobden, D.H., McEuen, P.L., Chopra, N.G., Zettl, A., Thess, A., Smalley, R. E.: Single-electron transport in ropes of carbon nanotubes. Science **1997**, 275 (1922)
19. Tans, S.J., Devoret, M.H., Dai, H., Thess, A., Smalley, R.E., Geerligs, L.J., Dekker, C.: Individual single-wall carbon nanotubes as quantum wires. Nature **386**, 474 (1997)
20. Tans, S.J., Verschueren, A.R.M., Dekker, C.: Room-temperature transistor based on a single carbon nanotube. Nature **393**, 49 (1998)
21. Hu, J., Ouyang, M., Yang, P., Lieber, C.M.: Growth and electrical properties of nanotube/nanowire heterojunctions. Nature, submitted for publication
22. Yang, P., Lieber, C.M.: Nanorod-superconductor composites: a pathway to high critical current density materials. Science **1996**, 273 (1836)
23. Yang, P., Lieber, C.M.: Columnar defect formation in nanorod/$Tl_2Ba_2Ca_2Cu_3O_z$ superconducting composites. Appl. Phys. Lett. **70**, 3158 (1997)
24. Yang, P., Lieber, C.M.: Nanostructured high-temperature superconductors: creation of strong-pinning columnar defects in nanorod/superconductor composites. J. Mater. Res. **12**, 2981 (1997)
25. Dai, H., Hafner, J.H., Rinzler, A.G., Colbert, D.T., Smalley, R.E.: Nanotubes as nanoprobes in scanning probe microscopy. Nature **384**, 147 (1996)
26. Wong, S.S., Harper, J.D., Lansbury, P.T., Lieber, C.M.: Carbon nanotube tips: high-resolution probes for imaging biological systems. J. Am. Chem. Soc. **120**, 603 (1998)
27. Wong, S.S., Joselevich, E., Woolley, A.T., Cheung, C., Lieber, C.M.: Covalently functionalized nanotubes as nanometer probes for chemistry and biology. Nature **394**, 52 (1998)
28. Wong, S.S., Woolley, A.T., Joselevich, E., Cheung, C., Lieber, C.M.: Covalently-functionalized single-walled carbon nanotube probe tips for chemical force microscopy. J. Am. Chem. Soc. **120**, 8557 (1998)
29. Wong, S.S., Woolley, A.T., Odom, T.W., Huang, J.-L., Kim, P., Vezenov, D.V., Lieber, C. M.: Single-walled carbon nanotube probes for high-resolution nanostructure imaging. Appl. Phys. Lett. **73**, 3465 (1998)
30. Collins, P.G., Bando, H., Zettl, A.: Nanotechnology **9**, 153 (1998)
31. Cui, Y., Lieber, M.: Science **291**, 851 (2001)
32. Dresselhaus, M.S., Lin, Y.-M., Cronin, S.B., Rabin, O., Black, M.R., Dresselhaus, G., Koga, T.: Semicond. Semimetals **71**, 1 (2001)
33. Wu, Y., Fan, R., Yang, P.: Int. J. Nanosci. **1**, 1 (2002)
34. Schwab, K., Henriksen, E.A., Worlock, J.M., Roukes, M.L.: Nature (London) **404**, 974 (2000)
35. Hone, J., Whitney, C., Piskoti, C., Zettl, A.: Phys. Rev. B **59**, 2514 (1999)
36. Balandin, A., Wang, K.: Phys. Rev. B **58**, 1544 (1998)
37. Khitun, A., Balandin, A., Wang, K.L.: Superlattices Microstruct. **26**, 181 (1999)
38. Volz, S., Chen, G.: Appl. Phys. Lett. **75**, 2056 (1999)
39. Cahill, D.G., Ford, W.K., Goodson, K.E., Mahan, G.D., Majumdar, A., Maris, H.J., Merlin, R., Phillpot, S.R.: J. Appl. Phys. **93**, 793 (2003)
40. Li, Deyu, Yiying, Wu, Kim, Philip, Shi, Li, Yang, Peidong, Majumdar, Arun: Thermal conductivity of individual silicon nanowires. Appl. Phys. Lett. **83**(4), 2934 (2003). https://doi.org/10.1063/1.1616981
41. Conibeer, G.: Mater. Today **10**, 42 (2007)
42. Daldosso, N., Pavesi, L.: Laser Photonics Rev. **3**, 508 (2009)
43. Rurali, R.: Rev. Mod. Phys. **82**, 427 (2010)
44. Khriachtchev, L., Ossicini, S., Iacona, F., Gourbilleau, F.: Int. J. Photoenergy **2012**, 1 (2012)

45. Govoni, M., Marri, I., Ossicini, S.: Nat. Photonics **6**, 672 (2012)
46. Abstreiter, G.: Thin Solid Films **183**, 1 (1989)
47. Gnutzmann, U., Clausecker, K.: Appl. Phys. **3**, 9 (1974)
48. Kasper, E., Herzog, H.-J., Kibbel, H.: Appl. Phys. **8**, 199 (1975)
49. Bean, J.C., Feldman, L.C., Fiory, A.T., Nakahara, S., Robinson, I.K.J.: Vac. Sci. Technol. A **2**, 436 (1984)
50. Kasper, E., Herzog, H.-J., Jorke, H., Abstreiter, G.: Superlattices Microstruct. **3**, 141 (1987)
51. Terashima, K., Tajima, M., Tatsumi, T.: Appl. Phys. Lett. **1990**, 57 (1925)
52. Sturm, J., Manoharan, H., Lenchyshyn, L.C., Thewalt, M.L.W., Roweel, N.L., Noël, J.P., Houghton, D.C.: Phys. Rev. Lett. **66**, 1362 (1991)
53. Whall, T.E., Parker, E.H.C.: J. Mater. Sci.: Mater. Electron. **6**, 249 (1995)
54. Patton, G.L., Comfort, J.H., Meyerson, B.S., Crabbé, E.F., Scilla, G.J., de Frésart, E., Stork, J.M.C., Sun, J.Y.-C., Harame, D.L., Burghartz, J.N.: IEEE Electron Device Lett. **11**, 171 (1990)
55. Lin, T.L., Maserjian, J.: Appl. Phys. Lett. **57**, 1422 (1990)
56. Xie, Y.-H., Monroe, D., Fitzgerald, E., Silverman, P., Thiel, F., Watson, G.: Appl. Phys. Lett. **63**, 2263 (1993)
57. Dingle, R., Störmer, H.L., Gossard, A.C., Wiegmann, W.: Appl. Phys. Lett. **33**, 665 (1978)
58. König, U., Schäffler, F.: IEEE Electron Device Lett. **14**, 205 (1993)
59. Schäffler, F.: Thin Solid Films **12**, 1515 (1997)
60. Whall, T.E., Parker, E.H.C.: Thin Sol. Films **369**, 297 (2000)
61. Paul, D.J.: Semicond. Sci. Technol. **9**, R75 (2004)
62. Lee, M.L., Fitzgerald, E.A., Bulsara, M.T., Currie, M.T., Lochtefeld, A.: J. Appl. Phys. **97**, 011101 (2005)
63. Brunner, K.: Rep. Prog. Phys. **65**, 27 (2002)
64. Berbezier, I., Ronda, A.: Surf. Sci. Rep. **64**, 47 (2009)
65. Aqua, J.-N., Berbezier, I., Favre, L., Frisch, T., Ronda, A.: Phys. Rep. **522**, 52 (2013)
66. Wagner, R.S., Ellis, W.C.: Appl. Phys. Lett. **4**, 89 (1964)
67. Morales, A.M., Lieber, C.M.: Science **279**, 208 (1998)
68. Duan, X., Lieber, C.M.: Adv. Mater. **12**, 298 (2000)
69. Wang, N., Tang, Y.H., Zhang, Y.F., Lee, C.S., Lee, S.T.: Phys. Rev. B **58**, R16024 (1998)
70. Holmes, J.D., Johnston, K.P., Doty, R.C., Korgel, B.A.: Science **287**, 1471 (2000)
71. Geyer, N., Huang, Z., Fuhrmann, B., Grimm, S., Reiche, M., Nguyen-Duc, T.-K., de Boor, J., Leipner, H.S., Werner, P., Gösele, U.: Nano Lett. **9**, 3106 (2009)
72. Non-VLS alloy SiGe NWs were previously reported. The nanowires were formed by pulsed UV laser induced epitaxy of Ge NWs deposited on a Si substrate 277
73. Lew, K.-K., Pan, L., Dickey, E., Redwing, J.: Adv. Mater. **15**, 2073 (2003)
74. Ma, J.W., Lee, W.J., Bae, J.M., Jeong, K.S., Kang, Y.S., Cho, M.H., Seo, J.H., Ahn, J.P., Chung, K.B., Song, J.Y.: Nano Lett. **13**, 1118 (2013)
75. Givan, U., Kwiat, M., Patolsky, F.: J. Phys. Chem. C **114**, 4331 (2010)
76. Dayeh, S.A., Wang, J., Li, N., Huang, J.Y., Gin, A.V., Picraux, S.T.: Other defects, such as twin boundaries and kinks, can also be found, though they are not specific of SiGe heterostructures and are also rather common in single-phase nanowires. Nano Lett. **11**, 4200 (2011)
77. Vincent, L., Boukhicha, R., Cherkashin, N., Reboh, S., Patriarche, G., Renard, C., Yam, V., Fossard, F., Bouchier, D.: Nanotechnology **23**, 395701 (2012)
78. Wu, Y., Fan, R., Yang, P.: Nano Lett. **2**, 83 (2002)
79. Givargizov, E., Sheftal, N.: J. Cryst. Growth **9**, 326 (1971)
80. Lauhon, L.J., Gudiksen, M.S., Wang, D., Lieber, C.M.: Nature **420**, 57 (2002)
81. Lu, W., Lieber, C.M.: J. Phys. D Appl. Phys. **39**, R387 (2006)
82. Fan, H.J., Werner, P., Zacharias, M.: Small **2**, 700 (2006)
83. Schmidt, V., Wittemann, J.V., Gösele, U.: Chem. Rev. **110**, 361 (2010)
84. Hochbaum, A.I., Yang, P.: Chem. Rev. **110**, 527 (2010)
85. Collins, G., Holmes, J.D.: J. Mater. Chem. **21**, 11052 (2011)

86. Chang, H.-Y., Tsybeskov, L., Sharma, S., Kamins, T.I., Wu, X., Lockwood, D.J.: Appl. Phys. Lett. **95**, 133120 (2009)
87. Lu, W., Xiang, J., Timko, B.P., Wu, Y., Lieber, C.M.: Proc. Natl. Acad. Sci. U.S.A. **102**, 10046 (2005)
88. Zhang, S., Lopez, F.J., Hyun, J.K., Lauhon, L.J.: Nano Lett. **10**, 4483 (2010)
89. Fukata, N., Mitome, M., Sekiguchi, T., Bando, Y., Kirkham, M., Hong, J.-I., Wang, Z.L., Snyder, R.L.: ACS Nano **6**, 8887 (2012)
90. Hochbaum, A.I., Chen, R., Delgado, R.D., Liang, W., Garnett, E.C., Najarian, M., Majumdar, A., Yang, P.: Nature **451**, 163 (2008)
91. Boukai, A.I., Bunimovich, Y., Tahir-Kheli, J., Yu, J.-K., Goddard III, W.A., Heath, J.R.: Nature **451**, 168 (2008)
92. Peressi, M., Binggeli, N., Baldereschi, A.: J. Phys. D Appl. Phys. **31**, 1273 (1998)
93. Zunger, A.: Curr. Opin. Solid State Mater. Sci. **3**, 32 (1998)
94. Van de Walle, C.G., Martin, R.M.J.: Vac. Sci. Technol. A **3**, 1256 (1985)
95. Van de Walle, C.G., Martin, R.M.: Phys. Rev. B **34**, 5621 (1986)
96. Ciraci, S., Batra, I.P.: Phys. Rev. B **1988**, 38 (1835)
97. Shi, L., Jiang, J., Zhang, G., Li, B.: Appl. Phys. Lett. **101**, 233114 (2012)
98. Liu, S., Xu, X., Xie, R., Zhang, G., Li, B.: Eur. Phys. J. B **85**, 1 (2012)
99. Palacios, T.: Nature **481**, 152 (2012)
100. Thelander, C., Agarwal, P., Brongersma, S., Eymery, J., Feiner, L., Forchel, A., Scheffler, M., Riess, W., Ohlsson, B., Gösele, U., Samuelson, L.: Mater. Today **9**, 28 (2006)
101. Gudiksen, M., et al.: Nature **415**, 617 (2002)
102. Wu, Y., et al.: Nature **430**, 61 (2004)

Chapter 14
Properties of Nanostructured Materials (NSMs) and Physicochemical Properties of (NPs)

Nanomaterials have the structural features in between of those of atoms and the bulk materials. While most microstructured materials have similar properties to the corresponding bulk materials, the properties of materials with nanometer dimensions are significantly different from those of atoms and bulks materials. This is mainly due to the nanometer size of the materials which render them: (i) large fraction of surface atoms; (ii) high surface energy; (iii) spatial confinement; (iv) reduced imperfections, which do not exist in the corresponding bulk materials.

Due to their small dimensions, nanomaterials have extremely large surface area to volume ratio, which makes a large to be the surface or interfacial atoms, resulting in more "surface" dependent material properties. Especially when the sizes of nanomaterials are comparable to length, the entire material will be affected by the surface properties of nanomaterials. This in turn may enhance or modify the properties of the bulk materials. For example, metallic nanoparticles can be used as very active catalysts. Chemical sensors from nanoparticles and nanowires enhanced the sensitivity and sensor selectivity. The nanometer feature sizes of nanomaterials also have spatial confinement effect on the materials, which bring the quantum effects.

L. H. Madkour, *Nanoelectronic Materials*, Advanced Structured Materials 116, https://doi.org/10.1007/978-3-030-21621-4_14

14.1 Properties of Nanoscale Matter

Nanoscale materials behave as surface matter.

Au-Cluster (55 atoms)

n	Atoms/Cluster	Percentage surface atoms	Cluster size
1	13	92 %	0.58 nm
2	55	76 %	1.4 nm
3	147	63 %	2.1 nm
4	309	52 %	2.8 nm
5	561	45 %	3.5 nm
7	1415	35 %	5 nm
9	2869	28 %	6.5 nm

Consequences for	solid state chemistry	catalysis	optical properties
Reactivity	+	+	-
Defect density	+	+	-
Band gap	-/+	-	-/+

14.2 Nanoscale Materials Show Quantum Confinement Effects

Very small nanoparticle show strong confinement of excitons (weakly bound electron-hole pairs) and thus dependence of absorption and luminescence on particle size.

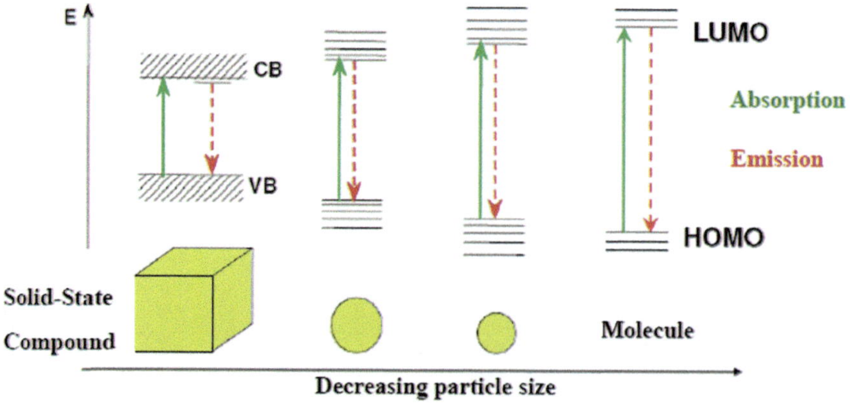

The energy band structure and charge carrier density in the materials can be modified quite differently from their bulk and in turn will modify the electronic and optical properties of the materials. For example, lasers and light emitting diodes (LED) from both of the quantum dots and quantum wires are very promising in the future optoelections. High density information storage using quantum dot devices is also a fast developing area. Reduced imperfections are also an important factor in determination of the properties of the nanomaterials. Nanostructures and Nanomaterials favors of a self-purification process in that the impurities and intrinsic material defects will move to near the surface upon thermal annealing. This increased materials perfection affects the properties of nanomaterials. For example, the chemical stability for certain nanomaterials may be enhanced, the mechanical properties of nanomaterials will be better than the bulk materials. The superior mechanical properties of carbon nanotubes are well known. Due to their nanometer size, nanomaterials are already known to have many novel properties. Many novel applications of the nanomaterials rose from these novel properties have also been proposed.

Nanocrystalline particles represent a state of matter in the transition region between bulk solid and single molecule. As a consequence, their physical and chemical properties gradually change from solid state to molecular behaviour with decreasing particle size. The reasons for this behaviour can be summarized as two basic phenomena.

First, as we have already discussed, owing to their small dimensions, the surface-to-volume ratio increases, and the number of surface atoms may be similar to or higher than those located in the crystalline lattice core, and the surface properties are no longer negligible. When no other molecules are adsorbed onto the nanocrystallites, the surface atoms are highly unsaturated and their electronic contribution to the behaviour of the particles is very different from that of the inner atoms. These effects may be even more marked when the surface atoms are ligated. This leads to different mechanical and electronic transport properties, which account for the catalytic properties of the nanocrystalline particles.

The second phenomenon, which only occurs in metal and semiconductor nanoparticles, is totally an electronic effect. The band structure gradually evolves with increasing particle size, i.e., and molecular orbital convert into delocalized band states. Figure 14.1 shows the *size quantization effect* responsible for the transition between a bulk metal or semiconductor, and cluster species. In a metal, the quasicontinuous density of states in the valence and the conduction bands splits into discrete electronic levels, the spacing between these levels and the band gap increasing with decreasing particle size.

In the case of semiconductors, the phenomenon is slightly different, since a band gap already exists in the bulk state. However, this band gap also increases when the particle size is decreased and the energy bands gradually convert into discrete molecular electronic levels. If the particle size is less than the De Broglie wavelength of the electrons, the charge carriers may be treated quantum mechanically as "particles in a box", where the size of the box is given by the dimensions of the crystallites. In semiconductors, the quantization effect that enhances the optical gap is routinely observed for clusters ranging from 1 nm to almost 10 nm. Metal particles consisting of 50–100 atoms with a diameter between one and two nm start to

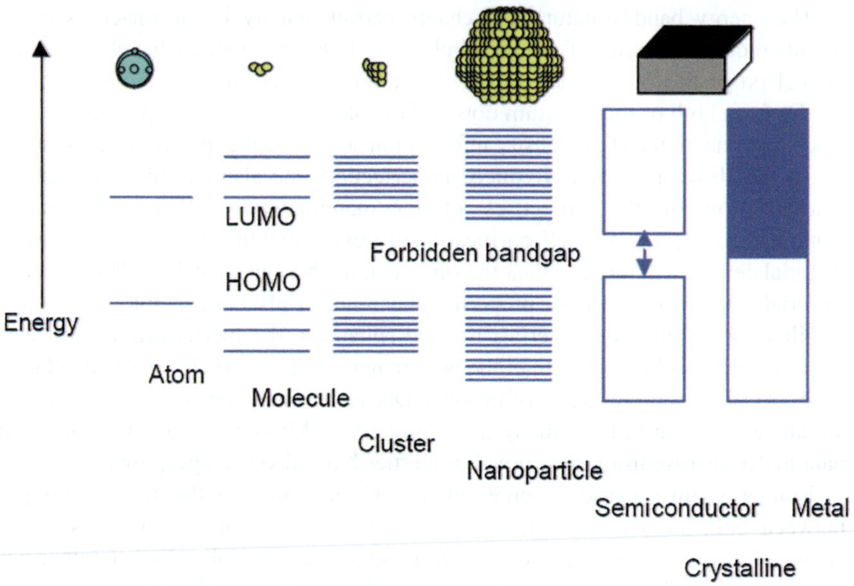

Fig. 14.1 Size quantization effect. Electronic state transition from bulk metal/semiconductor to small cluster

lose their metallic behaviour and tend to become semiconductors. Particles that show this *size quantization effect* are sometimes called *Q-particles* or *quantum dots*.

14.2.1 Nanoscale Luminescent Materials Are Mostly Less Efficient Than Microscale Materials

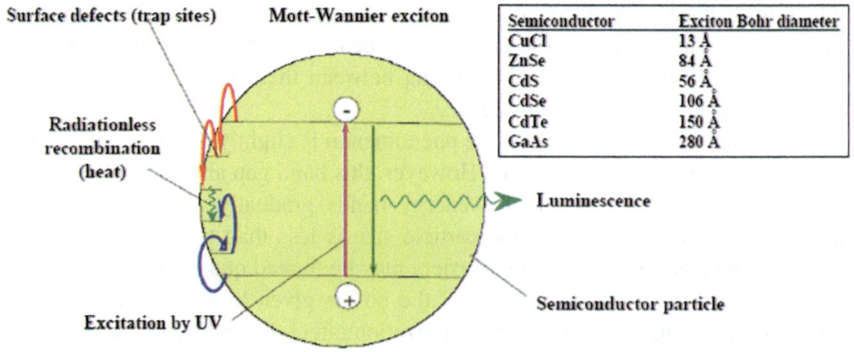

14.2.2 CdSe Nanocrystals

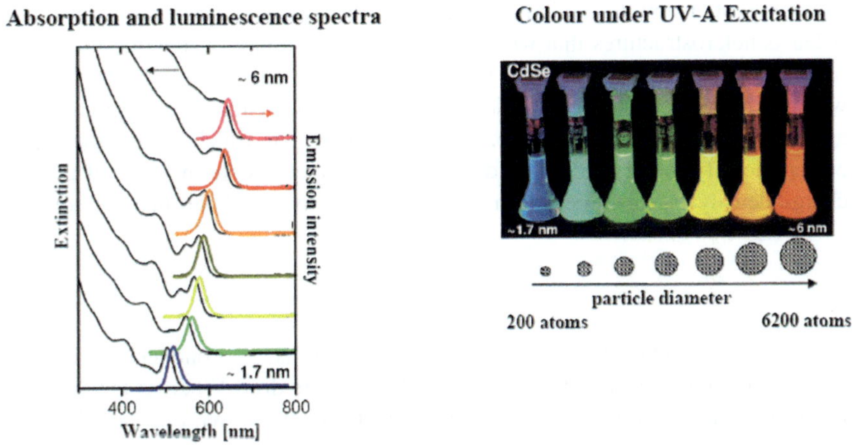

Certain particle sizes, determined by the so-called *magic numbers,* are more frequently observed than others. This is not only due to the thermodynamic stability of certain structures but is also related to the kinetics of particle growth. In semiconductor nanoparticles such as CdS, the growth of the initially formed smallest particles with an agglomeration number k occurs by combination of the particles.

Thus, particles so formed would have the agglomeration number of $2k$, $3k$ and so on. Metals have a cubic or hexagonal close-packed structure consisting of one central atom, which is surrounded in the first shell by 12 atoms, in the second shell by 42 atoms, or in principle by $10n2 + 2$ atoms in the nth shell. For example, one of the most famous ligand stabilized metal clusters is a gold particle with 55 atoms (Au_{55}) first reported by G. Schmid in 1981. In drive towards nanotechnology, there is tremendous interest in understanding the electric and thermal transport through nanowires. The questions that need to be answered pertain to how the electric and heat transport laws that govern the macroscopic systems get modified at the level of a few molecules or atoms. Wires and contacts having average radius **R**, in the range of the Fermi wavelength $\lambda\mathbf{F}$, can indeed be fabricated and have shown unusual mechanical and electronic properties. The effects of reduced size and dimensionality, and their quantum properties have been investigated actively in the last decade. Ultrathin nanowires and even monatomic chains suspended between two metal electrodes have been produced. The discovery of scanning tunneling microscopy (STM) and later atomic force microscopy (AFM) has sparked a tremendous recent research effort on nanoparticles because many properties have become 'directly' observable. The tunneling microscope has also made it possible to produce atomic size contacts and atomic wires only several angstroms in length. Electrical conduction and force have been measured concomitantly in the course of stretching of nanowires. Initially, much of the work was carried out on GaAs–AlGaAs heterostructures that were grown to contain a thin conducting layer at the interface. The conducting layer is treated as a two-dimensional (2D) electron gas in which a narrow constriction of desired width **w**, and length **l**, can be created by applying a negative gate voltage. The conductivity σ, which relates the electric current density to the electric field by $\mathbf{j} = \sigma\mathbf{E}$, is expressible in terms of the areal charge density ρ_S for 2D electron gas of effective mass **m*** through the equation:

$$\sigma = \frac{\rho_S e^2 \tau}{m^*}$$

The experimental quantity of interest, however, is the conductance **G = I/V**, which is the ratio of the total current **I** to the voltage drop **V** across the sample of length **l** in the direction of current flow. For 2D, since **I = wj**, one can also write.

$$G = \sigma \frac{w}{l}$$

For a 3D conductor, this relationship is valid provided that **w** is replaced by the cross sectional area **A** orthogonal to the current flow direction. Similar expressions are also valid for the thermal transport of energy. The thermal conductance related to the energy (or heat) current **Jx** through a sample between two reservoirs is given by $\mathbf{K_x} = \mathbf{J_x \Delta T}$, where **ΔT** is the temperature difference. Depending on whether the energy is carried by electrons (**x = e**) or by phonons (**x = p**) the thermal conductance is identified as electronic $\mathbf{K_e}$ or phononic $\mathbf{K_p}$. Here our focus is on quantum transport through materials of very small dimensions.

Novel size-dependent effects emerge as **w** and **l** are reduced towards atomic dimensions in the nanometre range. The relationship expressed by equation holds in the diffusive transport regime where both **w** and **l** are greater than the mean free path. As the width of the constriction decreases, there comes a point where quantum mechanics makes its presence known. The quantum confinement of a carrier in a strip of width **w** leads to the discretization of energy levels given by $\epsilon \mathbf{n} = \mathbf{n^2h^2/(8m * w^2)}$. The number of these **w**-dependent transverse modes, which are occupied, determines the conductance. Thus, rather than a simple linear dependence of **G** on **w**, quantum mechanics forces this 'indirect' dependence on **w**. As **w** is altered, the energy spectrum changes and so does the number of occupied modes below the Fermi energy and hence the conductance. Simple physical considerations show that the number of transverse occupied states, $\mathbf{N \sim 2w/\lambda F}$, increases with the width of the constriction. Since all these modes can contribute to the conductance, one still expects the conductance to increase linearly with **w** even in the nanodomain. This is almost the case, except with one important distinction. The width **w** can change continuously, at least in principle. But the number of modes, being an integer, can only vary in discrete steps. The concept becomes physically more transparent if we rewrite

$$N \sim n\sqrt{\frac{E_F}{\varepsilon_n}}$$

The highest occupied $\mathbf{\epsilon_n} = \mathbf{E_F}$ gives **N** from a simple counting of the number of discrete eigenvalues. Since it is a rare coincidence for an eigenvalue to exactly align with the Fermi energy, **N** is taken to be the integer, which corresponds to the highest occupied level just below $\mathbf{E_F}$. Thus, the effect of quantum mechanics due to the reduced dimension **w** is to cause the conductance to change in discrete steps in a staircase fashion. We have yet to find the step height. This simple view is necessarily modified when other factors explained later start to play a crucial role. Obviously, one then requires a more detailed analysis.

The characteristic length that comes into play is **λF**, since only those electrons having energy close to the Fermi energy carry current at low temperatures. If $\mathbf{w} \gg \lambda\mathbf{F}$, the number of conducting modes is very large and so is the conductance. This is typical in metals where **λF** ~ **0.2** nanometres is very small and consequently the observation of discrete conductance variation requires **w** ~ **λF**. That is why the discrete conductance behaviour was first observed in semiconductor heterojunctions having very low electron density and **λF** two orders of magnitude larger than in metals. The effect of reduced length **l** on the conductance is even more striking. If the ohmic regime were to hold in equation $G = \sigma\ W/l$, one would expect **G** to increase without limit (or resistance to reach zero) as **l** was reduced towards zero. We shall see that there would always be finite residual resistance. Heisenberg's uncertainty principle defines a natural characteristic length as the mean free path **le**. If **l** < **le**, carriers can propagate without losing their initial momentum, and this domain is referred to as the ballistic transport regime. Landauer pointed out that 'conduction is transmission'. If we follow the standard definition, that the conductance is a measure of current through a sample divided by the voltage difference, $G = I/\Delta V$.

The quantum of electrical conductance occurs naturally as a consequence of Heisenberg's uncertainty principle. The current is given by the rate of charge flow $\mathbf{I = \Delta Q/t_{transit}}$. Now charge is quantized in units of elementary charge, **e**. Hence in the extreme quantum limit, setting $\mathbf{\Delta Q = e}$ and recognizing that the transit time should then be at least in the range of time implied by Heisenberg's uncertainty principle, we get $I = e/\Delta t$.

Combining equation $G = I/\Delta V$ and equation $I = e/\Delta t$, and using the fact that the potential difference, **ΔV**, is equal to electrochemical potential difference **ΔE** divided by the electronic charge **e**, one gets

$$G = \frac{e^2}{\Delta E \Delta t}$$

Next invoking Heisenberg's uncertainty principle, $\mathbf{\Delta E \Delta t} \gg \mathbf{h}$, the expression for ballistic conductance including the spin degeneracy becomes $\mathbf{G = 2e^2/h}$ in the ideal case. It has a maximum value of $8 \times 10^{-5}\ \Omega^{-1}$. This is the step height, the conductance per transverse mode. The corresponding resistance $\mathbf{h/2e^2}$ has a value of 12.9 kΩ (it suffices to call it 12345 Ω for ease of remembrance) and is attributed to the resistance at the contacts where the conductor is attached to the electrodes or electron reservoirs. Thus classically $\mathbf{G} \propto \mathbf{w}$. Quantum mechanically it increases in discrete steps. It jumps by $\mathbf{2e^2/h}$ as **w** increases enough to permit one transverse mode to be occupied and hence available for conduction. More recently, the electronic and transport properties of metallic point contacts and wires having average dimension in the range of metallic λ**F** produced by STM and also by mechanical break junctions have displayed various quantum effects. The two-terminal electrical conductance, **G**, of the wire showed a stepwise variation with the stretch. The sudden jumps of conductance in the course of the stretch have been taken as the realization of quantized conductance at room temperature and thus have created much popular interest.

Metallic carbon nanotubes providing two conductance channels at the Fermi level are considered almost perfect one-dimensional (1D) conductors. These tubes, which can be grown to several micrometers in length with a variety of diameters and chirality, display unique electromechanical properties and dimensionality effects. They can be either semiconductor or metal depending on their chirality and diameter. The nanotubes are flexible, and at the same time are very strong with high yield strength. Their strength far exceeds that of any other fibre. These properties are being actively explored with a view to novel applications in nanotechnology. It has been demonstrated experimentally that an individual nanotube can sustain very high current density; first the current increases with the bias voltage and it eventually saturates at a value, which is very high for a nanowire. Recent studies have shown that the electronic properties of semiconducting carbon nanotubes can be modified variably and reversibly by radial deformation, a feature that may lead to some device applications. The thermal conductance may exhibit quantum features depending on the size of the nanostructure that transfers the phononic energy. Owing to the finite level spacing of vibrational frequencies, the phononic energy transfer through an electrically nonconducting nano-object (i.e. a molecule, atomic chain, or a single atom) between two reservoirs shows behaviour similar to the ballistic electron transport. Recent theoretical studies indicate that the thermal conductance of an acoustic branch at low temperature is independent of any material property, and is linearly dependent on temperature. At low temperature, the discrete vibrational frequency spectrum of a 'chain' gives rise to a sudden increase of the thermal conductance. Study of quantum effects in 1D conductors has seen a tremendous explosion because of growing interest in nanoscience and the quest for novel nanodevices.

14.3 The Physical Properties of Nanoclusters

14.3.1 The Morphology

- The γ plot

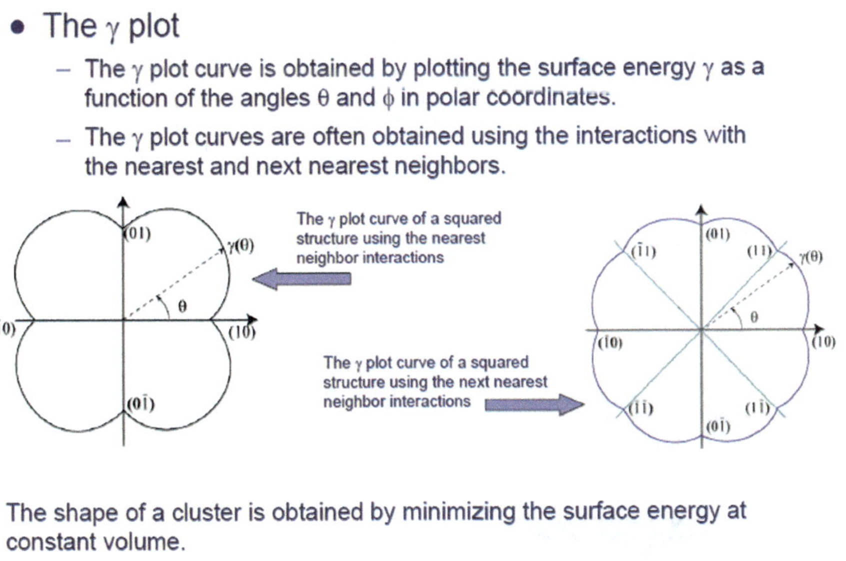

- The Wulff construction
  - To minimize the energy requires to use the Euler Lagrange theorem.
  - It has a rather simple geometrical solution. The Wulffdiagram is the pedal curve of the γ plot.

- **The Wulff construction**
  - To minimize the energy requires to use the Euler Lagrange theorem.
  - It has a rather simple geometrical solution. The Wulff diagram is the pedal curve of the γ plot.

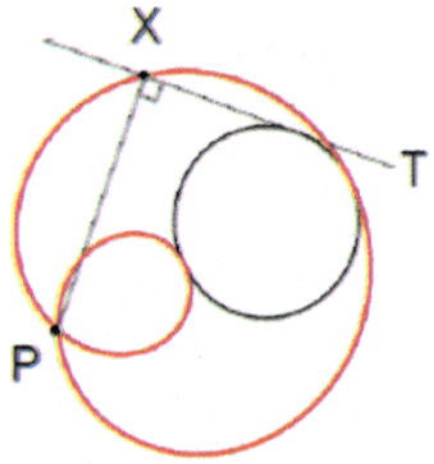

WIKIPEDIA: Pedal curve of a circle. Take a curve C and a fixed point *P* (called the **pedal point**). On any tangential line *T* there is a unique point *X* which is either *P* or forms with *P* a line perpendicular to *T*. **The pedal curve is the set of all *X* for which *T* is a tangent of the curve.**

For a parabola, the pedal point is the focus and the pedal curve is a line.

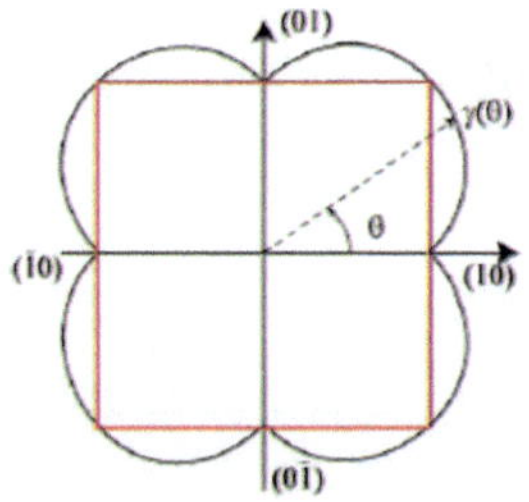

The Wulff construction of a squared structure using the nearest neighbor interactions

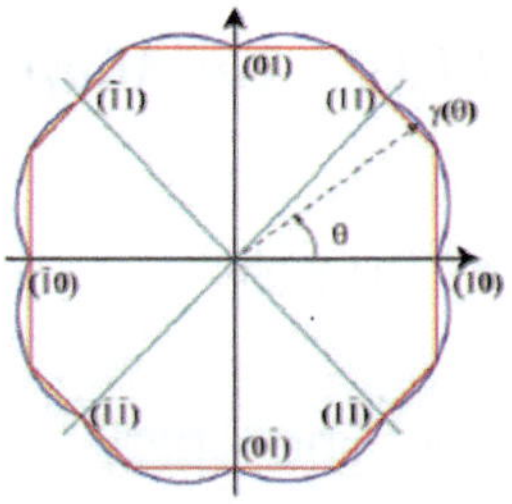

The Wulff construction of a squared structure using the next nearest neighbor interactions

- The Wulff formula (example of a hexagonal crystal)

- **The Wulff formula** (example of an hexagonal crystal)

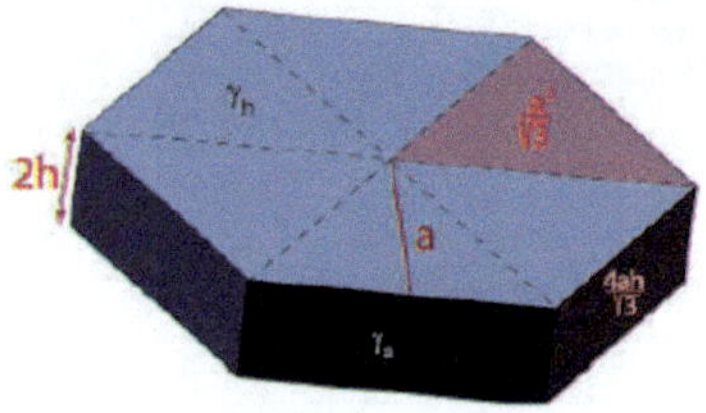

The formula is obtained by minimizing the surface energy at constant volume.

$$E=\sum_i \gamma_i E_i-\lambda(V_c-V_0)$$ where λ is the Lagrange multiplicator.

$$E=2\cdot 6\cdot\frac{a^2}{\sqrt{3}}\gamma_h+6\cdot\frac{4ah}{\sqrt{3}}\gamma_a-\lambda(2h.2\sqrt{3}a^2-V_0)$$

$$E=4\sqrt{3}a^2\gamma_h+8\sqrt{3}ah\gamma_a-\lambda(4\sqrt{3}a^2h-V_0)$$

Minimizing E with respect to a and h gives:

$$\frac{\partial E}{\partial a}=8\sqrt{3}a\gamma_h+8\sqrt{3}h\gamma_a-8\lambda\sqrt{3}ah=0 \quad (1)$$

$$\frac{\partial E}{\partial h}=8\sqrt{3}a\gamma_a-4\lambda\sqrt{3}a^2=0 \quad (2) \;\Rightarrow\; \lambda=\frac{2\gamma_a}{a} \quad (3)$$

Using (3) in (1) gives $\frac{\gamma_a}{a}=\frac{\gamma_h}{h}$

This result does not depend on the cluster shape: $\frac{\gamma_i}{h_i}=C^{te}$

- The morphology of clusters
  - The surface energy depends on the orientation of the surface.
  - The equilibrium shape is unique.
  - For a liquid, γ is constant and the equilibrium shape is a sphere in the absence of gravity.
  - For a solid, the total surface energy is $E_s = \sum_i \gamma_i dS_i$
  - The lowest energy is obtained for a polyhedra such that the distance from the center to the faces are proportional to their surface energy. $\frac{\gamma_i}{h_i} = C^{te}$

  - **The equilibrium shape for nanoclusters with size larger than 10 nm is the thermodynamic equilibrium shape.**
  - **For the smaller size, the γ_i are not constant and depend on the number of atoms on the edge.**

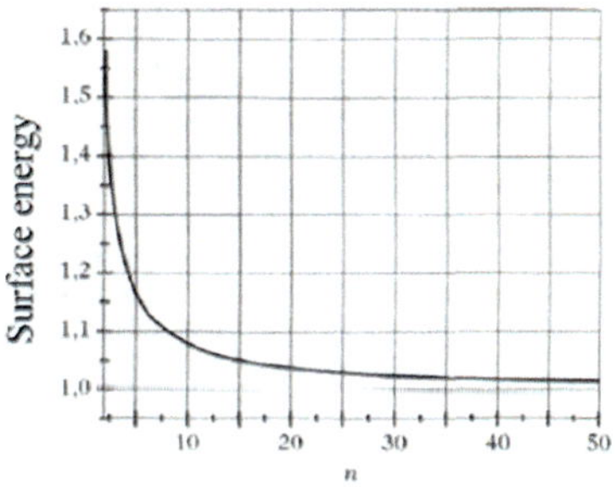

- The morphology of clusters: example of fcc materials
  - The smallest clusters have an icosahedral shape with the (111) face as the main faces, in order to decrease the energy.
  - An icosahedron is not an accepted equilibrium shape in crystallography since it has a 5 order symmetry. *The quasicrystals are another exception.*
  - As the cluster grows, it transforms to truncated octahedron with (111) and (100) faces.

Isocahedron

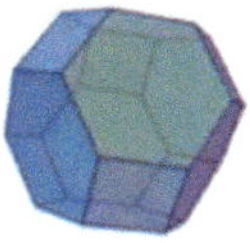

Truncated octahedron

- The equilibrium shape of supported clusters
  - There is an interaction between the cluster and the substrate.
  - This interaction is described using an energy of adhesion E_{adh}.
  - The equilibrium shape is a truncated Wulff polyhedra.

$$\frac{\Delta h_s}{h_i} = \frac{E_{adh}}{\gamma_i}$$

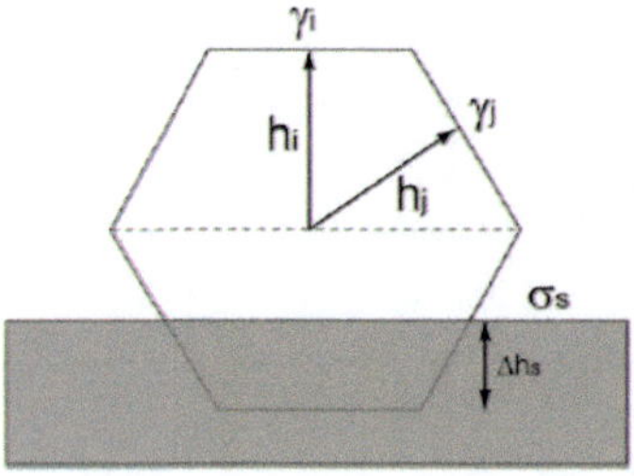

- If $E_{adh}=0$, then $\Delta h_s=0$.
- If E_{adh} is very large compare to the γi, the deposited material cluster perfectly wets the substrate and it becomes a film.

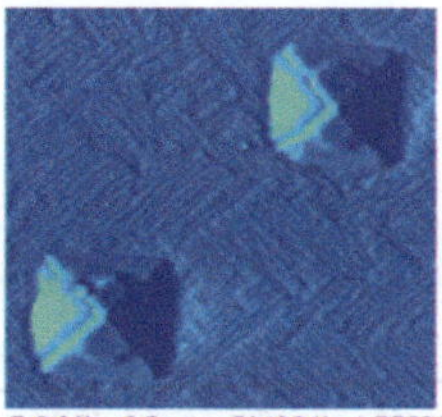

5.8 ML of Ge on Si (001) at 550°C

Appl. Phys. Letters 74, 994 (1999)

Usually, the substrate and the supported material have not the same lattice parameter and it leads to ***strain energy*** *that should be taken into account to calculate the equilibrium shape.*

- The equilibrium shape of supported clusters: Pd on MgO
  - It is a truncated octahedra.
  - It has (111) and (100) facets.

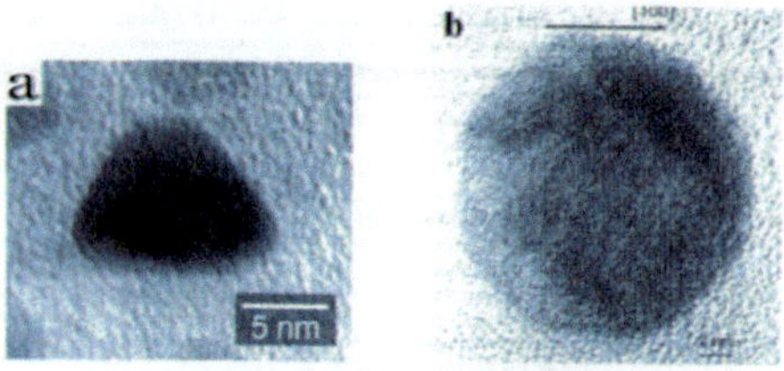

a. A view perpendicular to <110> direction

b. HRTEM. The distance between the (200) planes is 0.2 nm.

Graoui et al, **Surf. Sci 417**, 350, 1998

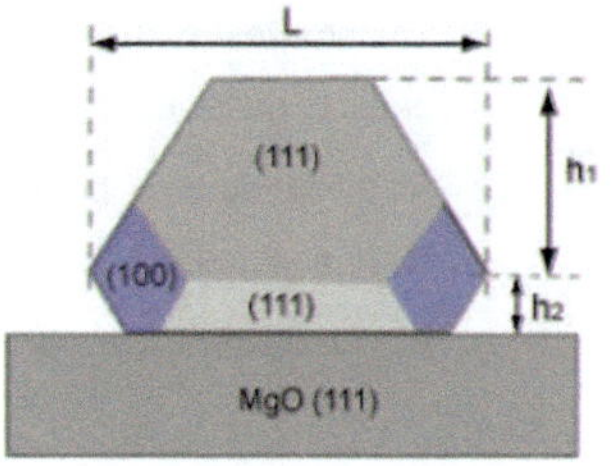

Scheme of a Pd cluster

$\Delta h_s = h_1 - h_2$

$$\frac{\Delta h_s}{h_1} = 0.55$$

$$\frac{\Delta h_s}{h_i} = \frac{E_{adh}}{\gamma_i}$$ → gives E_{adh}(111/100 MgO)=0.90 J.m^{-2}

In agreement with the calculations: E_{adh}(111/100 MgO)=**0.85** J.m^{-2}

- **The shape of embedded clusters: Xe bubbles in Aluminium**
  - The radius is about 1 nm.
  - The clusters exhibit (111) and (100) facets.
  - They can move and coalesce.
  - When two clusters coalesce, the resulting cluster rearrange its surface to minimize the energy.

With the courtesy of S.E. Donnelly, Salford University

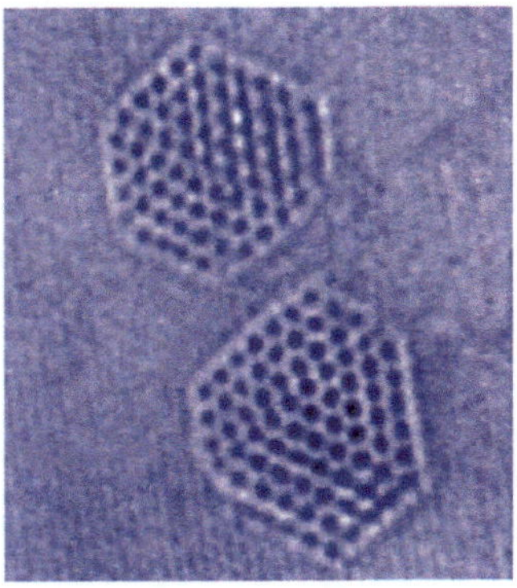

14.3.2 The Lattice Parameter

- The lattice parameter change
  - For a liquid sphere, the internal pressure is given by Laplace law: $P_{int} = P_{ext} + \frac{2\gamma}{r}$
  - For a solid, γ is replaced by the strain tensor σ. For a solid sphere of radius r and material with a constant value of γ, the strain tensor σ is: $\sigma = \gamma + \frac{d\gamma}{dS/S}$
  - The atomic volume is a^3, the total volume is $n.a^3$ and the compressibility is χ. $\chi = -\frac{1}{V}\frac{dV}{dP}$ $dP = \frac{2\sigma}{r}$
  - It leads to a decrease of the lattice parameter as the size of the particles increases (r^{-1} power law): $\frac{\Delta a}{a} = -\frac{2\chi\sigma}{3r}$

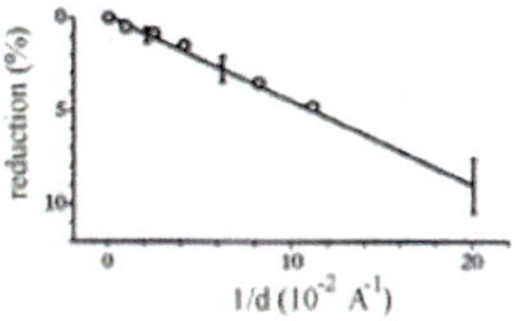

Cu clusters

The slope of the line gives the surface stress (3.35 GPa) and the pressure exerted on the nanocluster is terrific: 6.7 GPa.

*Apai et al, **Phys. Rev. Lettres 43**, 165, 1979*

- **The lattice parameter change: limitations**
  - The previous results are deduced from thermodynamics, i.e. when the numbers of particles in the studied system is very large.
  - Is thermodynamic still valid for nanoclusters?

- **Answer to this question:**
  - It is possible to numerically model the evolution of the lattice parameter using, not thermodynamics, but directly the interactions between particles, described by particular potentials.
  - The previous power law (R^{-1}) is valid till R=4 a_0 where a_0 is the bulk lattice parameter.
  - At lower size, the deviation is mainly due to the fact that the surface energy is no longer constant (more atoms on the edges).

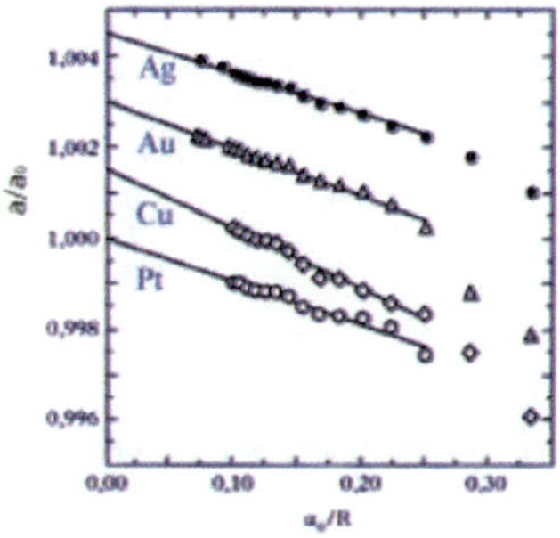

*Swaminarayan et al, **Surf. Sci. 306**, 367, 1994*

- **The equilibrium shape of supported clusters: Pd on MgO**
  - Due to the difference in lattice parameters between MgO and bulk Pd, it exists a strain energy.
  - This strain energy is responsible for a gradient of the lattice parameter in the Pd cluster.
  - The accommodation depth depends on the size of the cluster.

The **5 nm Pd cluster** has 12 atomic planes. The plane closer to the MgO substrate has a period equal to the MgO (100) planes (0.21 nm). The period of bulk Pd (0.195 nm) is reached for the third plane.

The **1.2 nm Pd cluster** possesses 4 planes. The period stays about the same as the MgO period.

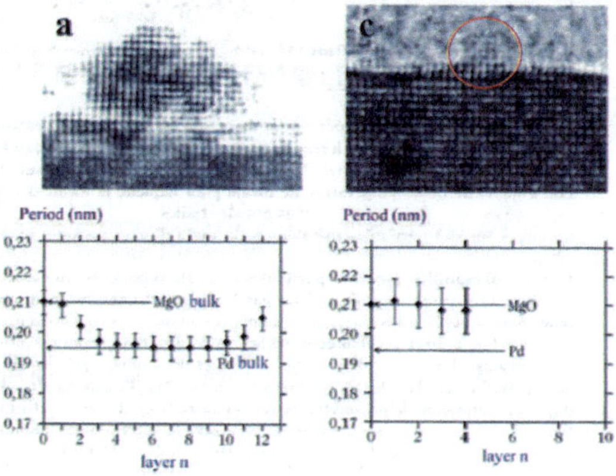

5 nm *1.2 nm*

*Giorgo et al, **Phil. Mag. B 67**, 773, 1993*

14.3.3 The Phase Changes

- The physical properties of nanomaterials depend on the grain size. Two types of effects:
  - Dimensional or confinement effect (the affected parameters evolutes non monotonously and exhibits extremes).
  - Surface effects that are described using the surface energy, as already mentioned. This surface energy evolutes monotonously and can be used in the frame of a thermodynamic point of view.
- The size of nanoclusters influences:
  - the shape,
  - the lattice parameter,
  - **and also the crystalline structure**.

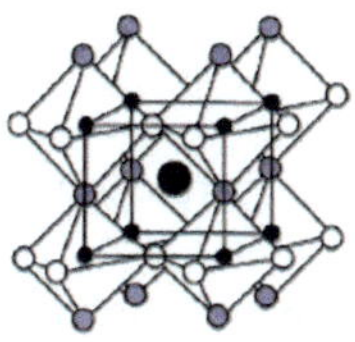

Tetragonal $BaTiO_3$

- Example of the Barium titanate $BaTiO_3$
  - a ferroelectric ceramic material
  - It has four phases as a solid: cubic, tetragonal, orthorhombic, and rhombohedric crystal structure.

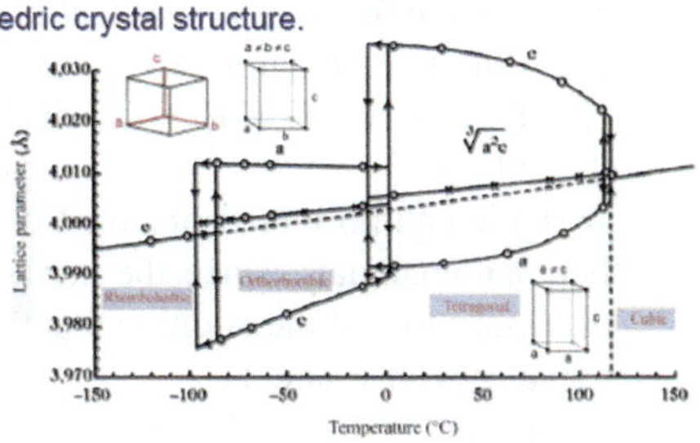

- Grains (powder) of $BaTiO_3$
  - If $d>d_c$ (80 nm), the structure is tetragonal. The ratio c/a decreases with the cluster size but the volume ca^2 is constant.
  - If $d<d_c$, the structure is cubic and the lattice parameter increases as the size decreases.

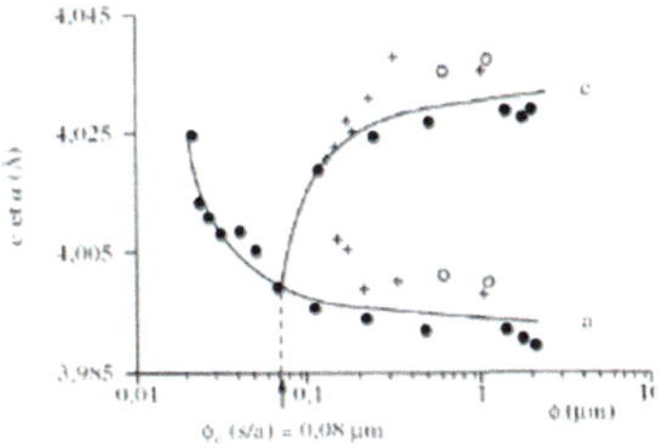

- Example of nickel
  - Bulk Ni is fcc, however, it has a metastable hcp phase.
  - Ni clusters have an hcp structure if the cluster diameter is less than 4 nm.

- Example of zirconia ZrO_2
  - Bulk zirconia is monoclinic at low temperature and becomes tetragonal at temperature higher than ~1100 °C.
  - Nanocrystals of zirconia (10nm) are tetragonal at room temperature.
  - If the cluster size is 7 nm, the transition from monoclinic to tetragonal occurs at 175 K.

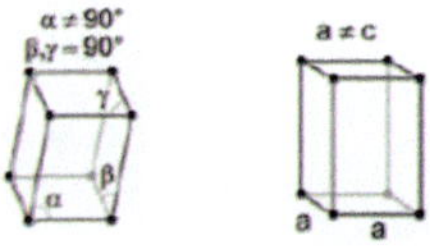

The size effect on the phase transition is a common characteristic of nanoclusters

- Elemental explanation using thermodynamics
  - The phase transitions at constant pressure P and temperature T are described using the free enthalpy G.
  - $H = E + PV$ and $G = H - TS$ where E is the energy, H is the enthalpy and S the entropy.
  - A stable equilibrium is obtained if G is minimal ($dG = 0$).
  - For nanomaterials, since the surface energy plays an important role, it is necessary to add the additional parameter D (diameter) or the curvature 2/D together with P and T.
  - Therefore, the free enthalpy becomes $G^* = G - (4\gamma/D)V$ and the stable state corresponds to $dG^* = 0$.
  - For a given (P, T) state, a phase transition occurs at a critical size D_c.
  - It is equivalent to say that the critical temperature T_c of the phase transition depends on the size of the nanoclusters. $T_c = f(P, D, \gamma\alpha, \gamma\beta)$ where $\gamma\alpha$ and $\gamma\beta$ are the interfacial energies of the two phases.

- Elemental explanation using thermodynamics
  - The phase transitions at constant pressure P and temperature T are described using the free enthalpy G.
  - H = E + PV and G=H-TS where E is the energy, H is the enthalpy and S the entropy.
  - A stable equilibrium is obtained if G is minimal (dG=0)
  - For nanomaterials, since the surface energy plays an important role, it is necessary to add the additional parameter D (diameter) or the curvature 2/D together with P and T.
  - Therefore, the free enthalpy becomes $G^*=G-(4\gamma/D)V$ and the stable state corresponds to $dG^*=0$.
  - For a given (P, T) state, a phase transition occurs at a critical size D_c.
  - It is equivalent to say that the critical temperature T_c of the phase transition depends on the size of the nanoclusters. $Tc=f(P, D, \gamma_\alpha, \gamma_\beta)$ where γ_α and γ_β are the interfacial energies of the two phases.

- Elemental explanation using thermodynamics
  - From a thermodynamical point of view, the size of the particles becomes a new intensive parameter, similarly to pressure and temperature.
  - $Tc = f(Pf(P, D, \gamma\alpha, \gamma\beta))$.
  - Application to the tetragonal → cubic phase transition of $BaTiO_3$.

- Elemental explanation using thermodynamics
  - From a thermodynamical point of view, the size of the particles becomes a new intensive parameter, similarly to pressure and temperature.
  - $Tc=f(P, D, \gamma_\alpha, \gamma_\beta)$
  - Application to the tetragonal →cubic phase transition of $BaTiO_3$.

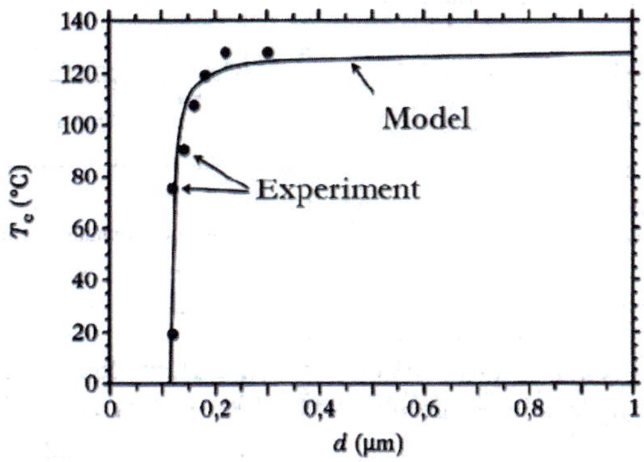

P. Labastie, F. Calvo, chapter 2, Les Nanosciences 2, Ed. Belin 2006

- Elemental explanation using thermodynamics
  - If the interfacial energies γ_α, γ_β plays a role; the temperature of transition should depend on the state of the interface.
  - A comparison of two types of $BaTiO_3$ grains demonstrates the role of these energies.

    Powder → interface solid-gas
    Ceramic → interface solid-solid.

- Elemental explanation using thermodynamics
  - If the interfacial energies γ_α, γ_β plays a role, the temperature of transition should depend on the state of the interface.
  - A comparison of two types of $BaTiO_3$ grains demonstrates the role of these energies.
    Powder → interface solid-gas
    Ceramic → interface solid-solid.
  - Application to the tetragonal →cubic phase transition of $BaTiO_3$.

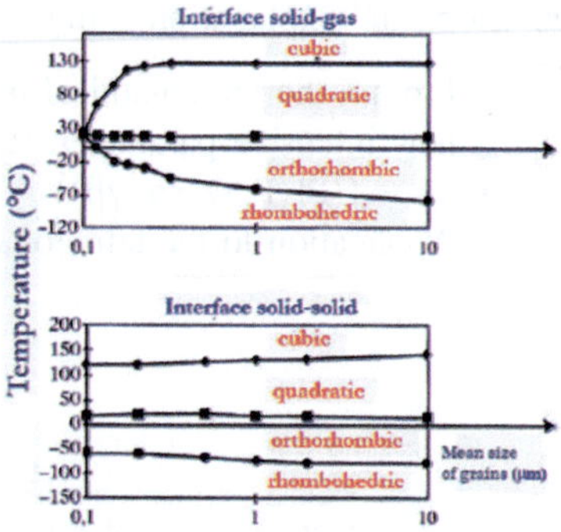

For given grain size and temperature, the stabilized phase depends on the nature of the interface.

The changes of the transition temperature with the grain size are rather high for the solid-gas interface. They are smaller for the interface solid-solid, where the interface energy is smaller.

Ref 13 in chapter 2, Les Nanosciences 2, Ed. Belin 2006

- The solid → liquid transition
  - Thermodynamics of nanosystems differs from the thermodynamics of macroscopic systems, where the number of particles tends to infinity.
  - In statistical mechanics, the partition function Z describes the statistical properties of a system in thermodynamic equilibrium. Most of the thermodynamic variables of the system, such as the total energy, free energy, entropy, and pressure, can be expressed in terms of the partition function or its derivatives.
  - There are actually several different types of partition functions, each corresponding to different types of statistical ensemble (or, equivalently, different types of free energy).

    The micro canonical ensemble: the system has no exchange with the environment and its energy is constant. All states are equiprobable.

The canonical ensemble: the number of particles is constant and the system can exchange energy with the environment (at a constant temperature).
The grand canonical ensemble: the system can exchange both heat and particles with the environment, at fixed temperature, volume, and chemical potential.

- The solid $\rightarrow$ liquid transition
  - Macroscopic level:
    The amount of energy in the form of heat released or absorbed by a substance during a phase change is the enthalpy of transformation (often called the Latent Heat L).
  - Microscopic level:

    The amplitude of vibration of atoms around their equilibrium position increases with temperature.
    f is the ratio of this amplitude and the distance between first neighbors.
    For the single crystalline solids, the solid $\rightarrow$ liquid transition occurs if f = 7% (it means that the first neighbor distance varies by 14%.

- The solid $\rightarrow$ liquid transition
  - Macroscopic level: The amount of energy in the form of heat released or absorbed by a substance during a phase change is the enthalpy of transformation (often called the Latent Heat L).
  - Microscopic level
    The amplitude of vibration of atoms around their equilibrium position increases with temperature.
    f is the ratio of this amplitude and the distance between first neighbors.
    For the single crystalline solids, the solid $\rightarrow$ liquid transition occurs if f=7% (it means that the first neighbor distance varies by 14%.

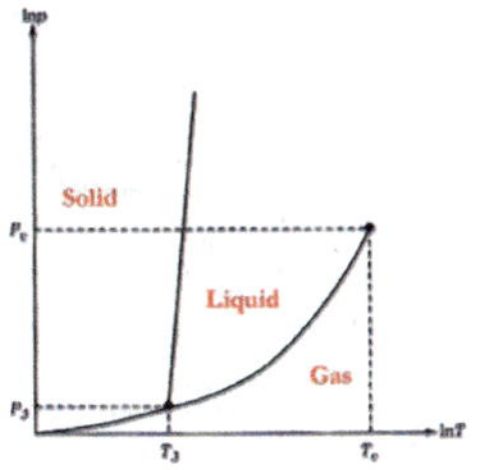

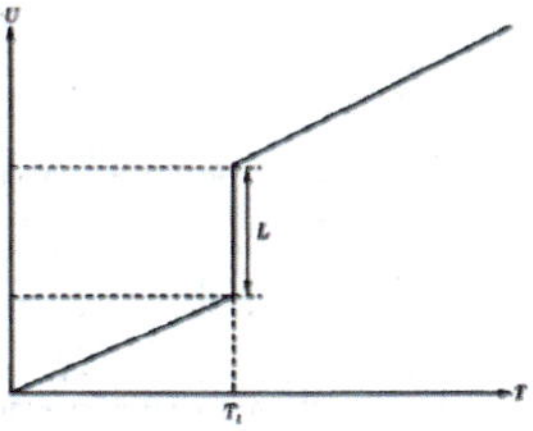

P. Labastie, F. Calvo, chapter 2, Les Nanosciences 2, Ed. Belin 2006

- Transition in a nanomaterial

- **Transition in a nanomaterial**
  - The temperature of the solid → liquid transition decreases as the size of the particles is reduced.
  - Classical approach

    A "Laplace" term is added to a Gibbs-Duheim equation.

$$1-\frac{T_f(r)}{T_f(\infty)}=\frac{2\alpha}{\rho_s L_\infty r}$$

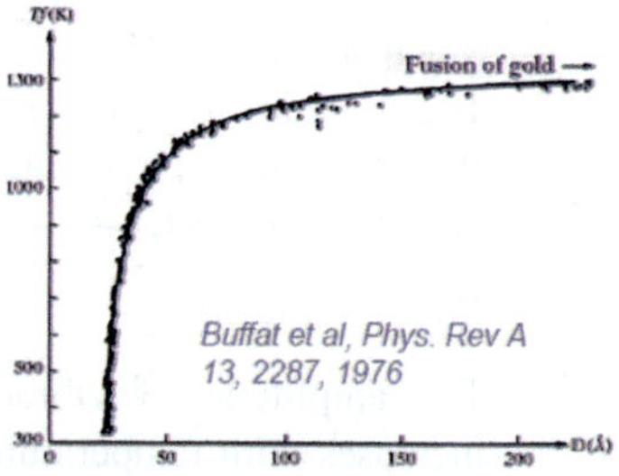

Buffat et al, Phys. Rev A 13, 2287, 1976

Where:
- $T_f(r)$ is the temperature of transition of the particle of radius r,
- $T_f(\infty)$ is the temperature of transition of the bulk material,
- ρ_s is the density of the bulk material,
- L_∞ is the enthalpy of transformation of the bulk material,
- α is a parameter which depends on the surface tensions γ_s and γ_L of the bulk solid and liquid material.

Assuming an equilibrium between the solid particles, the liquid and the vapor, it gives:

$$1-\frac{T_f(r)}{T_f(\infty)}=\frac{2\alpha}{\rho_s L(r) r_S}\left[\gamma_S-\gamma_L\left(\frac{\rho_S}{\rho_L}\right)^{\frac{2}{3}}\right]$$

- Where ρ_L is the density of the liquid and r_S the radius of the solid particles.
- It is in good agreement with the experimental values (solid line).

- **Transition in a nanomaterial**
  - It is experimentally observed that the transition occurs in an interval of temperature, contrary to the bulk.
  - As the size is reduced, the temperature of transition decreases, as well as the enthalpy of transition ΔH_m.

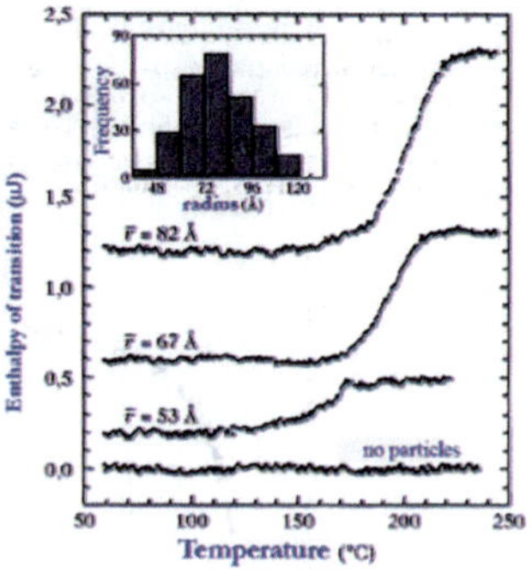

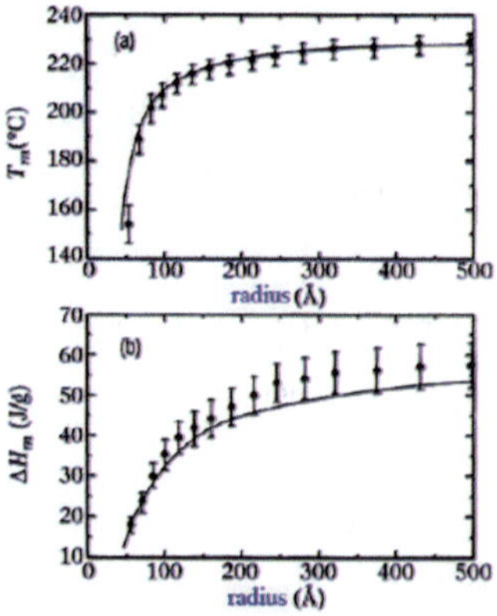

Solid → liquid transition of Sn nanoclusters.

Guo et al, phys; Rev. Letters 77, 99 1996.

- Transition in a nanomaterial
  - Assuming that the particles are either entirely liquid or entirely solid allows to express the enthalpy of transition as a function of the radius of the particles:

$$L(\infty)-L(r)=\frac{3}{r}\left(\frac{\gamma_S}{\rho_S}-\frac{\gamma_L}{\rho_L}\right)+\int_{T_f(r)}^{T_f(\infty)}(c_{P,L}(T)-c_{P,S}(T)\,dT$$

Where:
- $T_f(r)$ is the temperature of transition of the particle of radius r,
- $T_f(\infty)$ is the temperature of transition of the bulk material,
- ρ_s and ρ_L are the densities of solid and liquid bulk material,
- L_∞ is the enthalpy of transformation of the bulk material,
- L(r) is the enthalpy of a particle of radius r.
- γ_S and γ_L are the surface tensions of the bulk solid and liquid material.

  - The variation from the bulk value of the enthalpy of transition increases as the radius is decreased.
  - This is in rather good agreement with the experimental results (solid line).

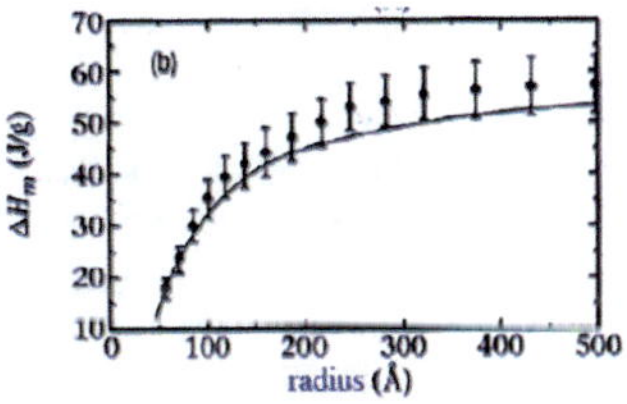

Solid → liquid transition of Sn nanoclusters.
Guo et al, phys; Rev. Letters 77, 99 1996.

- Transition in a nanomaterial
  - In small particles, the transition occurs in a range of temperature ΔT.
  - It has an intrinsic origin which results of the nanometric size.

Assuming a constant value of the enthalpy of formation and of the temperature of transition.

If N is the number of atoms in a particles, the probability for this particle to have an energy E at the temperature T is given by the Einstein equation:

$$p(E,T)=\frac{e^{\left(\frac{TS(E)-E}{kT}\right)}}{Z(T)}$$

Where Z(T) is the partition function and S(E) is the entropy.

$$Z(T)=\sum_i e^{-\frac{E_i}{kT}}$$

- s_s ans e_s are the entropy and energy per atom in a solid particle.
- s_ℓ and e_ℓ are the entropy and the energy per atom in a liquid particle.
- $\mathcal{L}$ being the enthalpy of formation per atom and T_f the temperature of transition, it means that $s_\ell = s_s + \mathcal{L}/T_f$ and $e_\ell = e_s + \mathcal{L}$.
- Therefore, the ratio of the probability to get a solid or liquid particle with an energy E at a temperature T is:

$$\frac{p_\ell}{p_s}=\frac{e^{\left(\frac{NTs_\ell-Ne_\ell}{kT}\right)}}{e^{\left(\frac{NTs_s-Ne_s}{kT}\right)}}=e^{\frac{N}{kT}\left[T(s_s+\frac{\mathcal{L}}{T_f}-s_s)-e_s-\mathcal{L}-e_s\right]}=e^{\frac{N}{kT}\left[\frac{T_\ell}{T_f}-\mathcal{L}\right]}=e^{\frac{N\mathcal{L}}{k}\left[\frac{1}{T_f}-\frac{1}{T}\right]}$$

- **Transition in a nanomaterial**
  - the ratio of the probability to get a solid or liquid particle with an energy E at a temperature T exhibits a huge variation on an interval ΔT such that $T=T_f \pm \Delta T$ with:

$$\frac{p_\ell}{p_s} = e^{\frac{N\mathcal{L}}{k}\left[\frac{1}{T_f}-\frac{1}{T}\right]}$$

$$\Delta T \approx k\frac{T_f^2}{N\mathcal{L}}$$

  - For a Sn particles with 20000 atoms, i.e. a size of 5.3 nm, it gives ΔT= 0.03K using $\mathcal{L}$= 4 10^{-21} J/at.
  - If the particle contains only 20 atoms, then ΔT= 30K.
  - In this size range, the evolution of the temperature of fusion and of the enthalpy of transition are no longer monotonous.
  - Complementary "interpretations" of ΔT:

    What is the meaning of fusion of a 20 sized cluster?

    More than half of the atoms are on the surface and classical thermodynamics is no longer valid.

    The stabilization of particular sizes could be due to a sort of quantum confinement.

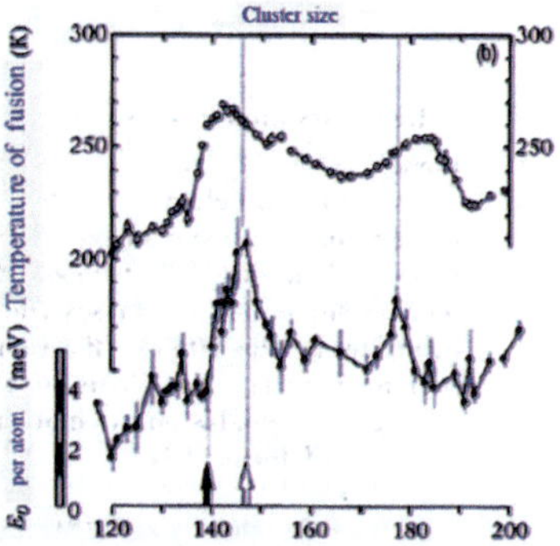

Solid → liquid transition of Na nanoclusters.
Schmidt et al, Phys. Rev. Letters 90, 103401, 2003

14.4 The Electronic Properties

- **The need of quantum mechanics**
  - The electronic properties of solids (energy levels, types of bonds between atoms, energy bands, energy gap, and Fermi level) are of primary importance.
  - The thermal, magnetic and optical properties of solids result from their electronic state.
  - The best description of the electronic properties is obtained using quantum mechanics, although classical descriptions can be used for macroscopic solids.
  - The basic steps of quantum mechanics studies for solids.
- The Schrödinger's equation.
- The energy levels and atomic orbitals for the hydrogen atom.
- Atoms with n electrons.
- Bonds between atoms, hybridization of atomic orbitals.
- Energy bands in solids.
- The Schrödinger's equation.
  - The probability to get the particle at location rat time t is given by $\Psi(r, t)\Psi^*(r, t)\, dV$ where $\Psi(r, t)$ is the wave function, $\Psi^*(r, t)$ the complex conjugate and $dV = r^2 \sin\theta\, dr\, d\theta\, d\phi$.

- Nowadays, the possible states are described using a ket vector IΨn > in the Hilbert space. The "conjugate" is the bra vector < ΨnI such that <ΨnIΨn> = 1.
- Therefore, the energy is a measurable quantity which is an eigenvalue of the operator Hamiltonien, known as the observable corresponding to the energy of the particle.
- Other observables are of primary importance to get the modern shape of the Schrödinger's equation:
- is the observable position.
- is the observable momentum.

$$\hat{H}|\Psi_n(t)\rangle = \frac{\hat{p}^2}{2m}|\Psi_n(t)\rangle + V(\hat{r},t)|\Psi_n(t)\rangle = i\hbar\frac{d}{dt}|\Psi_n(t)\rangle$$

The non-relativist time-dependent Schrödinger's equation.

- **The solution of the Schrödinger's equation**
  - The solution of the time-dependent Schrödinger's equation can be written as

- The solution of the Schrödinger's equation
  - The solution of the time-dependent Schrödinger's equation can be written as

$$|\Psi_n(t)\rangle = |\varphi_n\rangle e^{-i\frac{E_n t}{\hbar}}$$

  - The time-independent Schrödinger's equation is: $\hat{H}|\varphi_n\rangle = E_n|\varphi_n\rangle$

$$\hat{H}|\varphi_n\rangle = \frac{\hat{p}^2}{2m}|\varphi_n\rangle + V(\hat{r})|\varphi_n\rangle = E_n|\varphi_n\rangle$$

With $\hat{p} = -i\hbar\nabla$ where ∇ is the gradient operator.
V(r) is the potential energy operator.

  - The previous equation can be solved (quite rigorously) for the atom of hydrogen.

- **The solution of the Schrödinger's equation for the atom of hydrogen**
  - The potential energy is the interaction potential between the proton and the electron.

$$V(r) = -\frac{1}{4\pi\varepsilon_0}\frac{e^2}{r}$$

- The states Iϕn> can be described using quantic numbers n, l and m_l and can be written as In, l, ml>.
- n is the principal quantum number (n = 1, 2, 3, 4, …). It is an eigenvalue that denotes for the average distance between the electron and the nucleus (i.e. the electron shell).
- l is the orbital quantum number (l = 0, 1, …, n − 1). It corresponds to the eigenvalues l(l + 1) of the operator L^2 where L is the orbital angular momentum operator. The quantic number l specifies the shape of an atomic orbital and strongly influences chemical bonds and bond angles.

  l = 0 is called an s orbital.
  l = 1 is called a p orbital.
  l = 2 is called a d orbital.
  m_l is the orbital magnetic quantum number (m_l = −l, −l+1, …, 0, …, l−1, l). It is the eigenvalue of L_z, the z component of the orbital angular momentum operator.

- The solution of the Schrödinger's equation for the atom of hydrogen
  - The potential energy is the interaction potential between the proton and the electron.

$$V(r) = -\frac{1}{4\pi\varepsilon_0}\frac{e^2}{r}$$

  - The states Iφ_n> can be described using quantic numbers n, ℓ and m_ℓ and can be written as In,ℓ,m_ℓ>.
  - n is the principal quantum number (n = 1, 2, 3, 4 ...). It is an eigenvalue that denotes for the average distance between the electron and the nucleus (i.e. the electron shell).
  - ℓ is the orbital quantum number (ℓ = 0, 1 ... n−1). It corresponds to the eigenvalues $\ell(\ell+1)$ of the operator L^2 where L is the orbital angular momentum operator. The quantic number ℓ specifies the shape of an atomic orbital and strongly influences chemical bonds and bond angles.

    ℓ=0 is called an *s* orbital.

    ℓ=1 is called a *p* orbital.

    ℓ=2 is called a *d* orbital.

  - m_ℓ is the orbital magnetic quantum number (m_ℓ = − ℓ , − ℓ +1 ... 0 ... ℓ −1, ℓ). It is the eigenvalue of L_z, the z component of the orbital angular momentum operator.

- The solution of the Schrödinger's equation for the atom of hydrogen
  - The eigenvalues E are expressed as a function of the quantum number n.

$$E(r) = -\frac{E_0}{n^2} \text{ with } E_0 = \frac{1}{2\hbar_2}\frac{m_e e^2}{(4\pi\varepsilon_0)^2}$$

  - The "wavefunctions" or eigenstates $|n, \ell, m_\ell>$ have the form:

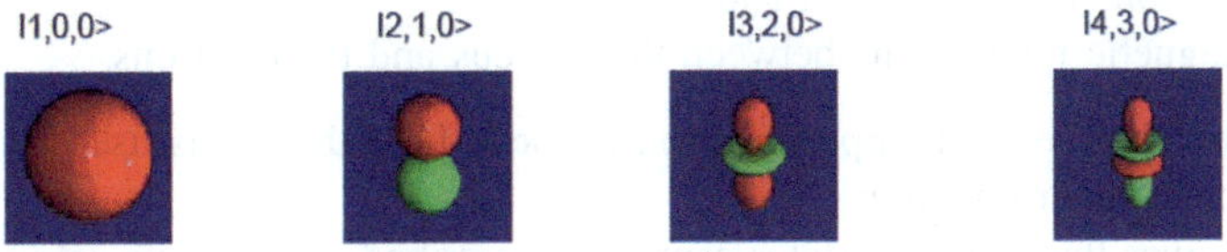

$$\left|\Psi_{n,\ell,m_\ell}(r,\theta,\varphi)\right\rangle = R_{n,\ell}(r)\,Y_\ell^{m_\ell}(\theta,\varphi)$$

where $Y_\ell^{m_\ell}(\theta,\varphi)$ are the spherical harmonics.

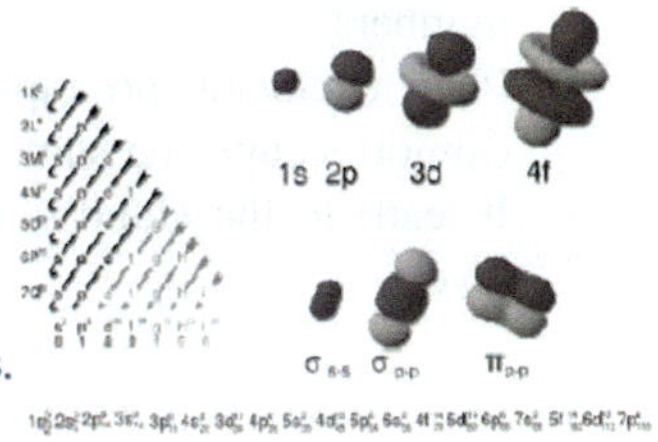

- The solution of the Schrödinger's equation for hydrogen
  - The Stern and Gerlach experiment has shown that an additional intrinsic angular momentum of the electron must be introduced to describe the electronic states of atoms.

  - This intrinsic angular momentum is called the spin.
  - The total angular momentum is J=L+S.
  - The spin of an electron is ½.
  - The eigenvalues of the operator S^2 are $s(s+1)$.
  - S_z is the component of the spin operator on Z axis. Its eigenvalues are m_s such that $(m_s = -s, -s+1 \ldots 0 \ldots s-1, s)$.
  - For an electron, $m_s = \pm½$.
  - The eigenstates can be written as $|n, \ell, m_\ell, s, m_s>$ or $|n, \ell, m_\ell, ½, m_s>$ for hydrogen.

- **Atoms with N electrons**
  - There are many interactions:

    Electrostatic interactions between electrons and the nucleus,
    Electrostatic interactions between electrons,
    Magnetic interactions between spin of electrons and orbital motion of electrons (coupling of L and S),
    Magnetic interactions between the nucleus and the electrons.

  - The Hartree-Fock approximation (a mean field theory) is used to solve the Schrödinger equation.
  - It gives that the eigenstates cannot be described by the same set of quantum numbers.
  - Pauli exclusion principle: two identical fermions (electrons are fermions) cannot occupy the same quantum state simultaneously.
  - It leads to the Hund's rules for the distribution of electrons on the energy levels.

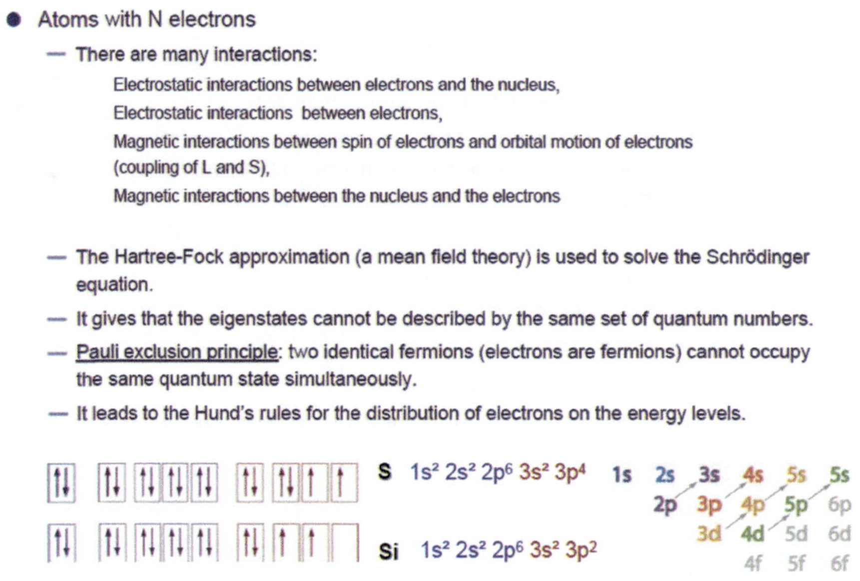

- Atomic bonding
  - Covalent bonding
    Sharing of one or more electrons between two atoms,
    The external energy shell of each atoms becomes saturated
    Example: CH_4
  - Ionic bonding
    Atoms looses or gains electrons to complete the external energy shell,
    They become ions. Example: NaCl
  - Polar covalent bond
    Due to the difference in electro negativity, the molecule is electrically polarized (Example: HCl)
  - Metallic bond
    Sharing of electrons between all atoms, interactions between ions and electrons.
    Example: sodium
  - Low intensity bonds
    In many molecules with covalent bonds, the distribution of charges is not homogenous and it leads to a weak permanent electrical dipole.
    The interactions between these dipoles are the origin of Van der Waals bond as well as the Hydrogen bond (a "bridge" between polar molecules).

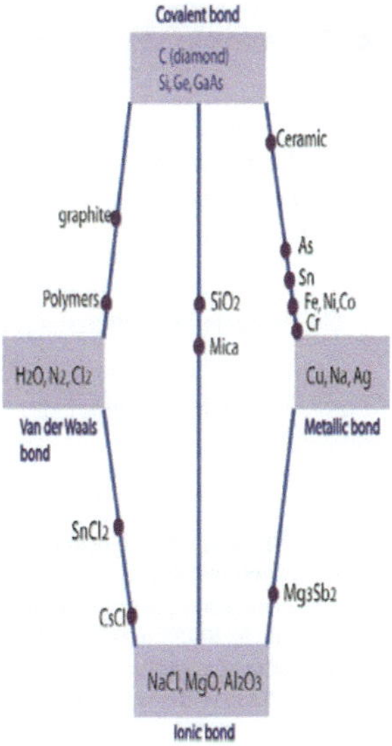

- Hybridization
  - The atoms combine their atomic orbitals to form new hybrid orbitals, so that the total energy is decreased.
  - The sp3 hybridization (diamond)

$$C \;\; \underset{1s}{\uparrow\downarrow}\; \underset{2s}{\uparrow\downarrow}\; \underset{2p_x}{\uparrow}\; \underset{2p_y}{\uparrow}\; \underset{2p_z}{\_\_} \;\Longrightarrow\; C^{*} \;\; \underset{1s}{\uparrow\downarrow}\; \underset{sp^3}{\uparrow}\; \underset{sp^3}{\uparrow}\; \underset{sp^3}{\uparrow}\; \underset{sp^3}{\uparrow}$$

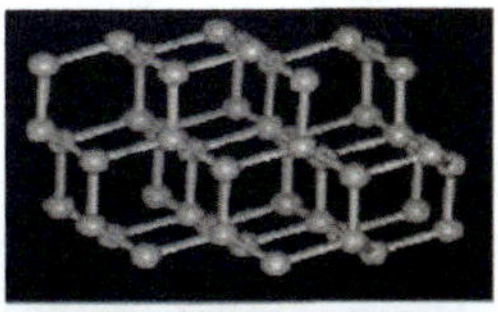

  - The sp2 hybridization (graphite)

$$C \;\; \underset{1s}{\uparrow\downarrow}\; \underset{2s}{\uparrow\downarrow}\; \underset{2p_x}{\uparrow}\; \underset{2p_y}{\uparrow}\; \underset{2p_z}{\_\_} \;\Longrightarrow\; C^{*} \;\; \underset{1s}{\uparrow\downarrow}\; \underset{sp^2}{\uparrow}\; \underset{sp^2}{\uparrow}\; \underset{sp^2}{\uparrow}\; \underset{p}{\uparrow}$$

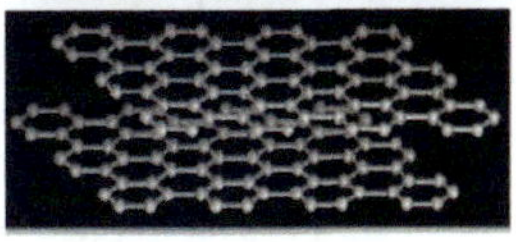

  - From the graphite to the fullerenes

  - From the fullerenes to the nanotubes

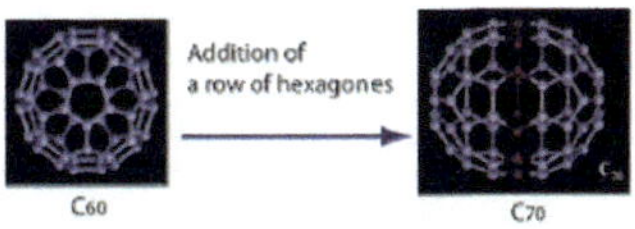

http://lesnanotubes.ibelgique.com/

- **Electrons in solids, the energy bands**

J.S. Blakemore, solid state physics, Saunders Comp. 1974

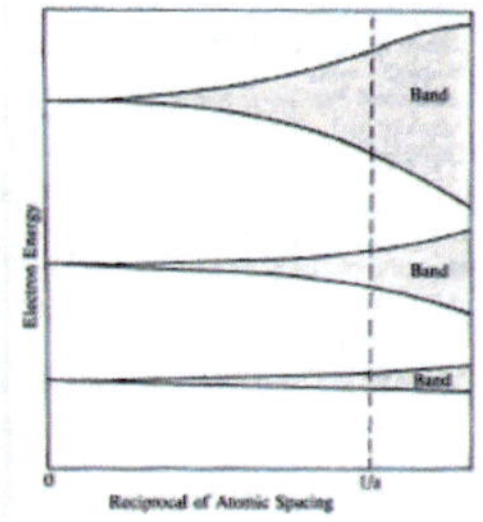

  - In an isolated atoms, the electrons are moving in the potential distributed on quantified energy levels, according to the Pauli principle.
  - In solids, the distance between atoms is low and the electron's orbitals interact. It leads to a broadening of the energy levels and to energy bands.

  - The inner-core electrons form narrow bands called internal bands.
  - The electrons from the external shells or valence shells form the valence bands.
  - The conduction bands corresponds to the excited states of the electrons.

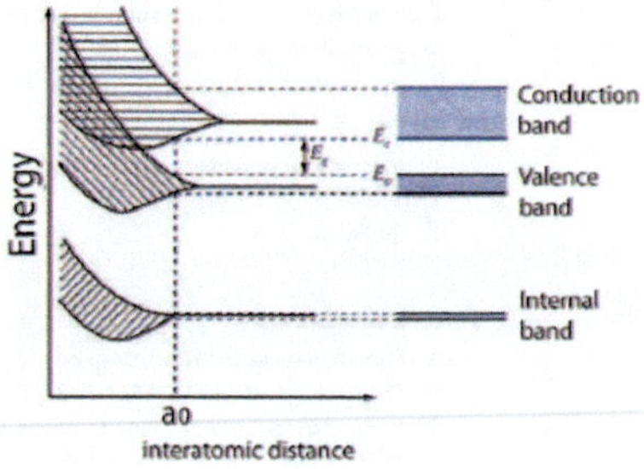

In « Les matériaux », JP Bailon, Montréal, 2000

- **Electrons in solids, the energy bands, the Fermi level**
  - The energy gap E_g is the difference between the upper value of the valence band and the lower value of the conduction band.

  - In solids with covalent and ionic bondings, E_g is large. E_g=1.12 eV for Si.
  - In most metals, the energy gap is equal to zero as the conduction band and the valence band overlap.

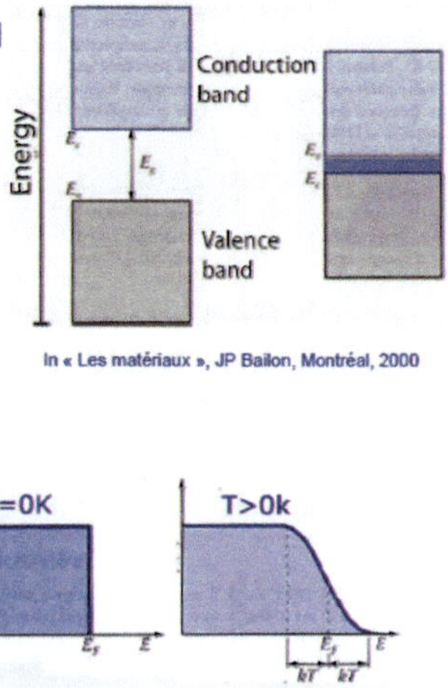

In « Les matériaux », JP Bailon, Montréal, 2000

  - The Fermi level or Fermi energy is the maximum energy for the electrons at 0K.
  - It plays an important role in physics as the physical properties of materials are mainly governed by the electrons that have an energy larger than this energy.
  - E_F value is a few eV.

    The Fermi function gives the probability to occupy a state of energy E.

    For metals, its value ranges from 3 to 7 eV.

    The population of electrons n(E)*p(E) depends upon the product of the Fermi function and the electron density of states.

$$p(E) = \frac{1}{1 + \exp^{\frac{E - E_F}{kT}}}$$

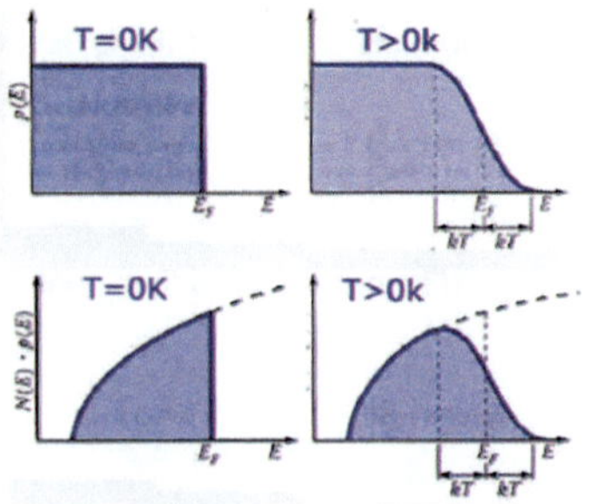

- **Electrons in solids, the energy bands, the Fermi level**
  - The Fermi level for insulators as Si semiconductors is halfway between the valence band and the conduction band. The valence band is completely filled and the sizable energy gap (1.12 eV for Si) prevents any conduction by electrons passing in the conduction band.
  - For intrinsic semiconductors, the Fermi level is halfway between the valence and conduction bands separated by a narrow energy gap (0.66 eV for Ge). Although no conduction occurs at 0 K, at higher temperatures a finite number of electrons can reach the conduction band and provide some current. In doped semiconductors, extra energy levels are added.
  - In conductors, the Fermi level is in the overlap region and electrons are present in the conduction band.

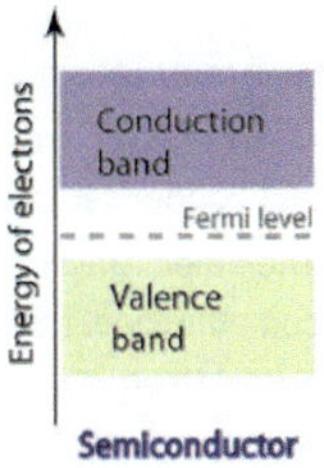

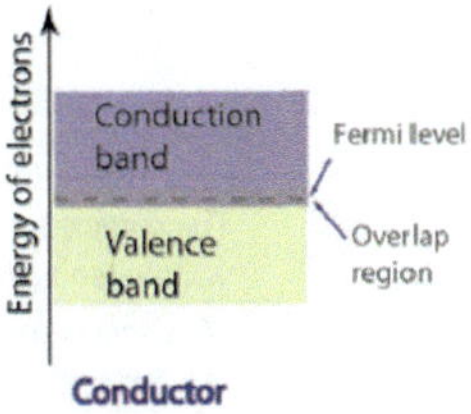

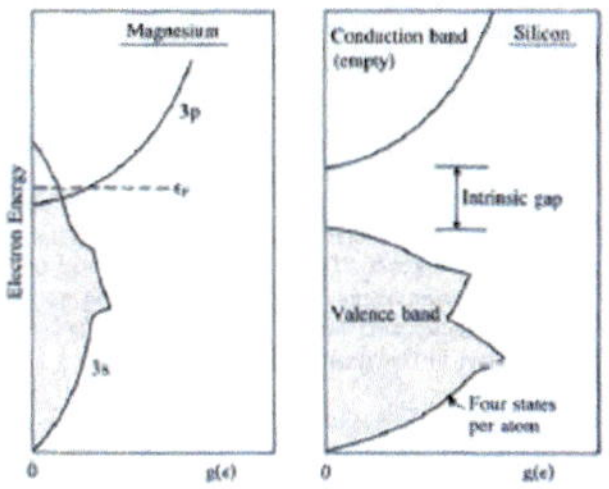

- **Electrons in crystalline solids, the Brillouin zones**
  - In crystalline solids, the electrons are moving in a periodic potential due to the nucleus and other electrons.
  - This potential has the periodicity of the array. The Bloch's Theorem states that the wavefunctions have the same periodicity.

$$\psi_k(\bar{r}) = u_k(\bar{r})\exp^{ik.\bar{r}} \qquad u_k(\bar{r}) = u_k(\bar{r} + \bar{T})$$

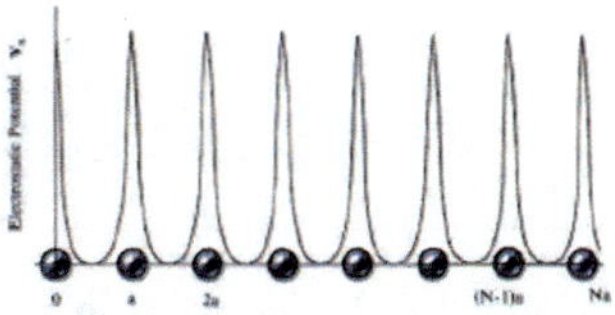

Periodic potential for a linear monoatomic lattice.
J.S. Blakemore, solid state physics, Saunders Comp. 1974

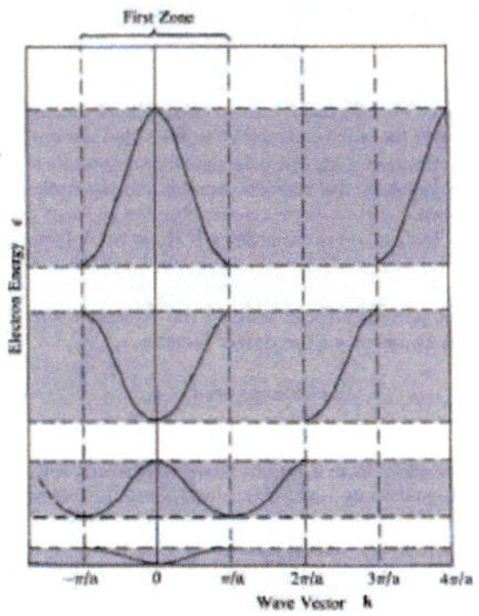

  - The periodicity in the reciprocal space allows to define shells which have the "dimension" of the reciprocal lattice.
  - The discontinuity of E(k) at k=G, where G is a vector of the reciprocal lattice, explains the apparition of forbidden bands.

- And in nanomaterials ?
  - Nanoparticles are a connection between isolated atoms and condensed matter.
  - Their electronic structure are deduced from computational methods starting from the electronic Hamiltonian and using several assumptions.

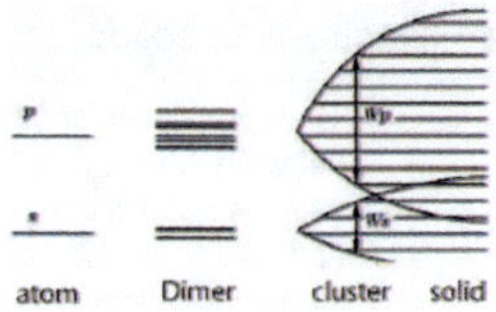

$$H_{el}(R_a, r_i) = -\sum_{i}^{N}\left(-\frac{1}{2}\Delta_i - \sum_{a}^{n}\frac{Z_a}{|r_i - R_a|}\right) + \sum_{i<j}^{N}\frac{Z_a Z_b}{|r_i - r_j|}$$

  - Metallic nanoclusters

    The confinement of the valence electrons, which are almost free electrons in a quasi spherical finite volume leads to the quantification of their energy, similarly to the energy of nucleons in nucleus.

    The liquid droplet model, where the nucleus is described as a fluid constituted of nucleons confined in a finite volume by the "Strong interactions ». It leads to discrete values of the energies of nucleons.

- Metallic nanoclusters
  - Structure of nucleus

    The nuclear droplet has an electrical charge and a surface.

    The binding energy per nucleon (the work required to put the nucleon outside the nucleus) depends on the mass number.

    The nucleus is stabilized when the A=2 Z.

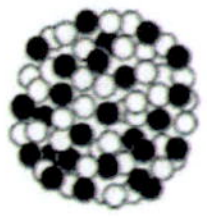

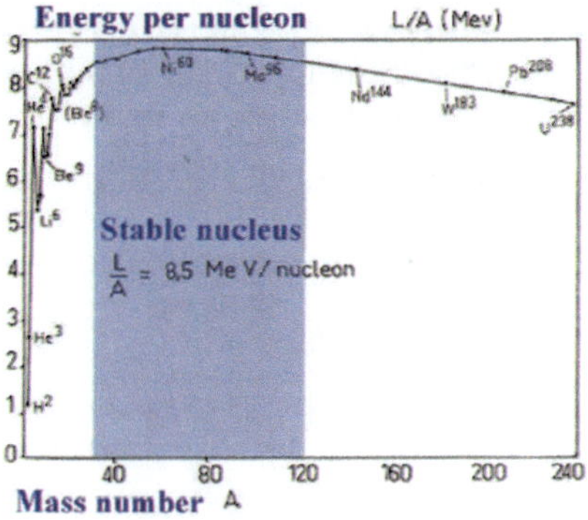

P. Morel, Physique quantique, Hermann, Paris 1969

- Metallic nanoclusters

  - Structure of metallic clusters

    In a Na cluster beam, cluster with **magic numbers** N=8, 20, 40, 58 are favored.

    Following the nucleus nomenclature, the shell model assume that all electrons in a given shell have the same radial quantum number n and angular momentum numbers ℓ. A shell can contain 2 (2ℓ+1) electrons.

    When the electrons in a cluster exactly fill an energy level, it gives a closed shell cluster.

    The closed shell clusters have a lower energy. It means that it is relatively more costly to add one more atom to a closed shell than to an open shell cluster.

    Extended theoretical models (ab initio methods) improve the electronic description.

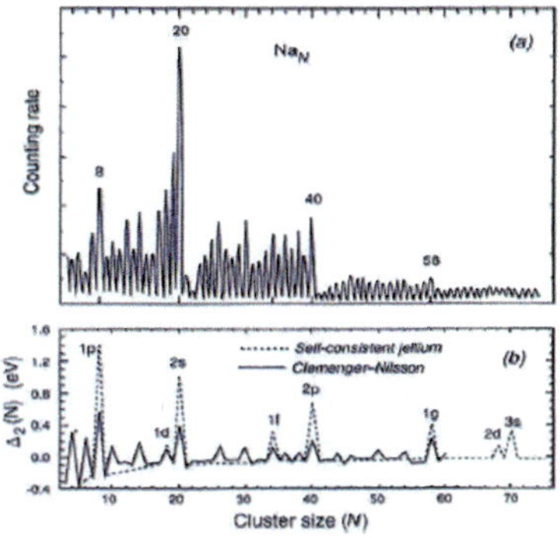

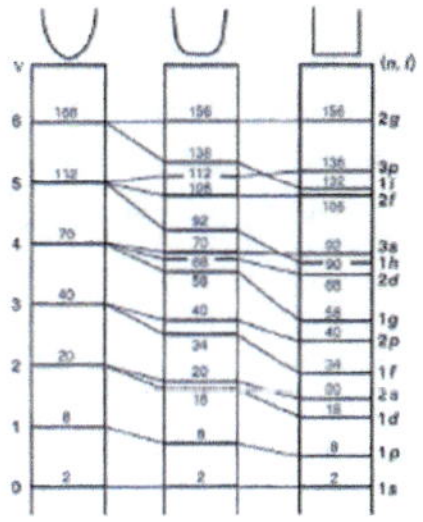

K.H. Meiwes-Broer, Metal clusters at surface, Springer 2000

- Quantum dots
  - In a quantum dot crystal, the energy gap is size dependent and can be altered to produce a range of energies between the valence and conduction band. Modeling using quantum mechanics gives that the gap of a quantum dot is enlarged.

  - In solids, energy levels in a band are separated by such small amount of energy that for some processes, they may be treated as if they were continuous. It works well for semiconductor crystals with large numbers of atoms and physical dimensions much greater than the exciton Bohr radius (a few nm, depending on the material).

  - In nanoclusters, the energy levels are well separated and the addition or subtraction of one atom or electron of the nanocrystal measurably change the electronic properties. It is the case of semiconductor crystals with a small number of atoms and physical dimensions on the order of 10 nanometers.

  - When a semiconductor crystal has discrete states, it can be defined as a quantum dot.
  - Quantum dots have useful, interesting and tunable properties.

***Exciton Bohr radius**: the natural physical separation in a crystal between an electron in the conduction band and the hole it leaves behind in the valence band. The size of this radius controls how large a crystal must be before its energy bands can be treated as continuous.*

- Quantum well
  - A few nm thick multilayers with different energy gaps. The material with the narrowest width constitute a potential well quite infinite in the layers plane and with nanometric size in the perpendicular direction (GaAs/Al_x Ga_{1-x} As).
  - It leads to narrower emission bands and it is used for laser diode.

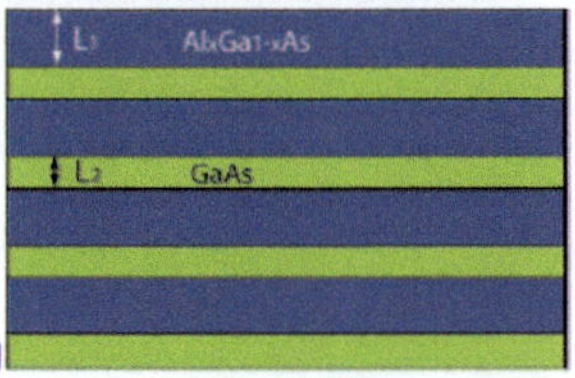

  - The discrete values of the energies can be tuned adjusting the dimension L.
  - If the confinement layers are without impurities, the mobility of electrons is high and these nanomaterials can be used as high frequency devices.
  - If the two layers have a periodicity L = L_1+ L_2, it leads to new Brillouin zones and eventually new forbidden bands. It is the origin of the "gap engineering" in microelectronics.

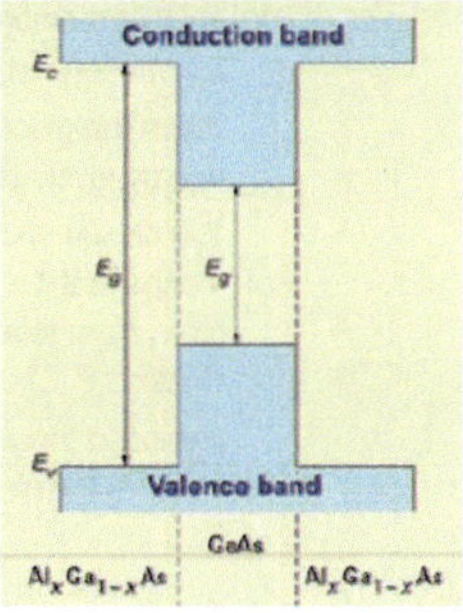

P. Costa, Nanomatériaux, Propriétés et applications Techniques de l'ingénieur

- Metallic nanoclusters in an insulator
  - The resistivity of such materials depends on the volumic concentration of the metallic material and the evolution changes as percolation occurs.
  - For x >x_0 , conducting path exist within the composite material and it has a metallic response with a low resistivity (about 10^{-6} Ω.m) which increases with temperature.
  - For x < x_0 , the conduction proceeds by jumps or tunneling effect within the insulator. The resistivity is drastically higher and it decreases with temperature.

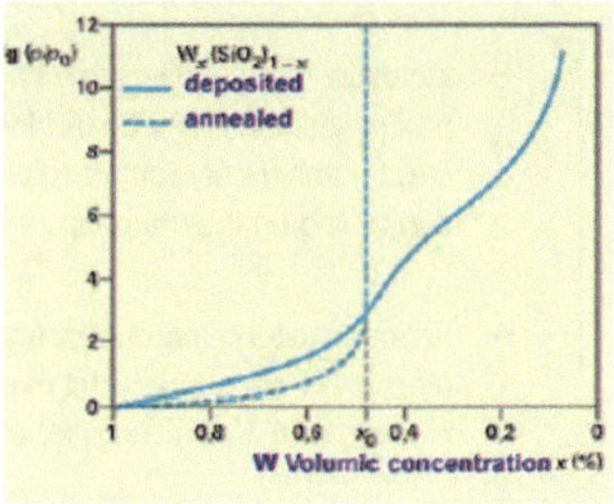

Abeles et al, Phys. Rev. Lett. vol. 35, p. 247 (1975).

14.5 The Magnetic Properties and Classifications of Magnetic Nanomaterials

The nanostructure may be two-phase, in which nanocrystallites are distinct phases from the matrix, or the ideal case in which both the crystallites and the matrix are made of the same material.

The magnetic behaviour of most experimental realizable systems of is result of contributions from both interaction and size effects. The correlation between nanostructure and magnetic properties suggests a classification of nanostructure morphologies. The following classification is designed to emphasize the physical mechanisms responsible for the magnetic behaviour. Figure 14.2 schematically represents the classification. At one extreme are systems of isolated particles with nanoscale diameters, which are denoted by type A. These non interacting systems derive their unique magnetic properties strictly from the reduced size from the components, with no contribution from interparticle interactions. At the other

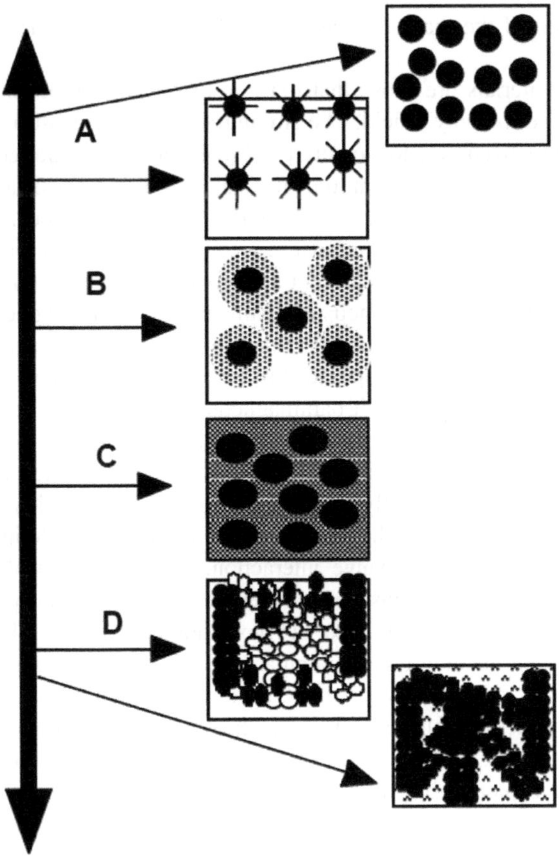

Fig. 14.2 Schematic representation of the different types of magnetic nanostructures. **Type-A** materials include the ideal ultrafine particle system, with interparticle spacing large enough to approximate the particles as noninteracting. **Ferrofluids**, in which magnetic particles are surrounded by a surfactant preventing interactions, are a subgroup of Type-A materials. **Type-B** materials are ultrafine particles with a core-shell morphology. **Type-C** nanocomposites are composed of small magnetic particles embedded in a chemically dissimilar matrix. The matrix may or may not be magnetoactive. **Type-D** materials consist of small crystallites dispersed in a noncrystalline matrix

extreme are bulk materials with nanoscale structure denoted by type D, in which a significant fraction (up to 50%) of the sample volume is composed of grain boundaries and interfaces. In contrast to type A systems, magnetic properties here are dominated by interactions. The length of the interactions can span many grains and is critically dependent on the character of interface. Due to this dominance of interaction and grain boundaries, the magnetic behaviour of type D nanostructures cannot be predicted simply by applying theories of polycrystalline materials with reduced length scale. In type B particles, the presence of a shell can help prevent particle interactions, but often at the cost of interaction between the core and the shell. In many cases the shells are formed via oxidation and may themselves be magnetic. In type C materials, the magnetic interactions are determined by the volume fraction of the magnetic particles and the character of the matrix they are embedded in.

Magnetic materials are those that exist in a state of permanent magnetization without the need to apply a field. The strength of a magnet is measured in terms of saturation magnetization and coercivity (H_c is the field required to reduce the magnetization to zero from saturation, and is applied in the opposite direction to the original saturating field) values. These values increase with a decrease in the grain size and an increase in the specific surface area (surface area per unit volume) of the grains. Therefore nanomaterials present also good properties in this field. There are three main categories of magnetism: diamagnetism, paramagnetism and ferromagnetism. Diamagnetism is a fundamental property of all atoms and the magnetization is very small and opposed to the applied magnetic fields direction. However, many materials exhibit paramagnetism, where a magnetization develops parallel to the applied magnet field as the field is increased from zero, but the strength of the magnetization is small. Ferromagnetism is the property of those materials which are intrinsically magnetically ordered and which develop spontaneous magnetization without the need to apply a field. The ordering mechanism is the quantum mechanical exchange interaction. Magnets made of nanocrystalline yttrium-samarium-cobalt grains possess very unusual magnetic properties due to their extremely large surface area. Typical applications for these high-power rare-earth magnets include quieter submarines, automobile alternators, land-based power generators, and motors for ships, ultra-sensitive analytical instruments, and magnetic resonance imaging (MRI) in medical diagnostics. Bulk gold and Pt are non-magnetic, but at the nano size they are magnetic. Surface atoms are not only different to bulk atoms, but they can also be modified by interaction with other chemical species, that is, by capping the nanoparticles. This phenomenon opens the possibility to modify the physical properties of the nanoparticles by capping them with appropriate molecules. Actually, it should be possible that non-ferromagnetic bulk materials exhibit ferromagnetic-like behavior when prepared in nano range. One can obtain magnetic nanoparticles of Pd, Pt and the surprising case of Au (that is diamagnetic in bulk) from non-magnetic bulk materials. In the case of Pt and Pd, the ferromagnetism arises from the structural changes associated with size effect (Fig. 14.3).

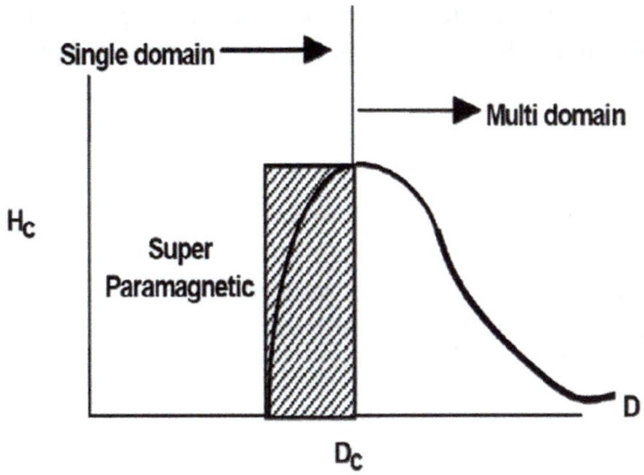

Fig. 14.3 Magnetic properties of nanostrucutred materials

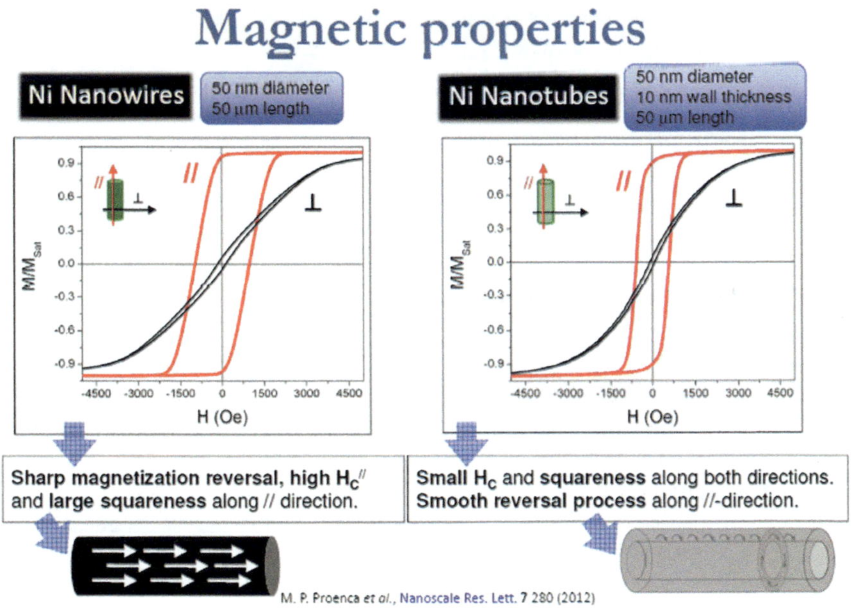

However, gold nanoparticles become ferromagnetic when they are capped with appropriate molecules: the charge localized at the particle surface gives rise to ferromagnetic-like behavior. Surface and the core of Au nanoparticles with 2 nm in diameter show ferromagnetic and paramagnetic character, respectively. The large spin-orbit coupling of these noble metals can yield to a large anisotropy and therefore exhibit high ordering temperatures. More surprisingly, permanent magnetism was observed up to room temperature for thiol-capped Au nanoparticles. For nanoparticles with sizes below 2 nm the localized carriers are in the 5d band. Bulk Au has an extremely low density of states and becomes diamagnetic, as is also the case for bare Au nanoparticles. This observation suggested that modification of the d band structure by chemical bonding can induce ferromagnetic like character in metallic clusters.

- Magnetism of solids
  - The magnetic momentum of an atom has two contributions:

    The magnetic orbital momentum $m_\ell = \mu_B \ell$ where μ_B is the Bohr magneton and ℓ is the orbital momentum. $\ell = 0, 1, 2$ and 3 for the electron shells *s*, *p*, *d* and *f*.

    The spin magnetic momentum for an electron is $m_s = -2\mu_B S$ where S is the spin quantum number which has the value ±1/2.

    $$\mu_B = \frac{e\hbar}{2m} = 9.274\ 10^{-24}\ \mathrm{J.K^{-1}}$$

  - The quantum description of magnetism

    It results from the partial filling of the energy levels. The manner in which electrons are assigned to various quantum states within a partially filled subshell is determined by the Pauli principle and the Hund's rule. **A magnetic atom has some unpaired electrons and therefore a non-vanishing permanent magnetic momentum.**

    It results also from the coupling between electrons from the various atoms.

    Magnetic properties of a solid are described by a magnetization $\mathcal{M}$ per unit volume.

    $$M = \mu_B(L + 2S) = g\ \mu_B\ J \qquad g = 1 + \frac{J(J+1) + S(S+1-L(L+1)}{2J(J+1)}$$

    J=L+S is the total angular momentum and g is the Lande splitting factor.

    The magnetic momentum associated with the nucleus can be neglected (except when looking at the hyperfine structure).

    $$\mu_N = \frac{e\hbar}{2M_p} = 5.051\ 10^{-27}\ \mathrm{J.K^{-1}}$$

- **The different types of magnetism**

- The different types of magnetism
  - The most interesting forms of magnetism are the strong forms which arise in solids having a permanent magnetization.
  - Diamagnetism: ↑↓ ↑↓ ↑↓ ↑↓

    For material in which all electrons are in paired states, there is <u>zero magnetic momentum</u>.

    If a magnetic field H is applied, the response is the appearance of a magnetization per unit volume $\mathcal{M}$ such that $\mathcal{M} = \chi_{dia}$ H (assuming a linear response). χ_{dia} is the diamagnetic susceptibility.

    According to Lentz law, χ_{dia} is negative.

    All materials exhibits diamagnetism but, since χ_{dia} is low, it it is hindered by the other type of magnetism.

    Copper is a diamagnetic material and $\chi_{dia} = -9.5\ 10^{-3}$.

  - Paramagnetism: ↑↓ ↑↓ ↑ ↑

    Atoms in paramagnetic materials have a <u>permanent magnetic momentum</u> M randomly oriented.

    In the absence of external magnetic field, the macroscopic magnetization $\mathcal{M}$ is equal to zero.

    When a magnetic field is applied, it induces a preferential orientation and $\mathcal{M} = \chi_m$ **H**.

    χ_m depends on temperature (proportional to 1/T in the weak field approximation).

    Aluminium is a paramagnetic material with $\chi_m = 2.2\ 10^{-3}$ at 300K.

- The different types of magnetism
  - Ferromagnetism

Ferromagnetic materials exhibit permanent magnetization (below a temperature known as the Curie point, 1043 K for Fe, 627K for Ni, 1388K for Co) which involves the cooperative alignment of permanent atomic moments.

The two conditions for a material to be ferromagnetic are:

- Atoms have a magnetic momentum.
- An internal interaction between the spins tends to align these spins through an exchange field B_E called the Weiss field. B_E can be as large as 10^3 T. **The range of these interactions is a few interatomic distances, up to 1-2 nm in rare earth elements.**

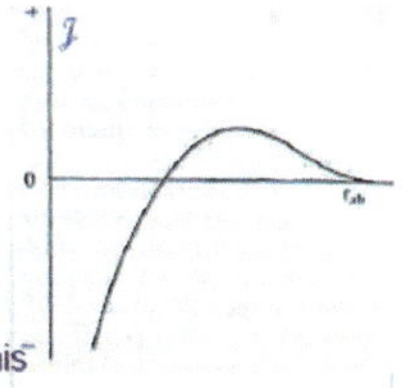

Using a mean field approximation, the exchange field is $B_E=\lambda \mathcal{M}$.
The interaction energy between two atoms 1 and 2 can be written as $\Delta E=-2\mathcal{J} S_1S_2$ where $\mathcal{J}$ is an exchange integral (corresponding to the overlapping of the 3d orbitals) and S_1 and S_2 are the spins of the two atoms.
Quantum mechanic shows that the exchange integral $\mathcal{J}$ is a negative quantity for a small distance between neighboring magnetic atoms. This makes parallel spins energetically unfavorable. However, when this distance is larger than a critical distance (about three times the nearest neighbor distance), $\mathcal{J}$ becomes positive (Fe, Co and Ni are ferromagnetic but not Cr).

Elements which are paramagnetic can form ferromagnetic alloys (Cu-Mn, Cu-Mn-Sn, Mn-Sb)

J.S. Blakemore, solid state physics, Saunders Comp. 1974

From the « KITTEL »

- The different types of magnetism
  - Ferromagnetism and Weiss domains

If all magnetic moments of the atoms in a magnetic material where aligned, the magnetic energy would be very large. When it is divided into domains, it lowers the total energy.

Thus, a ferromagnetic solid can appear non magnetized on a macroscopic scale.

The number of domains and the shapes and relationships between domains are determined by the crystal directions in which spin alignment can occur and by the requirement that energy is a minimum. Domain size is typically 1 μm.

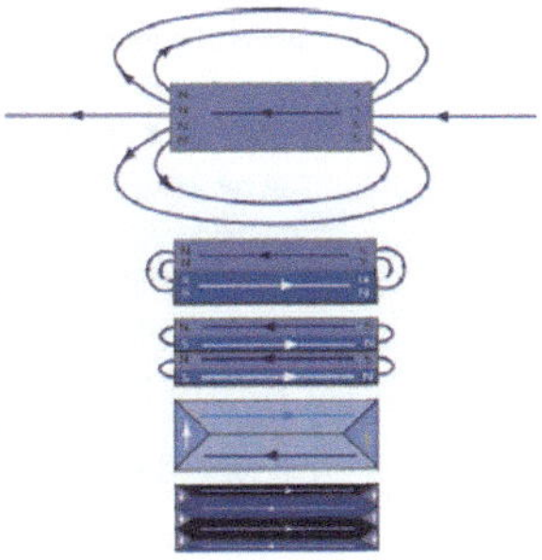

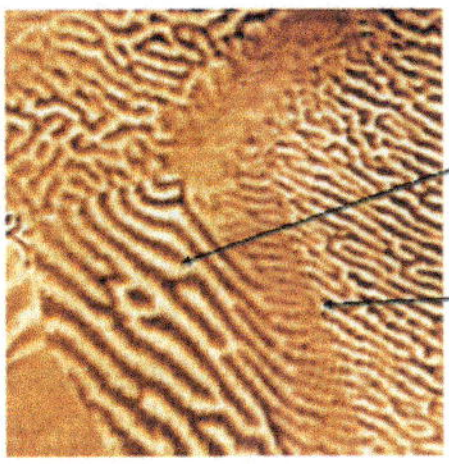

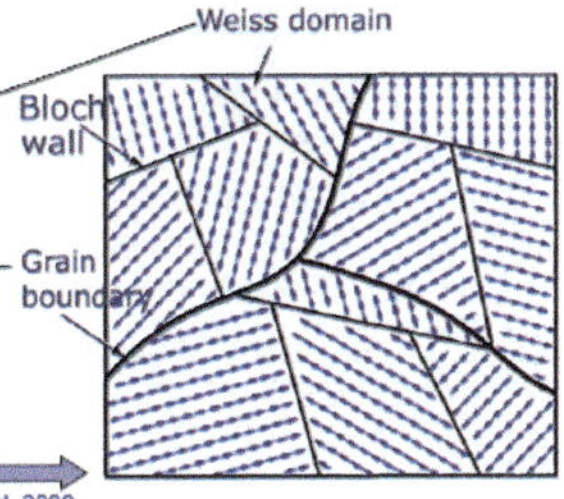

MFM image of nitrided ASS 316L, C. Tromas, LMP, Poitiers

In « Les matériaux », JP Bailon, Montréal, 2000

- The different types of magnetism
  - Ferromagnetism and Weiss domains

Each domain is separated from neighboring domains by a domain wall called Bloch wall. The thickness of a Bloch wall is about 300 nm for a « soft » material (easy magnetization) and 5 nm for a « hard » material (resistance to magnetization).

The Bloch wall width can be defined by $\delta = \mathcal{J}^{1/2} K^{-1/2}$ where $\mathcal{J}$ is an exchange integral and K an anisotropy constant.

The exchange length is $\lambda = \mathcal{J}^{1/2} / \mathcal{M}$ where $\mathcal{M}$ is the magnetization.

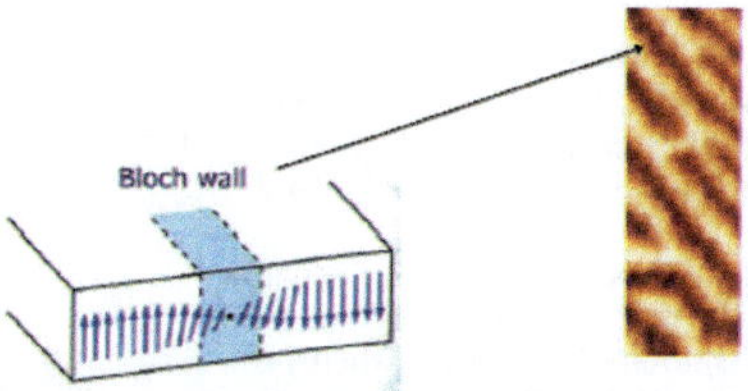

In « Les matériaux », JP Bailon, Montréal, 2000

- The different types of magnetism
  - Ferromagnetism and coercive field

When an external field H is applied, Bloch walls move in a manner that enlarges domains in which the magnetization $\mathcal{M}$ is close to the direction of this field.

The resulting field $B = \mu_0 (H + \mathcal{M})$ increases till a saturation value which involves wall movements (low applied field) and domain rotation (high applied field).

The motion of Bloch walls is partly reversible if the applied field is decreased and a remanent magnetization is observed.

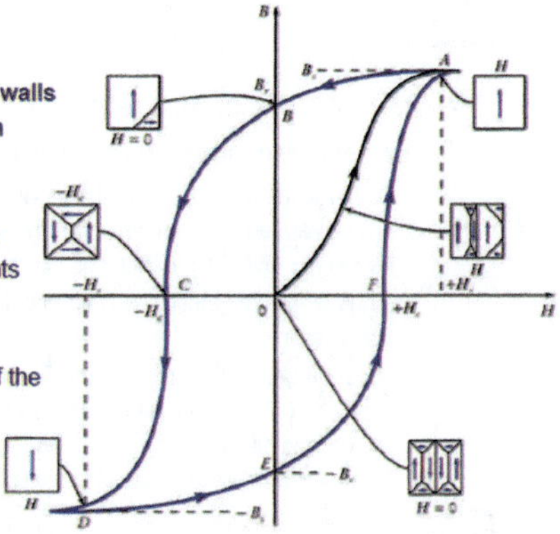

The magnetization goes to zero if an opposite field H_c is applied. **H_c is called the coercive field** ($H_c = 2K/M_s$ where M_s is the magnetization at saturation).

Magnetic anisotropy of Fe, Co and Ni is low and the Bloch walls easily move. These elements are "soft" materials. The rare earth materials ($Nd_2Fe_{14}B$) are "hard" since their anisotropy is large, giving a high coercive field.

An alternation between positive and negative applied fields produces the Hysteresis loop.

- The different types of magnetism
  - Antiferromagnetism

    There is no display of gross magnetization because neighboring atoms displays anti-parallels spins.

    The two directions originate from two magnetic sub-lattices that oppose the magnetization from each other.

    The antiferromagnetism breaks at a temperature called the Néel temperature.

    Example: MnO. The moments on a (111) plane are anti-parallels to the moments of the adjacent (111) planes.

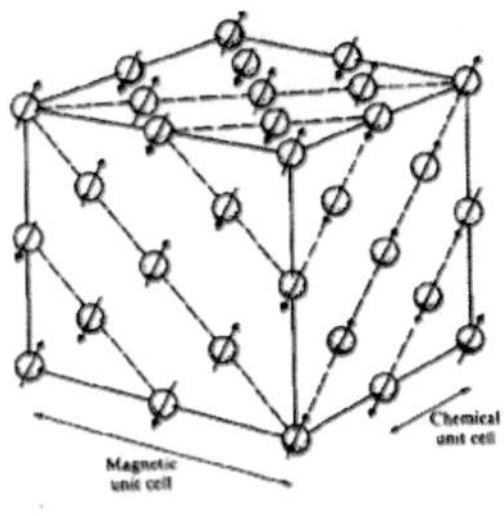

  - Ferrimagnetism

    The two groups of sub-lattices have unequal magnetic moments.

    There is a spontaneous magnetization at low temperature given by the difference between the magnetization of the two sub-lattices.

    Example: Ferrite Fe_2O_3.

- Magnetism in nanomaterials
  - Since the domain size is typically 1 μm, it is assumed that the nanoclusters are single domain and complications due to domain structure can be ignored.
  - When the system size is of the order of magnitude of λ and δ, the formation of domain walls requires too much energy. Thus, the magnetization remains in the single domain state. The magnetization reverse by uniform rotation.

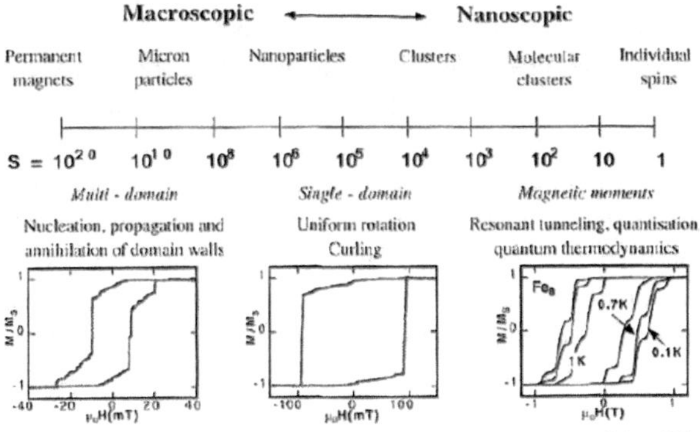

K.H. Meiwes-Broer, Metal clusters at surface, Springer 2000

- *At smallest size below λ and δ, one must take into account the band structure and its modifications at the particle's boundaries. When the size is so small that one must consider individual atoms, tunneling effects take place.*

- Magnetism in nanomaterials
  - The magnetization is larger in nanoclusters. For example, M=2.08 μ_B in Co nanoclusters and 1.72 μ_B in bulk Co.

    Due to the lowest number of atoms in nanoclusters, the band width is smaller and favor the jump of electrons from one band to another one, thus enhancing magnetization.

    However, the interatomic distance is sometimes reduced and the magnetization can be reduced.
  - Paramagnetic materials become ferromagnetic.

    Bulk rhodium (32 atoms) is not magnetic while nanoclusters of Rh have a magnetization M=1 μ_B. This is attributed to the shrinkage of the d band.
  - Soft ferromagnetic materials become harder

    Nanoclusters are more anisotropic.

    The rotation of the magnetic moments occurs in an entire grain, much lower than the width of the Bloch wall of the bulk soft material. It needs more energy and the coercive field is larger.

    Nanomaterials better withstand changes of magnetization.

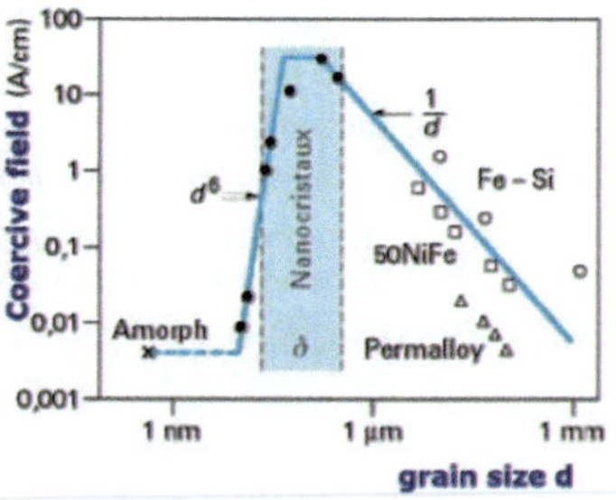

P. Costa, Nanomatériaux, Propriétés et applications Techniques de l'ingénieur

14.6 The Optical Properties

One of the most fascinating and useful aspects of nanomaterials is their optical properties. Applications based on optical properties of nanomaterials include optical detector, laser, sensor, imaging, phosphor, display, solar cell, photocatalysis, photo electrochemistry and biomedicine. The optical properties of nanomaterials depend on parameters such as feature size, shape, surface characteristics, and other variables including doping and interaction with the surrounding environment or other nanostructures. Likewise, shape can have dramatic influence on optical properties of metal nanostructures. Figure 14.4 exemplifies the difference in the optical properties of metal and semiconductor nanoparticles. With the CdSe semiconductor nanoparticles, a simple change in size alters the optical properties of the nanoparticles. When metal nanoparticles are enlarged, their optical properties change only slightly as observed for the different samples of gold nanospheres in Fig. 14.4. However, when an anisotropy is added to the nanoparticle, such as growth of nanorods, the optical properties of the nanoparticles change dramatically.

Nanocrystalline systems have attracted much interest for their novel optical properties, which differ remarkably from bulk crystals. Key contributory factors include quantum confinement of electrical carriers within nanoparticles, efficient energy and charge transfer over nanoscale distances and in many systems a highly

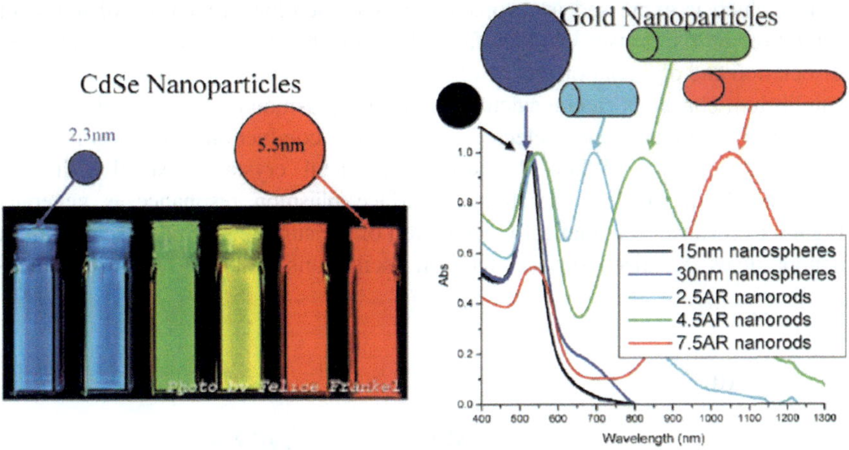

Fig. 14.4 Fluorescence emission of (CdSe) ZnS quantum dots of various sizes and absorption spectra of various sizes and shapes of gold nanoparticles (*Chem. Soc. Rev.*, 2006, 35, 209–217)

enhanced role of interfaces. The linear and nonlinear optical properties of such materials can be finely tailored by controlling the crystal dimensions, and the chemistry of their surfaces, fabrication technology becomes a key factor for the applications. Surface plasmons, also known as surface plasmon polaritons, are surface electromagnetic waves that propagate parallel along a metal/dielectric (or metal/vacuum) interface. Since the wave is on the boundary of the metal and the external medium (air or water for example), these oscillations are very sensitive to any change of this boundary, such as the adsorption of molecules to the metal surface. At an interface between two transparent media of different refractive index (glass and water), light coming from the side of higher refractive index is partly reflected and partly refracted. Above a certain critical angle of incidence, no light is refracted across the interface, and total internal reflection is observed. While incident light is totally reflected the electromagnetic field component penetrates a short (tens of nanometers) distance into a medium of a lower refractive index creating an exponentially detenuating evanescent wave. If the interface between the media is coated with a thin layer of metal (gold), and light is monochromatic and p-polarized, the intensity of the reflected light is reduced at a specific incident angle producing a sharp shadow (called surface plasmon resonance) due to the resonance energy transfer between evanescent wave and surface plasmons. Resonance conditions are influenced by the material adsorbed onto the thin metal film. Satisfactory linear relationship is found between resonance energy and mass concentration of biochemically relevant molecules such as proteins, sugars and DNA. The SPR signal which is expressed in resonance units is therefore a measure of mass concentration at the sensor chip surface. This means that the analyte and ligand association and dissociation can be observed and ultimately rate constants as well as equilibrium constants can be calculated. Surface plasmon resonance is the coherent excitation of all the "free" electrons within the conduction band, leading to an in-phase oscillation. When the size of a metal crystal is smaller than the wave-length of incident radiation, a surface plasmon resonance is generated. Figure 14.5 shows schematically generation of the surface plasmon oscillation. For nanoparticles, localized surface plasmon oscillations can give rise to the intense

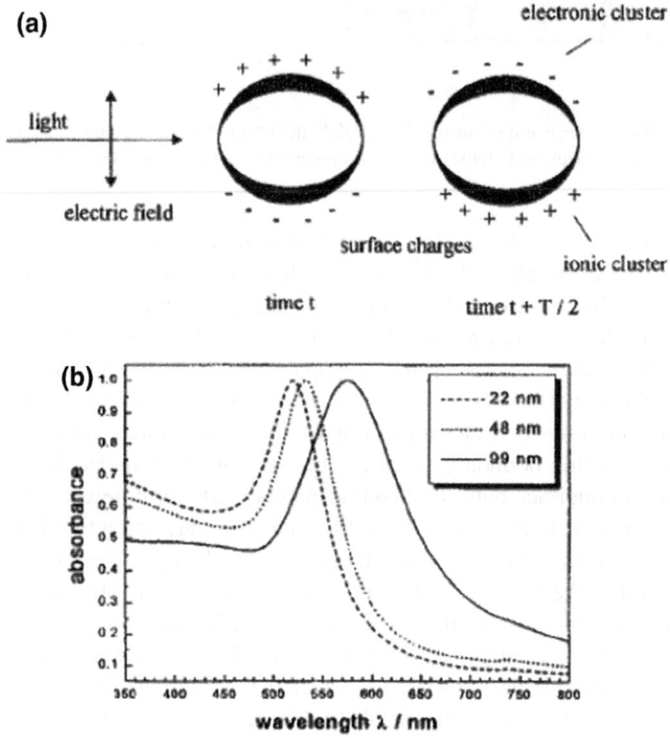

Fig. 14.5 Surface plasmon absorption of spherical nanoparticles and its size dependence. **a** A schematic illustrating the excitation of the dipole surface plasmon oscillation. The electric field of an incoming light wave induces a polarization of the (free) conduction electrons with respect to the much heavier ionic core of a spherical metal nanoparticle. A net charge difference is only felt at the nanoparticle surfaces, which in turn acts as a restoring force. In this way a dipolar oscillation of the electrons is created with period *T*. **b** Optical absorption spectra of 22, 48 and 99 nm spherical gold nanoparticles. The broad absorption band corresponds to the surface plasmon resonance (from S. Link, M.A. El-Sayed *Int. Rev. Phys. Chem.*, 2000, v. 19, 409)

colors of solutions of plasmon resonance nanoparticles and/or very intense scattering. Nanoparticles of noble metals exhibit strong ultraviolet—visible absorption bands that are not present in the bulk metal. Shifts in this resonance due to changes in the local index of refraction upon adsorption of biopolymers to the nanoparticles can also be used to detect biopolymers such as DNA or proteins. Related complimentary techniques include plasmon waveguide resonance, and Dual Polarization Interferometry.

Unique optical property of nanomaterials may also arise from another quantum size effect. When the size of a nanocrystal (i.e. a single crystal nanoparticle) is smaller than the de Broglie wavelength, electrons and holes are spatially confined and electric dipoles are formed, and discrete electronic energy level would be formed in all materials. Similar to a particle in a box, the energy separation between adjacent levels increases with decreasing dimensions. The electronic configurations of nanomaterials are significantly different from that of their bulk counterpart. These changes arise through systematic transformations in the density of electronic energy levels as a function of the size, and these changes result in strong variations in the optical and electrical properties with size. Nanocrystals lie in between the atomic and molecular limit of discrete density of electronic states and the extended crystalline limit of continuous band. In any material, there will be a size below which there is substantial variation of fundamental electrical and optical properties with size, when energy level spacing exceeds the temperature. For a given temperature, this occurs at a very large size (in nanometers) in semiconductors as compared with metals and insulators. In the case of metals, where the Fermi level lies in the center of a band and the relevant energy level spacing is very small, the electronic and optical properties more closely resemble those of continuum, even in relatively small sizes (tens or hundreds of atoms). In semiconductors, the Fermi level lies between two bands, so that the edges of the bands are dominating the low-energy optical and electrical behavior. Optical excitations across the gap depend strongly on the size, even for crystallites as large as 10,000 atoms. For insulators, the band gap between two bands is already too big in the bulk form. The quantum size effect is most pronounced for semiconductor nanoparticles, where the band gap increases with a decreasing size, resulting in the interband transition shifting to higher frequencies. In a semiconductor, the energy separation, i.e. the energy difference between the completely filled valence band and the empty conduction band is of the order of a few electrovolts and increases rapidly with a decreasing size. The same quantum size effect is also known for metal nanoparticles; however, in order to observe the localization of the energy levels, the size must be well below 2 nm, as the level spacing has to exceed the thermal energy (–26 meV). In a metal, the conduction band is half filled and the density of energy levels is so high that a noticeable separation in energy levels within the conduction band (intraband transition) is only observed when the nanoparticle is made up of 100 atoms. If the size of metal nanoparticle is made small enough, the continuous density of electronic states is broken up into discrete energy levels. Glues containing nanoparticles have optical properties that give rise to uses in optoelectronics. Casings, containing nanoparticles used in electronic devices, such as computers, offer improved shielding against electromagnetic interference. Electro chromic, devices are similar to liquid-crystal displays (LCD), are been developed with nanomaterials. The incorporation of nanomaterials in surface coatings can provide long-term abrasion resistance without significantly effecting optical clarity, gloss, colour or physical properties. The cosmetics industry has used nanomaterials as UV absorbers or sunscreens.

- **Photoluminescence of semiconductors**

  - This has been introduced in the electronic properties of nanomaterials.
  - Photoluminescence corresponds to the annihilation of an electron-hole pair. It corresponds to light emission in the visible range of wavelength.
  - For nanomaterials, this is shift towards higher frequencies → blue (lower wavelength).
  - This is due to the confinement effect when at least one dimension of the material becomes nanometric (quantum well).
  - Photoluminescence is used in optoelectronics to develop optical connections rather than electrical connections and to produce light-emitting diodes (LEDs).

- Photoluminescence of semiconductors

  - This has been introduced in the electronic properties of nanomaterials.
  - Photoluminescence corresponds to the annihilation of an electron-hole pair. It corresponds to light emission in the visible range of wavelenght.
  - For nanomaterials, this is shift towards higher frequencies → blue (lower wavelength).
  - This is due to the confinement effect when at least one dimension of the material becomes nanometric (quantum well).
  - Photoluminescence is used in optoelectronics to develop optical connections rather than electrical connections and to produce light-emitting diodes (LEDs).

- **Absorption bands in insulators.**
  - The interaction between an electromagnetic wave and an insulator (polarization) can be described using the dielectric constant ε which depends on the "frequency" ω of the wave.
  - The dielectric constant is complex, which gives a refractive index n_r and an absorption index n_i.

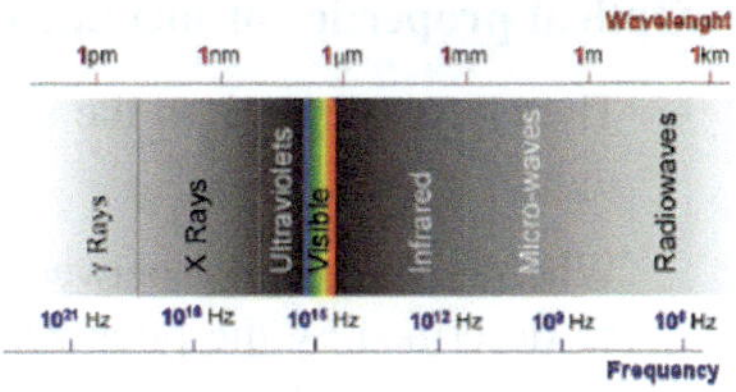

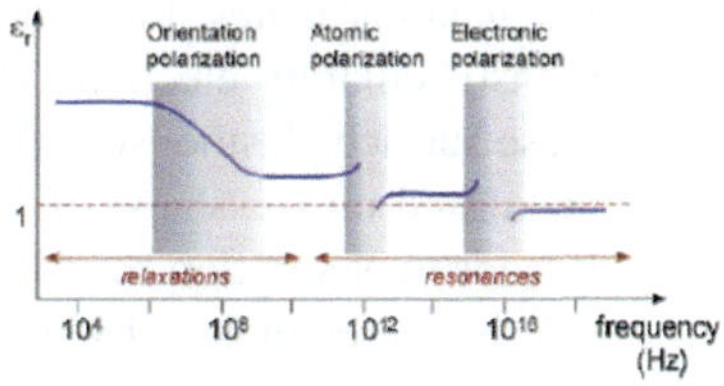

  - The Drude's model allows finding the expression of the electric susceptibility as a function of the frequency of the applied electric field. It gives a satisfactory description of the evolution of the permittivity in the frequency interval extending from infrareds to soft X-rays.

$$\varepsilon_r = 1 + \frac{\omega_p^2}{(\omega_0'^2 - \omega^2) + j\frac{\omega}{\tau}}$$

Where ω'_0 and τ are constants depending on the insulator.

- **Absorption bands** (filter for solar protection).
  - The study of the imaginary part shows that there is a forbidden band for propagation, which corresponds to absorption.
  - Insulators have a large forbidden band width which corresponds to ultraviolet absorption.
  - When their size is decreased towards the nanometric range, the absorption peaks are shifted towards the high frequencies.
  - For 20 nm TiO_2 nanograins, the absorption band starts at λ_0=310 nm; it allows reducing the UVB and keeping the UVA, which is suitable for solar protection.

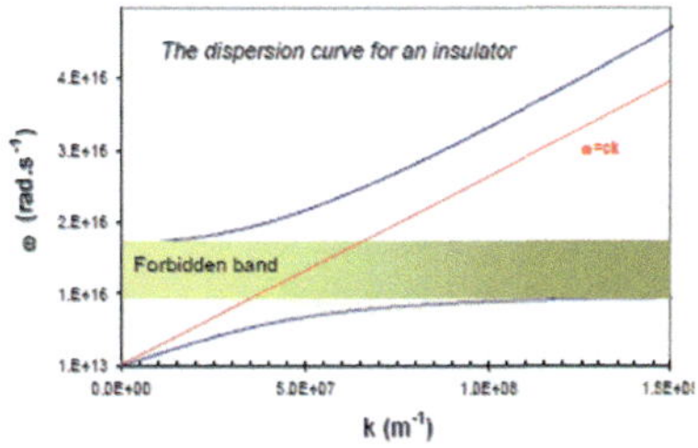

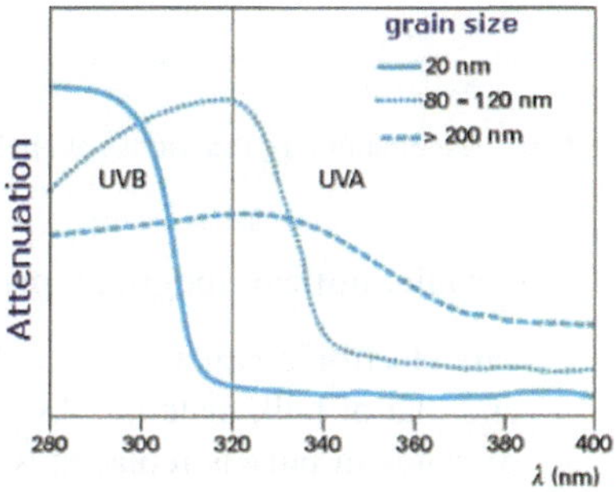

P. Costa, Nanomatériaux, Propriétés et applications
Techniques de l'ingénieur

- **Optical properties of metallic clusters in a matrix**
  - The effects depend on the size of the clusters and on the wavelength of the light.

    When the interaction between the electromagnetic wave and the material can be described by the polarization of the electron cloud, plasmonsare created (collective oscillations of the electrons).
    For nanoparticules, the confinement results in the creation of surface plasmons(oscillating electrical charges on their surface). The characteristics of these surface plasmonsdepend on the dielectric constant of the matrix and the metal and on the shape of the nanoclusters.
    It leads to absorption bands in the visible range and in infrared. Reducing the size increases drastically the absorption.

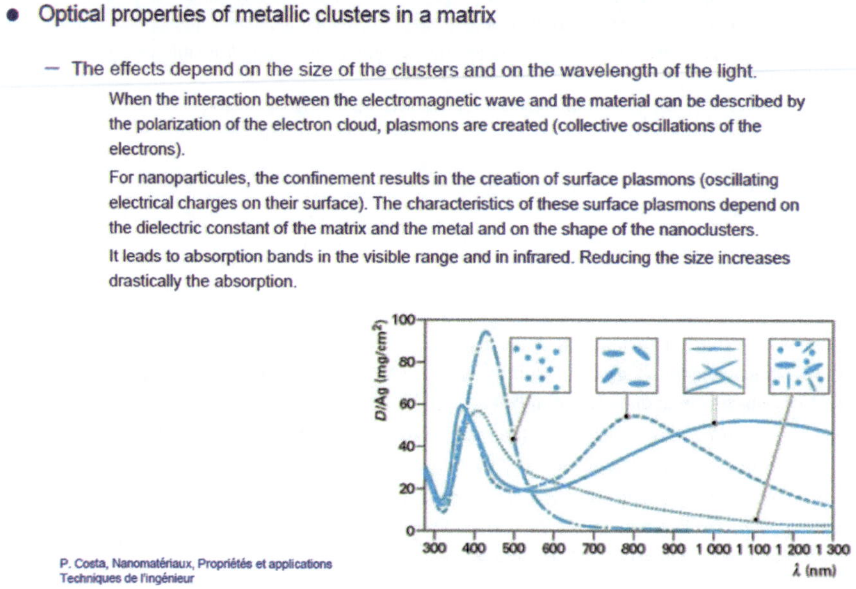

P. Costa, Nanomatériaux, Proprieties et applications Techniques de l'ingénieur

- **Electronic, optical, magnetic properties, reactivity**
  - The electronic structure of a cluster varies from both that of a single atom and that of bulk material. As particles shrink in size, the continuous density of states in bulk is replaced with a set of discrete energy levels.

- Electronic, optical, magnetic properties, reactivity

  – The electronic structure of a cluster varies from both that of a single atom and that of bulk material. As particles shrink in size, **the continuous density of states in bulk is replaced with a set of discrete energy levels**.

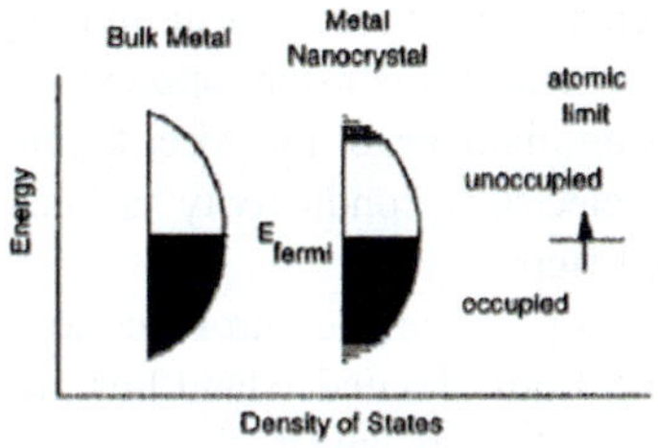

Nanomaterials For Environmental Applicatior
Roberto Comparelli, Thesis 2004
Universita di Bari

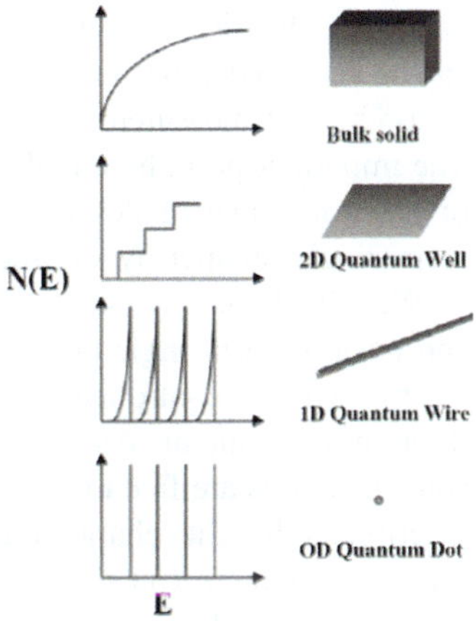

- Electronic, optical, magnetic properties, reactivity

  – This leads to interesting optical properties. In semiconductors the band gap will be increased as cluster size is decreased – a blue shift in the absorption spectrum of the semiconductor (good for tayloring lasers).

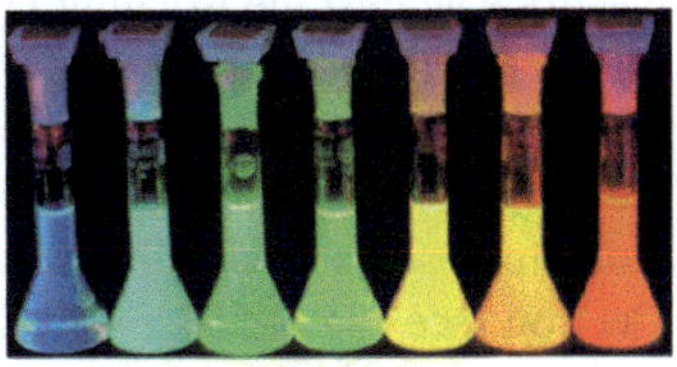

Photoluminescence color of colloidal solutions of CdSe nanocrystals passivated with a ZnS shell.
The smallest (~1.7 nm) CdSe particles emit blue, the largest (~5 nm) emit red.
Nanomaterial For Environmental Applications
Roberto Comparelli, Thesis 2004, Univeriista di Bari

  – Interesting magnetic properties are also observed in small clusters, the magnetic moment of each atom interacting with the moments of the other atoms. It can force all the moments to align in one direction with respect to some symmetry axis of the cluster and the cluster have a net magnetic moment and is magnetized. As cluster size decreases it therefore becomes **easier** for them to exhibit **ferromagnetic behaviour**.

  – Reactivity: since the electronic structure of nanoclusters depends on their size, their ability to react with other species should also depend on size. This has important applications for the design of catalytic agents.

14.7 The Electrical Properties

Electrical Properties of Nanoparticles discuss about fundamentals of electrical conductivity in nanotubes and nanorods, carbon nanotubes, photoconductivity of nanorods, electrical conductivity of nanocomposites. One interesting method which can be used to demonstrate the steps in conductance is the mechanical thinning of a nanowire and measurement of the electrical current at a constant applied voltage. The important point here is that, with decreasing diameter of the wire, the number of electron wave modes contributing to the electrical conductivity is becoming increasingly smaller by well-defined quantized steps.

We talk about single-electronics whenever it is possible to control the movement and position of a single or small number of electrons. To understand how a single electron can be controlled, one must understand the movement of electric charge through a conductor. An electric current can flow through the conductor because some electrons are free to move through the lattice of atomic nuclei. The current is determined by the charge transferred through the conductor. Surprisingly this transferred charge can have practically any value, in particular, a fraction of the charge of a single electron. Hence, it is not quantized. This, at first glance counterintuitive fact, is a consequence of the displacement of the electron cloud against the lattice of atoms. This shift can be changed continuously and thus the transferred charge is a continuous quantity (see left side of Fig. 14.6).

If a tunnel junction interrupts an ordinary conductor, electric charge will move through the system by both a continuous and discrete process. Since only discrete electrons can tunnel through junctions, charge will accumulate at the surface of the electrode against the isolating layer, until a high enough bias has built up across the tunnel junction (see right side of Fig. 14.7). Then one electron will be transferred. Likharev has coined the term 'dripping tap' as an analogy of this process. In other words, if a single tunnel junction is biased with a constant current **I**, the so called Coulomb oscillations will appear with frequency **f = I/e**, where **e** is the charge of an electron (see Fig. 14.7).

Charge continuously accumulates on the tunnel junction until it is energetically favorable for an electron to tunnel. This discharges the tunnel junction by an elementary charge **e**. Similar effects are observed in superconductors. There, charge

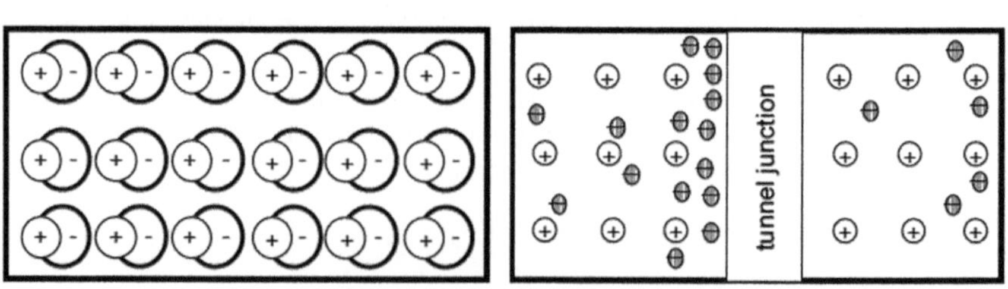

Fig. 14.6 The left side shows, that the electron cloud shift against the lattice of atoms is not quantized. The right side shows an accumulation of electrons at a tunnel junction

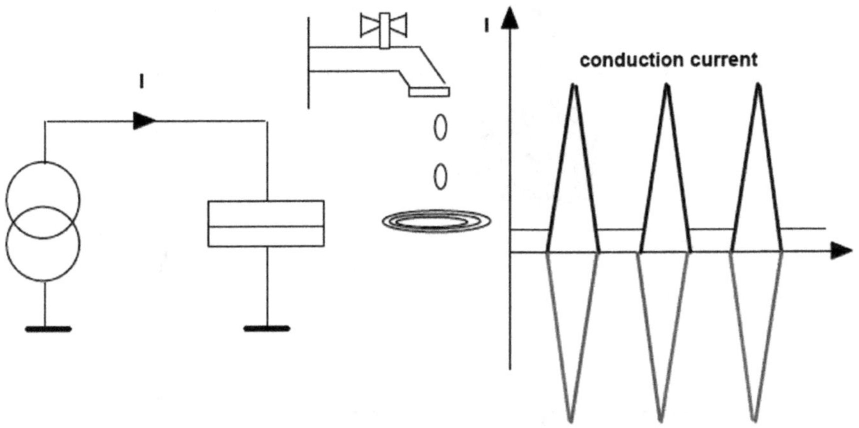

Fig. 14.7 Current biased tunnel junction showing Coulomb oscillations

carriers are Cooper pairs, and the characteristic frequency becomes **f = I/2c**, related to the so called Bloch oscillations.

The current biased tunnel junction is one very simple circuit that shows the controlled transfer of electrons. Another one is the electron-box (see Fig. 14.8). A particle is only on one side connected by a tunnel junction. On this side electrons can tunnel in and out. Imagine for instance a metal particle embedded in oxide, as shown in Fig. 14.9.

The top oxide layer is thin enough for electrons to tunnel through. To transfer one electron onto the particle, the Coulomb energy $\mathbf{E_C = e^2/2C}$, where **C** is the particles capacitance, is required. Neglecting thermal and other forms of energy, the only energy source available is the bias voltage **Vb**. As long as the bias voltage is small enough, smaller than a threshold **Vth = e/C**, no electron can tunnel, because not enough energy is available to charge the island. **This behavior is called the Coulomb blockade**. Raising the bias voltage will populate the particle with one, then two and so on electrons, leading to a staircase-like characteristic. It is easily

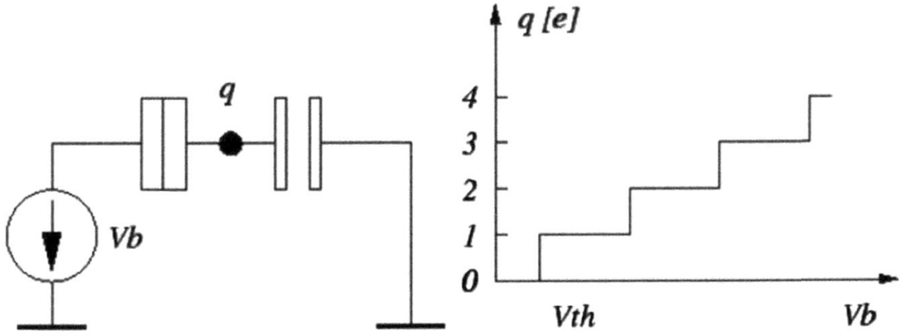

Fig. 14.8 The electron-box can be filled with a precise number of electrons

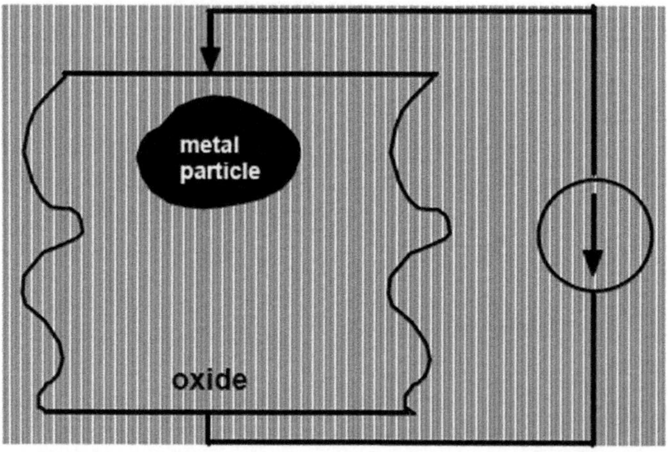

Fig. 14.9 Metal particle embedded in oxide. Tunneling is only possible through the thin top layer of oxide

understandable, that these single-electron phenomena, such **as Coulomb oscillations and Coulomb blockade**, only matter, if the Coulomb energy is bigger than the thermal energy. Otherwise, thermal fluctuations will disturb the motion of electrons and will wash out the quantization effects. The necessary condition is,

$$E_c = \frac{e^2}{2C} > k_B T$$

where $\mathbf{k_B}$ is Boltzmann's constant and **T** is the absolute temperature. This means that the capacitance **C** has to be smaller than 12 nF for the observation of charging effects at the temperature of liquid nitrogen and smaller than 3 nF for charging effects to appear at room temperature. A second condition for the observation of charging effects is, that quantum fluctuations of the number of electrons on an island must be negligible. Electrons need to be well localized on the islands. If electrons would not be localized on islands one would not observe charging effects, since islands would not be separate particles but rather one big uniform space. The charging of one island with an integer number of the elementary charge would be impossible, because one electron is shared by more than one island. The Coulomb blockade would vanish, since no longer would a lower limit of the charge, an island could be charged with, exist. This leads to the requirement that all tunnel junctions must be opaque enough for electrons in order to confine them on islands. The 'transparency' of a tunnel junction is given by its tunnel resistance $\mathbf{R_T}$ which must fulfill the following condition for observing discrete charging effects, where **h** is Planck's constant. This should be understood as an order-of-magnitude measure, rather than an exact threshold.

$$R_T > \frac{h}{e^2} = 25813\,\Omega$$

Therefore, these effects are experimentally verifiable only for very small high-resistance tunnel junctions, meaning small particles with small capacitances and/or very low temperatures. Advanced fabrication techniques, such as the production of granular films with particle sizes down to 1 nm, and deeper physical understanding allow today the study of many charging effects at room temperature.

Based on the Coulomb blockade many interesting devices are possible, such as precise current standard, very sensitive electrometers, logic gates and memories with ultra low power consumption, down-scalability to atomic dimensions, and high speed of operation. Altogether, single-electronics will bring new and novel devices and is a very promising candidate to partly replace MOS technology in the near future.

In electrically conducting carbon nanotubes, only one electron wave mode is observed which transport the electrical current. As the lengths and orientations of the carbon nanotubes are different, they touch the surface of the mercury at different times, which provides two sets of information: (i) the influence of carbon nanotube length on the resistance; and (ii) the resistances of the different nanotubes. As the nanotubes have different lengths, then with increasing protrusion of the fiber bundle an increasing number of carbon nanotubes will touch the surface of the mercury droplet and contribute to the electrical current transport (Fig. 14.10).

The effects of size on electrical conductivity of nanostructures and nanomaterials are complex, since they are based on distinct mechanisms. These mechanisms can be generally grouped into four categories: surface scattering including grain boundary scattering, quantized conduction including ballistic conduction, Coulomb charging and tunneling, and widening and discrete of band gap, and change of microstructures. In addition, increased perfection, such as reduced impurity,

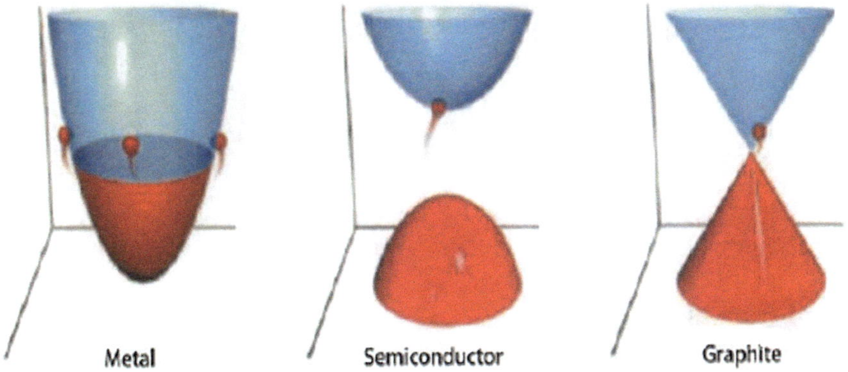

Fig. 14.10 Electrical behavior of naotubes (P. G. Collins and Ph. Avouris, *Scientific American*, 62, 2000, 283)

structural defects and dislocations, would affect the electrical conductivity of nanostructures and nanomaterials.

Nanomaterials can hold considerably more energy than conventional ones because of their large grain boundary area. They are materials in which an optical absorption band can be introduced, or an existing band can be altered by the passage of current through these materials, or by the application of an electric field. Tremendous efforts and progress have been made in the molecular electronics and nanoelectronics. In molecular electronics, single molecules are expected to be able to control electron transport, which offers the promise of exploring the vast variety of molecular functions for electronic devices, and molecules can now be crafted into a working circuit as shown schematically in Fig. 14.11 When the molecules are biologically active, bioelectronic devices could be developed. In molecular electronics, control over the electronic energy levels at the surface of conventional semiconductors and metals is achieved by assembling on the solid surfaces, poorly organized, partial monolayers of molecules instead of the more commonly used ideal ones. Once those surfaces become interfaces, these layers exert electrostatic rather than electrodynamic control over the resulting devices, based on both electrical monopole and dipole effects of the molecules. Thus electronic transport devices, incorporating organic molecules, can be constructed without current flow through the molecules.

The simplest molecular electronics are sensors that translate unique molecular properties into electrical signals. Sensors using a field effect transistor (FET) configuration with its gate displaced into a liquid electrolyte, and an active layer of molecules for molecular recognition were reported in early 70 s.

Fig. 14.11 Schematic of the molecules crafted into working circuit (from R.F. Service, *Science*, 2001, 293)

A selective membrane is inserted on the insulator surface of the FET, and this permits the diffusion of specific analyte ions and construction of a surface dipole layer at the insulator surface. Such a surface dipole changes the electric potential at the insulator surface and, thus, permits the current going through the device. Such devices are also known as ion-selective FET (ISFET) or chemical FET (CHEM-FET). Thin films attached to metal nanoparticles have been shown to change their electrical conductivity rapidly and reproducibly in the presence of organic vapors, and this has been exploited for the development of novel gas sensor. The monolayer on metal nanoparticles can reversibly adsorb and desorb the organic vapor, resulting in swelling and shrinking of the thickness of the monolayer, thus changing the distance between the metal cores. Since the electron hopping conductivity through the monolayers is sensitively dependent on the distance, the adsorption of organic vapor increases the distance and leads to a sharp decrease in electrical conductivity.

Conventional and rechargeable batteries are used in almost all applications that require electric power. The energy density (storage capacity) of these batteries is quite low requiring frequent recharging. Nanocrystalline materials are good candidates for separator plates in batteries because they can hold considerably more energy than conventional ones. Nickel-metal hydride batteries made of nanocrystalline nickel and metal hydrides are envisioned to require far less frequent recharging and to last much longer. An electrochromic device consists of materials in which an optical absorption band can be introduced, or an existing band can be altered by the passage of current through the materials or by the application of an electric field. They are similar to liquid-crystal displays (LCD) commonly used in calculators and watches and are primarily used in public billboards and ticker boards to convey information. The resolution, brightness, and contrast of these devices depend on the tungsten acid gel's grain size. Hence, nanomaterials, such as tungstic oxide gel, are being explored for this purpose. Many nanoscale electronic devices have been demonstrated: tunneling junctions, devices with negative differential electrically configurable switches, carbon nanotube transistors, and single molecular transistors. Devices have also been connected together to form circuits capable of performing single functions such as basic memory and logic functions. Ultrahigh density nanowires lattices and circuits with metal and semiconductor nanowires have also been demonstrated.

14.8 The Mechanical Properties of Nanomaterials

"Mechanical Properties of Nanoparticles" deals with bulk metallic and ceramic materials, influence of porosity, influence of grain size, superplasticity, filled polymer composites, particle-filled polymers, polymer-based nanocomposites filled with platelets, carbon nanotube-based composites. The discussion of mechanical properties of nanomaterials is, into some extent, only of quite basic interest, the reason being that it is problematic to produce macroscopic bodies with a high density and a grain size in the range of less than 100 nm. However, two materials, neither of which is produced by pressing and sintering, have attracted much greater interest as they will undoubtedly achieve industrial importance. These materials are polymers which contain nanoparticles or nanotubes to improve their mechanical behaviors, and severely plastic-deformed metals, which exhibit astonishing properties. However, because of their larger grain size, the latter are generally not accepted as nanomaterials. Experimental studies on the mechanical properties of bulk nanomaterials are generally impaired by major experimental problems in producing specimens with exactly defined grain sizes and porosities. Therefore, model calculations and molecular dynamic studies are of major importance for an understanding of the mechanical properties of these materials. Filling polymers with nanoparticles or nanorods and nanotubes, respectively, leads to significant improvements in their mechanical properties. Such improvements depend heavily on the type of the filler and the way in which the filling is conducted. The latter point is of special importance, as any specific advantages of a nanoparticulate filler may be lost if the filler forms aggregates, thereby mimicking the large particles. Particulate filled polymer-based nanocomposites exhibit a broad range of failure strengths and strains. This depends on the shape of the filler, particles or platelets, and on the degree of agglomeration. In this class of material, polymers filled with silicate platelets exhibit the best mechanical properties and are of the greatest economic relevance. The larger the particles of the filler or agglomerates, the poorer are the properties obtained. Although, potentially, the best composites are those filled with nanofibers or nanotubes, experience teaches that sometimes such composites have the least ductility. On the other hand, by using carbon nanotubes it is possible to produce composite fibers with extremely high strength and strain at rupture. Among the most exciting nanocomposites are the polymer ceramic nanocomposites, where the ceramic phase is platelet-shaped. This type of composite is preferred in nature, and is found in the structure of bones, where it consists of crystallized mineral platelets of a few nanometers thickness that are bound together with collagen as the matrix. Composites consisting of a polymer matrix and defoliated phyllosilicates exhibit excellent mechanical and thermal properties.

As a primary microstructural parameter, grain size imparts significant influence on the mechanical behaviour of materials. The drive for refining grain size comes from the demand for increased strength in many applications. There is the general observation that the yield strength of metals is proportional to the inverse square root of grain size (the Hall–Petch relation):

$$\sigma = \sigma_0 + \mathrm{kd}^{-\frac{1}{2}}$$

where d is the average grain size; σ_0 and k are constants. k may be defined as a slope of Hall-Petch plot. A number of mechanisms has been proposed to explain this phenomenon: pile-up of dislocations against the grain boundaries (GBs), GBs acting as dislocation sources and sinks, and presence of geometrically necessary dislocations in the vicinity of the GBs to account for the deformation compatibility of polycrystalline metals, etc.

Certainly, the grain size, or average grain diameter, influences the mechanical properties of a material. It is important to note in this context that the mechanisms of deformation and the properties of the NC material (NSM) not only depend on the average grain size, but are also strongly influenced by the grain size distribution and the grain boundary structure (e.g., low-angle versus high-angle grain boundaries). Some appealing characteristics of nc metals and alloys with potential significance for engineering applications include ultra-high yield and strengths, decreased elongation and toughness, superior wear resistance, and the promise of enhanced superplastic formability at lower temperatures and faster strain rates compared to their micro structured counterparts. The need for scratch, mar and/or abrasion resistance is well established in various markets, including fingernail polishes, flooring, plastic glazing, headlamp covers and other automotive parts, transportation windows and optical lenses, where clear scratch-resistant coatings are used. Because the nanosize, many of their mechanical properties of the materials is modified, among others, hardness and elastic modulus, fracture toughness, scratch resistance, fatigue strength, and hardness. Energy dissipation, mechanical coupling within arrays of components, and mechanical nonlinearities are influenced by structuring components at the nanometer scale. This includes also the interpretation of unusual mechanical behaviour (e.g., strengths approaching the theoretical limit) and the exploration of new ways to integrate diverse classes of mechanically functional materials on the nano-size.

14.8.1 Elasticity, Plasticity, Dislocations, Hardening, Twinning, Toughness …

- Unique mechanical properties
  - By reducing the grain size, lighter materials with enhanced mechanical properties can be produced. Nanograins are more resistant since they have no defects inside, contrary to micrograins (nanocrystalline copper is three times more resistant than usual copper). They are also more ductile since nanograin-nanograinglide is much easier.
  - The following mechanical properties will be presented.

    Elastic properties
    Plasticity
    Hardness
    Ductility
    Toughness.
- Understanding of these unique properties requires reminding:
  - Elements of elasticity
  - Elements of crystallography

- Dislocations
- Hardening.

Adjacent grains normally have different crystallographic orientations and common grain boundary. In coarse-grained materials, the grain boundary acts as a barrier to dislocation motion because of necessity in change of direction of motion in grain boundary regions and pile-up of dislocations at grain boundaries is envisioned as a key mechanistic process underlying an enhanced resistance to plastic flow from grain refinement. As the microstructure is refined from the micro and ultra-fine grained regime into the nc regime, this process invariably breaks down and the flow stress versus grain size relationship departs markedly from that seen at higher grain sizes, Fig. 14.12. With further grain refinement, the yield stress peaks in many cases at an average grain size value on the order of 10 nm or so. Further decrease in grain size can cause weakening of the metal. Experimental evidence indicates that below a grain size of ~ 10 nm, strength decreases with further grain refinement (the so-called "inverse Hall-Petch-type" relationship). Although there is a growing body of experimental evidence pointing to such unusual deformation responses in nc materials, the underlying mechanisms are not well understood yet. The mechanical properties of fully-dense facial cubic centered metals (Cu, Ni and Pd) with grain size less than 100 nm have been primarily derived from uniaxial tension/compression tests and micro- or nanoindentation. Typically, these

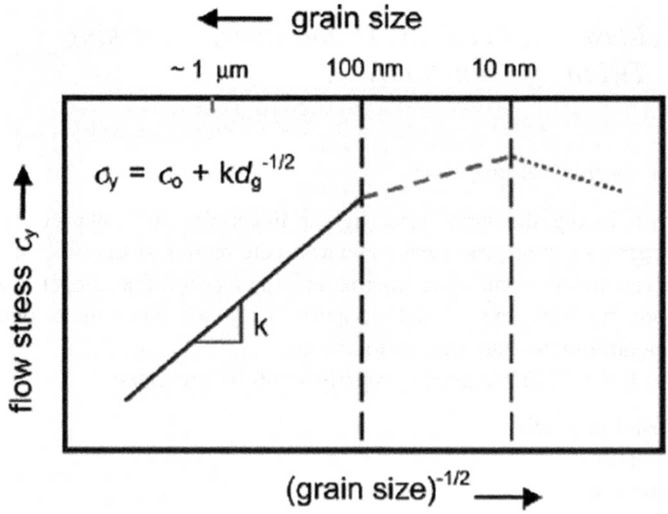

Fig. 14.12 Schematic representation of the variation of yield stress as a function of grain size (adapted from K. Kumar, et al., *Acta Materia*, 2003, v. 51)

nanocrystalline metals exhibit significantly higher yield strength, and reduced tensile elongation relative to their microcrystalline counterparts. Furthermore, hardness and yield strength have been found to increase with decreasing grain size in this regime (<100 nm) down to at least 15 nm. The reasons for this behaviour are still under debate as dislocation sources within grains are not expected to operate at these grain sizes. In addition, there is no documented evidence of dislocation pile-ups in deformed specimens, and any dislocation activity is primarily believed to originate and terminate at grain boundaries.

Among many excellent mechanical properties of nanomaterials, high hardness of nanocomposites systems is one of the most intriguing. A variety of superhard nanocomposites can be made of nitrides, borides and carbides by plasma-induced chemical and physical vapor deposition. In the appropriately synthesized binary systems, the hardness of the nanocomposite exceeds significantly that given by the rule of mixtures in bulk. For example, the hardness of nc-MnN/a–Si_3N_4 (M = Ti, W, V, …) nano-composites with the optimum content of amorphous Si3N4 matrix close to the percolation threshold reaches 50 GPa although that of the individual nitrides does not exceed 21 GPa. Elastic image forces are argued to require a very high stress to force dislocations to cut through the nanometer-sized nitride crystallites. This high stress may, however, not lead to fracture because any crack formed in one of the crystallites is suggested to be stopped by the ductile amorphous Si_3N_4 matrix surrounding the cracked crystallite. These superhard nanocomposites will have promising potential in hard protective coatings. Superhardness also may come from pure nanoparticles. For example, the nearly spherical, defect-free silicon nanospheres with diameters from 20 to 50 nm possess hardness of up to 50 GPa or four times greater than that of the bulk silicon. Since their discovery, carbon nanotubes have stimulated intensive research interests because of excellent mechanical properties. The strength of the carbon fibers increases with graphitization along the fibber axis. Carbon nanotubes, which are formed of seamless cylindrical graphene layers, represent the ideal carbon fiber and presumably have the best mechanical properties in the carbon fibers species, showing a high Young's modulus and high tensile strength. Theoretical simulations have predicted the high modulus of carbon nanotubes. Calculated Young's module Y of single wall carbon nanotubes is found to be in the range of 0.5–5.5 TPa, that is much higher than high strength steel ($Y \sim 200$ GPa). The experimental measurements of Young's modulus of multiwalled carbon nanotubes with the help of thermal vibrations studying by means of transmission electron microscopy (TEM), give the value of Young's modulus of 1.8 ± 0.9 TPa. Atomic force microscope (AFM) has also been employed to measure the modulus of the carbon nanotubes. This is realized by bending the anchored carbon nanotubes with AFM tip while simultaneously recording the force by the tube as a function of the displacement from its equilibrium position. The resultant Young's modulus was 1.28 ± 0.5 TPa. The values of Young's module measured from different ways were all in the range in theoretical prediction, proving the existence the high elastic modulus of the carbon nanotubes. The tensile strength of carbon nanotubes has also been under consideration. An individual multi wall carbon nanotube was mounted between two AFM tips, one on rigid cantilever and the other on soft cantilever. By recording the whole tensile loading experiment, both the deflection of the soft cantilever from which the force applied on the nanotube and the length change of the nanotube were simultaneously obtained. The carbon nanotubes broke in the outermost layer ("sword-in-sheath" failure), and the tensile strength of this layer ranged from 11 to 63 GPa and the measured strain at failure can be as high as 12%. For comparison, the tensile strength of high-strength steel is 1–2 GPa. Enhancement of mechanical properties of polymeric materials by nanofillers is one of the very active applications of nanomaterials. Micrometer size fillers used in traditional polymer composites show improvements in their mechanical properties such as the modulus, yield strength and glass transition temperature. However, these enhancements sacrifice the ductility and toughness of the materials. Comparably, polymer nanocomposites from nano size fillers result in unique mechanical properties at very low filler weight fractions. The dramatic improvements in the yield stress (30%) and Young's modulus (170%) in polypropylene filled with ultrafine SiO_2, as compared to micrometer filled polypropylene, has been reported. A rubbery polyurethane elastomer filled with 40wt% of 12 nm silica particles exhibits 6 times increase in the elongation-at-break and 3 times increase in the modulus as them compared to a micrometer-sized filler reinforced polymer.

- Elasticity theory deals with the small, continue and reversible deformations of isotropic elastic materials.
  - An elastic material has the three following properties:

    It deforms under stress and returns to its original shape when the stress disappears.
    It is homogenous and uniformly distributed in its occupied volume.
    It is isotropic.

- Elasticity of polycrystals
  - Materials are commonly not isotropic as they are polycrystalline, with grains having different shapes and orientations.
  - However, as the dimensions of the materials are very large compare to the mean grain, homogeneity and isotropy hypothesis are sometimes more or less fulfilled.
  - Thus, elasticity theory is also used for polycrystals.

- Stress and strain: Hooke law.

- Elasticity theory deals with the small, continue and reversible deformations of isotropic elastic materials.

- An elastic material has the three following properties:
  - It deforms under stress and returns to its original shape when the stress disappears.
  - It is homogenous and uniformly distributed in its occupied volume.
  - It is isotropic.

- Elasticity of polycrystals
  - Materials are commonly not isotropic as they are polycrystalline, with grains having different shapes and orientations.
  - However, as the dimensions of the materials are very large compare to the mean grain, homogeneity and isotropy hypothesis are sometimes more or less fulfilled.
  - Thus, elasticity theory is also used for polycrystals.

- Stress and strain: Hooke law.
  - When a material is submitted to an elongation force, its deformation $\varepsilon=\Delta\ell/\ell$ is proportional to the applied stress $\sigma=F/S$ for low deformation (elastic domain).
  - The coefficient of proportionality between the stress σ and the strain ε is the Young modulus E such that $\sigma=E\ \varepsilon$.
  - If the solid is considered isotropic, elongation is accompanied by a lateral contraction such that $\Delta r/r=-\nu\Delta\ell/\ell$ where ν is the Poisson's ratio.
  - When the elastic response is no longer sufficient to accommodate the stress, the material plastically deforms. It is called work hardening.
  - The **yield stress** is the maximum stress for elastic deformation (conventionally, the stress leaving 0.2% remanent strain).

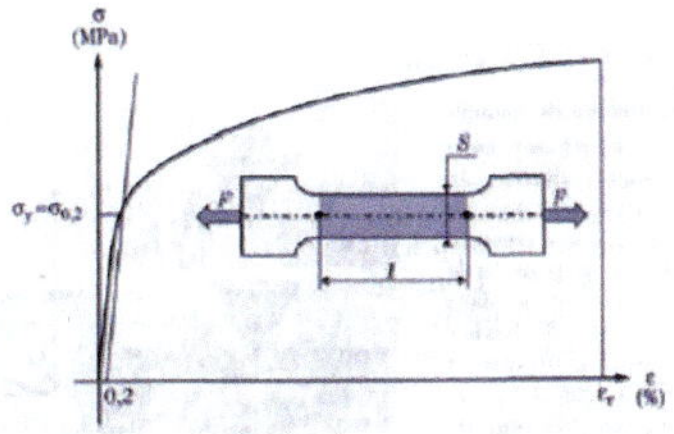

C. Tromas et al, chapter 8, Les Nanosciences 2, Ed. Belin 2006

- **Plasticity:** a non-reversible deformation in response to an applied stress.

- Stress-strain curve of a single crystal
  - The stress-strain curve exhibits stages.
  - This results from the microscopic nature of plasticity, which is a consequence of dislocations in the stressed material when stress is larger than the yield stress.

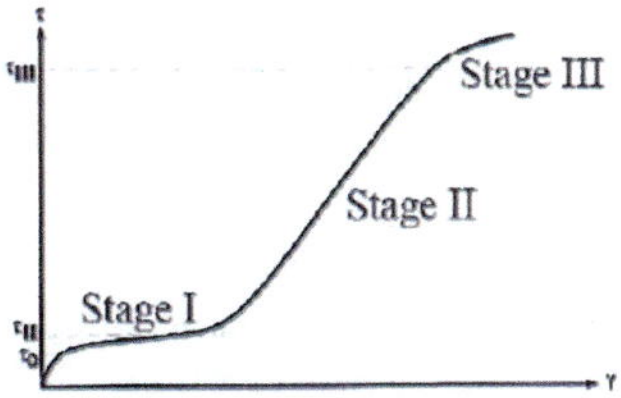

In Métallurgie, du minerai aux matériaux, J. Philibert et al, Masson Paris 1998.

- Plastic deformation of a single crystal
  - When a material plastically deforms, its surface exhibits slip bands.
  - These slip bands are the emergence on the surface of dislocations.
  - The orientations of these bands depends on the crystallographic structure of the material.
  - A part of the material is translated compare to the other part.
  - τ is the tangential component of the applied stress (**shear stress**). It is equal to F/S_0 multiplied by the **Schmid factor**.

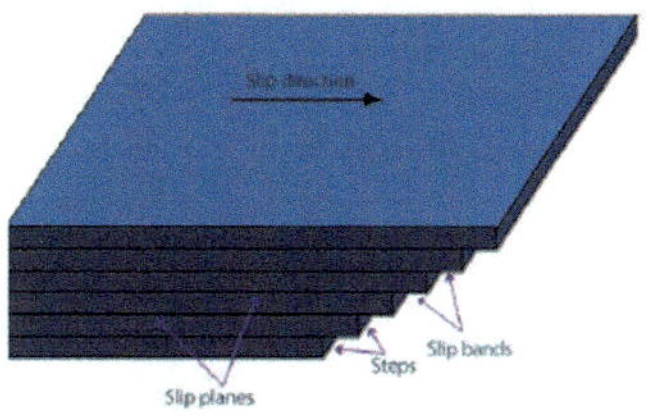

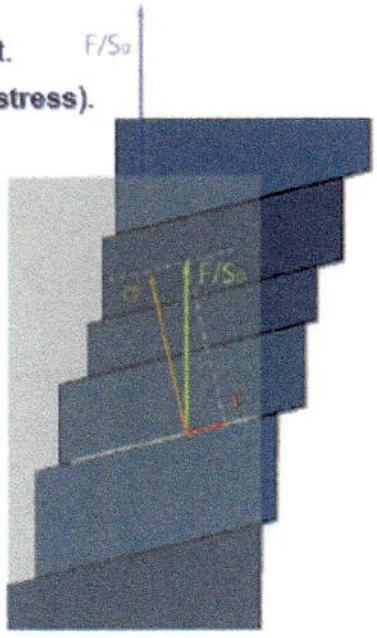

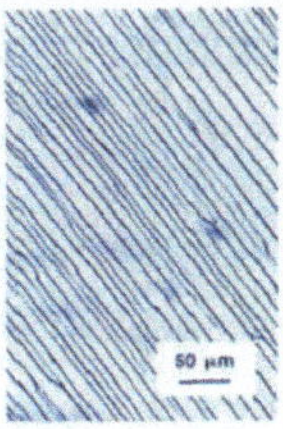

Slip bands in an Al single crystal
in La matière à l'état solide, Guisnier et al, Hachette Paris 1987

- **Plane-to-plane sliding?**
  - Using elasticity, it is possible to calculate the maximum value of the shear stress necessary to move all atoms of a plane.
  - For Cu, it gives 8000 MPa (about G/6, G is the shear modulus)
  - The experimental value is ~0.5 GPa.
  - It means that plasticity occurs through dislocations nucleation and glide.

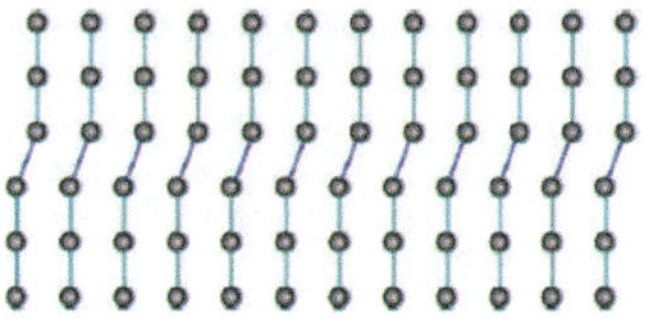

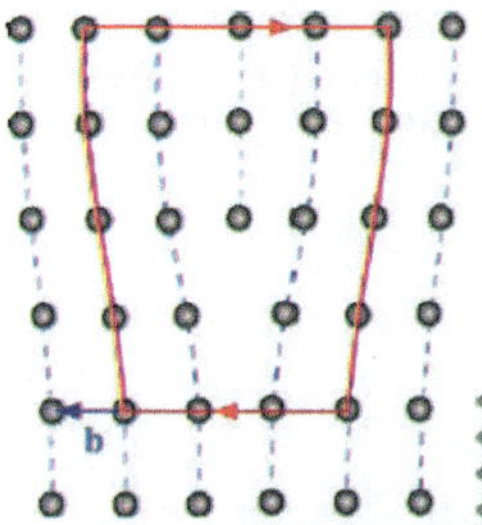

- **Dislocations**
  - Crystalline defects (edge and screw dislocations).
  - It costs less energy to move a dislocation.
  - The slip bands are the emergence on the surface of dislocations.

Edge dislocation

Dislocation glide
"caterpillar glide"

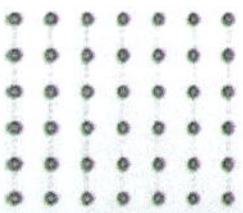

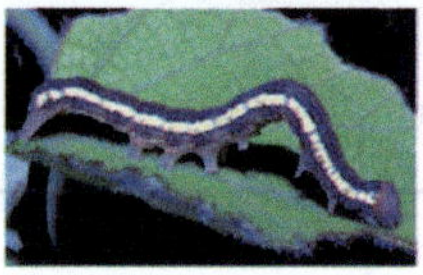

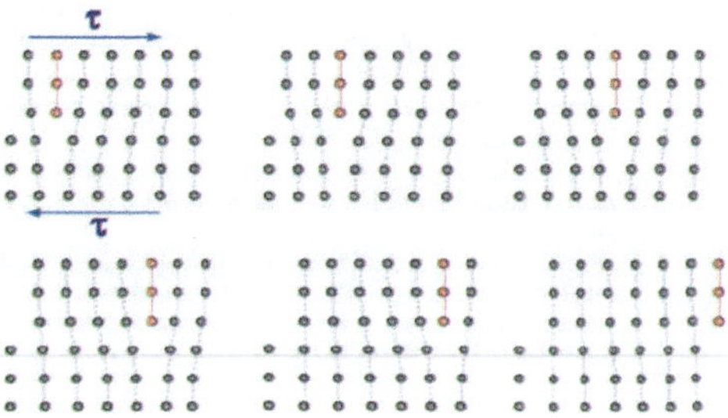

- **Crystalline materials**: *Some solids are crystalline, their atoms being periodically located in three dimensions on a lattice.*

- **Crystalline structure: the real space and the real lattice**
  - The Bravais lattice is a lattice in the real space, it is defined by three primitive vectors. (n_1, n_2 and n_3 are integers). $\mathbf{R} = n_1\,\mathbf{a_1} + n_2\,\mathbf{a_2} + n_3\,\mathbf{a_3}$

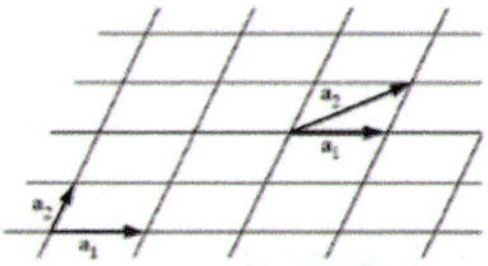

2D Bravais lattice

Fcc Wigner Seitz cell

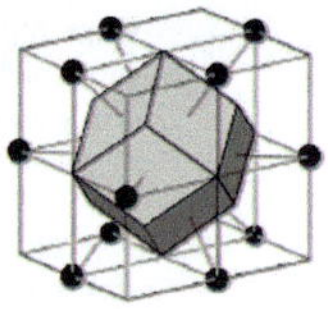

  - The unit cell (Wigner Seitz cell) is obtained from the Bravais lattice. $\mathbf{v} = \mathbf{a_1}.(\mathbf{a_2} \wedge \mathbf{a_3})$
  - 14 Bravais lattices, 32 crystal classes, 230 space groups.

  - A crystalline structure is a Bravais lattice associated with a base (a set of atoms attached to each node of the lattice).
  - The atomic planes (reticular planes) are characterized by their Miller's indexes. (h, k, ℓ are integers).
  - The ($hk\ell$) plane cut the directions $\mathbf{a_1}$, $\mathbf{a_2}$ and $\mathbf{a_3}$ on a_1/h, a_2/k *and* a_3/ℓ.

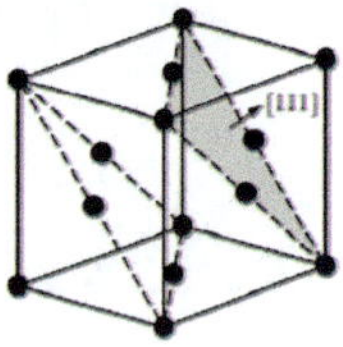

A (111) plane in a fcc lattice

- **Source and multiplication of dislocations**
  - Dislocations can be generated inside the material (homogenous nucleation) or at surfaces and interfaces (grain boundaries, precipitates…).
  - The best-known mechanism of dislocation generation is the Frank–Read source: a dislocation segment pinned at both ends and bowing out under an applied resolved shear stress.

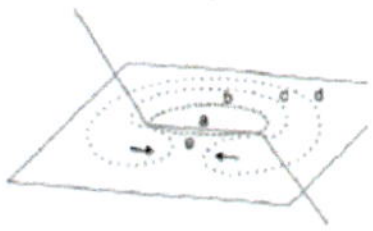

- **Slip planes**
  - Dislocation climb occurs in slip planes in the direction of the Burger's vector.
  - Usually, but not always, the slide planes are the denser planes since it corresponds to the planes with the lower "friction".

    FCC crystals: the glide directions are the <110> directions in the {111} planes

    CC crystals: the glide directions are the <111> directions in the {110}, {112} and {123} planes.

- **From single to polycrystal**
  - Single crystal: *During work hardening of single crystals, depending on the initial orientation of the sample, different "glide systems" can successively operate, corresponding to different stages.*
  - Polycrystal: *If the grain size is large enough (microscopic range), every grain is more or less homogenously deformed, proportionally to the macroscopic strain. It means that each grain simultaneously "activates" various "glide systems".*

- **Twinning: another mode of plastic deformation**
  - The two parts of a twin exhibit a mirror symmetry.
  - Only a slight misorientation exists between them, which results from a shear displacement parallel to the twinning plane. It requires from 5 to 35 GPa to nucleate twins in Zn.

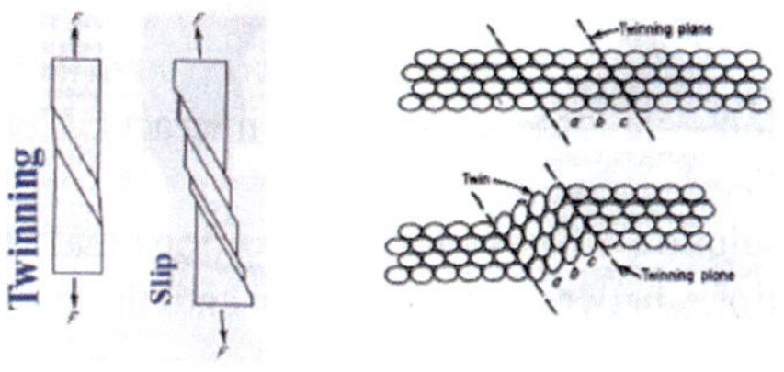

Physical Metallurgy Principles
R.E. Reed-Hill, PWS-Kent, Boston

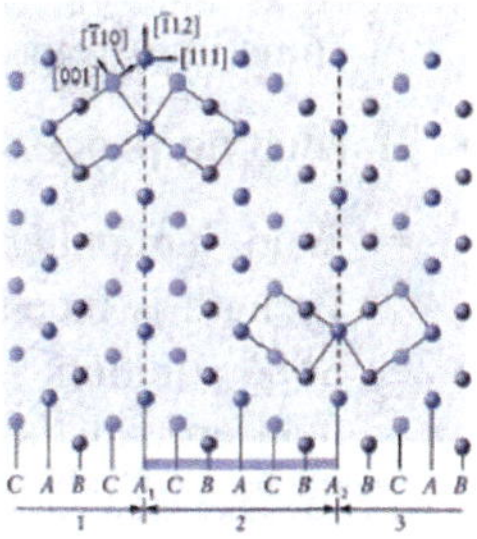

Twin scheme and twins in brass
In « Les matériaux », JP Bailon, Montréal, 2000

- Twinning: another mode of plastic deformation

  - Twins form primarily in metals that have suffered previous deformation by slip and nucleation centers are positions of highly localized strain in the lattice. Generally speaking, it means that twinning also involves slip.

  - The twins effectively reduce the grain size and raise the work hardening rate above what it would be if deformation were to occur by slip alone.

Twin in brass
In « Les matériaux », JP Bailon, Montréal, 2000

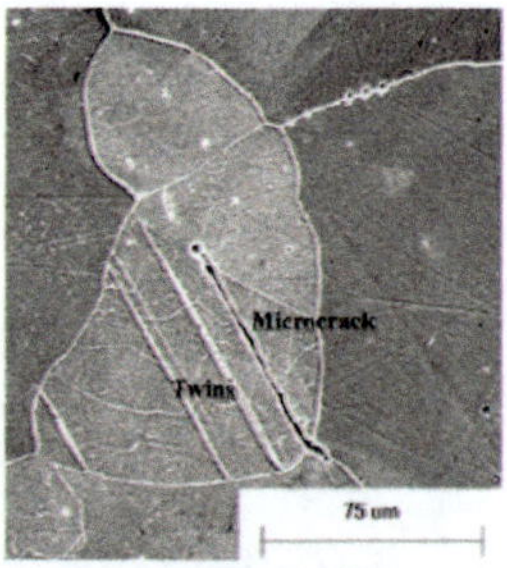

Twins in a ferritic steel
J. Bosansky et al, Materials Science and Engineering A323 (2002) 198–205

- **Hardening**
  - Hardening of a material occurs when the motion of dislocations becomes difficult, even impossible, in this materials.
  - Plastic deformation starts when dislocation climb becomes active under the applied stress.
  - Grain boundaries, precipitates, impurities or other dislocations are common obstacles for easy dislocation glide.
  - Over passing these obstacles requires higher stress and dislocations are sometimes trapped and their motion is stopped.
- **Hardening processes**
  - Work-hardening: New dislocations are produced which interacts. Tensile strength is increased.
  - Solid solution: Atoms in solid solution creates elastic distortions and stress field around them. The interactions between dislocations and these stress fields lead to a higher hardness.
  - Thermal treatment: Adjusting the conditions of cooling or annealing, one can get non-equilibrium phases with precipitates that interacts with the dislocations.
  - Grain refinement: Grain boundaries are major obstacles for the dislocations. They pile-up at these grain boundaries and it needs a larger stress to cross it and propagate in another grain. The smaller the grain size, the larger the surface per unit volume. This effect is described by the Hall-Petch relation:

where σ is the conventional tensile strength, σ_0 is a constant, **k** is the Hall-Petch coefficient which depends on the materials and **d** is the mean grain size.

- Hardening
  - Hardening of a material occurs when the motion of dislocations becomes difficult, even impossible, in this materials
  - Plastic deformation starts when dislocation climb becomes active under the applied stress.
  - Grain boundaries, precipitates, impurities or other dislocations are common obstacles for easy dislocation glide.
  - Over passing these obstacles requires higher stress and dislocations are sometimes trapped and their motion is stopped.

- Hardening processes
  - Work-hardening: New dislocations are produced which interacts. Tensile strenght is increased.
  - Solid solution: Atoms in solid solution creates elastic distortions and stress field around them. The interactions between dislocations and these stress fields lead to a higher hardness.
  - Thermal treatment: Adjusting the conditions of cooling or annealing, one can get non-equilibrium phases with precipitates that interacts with the dislocations.
  - Grain refinement: Grain boundaries are major obstacles for the dislocations. They pile-up at these grain boundaries and it needs a larger stress to cross it and propagate in an other grain. The smaller the grain size, the larger the surface per unit volume.
    This effect is described by the Hall-Petch relation:
    where σ is the conventional tensile strength, σ_0 is a constant, **k** is the Hall-Petch coefficient which depends on the materials and **d** is the mean grain size.

$$\sigma = \sigma_0 + k d^{-\frac{1}{2}}$$

- **Toughness**
  - The ability of material to absorb energy during deformation up to fracture.
  - A material is generally considered tough if it withstands high levels of loading (tensile, compressive, shear, etc.) and can dissipate strain energy without brittle fracture
- **Ductile fracture**
  - Extensive plastic deformation takes place before fracture.
  - It is related with the area under the stress-strain curve in tensile test.
- **Brittle failure**
  - It may occur before any detectable macroscopic plastic deformation for relatively high strength or hard materials at applied stress well below the yield strength.
  - Brittle fracture is the result of rapid crack growth.
  - K_{IC} (MPa.m^{½}) is the "fracture toughness".
  - For bulk materials, K_{IC} is determined from the load that causes crack propagation in pre-cracked sample under tensile opening loading mode.

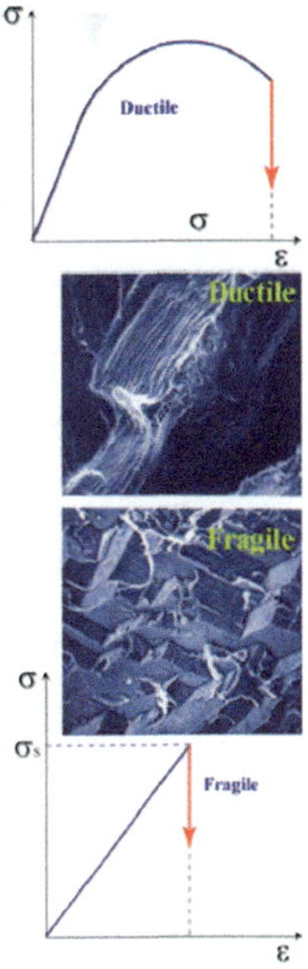

14.8.2 *The Role of Grain Boundaries*

- The properties of nanocrystalline (nc) materials are controlled not only by their small grain size but also by their interfaces, i.e. the grain boundaries.
- Structure of grain boundaries in nanomaterials
  - Two points of view:

    Amorphous like layer amorphous layer at G.B.: Less ordered material, atomic arrangements with lower density and possibility of void formation.
    Ordinary G.B. structure Ordinary structure: crystalline right up to the G.B.

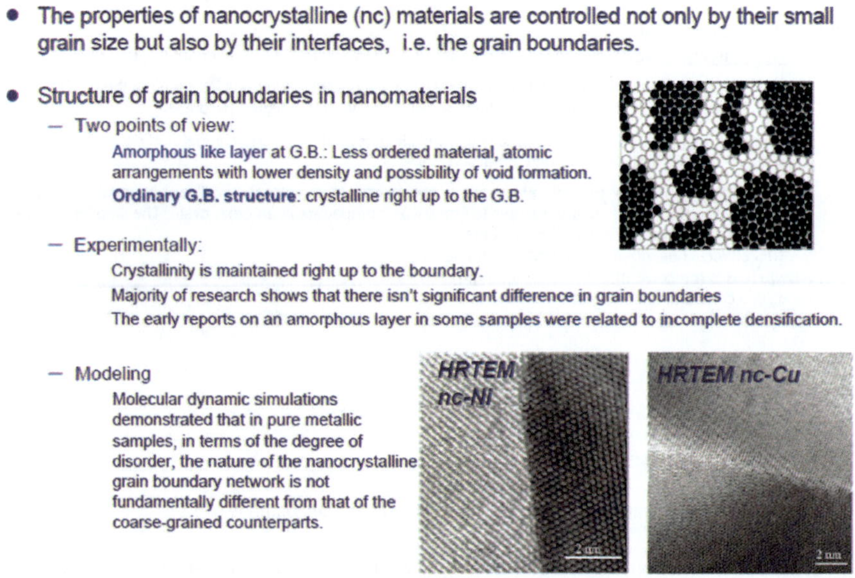

  - Experimentally:

    Crystallinity is maintained right up to the boundary.
    Majority of research shows that there isn't significant difference in grain boundaries
    The early reports on an amorphous layer in some samples were related to incomplete densification.

  - Modeling

    Molecular dynamic simulations demonstrated that in pure metallic samples, in terms of the degree of disorder, the nature of the nanocrystalline grain boundary network is not fundamentally different from that of the coarse-grained counterparts.

- Nanocrystalline materials are polycrystalline solids with an average crystallite size typically in the range 1 to 100 nm.
- Small grain sizes correspond to extremely numerous grain boundaries.

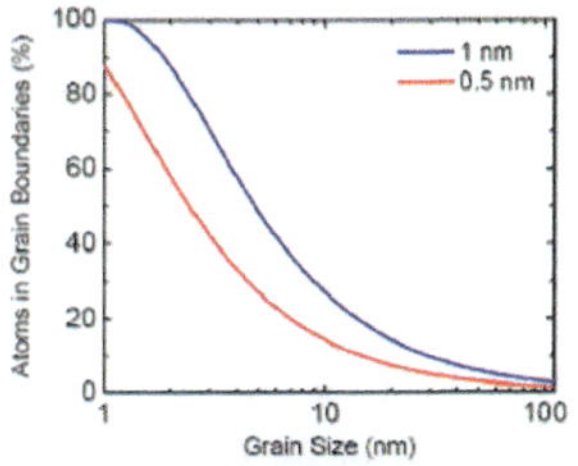

- Using a boundary thickness of 1 nm or 0.5 nm and assuming spherical/cubic grains, the volume fraction of atoms in grain boundaries can be as large as 30-50% for 5 nm, 15-30% for 10 nm and about a few % for 100 nm grain sizes.

- **In the nanometer range, it is generally accepted that grain boundaries act as both sources and sinks for dislocations.**

- Thus, the intrinsic plastic deformation behavior arises from the **interplay between dislocations and grain boundaries**.

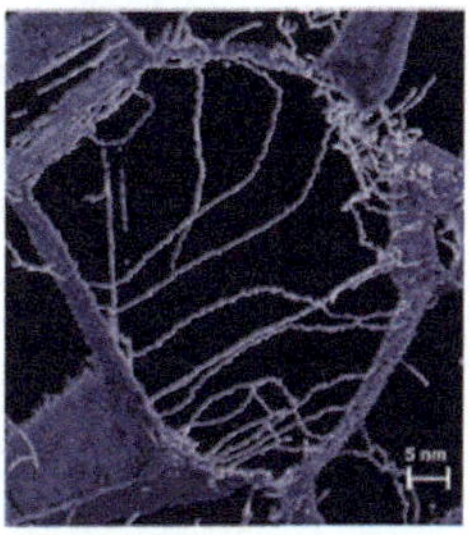

Pile-up of dislocations in a grain (model)
J. Schiøtz, Scripta Materialia 51 (2004) 837

14.8.3 Hardness, Ductility, Toughness of Nanomaterials

- In analogy with yield strength, Hardness can be related to the grain size by a relation deduced from the well-known empirical Hall-Petch equation:
  - $\mathbf{H = H_0 + k_{HP}\, d^{-1/2}}$ where H_0 and k_{HP} are constants.

- In conventional coarse grain materials, plastic deformation takes place mainly by dislocation activity, i.e. dislocation glide and dislocation generation by Frank-Read sources.

- There are several models for H-P effect. At least four different models have been put forward which explain the $\mathbf{d^{-1/2}}$ behavior of the yield stress and give expressions for $\mathbf{k_{HP}}$.
  - Dislocation pile up based explanation
  - Dislocation density based explanations
  - Slipped plane - shear crack modeling
  - Statistical modeling

- **Dislocation pile up based explanation of $H = H_0 + k_{HP}\, d^{-1/2}$**
- The original H-P analysis is based on the concept of dislocation pile-ups at grain boundaries, which generates the stress concentration and is responsible for nucleating slip in the neighboring grain.
- Eshelby et al. (1951) Number of dislocations in a pile up
  - n → number of dislocation in pile
  - α → constant
  - τ_s → average resolved shear stress in the slip plane
  - d → grain size
  - G → shear modulus
  - b → burgers vector

$$n = \frac{\alpha \tau_s d}{Gb}$$

- Stress acting on the lead dislocation
  - **$\tau = \sigma/M$** is the applied shear stress where M is the Taylor factor, which represents the average of the inverse of the Schmid factors for the various grains. In a non textured specimen, M~3.
  - τ_0 → intrinsic lattice resistance to dislocation motion.
  - τ_c → the resolved shear stress required to propagate slip across the grain boundary.

$$\tau_c = n\tau_s = \frac{\alpha \tau_s^2 d}{Gb} \quad \text{where} \quad \tau_s = \tau - \tau_0$$

- **Dislocation pile up based explanation of $H = H_0 + k_{HP}\, d^{-1/2}$**

$$\tau_c = \frac{\alpha(\tau - \tau_0)^2 d}{Gb} \Rightarrow \tau = \tau_0 + kd^{-1/2}$$

  - Thus $K_{H\text{-}P} = kM = M\left(\frac{Gb\tau_c}{\alpha}\right)^{1/2}$
  - Taking $K_{H\text{-}P}$ to represent the activation of slip in the neighboring grain, Armstrong and Smith have proposed the following expression where $M^* = 2.2$ is the Sachs orientation factor and υ is the Poisson's ratio.

$$K_{H\text{-}P} = M\left(\frac{\pi M^* Gb(2-\nu)\tau_c}{4(1-\nu)}\right)^{1/2}$$

R.W. Armstrong, in: T.N. Baker (Ed.), **Yield, Flow and Fracture of Polycrystals**, *Applied Science Publishers, UK, 1983, pp. 1–31.*
R.W. Armstrong, T.W. Smith, in: C. Suryanarayana, J. Singh, F. Fracs (Eds.), **Processing and Properties of Nanocrystalline Materials**, *TMS, Warrendale, PA, 1996, pp. 345–354.*

- **Dislocation density based explanations of $H = H_0 + k_{HP} d^{-1/2}$**
  - Assumptions:

    Grain boundaries act as dislocation sources.

    Dislocation density is inversely proportional to the grain size.

  - The basis of the dislocation density model for the effect of d on σ is the well-known relation between τ and the dislocation density ρ.

$$\tau = \tau_0 + \alpha G b \rho^{1/2} \qquad \sigma_y = \sigma_0 + M\alpha G b \rho^{1/2}$$

σ_0 is due to all defects other than dislocations

M → Taylor factor.

α → constant (1/2π for screw and 1/(2π (1-υ)) edge dislocations respectively),

G → shear modulus

b → Burgers vector.

  - The dislocation density ρ generally increases:

    with shear strain γ=Mε, where ε is the uniaxial strain

    as d decreases.

- **Dislocation density based explanations of $H = H_0 + k_{HP} d^{-1/2}$**
  - Three mechanisms for the increase in ρ as d decreases

    grain boundary ledges provide a source for dislocations,

    m is the ledge density (typically m = 0.02 – 0.2).

$$\rho_L = \frac{8m}{\pi}\frac{1}{d}$$

grain boundaries reduce the average free slip distance $\lambda_s = \alpha_s d$ (with $\alpha_s \leq 1$).

$$\rho_S = \frac{M\varepsilon}{\alpha_S b}\frac{1}{d}$$

ε is the tensile strain.

so-called “geometric dislocations” are needed to prevent voids and overlaps from occurring at the grain boundary due to the inhomogeneous deformation.

$$\rho_g = \frac{\varepsilon}{4b}\frac{1}{d}$$

  - The three dislocation mechanisms may act either independently or concurrently, and either sequentially or simultaneously.

- **Dislocation density based explanations of $H = H_0 + k_{HP}\, d^{-1/2}$. The Hansen analysis.**
  - The stress is divided into two parts.

$$\sigma(\varepsilon) = \sigma_0(\varepsilon) + k(\varepsilon) D_B^{-1/2}$$

$\sigma_0(\varepsilon)$ is independent of the grain size. It is the stress flow of the grain interior, related to the density of dislocations accumulated in dislocation boundaries.

$k(\varepsilon)\, D_B^{1/2}$ is grain-size dependent. (D_B is the average distance between grain boundaries). It is the strength contribution due the additional resistance to dislocation motion caused by the presence of grain boundaries.

The stress $\sigma_0(\varepsilon)$ and $k(\varepsilon)$ are constants at a given strain.

- **"Slip plane - shear crack" based explanations of $H = H_0 + k_{HP}\, d^{-1/2}$**
  - It treats a slip plane within a grain as a shear crack.
  - It gives:

$$K_{HP} = M\tau_c r_c$$

where τ_c is the critical stress required to operate a dislocation source at a distance r_c ahead of the "crack" and M is the Taylor factor.

- **Statistical modeling of $H = H_0 + k_{HP}\, d^{-1/2}$**
  - It gives:

$$K_{HP} = \frac{MKGb}{\ell}$$

  - Where:

K_{HP} depends on obstacle strength,

ℓ is the mean spacing between obstacles

  - Both of these constants are difficult to determine independently. Furthermore, if the obstacles are dislocations, as would often be the case, then this model requires the dislocation density ρ to be independent of the grain size, in direct conflict with the previous explanations.

- $H = H_0 + k_{HP} d^{-½}$
  - The presented "models" are capable of explaining the form of the Hall–Petch equation, but the meaning of the slope K_{HP} remains the object of some debate.
  - The Hall-Petch formulae gives unrealistic strength at nano size region/

    Example of a steel having a strength σ = 600 MPa at d=10 μm and σ_0 =400 Mpa.

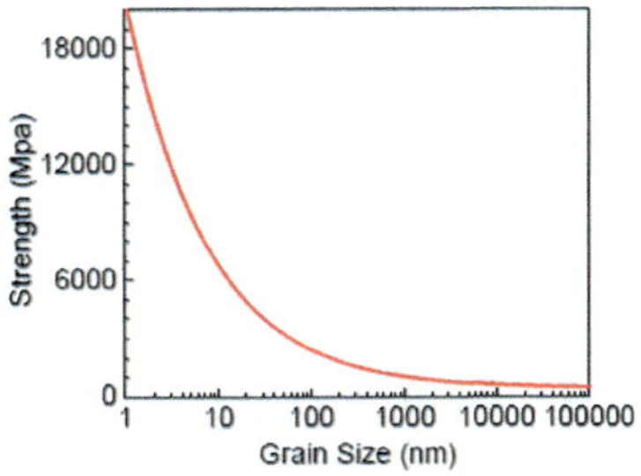

- **Experimental results**

$$\sigma_y = \sigma_0 + kd^{-1/2} \;\; ?$$

- σ_0 is the stress in the absence of grain boundaries, k is a constant.
- **d** is the grain size.

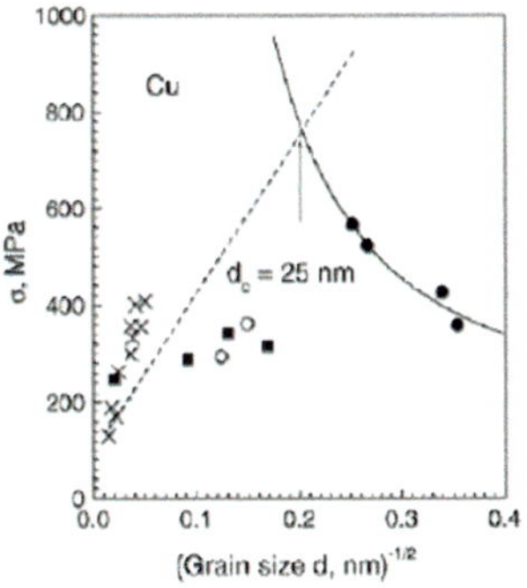

$$H = H_0 + k_{HP} d^{-1/2} \;\; ?$$

- H_0 is the hardness in the absence of grain boundaries, k_{HP} is the Hall-Petch constant.
- **d** is the grain size.

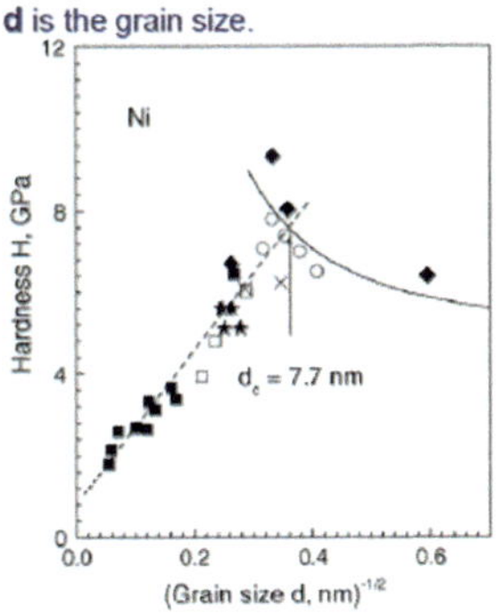

Yield strength in Cu nanograins *Hardness of Ni and Ni(W) nanograins*

GJ Fan et al, Materials Science and Engineering A 409 (2005) 243–248

There is a material softening (inverse Hall-Petch) with grain size for d< 20 nm

- **Experimental results**

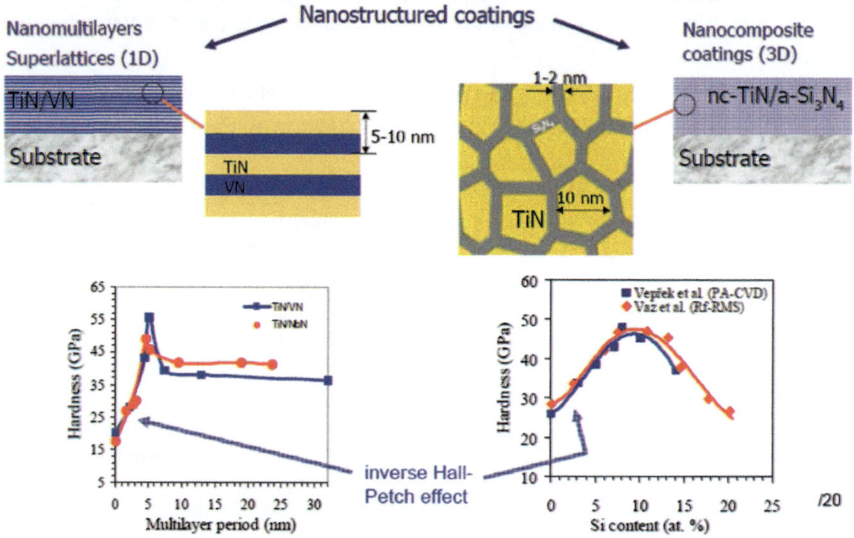

- **Inverse Hall Petch Effect**
  - Strengthening with grain refinement, rationalized by H-P mechanism, holds up for ultra fine crystalline (ufc) regime.
  - By decreasing grain size into the nanometer regime (<100 nm), this process invariably breaks down and hardness versus grain size relationship departs markedly from the H-P relation.
  - **A softening for the very small grain size** (i.e. less than 10-20 nm).

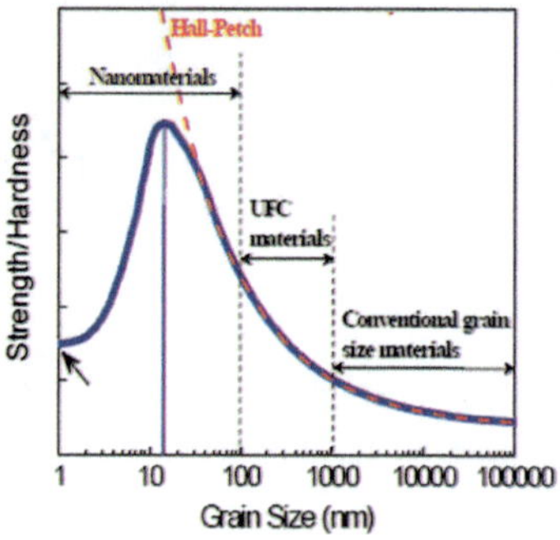

- **Inverse Hall Petch Effect**
  - A number of investigators have suggested that the inverse Hall–Petch relation can be **attributed to the increased grain boundary activity** due to:

    grain-boundary sliding

    and/or the **diffusional mass transfer via grain-boundary diffusion**.

  - Conventional mechanisms for the grain boundary diffusion are described using:

    The Nabarro-Herring creep, based on the vacancy flow through the crystal lattice. It predicts a steady-creep rate: $\dot{\varepsilon} = \frac{14 D_1 \Omega \sigma}{k_B T} d^{-2}$

    where D_1 is the lattice diffusivity and Ω is the atomic volume.

    The Coble creep is based on the vacancy flow along the grain boundaries. It predicts a steady-creep rate: $\dot{\varepsilon} = \frac{148 D_b \Omega e \sigma}{\pi k_B T} d^{-3}$

    where e is the thickness of the grain boundary and D_2 is the diffusivity along the grain boundary.

- **Inverse Hall Petch Effect:** *numerous models*
  - Most of the models proposed to explain the IHPE behavior fall into at least one of four categories:

    Dislocation-based models: *Dislocation motion is the primary agent for plastic flow in nanocrystalline and very small crystallite sizes affect the behavior of dislocations.*

    Diffusion-based models: *Various diffusion processes such as the Coble creep.*

    Grain-boundary-shearing models: *observed in various molecular dynamics simulations.*

    Two-phase-based models: *A nanocrystalline material is a composite material with at least two phases, namely a grain-boundary phase and a grain-interior phase.*

  - In the context of this lively and continuing discussion, a new model was recently suggested based on the nature of dislocation absorption by grain boundaries.

    P_{dis} is the probability of a dislocation being absorbed by a grain boundary.

$$\sigma = \sigma_0 + k(1 - P_{dis}) d^{-1/2}$$

- **Hardness of nanocrystalline materials**

Conventional grain size:
Dislocation mediated plasticity

↓

Nanocrystalline material (d<20nm):
Grain boundary mediated plasticity

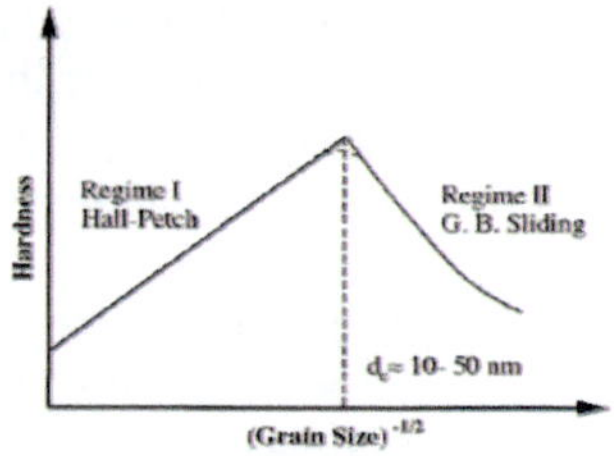

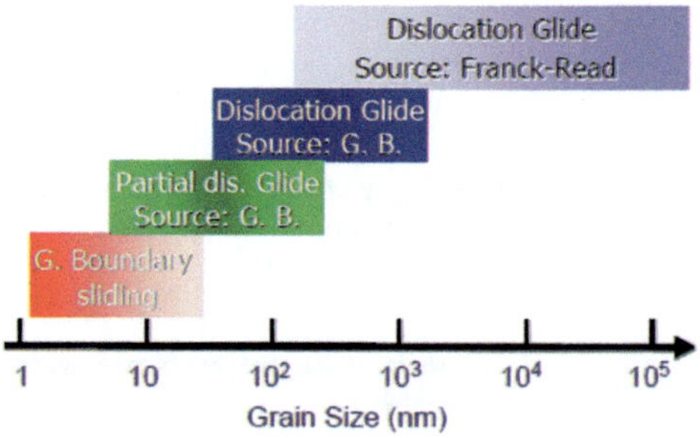

- **Elastic properties**

- Intrinsic elastic moduli of porosity-free nanostructured materials are **essentially the same as those of conventional grain size materials** until the grain size becomes very small, e.g. <5 nm.
- Below this grain size, as the number of atoms associated with the grain boundaries and triple junctions becomes very large, **elastic modulus decreases**.
- Such decrease in elastic modulus is linked to the increase of the Poisson's ratio.
- Modeling of nanocrystalline Cu.
  - At 0K, ~100 GPa in nc-Cu compare to ~150 GPa.
  - At 300K, ~80 GPa compare to the experimental value of 124 GPa for macroscopic Cu.

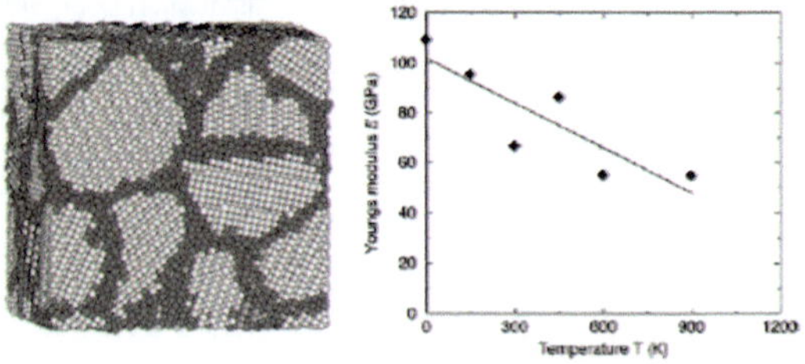

J. Schiotz et al, PHYSICAL REVIEW B VOLUME 60, NUMBER 17, 1999

- **Ductility** is defined as the percentage of elongation or uniform plastic flow prior to mechanical instability or fracture.
- In the micrometric range, metals are more ductile if their mean grain size is low (enhanced grain boundary diffusivity).
- However, if the grain size is lower than ~100 nm, metals become more fragile. It could be the result of the difficulty for dislocations to nucleate and to glide in such small grains.

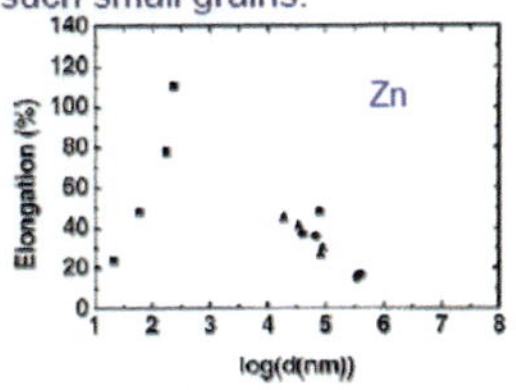

- Nanocrystalline materials exhibit no real improvement of the ductility, compare to micromaterials. In metals, the increase in ductility provided by diffusional mechanisms is apparently insignificant compared to the loss of dislocation-controlled ductility. *Typical ductility of less than 2 % was obtained for nc-metals with grain size less than 25 nm while the same metals, in conventional grain size regime, exhibit 40-60% tensile elongation (ductility)*

X. Zhang et al, App. Phys. Lett. 81, 5 2002

- However, **nanocomposite materials**, i.e. polyphase materials, **may exhibit a good ductility**. The ductility and therefore the toughness will be enhanced provided the particle size d becomes of the same size as the width of the shear localization s (s≈d).

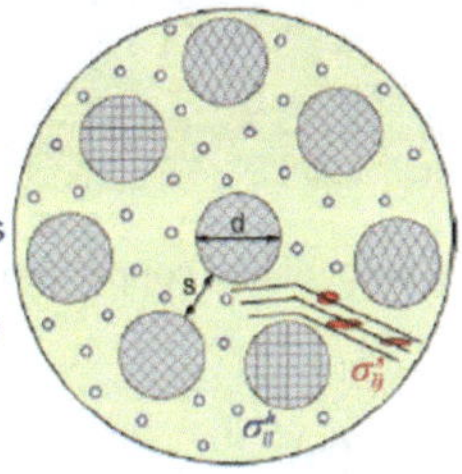

Sam Zhanget al, Surface & Coatings Technology 200 (2005) 1530 – 1534

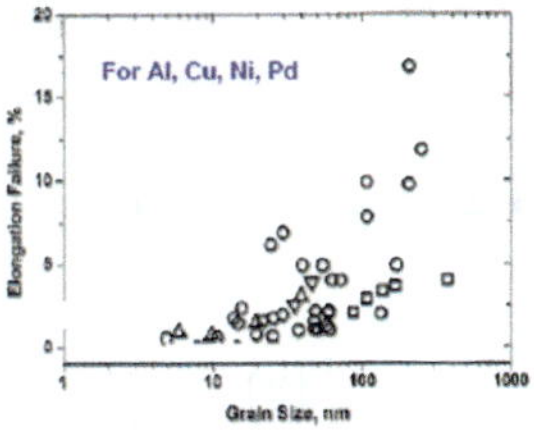

H. S. KIMA et al, App. Phys. Lett. 79, 25, 2001

Nc materials often exhibit low tensile ductility which essentially limits their practical use. The tensile elongation at fracture of nc metals is low relative to their conventional mc counterparts. There are three factors limiting the ductility: the presence of structural artefacts arising from processing, such as porosity and cracks; crack nucleation or propagation instability; plastic instability in tension. Defects obviously decrease a fracture strength as they serve as a crack initiation sites resulting in easy crack nucleation and growth, promoting brittle behavior in tension and causing failure before yielding has chance to occur. Plastic instability originates from the lack of an effective hardening mechanism and/or internal flaws; this instability manifests itself either as shear bands or through "early" necking. During deformation, the dislocation storage and annihilation at grain boundaries are two key competing mechanisms influencing the level of the flow stress. The absence of substantial macroscopic tensile ductility in nanocrystalline fcc metals together with the observation of dimpled rupture on fracture surfaces leads to the hypothesis that deformation is localized. Localized deformation is clearly manifested through the appearance of shear bands on deformed specimen surfaces as shown in Fig. 14.13.

Plastic flow localization in materials in the absence of strengthening leads to the macroscopic necking, Fig. 14.14, followed by the stress concentration in the neck region. Importance of the presence of nano-scale voids in the structure prior to deformation is now well understood. For example, the magnetron sputtered Ni, containing some grain boundary porosity, failed in a brittle manner via rapid coalescence of intergranular cracks whereas in the laser deposited film that

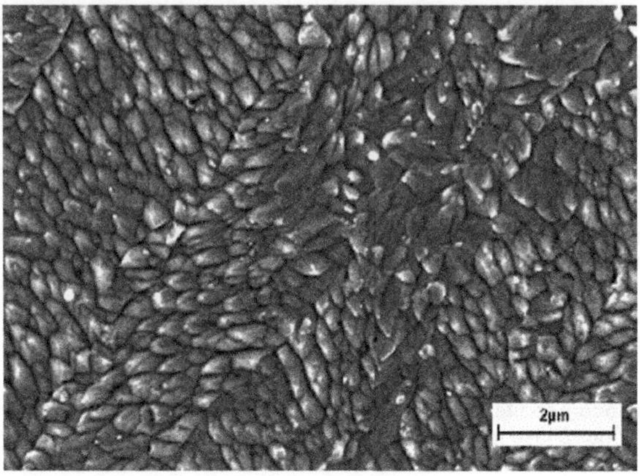

Fig. 14.13 Shear bands in ultra-fine grained copper produced by SPD method

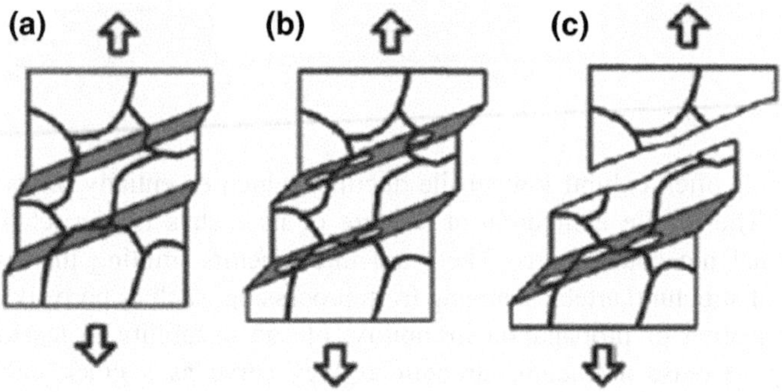

Fig. 14.14 Deformation and fracture of ultra-high-fine materials: **a** Plastic flow localization; **b** nanockrack nucleation; **c** final failure

contained no porosity the crack propagated slowly and is accompanied by continuous film thinning. Fracture surfaces resulting from tensile tests have frequently shown dimpled rupture in nc metals, Fig. 14.15. Further, it has been shown that the dimple size is significantly larger than the average grain size; in addition, a pair of mating fracture surfaces illustrates the presence a significant stretching of the ligaments between the dimples that indicates an appreciable local plasticity. When the grain size is reduced down to ~ 10 nm or less, the resulting fracture surface still shows the dimpled rupture. Only a difference is that the dimple diameter is finer in size (but still much larger than the grain size).

Based on the observations of dimpled rupture, dislocation activity at the crack tip, and the formation of voids at grain boundaries and triple junctions in the regions ahead of the advancing crack, a model for damage evolution and fracture was proposed and schematically depicted in Fig. 14.16. In the early stages of

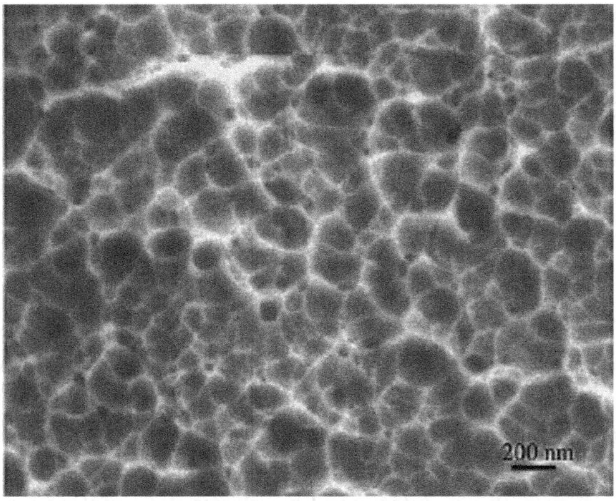

Fig. 14.15 Fracture surface of a 30 nm grain size electrodeposited Ni tensile specimen

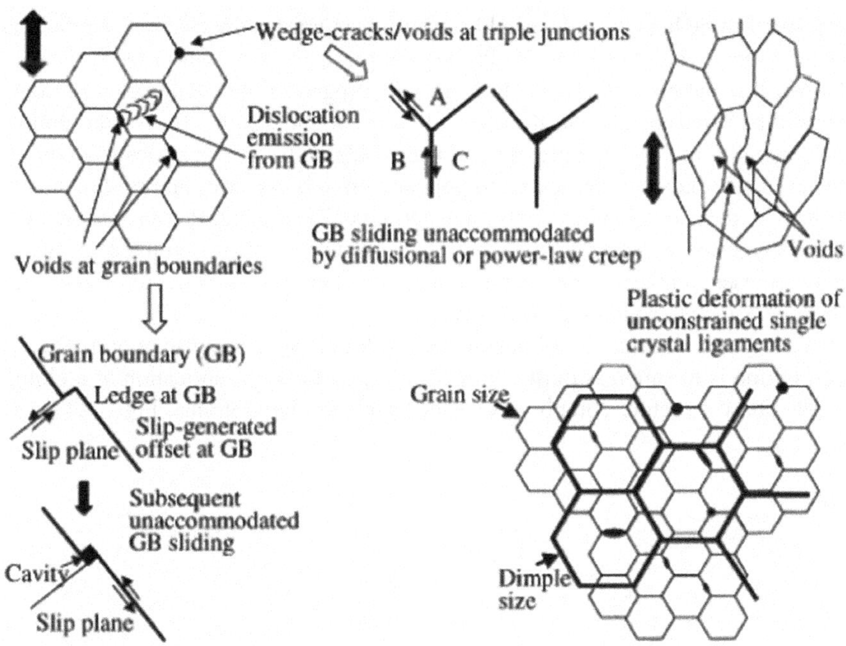

Fig. 14.16 A schematic illustration depicting how deformation evolves in nanocrystalline metals. Dislocation motion, void formation/growth at grain boundaries and triple junctions, the formation of partially unconstrained ligaments that deform plastically, and the interaction of these various features to produce the eventual fracture morphology are all synthesized in this figure (from Kumar et al., *Acta Materialia*, 2003, v. 51, 5743–5774)

deformation, dislocations are emitted from grain boundaries under the influence of an applied stress, when intragranular slip is coupled with unaccommodated grain boundary sliding to facilitate void formation at the grain boundaries. Such voids do not necessarily form at every boundary. Triple junction voids and wedge cracks can also result from grain boundary sliding if resulting displacements at the boundary are not accommodated by diffusional or power law creep. These grain boundary and triple junction voids then act as sites for nucleation of the dimples which are significantly larger than the individual grains and the rim of these dimples on the fracture surface do not necessarily coincide with grain boundaries. Thus, at a local level, the NSM demonstrates considerable plasticity and could represent localized deformation within a shear band. The deformation and fracture processes are closely related to the coupling of dislocation-mediated plasticity and formation and growth of voids. The fundamental difference between the two approaches is that the atomistic simulations reveal intergranular crack propagation, where the GBs that are chosen by the crack path are determined by plastic deformation processes, whereas the model illustrated in Fig. 14.16 proposes the formation of local ligaments with free surfaces as the voids evolve, stretching in concert and finally leading to transgranular fracture. Whatever the fracture mechanism, it is evident that the fracture will be heavily influenced by microstructural features such as the presence of nano-scale voids or even bubbles and the presence of grown-in twins. It is well known that bubbles filled with hydrogen (hydrogen bubbles) are often present in electrodeposited metals and these could serve as nucleation sites for the dimples. Additionally, in NSM the presence of nanovoids of 10–20 vacancies is a common feature, and voids are considered to populate grain boundaries and triple junctions. However, their location has not been experimentally verified. Presence of twins has been suggested as an inter-face control mechanism in coarse-grained metals and may represent a relevant microstructural feature that influences fracture, since many of the NS metals contain grown-in twins.

To improve the ductility of nanomaterials, four basic concepts are proposed. The first option is to suppress plastic flow localization through fabrication of a bimodal single-phase structure composed of nanograins and large grains, Fig. 14.17a. The

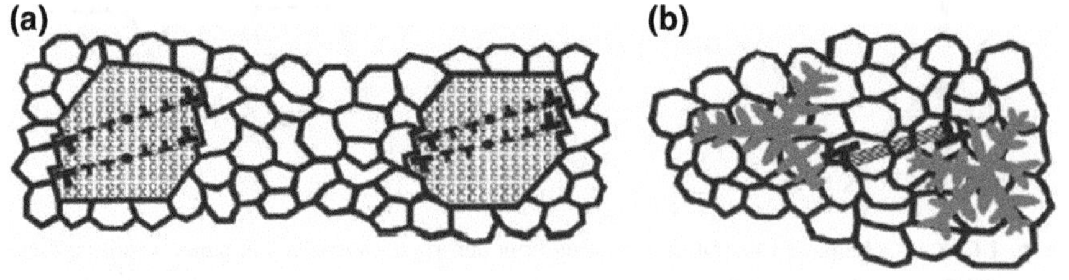

Fig. 14.17 NC materials with high ductility: **a** a bimodal single-phase structure composed of nanograins and large grains; and **b** nano-composite consisting of nanoscale grains and dendrite—like inclusions of the second phase (from I.A. Ovid'ko, *Rev. Adv. Mater. Sci.*, 2005, v. 10, 89–104)

second strategy is to suppress plastic flow localization through fabrication of a composite structure consisting of ductile second—phase inclusions embedded into a nc matrix, Fig. 14.17b. Localization of the plastic flow is effectively hampered due to the strengthening effects provided by large grains and inclusions, respectively. The next option is to deform materials at low temperatures. In this case, recovery or dislocation annihilation, processes at grain boundaries are suppressed and do not compensate the dislocation storage at the boundaries. As a result, a NC specimen shows a good ductility due to strengthening that prevents plastic flow localization.

The fourth option is using a positive strain rate sensitivity of the flow stress. The sensitivity means that a local increase of the plastic strain rate in the neck region leads to a local increase in a flow stress.

- **Toughness**

As the failure mechanism of high strength materials like nanomaterials is governed by brittle fracture, toughness is a key mechanical propertyused in material designing (design criteria), instead of strength or hardness alone.

- **Toughening mechanisms**

  - crack deflection,
  - crack tip shielding (by dislocations, microcracks, phase transformation and ductile phase),
  - crack-wake bridging (by grains, continuous-fibers, short whiskers, second ductile phase),
  - Multilayering.

- **In nanomaterials**

  - Indeed, the original design concept of superhard coatings aimed at obtaining superhardness by preventing all possible plastic deformation mechanisms. The zero plastic deformation means clearly zero ductility and reduced toughness and fracture toughness.

- **Toughness in nanomaterials**

- **Toughness in nanomaterials**
  - One way to improve nanomaterials toughness is to combine a hard nanocrystalline phase with a soft metal matrix (nanocomposites).
  - The concept of hard nanocrystalline carbide and amorphous diamond-like carbon (DLC) phases.
    - To maintain grain size at ~10 nm to restrict initial crack size and create a large volume of grain boundaries;
    - To separate the nanograins within an amorphous matrix thickness above 2 nm to prevent interaction of atomic planes in the adjusted nanocrystalline grains and facilitate grain boundary sliding, but less than 10 nm to restrict path of a straight crack;
    - To produce nanocrystalline grains with random orientation (high angle grain boundaries) to minimize grain incoherence strain and facilitate grain boundary sliding.
  - Toughening can be done using multilayer design.
    - The number and thickness of layers are combined to govern the resulted toughness.

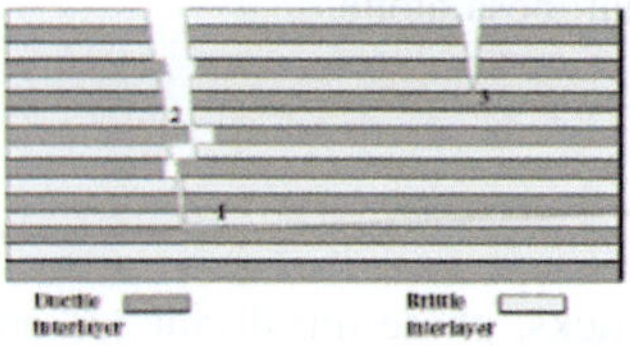

S. Zhang et al. / Surface & Coatings Technology 198 (2005) 2–8

Scratch testing: WC/DLC A.A. Voevodin, J.S. Zabinski / Thin Solid Films 370 (2000) 223±231

- **The mechanical properties depend on the structural parameters**

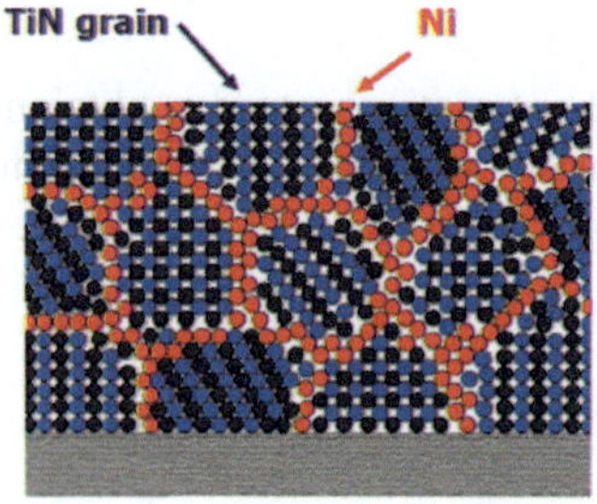

- **Nanocomposite nc-TiN/a-Ni**

  nanocrystalline TiN0.85 with grain size $\mathbf{d_{Ni}}$ in the range 4-12 nm → parameter d_{Ni} surrounded by amorphous like Ni with a thickness $\mathbf{T_{Ni}}$ about 1-2 monolayers.

- **The mechanical properties**

  Hardness, Toughness, Wear resistance.

  They depends on the two parameters $\mathbf{d_{Ni}}$ and $\mathbf{T_{Ni}}$ that characterize the microstructure of the nanocomposite

14.9 Thermal Properties of NSMs

Many properties of the nanoscale materials have been well studied, including the optical electrical, magnetic and mechanical properties. However, the thermal properties of nanomaterials have only shown a slower progress. This is partially due to the difficulties of experimental measuring and controlling the thermal transport in nano scale dimensions. Atomic force microscope (AFM) has been introduced to measure the thermal transport of nanostructures with nanometer-scale high spatial resolution, providing a promising way to probe the thermal properties. Moreover, the theoretical simulations and analysis of thermal transport in nanostructures are still in infancy. Available approaches including numerical solutions of Fourier's law, computational calculation based on Boltzmann transport equation and molecular-dynamics (MD) simulation. More important, as the dimensions go down into nanoscale, the definition of temperature becomes questionable. In non-metallic material system, the thermal energy is mainly carried by phonons, which have a wide variation in frequency and the mean free paths (mfp). The heat carrying photons often have large wave vectors and mfp in the order of nanometer range at room temperature, so that the dimensions of the nanostructures are comparable to the mfp and wavelengths of photons. However the general definition of temperature is based on the average energy of a material system in equilibrium. For macroscopic systems, the dimension is large enough to define a local temperature in each region within the materials and this local temperature will vary from region to region, so that one can study the thermal transport properties of the materials based on certain temperature distributions in the material. But for NC systems, the dimensions may be too small to define a local temperature. Moreover, it is also problematic to use the concept of temperature which is defined in equilibrium conditions for the non-equilibrium processes of thermal transport in nanomaterials posing difficulties for theoretical analysis of thermal transport in nano scales.

In spite of all the difficulties in both experimental and theoretical characterization the thermal properties of nanomaterials, recent advances in experiments have shown that certain nanomaterials have extraordinarily thermal properties compared to their macroscopic counterparts. For example, silicon nanowires have a much smaller thermal conductivities compared to bulk silicon. Because of tubular structures of carbon nanotubes, they have extreme high thermal conductivity in axial directions and high anisotropy in the heat transport over the specimen. Interfaces are also very important factor for determine the thermal properties of nanomaterials. Generally, the internal interfaces impede the flow of heat due to photon and phonon scattering. At interface or grain boundary between similar materials, the interface disorder scatters phonons, while as the differences in elastic properties and densities of vibration states affect the transfer of vibration energy across interfaces between dissimilar materials. As a result, the nc structures with high interfaces densities reduce the thermal conductivity of the materials. These interconnected factors joined together to determine the special thermal properties of the nanomaterials.

For instance, carbon nanotubes are carbon nanostructures relating to diamond and graphite, which are well known for their high thermal conductivities. The stiff *sp3* bonds in diamond structure result in high phonon velocity and consequently high thermal conductivity. In carbon nanotubes, the carbon atoms are held together by the even stronger *sp2* bonds, so that the nanotube structures, consisting of seamlessly joined graphitic cylinders are expected to have extraordinary high thermal conductivity. The rigidity of the these nanotubes, combined with virtual absence of atomic defects or coupling to soft photon modes of the embedding medium, should make isolated nanotubes very good candidates for efficient thermal conductors. On the other hand, one-dimensional nanowires may offer ultra-low thermal conductivities, quite different from that of carbon nanotubes. In nanowires, phonons behave differently from those in the corresponding bulk materials due to the quantum confinement in the one dimension structures. The nanowires surface can introduce surface phonon modes, resulting in many different phonons polarizations other than the two transverse and one longitudinal acoustic branch found in the bulk semiconductors. Those changes in the dispersion relation can modify the group velocity and the density of states of each branch. The phonon lifetime also changes due to the strong phonon–phonon interactions and the boundary scattering within the nanostructures. Thus the phonon transports and the thermal properties of the nanowires will be significantly different from that of the bulk materials.

The phonon transport of the semiconducting nanowires has been studied experimentally and theoretically. Measurement of the thermal conductivity of silicon nanowires using a micro fabricated suspended device over a temperature range of 20–320 K shows that although the nanowires had a well-defined crystalline order as in bulk materials, the observed thermal conductivity was more than two orders of magnitude smaller than that of bulk silicon, which also shows a strong dependence on the nanowires size. For a silicon nanowire with 22 nm diameter, the thermal conductivity was reduced to 10 W/m range. The appreciable change of the thermal conductivity compared to the bulk materials was ascribed to phonon–boundary scattering and the possible change in phonon dispersion due to confinement within the nanostructures. Besides the one dimensional nanowires, multi-layers and super lattices are another type of nc structures offering low thermal conductivity. Multi-layers and superlattices are thin films consisted of alternating layers of two or more different materials stacked upon each other. In multi-layer structure, the films can be either amorphous or polycrystalline while in superlattices the films are single crystal. There are many effects in the multi-layers or superlattices structures that affect the phonon transport properties. When alternating layers of materials are stacked together, many collective mode of phonon transport may appear besides the phonon modes in each single layer. This stacking effect will be more apparent when the coherence phonon length scales are much larger than that of a single layer. It is also a coupled behaviour resulting from interference of phonon waves reflected from multiple interfaces. When the mean free path of phonons span multiple interfaces, the phonon dispersion relation is modified and zone folding occurs, even resulting in multiple phonon band gaps. Moreover, due to the modification of the phonon dispersion, the phonon group velocities will be reduced significantly and the scattering rate will also be increased. The interfaces is also an important factor in determine the phonon transport properties due to the high density of interfaces in the multi-layers or superlattices structures, as well. For example, if two materials in the superlattice have large mismatch in the phonon dispersion relations, phonons in certain frequency range cannot propagate to the neighbouring layer unless there are mode conversions at the interface. The interfaces between two different materials with different lattice constants can contain dislocations and defects, which can also scatter phonons and reduce thermal conductivity. Physical roughness and alloying may also exist at the interface depending on the processing and affect the phonon transport. The overall effect in of these factors on the phonon transport is a general decrease of thermal conductivities.

For example, atomic layer deposition and magnetron sputter deposition were used to synthesize thin-film multi-layers of W/Al_2O_3. With individual layers only of a few nanometers thick, the high interface density produced a strong impediment to heat transfer, giving an ultra-low thermal conductivity of 0.6 W/mK.

Multi-layers and superlattice nanostructures have many potential applications. For example, multilayer thin films can be used as thermal barriers at high temperatures environments, such as in engines to improve their efficiencies; epitaxial superlattices of semiconductor films with low thermal conductivity can be used in thermoelectric power generation because of the ability to the control both electronic band structure and phonon transport simultaneously. The use of nanofluid to enhance the thermal transport is another promising application of the thermal properties of nanomaterials. Nanofluids are generally referred to the solid-liquid composite materials, which consist of nanomaterials of size in the range 1–100 nm suspended in a liquid. Nanofluids hold increasing attentions in both research and practical applications due to their greatly enhanced thermal properties compared to their base fluids. Many type of nanomaterials can be used in nanofluids including nanoparticles of oxides, nitrides, metals, metal carbides, and nanofibers such as single wall and multi wall carbon nanotubes, which can be dispersed into a variety of base liquid depending on the possible applications, such as water, ethylene glycol, and oils. The most important features of nanofluids are the significant increase of thermal conductivity compared with liquids without nanomaterials, which have been proved by many experimental works.

14.10 Chemical Properties of NSMs

One of the important factors for the chemical applications of nanomaterials is the increment of their surface area which increases the chemical activity of the material. Due to their enhanced chemical activity, nanostructure materials can be used as *catalysts* to react with such noxious and toxic gases as carbon monoxide and nitrogen oxide in automobile catalytic converters and power generation equipment to prevent environmental pollution arising from burning gasoline and coal. Bulk gold is chemically inert and thus considered to be not active or useful as a catalyst.

However, gold nanoparticles can have excellent catalytic properties. For example, gold nanoparticles with clean surface have demonstrated to be extremely active in the oxidation of carbon monoxide if deposited on partly reactive oxides, such as Fe_2O_3, NiO and MnO, alumina and titania and are also found to be reactive. Au nanoparticles also exhibit extraordinary high activity for partial oxidation of hydrocarbons, hydrogenation of unsaturated hydrocarbons and reduction of nitrogen oxides. Fuel cell technology is another important application of the noble metal nanoparticles relating the catalysis of the reactions. In the present, the fuel cell catalysts are based on platinum group metals (PGM). Pt and Pt-Ru alloys are some of the most frequently used catalysts from this group. In fact, the use of these metals is one major factor for cell costs, which has been one of the major drawbacks preventing it from growing into a more important technology. One possibility to produce economical catalysts is the use of bimetallic nanoparticles.

14.11 Physicochemical Properties of NPs

Various physicochemical properties such as large surface area, mechanically strong, optically active and chemically reactive make NPs unique and suitable applicants for various applications. Some of their important properties are discuss in the following section.

14.11.1 Electronic and Optical Properties

The optical and electronic properties of NPs are interdependent to greater extent. For instance, noble metals NPs have size dependent optical properties and exhibit a strong UV–visible extinction band that is not present in the spectrum of the bulk metal. This excitation band results when the incident photon frequency is constant with the collective excitation of the conduction electrons and is known as the localized surface plasma resonance (LSPR). LSPR excitation results in the wavelength selection absorption with extremely large molar excitation coefficient resonance Ray light scattering with efficiency equivalent to that of ten fluorophores and enhanced local electromagnetic fields near the surface of NPs that enhanced spectroscopies. It is well established that the peak wavelength of the LSPR spectrum is dependent upon the size, shape and interparticle spacing of the NPs as well as its own dielectric properties and those of its local environment including the substrate, solvents and adsorbates [1]. Gold colloidal NPs are accountable for the rusty colors seen in blemished glass door/windows, while Ag NPs are typically yellow. Actually, the free electrons on the surface in these NPs (d electrons in Ag and gold) are freely transportable through the nanomaterial. The mean free path for Ag and gold is $\sim$ 50 nm, which is more than the NPs size of these materials. Thus,

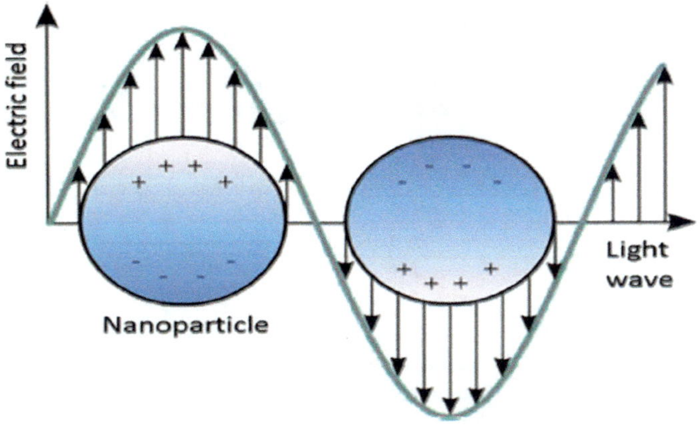

Fig. 14.18 Graphical illustration exemplifying the localized surface plasmon (LSPR) on nanoparticle outer surface [2, 3]

no scattering is expected from the bulk, upon light interaction, instead they set into a standing resonance conditions, which is responsible for LSPR in these NPs (Fig. 14.18) [2, 3].

14.11.2 Magnetic Properties

Magnetic NPs are of great curiosity for investigators from an eclectic range of disciplines, which include heterogenous and homogenous catalysis, biomedicine, magnetic fluids, data storage magnetic resonance imaging (MRI), and environmental remediation such as water decontamination. The literature revealed that NPs perform best when the size is < critical value i.e. 10–20 nm [4]. At such low scale the magnetic properties of NPs dominated effectively, which make these particle priceless and can be used in different applications [4–7]. The uneven electronic distribution in NPs leads to magnetic property. These properties are also dependent on the synthetic protocol and various synthetic methods such as solvothermal [8], coprecipitation, micro-emulsion, thermal decomposition, and flame spray synthesis can be used for their preparation [9].

14.11.3 Mechanical Properties

The distinct mechanical properties of NPs allow researchers to look for novel applications in many important fields such as tribology, surface engineering, nanofabrication and nanomanufacturing. Different mechanical parameters such as

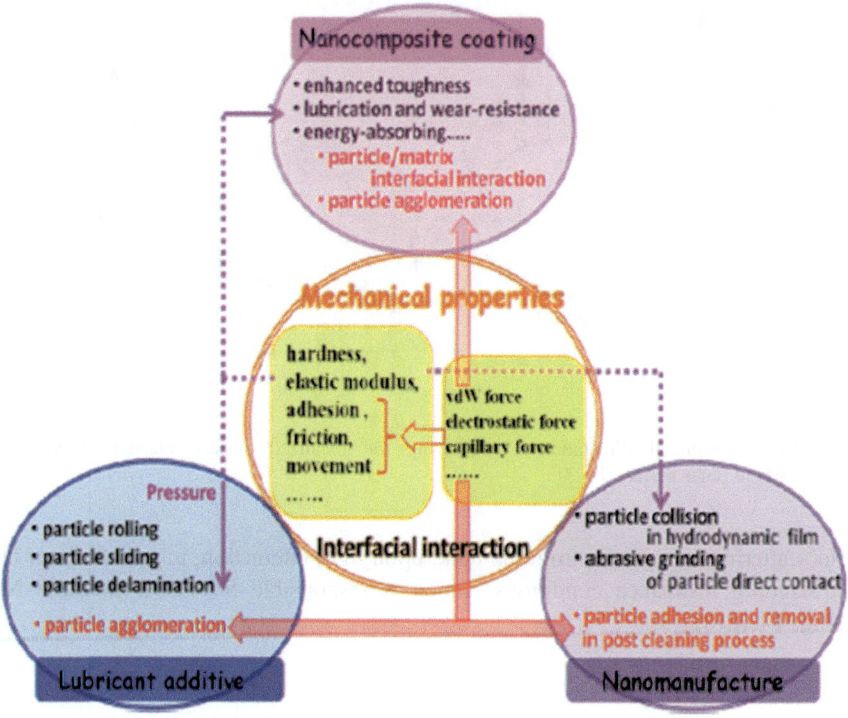

Fig. 14.19 Schematic view of the mechanical properties and their applications [10]

elastic modulus, hardness, stress and strain, adhesion and friction can be surveyed to know the exact mechanical nature of NPs. Beside these parameters surface coating, coagulation, and lubrication also aid to mechanical properties of NPs [10] (as shown in Fig. 14.19).

NPs show dissimilar mechanical properties as compared to microparticles and their bulk materials. Moreover, in a lubricated or greased contact, the contrast in the stiffness between NPs and the contacting external surface controls whether the NPs are indented into the plan surface or deformed when the pressure at contact is significantly large. This important information could divulge how the NPs perform in the contact situation. Decent controls over mechanical features of NPs and their interactions with any kind of surface are vital for enlightening the surface quality and elevating material removal. Fruitful outcomes in these fields generally need a deep insight into the basics of the mechanical properties of NPs, such as elastic modulus and hardness, movement law, friction and interfacial adhesion and their size dependent characteristics [10].

14.11.4 Thermal Properties

It is well-known fact that metals NPs have thermal conductivities higher than those of fluids in solid form. For example, the thermal conductivity of copper at room temperature is about 700 times greater than that of water and about 3000 times greater than that of engine oil. Even oxides such as alumina (Al_2O_3) have thermal conductivity higher than that of water. Therefore, the fluids containing suspended solid particles are expected to display significantly enhanced thermal conductivities relative to those of conventional heat transfer fluids. Nanofluids are produced by dispersing the nanometric scales solid particles into liquid such as water, ethylene glycol or oils. Nanofluids are expected to exhibit superior properties relative to those of conventional heat transfer fluids and fluids containing microscopic sized particles. Because the heat transfer takes place at the surface of the particles, it is desirable to use the particles with large total surface area. The large total surface area also increases the stability suspension [11]. Recently it has been demonstrated that the nanofluids consisting of CuO or Al_2O_3 NPs in water or ethylene exhibit advance thermal conductivity [12].

References

1. Eustis, S., El-Sayed, M.A.: Why gold nanoparticles are more precious than pretty gold: noble metal surface plasmon resonance and its enhancement of the radiative and nonradiative properties of nanocrystals of different shapes. Chem. Soc. Rev. **35**, 209–217 (2006). https://doi.org/10.1039/B514191E
2. Khlebtsov, N., Dykman, L.: Plasmonic nanoparticles. pp. 37–85 (2010). http://dx.doi.org/10.1201/9781439806296-c2
3. Khlebtsov, N.G., Dykman, L.A.: Optical properties and biomedical applications of plasmonic nanoparticles. J. Quant. Spectrosc. Radiat. Transf. **111**, 1–35 (2010). https://doi.org/10.1016/j.jqsrt.2009.07.012
4. Reiss, G., Hütten, A.: Magnetic nanoparticles: applications beyond data storage. Nat. Mater. **4**, 725–726 (2005). http://dx.doi.org/ https://doi.org/10.1038/nmat1494
5. Faivre, D., Bennet, M.: Materials science: magnetic nanoparticles line up. Nature **535**, 235–236 (2016). https://doi.org/10.1038/535235a
6. Priyadarshana, G., Kottegoda, N., Senaratne, A., de Alwis, A., Karunaratne, V., Priyadarshana, G., Kottegoda, N., Senaratne, A., de Alwis, A., Karunaratne, V.: Synthesis of magnetite nanoparticles by top-down approach from a high purity ore. J. Nanomater. **2015**, 1–8 (2015). https://doi.org/10.1155/2015/317312
7. Zhu, Y., Goodridge, A.G., Stapleton, S.R.: Zinc, vanadate and selenate inhibit the tri-iodothyronine-induced expression of fatty acid synthase and malic enzyme in chick-embryo hepatocytes in culture. Biochem. J. **303**, 213–216 (1994)
8. Qi, M., Zhang, K., Li, S., Wu, J., Pham-Huy, C., Diao, X., Xiao, D., He, H.: Superparamagnetic Fe3O4 nanoparticles: synthesis by a solvothermal process and functionalization for a magnetic targeted curcumin delivery system. New J. Chem. **4480**, 4480–4491 (2016). https://doi.org/10.1039/c5nj02441b
9. Wu, W., He, Q., Jiang, C.: Magnetic iron oxide nanoparticles: synthesis and surface functionalization strategies. Nanoscale Res. Lett. **3**, 397–415 (2008). https://doi.org/10.1007/s11671-008-9174-9

10. Guo, D., Xie, G., Luo, J.: Mechanical properties of nanoparticles: basics and applications. J. Phys. D Appl. Phys. **47**, 13001 (2014). https://doi.org/10.1088/0022-3727/47/1/013001
11. Lee, S., Choi, S.U.-S., Li, S., Eastman, J.A.: Measuring thermal conductivity of fluids containing oxide nanoparticles. J. Heat Transfer **121**, 280–285 (1999). https://doi.org/10.1115/1.2825978
12. Cao, Y.C.: Nanoparticles with Raman spectroscopic fingerprints for DNA and RNA detection. Science **80**(297), 1536–1540 (2002). https://doi.org/10.1126/science.297.5586.1536

Chapter 15
Applications of Nanomaterials and Nanoparticles

Considering the unique properties discussed in Chap. 14, NSMs and NPs can be used in variety of applications. Some important of these are given below.

It is evident that nanomaterials split their conventional counterparts because of their superior chemical, physical, and mechanical properties and of their exceptional formability.

Due to their incredible properties, nanoparticles have become significant in many representative applications in recent years such as: biomedical devices, drug-delivery, flat-panel displays, medical imaging, high energy batteries, sensors, paints and pigments, automobile components, formable ceramics, high-power magnets, smart structures, CMP slurries, KE penetrators/warheads, catalysts, phosphors, dielectrics, full cells, optical devices, cutting tools, cosmetics, aerospace components, health care, environment, agriculture and medicines, etc.

Figure 15.1a, b represents some types of metal nanoparticles and their applications in our biotechnology life.

15.1 Applications of NMs in Mechanical Industries

Enhanced mechanical properties of the nanomaterials have many potential applications both in nano scale such as mechanical nano resonators, mass sensors, microscope probe tips and nano tweezers for nano scale object manipulation, and in macro scale applications such as structural reinforcement of polymer materials, light weight high strength materials, flexible conductive coatings, wear resistance coatings, tougher and harder cutting tools etc. Cutting tools made of nanomaterials, such as tungsten carbide, tantalum carbide, and titanium carbide, are much harder, much more wear-resistant, erosion-resistant, and last longer than their conventional (large-grained) counterparts. Also, for the miniaturization of microelectronic circuits, the industry needs micro drills (drill bits with diameter less than the thickness of an average human hair or 100 μm) with enhanced edge retention and far better

L. H. Madkour, *Nanoelectronic Materials*, Advanced Structured Materials 116, https://doi.org/10.1007/978-3-030-21621-4_15

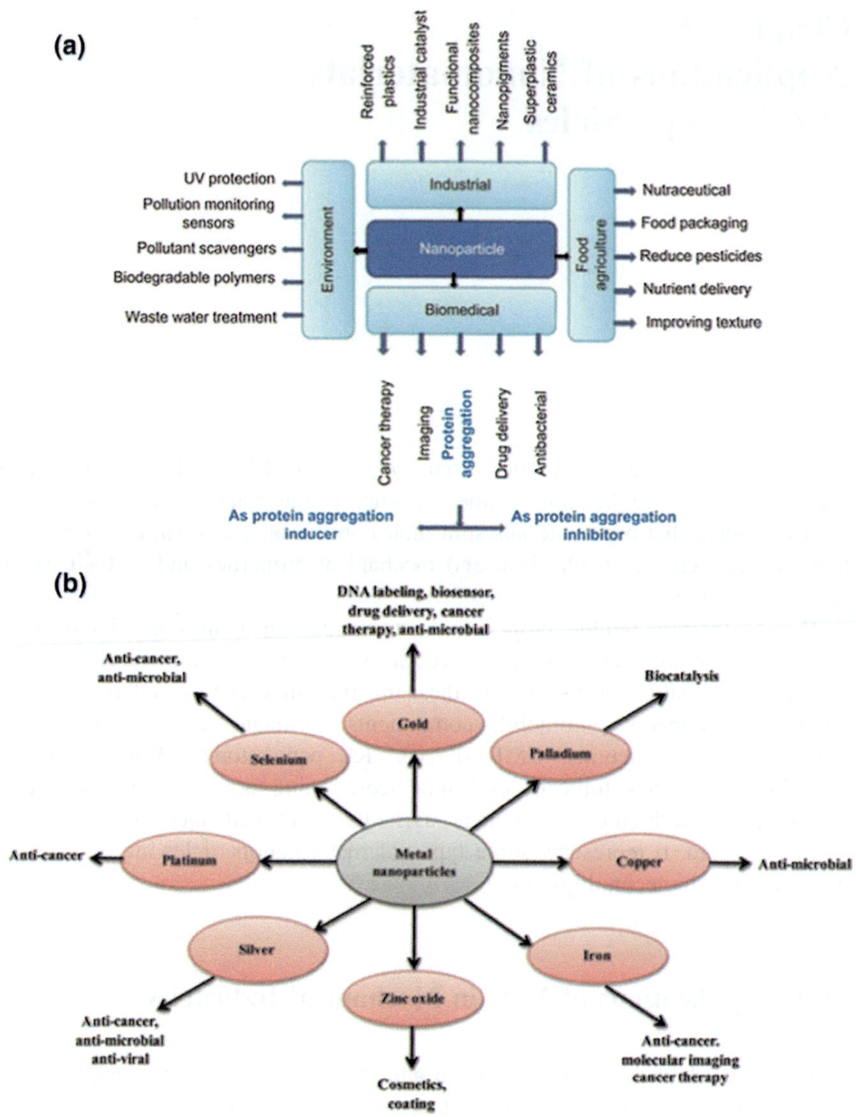

Fig. 15.1 **a** and **b** Types of metal nanoparticles and their applications in biotechnology

wear resistance. Since nanocrystalline, carbides are much stronger, harder, and wear-resistant; they are currently being used in these micro drills. In automobiles, nanomaterials are envisioned to be used in spark plugs. Also, automobiles waste significant amounts of energy by losing the thermal energy generated by the engine.

So, the engine cylinders are envisioned to be coated with nanocrystalline ceramics, such as zirconia and alumina, which retain heat much more efficiently that result in complete and efficient combustion of the fuel. One of the key properties required of the aircraft components is the fatigue strength, which decreases with the component's age. The fatigue strength increases with a reduction in the grain size of the material. Nanomaterials provide such a significant reduction in the grain size over conventional materials that the fatigue life is increased by an average of 200–300%. In spacecrafts, elevated-temperature strength of the material is crucial because the components (such as rocket engines, thrusters, and vectoring nozzles) operate at much higher temperatures than aircrafts and higher speeds. Nanomaterials are perfect candidates for spacecraft applications, as well. Ceramics are very hard, brittle, and hard to machine even at high temperatures. However, with a reduction in grain size, their properties change drastically. Nanocrystalline ceramics can be pressed and sintered into various shapes at significantly lower temperatures. Zirconia, for example, is a hard, brittle ceramic, has even been rendered superplastic, i.e., it can be deformed to great lengths (up to 300% of its original length). However, these ceramics must possess nanocrystalline grains to be superplastic. Ceramics based on silicon nitride (Si_3N_4) and silicon carbide (SiC), have been used in automotive applications as high-strength springs, ball bearings, and valve lifters, and because they possess good formability and machinability combined with excellent physical, chemical, and mechanical properties. They are also used as components in high-temperature furnaces. Aerogels are nanocrystalline porous and extremely lightweight materials and can withstand 100 times their weight. They are currently being used for insulation in offices, homes, etc. They are also being used as materials for "smart" windows, which darken when the sun is too bright and they lighten themselves otherwise.

As revealed from their mechanical properties through excellent young modulus, stress and strain properties, NPs can offer many applications in mechanical industries especially in coating, lubricants and adhesive applications. Besides, this property can be useful to achieve mechanically stronger nanodevices for various purposes. Tribological properties can be controlled at nanoscale level by embedding NPs in the metal and polymer matrix to increase their mechanical strengths. It is because, the rolling mode of NPs in the lubricated contact area could provide very low friction and wear. In addition, NPs offer good sliding and delamination properties, which could also effect in low friction and wear, and hence increase lubrication effect [1]. Coating can lead to various mechanically strong characteristics, as it improves toughness and wear resistance. Alumina, Titania and carbon based NPs successfully demonstrated to get the desirable mechanical properties in coatings [2–4].

15.1.1 Functional Coatings and Layers

Application areas

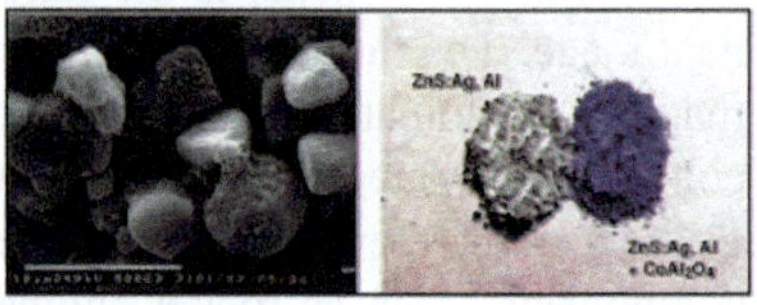

Nanoparticle coatings

- **Pigmentation of display phosphors, e.g. applied in CRTs or PDPs**
- **Protective coatings onto μ-scale pigments and μ-scale phosphors**
- **Dielectric coatings onto electroluminescence phosphors (EL)**

Nanoparticle layers

- **Colour filter, e.g. onto incandescent lamps (Philips Blue Vision)**
- **Interference filter onto display glass + light bulbs**
- **Transparent converter layer in LEDs**
- **Lamp production: Improvement of layer morphology and adhesion**

Coating materials

Requirements

- **(photo)chemical and thermal stability**
- **high transparency**

Protective materials		Band gap [eV]	PZC [pH]
• $LaPO_4$		8.7	7.8
• SiO_2		8.4	2-3
• MgO		8.0	12.0
• γ-Al_2O_3		7.5	9.0
• Y_2O_3		5.6	9.1
• La_2O_3		5.5	10.5
• C (diamond) "DLC"		5.4	?
• SnO_2		3.6	4.5
• ZnO		3.4	10.5
Colour filter			
• α-Fe_2O_3	rot	1.9	6.7
• $CoAl_2O_4$	blau	?	9 - 10

α-Fe_2O_3 Nanoparticles

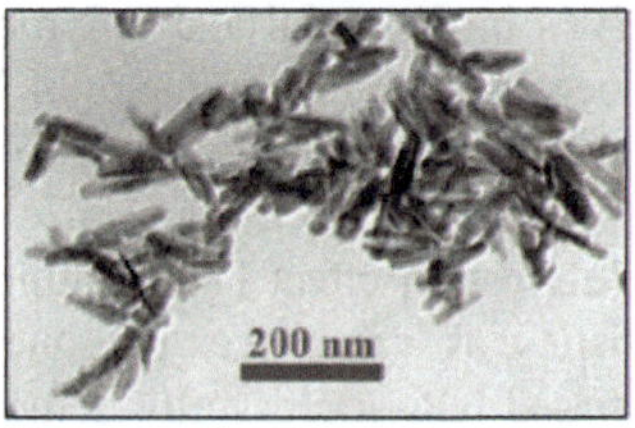

Protective MgO coating on $BaMgAl_{10}O_{17}$:Eu (BAM) "Blue Phosphor for PDPs"

Process

- $Mg(NO_3)_2$ + urea + BAM in H_2O
- Homogeneous pH value enhancement: $H_2N\text{-}CO\text{-}NH_2 + H_2O \rightarrow 2\,NH_3 + CO_2$
  $NH_3 + H_2O \rightleftharpoons NH_4^+ + OH^-$
  $Mg^{2+} + 2\,OH^- \rightarrow Mg(OH)_2$
- Sintering at 600°C: $Mg(OH)_2 \rightarrow MgO + H_2O \Rightarrow$ MgO nm particles onto BAM particles

SEM image $BaMgAl_{10}O_{17}$:Eu

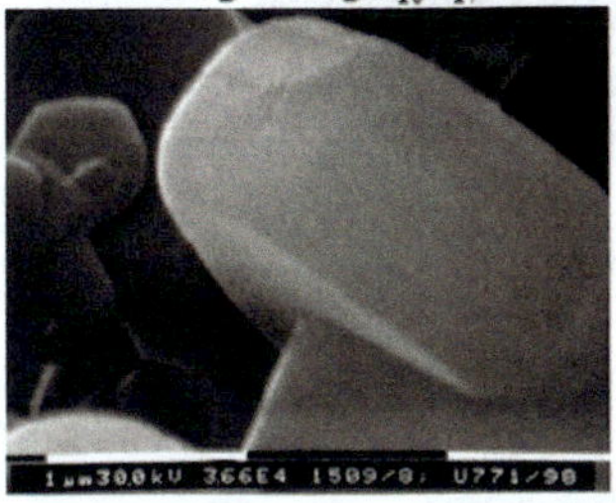

SEM image $BaMgAl_{10}O_{17}$:Eu (MgO)

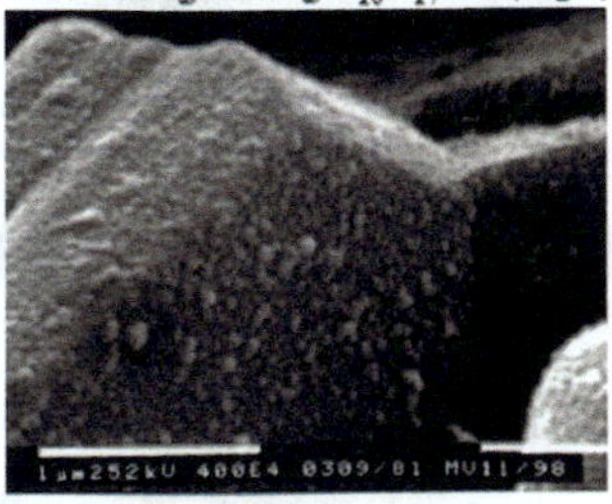

Protective La_2O_3 coating on $BaSi_2O_5$:Pb (BSP) "UV-A Phosphor for tanning lamps"

Application problem

BSP has a low PZC and thus shows fast Hg^+ take-up in fluorescent lamps, which results in strong light output decrease

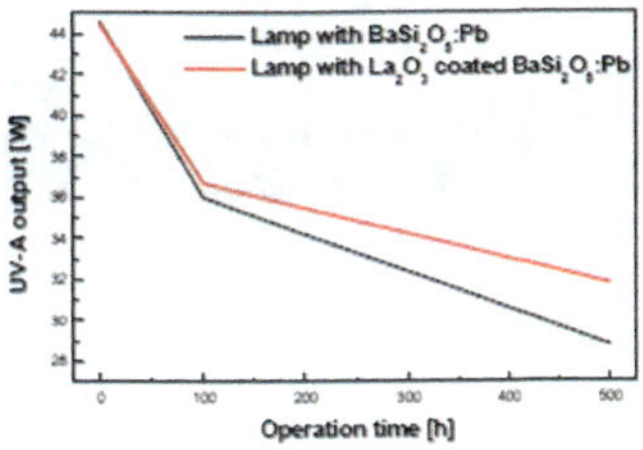

Solution

Coating by a high PZC material as e.g. La_2O_3 or Al_2O_3

Coating process

- Neutral to alkaline suspension pH 7 - 10
- Stabilisation of La^{3+} required, e.g. by complexation with EDTA

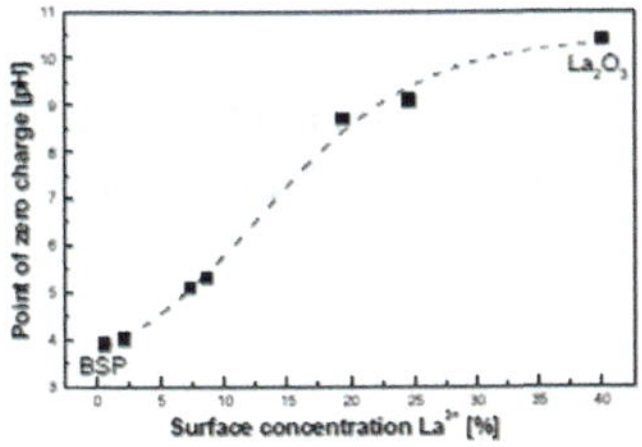

SEM images **TEM images**

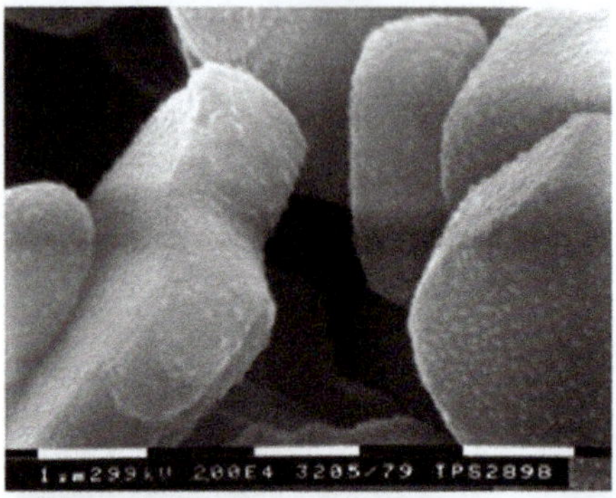

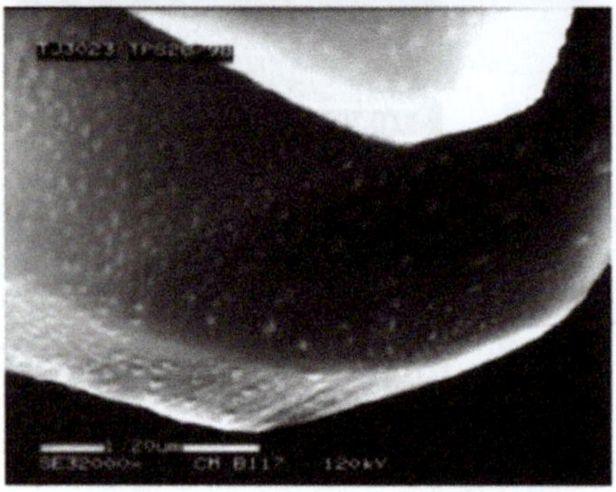

Protectice SiO_2 coating on (Ca,Sr)S:Eu "Red LED phosphor"

Application problem

- (Ca,Sr)S:Eu is highly sensitive towards H_2O and CO_2
- (Ca,Sr)S:Eu is highly refractive ($n > 2.0$)

Coating process

$Si(OEt)_4 + 2\ H_2O \rightarrow SiO_2 + 4\ EtOH$

"Hydrolysis of TEOS in EtOH"

SiO_2 coated (Ca,Sr)S:Eu

- Coating reduces diffusion of H_2O and CO_2 to the particle surface
- Coating is nano structured which results in refractive index gradient layer (anti-reflective surface)

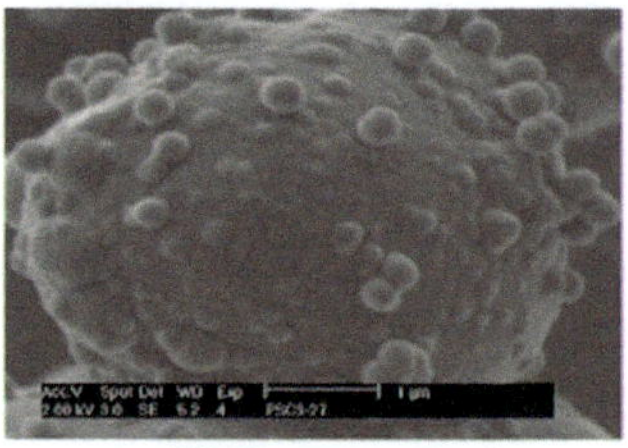

„Moth eye coating"

Pigmentation of ZnS:Ag by $CoAl_2O_4$

- **Co-prezipitation of Co^{2+} and Al^{3+} by Hydrolysis of $Co(CH_3COO)_2$ und $Al(CH_3COO)_3$ in aqueous solution:**

  $$Co^{2+} + Al^{3+} + 5\,OH^- \rightarrow Co(OH)_2 + Al(OH)_3$$

- **Sintering: Hydroxides $\rightarrow$ $CoAl_2O_4$**

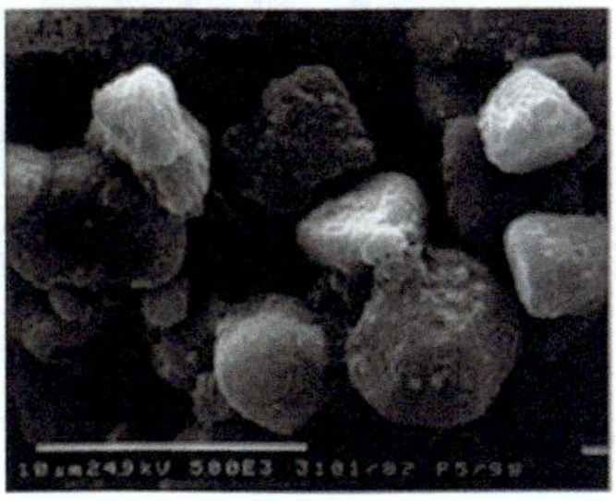

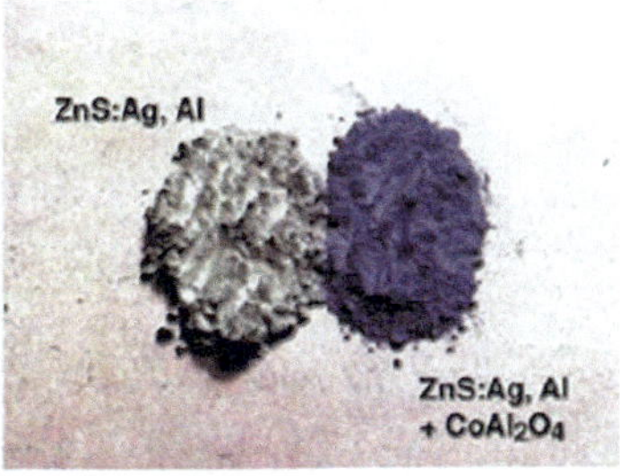

Interference layers

⇒ **Anti-reflective coatings (light outcoupling)**

- **stacks of layers (high and low index of refraction)**
- **monolayer of monodisperse spherical particles**

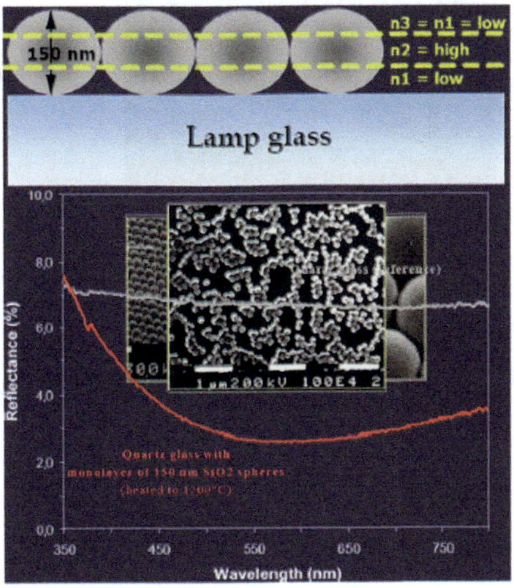

Interference layers

⇒ **Moth eye coatings (light incoupling)**

Fraunhofer Institute Freiburg

Solar cells with enhanced efficiency

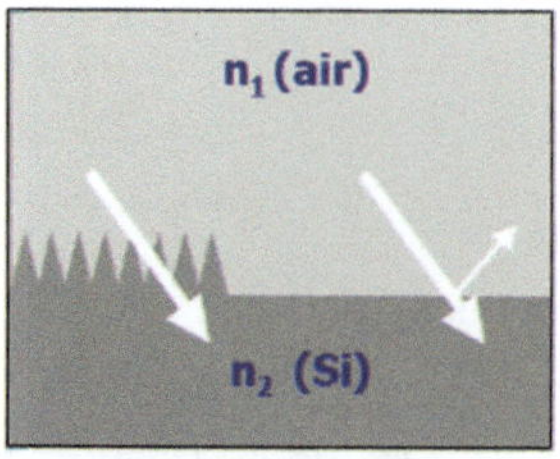

Porous refractive gradient layer
Periodic surface structure

Luminescent screen in fluorescent lamps

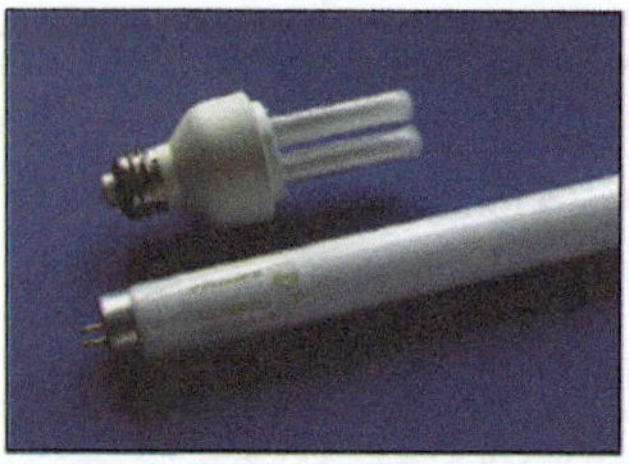

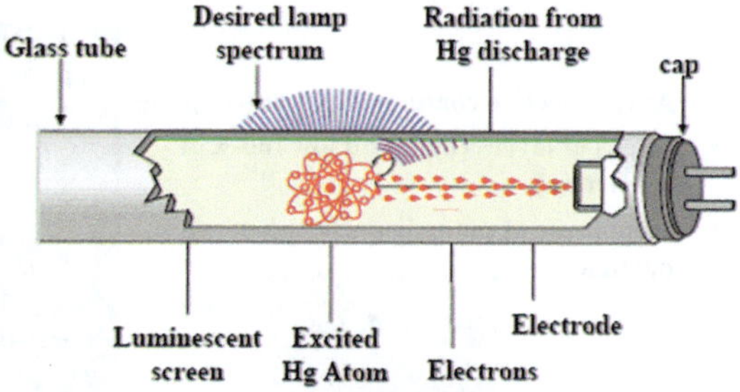

Hg discharge ⟶ **UV radiation** $\xrightarrow[\text{screen}]{\text{Luminescent}}$ **visible light**

UV radiation → **Purification**, **Disinfection**

visible light → **Illumination**

Nanoparticle filler (5 – 50 nm Al_2O_3 or $Ca_2P_2O_7$)

Effect of alon-c filler

- **Improves adhesion**
- **Improves lamp lifetime (reduces Hg take-up)**
- **Reduces initial light output due to absorption of 185 and 254 nm line**

Alternative high band gap materials

- Y_2O_3
- MgO
- $AlPO_4$
- $MgAl_2O_4$
- $AEMgAl_{10}O_{17}$ (AE = Ca, Sr, Ba)

Reflection spectra

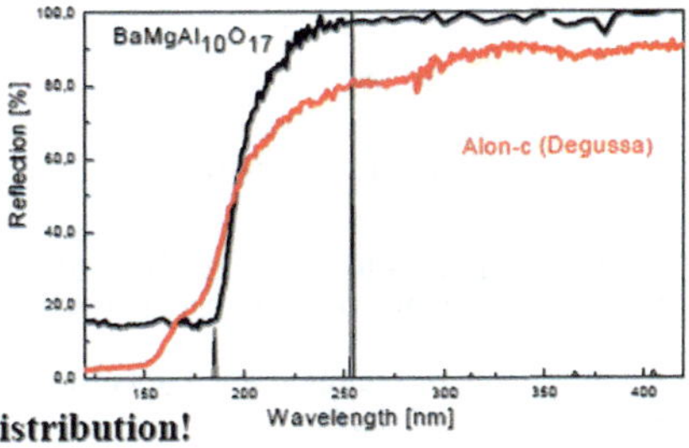

Scattering power depends on particle size distribution!

Precoating layers (20 - 50 nm Al_2O_3 or Y_2O_3)

Schematic build-up of a luminescent screen in fluorescent lamps

1. **Luminescent material + filler**
2. **Precoating**
3. **Glass substrate**

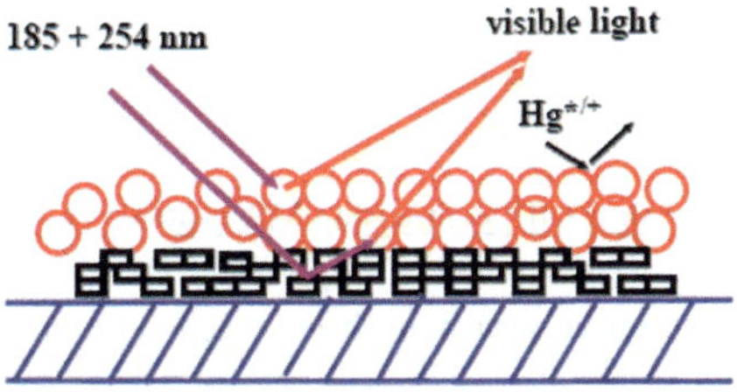

Functions of pre-coating layer

- **Protection of glass from Hg discharge**
- **Backscattering of transmitted UV photons** GE: "Star-coating"

Today: Nanoparticles of Al_2O_3 (alon-c)

15.1.2 *MR Contrast Enhancement and Hyperthermia*

Improvement of diagnostic methods, e.g. ^{1}H NMR (tomography)

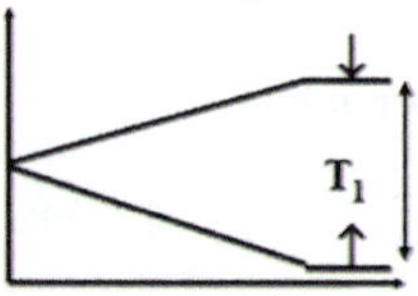

$$\Delta E = h\nu = \gamma \frac{h}{2\pi} B_0$$

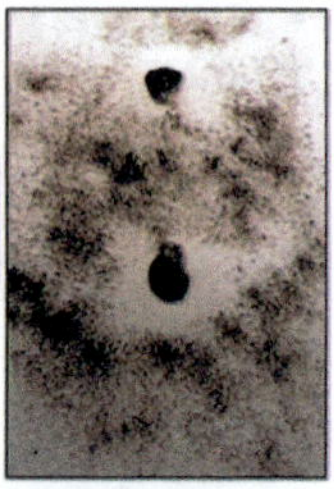

Physics
Protons ($^1H^+$) have a spin
⇒ I = 1/2
⇒ m_I = +1/2 and m_I = -1/2
⇒ energy separation ΔE by a magnetic field B
⇒ monitored are T_1 spin-lattice relaxation rates

Requirements on materials for MR contrast enhancement
- "Single-domain" magnetic nano particles, e g. γ-Fe_2O_3, Fe_3O_4, Fe, CrO_2
- Paramagnetic compounds, e.g. Gd-complexes or
- Gd^{3+} in nano vesicles
- Ø ~ 10 nm, spherical morphology

Fe_3O_4 Nanoparticles

Diagnostic advantages
- Higher image resolution
- Little impact by external magnetic fields
- Coupling of imaging with therapy

Improvement of diagnostic methods, e.g. ^{1}H NMR (tomography)

Physics: Gd^{3+} comprising MR contrast enhancement pharmaceuticals
Association of Gd^{3+} ([Xe]4f^7 S = 7/2) cations to the the ^{1}H nuclei (H_2O) results in accelerated relaxation of protons due to magnetic interaction

Approaches
(a) Application of Gd^{3+} complexes with coordination sites accessible towards H_2O molecules, e.g. [Gd(DTPA)]

(b) Application of Gd^{3+} comprising vesicles, e.g. $Gd_2HoN@C_{80}(OH)_n$ (Shinohara et al. have reported a significant increase (>20) in the ^{1}H MR T_1 spin-lattice relaxation rate)

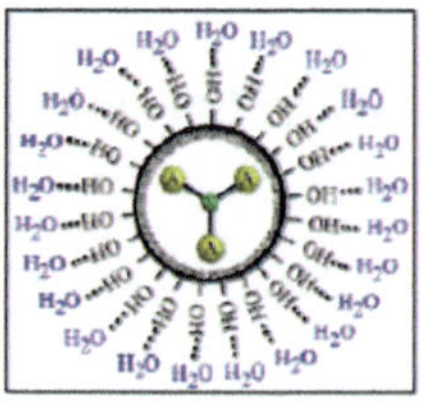

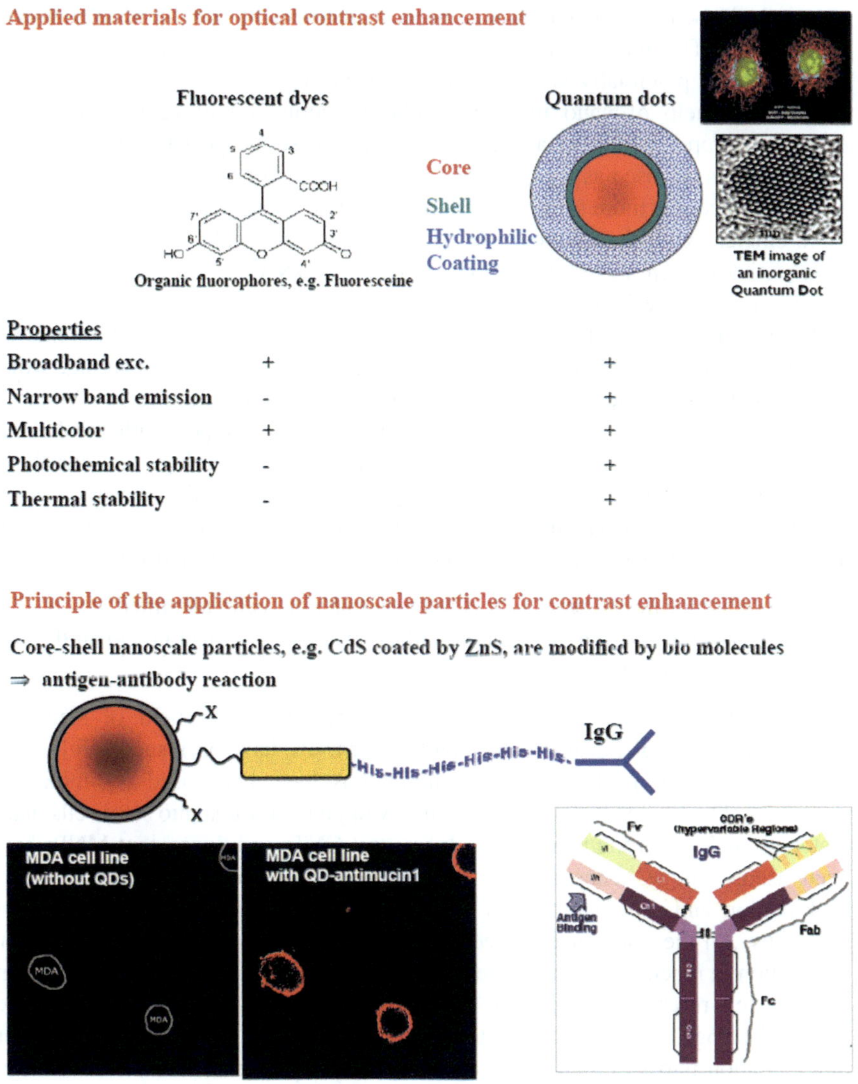

15.2 Applications of NMs in Health and Medical Therapy

Even today various disease like diabetes, cancer, Parkinson's disease, Alzheimer's disease, cardiovascular diseases and multiple sclerosis as well as different kinds of serious inflammatory or infectious diseases (e.g. HIV) constitute a high number of serious and complex illnesses which are posing a major problem for the mankind. Nano-medicine is an application of nanotechnology which works in the field of health and medicine. Nano-medicine makes use of nano materials, and nano electronic biosensors. In the future, nano medicine will benefit molecular nanotechnology. The medical area of nano science application has many projected benefits and is potentially valuable for all human races.

With the help of nano medicine early detection and prevention, improved diagnosis, proper treatment and follow-up of diseases is possible. Certain nano scale particles are used as tags and labels, biological can be performed quickly, the testing has become more sensitive and more flexible. Gene sequencing has become more efficient with the invention of nano devices like gold nano particles, these gold particles when tagged with short segments of DNA can be used for detection of genetic sequence in a sample.

With the help of nanotechnology, damaged tissue can be reproduced or repaired. These so called artificially stimulated cells are used in tissue engineering, which might revolutionize the transplantation of organs or artificial implants.

Advanced biosensors with novel features can be developed with the help of Carbon nano tubes. These biosensors can be used for astrobiology and can throw light on study origins of life. This technology is also being used to develop sensors for cancer diagnostics. Though CNT is inert, it can be functionalized at the tip with a probe molecule. Their study uses AFM as an experimental platform.

i. Probe molecule to serve as signature of leukemia cells identified.
ii. Current flow due to hybridization will be through CNT electrode to an IC chip.
iii. Prototype biosensors catheter development.

Nanotechnology has made excellent contribution in the field of stem cell research. For example, magnetic nanoparticles (MNPs) have been successfully used to isolate and group stem cells. Quantum dots have been used for molecular imaging and tracing of stem cells, for delivery of gene or drugs into stem cells, nano materials such as carbon nano tubes, fluorescent CNTs and fluorescent MNPs have been used. Unique nanostructures were designed for controllable regulation of proliferation and differentiation of stem cells is done by designed unique nano structures. All these advances speed up the development of stem cells toward the application in regenerative medicine [5]. The recent applications of nanotechnology in stem cell research promises to open new avenues in regenerative medicine. Nanotechnology can be a valuable tool to track and image stem cells, to drive their differentiation into specific cell lineage and ultimately to understand their biology. This will hopefully lead to stem cell-based therapeutics for the prevention, diagnosis and treatment of human diseases [6].

Nano devices can be used in stem cell research in tracking and imaging them. It has its applications for basic science as well as translational medicine. Stem cells can be modulated by mixing of nano carriers with biological molecules (Fig. 15.2). Nano devices can be used for intracellular access and also for intelligent delivery and sensing of biomolecules. These technologies have a great impact in stem cell microenvironment and tissue engineering studies and have a great potential for biomedical applications [7].

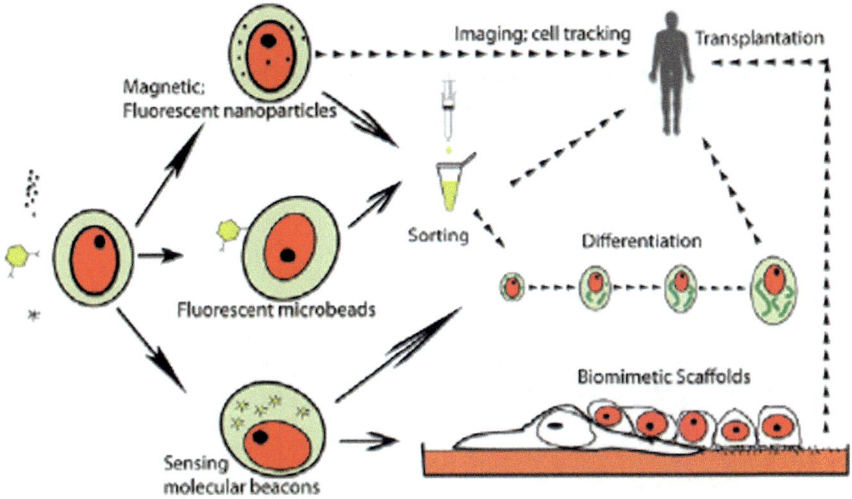

Fig. 15.2 Nanotechnology applications in stem cell biology and medicine

Nano-sized inorganic particles of either simple or complex nature, display unique, physical and chemical properties and represent an increasingly important material in the development of novel nanodevices which can be used in numerous physical, biological, biomedical and pharmaceutical applications [8–10].

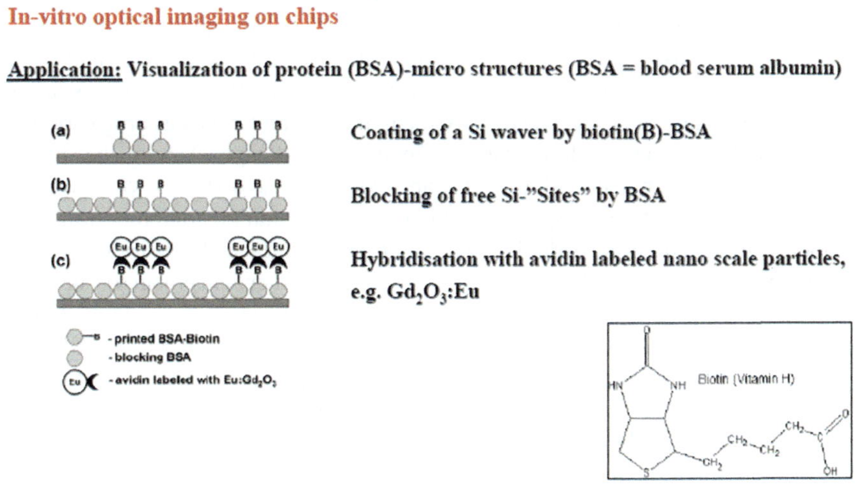

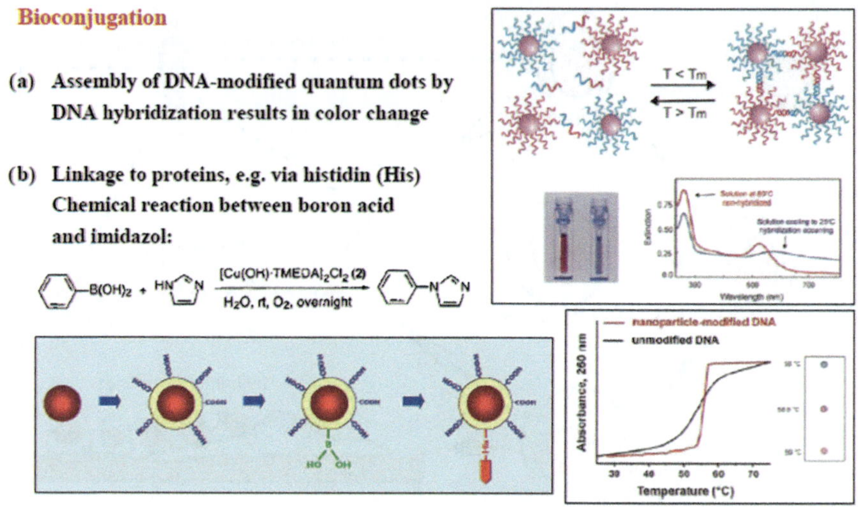

NPs have drawn increasing interest from every branch of medicine for their ability to deliver drugs in the optimum dosage range often resulting in increased therapeutic efficiency of the drugs, weakened side effects and improved patient compliance [11]. Iron oxide particles such as magnetite (Fe_3O_4) or its oxidized form maghemite (Fe_2O_3) are the most commonly employed for biomedical applications [12]. The selection of NPs for achieving efficient contrast for biological and cell imaging applications as well as for photo thermal therapeutic applications is based on the optical properties of NPs. Mie theory and discrete dipole approximation method can be used to calculate absorption and scattering efficiencies and optical resonance wavelength for the commonly used classes of NPs i.e. Au NPs, silica-Au NPs and Au nanorods [13]. The development of hydrophilic NPs as drug carrier has represented over the last few years an important challenge. Among the different approaches, polyethylene oxide (PEO) and polylactic acid (PLA) NPs have been revealed as very promising system for the intravenous administration of drugs [14]. Superparamagnetic iron oxide NPs with appropriate surface chemistry can be used for numerous in vivo applications such as MRI contrast enhancement, tissue repair, and immunoassay, detoxification of biological fluids hyperthermia, drugs delivery and cell separation. All of these biomedical applications require that the NPs have high magnetization value, a size smaller than 100 nm and a narrow particle size distribution [15]. The detection of analytes in tissue sections can be accomplished through antigen-antibody interactions using antibodies labeled with fluorescent dyes, enzymes, radioactive compounds or colloidal Au [16].

Hyperthermia - Thermal cancer therapy

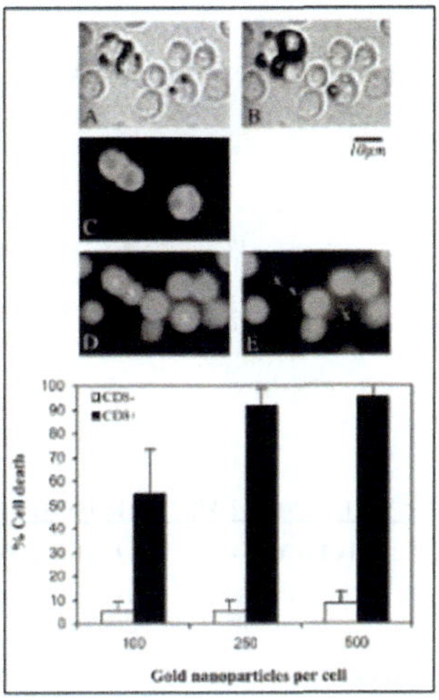

Idea

Cancer cell death induced by strong magnetic AC field or LASER irradiation

Method

- **Surface modification of magnetic nanoparticles (Fe_2O_3, FePt, or Au) to achieve selective up-take into cancer cells (antigen-antibody approach)**
- **Inject into blood or cancer tissue**
- **Apply AC field (oscillation) or LASER radiation (absorption) to heat up cancer cells**
- **Cell death for T > 44°C**

Radiative cancer therapy

Radiation therapy by application of radionuclides

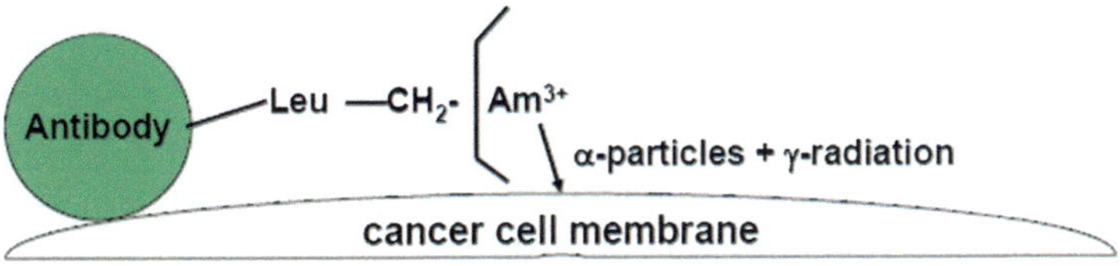

Coupling towards nanoparticle occurs e.g. by the application of the biotine-avidine system

References

- **Photogen Inc., US 6331286**
- **Light Sciences Limited Partnership, WO 99/52565**

Photodynamic cancer therapy

Idea: Application of UV-C or VUV emitting nanoscale materials

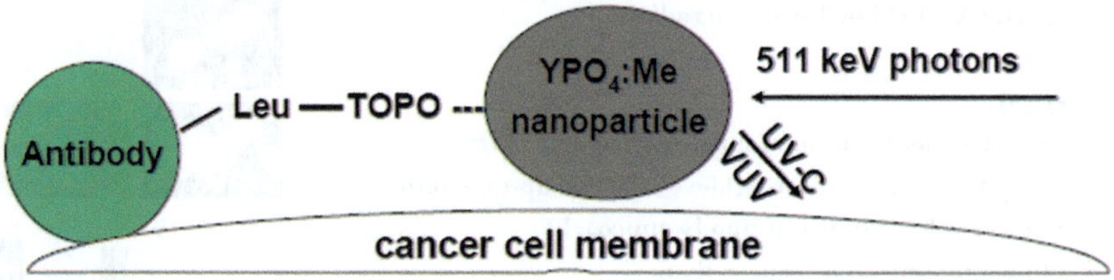

Excitation of the nanoscale material

- **Internally:** Doping with a (positron emitting) radionuclide
- **Externally:** Irradiation by x-rays, e.g. 511 keV (large contrast!)

Reference
Philips, EP 03047566

Radiative cancer therapy

1. **Irradiation by x-rays**

- **Application of keV to few MeV radiation**
- **Problems:**
  - **Low cross-section of absorbing material requires relatively high dose**
  - **No healthy/diseased tissue contrast**

2. **Application of radionuclides**

- $^{212}Bi \rightarrow \alpha + {}^{208}Tl$ **(half life ~ 1 h, 13.3 h for ^{123}I, 7 h for ^{212}At)**
- **To achieve high specificity to cancer cells, the radionuclide cations are chelated by organic moieties, e.g. edta, which is conjugated to an antibody with high specificity to cancer cells**
- **Problems:**
  - **Toxicity of the agents**
  - **Short half-life of useful radionuclides**

Photodynamic cancer therapy

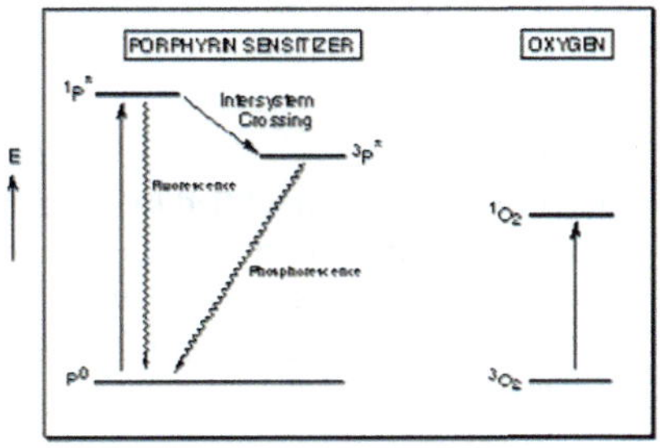

Principle

- **Administration of a photosensitive drug to an affected area (e.g. cancer tissue)**
- **Subsequent irradiation with light**
- **Light sources (50 mW/cm², 600 – 800 nm)**
  - **LASER**
  - **AlInGaP LEDs**
  - **Low-pressure discharge lamps**

Structure of a porphyrin sensitiser

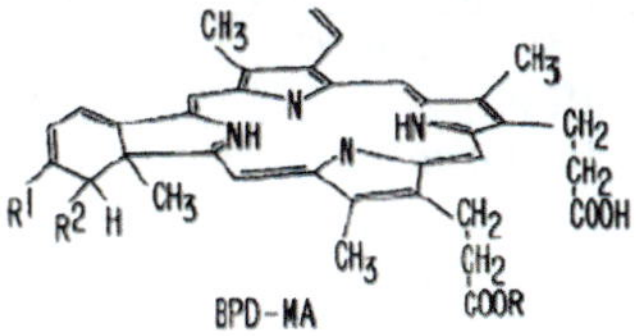

Application areas

- **Skin cancer treatment: Basal cell cancer, melanoma**
- **Blood cancer treatment: Leukemia**
- **Rheumatoid arthritis**
- **Bio stimulation: Wound healing**
- **Cosmetic skin treatment: Stain removal**

Drug delivery

Motivation

(a) **A major problem in pharmaceutical research is the formulation of the active ingredients. Substances have to be transported to the target cells (and only there) and release or activate the drug there in the desired concentration over time.**

Nanoparticle can function as a protective shell to prevent the immune system to destroy the drug, function as an envelope to ensure the correct delivery to the target cells or act as an ingredient deposit.

(b) **Nanoparticles and nanocrystalline materials are already commercialized as antimicrobial and antifungal agents. The health care industry needs for improved protection against bacteria in the face of growing antibiotic resistance.**

Some examples

- **Radiation therapy**
- **Photodynamic therapy**
- **Ag has antibiotic properties and is being used to made into crystalline nanoparticles, which increase solubility and potency**

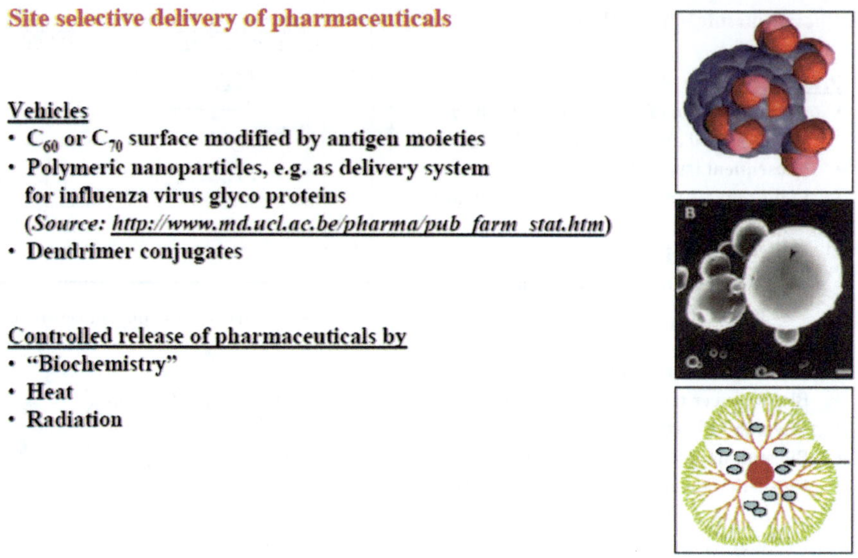

Over past few decades these has been considerable interest in developing biodegradable NPs as effective drug delivery devices [17]. Various polymers have been used in drug delivery research as they can effectively deliver the drugs to the target site thus increases the therapeutic benefit, while minimizing side effects. The controlled release of pharmacologically active drugs to the precise action site at the therapeutically optimum degree and dose regimen has been a major goal in designing such devices.

Liposomes have been used as a potential drug carrier instead of conventional dosage forms because of their unique advantages which include ability to protect drugs from degradation, target to the site of action and reduce the noxiousness and other side effects. However developmental work on liposome drugs has been restricted due to inherent health issues such as squat encapsulation efficiency, rapid water leakage in the commodity of blood components and very poor storage, and stability. On the other hand, polymeric NPs promise some critical advantages over these materials i.e. liposomes. For instance, NPs help to increase the ratability of drugs or problems and possess convenient controlled drug release properties.

Most of the semiconductor and metallic NPs have immense potential for cancer diagnosis and therapy on account of their surface plasmon resonance (SPR) enhanced light scattering and absorption. Au NPs efficiently convert the strong absorbed light into localized heat which can be exploited for the selective laser photo thermal therapy of cancer [18]. Beside this the antineoplastic effect of NPs is also effectively employed to inhibit the tumor growth. The multihydroxylated $[Gd@C_{82}(OH)_{22}]_n$ NPs showed antineoplastic activity with good efficiency and lower toxicity [19]. Ag NPs are being used increasingly in wound dressings, catheters and various households' products due to their antimicrobial activity [20].

Antimicrobial agents are extremely vital in textile, medicine, water disinfection and food packaging. Therefore, the antimicrobial characteristics of inorganic NPs add more potency to this important aspect, as compared to organic compounds, which are relatively toxic to the biological systems [21]. These NPs are functionalized with various groups to overcome the microbial species selectively. TiO_2, ZnO, $BiVO_4$, Cu- and Ni-based NPs have been utilized for this purpose due to their suitable antibacterial efficacies [22–25].

15.3 Applications in Manufacturing and Materials

Nanocrystalline materials provide very interesting substances for material science since their properties deviate from respective bulk material in a size dependent manner. Manufacture NPs display physicochemical characteristics that induce unique electrical, mechanical, optical and imaging properties that are extremely looked-for in certain applications within the medical, commercial, and ecological sectors [26–28]. NPs focus on the characterization, designing and engineering of biological as well as non-biological structures < than 100 nm, which show unique and novel functional properties. The potential benefits of nanotechnology have been documented by many manufacturer at high and low level and marketable products are already being mass-produced such as microelectronics, aerospace and pharmaceutical industries [29]. Among the nanotechnology consumer products to date, health fitness products from the largest category, followed by the electronic and computer category as well as home and garden category. Nanotechnology has been touted as the next revolution in many industries including food processing and packing. Resonant energy transfer (RET) system consisting of organic dye molecules and noble metals NPs have recently gamed considerable interest in bio photonics as well as in material science [30]. The presence of NPs in commercially available products is becoming more common.

Metals NPs such as noble metals, including Au and Ag have many colors in the visible region based on plasmon resonance, which is due to collective oscillations of the electrons at the surface of NPs [16, 31, 32]. The resonance wavelength strong depends on size and shape of NPs, the interparticle distance, and the dielectric property of the surrounding medium. The unique plasmon absorbance features of these noble metals NPs have been exploited for a wide variety of applications including chemical sensors and biosensors [32].

15.4 Applications in the Environment

Nanotechnology will play a critical role in coming 50 years by protecting the environment and providing sufficient energy for a growing world. The advanced techniques of nanotechnology can help storage of energy, its conversion into other forms, ecofriendly manufacturing of materials and by better enhanced renewable energy sources.

Nanoscale rare earth ortho-phosphates

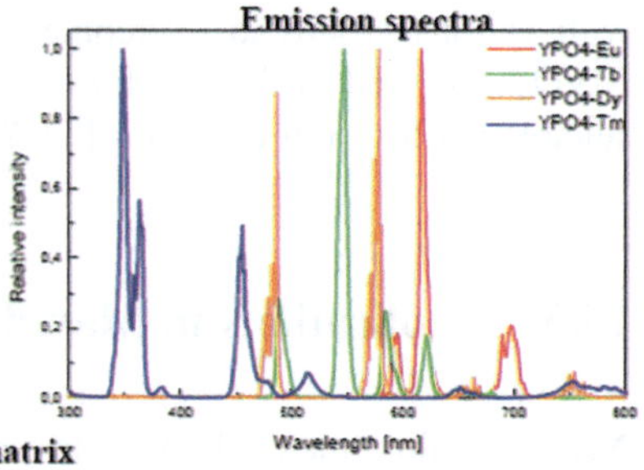

Optical properties

- Line emitters
- Emission spectra are independent of particle size
- Several emission lines can be combined e.g. Tb^{3+} with Eu^{3+}
- Slow decay (1 – 10 ms), i.e. easy discrimination from fluorescence of organic matrix

Chemical properties

- High photo stability
- Low solubility product: $K_L = c(Ln^{3+}) * c(PO_4^{3-}) \sim 10^{-30}$ mol²/l²
- Particle size (1 - 10 nm) comparable to organic dyes
- Buffer effect: $PO_4^{3-} \rightarrow HPO_4^{2-} \rightarrow H_2PO_4^{-}$

Nanoscale rare earth ortho-phosphates

Colloidal solutions of $(Y,Gd)PO_4$:RE upon UV-A excitation

Advantages

- narrow band emission
- many different colors
- up conversion
- high photo stability
- high chemical stability
- small size (1 – 10 nm)
- non or minor toxicity expected

Working principle

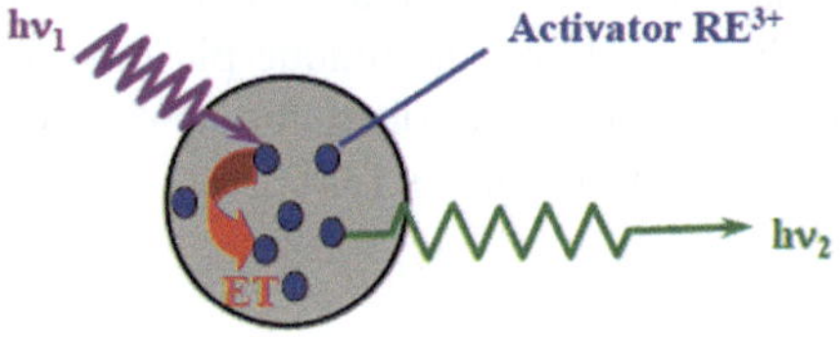

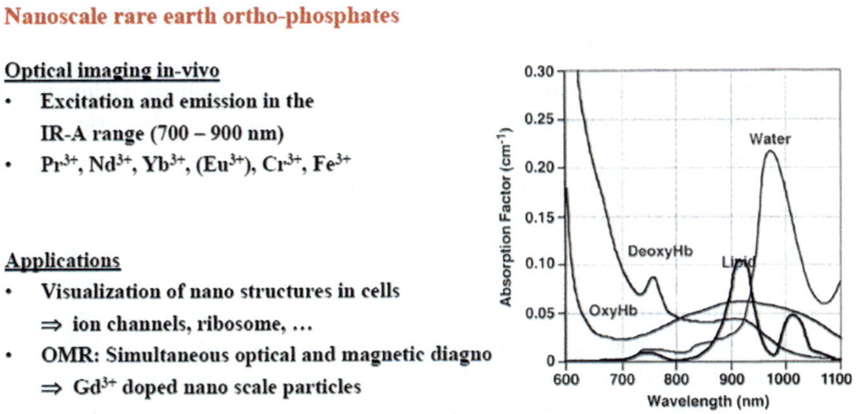

Nanotechnology can be used for less expensive energy production and for renewal energies, in solar technology, nano-catalysis, fuel cells and hydrogen technology. Carbon nano tube fuel cells are used for storage of hydrogen, thus finds application in power cars.

Nanotechnology is used on photovoltaic, for making them cheap, light weight and more efficient, which can reduce the combustion of engine pollutants by nano porous filters, and can clean the exhaust mechanically, with the help of catalytic converters made up of nano scale noble metal particles and by catalytic coatings on cylinder walls and catalytic nanoparticles as additive for fuels.

Nanotechnology can help in developing new ecofriendly and green technologies that can minimize undesirable pollution. Solid state lightening can reduce total electricity consumption. Nano technological approaches can lead to a strong reduction of energy consumption for illumination.

The increasing area of engineered NPs in industrial and household applications leads to the release of such materials into the environment. Assessing the risk of these NPs in the environment requires on understanding of their mobility, reactivity, Eco toxicity and persistency [33, 34]. The engineering material applications can increase the concentration of NPs in groundwater and soil which presents the most significant exposure avenues for assessing environmental risks [35, 36]. Due to high surface to mass ratio natural NPs play an important role in the solid/water partitioning of contaminants can be absorbed to the surface of NPs, co-precipitated during the formation of natural NPs or trapped by aggregation of NPs which had contaminants adsorbed to their surface. The interaction of contaminants with NPs is dependent on the NPs characteristics, such as size, composition, morphology, porosity, aggregation/disaggregation and aggregate structure. The luminophores are not safe in the environment and are protected from the environmental oxygen when they are doped inside the silica network [37].

Most of environmental applications of nanotechnology fall into three categories:

1. Environmentally benign sustainable products (e.g. green chemistry or pollution prevention).

2. Remediation of materials contaminated with hazardous substances and
3. Sensors for environmental stages [38].

The removal of heavy metals such as mercury, lead, thallium, cadmium and arsenic from natural water has attracted considerable attention because of their adverse effects on environmental and human health. Superparamagnetic iron oxide NPs are an effective sorbent material for this toxic soft material. So, for no measurements of engineered NPs in the environment have been available due to the absence of analytical methods, able to quantify trace concentration of NPs [39]. Photodegradation by NPs is also very common practice and many nanomaterials are utilized for this purpose. Rogozea et al. used NiO/ZnO NPs modified silica in the tandem fashion for photodegradation purpose. The high surface area of NPs due to very small size (<10 nm), facilitated the efficient photodegradation reaction [40]. The same group has reported the synthesis of variety of NPs and reported their optical, florescence and degradation applications [41–43].

15.5 Applications in the Electronics

There has been growing interest in the development of printed electronics in last few years because printed electronics offer attractive to traditional silicon techniques and the potential for low cost, large area electronics for flexible displays, sensors. Printed electronics with various functional inks containing NPs such as metallic NPs, organic electronic molecules, CNTs and ceramics NPs have been expected to flow rapidly as a mass production process for new types of electronic equipment [44].

Applications of Environmental Sensors:

- Detection of gaseous molecules
- NO
- CO
- ethylene (plant stress signal)
- cis-3-hexen-ol (plant odorous substance)
- α-pyrene, 3-carene, 2-methoxyphenol (early fire detection by electro antennographic detector (antenna of the jewel beetle)
- Detection of poisonous substances in soil
- 2,4-Dinitrophenol
- Pentachlorphenol
- FCCP →

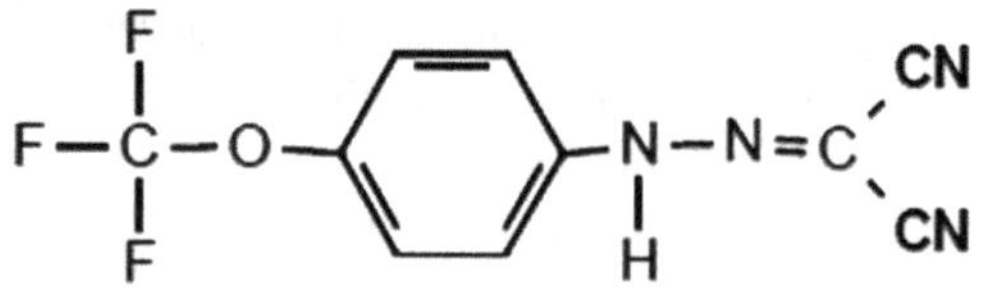

Unique structural, optical and electrical properties of one dimensional semiconductor and metals make them the key structural block for a new generation of electronic, sensors and photonic materials [45–47].

The good example of the synergism between scientific discovery and technological development is the electronic industry, where discoveries of new semiconducting materials resulted in the revolution from vacuumed tubes to diodes and transistors, and eventually to miniature chips [48].

The important characteristics of NPs are facile manipulation and reversible assembly which allow for the possibility of incorporation of NPs in electric, electronic or optical devices such as "bottom up" or "self-assembly" approaches are the bench mark of nanotechnology [49].

15.6 Applications in Energy Harvesting

Recent studies warned us about the limitations and scarcity of fossil fuels in coming years due to their nonrenewable nature. Therefore, scientists shifting their research strategies to generate renewable energies from easily available resources at cheap cost. They found that NPs are the best candidate for this purpose due to their, large surface area, optical behavior and catalytic nature. Especially in photocatalytic applications, NPs are widely used to generate energy from photoelectrochemical (PEC) and electrochemical water splitting [39, 50, 51]. Beside water splitting, electrochemical CO_2 reduction to fuels precursors, solar cells and piezoelectric generators also offered advance options to generate energy [30, 52–58]. NPs also use in energy storage applications to reserve the energy into different forms at nanoscale level [56, 59–62]. Recently, nanogenerators are created, which can convert the mechanical energy into electricity using piezoelectric, which is an unconventional approach to generate energy [63]. Figure 15.3 shows some energy generating devices, and uses NPs.

- **Quantum dots**

Inorganic nanostructures measuring only a few nanometers (quantum dots) have unique optical properties.

Material properties change dramatically because quantum effects arise from the confinement of electrons and "holes" in the material (a hole is the absence of an electron; the hole behaves as though it were a positively charged particle).

Size changes other material properties such as the electrical and nonlinear optical properties of a material, making them very different from those of the material's bulk form. If a dot is excited, the smaller the dot, the higher the energy and intensity of its emitted light. These very small, semiconducting quantum dots are gateways to an enormous array of possible applications and new technologies.

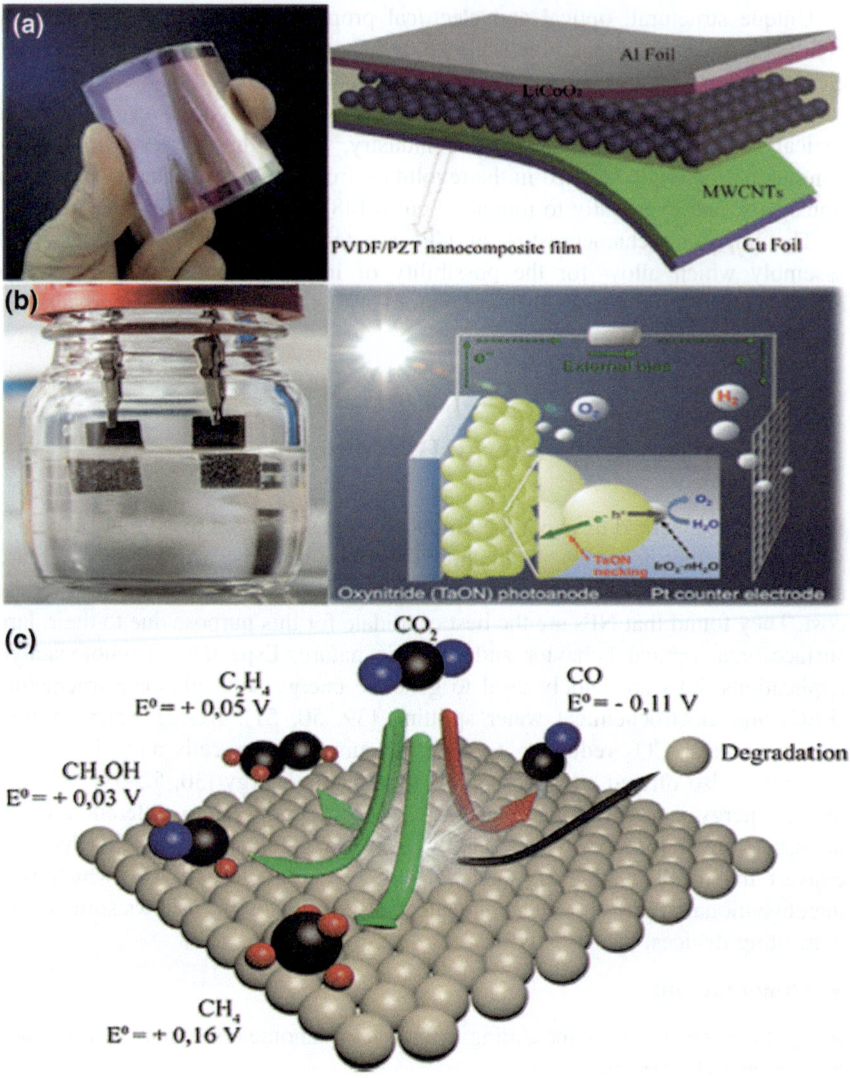

Fig. 15.3 Energy generation approaches from **a** Piezoelectrics actuators, **b** water splitting, **c** CO_2 reduction

- Quantum dots

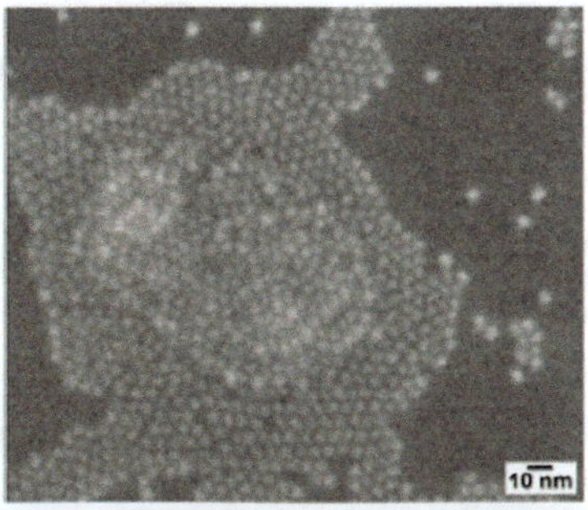

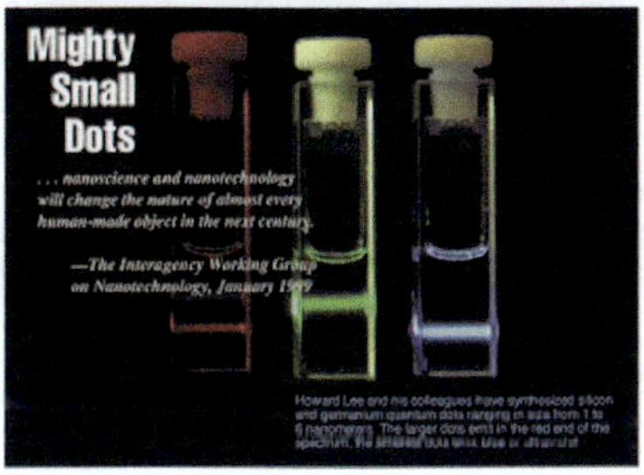

Inorganic nanostructures measuring only a few nanometers (quantum dots) have unique optical properties.

Material properties change dramatically because quantum effects arise from the confinement of electrons and "holes" in the material (a hole is the absence of an electron; the hole behaves as though it were a positively charged particle).

Size changes other material properties such as the electrical and nonlinear optical properties of a material, making them very different from those of the material's bulk form. If a dot is excited, the smaller the dot, the higher the energy and intensity of its emitted light.

These very small, semiconducting quantum dots are gateways to an enormous array of possible applications and new technologies.

15.7 Current and Future Trends

Here we list some of the present and future applications of nanomaterials that has been reported in recent Scheme 15.1: (Table 15.1).

15.8 Examples of Nanomaterials' Applications

15.8.1 Fuel Cells

A fuel cell is an electrochemical energy conversion device that converts the chemical energy from fuel (on the anode side) and oxidant (on the cathode side) directly into electricity. The heart of fuel cell is the electrodes. The performance of a fuel cell electrode can be optimized in two ways; by improving the physical structure and by using more active electro catalyst. A good structure of electrode must provide ample surface area, provide maximum contact of catalyst, reactant gas and electrolyte, facilitate gas transport and provide good electronic conductance. In this fashion the structure should be able to minimize losses (Fig. 15.4).

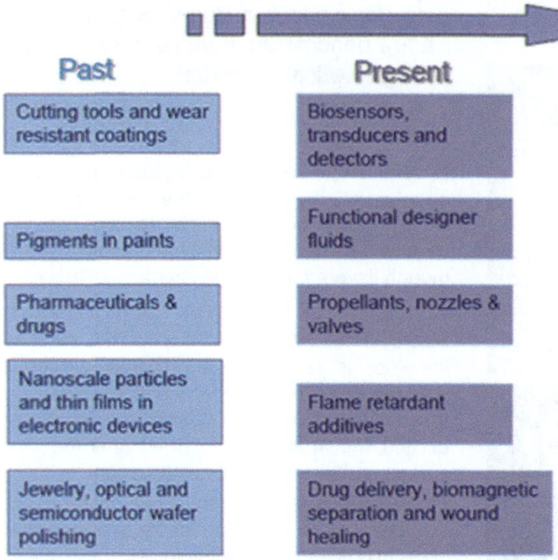

Scheme 15.1 Current applications of nanotechnology and time line for anticipated advances

Carbon nanotubes—Microbial fuel cell

Microbial fuel cell is a device in which bacteria consume water-soluble waste such as sugar, starch and alcohols and produces electricity plus clean water. This technology will make it possible to generate electricity while treating domestic or industrial wastewater. Microbial fuel cell can turn different carbohydrates and complex substrates present in wastewaters into a source of electricity. The efficient electron transfer between the microorganism and the anode of the microbial fuel cell plays a major role in the performance of the fuel cell. The organic molecules present in the wastewater possess a certain amount of chemical energy, which is released when converting them to simpler molecules like CO_2. The microbial fuel cell is thus a device that converts the chemical energy present in water-soluble waste into electrical energy by the catalytic reaction of microorganisms. Carbon nanotubes (CNTs) have chemical stability, good mechanical properties and high surface area, making them ideal for the design of sensors and provide very high surface area due to its structural network. Since carbon nanotubes are also suitable supports for cell growth, electrodes of microbial fuel cells can be built using of CNT. Due to three-dimensional architectures and enlarged electrode surface area for the entry of growth medium, bacteria can grow and proliferate and get immobilized. Multi walled CNT scaffolds could offer self-supported structure with large surface area through which hydrogen producing bacteria (e.g., E. coli) can eventually grow and proliferate. In addition, CNTs and MWCNTs have been reported to be biocompatible for different eukaryotic cells. The efficient proliferation of hydrogen producing bacteria throughout an electron-conducting scaffold of CNT

Table 15.1 Some examples of present and potential applications with significant technological impact

Technology	Present Impact	Potential Impact
Dispersions and coatings	Thermal barriers	Enhanced thermal barriers
	Optical (visible and UV) barriers	Multifunctional nanocoatings
	Imaging enhancement	Fine particle structure
	Ink-jet materials	Super absorbent materials (Ilford paper)
	Coated abrasive slurries	Higher efficiency and lower contamination
	Information-recording layers	Higher density information storage
High surface area	Molecular sieves	Molecule-specific sensors
	Drug delivery	Particle induced delivery
	Tailored catalysts	Energy storage (fuel cells, batteries)
	Absorption/desorption materials	Gratzel-type solar cells, gas sensors
Consolidated materials	Low-loss soft magnetic materials	Superplastic forming of ceramics materials
	High hardness, tough WC/Co cutting tools	Ultrahigh-strength, tough structural materials
	Nanocomposite cements	Magnetic refrigerants nanofilled polymer composites Ductile cements
Bio-medical aspects	Functionalised nanoparticles	Cell labelling by fluorescent nanoparticles
		Local heating by magnetic nanoparticles
Nanodevices	GMR read heads	Terabit memory and microprocessing
		Single molecule DNA sizing and sequencing
		Biomedical sensors
		Low noise, low threshold lasers

can form the basis for the potential application as electrodes in MFCs leading to efficient performance.

15.8.2 *Catalysis*

Higher surface area available with the nanomaterial counterparts, nano-catalysts tend to have exceptional surface activity. For example, reaction rate at nano-aluminum can go so high, that it is utilized as a solid-fuel in rocket propulsion, whereas the bulk aluminum is widely used in utensils. Nano-aluminum becomes highly reactive and supplies the required thrust to send off pay loads in

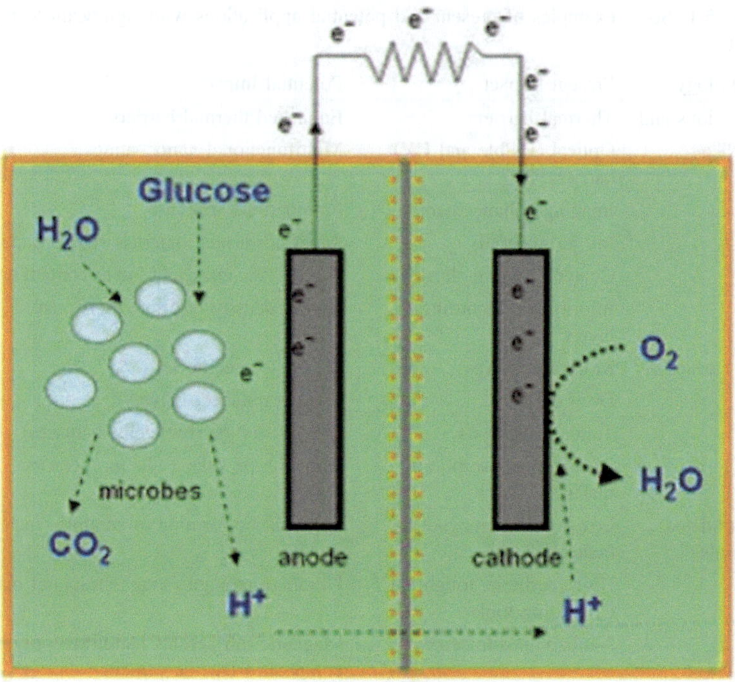

Fig. 15.4 Schematic representation of microbial fuel cell

space. Similarly, catalysts assisting or retarding the reaction rates are dependent on the surface activity, and can very well be utilized in manipulating the rate-controlling step.

15.8.3 Phosphors for High-Definition TV

The resolution of a television, or a monitor, depends greatly on the size of the pixel. These pixels are essentially made of materials called "phosphors," which glow when struck by a stream of electrons inside the cathode ray tube (CRT). The resolution improves with a reduction in the size of the pixel, or the phosphors. Nanocrystalline zinc selenide, zinc sulfide, cadmium sulfide, and lead telluride synthesized by the sol-gel techniques are candidates for improving the resolution of monitors. The use of nanophosphors is envisioned to reduce the cost of these displays so as to render high definition televisions (HDTVs) and personal computers affordable to be purchase.

15.8.4 Next-Generation Computer Chips

The microelectronics industry has been emphasizing miniaturization, whereby the circuits, such as transistors, resistors, and capacitors, are reduced in size. By achieving a significant reduction in their size, the microprocessors, which contain these components, can run much faster, thereby enabling computations at far greater speeds. However, there are several technological impediments to these advancements, including lack of the ultrafine precursors to manufacture these components; poor dissipation of tremendous amount of heat generated by these microprocessors due to faster speeds; short mean time to failures (poor reliability), etc. Nanomaterials help the industry break these barriers down by providing the manufacturers with nanocrystalline starting materials, ultra-high purity materials, materials with better thermal conductivity, and longer-lasting, durable interconnections (connections between various components in the microprocessors).

Example: Nanowires for junctionless transistors Transistors are made so tiny to reduce the size of sub-assemblies of electronic systems and make smaller and smaller devices, but it is difficult to create high-quality junctions. In particular, it is very difficult to change the doping concentration of a material over distances shorter than about 10 nm. Researchers have succeeded in making the junctionless transistor having nearly ideal electrical properties. It could potentially operate faster and use less power than any conventional transistor on the market today.

The device consists of a silicon nanowire in which current flow is perfectly controlled by a silicon gate that is separated from the nanowire by a thin insulating layer. The entire silicon nanowire is heavily n-doped, making it an excellent conductor. However, the gate is p-doped and its presence has the effect of depleting the number of electrons in the region of the nanowire under the gate. The device also has near-ideal electrical properties and behaves like the most perfect of transistors without suffering from current leakage like conventional devices and operates faster and using less energy (Fig. 15.5).

15.8.5 Elimination of Pollutants

Nanomaterials possess extremely large grain boundaries relative to their grain size. Hence, they are very active in terms of their chemical, physical, and mechanical properties. Due to their enhanced chemical activity, nanomaterials can be used as catalysts to react with such noxious and toxic gases as carbon monoxide and nitrogen oxide in automobile catalytic converters and power generation equipment to prevent environmental pollution arising from burning gasoline and coal.

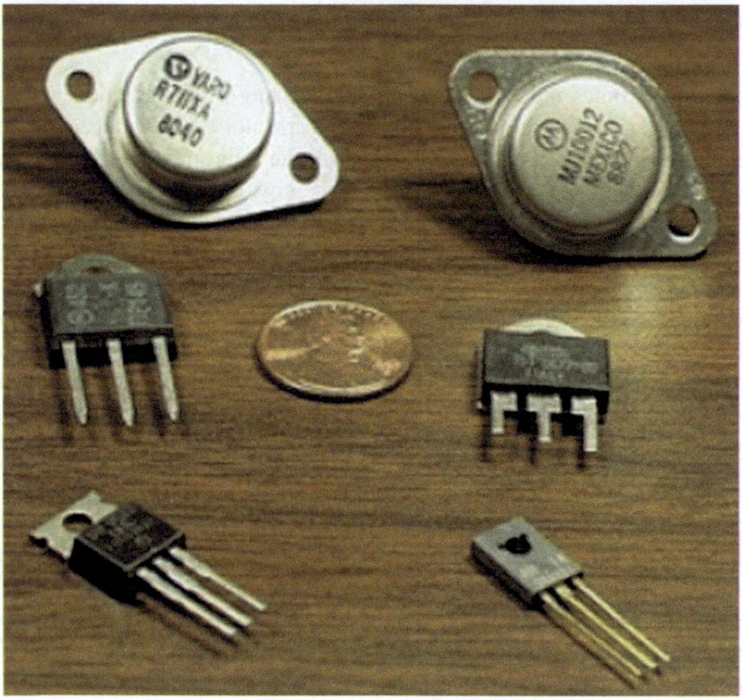

Fig. 15.5 Silicon nanowires in junction less transistors

15.8.6 *Sun-Screen Lotion*

Prolonged UV exposure causes skin-burns and cancer. Sun-screen lotions containing nano-TiO_2 provide enhanced sun protection factor (SPF) while eliminating stickiness. The added advantage of nano skin blocks (ZnO and TiO_2) arises as they protect the skin by sitting onto it rather than penetrating into the skin. Thus they block UV radiation effectively for prolonged duration. Additionally, they are transparent, thus retain natural skin color while working better than conventional skin-lotions.

15.8.7 *Sensors*

Sensors rely on the highly active surface to initiate a response with minute change in the concentration of the species to be detected. Engineered monolayers (few Angstroms thick) on the sensor surface are exposed to the environment and the peculiar functionality (such as change in potential as the CO/anthrax level is detected) is utilized in sensing.

15.8.8 Tools

Nowadays surface engineered materials are developed so as to tailor desired material properties for a specific application: A large variety of functional properties can be optimized for the bulk material and surface by applying an appropriate coating. The incorporation of nanomaterials in surface coatings can provide long-term abrasion resistance without significantly effecting optical clarity, gloss, color or physical properties.

- Tools

Nowadays surface engineered materials are developed so as to tailor desired material properties for a specific application: A large variety of functional properties can be optimized for the bulk material and surface by applying an appropriate coating. The incorporation of nanomaterials in surface coatings can provide long-term abrasion resistance without significantly effecting optical clarity, gloss, color or physical properties.

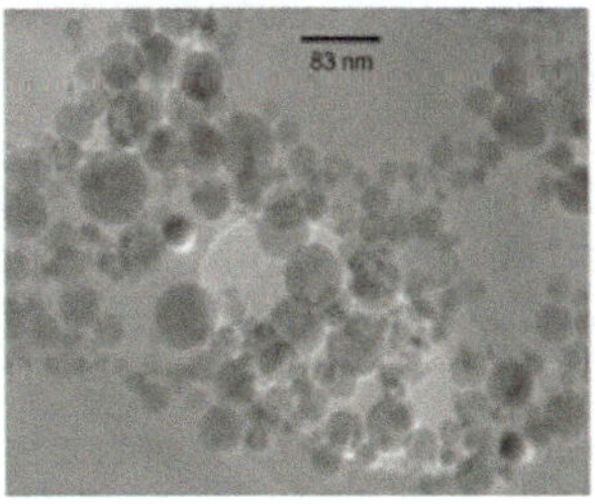

NanoDur® Aluminum Oxide is designed for coatings where transparency is of utmost importance. NanoDur® Aluminum Oxide is especially suited for incorporation into coating systems, greatly enhancing abrasion resistance with minimal effects on clarity, gloss and physical properties of the coatings.

15.8.9 Nanomedicine

Nanomedicine is the application of nanotechnology (the engineering of tiny machines) to the prevention and treatment of disease in the human body. This discipline is in its infancy. It has the potential to change medical science dramatically in the 21st century.

- **Diagnose illness**: to monitor or to report on the internal chemistry of the body.
- **Nanodevices**: to dispense drugs or hormones as needed in people with chronic imbalance or deficiency states.
- **Nanosurgeons**

http://www.foresight.org/Nanomedicine/

- Nanomedicine

Nanomedicine is the application of nanotechnology (the engineering of tiny machines) to the prevention and treatment of disease in the human body. This discipline is in its infancy. It has the potential to change medical science dramatically in the 21st century.

Diagnose illness: to monitor or to report on the internal chemistry of the body.

Nanodevices: to dispense drugs or hormones as needed in people with chronic imbalance or deficiency states.

This image shows an hypothetical device, attached to a red blood cell and releasing its payload molecules.

©Copyright 2002 Rutgers University Bio-Nano Project.

Nanosurgeons

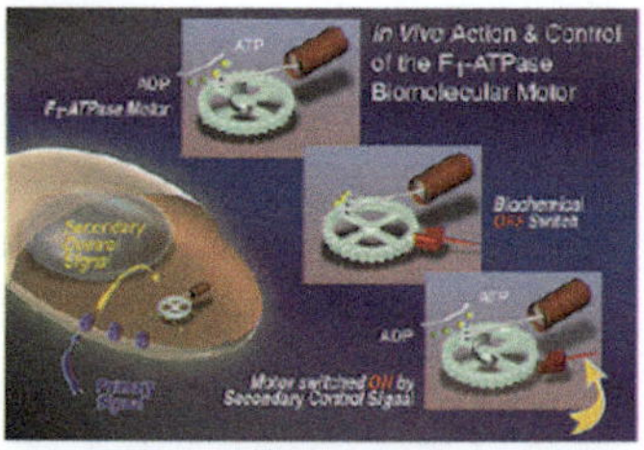

Nanoscale Biological Engineering and Transport Group/Cornell University, Copyright © Cornell University

This image shows a hypothetical device, attached to a red blood cell and releasing its payload molecules. © Copyright 2002 Rutgers University Bio-NanoProject. Nanoscale Biological Engineering and Transport Group/Cornell University, Copyright © Cornell University.

Various pharmaceutical applications for material nanoparticles MNPs are shown in Fig. 15.6.

15.8.10 Paint, Ink

In paint and coatings nanomaterials fulfil decorative needs (e.g. color and gloss), functional purposes (e.g. conductivity, microbial inactivation) and improve protection (e.g. scratch resistance, UV stability) of paints and coatings. In particular nano-size metal-oxides, such as TiO_2 and ZnO or Alumina, Ceria and Silica and nano-size pigments find application in new paint and coating formulations.

When matter is reduced in size it changes its characteristics, such as color and interaction with other matter such as chemical reactivity. The change in the characteristics is caused by the change of the electronic properties.

Nano-meter sized molecules penetrate into the smallest inclusions in painted metal, fiberglass, wood and ceramic, protecting them from harsh elements.

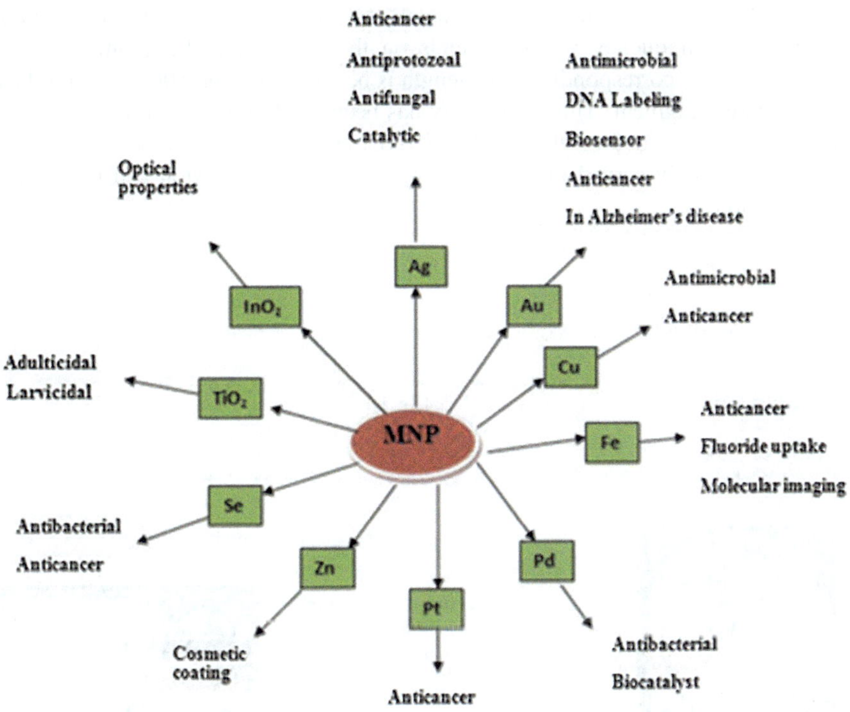

Fig. 15.6 Various pharmaceutical applications of MNPs

Nanomaterials exhibit uncommon characteristics like superior scratch resistance, clarity, UV protection, hydrophobicity and adhesion. Resistant self-cleaning effect.

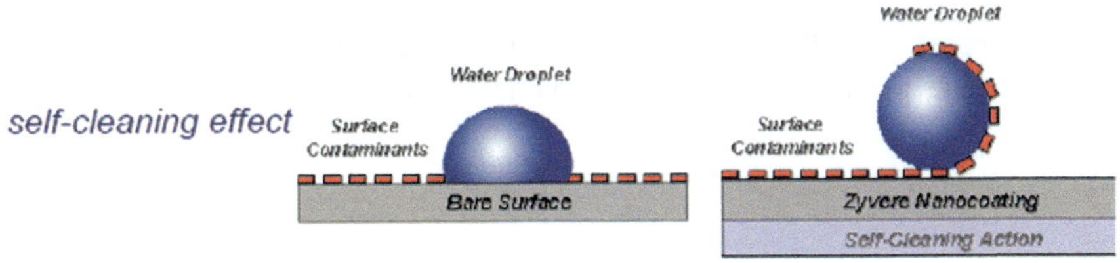

15.8.11 Nanoinclusions

Metallic nanoinclusions in a dielectric matrix affect its optical properties. A part of the visible spectra is absorbed due to surface Plasmon. A Plasmon is a collective oscillation of electric charges submitted to an electromagnetic wave. It appears in gaseous plasma as well as in metals like silver, gold and copper. If the electrons from the metal are confined in a very small inclusion, resonance takes place with a characteristic frequency which depends on the nature and the geometry of the inclusion. If the corresponding wavelength is between 400 and 800 nm, it produces a color like a pigment. This technology has been used during the roman era. Gold and silver were added to glasses. The **Lycurgus cup** is from the 4th century AD and was probably carved in Rome. It appears RED by reflexion and GREEN by transmission.

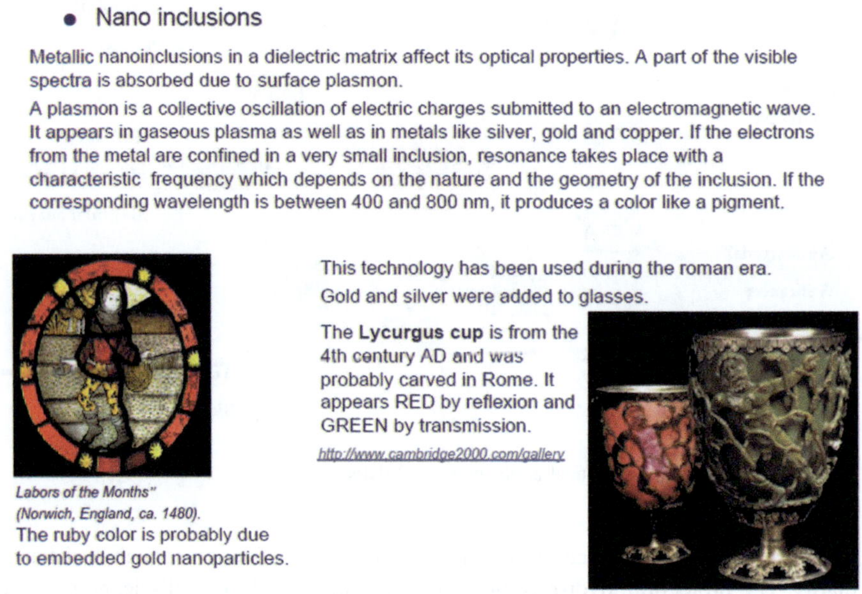

"Labors of the Months" (Norwich, England, ca. 1480). The ruby color is probably due to embedded gold nanoparticles. http://www.cambridge2000.com/gallery.

15.8.12 *Deodorant/Antiperspirant (Shaving/Depilatory Products)*

Nanomaterials can impart antibacterial and anti-odour functionality on human skin in powder (Foot Powder), gel, stick or spray underarm products (Oral care). It has also antimicrobial and anti-irritant properties.

- Deodorant/Antiperspirant
- Shaving/Depilatory Products
- Foot Powder
- Oral care

Nanomaterials can impart antibacterial and anti-odour functionality on human skin in powder, gel, stick or spray underarm products. It has also antimicrobial and anti-irritant properties.

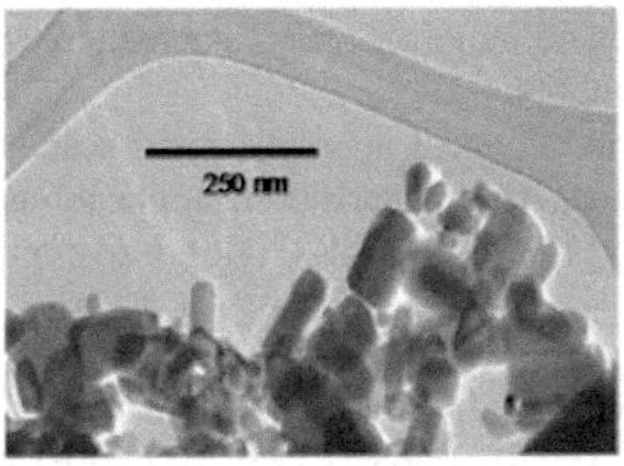

NanoGard® zinc oxide or silver doped zinc oxide appear to be naturals for this application. It can provide effective odour and bacterial control while maintaining clarity in silky-smooth, non-sticky formulations. The anti-irritant properties of zinc oxide are an added benefit! Silver doped zinc oxide may offer an enhancement in odor fighting antibacterial properties.

NanoGard® zinc oxide or silver doped zinc oxide appear to be naturals for this application. It can provide effective odour and bacterial control while maintaining clarity in silky-smooth, non-sticky formulations. The anti-irritant properties of zinc oxide are an added benefit! Silver doped zinc oxide may offer an enhancement in odor fighting antibacterial properties.

References

1. Guo, D., Xie, G., Luo, J.: Mechanical properties of nanoparticles: basics and applications. J. Phys. D Appl. Phys. **47**, 13001 (2014). https://doi.org/10.1088/0022-3727/47/1/013001
2. Kot, M., Major, Ł., Lackner, J.M., Chronowska-Przywara, K., Janusz, M., Rakowski, W.: Mechanical and tribological properties of carbon-based graded coatings. J. Nanomater. **2016**, 1–14 (2016). https://doi.org/10.1155/2016/8306345
3. Mallakpour, S., Sirous, F.: Surface coating of a-Al_2O_3 nanoparticles with poly(vinyl alcohol) as biocompatible coupling agent for improving properties of bio-active poly(amide-imide) based nanocomposites having l-phenylalanine linkages. Prog. Org. Coat. **85**, 138–145 (2015). https://doi.org/10.1016/j.porgcoat.2015.03.021
4. Shao, W., Nabb, D., Renevier, N., Sherrington, I., Luo, J.K.: Mechanical and corrosion resistance properties of TiO_2 nanoparticles reinforced Ni coating by electrodeposition. IOP Conf. Ser. Mater. Sci. Eng. **40**, 12043 (2012). https://doi.org/10.1088/1757-899X/40/1/012043
5. Wang, Z., Ruan, J., Cui, D.: Advances and prospect of nanotechnology in stem cells. Nanoscale Res. Lett. **4**, 593–605 (2009)
6. Ricardo e Lino F, P.N.: Stem cell research meets nanotechnology. Revista Da Sociedade Portuguesa D Bioquimica, CanalBQ **7**, 38–46 (2010)
7. Deb, K.D., Griffith, M., Muinck, E.D., Rafat, M.: Nanotechnology in stem cells research: advances and applications. Front Biosci (Landmark Ed) **17**, 1747–1760 (2012)

8. Loureiro, A., Azoia, N.G., Gomes, A.C., Cavaco-Paulo, A.: Albumin-based nanodevices as drug carriers. Curr. Pharm. Des. **22**, 1371–1390 (2016)
9. Martis, E., Badve, R., Degwekar, M.: Nanotechnology based devices and applications in medicine: an overview. Chron. Young Sci. **3**, 68 (2012). https://doi.org/10.4103/2229-5186.94320
10. Nikalje, A.P., 2015. Nanotechnology and its applications in medicine. Med Chem 5. http://dx.doi.org/10.4172/2161-0444.1000247
11. Alexis, F., Pridgen, E., Molnar, L.K., Farokhzad, O.C.: Factors affecting the clearance and biodistribution of polymeric nanoparticles. Mol. Pharm. **5**, 505–515 (2008). https://doi.org/10.1021/mp800051m
12. Ali, A., Zafar, H., Zia, M., Ul Haq, I., Phull, A.R., Ali, J.S., Hussain, A.: Synthesis, characterization, applications, and challenges of iron oxide nanoparticles. Nanotechnol. Sci. Appl. **9**, 49–67 (2016). https://doi.org/10.2147/NSA.S99986
13. Jain, P.K., Lee, K.S., El-Sayed, I.H., El-Sayed, M.A.: Calculated absorption and scattering properties of gold nanoparticles of different size, shape, and composition: applications in biological imaging and biomedicine. J. Phys. Chem. B **110**, 7238–7248 (2006). https://pubs.acs.org/doi/10.1021/jp057170o
14. Calvo, P., Remuoon-Lopez, C., Vila-Jato, J.L., Alonso, M.J.: Novel hydrophilic chitosan-polyethylene oxide nanoparticles as protein carriers. J. Appl. Polym. Sci. **63**, 125–132 (1997). http://dx.doi.org/10.1002/(SICI)1097-4628(19970103)63:1<125::AID-APP13>3.0.CO;2-4
15. Laurent, S., Forge, D., Port, M., Roch, A., Robic, C., Vander Elst, L., Muller, R.N.: Magnetic iron oxide nanoparticles: synthesis, stabilization, vectorization, physicochemical characterizations, and biological applications. Chem. Rev. **110**, 2574–2574 (2010). http://dx.doi.org/10.1021/cr900197g
16. Khlebtsov, N.G., Dykman, L.A.: Optical properties and biomedical applications of plasmonic nanoparticles. J. Quant. Spectrosc. Radiat. Transf. **111**, 1–35 (2010). https://doi.org/10.1016/j.jqsrt.2009.07.012
17. Zhang, J., Saltzman, M.: Engineering biodegradable nanoparticles for drug and gene delivery. Chem. Eng. Prog. **109**, 25–30 (2013)
18. Prashant, K.J., Ivan, H.S.: Au NPs target cancer. Nanotoday **2**, 19–29 (2007)
19. Chen, C., Xing, G., Wang, J., Zhao, Y., Li, B., Tang, J., Jia, G., Wang, T., Sun, J., Xing, L., Yuan, H., Gao, Y., Meng, H., Chen, Z., Zhao, F., Chai, Z., Fang, X.: Multihydroxylated [Gd@C82(OH)22]n nanoparticles: antineoplastic activity of high efficiency and low toxicity. Nano Lett. **5**, 2050–2057 (2005). http://dx.doi.org/10.1021/nl051624b
20. AshaRani, P.V., Low Kah Mun, G., Hande, M.P., Valiyaveettil, S.: Cytotoxicity and genotoxicity of silver nanoparticles in human cells. ACS Nano. **3**, 279–290 (2009). https://doi.org/10.1021/nn800596w
21. Hajipour, M.J., Fromm, K.M., Ashkarran, A. Akbar, de Aberasturi, D. Jimenez, de Larramendi, I.R., Rojo, T., Serpooshan, V., Parak, W.J., Mahmoudi, M.: Antibacterial properties of nanoparticles. Trends Biotechnol. **30**, 499–511 (2012). http://dx.doi.org/10.1016/j.tibtech.2012.06.004
22. Akhavan, O., Azimirad, R., Safa, S., Hasani, E.: CuO/Cu$(OH)_2$ hierarchical nanostructures as bactericidal photocatalysts. J. Mater. Chem. **21**, 9634 (2011). https://doi.org/10.1039/c0jm04364h
23. Pant, H.R., Pant, B., Sharma, R.K., Amarjargal, A., Kim, H.J., Park, C.H., Tijing, L.D., Kim, C.S.: Antibacterial and photocatalytic properties of Ag/TiO_2/ZnO nano-flowers prepared by facile one-pot hydrothermal process. Ceram. Int. **39**, 1503–1510 (2013). https://doi.org/10.1016/j.ceramint.2012.07.097
24. Qu, Z., Liu, P., Yang, X., Wang, F., Zhang, W., Fei, C.: Microstructure and characteristic of $BiVO_4$ prepared under different pH values: photocatalytic efficiency and antibacterial activity. Materials **9**, 129 (2016). https://doi.org/10.3390/ma9030129

25. Yin, Q., Wu, W., Qiao, R., Ke, X., Hu, Y., Li, Z.: Glucoseassisted transformation of Ni-doped-ZnO@carbon to a Ni-doped-ZnO@void@SiO_2 core–shell nanocomposite photocatalyst. RSC Adv. **6**, 38653–38661 (2016). https://doi.org/10.1039/C5RA26631A
26. Dong, H., Wen, B., Melnik, R.: Relative importance of grain boundaries and size effects in thermal conductivity of nanocrystalline materials. Sci. Rep. **4**, 7037 (2014). https://doi.org/10.1038/srep07037
27. Ma, E.: Nanocrystalline materials: controlling plastic instability. Nat. Mater. **2**, 7–8 (2003). https://doi.org/10.1038/nmat797
28. Todescato, F., Fortunati, I., Minotto, A., Signorini, R., Jasieniak, J., Bozio, R.: Engineering of semiconductor nanocrystals for light emitting applications. Materials **9**, 672 (2016). http://dx.doi.org/10.3390/ma9080672
29. Weiss, J., Takhistov, P., McClements, D.J.: Functional materials in food nanotechnology. J. Food Sci. **71**, R107–R116 (2006). https://doi.org/10.1111/j.1750-3841.2006.00195.x
30. Lei, Y.-M., Huang, W.-X., Zhao, M., Chai, Y.-Q., Yuan, R., Zhuo, Y.: Electrochemiluminescence resonance energy transfer system: mechanism and application in ratiometric aptasensor for lead ion. Anal. Chem. **87**, 7787–7794 (2015). https://doi.org/10.1021/acs.analchem.5b01445
31. Khlebtsov, N., Dykman, L.: Plasmonic nanoparticles, pp. 37–85 (2010). http://dx.doi.org/10.1201/9781439806296-c2
32. Unser, S., Bruzas, I., He, J., Sagle, L.: Localized surface plasmon resonance biosensing: current challenges and approaches. Sensors **15**, 15684–15716 (2015). https://doi.org/10.3390/s150715684
33. Ripp, S., Henry, T.B. (eds.): Biotechnology and Nanotechnology Risk Assessment: Minding and Managing the Potential Threats around Us, ACS Symposium Series. American Chemical Society, Washington DC (2011). http://dx.doi.org/10.1021/bk-2011-1079
34. Zhuang, J., Gentry, R.W.: Environmental application and risks of nanotechnology: a balanced view, pp. 41–67 (2011). http://dx.doi.org/10.1021/bk-2011-1079.ch003
35. Golobič, M., Jemec, A., Drobne, D., Romih, T., Kasemets, K., Kahru, A.: Upon exposure to Cu nanoparticles, accumulation of copper in the isopod Porcellio scaber is due to the dissolved Cu ions inside the digestive tract. Environ. Sci. Technol. **46**, 12112–12119 (2012). https://doi.org/10.1021/es3022182
36. Masciangioli, T., Zhang, W.-X.: Peer reviewed: environmental technologies at the nanoscale. Environ. Sci. Technol. **37**, 102A–108A (2003). https://doi.org/10.1021/es0323998
37. Swadeshmukul, S., Peng, Z., Kemin, W., Rovelyn, T., Weihong, T.: Conjugation of biomolecules with luminophore-doped silica nanoparticles for photostable biomarkers. Anal. Chem. **73**, 4988–4993 (2001). https://doi.org/10.1021/AC010406+
38. Tratnyek, P.G., Johnson, R.L.: Nanotechnologies for environmental cleanup. Nano Today **1**, 44–48 (2006). https://doi.org/10.1016/S1748-0132(06),70048-2
39. Mueller, N.C., Nowack, B.: Exposure modeling of engineered nanoparticles in the environment. Environ. Sci. Technol. **42**, 4447–4453 (2008). https://doi.org/10.1021/es7029637
40. Rogozea, E.A., Petcu, A.R., Olteanu, N.L., Lazăr, C.A., Cadar, D., Mihaly, M.: Tandem adsorption-photodegradation activity induced by light on NiO-ZnO p–n couple modified silica nanomaterials. Mater. Sci. Semicond. Process. **57**, 1–11 (2017). http://dx.doi.org/10.1016/j.mssp.2016.10.006
41. Olteanu, N.L., Lazăr, C.A., Petcu, A.R., Meghea, A., Rogozea, E.A., Mihaly, M.: "One-pot" synthesis of fluorescent Au@SiO_2 and SiO_2@Au nanoparticles. Arab. J. Chem. **9**, 854–864 (2016). http://dx.doi.org/10.1016/j.arabjc.2015.12.014
42. Olteanu, N.L., Rogozea, E.A., Popescu, S.A., Petcu, A.R., Lazăr, C. A., Meghea, A., Mihaly, M.: "One-pot" synthesis of Au–ZnO–SiO_2 nanostructures for sunlight photodegradation. J. Mol. Catal. A: Chem. **414**, 148–159 (2016). http://dx.doi.org/10.1016/j.molcata.2016.01.007

43. Rogozea, E.A., Olteanu, N.L., Petcu, A.R., Lazăr, C.A., Meghea, A., Mihaly, M.: Extension of optical properties of ZnO/SiO_2 materials induced by incorporation of Au or NiO nanoparticles. Opt. Mater. **56**, 45–48 (2016). https://doi.org/10.1016/j.optmat.2015.12.020
44. Kosmala, A., Wright, R., Zhang, Q., Kirby, P.: Synthesis of silver nano particles and fabrication of aqueous Ag inks for inkjet printing. Mater. Chem. Phys. **129**, 1075–1080 (2011). http://dx.doi.org/10.1016/j.matchemphys.2011.05.064
45. Holzinger, M., Le Goff, A., Cosnier, S.: Nanomaterials for biosensing applications: a review. Front. Chem. **2**, 63 (2014). http://dx.doi.org/10.3389/fchem.2014.00063
46. Millstone, J.E., Kavulak, D.F.J., Woo, C.H., Holcombe, T.W., Westling, E.J., Briseno, A.L., Toney, M.F., Fre´chet, J.M.J.: Synthesis, properties, and electronic applications of size-controlled poly (3-hexylthiophene) nanoparticles. Langmuir **26**, 13056–13061 (2010). https://doi.org/10.1021/la1022938
47. Shaalan, M., Saleh, M., El-Mahdy, M., El-Matbouli, M.: Recent progress in applications of nanoparticles in fish medicine: a review. Nanomed. Nanotechnol. Biol. Med. **12**, 701–710 (2016). http://dx.doi.org/10.1016/j.nano.2015.11.005
48. Cushing, B.L., Kolesnichenko, V.L., O'Connor, C.J.: Recent advances in the liquid-phase syntheses of inorganic nanoparticles. Chem. Rev. **104**, 3893–3946 (2004). https://doi.org/10.1021/cr030027b
49. O'Brien, S., Brus, L., Murray, C.B.: Synthesis of monodisperse nanoparticles of barium titanate: toward a generalized strategy of oxide nanoparticle synthesis. J. Am. Chem. Soc. **123**, 12085–12086 (2001). https://doi.org/10.1021/ja011414a
50. Avasare, V., Zhang, Z., Avasare, D., Khan, I., Qurashi, A.: Room-temperature synthesis of TiO_2 nanospheres and their solar driven photoelectrochemical hydrogen production. Int. J. Energy Res. **39**, 1714–1719 (2015). https://doi.org/10.1002/er.3372
51. Ning, F., Shao, M., Xu, S., Fu, Y., Zhang, R., Wei, M., Evans, D.G., Duan, X.: TiO_2/graphene/NiFe-layered double hydroxide nanorod array photoanodes for efficient photoelectrochemical water splitting. Energy Environ. Sci. **9**, 2633–2643 (2016). http://dx.doi.org/10.1039/C6EE01092J
52. Fang, X.-Q., Liu, J.-X., Gupta, V.: Fundamental formulations and recent achievements in piezoelectric nano-structures: a review. Nanoscale **5**, 1716 (2013). https://doi.org/10.1039/c2nr33531j
53. Gawande, M.B., Goswami, A., Felpin, F.-X., Asefa, T., Huang, X., Silva, R., Zou, X., Zboril, R., Varma, R.S.: Cu and Cu-based nanoparticles: synthesis and applications in catalysis. Chem. Rev. **116**, 3722–3811 (2016). https://doi.org/10.1021/acs.chemrev.5b00482
54. Li, D., Baydoun, H., Verani, C.N., Brock, S.L.: Efficient water oxidation using CoMnP nanoparticles. J. Am. Chem. Soc. **138**, 4006–4009 (2016). https://doi.org/10.1021/jacs.6b01543
55. Nagarajan, P.K., Subramani, J., Suyambazhahan, S., Sathyamurthy, R.: Nanofluids for solar collector applications: a review. Energy Procedia **61**, 2416–2434 (2014). https://doi.org/10.1016/j.egypro.2014.12.017
56. Sagadevan, S.: A review on role of nanofluids for solar energy applications, vol. 3, p. 53 (2015). http://www.sciencepublishinggroup.com, http://dx.doi.org/10.11648/J.NANO.20150303.14
57. Young, K.J., Martini, L.A., Milot, R.L., Snoeberger, R.C., Batista, V.S., Schmuttenmaer, C.A., Crabtree, R.H., Brudvig, G.W.: Light-driven water oxidation for solar fuels. Coord. Chem. Rev. (2012). http://dx.doi.org/10.1016/j.ccr.2012.03.031
58. Zhou, Y., Dong, C.-K., Han, L., Yang, J., Du, X.-W.: Topdown preparation of active cobalt oxide catalyst. ACS Catal. **6**, 6699–6703 (2016). https://doi.org/10.1021/acscatal.6b02416
59. Greeley, J., Markovic, N.M.: The road from animal electricity to green energy: combining experiment and theory in electrocatalysis. Energy Environ. Sci. **5** (2012). http://dx.doi.org/10.1039/c2ee21754f

60. Liu, D., Li, C., Zhou, F., Zhang, T., Zhang, H., Li, X., Duan, G., Cai, W., Li, Y.: Rapid synthesis of monodisperse Au nanospheres through a laser irradiation-induced shape conversion, self-assembly and their electromagnetic coupling SERS enhancement. Sci. Rep. **5**, 7686 (2015). https://doi.org/10.1038/srep07686
61. Liu, J., Liu, Y., Liu, N., Han, Y., Zhang, X., Huang, H., Lifshitz, Y., Lee, S.-T., Zhong, J., Kang, Z.: Metal-free efficient photocatalyst for stable visible water splitting via a two-electron pathway. Science **80**(347), 970–974 (2015). http://dx.doi.org/10.1126/science.aaa3145
62. Wang, D.-W., Su, D.: Heterogeneous nanocarbon materials for oxygen reduction reaction. Energy Environ. Sci. **7**, 576 (2014). http://dx.doi.org/10.1039/c3ee43463j
63. Wang, Z., Pan, X., He, Y., Hu, Y., Gu, H., Wang, Y., Wang, Z., Pan, X., He, Y., Hu, Y., Gu, H., Wang, Y.: Piezoelectric nanowires in energy harvesting applications. Adv. Mater. Sci. Eng. **2015**, 1–21 (2015). https://doi.org/10.1155/2015/165631

Chapter 16
Environmental Impact of Nanotechnology and Novel Applications of Nano Materials and Nano Devices

Nanomaterial comprises several hundred atoms and molecules, which a lot of complicated and interesting interactions occur in it. Characteristics in nanoscale, such as transmittance, electrical conductivity, thermal conductivity, magnetic, corrosion, and other effects.

Furthermore, nanomaterial manufacturing technology is applied on all kinds of nanomaterial devices, such as photo device, FET, and biosensor, and enhances electronic, optical, and mechanical properties. Most importantly, taking advantage of nanotechnology, the cost of devices will go down with better quality. This chapter is about novel nanomaterials device and applications and other related fields. This chapter enables interdisciplinary collaboration between material science and engineering technologists in the academic and industrial fields.

Novel applications of 1D nanomaterials are the core of the current research [1]. Guozhen Shen and Chongwu Zhou's groups have presented devices and chemical sensing applications of metal oxide nanowires. Kuei-Hsien Chen and Li-Chyong Chen's groups reported functionalized GaN nanowire-based electrodes for direct label-free volt ammetric detection of DNA hybridization. Younan Xia's group has reported synthesis and application of $RuSe_2$ nanotubes as a methanol tolerant electrocatalyst for the oxygen reduction reaction. Guozhong Cao's group reported carbon monoxide annealed TiO_2 nanotube array electrodes for efficient biosensor applications. Moon-Ho Jo and Seunghun Hong's groups gave a review about three promising strategies for the massive assembly of inorganic nanowires on solid substrates for their device applications. This is an important step towards nanomanufacturing. Huibiao Liu, Yuliang Li and colleagues reported the enhanced field emission properties of ZnO nanoparticles and CuTCNQ (copper 7,7,8,8-tetracyanoquinodimethane) nanowires.

Nanotubes, nanowires and nanobelts are important one-dimensional (1D) nanomaterials that are the foundation for nanoscience and nanotechnology. 1D nanomaterials have important applications in electronics, optoelectronics, sensors, biomedical science, environmental science, green energy and advanced defence technologies.

L. H. Madkour, *Nanoelectronic Materials*, Advanced Structured Materials 116, https://doi.org/10.1007/978-3-030-21621-4_16

Beyond binary oxides, synthesis of 1D nanostructures of complex oxides and oxides with unique crystallographic structures is a new trend in nanomaterials synthesis. Hong Jin Fan et al. gave a review on the synthesis of 1D nanomaterials of spinel structured materials, such as Al_2O_3, Ga_2O_3, Fe_2O_3, Sb_2O_3, TiO_2, and SnO_2. Using ZnO nanowire arrays as sacrificial templates, Kuang and Xie reported the fabrication of aligned SnO_2 nanotube arrays, which are great candidates for gas and chemical sensors. Shihe Yang's group described a general surfactant-free synthesis of $MTiO_3$ (M = Ba, Sr, Pb) perovskite nanostrips. Xiaodong Li's group introduced low temperature, organic-free synthesis of 1D nanostructures of complex hydroxides and oxides. Pu-Xian Gao et al. described synthesis, characterization, and photocatalytic properties of ZnO/(La, Sr)CoO_3 composite nanorod arrays.

Besides synthesis and structure analysis, in situ characterization of the physical properties of individual nanostructures is of vital interest for understanding their unique performance. Golberg and Bando introduced properties and engineering of individual inorganic nanotubes using in situ transmission electron microscopy, which provides a direct correlation between the measured property and the observed structure. X. D. Bai and E. G. Wang's group studied the bending effect of individual zinc oxide nanowires on photoconducting behavior of individual ZnO nanowires, and they found an improved photoresponse performance of the nanowires as a result of bending. The result supports the mechanism proposed for piezotronics.

The property of the 1-D nanostructures provides a sensing modality for label-free and direct electrical readout when the nanostructure is used as a semiconducting channel of a chemiresistor or field-effect transistor. The methods used to fabricate 1-D nanostructures such as silicon nanowires (SiNWs), carbon nanotubes (CNTs), and conducting polymer nanowires (CP NWs) predict the direction toward which future research in nanotechnology might be directed.

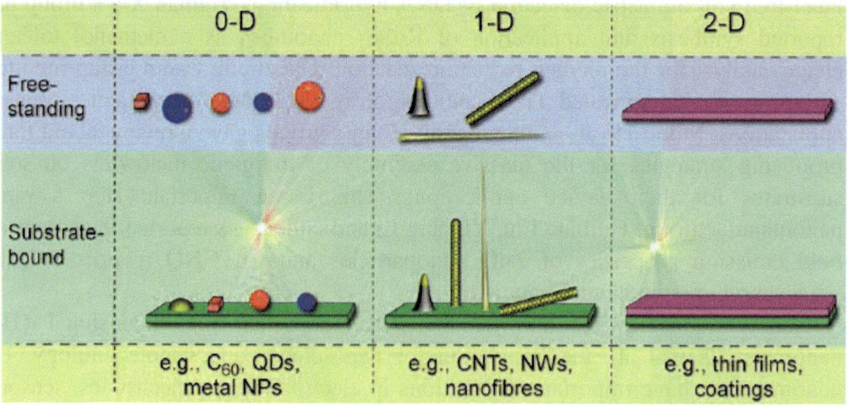

Nanowires are promising materials for many novel applications for their unique geometry and unique physical properties (electrical, magnetic, optical and mechanical). The properties of nanowires have caused researchers and companies to consider using this material in several fields.

- Nanowires can be precisely controlled during:
  - Synthesis
  - Chemical composition
  - Length
  - Diameter
  - Doping/electronic properties
- Enables the assembling of building blocks into increasingly complex structures.
- Low device-to-device reproducibility.

16.1 From Microelectronics to Nanoelectronics and Molecular Electronics

In 1965, Intel co-founder Gordon Moore forecasted the rapid pace of technology innovation. His prediction, popularly known as "Moore's Law," states that transistor density, that is the number of transistors in an integrated circuit or chip on integrated circuits, doubles about every two years (Fig. 16.1). The first microprocessor was introduced by Intel in 1971 (4004) and contained 2,300 transistors. In 2004, Intel's fastest processor (Intel® Itanium® 2 processor, 9 MB cache) contained 592,000,000 transistors. In 2010 Intel's processor exceeded 2,000,000,000 transistors [2]. However, this development is now reaching a wall so that smaller is no

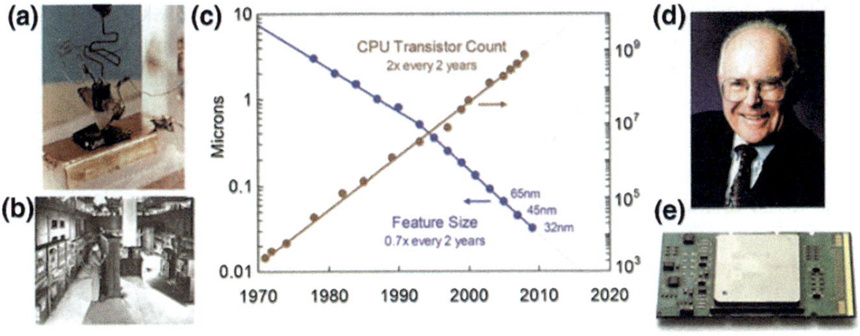

Fig. 16.1 **a** The first, point-contact transistor invented by John Bardeen and Walter Brattain in December 1947. Photo courtesy of Alcatel-Lucent. **b** Electronic Numerical Integrator and Computer, ENIAC conceived and designed by John Mauchly and J. Presper Eckert in 1946. **c** Transistors per microprocessor history [278]. **d** Gordon Moore. **e** Intel Itanium 2 processor

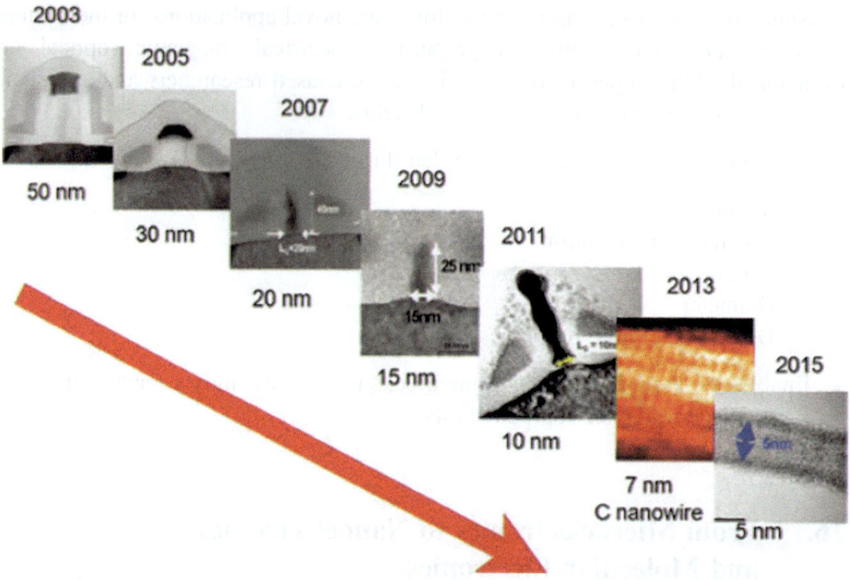

Fig. 16.2 Scaling history of the transistor channel length. The channel length was reduced from 50 nm in 2003 to 10 nm today. Future perspectives incorporate the use of nanowires in OFETs [2]

longer any faster. The prime reason for the limitation the semiconductor electronics experiences is its power dissipation and thus heat [3].

Figure 16.2 shows the evolution miniaturization of the conducting channel between the two other contacts, the source and the drain of a transistor. The channel length which is made of n- or p-doped silicon was reduced from 50 nm in 2003 to 10 nm today (2011).

However technical factors limit the top-down development of microelectronics, the non-scalability of the MOS transistor below a critical size L_{phys} (the physical limit) or the impossibility of batch defining, using proximity masks, feature sizes below a critical one L_{litho} (the lithographic limit) [4]. At the IC (integrated circuit) level, the channel length (L) of the transistor is very critical because as this length decreases we have: (a) increase in the number of the transistors of the IC, thus then more "logic gates", so more "processing" power and (b) decrease in the "responding time" of the "logical operations". At the transistor level, the channel length (L) of the transistor is included in the basic equation of the gain factor β.

$$\beta = \frac{\mu_{\varepsilon}}{t_{\text{OX}}}\left(\frac{W}{L}\right).$$

That means, when the length L decreases, we have: (a) increase in the gain factor β of the transistor and (b) better directivity of electrons in the channel path [5].

Organic field-effect transistors (OFETs) are an alternative technology with high technological potential due to the possibility of low-cost and large-area manufacturing processes. An OFET uses an organic semiconductor in its channel and can be prepared either by vacuum evaporation of small molecules or by solution casting of polymers or small molecules. It has been demonstrated that single-walled carbon nanotubes (SWNTs) can be used as quasi-one-dimensional (1D) electrodes to construct organic FETs with molecular-scale width ($\sim$2 nm) and channel length (down to 1–3 nm) [6].

Ultra-dense integrated circuits with features smaller than 10 nm would provide enormous benefits for all information technologies, including computing, networking, and signal processing. The top-down route of the silicon technology has indeed been relatively easy to run until its basic step (optical lithography) has met its physical limits (minimum feature size around the light wavelength). To overcome this limit, shorter wavelengths are required, such as in extreme ultraviolet lithography. X-ray lithography is currently the leading technology in the drive to replace photolithography as a large-scale production tool because it uses masks, which are suited to high-volume production.

Nanoelectronics research is currently looking not only for the successor to CMOS processing but also for a replacement for the transistor itself. On the scale of 10 nm dimensions, components have a wavelength comparable to that of an electron at the Fermi energy. The confinement and coherence of the electron gives rise to gross deviations from the classical charge transport found in conventional devices. Quantum-mechanical laws become increasingly dominant on the nanoscale, and it is probable that nanoelectronics will operate on quantum principles [7].

Molecular electronics, i.e., the information processing at the molecular-scale, becomes more and more investigated and envisioned as a promising candidate for the nanoelectronics of the future. More than a possible answer to ultimate miniaturization problem in nanoelectronics, molecular electronics is foreseen as a possible way to assemble a large number of nanoscale objects (molecules, nanoparticles, nanotubes, and nanowires) to form new devices and circuit architectures [8].

The difference between molecular- (nano) and micro-electronics is not the size (dimensionality), but the profoundly different device- and system-level solutions, the device physics, and the phenomena, fabrication, and topologies/organizations/architectures. Three-dimensional topology molecular and nanoelectronic devices, engineered from atomic aggregates and synthesized utilizing *bottom-up* fabrication, exhibit quantum phenomena and electrochemomechanical effects that should be uniquely utilized. Given technological advancements, molecular electronics proponents believe purposeful bottom-up design will be more efficient than the top-down method, and that the incredible structural diversity available to the chemist will lead to more effective molecules, thus approaching optional functionality for each application. A single mole of molecular switches, weighing about 450 g and synthesized in small reactors (a 22-L flask might suffice for most steps of the synthesis), contains 6×10^{23} molecules—a number greater than all the transistors ever made. While we do not expect to build a circuit in which each single molecule is both addressable and connected to a power supply (at least not in the first few generations), the extremely large numbers of switches available in a small mass illustrate one reason molecular electronics can be a powerful tool for future computing development [9].

16.2 Nano in Energy and Clean Energy

Energy is one of the most challenging needs of humanity, and is highest on the list of priorities and requisites for human welfare [10]. According to the International Energy Agency (IEA), World's primary energy demand will increase by 36% between 2008 and 2035. Electricity demand is expected to grow by 2.2% per year between 2008 and 2035. Taking in account the CO_2 emissions and the global climate change impact on life and the health of the planet renewable energy sources will have to play a central role in moving the world onto a more secure, reliable, and sustainable energy path [11].

Solar energy is the most abundant, inexhaustible and clean of all the renewable energy resources till date. The power from sun intercepted by the earth is about 1.8×10^{11} MW, which is many times larger than the present rate of all the energy consumption. Photovoltaic technology is one of the finest ways to harness the solar power [12].

Figure 16.3 shows the history of confirmed "champion" laboratory cell efficiencies. The performance of conventional solar cells is approaching a plateau; only incremental improvements have been accomplished in the last decade despite dedicated R&D effort.

Tandem solar cells based on III–V materials have achieved the highest efficiencies of any present photovoltaic device exceeding 40% recently However, the cost of these devices is very high, limiting their application to space applications [13]. The efficiencies reached with commercial solar cell modules are significantly lower than those of the best laboratory cells due to losses incurred during scaleup. The typical size of "champion laboratory cells" is in the square centimeter range or even below, facilitating the collection of photocurrent. High efficiency multijunction solar cells with finely-tuned quantum wells are investigated by Varonides.

Solar energy has been greatly cared as a renewable energy source. Organic photovoltaic devices are attractive low-cost alternative compared with thin-film solar cells (single-polycrystalline and amorphous thin films) and compound-based solar cells (copper indiumgallium(di)selenide (CIGS), GaAs, CdTe). They are easily produced and large-area modules can be made with roll-to-roll process.

Silicon nanoparticles doped poly(3-hexylthiophene) and [6,6]-phenyl C61-butyric acid methyl ester blends (P3HT:PCBM:Si NP) have been produced as the photoactive layer of organic photovoltaic devices (OPVs). The silicon nanoparticles' size is between 80 and 100 nm checked by transmission electron microscope (TEM). The 0.35 wt% Si NP doping OPVs exhibit higher power

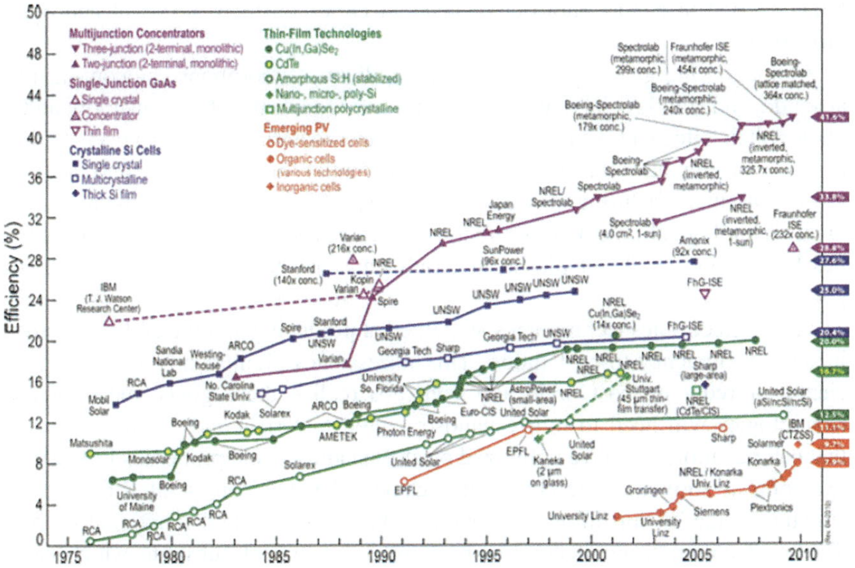

Fig. 16.3 Historic summary of champion cell efficiencies for various PV technologies. Tandem solar cells based on III–V materials present the highest efficiencies of any present photovoltaic device exceeding 40% [279]

conversion efficiency (PCE) than other OPVs. The PCE of the OPVs increases from 3.01 to 3.38% mainly due to increasing short-circuit current density from 8.38 to 9.48 mA/cm^2, while the open circuit voltage remains the same. The Si NP can provide extra exciton separation and electron pathways in hybrid solar cells. The PCE of OPVs with poly(3-hexylthiophene) (P3HT) and [6,6]-phenyl C61-butyric acid methyl ester (PCBM) blend as the photoactive layer is high compared with that of other OPVs so far. The P3HT provides the pi-orbit electron excitation during the photovoltaic process [14]. The PCBM usually acts an electron transport material in polymer solar cells.

The PCE of OPVs under AM 1.5 conditions has exceeded 7.4% [15]. The PCE of OPVs can be improved by doping nanomaterials like single-walled carbon nanotubes [16], multiwalled carbon nanotubes [17], nanodiamond [18], Fe_3O_4 magnetic nanoparticles [19], and ZnO nanorods [20]. Nanomaterials, which can be used to make superstrong metals [21], have become popular as semiconductor nanocrystals and solar cell materials due to their high surface to volume ratio, quantum confinement [22], quantum size [23], and small size effect [18].

Silicon, which has been successfully applied in solar cell and microelectronics industry, is a potential candidate for polymer-nanomaterials solar cells owing to its abundance, nontoxicity, and strong ultraviolet absorption. Doping Si NP into polymer materials acting as the photoactive layer of OPVs may increase the electron collection rate in the mixture. This work investigates the effect of employing

P3HT:PCBM:Si NP as photoactive layer in OPVs on their electrical and optical properties.

Although inorganic semiconductors (silicon, amorphous silicon, gallium arsenide, and sulfide salts) have been the primary focus, the photosensitivity and the photovoltaic effects in devices made with organic materials have also been explored because of the advantages such as the potential of large-scale and low cost roll-to-roll manufacturing processes. More specific, organic photovoltaics offer great technological potential as a renewable energy source due to their mechanical flexibility, low weight of plastic materials and easy thin-film casting technology [24–28].

Plastic electronics technologies are aimed at producing significant improvements in device efficiency-to-cost ratios. This necessitates significantly improving efficiency or reducing cost or ideally both. To realize these goals, many of these technologies will need to utilize nanostructured materials and composite systems that can be tailored to have optimized electronic and optical properties [29, 30]. For example, as shown in Fig. 16.4, an organic solar cell consist of a multilayered structure made of thin films each one of which has a certain functional property. The most common architecture consists of a transparent substrate which can be either glass or a polyester film such as poly(ethylene terephthalate) PET or poly (ethylene naphthalate) PEN. A high work function metal electrode such as indium tin oxide (ITO) serves as an anode for collecting holes and a low work function metal such as aluminum serves as the cathode collecting the electrons which are produced in the active layer. Additional buffer layers such as hole transport layers (PEDOT:PSS) or electron transport layers (Ca, LiF, and TiO_x) are placed between the electrodes and the active layer to provide better energy level alignment and better ohmic contact between the organic layer and the metal electrodes.

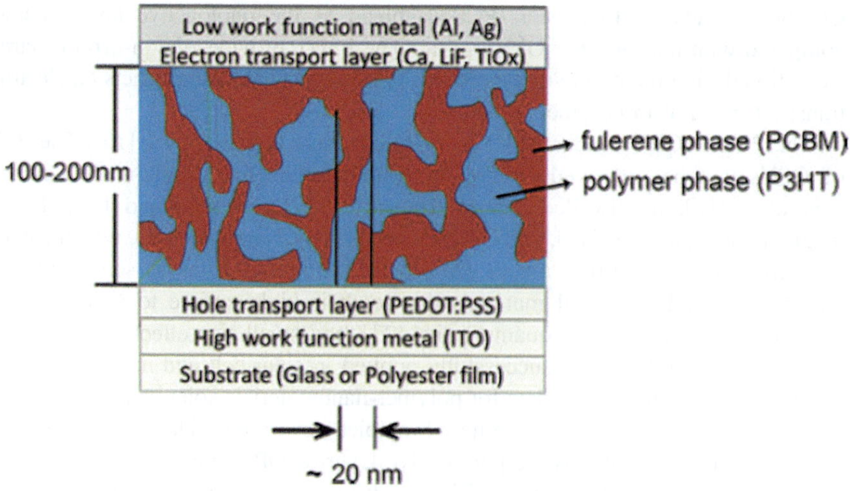

Fig. 16.4 Schematic structure of a typical organic solar cell device

The most successful active layer up to date consists of a bulk heterojunction (BHJ) that is formed by a p-type semiconductor (electron donor), such as poly (3-hexylthiophene) (P3HT) with an n-type semiconductor (electron acceptor), such as methanofullerene derivatives (PCBM). Due to the low dielectric constant in organic components (~ 3) photoexcitation leads to a strongly bound exciton, which needs to be dissociated into free carriers. This dissociation can take place in a strong electric field or at the donor–acceptor interface. Then, free carriers need to be transported to the corresponding electrodes via drift and diffusion processes, where they are collected, giving rise to an electric current. The morphology of the active donor–acceptor film is critical for charge generation and transport, strongly influencing device performance. Despite the rapid increase in device performance observed recently, much effort is still required to understand the fundamental processes of photovoltaic energy generation, in particular to elucidate the complex relationship between nanoscale morphology/electronic properties and device performance and to further develop the appropriate nanometrology needed to address this interplay [31].

Owing to the advantages consisting of low cost, easy fabrication, and environmental friendliness, silicon (Si) is a very promising material for the photovoltaic applications [32, 33]. Thin-film solar cells based on amorphous silicon (a-Si) or microcrystalline silicon (μc-Si) are the most popular products applied to the building-integrated photovoltaics and consumer electronics. Transparent conductive oxide (TCO) films are usually used as the front electrode of thin-film solar cells. For a transparent electrode, the requirements of TCO films are a low sheet resistance (R_{sheet}) to minimize the current loss, a low contact resistance to semiconductor layers, and a high transmission of incident light. To completely utilize the incident light, an important technique of the so-called light trapping has been developed using a TCO with suitable surface texture.

Transparent electrodes of tin dioxide (SnO_2) on glasses were further wet-etched in the diluted HCl:Cr solution to obtain larger surface roughness and better light-scattering characteristic for thin-film solar cell applications. The process parameters in terms of HCl/Cr mixture ratio, etching temperature, and etching time have been investigated. After etching process, the surface roughness, transmission haze, and sheet resistance of SnO_2 glasses were measured. It was found that the etching rate was increased with the additions in etchant concentration of Cr and etching temperature. The optimum texture-etching parameters were 0.15 wt% Cr in 49%HCl, temperature of 90 °C, and time of 30 s. Moreover, silicon thin-film solar cells with the p-i-n structure were fabricated on the textured SnO_2 glasses using hot-wire chemical vapor deposition. By optimizing the texture-etching process, the cell efficiency was increased from 4.04 to 4.39%, resulting from the increment of short-circuit current density from 14.14 to 15.58 mA/cm^2. This improvement in cell performances can be ascribed to the light-scattering effect induced by surface texturization of SnO_2.

It was indicated that a specific design of TCO films plays an important role in enhancing the performances of thin-film solar cells. A surface-textured TCO can scatter the light greatly and increase the effective optical path length within the active layers [34–37]. A rough TCO film also ensures that the roughness is copied by the film deposited on it, so that the back metal electrode produces an increased scattering of reflected light. Therefore, this would improve both the optical absorption and current density in thin-film solar cells. After that, various randomly textured TCO substrates have been proposed to increase the light scattering. As-deposited and postchemical textured TCOs are both useful to achieve the rough substrates. The common TCO materials applied to thin-film solar cells consist of tin dioxide (SnO_2) [37] and zinc oxide (ZnO) [38, 39]. The great majority of glass based Si thin-film modules are prepared on the fluorine doped SnO_2 (SnO_2:F) glasses due to their intrinsic rough surface [35, 36]. Moreover, the directly deposited or texture etched ZnO is another attractive method for the fabrication of rough TCO substrate [40–43]. By using the texture-etched ZnO, the quantum efficiency of thin-film solar cells over the whole spectral range can be increased [42, 43]. A further improvement in the reduction of free carrier absorption losses in the red/IR and absorption around the optical edge of ZnO in the blue/UV region can be achieved [44].

Magnesium (Mg) was used as a negative electrode in lithium (Li) ion secondary batteries due to its high theoretic capacity [45]. Magnesium-nickel (Mg–Ni) powders are used as the anode materials for secondary lithium (Li) ion batteries. Mg–Ni powders with ratios of 1:1 (Mg:Ni) are prepared and their structure and electrochemical behavior at room temperature and 55 °C are investigated. The results show that adding Ni powders to Mg powders can reduce the charge-discharge capacities and improve cycling life. In charge-discharge cycle testing at 55 °C, the Li ion concentration gradually increased with increasing the duration of electrochemical reactions, indicating that the charge-discharge capacities increase with increment of cycling number. The formation of a solid electrolyte interface (SEI) layer restrains Mg ions from dissolving into the electrolyte and thus improves the charge-discharge capacities at high temperature. Mg–Ni material is a typical active/nonactive alloy system that can effectively suppress the volume expansion and improve battery cycling life. Many researchers have prepared Mg–Ni alloys using the mechanical grinding method [46, 47]. However, the proportion of Mg to Ni is not easy to control, and Mg–Ni alloys with excessive Ni have reduced charge discharge capacity [48]. In addition, Mg easily oxidized to form MgO phases during mechanical milling [49]. The formation of MgO phases may degrade the cycling performance of batteries. Mg powder was directly mixed with Ni powder in a glove box. Because the particle size of Mg powder and that of Ni powder is very different, alloying effect prevents oxidation problems [50].

Recently, nonvolatile memory (NVM) has become an important device for portable products. Several novel technologies have been proposed to develop an advanced device as the next generation of NVM. Resistive random access memory (RRAM) has attracted significant interest due to its simple structure, high switching speed, and low operating voltage [51]. RRAM devices can be reversibly switched between a high resistance state (HRS) and a low resistance state (LRS) by DC voltages or voltage pulses. The resistance change can be used to store information for NVM applications. The resistive switching behavior is primarily influenced by the material groups [52, 53], electrodes [54], and interface status [55, 56]. In general, the switching mechanisms of these resistive switching behaviors can be divided into three groups: thermochemical reactions [57], electrochemical reactions [58–60], and the valence change effect [61]. However, a detailed switching mechanism is still lacking. Several studies [62–64] proposed the surface effect on the resistive switching behaviors, which indicated that oxygen or humidity may influence the redox reaction in the interface or resistive layer. Although a RRAM device has many promising advantages, the resistive switching behavior is unstable and may cause operation failure. Therefore, several methods such as nanoparticle [65], stacked layers [66], and process optimization [67] have been proposed to improve the switching dispersion. Plasma treatments have also been proposed to improve switching dispersion [68, 69]. Wang et al. proposed the CF_4 plasma treatment to improve the resistive switching properties with the valence change effect [69]. The CF_4 plasma treatment changed the Schottky barrier height to reduce the operating voltages. However, the effect of plasma treatment on an electrochemical-based RRAM has not been studied.

A 20 nm SiO_x layer is deposited using radio-frequency sputtering to form the resistive switching layer of a Cu/SiO_x/Pt memory device. The SiO_x-based device demonstrates the resistive switching characteristics with an electrochemical reaction. CF_4 plasma treatment was used to modify the SiO_x layer and incorporate fluorine atoms into the SiO_x layer. The bombardment damage and fluorine incorporation caused the SiO_x film to form a stack-like structure. This reduced the operating voltage and improved switching dispersion. The fluorine repaired the Cu/SiO_x interface, thus increasing the barrier height of the Cu/SiO_x interface and the resistance of the high resistance state. A statistical analysis of the conducting filament formation was performed in order to evaluate the number of formation/rupture sites. The resistive switching of the CF_4-treated sample had higher possibility to use the same filament sites; thus, the CF_4-treated sample had stable resistive switching behavior.

Low-frequency noise is an important issue for analog, digital, mixed signal, and RF application. Nowadays, complementary metal-oxide-semiconductor (CMOS) technology has intruded into RF and/or analog circuits, and hence the excessive low-frequency noise will lead to a limitation of the functionality for related circuits [70, 71]. Low-frequency noise, including flicker ($1/f$) noise and random telegraph noise (RTN), is increasingly attracting much interest in the CMOS device and technology community. In metal-oxide semiconductor field-effect transistor (MOSFET) devices, the origin of $1/f$ noise and RTN is considered stemming from the carrier behaviors related to the oxide traps. $1/f$ noise is most often due to fluctuations in carrier number and/or carrier scattering [72, 73]. RTN is another special kind of noise, which also appears in a low-frequency spectrum and originates from the trapping/detrapping behavior of a single or few traps [74, 75].

On the other hand, the continued shrinking of conventional CMOS to enhance device performance has revealed limitations. The mobility enhancement has emerged as a key technology for improving drive current [76]. Ways of optimizing channel mobility had been proposed to overcome the limitations on the scaling down of devices and to further improve the speed of CMOS circuits. The introduction of channel strain engineering in the state-of-the-art CMOS technology is recognized as an indispensable performance booster in producing next generation CMOS devices [77, 78]. For p-type MOSFET (pMOSFET), using embedded SiGe in the recessed source/drain (S/D) region can efficiently provide uniaxial compressive strain in the channel and improve hole mobility, bringing the enhancement of device performance [79–81]. In addition, high-k (HK) materials are also adopted into advanced CMOS process for solving the increased gate leakage current. This is because HK dielectrics are the promising candidate for gate insulator to achieve low equivalent oxide thickness as required for the advanced CMOS technology nodes [82]. However, a fabricated pMOSFET with SiGe S/D is possibly accompanied by the extra amount of process-induced defects, and a gate insulator that is replaced from conventional SiO_2 to HK materials usually causes the changes of trap properties in MOSFET, resulting in the influence on low-frequency noise characterizations. Though low-frequency noise characterizations in pMOSFET with SiGe S/D had been reported [81, 83], low-frequency noise in HK pMOSFET with SiGe S/D is still unclear and needs to be addressed to improve the understanding.

Recently, the low-frequency noise characterizations in 28-nm high-k (HK) pMOSFET with embedded SiGe source/drain (S/D) through $1/f$ noise and random telegraph noise measurements simultaneously has been studied. It is found that uniaxial compressive strain really existed in HK pMOSFET with embedded SiGe S/D. The compressive strain induced the decrease in the tunneling attenuation length reflecting in the oxide trap depth from Si/SiO_2 interface to the HK layer, so that the oxide traps at a distance from insulator/semiconductor interface cannot capture carrier in the channel. Consequently, lower $1/f$ noise level in HK pMOSFET with embedded SiGe S/D is observed, thanks to the less carrier fluctuations from trapping/detrapping behaviors. This result represents an intrinsic benefit of HK pMOSFET using embedded SiGe S/D in low-frequency noise characteristics.

Researchers at MIT have developed a solar cell using graphene coated with zinc oxide nanowires. The researchers believe that this method will allow the production of low cost flexible solar cells at high enough efficiency to be completive. Sensors powered by electricity generated by piezoelectric zinc oxide nanowires. This could allow small, self-contained, sensors powered by mechanical energy such as tides or wind. Researchers are using a method called Aerotaxy to grow semiconducting nanowires on gold nanoparticles. They plan to use self-assembly techniques to align the nanowires on a substrate; forming a solar cell or other electrical devices. The gold nanoparticles replace the silicon substrate on which conventional semiconductor based solar cells are built. Researchers at the Nies Bohr Institute have determined that sunlight can be concentrated in nanowires due to a resonance effect. This effect can result in more efficient solar cells, allowing more of the energy from the sun to be converted to electricity. Using light absorbing nanowires embedded in a flexible polymer film is another method being developed to produce low cost flexible solar panels. Researchers at Lawrence Berkeley have demonstrated an inexpensive process for making solar cells. These solar cells are composed of cadmium sulfide nanowires coated with copper sulfide. Researchers at Stanford

University have grown silicon nanowires on a stainless steel substrate and demonstrated that batteries using these anodes could have up to 10 times the power density of conventional lithium ion batteries. Using silicon nanowires, instead of bulk silicon fixes a problem of the silicon cracking that has been seen on electrodes using bulk silicon. The cracking is caused because the silicon swells it absorbs lithium ions while being recharged, and contracts as the battery is discharged and the lithium ions leave the silicon. However the researchers found that while the silicon nanowires swell as lithium ions are absorbed during discharge of the battery and contract as the lithium ions leave during recharge of the battery the nanowires do not crack, unlike anodes that used bulk silicon.

16.3 The Environmental Impact of Nanotechnology

The world is facing significant environmental challenges such as improving the quality of air, soil, and water. Currently, industry is focusing on detecting pollutants (from chemical spills, fertilizer and pesticide run-off), improving industrial and mining sites, treating contaminants and stopping further pollution.

Image Credits shutterstock.com/g/Irina+Chernysh

A potential solution to these problems is to use nanomaterials [84–89]. Nanomaterials can be used to assist with cleaning the environment and even provide efficient energy solutions, such as nanomaterial based solar cells. In addition to this, nanomaterials help to improve the quality and performance of many consumer products. As a result of this, the exposure to manufactured nanomaterials is increasing day-by-day. However, there are both positive and negative impacts on the environment due to nanotechnology.

16.3.1 Positive Impacts

With the help of nanotechnology, water quality can be improved. Some of the nanomaterials that can be used for remediation of water are carbon nanotubes (CNTs), zeolites, nanoparticles of zero valent iron (ZVI), silver nanoparticles, etc. Other nanomaterials like zinc oxide (ZnO), titanium dioxide (TiO_2), tungsten oxide, serve as a photocatalyst. These photocatalysts can oxidize organic pollutants into harmless materials. TiO is the most preferred material as it has high photostability, high photoconductivity, easily available, inexpensive and nontoxic. Silver nanoparticles have antimicrobial effect. Also, many polymeric nanoparticles are being used for wastewater treatment.

Another new technology is known as nanofiltration which can be used in water treatment in homes, offices, and industries. Molybdenum disulphide (MoS_2) non-porous membrane is used for energy efficient desalination of water which filters five times more than the conventional ones. To clean oil spills in the water bodies, a nanofabric paper towel has been developed which are woven from tiny wires of potassium manganese oxide that can absorb oil 20 times its weight. Thus, nanotechnology provides a solution to clean the contaminated water and prevent new pollution.

With the help of nanotechnology, toxic gases in the air can be cleaned. But first, we have to detect the pollutants at the molecular level using precise sensors. A sensor called nanocontact sensor has been developed which can detect the heavy metal ions and radioactive elements. These sensors have a small size, are inexpensive and are easy to use onsite. Currently, single walled nanotubes (SWNTs) are being used for the detection of NO_2 and NH_3 gases. Also, SWNTs sensor can accomplish high sensing activity at room temperature when compared to the conventional sensors which work at 200 to 600 C. Cantilever sensors have been developed to sense VOCs, heavy metals and pesticides. A mixture of CNTs with gold particles helps adsorb toxic gases like NO_X, SO_2 and CO_2. Another porous nanomaterial manganese oxide has better adsorption of toxic gases due to its large surface area.

Therefore, by detecting pollutants by specific sensors, we can help protect the sustainability of human health and the environment. Thus, nanotechnology provides us with a new approach to cut down the waste production, reduce the emission of greenhouse gases and discharge of hazardous chemicals in water bodies.

16.3.2 *Negative Impacts*

Nanomaterials can also have a negative impact. Currently there is very little information describing the relative environmental risk of the manufactured nanomaterials. Only a few studies have been conducted with the aim of discovering the direct and indirect exposure to nanomaterials and there are no clear guidelines to quantify the effects.

Recently, a workshop, conducted by the National Science Foundation and the US Environmental Protection Agency, was undertaken to identify the critical risk issues concerning nanomaterials. Specifically, the workshop aimed at determining the exposure and toxicity of manufactured nanoparticles, ability to extrapolate manufactured nanoparticles toxicity using existing particle and fiber toxicology database, and recyclability and overall sustainability of the manufactured nanomaterials.

16.3.3 *Green Technology*

A remedial solution is the use of green technology or green manufacturing. This is an environmentally friendly technology which is developed and used to conserve natural resources. This technology aims at producing nanomaterials with lesser raw materials, minimum energy consumption, and minimum waste production. It is known that any manufacturing process is accompanied by a large amount of waste production. This is minimized by green manufacturing, which uses green chemicals which are less harmful to the environment and energy efficient processes. Microemulsions, which are used instead of VOCs in the cleaning industry, are an example of green technology.

Thus, scientific authorities are monitoring various nanoparticles produced and used, as well as their subsequent impact. This is in order to balance the technology's benefits and possible unintended consequences.

16.4 AI and Nanotechnology How Do They Work Together?

Artificial Intelligence (AI) technology [90] has been developing for many years now; not only as an area of technology in and of itself, but also in the various spaces and industries where it can now be found.

Image Credits shutterstock.com/g/peshkova

Technology that operates on the nanometre scale often involves intricate systems that are not always suited to the various facets of AI. However, there are some growing areas where AI converges with nanotechnology. In addition to merging the two technologies, combined work in nanotechnology and AI can also boost study in each discipline, possibly leading to all kinds of new tools for gaining insights and communication technologies.

Consider the following areas where AI and nanotechnology are working together.

16.4.1 Microscopy

Although atomic force microscopy (AFM) has seen many significant advances in recent years, it can still be a challenge to get high quality signals from these imaging devices. The predominant problem is that many of the tip-sample interactions these microscopes rely on are complex, varied and therefore not easy to decipher. AI can be very useful in dealing with these kinds of signal-related issues.

An AI approach known as functional recognition imaging (FRSPM) looks to address this issue through the direct identification of local actions from measured spectroscopic reactions. This process brings together the use of artificial neural networks (ANNs) with principal component analysis (PCA), utilized to streamline the input data to the neural network.

In another imaging development, researchers from The University of Texas Rio Grande recently announced the creation of a microfluidic channel with a removable nanotextured surface that binds specifically to breast cancer cells. Once linked, it can be extracted and imaged. The imaging is segmented and combined with an AI algorithm that automatically determines if a cell is cancerous based on historical existing cell data. The novel imaging system can contrast historical samples to the cells being evaluated in real-time.

16.4.2 Chemical Modelling

Algorithms are already being used to illustrate the frameworks of a molecules and materials in order to figure out various qualities and how they may interact in various environments. A natural progression has led to the incorporation of AI and the use of complex machine learning algorithms.

From a modelling point-of-view, there are numerous types of parameters that must be correlated to generate either an image or a dynamic depiction of a chemical system. As with some imaging techniques, AI can better analyse information and learn from the past to create a more precise representation of the system under study. For instance, AI can minimise the degree of error related to the geometry or size of a system or particle. This is especially useful for nanomaterials as the several effects and phenomena seen with materials like graphene can often be difficult to recreate.

Portrayal of the structural qualities of nanomaterials has also been resolved by the usage of ANNs. For instance, these algorithms have been used to figure out the configuration of carbon nanotube structures by quantifying structural qualities like alignment and curvature. Furthermore, the portrayal of several qualities of thin films is an issue that has been broadly addressed by using AI, machine learning and neural networks.

16.4.3 Nanocomputing

Unsurprisingly, AI is also extremely useful with respect to the future of nanocomputing, which is computing conducted through nanoscale mechanisms. Currently, there are a lot of ways nanocomputing devices can execute a function, and these can cover anything from the physical operations to computational methods. Due to a great deal of these devices depending on intricate physical systems to allow for intricate computational algorithms, machine learning procedures can be used to generate novel information representations for a broad range of uses.

16.5 Novel Nanotubes and Encapsulated Nanowires

Carbon nanotubes, with or without encapsulated material, generated by arc discharge and electrolytic techniques have been studied. Microcrystals of refractory carbides (i.e. NbC, TaC, MoC), contained in nanotubes and polyhedral particles, produced by arcing electrodes of graphite/metal mixtures, were analysed by high resolution transmission electron microscopy (HRTEM) and X-ray powder diffraction. Encapsulation of MoC was found to give rise to an unusual stable form, namely face-centered-cubic MoC. SQUID measurements indicate that the encapsulated carbides exhibit superconducting transitions at about 10–12 K, thus they differ from carbon nanotubes/nanoparticles which do not superconduct. Four-probe and microwave (contactless) conductivity measurements indicate that most of the analysed samples behave as semiconductors. However, metallic transport was observed in specimens containing single conglomerated carbon nanotube bundles and boron-doped carbon nanotubes. Novel *metallic* b-Sn nanowires were produced by electrolysis of graphite electrodes immersed in molten LiCl = $SnCl_2$ mixtures. Prolonged electron irradiation of these nanowires leads to axial growth and to dynamic transformations. These observations suggest ways in which materials may be modified by microencapsulation and irradiation.

Soon after carbon nanotubes were prepared in bulk by the arc discharge technique [91], the suggestion was made that it might be possible to encapsulate metals within the inner core of the nanotubes. Such a proposal followed logically from the successful preparation of fullerenes containing endohedral metals [92, 93]. Elements such as Pb, Bi, Cs, S, and Se were first introduced by heating the elements with openended nanotubes [94–96] (sometimes in the presence of oxygen). Capillarity and wetting plays an important role in the process. Generally, only *low-surface-tension substances* can be introduced into nanotubes [96] in this way.

In subsequent experiments, a graphite anode was drilled and packed with a metal. Arcing then led to the formation at the cathode of nanotubes and polyhedral particles, containing both encapsulated metals and carbides [97–102]. A problem arises with this technique in that the nanotubes are only partly filled, a fact which may be disadvantageous if nanoscale applications (e.g. conduction) are to be considered. Recently, Loisseau and Pascard described optimal conditions for more or-less complete encapsulation of Se, S, Sb, and Ge by using arc discharge techniques [103]. It was concluded from allied experiments [104] that elements with incomplete electronic shells stood the best chance of being encapsulated. However, the growth mechanism associated with these filled nanotubes (termed nanowires) is not well understood. To date, it has proved possible to produce encapsulated LaC_2 [105], YC_2 [98, 106], LaC_2 [107], CeC_2 [102, 107], Gd_2C_3 [104, 108], TiC [104, 109], V_4C_3 [110], ZrC [110], TaC [102, 105], HfC [111], Fe_3C [104], Ni_3C [104], and carbides of Cr [104], Dy [104], and Yb [104]. Only a few metals such as Mn, Co, and Cu and semiconducting elements such as Se, Ge, and Sb have been inserted into nanotubes without carbide formation by using plasma arcs.

Chemical techniques have also been used to prepare nanotube encapsulated-materials. In this context, carbon nanotubes were oxidatively opened by boiling with aqueous nitric acid, then filled with metal oxides (e.g. NiO, UO_2) [112] and metals (e.g. Pd) [113] by wetting techniques.

Carbon nanotubes react with metal oxides above 1200 °C to yield carbide nanorods [114] (i.e. filaments consisting of pure metal carbides which are not encapsulated: TiC, NbC, Fe_3C, SiC, BC_x). Ajayan and co-workers reported that it is also possible to coat carbon nanotubes with a single layer of metal oxide (e.g. V_2O_5) in addition to inserting the oxides into the core [115]. These authors produced so-called nanocomposites by annealing partly oxidised nanotubes with oxide mixtures in air above the melting point of the corresponding oxide [115].

Encapsulation of metals (e.g. Li) within fullerenes was also achieved by passing a current between graphite electrodes in molten mixtures of ionic salts (e.g. LiCl, LiBr) in an argon atmosphere [116]. This technique has been successfully used to introduce b-Sn (using mixtures of LiCl and $SnCl_2$) into nanotubes.

For encapsulated ferromagnetic materials (e.g. Fe, Ni, Co), it may prove possible to develop magnetic data storage devices and toners and inks for xerography [117]. Refractory metal carbides such as TaC, NbC, and MoC exhibit superconducting transitions at low temperatures (10–14 K). Therefore, carbide nanorods or nanotubes containing encapsulated metal carbides [107, 118] may be useful in constructing superconducting nanoscale circuits.

This account describes the encapsulation of TaC, MoC and NbC in carbon nanotubes and/or nanoparticles generated by arc techniques. TEM, HRTEM, SQUID, and transport measurements reveal fascinating features (see below) associated with these nanostructures. For example, molybdenum carbide, which usually occurs as a hexagonal phase, is found as an fcc phase when encapsulated. SQUID measurements show that superconducting transitions occur (for the above encapsulated carbides). X-ray powder diffraction studies reveal that slight changes in lattice parameters may occur, thus altering the T_c for the corresponding carbides (e.g. with MoC from 14.3 to 10.5 K).

Additionally, The Generation and characterisation of tin nanowires by condensed phase electrolysis in molten mixtures of ionic salts has been investigated.

16.5.1 Carbon Nanotube Sensors—Applications and Advantages

What Are Sensors?

Sensors are devices designed to detect or quantify physical-chemical characteristics, such as pressure, temperature, concentration, sound level or light. The process is typically mediated by the conversion of the measured factor into an electric current.

A sensor is assessed by its sensitivity to the signal, the speed of the response, the cost-effectiveness of the device, the capacity for high volume production and its reliability.

Sensors find applications in a vast range of fields, from automotive sensors to biomedical devices. This growing need for sensors has spurred intensive research on novel technologies [119–124] and materials, such as the nanotube sensor.

Image Credit Mopic/Shutterstock

The Advantages

Nanotechnology enables the production of miniaturized sensors that are not just tiny but incredibly sensitive to minute signals. As a result, they add very little weight to the device, use limited power and are cost effective. The initial nanosensors were made from inorganic semiconductors, but the emergence of carbon nanotubes (CNTs) has transformed the world of sensor research.

CNTs are unique among the newer sensors because they are inherently light and small, very strong, with high conductivity for both heat and electrical energy, and with a high specific surface area (ratio of surface area to mass).

CNTs are layers of graphite that have been rolled up to form a cylindrical tube. Both Single-walled and multi-walled nanotubes (SWNTs and MWNTs respectively) have been synthesized.

The value of metallic CNTs lies in the ballistic electron transport within them. This means that electrons shoot over long lengths of the tube without scattering. In other words, MWNTs can transport high currents of up to 10 A/cm with minimal heating, change in resistance or alteration in form, for long durations, at temperatures as high as 250 °C.

Applications

The use of CNT sensors is based upon the changes that occur in their electrical properties in response to any change in other parameters such as temperature or gas adsorption, coupled with their unique characteristics.

Temperature Sensors

They have been used as miniaturized temperature sensors, which respond to changes in temperature by changes in their electric characteristics.

Gas Sensors

Similarly, gas adsorption causes extremely rapid and sensitive changes in electrical properties of the CNTs, enabling their use as nanoscale sensors for the detection of minute amounts of chemical vapors. The response times are at least one order of magnitude higher when compared to conventional solid-state sensors.

These sensors are very small and can be held at room temperature. They could be useful in drug or explosive detection, for instance, as small cheap sensors that can be mounted throughout sensitive facilities or crowded but security intensive places like airports.

However, such sensors also experience interference as a result of changes in moisture, temperature, and the velocity of flow. Poorly diffusible gases are not detected well. Also, they cannot accurately differentiate gases from mixtures.

These obstacles are overcome by using gas ionization sensors which break down gases at the tips of the CNTs. These are very sensitive and can be used for rapid chemical sensing based on gas adsorption, at low power.

Piezoresistive Sensors

Piezoresistive sensors have been designed to make use of isotropic CNTs, detecting strain based on changes in the electronic properties of the CNT when under strain, at different locations, and in different directions.

Pressure Sensors

Tunable pressure sensors are based on the transition from metal to semiconductor under pressure in an SWNT.

Flow Sensors

Flow sensors take advantage of the change in electrical signal in response to the flow of fluid because of the polar particles in the fluid stream cause fluctuations in the coulombic fields, which in turn causes direct scattering of the free carriers.

Biosensors

Biosensors have been developed from MWNTs which can be used to monitor body temperature, pulse, blood glucose levels and to diagnose illnesses, detect inflammation or oncogenes, repair damaged cells, and even to act as a pacemaker.

Such sensors avoid invasive procedures and do not cause inflammation.

Environmental Gas Sensors

The use of such sensors could enable tracking air pollution to the source as well as evaluation of air quality.

Food Spoilage Sensors

Sensors to detect food spoilage have also been visualized, by spraying CNTs on a flexible plastic surface such as plastic food wrapping film. These nanotubes then act as sensors to detect the chemicals that indicate rotting. This helps control food quality.

Other Types of Sensor

CNTs are also capable of being used as optical, stress, strain, and position sensors.

Other areas of application include the automotive industry, agriculture (such as monitoring relative humidity in a greenhouse or application of liquid chemicals to plants), and the fishing industry (maintain pH and water quality in the fish tanks).

Many challenges remain to be resolved in the application of nanotube sensors, including the production of CNTs in pure form, perfecting the growth mode and controlling the surface growth. Solutions to these production parameter issues and growth problems, avoiding structural instability such as excessive kinking and buckling, and above all, confirming their safety in animals.

16.5.2 Carbon Nanotube Optics and Their Uses

Among its well know mechanical and electronic properties, carbon nanotubes also showcase a series of unique optical properties [125–133]. Whilst a lot of the focus for carbon nanotubes has been on structural materials, they can also be used in specialist optical applications. In this article, we look at the optical properties of carbon nanotubes, as well as looking at some of the areas of optics that they can be used in.

© nobeastsofierce/Shutterstock.com

Optical Properties of Carbon Nanotubes

Carbon nanotubes are often known for their high aspect ratio and mechanical strength properties; however, carbon nanotubes also possess a series of beneficial optical properties that mean they can be used in specialist optic applications.

Carbon nanotubes have unique optical absorption properties and differ from those found in bulk materials. However, given the high aspect ratio of nanotubes, many of their properties are only applicable to one direction along the nanotube. These unique absorption properties also arise because carbon nanotubes can often be classified as a metal or a semiconductor, depending on its orientation. Light will elastically scatter through a carbon nanotube, but it will scatter in an anisotropic fashion. This is attributed to carbon nanotubes possessing non-linear refractive properties (third order nonlinearity). These nonlinear optical properties arise because the delocalized π-electrons only have a 1D motion within a fixed lattice configuration. These third order nonlinear optical properties can also give rise to other phenomenon, such as third harmonic generation (THG), optical Kerr effect, self-focusing and phase conjugation.

Carbon nanotubes also have saturable absorption properties. So, when light comes towards the nanotube, it linearly absorbs the light until it reaches a saturation point. Once the amount of light absorbed is greater than the saturation limit, the nanotubes become transparent.

The absorption of light by a carbon nanotube is also known to produce excitons, which is a charge separation of electrons and holes, where one of each makes up an exciton. Additionally, it has been shown that carbon nanotubes can also form biexciton complexes (4 particle systems of two holes and two electrons) and remain stable. Because these excitation states can be created, carbon nanotubes can also exhibit fluorescence, photoluminescence and electroluminescence properties. Carbon nanotubes generally emit light in the near infrared range and possess a high photostability.

Applications

Given that carbon nanotubes have unique optical properties, there are various applications that they can be used in. Some examples include various types of biosensors, optical rectennas and fiberoptic anemometers.

Optical Biosensors

Carbon nanotube can be employed as part of a sensor to detect various biomolecules, such as those that detect DNA, glucose, hydrogen peroxide, adenosine triphosphate (ATP), nitrogen monoxide and various proteins. To make them more biocompatible, carbon nanotubes are often functionalized with organic molecules, or are composited into a biocompatible matrix. There are various types of carbon nanotube optical sensors to date, and it is perhaps the widest application area that exploits the optical properties of carbon nanotubes.

There are three common sensing mechanisms of carbon nanotubes in these devices, and these are changes in the electrooptical properties, electrochemical fluorescence and quenching mechanisms; and these provide a measurable response that enables the monitoring of exocytosis of biomolecules, the realtime monitoring of the above small molecules, for detecting reactive/radical species, detecting polymorphism in nucleotides, detecting the presence of biomarkers and for detecting the levels of hybridization, sequencing and conformational polymorphism in DNA.

Optical Rectenna

Carbon nanotubes can be used as part of an optical rectenna—which is a device that converts freely propagating electromagnetic waves (at defined optical frequencies) into a direct current (DC). This is a process known as rectification.

In this application, carbon nanotubes can be used as the antenna material and is coupled to PHz-scale metal–insulator–metal tunnel diodes. To act as the antenna, the carbon nanotubes are placed vertically and when they are irradiated with visible and infrared light, a DC open circuit voltage and open-circuit current can be generated. These rectennas have be shown to also work with solar light without there being any change in the performance of the rectenna.

Fiber-Optic Anemometer

Carbon nanotubes can also be used within anemometers, i.e. a type of sensor that measures the speed of wind or gas. Carbon nanotubes can be used to coat tilted fiber Bragg gratings (TFBGs) in hotwire anemometers. These Bragg gratings can be implemented in the sensors to act as both the light coupling and temperature sensing elements of the device. Whilst these Bragg gratings are often used without the carbon nanotubes (usually utilizing a metallic coating), the incorporation of these carbon nanotube coatings have been proven to the increase the light absorption efficiency (up to 93%) of these gratings.

This is because, when the light enters the sensor, it travels down the optical fiber, where it propagates. But because it can't escape the cladding of the fiber, the carbon nanotubes absorb a significantly high amount of light.

16.5.3 Graphene as a Renewable Energy [134]

There is a huge demand to find renewable alternatives to our traditional, finite energy sources. The elements are already playing an important role—only this year wind and solar energy generated more energy over a three-month period than coal—but other substitutes are still required.

Graphene is a single layer of carbon atoms arranged in a honeycomb lattice. It is a unique material with a range of desirable properties including high carrier mobility—meaning electrons move quickly through the material optical transparency and extraordinary mechanical and electronic properties. Ever since its isolation using sticky tape by Andre Geim and Konstantin Novoselov in Manchester in 2004, graphene has become a research hot topic. The versatile substance is capable of revolutionizing a number of technologically important areas-including the energy sector.

This novel material could be used for electrical components, sensors, batteries, composites and ionexchange membranes to give just a few examples. Graphene has been shown to make excellent battery electrodes thanks to its high surface area capable of storing lithium ions, while polymer matrices laced with graphene flakes make strong but workable composites.

Better Batteries

In 2011, engineers at Northwestern University discovered that graphene anodes could hold up to 10 times more power than graphite ones, charged 10 times faster and lasted longer. The anode contains sheets of graphene riddled with millions of holes 10–20 nm in diameter which allow the lithium ions to jump through the nanoholes instead of traveling around the edge of the graphene layer.

Shutterstock/OliveTree

Such batteries could have potential use in electric vehicles: by reducing the weight of batteries, graphene could be used to boost the efficiency of such vehicles, helping to extend their range and overcome a major concern preventing their wider uptake.

Two years later in 2013, researchers from Rice University found graphene laced with boron could be used to produce an ultrathin flexible anode for lithiumion batteries. The boron—which replaces a quarter of the carbon atoms helps the lithium ions stick to the graphene, which results in faster charging. The same team also discovered that graphene mixed with vanadium oxide could be utilized in high-performance, cost-effective cathodes that can be recharged in 20 s and hold more than 90% of their capacity after extensive use.

In addition to being used in electric vehicles, graphene in batteries could also boost the energy storage sector. The University of Manchester, the home of graphene, has already trialed a gridscale battery and converter system on campus. The aim is to develop high capacity electrical storage capable of supplying the National Grid. The system is being used to study methods to control the flow of electricity and settle differences between power generation and local demand.

Renewable Fuels
Researchers at the University of Manchester have recently discovered that graphene membranes, when illuminated with sunlight, can conduct protons at an increased rate. This photoproton effect could be employed to artificially mimic photosynthesis and directly harvest solar energy to produce hydrogen gas. This gas, could then be used in fuel cells in electric vehicles. Such a move would help lower the cost of renewable hydrogen fuel and make hydrogen fuel stations a more common sight along the roads as the cost of processing drops.

Photovoltaic Cells
Eight thousand times more solar energy is produced each year than is consumed worldwide; while some of the power reaching the Earth is harvested and utilized, a massive percentage is not. As energy needs increase, solar power is becoming an attractive alternative, but its efficiency is still lagging. Furthermore, commercially available silicon based photovoltaic cells are expensive to produce and install.

Graphene [134] could have an important role to play here too, in antireflection coatings for solar cells. Researchers in India found that such cells lower reflectance near the ultraviolet part of the spectrum from 35 to 15%.

Graphene incorporated within the photovoltaic cells can, say a team at MIT, deliver a higher power conversion efficiency. Hybrid cells based on graphene are less expensive and more effective as a conducting electrode than the traditionally used indium tin oxide. It also provided flexibility, is light weight, and imparts mechanical strength and a chemical robustness to the cell.

While graphene has the potential to revolutionize the energy sector, there is still much work to be done before graphene-based batteries and hydrogen fuel cells are powering cars on our roads.

16.6 Novel Applications of Nanowires and Nanotubes

Silver chloride nanowires are used as a photocatalysis to decompose organic molecules in polluted water. Using an electrified filter composed of silver nanowires, carbon nanotubes and cotton to kill bacteria in water. Using nanowire mats to absorb oil spill. Nanotubes are applied for Neural Networks.

- Nanotubes for neural networks

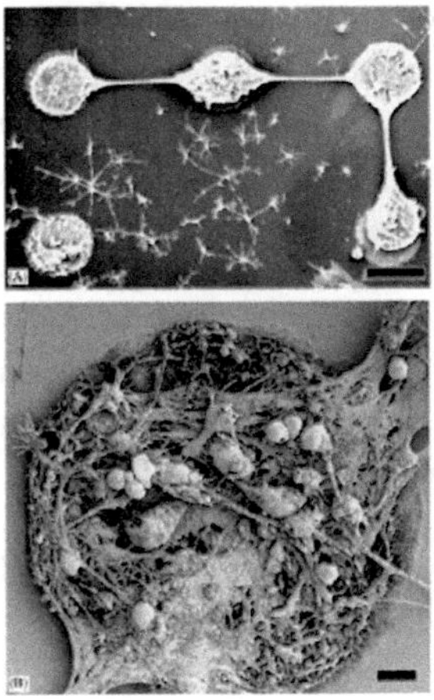

A. Microscopic images of neuron bridging an array of carbon nanotubes thereby creating neural networks.

The nanotube coating consists of a pristine carbon nanotube film, which is grown using a chemical vapor deposition process. Dissociated cortical cells from 1-dayold Charles River rats were plated for several days onto patterned quartz or oxidized silicon substrates at a density of 1103 cells/mm^2. The bar represents 100 mm.

B. Increased magnification of one nanotube pad

Neurons are generally confined to the nanotubes surface except where neurons extend to adjacent nanotube pads. Bar represents 10 mm.

Annual Review of Physical Chemistry
Vol. 55: 613-649 (Volume publication date June 2004)
First published online as a Review in Advance on February 26, 2004

Annual Review of Physical Chemistry, Vol. 55: 613–649 (Volume publication date June 2004)
First published online as a Review in Advance on February 26, 2004

(A) Microscopic images of neuron bridging an array of carbon nanotubes thereby creating neural networks. The nanotube coating consists of a pristine carbon nanotube film, which is grown using a chemical vapor deposition process. Dissociated cortical cells from 1-day old Charles River rats were plated for several days onto patterned quartz or oxidized silicon substrates at a density of 1103 cells/mm^2. The bar represents 100 mm.

(B) Increased magnification of one nanotube pad Neurons are generally confined to the nanotubes surface except where neurons extend to adjacent nanotube pads. Bar represents 10 mm.

16.6.1 Novel Photodetectors

Ultraviolet (UV) photodetectors have received more attention due to their wide applications, such as flame detection, chemical agent detection, space communications, and solar astronomy [135–137]. Different kinds of UV photodetectors have been developed to fit the requirements for various applications.

Enhanced performance of $Mg_{0.1}Zn_{0.9}O$ UV photodetectors using photoelectrochemical treatment and silica nanospheres are investigated. The $Mg_{0.1}Zn_{0.9}O$ films were grown using atomic layer deposition (ALD) system and applied to metal-semiconductor-metal ultraviolet photodetectors (MSM-UPDs) as an active layer. To suppress the dangling bonds on the $Mg_{0.1}Zn_{0.9}O$ surface, the photoelectrochemical (PEC) treatment was used to passivate the $Mg_{0.1}Zn_{0.9}O$ surface, which could reduce the dark current of the MSM-UPDs about one order. Beside, to increase more incident light into the $Mg_{0.1}Zn_{0.9}O$ active layer of the MSM-UPDs, the 500-nm-diameter silica nanospheres were spin-coated on the $Mg_{0.1}Zn_{0.9}O$ active layer to improve the antireflection capability at the wavelength of 340 nm. The reflectivity of the $Mg_{0.1}Zn_{0.9}O$ films with silica nanospheres antireflection layer decreased about 7.0% in comparison with the $Mg_{0.1}Zn_{0.9}O$ films without silica nanospheres. The photocurrent and UV-visible ratio of the passivated $Mg_{0.1}Zn_{0.9}O$ MSM-UPDs with antireflection layer were enhanced to 5.85 μA and 1.44×10^4, respectively, at the bias voltage of 5 V. Moreover, the noise equivalent power and the specific detectivity of the passivated $Mg_{0.1}Zn_{0.9}O$ MSM-UPDs with antireflection layer were decreased to 2.60×10^{-13}W and increased to 1.21×10^{12} cm $Hz^{1/2}$ W^{-1}, respectively, at the bias voltage of 5 V. According to the above mentions, the PEC treatment and silica nanospheres antireflection layer could effectively enhance the performance of $Mg_{0.1}Zn_{0.9}O$ MSM-UPDs.

16.6.2 White Light-Emitting Diodes

Recently, people have paid more and more attention to the problems of energy shortage and environment problems. Many research teams actively investigated how to improve energy saving and how to produce renewable energy. The white light-emitting diodes (WLEDs) are expected to be the promising green lighting sources and have been widely used in solid-state lighting and taken as the next generation lighting sources to replace the compact fluorescent lamp (CFL) and incandescent light. The WLEDs exhibit many superior advantages such as the long lifetime, high luminous intensity, energy saving, fast response, low heat dissipation, and high reliability. In general, there are three approaches that can be used to fabricate WLEDs: (1) using the blue light emitting diodes (LEDs) to excite the yellow phosphor [138, 139], (2) using an ultraviolet (UV) LEDs to excite red, green, and blue phosphors [140, 141], and (3) mixing three primary colors of red, green, and blue LEDs [142]. The commercial WLEDs were commonly integrated by the GaN-based blue LEDs with yellow phosphor ($Y_3Al_5O_{12}:Ce_3^+$ or $YAG:Ce_3^+$) layer. To augment the application of WLEDs, the development of high performance WLEDs is immediately needed.

The GaN-based flip-chip white light-emitting diodes (FCWLEDs) with diffused ZnO nanorod reflector and with ZnO nanorod antireflection layer were fabricated. The ZnO nanorod array grown using an aqueous solution method was combined with Al metal to form the diffused ZnO nanorod reflector. It could avoid the blue light emitted out from the Mg-doped GaN layer of the FCWLEDs, which caused more blue light emitted out from the sapphire substrate to pump the phosphor. Moreover, the ZnO nanorod array was utilized as the antireflection layer of the FCWLEDs to reduce the total reflection loss. The light output power and the phosphor conversion efficiency of the FCWLEDs with diffused nanorod reflector and 250 nm long ZnO nanorod antireflection layer were improved from 21.15 mW to 23.90 mW and from 77.6% to 80.1% in comparison with the FCWLEDs with diffused nanorod reflector and without ZnO nanorod antireflection layer, respectively.

16.6.3 Nanowire Applications in Electronics

Using electrodes made from nanowires that would enable flat panel displays to be flexible as well as thinner than current flat panel displays. Using nanowires to build transistors without p-n junctions. Using nanowires made of an alloy of iron and nickel to create dense memory devices. By applying a current magnetized sections along the length of the wire. As the magnetized sections move along the wire, the data is read by a stationary sensor. This method is called race track memory. Using silver nanowires embedded in a polymer to make conductive layers that can flex,

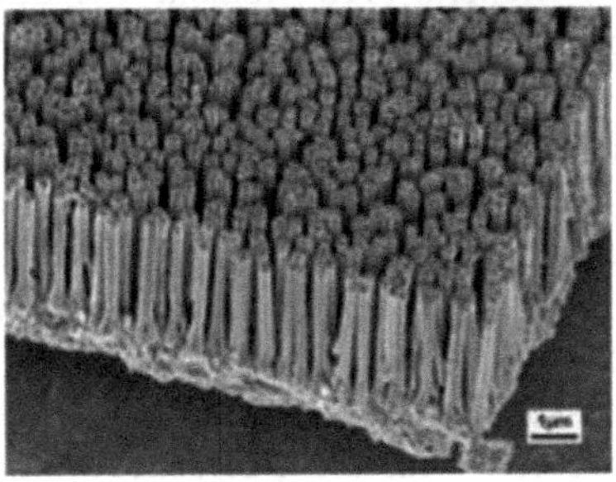

A forest of copper rods about 100 nm in diameter create much more surface area for high-capacity battery electrodes

without damaging the conductor. Sensors using zinc oxide nano-wire detection elements capable of detecting a range of chemical vapors.

In the past few decades, continuous miniaturization of semiconductor devices has led to high-speed operations and large-scale integration of electronics. A single silicon chip can now contain more than a billion transistors and operate at frequencies higher than 1 GHz. The operation frequency can be further increased by utilizing a planar architecture. In planar devices, electrodes are connected to the sides of the active semiconductor layer, rather than being placed on its top and bottom surface, as in conventional multilayered vertical-structured devices. As a result, very low parasitic capacitances are obtainable, leading to very high operating speeds. Planar nanodevices based on 2DEGs in semiconductor heterostructures have been demonstrated to operate at tens of GHz or more [143–145]. In particular, self-switching diodes (SSDs) have been shown to possess zero threshold voltages and be able to work as high-frequency detectors. At room temperature, they can operate at frequencies up to 1.5 THz [146] at room temperature and up to 2.5 THz [147] at a temperature of 150 K.

Apart from improving the operational frequency of the devices, planar architecture can also lead to some additional mechanisms. It has been shown that when the thickness of the conducting layer is lower than 10 nm, 2D plasmonic noise is dominant, resulting in a geometric dependence of noise characteristics [148]. Moreover, it has been proved that in a submicron field-effect transistor (FET) with asymmetric boundary conditions, 2D plasma waves can be excited simply by a DC current [149]. This discovery has prompted intensive investigations of nanometer FETs as THz sources and detectors, resulting in substantial progress in this field [150–155].

The physics of plasma wave based collective electron transport is vastly different from conventional individual (uncorrelated) electron transport. However, the studies so far have been mostly limited to plasma waves occurring in submicron FET structures.

Recently, using a combined two-dimensional-three-dimensional (2D-3D) ensemble Monte Carlo (EMC) model, the performance of a planar nanochannel device is studied at the terahertz (THz) region. The device is based on a GaN/AlGaN heterostructure in which a two-dimensional electron gas (2DEG) forms at the interface. Simulation results reveal that, at low working frequencies, the performance of the device is almost frequency independent. However, when the working frequency is higher than 0.5 THz, obvious enhancements in the device performance have been observed. The enhancements are characterized by two resonant peaks at frequencies of about 4 THz and 8 THz. Also, the frequency-dependent performance exhibits nonmonotonicity. Further studies show that the performance enhancements can be attributed to the excitations of 2D plasma waves in the device, with the emergence of the above resonant peaks corresponding to the formation of standing plasma waves. Moreover, simulation results show that the device performance increases monotonically with signal amplitude, when the device is unbiased. However, when a DC bias is applied, the performance remains almost unchanged for large signals but is significantly enhanced for small signals. Therefore, the device performance shows a strong nonmonotonic dependence on signal amplitude, and its minimal value occurs when the signal amplitude is only about $\sqrt{2}$ times the DC bias.

16.6.4 Devices and Applications of SiGe Nanostructures

The potential of SiGe NWs as building blocks for next generation technological devices is enormous thanks to their remarkable electronic, thermal, and optical features. Moreover, as already stated at the end of the section on the chemical and physical properties, they offer the unique opportunity to be implemented on a semiconductor infrastructure which is easily integrated with current CMOS technology. However, despite some preliminary ongoing innovations and breakthroughs, there is no doubt that research in this field is still in a premature state and further challenges have to be overcome before SiGe nanowires are implemented in a high-end product. At this stage it seems that the great potential of these nanostructures can be exploited in the following applied research fields: (i) nanoelectronics, (ii) energy conversion, and (iii) spintronics.

16.6.5 High-Performance Nanoelectronic Components

The interest in alloy NWs, for instance, derives from the use of a material with enhanced properties, such as a tunable band gap. Core—shell NWs, on the other hand, are expected to outperform conventional Si or Ge channel FET mostly due to a design reason: the shell effectively confine carriers in the core, thus reducing surface scattering and increasing the mobility.

16.6.6 $Si_{1-x}Ge_x$ Alloy Nanowire Transistor

$Si_{1-x}Ge_x$ alloy NWs are interesting as material for FET channels mainly because adding some Ge to a Si NW should lead to an increase in the mobility, which is larger in Ge than in Si for both electron and holes.

16.6.7 Si-Shell Ge-Core Nanowire Transistor

Unoptimized devices based on core—shell Si/Ge NWs had already been reported to be comparable to carbon nanotube FET, but it was the observation of a one-dimensional hole-gas that opened up the way to a new class of nanowire based devices.

16.6.8 From Quantum Transport to Superconductivity: SiGe Nanowires as Platforms for Fundamental Physics Studies

SiGe NWs have been used to address fundamental physics topic, rather than to implement (nano) electronic devices. The growth of quantum dots with controlled shape and distribution has provided the ideal test-beds for several fundamental quantum transport studies.

16.7 Nanowire-Based Transistors (Nanotube Field-Effect Transistor)

The electrical characteristics of a nanowire-based field-effect transistor (NW-FET) denotes a three-terminal device where the transverse electric-field imparted by the gate electrode controls the resistance of a semiconductor channel between the source and drain terminals. It is important to note that the NW-FET structure possesses some differences as compared to the widely studied MOSFET (metal oxide semiconductor) devices. In a NW-FET, the source and drain regions (which are in contact with the channel) are comprised of metal electrodes. This differs from the MOSFET case, where the source and drain regions are made up of degenerately doped semiconductor. This leads to the possibility of am bipolar conduction. Ambipolar behavior in FETs is defined as both n- and p-type conduction in a single device at the appropriate gate bias conditions.

Nanowires can be used for transistors. Transistors are used widely as fundamental building element in today's electronic circuits. As predicted by Moore's law, the dimension of transistors is shrinking smaller and smaller into nanoscale. One of the key challenges of building future nanoscale transistors is ensuring good gate control over the channel. Due to the high aspect ratio, if the gate dielectric is wrapped around the nanowire channel, we can get good control of channel electrostatic potential, thereby turning the transistor on and off efficiently. Transistors are the basic building blocks of integrated circuits. IBM has successfully fabricated and tested nanotube transistors using individual multi-wall or single-wall nanotubes as the channel of a field-effect transistor (FET).

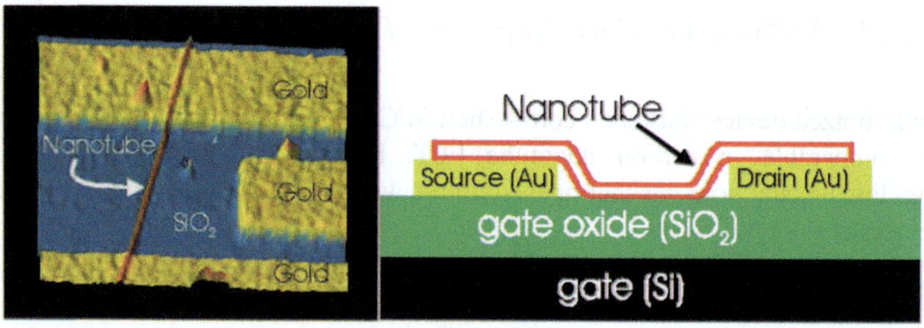

To create active electronic elements, the first key step was to chemically dope a semiconductor nanowire. This has already been done to individual nanowires to create p-type and n-type semiconductors.

The next step was to find a way to create a p-n junction, one of the simplest electronic devices. This was achieved in two ways. The first way was to physically cross a p-type wire over an n-type wire. The second method involved chemically doping a single wire with different dopants along the length. This method created a p-n junction with only one wire. After p-n junctions were built with nanowires, the next logical step was to build logic gates. By connecting several p-n junctions together, researchers have been able to create the basis of all logic circuits: the AND, OR, and NOT gates have all been built from semiconductor nanowire crossings. For practical circuit applications, both p- and n-type transistors are required. The nascent field of nanowire (NW) research has made a great deal of progress in broadening the field of knowledge related to nanowire growth mechanisms and their associated electrical properties. A practical NW-based electronic device must be able to be fabricated in sizable quantities and in an affordable fashion while maintaining a good uniformity in performance among different devices. The bright prospects related to NW-based electronics can be attributed to the unique physical properties of nanowires. As a result of their growth mechanism, nanowires may be synthesized in an independent step and later be transferred (post-growth) to device substrates of interests. Moreover, the nanowires to be discussed also possess a single-crystalline structure which allows for higher speed electronic carrier transport as compared to poly-crystalline or amorphous materials. The device structure to be presented that can take advantage of these properties in a practical manner is the nanowire-based thin-film transistor (NW-TFT). The main novelty of a NW-TFT as compared to other electronic devices is that the switching element, i.e. semiconductor channel material, is made up of an array of a large number of aligned nanowires. Such NW arrays may be transferred onto a wide variety of substrates, like glass, in a controlled fashion by a contact printing method. Following the nanowire transfer to the substrate of interest, device fabrication can be carried out using established semiconductor processing methods. The fabrication of NW-TFTs on transparent substrates like glass opens up the possibility of achieving transparent electronic devices. The semiconductor SnO_2 was utilized as the n-type material, while Ge-based nanowires were explored as the p-type material. A future goal will be to fabricate both structures on the same chip with the aim of creating low-power, complementary logic circuits. Although not a nanowire-based device, the ZnO TFT shares many structural similarities with NW-TFTs, and serves as a useful performance comparison.

16.7.1 Nanowire Based Field Effect Transistors

- The crossed NW structure can be configured into a variety of devices, such as diodes and transistors
- A p-n diode can be obtained by simply crossing p- and n-type NW
- Crossed NW p-n diodes and NW-FETs enable more complex circuits, such as logic gates to be produced
- Nano-logic gates.

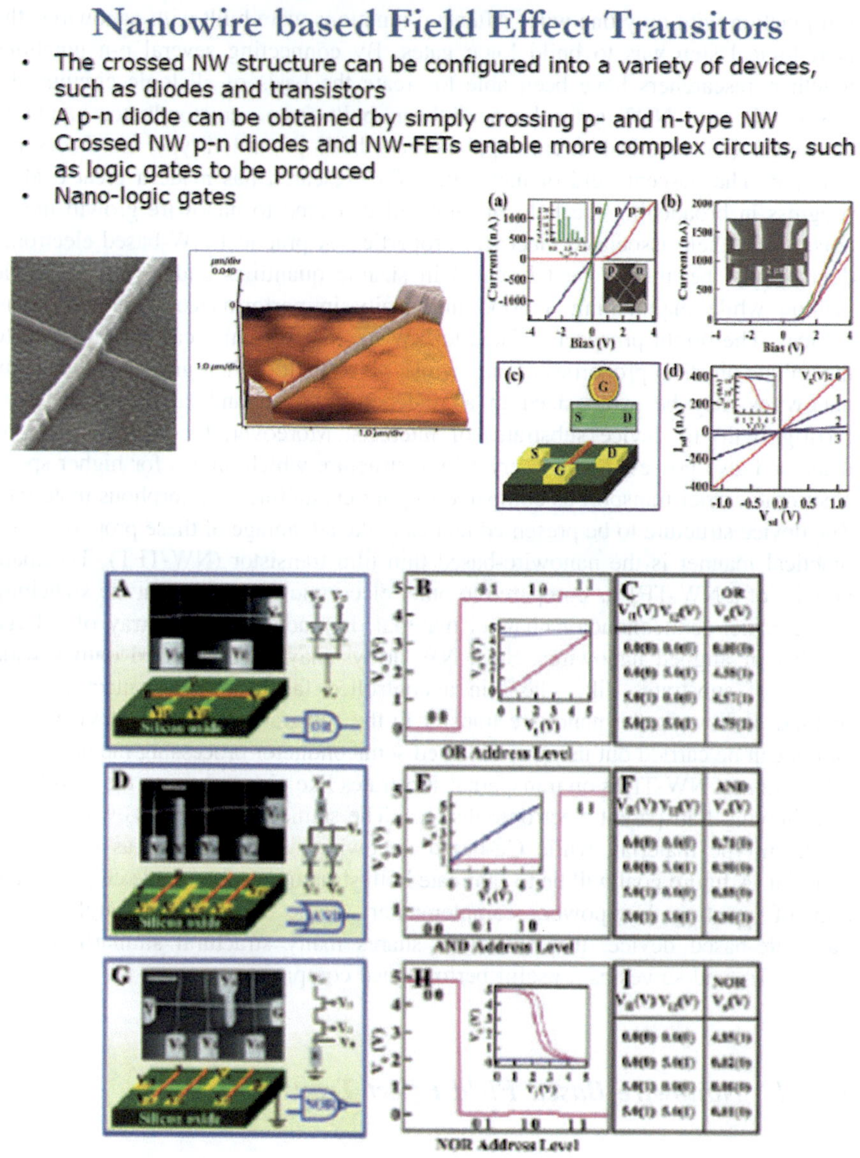

16.7.2 *Sensing of Proteins and Chemicals Using Semiconductor Nanowires*

In an analogous way to FET devices in which the modulation of conductance (flow of electrons/holes) in the semiconductor, between the input (source) and the output (drain) terminals, is controlled by electrostatic potential variation (gate-electrode) of the charge carriers in the device conduction channel, the methodology of a Bio/

Chem-FET is based on the detection of the local change in charge density, or so-called "field effect", that characterizes the recognition event between a target molecule and the surface receptor. This change in the surface potential influences the Chem-FET device exactly as a 'gate' voltage does, leading to a detectable and measurable change in the device conduction. When these devices are fabricated using semiconductor nanowires as the transistor element the binding of a chemical or biological species to the surface of the sensor can lead to the depletion or accumulation of charge carriers in the "bulk" of the nanometer diameter nanowire i.e. (small cross section available for conduction channels). Moreover, the wire, which serves as a tunable conducting channel, is in close contact with the sensing environment of the target, leading to a short response time, along with orders of magnitude increase in the sensitivity of the device as a result of the huge S/V ratio of the nanowires. While several inorganic semiconducting materials such as Si, Ge, or metal oxides (e.g. In_2O_3, SnO_2, ZnO, etc.) have been used for the preparation of nanowires. Silicon nanowires are usually the material of choice when fabricating nanowire FET-based chemo/biosensors. Several examples of the use of silicon nanowire sensing devices include the ultra sensitive, real-time sensing of biomarker proteins for cancer, detection of single virus particles, and the detection of nitro-aromatic explosive materials such as 2,4,6 Tri-nitro toluene (TNT) in sensitive superior to these of canines. Silicon nanowires could also be used in their twisted form, as electromechanical devices, to measure intermolecular forces with great precision.

16.7.3 Limitations of Sensing with Silicon Nanowire FET Devices

Generally, the charges on dissolved molecules and macromolecules are screened by dissolved counterions, since in most cases molecules bound to the devices are separated from the sensor surface by approximately 2–12 nm (the size of the receptor proteins or DNA linkers bound to the sensor surface). As a result of the screening, the electrostatic potential that arises from charges on the analyte molecule decays exponentially toward zero with distance. Thus, for optimal sensing, the Debye length must be carefully selected for nanowire FET measurements. One approach of overcoming this limitation employs fragmentation of the antibody-capturing units and control over surface receptor density, allowing more intimate binding to the nanowire of the target protein. This approach proved useful for dramatically enhancing the sensitivity of cardiac biomarkers (e.g. Troponin) detection directly from serum for the diagnosis of acute myocardial infarction.

16.7.4 Field Emitting Transistor (FET) Based on C-NTs

Transistor (triode) is the electronic device the current in which between two electrodes under bias voltage V_{bias} depends strongly on gate voltage V_{gate} on third gate

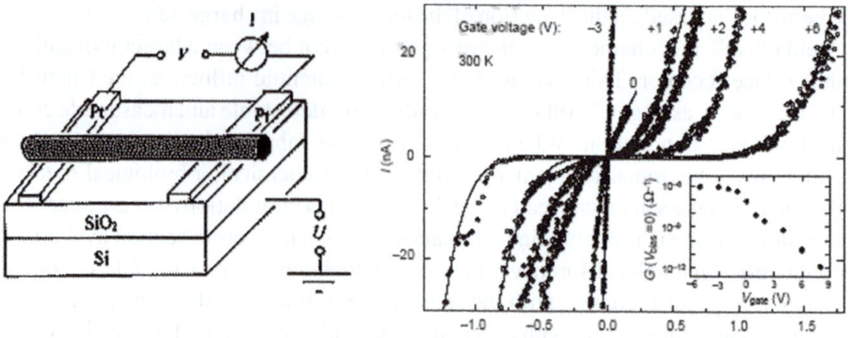

Fig. 16.5 Set-up of the first field emitting transistor on base of C-NT and it current voltage curve, the I–V_{bias} dependence on Vgate ~−3 to 8 V. Nanotube lie between two platinum nanoelectrodes on nonconducting quartz substrate. Silicon layer plays the role of gate electrode. Under positive Vgate the current drops while under negative it is increased (S. J. Tans, et al. Nature, Vol. 393, 49–51 (1998))

electrode. Tans was the first who fabricated the transistor on base of semiconducting carbon nanotube, shown in Fig. 16.5.

Carbon Nano Tubes: Application in Field Emission Displays (FEDs)

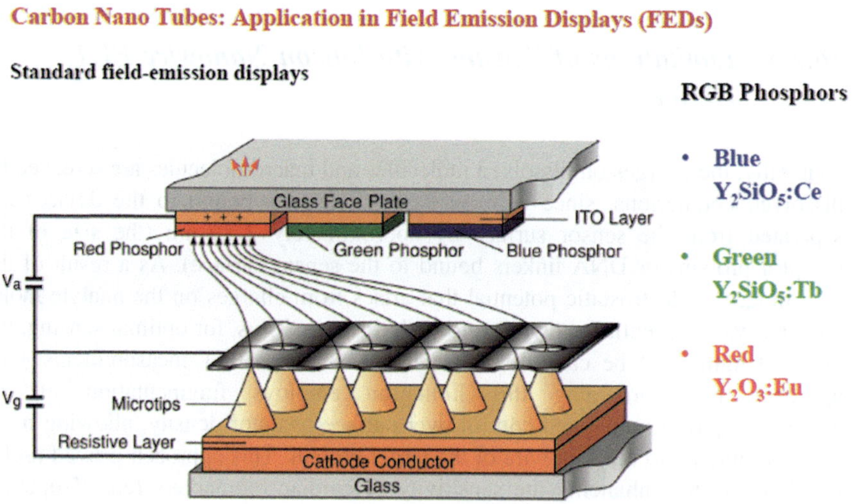

Semiconducting C-NT in general have the strong resistance at V_{bias} due to existence of band gap. However, applied gate voltage V_{gate} induces an electric field in C-NT, which bends energetic bands that in turns changes the current. Negative potential leads to increase while positive to decrease of the current through C-NT.

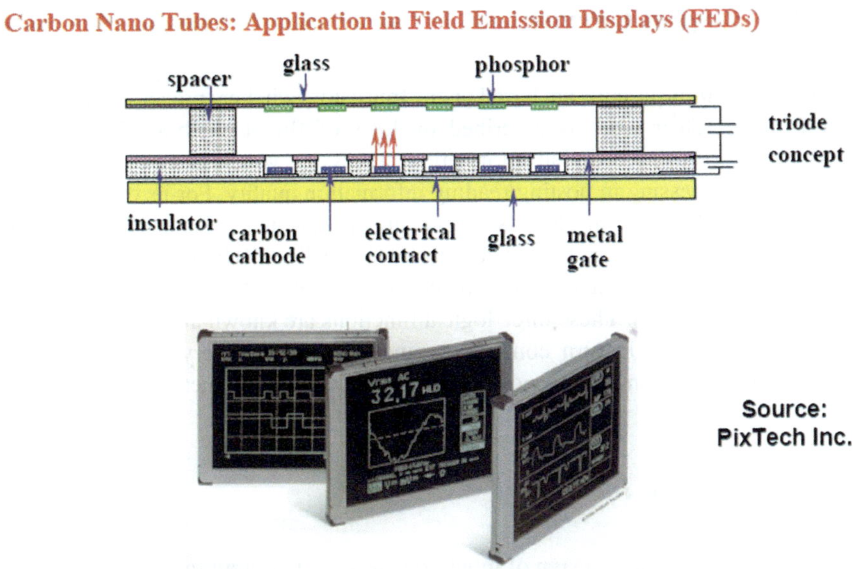

It is evident of the hole conductivity as being the main charge carriers in the C-NT transistor. Concentration of holes in C-NT is estimated to be one per 250 carbon atoms, that in order of magnitude greater than in graphite (one hole per 10^4 carbon atoms).

Carbon Nano Tubes: Application in Electron Guns of Cathode Ray Tubes (CRTs)

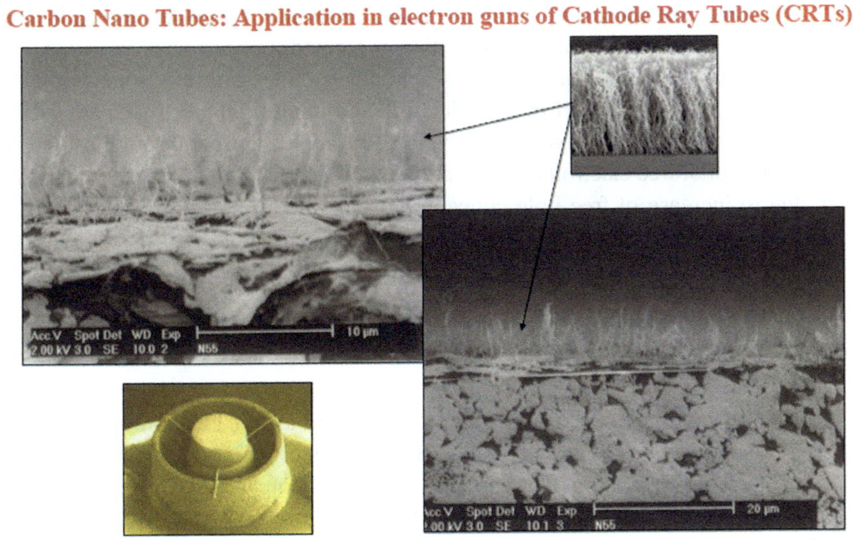

16.7.5 Logical Circuits

Modern electronic circuits are based on the transistor switchers processing in the ON/OUT mode. Their work is described by lows of Bullet algebra. This gives the opportunity of using a digital information processing which in turn overcomes an analogous processing in posting/reading information quality. For its electronic realization a calculus scale on 2 base is ideally suitable in which ON and OUT correspond to 0 or 1 of binary integers, that relate in turn to logical 1 (TRUE) or 0 (FALSE). Three main logical functions have been designed called as OR, AND, NOT on base of these elementary functions. These three logical functions are known all requested to build digital circuits and in turn computers, since in essence every processor is well known to present complex branchy combination of these 3 functions.

16.7.6 Voltage Inverter

Voltage inverter change a sign of input voltage V_{in} into output reversible sign $V_{out} = - V_{in}$. If in logical circuit its positive value will adopted as logical 1 while negative as 0 then the inverter becomes to be logical NOT. Such inverter on base of C-NTs have become to be the first example of logical nanocircuit on single molecule. Derycle has developed an inverter on base of semiconducting NT of 1.4 nm in diameter with band gap 0.6 eV. Nanotube is replaced on dielectric silicon (Si) substrate covered by thin silica (SiO_2) layer in contact with three gold electrodes. The inverter consist of two kinds of field emitting transistors, namely, with electron (*n*-FET) and hole (*p*-FET) conductivity (Fig. 16.6). Usually the C-NT have the hole type of conductivity but it is changed on electron conductivity when doped by potassium. If one half of *p*-type NT will be doped by potassium with intermediate layer of PMMA than *pn* transition is formed.

In such inverter the voltage V_{in} is applied at outer gate electrode while the potential V_{out} is killed out of middle electrode. Left and right electrodes are used to apply bias voltage *V*. Positive gate V_g voltage promotes current through n-type NT due to growth of free electron concentration in n-FET and suppress current in p-type NT due to increase of free hole concentration in P-FET. Negative voltage acts in opposite manner.

16.7.7 Chips with Logical Elements

To enhance the affectivity of electronic circuits on base of FETs, they have to be packed at substrate as dense as possible. Bachtold takes a first step in the way of the logical elements integration, fabricating a smallest chip on base of C-NTs arranged on alumina Al_2O_3 film with aluminums gate below. Figure 16.7 show one of this chip FET.

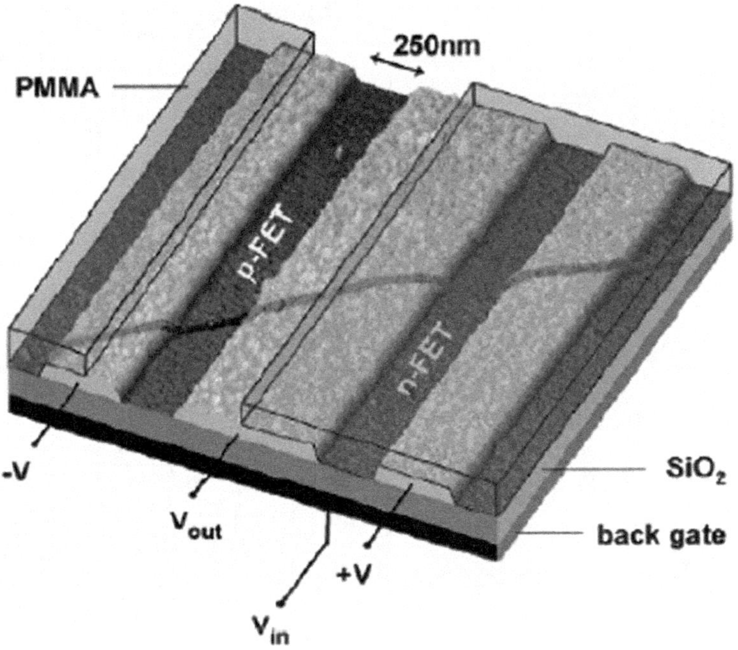

Fig. 16.6 Voltage inverter on base of single C-NT with conductivity of *p*-type (left half) and *n*-type (right half) replaced on three electrodes (A. Derycke, et al. Nano Lett. 1, 453 (2001)

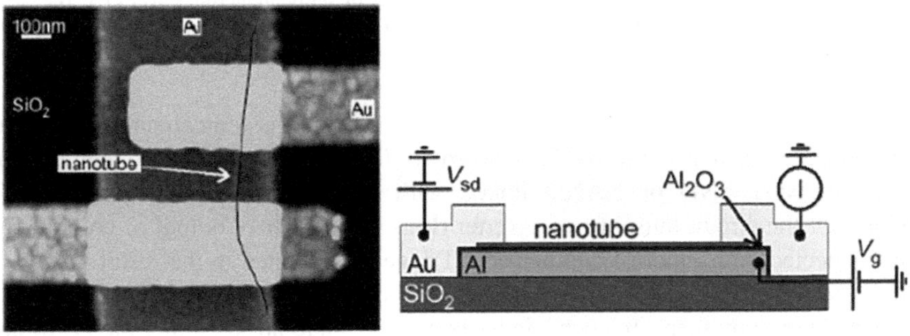

Fig. 16.7 Field transistor based on C-NT as one element of smallest chip (A. Bachtold, et al. Science, Vol. 294, 1317 (2001))

Figure 16.8 demonstrate the example of logical elements circuits, b is the two-transistor circuit with two inputs V_{in1} and V_{in2}, obtained by change of one inverter on two ones linked in parallel. If one or both transistors are in state of logical unity then one NT at least is in conductive state and $V_{out} = 0$ (logical zero). If both NTs are in nonconductive state then a logical unity is in output. Thus for these two-transistor logical circuit $Y(0, 0) = 1$, $Y(1, 0) = 0$, $Y(0, 1) = 0$, $Y(1, 1) = 0$,

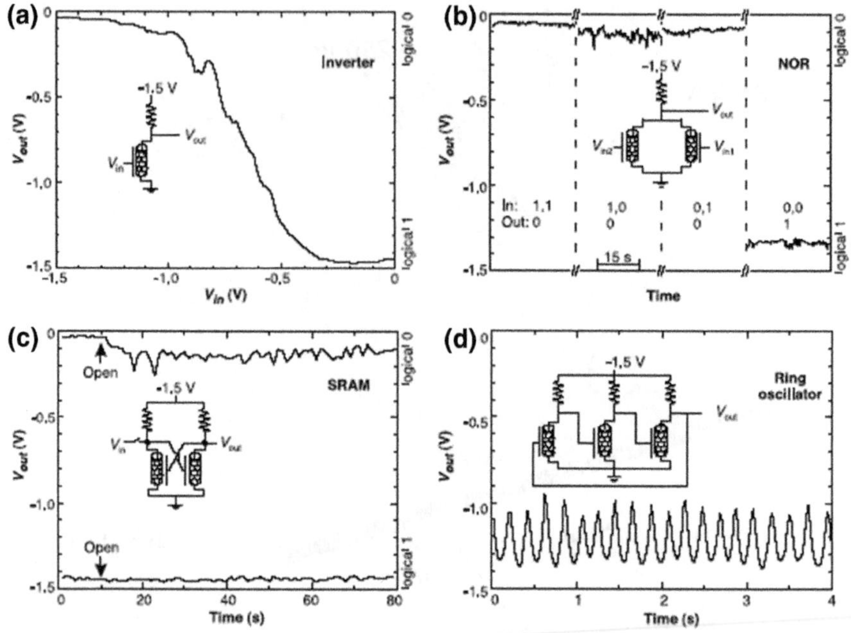

Fig. 16.8 Electrical characteristics of one-, two-, and three-transistor circuits on base of C-NTs. **a** One-transistor inverter and its dependence of output voltage V_{out} on input V_{in}. **b** Two transistors logical unit NOT-OR (NOR) and its dependence of output voltage V_{out} on possible input states (0, 0), (0, 1), (1, 0) and (1, 1). **c** Static random access memory (SRAM) on base of two-transistor trigger. **d** Three-transistors circuit ring oscillator on base of semiconducting C-NTs (A. Bachtold, P. Hadley, T. Nakanishi, C. Dekker, Logic circuits with carbon nanotube transistors, Science, 294, 1317 (2001))

that correspond to the logical function $Y = A + B$, that is logical summation with following logical negator NOT, denoted as OR-NOT or NOR. Nonconjunction element NOT-AND or NAND linked with element NOT gives element OR. Consider the circuit built of two inverter (Fig. 16.8c), where output of one inverter is connected with input of another one. Trigger is obtained in the result known as the flip flop circuit that is the elementary cell of the static random access memory (SRAM) in computers. In such trigger two stable state are possible, the (1, 0) and (0, 1), corresponding to a current only left or only through right nanotube. This pair of states is used for design of logical unity 1 or zero. If input voltage is applied on gate of right transistor $V_{in} = 0$, then no current in right nanotube exists and the voltage on left gate become to be $V_{out} = -1.5$ V inspiring the current in left nanotube. If here input V_{in} contact is broken than the system comes in this (1, 0) state during long time (70 s on Fig. 16.8c). If than gate voltage $V_{in} = -1.5$ V is applied to right transistor, then a resistance of right NT will drop while voltage on left gate will increase to $V_{out} = 0$ that correspond to the state (0, 1) of the system. In the result the trigger "remember" the last voltage V_{in} so the information may be read measuring output V_{out} voltage.

16.8 Sensing Devices

- Nanowire sensors will potentially be smaller, more sensitive, demand less power, and react faster.
- pH sensors, single virus detection, protein binding, gas sensors.

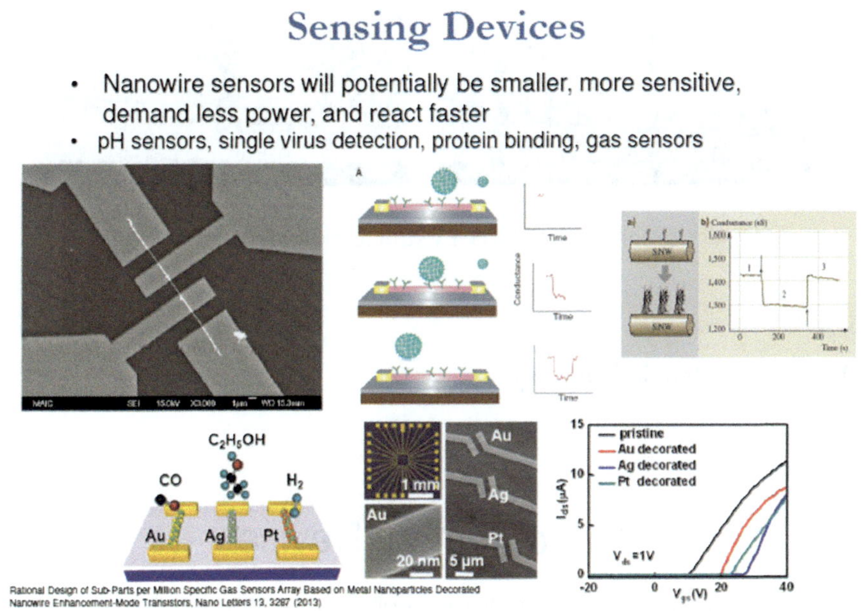

Rational Design of Sub-Parts per Million Specific Gas Sensors Array Based on Metal Nanoparticles Decorated Nanowire Enhancement-Mode Transistors, Nano Letters 13, 3287 (2013)

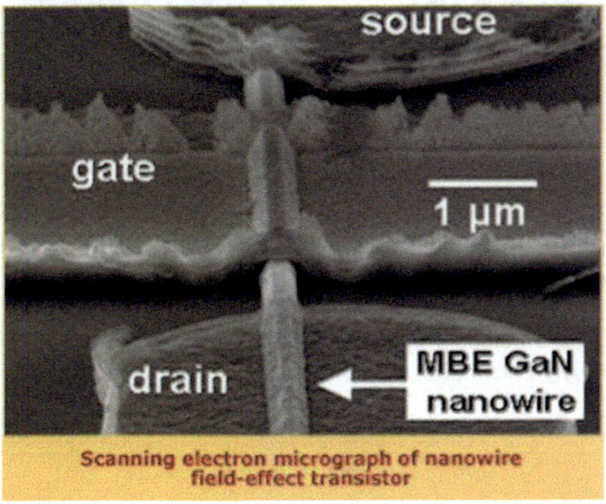

http://www.boulder.nist.gov/div815/OMM_Project/Accomplishments.htm

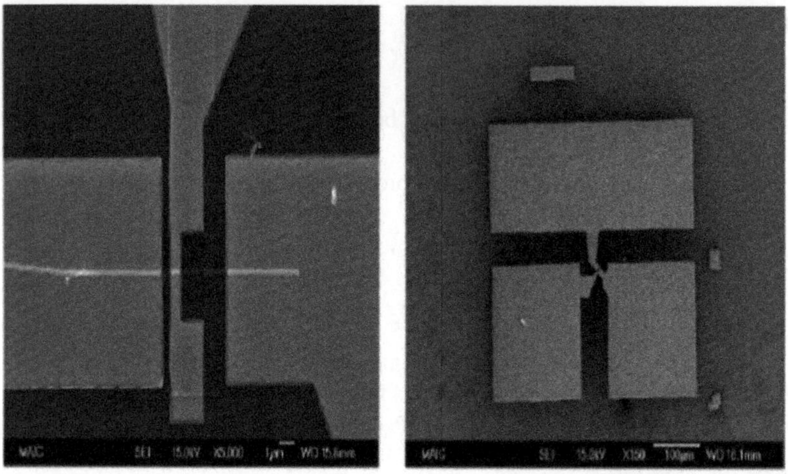

SEM micrographs of ZnO MOSFET structure(crossed nanowire device)

16.9 Racetrack Memory

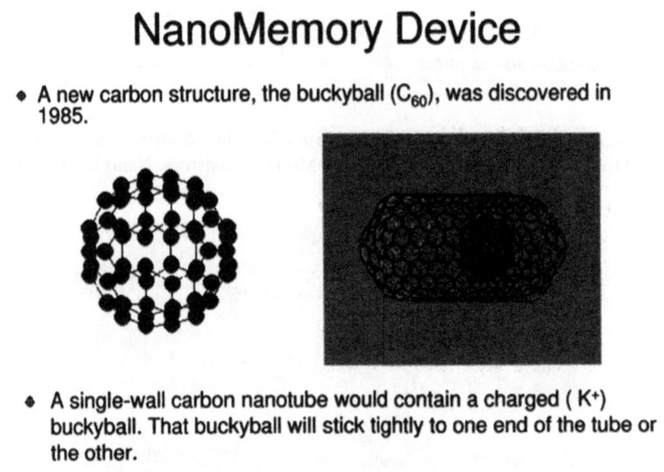

Source M. Brehob "The Potential of Carbon-Based Memory Systems", IEEE (1999)
Source M. Brehob "The Potential of Carbon-Based Memory Systems", IEEE (1999)

NanoMemory Device

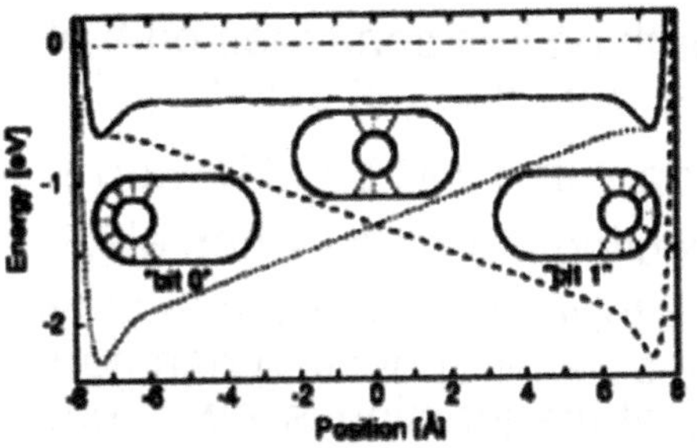

Potential energy of the shuttle at different locations in the capsule. The solid line is when no electric field is applied. The dashed lines are the potential energy when the two-volt potential difference is applied.

- Assign the bit value of the device depending on which side of the tube the ball is. The result is a high-speed, non-volatile bit of memory.

Source M. Brehob "The Potential of Carbon-Based Memory Systems", IEEE (1999)

NanoMemory Device

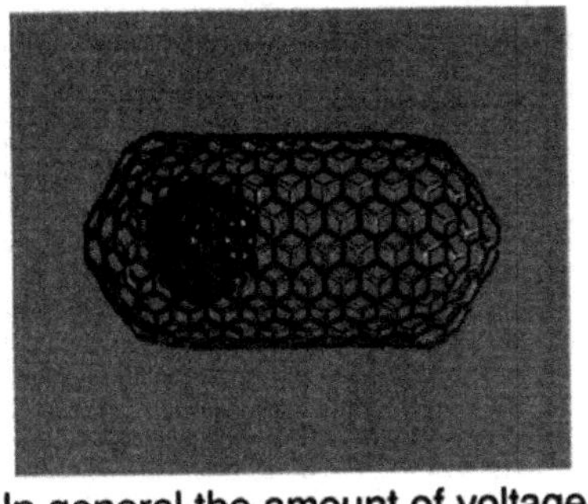

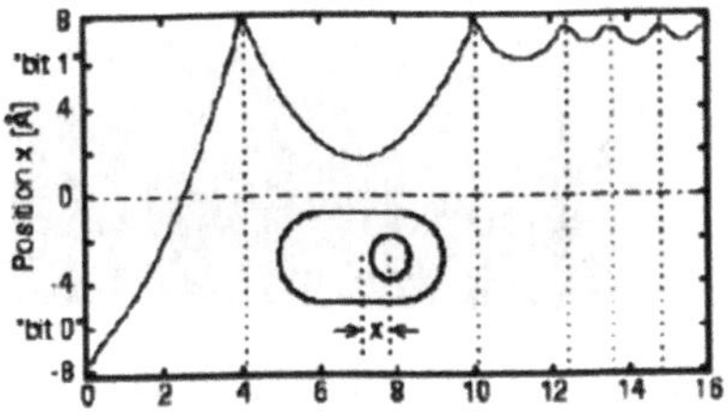

Location vs. time of the shuttle as a write is preformed. Time is in picoseconds.

- In general the amount of voltage which needs to be applied depends upon the length of the capsule.
- A field of 0.1 volts/cm is sufficient to move the shuttle from one side of the tube to the other.
- Write speed: 20 picoseconds

Source M. Brehob "The Potential of Carbon-Based Memory Systems", IEEE (1999)

Bi-layer CNT RAM

- The clever thing is it combines both electronic and mechanical properties of single-wall nanotubes.
- Metallic nanotubes will bend toward a perpendicular semiconducting nanotube when electrically charged.
- When a metallic nanotube is one to two nanometers away from a semiconducting nanotube, the electrical resistance at the junction is low, creating an ON state. When the nanotubes are apart the resistance is much higher, creating an OFF state.

Working Principle

Bistable at NT crossing:
- Top NT Suspended: potential energy minimum
- Top NT contacting lower NT: van der Waals attraction

Source Thomas Rueckes, et al., "Carbon Nanotube Based Nonvolatile Random Access Memory for Molecular Computing", Science, Vol. 289, 7 July 2000

I-V Characteristic

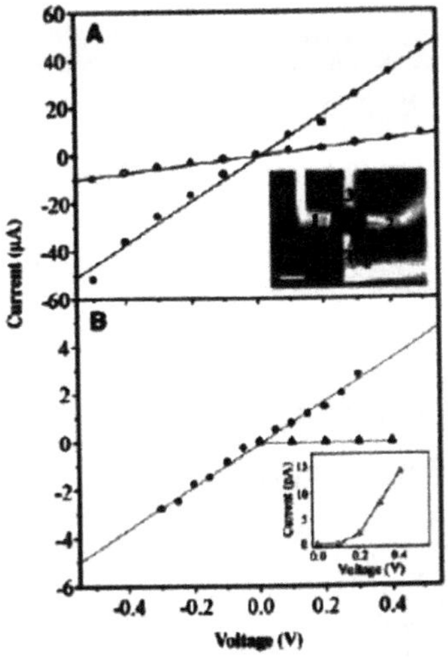

- The touching of two nanotubes decreases resistance between the two wires dramatically, yielding different I-V characteristics.
- Experimental results show 10X higher resistance for off state
- Bit value can be sensed by determining resistance with low voltage applied at electrodes
- Once a bend is made, it will remain until opposite charges are placed at the intersection.

Source Thomas Rueckes, et al., "Carbon Nanotube Based Nonvolatile Random Access Memory for Molecular Computing", Science, Vol. 289, 7 July 2000

NRAM™ by Nantero

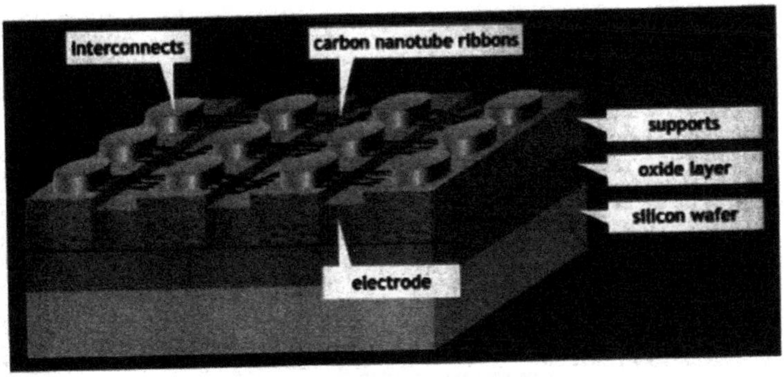

- Applied charge make CNT ribbons bend down to touch the substrate or bend up back to its original state.
- Ribbon-up gives 'zero' and ribbon-down is 'one'.

Source http://www.nantero.com/nram.html

Racetrack memory

- Need deep trench with notches to "pin" domains
- Need sensitive sensors to "read" presence of domains
- Must insure a moderate current pulse moves every domain one and only one notch

Promise (10-100 bits/F^2) is enormous...

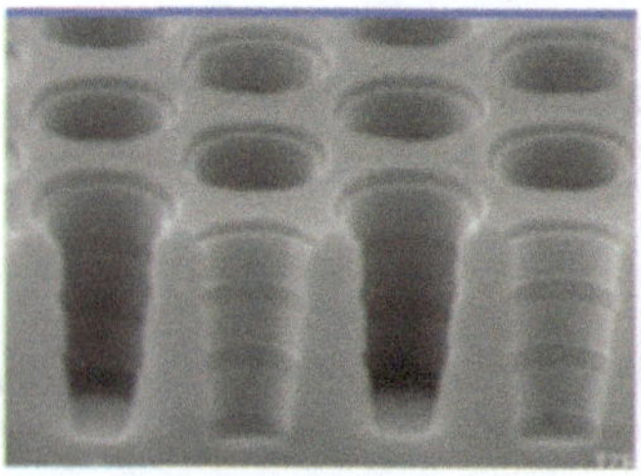

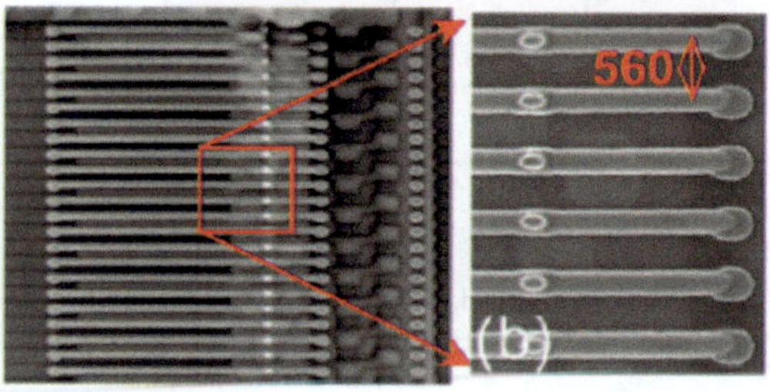

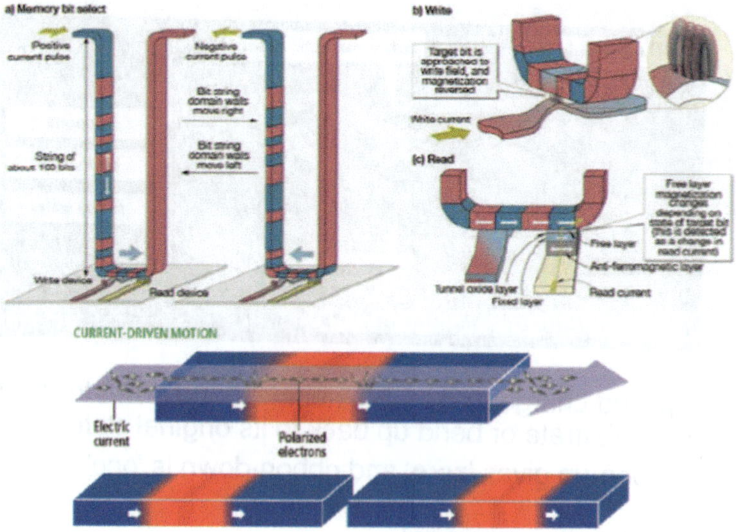

In August 2012, researchers reported constructing the first NAND gate from undoped silicon nanowires. This avoids the problem of how to achieve precision doping of complementary nanocircuits, which is unsolved. They were able to control the Schottky barrier to achieve low-resistance contacts by placing a silicide layer in the metal-silicon interface. It is possible that semiconductor nanowire crossings will be important to the future of digital computing. Though there are other uses for nanowires beyond these, the only ones that actually take advantage of physics in the nanometer regime are electronic. In addition, nanowires are also being studied for use as photon ballistic waveguides as interconnects in quantum dot/quantum effect well photon logic arrays. Photons travel inside the tube, electrons travel on the outside shell. When two nanowires acting as photon waveguides cross each other the juncture acts as a quantum dot. Conducting nanowires offer the possibility of connecting molecular-scale entities in a molecular computer. Dispersions of conducting nanowires in different polymers are being investigated for use as transparent electrodes for flexible flat-screen displays. Because of their high Young's moduli, their use in mechanically enhancing composites is being investigated. Because nanowires appear in bundles, they may be used as tribological additives to improve friction characteristics and reliability of electronic transducers and actuators.

16.10 Nanowire-Based Metamaterials

- Engineered composites in which the dielectric permittivity and magnetic permeability are artificially manipulated
- Periodical structures that in some range of frequencies react to the electric and magnetic fields as medium with negative ε or μ.
- Negative refractive index materials can make an object invisible (cloaking applications; most have been made for microwaves). For visible light requires structures small compared with the corresponding wavelength.

Metamaterials

- Engineered composites in which the dielectric permittivity and magnetic permeability are artificially manipulated
- Periodical structures that in some range of frequencies react to the electric and magnetic fields as medium with negative ε or μ.
- Negative refractive index materials can make an object invisible (cloaking applications; most have been made for microwaves). For visible light requires structures small compared with the corresponding wavelength.

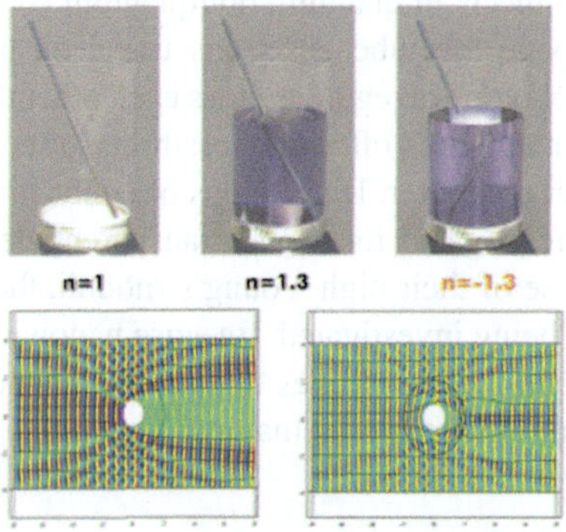

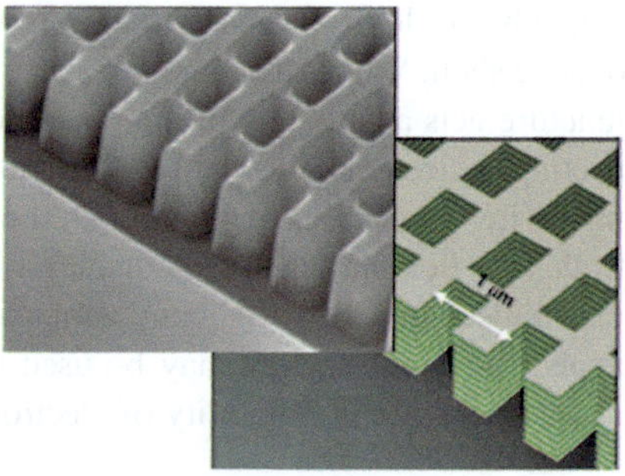

Nanowire-based Metamaterials

Optical properties and applications:

- *Subdifraction imaging (hyperlens)*
- *Freedom of refraction up to 180º*
- *Superfine spatial polarizer*
- *Ultra sensitive Raman sensors*
- *Nanoplasmonics devices*

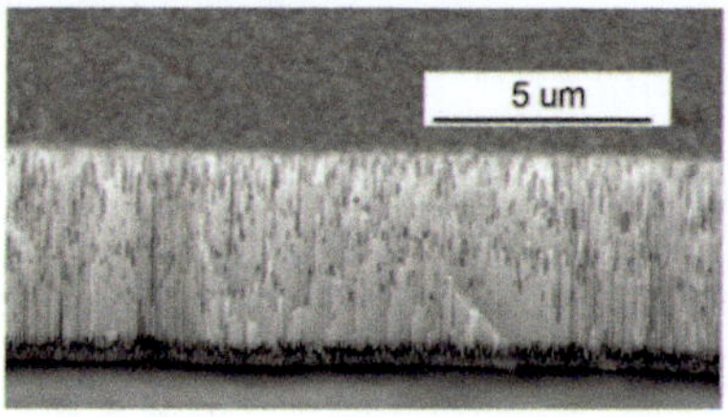

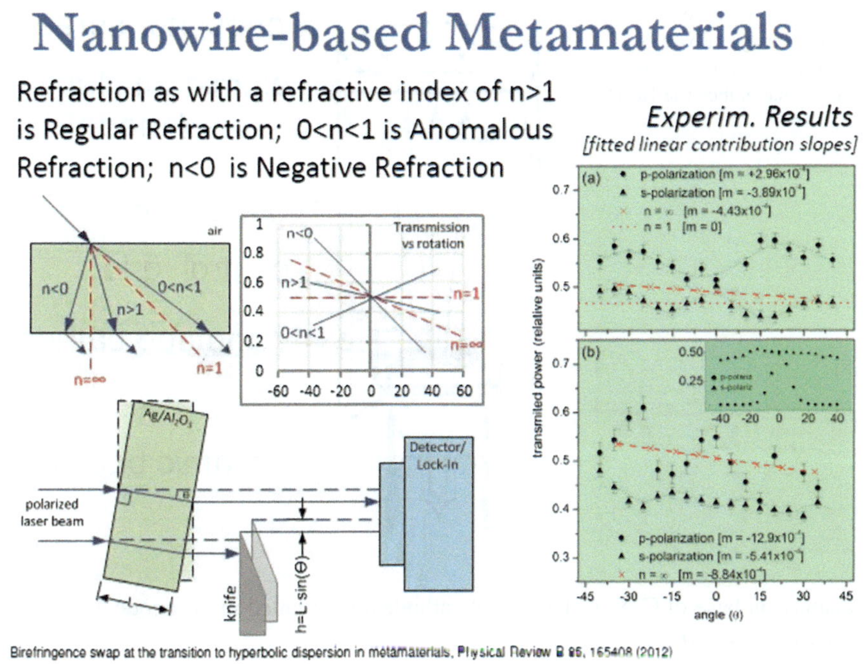

Birefringence swap at the transition to hyperbolic dispersion in metamaterials, Physical Review B 85, 165408 (2012).

Because of their high aspect ratio, nanowires are also uniquely suited to dielectrophoretic manipulation, which offers a low-cost, bottom-up approach to integrating suspended dielectric metal oxide nanowires in electronic devices such as UV, water vapor, and ethanol sensors.

16.11 Indicators and Flat Displays

Thin carbon NTs have ideal geometry for electron emission due to their smallest atomistic tip radius that result in turn to a high electric field strength, in order of magnitude greater than a work function of electrons. Moreover C-NTs are known to exhibit the high strength, melting temperature, resistance to aggressive media as graphite, and are able to work in technical vacuum. Cold emitter on base of C-NTs was proposed to be a key element for the future flat panel TV set as being the alternative for hot emitters on base of electron-beam tube, that make possible to avoid high dangerous asseverated 20–30 kV voltages. At ambient temperature the C-NT are able to emit electrons and current of the same density as the standard tungsten cathode operating at high temperatures and voltages. Set-up of light panel

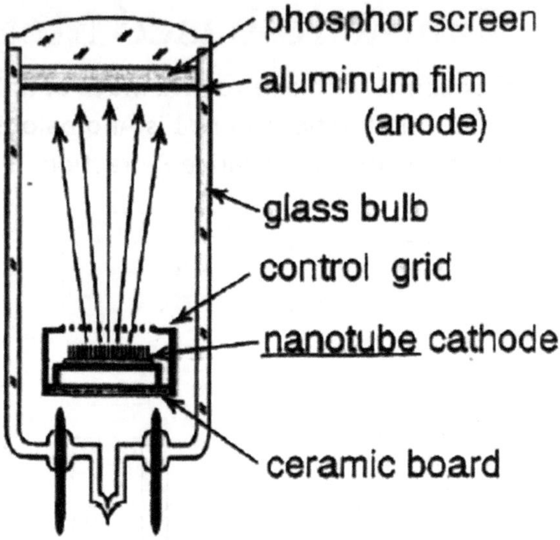

Fig. 16.9 Set-up of the light panel display on base of autoelectron emission of carbon nanotubes (de Heer)

display on base of C-NTs clamped at cathode and oriented in the anode direction is shown in Fig. 16.9.

To obtain the image a phosphor is covered at aluminum anode. Electron shock excites phosphor's molecules which phosphorescence the light photons under transition into ground state. For instance, the phosphor on base of zinc sulphide with copper and aluminum doping shines in green range while silver doped in the blue range. Red light gives phosphor on base of yttrium oxide Y_2O_3 doped by europium. Obviously to fabricate such the panels the high ordered arrays of nanotubes on large area are requested. They can be grown by CVD template technique. In kind of template the zeolite like alumina Al_2O_3 or silica SiO_2 are used often because they contain ordered arrays of cylindrical nanoporous. The arrays of pores of typical diameter 40 nm and interpore distance 100 nm are grown by electrochemical etching of aluminum in acids. Preliminary cobalt is deposited at the pores bottom that serves as catalysist for growth of nanotubes. They growth inside pores under pyrolysis of acetylene in nitrogen atmosphere at 700 °C. Average density of nanotubes is $\sim 10^{10}$ cm^{-2}, a wall thickness is near 10 graphene layers, and a substrate area may be of several square meters.

Today such high-bright light source on base of C-NTs have been fabricated (Fig. 16.10). It consist if cathode, grid and anode inside glass tube of 32 mm in diameter At it inner surface the doped phosphors giving the blue, green and red lights are deposited. At it external surface the lenses are clamped which focuses light in center of glass tube. Electronic circuits are clamped at the lamp-cap.

This lamp with cold cathode is very efficient, the light of which is visible at very large distances under 70° view angle.

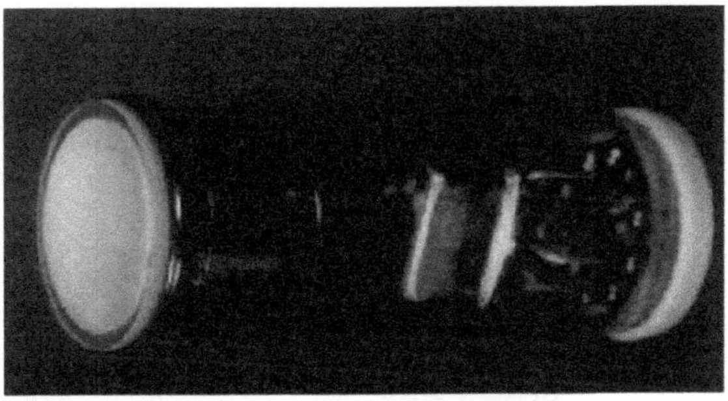

Fig. 16.10 Power bright mini-lamp with cold cathode on base of CNT (NS027A, *Noritake*)

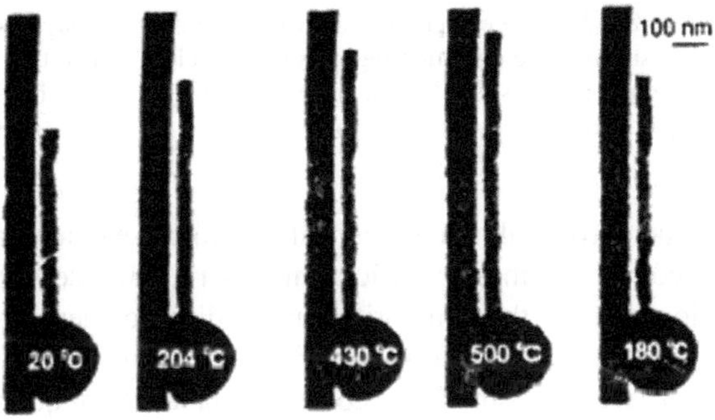

Fig. 16.11 Thermometer on base of MgO nanotube, fabricated in NIMS, Japan

16.11.1 Thermometer

The least thermometer have been fabricated on base MgO nanotube playing the role of capillary shown in Fig. 16.11.

16.12 Nanowire Photovoltaics

The unique properties of semiconductor nanowires make them promising candidates for efficient and inexpensive solar cells. Nanowire array solar cells are the natural heir to second-generation thin-film photovoltaics (PV) for reducing processing costs while maintaining practical efficiencies. The potential of nanowires lies primarily in their tunable geometric effects rather than in any new physical phenomena, as is the case for third-generation quantum-dot solar cells. Nanowire solar cells typically use large nanowires that are well beyond quantum confinement effects so that conventional semiconductor physics applies.

One of the earliest-recognized advantages of nanowires is their potential for improved separation and collection of photogenerated charges. Radial p-n junctions decouple the length scales for absorption and separation of charges: absorption occurs throughout the length of the wire, while the photogenerated minority carriers

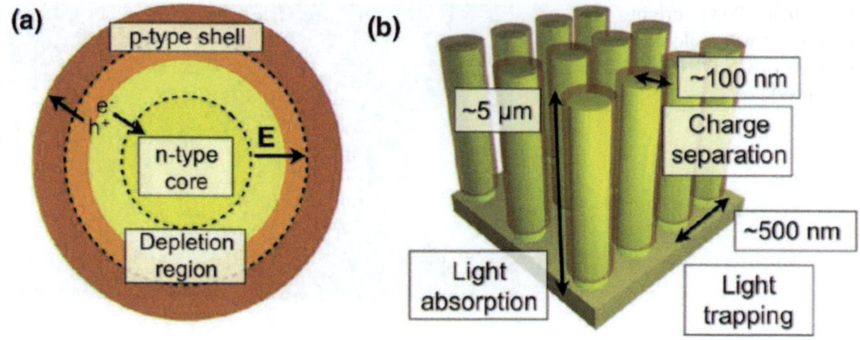

Fig. 16.12 The geometry of nanowire solar cells. **a** Diagram of a radial core-shell nanowire solar cell showing the depletion region's built-in electric field that separates charges between an n-type core and p-type shell. **b** Schematic of the length scales that improve light absorption and charge separation in nanowire array solar cells

must travel only the radial distance to be separated and collected (Fig. 16.12). This geometry is most beneficial in materials whose optical absorption length is much longer than their minority carrier diffusion length; consequently, nanowire solar cells can be made from materials that are unsuitable for planar thin-fi lm solar cells because the core shell geometry relaxes the requirements on minority carrier diffusion length [156]. Also, less expensive and lower purity materials can be used. Interesting candidates include sulfides, [157] oxides [158], and low-purity [159] or heavily doped silicon [160], which exhibit acceptable light absorption but low minority carrier diffusion lengths. This concept has even been applied to dye sensitized solar cells [161] and semiconductor-organic hybrid solar cells [162] to improve charge collection from the dye or low mobility organic phase.

Nanowire solar cell absorption, charge transfer, and carrier collection efficiency can be measured using standard techniques for bulk solar cells. The nanowire geometry provides potential advantages over planar wafer based or thin-film solar cells in every step of the photoconversion process. These advantages include reduced reflection, extreme light trapping, improved band gap tuning, facile strain relaxation, and increased defect tolerance. These benefits are not expected to increase the maximum efficiency above standard limits; instead, they reduce the quantity and quality of material necessary to approach those limits, allowing for substantial cost reductions. Additionally, nanowires provide opportunities to fabricate complex single-crystalline semiconductor devices directly on low-cost substrates and electrodes such as aluminum foil, stainless steel, and conductive glass, addressing another major cost in current photovoltaic technology.

Over the past decade, both solar cell research and nanowire research have become hot topics within science and engineering. The need for higher solar cell efficiencies at lower cost has become apparent, and at the same time synthetic control in nanoscience has improved such that high-performance electronic devices are becoming possible. Nanowire solar cells have some potential benefits over

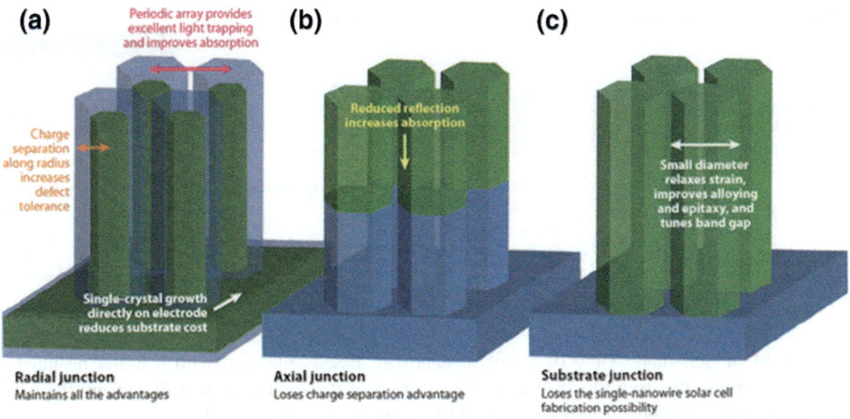

Fig. 16.13 Benefits of the nanowire geometry. **a** Periodic arrays of nanowires with radial junctions maintain all the advantages, including reduced reflection, extreme light trapping, radial charge separation, relaxed interfacial strain, and single-crystalline synthesis on nonepitaxial substrates. **b** Axial junctions lose the radial charge separation benefit but keep the others. **c** Substrate junctions lack the radial charge separation benefit and cannot be removed from the substrate to be tested as single-nanowire solar cells

traditional wafer-based or thin-film devices related to optical, electrical, and strain relaxation effects; new charge separation mechanisms; and cost. Ordered arrays of vertical nanowires with radial junctions take advantage of all these effects, although solar cells made using axial junctions or random arrays can still have some benefits over planar cells, as shown schematically in Fig. 16.13.

16.12.1 Silicon Nanowire Based Solar Cells and Anodes for Li-Ion Batteries

Energy generation from renewable sources and storage of electrical energy are among the most pressing technical challenges for our society. These fields can benefit from silicon nanowires [163] as well, as they have the potential to increase the optical absorption and collection efficiency in solar cells [164–166]. Single nanowire solar cells can be used to study the parameters that influence the performance of nanowire based photovoltaics [166–169]. Additionally, they allow a seamless integration with nanowire based electronics and sensors [167, 168]. For doing general purpose solar cells, however, the single wire elements have to be arranged in large arrays [164, 170]. The potential of the single element to enable such a large scale assembly is therefore a strong differentiator for the different

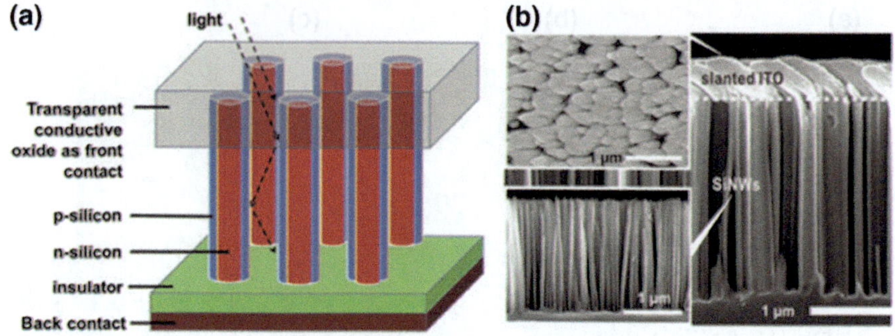

Fig. 16.14 **a** Schematic view of a silicon nanowire based solar cell with a radial pn-junction. **b** SEM micrographs from top view and cross section of the solar cell, as well as from the silicon nanowire array. © Optics Express, reprinted with permission from [280]

concepts. Therefore, the low cost bottom-up growth, which is compatible with different types of substrates, is the fabrication method of choice for such devices. In order to accomplish all the requirements, a coaxial structure is beneficial. First of all, the possibility to fabricate radial pnjunctions has shown promising results [171]. Figure 16.14 shows a schematic view of a silicon nanowire array based solar cell using this concept. Moreover, the flexibility of combining different materials in the nanowire arrangement is utilized as can be seen, for example, in [168]. Also classical heterojunction concepts using crystalline and amorphous silicon can easily be implemented [170].

16.12.2 Dye-Sensitized Solar Cells

- Dye-sensitized solar cell—could offer a cheaper alternative to silicon-based photovoltaics.
- A dye produces excited electrons after absorbing photons from the sunlight and injects them in the conduction band of a TiO_2 thin film.
- Electrons flow through the external load and return to the electrolyte through the counter-electrode.
- Dye molecule is regenerated by the redox system.

Dye-Sensitized Solar Cells

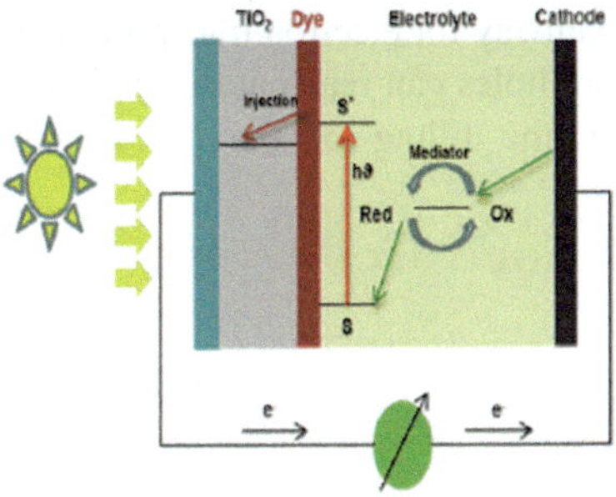

- Dye-sensitized solar cell - could offer a cheaper alternative to silicon-based photovoltaics
- A dye produces excited electrons after absorbing photons from the sunlight and injects them in the conduction band of a TiO2 thin film
- Electrons flow through the external load and return to the electrolyte through the counter-electrode
- Dye molecule is regenerated by the redox system

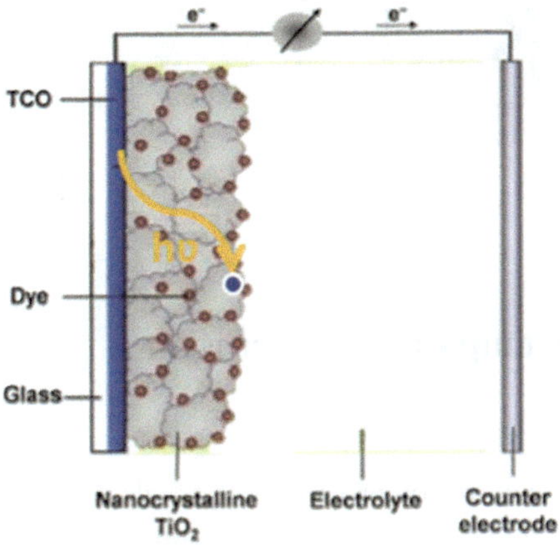

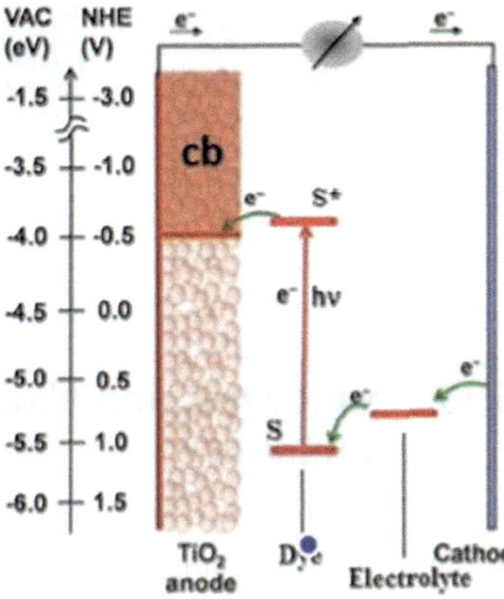

Nanoparticles—higher surface area (efficiency 10–11%)
Nanowires and **Nanorods**—reduce the electron recombination but have a smaller surface area
Nanotubes—have a bigger surface area than nanorods and nanowires (efficiency 7%).

- Current approaches use mesoporous thin films, random nanoparticle network
- Problems: slow electron transport, difficult to fill with electrolyte.
- The use of aligned nanotube arrays allows the enhancement of the absorption coefficient (larger surface area), a direct pathway for electron transport, reducing probability of electron-hole recombination (holes can reach at the surface faster than other architectures) and facilitates the filling of the pores with the

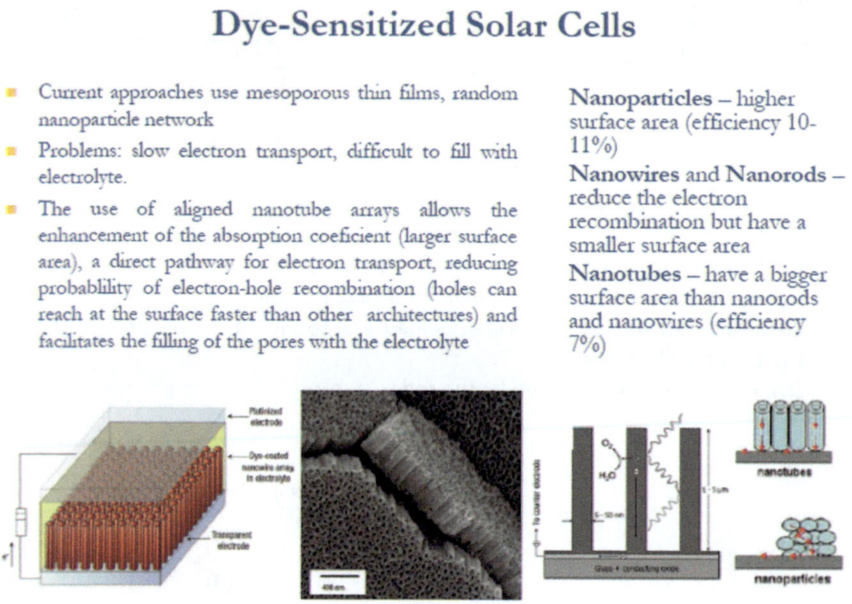

16.13 Nanowires and Nano-Composite as Corrosion Inhibitors

The application of nanotechnology in the corrosion protection of metals has recently gained momentum; Environmental impact can improved by utilizing nanostructure particulates in corrosion inhibition, Nano composites have also proven to be an effective alternative to other hazardous and toxic compounds. The good inhibitor has many advantages such as high inhibition efficiency, low price, low toxicity and easy production. Nanocomposite exhibit such advantages with addition of improvement of the environmental impact. In the last two decades, nanotechnology has been playing an increasing important role in supporting innovative technological advances to manage the corrosion of metals. The application of nanotechnology in the corrosion protection of metals has recently gained momentum. A polymer Nano composite coating can effectively combine the benefits of organic polymers, such as elasticity and water resistance, to that of advanced inorganic materials, such as hardness and permeability. Environmental impact can also be improved by utilizing nanostructure particulates in corrosion inhibition, coating, and eliminating the requirement of toxic solvents. Nano composites have also proven to be an effective alternative to phosphate-chromate pretreatment of metallic substrate, which is hazardous due to the presence of toxic hexavalent chromium.

Nanotechnology has been employed to enhance the inherent corrosion resistance and performance of the metal itself, by achieving the desirable finely crystalline microstructure of metal (e.g., Nano-crystallization) or by modifying its chemical composition at the nanometer scale (e.g., formation of copper nanoparticles at the steel grain boundaries). Metallurgy approaches to the production of high-performance steel with a fine-grain structure and/or self-organization of strengthening Nano phases (carbides, nitrides, carbonitrides, intermetallides) have been burgeoning under the guide of Nano technological principles, including Nano processes for steel smelting and micro alloying, mechanical pressure treatment (e.g., intense plastic deformation), and heat treatment (e.g., superfast quenching of melts). One such technology commercialized in the U.S. produces high-performance carbon steels that feature a "three-phase microstructure consisting of grains of ferrite fused with grains that contain dislocated lath structures in which laths of marten site alternate with thin films of austenite".

Nanotechnology has been employed to reduce the impact of corrosive environments through the alternation of the steel/electrolyte interface (e.g., formation of Nano-composite thin film coatings on steel). Significant improvements in the corrosion protection of steel have been reported through the co-deposition of Ni–SiC or Ni–Al_2O_3 Nano composite coatings on mild steel and the application of TiO_2-naoparticle sol-gel coatings or multilayer polyelectrolyte Nano films on 316L stainless steel. The incorporation of Nano-sized particles (e.g., polyaniline/ferrite, ZnO, Fe_2O_3, hallo site clay, and other nanoparticles) into conventional polymer coatings also significantly enhanced the anti-corrosive performance of such coatings on steel substrates.

Intrinsically conducting polymer metal Nano composites were synthesized by polymerizing anthranilic acid (PANA) with metal salts like ferric chloride, Zinc oxide and Magnesium oxide by chemical oxidation method. Polyanthranilic acid iron Nano composite (PANA-Fe), Polyanthranilic acid-Zinc Nano composite (PANA-Zn) and Polyanthranilic acid-magnesium Nanocomposite (PANA-Mg) ware synthesized BY Sophia et al. Cyclic voltammetric studies revealed the presence of an adherent polymer film on the glassy carbon electrode and showed redox behavior of the polymer metal Nano composites. The XRD (X Ray Diffraction) studies showed a rather more crystalline behavior of the Nano composites and the grain size was calculated using Scherrer's formula and it was found to be in Nano range. SEM (Scanning Electron Microscope) analysis showed a rather mixed crystalline and amorphous behavior. EDAX (Energy Dispersive X Ray Spectroscopy) confirms the incorporation of the metals iron. Zinc and Magnesium in the polymer metal Nano composites.

Composite materials have various properties such as dispersion hardening, self-lubricity, high-temperature oxidation resistance, excellent wear, and corrosion resistance. Because of their importance in many fields, the newer composite materials are synthesized through different existing methods. Among these methods, the electrodeposition is considered to be one of the most important techniques for producing composites, owing to precisely controlled near room temperature operation, rapid deposition rates, and low cost. A number of the literatures appear in scientific journals connected to the code position of SiC, ZrO_2, Al_2O_3, TiO_2, and PTFE with single metal and alloy electrodeposition.

So, generated composite coatings on steel exhibited excellent atmospheric corrosion resistance property and thus reducing or eliminating its chromium passivation. Further the corrosion resistance property is enhanced by code position of nano materials like CNT, MWCNTs, TiO_2, Fe_2O_3, and so forth, with metals. The size of nano structural materials ranging from 1–100 nm, are used widely in electrodeposition. Thus, these materials enhance the mechanical and physical properties of the coatings due to their extremely small size. Researches adopted pulse deposition or electrodeposition methods for preparing Zn–TiO_2 and Zn–Ni–TiO_2 composites.

16.13.1 Corrosion Resistant of ZnO Nanowires Coatings

Synthesis of ZnO Nanowires

There are many ways to synthesized ZnO nanowires, we will illustrate the two easily and better ways to synthesize ZnO:

Hydrothermal Method *(Yangyang Zhang, et al., 2012).*

Generally, solution phase synthesis is carried out in an aqueous solution, and the process is then referred to as the hydrothermal growth method. Hydrothermal methods have received a lot of attention and have been widely used for synthesis of 1D nanomaterials. In addition, hydrothermally grown ZnO nanowires have more crystalline defects than others primarily due to oxygen vacancies. Nanowires with inherent defects are capable of exhibiting visible light photocatalysis even without doping with transition metals. The general process for vertically aligned ZnO nanowires grown on a substrate by the hydrothermal method is the following.

(a) A thin layer of ZnO nanoparticles is seeded on a certain substrate. The seeding layer promotes nucleation for the growth of nanowires due to the lowering of the thermodynamic barrier.
(b) An alkaline reagent (such as NaOH or hexamethylenetetramine) and Zn^{2+} salt ($Zn(NO_3)_2$, $ZnCl_2$, etc.) mixture aqueous solution is used as a precursor (or growth solution).
(c) The ZnO seeded substrate is kept in the growth solution at a certain temperature and a certain period of time.

Fig. 16.15 SEM images of the ZnO nanowires prepared by the oxidation of Zn thin films in air for 1 h at temperatures of **a** 450 °C, **b** 500 °C, **c** 550 °C, **d** 600 °C, **e** 650 °C

(d) The resultant substrate and growth layer is washed and dried. When hexamethylenetetramine ($(CH_2)_6N_4$, or HTMA) and $Zn(NO_3)_2$ are chosen as precursor, the chemical reactions can be summarized in the following equations.

$$\text{Decomposition reaction: } (CH_2)_6N_4 + 6\,H_2O \rightarrow 6\,HCHO + 4NH_3 \tag{16.1}$$

$$\text{Hydroxyl supply reaction: } NH_3 + H_2O \leftrightarrow NH_4^+ + OH^- \tag{16.2}$$

$$\text{Super saturation reaction: } 2OH^- + Zn^{2+} \rightarrow Zn(OH)_2 \tag{16.3}$$

$$\text{Znb nanowire growth reaction : } Zn\,(OH)_2 \rightarrow ZnO + H_2O \tag{16.4}$$

One of the key parameters for the growth of ZnO nanowires is controlling the supersaturating of the reactants. It is believed that high super saturation levels favor nucleation and low supersaturation levels favor crystal growth. Thus the concentration of OH^- should be controlled in the solution to maintain low supersaturation levels during the whole nanowire growth process (Fig. 16.15).

Growth of ZnO Nanowires on Silicon

The p-type, (100) oriented Si wafers were used as the substrates for nanowires growth. After standard RCA cleaning, a layer of 70 Å thick Au thin film was deposited on silicon substrate by rf-magnetron sputtering. The substrates with Au thin films were annealed at 500 °C for 30 min in vacuum in order to form Au nanoparticles on the substrates. Equal amounts of ZnO powder (Alfa Aesar, 99.999%) and graphite powder (Alfa Aesar, 99%) was milled, placed in an alumina boat, and then loaded to the center of a quartz tube. The annealed Au/Si substrate was placed in the same boat with mixed powder. The ZnO nanowires were grown on the annealed Au-coated Si substrates at temperatures from 850 to 950 °C under 5–30 sccm argon gas flow for 1–30 min. Atomic force microscopy (AFM, DI-3100) in tapping mode under ambient conditions was performed to observe the surface morphology of annealed Au/Si substrate. The phase characterization of the nanowires was studied by X-ray diffraction (XRD, MAC Science, and MXP18) with Cu Kα radiation. The chemical composition of the nanowires was analyzed by X-ray photoelectron spectroscopy (XPS, VG Scientific ESCALAB 250) and energy dispersive X-Ray spectrometer (EDX, JEOL, JEM-2010). The morphology and size distribution of the nanowires were characterized using scanning electron microscopy (SEM, Hitachi, S-4700I) and transmission electron microscopy (TEM, Philip, tecnai-20). A photoluminescence spectrophotometer (PL, Hitachi, F-4500) with a Xe lamp as an excited source was used for the luminescence studies at room temperature. Figure 16.16 shows proposed mechanism of the VLS growth of the ZnO nanowires. There are at least four different stages involved in the formation of ZnO nanowires: Au film deposition, alloy nanoparticles formation, nucleation of ZnO and epitaxial growth of ZnO nanowires. The Au thin film on Si substrate annealed at 500 °C for 30 min was expected to form nanosized Au–Si alloy droplets as the eutectic point of Au–Si binary system is about 363 °C on the basis of their phase diagram. An AFM image of the annealed Au/Si substrate indicates that the annealing process leads to the formation of the Au–Si alloy nanoparticles having a mean diameter of 25 nm. The ZnO nanowires might be nucleated on the Au/Si alloy nanoparticles. As the temperature increased to ~900 °C, the ZnO was reduced by graphite and CO (g). The corresponding chemical reaction can be expressed as:

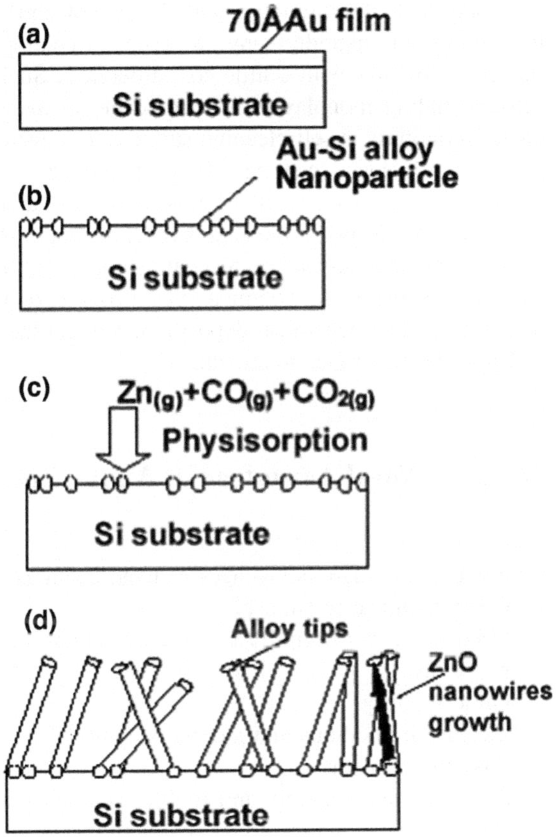

Fig. 16.16 Schematic illustration of vapor-liquid-solid nanowire growth mechanism **a** Au film deposition; **b** alloy nanoparticles formation; **c** absorption and nucleation; **d** epitaxy growth

$$ZnO(s) + C(s) \rightarrow Zn(g) + CO(g) \tag{16.5}$$

$$CO(g) + ZnO(s) \rightarrow CO_2(g) + Zn(g) \tag{16.6}$$

16.13.2 Corrosion Resistance of Nanoparticle—Incorporated Nano Coatings

Nano coatings: are materials that are produced by shrinking the material at the molecular level to form a denser product. Nano coatings have several advantages such as better surface appearance, good chemical resistance, better thermal and electrical conductivity and better corrosion resistance. Incorporation of nanoparticles enhance the corrosion resistance of zeolite coatings, epoxy coatings and antimicrobial coatings. The appearance and usefulness of nanoparticles brings many advantages and opportunities to paint and coating industry. Coating industry is among the first to tap the potential of nanotechnology. Addition of nanoparticles to coatings can upgrade many properties of coating system and can produce multipurpose coatings with a little cost difference. Such coatings, sometimes made of self-assembling monolayers are applicable in many ways: from scratch resistant coatings on glass to self-cleaning surfaces to moisture absorbing to anti-graffiti and corrosion resistance coatings. Further, unique composition, better strength and flexibility along with excellent gloss and transparency makes nano-coatings even more effective. Many of the nanoparticls like nano–ZnO are non-toxic in nature and thus add another advantage to coating industry. Nano-coating can be applied in many ways including chemical vapor phase deposition, physical vapor phase deposition. Electrochemical deposition, Sol–gel methods, electro-spark deposition, and laser beam surface treatment.

16.13.3 Novel Advantage of Nano-Coatings [172]

- Some of the main advantages of nano-coatings are better surface appearance.
- Good chemical resistance.
- Decrease in permeability to corrosive environment and hence better corrosion properties.
- Optical clarity.
- Increase in modulus and thermal stability.
- Easy to clean surface.
- Anti-skid, anti-fogging, anti-fouling and anti-graffiti properties.
- Better thermal and electrical conductivity.
- Better retention of gloss and other mechanical properties like scratch resistance.
- Anti-reflective in nature.
- Chromate and lead free.
- Good adherence on different type of materials.

16.13.4 Nanoparticle—Based Coatings for Magnesium Alloys with Thermal and Mechanical Stability

Magnesium alloys have an excellent strength to weight ratio and are attractive materials in cars, aircrafts and high end portable devices. Nevertheless they are also very reactive and require excellent corrosion protection. Inorganic coatings offer good protection especially under mechanical and thermal strain. However, due to the low temperature stability of most magnesium alloys, dense inorganic coatings cannot be applied using conventional techniques.

Purely inorganic, transparent coatings have been developed for the corrosion protection of magnesium alloys, based on commercially available, aqueous SiO_2 nanodispersions. These coatings can be processed under mild conditions, but are permeable for aggressive anions. The permeability can be decreased by sealing with inorganic polymeric sols. Despite their low thickness the resulting coatings show remarkable protective properties.

With decreasing particle size the sintering temperature decreases. Commercial water based silica dispersion with suitable sintering additives can be applied onto AZ31 and AZ91 (grinded, degreased with acetone and activated with Nital) by dip coating. Small particles, slow withdrawal and low dispersion viscosity result in thin (−300 nm) defect-free and transparent coatings. Thicker crack-free coatings can be achieved by multiple coating application (up to 5 layers = 1.5 μm). The coating resistance (Rpo) increases with each additional layer and localized corrosion becomes inhibited.

16.13.5 Corrosion Resistant Zeolite Coatings

Incorporation nanoparticles enhance the corrosion protection efficiency of zeolite coatings, epoxy coatings and biocidal coatings. Zeolites are microporous crystalline aluminosilicates minerals with uniform and molecular sized pores. They are non-toxic and have been explored for medical applications, commonly used as commercial adsorbents.

The term zeolite was originally coined by Swedish mineralogist Axel Fredrik Cronstedt, who observed that upon rapidly heating the material stilbite, it produced large amounts of steam from water that had been adsorbed by the material. Based on this, he called the material zeolite, from the Greek meaning boiling stone.

About 194 unique zeolite frameworks have been identified, and over 40 naturally occurring zeolite frameworks are known. Zeolites are widely used in industry for water purification, as catalysts, and in nuclear reprocessing. Their biggest use is in the production of laundry detergents. They are also used in medicine and in agriculture.

Zeolites have a porous structure that can accommodate a wide variety of cations, such as Na^+, K^+, Ca^{2+}, Mg^{2+} and others. These positive ions are rather loosely held and can readily be exchanged for others in a contact solution. Some of the more common mineral zeolites are analcime, chabazite, clinoptilolite, heulandite, natrolite, phillipsite, and stilbite. An example mineral formula is: $Na_2Al_2Si_3O_{10}{\cdot}2H_2O$, the formula for natrolite.

Zeolites are a class of crystalline oxides with uniform channels and/or cages ranging from 0.3 to 2 nm. They are extensively used as adsorbents, catalysts, and ion exchange agents [173–175]. The idea of corrosion-resistant zeolite coatings takes advantages of several well-known facts about zeolite.

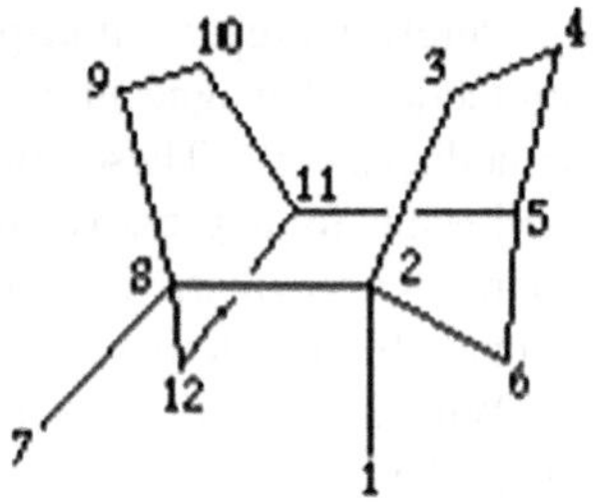

Fig. 16.17 Secondary building units (SBU) in pentasil zeolites

(i) Most zeolites are non-toxic, and some compositions are even used in food and drug applications.
(ii) High-silica and pure-silica zeolites are thermally and chemically stable. For example, many high silica zeolites (HSZ) are thermally stable up to 1000 °C, and pure-silica zeolites are stable in all mineral acids except hydrofluoric acid. Thus, the framework of zeolites contains the needed corrosion resistance. (As used herein, HSZs have a Si:Al ratio greater than 50, while the ratio for low-silica zeolite (LSZ) is less than 2.)
(iii) Most HSZ syntheses use structure-directing agent (SDA) molecules, which are eventually trapped inside the zeolite pores.

Like most silicates—the zeolites are based on TO_4 tetrahedra, where T is an aluminium or silicon atom (phosphorus in aluminophosphates). The vast 3-dimensional networks are a result of all four corners for the tetrahedra being shared, producing low density microporous materials. Zeolite structures can be thought to exist of finite or infinite (chains, layers etc.) component units. An important class of materials is the pentasil zeolites, named so because they are constructed of five membered rings. The most important example is that of ZSM-5—which is shown in Fig. 16.17. ZSM-5 is used in a variety of catalytic applications, including the catalysis of methanol into fuel hydrocarbons. The individual pentasil units are shown in Fig. 16.17. They combine to form long chains, (Fig. 16.18), which then join together to form layers, (Fig. 16.19).

Zeolite molecular sieves are characterized by the following properties:

- Selective adsorption due to the uniform pore size of the zeolite structure.
- High adsorption capacity for polar substances at low concentrations.

Zeolite molecular sieves are crystalline, highly porous materials, which belong to the class of alumino silicates. These crystals are characterised by a three-dimensional pore system, with pores of precisely defined diameter. The corresponding crystallographic structure is formed by tetrahedras of (AlO_4) and (SiO_4). These tetrahedras are the basic building blocks for various zeolite structures, such as zeolites (Fig. 16.20 **A** and **X**), the most common commercial adsorbents.

Due to the presence of alumina, zeolites exhibit a negatively charged framework, which is counter-balanced by positive cations resulting in a strong electrostatic field

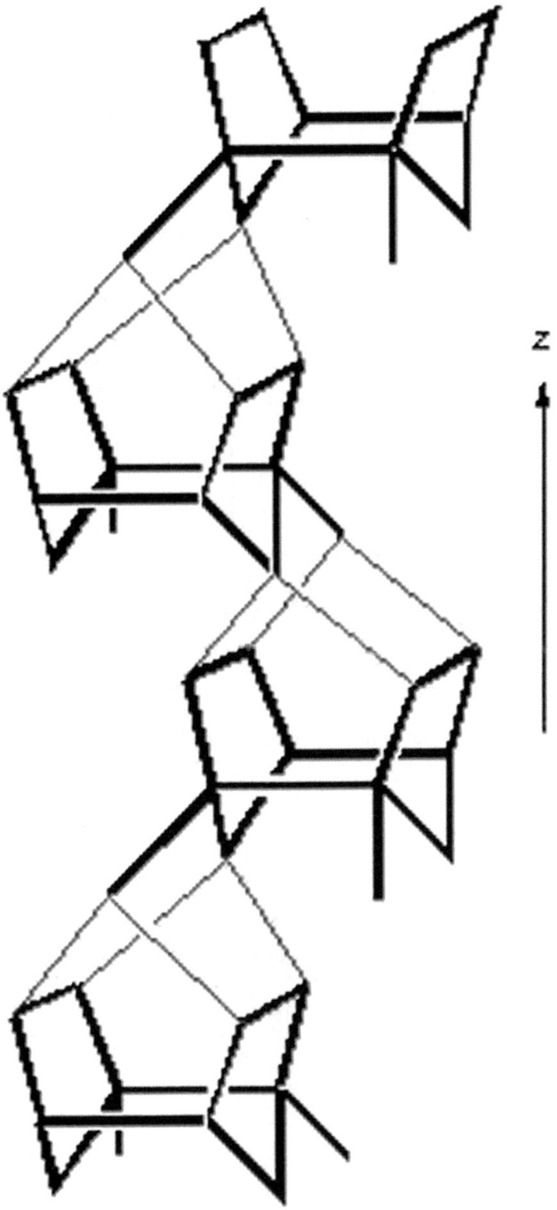

Fig. 16.18 SBU linked chains

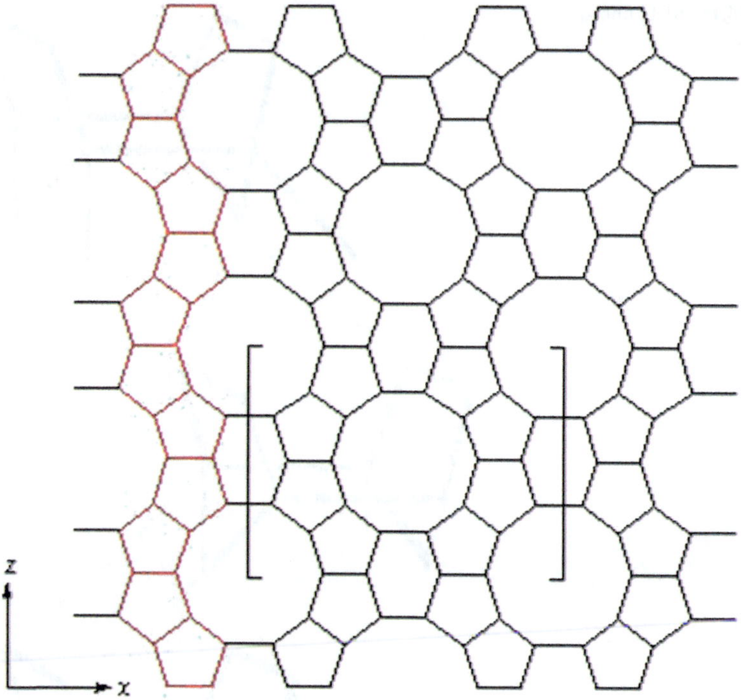

Fig. 16.19 Layer formed from linked chains in ZSM-5 structure

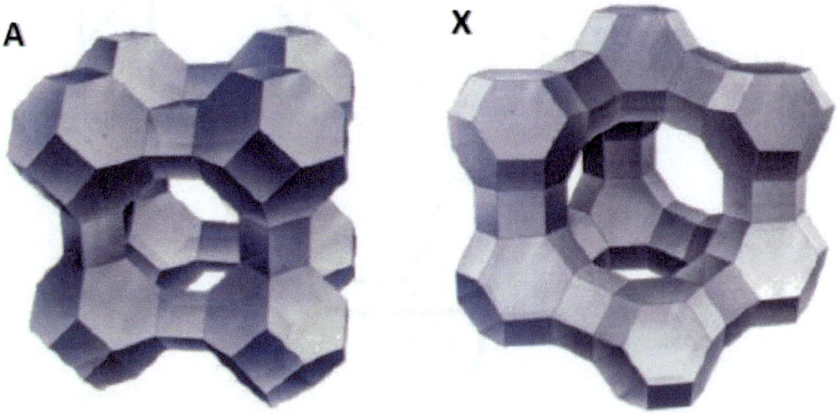

Fig. 16.20 Molecular sieve types **A** and **X**

on the internal surface. These cations can be exchanged to finetune the pore size or the adsorption characteristics. For instance, the sodium form of zeolite A has a pore opening of approximately 4 Ångstrom (4×10^{-10} m), called 4 Å molecular sieve. Zeolites are microporous crystalline aluminosilicates with uniform and molecular sized pores. They are non-toxic and have been explored for medical applications. The application of zeolite films as chromium-free protective coatings on aluminum alloys have been investigated. Recent technological advances of Zeolite corrosion resistant coatings have been reported [176–180].

16.13.6 Epoxy Coatings-Influence of Nanoparticles on the Anti-corrosion and Mechanical Properties of Epoxy Coatings

Epoxy has been widely used as a coating material to protect the steel reinforcement in concrete structures [181–183], because of its outstanding process ability, excellent chemical resistance, good electrical insulating properties, and strong adhesion/affinity to heterogeneous materials. Epoxy coatings generally reduce the corrosion of a metallic substrate subject to an electrolyte in two ways. First, they act as a physical barrier layer to control the ingress of deleterious species. Second, they can serve as a reservoir for corrosion inhibitors to aid the steel surface in resisting attack by aggressive species such as chloride anions. Nonetheless, the successful application of epoxy coatings is often hampered by their susceptibility to damage by surface abrasion wear [184, 185]. They also show poor resistance to the initiation ad propagation of cracks [186] such processes introduce localized defects in the coating and impair their appearance and mechanical strength. The defects can also act as pathways accelerating the ingress of water, oxygen and aggressive species onto the metallic substrate, resulting in its localized corrosion. Furthermore, being hydrophilic in nature, epoxy coatings experience large volume shrinkage upon curing and can absorb water from surroundings [187, 188]. The pores in the cured epoxy coating can assist in the migration of absorbed water and other species to the epoxy metal interface, leading to the initiation of corrosion of the metallic substrate and to delamination of the coating. The barrier performance of epoxy coatings can be enhanced by the incorporation of a second phase that is miscible with the epoxy polymer, by decreasing the porosity and zig zagging the diffusion path for deleterious species. For instance, inorganic filler particles at nanometer scale can be dispersed within the epoxy resin matrix to form an epoxy nano composite. The incorporation of nanoparticles into epoxy resins offers environmentally benign solutions to enhancing the integrity and durability of coatings, since the fine particles dispersed in coatings can fill cavities [189–191] and cause crack bridging, crack deflection and crack bowing [192]. Nanoparticles can also prevent epoxy disaggregation during curing, resulting in a more homogenous coating. Nanoparticles tend to occupy small hole defects formed from local shrinkage during curing of the epoxy resin and act as a bridge interconnecting more molecules. This results in a reduced total free volume as well as an increase in the cross-linking density [193, 194]. In addition, epoxy coatings containing nanoparticles offer significant barrier properties for corrosion protection [195, 196] and reduce the trend for the coating to blister or delaminate.

Homogeneous epoxy coatings containing nanoparticles of SiO_2, Zn, Fe_2O_3 and halloysite clay were successfully synthesized on steel substrates by room-temperature curing of a fully mixed epoxy slurry diluted by acetone. The surface morphology and mechanical properties of these coatings were characterized by scanning electron microscopy and atomic force microscopy, respectively. The effect of incorporating various nanoparticles on the corrosion resistance of epoxy-coated steel was investigated by potentiodynamic polarization and electrochemical impedance spectroscopy. The electrochemical monitoring of the coated steel over 28 days of immersion in both 0.3 and 3 wt% NaCl solutions suggested the beneficial role of nanoparticles in significantly improving the corrosion resistance of the coated steel, with the Fe_2O_3 and halloysite clay nanoparticles being the best. The SiO_2 nanoparticles were found to significantly improve the microstructure of the coating matrix and thus enhanced both the anticorrosive performance and Young's modulus of the epoxy coating. In addition to enhancing the coating barrier performance, at least another mechanism was at work to account for the role of the nanoparticles in improving the anticorrosive performance of these epoxy coatings [197].

16.13.7 Nano Particle Incorporated Self-cleaning Paints and Biocidal Coatings

There is a great interest in the design and development of surfaces that not only provide biocidal activity but are also easy to clean and even self-cleaning. Most of such coatings acquire their biocidal/self-cleaning capacity by incorporating specific nanoparticles: basically silver (Ag) and titanium oxide ($TiCO_2$) [198, 199]. Nano TiO_2 is used for developing anti-UV, anti-bacterial and self-cleaning paints. This possesses self-cleaning hydrophobic properties, which causes water droplets to bead-off of a fully cured surface picking up dirt and other surface contaminants along the way. This self-cleaning action helps clean and maintain important surfaces and to accelerate drying, leaving the surface with minimal spotting. A recent study by Cai et al. [200] utilizes corona treatment technique, inert sol-gel coating and anatase TiO_2 layer. With the corona treatment, an organic surface was activated to allow a uniform TiO_2 sol-gel coating. Nanoparticles of surface-treated $A1_2O_3$ molecules help increase hydrophobicity and increase scratch resistance.

Microbial evolution on a wide variety of surfaces can cause corrosion, dirt, bad odor and even serious hygiene and health problems. AMBIO (Advanced

Nanostructured Surfaces for the Control of Biofouling), a European Union research project is investigating how to prevent the build-up of organisms on surfaces under marine conditions to avoid biofouling. The project aims to use nanostructuring to significantly reduce the adhesion of organisms to surfaces in aquatic environments, and thus control the fouling process without the use of toxic biocides such as copper and organotin compounds that prevent fouling by killing organisms. Nanostructuring of the surface alters the wetting properties and is intended to signal that the site is not suitable for the organisms to settle. The project aims to synthesize new nanostructured polymers that are stable under marine conditions. Although no alternatives to the use of biocides are available at present, creation of nanostructured surfaces could offer an innovative and environment-friendly solution to the problem of biofouling [201]. Research has developed new biocidal coating systems that prolong biocidal activity by immobilizing such additives on nanoparticles; the embedded biocides are designed to be released into the environment only when needed, thus extending the lifetime of the biocidal activity [202].

16.13.8 Nanoparticle Based Antimicrobial Corrosion Coatings

Marine microbiological corrosions are responsible for considerable damages to all devices and vessels immersed in seawater, and this induces serious economic problem to maritime activities [203–205]. Coating is the most convenient method for preventing the corrosion/fouling of steel surfaces, storage tanks, ship bottoms and their super structures. Employing effective antifouling marine paints, containing booster biocides at non-toxic levels is one approach to solve the issue of fouling [206–208]. Copper and its oxides are common biocides in a large number of commercial antifouling marine paints. Commercially available antifouling coatings have cuprous oxide (Cu_2O) as the primary biocide, which typically varies from 20 to 76% copper content [209]. However, recent reports have shown that 26% sampling locations/year in UK exceeded the concentration of current Environmental Quality Standard (5 μg/L) [210] for copper [211] and marine microalgae were able to take up copper very quickly in natural coastal seawater [212]. These data suggest that monitoring of concentrations of copper is important so as not to cause ecological damage in marinas. Besides Cu, other metals such as zinc and silver exhibit antifungal and antibacterial properties, which have been widely utilized in advanced coating technologies [213–217]. Especially, silver has long been known to have strong inhibitory and bactericidal effects as well as a broad spectrum of antimicrobial activities, even at low concentrations in the ppb range [218]. Meanwhile, silver has a remarkably low toxicity compared to other heavy metal ions [219], which make it possible to be more environmentally friendly antimicrobial painting additives instead of copper.

One important function of paints containing biocides or inhibitors is to obtain optimal release rate of the actual active substance into the sea. The leaching rate of biocides should not be too fast, resulting in rapid and premature depletion of the antifouling activity of marine coatings and unnecessarily high concentration in the sea. However, the release rate should not be too slow since this would undoubtedly result in fouling [220]. In order to deal with both issues, application of core–shell structured materials should be one of the best alternatives since the shells offer protection to the cores and introducing new properties to the hybrid structures [221–223].

Up to date, there are only some reports exploring the marine corrosion protection based on core–shell structures. White et al. demonstrated the application of micrometer scale capsules loaded with oligomers for self-healing of the cracks in polymer coatings [224, 225] and Shchukin et al. developed corrosion protection coatings with halloysite nanotubes as nanocontainers loaded with the corrosion inhibitors [226]. However, there are still challenges such as stability of core–shell particles, uniform dispersity in a matrix, and control of the release rate. Ke et al., have successfully prepared monodispersed Ag/SiO_2 core–shell nanoparticles with facile pathway [227].

Ag/SiO_2 core–shell nanoparticles within the size of 60 nm were applied in marine antimicrobial corrosion coatings. TEM and XRD results indicated silver was completely covered by silica and its crystal form was not affected after coated by silica. Effects of Ag/SiO_2 nanoparticles in the microbial corrosion of acrylic-coated steel were studied by means of electrochemical noise analysis (ENA). ENA results reveal that 1 wt% Ag/SiO_2 nanoparticles exhibit better antimicrobial corrosion activity than that of conventional 40 wt% Cu_2O biocides. Inductively coupled plasmaoptical emission spectrometry (ICP) showed very lower amount of Ag ions leaching from the matrix resin, whereas a higher amount of copper ion leaching from the resin in the same period. It was concluded Ag/SiO_2 core–shell nanoparticles could enhance long-term corrosion protection in comparison with the copper biocides. This composite has a strong potential use in environmentally friendly antimicrobial coatings. Thus, Incorporation nanoparticles enhance the corrosion protection efficiency of zeolite coatings, epoxy coatings and biocidal coatings [228].

Finally, *scanning probe microscopes (SPMs)* are standard instruments at scientific and industrial laboratories that allow imaging, modifications, and manipulations with the nanoobjects. They permit imaging of a surface topography and correlation with different physical properties within a very broad range of magnifications, from millimeter to nanometer-scale range. Atomic force microscopy (AFM) and AFM related techniques (e.g., scanning near-field optical microscopy—SNOM) have become sophisticated tools, not only to image surfaces of molecules but also to measure molecular forces between molecules. This is substantially increasing our knowledge of molecular interactions. Electrical scanning probe microscopy (SPM) techniques have already made a number of important contributions to the field of organic electronics. Conductive atomic force microscopy (c-AFM), electrostatic force microscopy (EFM), scanning Kelvin probe microscopy

(SKPM), and similar variants were successfully used to elucidate charge injection/ extraction, transport, trapping, and generation/recombination in organic devices [229].

16.14 Superconducting Nanowire Single-Photon Detectors (SNSPDs)

Over the past decade superconducting nanowire single-photon detector (SNSPD) technology has advanced apace. The basic device performance has improved dramatically, offering high η_{sde}, low DCR and low Δt at infrared wavelengths, outperforming the best available semiconductor based single-photon detectors. Improvements in cooling technology have aided the adoption of SNSPDs in a wide range of applications. As this section illustrates, high impact photon-counting demonstrations using SNSPDs have been carried out in diverse fields, from quantum information science to atmospheric time-of-flight ranging. There is no question that over the coming decade SNSPDs will have an important role to play in many further scientific advances and applications.

Single-photon detectors based on superconducting nanowires have rapidly emerged as a highly promising photon counting technology for infrared wavelengths. These devices offer high efficiency, low dark counts and excellent timing resolution. This type of device, known as the superconducting single-photon detector (SSPD) or superconducting nanowire single-photon detector (SNSPD), is single-photon sensitive at visible and infrared wavelengths, with recovery times and timing precision orders of magnitude faster than existing single-photon detectors based on superconducting materials. SNSPDs have excellent potential for time-correlated single-photon counting (TCSPC) [230] in the infrared wavelength regime, where important new applications are emerging. The main competitors in this arena are solid-state single photon avalanche photodiodes (SPADs), which have superseded bulky photomultipliers (PMTs) [231]. The long wavelength sensitivity of the SNSPD extends far beyond that of the Si single photon avalanche photodiode (SPAD) [232] and the SNSPD is superior to the InGaAs SPAD [233] in terms of signal-to-noise ratio. SNSPDs have been the subject of intense interest of over the past decade and many research groups around the world have contributed to device development.

An ideal single-photon detector (SPD) generates an electrical signal only upon absorption of a photon. The signal level is well defined above the noise; in the absence of illumination no electrical signal is returned. In practice, real SPDs have many non-ideal characteristics and performance metrics must be carefully defined. The most obvious performance metric in a SPD is the detection efficiency (η)—the probability that an output signal is registered when a photon is incident on the detector. In practice η is lower than 100% and is likely to depend strongly on the wavelength λ of the incident photons. In any real photon-counting experiment, photons can be lost before reaching the detector due to absorption, scattering or reflection within the experimental environment—this loss can be defined as the *coupling efficiency* (η_{coupling}). The detector material and geometry determines the *absorption efficiency* ($\eta_{\text{absorption}}$). Finally, there may be a non-unity probability that the detector generates an output electrical signal after photon absorption—we define this as the *registering probability* ($\eta_{\text{registering}}$). Taking all these contributions into account, the overall *system detection efficiency* (η_{sde}) is

$$\eta_{\text{sde}} = \eta_{\text{coupling}} \times \eta_{\text{absorption}} \times \eta_{\text{registering}} \tag{16.7}$$

The intrinsic device detection efficiency (η_{dde}) is defined as

$$\eta_{\text{dde}} = \eta_{\text{absorption}} \times \eta_{\text{registering}} \tag{16.8}$$

Therefore η_{sde} and ηdde are only equal for perfect optical coupling i.e. $\eta_{\text{coupling}} = 1$. The device detection efficiency, η_{dde}, is often reported as 'quantum efficiency' in the literature. However this term is used with different meanings for different technologies, so 'device detection efficiency' will be used in this text for clarity. Readers should also heed the following caveat: unfortunately across the published SPD literature there is no uniform consensus on whether to report η_{sde} or η_{dde}. In general early SPD device demonstrations report η_{dde}, whereas as when SPDs are implemented in actual photon-counting experiments, η_{sde} is more likely to be reported.

Other performance metrics quantify limitations in SPD performance. Stray light and electrical noise can also potentially mimic the optical signal. These false detection events are called *dark counts,* usually quantified in terms of a dark count rate (DCR). The timing uncertainty between the arrival of the photon at the SPD (which can be known extremely precisely with modern optical sources) and the generation of the output pulse from the SPD, may set the timing resolution. This is known as the *jitter* (Δt) of the SPD. In addition, a real SPD will have a finite *recovery time* (τ) before it is ready to register a subsequent photon—this sets a limit on the theoretical maximum count rate of the SPD (which in practice may be limited by other factors, such as the readout electronics).

16.14.1 Origins of Device Concept

In 1971, a laser was used for the first time to disrupt the superconductivity of Pb films, forming a resistive state that could not be explained just by heating effects [234]. This experiment showed that the energy of the absorbed photons resulted in a nonequilibrium state, with hot excited quasiparticles at a higher temperature than the Cooper pairs in the superconductor. Equilibrium is reached via a sequence of relaxation processes: (i) inelastic scattering of quasiparticles by electron-electron interactions and electron-phonon interactions, (ii) generation of quasiparticles by phonons, (iii) quasiparticle recombination and (iv) energy dissipation into the substrate by the phonon in the superconductor. The temporal characteristics of the relaxation depend on the time constants associated with each of these processes. Twenty years later, an elegant approach to understand the nonequilibrium relaxation was developed by studying transient photoimpedence response (TPR) of superconducting films [235]. In a TPR measurement the superconducting film is approximated as a changing kinetic inductance and the film in the resistive state is approximated as a photoresistor. A two-temperature model is used to quantitatively extract the time constants from the photoresponse of a superconductor film to an optical pulse [236]. In this model, the time-dependent temperature of the electron subsystem (T_e) and phonon subsystem (T_p) are obtained by solving the following coupled linear heat-balance equations:

$$
\begin{aligned}
c_e \frac{dT_e}{dt} &= -\frac{c_e}{\tau_{e-p}}\left(T_e - T_p\right) + P(t) \\
c_p \frac{dT_p}{dt} &= -\frac{c_e}{\tau_{e-p}}\left(T_e - T_p\right) - \frac{c_p}{\tau_{es}}\left(T_p - T_0\right)
\end{aligned}
\qquad (16.9)
$$

where c_e and c_p are the electron and phonon specific heat, respectively, T_0 is the substrate temperature, $\tau_{e\text{-}p}$ is the electron-phonon interaction time, τ_{es} is the phonon escape time and $P(t)$ is the time dependent power of radiation absorbed in the unit volume of the film. This two-temperature model ignores any Joule heating via the bias current. This hot-electron and phonon-heatsink scheme predicts a picosecond photoresponse from the superconductor [236]. This was experimentally verified in NbN [237] and $YBa_2Cu_3O_{7\text{-}\delta}$ [238] microbridges. The region of hot electrons or quasiparticles forming the resistive region in the superconductor is known as the 'hotspot'. The hotspot concept was originally developed by Skocpol et al. to study self-heating effects in superconducting microbridges [239]. In 1996, Kadin et al. [240] approximated the two-temperature model equations as a heat flow equation:

$$
Cd\frac{\partial T}{\partial t} = \kappa d \nabla^2 T + \alpha (T_0 - T) \qquad (16.10)
$$

where d is the thickness of the nanowire, κ is the thermal conductivity, C [$C = c_e + c_p$] is the heat capacity of the superconducting film, α is the thermal boundary conductance between the film and the substrate, T_0 is the substrate temperature (since the nanowire is thin enough) and the temperature of the nanowire is approximated to one temperature T. The solutions to Eq. 16.10 predict that the photon-induced hotspot is on the few nanometre to tens of nanometre scale. Therefore it was proposed that patterning the superconducting film into a device with comparable dimensions to the size of a hotspot would result in an infrared sensitive SPD [240].

In 2001, Gregory Gol'tsman and coworkers [241] demonstrated the single-photon sensitivity ($\lambda = 790$ nm) of a current-biased NbN superconducting nanowire (200 nm wide, 5 nm thick, 1 μm long) maintained at 4.2 K. This device is known as the superconducting single-photon detector (SSPD) or superconducting nanowire single-photon detector (SNSPD).

16.14.2 SNSPD Device Physics

The basic device operation of the Superconducting Nanowire Single-Photon Detector (SSPD/SNSPD) can be described as follows [241–243]: the NbN nanowire is maintained well below its superconducting critical temperature Tc and direct current biased just below its critical current (Fig. 16.21a(i)). Individual infrared photons have enough energy to disrupt hundreds of Cooper pairs in a superconductor thereby forming a hotspot (Fig. 16.21a(ii)). The hotspot itself is not large enough to span the width of the ~100 nm nanowire. The hotspot region forces the supercurrent to flow around the resistive region (Fig. 16.21a(iii)). The local current density in the sidewalks increases beyond the critical current density and forms a resistive barrier across the width of the nanowire. The sudden increase in resistance from zero to a finite value generates a measurable output voltage pulse across the nanowire (Fig. 16.21a(iv)).

Growth of the resistive region due to Joule heating accounts for the high resistance of the nanowire (Fig. 16.21a(v)) [243]. Equation 16.10 can be modified to include Joule heating as follows:

$$Cd\frac{\partial T}{\partial t} = J^2 d\rho + \kappa d\nabla^2 T + \alpha(T_0 - T) \tag{16.11}$$

where J is the current density through the wire and ρ is the electrical resistivity. The hotspot cools down by coupling the energy of the excited electrons to the phonons via electron-phonon scattering, with a time constant $\tau_{\text{e-p}}$ (~10 ps). The phonon-phonon scattering then couples the energy into the substrate with a time constant $\tau_{\text{p-sub}}$. A fraction of the energy is backscattered into the electron system owing to mechanisms such as the lattice mismatch between the superconducting nanowire and the substrate. The substrate acts as a heat sink at temperature T_0 [244]. Then the nanowire recovers its superconducting state (Fig. 16.21a(vi)). The energy of the incident photon is negligible compared to the energy stored in the kinetic inductance; therefore energy or photon-number resolution is lost due to the Joule heating of the hotspot [242, 243].

A simple but elegant phenomenological model can be used to simulate the measurable output voltage pulse from an SNSPD biased at I_{bias} as shown in Fig. 16.21b [245, 246]. The simulated voltage pulse is shown in Fig. 16.21c. As the current flowing through the nanowire decays to a return current (I_r), the electrothermal feedback reduces the Joule heating of the resistive region by the bias current in the nanowire. The current recovers from I_r to I_{bias} with a longer time constant due to the reduction of total resistance in the circuit (dashed red curve in

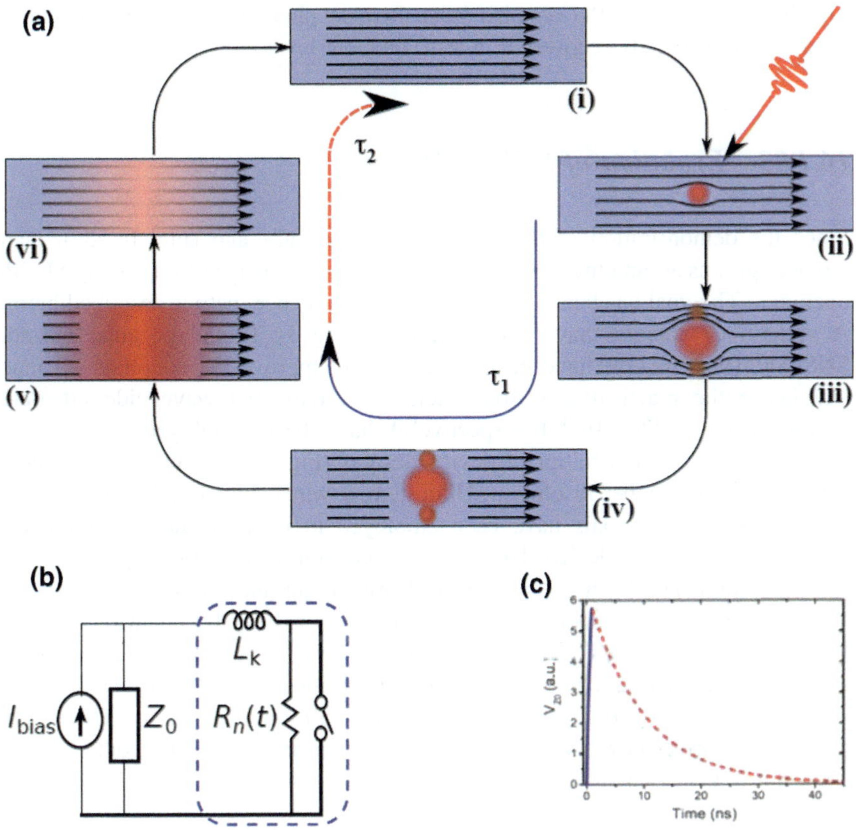

Fig. 16.21 The basic operation principle of the superconducting nanowire single-photon detector (SNSPD) (after Gol'tsman et al. [241], Semenov et al. [242] and Yang et al. [243]): **a** A schematic illustrating the detection cycle. (i) The superconducting nanowire maintained well below the critical temperature is direct current (DC) biased just below the critical current. (ii) When a photon is absorbed by the nanowire creating a small resistive hotspot. (iii) The supercurrent is forced to flow along the periphery of the hotspot. Since the NbN nanowires are narrow, the local current density around the hotspot increases, exceeding the superconducting critical current density. (iv) This in turn leads to the formation of a resistive barrier across the width of the nanowire [241]. (v) Joule heating (via the DC bias) aids the growth of resistive region along the axis of the nanowire [243] until the current flow is blocked and the bias current is shunted by the external circuit. (vi) This allows the resistive region to subside and the wire becomes fully superconducting again. The bias current through the nanowire returns to the original value (i). **b** A simple electrical equivalent circuit of a SNSPD. L_k is the kinetic inductance of the superconducting nanowire and *R*n is the hotspot resistance of the SNSPD. The SNSPD is current biased at I_{bias}. Opening and closing the switch simulates the absorption of a photon. An output pulse is measured across the load resistor Z_0 [245, 246]. **c** A simulation of the output voltage pulse of the SNSPD (approximating the pulse shape typically observed on an oscilloscope after amplification). Values of $L_k = 500$ nH and $R_n = 500\ \Omega$ have been used for this simulation (for simplicity the R_n is assumed fixed, although a more detailed treatment [243] shows $R_n(t)$). The solid blue line is the leading edge of the SNSPD output pulse, whilst the dotted red line is the trailing edge of the output pulse. The time constants relate to the phases of the detection cycle in (**a**)

Fig. 16.21c). The SNSPD remains insensitive to photons while the supercurrent decays and eventually recovers to a re-triggerable level.

16.14.3 Evolution of SNSPD Devices

Since the demonstration of the first SNSPD a decade ago (Fig. 16.22a), many research groups around the world have taken up the challenge of improving SNSPD operation. The goal has been to improve the performance metrics described before. A variety of strategies have been employed (Fig. 16.22). Large area 'meander SNSPDs' (Fig. 16.22b) have been developed to improve the coupling efficiency and hence the practical detection efficiency. Cavity and waveguide integrated designs (Fig. 16.22c, 16.22f respectively) have been employed to boost the absorption efficiency. Multipixel device designs (Fig. 16.22d) have been used to achieve photon number resolution. Ultra narrow wires (Fig. 16.22e) and alternate superconducting materials have been employed to improve the registering efficiency. As new device designs have been evaluated, important limiting factors such as constrictions and latching have been identified and understood.

Focusing an optical spot at λ = 1550 nm on to a 150–200 nm wide nanowire [247] (Fig. 16.22a) with η_{coupling} approaching unity is not feasible. In order to enhance ηcoupling, a nanowire meander is written typically across a 10 μm 10 μm [248, 249] (Fig. 16.22b) or 20 μm × 20 μm area [250]. These NbN meander devices show system detection efficiencies (η_{sde}) of about 2–3% at 1550 nm, 1 kHz DCR at T ~3 K (Fig. 16.23a).

16.14.4 Noise Mechanisms in SNSPDs: Dark Counts and Timing Jitter

As illustrated in Fig. 16.23a, in general for SNSPDs ηsde decreases as the energy of the photon decreases (i.e. at longer wavelength)—pointing to a decrease in ηregistering [248, 251]. Figure 16.23c shows typical timing jitter (60 ps FWHM) for a large area SNSPD measured with TCSPC electronics and a femtosecond laser. The signal-to-noise ratio, which can be achieved with an SNSPD, is governed not just by the practical efficiency (η_{sde}) and the dark count rate (DCR) but also timing jitter (Δt) [231]. The DCR of the SNSPD is a crucial factor in the device performance. Empirically, the DCR rate rises exponentially as I_{bias} approaches I_c (Fig. 16.23a) [250]. It is clear from these data (from a device with a large area meander with no cavity) at the shorter illumination wavelength (λ = 830 nm) relatively high efficiency can be achieved at negligible dark count rates. The origin

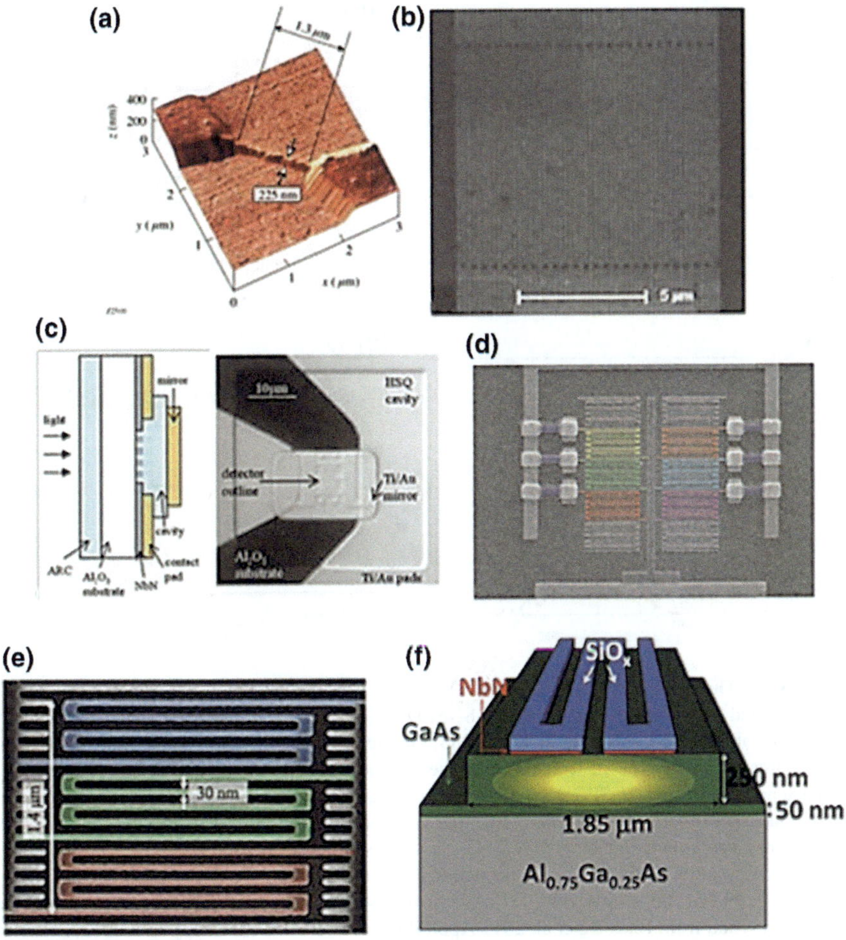

Fig. 16.22 The evolution of SNSPD design. **a** An atomic force microscopy (AFM) image of an early 1.3 µm × 225 nm NbN nanowire (after [247]), **b** A scanning electron micrograph (SEM) of NbN meander SNSPD covering a 10 µm × 10 µm area (of the type first presented in [248]). **c** A 3 µm × 3 µm NbN meander SNSPD embedded in an optical cavity, designed for optical illumination via the substrate [264]. **d** Multiple nanowire elements are biased in parallel via independent resistors resulting in a photon-number-resolving SNSPD (PNRSNSPD) [281]. **e** Ultrathin nanowires (30 nm wide) are connected in parallel to improve the sensitivity (registering efficiency) of the SNSPD—this device is known as a superconducting nanowire avalanche photodetector (SNAP) [282]. **f** A SNSPD fabricated on an optical waveguide structure to improve the optical coupling efficiency (after [283]). Figure **a** reproduced with permission from IEEE [247]. **b** Reproduced with permission from Nature Publishing Group [284]. **c** Reproduced with permission from The Optical Society of America [264]. **d** Reproduced with permission from Nature Publishing Group [281]. **e** Reprinted with permission from Marsili F., et al. 2011 Nano Lett. **11** 2048. Copyright 2011 American Chemical Society [282]. **f** Reprinted with permission from Sprengers J. P., et al. 2011 Appl. Phys. Lett. **99** 181110. Copyright 2011, American Institute of Physics [283]

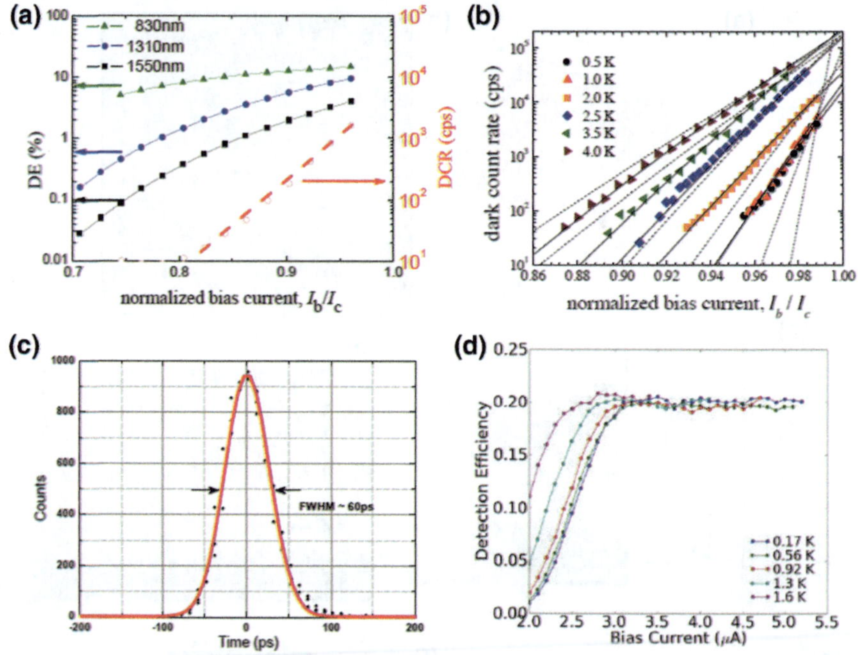

Fig. 16.23 The characteristics of SNSPDs **a** the system detection efficiency (ηsde) of a fibre-coupled 20 μm × 20 μm area NbN SNSPD at ~3 K for λ = 830, 1310 and 1550 nm. This device is of the type reported in [250, 285], **b** the variation in dark count rate (DCR) with temperature [255], **c** The instrument response function of a NbTiN SNSPD measured using a femtosecond laser and TCSPC card at λ = 1550 nm. The Δt of 60 ps FWHM (determined via a fitted Gaussian function) is quoted as the jitter of the device in [286], **d** A SNSPD fabricated in a low energy gap superconductor, a-W_xSi_{1-x}. A clear saturation of ηsde (λ = 1550 nm) occurs at higher bias currents [256]. Figure **b** reprinted with permission from Yamashita T., et al. 2011 Appl. Phys. Lett. **99** 161105. Copyright 2011, American Institute of Physics [255]. Figure **d** reproduced with permission from Baek B., et al. 2011 Appl. Phys. Lett. **98** 251105. Copyright 2011, American Institute of Physics [256]

of the exponential behaviour of dark count rate with bias is poorly understood. Theoretical explanations centre on the possible transit of flux vortices across the current-carrying superconducting nanowire. Several mechanisms have been put forward: quantum phase-slips, single-vortex crossings and vortex-antivortex pair nucleation [252].

Experimental tests have been far from conclusive [253, 254], owing to the difficulty of differentiating these fundamental mechanisms from prosaic environmental noise owing to the bias and readout configuration. Recent studies of dark counts over a wide temperature range (0.5–4 K) provide valuable fresh data [255] (see Fig. 16.23b). The registering probability saturation is also observed when smaller superconducting gap energy materials such as a-W_xSi_{1-x} [256] (Fig. 16.23d) are used for fabrication of SNSPDs. These devices exhibit η_{dde}

saturation (and near unity $\eta_{registering}$) at λ = 1550 nm even when the nanowires are 150 nm wide with 4.5 nm film thickness. This behaviour is most likely a consequence of the smaller size of superconducting energy gap in this material leading to an increased hotspot size for a given photon energy.

Furthermore, new theoretical studies also indicate that the device geometry (in particular, tight bends in the meander structure) may play a crucial role in triggering dark counts [257].

The very low timing jitter Δt of SNSPDs makes these detectors very attractive for TCSPC applications. When the arrival time of the photon is known extremely well (using modern ultrafast optical sources) the main timing uncertainty arises from the jitter of the SPD. The lowest timing jitter reported is 18 ps [258], but this result has not been widely reproduced and the dimensions of the device used and photon flux were unclear. 29 ps FWHM has been demonstrated in small area (4 μm × 4.2 μm) single-and multi-pixel SNSPDs [259]. Larger area SNSPDs typically give larger timing jitter when measured with femtosecond laser sources and state-of-the-art TCSPC electronics: for example 10 μm × 10 μm meanders of the type first reported by Verevkin et al. [248] give 68 ps FWHM [260] and 20 μm × 20 μm meander devices [250] give 60 ps FWHM (Fig. 16.23c). These results indicate that a highly uniform nanowire (most easily achieved with a short wire length) will give lower timing jitter. Higher critical currents have been observed to improve jitter in uniform devices [261] allowing 40 ps FWHM to be demonstrated in 15 μm × 15 μm meander devices [262]. Recently studies using nanooptical techniques to study the local efficiency and timing properties of SNSPDs [263] indicate that different parts of a non-uniform nanowire give different hotspot resistances yielding different pulse rise times, broadening the overall timing jitter.

16.14.5 Cooling, Optical Coupling and Device Readout

The most common method for cooling a SNSPD is to immerse it in liquid helium (He) [241] or to mount the device in a cryogenic probe station [264] and reach 4.2 K. Pumping on the He bath [265, 266] can reduce the temperature further (see Fig. 16.24a(i)). Liquid He is expensive, hazardous and demands trained personnel for correct use. This technique is satisfactory for testing superconducting devices in a low temperature physics laboratory; however if the ultimate goal is to provide a working device for users in other scientific fields or in industrial applications, alternative cooling methods must be sought. Operating SNSPDs in a closed-cycle refrigerator [267, 268] offers a solution to this problem. The circulating fluid is high pressure, high purity He gas which is enclosed inside the refrigerator allowing continuous operation and eliminating repeated cryogenic handling. A recent version of a SNSPD system operating on this principle is shown in Fig. 16.24a(ii).

The most obvious way to couple photons to a detector inside a cryostat is via an optical window [241]. This technique has the following disadvantages: (i) it is

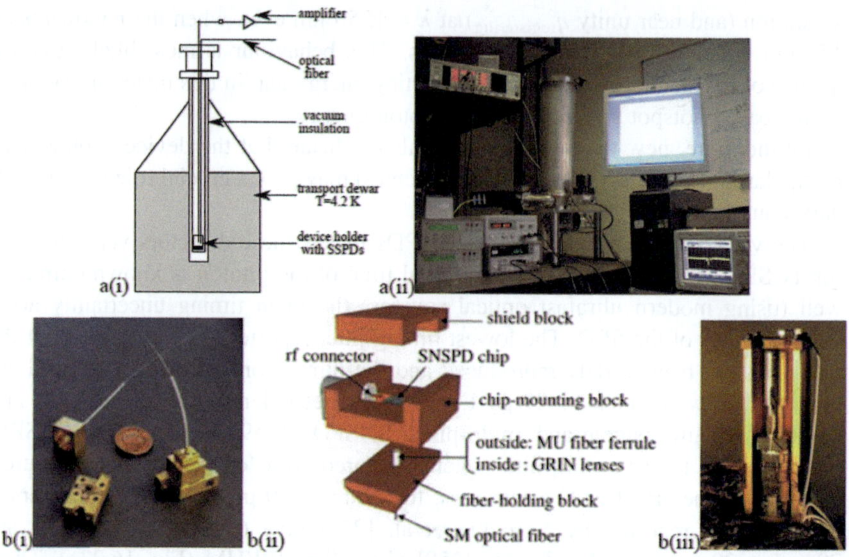

Fig. 16.24 **a** Cooling techniques for SNSPDs (i) Liquid Helium dewar [266] (ii) Gifford-McMahon (GM) closed-cycle refrigerator. **b** Optical-coupling techniques (i) fibre-coupling (as used in [285, 286]) (ii) GRIN lens fibre-coupling [269] and (iii) confocal microscopy setup in a GM cryocooler (as used in [263]). Figure **a** (i) reproduced with permission from IEEE [266], **b** (ii) reproduced with permission from The Optical Society of America [269]

difficult to focus the optical spot on the 10 μm × 10 μm or 20 μm × 20 μm nanowire meander area, (ii) vibrations in the refrigerator and ambient temperature fluctuations outside the cryostat result in poor alignment stability, (iii) operating multiple SNSPDs in parallel will require separate alignment optics, this will eventually limit the number of simultaneous operating detectors and (iv) the DCR will be higher due to the increase in blackbody radiation and stray light coupling to the device. Delivering the photons through a standard single mode optical fibre [250, 265, 266, 268] (package shown in Fig. 16.24b(i)) or lensed fibre [269, 270] (package shown in Fig. 16.24b(ii)) overcomes these issues, and removes the need for an optical window in the cryostat. The fibre coupled configuration makes the operating environment more favourable for low noise SNSPD operation at low DCR, due to reduction in blackbody radiation and stray light reaching the detector. However, this technique can potentially suffer from optical misalignment, since the fibre is typically aligned to the active area of the SNSPD at room temperature and subsequently cooled down to cryogenic temperatures. New 'self-alignment' schemes [271] cleverly constrain the position of the device relative to the fibre and can be applied to SNSPDs on silicon substrates [272]. Cryogenic piezoelectric nano-positioners have been employed to provide precision in situ optical alignment at low temperatures. This method has helped to improve the system detection efficiency for smaller active area devices [270], but the high cost of the nanopositioners is a barrier to widespread adoption. Cryogenic nano-positioners are also of great benefit in carrying out nanooptical studies of local device properties. Local sections within the SNSPD have been examined with sub-micrometre resolution with the help of a confocal microscopy setup (shown in Fig. 16.24b(iii)), which can be implemented either in a liquid-He cryostat [273] or a closed-cycle refrigerator [263].

16.15 Superconducting Nanowire Photodetector Arrays

Past successful NASA space exploration missions have relied on radio-communications links for interplanetary spacecraft as well as for Earth-orbiting satellites. However, this modality will not support future missions whose instruments demand much larger data rates. Optical networks represent the future of deep-space communications and will make it possible to collect ambitious data products from more distant destinations. Highly sensitive single photon detectors are a key enabler for this next generation of optical communication technologies.

Advanced photon detectors are also enabling new optical quantum information applications, which may revolutionalize communication and computation. One such application is quantum key distribution (QKD), which takes advantage of the information properties of quantum systems to provide secure communication that cannot be intercepted without being corrupted.

These advancements in communication technologies would not be possible without significant improvements to the detection efficiency, signal-to-noise ratio, spectral range, and resolving power of single-photon detectors. Understanding the basic physical limitations on individual detector speed led to a collaboration

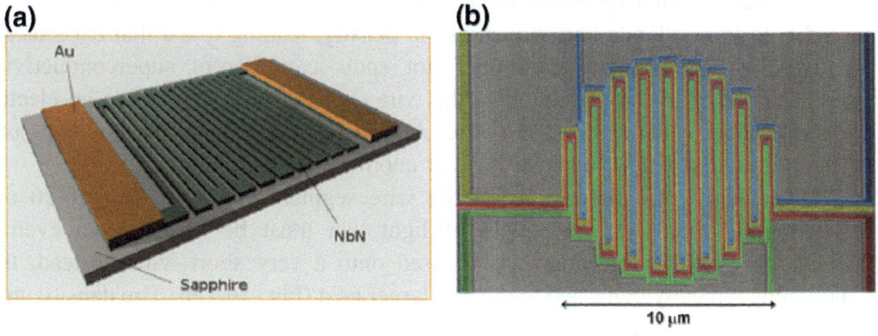

Fig. 16.25 **a** A schematic of a single SNPD. The nanowire is typically 4 nm to 6 nm thick and 50 nm to 100 nm in width. **b** A false-color scanning electron micrograph of a four-element SNPD array

between MIT Lincoln Laboratory and MIT Research Laboratory for Electronics to build a superconducting nanowire photodetector (SNPD, Fig. 16.25a). These detectors enable broadband single-photon detection with high efficiency and low noise at record-high rates exceeding one billion photons per second. An interleaved, four-element detector (Fig. 16.25b) provides a new capability for ultrafast, ultra-sensitive laser communications, as well as other applications and scientific measurements. The SNPD array utilizes a spatially interleaved array of up to eight serpentine detectors and occupies a region 7–20 μm in diameter.

An SNPD array can take advantage of a spatially multiplexed approach that allows detection of photons at a higher rate than a single SNPD would support without requiring complex optical coupling techniques typical of other photon-counting arrays. When illuminated by a single photon, only one of the SNPDs loses its superconductive properties. The remaining SNPDs remain available to detect the next photon, while the first SNPD recovers its superconductive properties.

16.15.1 Theory of Operation

Each SNPD consists of a superconducting film ~ 5 nm (nm) thick that has been lithographically patterned into a wire whose width is less than 100 nm. The wire is cryogenically cooled below 4 K, which is cool enough to ensure superconducting properties. When a photon is absorbed by the wire (Fig. 16.26a), a small, localized region known as a *hot spot* is created. Within such a hot spot, the temperature of the electrons is sufficiently elevated to disrupt the superconductivity and return the material to its normal, highly resistive state, causing the supercurrent to divert its path (Fig. 16.26b). Although this region is very small (e.g., on the order of 10 nm in diameter for a near-infrared photon [274]), its presence can be detected electrically if the wire is sufficiently narrow (e.g., less than 100 nm in diameter) and the initial DC current density J is close enough to the critical value JC, above which a material's superconductivity breaks down. The occurrence of the initial hot spot can result in additional hot spots because the charge carriers moving around the spot are forced both to accelerate and to increase in density, leading to a J that exceeds JC locally (Fig. 16.26c). The additional hot spots can disrupt superconductivity throughout an entire cross section of the wire (Fig. 16.26d), producing an electrically detectable voltage across the wire. After a relaxation time, the hot spot subsides and the detector is ready to register another photon.

Unfortunately, the dimensions of the wire segment shown in Fig. 16.26 are much smaller than the wavelength of light that must be detected, so even a diffraction-limited beam cannot be focused onto a very short wire. Instead, the nanowire is typically patterned to cover a larger area (Fig. 16.25b) The density with which wires are packed in this structure affects the absorptance of the detector, with a more tightly packed structure increasing the absorptance. Consequently, a longer nanowire increases the probability of a detector absorbing photons, thus producing a higher-efficiency communication channel.

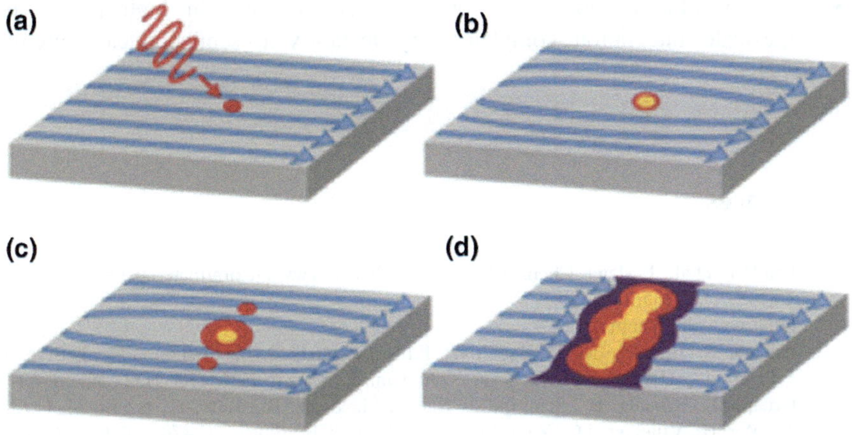

Fig. 16.26 **a** An incident photon heats a nanowire, forming a localized hot spot **b** where the superconductivity is disrupted. The supercurrent diverts around this spot. **c** When the local current density on either side of the hot spot exceeds the critical current density, a resistive region spanning the entire cross section of the nanowire is formed (**d**)

Although a tradeoff clearly exists in choosing nanowire length for a single detector (speed and yield improve for shorter nanowires while absorptance improves for longer nanowires), the SNPD array simultaneously achieves many of the advantages of short and long nanowires. One realization of this invention (Fig. 16.25b) interleaves four independent nanowires to form a single optical active area. In this case, each nanowire is relatively short, compared to the four-times-longer wire that would be needed if only a single nanowire had been used. Therefore, each of the four shorter nanowires is faster and can be produced with high yield, while the combination of elements covers a large area with tightly packed wires. Although two- and four element SNPD arrays have been investigated in the most detail, other arrangements can be highly customized for specific applications by implementing simple changes to the lithography pattern, without changing the fabrication processes themselves.

16.15.2 Example Applications of the SNPD Array Include

- a Moon-to-Earth laser-communications link that will be part of NASA's Moon-orbiting Lunar
- Atmosphere and Dust Environment Explorer satellite, scheduled for launch in 2012 [275];
- a single-photon-counting detector for QKD systems operating at rates 100 times faster than current rates (for a fixed fiber optical cable length) [276]; and

- a high-resolution, noncontact probe to detect flaws in high-speed, very large-scale integration complementary metal-oxide semiconductor circuitry [277].

References

1. Fan H.J., et al.: J. Mater. Chem. **19**, 826–827 (2009). www.rsc.org/materials, https://doi.org/10.1039/b900135m
2. http://www.intel.com
3. Anders, S, Blamire, M.G., Buchholz, F.I.M., Crétéd, D.G., Cristiano, R., Febvre, P., Fritzsch, L., Herr, A., Il'ichev, E., Kohlmann J., Kunert J., Meyer H.G., Niemeyer J., Ortlepp T., Rogalla, H., Schurig, T., Siegel, M., Stolz, R., Tarte, E., ter Brake, H.J.M., Toepfer, H., Villegier, J.C., Zagoskin, A.M., Zorin, A.B.: Physica C **470**, 2079–2126 (2010)
4. Cerofolini, G.F., et al.: Microelectron. Eng. **81**, 405–419 (2005)
5. Weste, N., Eshraghian, K.: Principles of CMOS VLSI Design: a Systems Perspectives. Addison-Wesley, Boston (1993)
6. Qi, P., Javey, A., Rolandi, M., Wang, Q., Yenilmez, E., Dai, H.: J. Am. Chem. Soc. **126**, 11774–11775 (2004)
7. Gimzewski, J.K.: Nanoelectronics. In: McGraw-Hill Yearbook of Science and Technology 2000. McGraw-Hill, New York, pp. 274–278 (1999)
8. Vuillaume, D.: Molecular nanoelectronics. Proc. IEEE **98**(12), 2111–2123 (2010)
9. Lyshevski, S.E.: Nano and Molecular Electronics Handbook. CRC, New York (2007)
10. Ghoniem, A.F.: Prog. Energy Combust. Sci. **37**, 15–51 (2011)
11. The International Energy Agency, World Energy Outlook, http://www.worldenergyoutlook.org/
12. Parida, B., Iniyan, S., Goic, R.: Renew. Sustain. Energy Rev. **15**, 1625–1636 (2011)
13. Grätzel, M., Phil. Trans. R. Soc. A **365**, 993–1005 (2007)
14. Nalwa, H.S.: Handbook of Organic Conductive Molecules and Polymers, Volume 3, Conductive Polymers: Spectroscopy and Physical Properties. Wiley (1997)
15. Liang, Y., Xu, Z., Xia, J., et al.: For the bright future-bulk heterojunction polymer solar cells with power conversion efficiency of 7.4%. Adv. Mater. **22**(20), E135–E138 (2010)
16. Singh, R.K., Kumar, J., Kumar, A., Kumar, V., Kant, R., Singh, R.: Poly(3-hexylthiophene): functionalized single-walled carbon nanotubes: (6,6)-phenyl-C61-butyric acid methyl ester composites for photovoltaic cell at ambient condition. Sol. Energy Mater. Sol. Cells **94**(12), 2386–2394 (2010)
17. Wu, M., Lin, Y., Chen, S., et al.: Enhancing light absorption and carrier transport of P3HT by doping multi-wall carbon nanotubes. Chem. Phys. Lett. **468**(1–3), 64–68 (2009)
18. Hsiao, Y., Fang, T., Ji, L., Lee, Y., Dai, B.: Size effect of nanodiamonds on P3HT:PCBM heterojunction solar cells. Electrochem. Commun. **18**(1), 4–7 (2012)
19. Zhang, W., Xu, Y., Wang, H., Xu, C., Yang, S.: Fe3O4 nanoparticles induced magnetic field effect on efficiency enhancement of P3HT:PCBM bulk heterojunction polymer solar cells. Sol. Energy Mater. Sol. Cells **95**(10), 2880–2885 (2011)
20. Bhat, S.V., Govindaraj, A., Rao, C.N.R.: Hybrid solar cell based on P3HT-ZnO nanoparticle blend in the inverted device configuration. Sol. Energy Mater. Sol. Cells **95**(8), 2318–2321 (2011)
21. Valiev, R.: Materials science: nanomaterial advantage. Nature **419**(6910), 887–889 (2002)
22. Feng, S.Q., Yu, D.P., Zhang, H.Z., Bai, Z.G., Ding, Y.: The growth mechanism of silicon nanowires and their quantum confinement effect. J. Cryst. Growth **209**(2–3), 513–517 (2000)

23. Varonides, A.C.: Quantum size effects in amorphous Si superlattice solar cells. Renew. Energy **33**(2), 273–276 (2008)
24. Yu, G., Gao, J., Hummelen, J.C., Wudl, F., Heeger, A.J.: Polymer photovoltaic cells: enhanced efficiencies via a network of internal donor–acceptor heterojunctions. Science **270**, 1789–1795 (1995)
25. Logothetidis, S.: Mater. Sci. Eng., B **152**, 96–104 (2008)
26. Hamers, R.J.: Flexible electronic futures. Nature **412**, 489 (2001)
27. Padinger, F., Rittberger, R.S., Sariciftci, N.S.: Adv. Funct. Mater. **13**(2) (2003)
28. Forrest, S.R.: The path to ubiquitous and low cost organic electronic appliances on plastic. Nature **428**, 911 (2004)
29. Shaheen, S.E., Ginley D.S.: Photovoltaics for the Next Generation: Organic-Based Solar Cells, Dekker Encyclopedia of Nanoscience and Nanotechnology (Schwarz, Contescu, and Putyera, eds.), pp. 2879–2895. Marcel Dekker, Inc., New York (2004)
30. Contescu, C., Putyera, K.: Dekker Encyclopedia of Nanoscience and Nanotechnology, 2nd edn. CRC, New York (2009)
31. Nicholson, P.G., Castro, F.A.: Nanotechnology **21**, 492001 (2010)
32. Lee, C.M., Chang, S.P., Chang, S.J., Wu, C.I.: High efficiency Si solar cell fabricated by ion implantation and inline backside rounding process. Int. J. Photoenergy, 7 pp. (2012). Article ID 670981
33. Lee, C.M., Chang, S.P., Chang, S.J., Wu, C.I.: p-type quasimono silicon solar cell fabricated by ion implantation. Int. J. Photoenergy, 8 pp. (2013). Article ID 171390
34. Rech, B., Repmann, T., van den Donker, M.N., et al.: Challenges in microcrystalline silicon based solar cell technology. Thin Solid Films **511–512**, 548–555 (2006)
35. Müller, J., Rech, B., Springer, J., Vanecek, M.: TCO and light trapping in silicon thin film solar cells. Sol. Energy **77**(6), 917–930 (2004)
36. Yamamoto, K., Nakajima, A., Yoshimi, M., et al.: A high efficiency thin film silicon solar cell and module. Sol. Energy **77**(6), 939–949 (2004)
37. Matsui, T., Tsukiji, M., Saika, H., Toyama, T., Okamoto, H.: Influence of substrate texture on microstructure and photovoltaic performances of thin film polycrystalline silicon solar cells. J. Non-Cryst. Solids **299–302**(2), 1152–1156 (2002)
38. Söderström, T., Haug, F.-J., Terrazzoni-Daudrix, V., Ballif, C.: Optimization of amorphous silicon thin film solar cells for flexible photovoltaics. J. Appl. Phys. **103**(11) (2008). Article ID 114509
39. Berginski, M., Hüpkes, J., Schulte, M., et al.: The effect of front ZnO:Al surface texture and optical transparency on efficient light trapping in silicon thin-film solar cells. J. Appl. Phys. **101**(7) (2007). Article ID 074903
40. Müller, J., Schöpe, G., Kluth, O., et al.: Upscaling of textureetched zinc oxide substrates for silicon thin film solar cells. Thin Solid Films **392**(2), 327–333 (2001)
41. Lin, Y.C., Jian, Y.C., Jiang, J.H.: A study on the wet etching behavior of AZO (ZnO:Al) transparent conducting film. Appl. Surf. Sci. **254**(9), 2671–2677 (2008)
42. Kluth, O., Rech, B., Houben, L., et al.: Texture etched ZnO: Al coated glass substrates for silicon based thin film solar cells. Thin Solid Films **351**(1–2), 247–253 (1999)
43. Wang, Y., Zhang, X., Bai, L., Huang, Q., Wei, C., Zhao, Y.: Effective light trapping in thin film silicon solar cells from textured Al doped ZnO substrates with broad surface feature distributions. Appl. Phys. Lett. **100**(26) (2012). Article ID 263508
44. Springer, J., Rech, B., Reetz, W., Müller, J., Vanecek, M.: Light trapping and optical losses in microcrystalline silicon pin solar cells deposited on surface-textured glass/ZnO substrates. Sol. Energy Mater. Sol. Cells **85**(1), 1–11 (2005)
45. Zhang, W.-J.: A review of the electrochemical performance of alloy anodes for lithium-ion batteries. J. Power Sources **196**(1), 13–24 (2011)
46. Inoue, H., Ueda, T., Nohara, S., Fujita, N., Iwakura, C.: Effect of ball-milling on electrochemical and physicochemical characteristics of crystalline Mg2Ni alloy. Electrochim. Acta **43**(14–15), 2215–2219 (1998)

47. Varin, R.A., Czujko, T., Mizera, J.: Microstructural evolution during controlled ball milling of (Mg2Ni + MgNi2) intermetallic alloy. J. Alloy. Compd. **350**(1–2), 332–339 (2003)
48. Zhang, S.G., Hara, Y., Morikawa, T., Inoue, H., Iwakura, C.: Electrochemical and structural characteristics of amorphous MgNix ($x >= 1$) alloys prepared by mechanical alloying. J. Alloy. Compd. **293**, 552–555 (1999)
49. Liu, G., Xi, S., Ran, G., Zuo, K., Li, P., Zhou, J.: A new phase $MgNi_3$ synthesized by mechanical alloying. J. Alloy. Compd. **448**(1–2), 206–209 (2008)
50. Anik, M., Özdemir, G., Küükdeveci, N., Baksan, B.: Effect of Al, B, Ti and Zr additive elements on the electrochemical hydrogen storage performance of MgNi alloy. Int. J. Hydrog. Energy **36**(2), 1568–1577 (2011)
51. Beck, A., Bednorz, J.G., Gerber, C., Rossel, C., Widmer, D.: Reproducible switching effect in thin oxide films for memory applications. Appl. Phys. Lett. **77**(1), 139–141 (2000)
52. Sawa, A.: Resistive switching in transition metal oxides. Mater. Today **11**(6), 28–36 (2008)
53. Lai, C.-H., Tseng, T.-Y.: Preparation and properties of perovskite thin films for resistive nonvolatile memory applications. Ferroelectrics **357**(1), 17–27 (2007)
54. Goux, L., Opsomer, K., Degraeve, R., et al.: Influence of the Cu-Te composition and microstructure on the resistive switching of Cu-Te/Al_2O_3/Si cells. Appl. Phys. Lett. **99**(5) (2011). Article ID 053502
55. Zhou, P., Yin, M., Wan, H.J., Lu, H.B., Tang, T.A., Lin, Y.Y.: Role of TaON interface for Cu*x*O resistive switching memory based on a combined model. Appl. Phys. Lett. **94**(5) (2009). Article ID 053510
56. Lai, C.H., Chen, C.H., Tseng, T.Y.: Resistive switching behavior of sol-gel deposited TiO_2 thin films under different heating ambience. Surf. Coat. Technol. **231**, 399–402 (2013)
57. Cagli, C., Nardi, F., Ielmini, D.: Modeling of set/reset operations in NiO-based resistive-switching memory devices. IEEE Trans. Electron Devices **56**(8), 1712–1720 (2009)
58. Liu, C.-Y., Huang, Y.-H., Ho, J.-Y., Huang, C.-C.: Retention mechanism of Cu-doped SiO_2 based resistive memory. J. Phys. D **44**(20) (2011). Article ID 205103
59. Valov, I., Waser, R., Jameson, J.R., Kozicki, M.N.: Electrochemical metallization memories: fundamentals, applications, prospects. Nanotechnology **22**(25) (2011). Article ID 254003
60. Yu, S., Wong, H.-S.P.: Modeling the switching dynamics of Programmable-Metallization-Cell (PMC) memory and its application as synapse device for a neuromorphic computation system. In: Proceedings of the IEEE International Electron Devices Meeting (IEDM '10), pp. 520–523, Dec 2010
61. Waser, R., Dittmann, R., Staikov, C., Szot, K.: Redox-based resistive switching memories nanoionic mechanisms, prospects, and challenges. Adv. Mater. **21**(25–26), 2632–2663 (2009)
62. Yang, P.K., Chang, W.Y., Teng, P.Y., Jeng, S.F.: Fully transparent resistive memory employing graphene electrodes for eliminating undesired surface effects. Proc. IEEE **101**(7), 1732–1739 (2013)
63. Ke, J.J., Liu, Z.J., Kang, C.F., Lin, S.J.: Surface effect on resistive switching behaviors of ZnO. Apply Phys. Lett. **99**, 3 pp. (2011). Article ID 192106
64. Huang, T.H., Yang, P.K., Chang, W.Y., et al.: Eliminating surface effects via employing nitrogen doping to significantly improve the stability and reliability of ZnO resistive memory. J. Mater. Chem. C **1**(45), 7593–7597 (2013)
65. Tsai, Y.-T., Chang, T.-C., Lin, C.-C., et al.: Influence of nanocrystals on resistive switching characteristic in binary metal oxides memory devices. Electrochem. Solid-State Lett. **14**(3), H135–H138 (2011)
66. Sun, Q.-Q., Gu, J.-J., Chen, L., et al.: Controllable filament with electric field engineering for resistive switching uniformity. IEEE Electron Device Lett. **32**(9), 1167–1169 (2011)
67. Jung, R., Lee, M.-J., Seo, S., et al.: Decrease in switching voltage fluctuation of PtNi O*x* Pt structure by process control. Appl. Phys. Lett. **91**(2) (2007). Article ID022112
68. Vallée, C., Gonon, P., Mannequin, C., et al.: Plasma treatment of HfO_2-based metal insulator metal resistive memories. J. Vac. Sci. Technol. A **29**(4) (2011). Article ID 041512

69. Wang, J.C., Ye, Y.R., Lai, C.S., et al.: Characterization of gadolinium oxide thin films with CF_4 plasma treatment for resistive switching memory applications. Appl. Surf. Sci. **276**, 497–501 (2013)
70. Nemirovsky, Y., Brouk, I., Jakobson, C.G.: 1/f noise in CMOS transistors for analog applications. IEEE Trans. Electron Devices **48**(5), 921–927 (2001)
71. Yasuda, Y., Hu, C.: Effect of fluorine incorporation on 1/f noise of HfSiON FETs for future mixed-signal CMOS. In: Proceedings of the International Electron Devices Meeting (IEDM '06), pp. 1–4, 2006
72. Hung, K.K., Ko, P.K., Hu, C., Cheng, Y.C.: Unified model for the flicker noise in metal-oxide-semiconductor field-effect transistors. IEEE Trans. Electron Devices **37**(3), 654–665 (1990)
73. Vandamme, L.K.J., Li, X., Rigaud, D.: 1/f noise in MOS devices, mobility or number fluctuations? IEEE Trans. Electron Devices **41**(11), 1936–1945 (1994)
74. Ghibaudo, G., Boutchacha, T.: Electrical noise and RTS fluctuations in advanced CMOS devices. Microelectron. Reliab. **42**(4–5), 573–582 (2002)
75. Vandamme, L.K.J., Hooge, F.N.: What do we certainly know about 1/f noise inMOSTs? IEEE Trans. Electron Devices **55**(11), 3070–3085 (2008)
76. Hu, C.: Device challenges and opportunities. Dig. Tech. Pap. Symp. VLSI Technology, 4 (2004)
77. Wang, T.-J., Ko, C.-H., Lin, H.-N., et al.: Investigation of metallized source/drain extension for high-performance strained NMOSFETs. IEEE Trans. Electron Devices **55**(11), 3221–3226 (2008)
78. Thompson, S.E., Sun, G., Choi, Y.S., Nishida, T.: Uniaxial process-induced Strained-Si: extending the CMOS roadmap. IEEE Trans. Electron Devices **53**(5), 1010–1020 (2006)
79. Simoen, E., Verheyen, P., Shickova, A., Loo, R., Claeys, C.: On the low-frequency noise of pMOSFETs with embedded SiGe source/drain and fully silicided metal gate. IEEE Electron Device Lett. **28**(11), 987–989 (2007)
80. Kuo, C.W., Wu, S.L., Chang, S.J., Lin, H.Y., Wang, Y.P., Hung, S.C.: Investigation of interface characteristics in strained-Si nMOSFETs. Solid-State Electron. **53**(8), 897–900 (2009)
81. Wang, B.C., Wu, S.L., Huang, C.W., et al.: Correlation between random telegraph noise and 1/f noise parameters in 28-nm pMOSFETs with tip-shaped SiGe source/drain. IEEE Electron Device Lett. **33**(7), 928–930 (2012)
82. Jung, H.-S., Lee, J.-H., Han, S.K., et al.: A highly manufacturable MIPS (Metal Inserted Poly-Si Stack) technology with novel threshold voltage control. In: Proceedings of the Symposium on VLSI Technology, pp. 232–233, June 2005
83. Wang, B.C., Wu, S.L., Huang, C.W., et al.: Characterization of oxide tarps in 28 nm p-type metal-oxide-semiconductor field effect transistors with tip-shaped SiGe source/drain based on random telegraph noise. Jpn. J. Appl. Phys. **51** (2012). Article ID 02BC11
84. Ray, U: The Environmental Impact of Nanotechnology, 4 Jan 2019. https://www.azonano.com/article.aspx?ArticleID=5114
85. Yunus, I.S., Harwin, Kurniawan, A., Adityawarman, D., Indarto, A.: Nanotechnologies in water and air pollution treatment. Environ. Technol. Rev. **1**(1), 136148 (2012). https://doi.org/10.1080/21622515.2012.733966
86. Rani, K., Sridevi, V.: An overview on role of nanotechnology in green and clean technology. Austin Environ Sci. **2**(3), 1026 (2017)
87. https://www.nano.gov/you/nanotechnologybenefits
88. https://www.asme.org/engineeringtopics/articles/technologyandsociety/10waysnanotechnologyimpactslives
89. http://www.academia.edu/7099463/impact_of_nanotechnology_on_environment
90. Smith, B.: AI and nanotechnology—how do they work together? 4 Jan 2019. https://www.azonano.com/article.aspx?ArticleID=5116
91. Ebbesen, T.W., Ajayan, P.M.: Nature (London) **358**, 220 (1992)

92. Heath, J.R., O'Brien, S.C., Zhang, Q., Liu, Y., Curl, R.F., Kroto, H.W., Tittel, F.K., Smalley, R.E.: J. Am. Chem. Soc. **107**, 7779 (1985)
93. O'Brien, S.C., Heath, J.R., Curl, R.F., Smalley, R.E.: J. Chem. Phys. **88**, 220 (1988)
94. Ajayan, P.M., Iijima, S.: Nature (London) **361**, 333 (1993)
95. Ajayan, P.M., Ebbesen, T.W., Ichihashi, T., Iijima, S., Tanigaki, K.: Nature (London) **362**, 522 (1993)
96. Dujardin, E., Ebbesen, T.W., Hiura, H., Tanigaki, K.: Science **265**, 1850 (1994)
97. Subramoney, S., Kavelaar, V., Ruoff, R.S., Lorents, D.C., Malhotra, R., Kazmer, A.J.: Chemistry and Physics of Fullerenes and Related Materials (Kadish, K.M., Ruoff, R.S., eds.), PV 94–24, p. 1498. The Electrochemical Society, Pennington, NJ (1994)
98. Ruoff, R.S., Lorents, D.C., Chan, B.C., Malhotra, R., Subramoney, S.: Science **259**, 346 (1993)
99. Liu, M., Cowley, J.M.: Carbon **33**, 225 (1995)
100. Seraphin, S.: Chemistry and Physics of Fullerenes and Related Materials (Kadish, K.M., Ruoff, R.S., eds.), PV 94–24, p. 1433. The Electrochemical Society, Pennington, NJ (1994)
101. Subramoney, S., Ruoff, R.S., Lorents, D.C., Chan, B.C., Malhotra, R., Dyer, R.M.J., Parvin, K.: Carbon **32**, 507 (1994)
102. Yosida, Y.: Appl. Phys. Lett. **62**, 3447 (1993)
103. Loisseau, A., Pascard, H.: Chem. Phys. Lett. **256**, 246 (1996)
104. Guerret-Piécourt, C., Le Bouar, Y., Loisseau, A., Pascard, H.: Nature (London) **372**, 761 (1994)
105. Tomita, M., Saito, Y., Hayashi, T.: Jpn. J. Appl. Phys. **32**, L280 (1993)
106. Saito, Y., Yoshikawa, T., Okuda, M., Ohkohchi, M., Ando, Y., Kasuya, A., Nishina, Y.: Chem. Phys. Lett. **209**, 72 (1993)
107. Murakami, Y., Shibata, T., Okuyama, K., Arai, T., Suematsu, H., Yoshida, Y.: J. Phys. Chem. Solids **54**, 1861 (1994)
108. Majetich, S.A., Artman, J.O., McHenry, M.E., Nuhfer, N.T., Staley, S.W.: Phys. Rev. B **48**, 16845 (1993)
109. Seraphin, S., Zhou, D., Jiao, J., Withers, J.C., Loufty, R.: Appl. Phys. Lett. **63**, 2073 (1993)
110. Bandow, S., Saito, Y.: Jpn. J. Appl. Phys. **32**, L1677 (1994)
111. Ata, M., Yamaura, K., Hudson, A.J.: J. Adv. Mater. **7**, 286 (1995)
112. Tsang, S.C., Chen, Y.K., Harris, P.J.F., Green, M.L.H.: Nature (London) **372**, 159 (1994)
113. Lago, R.M., Tsang, S.C., Lu, K.L., Chen, Y.K., Green, M.L.H.: J. Chem. Soc., Chem. Commun. 1355 (1995)
114. Dai, H., Wong, E.W., Lu, Y.Z., Fan, S., Lieber, C.M.: Nature (London) **375**, 769 (1995)
115. Ajayan, P.M., Stephan, O., Redlich, Ph, Colliex, C.: Nature (London) **375**, 564 (1995)
116. Hsu, W.K., Terrones, M.: Unpublished results (1995)
117. McHenry, M.E., Nakamura, Y., Kirkpatrick, S., Johnson, F., Curtin, S., DeGraef, M., Uhfer, N.T., Majetich, S.A., Brunsman, E.M.: Chemistry and Physics of Fullerenes and Related Materials (Kadish, K.M., Ruoff, R.S., eds.), PV 95–10, p. 1463. The Electrochemical Society, Pennington, NJ (1995)
118. Hare, J.P., Hsu, W.K., Kroto, H.W., Lappas, A., Prassides, K., Terrones, M., Walton, D.R. M.: Chem. Mater. **8**, 6 (1996)
119. Thomas, L.: Carbon nanotube sensors—applications and advantages, 19 Dec (2018). https://www.azosensors.com/article.aspx?ArticleID=1536
120. http://www.understandingnano.com/sensor.html
121. https://3c1703fe8d.site.internapcdn.net/newman/gfx/news/hires/2013/withcarbonna.jpg
122. http://biomems.uwaterloo.ca/papers/JNN.pdf
123. https://www.nasa.gov/centers/ames/research/technologyonepagers/ gas_detection.html
124. https://www.hindawi.com/journals/js/2009/493904/
125. Critchley, L.: Carbon nanotube optics and their uses, 4 Jan 2019 https://www.azooptics.com/Article.aspx?ArticleID=1507
126. Aalborg University: http://homes.nano.aau.dk/tgp/nanotube%20research.pdf

127. Wang, F., et al.: The optical resonances in carbon nanotubes arise from excitons. Science (2005). https://doi.org/10.1126/science.1110265
128. Cola, B.A., et al.: A carbon nanotube optical rectenna. Nat. Nanotechnol. (2015). https://doi.org/10.1038/nnano.2015.220
129. Strano, M.S., et al.: Carbon nanotubes as optical biomedical sensors. Adv. Drug Deliv. Rev. (2013). https://doi.org/10.1016/j.addr.2013.07.015
130. Feliu, N., et al.: Carbon nanotubes as optical sensors in biomedicine. ACS Nano (2017). https://doi.org/10.1021/acsnano.7b06701
131. Voisin, C., et al.: Carbon nanotubes as emerging quantum light sources. Nat. Mater. (2018). https://doi.org/10.1038/s4156301801092
132. Peng, W., et al.: Fiberoptic anemometer based on single walled carbon nanotube coated tilted fiber Bragg grating. Opt. Express (2017). https://doi.org/10.1364/oe.25.024521
133. Martinez, A., Yamashita, S.: Carbon nanotube based photonic devices: applications in nonlinear optics. In: Marulanda, J.M. (eds) Carbon Nanotubes (2011). https://doi.org/10.5772/16527
134. Taylor-Smith, K.: Graphene as a renewable energy, 26 Nov 2018. https://www.azocleantech.com/article.aspx?ArticleID=814
135. Zheng, Q., Huang, F., Ding, K., et al.: MgZnO-based metal semiconductor-metal solar-blind photodetectors on ZnO substrates. Appl. Phys. Lett. **98**(22), 221112-1–221112-3 (2011). Article ID221112
136. Liao, M., Koide, Y.: High-performance metal semiconductor-metal deep-ultraviolet photodetectors based on homoepitaxial diamond thin film. App. Phys. Lett. **89**(11), 113509-1–113509-3 (2006). Article ID 113509
137. Sun, L., Chen, J., Li, J., Jiang, H.: AlGaN solar-blind avalanche photodiodes with high multiplication gain. App. Phys. Lett. **97**(19), 191103-1–191103-3 (2010). Article ID 191103
138. Rao, H., Wang, W., Wan, X. L., et al.: An improved slurry method of self-adaptive phosphor coating for white pc-LED packaging. J. Disp. Technol. **9**(6), 453–458 (2013)
139. Lee, C.T., Wu, T.J.: Light distribution and light extraction improvement mechanisms of remote GaN-based white light emitting-diodes using ZnO nanorod array. J. Lumin. **137**(5), 143–147 (2013)
140. Mao, Z.Y., Zhu, Y.C., Gan, L., et al.: Tricolor emission $Ca_3Si_2O_7$: Ln (Ln = Ce, Tb, Eu) phosphors for near-UV white light emitting-diode. J. Lumin. **134**(2), 148–153 (2013)
141. Park, S.H., Lee, K.H., Unithrattil, S., Yoon, H.S., Jang, H.G., Im, W.B.: Melilite-structure $CaYAl_3O_7$: Eu^{3+} phosphor: structural and optical characteristics for near-UV LED based white light. J. Phys. Chem. **116**(51), 26850–26856 (2012)
142. Muthu, S., Schuurmans, F.J.P., Pashley, M.D.: Red, green, and blue LEDs for white light illumination. IEEE J. Sel. Top. Quantum Electron. **8**(2), 333–338 (2002)
143. Bisotto, I., Kannan, E.S., Sassine, S., et al.: Microwave based nanogenerator using the ratchet effect in Si/SiGe heterostructures. Nanotechnology **22**(24) (2011). Article ID 245401
144. Irie, H., Sobolewski, R.: Terahertz electrical response of nanoscale three-branch junctions. J. Appl. Phys. **107**(8) (2010). Article ID 084315
145. Íñiguez-de-la-Torre, I., Purohit, S., Kaushal, V., et al.: Exploring digital logic design using ballistic deflection transistors through monte carlo simulations. IEEE Trans. Nanotechnol. **10** (6), 1337–1346 (2011)
146. Balocco, C., Kasjoo, S.R., Lu, X.F., et al.: Room-temperature operation of a unipolar nanodiode at Terahertz frequencies. App. Phys. Lett. **98**(22) (2011). Article ID 223501
147. Balocco, C., Halsall, M., Vinh, N.Q., Song, A.M.: THz operation of asymmetric-nanochannel devices. J. Phys. Condens. Matter **20**(38) (2008). Article ID 384203
148. Millithaler, J.-F., Reggiani, L., Pousset, J., et al.: A Monte Carlo investigation of plasmonic noise in nanometric n-In0.53Ga0.47As channels. J. Stat. Mech.: Theory Exp. (1) (2009). Article IDP01040

149. Dyakonov, M., Shur, M.: Shallow water analogy for a ballistic field effect transistor: new mechanism of plasma wave generation by dc current. Phys. Rev. Lett. **71**(15), 2465–2468 (1993)
150. El Fatimy, A., Dyakonova, N., Meziani, Y., et al.: AlGaN/GaN high electron mobility transistors as a voltage-tunable room temperature Terahertz sources. J. Appl. Phys. **107**(2) (2010). Article ID 024504
151. Yu, V., Kachorovskii, Shur, M.S.: Current-induced terahertz oscillations in plasmonic crystal. Appl. Phys. Lett. **100**(23) (2012). Article ID 232108
152. Nagatsuma, T.: Terahertz technologies: present and future. IEICE Electron. Express **8**(14), 1127–1142 (2011)
153. Chen, Y., He, J., Liang, H.L., et al.: Model-based prediction of the plasma oscillation excitation response characteristics of a high-electron mobility transistor-based terahertz photomixer with the cap region. J. Comput. Theor. Nanosci. **9**(4), 549–554 (2012)
154. Wang, L., Hu, W., Wang, J., et al.: Plasmon resonant excitation in grating-gated AlN barrier transistors at Terahertz frequency. Appl. Phys. Lett. **100**(12) (2012). Article ID123501
155. Zhou, Y., Sun, J., Sun, Y., et al.: Characterization of a room temperature Terahertz detector based on a GaN/AlGaN HEMT. J. Semicond. **32**(6) (2011). Article ID064005
156. Kayes, B.M., Atwater, H.A., Lewis, N.S.: J. Appl. Phys. **97**, 114302 (2005)
157. Tang, J., Huo, Z., Brittman, S., Gao, H., Yang, P.: Nat. Nanotechnol. **6**, 568 (2011)
158. Musselman, K.P., Marin, A., Schmidt-Mende, L., MacManus-Driscoll, J.L.: Adv. Funct. Mater. **22**, 2202 (2012)
159. Boukai, A., Haney, P., Katzenmeyer, A., Gallatin, G.M., Talin, A.A., Yang, P.: Chem. Phys. Lett. **501**, 153 (2011)
160. Yoon, H.P., Yuwen, Y.A., Kendrick, C.E., Barber, G.D., Podraza, N.J., Redwing, J.M., Mallouk, T.E., Wronski, C.R., Mayer, T.S.: Appl. Phys. Lett. **96**, 213503 (2010)
161. (a) Law, M., Greene, L.E., Johnson, J.C., Saykally, R., Yang, P.: Nat. Mater. 4, 455 (2005); (b) Peters, C.H., Guichard, A.R., Hryciw, A.C., Brongersma, M.L., McGehee, M.D.: J. Appl. Phys. 105, 124509 (2009); (c) Weintraub, B., Wei, Y., Wang, Z.L.: Angew. Chem. Int. Ed. 48, 8981 (2009)
162. Briseno, A.L., Holcombe, T.W., Boukai, A.I., Garnett, E.C., Shelton, S.W., Frechet, J.J., Yang, P.: Nano Lett. **10**, 334 (2010)
163. Peng, K.-Q., Wang, X., Li, L., Hu, Y., Lee, S.-T.: Silicon nanowires for advanced energy conversion and storage. Nano Today **8**, 75–97 (2013)
164. Kelzenberg, M.D., Boettcher, S.W., Petykiewicz, J.A., Turner-Evans, D.B., Putnam, M.C., Warren, E.L., Spurgeon, J.M., Briggs, R.M., Lewis, N.S., Atwater, H.A.: Enhanced absorption and carrier collection in Si wire arrays for photovoltaic applications. Nat. Mater. **9**, 239–244 (2010)
165. Tsakalakos, L., Balch, J., Fronheiser, J., Korevaar, B.A., Sulima, O., Rand, J.: Silicon nanowire solar cells. Appl. Phys. Lett. **91**, 233117 (2007)
166. Kelzenberg, M.D., Turner-Evans, D.B., Kayes, B.M., Filler, M.A., Putnam, M.C., Lewis, N. S., Atwater, H.A.: Photovoltaic measurements in single-nanowire silicon solar cells. Nano Lett. **8**, 710–714 (2008)
167. Tian, B., Zheng, X., Kempa, T.J., Fang, Y., Yu, N., Yu, G., Huang, J., Lieber, C.M.: Coaxial silicon nanowires as solar cells and nanoelectronic power sources. Nature **449**, 885–889 (2007)
168. Kempa, T.J., et al.: Semiconductor nanowires: a platform for exploring limits and concepts for nano-enabled solar cells. Energy Environ. Sci. **6**, 719–733 (2013)
169. Tian, B.Z., Kempa, T.J., Lieber, C.M.: Single nanowire photovoltaics. Chem. Soc. Rev. **38**, 16–24 (2009)
170. Hua, X., Zeng, Y., Wang, W., Shen, W.: Light absorption mechanism of c-Si/a-Si Half-coaxial nanowire arrays for nanostructured heterojunction photovoltaics. IEEE Trans. Electron Devices **61**, 4007–4013 (2014)
171. Garnett, E.C., Yang, P.: Silicon nanowire radial p-n junction solar cells. J. Am. Chem. Soc. **130**, 9224–9225 (2008)

172. Kanna, A.S., Asian, J.: Exp. Sci. **21**(2), 25–32 (2008)
173. Frankel, G.S.: J. Corr. Sci. Eng., Proc. "Corrosion Science in the 21st Century" **l6**(28). UMIST Manchester, England (2003)
174. Kendig, M.W., Buchheit, R.G.: Corros. **59**, 379 (2003)
175. Kendig, M.W.: International Workshop on Advanced Research & Development of Coatings for Corrosion Protection Biloxi, Mississippi (2004)
176. Beving, D.E., McDonnell, A.M.P., Yang, W.S., Yan, Y.S.: J. Electrochem. Soc. **153**, 325 (2006)
177. Cai, R., Yan, Y.S.: Corros. **64**, 271 (2008)
178. Cai, R., Sun, M.W., Chen, Z.W., Munoz, R., O'Neill, C., Beving, D., Yan, Y.S.: Angew. Chem. Int. Ed. **47**, 525 (2008)
179. Choi, J.Y., Lai, Z.P., Ghosh, S., Beving, D.E., Yan, Y.S., Tsapatsis, M.: Ind. Eng. Chem. Res. **46**, 7096 (2007)
180. Beving, D.E., O'Neill, C.R., Yan, Y.: Microporous Mesoporous Mater. **108**, 77 (2008)
181. Galliano, F., Landolt, D.: Prog. Org. Coat. **44**, 217 (2002)
182. Talo, A., Forsen, O., Ylasaari, S.: Synth. Methods **102**, 1394 (1999)
183. Miskovic-Stankovic, V.B., Stanic, M.R., Drazic, D.M.: Prog. Org. Coat. **36**, 53 (1999)
184. Wetzel, B., Haupert, F., Zhang, M.Q.: Comp. Sci. Technol. **63**, 2055 (2003)
185. Zhang, M.Q., Rong, M.Z., Yu, S.I., Wetzel, B., Friedrich, K.: Macromol. Mater. Eng. **287**, 111 (2002)
186. Yamini, S., Young, R.J.: Polymer **18**, 1075 (1977)
187. Perreux, D., Suri, C.: Comp. Sci. Technol. **57**, 1403 (1997)
188. Loos, C., Springer, G.S.: J. Comp. Mater. **13**, 131 (1979)
189. Lam, K., Lau, K.T.: Comp. Struct. **75**, 553 (2006)
190. Shi, G., Zhang, M.Q., Rong, M.Z., Wetzel, B., Friedrich, K.: Wear **254**, 784 (2003)
191. Hartwig, A., Sebald, M., Putz, D., Aberle, L.: Macromol. Symp. **221**, 127 (2005)
192. Dietsche, F., Thomann, Y., Thomann, R., Mulhaupt, R.: J. Appl. Polym. Sci. **75**, 396 (2000)
193. Huong, N.: Improvement of bearing strength of laminated composites by nanoclay and Z-pin reinforcement. Ph.D. Dissertation, University of New South Wales, Australia (2006)
194. Becker, O., Varley, R., Simon, G.: Polymer **43**(16), 4365 (2002)
195. Yang, L.H., Liu, F.C., Han, E.H.: Prog. Org. Coat. **53**, 91 (2005)
196. Umaka, S.V., Zheludkevich, M.L., Yasakau, K.A., Serra, R., Poznyak, S.K., Ferreira, M.G. S.: Prog. Org. Coat. **58**, 127 (2007)
197. Shi, X., Nguyen, T.A., Suo, Z., Liu, Y., Avic, R.: Surf. Coat. Tech. **204**, 237 (2009)
198. Li, R., Chen, L.: A paint containing nano titanium oxide and nano silver, and its preparation method. CN 10027622 (2005)
199. Morrow, W.H., McLean, L.J.: Self-cleaning UV reflective coating, its applying methods, and UV irradiating device prepared therefrom. US 2003059549 (2003)
200. Cai, R., Van, G.M., Aw, P.K., Itoh, K.: Solar-driven selfcleaning coating for a painted surface. C. R. Chim. **9**, 829–835 (2006)
201. www.ambio.bham.ac.uk
202. www.tda.com/Library/docs/Nanomatcrials%20for%20Coatings%205-l7-04.pdf
203. Townsin, R.: Biofouling **19**, 9 (2003)
204. Kiil, S., Weinell, C.E., Pedersen, M.S., Dam-Johanssen, K.: Ind. Eng. Chem. Res. **40**, 3906 (2001)
205. Adkins, J.D., Mera, A.E., Roe-Short, M.A., Pawlikowski, G.T., Brady, R.F.: Prog. Org. Coat. **29**, 1 (1996)
206. Omae, I.: Chem. Rev. **103**, 3431 (2003)
207. Yebra, D.M., Kiil, S., Johansen, K.D.: Prog. Org. Coat. **50**, 75 (2004)
208. Iguerb, O., Poleunis, C., Mazeas, F., Compere, C., Bertrand, P.: Langmuir **24**, 12272 (2008)
209. Schiff, K., Diehl, D., Valkirs, A.: Marine Pollut. Bull. **48**, 371 (2004)
210. Voulvoulis, N., Scrimshaw, M.D., Lester, J.N.: Appl. Organomet. Chem. **13**, 135 (1999)
211. Matthiessen, P., Reed, J., Johnson, M.: Marine Pollut. Bull. **38**, 908 (1999)
212. Vasconcelos, M.T.S.D., Leal, M.F.C.: Environ. Sci. Technol. **35**, 508 (2001)

213. Mann, E.L., Nathan, A., James, W.M., Sallie, W.C.: Limnol. Oceanogr. **47**, 976 (2002)
214. Avery, S.V., Howlett, N.G., Radice, S.A.: Appl. Environ. Microbiol. **62**, 3960 (1996)
215. Antonietta, Z.M., Stefania, Z., Rebecca, P., Riccardo, B.J.: Inorg. Biochem. **35**, 291 (1996)
216. Stoimenov, P.K., Klinger, R.L., Marchin, G.L., Klabunde, K.J.: Langmuir **18**, 6679 (2002)
217. Yoichi, Y., Hiroshi, Y., Chikara, K., Kei, I.: Prog. Org. Coat. **42**, 150 (2001)
218. Kawashita, M., Toda, S., Kim, H.M., Kokubo, T., Masuda, N.: J. Biomed. Mater. Res. **66**, 266 (2003)
219. Shah, M.S.A.S., Kalagara, T., Singh, S., Manorama, S.: Chem. Mater. **20**, 2455 (2008)
220. Shtykova, L., Fant, C., Handa, P.: Prog. Org. Coat., 64 (2009)
221. Tovmachenko, O.G., Graf, C., Heuvel, D.J.: Adv. Mater. **18**, 91 (2006)
222. Sakai, H., Kanda, T., Shibata, H.: J. Am. Chem. Soc. **128**, 4944 (2006)
223. Tom, R.T., Nair, A.S., Singh, N.: Langmuir **19**, 3439 (2003)
224. White, S.R., Sottos, N.R., Geubelle, P.H.: Nature **409**, 794 (2001)
225. Cho, S.H., Andersson, H.M., White, S.R.: Adv. Mater. **18**, 99 (2006)
226. Shchukin, D.G., Lamaka, S.V., Yasakau, K.A.: J. Phys. Chem. **112**, 958 (2008)
227. Ke, X., Wang, J.X., Kang, X.L., Chen, J.F.: Mater. Lett. **63**, 31 (2009)
228. Le, Y., Hou, P., Wang, J., Chen, J.-F.: Mater. Chem. Phys. **120**(2–3), 351–355 (2010)
229. Pingree, L.S.C., Reid, O.G. Ginger, D.S.: Adv. Mater. **21**, 19–28 (2009)
230. Becker, W.: Advanced Time-Correlated Single Photon Counting Techniques, vol. 81. Springer, Berlin (2005)
231. Hadfield, R.H.: Single-photon detectors for optical quantum information applications. Nat. Photon. **3**, 696–705 (2009)
232. Cova, S., Ghioni, M., Lotito, A., Rech, I., Zappa, F.: Evolution and prospects for single-photon avalanche diodes and quenching circuits. J. Mod. Optic. **51**, 1267–1288 (2004)
233. Itzler, M.A., Jiang, X.D., Entwistle, M., Slomkowski, K., Tosi, A., Acerbi, F., Zappa, F., Cova, S.: Advances in InGaAsP-based avalanche diode single photon detectors. J. Mod. Optic. **58**, 174–200 (2011)
234. Testardi, L.R.: Destruction of superconductivity by laser light. Phys. Rev. B **4**, 2189–2196 (1971)
235. Bluzer, N.: Temporal relaxation measurements of photoinduced nonequilibrium in superconductors. J. Appl. Phys. **71**, 1336–1348 (1992)
236. Semenov, A.D., Nebosis, R.S., Gousev, Y.P., Heusinger, M.A., Renk, K.F.: Analysis of the nonequilibrium photoresponse of superconducting films to pulsed radiation by use of a two temperature model. Phys. Rev. B **52**, 581–590 (1995)
237. Lindgren, M., Currie, M., Zeng, W.S., Sobolewski, R., Cherednichenko, S., Voronov, B., Gol'tsman, G.N.: Picosecond response of a superconducting hot-electron NbN photodetector. Appl. Supercond. **6**, 423–428 (1998)
238. Lindgren, M., Currie, M., Williams, C., Hsiang, T.Y., Fauchet, P.M., Sobolewski, R., Moffat, S.H., Hughes, R.A., Preston, J.S., Hegmann, F.A.: Intrinsic picosecond response times of Y-Ba-Cu-O superconducting photodetectors. Appl. Phys. Lett. **74**, 853–855 (1999)
239. Skocpol, W.J., Beasley, M.R., Tinkham, M.: Self-heating hotspots in superconducting thin-film microbridges. J. Appl. Phys. **45**, 4054–4066 (1974)
240. Kadin, A.M., Johnson, M.W.: Nonequilibrium photon-induced hotspot: a new mechanism for photodetection in ultrathin metallic films. Appl. Phys. Lett. **69**, 3938–3940 (1996)
241. Gol'tsman, G.N., Okunev, O., Chulkova, G., Lipatov, A., Semenov, A., Smirnov, K., Voronov, B., Dzardanov, A., Williams, C., Sobolewski, R.: Picosecond superconducting single-photon optical detector. Appl. Phys. Lett. **79**, 705–707 (2001)
242. Semenov, A.D., Gol'tsman, G.N., Korneev, A.A.: Quantum detection by current carrying superconducting film. Physica C **351**, 349–356 (2001)
243. Yang, J.K.W., Kerman, A.J., Dauler, E.A., Anant, V., Rosfjord, K.M., Berggren, K.K.: Modeling the electrical and thermal response of superconducting nanowire single-photon detectors. IEEE T. Appl. Supercon. **17**, 581–585 (2007)

244. Il'in, K.S., Lindgren, M., Currie, M., Semenov, A.D., Gol'tsman, G.N., Sobolewski, R., Cherednichenko, S.I., Gershenzon, E.M.: Picosecond hot-electron energy relaxation in NbN superconducting photodetectors. Appl. Phys. Lett. **76**, 2752–2754 (2000)
245. Hadfield, R.H., Miller, A.J., Nam, S.W., Kautz, R.L., Schwall, R.E.: Low-frequency phase locking in high-inductance superconducting nanowires. Appl. Phys. Lett. **87**, 203505-3 (2005)
246. Kerman, A.J., Dauler, E.A., Keicher, W.E., Yang, J.K.W., Berggren, K.K., Gol'tsman, G., Voronov, B.: Kinetic-inductance-limited reset time of superconducting nanowire photon counters. Appl. Phys. Lett. **88**, 111116 (2006)
247. Gol'tsman, G., Okunev, O., Chulkova, G., Lipatov, A., Dzardanov, A., Smirnov, K., Semenov, A., Voronov, B., Williams, C., Sobolewski, R.: Fabrication and properties of an ultrafast NbN hot-electron single photon detector. IEEE T. Appl. Supercon. **11**, 574–577 (2001)
248. Verevkin, A., Zhang, J., Sobolewski, R., Lipatov, A., Okunev, O., Chulkova, G., Korneev, A., Smirnov, K., Gol'tsman, G.N., Semenov, A.: Detection efficiency of large-active-area NbN single-photon superconducting detectors in the ultraviolet to near-infrared range. Appl. Phys. Lett. **80**, 4687–4689 (2002)
249. Gol'tsman, G.N., Smirnov, K., Kouminov, P., Voronov, B., Kaurova, N., Drakinsky, V., Zhang, J., Verevkin, A., Sobolewski, R.: Fabrication of nanostructured superconducting single-photon detectors. IEEE T. Appl. Supercon. **13**, 192–195 (2003)
250. Miki, S., Fujiwara, M., Sasaki, M., Baek, B., Miller, A.J., Hadfield, R.H., Nam, S.W., Wang, Z.: Large sensitive-area NbN nanowire superconducting single-photon detectors fabricated on single-crystal MgO substrates. Appl. Phys. Lett. **92**, 061116 (2008)
251. Gol'tsman, G., Korneev, A., Tarkhov, M., Seleznev, V., Divochiy, A., Minaeva, O., Kaurova, N., Voronov, B., Okunev, O., Chulkova, G., Milostnaya, I., Smirnov, K.: Middle-infrared ultrafast superconducting single photon detector. In: Joint 32nd International Conference on Infrared and Millimeter Waves, 2007 and the 2007 15th International Conference on Terahertz Electronics. IRMMW-THz, pp. 115–116 (2007)
252. Bulaevskii, L.N., Graf, M.J., Batista, C.D., Kogan, V.G.: Vortex-induced dissipation in narrow current-biased thin-film superconducting strips. Phys. Rev. B **83**, 144526 (2011)
253. Kitaygorsky, J., Komissarov, I., Jukna, A., Pan, D., Minaeva, O., Kaurova, N., Divochiy, A., Korneev, A., Tarkhov, M., Voronov, B., Milostnaya, I., Gol'tsman, G., Sobolewski, R.R.: Dark counts in nanostructured NbN superconducting single-photon detectors and bridges. IEEE T. Appl. Supercon. **17**, 275–278 (2007)
254. Bartolf, H., Engel, A., Schilling, A., Il'In K., Siegel, M., Hübers, H.W., Semenov, A.: Current-assisted thermally activated flux liberation in ultrathin nanopatterned NbN superconducting meander structures. Phys. Rev. B **81**, 024502 (2010)
255. Yamashita, T., Miki, S., Makise, K., Qiu, W., Terai, H., Fujiwara, M., Sasaki, M., Wang, Z.: Origin of intrinsic dark count in superconducting nanowire single-photon detectors. Appl. Phys. Lett. **99**, 161105-3 (2011)
256. Baek, B., Lita, A.E., Verma, V., Nam, S.W.: Superconducting a-WxSi1 − x nanowire single-photon detector with saturated internal quantum efficiency from visible to 1850 nm. Appl. Phys. Lett. **98**, 251105 (2011)
257. Clem, J.R., Berggren, K.K.: Geometry-dependent critical currents in superconducting nanocircuits. Phys. Rev. B **84**, 174510 (2011)
258. Pearlman, A., Cross, A., Slysz, W., Zhang, J., Verevkin, A., Currie, M., Korneev, A., Kouminov, P., Smirnov, K., Voronov, B., Gol'tsman, G., Sobolewski, R.: Gigahertz counting rates of NbN single-photon detectors for quantum communications. IEEE T. Appl. Supercon. **15**, 579–582 (2005)
259. Dauler, E.A., Kerman, A.J., Robinson, B.S., Yang, J.K.W., Voronov, B., Goltsman, G., Hamilton, S.A., Berggren, K.K.: Photon-number-resolution with sub-30-ps timing using multi-element superconducting nanowire single photon detectors. J. Mod. Optic. **56**, 364–373 (2009)

260. Stevens, M.J., Hadfield, R.H., Schwall, R.E., Nam, S.W., Mirin, R.P., Gupta, J.A.: Fast lifetime measurements of infrared emitters using a low-jitter superconducting single-photon detector. Appl. Phys. Lett. **89**, 031109-3 (2006)
261. Yang, X., Yamashita, T., Miki, S., Fujiwara, M., Sasaki, M., You, L., Wang, Z.: Performances of superconducting nanowire single-photon detectors with different designs. In: Superconducting centennial conference EUCAS. The Hague, Netherlands (2011)
262. Miki, S., Terai, H., Yamashita, T., Makise, K., Fujiwara, M., Sasaki, M., Wang, Z.: Superconducting single photon detectors integrated with single flux quantum readout circuits in a cryocooler. Appl. Phys. Lett. **99**, 111108 (2011)
263. O'Connor, J.A., Tanner, M.G., Natarajan, C.M., Buller, G.S., Warburton, R.J., Miki, S., Wang, Z., Nam, S.W., Hadfield, R.H.: Spatial dependence of output pulse delay in a niobium nitride nanowire superconducting single-photon detector. Appl. Phys. Lett. **98**, 201116-3 (2011)
264. Rosfjord, K.M., Yang, J.K.W., Dauler, E.A., Kerman, A.J., Anant, V., Voronov, B.M., Gol'tsman, G.N., Berggren, K.K.: Nanowire Single-photon detector with an integrated optical cavity and antireflection coating. Opt. Express **14**, 527–534 (2006)
265. Slysz, W., Wegrzecki, M., Bar, J., Grabiec, P., Gorska, M., Zwiller, V., Latta, C., Bohi, P., Milostnaya, I., Minaeva, O., Antipov, A., Okunev, O., Korneev, A., Smirnov, K., Voronov, B., Kaurova, N., Gol'tsman, G., Pearlman, A., Cross, A., Komissarov, I., Verevkin, A., Sobolewski, R.: Fiber-coupled single-photon detectors based on NbN superconducting nanostructures for practical quantum cryptography and photon-correlation studies. Appl. Phys. Lett. **88**, 261113-3 (2006)
266. Korneev, A., Vachtomin, Y., Minaeva, O., Divochiy, A., Smirnov, K., Okunev, O., Gol'tsman, G., Zinoni, C., Chauvin, N., Balet, L., Marsili, F., Bitauld, D., Alloing, B., Lianhe, L., Fiore, A., Lunghi, L., Gerardino, A., Halder, M., Jorel, C., Zbinden, H.: Single-photon detection system for quantum optics applications. IEEE J. Sel. Top. Quant. **13**, 944–951 (2007)
267. Radebaugh, R.: Refrigeration for superconductors. P. IEEE **92**, 1719–1734 (2004)
268. Hadfield, R.H., Stevens, M.J., Gruber, S.S., Miller, A.J., Schwall, R.E., Mirin, R.P., Nam, S. W.: Single photon source characterization with a superconducting single photon detector. Opt. Express **13**, 10846–10853 (2005)
269. Miki, S., Yamashita, T., Fujiwara, M., Sasaki, M., Wang, Z.: Multichannel SNSPD system with high detection efficiency at telecommunication wavelength. Opt. Lett. **35**, 2133–2135 (2010)
270. Hu, X., Zhong, T., White, J.E., Dauler, E.A., Najafi, F., Herder, C.H., Wong, F.N.C., Berggren, K.K.: Fiber-coupled nanowire photon counter at 1550 nm with 24% system detection efficiency. Opt. Lett. **34**, 3607–3609 (2009)
271. Miller, A.J., Lita, A.E., Calkins, B., Vayshenker, I., Gruber, S.M., Nam, S.W.: Compact cryogenic self-aligning fiber-to-detector coupling with losses below one percent. Opt. Express **19**, 9102–9110 (2011)
272. Dorenbos, S.N., Heeres, R.W., Driessen, E.F.C., Zwiller, V.: Efficient and robust fiber coupling of superconducting single photon detectors. arXiv:1109.5809v1 (2011)
273. Hadfield, R.H., Dalgarno, P.A., O'Connor, J.A., Ramsay, E., Warburton, R.J., Gansen, E.J., Baek, B., Stevens, M.J., Mirin, R.P., Nam, S.W.: Submicrometer photoresponse mapping of nanowire superconducting single-photon detectors. Appl. Phys. Lett. **91**, 241108 (2007)
274. Korneev, A., et al.: Sensitivity and gigahertz counting performance of NbN superconducting single-photon detectors. Appl. Phys. Lett. **84** (2004)
275. Boroson, D.M., Scozzafava, J.J., Murphy, D.V., Robinson, B.S.: The lunar laser communications demonstration (LLCD). In: Third IEEE International Conference on Space Mission Challenges for Information Technology (2009)

276. Dauler, E.A.: High-rate quantum key distribution with superconducting nanowire single photon detectors. In: Workshop on Single and Entangled Photons: Sources, Detectors, Components, and Applications, Boulder, Col. (2009)
277. Zhang, J., et al.: A superconducting single-photon detector for CMOS IC Probing. In: Eleventh IEEE International Conference on Requirements Engineering (2003)
278. Moore's Law past 32 nm: future challenges in device scaling. Kelin Kuhn/IWCE/Beijing (2009)
279. National Renewable Energy Laboratory, NREL, http://www.nrel.gov/solar/
280. Lee, Y.-J., Yao, Y.-C., Yang, C.-H.: Direct electrical contact of slanted ITO film on axial p-n junction silicon nanowire solar cells. Opt. Express **21**, A7–A14 (2012)
281. Divochiy, A., Marsili, F., Bitauld, D., Gaggero, A., Leoni, R., Mattioli, F., Korneev, A., Seleznev, V., Kaurova, N., Minaeva, O., Gol'tsman, G., Lagoudakis, K.G., Benkhaoul, M., Levy, F., Fiore, A.: Superconducting nanowire photon-number-resolving detector at telecommunication wavelengths. Nat. Photon. **2**, 302–306 (2008)
282. Marsili, F., Najafi, F., Dauler, E., Bellei, F., Hu, X., Csete, M., Molnar, R.J., Berggren, K.K.: Single-photon detectors based on ultra narrow superconducting nanowires. Nano Lett. **11**, 2048–2053 (2011)
283. Sprengers, J.P., Gaggero, A., Sahin, D., Jahanmirinejad, S., Frucci, G., Mattioli, F., Leoni, R., Beetz, J., Lermer, M., Kamp, M., Hofling, S., Sanjines, R., Fiore, A.: Waveguide superconducting single-photon detectors for integrated quantum photonic circuits. Appl. Phys. Lett. **99**, 181110 (2011)
284. Takesue, H., Nam, S.W., Zhang, Q., Hadfield, R.H., Honjo, T., Tamaki, K., Yamamoto, Y.: Quantum key distribution over a 40-dB channel loss using superconducting single-photon detectors. Nat. Photon. **1**, 343–348 (2007)
285. Natarajan, C.M., Peruzzo, A., Miki, S., Sasaki, M., Wang, Z., Baek, B., Nam, S., Hadfield, R.H., O'Brien, J.L.: Operating quantum waveguide circuits with superconducting single-photon detectors. Appl. Phys. Lett. **96**, 211101 (2010)
286. Tanner, M.G., Natarajan, C.M., Pottapenjara, V.K., O'Connor, J.A., Warburton, R.J., Hadfield, R.H., Baek, B., Nam, S., Dorenbos, S.N., Urena, E.B., Zijlstra, T., Klapwijk, T.M., Zwiller, V.: Enhanced telecom wavelength single-photon detection with NbTiN superconducting nanowires on oxidized silicon. Appl. Phys. Lett. **96**, 221109 (2010)

Chapter 17
Interfacing Biology Systems with Nanoelectronics for Nanodevices

The interface between nanoscale electronic devices and biological systems enables interactions at length scales natural to biology, and thus should maximize communication between these two diverse yet complementary systems. Moreover, nanostructures and nano–structured substrates show enhanced coupling to artificial membranes, cells, and tissue. Such nano–bio interfaces offer better sensitivity and spatial resolution as compared to conventional planar structures. The exceptional synthetic control and flexible assembly of nanowires (NWs) provides powerful tools for fundamental studies and applications in life science, and opens up the potential of merging active transistors with cells such that the distinction between nonliving and living systems is blurred.

17.1 Nanoelectronic-Biological Interfaces Enable

- Diagnostic devices for disease detection
- General detection and kinetics platform
- New tool for single-molecule detection/biophysics
- Powerful devices for electronic and chem/bio recording from cells, tissue and organs
- Potential implants for highly functional and powerful prosthetics, as well as hybrid biomaterials enabling new opportunities.

Evaluating Research Motivation: Progress?

- Synthetic challenge of controlling structure and composition on many length scales
- Fundamental scientific questions in 1-dimensional systems
- Central importance of nanoscale wires in integrated nanosystems

L. H. Madkour, *Nanoelectronic Materials*, Advanced Structured Materials 116, https://doi.org/10.1007/978-3-030-21621-4_17

- New/novel materials can make revolutionary versus evolutionary changes in science and technology!
  - Many fundamental scientific questions remain, and will require bold researchers to address.
  - Pushing ourselves to identify and tackle these 'big' challenges, while difficult, offers the best opportunity to make revolutionary advances and benefit society!

Nanotechnology is the study of extremely small structures, having size of 0.1–100 nm. Nano medicine is a relatively new field of science and technology. Brief explanation of various types of pharmaceutical nano systems is given. Classification of nano materials based on their dimensions is given. An application of Nanotechnology in various fields such as health and medicine, electronics, energy and environment, is discussed in detail. Applications of nano particles in drug delivery, protein and peptide delivery, cancer are explained. Applications of various nano systems in cancer therapy such as carbon nano tube, dendrimers, nano crystal, nano wire, nano shells etc. are given. The advancement in nano technology helps in the treatment of neuro degenerative disorders such as Parkinson's disease and Alzheimer's disease. Applications of nano technology in tuberculosis treatment, the clinical application of nanotechnology in operative dentistry, in ophthalmology, in surgery, visualization, tissue engineering, antibiotic resistance, immune response are discussed in this chapter. Nano pharmaceuticals can be used to detect diseases at much earlier stages.

Advancement in the field of nanotechnology and its applications to the field of medicines and pharmaceuticals has revolutionized the twentieth century. Nanotechnology [1] is the study of extremely small structures. The prefix "nano" is a Greek word which means "dwarf". The word "nano" means very small or miniature size. Nanotechnology is the treatment of individual atoms, molecules, or compounds into structures to produce materials and devices with special properties. Nanotechnology involve work from top down i.e. reducing the size of large structures to smallest structure e.g. photonics applications in nano electronics and nano engineering, top-down or the bottom up, which involves changing individual atoms and molecules into nanostructures and more closely resembles chemistry biology.

Since nanomaterials are increasingly used in industrial and consumer products, a science-based risk assessment is highly required. Metallic nanoparticles exhibit unique physical and chemical properties that make them valuable for environmental biotechnological applications. The growing interest in the use of superparamagnetic nanoparticles in environmental bioremediation may become of great importance in the development of new efficient technologies with the possibility to overcome the weakness of conventional nanoparticles. Overall, there is still much to do for the improvement of the nanotechnology systems before its application in bioremediation and continuous research is needed. The synthesis of magnetic nanoparticles, covering a wide range of compositions and sizes, has made substantial progress, especially over the past decade. Nevertheless, an understanding regarding the mechanism of nanoparticle formation is in its infancy. There is a need to control the size, shape and dispersity, which would be easier to manipulate if the exact mechanisms of the formation of nanoparticles are better known. High reproducibility of biomaterials is also a crucial point to be considered. Up to now, the formation of mainly Ag, Au, Cd and Pd nanoparticles has been studied, while knowledge of the synthesis of other metal nanoparticles is lacking. There is also a

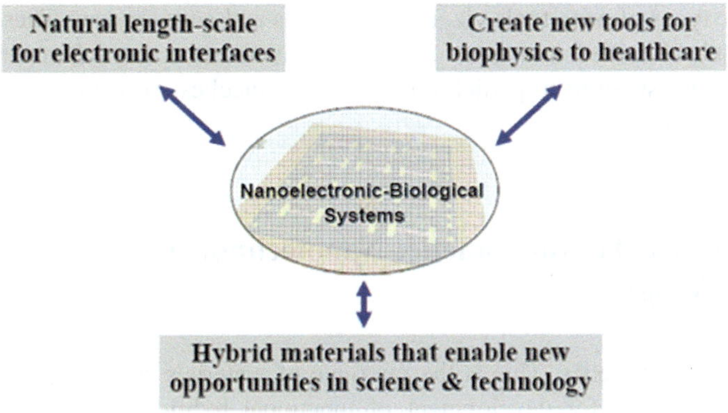

Fig. 17.1 Molecular biomimetics: nanotechnology through biology

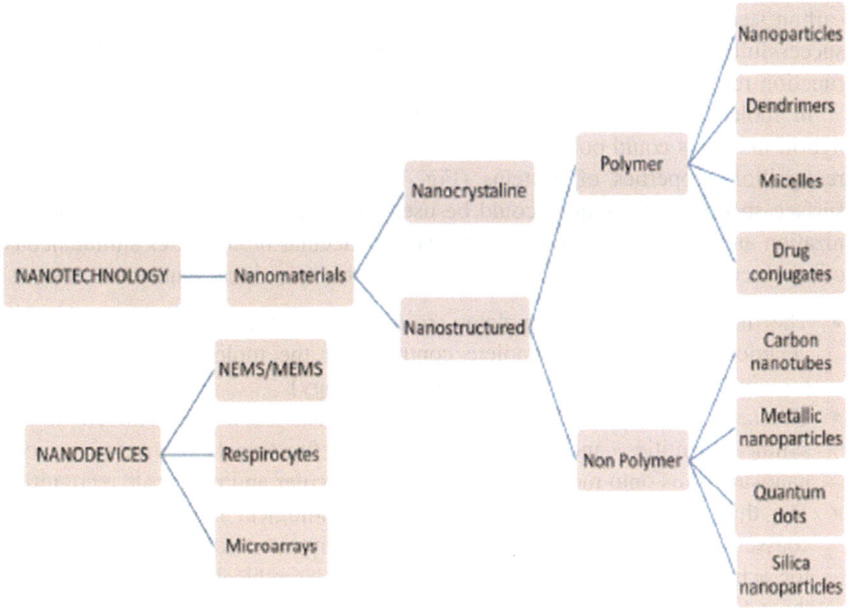

Fig. 17.2 Schematic diagram of various types of pharmaceutical nano systems

need to explore the microbial diversity to obtain novel microorganisms on nanoparticles synthesis and to expand the range of oxide nanoparticles formed by microorganisms. Currently, neither the exact biochemical mechanism nor the genes and proteins involved are known. Modern techniques, like differential proteomics combined with genomics could be helpful in identifying the key proteins involved in the nanoparticle synthesis. Figure 17.1 show nanotechnology through biological systems.

The brief explanation of pharmaceutical nano system is as follows: As shown in the schematic diagram (Fig. 17.2), pharmaceutical nanotechnology is divided in two basic types of nano tools viz. nano materials and nano devices. These materials can be sub classified into nano crystalline and nano structured materials. Nano structure consists of nano particles, dendrimers, micelles, drug conjugates, metallic nano particles etc.

17.2 Molecular Biomimetic: Nanotechnology Through Biology

Molecular biomimetics. This is the marriage of materials science engineering and molecular biology for development of functional hybrid systems, composed of inorganics and inorganic-binding proteins. The new approach takes advantage of DNA-based design, recognition, and self-assembly characteristics of biomolecules. Traditional materials science engineering produces materials (for example, medium carbon steels depicted in the bright- and dark-field TEM images), that have been successfully used over the last century. Molecular biology focuses on structure—function relations in biomacromolecules, for example, proteins.

In molecular biomimetics, a marriage of the physical and biological fields, hybrid materials could potentially be assembled from the molecular level using the recognition properties of proteins (Fig. 17.3) under the premise that inorganic surface-specific polypeptides could be used as binding agents to control the organization and specific functions of materials. Molecular biomimetics simultaneously offers three solutions to the development of heterofunctional nanostructures.

- The first is that protein templates are designed at the molecular level through genetics. This ensures complete control over the molecular structure of the protein template (that is, DNA-based technology).
- The second is that surface-specific proteins can be used as linkers to bind synthetic entities, including nanoparticles, functional polymers, or other nanostructures onto molecular templates (molecular and nanoscale recognition).
- The third solution harnesses the ability of biological molecules to self- and coassemble into ordered nanostructures. This ensures a robust assembly process for achieving complex nano-, and possibly hierarchical structures, similar to those found in nature (self-assembly). The current knowledge of protein-folding predictions and surface-binding chemistries does not provide sufficiently

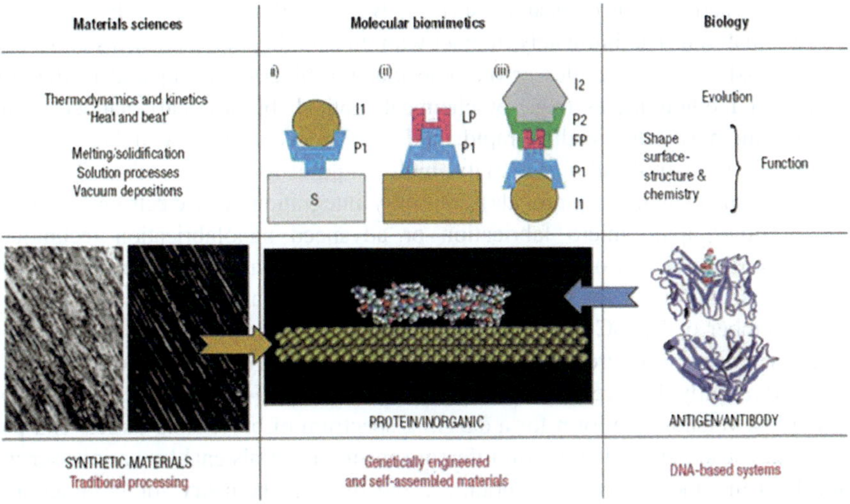

Fig. 17.3 In molecular biomimetics, inorganic-binding proteins could potentially be used as (i) linkers for nanoparticle immobilization; (ii) functional molecules assembled on specific substrates; and (iii) heterobifunctional linkers involving two (or more) binding proteins linking several nanoinorganic units (I1: inorganic-1, I2: Inorganic-2, P1 and P2: inorganic specific proteins, LP: linker protein, FP: fusion protein) (Sarikaya, C. Tamerler, A.K.Y. Jen, K. Schulten F. Baneyx, Molecular biomimetics: nanotechnology through biology, Nature Materials, vol 2, 2003, pp. 577–585)

detailed information to perform rational design of proteins. To circumvent this problem, massive libraries of randomly generated peptides can be screened for binding activity to inorganic surfaces using phage and cell-surface display techniques. It may ultimately be possible to construct a 'molecular erector' set, in which different types of proteins, each designed to bind to a specific inorganic surface, could assemble into intricate, hybrid structures composed of inorganics and proteins. This would be a significant leap towards realizing molecularly designed, genetically engineered technological materials.

Biological processes are spatiotemporally mediated by the dynamic actions of molecular machineries which are often organized in nanoscale domains and exist in extremely low abundance. Therefore, there is a pressing need to develop biosensors that can offer high enough sensitivity to detect trace amount of target molecules or minute biological signals, and high enough spatial and temporal resolution to resolve the subtle biological activities. The emerging nanoelectronic biosensors provide unprecedented possibilities towards such ambition which is unattainable by the conventional bioanalytical methods.

The active conducting channel (sensing element) of a nanoelectronic biosensor is made of nanostructured semiconducting (NS-SC) materials whose current flow is completely or largely confined on the surface, and therefore, highly sensitive to minute electrical perturbations imposed by interacting biomolecules or electrogenic biological activities at or near its surface. Because of their nanoscale dimension, NS-SC materials can intimately interact with, thus highly and locally responsive to, similarly-sized biomolecules or biomolecular machineries. Compared to the conventional methodologies (e.g., biochemical, optical, biophysical), nanoelectronic biosensing may offer readily, rapid, and label-free detection with high or even ultimate (single-molecule) sensitivity, high temporal resolution (sub-millisecond), high spatial resolution (nanoscale), and easy integration with electronic readouts. Using lithographic micro fabrication or advanced nanofabrication techniques, nanoelectronic biosensors can be miniaturized for lab-on-a-chip developments and made in array format for parallel detection of multiple analytes.

A number of NS-SC materials can be employed for nanoelectronic biosensing, such as, silicon nanowires (SiNWs) [2], single-walled carbon nanotubes (SWCNTs) [3], and recently discovered graphene and its derivatives [4]. Nanoelectronic sensors provide a versatile platform for a dazzling spectrum of biosensing targets and purposes because the availability of different sensing materials enables various sensing mechanisms and device configurations, and diverse chemistry for biofunctionalization. Each NS-SC material possesses unique advantages and limitations while being applied to a particular application. The choice should be made based on balanced consideration on specific requirements on the performance, detection mechanisms to be utilized, ease of fabrication and recording, cost, and so on.

Nanoelectronic sensors have been used to detect the presence of biomolecules, cells, or microbes at extremely low concentration. Remarkably, SiNW transistors have demonstrated the ability to detect single virus [5]; and single molecule detection of DNA-hybridization kinetics has been achieved by SWCNT transistors [6]. The promising and unique potential of nanoelectronic sensors for molecular diagnosis or studies are clearly evidenced by many studies in the past two decades.

Recently, several groups have attempted to interface nanoelectronic devices with live cells for detection of dynamic cellular activities, such as, secretion of signaling molecules and generation of bioelectrical signals [7, 8]. In their landmark work published in 2006, Charles Lieber's group of Harvard University demonstrated the use of array of nanowire transistors for spatially resolved detection, stimulation or inhibition of nerve impulses (electrical spike or action potential produced by ion channel activities in a neuron) and their propagation along individual neurites (slender projection from the cell body of a neuron) [9]. In contrast to the invasive intracellular recording by patch-clamp technique, this nanowire approach provides noninvasive extracellular measurement; as compared to the extracellular recording by metal microelectrode, the nanowire approach offers unraveled spatial resolution. Later, the same group made a nano-transistor device at the tip of a bent silicon nanowire, and used it, as a novel alternative to patch-clamp yet with minimum invasiveness, to intracellularly record action potentials [10]. Very recently, Lieber group demonstrated a new type of SiNW transistor with a branched hollow nanotube to establish connection with the intracellular space for the recording of action potentials [11]. Nanoelectronic sensors can probe at single cell or sub cellular level (even nano-domains). With an array of them, signal distribution and propagation in cell networks could be resolved, e.g., in a brain slice or heart tissue. It is envisaged that nanoelectronics-cell interface would be instrumental or revolutionary to the study of cell functions as well as to the discovery of drugs targeting on these functions [12].

Exciting progress has been made in interfacing nanoelectronics with biomolecules, cells, and tissues/organs. No doubt that the convergence of nanoelectronics and biological systems will bring profound impacts on fundamental biological studies, diagnosis, drug discovery, and development of human-machine interface. But currently, nanoelectronic sensing still remains as a scientist niche. To transform it to practical and widely accessible techniques, tremendous and interdisciplinary efforts are still needed, for examples, to develop facile and scalable approach to make these fine devices, and to understand better and optimally engineer the interaction/communication between the bio-system and the nanodevice.

High spatio-temporal resolution interfacing between electrical sensors and biological systems, from single live cells to tissues, is crucial for many areas, including fundamental biophysical studies as well as medical monitoring and intervention. Recording electrical signals from cells and tissues, such as action potentials in the nervous system, is central to areas ranging from fundamental electrophysiological studies to brain activity mapping and biomedical prosthetic applications [13]. Conventional methods based on glass micropipette electrodes have been widely used for intracellular action potential recording, and have shown excellent signal-to-noise ratio (SNR) and temporal resolution. However, these methods have constraints [13b, c, 14] that limit their applicability to simultaneous measurements from large numbers of cells with single-cell resolution. In addition, the typical micrometer size of these probes poses a challenge to recording from small subcellular structures and also results in invasiveness and toxicity to cells [13b, 14]. On the other hand, methods using microfabricated metal electrodes and arrays (MEAs) have made possible large-scale multi-site recording, although the size of these electrodes remains micrometer scale in order to meet electrode/electrolyte interface impedance conditions necessary to achieve usable SNR. This size restriction precludes subcellular resolution needed for many important studies [13d, e, 15]. In the case of tissue level electrical measurements, implementing electronic sensors in three dimensions (3D) and the capabilities for monitoring cells throughout the 3D micro-environment of tissues is critical for functional neural activity mapping and understanding physicochemical changes relevant to living organisms [13a, 16]. Most work has, however, focused on coupling electronics to the surface of tissues or artificial tissue constructs, including recently reported studies which use flexible and/or stretchable planar devices that conform to tissue surfaces [17].

The nanometer size of nanoelectronic devices makes them advantageous for realizing high resolution and minimally invasive cellular- and subcellular-level interfaces between recording probes and biological systems, and making such interfaces on a large-scale and with high-density is of significant importance for mapping activity in the brain and other excitable biological systems [13a, 18]. Moreover, the bottom-up paradigm used for nanodevices fabrication that we have pioneered [19] enables the preparation of 3D free-standing, macroporous device arrays that can be utilized as the scaffolding for synthetic tissue constructs, and thus realize monitoring of cellular activity throughout 3D cellular Networks [20]. In this focus review, we will talk about the development of novel nanoscale devices for intracellular action potential recording and macroporous nanoelectronic scaffolds for 3D interfacing with synthetic tissue constructs. We will focus on field-effect transistor (FET) devices from semiconducting nanowires or nanotubes, where the active nanowire or nanotube channel serves as the voltage sensing element. There are several reasons that make this nanoscale FET approach attractive. First, there is no dependence on device/electrolyte interface impedance for an FET voltage sensor [21], which enables dramatic miniaturization of recording probes (compared to passive metallic electrodes); this miniaturization of the sensors facilitates both subcellular level resolution and high-density recording. Second, the structure, morphology, physical properties and corresponding functions of semiconducting nanowires and nanotubes can be well controlled by encoded synthesis, which makes them ideally suited for hierarchical design of devices [22].

Research at the interface between nanoscience and biology has the potential to produce breakthroughs in fundamental science and lead to revolutionary technologies [23, 24]. In this chapter, we focus on nanoelectronic/biological interfaces. First, we discuss nanoscale field effect transistors (nanoFETs) as probes to study cellular systems, including the realization of nanoFET comparable in size to biological nanostructures involved in communication using synthesized nanowires. Second, we overview current progress in multiplexed extracellular sensing using planar nanoFET arrays. Third, we describe the design and implementation of three distinct nanoFETs used to realize the first intracellular electrical recording from single cells. Fourth, we present recent progress in merging electronic and biological systems at the 3D tissue level by using macroporous nanoelectronic scaffolds. Finally, we discuss future development in this research area, the unique challenges and opportunities, and the tremendous impact these nanoFET based technologies might have in advancing biology and medical sciences.

There are two types of nanowire-based platforms in biomedical sciences: basic platforms that can be readily adapted to address biomedical questions; and advanced platforms that are specifically designed to push the frontiers of what is possible by, for example, enabling a new measurement tool. The basic platforms use conventional nanowire material and device systems with well-exploited physical or chemical properties, and they also have wideranging applications in many other fields, such as energy scavenging systems [25–32] or components for integrated circuit [33, 34]. These basic platforms, such as planar nanowire field effect transistors [33–37] or vertical nanowire arrays [26–29, 31, 32], have been used in biomolecular sensing [23, 24], extracellular recording [23, 24], drug delivery [38–40] and localized cellular imaging [41]. On the other side, the advanced platforms have been designed to address some intrinsic complexity in biology and medical sciences in way simply not possible previously. They allow new types or new scales of interact and measurements with their target systems [10, 42–44], and in so doing, open up completely new opportunities in science and technology. Examples of advanced platforms include recent intracellular field effect transistor probes [10, 43–45] and nanoelectronics-innervated synthetic tissues [42].

17.3 Fundamentals of NanoFET in Biology and Medicine

The ability to make electrical measurements inside single cells or throughout the entire 3D space of the tissue can have many important impacts in electrophysiology and biomedical sciences. The patch clamp technique, in which a pulled glass micropipette filled with electrolyte is inserted into a cell, offers intracellular electrical measurements with high signal-to-noise ratio (S/N) and single ion channel recording capability [46]. Ideally, the micropipette should be as small as possible to increase the spatial resolution and reduce the invasiveness of the measurement, and ideally, allow for recording from subcellular structures. However, the overall performance of the technique also depends on the impedance of the interface between the micropipette and the cell interior (i.e., the smaller the probe tip size, the larger the junction impedance), which sets limits on the temporal resolution and S/N of the micropipette-based electrical probes [10, 47]. Advanced techniques that involve inserting metal or carbon microelectrodes or nanoelectrodes into cells or tissues could be subject to similar dilemma, because all these tools are single terminal devices and electrochemical thermodynamics and kinetics must be considered for device operation [48–55].

In integrated circuits, the basic device element is a multi-terminal FET that uses either electrons or holes as the charge carriers [56] (Fig. 17.4a). Although the charge carriers are ions in biological systems, there are many biophysical links that connect ions to electrons and holes in a FET. For example, the dynamic flow of ions in biological system can generate spatially defined field potential [57]. The Poisson equation [58] links such potentials directly to the ionic current sources and sinks that produce them. The Goldman-Hodgkin-Katz voltage equation [58] has also been used in cell membrane physiology to determine the equilibrium potential across a cell's membrane, where it takes into account all of the ions that permeate through that membrane. The potentials, generated by ion flows and gradients, can function as the gate signals to modulate the electrical output in FET devices (Fig. 17.4b, c). The sensitivity of a FET or how well the transistor can receive and amplify the gate signal is usually defined as transconductance (Gm) [23, 24, 56, 59], which is inversely proportional to the dimension (L) of the active device [59]. This fact implies that the use of nanoelectronics would have improved sensitivity compared to its bulk and planar counterparts. As shown in the following sections, nanoFETs have shown to be able to record electric potentials inside cells [10, 43–45] and from the internal regions of synthetic tissues [42], and because their performance does not depend on impedance, they can be made much smaller than micropipettes and microelectrodes. Moreover, nanoFET arrays are better suited for multiplexed measurements [43, 44].

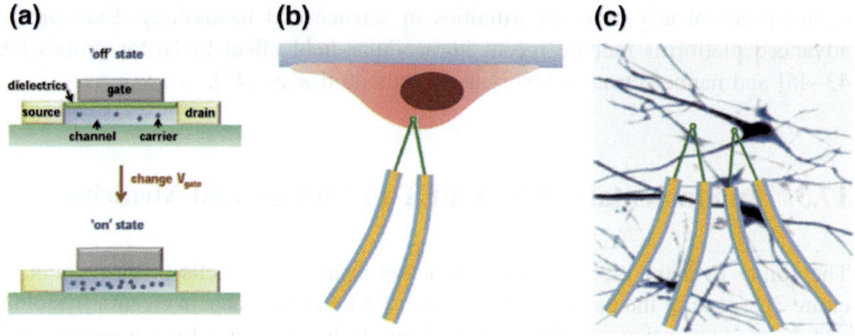

Fig. 17.4 FET basics and electrical interfaces between nanoFET and biological systems. **a** Schematic of a planar FET device. In FET, current flows along a semiconductor path called the channel. At one end of the channel, there is an electrode called the source. At the other end of the channel, there is an electrode called the drain. The third electrode that applies a voltage to the channel is called gate, which modulates the electron/hole carrier density and the output of the FET devices. A small voltage change in gate signal can cause a large variation in the current from the source to the drain. This is how FET works and in particular, amplifies signals. **b–c** Schematics of electrically based cellular sensing using a kinked nanoFET, where intracellular potentials (**b**) or extracellular field potentials (**c**) can be used to change the nanoFET conductance, analogous to applying a voltage using a gate electrode

17.3.1 Chemical Synthesis of NanoFETs

Three distinct classes of de novo design and synthesis have been used to yield nanoFETs building blocks, covering structural motifs in one-dimension (1D), 2D and 3D (Fig. 17.5). The basic semiconductor nanowire structure (Fig. 17.5a, I) consists of a uniform composition, 1D structure with a diameter typically in the range of 3–500 nm. In the growth process, which builds upon earlier work showing vapor-liquid-solid (VLS) growth of micrometer to millimeter diameter wires [60, 61], the nanocluster catalyst (typically gold nanoparticles) forms a liquid solution with nanowire reactant component(s), and when supersaturated, acts as the nucleation site for crystallization and preferential 1D growth [62, 63]. Other growth mechanisms, such as vapor-solid-solid (VSS) and vapor-solid (VS) [64], can also be explored to yield high quality semiconductor nanowires. Within this framework, it is straightforward to synthesize nanowires with different compositions, such as groups III–V, IV and II–VI semiconductors [64–67], using the appropriate nanocluster catalysts and growth temperatures/pressures. Additionally, nanowire structures in which the composition, dopant and even growth mechanisms (e.g., VLS, VSS) are modulated along axial [68–72] (Fig. 17.5b) or radial directions [73–75] have also been widely exploited. These axial and radial nanowire heterostructures provide a number of advantages compared to homogeneous semiconductor nanowires, and they have proven exceptionally powerful for a broad range of electronic, photonic and optoelectronic device applications [76]. For example, germanium/silicon core/shell nanowires have been chemically

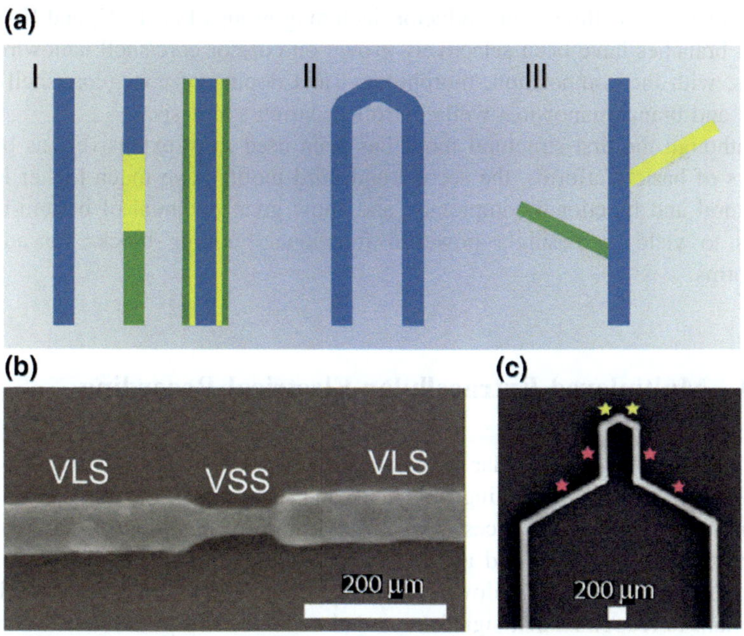

Fig. 17.5 Semiconductor nanowire structural motifs for nanoFETs. **a** Schematics of 1D (I), 2D (II) and 3D (III) motifs. 1D motif (I) can have uniform composition and doping (I, left) or axially (I, middle) or radially (I, right) modulated. A kinked nanowire with structurally coherent "kinks" introduced in a controlled manner during axial elongation represents an example of 2D motif (II). Heterobranched nanowires yield 3D structure (III) and the branch junction (e.g., blue/yellow segment junction) can be exploited for localized sensing. **b** An axial nanowire heterostructure made by modulation in VLS/VSS growth mechanisms. **c** A multiply kinked nanowire showing a probe structure. Yellow and magenta stars denote cis- and trans-conformations, respectively

synthesized for high mobility nanowire FETs due to quantum confinement of carriers within the germanium core by the larger band-gap silicon shell [77–81].

The second structural motif was recently demonstrated by an approach in which topological centers are synthetically introduced in a controlled manner in linear 1D structures (Fig. 17.5a, II) [10, 82]. In this area, we demonstrated that iterative control over nucleation and growth leads to kinked nanowires, in which the straight sections are separated by triangular joints and where doping can be varied at these topologically defined points (Fig. 17.5c). Moreover, new work has shown that it is possible to control the stereochemistry of adjacent kinks in a manner that allows the synthesis of increasingly complex two- and three-dimensional structures akin to organic chemistry, thus opening up a great opportunity for the future in terms of designed synthesis [10].

A third basic motif involves the synthesis of branched or tree-like nanowire structures (Fig. 17.5a, III) [83–85]. To this end, we reported a rational, multistep approach toward the general synthesis of 3D branched nanowire heterostructures

[83]. Single-crystalline semiconductor, including groups IV, III–V, and II–VI, and metal branches have been selectively grown on core or core/shell nanowire backbones, with the composition, morphology, and doping of core (core/shell) nanowires and branch nanowires well controlled during synthesis.

Although the first structural motif has been used most extensively as building blocks of basic platforms, the second and third motifs have much higher level of structural and functional complexity, and show great potential of bottom-up synthesis to yield increasingly powerful functional building blocks for advanced platforms.

17.4 Multiplexed Extracellular Electrical Recording

Natural and synthetic cellular assemblies are usually organized into 2D or 3D hierarchical networks operating on spatial and temporal scales that span multiple orders of magnitude. Advances in microfabrication of high-density passive multi-electrode arrays (MEAs) and active transistor arrays on silicon substrates enable direct electrical recording down to ca. 10 Âμm length scales, although it is important to recognize that signals recorded within ~100 μm are often correlated [59, 77, 86], and moreover, it has been difficult to to resolve the cellular signals at the single cell level. As mentioned above, simply reducing the size of individual metal electrodes to achieve more localized detection is not viable due to corresponding increases in their impedance [65, 87], which intrinsically limits the resolution of such passive recording devices.

Silicon nanowire nanoFET arrays have several features that make them unique for high resolution multiplexed extracellular recording from cellular systems. First, previous studies have shown that nanowire nanoFETs can exhibit ultra-high sensitivity detection of charged biomolecules, including detection of single particles [24]. Second, bottom-up fabrication of nanoFETs yields devices that have nanoscale protrusions from the substrate surface [24, 88]. This can reduce device to cell/tissue separation and promote enhanced cell-nanostructure interaction and has resulted in high S/N extracellular recording of field potentials from cultured cells and cardiac tissue with signals improved compared to planar FETs. Third, the bottom-up approach also enables high-performance nanoFET fabrication on transparent, flexible and stretchable substrates [33, 36, 89, 90]. The freedom to design device structures and arrays on substrates adapted to specific biological applications also opens up new possibilities for interfacing with living tissues, for example, bio-resorbable and implantable devices [91–94]. This freedom also allows other measurements or manipulations to be performed in conjunction with nanoFET recordings, such as high-resolution optical imaging. Fourth, the active junction area of typical nanoFETs, 0.01–0.1 μm^2, is much smaller and can provide better spatial resolution of signals compared to MEA and planar FETs that are 10^2–10^5 times larger in active area [47]. Last, nanoFET detectors provide fast intrinsic response time which is critical for high temporal resolution recordings [81, 95].

17.4.1 Electrical Interfacing with Cultured Neurons

An early example of multiplexed nanoFET recording layout consists of a neonatal rat cortical neuron and four peripheral silicon nanoFETs that are arranged at the corners of a rectangle, where polylysine patterning was used to promote axon and dendrites growth across single nanoFETs [9] (Fig. 17.6a). This multiplexed

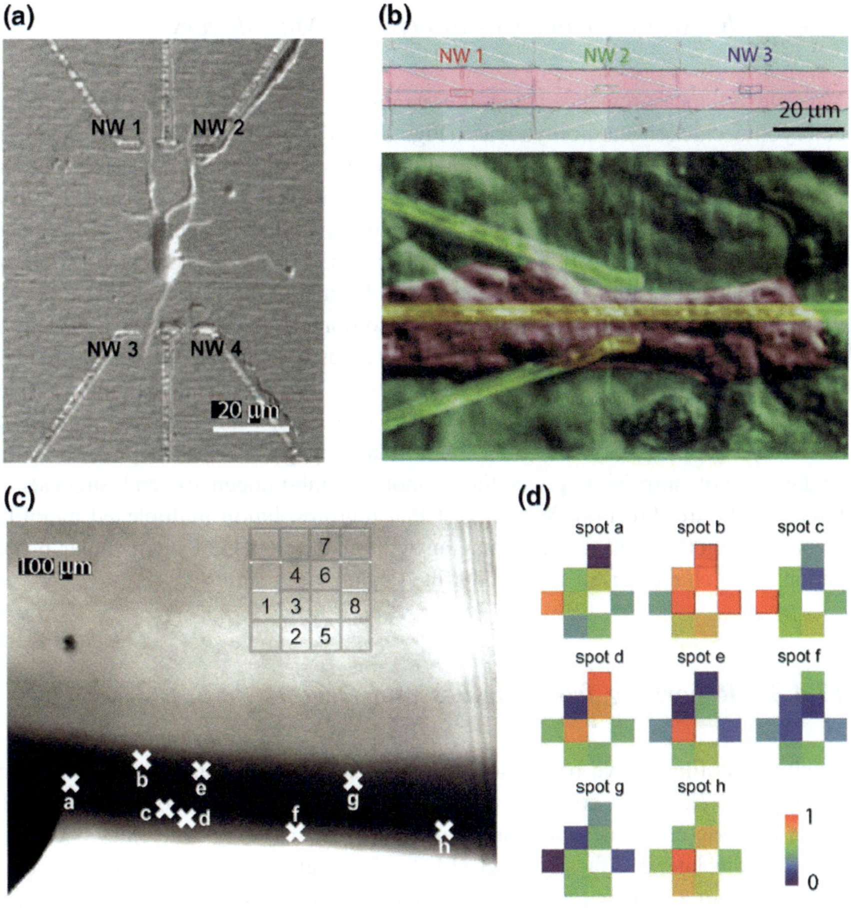

Fig. 17.6 Multiplexed extracellular electrical recordings using nano FETs. a Optical image of a cortical neuron interfaced to three of the four functional nanoFETs in an array. **b** Upper panel, optical micrograph showing three nanoFET devices (NW1, NW2, and NW3) in a linear array, where pink indicates the area with exposed NW devices. Lower panel, a differential interference contrast bright field image showing individual cardiomyocytes (purple) and single nanoFETs (yellow). **c** Optical image of an acute slice over a 4 × 4 NWFET array. Signals were recorded simultaneously from the eight devices indicated on the image. Crosses along the LOT fiber region of the slice mark the stimulation spots **a–h**. The stimulator insertion depth was not controlled precisely in these experiments. **d** Maps of the relative signal intensity or activity for devices 1–8

nanoFET/neurite hybrid was used to study spike propagation with NW1 as a local input to elicit action potential spikes. After stimulation with a biphasic pulse sequence, back propagation of the elicited action potential was detected in the two dendrites crossing elements NW2 and NW3. The lack of observed signal from NW4 demonstrates the absence of crosstalk in the hybrid device array, and thus the capability for multiplexed subcellular resolution detection.

17.4.2 Recording from Cardiomyocyte Monolayers

Multiplexed measurements using the nanoFET arrays interfaced with cultured embryonic chicken cardiomyocytes (Fig. 17.6b) [96]. The nanoFETs were patterned in a linear array with an average spacing of 300 μm so that signal propagation within cardiomyocyte monolayers could be characterized. Recording from multiple nanoFETs in contact with spontaneously beating monolayer yielded very stable and high S/N (>10) field potential spikes. In this experiment, the relative large signal magnitude confirmed that a good junction is formed between each of the nanoFETs and PDMS/cell substrate. Additionally, a cross-correlation method was used to determine robustly the time differences between the signals recorded by the devices. The time shifts between devices and device separations yielded propagation speeds of 0.07–0.21 m/s that are consistent with other measurement on cardiomyocyte monolayers. The variation in propagation speeds in these initial studies is not surprising given the monolayer inhomogeneity and suggests an important future direction. We suggest that high-resolution multiplexed nanoFET recording together with optical imaging will enable details of intercellular propagation to be characterized for well-defined cellular structures.

17.4.3 Recording from Tissues and Organs

Finally, nanoFETs have been used to probe electrical activities from tissues and organs [47, 97]. To this end, we have studied the activity patterns of layer II/III cells in the piriform cortex of acute rat brain slices by stimulating different sets of axon fibers in the lateral olfactory tract (LOT). In a representative experiment, eight devices within a four-by four 2D array oriented under the pyramidal cell layer of an acute slice were simultaneously monitored following stimulation at eight different spots (a–h) in the LOT [47] (Fig. 17.6c). Strong stimulation of all axons fibers in the LOT yielded similar response by nanoFETs 1–8 with clear population spike signals (postsynaptic activities) regardless of stimulation positions. Reduced stimulation intensity was also used so that at each spot only a subgroup of fibers was activated. Notably, visual inspection of 2D activity maps for each of the eight stimulation positions demonstrates clearly how heterogeneous activity can be resolved (Fig. 17.6d), and thus define a complex functional connectivity in the piriform cortex.

17.4.4 Challenges and Promises

Although great progress has been made in the extracellular electrical recordings using nanowire nanoFETs, many challenges remain. For example, there is still a pressing need to further enhance the nanoFET S/N so that very weak endogenous biological signals, with the amplitude of $\sim 100\ \mu V$, can be readily resolved. We can potentially achieve this goal by [98] new chemical design and synthesis of high mobility nanowire building blocks for nanoFET, [99] nanoscale engineering of nanowire materials to reduce nanoFET noise by, for example, thermal annealing and/or surface passivation. It is also important to note that the high input impedance of the nanoFETs circumvents the common challenges confronted by implanted microelectrodes, where gradual increases of single terminal device impedance due to, for example, absorption of proteins, leads to degraded S/N over time [47, 100, 101]. This feature makes nanoFETs very promising for multiplexed, in vivo chronic recordings. This is particularly true considering the facts that (i) nanoscale device feature size allows integration of multiple nanoFETs on minimally invasive and movable electrophysiological probes [44], (ii) bottom-up fabrication makes it possible to choose biocompatible or even biodegradable materials as substrates to reduce mechanical mismatch and to minimize inflammatory tissue response [10, 42, 44, 91–94], and (iii) the nanoscale topology could be arbitrarily designed de novo to promote better attachment of single cells or even intracellular contacts. Therefore, nanoFETs should bring many exciting opportunities to interfacing living tissue and organs with electronics for biomedical applications (e.g., diagnostic devices for brain trauma and surgical tools for cardiac therapy), and even new cybernetic biosystems for hybrid information processing.

17.5 Intracellular Electrical Recording

As the key cellular component, lipid membranes represent important structural and protective elements of the cell that form a stable, self-healing, and virtually impenetrable barrier to the ions and small molecules [102]. Since these membranes have resistance (R) and capacitance (C), the membrane RC circuit also behaves as an electrical barrier and would attenuate and even distort the intracellular signals as they are detected by extracellular sensors. More importantly, although cellular signal transduction often starts with an extracellular signaling molecules activating a cell surface receptor, it is the subsequent intracellular processing that eventually creates a cellular response. Deciphering of such intracellular signal transmission and amplification processes is critical to the understanding of cellular information flow and cell physiology. Therefore, it is highly desirable to deliver nanoFETs into the cell and directly record intracellular electrical activities, which can provide much more detailed understanding of the inner workings of cells.

Although nanoFETs have been exploited for ultrasensitive detection of biological markers and high-resolution extracellular recording from cells [24], localized and tunable intracellular sensing and recording had not been demonstrated prior to our work because all FET and nanoFET devices were created on planar substrates—using the basic nanoFET platform. Ideally, rather than force the cell to conform to the substrate, a movable and 3D nanoFET with the necessary source (S) and drain (D) electrical connections could move into contact with the cell and probe within the cell membrane. However, minimally invasive insertion of a nanoFET into the confined 3D space of single cells, or even 3D cellular networks, was still a major challenge before year 2010 because the S and D typically dominated the overall device size and defined a planar and rigid structure, regardless of whether the nanoFET was on or suspended above a substrate. An advanced nanoFET platform that is designed specifically for intracellular measurement is needed to meet this requirement [43–45, 82]. Three distinct examples that we have recently introduced to address this central challenge are shown schematically in Fig. 17.7a, and include (1) kinked nanowire nanoFET, (2) branched-intracellular nanotube nanoFET, and (3) active nanotube nanoFET devices.

Existing probes capable of intracellular sensing and recording include voltage-sensitive optical dyes or proteins [103–106], and single-terminal glass or carbon microelectrodes as mentioned briefly in prior section [46, 49] (Fig. 17.8). Voltage-sensitive dyes can readily be used to interrogate action potentials with high spatial resolution, but they still have limitations in terms of signal-to-noise (S/N) ratio, pharmacological side effects, phototoxicity, and difficulty in differentiating single spikes [104]. For electrical probes (Fig. 17.8), the single electrical connection facilitates design and mechanical insertion into cells, but the requirement of direct ionic and/or electrical junctions between probe tips and cytosol also introduce several limitations. First, the tip size of these probes ($\sim$0.2 to 5 μm) is a compromise between being small enough (<5 μm) to penetrate or rupture the cell membrane with minimum damage and large enough (>0.2 μm) to yield a junction impedance that is sufficiently low so that small cellular signals can be discerned from thermal noise. Second, direct exposure of intracellular species to extraneous probe surfaces or electrolytes in probe lumen, especially for larger glass micropipettes, might induce irreversible changes to cells and, thus, prevent long-term and noninvasive cellular recordings. Finally, these probe techniques are intrinsically passive and are not capable of built-in signal processing and facile integration with other circuitries, especially given the emerging need to enable a cell-machine communication [107–110].

NanoFETs can function in a sub-10-nm-size regime [99]. In principle, their exceptionally small size enables them to function as mechanically noninvasive probes capable of entering cells through endocytic pathways, as can occur with nanoparticles [111–114]. Moreover, when interfacing with cells, nanoFETs process

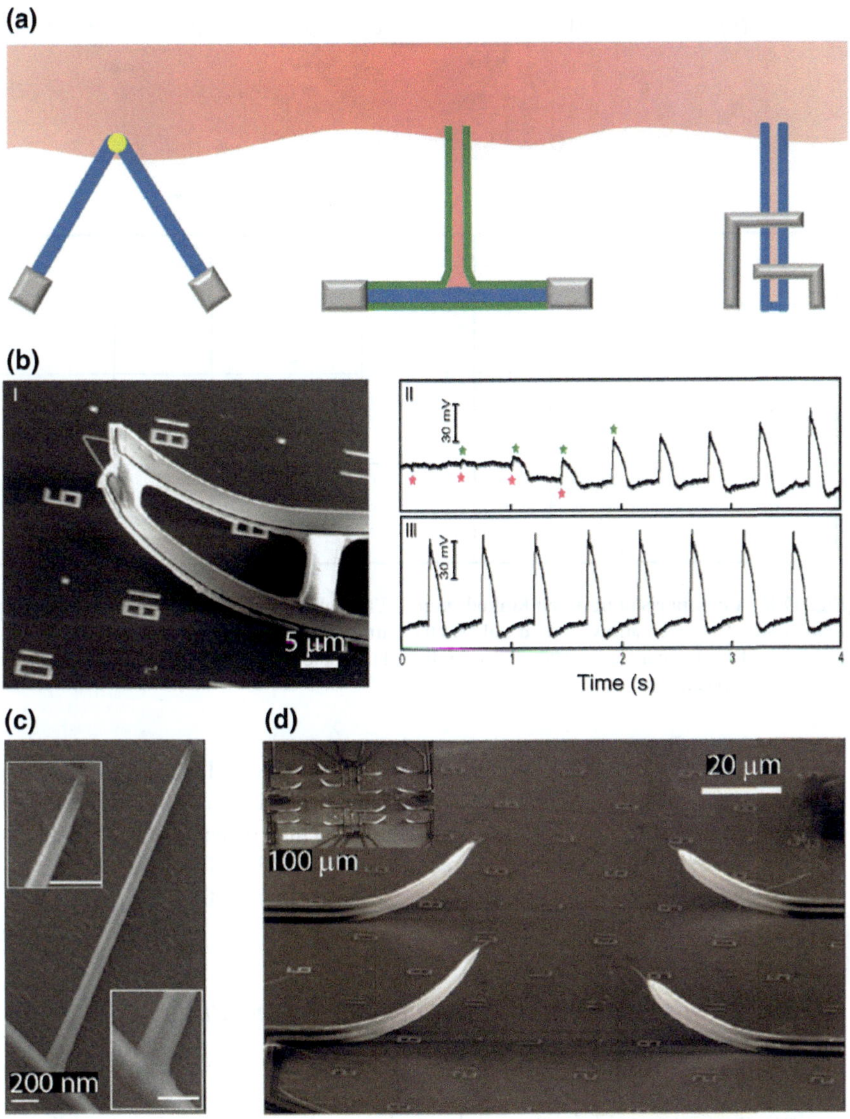

Fig. 17.7 Intracellular electrical recordings using nanoFETs. a Schematics of kinked nanoFET (left), BIT-FET (middle) and ANTT (right) probes. **b** SEM image of a kinked nanoFET probe (I) and its intracellular electrical recordings (II, III) from spontaneously beating cardiomyocytes. **c** SEM of a BIT-FET probe, insets highlight the tip and root parts of the hollow branch. **d** SEM image of ANTT probe array

input/output information without the need for direct exchange with cellular ions; thus, the issues of interfacial impedance and biochemical invasiveness to cells can be ignored or minimized (Fig. 17.8). In addition, because signals are transduced by

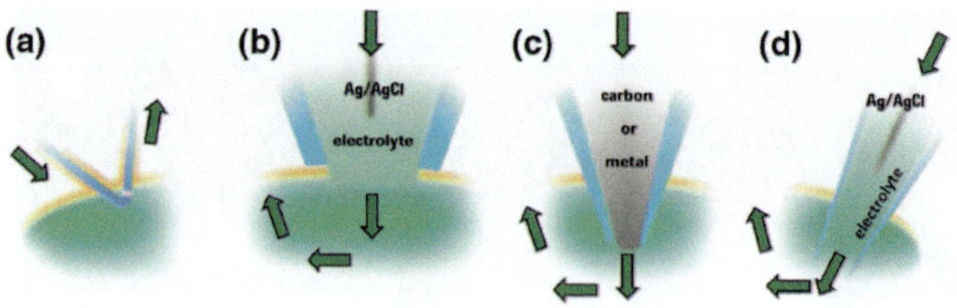

IC technique	Equivalent circuit	Size (nm)	Calibrations	Capabilities	Invasiveness	Cellular entrance
Glass micropipette (b and d)	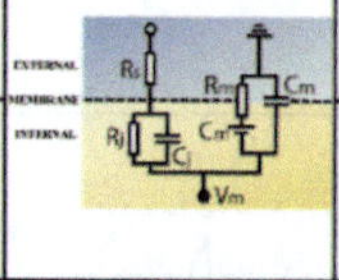	~ 50-5000 Impedance limited	Both amplitude and shape	Can record both current and voltage, Single ion channel to whole cell recording	Electrochemical and mechanical	Mechanical or electrical
Carbon or metal micro-/nano-electrode (c)		~500-1000 Impedance limited	Both amplitude and shape	Can record both current and voltage, Whole cell recording	Electrochemical and mechanical	Mechanical or electrical
nanoFET (a)	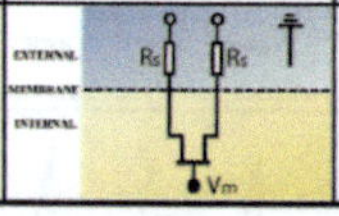	~10-100	Amplitude	Can only record voltage, Whole cell recording, Multiplexing is scalable, High spatiotemporal resolutions	Minimal	Biological

Fig. 17.8 A comparison between kinked nano FET probe (a) and conventional intracellular tools (b–d). The green arrows in **a–d** indicate the current flows. Rs, series resistance; Rj, junction resistance; Rm, membrane resistance; Vm, intracellular potential; Cj, junction capacitance; Cm, membrane capacitance

change in field/potential at well-isolated surfaces, nanoFETs can detect cellular potential, as well as biological macromolecules, and could be integrated for potential multiplexed intracellular measurements. Until recently, these advantages could not be exploited, although our recent work [10, 43–45] (Fig. 17.7a) has now shown three solutions that open up these exciting opportunities.

17.5.1 Designs and Implementation of Intracellular NanoFET Probes

In 2010, the first nanoFET intracellular probes were designed and chemically synthesized without lithography to encode a ~100 nm FET device at the apex of a kinked nanowire [10] (Fig. 17.7a, b). This was achieved through control over cis-/trans-conformations and modulation doping during the silicon nanowire synthesis [10, 82]. Subsequently, the free arms of such kinked nanowires were electrically contacted to free-standing and flexible electrodes. Electrical characterization of the 3D nanowire probes showed they were robust to mechanical deformation, recorded solution pH changes with high-resolution, and, when modified with phospholipid bilayers, recorded the intracellular potential of single cells. Significantly, electrical recordings of spontaneously beating cardiomyocytes demonstrated that the 3D nanoFET probes continuously monitored extra- to intracellular signals during cellular uptake for the first time. The nanometer size, biomimetic surface coating, and flexible 3D device geometry render these active semiconductor nanoprobes a new and powerful tool for intracellular electrophysiology.

The kinked nanoFET based intracellular recording represents the first example of interfacing semiconductor devices with cells intracellularly, but the kink configuration and device design also place certain limits on the probe size and the potential for multiplexing. To address these issues, we reported a new device platform in which a branched SiO_2 nanotube was synthetically integrated on top of a nanoFET (BIT-FET) [43] (Fig. 17.7a, c). This branched nanotube penetrated the cell membrane, bringing the cell cytosol into contact with the extracellular FET, thus allowing intracellular recording of transmembrane potential. Studies of cardiomyocyte cells demonstrated that when phospholipid-modified BIT-FETs are brought close to cells, the nanotubes spontaneously penetrate the cell membrane and yield full-amplitude intracellular action potentials, thus showing that a stable and tight seal forms between the nanotube and cell membrane. Significantly, we also showed that multiple BIT-FETs can be used for multiplexed intracellular electrical recordings from both single cells and networks of cells.

Recently, we also demonstrated a conceptually new and practically simple nanoFET probe that consists of a single semiconductor nanotube [44] (Fig. 17.7a, d). The fabrication of the active nanotube transistor (ANTT) intracellular probe involves the fabrication of S/D contacts to one end of a silicon or other semiconductor nanotube, and electrical isolation of these S/D contacts from surrounding medium. Then the solution filling the interior of the nanotube can gate the transistor and the variation of interior electrochemical potential is recorded as a change in device conductance. In experiments, the free end of ANTT probes were inserted into cardiomyocyte cells, and the time-dependent changes associated with action potential spikes were recorded by this nanoFET probe. As expected, if a similarly configured solid nanowire nanoFET was inserted into the cell, no signal was observed since it would not be possible to "gate" the nanoFET. Finally, the straightforward fabrication of ANTT devices was exploited to prepare multiple ANTTs at the end of single probes, which enabled multiplexed recording of full-amplitude intracellular action potentials from single cells, and multiplexed arrays of single ANTT device probes (Fig. 17.7d).

17.5.2 Challenges and Promises

Despite these advances, additional work remains to advance further the nanoFET-based intracellular measurement techniques (Fig. 17.8). For example, the S/N is, at current stage, not better than that from glass micropipette recordings although spatial resolution is much higher. The current designs of nanoFETs only enable potential recordings, but measurement of ionic currents is also possible if other signal transduction mechanisms are combined with nanoFET. Moreover, the capability for cell stimulation in addition to recording is still lacking. Nevertheless, we believe that the advantages of the nanoFET intracellular probes already demonstrated in our work, including the capability to realize sub-10 nm probes, ease of operations (e.g., there is no need to compensate/calibrate the probe junction potential and capacitance, etc.), the biomimetic cellular entrance, minimal mechanical and biochemical invasiveness, and the potential for large-scale, high-density, multiplexed recording, make them very attractive new measurement tools that will extend substantially the scope of fundamental and applied electrophysiology studies to regimes hard to access by current methods. For example, an exciting future application of these nanoFET probes will be measuring membrane potentials directly from cellular organelles, a Holy Grail in intracellular electrophysiology.

17.6 Nanoelectronics Innervated Synthetic Tissues

The development of synthetic 3D macroporous biomaterials as extracellular matrices (ECMs) represents a key area because (i) functionalized 3D biomaterials allow for studies of cell/tissue development in the presence of spatiotemporal biochemical stimulants [115, 116], and (ii) the understanding of pharmacological response of cells within synthetic tissues [117–119] is expected to provide a more robust link to in vivo disease treatment than that from 2D cell cultures. Advancing further such biomaterials requires capabilities for monitoring cells throughout the 3D microenvironment. While electrical sensors are attractive tools, it has not been possible to integrate such elements with porous 3D scaffolds for localized real-time monitoring of cellular activities and physicochemical changes.

Recent efforts in coupling electronics and tissues have focused on flexible, stretchable planar arrays that conform to tissue surfaces [24, 91–94, 97, 120], or implantable microfabricated probes [121]. These approaches have been used to probe electrical activities near surfaces of the heart, brain and skin, and they have shown translational potential. However, these new electronic tools are currently limited in merging electronics with tissues throughout 3D space while minimizing tissue disruption, because of the 2D support structures and the electronic sensors are generally much larger scale than the extracellular matrix (ECM) and cells. Studies using nanoFETs have shown that electronic devices with nanoscopic features were able detect extra- and intracellular potentials from single cells but had also been limited to surface or near surface recording from tissue and organs [24, 97]. Merging electronics seamlessly throughout tissues (Fig. 17.9a) had remained a major challenge. To address this challenge we recently set-forth the key constraints [42] include: (1) The electronic structures must be macroporous, not planar, to enable 3D interpenetration with biomaterials; (2) the electronic network should have nanometer to micrometer scale features comparable to biomaterial scaffolds; and (3) the electronic network must have 3D interconnectivity and mechanical properties similar to biomaterials (Fig. 17.9b).

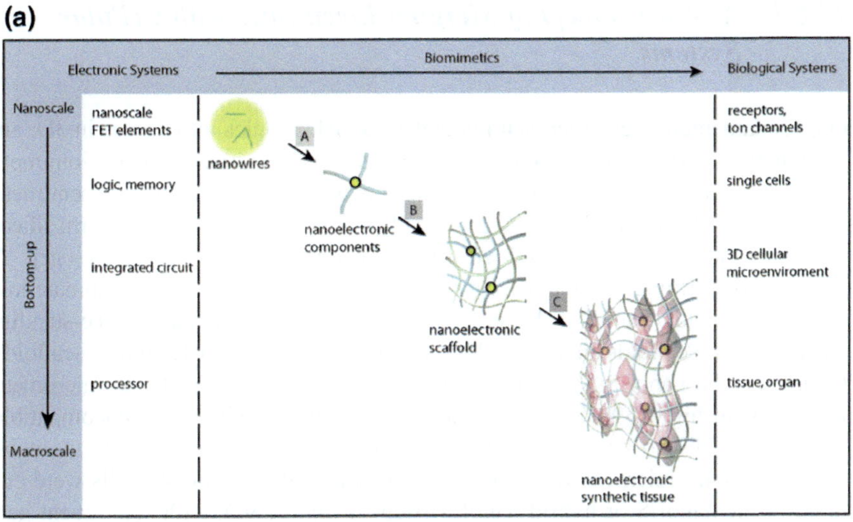

(b)

Biomimetics: Targets	Approaches	Implementation Examples	Steps		
Composition	+ Synthetic polymer	Deposit PLGA, alginate	A	B	
	+ Natural polymer	Add or secrete collagen, polysaccharides		B	C
	+ Cellular species	Biosynthesize proteins and lipids			C
	- Electronic materials	Reduce silicon, metals, dielectrics	A	B	
Structure	+ Porosity	Increase to > 90%	A	B	
	- Feature size	Decrease to sub-micrometer	A	B	
Mechanics	+ Flexibility	Decrease bending stiffness to < 10 nN-m	A	B	C
	+ Stretchability	Use serpentine device designs	A	B	C
Bio-functions	+ Morphogenesis	Enable cell growth and proliferation			C
	+ Other Signalings	Activate ECM cues		B	C
	- Cytotoxicities	Use biocompatible materials	A	B	

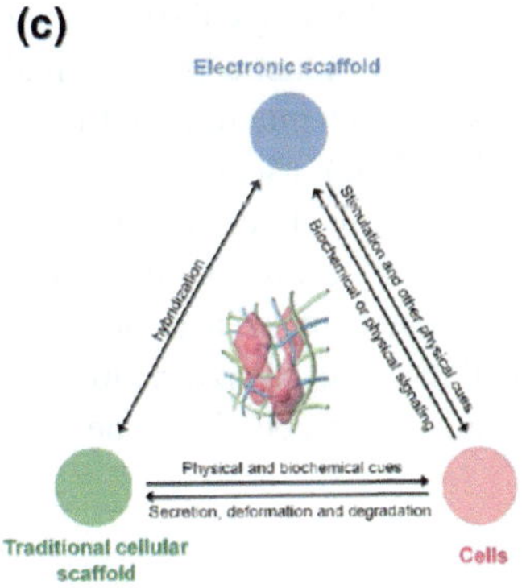

Fig. 17.9 Integrating nanoelectronics with cells and tissue. Conventional bulk electronics are distinct from biological systems in composition, structural hierarchy, mechanics and function. Their electrical coupling at the tissue/organ level is usually limited to the tissue surface, where only boundary or global information can be gleaned unless invasive approaches are used. **a** A new concept was introduced where an integrated system can be created from discrete electronic and biological building blocks (for example, semiconductor nanowires, molecular precursors of polymers and single cells). Three biomimetic and bottom-up steps have been designed: step A, patterning, metallization and epoxy passivation for single-nanowire FETs; step B, forming 3D nanowire FET matrices (nanoelectric scaffolds) by self or manual organization and hybridization with traditional ECMs; step C, incorporation of cells and growth of synthetic tissue through biological processes. Yellow dots: nanowire components; blue ribbons: metal and epoxy interconnects; green ribbons: traditional ECMs; pink: cells. **b** Rationale and approaches for biomimetic implementation of nanoelectronics innervated synthetic tissues. A, B and C are the same steps used in **a**. **c** The new electronic scaffold component in synthetic tissues enables additional interactions with traditional cellular scaffold and cells

17.6.1 A New Concept of Merging Electronics with Cellular Systems

Our fundamentally new approach integrates nanoelectronics into tissues in 3D, and the integrative synthetic approach involved stepwise incorporation of biomimetic and biological elements into nanoelectronic networks across nanometer to centimeter size scales [42] (Fig. 17.9a). First, chemically synthesized kinked or uniform silicon nanowires were registered and electrically connected to yield FETs (step A, Fig. 17.9a), forming the nanoelectronic sensor elements for hybrid biomaterials. Second, individual nanoFET devices were arranged and integrated into free-standing macroporous scaffolds (step B, Fig. 17.9a), termed 'nanoelectronic scaffolds' (nanoES). The nanoES were tailored to be 3D, to have nanometer to micrometer features with high (>99%) porosity, and to be highly flexible and biocompatible. NanoES could also be hybridized with biodegradable synthetic ECMs to enable suitable cellular microenvironments prior to tissue culture. Finally, cells were cultured inside nanoES or hybrid nanoES (step C, Fig. 17.9a), with subsequent generation of biological species and the merging of cells with nanoelectronics in 3D. The entire biomimetic process make a natural transition from electronic to biological systems by integrating the third component, nanoES, into the synthetic tissues (Fig. 17.9c). Metal electrode or carbon nanotube/nanofiber based passive detectors are not considered in our work because impedance limitations (i.e., signal/noise and temporal resolution degrade as the area of the metal or carbon electrodes is decreased) make it difficult to reduce the size of individual electrodes to the subcellular level, a size regime necessary to achieve noninvasive 3D interface of electronics with cells in tissue.

17.6.2 Designs and Preparation of Synthetic Tissues

We have designed two types of 3D macroporous nanoES (reticular- and mesh-nanoES) to mimic the structure of natural tissue scaffolds (Fig. 17.10) [42]. These nanoES were formed by self-organization of coplanar reticular networks with built-in strain (Fig. 17.10a) and by manual manipulation of 2D mesh matrices (Fig. 17.10b). We showed that nanoES exhibited robust electronic properties and could be used alone or seamlessly merged with other biomaterials as biocompatible extracellular scaffolds for efficient 3D culture of neurons, cardiomyocytes and smooth muscle cells (Fig. 17.10c, d). Significantly, we have demonstrated multiplexed electrical recordings of extracellular field potentials from 3D nanoelectronic innervated cardiac patches, including the effects of drugs (Fig. 17.10e, f). The results suggested the feasibility of continuous electrical monitoring of engineered tissue in 3D for in vitro therapeutic assays. Finally, we have used 3D distributed

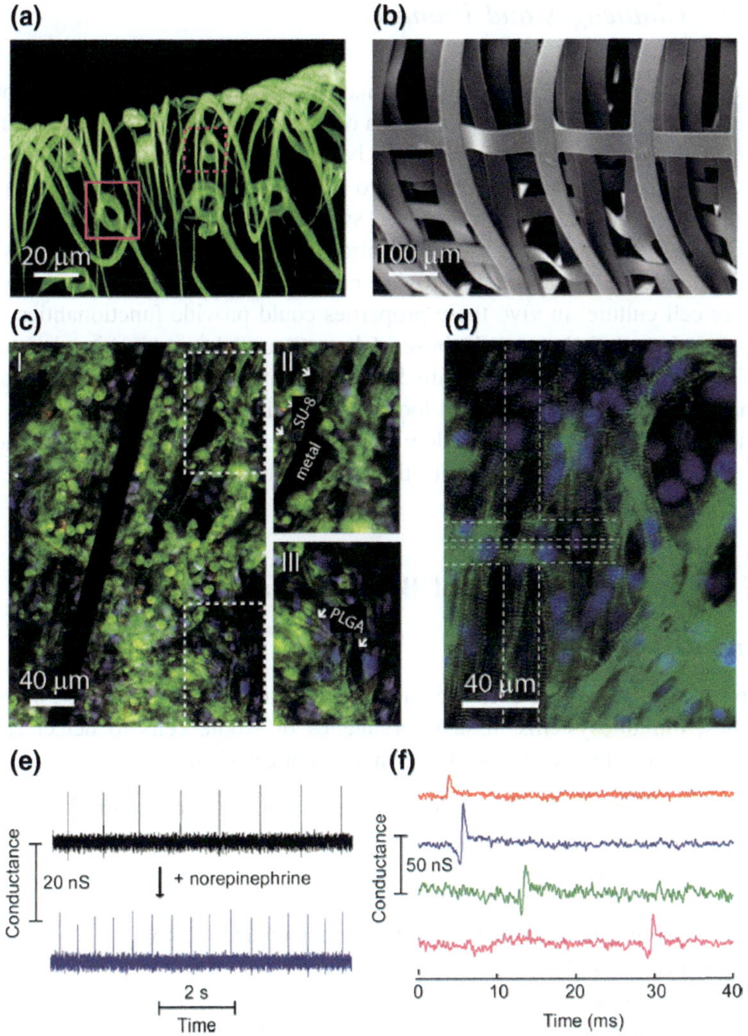

Fig. 17.10 NanoES exhibited robust electronic properties recorded from a single-nanowire FET before (black) and after (blue) applying noradrenaline. **f** Multiplex electrical recording of extracellular field potentials from four nanowire FETs in a mesh nanoES. Data are conductance versus time traces of a single spike recorded at each nanowire FET

nanoelectronic devices for simultaneous monitoring of pH inside and outside an engineered tubular vascular construct that was developed from the nanoelectronic scaffold, suggesting the potential of a multifunctional prosthetics.

17.6.3 Challenges and Promises

The results open up a new field whereby nanoelectronics are merged with biological systems in 3D, and as in any nascent area opportunities and challenges abound. For example, the sensing capabilities could be broadened to address various disease states, in vitro (organ-on-a-chip) or in vivo [122] by exploiting the diverse nanowire building blocks available from designed synthesis. Cell or tissue interactions with nanoES could be fine tuned by modification with cell growth determinants [117]. NanoES could be enhanced to provide electrical and mechanical stimulation to enhance cell culture; in vivo these properties could provide functionalities such as pacing, and moduli that match those of host tissues. Long-term in vivo biocompatibility of nanoES should be studied. One can envision nanoES-based tissues that are hard-wired to provide closed-loop systems that sense and treat, that enable telemetric monitoring of physiological processes, or that provide connections between engineered constructs with the host nervous system.

17.7 Application Areas of Biosensors and -Assays

Definitions

Biosensor: A device that uses specific biochemical reactions mediated by isolated enzymes, immunosystems, tissues, organelles or whole cells to detect chemical compounds, usually by electrical, thermal or optical signals.

Bioassay: A bioassay is a procedure for determining the concentration, purity, or biological activity of a substance by measuring the biological response that it produces compared to a standard

- Deoxyribonucleic acid (DNA) and ribonucleic acid (RNA)
- Hormones (steroids)
- Proteins (polypeptides)
- Immune globulins IgG, IgM, IgA, IgD, IgE $\Rightarrow$ immunoassays (antibody-antigen reaction)

Determination of:

- a single analyst $\Rightarrow$ Single analyst assays
- several analysts $\Rightarrow$ Multi analyst assays.

Biosensors can be applied in detection of:

- Glucose in blood
- Cancer markers in blood
- Penicillin in fungi bioreactors
- Urea in urine.

Principle of operation

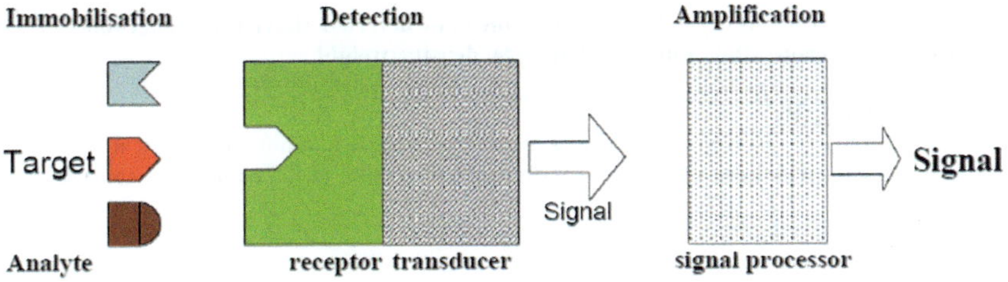

Transducer

A device that converts the physical or chemical changes due to analyte receptor reaction to another form of physical signal (in general, electronic signals) whose magnitude is proportional to the amount of the analyte

Electrochemical detection	Potentiometric, Voltammetric, Conductimetric
Optical detection	Fluorescence, Absorbance, Light scattering, Refractive index
Electrical detection	Field effect transistor (FET)

Mechanical, Thermal, Piezoelectric, Surface acoustic waves, Magnetical, ...

Principle of operation

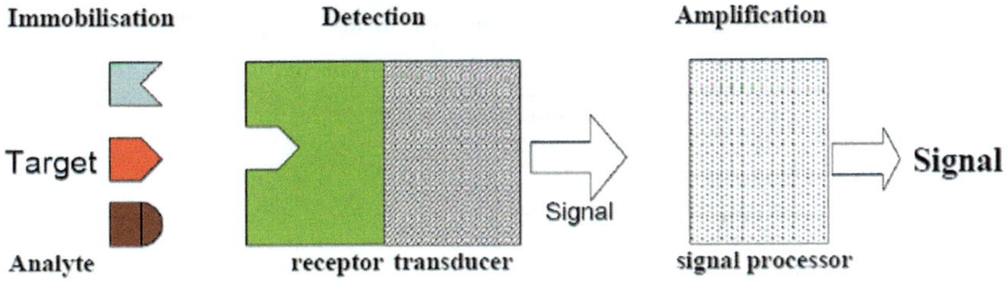

Analyte

- The substances to be measured
- Small molecules: Sugars, urea, cholesterol, glutamic acid, phosphate, ..
- Macro molecules: Nucleic acids (DNA, RNA), poly peptides (protein, antibody, enzyme)

Receptor

- A sensing element that responds to the substances being measured
- The interaction must be highly selective ⇒ Enzyme, Antibody, Nucleic acids, Cells

Performance factors

Sensitivity

- Minimum amount of analyte that are able to be detected above the background
- Units: Concentration, number of analyte, density, weight

Specificity/Selectivity

- The ability to discriminate between substrates. This is function of biological component, principle, although sometimes the operation of the transducer contributes to selectivity
- Molecular recognition
- Separation scheme
- Signal overlap

Speed/Response Time

- Sample preparation + Biological/Chemical reaction + Signal Processing
- Bench process : hours to weeks
- Chip process: minutes to hours
- Ultra-high temporal resolution, 10 ns, for real-time measurement of molecular kinetics

Moreover: Accuracy, Simplicity, Cost, Life time, …

Examples

Conversion of bio molecules by an enzyme bound to a surface (e.g. polyaniline)

Enzyme	Reaction
GOD	glucose + O_2 → gluconic acid + H_2O_2
Urease	$(NH_2)_2CO + H_2O \rightarrow 2\,NH_3 + CO_2$
Catalase	$H_2O_2 \rightarrow H_2O + O_2$
Trypsin	polypeptide → amino acids
Amylase	starch → glucose
Uricase	uric acid + H_2O → $NH_3 + CO_2$

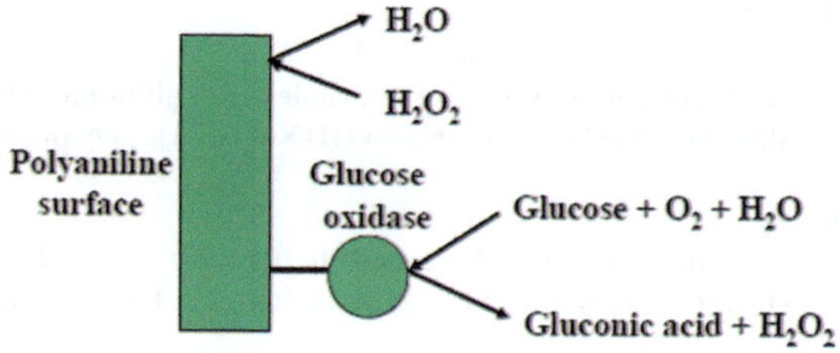

Biosensing of macromolecules

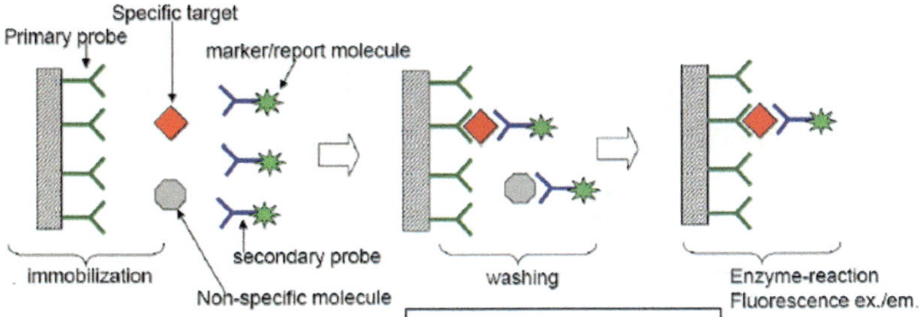

Steps

(1) Immobilization of primary probes
(2) Mixing/incubation of the mixtures
(3) Washing the non-specific bindings
(4) Signal transduction

Markers/Report molecules

(1) Enzymes
(2) Fluorescence tags
nanoparticles
fluorescent dyes
(3) Radioactive tags

Molecular recognition

(1) Watson-Crick base pairing ATTGGC (target) + TAACCG (probe) → ATTGGC / TAACCG
(2) Antibody-antigen binding Ab + Ag → Ab-Ag

Nanowire nanosensor

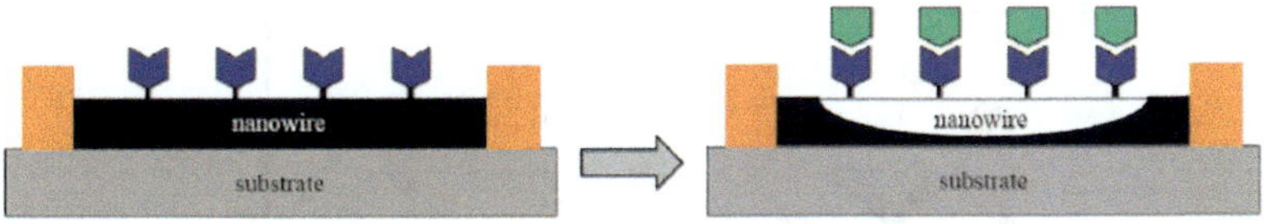

Binding of chemical or biological species to the surface of a nanowire will result in depletion or accumulation of carriers.
The change in carrier concentration due to binding can be directly monitored by measuring the nanowire conductance.

A solid state FET, whose conductance is modulated by an applied gate, is transformed into a nanosensor by modifying the silicon oxide surface. The conductance of modified Si-NWs increases stepwise with discrete changes in pH from 2 to 9. Changes in the surface charge can chemically-gate the Si-NW.

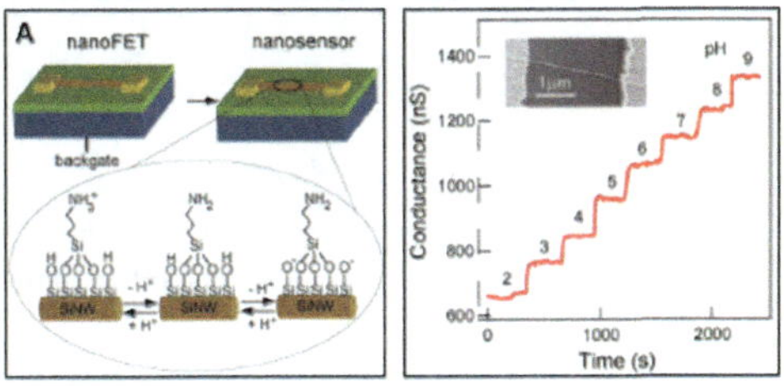

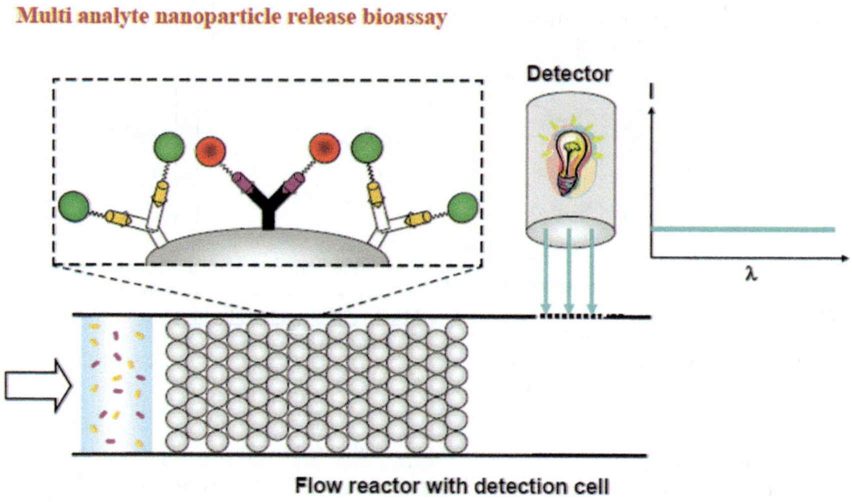
Multi analyte nanoparticle release bioassay
Detector
I
λ
Flow reactor with detection cell

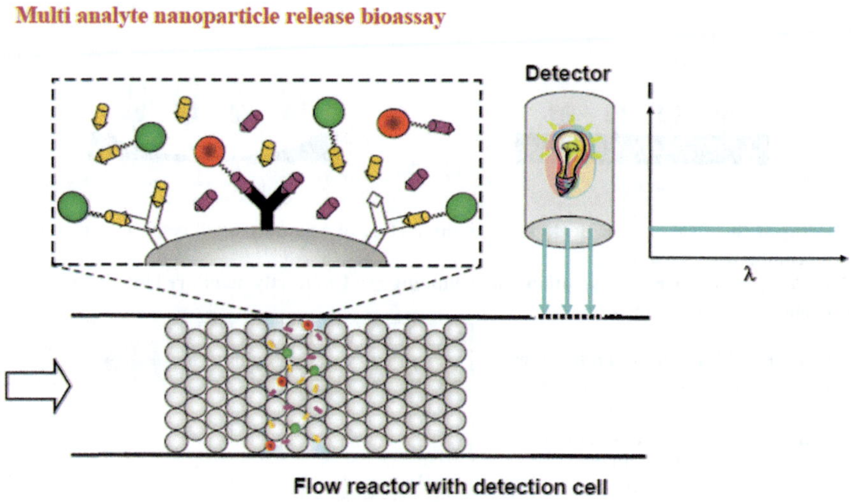
Multi analyte nanoparticle release bioassay
Detector
I
λ
Flow reactor with detection cell

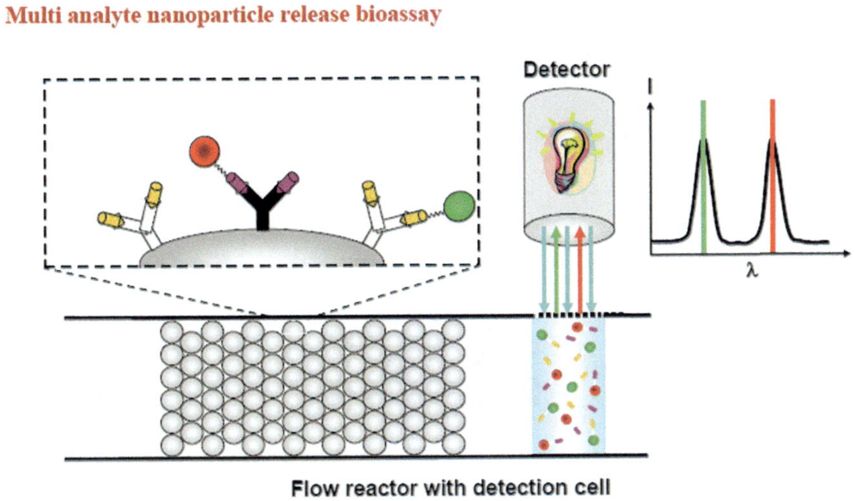

17.8 Selection of Inorganic-Binding Proteins Through Display Technologies

There are several possible ways of obtaining polypeptide sequences with specific affinity to inorganics. A number of proteins may fortuitously bind to inorganics, although they are rarely tested for this purpose. Inorganic-binding peptides may be designed using a theoretical molecular approach similar to that used for pharmaceutical drugs. This is currently impractical because it is time consuming and expensive. Another possibility would be to extract biomineralizing proteins from hard tissues followed by their isolation, purification and cloning. Several such proteins have been used as nucleators, growth modifiers, or enzymes in the synthesis of certain inorganics. One of the major limitations of this approach is that a given hard tissue usually contains many proteins, not just one, all differently active in biomineralization and each distributed spatially and temporally in complex ways. Furthermore, tissue-extracted proteins may only be used for the regeneration of the inorganics that they are originally associated with, and would be of limited practical use. The preferred route, therefore, is to use combinatorial biology techniques. Here, a large random library of peptides with the same number of amino acids, but of different sequences, is used to mine specific sequences that strongly bind to a chosen inorganic surface. Since their inception, well-established in vivo combinatorial biology protocols (for example, phage display (PD) and cell-surface display (CSD)) have been used to identify biological ligands and to map the epitope (molecular recognition site) of antibodies. Libraries have also been screened for various biological activities, such as catalytic properties or altered affinity and specificity to target molecules in many applications including the design of new drugs, enzymes, antibodies, DNA-binding proteins and diagnostic agents. The power of display technologies relies on the fact

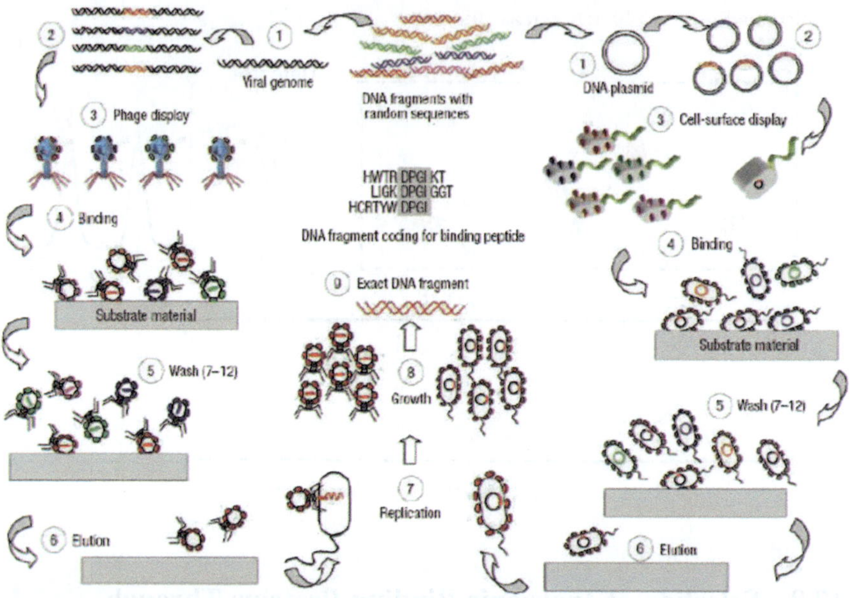

Fig. 17.11 Phage display and cell-surface display. Principles of the protocols used for selecting polypeptide sequences that have binding affinity to given inorganic substrates (Sarikaya, C. Tamerler, A.K.Y. Jen, K. Schulten F. Baneyx, Molecular biomimetics: nanotechnology through biology, Nature Materials, vol 2, 2003, pp. 577–585)

that an a priori knowledge of the desired amino acid sequence is not necessary, as it can simply be selected and enriched if a large enough population of random sequences is available. In vitro methods, such as ribosomal and messenger RNA display technologies, have been developed for increased library size (10^{15}) compared to those of in vivo systems (10^{7-10}). Combinatorial biology protocols can be followed in molecular biomimetics to select polypeptide sequences that preferentially bind to the surfaces of inorganic compounds chosen for their unique physical properties in nano- and biotechnology. Libraries are generated by inserting randomized oligonucleotides within certain genes encoded on phage genomes or on bacterial plasmids (step 1 in Fig. 17.11).

This leads to the incorporation of a random polypeptide sequence within a protein residing on the surface of the organism (for example, the coat protein of a phage or an outer membrane or flagellar protein of a cell; step 2). The eventual result is that each phage or cell produces and displays a different, but random peptide (step 3). At this stage, a heterogeneous mixture of recombinant cells or phages are contacted with the inorganic substrate (step 4). Several washing cycles of the phages or the cells eliminate non-binders by disrupting weak interactions with the substrate (step 5). Bound phages or cells are next eluted from the surfaces (step 6). In PD, the eluted phages are amplified by reinfecting the host (step 7). Similarly in CSD, cells are allowed to grow (steps 7, 8). This step completes a round of biopanning. Generally, three to five cycles of biopanning are repeated to

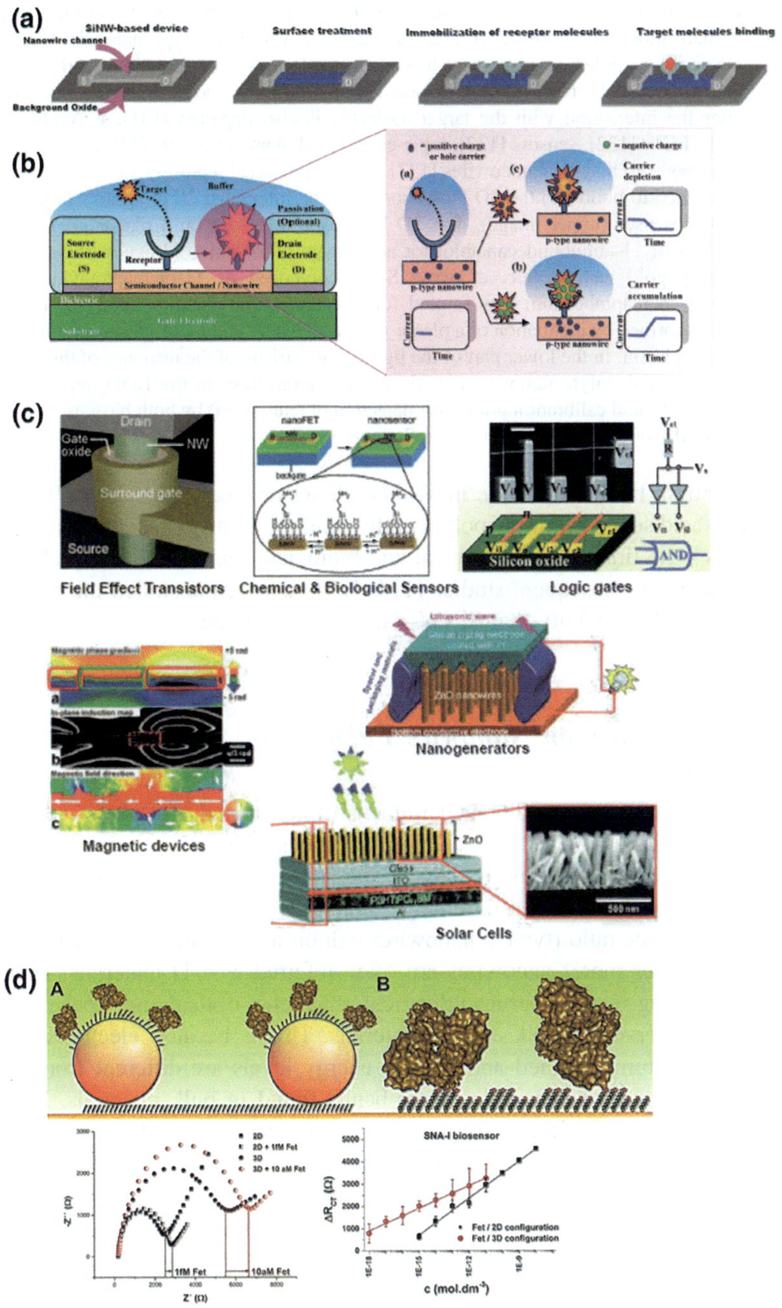
(a)
SiNW-based device
Nanowire channel
Background Oxide
Surface treatment
Immobilization of receptor molecules
Target molecules binding
(b)
Target
Buffer
Passivation (Optional)
Source Electrode (S)
Receptor
Drain Electrode (D)
Semiconductor Channel / Nanowire
Dielectric
Gate Electrode
Substrate
= positive charge or hole carrier
= negative charge
Carrier depletion
Carrier accumulation
p-type nanowire
Current
Time
(c)
Drain
Gate oxide
NW
Surround gate
Source
nanoFET
nanosensor
backgate
Silicon oxide
AND
Field Effect Transistors
Chemical & Biological Sensors
Logic gates
Nanogenerators
ZnO nanowires
Magnetic devices
ZnO
Glass
ITO
Solar Cells
500 nm
(d)
A
B
2D
2D + 1fM Fet
3D
3D + 10 aM Fet
1fM Fet
10aM Fet
-Z'' (Ω)
Z' (Ω)
SNA-I biosensor
Fet / 2D configuration
Fet / 3D configuration
ΔR_CT (Ω)
c (mol.dm^-3)

◂**Fig. 17.12** Schematic illustrating a Si nanowire-based device subjected to surface treatment, bio-functionalization and exposure to target molecules [130]. **a** A p-type nanowire-base sensor, along with receptor molecule attachment to the surface for specific recognition and capture of a target molecule from a buffer solution. The detailed mechanism to modulate the conductance prior to and after the interaction with the target molecule is also depicted [131]. **b** Applications of nanowires as FETs [132], sensors [133], logic gates [134], nanogenerators [135], solar cells [136] and applications with magnetic nanowires [137] **c** and A graphical representation drawn to scale of interfaces applied to build (A) the 3D biosensor based on integrated 20 nm gold nanoparticles or (B) the 2D biosensor (upper image). The 3D biosensor was built on a planar gold surface by chemisorptions of 11-aminoundecanethiol for attachment of 20 nm gold nanoparticles (spheres). (A) On every gold nanoparticle, a mixed SAM composed of 11-mercaptoundecanoic acid (MUA) and 6-mercaptohexanol was formed for covalent immobilization oflectin. (B) The 2D biosensor was formed by incubation of a planar gold with MUA and mercaptohexanol for covalent attachment of a lectin. In the lower part of the figure, comparison of the response of the 2D and the 3D biosensor to its analyte fetuin (Fet) with concentration close to the LOD, represented in a Nyquist plot (left), and calibration graphs for detection of fetuin (Fet) by both biosensors (right) are shown [138]. **d** Copyright 2014, Enterprise Strategy Group

enrich for tight binders. Finally, individual clones are sequenced (step 9) to obtain the amino acid sequence of the polypeptides binding to the target substrate material.

SiNWs were first used as chemical and biological sensors by Cui et al. in 2001 [123], and many subsequent studies [124–129] have demonstrated the robustness, sensitivity, and specificity that SiNW sensors can provide.

17.8.1 Overview on Nanowire Fabrication

Nanowire structures have attracted immense interest and have been emerged as promising building blocks for many applications (Fig. 17.12), implemented as nanoscale electronic devices, or having the role of interconnections and wiring, linking components into extremely small circuits. Thank to their high surface-to-volume ratio (typical nanowires exhibit aspect ratios i.e. length-to-width ratio, of 1000 or more) nanowires are often referred as 1-D materials, exhibiting many interesting new properties (electrical, magnetic, optical and mechanical) not previously observed in bulk or 3-D materials. This is because electrons in nanowires are quantum confined and occupy energy levels are different from the traditional continuum of energy levels or bands found in bulk materials. Nanowire structures exhibit tunable electron transport properties, that lead to a highly sensitive electrical response and consequently their electrical properties are also strongly influenced by even minor perturbations.

Nanowire structures may consist of inorganic materials or of organic materials. Many different types of nanowires exist, including metallic (i.e., Ni, Pt, Au), semiconducting (i.e., Si, InP, Si, GaN), insulating (i.e., SiO_2, TiO_2), multi-segment and molecular (i.e., DNA, MoSI) that involve repeating organic or inorganic molecular units. The synthesis of nanowire structures can be performed (a) via spontaneous growth, a growth driven by the reduction of Gibbs free energy or chemical potential, that involves evaporation (dissolution) condensation, vapor (or solid) liquid-solid growth and stress induced re-crystallization, (b) via template based synthesis that includes electrochemical deposition, electrophoretic deposition, colloid dispersion, melt or solution filling and conversion with chemical reactions, (c) via electro-spinning and (d) by lithographic techniques followed by etching and others techniques.

In the fabrication of nanowires, there are two main approaches that are widely applied: the "top-down" method and the "bottom-up" method. In the top-down methodology, the structures are etched out of a substrate whereas in the bottom-up the structures are grown and subsequently assembled. Bottom-up methods include a large variety of techniques, allowing the growth of nanowires or nanowire-heterostructures of different lengths, crosssectional profiles, materials and properties.

Self-assembly, the coordinated action of independent entities to produce larger, is the principle behind bottom-up processing. A typical growth-method involves the presence of catalyst nanoparticles placed in a reactive atmosphere containing precursor species. The precursor reacts with the surface of the nanoparticle and a wire grows in a certain direction from the particle.

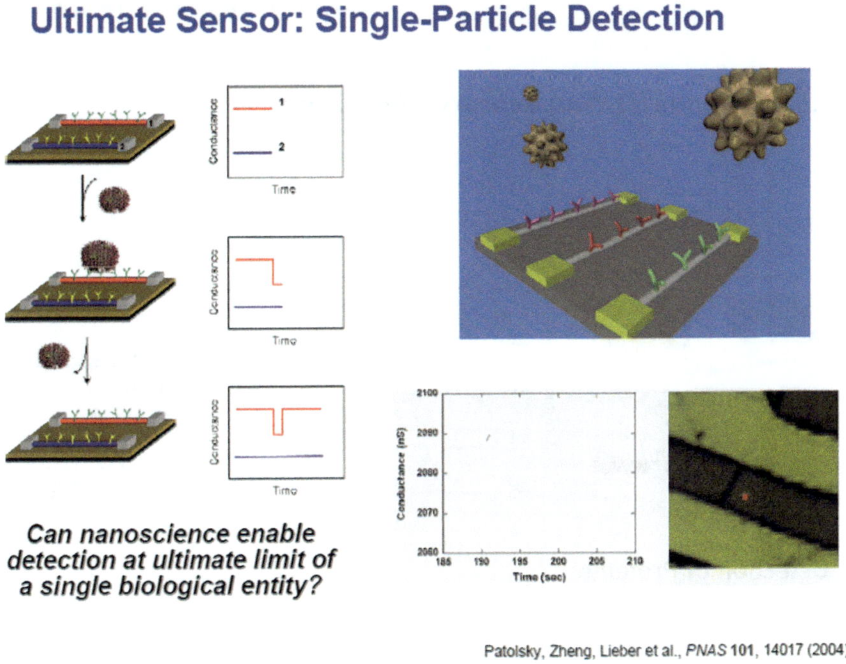

17.8.2 Bio-nanowire Device Interface

Because the sizes of biological macromolecules are comparable to nanowire building blocks, these structures represent natural transducers for ultra-sensitive detection.

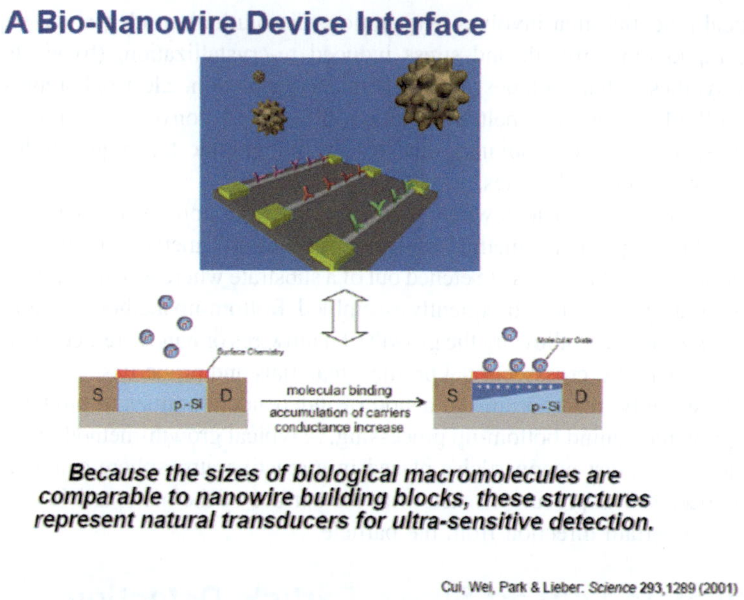

17.8.3 Nanowire Nanosensors: Beginning

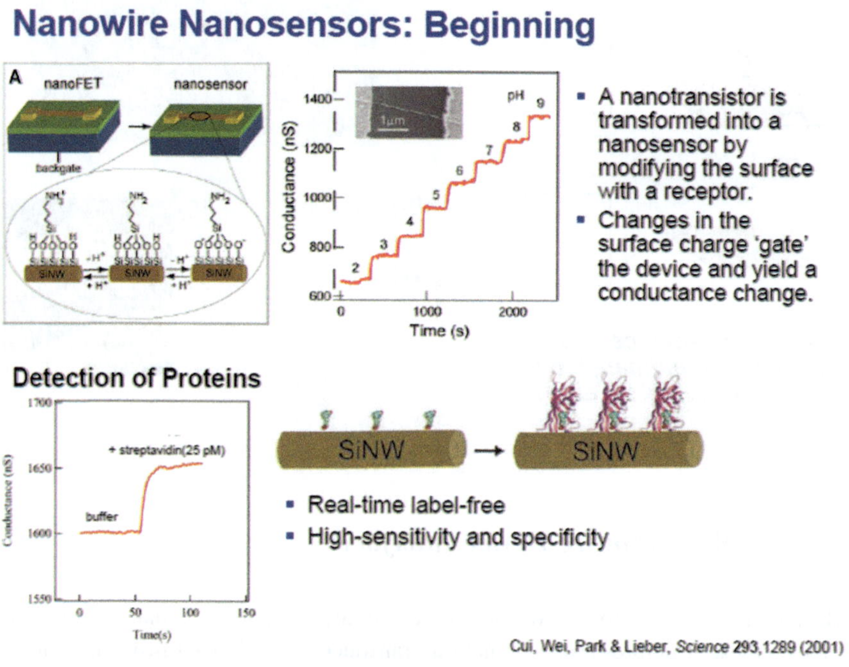

Cui, Wei, Park & Lieber: *Science* **293**, 1289 (2001)

17.8.4 Multiplexed Cancer Marker Detection

Due to the small size of nano particles can be of great use in oncology, particularly in imaging. Nano particles, such as quantum dots, with quantum confinement properties, such as size-tunable light emission, can be used in conjunction with magnetic resonance imaging, to produce exceptional images of tumor sites. As compared to organic dyes, nano particles are much brighter and need one light source for excitation. Thus the use of fluorescent quantum dots could produce a higher contrast image and at a lower cost than organic dyes used as contrast media. But quantum dots are usually made of quite toxic elements.

Nano particles have a special property of high surface area to volume ratio, which allows various functional groups to get attached to a nano particle and thus bind to certain tumor cells. Furthermore, the 10–100 nm small size of nanoparticles, allows them to preferentially accumulate at tumor sites as tumors lack an effective lymphatic drainage system. Multifunctional nano particles can be manufactured that would detect, image, and then treat a tumor in future cancer treatment [139]. Kanzius RF therapy attaches microscopic nano particles to cancer cells and then "cooks" tumors inside the body with radio waves that heat only the nanoparticles and the adjacent (cancerous) cells.

Nano wires are used to prepare sensor test chips, which can detect proteins and other biomarkers left behind by cancer cells, and detect and make diagnosis of cancer possible in the early stages from a single drops of a patient's blood [126].

Nano technology based drug delivery is based upon three facts: (i) efficient encapsulation of the drugs, (ii) successful delivery of said drugs to the targeted region of the body, and (iii) successful release of that drug there.

Nano shells of 120 nm diameter, coated with gold were used to kill cancer tumors in mice by Prof. Jennifer at Rice University. These nano shells are targeted to bond to cancerous cells by conjugating antibodies or peptides to the nano shell surface. Area of the tumor is irradiated with an infrared laser, which heats the gold sufficiently and kills the cancer cells [140].

Cadmium selenide nano particles in the form of quantum dots are used in detection of cancer tumors because when exposed to ultraviolet light, they glow. The surgeon injects these quantum dots into cancer tumors and can see the glowing tumor, thus the tumor can easily be removed.

Nano particles are used in cancer photodynamic therapy, wherein the particle is inserted within the tumor in the body and is illuminated with photo light from the outside. The particle absorbs light and if it is of metal, it will get heated due to energy from the light. High energy oxygen molecules are produced due to light which chemically react with and destroy tumors cell, without reacting with other body cells. Photodynamic therapy has gained importance as a noninvasive technique for dealing with tumors.

The applications of various nano systems in cancer therapy [141] are summarized as:

Carbon nano tubes, 0.5–3 nm in diameter and 20–1000 nm length, are used for detection of DNA mutation and for detection of disease protein biomarker.

Dendrimers, less than 10 nm in size are useful for controlled release drug delivery, and as image contrast agents.

Nano crystals, of 2–9.5 nm size cause improved formulation for poorly-soluble drugs, labeling of breast cancer marker HeR2 surface of cancer cells.

Nano particles are of 10–1000 nm size and are used in MRI and ultrasound image contrast agents and for targeted drug delivery, as permeation enhancers and as reporters of apoptosis, angiogenesis.

Nano shells find application in tumor-specific imaging, deep tissue thermal ablation.

Nano wires are useful for disease protein biomarker detection, DNA mutation detection and for gene expression detection.

Quantum dots, 2–9.5 nm in size, can help in optical detection of genes and proteins in animal models and cell assays, tumor and lymph node visualization.

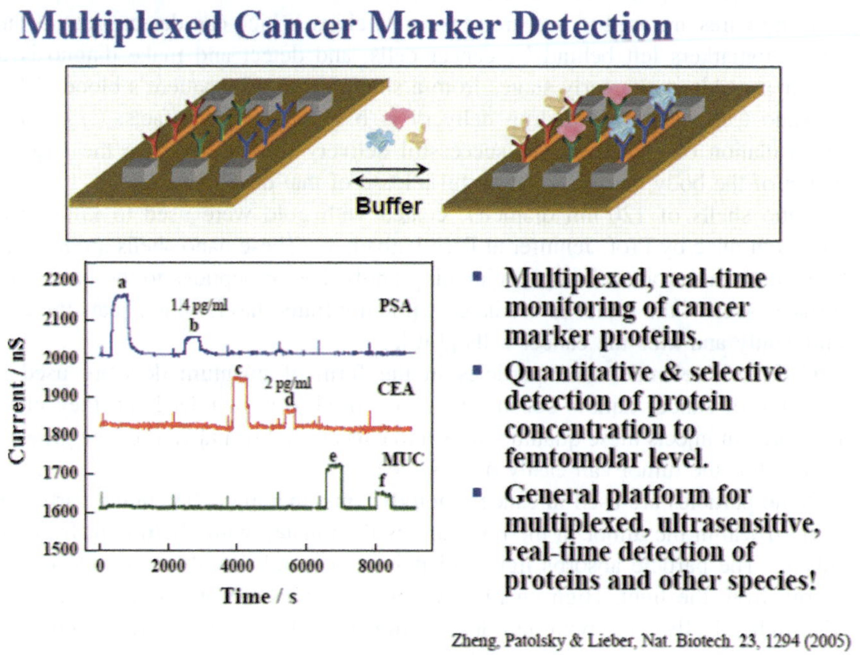

Zheng, Patolsky & Lieber, Nat. Biotech. **23**, 1294 (2005).

17.8.5 *Undiluted Blood Serum Analysis*

- Serum samples are characterized after single step ‘desalting’ purification.
- (1) Buffer; (2) *Donkey Serum* (*DS*), 59 mg/ml total protein; (3) DS + 2.5 pM PSA; (4) DS + 25 pM PSA
- (1) DS + 0.9 pg/ml; (2) DS.

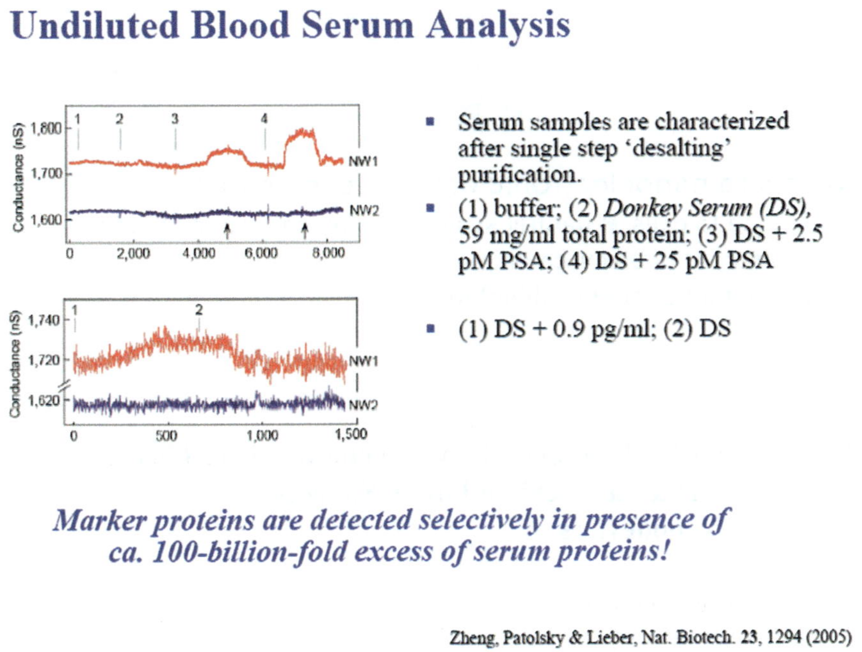

Marker proteins are detected selectively in presence of ca. 100-billion-fold excess of serum proteins!

Zheng, Patolsky & Lieber, Nat. Biotech. **23**, 1294 (2005).

17.8.6 *Nanoelectronic-Cell Interfaces*

Nanowire nanoelectronic devices can enable:

- Interface to cells at natural scale of biological communication
- Input/output of electrical signals
- Input/output of chemical/biological signals.

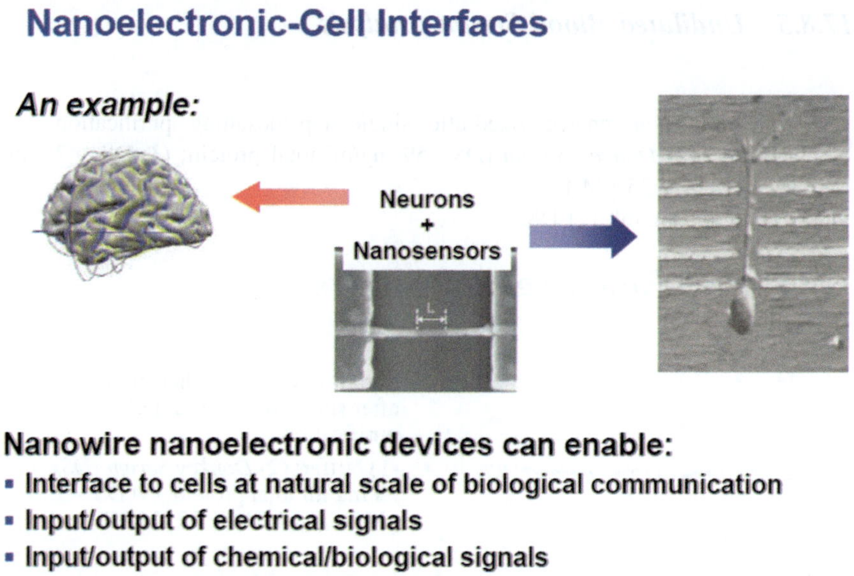

17.9 Nanowire Piezoelectric Nanogenerators on Plastic Substrates as Flexible Power Sources for Nanodevices

Research applications in biomedical science and technology usually require various portable, wearable, easy-to-use, and/or implantable devices that can interface with biological systems [142, 143]. Organic or hybrid organic–inorganic microelectronics and nanoelectronics have long been a possibility [144–148]. However, these devices require a power source, such as electrochemical cells [149] or piezoelectric [150], thermoelectric, [151] and pyroelectric transducers, [152] to generate or store the electrical energy created through chemical, mechanical, or thermal processes. Finding a suitable power source has remained major challenge for many devices in bioengineering and medical fields.

ZnO is a typical piezoelectric and pyroelectric inorganic semiconducting material used for electromechanical and thermoelectrical energy conversion. Nanostructures of ZnO, [153] such as nanowires (NWs) [154, 155], nanobelts (NBs) [156], nanotubes [157–159], nanorings [160], nanosprings [161, 162], and nanohelices [163], have attracted extensive research interest because of their potential applications as nanoscale sensors and actuators. While most of the current applications focus on its semiconducting properties, only a few efforts have utilized the nanometer scale piezoelectric properties of ZnO. Using ZnO NW arrays grown on a single-crystal sapphire substrate, it has been successfully converted mechanical energy into electrical energy at the nanoscale [164]. A conductive atomic force microscopy (AFM) tip was used in contact mode to deflect the aligned NWs. The coupling of piezoelectric and semiconducting properties in ZnO creates a strain field and charge separation across the NWs as a result of their bending. The rectifying characteristic of the Schottky barrier formed between the metal tip and the NW leads to electrical current generation. This is the principle behind piezoelectric nanogenerators.

The ceramic and semiconducting substrates used for growing ZnO NWs are hard and brittle and cannot be used in applications that require a foldable or flexible power source, such as implantable biosensors. In this Communication, by using ZnO NW arrays grown on a flexible plastic substrate. It was demonstrate the first successful flexible power source built on conducting-polymer films. This approach has two specific advantages: it uses a cost-effective, large-scale, wet-chemistry strategy to grow ZnO NW arrays at temperatures lower than 80 °C, and the growth of aligned ZnO NW arrays can occur on a large assortment of flexible plastic substrates. The latter advantage could play an important role in the flexible and portable electronics industry. Various dimensions, shapes, and orientations of ZnO NWs and microwires on flexible plastic substrates have been shown to be capable of producing piezoelectric voltage output, giving a real advantage for energy harvesting using large-scale ZnO NW arrays. The voltage generated from a single NW can be as high as 50 mV, which is large enough to power many nanoscale devices.

The ZnO NWs were grown in solution using a synthetic chemistry approach. Figure 17.13 shows a series of scanning electron microscopy (SEM) and transmission electron microscopy (TEM) images of typical ZnO NW arrays grown on a conductive plastic substrate. Figure 17.13a shows a low-resolution, topdown view of the densely aligned NW arrays. The aligned NWs have a uniform diameter of 200–300 nm and a hexagonal cross section, as indicated by the magnified SEM image in Fig. 17.13b. The SEM image in Fig. 17.13c displays ZnO NWs ca. 2 μm long aligned perpendicular to the plastic substrate. Figure 17.13d is a typical TEM image of a ZnO NW that has a diameter of ca. 300 nm. The corresponding electron diffraction pattern indicates that the NW growth direction is [0001] and its side surfaces are $\{2\bar{1}\bar{1}0\}$.

It is worth noting that by controlling reaction conditions such as temperature, concentration, pH value, reaction time, and plastic-substrate surface quality, NWs of different density distributions, dimensionality, and alignment have been fabricated for electrical measurements. Figure 17.14a, b shows low resolution and magnified SEM images, respectively, depicting sparsely grown ZnO NW arrays on a plastic substrate. The NWs have a density of ca. 1 μm^{-2} on a Au-coated plastic substrate. The NWs are typically 100–350 nm wide and ca. 1 μm long. The SEM image shown in the inset of Fig. 17.14b is a side view of the NWs aligned perpendicularly to the plastic substrate.

Figure 17.14c is a schematic depicting the experimental setup used for measuring the mechanically induced piezoelectric discharge from individual NWs [164]. A conductive Si tip coated with a Pt film with a cone angle of 70° was used for AFM measurements. The rectangular cantilever had a calibrated normal spring

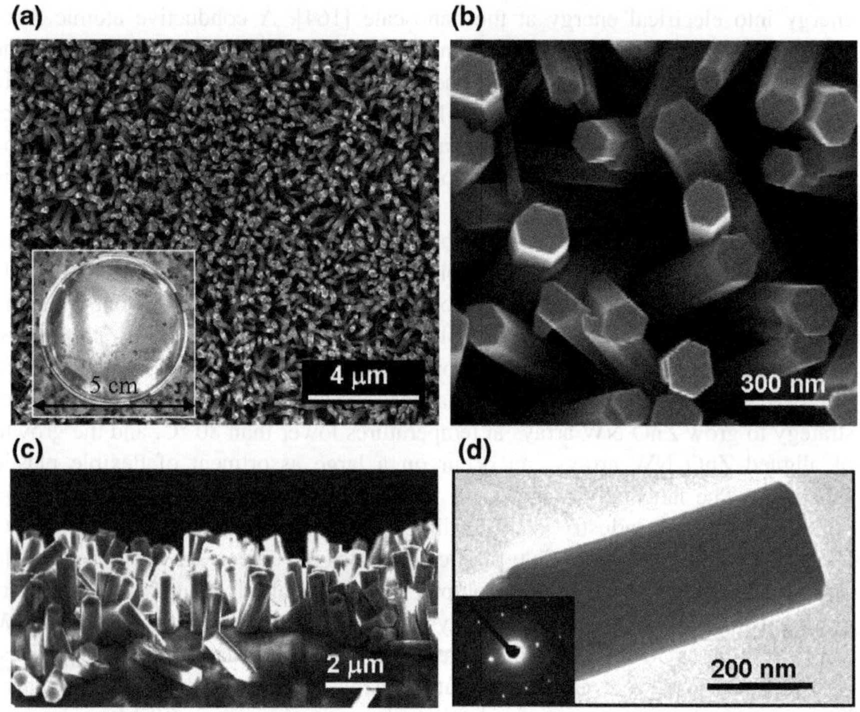

Fig. 17.13 A series of SEM and TEM images of typical ZnO NW arrays grown on a plastic substrate. **a** A low-magnification, top-down view of the densely aligned NW arrays and **b** a corresponding higher-resolution SEM image. The inset in **a** is an optical image of a sample of NWs on a polymer surface. **c** An SEM image of a side view of the sample displaying ZnO NWs aligned normal to the plastic substrate. **d** A typical TEM image of an individual ZnO NW; inset is the corresponding electron diffraction pattern of the selected area

constant of 1.857 Nm^{-1}. In the AFM contact mode, a constant normal force of 5 nN was maintained between the tip and sample surface. When the tip was scanned over the top of the ZnO NWs, the tip height was adjusted according to the surface morphology and local contacting force. For the electric contact at the bottom of the NWs, Ag paste was applied to connect the Au film on the plastic substrate surface to the measurement circuit. Connecting Ag and ZnO produces an Ohmic contact [165]. The output voltage across an external load of resistance R_L = 500 MΩ was continuously monitored as the tip was scanned over the NWs. In contact mode, as the tip was scanned over the vertically aligned NWs, the NWs were bent consecutively. The tip forced the elastic deflection of the oriented ZnO NWs and produced a charge separation and a voltage drop across the diameter of the NWs, with the stretched and compressed sides having positive and negative piezoelectric potentials, respectively. The center axis of the NW, as indicated by the dotted line, remained neutral.

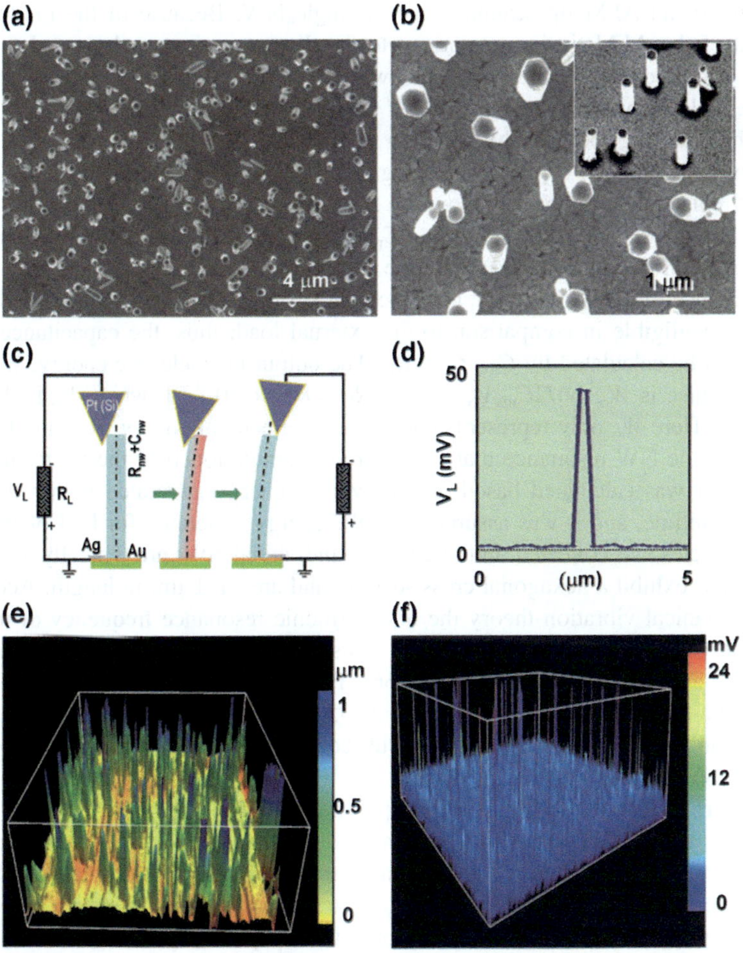

Fig. 17.14 **a** A low-magnification and **b** a higher-resolution SEM image showing a sparsely grown ZnO NW array on a plastic substrate. **c** Schematic illustrating the experimental setup for the conductive AFM measurement of the NW array. Note the polarity of the defined output voltage on the external load. **d** A typical voltage output profile for the AFM tip scanning over a single NW on the plastic substrate. **e** A 3D plot of the AFM topography image and **f** a voltage output profile obtained by scanning the AFM tip over a 40 μm × 40 μm area of aligned ZnO NWs on a plastic substrate

As the conductive tip was scanned across the neutral axis, a discharge occured when the tip touched the compressed side of the NW. A single NW that has a diameter of 300 nm can produce an output voltage discharge of ca. 45 mV (Fig. 17.14d), which is the voltage drop across an external resistor converted using the measured electrical current. In fact, the true output voltage should be higher if we consider the inner resistance of the NW. Figure 17.14d is a typical voltage output for an AFM tip scanning over a single NW. Because of the limited scan speed of the AFM tip in comparison to the discharge time, only two data points were captured around the peak area, which suggests that the true peak could be twice as large as the measured 45 mV shown in this profile.

The scanning speed of the AFM tip was 53.51 μms^{-1}, and the full width at half maximum (FWHM) of the peak in Fig. 17.14d is 234 nm. Therefore, the lifetime decay constant of the circuit shown in Fig. 17.14c is estimated to be $\tau_c = 4.4$ ms, based on the equation $\tau_c = (R_L + R_{nw})C_{nw}$, where R_L and R_{nw} are the resistances for the external load (500 MΩ) and the ZnO NW, respectively, and C_{nw} is the capacitance of the NW and the measurement system. As previously reported [164], R_{nw} is negligible in comparison to the external load; thus, the capacitance of the NW can be calculated by $C_{nw} \approx \tau_c/R_L$. The output piezoelectric energy, W_o, for a single pulse is $W_o = 1/2C_{nw}V_p^2 = \tau_c V_p^2/2R_L = 8.9 \times 10^{-15}$ J, where V_p is the peak voltage. Here W_o only represents the harvested electrical energy from the first half cycle of the NW resonance that results from a single touch of the tip to the NW.

Power was calculated based on the average energy generated within the resonance lifetime, and it was assumed that the energy generated for half of the resonance cycles was collected. In this case, the individual NWs are typically 300 nm in diameter, exhibit a hexagonal cross section, and are ca. 1 μm in length. According to mechanical vibration theory the first harmonic resonance frequency of the NW can be calculated to be ca. 70 MHz; considering the damping effect in air, the lifetime of a single NW is ca. 50 μs [166]. Therefore, the effective number of cycles required for energy harvesting is about 3500. By considering the decay of the vibration amplitude of a NW for the subsequent resonance, the output power of a single NWis 10–20 pW.

Figure 17.14e is a 3D plot of a topographic AFM scan of a 40 μm × 40 μm area of the ZnO NW array. The scanning direction is from the front to the back, which can be seen from the raised linear traces on the flexible plastic substrate. Because of the flexibility of the plastic substrate and in spite of the firm adhesion to the Ag paste, the substrate surface profile under tip contact was somewhat wavy. Relative to the plastic substrate, the NW array was revealed to have a height distribution of 0.5–2.0 μm. The corresponding 3D plot of the voltage output image is shown in Fig. 17.14f. There are a number of sharp peaks ranging from 15 to 25 mVthat represent voltage outputs. By counting the pulse numbers shown here with respect to the number of NWs in the topography profile in Fig. 17.14e, the ratio of voltage peaks to the number of available NWs is 90:150 or about 60%, which suggests that the discharge events of the NWs captured by the AFM tip correspond to at least 50% of the NWs. In fact, because of the limitation in the data-collecting speed of the atomic force microscope, which has a data step size of ca. 156.2 nm in this measurement, the discharge peaks of some NWs were missed, possibly due to poor contact between the tip and the NWs and/or due to multiple contacts with neighboring NWs. It is suggested that both suitable bonding strength between the ZnO NWs and the polymer substrate and a uniform density distribution of NWs in the array might be very important in terms of improving the piezoelectric discharge efficiency. These two issues, together with the power output calculations, will be explored in the following discussions.

To calculate the total number of NWs that effectively produced electrical energy output, it was assumed an average voltage peak height of ca. 20 mV, and the density of ZnO NWs on the plastic substrate was very conservatively estimated to be 1 μm^{-2}. The power density per unit substrate area is ca. 1–2 pWμm^{-2}, that is, 0.1–0.2 mWcm^{-2}, which is large enough to power a variety of devices that operate at lower power consumption, such as microelectromechanical systems (MEMS), nanoelectromechanical systems (NEMS), and other nanoscale devices.

It was found that, NWs grown on a polymer substrate with diameters of 300 nm and lengths of 1 µm give rise to an output power of ca. 5 pWat 45 mV. This value has improved by an order of magnitude.

To evaluate the power-generating capability of ZnO NW arrays grown on plastic substrates, NW arrays configured differently in terms of dimensions and orientation have been tested to demonstrate the performance of the nanogenerator, and these arrays are shown in Figs. 17.15 and 17.16. Figure 17.15a is a topdown view of an as-grown, densely packed, vertically aligned ZnO *microwire* array on a plastic substrate. The microwires are uniform with a width of 1–2 µm and a height of 5–10 µm.

Finally, the density of the wires could be so high that there is not much room for the wires to be deflected by the AFM tip without touching neighboring wires; thus, the piezoelectric current, if any, may slowly leak out, resulting in a gradual discharge signal.

To improve the bonding between the wires and the substrate, a poly(methyl methacrylate) (PMMA) coating approximately 200 nm thick was introduced on the substrate after wire growth. Figure 17.15c shows the ZnO microwire array on a plastic substrate after the PMMA coating. It has been clearly see that the bottom portion of the wire array has been encapsulated by the PMMA coating (inset). The top portion of the wire array is clean after 10 min of UV irradiation following a spin coating and soft-baking process. In Fig. 17.15c, the shorter wires have been buried by the PMMA coating, while most of the long microwires are standing on the surface. The result from AFM measurements is shown in Fig. 17.15d. The number of voltage peaks in an area of 20 µm $\times$ 20 µm has increased to ca. 50. The peak voltage output amplitude remains in the range of 6–12 mV. The inset is an AFM topography image of the corresponding microwire array affixed by PMMA. It is clearly seen that the microwires are firmly held onto the plastic substrate during the tip scan, as shown in the topography image, suggesting that by using a reinforcing layer of polymer coating, the NW/substrate connection can be significantly improved.

To determine the effect of NW shape and orientation on power generation, cone-shaped ZnO NWs and randomly oriented ZnO NWs on plastic substrates were grown and used for AFM measurements. Figure 17.16a shows as-grown, cone-shaped ZnO hexagonal NWs with ca. 50% vertical alignment on a plastic substrate. The nanocones typically have tip diameters of 100–200 nm, base diameters of 300–500 nm, and lengths of 3–5 µm (inset). Under the conductive tip

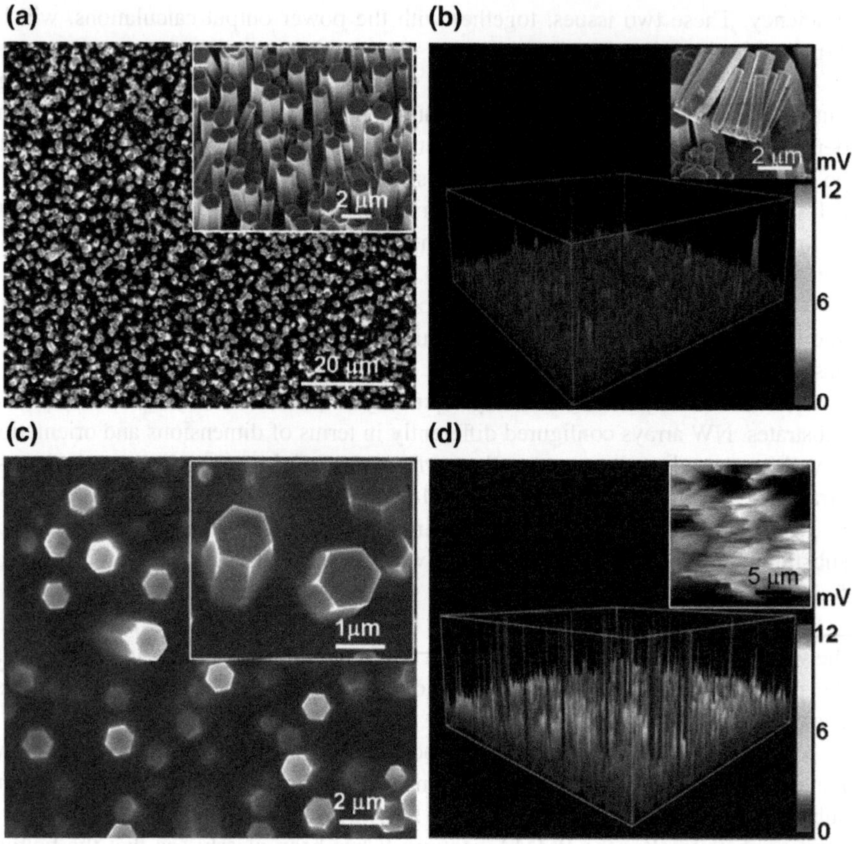

Fig. 17.15 **a** An SEM image showing a top-down view of the as-grown, densely packed, vertically aligned ZnO microwire array (inset: a side view showing the alignment of the microwires) on a plastic substrate and **b** the corresponding AFM voltage output profile (20 μm × 20 μm). The inset in **b** is an SEM image showing a portion of the microwires pushed down by the AFM tip. **c** An SEM image showing a top-down view of ZnO wires immersed in a PMMA coating on a plastic substrate (inset: an enlarged view of the PMMA encapsulated microwires) and **d** the corresponding AFM voltage output profile from AFM measurements (inset: topography image of the 20 μm × 20 μm area)

deflection, only a few voltage output peaks in the range of 6–12 mV are observed (Fig. 17.16 b). Figure 17.16c is a typical SEM image of the as-grown ZnO NWs oriented randomly on the plastic substrate. The SEM image in the inset reveals the tilted hexagonal NWs rooted at the plastic substrate with dimensions of 200–600 nm in width and 1–2 μm in length. The AFM measurement result is shown in Fig. 17.16d, and it reveals an increased number of voltage output peaks of 6–12 mV, suggesting that as long as there is a way to bend the wires, a voltage/current signal is likely to be generated no matter what kind of orientation the wires adopt with respect to the plastic substrate. This result offers another advantage to

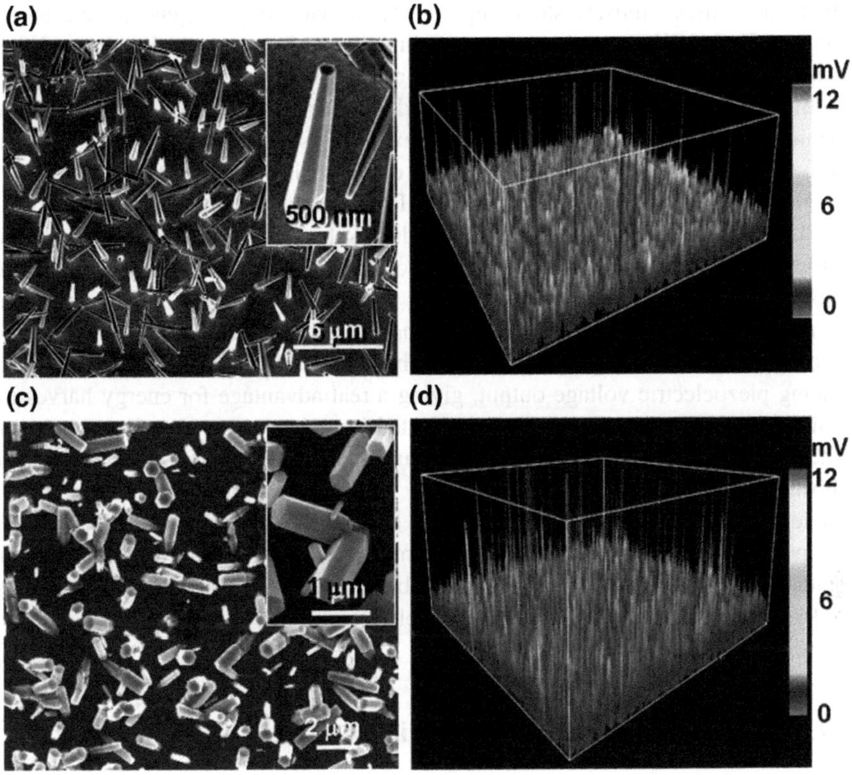

Fig. 17.16 **a** A typical SEM image of sparsely grown ZnO nanocones (inset: a magnified view of a single nanocone) and **b** the corresponding AFM voltage output profile (20 μm × 20 μm). **c** An SEM image of a typical top-down view of a randomly oriented ZnO NW array on a plastic substrate (inset: an enlarged view showing the randomly tilted NWs on a plastic substrate) and **d** the corresponding AFM voltage output profile (20 μm × 20 μm)

energy harvesting utilizing large-scale NW arrays made by using solution based chemistry.

The lead zirconium titania (PZT) ceramic is the most typical ceramic material for piezoelectric applications, and it has been used for producing output voltage [167]. In comparison to PZT, a NW array grown on a polymer substrate has a number of advantages for generating electricity. First, the NW-based nanogenerators can be subjected to extremely large deformations, both on the NWs and the substrate, indicating that they can be used for flexible electronics as flexible/foldable power sources. Second, the large degree of deformation that the NWs can bear is likely to afford a much larger volume density of power output. Third, in contrast to PZT, ZnO is a biocompatible, "biosafe" material, suggesting that it has a greater potential for use as an implantable power source in the human body. Fourth, the flexibility of the polymer substrate used for growing ZnO NWs makes it feasible to accommodate the flexibility of human muscles so that we can use the mechanical energy (body movement, muscle stretching) in the human body to generate electricity. Finally, ZnO NW nanogenerators can directly produce current because of their enhanced conductivity in the presence of oxygen vacancies.

In conclusion, it has been successfully demonstrated that arrays of piezoelectric, semiconducting ZnO NWs grown on flexible plastic substrates can be used to convert mechanical energy into electrical energy using a conductive atomic force microscope. The output discharge voltage from a single ZnONW can be as large as ca. 45 mV. The estimated piezoelectric induced electrical power density of the NW array upon deflection of an AFM tip was on the order of milliwatts per centimeter squared, which is large enough to power a variety of MEMS, NEMS, and other nanoscale devices. Various dimensions, shapes, and orientations of ZnO NWs and microwires on flexible plastic substrates have been shown to be capable of producing piezoelectric voltage output, giving a real advantage for energy harvesting using large-scale ZnO NW arrays fabricated by using a wet chemistry approach. The piezoelectric power generators that use ZnO NW arrays on flexible plastic substrates may be able to harvest energy from their environment, such as body movements (e.g., gestures, respiration, or locomotion) and mechanical vibrations, for powering nanodevices and nanosystems. This flexible power source can find potential applications in implantable biosensors and biodetection, wireless self-powered sensors, and self-powering electronic devices.

17.10 Future Vision for Life Sciences

The challenges associated with nanotechnology applications in biomedical sciences are numerous, but the impact on understanding how the cardiac or nervous systems work, how they fails in disease and how we can intervene at a nanoscopic or even a molecular level is significant. For example, neural developmental factors, such as the cadherins, laminins and bone morphometric protein families, as well as their receptors, could be manipulated in new ways [168]. The bottom-up nanowire nanotechnology offers the capacity to explore the functional specificity of these molecules by incorporating them into pre-defined locations in nanowire devices to have highly targeted effects towards single cells.

The merging of nanoelectronics or nanoscience in general with the entire fields of synthetic biology and/or system biology [169, 170] is also tempting and could be highly rewarding. This would be one of the next big leaps in materials sciences and biological sciences. It is especially true given that there's a whole toolbox of nanoelectronic and nanophotonic devices that one can think about building into cellular circuitry and merging them with biological information processing systems, and the fact that we have already achieved the intracellular interrogation [10] and the 3D electrical innervation of tissues [42] with semiconductor nanoelectronics!

In recent years, interest has been generated in the capability of MNPs to bind a wide range of organic molecules, their low toxicity and their strong and tunable absorption. Unique chemical, physical and photo-physical properties of MNPs [171–182] paved innovative ways in drug delivery systems to achieve controlled transport, controlled release and specific targeting of drugs. It has been shown that conjugates of MNPs with antibiotics provide promising results in antimicrobial therapy. Combination of antibiotic with MNPs would be helpful to improve antibiotic efficacy [183–196].

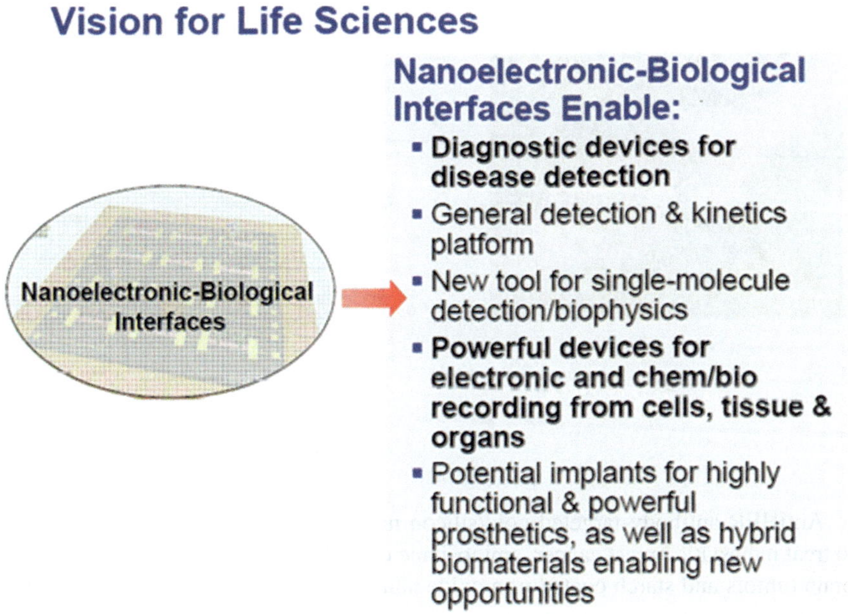

This conjugation can be via covalent, ionic or physical absorption. Cisplatin conjugation to MNPs has shown a significant cytotoxic effect, which is seven times higher than that of cisplatin alone. Conjugation of methotrexate to AuNPs, which involves the interaction of carboxylic group of drug with the metal surface found to have high concentrations of drug in Lewis lung carcinoma cells. Conjugation of tamoxifen and AuNPs has been reported. MNPs surfaces have to be modified in order to avoid aggregation and to improve the efficiency of MNPs drug delivery systems. Doxorubicin in combination with surface modified AuNPs reported to have significant cytotoxicity when compared to the free form of doxorubicin. Polyethylene glycol (PEG) can be used to modify the surface of MNPs so that the cellular uptake of nanoparticles can be improved.

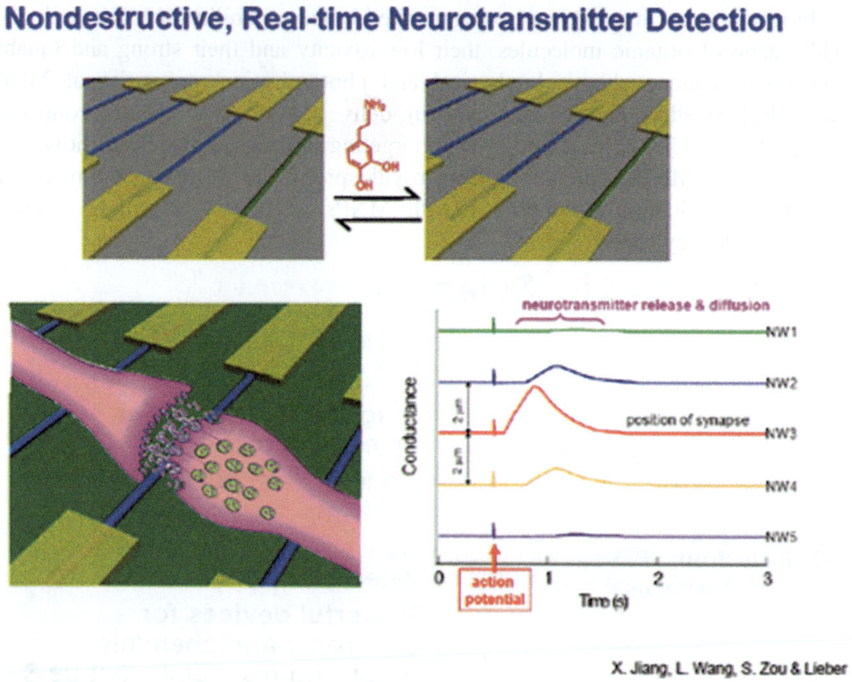

AntiHER antibody-targeted gold/silicon nanoparticles in the form of nanoshells to treat metastatic breast cancer, aminosilane coated iron oxide nanoparticles to treat brain tumors and starch coated iron oxide nanoparticles in the form of magnetically guided mitoxantrone to treat tumor angiogenesis have been reported. Calcium phosphate nanoparticles as vaccine adjuvant (Biovant) is developed by Biosante for subcutaneous administration has entered into phase I trials. Author previously reported a simple, cost effective and eco-friendly method for the synthesis of water soluble AgNPs using root bark extract of *Azadirachta indica* (RBAI). Clinical ultrasound gel prepared from these NPs proved to be effective with significant antibacterial activity. Chronic exposure to silver NPs causes adverse effects like argyria and argyrosis, soluble silver may cause organ damage.

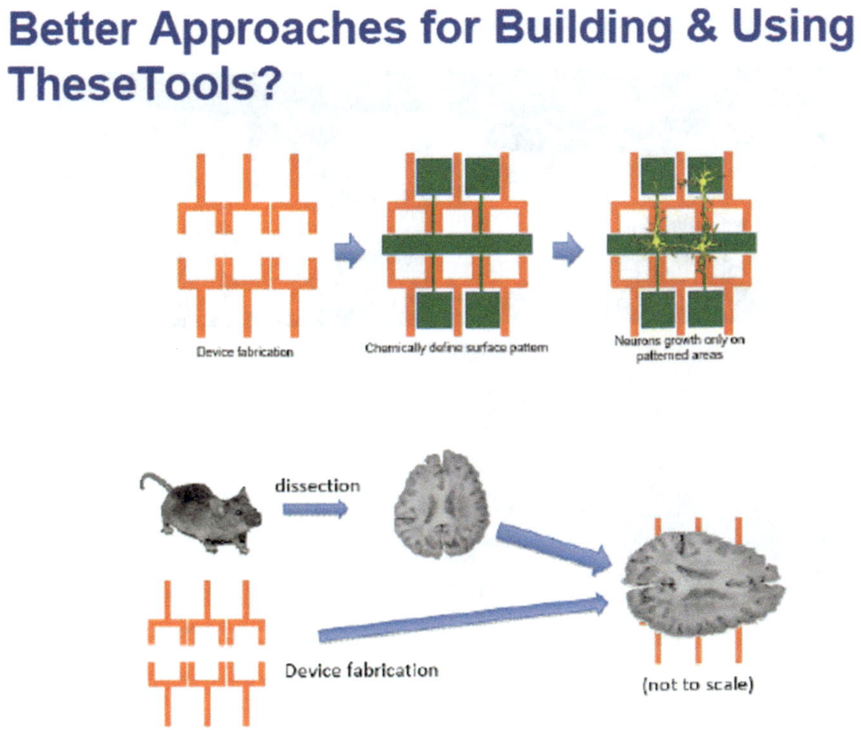

Biological synthesis of MNPs has been always beneficial. Especially green synthesis of MNPs using plants and their extracts is more economical, energy efficient and eco-friendly approach, which is free of toxic contaminates as required in therapeutic applications. Microbes are regarded as potential biofactories for MNPs synthesis and serve a new generation antimicrobial agents with their unique physicochemical properties. The MNPs have found diverse applications in the field of pharmacy as direct therapeutic agents to treat ailments and also as carriers for drug delivery systems. In both the cases, stability and surface activity of MNPs are the vital areas where researchers have to concentrate. In particular, the development of rational protocol for green synthesis of MNPs in keeping view of the advantages of this approach; application of these particles to resolve problem of antibiotic resistance and in the target specific drug delivery systems to treat microbial infections including HIV and cancer should be highly focused in the future. For the applicability of nanoparticles during bioremediation, it is important to develop effective technologies to retain the nanoparticles in the reactors and to separate and remove them from the reaction medium after treatment. Future studies should also aim to extend the number of applications exploring the maximal characteristics of these new materials.

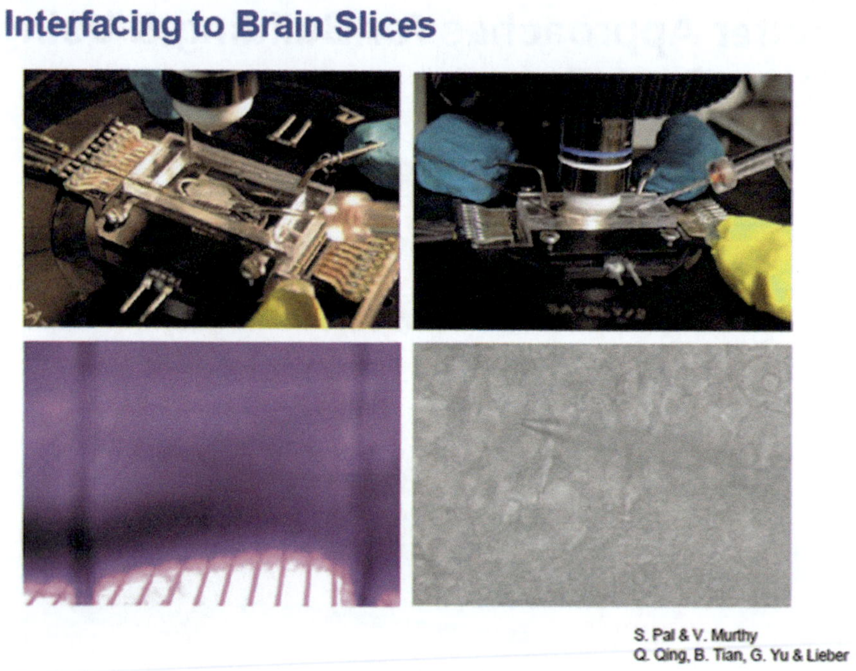

The studies on bioavailability and biodegradation of nanoparticles (except iron nanoparticles) are lacking and need more attention in future. There is dire need for detailed experiments to determine the extent of nanoparticles biodegradability under different environmental conditions (oxic/anoxic, at different pH and temperatures). Except for one study, others have shown that nanoparticles are more bioavailable and degraded faster than the bulk particles. Bacterial enzymes, like various mono-or dioxygenases including cytochrome P450 oxygenases, oxidases or peroxidases can oxidize nanomaterials in aerobic environments and some reductases might be able to reduce them in the anaerobic environments. Bioavailability is a corner stone in further assessing risks for ecosystems and human health, and is therefore an important topic for future nanoparticles research and their application.

References

1. Yousaf, S.A., Salamat, A.: Effect of heating environment on fluorine doped tin oxide (f: SnO/sub 2/) thin films for solar cell applications. Faculty of Engineering & Technology, Islamabad (2008)
2. Chen, K.I., Li, B.R., Chen, Y.T.: Silicon nanowire field-effect transistor-based biosensors for biomedical diagnosis and cellular recording investigation. Nano Today **6**, 131–154 (2011)
3. Allen, B.L., Kichambare, P.D., Star, A.: Carbon nanotube field-effect-transistor-based biosensors. Adv. Mater. **19**, 1439–1451 (2007)

4. Liu, Y., Dong, X., Chen, P.: Biological and chemical sensors based on graphene materials. Chem. Soc. Rev. **41**, 2283–2307 (2012)
5. Patolsky, F., Zheng, G., Hayden, O., Lakadamyali, M., Zhuang, X., et al.: Electrical detection of single viruses. Proc. Natl. Acad. Sci. USA **101**, 14017–14022 (2004)
6. Sorgenfrei, S., Chiu, C.Y., Gonzalez Jr., R.L., Yu, Y.J., Kim, P., et al.: Label-free single-molecule detection of DNA-hybridization kinetics with a carbon nanotube field-effect transistor. Nat. Nanotechnol. **6**, 126–132 (2011)
7. Huang, Y., Cai, D., Chen, P.: Micro and nanotechnologies for study of cell secretion. Anal. Chem. **83**, 4393–4406 (2011)
8. Huang, Y., Chen, P.: Nanoelectronic biosensing of dynamic cellular activities based on nanostructured materials. Adv. Mater. **22**, 2818–2823 (2010)
9. Patolsky, F., Timko, B.P., Yu, G., Fang, Y., Greytak, A.B., et al.: Detection, stimulation, and inhibition of neuronal signals with high-density nanowire transistor arrays. Science **313**, 1100–1104 (2006)
10. Tian, B., Cohen-karni, T., Qing, Q., Duan, X., Xie, P., et al.: Three-dimensional, flexible nanoscale field-effect transistors as localized bioprobes. Science **329**, 830–834 (2010)
11. Duan, X., Gao, R., Xie, P., Cohen-karni, T., Qing, Q., et al.: Intracellular recordings of action potentials by an extracellular nanoscale field-effect transistor. Nat. Nanotechnol. **7**, 174–179 (2011)
12. Pui, T.S., Sudibya, H.G., Luan, X., Zhang, Q., Ye, F., et al.: Non-invasive detection of cellular bioelectricity based on carbon nanotube devices for high-throughput drug screening. Adv. Mater. **22**, 3199–3203 (2010)
13. (a) Koch, C., Reid, R.C.: Nature **483**, 397–398 (2012). (b) Hille, B.: Ion Channels of Excitable Membranes, 3rd edn. Sinauer Associates Inc; Sunderland (2001). (c) Dunlop, J., Bowlby, M., Peri, R., Vasilyev, D., Arias, R.: Nature Rev. Drug Discov. **7**, 358–368 (2008). (d) Meyer, T., Boven, K.H., Gunther, E., Fejtl, M.: Drug Saf. **27**, 763–772 (2004); (e) Dhein, S., Mohr, F.W., Delmar, M.: Practical Methods in Cardiovascular Research. Springer, Berlin, pp. 215–453 (2005). (f) Mark Wightman, R.: Science **311**, 1570–1574 (2006). (g) Rutten, W.L.C.: Annu. Rev. Biomed. Engl. **4**, 407–452 (2002)
14. (a) Molleman, A.: Patch Clamping: An Introductory Guide to Patch Clamp Electrophysiology. Wiley (2003). (b) Purves, R.D.: Microelectrode Methods for Intracellular Recording and Ionophoresis. Academic Press Inc., Burlington (1981)
15. (a) Erickson, J., Tooker, A., Tai, Y.C., Pine, J.J.: Neurosci. Method. **175**, 1–16 (2008). (b) Hai, A., Shappir, J., Spira, M.E.: Nat. Methods. **7**, 200–202 (2010). (c) Xie, C., Lin, Z., Hanson, L., Cui, Y., Cui, B.: Nat. Nanotech. **7**, 185–190 (2012). (d) Robinson, J.T., Jorgolli, M., Shalek, A.K., Yoon, M.H., Gertner, R.S., Park, H.: Nat. Nanotech. **7**, 180–184 (2012)
16. Dvir, T., Timko, B.P., Kohane, D.S., Langer, R.: Nature Nanotech. **6**, 13–22 (2011)
17. (a) Timko, B.P., Cohen-Karni, T., Yu, G., Qing, Q., Tian, B., Lieber, C.M.: Nano Lett. **9**, 914–918 (2009). (b) Viventi, J., Kim, D.H., Vigeland, L., Frechette, E.S., Blanco, J.A., Kim, Y.S., et al.: Nat. Neurosci. **14**, 1599–1605 (2011). (c) Kim, D.H., Lu, N., Ma, R., Kim, Y.S., Kim, R.H., Wang, S., et al.: Science **333**, 838–843 (2011)
18. (a) Scanziani, M., Hausser, M.: Nature **461**, 930–939 (2009). (b) Tian, B., Cohen-Karni, T., Qing, Q., Duan, X., Xie, P., Lieber, C.M.: Science **329**, 831–834 (2010). (c) Duan, X., Gao, R., Xie, P., Cohen-Karni, T., Qing, Q., Choe, H.S., Tian, B., Jiang, X., Lieber, C.M.: Nat Nanotech. **7**, 174–179 (2012). (d) Jiang, Z., Qing, Q., Xie, P., Gao, R., Lieber, C.M.: Nano Lett. **12**, 1711–1716 (2012). (e) Gao, R., Strehle, S., Tian, B., Cohen-Karni, T., Xie, P., Duan, X., Quan, Q., Lieber, C.M.: Nano Lett. **12**, 3329–3333 (2012). (f) Patolsky, F., Timko, B.P., Yu, G., Fang, Y., Greytak, A.B., Zheng, G., Lieber, C.M.: Science **313**, 1100–1104 (2006). (g) Cohen-Karni, T., Qing, Q., Li, Q., Fang, Y., Lieber, C.M.: Nano Lett. **10**, 1098–1102 (2010). (h) Timko, B.P., Cohen-Karni, T., Qing, Q., Tian, B., Lieber, C.M.: IEEE Trans. Nanotechnol. **9**, 269–280 (2010). (i) Cohen-Karni, T., Timko, B.P., Weiss, L. E., Lieber, C.M.: Proc. Natl. Acad. Sci. USA **106**, 7309–7313 (2009). (j) Cohen-Karni, T., Casanova, D., Cahoon, J., Qing, Q., Bell, D., Lieber, C.M.: Nano Lett. **12**, 2639–2644

(2012). (k) Qing, Q., Pal, S.K., Tian, B., Duan, X., Timko, B.P., Cohen-Karni, T., Murthy, V.N., Lieber, C.M.: Proc. Natl. Acad. Sci. USA **107**, 1882–1887 (2010)
19. Lu, W., Lieber, C.M.: Nature Mater. **6**, 841–850 (2007)
20. https://doi.org/10.1073/pnas.1305209110
21. (a) Prohaska, O.J., Olcaytug, F., Pfundner, P., Dragaun, H.: IEEE Trans. Biomed. Eng. **33**, 223–229 (1986) (PubMed: 3957371). (b) Patolsky, F., Zheng, G., Lieber, C.M.: Anal. Chem. **78**, 4260–4269 (2006). (c) Sze, S.M., Ng, K.K.: Physics of Semiconductor Devices, 3rd edn. Wiley Interscience (2006)
22. (a) Jiang, X., Tian, B., Xiang, J., Qian, F., Zheng, G., Wang, H., Mai, L., Lieber, C.M.: Proc. Natl. Acad. Sci. USA **108**, 12212–12216 (2011). (b) Yang, C., Zhong, Z., Lieber, C.M.: Science **310**, 1304–1307 (2005) (PubMed: 16311329). (c) Tian, B., Xie, P., Kempa, T.J., Bell, D.C., Lieber, C.M.: Nat. Nanotech. **4**, 824–829 (2009). (d) Pan, Z., Dai, Z., Wang, Z.: Science **291**, 1947–1949 (2001). (e) Wang, X., Summers, C.J., Wang, Z.: Nano Lett. **4**, 423–426 (2004)
23. Patolsky, F., Timko, B.P., Zheng, G., Lieber, C.M.: Nanowire-based nanoelectronic devices in the life sciences. MRS Bull. **32**, 142–149 (2007)
24. Timko, B.P., Cohen-Karni, T., Qing, Q., Tian, B.Z., Lieber, C.M.: Design and implementation of functional nanoelectronic interfaces with biomolecules, cells, and tissue using nanowire device arrays. IEEE Trans. Nanotechnol. **9**, 269–280 (2010)
25. Fan, Z.Y., Razavi, H., Do, J.W., Moriwaki, A., Ergen, O., et al.: Three-dimensional nanopillar-array photovoltaics on low-cost and flexible substrates. Nat. Mater. **8**, 648–653 (2009)
26. Boettcher, S.W., Spurgeon, J.M., Putnam, M.C., Warren, E.L., Turner-Evans, D.B., et al.: Energy-conversion properties of vapor-liquid-solid-grown silicon wire-array photocathodes. Science **327**, 185–187 (2010)
27. Kelzenberg, M.D., Boettcher, S.W., Petykiewicz, J.A., Turner-Evans, D.B., Putnam, M.C., et al.: Enhanced absorption and carrier collection in Si wire arrays for photovoltaic applications. Nat. Mater. **9**, 239–244 (2010)
28. Garnett, E., Yang, P.D.: Light trapping in silicon nanowire solar cells. Nano Lett. **10**, 1082–1087 (2010)
29. Qin, Y., Wang, X.D., Wang, Z.L.: Microfibre-nanowire hybrid structure for energy scavenging. Nature **451**, U809–U805 (2008)
30. Xu, S., Qin, Y., Xu, C., Wei, Y.G., Yang, R.S., Wang, Z.L.: Self-powered nanowire devices. Nat. Nanotechnol. **5**, 366–373 (2010)
31. Wang, X.D., Song, J.H., Liu, J., Wang, Z.L.: Direct-current nanogenerator driven by ultrasonic waves. Science **316**, 102–105 (2007)
32. Wang, Z.L., Song, J.H.: Piezoelectric Nanogenerators Based on Zinc Oxide Nanowire Arrays. Science **312**, 242–246 (2006)
33. Fan, Z., Ho, J.C., Jacobson, Z.A., Razavi, H., Javey, A.: Large-scale, heterogeneous integration of nanowire arrays for image sensor circuitry. Proc. Natl. Acad. Sci. USA **105**, 11066–11070 (2008)
34. Fan, Z., Ho, J.C., Takahashi, T., Yerushalmi, R., Takei, K., et al.: Toward the development of printable nanowire electronics and sensors. Adv. Mater. **21**, 3730–3743 (2009)
35. Javey, A., Nam, S., Friedman, R.S., Yan, H., Lieber, C.M.: Layer-by-layer assembly of nanowires for three-dimensional, multifunctional electronics. Nano Lett. **7**, 773–777 (2007)
36. Nam, S., Jiang, X.C., Xiong, Q.H., Ham, D., Lieber, C.M.: Vertically integrated, three-dimensional nanowire complementary metal-oxide-semiconductor circuits. Proc. Natl. Acad. Sci. USA **106**, 21035–21038 (2009)
37. Whang, D., Jin, S., Wu, Y., Lieber, C.M.: Large-scale hierarchical organization of nanowire arrays for integrated nanosystems. Nano Lett. **3**, 1255–1259 (2003)
38. Kim, W., Ng, J.K., Kunitake, M.E., Conklin, B.R., Yang, P.D.: Interfacing silicon nanowires with mammalian cells. J. Am. Chem. Soc. **129**, 7228 (2007)

39. Chevrier, N., Mertins, P., Artyomov, M.N., Shalek, A.K., Iannacone, M., et al.: Systematic discovery of TLR signaling components delineates viral-sensing circuits. Cell **147**, 853–867 (2011)
40. Shalek, A.K., Robinson, J.T., Karp, E.S., Lee, J.S., Ahn, D.R., et al.: Vertical silicon nanowires as a universal platform for delivering biomolecules into living cells. Proc. Natl. Acad. Sci. USA **107**, 1870–1875 (2010)
41. Xie, C., Hanson, L., Cui, Y., Cui, B.X.: Vertical nanopillars for highly localized fluorescence imaging. Proc. Natl. Acad. Sci. USA **108**, 3894–3899 (2011)
42. Tian, B.Z., Liu, J., Dvir, T., Jin, L.H., Tsui, J.H., et al.: Macroporous nanowire nanoelectronic scaffolds for synthetic tissues. Nat. Mater. (2012) (Published online)
43. Duan, X.J., Gao, R.X., Xie, P., Cohen-Karni, T., Qing, Q., et al.: Intracellular recordings of action potentials by an extracellular nanoscale field-effect transistor. Nat. Nanotechnol. **7**, 174–179 (2012)
44. Gao, R.X., Strehle, S., Tian, B.Z., Cohen-Karni, T., Xie, P., et al.: Outside looking in: nanotube transistor intracellular sensors. Nano Lett. **12**, 3329–3333 (2012)
45. Jiang, Z., Qing, Q., Xie, P., Gao, R.X., Lieber, C.M.: Kinked p-n junction nanowire probes for high spatial resolution sensing and intracellular recording. Nano Lett. **12**, 1711–1716 (2012)
46. Sakmann, B., Neher, E.: Patch clamp techniques for studying ionic channels in excitable membranes. Annu. Rev. Physiol. **46**, 455–472 (1984)
47. Qing, Q., Pal, S.K., Tian, B., Duan, X., Timko, B.P., et al.: Nanowire transistor arrays for mapping neural circuits in acute brain slices. Proc. Natl. Acad. Sci. USA **107**, 1882–1887 (2010)
48. Robinson, J.T., Jorgolli, M., Shalek, A.K., Yoon, M.H., Gertner, R.S., Park, H.: Vertical nanowire electrode arrays as a scalable platform for intracellular interfacing to neuronal circuits. Nat. Nanotechnol. **7**, 180–184 (2012)
49. Ewing, A.G., Strein, T.G., Lau, Y.Y.: Analytical chemistry in microenvironments-single nerve cells. Acc. Chem. Res. **25**, 440–447 (1992)
50. Schrlau, M.G., Dun, N.J., Bau, H.H.: Cell electrophysiology with carbon nanopipettes. Acs Nano. **3**, 563–568 (2009)
51. Xie, C., Lin, Z.L., Hanson, L., Cui, Y., Cui, B.X.: Intracellular recording of action potentials by nanopillar electroporation. Nat. Nanotechnol. **7**, 185–190 (2012)
52. Bohn, P.W.: Nanoscale control and manipulation of molecular transport in chemical analysis. Annu. Rev. Anal. Chem. 279–296 (2009)
53. Henstridge, M.C., Compton, R.G.: Mass Transport to micro- and nanoelectrodes and their arrays: a review. Chem. Rec. **12**, 63–71 (2012)
54. Walsh, D.A., Lovelock, K.R.J., Licence, P.: Ultramicroelectrode voltammetry and scanning electrochemical microscopy in room-temperature ionic liquid electrolytes. Chem. Soc. Rev. **39**, 4185–4194 (2010)
55. Yeh, J.I., Shi, H.B.: Nanoelectrodes for biological measurements. Wiley Interdisc. Rev. Nanomed. Nanobiotechnol. **2**, 176–188 (2010)
56. Sze, S.M.: Physics of Semiconductor Devices, 2nd edn. Wiley-Interscience, p. 880 (1981)
57. Buzsaki, G., Anastassiou, C.A., Koch, C.: The origin of extracellular fields and currents—EEG, ECoG, LFP and spikes. Nat. Rev. Neurosci. **13**, 407–420 (2012)
58. Plonsey, R., Barr, R.C.: Bioelectricity—A Quantitative Approach. 2nd edn. Kluwer Academic/Plenum Publishers (2000)
59. Lu, W., Xie, P., Lieber, C.M.: Nanowire transistor performance limits and applications. IEEE Trans. Electron. Dev. **55**, 2859–2876 (2008)
60. Givargizov, E.I.: Fundamental aspects of VLS growth. J. Cryst. Growth **31**, 20–30 (1975)
61. Wagner, R.S., Ellis, W.C.: Vapor-liquid-solid mechanism of single crystal growth. Appl. Phys. Lett. **4**, 89 (1964)
62. Cui, Y., Lauhon, L.J., Gudiksen, M.S., Wang, J.F., Lieber, C.M.: Diameter-controlled synthesis of singlecrystal silicon nanowires. Appl. Phys. Lett. **78**, 2214–2216 (2001)

63. Morales, A.M., Lieber, C.M.: A laser ablation method for the synthesis of crystalline semiconductor nanowires. Science **279**, 208–211 (1998)
64. Law, M., Goldberger, J., Yang, P.D.: Semiconductor nanowires and nanotubes. Annu. Rev. Mater. Res. **34**, 83–122 (2004)
65. Duan, X.F., Huang, Y., Cui, Y., Wang, J.F., Lieber, C.M.: Indium phosphide nanowires as building blocks for nanoscale electronic and optoelectronic devices. Nature **409**, 66–69 (2001)
66. Duan, X.F., Lieber, C.M.: General Synthesis of compound semiconductor nanowires. Adv. Mater. **12**, 298–302 (2000)
67. Duan, X.F., Lieber, C.M.: Laser-assisted catalytic growth of single crystal GaN nanowires. J. Am. Chem. Soc. **122**, 188–189 (2000)
68. Bjork, M.T., Ohlsson, B.J., Sass, T., Persson, A.I., Thelander, C., et al.: One-dimensional heterostructures in semiconductor nanowhiskers. Appl. Phys. Lett. **80**, 1058–1060 (2002)
69. Gudiksen, M.S., Lauhon, L.J., Wang, J., Smith, D.C., Lieber, C.M.: Growth of nanowire superlattice structures for nanoscale photonics and electronics. Nature **415**, 617–620 (2002)
70. Cohen-Karni, T., Casanova, D., Cahoon, J.F., Qing, Q., Bell, D.C., Lieber, C.M.: Synthetically encoded ultrashort-channel nanowire transistors for fast, pointlike cellular signal detection. Nano Lett. **12**, 2639–2644 (2012)
71. Lieber, C.M.: Nanowire superlattices. Nano Lett. **2**, 81–82 (2002)
72. Wu, Y.Y., Fan, R., Yang, P.D.: Block-by-block growth of single-crystalline Si/SiGe superlattice nanowires. Nano Lett. **2**, 83–86 (2002)
73. Lauhon, L.J., Gudiksen, M.S., Wang, C.L., Lieber, C.M.: Epitaxial core-shell and core-multishell nanowire heterostructures. Nature **420**, 57–61 (2002)
74. Tian, B.Z., Zheng, X.L., Kempa, T.J., Fang, Y., Yu, N.F., et al.: Coaxial silicon nanowires as solar cells and nanoelectronic power sources. Nature **449**, U885–U888 (2007)
75. Qian, F., Gradecak, S., Li, Y., Wen, C.Y., Lieber, C.M.: Core/multishell nanowire heterostructures as multicolor, high-efficiency light-emitting diodes. Nano Lett. **5**, 2287–2291 (2005)
76. Li, Y., Qian, F., Xiang, J., Lieber, C.M.: Nanowire electronic and optoelectronic devices. Mater. Today **9**, 18–27 (2006)
77. Lu, W., Lieber, C.M.: Nanoelectronics from the bottom up. Nat. Mater. **6**, 841–850 (2007)
78. Hu, Y.J., Churchill, H.O.H., Reilly, D.J., Xiang, J., Lieber, C.M., Marcus, C.M.: A Ge/Si heterostructure nanowire-based double quantum dot with integrated charge sensor. Nat. Nanotechnol. **2**, 622–625 (2007)
79. Hu, Y.J., Kuemmeth, F., Lieber, C.M., Marcus, C.M.: Hole spin relaxation in Ge-Si core-shell nanowire qubits. Nat. Nanotechnol. **7**, 47–50 (2012)
80. Lu, W., Xiang, J., Timko, B.P., Wu, Y., Lieber, C.M.: One-dimensional hole gas in germanium/silicon nanowire heterostructures. Proc. Natl. Acad. Sci. USA **102**, 10046–10051 (2005)
81. Xiang, J., Lu, W., Hu, Y.J., Wu, Y., Yan, H., Lieber, C.M.: Ge/Si nanowire heterostructures as highperformance field-effect transistors. Nature **441**, 489–493 (2006)
82. Tian, B.Z., Xie, P., Kempa, T.J., Bell, D.C., Lieber, C.M.: Single-crystalline kinked semiconductor nanowire superstructures. Nat. Nanotechnol. **4**, 824–829 (2009)
83. Jiang, X.C., Tian, B.Z., Xiang, J., Qian, F., Zheng, G.F., et al.: Rational growth of branched nanowire heterostructures with synthetically encoded properties and function. Proc. Natl. Acad. Sci. USA **108**, 12212–12216 (2011)
84. Dick, K.A., Deppert, K., Larsson, M.W., Martensson, T., Seifert, W., et al.: Synthesis of branched 'nanotrees' by controlled seeding of multiple branching events. Nat. Mater. **3**, 380–384 (2004)
85. Wang, D., Qian, F., Yang, C., Zhong, Z.H., Lieber, C.M.: Rational growth of branched and hyperbranched nanowire structures. Nano Lett. **4**, 871–874 (2004)
86. Lu, W., Lieber, C.M.: Semiconductor nanowires. J. Phys. D-Appl. Phys. **39**, R387–R406 (2006)

87. Cui, Y., Lieber, C.M.: Functional nanoscale electronic devices assembled using silicon nanowire building blocks. Science **291**, 851–853 (2001)
88. Zhou, X., Moran-Mirabal, J.M., Craighead, H.G., McEuen, P.L.: Supported lipid bilayer/carbon nanotube hybrids. Nat. Nanotechnol. **2**, 185–190 (2007)
89. McAlpine, M.C., Friedman, R.S., Jin, S., Lin, K.H., Wang, W.U., Lieber, C.M.: High-performance nanowire electronics and photonics on glass and plastic substrates. Nano Lett. **3**, 1531–1535 (2003)
90. McAlpine, M.C., Friedman, R.S., Lieber, C.M.: High-performance nanowire electronics and photonics and nanoscale patterning on flexible plastic substrates. Proc. IEEE **93**, 1357–1363 (2005)
91. Kim, D.H., Lu, N., Ghaffari, R., Kim, Y.S., Lee, S.P., et al.: Materials for multifunctional balloon catheters with capabilities in cardiac electrophysiological mapping and ablation therapy. Nat. Mater. **10**, 316–323 (2011)
92. Viventi, J., Kim, D.H., Moss, J.D., Kim, Y.S., Blanco, J.A., et al.: A conformal, bio-interfaced class of silicon electronics for mapping cardiac electrophysiology. Sci. Transl. Med. **2** (2010)
93. Viventi, J., Kim, D.H., Vigeland, L., Frechette, E.S., Blanco, J.A., et al.: Flexible, foldable, actively multiplexed, high-density electrode array for mapping brain activity in vivo. Nat. Neurosci. **14**, U1599–U1138 (2011)
94. Kim, D.H., Viventi, J., Amsden, J.J., Xiao, J.L., Vigeland, L., et al.: Dissolvable films of silk fibroin for ultrathin conformal bio-integrated electronics. Nat. Mater. **9**, 511–517 (2010)
95. Hu, Y.J., Xiang, J., Liang, G.C., Yan, H., Lieber, C.M.: Sub-100 nanometer channel length Ge/Si nanowire transistors with potential for 2 THz switching speed. Nano Lett. **8**, 925–930 (2008)
96. Cohen-Karni, T., Timko, B.P., Weiss, L.E., Lieber, C.M.: Flexible electrical recording from cells using nanowire transistor arrays. Proc. Natl. Acad. Sci. USA **106**, 7309–7313 (2009)
97. Timko, B.P., Cohen-Karni, T., Yu, G.H., Qing, Q., Tian, B.Z., Lieber, C.M.: Electrical recording from hearts with flexible nanowire device arrays. Nano Lett. **9**, 914–918 (2009)
98. http://nano.gov/sites/default/files/pub_resource/nni_siginit_nanoelectronics_jul_2010.pdf
99. Ieong, M., Doris, B., Kedzierski, J., Rim, K., Yang, M.: Silicon device scaling to the sub-10-nm regime. Science (New York, N.Y.) **306**, 2057–2060 (2004)
100. Mercanzini, A., Colin, P., Bensadoun, J.C., Bertsch, A., Renaud, P.: In vivo electrical impedance spectroscopy of tissue reaction to microelectrode arraysl. IEEE Trans. Biomed. Eng. **56**, 1909–1918 (2009)
101. Patrick, E., Orazem, M.E., Sanchez, J.C., Nishida, T.: Corrosion of tungsten microelectrodes used in neural recording applications. J. Neurosci. Methods **198**, 158–171 (2011)
102. Chernomordik, L.V., Kozlov, M.M.: Mechanics of membrane fusion. Nat. Struct. Mol. Biol. **15**, 675–683 (2008)
103. Kauer, J.S., White, J.: Imaging and coding in the olfactory system. Annu. Rev. Neurosci. **24**, 963–979 (2001)
104. Grinvald, A., Hildesheim, R.: VSDI: A new era in functional imaging of cortical dynamics. Nat. Rev. Neurosci. **5**, 874–885 (2004)
105. Kralj, J.M., Douglass, A.D., Hochbaum, D.R., Maclaurin, D., Cohen, A.E.: Optical recording of action potentials in mammalian neurons using a microbial rhodopsin. Nat. Methods **9**, U90–U130 (2012)
106. Kralj, J.M., Hochbaum, D.R., Douglass, A.D., Cohen, A.E.: Electrical Spiking in escherichia coli probed with a fluorescent voltage-indicating protein. Science **333**, 345–348 (2011)
107. Hochberg, L.R., Bacher, D., Jarosiewicz, B., Masse, N.Y., Simeral, J.D., et al.: Reach and grasp by people with tetraplegia using a neurally controlled robotic arm. Nature **485**, U372–U121 (2012)
108. Hochberg, L.R., Serruya, M.D., Friehs, G.M., Mukand, J.A., Saleh, M., et al.: Neuronal ensemble control of prosthetic devices by a human with tetraplegia. Nature **442**, 164–171 (2006)

109. Serruya, M.D., Hatsopoulos, N.G., Paninski, L., Fellows, M.R., Donoghue, J.P.: Instant neural control of a movement signal. Nature **416**, 141–142 (2002)
110. Truccolo, W., Hochberg, L.R., Donoghue, J.P.: Collective dynamics in human and monkey sensorimotor cortex: predicting single neuron spikes. Nat. Neurosci. **13**, U105–U275 (2010)
111. Ferrari, M.: Beyond drug delivery. Nat. Nanotechnol. **3**, 131–132 (2008)
112. Nel, A.E., Madler, L., Velegol, D., Xia, T., Hoek, E.M.V., et al.: Understanding biophysicochemical interactions at the nano-bio interface. Nat. Mater. **8**, 543–557 (2009)
113. Rajendran, L., Knolker, H.J., Simons, K.: Subcellular targeting strategies for drug design and delivery. Nat. Rev. Drug Discov. **9**, 29–42 (2010)
114. Summers, H.D., Rees, P., Holton, M.D., Brown, M.R., Chappell, S.C., et al.: Statistical analysis of nanoparticle dosing in a dynamic cellular system. Nat. Nanotechnol. **6**, 170–174 (2011)
115. Wylie, R.G., Ahsan, S., Aizawa, Y., Maxwell, K.L., Morshead, C.M., Shoichet, M.S.: Spatially controlled simultaneous patterning of multiple growth factors in three-dimensional hydrogels. Nat. Mater. **10**, 799–806 (2011)
116. Kloxin, A.M., Kasko, A.M., Salinas, C.N., Anseth, K.S.: Photodegradable hydrogels for dynamic tuning of physical and chemical properties. Science **324**, 59–63 (2009). (PubMed: 19342581)
117. Dvir, T., Timko, B.P., Kohane, D.S., Langer, R.: Nanotechnological strategies for engineering complex tissues. Nat. Nanotechnol. **6**, 13–22 (2011)
118. Hutmacher, D.W.: Biomaterials offer cancer research the third dimension. Nat. Mater. **9**, 90–93 (2011)
119. Prestwich, G.D.: Evaluating drug efficacy and toxicology in three dimensions: using synthetic extracellular matrices in drug discovery. Acc. Chem. Res. **41**, 139–148 (2008)
120. Rogers, J.A., Lagally, M.G., Nuzzo, R.G.: Synthesis, assembly and applications of semiconductor nanomembranes. Nature **477**, 45–53 (2011)
121. Prohaska, O.J., Olcaytug, F., Pfundner, P., Dragaun, H.: Thin-film multiple electrode probes possibilities and limitations. IEEE Trans. Biomed. Eng. **33**, 223–229 (1986)
122. Huh, D., Matthews, B.D., Mammoto, A., Montoya-Zavala, M., Hsin, H.Y., Ingber, D.E.: Reconstituting organ-level lung functions on a chip. Science **328**, 1662–1669 (2010)
123. Cui, Y., Wei, Q., Park, H., Lieber, C.M.: Science **293** (2001)
124. Hahm, J.-I., Lieber, C.M.: Nano Lett. **4**, 51 (2004)
125. Patolsky, F., Zheng, G., Hayden, O., Lakadamyali, M., Zhuang, X., Lieber, C.M.: Proc. Natl. Acad. Sci. USA **101**, 14017 (2004)
126. Zheng, G., Patolsky, F., Cui, Y., Wang, W.U., Lieber, C.M.: Nat. Biotechnol. **23**, 1294 (2005)
127. Kim, A., Ah, C.S., Yu, H.Y., Yang, J.H., Baek, I.B., Ahn, C.G., Park, C.W., Jun, M.S., Lee, S.: Appl. Phys. Lett. **91**, 103901 (2007)
128. Patolsky, F., Zheng, G., Lieber, C.M.: Nat. Protoc. **1**, 1711 (2006)
129. Stern, E., Klemic, J.F., Routenberg, D.A., Wyrembak, P.N., Turner-Evans, D.B., Hamilton, A.D., LaVan, D.A., Fahmy, T.M., Reed, M.A.: Nature **445**, 519 (2007)
130. de Smet, L.C.P.M., et al.: Organic surface modification of silicon nanowire-based sensor devices. InTech (2011)
131. Curreli, M., Zhang, R., Ishikawa, F.N., Chang, H.K., Cote, R.J., Zhou, C., Thompson, M.E.: Real-time, label-free detection of biological entities using nanowire-based fets. IEEE Trans. Nanotechnol. **7**(6), 651–667 (2008)
132. Ng, H.T., Han, J., Yamada, T., Nguyen, P., Chen, Y.P., Meyyappan, M.: Single crystal nanowire vertical surround-gate field-effect transistor. Nano Lett. **4**(7), 1247–1252 (2004)
133. Cui, Y., Wei, Q., Park, H., Lieber, C.M.: Nanowire nanosensors for highly sensitive and selective detection of biological and chemical species. Science **293**(5533), 1289–1292 (2001)
134. Huang, Y., Duan, X., Cui, Y., Lauhon, L.J., Kim, K.-H., Lieber, C.M.: Logic gates and computation from assembled nanowire building blocks. Science **294**(5545), 1313–1317 (2001)

135. Wang, X., Song, J., Liu, J., Wang, Z.L.: Direct-current nanogenerator driven by ultrasonic waves. Science **316**(5821), 102–105 (2007)
136. Hiralal, P., Chien, C., Lal, N.N., Abeygunasekara, W., Kumar, A., Butt, H., Zhou, H., Unalan, H.E., Baumberg, J.J., Amaratunga, G.A.J.: Nanowire-based multifunctional antireflection coatings for solar cells. Nanoscale **6**, 14555–14562 (2014)
137. Ivanov, Y.P., Chuvilin, A., Lopatin, S., Kosel, J.: Modulated magnetic nanowires for controlling domain wall motion: toward 3d magnetic memories. ACS Nano **10**(5), 5326–5332 (2016)
138. Bertok, T., Sediva, A., Vikartovska, A., Tkac, J.: Int. J. Electrochem. Sci. **9**, 890 (2014)
139. Nie, S., Xing, Y., Kim, G.J., Simons, J.W.: Nanotechnology applications in cancer. Annu. Rev. Biomed. Eng. **9**, 257–288 (2007)
140. Loo, C., Lin, A., Hirsch, L., Lee, M.H., Barton, J., et al.: Nanoshell-enabled photonics-based imaging and therapy of cancer. Technol. Cancer Res. Treat. **3**, 33–40 (2004)
141. Nahar, M., Dutta, T., Murugesan, S., Asthana, A., Mishra, D., et al.: Functional polymeric nanoparticles: an efficient and promising tool for active delivery of bioactives. Crit. Rev. Ther. Drug Carrier Syst. **23**, 259–318 (2006)
142. Billinghurst, M., Starner, T.: IEEE Comput. **32**, 57 (1999)
143. De Rossi, D., Della Santa, A., Mazzoldi, A.: Mater. Sci. Eng. C **7**, 31 (1999)
144. R. F. Service: Science **301**, 909 (2003)
145. Zhang, Q.M., Bharti, V., Zhao, X.: Science **280**, 2101 (1998)
146. Heeger, A.J.: J. Phys. Chem. B **105**, 8475 (2001)
147. Ashley, S.: Sci. Am. **10**, 52 (2003)
148. Zhang, Q.M., Li, H., Poh, M., Xia, F., Cheng, Z.Y., Xu, H., Huang, C.: Nature **419**, 284 (2002)
149. Yu, X., Bates, J.B., Jellison Jr., G.E., Hart, F.X.: J. Electrochem. Soc. **144**, 524 (1997)
150. Shenck, N.S., Paradiso, J.A.: IEEE Micro. **21**, 30 (2001)
151. Venkatasubramanian, R., Siivola, E., Colpitts, T., O'Quinn, B.: Nature **413**, 597 (2001)
152. Setiadi, D., Weller, H., Binnie, T.D.: Sens. Actuators, A **76**, 145 (1999)
153. Wang, Z.L., Kong, X.Y., Ding, Y., Gao, P.X., Hughes, W.L., Yang, R., Zhang, Y.: Adv. Funct. Mater. **14**, 943 (2004)
154. Xia, Y.N., Yang, P.D., Sun, Y.G., Wu, Y.Y., Mayers, B., Gates, B., Yin, Y.D., Kim, F., Yan, Y.Q.: Adv. Mater. **15**, 353 (2003)
155. Huang, M.H., Wu, Y.Y., Feick, H., Tran, N., Weber, E., Yang, P.D.: Adv. Mater. **13**, 113 (2003)
156. Pan, Z.W., Dai, Z.R., Wang, Z.L.: Science **2001**, 291 (1947)
157. Wu, J.J., Liu, S.C., Wu, C.T., Chen, K.H., Chen, L.C.: Appl. Phys. Lett. **81**, 1312 (2002)
158. Xing, Y.J., Xi, Z.H., Xue, Z.Q., Zhang, X.D., Song, J.H., Wang, R.M., Xu, J., Song, Y., Zhang, S.L., Yu, D.P.: Appl. Phys. Lett. **83**, 1689 (2004)
159. Gao, P.X., Lao, C.S., Ding, Y., Wang, Z.L.: Adv. Funct. Mater. **16**, 53 (2006)
160. Kong, X.Y., Ding, Y., Yang, R.S., Wang, Z.L.: Science **303**, 1348 (2004)
161. Kong, X.Y., Wang, Z.L.: Nano Lett. **3**, 1625 (2003)
162. Gao, P.X., Wang, Z.L.: Small **1**, 945 (2005)
163. Gao, P.X., Ding, Y., Mai, W.J., Hughes, W.L., Lao, C.S., Wang, Z.L.: Science **309**, 1700 (2005)
164. Wang, Z.L., Song, J.H.: Science **312**, 242 (2006)
165. Song, J.H., Zhou, J., Wang, Z.L.: Nano Lett. **6**, 1656 (2006)
166. Zhou, J., Lao, C.S., Gao, P.X., Mai, W.J., Wang, Z.L., Xu, N.S.: Solid State Commun. **139**, 222 (2006)
167. Platt, S.R., Farritor, S., Haider, H.: IEEE/ASME Trans. Mechatron. **10**, 240 (2005)
168. Silva, G.A.: Neuroscience nanotechnology: progress, opportunities and challenges. Nat. Rev. Neurosci. **7**, 65–74 (2006)
169. Nandagopal, N., Elowitz, M.B.: Synthetic biology: integrated gene circuits. Science **333**, 1244–1248 (2011)

170. Ruder, W.C., Lu, T., Collins, J.J.: Synthetic biology moving into the clinic. Science **333**, 1248–1252 (2011)
171. Guo, L., Gao, Y., Xu, Y., Zhang, R., Madkour, L.H., Yang, Y.: Understanding the corrosion behavior of amorphous multiple-layer carbon coating. In: Advances In Materials, Machinery, Electronics II: Proceedings of the 2nd International Conference on Advances in Materials, Machinery, Electronics (AMME 2018), vol. 1955, 18 Apr 2018. AIP Conference Proceedings 1955, 020001 (2018). https://doi.org/10.1063/1.5033573, https://aip.scitation.org/doi/pdf/10.1063/1.5033573
172. Madkour, L.H.: Applications of gold nanoparticles in medicine and therapy. Pharm. Pharmacol. Int. J. **6**(3), 157–174. https://doi.org/10.15406/ppij.2018.06.00172, http://medcraveonline.com/PPIJ/PPIJ-06–00172.pdf (2018)
173. Madkour, L.H.: Toxic effects of environmental heavy metals on cardiovascular pathophysiology and heart health function: chelation therapeutics. UPI J. Pharm. Med. Health Sci. (UPI-JPMHS) **1**(1), 19–62. https://uniquepubinternational.com/wp-content/uploads/2018/03/UPI-JPMHS-2018-7.pdf (2018)
174. Madkour, L.H.: Biogenic–biosynthesis metallic nanoparticles (MNPs) for pharmacological, biomedical and environmental nanobiotechnological applications. Chron. Pharm. Sci. J. **2** (1), 384–444. https://scientiaricerca.com/srcops/SRCOPS-02-00038.php (2018)
175. Madkour, L.H.: Ecofriendly green biosynthesized of metallic nanoparticles: Bio-reduction mechanism, characterization and pharmaceutical applications in biotechnology industry. Glob. Drugs Ther. **3**(1), 1–11. http://www.oatext.com/ecofriendly-green-biosynthesized-of-metallic-nanoparticles-bio-reduction-mechanism-characterization-and-pharmaceutical-applications-in-biotechnology-industry.php (2018)
176. Madkour, L.H.: Review Article: Advanced AuNMs as nanomedicine's central goals capable of active targeting in both imaging and therapy in biomolecules. Glob. Drugs Ther. **2**(6), 1–12. http://www.oatext.com/advanced-aunms-as-nanomedicines-central-goals-capable-of-active-targeting-in-both-imaging-and-therapy-in-biomolecules.php (2017)
177. Madkour, L.H.: Biotechnology of nucleic acids medicines as gene therapeutics and their drug complexes. Chron. Pharm. Sci. J. **1**(4), 204–253. https://scientiaricerca.com/srcops/pdf/SRCOPS-01–00023.pdf (2017)
178. Madkour, L.H.: Advanced AuNMs as nanomedicine's central goals capable of active targeting in both imaging and therapy in biomolecules. Bio Accent. Online BAOJ Nanotechnol. **3**(1); 015, 1–18. https://bioaccent.org/nanotechnology/nanotechnology15.pdf (2017)
179. Madkour, L.H.: Vision for life sciences: interfaces between nanoelectronic and biological systems. Glob. Drugs Ther. **2**(4), 1–4. https://doi.org/10.15761/gdt.1000126, https://oatext.com/Vision-for-life-sciences-interfaces-between-nanoelectronic-and-biological-systems.php (2017)
180. Foundation CaDR.: One Degree of Separation: Paralysis and Spinal Cord Injury in the United States (2010)
181. Singh, A., Tetrault, L., Kalsi-Ryan, S., Nouri, A., Fehlings, M.G.: Global prevalence and incidence of traumatic spinal cord injury. Clin. Epidemiol. **6**, 209–331 (2014)
182. "Global Hearts", a new initiative from the World Health Organization (WHO) (2016). New initiative launched to tackle cardiovascular disease, the world's number one killer. Cardiovasc. Dis. http://www.who.int/cardiovascular_diseases/en/
183. WHO Media Centre (2017) Cancer. http://www.who.int/mediacentre/factsheets/fs297/en/
184. http://www.cancerresearchuk.org/
185. Fujiwara, A., Hoshino, T., Westley, J.M.: Anthracycline antibiotics. Crit. Rev. Biotechnol. Anthracycline **3**, 133 (1985)
186. Mele, D., Nardozza, M., Spallarossa, P., Frassoldati, A., Tocchetti, C.G., et al.: Current views on anthracycline cardiotoxicity. Heart Fail. Rev. **21**, 621–634 (2016)
187. Kucharska, W., Negrusz-kawecka, M., Gromkowska, M.: Cardiotoxicity of oncological treatment in children. Adv. Clin. Exp. Med. **21**, 281–288 (2012)

188. Hagen, E.M., Rekand, T., Gronning, M., Faerestrand, S.: Cardiovascular complications of spinal cord injury. Tidsskr. Nor. Laegeforen. **132**, 1115–1120 (2012)
189. Kalisvaart, J.F., Katsumi, H.K., Ronningen, L.D., Hovey, R.M.: Bladder cancer in spinal cord injury patients. Spinal Cord **48**, 257–261 (2010)
190. Kao, C.H., Sun, L.M., Chen, Y.S., Lin, C.L., Weng, M.W.: Risk of nongenitourinary cancers in patients with spinal cord injury—a population-based cohort study. Medicine **95**, e2462 (2016)
191. Chen, J.J., Wu, P.T., Middlekauff, H.R., Nguyen, K.L.: Aerobic exercise in anthracycline-induced cardiotoxicity: a systematic review of current evidence and future directions. Am. J. Physiol. Heart Circ. Physiol. (2016) (ahead of print)
192. Smith, A.E., Molton, I.R., Jensen, M.P.: Self-reported incidence and age of onset of chronic comorbid medical conditions in adults aging with long-term physical disability. Disabil. Health J. **9**, 533–538 (2016)
193. Guertin, P.A., Ung, R.V., Rouleau, P., Steuer, I.: Effects on locomotion, muscle, bone, and blood induced by a combination therapy eliciting weight-bearing stepping in nonassisted spinal cord-transected mice. Neurorehabil. Neural Repair **25**, 234–242 (2011)
194. Radhakrishna, M., Steuer, I., Prince, F., Roberts, M., Mongeon, D., et al.: Double-blind, placebo-controlled, randomized phase I/IIa study (safety and efficacy) with buspirone/levodopa/carbidopa (*Spinalon*) in subjects with complete AIS A or motor-complete AIS B spinal cord injury. Curr. Pharm. Des. (2016) (Ahead of print)
195. Hofstoetter, U.S., Knikou, M., Guertin, P.A., Minassian, K.: Probing the human spinal locomotor circuits by phasic step-induced feedback and by tonic electrical and pharmacological neuromodulation. Curr. Pharm. Des. (2016) (Ahead of print)
196. Henninger, C., Fritz, G.: Statins in anthracycline-induced cardiotoxicity: Rac and Rho, and the heartbreakers. Cell Death Dis. **8**, e2564 (2017)

Future Perspectives

Nanotechnology is distinguished by its interdisciplinary nature. As investigations at the nanolevel are occurring in a variety of fields, it is expected that the results of this research are going to have a significant impact on a broad range of applications Nanomaterials with tailored unique properties have limitless possibilities in materials science Products where the addition of a relatively small amount of functionalized nanoparticles or carbon nanotubes leads to a major change in the properties are going to revolutionize many commercial technologies.

Nowadays: the science of small will have a HUGE impact in society. Nanoparticles are being used for diverse purposes, from medical treatments, using in various branches of industry production such as solar and oxide fuel batteries for energy storage, to wide incorporation into diverse materials of everyday use such as cosmetics or clothes, optical devices, catalytic, bactericidal, electronic, sensor technology, biological labelling and treatment of some cancers. Due to their exceptional properties including antibacterial activity, high resistance to oxidation and high thermal conductivity, nanoparticles have attracted considerable attention in recent years. Nanoparticles can be synthesized chemically or biologically. Metallic nanoparticles that have immense applications in industries are of different types, namely, Gold, Silver, Alloy, magnetic etc. This study aims to present an overview of nanoparticles, with special reference to their mechanism of biosynthesis and types.

Besides technological prospects, nano-scale devices also offer a convenient platform to explore the fundamental physics of electrons in solids and molecules. In this ongoing journey of understanding the nature of materials, researchers around the world have developed many theories, among which the density functional theory (DFT) has emerged as one of the most powerful tools. One of the most active areas of computational materials research in recent years, yet also one of the numerically most costly, is the combination of DFT with modern theories of quantum transport in the attempt to study electronic transport through nano-scale devices under external bias. Among the variety of schemes developed, a special place is occupied by the application of the nonequilibrium Green's function

L. H. Madkour, *Nanoelectronic Materials*, Advanced Structured Materials 116, https://doi.org/10.1007/978-3-030-21621-4

(NEGF) method within the DFT approach. The sizes of the nanoscale systems considered in calculations have been much smaller than realistic experimental configurations. Usually the number of atoms is no more than a few hundred. NEGF-DFT has become a cost-effective first-principles method. Role of the computer simulations, stretchable and wearable nanoelectronics are discussed in this book.

Unavoidably, the continuation of this trend, well known as Moore's law, will reach a domain where the feature size becomes comparable to the wave length of the electrons in the circuits. In that case, quantum mechanical effects can arise and the classical Ohm's law of electronic transport breaks down. On the other hand, such quantum effects are not necessarily harmful if we have the basic understanding of them, as they can be used to design radically new types of electronic components. The development and manufacturing of such components at the scale of nanometers is the key aspect and idea of nanotechnology.

It is believed that nanotechnology can greatly contribute to the evolution of modern medical approaches and practices. Nanoscale constructs are already used in therapeutical applications against cancer and pathogens mostly by acting as drug carriers. Also either in-vivo or ex-vivo engineered scaffolds and tissues are implanted in patients whose own organs and tissues are damaged or lost. Furthermore, specific nanostructures are widely used as imaging and detection agents in diagnostic procedures. The ideal goal is to improve health by enhancing the efficacy and safety of nanosystems and nanodevices while at the same time use nanomedicine in order to cure diseases that remain incurable or the conventional therapeutical approaches against them are either expensive or inefficient.

The advances in fundamental nanosciences, the design of new nanomaterials, and ultimately the manufacturing of new nanoscale products and devices all depend to some degree on the ability to accurately and reproducibly measure their properties and performance at the nanoscale. Therefore, nanometrology tools and techniques are both integral to the emerging nanotechnology enterprise and are two of the main areas critical to the success of nanotechnology. Decades of nanoscience research have led to remarkable progress in nanotechnology as well as an evolution of instrumentation and metrology suitable for some nanoscale measurements. Consequently, today's suite of metrology tools has been designed to meet the needs of exploratory nanoscale research. New techniques, tools, instruments and infrastructure will be needed to support a successful nanomanufacturing industry.

The currently available metrology tools are also beginning to reach the limits of resolution and accuracy and are not expected to meet future requirements for nanotechnology or nanomanufacturing. Novel methods and combinations, such as the TERS technique, achieve much higher resolution values since they provide a significant increase in the Raman signal and in the lateral resolution by up to nine orders of magnitude. This combination overcomes the difficulties that originate from low signal since the Raman systems have limit in lateral resolution of 300 μm and require high laser power for surface investigation because the measured Raman intensity is six orders of magnitude lower than the excitation power. Thus, TERS is a promising technique and we can see it in the near future to be used for probing the chemical analysis of very small areas and for the imaging of nanostructures and biomolecules such as proteins. New approaches have to be developed and existing ones based on XPS, X-ray absorption spectroscopy, SPM and SIMS have to be improved in terms of better spectral and spatial resolution, better contrast and better sensitivity for elements and molecular species. Ideally new methods should have capabilities to work in situ, at ambient air and/or in liquid surroundings.

However, clever new approaches need to be developed. For this, it is required to understand the fundamental mechanisms by which the probes of the nanometrology measuring systems interact with the materials and objects that are being measured. Also, it is important to develop standard samples and to construct standardized procedures for measurements in nanometer scale, which enable the transfer of the properties and response of the unit from the nanometer to macroscopic scale without any appreciable loss of accuracy, for certifying, calibrating, and checking nanometrology instruments. Finally, even with the vast array of current tools available, the important question is whether or not they are providing the required information or reams of inconsequential data. Revolutionary approaches to the nanometrology needed may be required in the near future and therefore, revolutionary and not just evolutionary instrumentation and metrology are needed. In the future, develop new computational methods for the modeling of the transport properties of nano-scale devices, in order to treat systems with thousands of atoms and thus reduce the gab between simulations and realistic present day experiments.

Conclusions

Nanotechnology is an emerging technology with applications in several scientific and research fields, such as information and communication technology, electronics, energy, biology, medical technology, etc. Novel nano- and biomaterials, and nanodevices are fabricated and controlled by nanotechnology tools and techniques, which investigate and tune the properties, responses and functions of living and non-living matter, at sizes below 100 nm.

Nanotechnology is a science with huge potential and great expectations. The daily announcements of new discoveries and breakthroughs are going to influence all aspects of human society. Nanomaterials bring new possibilities by tailoring the optical, the electronic the mechanical, the chemical, and the magnetic properties. In the last few years there was a rapid progress in the fabrication and processing of nanostructures. As a result nanophase materials and applications are already in the market and a large volume of new applications is expected over the next several years. However, the development and commercialization of products containing nanomaterials raises many of the same issues as with introduction of any new technology, including concerns about the toxicity and environmental impact of nanomaterial exposures. Despite the extensive research of the last decade the literature on toxicological risks of the application of nanotechnology in medical technology is scarce.

In conclusion, risk assessment of nanomaterials requires the same information as needed for microscale materials, and also requires comprehensive characterization of the particles under investigation, including agglomeration/aggregation in relevant biological fluids. At present, there are increasing numbers of publications dealing with distinct aspects of the toxicity of nanomaterials. However, only during the very last years has attention been given to a careful characterization of the particles and also to the role of the protein corona around the particles, which appears to be critical for translocation and distribution. Since it seems to be neither feasible nor desirable to investigate every single particle type, common criteria for risk assessment should be defined for groups of nanoparticles, including comparison with microscale particles of the same material To establish science-based

L. H. Madkour, *Nanoelectronic Materials*, Advanced
Structured Materials 116, https://doi.org/10.1007/978-3-030-21621-4

occupational exposure limits, especially dose-response experiments under low exposure conditions are needed to identify relevant target organs and critical cellular reactions including primary and secondary genotoxicity, and to derive no observed effect levels (NOELs) and NOAELs after inhalation at realistic exposure levels.

- Nanoparticles can be defined as objects ranging in size from 1 to 100 nm that due to their size may differ from the bulk material.
- Presently, different metallic nanomaterials are being produced using copper, zinc, titanium, magnesium, gold, alginate and silver.
- A key property of the nanostructured materials (NSM) is their high surface-to-volume ratio, and that one or more of the physical dimensions are less than or comparable to the charge screening length, i.e., the Debye length. Therefore, these nanomaterials very often exhibit superior sensitivity in chemical surface processes, chemical industry, environmental pollution monitoring and the human health fields.
- Nanoparticles are being used for diverse purposes, from medical treatments, using in various branches of industry production such as solar and oxide fuel batteries for energy storage, to wide incorporation into diverse materials of everyday use such as cosmetics or clothes
- Nanoparticle technologies have great potentials, being able to convert poorly soluble, poorly absorbed and labile biologically active substance into promising deliverable substances.
- Nanoparticles are applied as delivery systems, e.g. for drugs or bioactive food ingredients. They are designed to target drugs to specific organs or to increase the bioavailability of bioactive food ingredients that may have a health impact.
- Differences between microscale particles and nanoscale particles depend on particle size and further particle characteristics, granular biopersistent particles (GBP) with no or little intrinsic chemical toxicity (inert particles). Relevant toxic effects are inflammation, oxidative stress and secondary genotoxicity at the lung as the target organ after inhalation.
- Another group of nanomaterials is fibre-like nanomaterials, e.g. nanotubes. They may exert fibre-like toxicity, depending on shape and chemical and physical characteristics. Risk assessment should be performed taking the dimensions and biopersistence of the respective materials into account. This requires a case-by-case evaluation.
- Nanomaterials currently possesses the technologies to deliver nanomaterials to the marketplace.
- Nanomaterials is currently selling numerous nanomaterials for a wide variety of proprietary developmental applications.
- Nanomaterials is currently in discussions with domestic and international investors for spinning off certain technologies into separate companies.
- Nanotechnology, from a business perspective, will create better and entirely new materials, devices, and systems. It means new jobs due to new markets.

- It is projected that by the year 2010, the demand for nanotechnology products and services in the U.S. alone will hit the $1 trillion dollar amount.
- Nanotechnology applications can be summarized into several basic areas: Smart Materials, Sensors, Nanoscale Biostructures, Energy Capture and Storage, Magnets, Fabrication, Electronics.
- Nanotechnology has potential to revolutionize a wide array of tools in biotechnology so that they are more personalized, portable, cheaper, safer, and easier to administer.
- Nanotechnology has been playing an increasing important role in supporting innovative technological advances.
- Few of the disadvantages of MNPs are the exact mechanism for synthesis of NPs needs to be elucidated, limitations to scale up production processes and reproducibility of the processes.

In order to investigate in depth the complex nanosystems, highly sophisticated nanoscale precision metrology tools are required. The advances in nanomaterials necessitate parallel progress of the nanometrology tools and techniques. Examples of important nanometrology tools as they have been discussed above include: ellipsometry, highly focused X-ray sources and related techniques, nanoindentation and scanning probe microscopies. The above described nanometrology methods contribute towards the understanding of several aspects of the state-of-the-art nanomaterials in terms of their optical, structural, and nanomechanical properties. The nanoscale precision and the detailed investigation that these nanometrology techniques offer, give them an enormous potential for even more advanced applications for the improvement of the quality of research and of the everyday life.

Bibliography

1. Thassu, D., Deleers, M., Pathak, Y.V.: Nanoparticulate Drug Delivery Systems, pp. 1–7. Informa Health Care, New York (2009)
2. http://easyayurveda.com/2013/10/21/charaka-samhita-sutrasthana-chapter-1-quest-longevity/
3. Saper, R.B., Kales, S.N., Paquin, J., Burns, M.J., Eisenberg, D.M., Davis, R.B.: Heavy metal content of ayurvedic herbal medicine products. JAMA **292**, 2868–2873 (2004)
4. Saper, R.B., Phillips, R.S., Schgal, A., Khouri, N., Davis, R.B., Paquin, J., et al.: Lead, mercury and arsenic in US- and Indian-manufactured ayurvedic medicines sold via the internet. JAMA **300**, 915–923 (2008)
5. Kales, S.N., Saper, R.B.: Ayurvedic lead poisoning: An under recognized international problem. Int. J. Med. Sci. **63**, 379–381 (2009)
6. Faraday, M.: Experimental relations of gold to light, p. 1475. Physiological Transactions of Royal Society, London (1857)
7. Mei, G.: Beiträge zur Optik trüber Medien, speziell kolloidaler Metallösungen. Ann. Phys. **330**, 377–445 (1908)
8. Beecroft, L.L., Ober, C.K.: Nanocomposite material for optical applications. Chem. Mater. **9**, 1302–1317 (1997)
9. Fendler, J.H.: Atomic and molecular clusters in membrane mimetic chemistry. Chem. Rev. **87**, 877–899 (1987)
10. Gates, B.C.: Supported metal clusters: synthesis, structure and catalysis. Chem. Rev. **95**, 511–522 (1995)
11. Kamat, P.V.: Photochemistry on non-reactive and reactive (semiconductor) surfaces. Chem. Rev. **93**, 267–300 (1993)
12. Madkour, L.H.: Applications of gold nanoparticles in medicine and therapy. Pharm. Pharmacol. Int. J. **6**(3), 157–174 (2018)
13. Madkour, L.H.: Biogenic–biosynthesis metallic nanoparticles (MNPs) for pharmacological, biomedical and environmental nanobiotechnological applications. Chron. Pharm. Sci. J. **2** (1), 384–444 (2018)
14. Madkour, L.H.: Ecofriendly green biosynthesized of metallic nanoparticles: bio-reduction mechanism, characterization and pharmaceutical applications in biotechnology industry. Glob. Drugs Ther. **3**(1), 1–11 (2018)
15. Madkour, L.H.: Advanced AuNMs as nanomedicine's central goals capable of active targeting in both imaging and therapy in biomolecules. Glob. Drugs Ther. **2**(6), 1–12 (2017)
16. Madkour, L.H.: Biotechnology of nucleic acids medicines as gene therapeutics and their drug complexes. Chron. Pharm. Sci. J. **1**(4), 204–253 (2017)

L. H. Madkour, *Nanoelectronic Materials*, Advanced Structured Materials 116, https://doi.org/10.1007/978-3-030-21621-4

17. Madkour, L.H.: Advanced AuNMs as nanomedicine'scentral goals capable of active targeting in both imaging and therapy in biomolecules. BAOJ Nanotechnol. **3**(1), 1–18 (2017)
18. Madkour, L.H.: Vision for life sciences: interfaces between nanoelectronic and biological systems. Glob. Drugs Ther. **2**(4), 1–4 (2017)
19. Madkour, L.H.: Toxic effects of environmental heavy metals on cardiovascular pathophysiology and heart health function: chelation therapeutics. UPI J. Pharm. Med. Health Sci. (UPI-JPMHS) **1**(1), 19–62 (2018). https://uniquepubinternational.com/upi-journals/upi, https://uniquepubinternational.com/wp-content/uploads/2018/03/UPI-JPMHS-2018-7.pdf
20. Foundation CaDR: One degree of separation: paralysis and spinal cord injury in the United States (2010)
21. Singh, A., Tetrault, L., Kalsi-Ryan, S., Nouri, A., Fehlings, M.G.: Global prevalence and incidence of traumatic spinal cord injury. Clin. Epidemiol. **6**, 209–331 (2014)
22. "Global Hearts", a new initiative from the World Health Organization (WHO). New initiative launched to tackle cardiovascular disease, the world's number one killer. Cardiovascular disease, Sept 2016. http://www.who.int/cardiovascular_diseases/en/
23. WHO Media centre: Cancer (2017). http://www.who.int/mediacentre/factsheets/fs297/en/
24. http://www.cancerresearchuk.org/
25. Fujiwara, A., Hoshino, T., Westley, J.M.: Anthracycline antibiotics. Crit. Rev. Biotechnol. Anthracycline **3**, 133 (1985)
26. Mele, D., Nardozza, M., Spallarossa, P., Frassoldati, A., Tocchetti, C.G., et al.: Current views on anthracycline cardiotoxicity. Heart Fail Rev. **21**, 621–634 (2016)
27. Kucharska, W., Negrusz-kawecka, M., Gromkowska, M.: Cardiotoxicity of oncological treatment in children. Adv. Clin. Exp. Med. **21**, 281–288 (2012)
28. Hagen, E.M., Rekand, T., Gronning, M., Faerestrand, S.: Cardiovascular complications of spinal cord injury. Tidsskr Nor Laegeforen **132**, 1115–1120 (2012)
29. Kalisvaart, J.F., Katsumi, H.K., Ronningen, L.D., Hovey, R.M.: Bladder cancer in spinal cord injury patients. Spin. Cord **48**, 257–261 (2010)
30. Kao, C.H., Sun, L.M., Chen, Y.S., Lin, C.L., Weng, M.W.: Risk of nongenitourinary cancers in patients with spinal cord injury—a population-based cohort study. Medicine **95**, e2462 (2016)
31. Chen, J.J., Wu, P.T., Middlekauff, H.R., Nguyen. K.L.: Aerobic exercise in anthracycline-induced cardiotoxicity: a systematic review of current evidence and future directions. Am. J. Physiol. Heart Circ. Physiol. **6** (2016) (ahead of print)
32. Smith, A.E., Molton, I.R., Jensen, M.P.: Self-reported incidence and age of onset of chronic comorbid medical conditions in adults aging with long-term physical disability. Disabil. Health J. **9**, 533–538 (2016)
33. Guertin, P.A., Ung, R.V., Rouleau, P., Steuer, I.: Effects on locomotion, muscle, bone, and blood induced by a combination therapy eliciting weight-bearing stepping in nonassisted spinal cord-transected mice. Neurorehabil. Neural Repair **25**, 234–242 (2011)
34. Radhakrishna, M., Steuer, I., Prince, F., Roberts, M., Mongeon, D. et al.: Double-blind, placebo-controlled, randomized phase I/IIa study (safety and efficacy) with buspirone/levodopa/carbidopa (*Spinalon*) in subjects with complete AIS A or motor-complete AIS B spinal cord injury. Curr. Pharm. Des. **27** (2016). (Ahead of print)
35. Hofstoetter, U.S., Knikou, M., Guertin, P.A., Minassian, K.: Probing the human spinal locomotor circuits by phasic step-induced feedback and by tonic electrical and pharmacological neuromodulation. Curr. Pharm. Des. **27** (2016) (Ahead of print)
36. Henninger, C., Fritz, G.: Statins in anthracycline-induced cardiotoxicity: Rac and Rho, and the heartbreakers. Cell Death Dis. **8**, e2564 (2017)
37. Gong, P., Li, H., He, X., Wang, K., Hu, J., Tan, W., Tan, S., Zhang, X.Y.: Preparation and antibacterial activity of Fe_3O_4@Ag nanoparticles. Nanotechnology **18**, 604–611 (2007)
38. Rai, M., Yadav, A., Gade, A.: Biotech. Adv. **27**(1), 76–83 (2009)
39. Rai, M., Yadav, A., Gade, A.: Biotech. Adv. **27**, 76–83 (2009)

40. Sharma, V.K., Ria, A.Y., Lin, Y.: Adv. Colloid Interface Sci. **145**, 83–96 (2009)
41. Shankar, S.S., Rai, A., Ankamwar, B., Singh, A., Ahmad, A., Sastry, M.: Nat. Mater. **3**, 482–488 (2004)
42. Bar, H., Bhui, D.K., Sahoo, G.P., Sarkar, P., De, S.P., Misra, A.: Colloids and surfaces A. Physicochem. Eng. Asp. **339**, 134–139 (2009)
43. Jha, A.K., Prasad, K.: Int. J. Green Nanotechnol. Phys. Chem. **1**, 110–117 (2010)
44. Baban, D., Seymour, L.W.: Control of tumour vascular permeability. Adv. Drug Deliv. Rev. **34**, 109–119 (1998)
45. Tomar, A., Garg, G.: Short review on application of gold nanoparticles. Glob. J. Pharmacol. **7**(1), 34–38 (2013)
46. Ceylan, A., Jastrzembski, K., Shah, S.I.: Metall. Mater. Trans. A **37**, 2033 (2006)
47. Junggwon, Y., Kyoungah, C., Byoungjun, P., Ho-Chul, K., Byeong-Kwon, J., Sangsig, K. J.: J. Appl. Phys. **47**, 5070 (2008)
48. Mohl, M., Dobo, D., Kukovecz, A., Konya, Z., Kordas, K., Wei, J., Vajtai, R., Ajayan, P. M.: J. Phys. Chem. C **115**, 9403 (2011)
49. Fan, T.X., Chow, S.K., Zhang, D.: Biomorphic mineralization: from biology to materials. Prog. Mater. Sci. **54**(5), 542–659 (2009)
50. Harris, P.J.: Carbon Nanotubes and Related Structures. Cambridge University Press (1999)
51. He, X., Du, C., Shen, B., Chen, C., Xu, X., Wang, Y., Zuo, P., Ma, Y., Cheng, X., Yin, G.: Electronically conductive Sb-doped SnO_2 nanoparticles coated $LiNi_{0.8}Co_{0.15}Al_{0.0}5O_2$ cathode material with enhanced electrochemical properties for Li-ion batteries. Electrochim. Acta **236**, 273–279 (2017)
52. Wang, J., Sun, X.: Understanding and recent development of carbon coating on $LiFePO_4$ cathode materials for lithium-ion batteries. Energy Environ. Sci. **5**, 5163–5185 (2012)
53. Zhao, D., Feng, Y., Wang, Y., Xia, Y.: Electrochemical performance comparison of $LiFePO_4$ Supported by various carbon materials. Electrochim. Acta. **88**, 632–638 (2013)
54. Liu, H., Yang, G., Zhang, X., Gao, P., Wang, L., Fang, J., Pinto, J., Jiang, X.: Kinetics of Conventional carbon coated-$Li_3V_2(PO_4)_3$ and nanocomposite $Li_3V_2(PO_4)_3$/graphene as cathode materials for lithium ion batteries. J. Mater. Chem. **22**, 11039–11047 (2012)
55. Wang, X., Lv, L., Cheng, Z., Gao, J., Dong, L., Hu, C., Qu, L.: High-density monolith of N-doped holey graphene for ultrahigh volumetric capacity of li-ion batteries. Adv. Energy Mater. **6**, 1502100 (2016)
56. Zhao, Y., Li, X., Yan, B., Li, D., Lawes, S., Sun, X.: Significant impact of 2D graphene nanosheets on large volume change tin-based anodes in lithium-ion batteries: a review. J. Power Sour. **274**, 869–884 (2015)
57. Moore, M.N.: Do nanoparticles present ecotoxicological risks for the health of the aquatic environment? Environ. Int. **32**(8), 967–976 (2006)
58. Sergeev, G.B., Shabatina, T.I.: Cryochemistry of nanometals. Colloids Surf. A Physicochem. Eng. Asp. (2008)
59. Xue, Chenming, et al.: Size- and shape-dependent fluorescence quenching of gold nanoparticles on perylene dye. Adv. Opt. Mater. **1**(8), 581–587 (2013)
60. Aljerf, L., Nadra, R.: Developed greener method based on MW implementation in manufacturing CNFs. Int. J. Nanomanuf. (2018)
61. Nomura, Toshiyuki, et al.: Shape and size control of barium chromate nanoparticles using reverse micelle. Adv. Powder Technol. **20**(1), 101–105 (2009)
62. Katyal, J., Soni, R.K.: Size- and shape-dependent plasmonic properties of aluminum nanoparticles for nanosensing applications. J. Mod. Opt. **60**(20), 1717–1728 (2013)
63. Kaushik, N. et al.: Biological synthesis of metallic nanoparticles. Nanomed. Nanotechnol. Biol. Med. **6**(2), 257–262 (2010)
64. Rai, M., et al.: Silver nanoparticles as a new generation of antimicrobials. Biotechnol. Adv. **27**(1), 76–83 (2009)
65. Lu, K.: Nanoparticulate Materials: Synthesis, Characterization, and Processing (2013)

66. Porter, A., Youtie, J.: How interdisciplinary is nanotechnology. J. Nanopart. Res. **11**(5), 1023–1041 (2009)
67. Evans, J. et al.: Nanotechnology tackles chemotherapy. In: Proceedings of the National Academy of Sciences (2006)
68. Berger, M.: Carbon nanotubes could make t-shirts bullet proof. In: Nanotechnology Spotlight (2006)
69. Ju-Nam, Y., Lead, J.R.: Sci. Total Environ. **1**, 400
70. De, M., Ghosh, P.S., Rotello, V.M.: Adv. Mater. **20,** 4225 (2008)
71. Lu, A.-H., Salabas, E.L.: Ferdi Schüth Angewandte. Chemie Int. Edn. **46**, 1222 (2007)
72. Ginley, D., Green, M., Collins, R.: Solar energy conversion toward 1 terawatt. MRS Bull. **33**(355), 73 (2008)
73. Hochbaum, A.I., Yang, P.: Semiconductor nanowires for energy conversion. Chem. Rev. **110**(1), 527–546 (2010)
74. Law, M., Goldberger, J., Yang, P.: Semiconductor nanowires and nanotubes. Annu. Rev. Mater. Res. **34**(1), 83–122 (2004)
75. Tian, B., Kempa, T.J., Lieber, C.M.: Single nanowire photovoltaics. Chem. Soc. Rev. **38** (1), 16–24 (2009)
76. Fan, Z., Ruebusch, D.J., Rathore, A.A., Kapadia, R., Ergen, O., et al.: Challenges and prospects of nanopillar-based solar cells. Nano Res. **2**(11), 829–843 (2009)
77. Yu, K., Chen, J.: Enhancing solar cell efficiencies through 1-D nanostructures. Nanoscale Res. Lett. **4**(1), 1–10 (2008)
78. Cao, L., Park, J., Fan, P., Clemens, B., Brongersma, M.L.: Resonant germanium nanoantenna photodetectors. Nano Lett. **10**(4), 1229–1233 (2010)
79. Gunawan, O., Sekaric, L., Majumdar, A., Rooks, M., Appenzeller, J., et al.: Measurement of carrier mobility in silicon nanowires. Nano Lett. **8**(6), 1566–1571 (2008)
80. Ford, A.C., Ho, J.C., Chueh, Y., Tseng, Y., Fan, Z., et al.: Diameter-dependent electron mobility of InAs nanowires. Nano Lett. **9**(1), 360–365 (2009)
81. Xiang, J., Lu, W., Hu, Y., Wu, Y., Yan, H., Lieber, C.M.: Ge/Si nanowire heterostructures as highperformance field-effect transistors. Nature **441**(7092), 489–493 (2006)
82. Garnett, E., Yang, P.: Light trapping in silicon nanowire solar cells. Nano Lett. **10**(3), 1082–1087 (2010)
83. Fan, Z., Razavi, H., Do, J., Moriwaki, A., Ergen, O., et al.: Three-dimensional nanopillar-array photovoltaics on low-cost and flexible substrates. Nature Mater. **8**(8), 648–653 (2009)
84. Schmid, H., Björk, M.T., Knoch, J., Riel, H., Riess, W. et al.: Patterned epitaxial vapor-liquid-solid growth of silicon nanowires on Si(111) using silane. J. Appl. Phys. **103** (2), 024304 (2008)
85. Mårtensson, T., Carlberg, P., Borgström, M., Montelius, L., Seifert, W., Samuelson, L.: Nanowire arrays defined by nanoimprint lithography. Nano Lett. **4**(4), 699–702 (2004)
86. Zschech, D., Kim, D.H., Milenin, A.P., Scholz, R., Hillebrand, R. et al.: Ordered arrays of<100>-oriented silicon nanorods by CMOS-compatible block copolymer lithography. Nano Lett. **7**(6), 1516–1520 (2007)
87. Huang, Z., Fang, H., Zhu, J.: Fabrication of silicon nanowire arrays with controlled diameter, length, and density. Adv. Mater. **19**(5), 744–748 (2007)
88. Garnett, E.C., Yang, P.: Silicon nanowire radial *p-n* junction solar cells. J. Am. Chem. Soc. **130**(29), 9224–9225 (2008)
89. HsuC, Connor S.T., Tang, M.X., Cui, Y.: Wafer-scale silicon nanopillars and nanocones byLangmuir–Blodgett assembly and etching. Appl. Phys. Lett. **93**(13), 133109 (2008)
90. Huang, Z., Zhang, X., Reiche, M., Liu, L., Lee, W., et al.: Extended arrays of vertically aligned sub-10 nm diameter [100] Si nanowires by metal-assisted chemical etching. Nano Lett. **8**(9), 3046–3051 (2008)
91. Bogush, G.H., Tracy, M.A., Zukoski IV, C.F.: Preparation of monodisperse silica particles: control of size and mass fraction. J. Non-Cryst. Solids **104**, 95–106 (1988)

92. Jeong, S., Hu, L., Lee, H.R., Garnett, E., Choi, J.W., Cui, Y.: Fast and scalable printing of large area monolayer nanoparticles for nanotexturing applications. Nano Lett. **10**, 2989–2994 (2010)
93. Law, M., Greene, L.E., Johnson, J.C., Saykally, R., Yang, P.: Nanowire dye-sensitized solar cells. Nat. Mater. **4**(6), 455–459 (2005)
94. Wang, K., Chen, J.J., Zeng, Z.M., Tarr, J., Zhou, W.L., et al.: Synthesis and photovoltaic effect of vertically aligned ZnO/ZnS core/shell nanowire arrays. Appl. Phys. Lett. **96**(12), 123105 (2010)
95. Fan, Z., Razavi, H., Do, J., Moriwaki, A., Ergen, O., et al.: Three-dimensional nanopillar-array photovoltaics on low-cost and flexible substrates. Nature Mater. **8**(8), 648–653 (2009)
96. Lévy-Clément, C., Tena-Zaera, R., Ryan, M.A., Katty, A., Hodes, G.: CdSe-sensitized *p*-CuSCN/nanowire *n*-ZnO heterojunctions. Adv. Mater. **17**(12):1512–1515 (2005)
97. Takanezawa, K., Tajima, K., Hashimoto, K.: Efficiency enhancement of polymer photovoltaic devices hybridized with ZnO nanorod arrays by the introduction of a vanadium oxide buffer layer. Appl. Phys. Lett. **93**(6), 063308 (2008)
98. Dong, Y., Tian, B., Kempa, T.J., Lieber, C.M.: Coaxial group III-nitride nanowire photovoltaics. Nano Lett. **9**(5), 2183–2187 (2009)
99. Kayes, B.M., Atwater, H.A., Lewis, N.S.: Comparison of the device physics principles of planar and radial *p-n* junction nanorod solar cells. J. Appl. Phys. **97**(11), 114302 (2005)
100. Hecht, E.: Optics, p. 698. Pearson, San Francisco (2002)
101. Zhao, J., Green, M.: Optimized antireflection coatings for high-efficiency silicon solar cells. IEEE Trans. Electron Devices **38**(8), 1925–1934 (1991)
102. Zhu, J., Yu, Z., Burkhard, G.F., Hsu, C., Connor, S.T., et al.: Optical absorption enhancement in amorphous silicon nanowire and nanocone arrays. Nano Lett. **9**(1), 279–282 (2009)
103. Wu, X.: High-efficiency polycrystalline CdTe thin-film solar cells. Solar Energy **77**(6), 803–814 (2004)
104. Ramanathan, K., Contreras, M.A., Perkins, C.L., Asher, S., Hasoon, F.S., et al.: Properties of 19.2% efficiency ZnO/CdS/$CuInGaSe_2$ thin-film solar cells. Prog. Photovolt. Res. Appl. **11**(4), 225–230 (2003)
105. Todorov, T.K., Reuter, K.B., Mitzi, D.B.: High-efficiency solar cell with earth-abundant liquidprocessed absorber. Adv. Mater. **22**(20), E156–E159 (2010)
106. Hains, A.W., Liang, Z., Woodhouse, M.A., Gregg, B.A.: Molecular semiconductors in organic photovoltaic cells. Chem. Rev. **110**(11), 6689–6735 (2010)
107. Hibberd, C.J., Chassaing, E., Liu, W., Mitzi, D.B., Lincot, D., Tiwari, A.N.: Non-vacuum methods for formation of Cu(In, Ga)(Se, S)2 thin film photovoltaic absorbers. Prog. Photovolt. Res. Appl. **18**(6), 434–452 (2009)
108. Fan, Z., Kapadia, R., Leu, P.W., Zhang, X., Chueh, Y., et al.: Ordered arrays of dual-diameter nanopillars for maximized optical absorption. Nano Lett. **10**(10), 3823–3827 (2010)
109. Tian, B., Zheng, X., Kempa, T.J., Fang, Y., Yu, N., et al.: Coaxial silicon nanowires as solar cells and nanoelectronic power sources. Nature **449**(7164), 885–889 (2007)
110. Wu, Z., Neaton, J.B., Grossman, J.C.: Quantum confinement and electronic properties of tapered silicon nanowires. Phys. Rev. Lett. **100**, 246804 (2008)
111. Takanezawa, K., Hirota, K., Wei, Q., Tajima, K., Hashimoto, K.: Efficient charge collection with ZnO nanorod array in hybrid photovoltaic devices. J. Phys. Chem. C **111**(19), 7218–7223 (2007)
112. Kempa, T.J., Tian, B., Kim, D.R., Hu, J., Zheng, X., Lieber, C.M.: Single and tandem axial *p-i-n* nanowire photovoltaic devices. Nano Lett. **8**(10), 3456–3460 (2008)
113. Chan, C.K., Peng, H., Liu, G., McIlwrath, K., Zhang, X.F., et al.: High-performance lithium battery anodes using silicon nanowires. Nat. Nanotechnol. **3**(1), 31–35 (2008)

114. Green, M., Basore, P., Chang, N., Clugston, D., Egan, R., et al.: Crystalline silicon on glass (CSG) thin-film solar cell modules. Solar Energy **77**(6), 857–863 (2004)
115. Cao, L., White, J.S., Park, J., Schuller, J.A., Clemens, B.M., Brongersma, M.L.: Engineering light absorption in semiconductor nanowire devices. Nature Mater. **8**(8), 643–647 (2009)
116. Cao, L., Fan, P., Vasudev, A.P., White, J.S., Yu, Z., et al.: Semiconductor nanowire optical antenna solar absorbers. Nano Lett. **10**(2), 439–445 (2010)
117. Cao, L., Fan, P., Barnard, E.S., Brown, A.M., Brongersma, M.L.: Tuning the color of silicon nanostructures. Nano Lett. **10**(7), 2649–2654 (2010)
118. Wang, Z.L., Liu, Y., Zhang, Z. (eds.): Handbook of Nanophase and Nanostructured Materials. V.1. Synthesis; V.2. Characterization; V.3. Materials Systems and Applications I; V.4. Materials Systems and Applications II. Kluwer Academic/Plenum Publishers (2003)
119. Cao, G.: Nanostructures and Nanomaterials. Synthesis, Properties and Applications. Imperial College Press (2004)
120. Kelsall, R.W., Hamley, I.W., Geoghegan, M. (eds.): Nanoscale Science and Technology. Wiley (2005)
121. Harrison, P.: Quantum Wells, Wires and Dots. Wiley (2005)
122. Yu, G. (ed.): Nanomaterials Handbook. Taylor & Francis Group, LLC (2006)
123. Dresselhaus, M.S., Eklund, P.C.: Phonons in carbon nanotubes. Adv. Phys. **49**(6), 705–814 (2000)
124. Gleiter, H.: Nanostructured materials: basic concepts and microstructures. Acta Metall. **48** (1), 1–29 (2000)
125. Joannopoulos, J.D., Villeneuve, P.R., Fan, S.: Photonic crystals: putting a new twist on light. Nature **386**, 143–149 (1997)
126. Pokropivny, V.V.: Noncarbon nanotubes. Rev. Powder Metall. Metal Ceram. I. Methods of synthesis **40**(9/10), 485–496 (2001); II. Types and structures **40**(11/12), 582–594 (2001); III. Properties and promising applications **41**(3/4), 123–135 (2002)
127. Pokropivny, V.V.: Two-dimensional nanocomposites. Rev. Powder Metall. Metal Ceram. I. Types and fabrication **41**(5/6), 264–272 (2002). II. Properties and applications **41**(7/8), 369–381 (2002)
128. Viswanathan, B.: Nanomaterials. Narosa Publishing House
129. Zhang, J.Z.: Optical Properties and Spectroscopy of Nanomaterials. World Scientific Publishing Co. Pte. Ltd.
130. Sajanlal, P.R., Sreeprasad, T.S., Samal, A.K., Pradeep, T.: Anisotropic Nanomaterials: Structure, Growth, Assembly, and Functions. Nano Rev. **2** (2011)
131. Lue, J.T.: Physical properties of nanomaterials. In: Encyclopedia of Nanoscience and Nanotechnology, vol. X, pp. 1–46
132. Nanomaterials—An introduction to synthesis, properties and application. Environ. Eng. Manag. J. **7**(6), 865–870 (2008)
133. Sarikaya, M., Tamerler, C., Jen, A.K.Y., Schulten, K., Baneyx, F.: Molecular biomimetics: nanotechnology through biology. Nature Mater. **2**, 577–585 (2003)
134. Ikkala, O., Brinke, G.T.: Functional materials based on self-assembly of polymeric supramolecules. Science, New Series, **295** (5564), 2407–2409 (2002)
135. Rao, C.N.R., Nath, M.: Inorganic nanotubes. Dalton Trans. 1–24 (2003)
136. Boston, R., Schnepp, Z., Nemoto, Y., Sakka, Y., Hall, S.R.: In Situ TEM observation of a microcrucible mechanism of nanowire growth. Science **344**(6184), 623. Bibcode: 2014Sci…344…623B. https://doi.org/10.1126/science.1251594. PMID 24812400
137. Carter, R., Suyetin, M., Lister, S., Dyson, M.A., Trewhitt, H., Goel, S., Liu, Z., Suenaga, K., Giusca, C., Kashtiban, R.J., Hutchison, J.L., Dore, J.C., Bell, G.R., Bichoutskaia, E., Sloan, J.: Band gap expansion, shear inversion phase change behaviour and low-voltage induced crystal oscillation in low-dimensional tin selenide crystals. Dalton Trans. **43**(20), 7391 (2014). https://doi.org/10.1039/c4dt00185k. PMID 24637546

138. Spencer, J., Nesbitt, J., Trewhitt, H., Kashtiban, R., Bell, G., Ivanov, V., Faulques, E., Smith, D.: Raman spectroscopy of optical transitions and vibrational energies of ~1 nm HgTe extreme nanowires within single walled carbon nanotubes. ACS Nano **8**(9), 9044 (2014). https://doi.org/10.1021/nn5023632. PMID 25163005
139. von Klitzing constant. physics.nist.gov
140. Holmes, J.D., Johnston, K.P., Doty, R.C., Korgel, B.A.: Control of thickness and orientation of solution-grown silicon nanowires. Science **287**(5457), 1471–1473 (2000). Bibcode: 2000Sci...287.1471H. https://doi.org/10.1126/science.287.5457.1471. PMID 10688792
141. Heitsch, A.T., Akhavan, V.A., Korgel, B.A.: Rapid SFLS synthesis of Si nanowires using trisilane with in situ alkyl-amine passivation. Chem. Mater. **23**(11), 2697 (2011). https://doi.org/10.1021/cm2007704
142. Hanrath, T., Korgel, B.A.: Supercritical Fluid–Liquid–Solid (SFLS) synthesis of Si and Ge nanowires seeded by colloidal metal nanocrystals. Adv. Mater. **15**(5), 437 (2003). https://doi.org/10.1002/adma.200390101
143. Rackauskas, S., Nasibulin, A.G., Jiang, H., Tian, Y., Kleshch, V.I., Sainio, J., Obraztsova, E.D., Bokova, S.N., Obraztsov, A.N., Kauppinen, E.I.: A novel method for metal oxide nanowire synthesis. Nanotechnology **20**, 165603 (2010). https://doi.org/10.1088/0957-4484/20/16/165603
144. Sears, G.W.: A growth mechanism for mercury whiskers. Acta Metall. **3**, 361–366 (1955). https://doi.org/10.1016/0001-6160(55)90041-9
145. Frank, F.C.: The influence of dislocations on crystal growth. Discuss. Faraday Soc. **5**, 48 (1949). https://doi.org/10.1039/df9490500048
146. Burton, W.K., Cabrera, N., Frank, F.C.: The growth of crystals and the equilibrium structure of their surfaces. Philos. Trans. R. Soc. Lond. Ser. A, Math. Phys. Sci. **243**, 299–358 (1951). http://dx.doi.org/10.1098/rsta.1951.0006
147. Morin, S.A., Bierman, M.J., Tong, J., Jin, S.: Mechanism and kinetics of spontaneous nanotube growth driven by screw dislocations. Science **328**, 476–480 (2010). https://doi.org/10.1126/science.1182977
148. Bierman, M.J., Lau, Y.K.A., Kvit, A.V., Schmitt, A.L., Jin, S.: Dislocation-driven nanowire growth and Eshelby twist. Science **320**, 1060–1063 (2008). https://doi.org/10.1126/science.1157131
149. Rackauskas, S., Jiang, H., Wagner, J.B., Shandakov, S.D., Hansen, T.W., Kauppinen, E.I., Nasibulin, A.G.: In Situ study of noncatalytic metal oxide nanowire growth. Nano Lett. **14**, 5810–5813 (2014). https://doi.org/10.1021/nl502687s
150. Tilke, A.T., Simmel, F.C., Lorenz, H., Blick, R.H., Kotthaus, J.P.: Quantum interference in a one-dimensional silicon nanowire. Phys. Rev. B **68**(7), 075311 (2003). Bibcode: 2003PhRvB.68g5311T. https://doi.org/10.1103/physrevb.68.075311
151. Halford, Bethany: Wee welding with nanosolder. Chem. Eng. News **86**(51), 35 (2008)
152. Lu, Y., Huang, J.Y., Wang, C., Sun, S., Lou, J.: Cold welding of ultrathin gold nanowires. Nature Nanotechnol. **5**(3), 218 (2010). Bibcode: 2010NatNa...5..218L. https://doi.org/10.1038/nnano.2010.4. PMID: 20154688
153. Zhong, Z., Wang, D., Cui, Y., Bockrath, M.W., Lieber, C.M.: Nanowire crossbar arrays as address decoders for integrated nanosystems. Science **302**(5649), 1377–1379 (2003). Bibcode: 2003Sci...302.1377Z. https://doi.org/10.1126/science.1090899. PMID: 14631034
154. Huo, F., Zheng, Z., Zheng, G., Giam, L.R., Zhang, H., Mirkin, C.A.: Polymer pen lithography. Science **321**(5896), 1658 (2008). Bibcode: 2008Sci...321.1658H. https://doi.org/10.1126/science.1162193. PMID: 18703709
155. Mongillo, M., Spathis, P., Katsaros, G., Gentile, P., De Franceschi, S.: Multifunctional devices and logic gates with undoped silicon nanowires. Nano Lett. **12**(6), 3074 (2012). https://arxiv.org/abs/1208.1465. https://doi.org/10.1021/nl300930m. PMID: 22594644

156. Appenzeller, J., Knoch, J., Bjork, M.T., Riel, H., Schmid, H., Riess, W.: Toward nanowire electronics. IEEE Trans. Electron Devices **55**(11), 2827 (2008). Bibcode: 2008ITED... 55.2827A. https://doi.org/10.1109/ted.2008.2008011
157. Wissner-Gross, A.D.: Dielectrophoretic reconfiguration of nanowire interconnects (PDF). Nanotechnology **17**(19), 4986–4990. Bibcode: 2006Nanot..17.4986W. https://doi.org/10.1088/0957-4484/17/19/035
158. Nanowires get reconfigured. Nanotechweb.org, 19 Oct 2006
159. Grange, R., Choi, J.W., Hsieh, C.L., Pu, Y., Magrez, A., Smajda, R., Forro, L., Psaltis, D.: Lithium niobate nanowires: synthesis, optical properties and manipulation. Appl. Phys. Lett. **95**(14),143105 (2009). Bibcode: 2009ApPhL..95n3105G. https://doi.org/10.1063/1.3236777
160. Vizcaíno, J.L.P., Núñez, C.G.A.: Fast, effective manipulation of nanowires for electronic devices. SPIE Newsroom (2013). https://doi.org/10.1117/2.1201312.005260
161. Engel, Y., Elnathan, R., Pevzner, A., Davidi, G., Flaxer, E., Patolsky, F.: Supersensitive detection of explosives by silicon nanowire arrays. Angew. Chemie Int. Ed. **49**(38), 6830–6835 (2010). https://doi.org/10.1002/anie.201000847. Cite uses deprecated parameter | coauthors= (help)
162. Garcia, J.C., Justo, J.F.: Twisted ultrathin silicon nanowires: a possible torsion electromechanical nanodevice. Europhys. Lett. **108**(3), 36006 (2014). Bibcode: 2014EL.... 10836006G. https://doi.org/10.1209/0295-5075/108/36006
163. Elnathan, R., Kwiat, M., Pevzner, A., Engel, Y., Burstein, L., Khatchtourints, A., Lichtenstein, A., Kantaev, R., Patolsky, F.: Biorecognition layer engineering: overcoming screening limitations of nanowire-based FET devices. Nano Lett. **12**(10), 5245–5254 (2012). Bibcode: 2012NanoL.12.5245E. https://doi.org/10.1021/nl302434w
164. Lu, W., Lieber, C.M.: Semiconductor nanowires. J. Phys. D Appl. Phys. R387 (2006)
165. Lu, J.G., Chang, P., Fan, Z.: Quasi-one-dimensional metal oxide materials—synthesis, properties and applications. Math. Sci. Eng. R-Rep. **52**, 49–91 (2006)
166. Xia, Y., Yang, P., Sun, Y., Wu, Y., Mayers, B., Gates, B., Yin, Y., Kim, F., Yan, H.: One-dimensional nanostructures: synthesis, characterization, and applications. Adv. Mater. **15**, 353–389 (2003)
167. Wagner, R.S., Ellis, W.C.: Vapor–Liquid–Solid mechanism of single crystal growth. Appl. Phys. Lett. **4**, 89–90 (1964)
168. Hochbaum, A.I., Fan, R., He, R., Yang, P.: Controlled growth of Si nanowire arrays for device integration. Nano Lett. **5**, 457–460 (2005)
169. Kodambaka, S., Tersoff, J., Reuter, M.C., Ross, F.M.: Germanium nanowire growth below the eutectic temperature. Science **316**, 729–732 (2007)
170. Tian, B., Xie, P., Kempa, T.J., Bell, D.C., Lieber, C.M.: Single-crystalline kinked semiconductor nanowire superstructures. Nat. Nano, vol. advance online publication (2009)
171. Lieber, C.M.: Nanoscale science and technology: building a big future from small things. Bulletin **28**, 486–491 (2003)
172. Plummer, J.D., Deal, M.D., Griffin, P.B.: Silicon VLSI Technology: Fundamentals, Practice, and Modelling. Prentice Hall (2000)
173. Wang, Z.L.: Functional oxide nanobelts: materials, properties and potential applications inanosystems and biotechnology. Annu. Rev. Phys. Chem. **55**, 159–196 (2004)
174. Hejazi, S.R., Hosseini, H.R.M., Ghamsari, M.S.: The role of reactants and droplet interfaon nucleation and growth of ZnO nanorods synthesized by vapor-liquid-solid (VLS) mechanism. J. Alloys Compd. **455**, 353–357 (2008)
175. Park, J., Choi, H.-H., Siebein, K., Singh, R.K.: Two-step evaporation process for formation of aligned zinc oxide nanowires. J. Cryst. Growth **258**, 342–348 (2003)
176. Masumura, R.A., Hazzledine, P.M., Pande, C.S.: Yield stress of fine grained materials. Acta Mater. (1998)

177. Arzt, E.: Size effects in materials due to microstuctural and dimensional constraints: a comparative review. Acta Mater. (1998); Chokshi, A.H., Rosen, A., Karch, J., Gleiter, H.: Scripta Metall. **23**, 1679 (1989)
178. Van Swygenhoven, H., Caro, A., Farkas, D.: A molecular dynamics study of polycrystalline FCC metals at the nanoscale grain boundary structure and its influence on plastic deformation. Mater. Sci. Eng. A (2001)
179. Van Swygenhoven, H., Spaczer, M., Caro, A.: Microscopic description of plasticity in computer generated metallic nanophase samples: a comparison between Cu and Ni. Acta Mater. (1999)
180. Konstantinidis, D.A., Aifantis, E.C.: On the 'anomalous' hardness of nanocrystalline materials. Nanostruct. Mater. (1998)
181. Klemens, P.G., Gell, M.: Thermal conductivity of thermal barrier coatings. Mater. Sci. Eng. A **245**, 143–149 (1998)
182. Braginsky, L., Shklover, V., Hofmann, H., Bowen, P.: High-temperature thermal conductivity of porous Al_2O_3 nanostructures. Phys. Rev. B **70**, 134201 (2004)
183. Alivisatos, A.P.: Perspectives on the physical chemistry of semiconductor nanocrystals. J. Phys. Chem. (1996)
184. Ciraci, S., Buldum, A., Batra, I.P.: Quantum effects in electrical and thermal transport through nanowires. J. Phys. Condens. Matter (2001)
185. Eustis, S., El-Sayed, M.A.: Why gold nanoparticles are more precious than pretty gold: noble metal surface plasmon resonance and its enhancement of the radiative and nonradiative properties of nanocrystals of different shapes. R. Soc. Chem. (2005)
186. Kelly, K.L., Coronado, E., Zhao, L.L., Schatz, G.C.: The optical properties of metal nanoparticles: the influence of size, shape, and dielectric environment. J. Phys. Chem. (2003)
187. Chen, J.Y., Pan, C.J., Tsao, F.C., Kuo, C.H., Chi, G.C., Pong, B.J., Chang, C.Y., Norton, D.P., Pearton, S.J.: Characterization of ZnO nanowires grown on Si (100) with and without Au catalyst. Vacuum **84**, 1076–1079 (2009)
188. Huang, M.H., Wu, Y., Feick, H., Tran, N., Weber, E., Yang, P.: Catalytic growth of zinc oxide nanowires by vapor transport. Adv. Mater. **13**, 113–116 (2001)
189. Li, S.Y., Lee, C.Y., Tseng, T.Y.: Copper-catalyzed ZnO nanowires on silicon (100) grownby vapor-liquid-solid process. J. Cryst. Growth **247**, 357–362 (2003)
190. Wang, J.X., Liu, D.F., Yan, X.Q., Yuan, H.J., Ci, L.J., Zhou, Z.P., Gao, Y., Song, L., Liu, L.F., Zhou, W.Y., Wang, G., Xie, S.S.: Growth of SnO_2 nanowires with uniform branched structures. Solid State Commun. **130**, 89–94 (2004)
191. Chen, J.Y., Pan, C.J., Tsao, F.C., Kuo, C.H., Chi, G.C., Pong, B.J., Chang, C.Y., NortonS, D.P., Pearton, J.: Characterization of ZnO nanowires grown on Si (100) with and without Au catalyst. Vacuum **83**, 1076–1079 (2009)
192. Johnson, M.C., Aloni, S., McCready, D.E., Bourret-Courchesne, E.D.: Controlled vapor −liquid−solid growth of indium, gallium, and tin oxide nanowires via chemical vapor transport. Cryst. Growth Des. **6**, 1936–1941 (2006)
193. Kirkham, M., Wang, X., Wang, Z.L., Snyder, R.L.: Solid Au nanoparticles as a catalyst for growing aligned ZnO nanowires: a new understanding of the vapour–liquid–solid process. Nanotechnology **18**, 365304 (2007)
194. Law, J.B.K., Boothroyd, C.B., Thong, J.T.L.: Site-specific growth of ZnO nanowires from patterned Zn via compatible semiconductor processing. J. Cryst. Growth **310**, 2485–2492 (2008)
195. Fan, H.J., Scholz, R., Kolb, F.M., Zacharias, M.: Two-dimensional dendritic ZnO nanowiresfrom oxidation of Zn microcrystals. Appl. Phys. Lett. **85**, 4142–4144 (2004)
196. Dang, H.Y., Wang, J., Fan, S.S.: The synthesis of metal oxide nanowires by directly heat in metal samples in appropriate oxygen atmospheres. Nanotechnology, 738 (2003)
197. Wan, Q., Lin, C.L., Yu, X.B., Wang, T.H.: Room-temperature hydrogen storage characteristics of ZnO nanowires. Appl. Phys. Lett. **84**, 124–126 (2004)

198. Chen, Y.Q., Jiang, J., Wang, B., Hou, J.G.: Synthesis of tin-doped indium oxide nanowires by self-catalytic VLS growth. J. Phys. D Appl. Phys. 3319 (2004)
199. Orlandi, M.O., Leite, E.R., Aguiar, R., Bettini, J., Longo, E.: Growth of SnO nanobelts and dendrites by a self-catalytic VLS process. J. Phys. Chem. B **110**, 6621–6625 (2006)
200. Dai, Z.R., Pan, Z.W., Wang, Z.L.: Novel nanostructures of functional oxides synthesizedby thermal evaporation. Adv. Funct. Mater. **13**, 9–24 (2003)
201. Pan, Z.W., Dai, Z.R., Wang, Z.L.: Nanobelts of semiconducting oxides. Science **291**, 1947–1949 (2001)
202. Edwards, J.R., Marr, G.: Depletion-mode IGFET made by deep ion implantation. IEEE Trans. Electron Devices **20**, 283–289 (1973)
203. Streetman, B.G., Banerjee, S.: Solid State Electronic Devices, 5th edn. Prentice-Hall Inc. (2000)
204. Muller, R.S., Kamins, T.I.: Device Electronics for Integrated Circuits, 3rd edn. Wiley & Sons. Inc, (2003)
205. Byon, K., Tham, D., Fischer, J.E., Johnson, A.T.: Systematic study of contact annealing: ambipolar silicon nanowire transistor with improved performance. Appl. Phys. Lett. **90**, 143513 (2007)
206. Zheng, G., Lu, W., Jin, S., Lieber, C.M.: Synthesis and fabrication of high-performance n-type silicon nanowire transistors. Adv. Mater. **16**, 1890–1893 (2004)
207. Wunnicke, O.: Gate capacitance of back-gated nanowire field-effect transistors. Appl. Phys. Lett. **89**, 083102–083103 (2006)
208. Fan, Z., Wang, D., Chang, P.-C., Tseng, W.-Y., Lu, J.G.: ZnO nanowire field-effect transistor and oxygen sensing property. Appl. Phys. Lett. **85**, 5923–5925 (2004)
209. Alexander, C.K., Sadiku, M.N.O.: Fundamentals of Electric Circuits, 2nd edn. Tata McGrHill (2004)
210. Sedra, A.S., Smith, K.C.: Microelectronic Circuits, 5th edn. Oxford University Press (2004); Hejhall, R.: RF small signal design using two-port parameters (1993)
211. Fan, Z., Ho, J.C., Takahashi, T., Yerushalmi, R., Takei, K., Ford, A.C., Chueh, Y.-L., Javey, A.: Toward the development of printable nanowire electronics and sensors. Adv. Mater. **21**, 3730–3743 (2009)
212. Desai, A.V., Haque, M.A.: Sliding of zinc oxide nanowires on silicon substrate. Appl. Phys. Lett. **90**, 033102–033103 (2007)
213. Pirro, D.M., Wessol, A.A., Wills, J.G.: Lubrication Fundamentals, 2nd edn. CRC Press (2001)
214. Liu, Z., Zhang, D., Han, S., Li, C., Tang, T., Jin, W., Liu, X., Lei, B., Zhou, C.: Laser ablation synthesis and electron transport studies of tin oxide nanowires. Adv. Mater. **15**, 1754–1757 (2003)
215. Kofstad, P.: Defects and transport properties of metal oxides. Oxid. Metals **44**, 3–27 (1995)
216. Van de Walle, C.G.: Hydrogen as a cause of doping in zinc oxide. Phys. Rev. Lett. **85**, 1012 (2000)
217. Anderson, J., Chris, G.V.D.W.: Fundamentals of zinc oxide as a semiconductor. Rep. Prog. Phys. **72**, 126501 (2009)
218. Liu, W., Cao, X., Zhu, Y., Cao, L.: The effect of dopants on the electronic structure of SnO_2 chemical, **66**, 219–221 (2000)
219. Toyosaki, H., Kawasaki, M., Tokura, Y.: Electrical properties of Ta-doped SnO[sub 2] thin films epitaxially grown on TiO[sub 2] substrate. Appl. Phys. Lett. **93**, 132109–132113 (2008)
220. Wang, C., Wang, J., Chen, H., Wang, W., Su, W., Zang, G., Qi, P.: Effects of Ta_2O_5 on the grain size and electrical properties of SnO_2-based varistors. J. Phys. D Appl. Phys. **36**, 3069 (2003)
221. Gordon, R.G.: Criteria for choosing transparent conductors. Mrs Bulletin **25**, 52–57 (2000)
222. Tahar, R.B.H., Ban, T., Ohya, Y., Takahashi, Y.: Tin doped indium oxide thin films: electrical properties. J. Appl. Phys. **83**, 2631–2645 (1998)

223. Wan, Q., Song, Z.T., Feng, S.L., Wang, T.H.: Single-crystalline tin-doped indium oxide whiskers: synthesis and characterization. Appl. Phys. Lett. **85**, 4759–4761 (2004)
224. Law, M., Greene, L.E., Johnson, J.C., Saykally, R., Yang, P.: Nanowire dye-sensitized solar cells. Nat. Mater. **4**, 455–459 (2005)
225. Banerjee, D., Jo, S.H., Ren, Z.F.: Enhanced field emission of ZnO nanowires. Adv. Mater. **16**, 2028–2032 (2004)
226. Dayeh, S.A., Soci, C., Yu, P.K.L., Yu, E.T., Wang, D.: Influence of surface states on the extraction of transport parameters from InAs nanowire field effect transistors. Appl. Phys. Lett. **90**, 162112–162113 (2007)
227. Kelley, T.W., Baude, P.F., Gerlach, C., Ender, D.E., Muyres, D., Haase, M.A., Vogel, D.E., Theiss, S.D.: Recent progress in organic electronics: materials, devices, and processes. Chem. Mater. **16**, 4413–4422 (2004)
228. Wang, D., Chang, Y.-L., Wang, Q., Cao, J., Farmer, D.B., Gordon, R.G., Dai, H.: Surface chemistry and electrical properties of germanium nanowires. J. Am. Chem. Soc. **126**, 11602–11611 (2004)
229. Rosaye, J.Y., Yasuda, Y., Sakai, A., Mialhe, P., Charles, J.P., Watanabe, Y.: C-V and C-P characterization sensitivities for fast and slow-state traps in very thin oxide MOSFETs. In: 2004 IEEE International Reliability Physics Symposium Proceedings 42nd Annual, pp. 353–356 (2004)
230. Batey, J., Tierney, E.: Low-temperature deposition of high-quality silicon dioxide by plasma-enhanced chemical vapor deposition. J. Appl. Phys. **60**, 3136–3145 (1986)
231. Lu, W., Xiang, J., Timko, B.P., Wu, Y., Lieber, C.M.: One-dimensional hole gas in germanium/silicon nanowire heterostructures. Proc. Nat. Acad. Sciences U. S. A. **102**, 10046–10051 (2005)
232. Callister, W.D.: Science et genie des matériaux, Dunod (2003)
233. Klabunde, K.J.: Nanoscale Materials in Chemistry, Wiley-Interscience (2001)
234. Batle, X., Labarta, A.: Finite-size effects in fine particles: magnetic and transport properties. J. Phys. R15–R42 (2002)
235. Zutic, I., Fabian, J., Das Sarma, S.: Spintronics: fundamentals and applications. Rev. Mod. Phys. **76**, 323–410 (2004)
236. Abdel Hameed, R.S., Abu-Nawwas, A.-A.H., Shehataa, H.A.: Nano-composite as corrosion inhibitors for steel alloys in different corrosive media. Adv. Appl. Science Res. **4**(3), 126–129 (2013)
237. Wanekaya, A.K., Chen, W., Myung, N.V., Mulchandani, A.: Nanowire-based electrochemical biosensors. Electroanalysis **18**(6), 533–550 (2006)
238. Praveen, B.M., Venkatesha, T.V.: Electrodeposition and corrosion resistanceproperties of Zn–Ni/TiO_2 nano composite coatings. Int. J. Electrochem. **2011**, Article ID 261407, 4 pp.
239. Joseph Rathish, R., Dorothy, R., Joany, R.M., Pandiarajan, M., Rajendran, S.: Corrosion resistance of nanoparticle incorporated nanocoatings. Eur. Chem. Bull. **2**(12), 965–970 (2013)
240. Galliano, F., Landolt, D.: Prog. Org. Coat. **44**, 217 (2002)
241. Talo, A., Forsen, O., Ylasaari, S.: Synth. Methods **102**, 1394 (1999)
242. Miskovic-Stankovic, V.B., Stanic, M.R., Drazic, D.M.: Prog. Org. Coat. **36**, 53 (1999)
243. Wetzel, B., Haupert, F., Zhang, M.Q.: Comp. Sci. Technol. **63**, 2055 (2003)
244. Zhang, M.Q., Rong, M.Z., Yu, S.I., Wetzel, B., Friedrich, K.: Macromol. Mater. Eng. **287**, 111 (2002)
245. Yamini, S., Young, R.J.: Polymer **18**, 1075 (1977)
246. Perreux, D., Suri, C.: Comp. Sci. Technol. **57**, 1403 (1997)
247. Loos, C., Springer, G.S.: J. Comp. Mater. **13**, 131 (1979)
248. Lam, K., Lau, K.T.: Comp. Struct. **75**, 553 (2006)
249. Shi, G., Zhang, M.Q., Rong, M.Z., Wetzel, B., Friedrich, K.: Wear **254**, 784 (2003)
250. Hartwig, M.S., Putz, D., Aberle, L.: Macromol. Symp. **221**, 127 (2005)

251. Dietsche, F., Thomann, Y., Thomann, R., Mulhaupt, R.: J. Appl. Polym. Sci. **75**, 396 (2000)
252. Huong, N.: Improvement of bearing strength of laminated composites by nanoclay and Z-pin reinforcement. Ph.D. Dissertation, University of New South Wales, Australia (2006)
253. Becker, O., Varley, R., Simon, G.: Polymer **43**(16), 4365 (2002)
254. Yang, L.H., Liu, F.C., Han, E.H.: Prog. Org. Coat. **53**, 91 (2005)
255. Umaka, S.V., Zheludkevich, M.L., Yasakau, K.A., Serra, R., Poznyak, S.K., Ferreira, M.G. S.: Prog. Org. Coat. **58**, 127 (2007)
256. Shi, X., Nguyen, T.A., Suo, Z., Liu, Y., Avic, R.: Surf. Coat. Tech. **204**, 237 (2009)
257. Li, R., Chen, L.: A paint containing nano titanium oxide and nano silver, and its preparation method. CN 10027622 (2005)
258. Morrow, W.H., McLean, L.J.: Self-cleaning UV reflective coating, its applying methods, and UV irradiating device prepared therefrom. US 2003059549 (2003)
259. Cai, R., Van, G.M., Aw, P.K., Itoh, K.: Solar-driven selfcleaning coating for a painted surface. C. R. Chim. **9**, 829–835 (2006)
260. www.ambio.bham.ac.uk
261. www.tda.com/Library/docs/Nanomatcrials%20for%20Coatings%205-l7-04.pdf
262. Townsin, R.: Biofouling **19**, 9 (2003)
263. Kiil, S., Weinell, C.E., Pedersen, M.S., Dam-Johanssen, K.: Ind. Eng. Chem. Res. **40**, 3906 (2001)
264. Adkins, J.D., Mera, A.E., Roe-Short, M.A., Pawlikowski, G.T., Brady, R.F.: Prog. Org. Coat. **29**, 1 (1996)
265. Omae, I.: Chem. Rev. **103**, 3431 (2003)
266. Yebra, D.M., Kiil, S., Johansen, K.D.: Prog. Org. Coat. **50**, 75 (2004)
267. Iguerb, O., Poleunis, C., Mazeas, F., Compere, C., Bertrand, P.: Langmuir **24**, 12272 (2008)
268. Schiff, K., Diehl, D., Valkirs, A.: Marine Pollut. Bull. **48**, 371 (2004)
269. Voulvoulis, N., Scrimshaw, M.D., Lester, J.N.: Appl. Organomet. Chem. **13**, 135 (1999)
270. Matthiessen, P., Reed, J., Johnson, M.: Mar. Pollut. Bull. **38**, 908 (1999)
271. Vasconcelos, M.T.S.D., Leal, M.F.C.: Environ. Sci. Technol. **35**, 508 (2001)
272. Mann, E.L., Nathan, A., James, W.M., Sallie, W.C.: Limnol. Oceanogr. **47**, 976 (2002)
273. Avery, S.V., Howlett, N.G., Radice, S.A.: Appl. Environ. Microbiol. **62**, 3960 (1996)
274. Antonietta, Z.M., Stefania, Z., Rebecca, P., Riccardo, B.J.: Inorg. Biochem. **35**, 291 (1996)
275. Stoimenov, P.K., Klinger, R.L., Marchin, G.L., Klabunde, K.J.: Langmuir **18**, 6679 (2002)
276. Yoichi, Y., Hiroshi, Y., Chikara, K., Kei, I.: Prog. Org. Coat. **42**, 150 (2001)
277. Kawashita, M., Toda, S., Kim, H.M., Kokubo, T., Masuda, N.: J. Biomed. Mater. Res. **66**, 266 (2003)
278. Shah, M.S.A.S., Kalagara, T., Singh, S., Manorama, S.: Chem. Mater. **20**, 2455 (2008)
279. Shtykova, L., Fant, C., Handa, P.: Prog. Org. Coat. **64** (2009).
280. Tovmachenko, O.G., Graf, C., Heuvel, D.J.: Adv. Mater. **18**, 91 (2006)
281. Sakai, H., Kanda, T., Shibata, H.: J. Am. Chem. Soc. **128**, 4944 (2006)
282. Tom, R.T., Nair, A.S., Singh, N.: Langmuir **19**, 3439 (2003)
283. White, S.R., Sottos, N.R., Geubelle, P.H.: Nature **409**, 794 (2001)
284. Cho, S.H., Andersson, H.M., White, S.R.: Adv. Mater. **18**, 99 (2006)
285. Shchukin, D.G., Lamaka, S.V., Yasakau, K.A.: J. Phys. Chem. **112**, 958 (2008)
286. Ke, X., Wang, J.X., Kang, X.L., Chen, J.F.: Mater. Lett. **63**, 31 (2009)
287. Le, Yuan, Hou, Pengtao, Wang, Jiexin, Chen, Jian-Feng: Mater. Chem. Phys. **120**(2–3), 351–355 (2010)
288. Gebel, T.: Arch. Toxicol. (2011). https://doi.org/10.1007/s00204-012-0835-1
289. Sharma, V.K., Ria, A.Y., Lin, Y.: Adv. Colloid Interface Sci. **145**, 83–96 (2009)
290. Shankar, S.S., Rai, A., Ankamwar, B., Singh, A., Ahmad, A., Sastry, M.: Nat. Mater. **3**, 482–488 (2004)
291. Bar, H., Bhui, D.K., Sahoo, G.P., Sarkar, P., De, S.P., Misra, A.: Colloids and surfaces A. Physicochem. Eng. Asp. **339**, 134–139 (2009)

292. Jha, A.K., Prasad, K.: Int. J. Green Nanotechnol. Phys. Chem. **1**, 110–117 (2010)
293. Baban, D., Seymour, L.W.: Control of tumour vascular permeability. Adv. Drug Deliv. Rev. **34**, 109–119 (1998)
294. Tomar, A., Garg, G.: Short review on application of gold nanoparticles. Glob. J. Pharmacol. **7**(1), 34–38 (2013)
295. Ceylan, A., Jastrzembski, K., Shah, S.I.: Metall. Mater. Trans. A **37**, 2033 (2006)
296. Junggwon, Y., Kyoungah, C., Byoungjun, P., Ho-Chul, K., Byeong-Kwon, J., Sangsig, K. J.: J. Appl. Phys. **47**, 5070 (2008)
297. Mohl, M., Dobo, D., Kukovecz, A., Konya, Z., Kordas, K., Wei, J., Vajtai, R., Ajayan, P. M.: J. Phys. Chem. C **115**, 9403 (2011)
298. Fan, T.X., Chow, S.K., Zhang, D.: Biomorphic mineralization: from biology to materials. Prog. Mater. Sci. **54**(5), 542–659 (2009)
299. Klaus, T., Joerger, R., Olsson, E., Granqvist, C.G.: Silver-based crystalline nanoparticles, microbially fabricated. J. Proc. Natl. Acad. Sci. USA **96**, 13611–13614 (1999)
300. Konishi, Y., Uruga, T.: Bioreductive deposition of platinum nanoparticles on the bacterium Shewanella algae. J. Biotechnol. **128**, 648–653 (2007)
301. Willner, I., Baron, R., Willner, B.: Growing metal nanoparticles by enzymes. J. Adv. Mater. **18**, 1109–1120 (2006)
302. Vigneshwaran, N., Ashtaputre, N.M., Varadarajan, P.V., Nachane, R.P., Paralikar, K.M., Balasubramanya, R.H.: Mater. Lett. **61**(6), 1413–1418 (2007)
303. Shankar, S.S., Ahmed, A., Akkamwar, B., Sastry, M., Rai, A., Singh, A.: Biological synthesis of triangular gold nanoprism. Nature **3**, 482 (2004)
304. Ahmad, N., Sharma, S., Singh, V.N., Shamsi, S.F., Fatma, A., Mehta, B.R.: Biosynthesis of silver nanoparticles from *Desmodium triflorum* : a novel approach towards weed utilization. Biotechnol. Res. Int. **454090**, 1–8 (2011)
305. Armendariz, V., Gardea-Torresdey, J.L., Jose Yacaman, M., Gonzalez, J., Herrera, I., Parsons, J.G.: Proceedings of Conference on Application of Waste Remediation Technologies to Agricultural Contamination of Water Resources, Kansas City, MO, USA (2002)
306. Kim, B.Y., Rutka, J.T., Chan, W.C.: Nanomedicine. N. Engl. J. Med. **363**(25), 2434–2443 (2010)
307. Kyriacou, S.V., Brownlow, W.J., Xu, X.N.: Using nanoparticle opti chitosan assay for direct observation of the function of antimicrobial agents in single live bacterial cells. Biochemistry **43**, 140–147 (2004)
308. Anil Kumar, S., Khan, M.I.: Hetero functional nanomaterials: fabrication, properties and applications in nanobiotechnology. J. Nanosci. Nanotechnol. **10**, 4124 (2010)
309. Skirtach, A.G., Javier, A.M., Kreft, O., Köhler, K., Alberola, A.P., Möhwald, H., et al.: Laser-induced release of encapsulated materials inside living cells. Angew Chem. **118**, 4728 (2006)
310. Sershen, S.R., Westcott, S.L., Halas, N.J., West, J.L.: Temperaturesensitive polymer-nanoshell composites for photothermally modulated drug delivery. J. Biomed. Mater. Res. **51**, 293 (2000)
311. Gu, H., Ho, P.L., Tong, E., Wang, L., Xu, B.: Presenting vancomycin on nanoparticles to enhance antimicrobial activities. Nano Lett. **3**, 1261 (2003)
312. Estella-Hermoso de Mendoza, A., Campanero, M.A., Mollinedo, F., Blanco-Prieto, M.J.: Lipid nanomedicines for anticancer drug therapy. J. Biomed. Nanotechnol. **5**, 323 (2009)
313. Saravanakumar, G., Kim, K., Park, J.H., Rhee, K., Kwon, I.C.: Current status of nanoparticle based imaging agents for early diagnosis of cancer and atherosclerosis. J. Biomed. Nanotechnol. **5**, 20 (2009)
314. Chen, Y.H., Tsai, C.Y., Huang, P.Y., Chang, M.Y., Cheng, P.C., Chou, C.H., et al.: Methotrexate conjugated to gold nanoparticles inhibits tumor growth in a syngeneic lung tumor model. Mol. Pharm. **4**, 713 (2007)

315. Tseng, Y.C., Huang, L.: Self-assembled lipid nanomedicines for siRNA tumor targeting. J. Biomed. Nanotechnol. **5**, 351 (2009)
316. Andersen, A.J., Hashemi, S.H., Andresen, T.L., Hunter, A.C., Moghimi, S.M.: Complement: alive and kicking nanomedicines. J. Biomed. Nanotechnol. **5**, 364 (2009)
317. Asadishad, B., Vossoughi, M., Alemzadeh, I.: Folate-receptortargeted delivery of doxorubicin using polyethylene glycolfunctionalized gold nanoparticles. Ind. Eng. Chem. Res. **49**, 1958 (2010)
318. Erdogan, S.: Liposomal nanocarriers for tumor imaging. J. Biomed. Nanotechnol. **5**, 141 (2009)
319. Hirsch, L.R., Stafford, R.J., Bankson, J.A., Sershen, S.R., Rivera, B., Price, R.E., et al.: Nanoshell-mediated near-IR thermal therapy of tumors under magnetic resonance guidance. Proc. Natl. Acad. Sci. USA **100**, 13549–13554 (2003)
320. Jordan, A., Scholz, R., Maier-Hauff, K., van Landeghem, F.K., Waldoefner, N., Teichgraeber, U., et al.: The effect of thermotherapy using magnetic nanopartcles on rat malignant glioma. J. Neurooncol. **78**, 7–14 (2006)
321. Jurgons, R., Seliger, C., Hilpert, A., Trahms, L., Odenbach, S., Alexiou, C.: Drug loaded magnetic nanopartcles for cancer therapy. J. Phys. Condens. Matter **18**, S2893–S2902 (2006)
322. Zhang, L., Gu, F.X., Chan, J.M., Wang, A.Z., Langer, R.S., Farokhzad, O.C.: Nanoparticles in medicine: therapeutic applications and developments. Clin. Pharmacol. Ther. **83**, 761–769 (2008)
323. Aravind, K., Tasneem, M., Kiranmai, M.: Rapid green synthesis of silver nanopartcles from root bark extract of Azadirachta indica A. Juss and their application in clinical ultra sound gel. Ann. Drug Discov. Biomed. Res. **1**, 144–154 (2014)
324. Seabra, A.B., Duran, N.: Nanotoxicity of metal oxide nanoparticles. Metals **5**, 934–975 (2015)
325. Reddy, P.N., Eladia, M.P., Havel, J.: Silver or silver nanoparticles: a hazardous threat to the environment and human health? J. Appl. Biomed. **6**, 117–129 (2008)

Further Readings

1. Nanohedron.com | Nano Image Gallery several images of nanowires are included in the galleries
2. Stanford's nanowire battery holds 10 times the charge of existing ones
3. Von Klitzing, K., Dorda, G., Pepper, M.: Quantum hall effect. Phys. Rev. Lett. **45**, 494–497 (1980)
4. Strongest theoretical nanowire produced at Australia's University of Melbourne
5. Penn Engineers Design Electronic Computer Memory in Nanoscale Form That Retrieves Data 1,000 Times Faster
6. One atom thick, hundreds of nanometers long Pt-nanowires are one of the best examples of self-assembly (University of Twente)
7. http://en.wikipedia.org/wiki/Potential_well
8. http://phycomp.technion.ac.il/~anastasy/thesis/node10.html
9. http://mrsec.wisc.edu/Edetc/cineplex/nanotube/index.html
10. http://www.chm.bris.ac.uk/webprojects2001/andrews/nanotubes-synthesis.html
11. http://students.chem.tue.nl/ifp03/synthesis.html
12. http://www.fy.chalmers.se/atom/research/nanotubes/experimental.xml
13. http://nanogloss.com/nanotubes/what-is-carbon-nanotube-synthesis/#axzz1NJyWPECS
14. http://www.personal.reading.ac.uk/~scsharip/tubes.htm
15. http://www.research.ibm.com/nanoscience/manipulation.html
16. http://eeesrikar.blogspot.com/p/ppts.html
17. http://www.nature.com/nnano/journal/v5/n1/fig_tab/nnano.2009.319_F1.html

18. http://www.ioap.tu-berlin.de/fileadmin/i5/AG_Moeller/Papers/anderes/QC_in_Ge_nanocrystals_-_APL_84__4056__2004_.pdf
19. http://www.nano.ugent.be/index.php?Itemid=89&id=31&option=com_content&sectionid=2&task=category
20. http://www.nanotech-now.com/nanotube-buckyball-sites.htm
21. http://www.ece.rochester.edu/courses/ECE580/docs/Quantum_Dot_Lasers.pdf
22. http://physics.schooltool.nl/quantumoptics/qd.php
23. http://www.chm.bris.ac.uk/webprojects2001/andrews/nanotubes-synthesis.html
24. http://students.chem.tue.nl/ifp03/synthesis.html
25. http://www.pharmainfo.net/reviews/methods-carbon-nanotube-and-nanohorn-synthesis-review
26. http://www.engr.utk.edu/mse/Textiles/Nanofiber%20Nonwovens.htm
27. http://www.ece.gatech.edu/research/labs/vc/theory/photolith.html
28. http://en.wikipedia.org/wiki/Photolithography
29. http://www.physicscentral.com/explore/action/lithium-links.cfm
30. http://focus.aps.org/story/v11/st15
31. http://sundoc.bibliothek.uni-halle.de/diss-online/07/07H121/prom.pdf
32. http://www.engr.colostate.edu/ECE581/fall07/Nanowires%20and%20Nanorods.pdf
33. https://www.ncbi.nlm.nih.gov/pubmed/29369820
34. https://www.ncbi.nlm.nih.gov/pubmed/23782671
35. https://en.wikipedia.org/wiki/Colloidal_gold
36. www.understandingnano.com
37. http://En.m.wikipedia.org
38. www.nano.gov

9783030216207